Molecular Mechanism of Crucifer's Host-Resistance

Govind Singh Saharan • Naresh K. Mehta •
Prabhu Dayal Meena

Molecular Mechanism of Crucifer's Host-Resistance

Springer

Govind Singh Saharan
Department of Plant Pathology
CCS Haryana Agricultural University
Hisar, Haryana, India

Naresh K. Mehta
Department of Plant Pathology
CCS Haryana Agricultural University
Hisar, India

Prabhu Dayal Meena
Crop Protection Unit
ICAR-Directorate of Rapeseed-Mustard Research
Bharatpur, Rajasthan, India

ISBN 978-981-16-1973-1 ISBN 978-981-16-1974-8 (eBook)
https://doi.org/10.1007/978-981-16-1974-8

This Springer imprint is published by the registered company Springer Nature Singapore Pte Ltd.
The registered company address is: 152 Beach Road, #21-01/04 Gateway East, Singapore 189721, Singapore

Foreword

Plants are attacked by a wide variety of potential plant pathogens within their environment such as fungi, oomycetes, bacteria, phytoplasmas, viruses, and viroids that cause harmful and economically important diseases across a very broad range of plant species worldwide. The most promising way to manage the losses is to generate disease-resistant varieties, likely involving manipulation of target genes implicated in the induced resistance to pathogens or the signal transduction pathways controlling the expression of the defense-related genes. Plant immune/defense systems rely on their ability to recognize the presence of the pathogen, to carry out signal transduction, and to respond defensively through pathways involving many genes and/or their products. Pathogens actively attempt to evade and interfere with response pathways, selecting for a decentralized, multicomponent immune system. Recent advances in molecular techniques have greatly expanded our understanding of plant resistance and immunity, largely driven by potential application to agricultural systems. The plant immune systems have provided both increased evolutionary opportunity for pathogen resistance and additional mechanisms for pathogen inhibition of such defense responses. The advances in bioinformatics and molecular biology are driving an explosion of information that will advance agricultural production and illustrate the complex molecular interactions involved in disease resistance.

The present book entitled "*Molecular Mechanisms of Crucifer's Host-Resistance*" authored by Prof. (Dr.) G. S. Saharan, Prof. (Dr.) Naresh K. Mehta, and Dr. P. D. Meena is a comprehensive treatise on molecular aspects of host–pathogen interactions, including induction of signal molecules to express defense genes, against major pathogens of crucifers that have adverse impact at a global level. The book illustrates how the different mechanisms play their roles in imparting disease resistance against the important pathogens of crucifers.

I am confident that this book will serve as a reference for future research work by Brassica scientists, including breeders, molecular biologists, and pathologists. It will also be immensely useful for science teachers and students. In particular, it provides a solid foundation of knowledge and understanding for further investigations concerning the molecular mechanisms of disease resistance. It is highly relevant to all wanting to see profitable cruciferous crops maintained and improved across the world.

My heartiest congratulations to the authors for combining their lifelong professional interests and expertise in diseases of crucifers to make this prestigious book possible for the benefit of all those interested in the research, development, and improvement of cruciferous crops.

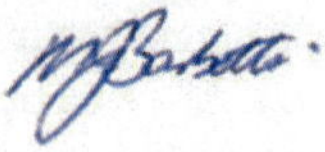

School of Agriculture and Environment
The University of Western Australia
Crawley, WA, Australia

Martin J. Barbetti

Preface

In the galaxy of plant biodiversity at a global level, crucifers share a significant role in feeding human and animal populations with vegetables, edible oil, and fodder crops of rich sources of nutrition and other amenities. Crucifers encompass a very large family Brassicaceae consisting of 325 genera and 3740 species with the genus *Brassica* having 75 accepted species. The gold mine of genomics studies *Arabidopsis thaliana* is also a part of the *Brassica* family which has transformed traditional sciences like Genetics, Plant Breeding, and Plant Pathology into Molecular Genetics, Molecular Breeding, and Molecular Plant Pathology by exploitation of *Arabidopsis* host-pathosystem to reveal molecular mechanisms of host–pathogen interactions. Crucifers are challenged by a large number of biotic and abiotic stresses, but 16 pathogens have received the attention of scientists to reveal molecular mechanisms of host–pathogen interactions and expression of molecular defense mechanisms. During the last three decades, substantial progress has been made on the investigations of molecular mechanisms of host resistance using *Brassica* host pathosystem as a model. After scanning more than 1500 publication, we have compiled the information in the form of a book entitled "*Molecular Mechanisms of Crucifer's Host-Resistance*" for the benefit of students, teachers, research scientists, and all other who are interested in amelioration of *Brassica* crops production and productivity. The book is a comprehensive treatise on the molecular aspects of host–pathogen interactions and induction of signal molecules to express defense genes against different pathogens of crucifers which have impact at a global level. The information is arranged in nine chapters with proper headings and subheadings, which have been arranged in numbered series to make the subject matter contiguous along with sections of references to consult original publications. It is the eighth book on the diseases of cruciferous crops series after *Sclerotinia, Albugo, Alternaria, Hyaloperonospora, Erysiphe, Plasmodiophora,* and "*Genomics of Crucifers Host-Resistance*" published by Springer Nature. The chapter-wise sections include information on, viz., molecular mechanisms of disease resistance, molecular mechanisms of host resistance to biotrophs, molecular mechanisms of host resistance to hemi-biotrophs and necrotrophs, biometabolomics of disease resistance to biotrophs, biometabolomics of host resistance to hemi-biotrophs and necrotrophs, and glimpses of host resistance genomics. A chapter on standardized, reproducible techniques has been included for researchers of cruciferous crops for studying the molecular mechanisms of resistant cultivars development. The last section deals with the gaps in understanding and knowledge of molecular mechanisms of disease resistance and offers suggestions for future research priorities in order to initiate advanced research programs on disease resistance. The subject matter has been vividly illustrated with photographs, graphs, figures, histogram, tables, and colored plates, which make it stimulating, effective, and easy to comprehend by readers.

We believe that this book will be immensely useful for researchers, especially *Brassica* breeders, molecular biologists, plant pathologists, teachers, extension specialists, students, industrialists, farmers, planners, and all others who are interested in growing healthy and profitable cruciferous crops all over the world. Suggestions by readers are always a source of inspiration for the authors. Any shortcomings, lacunae, and flaws in the book are the responsibility of the authors, and suggestions for its improvement are most welcome.

Hisar, Haryana, India — Govind Singh Saharan
Hisar, India — Naresh K. Mehta
Bharatpur, Rajasthan, India — Prabhu Dayal Meena

Acknowledgments

The authors are highly grateful to all the persons, scientists, publishers, societies, journals, institutes, websites, and others whose valuable materials such as photographs (macroscopic, microscopic, electron micrographs, scanning electron micrographs), drawings, figures, histograms, graphs, tables, and flow charts have been used through reproduction in the present document. The address of the author (s)/source (s) from where material was adopted whether institutes, websites, societies, etc. can be obtained from the reference which has been listed in the reference section of the book in each chapter. We have duly acknowledged all the authors for their excellent and valuable publications/articles in this manuscript by citing their names against each tables, figures, plates, dendrograms, etc. We have sought the necessary permission to reproduce their material in our document, and majority of them have granted the permission to reuse them. The authors are sincerely thankful to all the scientists, publishers, journals, institutes, societies, and websites whose materials have been reproduced in one or the other form in the present manuscript but forgot to acknowledge their name(s) inadvertently. Their prompt help in providing permission, information, photographs, figures, and consent for reproduction in the present manuscript has been duly acknowledged, and the authors are grateful to all of them. There might be some errors, mistakes, and shortcomings in this manuscript. We would greatly appreciate healthy criticism and suggestions for the improvement of this publication in future.

Govind Singh Saharan
Naresh K. Mehta
Prabhu Dayal Meena

Contents

About the Authors

Govind Singh Saharan Ex. Professor and Head, Department of Plant Pathology, CCS HAU, Hisar. Dr. Saharan has conducted research in diverse fields of Plant Pathology and had 250 articles in National and International Journals. He has been editor/author of several books, monographs, and Crop Production Compendium. He is on the panel of Experts of SAU, ICAR, IARI, CSIR, UGC, and Department of Biotechnology. Dr. Saharan has been a visiting Professor at University of Alberta, Edmonton, Canada; Agriculture and Agri-Food Canada, Saskatoon, Canada, and Rothamsted Research, IACR, Harpenden, UK. He has been President (NZ) of the Indian Phytopathological Society, Editor-in-Chief, Journal of Mycology and Plant Pathology, Journal of Oilseed Brassica, and President of the Indian Society of Mycology and Plant Pathology. He has been awarded with Y. L. Nene, Outstanding Plant Pathology Teacher by ISMPP, Udaipur, and Life Time Achievement Award by the Society for Rapeseed-Mustard Research, Bharatpur, India.

Naresh Kumar Mehta Former Associate Dean, Professor and Consultant Faculty, Department of Plant Pathology, CCS HAU, Hisar. He has been teaching Plant Pathology courses to UG and PG students and conducted research in diverse fields of Plant Pathology especially on rapeseed-mustard. He has guided several M.Sc. and Ph.D. students. He has published about 200 research articles and editor/author of several books, book chapters, review articles and teaching manuals. Dr. Mehta has been President of INSOPP, Ludhiana and Editor-in-Chief ISMPP and served as member of Editorial board of various phytopathological societies in India. Dr. Mehta has been awarded Y. L. Nene, Outstanding Plant Pathology Teacher by ISMPP, Udaipur and Ms. Manju Utereja Memorial Gold Medal, HAU. He is on the panel of Experts of SAU, ICAR, UGC, and member of various National and International committees. Dr. Mehta has been a visiting scientist to University of Alberta, Edmonton, Canada.

Prabhu Dayal Meena Principal Scientist (Plant Pathology) at ICAR-Directorate of Rapeseed-Mustard Research, Bharatpur, Rajasthan, India. He has conducted research on different aspects of rapeseed-mustard diseases including host resistance, management, epidemiology and biology of crucifer's pathogen. He has published several research papers, reviews, and book chapters in reputed journals, and number of books. Dr. Meena has been honored with Fellow of Indian Society of Mycology and Plant Pathology, Plant Protection Association of India, Indian Society of Oilseed Research, Society for Rapeseed-Mustard Research, Indian Phytopathological Society, and awarded Dr. P.R. Kumar Outstanding *Brassica* Scientist Award, Society for Rapeseed-Mustard Research. He is founder Secretary and managing editor of Society for Rapeseed-Mustard Research. He visited UK during 2007 under Indo-UK Collaborative Research on Oilseed Brassica crops. He has guided several M.Sc. and Ph.D. students.

List of Abbreviations

AAFC	Agriculture and Agri-Food Canada
AB	Alternaria blight
ABA	Abscisic acid
ABC	ATP-binding cassette
ABP1	Auxin-binding protein 1
AFLP	Amplified fragment length polymorphism
AG	Anastomosis groups
AIP	Aminoindan phosphonic acid
AM	Arbuscular mycorrhizal
ANOVA	Analysis of variance
AOC3	*ALLENE OXIDE CYCLASE 3*
AOS	*ALLENE OXIDE SYNTHASE*
ap	Appressorium
APX	Ascorbate peroxidase
ARF-GEF	ADP ribosylation factor-GTP exchange factor
AT	Associative transcriptomics
AtSTP4	*Arabidopsis sugar transport protein 4*
AUDPC	Area under disease progress curve
Avr	Avirulence
BA	Benzoic acid
BABA	β-amino butyric acid
BC	Backcross
Bgt	*Blumeria graminis* f. sp. *tritici*
BiFC	Bimolecular fluorescence complement
BLAST	Basic Local Alignment Search Tool
BLAT	BLAST-like alignment tool
BLUPs	Best linear unbiased predictions
BSA	Bulked segregant analysis
BSMV	Barley stripe mosaic virus
CA	Constitutively activated
CAC	Clathrin adaptor complex
CAD	Cinnamyl alcohol dehydrogenase
CaM	Calmodulin

CaMV	Cauliflower mosaic virus
CAT	Catalase
CBD	Chitin-binding domain
CBls	Calcineurin-B-like protein
CC	Coiled coil
CDPKs	Calcium-dependent protein kinases
CF	Culture filtrate
ChIP	Chromatin immune precipitation
CK	Cytokinins
cM	centiMorgans
CMLs	Calmodulin-like protein
CNGCs	Cyclic nucleotide-gated ion channels
COIP	Co-immune precipitation
CW	Cell wall
CWAs	Cell wall apposition
CWDE	Cell wall-degrading enzymes
DAI	Days after inoculation
DAMPs	Damage-associated molecular patterns
DAPC	Discriminant analysis of principal components
DAS	Days after sowing
DEGs	Differentially expressed genes
DGE	Differential gene expression
DH	Doubled haploid
DI	Disease indices
DM	Downy mildew
DNA	Deoxyribonucleic acid
dpi	Days post-inoculation
DTT	Dithiothreitol
Ec	*Erysiphe cruciferarum*
ECD	European clubroot differential
EDS1	ENHANCED DISEASE SUSCEPTIBILITY 1
EHM	Extra-haustorial membrane
ELISAs	Enzyme-linked immunosorbent assays
EMS	Ethyl methane-sulfonate
ENU	Ethyl nitrosourea
Ep	*Erysiphe pisi*
e-PCR	Electronic PCR
ERK	Extracellular signal-regulated kinase
ESTs	Expressed sequence tags
ET	Ethylene
ETI	Effector-triggered immunity
ETS	Effector-triggered susceptibility
f. sp.	Fungal forma specialis
FCF	Fungal culture filtrate
FDR	False discovery rate

Fig.	Figure
FPKM	Fragments per kilobase of transcript per million mapped
FTIR	Fourier-transform infrared spectroscopy
GA	Gibberellins
GBS	Genotyping by sequencing
GC	*Golovinomyces cichoracearum*
GC-MS	Gas chromatography-mass spectrometry
GCRMA	Guanine Cytosine Robust Multi-Array Analysis
GDIs	Guanine nucleotide dissociation inhibitors
GDP	Guaiacol-dependent peroxidase
GFP	Green fluorescent protein
GISH	Genomic in situ hybridization
GLM	Generalized linear models
GM	Genetically modified
GO	Gene ontology
Go	*Golovinomyces orontii*
GSS	Genomic survey sequences
GWAS	Genome-wide association analysis
HGP	Human Genome Project
HIGS	Host-induced gene silencing
HMM	Hidden Markov Model
HNRT	Homeologous non-reciprocal transposition
Hp	*Hyaloperonospora parasitica*
Hpa	*Hyaloperonospora arabidopsidis*
hpi	Hours post-inoculation
HR	Hypersensitive response
HRMS	High-resolution mass spectrometry
HS	Highly susceptible
HSPs	Heat shock proteins
HTGs	High-throughput genome sequences
HTS	Host targeting signal
IAA	Indole-3-acetic acid
IAN	Indole-3-acetonitrile
IAOx	Indole-3-acetaldoxime
IC	Isochorismate
ICAR	Indian Council of Agricultural Research
ICIM	Inclusive composite interval mapping
ICM	Composite interval mapping
ICS	Isochorismate synthases
iGS	Indole glucosinolates
ILs	Introgression lines
IM	Interval mapping
INA	Isonicotinic acid
IP	Intron polymorphic
IPG	Immobilized pH gradient strips

ISSRs	Inter-simple sequence repeats
ITCs	Isothiocyanates
ITS	Internal transcribed spacer
JA	Jasmonic acid
KASP	Kompetitive Allele-Specific PCR
LD	Linkage disequilibrium
LGs	Linkage groups
LIF	Lignification inducing factor
Lm	*Leptosphaeria maculans*
LOD	Logarithm of odds difference
LRR	Leucine-rich repeat
LRR-RLKs	Leucine-rich repeat receptor-like kinase
LRR-RLPs	Leucine-rich repeat receptors-like protein
LRRs	Leucine-rich repeats
LysM	Lysine motif
LZ	Leucine zipper
MAB	Marker-assisted backcross breeding
MAMPs	Microbe-associated molecular patterns
MAP	Mitogen-activated protein
MAPK	Mitogen-activated protein kinase
MAPKKK	MAP kinase kinase kinase
MAS	Marker-assisted selection
MeSA	Methyl salicylate
MET	Multi-Environment Trials
MKK	MAP kinase kinase
ML	Maximum likelihood
MLM	Mixed linear model
MLO	Mildew resistance locus O
MPK	MAP kinase
MPTO	Methylthiopentanaldoxime
MQM	Multiple QTL mapping
MR	Moderately resistant
MS	Mass spectrometry
MSA	Multiple sequence alignment
MT	Microtubule
MTAs	Marker trait associations
MYA	Million years ago
MYB	Myeloblastosis
NBS	Nucleotide-binding site
NBS-LRR	Nucleotide-binding site leucine-rich repeat
NBT	Nitro blue tetrazolium
NC	Nucleotidyl cyclase
NCBI	National Centre for Biotechnology Information
NGS	Next generation sequencing
NHR	Non-host resistance

NILs	Near isogenic lines
NIRS	Near-infrared reflectance spectroscopy
NLRs	Nucleotide-binding site leucine-rich repeats
NMR	Nuclear magnetic resonance
NO	Nitric oxide
NPR1	Non-expressor of Pathogenesis-Related Genes 1
NWCVT	National Winter Canola Variety Trials
OA	Oxalic acid
OD	Optical density
OGs	Oligogalaturonides
On	*Oidium neolycopersici*
OST1	Open stomata 1
PA	Phosphatidic acid
PAD4	Phytoalexin-deficient 4
PAL	Phenylalanine ammonia lyase
PAMP	Pathogen-associated molecular pattern
PCA	Principal component analysis
PCD	Programmed cell death
PCR	Polymerase chain reaction
PDA	Potato dextrose agar
PDI	Per cent Disease Index
PDR	Pleiotropic drug resistance
PEIs	Pectinesterase inhibitors
PEN2	Peroxisome-associated myrosinase penetration 2
PFK	Phosphofructokinase
PGA	Polygalacturonase
PGIPs	Polygalacturonase inhibitor proteins
PIC	Polymorphic information content
PIPs	Phosphatidylinositol phosphates
PM	Plasma membrane
PM	Powdery mildew
PMEI	Pectin methylesterase inhibitors
PMSF	Phenylmethylsulfonyl fluoride
PO	Peroxidase
pp	Penetration peg formation
PPO	Polyphenol oxidase
PPT	Phosphinothricin
PR	Pathogenesis related
PRRs	Pattern recognition receptors
PTI	Pathogen-associated molecular patterns (PAMPs)-triggered immunity or pattern-triggered immunity
pv.	Pathovar
QDR	Quantitative disease resistance
qRT-PCR	Real-time quantitative-PCR
QTL	Quantitative trait locus

R	Resistance
RAPD	Random amplification of polymorphic DNA reaction
RFLP	Restriction fragment length polymorphism
RGL	Resistant genes like
RIN	RNA integrity numbers
RLCK	Receptor-like cytoplasmic kinase
RLKs	Receptor-like kinases
RLPs	Receptor-like proteins
RNA	Ribonucleic acid
RNAi	RNA interference
ROI	Reactive oxygen intermediates
ROP	Rho of plants
ROS	Reactive oxygen species
RT-PCR	Reverse transcription and quantitative reverse transcription-polymerase chain
SA	Salicylic acid
SAG	Salicylic acid glycoside
SAG101	Senescence-associated gene 101
SAM	*S*-adenosine-L-methionine
SAR	Systemic acquired resistance
SARF	Sum of adjacent recombination fractions
SCAR	Sequence characterized amplified region
SD	Standard deviation
Si	Silicon
SMA	Single marker analysis
SMRT	Single-molecule real-time
SNAP	Soluble *N*-ethylmaleimide-sensitive factor adaptor protein
SNP	Single nucleotide polymorphism
SOD	Superoxide dismutase
SR	Sclerotinia rot
SRAP	Sequence-related amplified polymorphism
SSR	Simple sequence repeat
STK	Serine-threonine kinase
STS	Sequence tagged sites
TAIR	The *Arabidopsis* Information Resource
TBIAs	Tissue blot immunoassays
TBS	Tris-buffered saline
TDFs	Transcript-derived fragments
TFs	Transcription factors
TGS	TRIS-glycine-SDS
TIGS	Transient-induced gene silencing
TIR	Toll/interleukin-1 receptor
TM	Trans-membrane
TMM	Trimmed mean of means
TNL	TIR-NBS-LRR

TuMV	Turnip mosaic virus
TuYV	Turnip yellows virus
UPGMA	Unweighted pair group method with arithmetic mean
UPS	Ubiquitin proteasome system
USA	United States of America
USDA	United States Department of Agriculture
UTR	Untranslated region
VAMPs	Vesicle-associated membrane proteins
VIGS	Virus-induced gene silencing
WAKL	Wall-associated kinase-like
WGT	Whole genome triplication

Symbols

$	dollar
%	percent
≈	is approximately equal to
/	per
:	ratio
~	approximately
£	pound
+	plus
<	less than
=	is equal to
>	more than
±	plus-minus sign
×	multiply
≤	less than or equal to
≥	more than or equal to
β	Beta
μg	microgram
μL	microliter
μm	micrometer
μmol	micromole
μM	micrometers
Σ	sigma means "sum up"
°C	degree Celsius
AU$	Australian dollar
B	boron
bp	base pair
Ca	calcium
Ca^{2+}	calcium cation
CFU	colony forming unit
cM	centiMorgans
cm	centimeter

cm^2	square centimeter
cm^{-3}	per cubic centimeters
cvs.	cultivars
d	day
dpi	days post-inoculation
dS/m	deciSiemens per meter
e.g.	for example
et al.	"et alia" meaning "and others"
g	gram
g^{-1}	per gram
Gy	Gray
h	hours
h^2	heritability
H_2O_2	hydrogen peroxide
i.e.	id est, meaning "that is"
kb	kilobase
KCl	potassium chloride
kV	kilovolt
L	liter
L^{-1}	per liter
Log	logarithm
M	million
M	molar
m	meter
m^2	square meter
Mb	million bases
MB	megabyte
mg	milligram
Mg	magnesium
$MgCl_2$	magnesium chloride
Mha	million hectare
min	minute
mL	milliliter
mL^{-1}	per milliliter
mM	millimolar
mm	millimeter
mmol	millimoles
mmt	metric million tons
MW	molecular weight
N	nitrogen
nm	nanometer
nmol	nanomole
P	potassium
pH	potential of hydrogen
rpm	revolutions per minute

S	sulfur
S	susceptible
s	second
s^{-1}	per second
sdH_2O	sterile de-ionized water
SE	standard error
Subsp.	subspecies
V	volts
v/v	volume by volume
v:w	volume equal to weight
var.	variety
viz.	videlicet which means namely
viz.	videlicet meaning "namely" or "which is"
w/v	weight by volume

1 Molecular Mechanisms of Disease Resistance

Abstract

Cruciferous plants evolve R-genes in response to pathogen attack which encodes receptor proteins that detect and respond to pathogen effector protein leading to the activation of effector-triggered host immunity (ETI). Effectors that trigger ETI are referred to as avirulence gene (*Avr*). Mitogen-activated protein (MAP) kinase cascades are highly conserved signaling molecules in eukaryotes and have a central role in plant immunity against pathogen attack. Plants are very often exposed to a variety of biotic stresses and thus have evolved multidimensional defense approaches to survive or retain their fitness. Plant use both preformed and inducible defense mechanisms to overcome pathogen challenges. Much stronger and long-lasting is inducible defense response such as systemic acquired resistance (SAR), which confers enhanced disease resistance to broad range of pathogens. *NPR1* gene in *Brassica* plants mediates the SA-induced expression of pathogenesis-related (PR) genes and SAR. Overexpression of NPR1 results in the increased transcript levels of antifungal genes like PR1, PR2 (glucanase), and PR5 (thaumatin). Many of the underlying resistance genes have been cloned, and their downstream signaling pathways have been characterized using *Arabidopsis* host-pathosystem in molecular mechanisms of host resistance. Plants have evolved intracellular receptors with nucleotide-binding, leucine-rich repeat domains (NLRs) to detect fungal effectors, or their activity to trigger immunity. The high polymorphic information content (PIC) and number of phylogenetically informative bands established RAPD as a useful tool for phylogenetic reconstruction, quantification of genetic diversity for conservation, cultivar classification, and molecular disease resistance cultivars breeding in *Brassica*. The clustering and duplication of R-genes can cause problems in R-genes identification as these genes tend to collapse in genome sequence assemblies. The outcome of host–pathogen interactions is determined by R-genes that enable host to recognize invading pathogens and activate inducible defense mechanisms. A typical

G. S. Saharan et al., *Molecular Mechanism of Crucifer's Host-Resistance*,
https://doi.org/10.1007/978-981-16-1974-8_1

R-genes allele encodes “race-specific” resistance to only one or a few strains of a single pathogen species. R-genes loci are functionally polymorphic within a plant species and encode alternate alleles that either recognize different strains of the same pathogen or do not recognize any tested pathogen. The molecular studies have revealed that R-genes often reside in complex loci consisting of R genes and lightly linked homologs. The cloning of R-genes against diverse pathogen provides the tools for comparative analysis of R-genes alleles within and between species, which will provide insight into the evolutionary history of R-genes. The functional polymorphism at R-genes loci can arise from gene deletions. The majority of plant pattern recognition receptors (PRRs) identified belongs to the family of leucine-rich repeat receptor-like kinase (LRR-RLKs) or receptor-like protein (LRR-RLPs) that lacks a C-terminal kinase domain and interacts with RLKs for transducing signals. Several number of genes have been well documented that they are differentially expressed against different biotic stresses. Several techniques and approaches have been used to reveals molecular mechanisms of crucifers' host resistance. These techniques include genomics, bioinformatics, epigenetics, transcriptomics, proteomics, metabolomics, metagenomics, pathogenomics, phylogenomics, and pangenomics.

Keywords

Major diseases resistance in *Brassica* · *Arabidopsis* model and host pathosystem of cruciferous diseases · Intracellular receptors · Phylogenetic relationship · NBS encoding · *Brassica* genome complexicity · Molecular basis of R-genes · Molecular basis of pathogen recognition · Differential expression genes · Approaches for molecular mechanisms and molecular events · Classification of crucifers pathogen for molecular mechanisms · Molecular markers application in disease resistance · Omics approaches in disease resistance

1.1 Introduction

Biotic and abiotic stresses of crucifer crop plants cause significant yield losses and affect global food security, particularly in the face of increased demand due to a growing world population (Fisher et al. 2012). Diverse lifestyle, genome plasticity and evolution, prolific reproduction, and longevity of spores under harsh environment impede the efficient control of pathogen causing crucifer crop diseases. Bacterial and fungal pathogens secrete small proteins, known as effectors, which counter plant defense and modulate plant physiology to promote pathogen growth and reproduction (Presti et al. 2015; Stergiopoulos de Wit 2009). Accumulating evidences show that the secreted effectors are key players in suppressing pathogen-associated molecular patterns (PAMPs)-triggered immunity (PTI) induced upon fungal recognition by plants (Presti et al. 2015). The lysine motif (LysM) effector ECP6 secreted by the tomato pathogen *Cladosporium fulvum* suppresses chitin-triggered immunity. ECP6 binds chitin via the intramolecular chitin-binding groove formed by the LysM domains, resulting in the sequestering of chitin released from the cell walls of invading hyphae (de Jonge et al. 2010; Sanchez-Vallet et al. 2013). Plants, on the other hand, have evolved disease resistance (R)-genes encoding receptor proteins that detect and respond to pathogen effector proteins, leading to the activation of effector-triggered immunity (ETI) (Jones and Dangl 2006). Effectors that trigger ETI are referred to as avirulence (*Avr*) genes. Discovery of the host targets of effectors is essential for our understanding of effector activity and molecular mechanisms of plant defense (Win et al. 2012). There are many examples of plant proteins that are targeted by bacterial effectors secreted by the type III secretion system (Block and Alfano 2011; Deslandes and Rivas 2012). However, despite the economic importance of diseases caused by the fungal plant pathogens, for most fungal effectors, the host targets and underlying molecular mechanisms remain unknown. Mitogen-activated protein (MAP) kinase cascades are highly conserved signaling modules in eukaryotes and have a central role in plant immunity against pathogen attack (Meng and Zhang 2013; Pitzschke et al. 2009). A MAP kinase cascade is commonly composed of three types of kinases: a MAP kinase kinase kinase (MAPKKK), a MAP kinase kinase (MKK), and a MAP kinase (MPK). In *Arabidopsis*, pathogen-responsive MPK signaling has been reported to be involved in both PTI and ETI, in which the best characterized MPKs are MPK3, MPK4, and MPK6 (Asai et al. 2002; Rasmussen et al. 2012). PAMPs, such as flg22 (a conserved 22-amino acid flagellin peptide) and elf18 (elongation factor-Tu peptide), activate the kinase signaling cascades involving MPK3, MPK4, and MPK6 (Felix et al. 1999; Gao et al. 2008; Zipfel et al. 2006). In contrast, plant pathogens have evolved mechanisms to target the MPK signaling pathways to enhance plant susceptibility. *Pseudomonas syringae* HopAI1 targets MPK3 and MPK6 and inactivates their kinase function to suppress plant defense responses (Zhang et al. 2007). *P. syringae* HopF2 targets MKK5 and can inactivate MKK5 via ADP-ribosylation of the C terminus of MKK5 in vitro (Wang et al. 2010). AvrB is a *P. syringae* effector that interacts with MPK4 to perturb hormone signaling and promote infection (Cui et al. 2010). *Phytophthora infestans* RXLR effector PexRD2 as a virulence factor interacts with the kinase domain of the host MAPKKKε to suppress MAPKKKε-dependent phosphorylation of MPKs to modulate plant immunity (King et al. 2014). These findings highlight the importance of MPK pathways in plant immunity and as targets of bacterial and biotrophs pathogen effectors. However, to date no MAPKs have been identified as targets of effectors from fungal plant pathogens (Ma et al. 2018).

Plant MPKs can be classified into four groups (A, B, C, D) based on the conserved amino acid sequences of the TxY motif present in the activation loop of MPKs (Ichimura et al. 2002). Among the best studied MPKs, MPK3 and MPK6 are in group A and MPK4 belongs to group B. However, MPK9 belongs to the D group with the plant-specific TDY phosphorylation motif and is distinct from the A, B, and C groups. In *Arabidopsis*, studies have reported that AtMPK9, together with AtMPK12, is involved in stomatal closure (Lee et al. 2016). Both abscisic acid (ABA)- and methyl jasmonate-induced stomatal closures are impaired in *Arabidopsis* double mpk9/mpk12 mutants, but not in mpk9 or mpk12 single mutants, indicating a functional redundancy (Jammes et al. 2009; Khokon et al. 2015). The mpk9/mpk12 double mutant is highly susceptible to *P. syringae* pv. tomato (Pst) DC3000 infection (Jammes et al. 2011). However, Montillet et al. (2013) reported that ABA signaling, including open stomata 1 (OST1) protein kinase, AtMPK9, or AtMPK12, plays a limited role in response to Pst early infection. This is based on the observations that there is no difference between double mpk9/mpk12 mutants and wild-type *Arabidopsis* Col in response to flg22-induced stomatal closure and expression of ABA-specific marker genes is not affected by Pst treatment (Montillet et al. 2013). Thus it is unclear whether MPK9 acts in ABA-mediated guard cell immune signaling in response to biotic stresses. A study by Nagy et al. (2015) revealed that AtMPK9 is activated through intramolecular autophosphorylation independent of any upstream MAPKKs, which is similar to MAPKK-independent activation mechanisms reported for the mammalian atypical MAPKs, such as extracellular signal-regulated kinase (ERK)7/8 (Klevernic et al. 2006; Nagy et al. 2015).

Host resistance in crucifers to biotic and abiotic stresses is multilayered and multicomponent. The perception of defense response and molecular mechanisms in a host against the pathogens are of great significance in the breeding of resistant cultivars. The sequencing of host and pathogen genome during the last decade has facilitated the researchers using modern omics approaches to reveal molecular mechanisms of host–pathogen interaction and induction of signal molecules for expression of R-genes in the host at different levels of host growth. The intricate immune responses are evolved through accumulation of ROS, H_2O_2, deposition of callose, pectin, cellulose, waxes, silicon, ion-influxes, formation of papilla, cell wall apposition, phenolic compounds, overexpression of R-genes, PR proteins, protein phosphorylation, biosynthesis of phytoalexins, fungal enzymes inhibitors, triggering of HR, induction of SAR, nonhost resistance mechanisms, the production of secondary metabolites, calcium and hormones signaling and homeostasis, and the induction of transcriptional factors and mitogen-activated protein kinases. Resistance is activated through SA signaling and simultaneous perception of ethylene and jasmonic acid depending on biotrophic and necrotrophic host–pathogen interactions. The majority of plant R-genes encode nucleotide-binding site leucine-rich repeat (NB-LRR)–type protein, which can be further, grouped into two subclasses based on their N-terminal sequence. First, those containing coiled coil (CC) domain (CC-NB-LLR) and second those containing a domain with interleukin-1 receptor (TIR) (TIR-NB-LRR). The NBS-LRR proteins are thought to recognize effectors or effector-specific signals from the pathogen and activate the host immune system to provide hypersensitive responses.

1.2 Major Disease Resistance Molecular Mechanisms and Events Operating in *Brassica*

Plants are very often exposed to a variety of biotic stresses and thus have evolved multidimensional defense approaches to survive or retain their fitness (Roux et al. 2014). The plants display both preformed and inducible defense mechanisms to overcome pathogen challenges. However, much stronger and long-lasting is inducible defense response such as systemic acquired resistance (SAR), which confers enhanced disease resistance to broad range of phytopathogens (Durrant and Dong 2004). In plants, SAR is generally activated by local infections and immunizes the whole plant to subsequent infectious diseases (Fu and Dong 2013; Shah and Zeier 2013). Realistic evidences have shown that activation of SAR is reliant on the higher levels of the endogenous salicylic acid (SA), and activation of a battery of pathogenesis-related (PR) genes. Most of these PR proteins such as glucanases, chitinases, thaumatins, and defensins possess antifungal activities and are known to play important role in disease resistance. Exogenous application of SA or its analogs has been also revealed to activate SAR pathway in plants (Durrant and Dong 2004; Makandar et al. 2006). Conversely, *Arabidopsis thaliana* plants expressing NahG transgene which codes for salicylate hydroxylase (SA-degrading enzyme) were deficient in accumulating SA, and hence failed to activate SAR (Delaney et al. 1995). In addition to SA, a group of heterogeneous proteins is crucial for the activation of SAR. Among them is PR1 protein, a key regulator in the SA-mediated SAR signal transduction pathway. The quest to discover the SA receptor led to the discovery of a regulatory or transcription cofactor protein NPR1 (Cao et al. 1994). However, many studies have revealed that NPR1 is linked to SA signaling; however, its role as SA receptor remains largely unknown. In this context, Wu et al. (2012) have reported that NPR1 is the receptor for SA pathway in *Arabidopsis*. In addition, two NPR1 paralogs, namely, NPR3 and NPR4, bind to SA and control the proteasome-mediated degradation of NPR1 protein through their interaction with NPR1 (Fu et al. 2012). After pathogen assault, plants produce a variety of phytohormones; their composition, quantity, and timing significantly vary among plant species and depend mainly on the pathogens' lifestyle and their mode of infection (De-Vos et al. 2005). SA pathway generally provides resistance to biotrophic pathogens, whereas jasmonic acid/ethylene (JA/ET) pathways are commonly associated with resistance to necrotrophic pathogens and to herbivorous pests (Glazebrook 2005; Bari and Jones 2009). Generally, SA and JA signaling pathways operate antagonistically, and thus, elevated resistance against biotrophs is often related with increased susceptibility to necrotrophs and vice versa (Grant and Lamb 2006). Many regulatory components involved in SA/JA crosstalk have been identified; among them is NPR1 which plays a crucial role in regulating SA-mediated suppression of the JA pathway (Spoel et al. 2003; Pieterse et al. 2012; Thaler et al. 2012; Van der Does et al. 2013). Furthermore, SA/JA antagonism is commonly found in many plant species under various taxonomic groups and therefore seems to be evolutionary conserved (Thaler et al. 2012).

As first discovered in *Arabidopsis*, various AtNPR1 homologs have been isolated thereafter in many agriculturally important crops (Chen et al. 2013; Zhong et al. 2015). *NPR1* is a multigene family in *Arabidopsis* with multifaceted functions. For example, *AtNPR1* and *AtNPR2* are notably considered as key regulators of SAR (Cao et al. 1997, 1998; Zhang et al. 2003), while *AtNPR3* and *AtNPR4* are known as negative regulators of SAR (Fu et al. 2012). Moreover, another group of *AtNPR1* homologs are *AtBOP1*, and *AtBOP2*, which are related with lateral organ development (Hepworth et al. 2005). However, most of the studies were carried out on *Arabidopsis NPR1* (*AtNPR1*). Structurally, *AtNPR1* and its homologs contain an ankyrin repeat, N-terminal BTB/POZ broad-complex, Tramtrack, Bric a brac/poxvirus, and zinc finger domains, respectively (Cao et al. 1997; Aravind and Koonin 1999). In non-induced state, *NPR1* exists as an inactive oligomer form in the cytosolic region. However, after SA accumulation, the redox status of the cell changes which leads to dissociation of the inactive oligomer *NPR1* to active monomers and their translocation to the nucleus where it binds to TGA factors thereby inducing PR genes (Mou et al. 2003; Tada et al. 2008).

Transcriptional studies have shown that *NPR1* is expressed at low levels in mock plants but is induced significantly after microbial attack or treatment with SA or its biologically active analogs. Many studies have revealed *NPR1* mutant (*NPR1*) plants are more prone to diseases and also show altered expression of defense marker PR genes compared to *NPR1*-expressing plants (Glazebrook et al. 1996; Cao et al. 1997). Furthermore, *NPR1* also plays a role in cross talk of SA/JA signaling pathways and in antagonistic effect of SA on JA signaling (Spoel et al. 2003). Hence, *NPR1* is considered as the positive regulator of SA-mediated plant immune responses. To explore the defense role of *NPR1* against bacterial, viral, and fungal pathogens, various overexpression studies have been carried out in both model and crop plant systems. *NPR1* mediates the SA-induced expression of pathogenesis-related (PR) genes and SAR. Overexpression of *NPR1* results in the increase of transcript levels of antifungal genes like *PR1*, *PR2* (glucanase), and *PR5* (thaumatin) which are universally known to have antifungal activity. Many studies have revealed the potential antifungal activity of these PR genes against wide range of fungal pathogens. PR gene activity is regulated at the level of redox-dependent nuclear transport of *NPR1*. For example, overexpression of *NPR1* in *Arabidopsis* plants confers enhanced disease resistance to bacterial and fungal infections (Cao et al. 1998; Friedrich et al. 2001).

Transgenic carrot plants overexpressing *AtNPR1* exhibit high disease resistance not only to biotrophs (*Erysiphe hordei*) but also to necrotrophic fungal pathogen (*Botrytis cinerea*, *Sclerotinia sclerotiorum*, and *Alternaria radicina*), respectively (Wally et al. 2009). Parkhi et al. (2010) also reported that cotton transgenic plants expressing *AtNPR1* exhibited broad spectrum of disease resistance not only to fungal pathogens but also to nematodes. Additionally, tobacco plants overexpressing *Malus hupehensis NPR1* confers resistance to *Botrytis cinerea* as well as induces a battery of pathogen-related genes. Furthermore, studies have revealed that rice and wheat plants overexpressing *NPR1* gene confers broad spectrum of disease resistance against most disastrous pathogens *Magnaporthe oryzae*, *Fusarium verticillioides*, and *Fusarium oxysporum*, respectively (Makandar et al. 2006; Quilis et al. 2008). *NPR1* overexpression in *A. thaliana* was reported to result in an increase in the transcript levels of PR genes and hence proves that NPR1-dependent PR gene–mediated disease resistance (Cao et al. 1997; Friedrich et al. 2001). The overexpression of NPR1 was also revealed to confer disease resistance against broad range of pathogens in different crops (Dutt et al. 2015; Sundaresha et al. 2016). These results revealed that NPR1 is a potential candidate gene for developing disease-resistant transgenic crops against multiple pathogens.

Studies on the mechanisms of resistance to Alternaria blight have implicated polygenes (Tripathi et al. 1980; Medhi 1985; Zhang et al. 1996; Krishnia et al. 2000), while some studies have attributed it to dominant nuclear genes (Tripathi et al. 1978; Subudhi and Raut 1994; Zhang et al. 1997). Thus far, different sources of resistance to *A. Brassicae* have been identified from host germplasm, including wild and tolerant species of *Brassica*. The resistance in the host germplasm has different components, which mainly include structural components such as epicuticular waxes, as well as biochemical components like phenols and phytoalexins. Epicuticular waxes form a direct physical barrier in the plant, providing resistance to *Alternaria* (Conn 1986; Conn and Tewari 1989). High deposits of epicuticular wax, forming a protective hydrophobic coating on the leaf surface, reduce the adherence of inoculum, conidia germination, and germ tube formation in the plant (Saharan 1992). Some Brassicaceae members, including *B. napus*, *B. carinata*, and *Sinapis alba*, have been reported to have higher epicuticular wax compared to *B. rapa* and *B. juncea*, and therefore, the former are less susceptible to Alternaria blight infection (Conn et al. 1984). High quantities of epicuticular wax have been observed in the progeny of interspecific crosses between *B. napus* and *B. juncea* (Singh et al. 1999).

Alternaria blight-resistant varieties accumulate high amounts of biochemical compounds such as phenols. After infection, blight-tolerant species of *Brassica* such as *B. carinata* and *B. napus* have been observed to accumulate a higher amount of total phenols compared to the susceptible species *B. juncea* and *B. rapa* (Gupta et al. 1990, 1995; Gupta and Kaushik 2002). Meanwhile, the levels of soluble and reducing sugars and soluble nitrogen have been found to be lower in resistant species. In another study, resistance to the disease was reported to be associated with increased levels of antioxidant enzymes of the phenolic pathway, such as polyphenol oxidase, peroxidase, and catalase (Singh et al. 2009). Furthermore, phytoalexin accumulation in response to pathogen infection and its role in disease resistance have been well studied in *Brassica*.

Phytoalexins are low-molecular-weight antimicrobial compounds produced by plants after pathogen infection. Cultivars of both resistant (*B. napus*, *Camelina sativa*, Eruca sativa) and susceptible (*B. rapa*) species show the elicitation of phytoalexins (Tewari et al. 1987, 1988; Conn et al. 1988, 1991). The highly resistant *C. sativa* produces a large number of phytoalexins that are involved in regulating resistance to *A. Brassicae*. Rapid accumulation of phytoalexins in *C. sativa* after pathogen infection has been found to inhibit fungal growth on the leaf surface (Jejelowo et al. 1991).

The *Alternaria* pathogen has also been found to secrete both host-specific and nonhost-specific toxic metabolites, enabling a wide range of infection symptoms (Nishimura and Kohmoto 1983; Kohmoto et al. 1995; Agrios 2005). These toxins have been shown to alter the permeability and functioning of the cell membrane and organelles, thereby inhibiting various physiological processes of the host plant (Mathur and Chand 1991). Destruxin B, a host-specific *Alternaria* phytotoxin, acts as a pathogenicity factor responsible for its aggressiveness and for the susceptibility of the host plant. The *Alternaria*-tolerant species *S. alba* detoxifies destruxin B, and this is followed by simultaneous phytoalexin formation and elicitation. Together, these processes constitute the resistance mechanism of *S. alba* against *A. Brassicae* (Pedras and Smith 1997; Pedras et al. 2001). Some compounds related to camalexin and 6-methoxycamalex have also been found to confer toxicity to *A. Brassicae* (Dzurilla et al. 1998; Fatima et al. 2019).

1.3 Role of *Arabidopsis* Model Host-Patho System in Molecular Resistance

In nature, *Arabidopsis thaliana* is commonly infected by the downy mildew pathogen *Hyaloperonospora arabidopsidis* and two species of the white rust pathogen *Albugo* (*A. laibachii* and *A. candida*). These naturally occurring pathogens have imposed evolutionary pressures on *Arabidopsis* populations that resulted in a high level of intraspecific variation in resistance (Holub and Beynon 1996). Many of the underlying resistance genes have been cloned and their downstream signaling pathways have been characterized.

1.3.1 *Arabidopsis Hyaloperonospora* Pathosystem

This species was the first eukaryotic pathogen of *Arabidopsis* to be documented (Koch and Slusarenko 1990). It was initially described as *Peronospora parasitica* (Koch and Slusarenko 1900) and was later renamed as *Hyaloperonospora parasitica* (Constantinescu and Fatehi 2002) and currently *Hyaloperonospora arabidopsidis* (Voglmayr et al. 2004; Goker et al. 2009). This species is a frequently occurring pathogen in natural *Arabidopsis* populations (Holub and Beynon 1996; Holub 2008). Moreover, the interactions are typified by abundant genetic polymorphism in the host and the pathogen (Holub et al. 1994b). For these reasons, *H. parasitica* was adopted as a reference pathogen during the early days of developing *Arabidopsis* as a system for molecular plant–microbe interactions (Dangl et al. 1992; Crute et al. 1994). These pioneering efforts substantially broadened the impact of *Arabidopsis* as model system. Like other phytopathogenic oomycetes, downy mildew species can quickly overcome host resistance genes and develop resistance to chemical control agents (Lucas et al. 1995). Thus, *H. parasitica* is a reference species for a group of pathogens with significant economic impact.

A significant advantage of *H. parasitica* is that it is a bonafide pathogen of *Arabidopsis* in the natural world, and has been coevolving with its host (Holub 2001). The polymorphisms that have evolved in the host and pathogen were used to molecularly clone *Arabidopsis* disease resistance genes ("R-genes") and the corresponding effector/avirulence genes from the pathogen (Table 1.1). The studies have demonstrated additional natural variation in effector gene repertoires in *H. parasitica*, and in the responses of different *Arabidopsis* accessions to *Hpa* effectors, suggesting functional variability in the *Arabidopsis* proteins that are targeted by *Hpa* effectors (Fabro et al. 2011; Asai et al. 2014). Thus, natural genetic variation in the *Arabidopsis*/*H. parasitica* interaction can be further exploited for mechanistic insight into plant–pathogen interaction and coevolution. Furthermore, because of the experimental advantages of *Arabidopsis*, *H. parasitica* is also one of the best reference organisms for investigating obligate biotrophy (McDowell 2011). A final advantage of *Hpa* is the availability of a reference genome sequence (Baxter et al. 2010).

1.3.2 *Arabidopsis Albugo* Pathosystem

White rust pathogens belonging to the genus *Albugo* are naturally occurring, obligate biotrophic pathogens of *Arabidopsis*, and their attributes will be summarized in comparison with *Hpa*. Like *Hpa*, *Albugo* is commonly found on *Arabidopsis* in the wild, and their interactions display natural variability, including gene-for-gene interactions (Holub et al. 1994a; Holub and Beynon 1996; Borhan et al. 2001; Holub 2008). In addition, *Albugo* species cause important crop diseases on cultivated *Brassica* species (Saharan et al. 2014).

Molecular phylogenies clearly illustrate that the Albugo and downy mildew lineages are distinct (Thines 2014; Ascunce et al. 2017). Thus, *Albugo* and *H. parasitica* lineages have independently evolved to an obligate lifestyle and to compatibility with *Arabidopsis* (Kemen and Jones 2012). Interestingly, *A. laibachii* and *A. candida* display different host ranges: the former, like *H. parasitica*, is restricted to *A. thaliana*, while host range of *A. candida* encompasses 63 genera and 241 species within the Brassicaceae (Thines et al. 2009; Saharan et al. 2014). Thus, these two species provide an opportunity for comparative studies to understand the factors that control host range (McMullan et al. 2015; Jouet et al. 2018). As with *H. parasitica* intraspecific variation for resistance/ susceptibility to *Albugo* is prevalent in *Arabidopsis* and has been used to clone several *Arabidopsis* genes for resistance to *Albugo* (Table 1.1). Perhaps most interestingly, *Albugo* species display a remarkable capacity to suppress defense in *Arabidopsis* (Cooper et al. 2008). Co-infection with *Albugo* can render *Arabidopsis* susceptible to otherwise incompatible pathogens, including the late blight pathogen *Phytophthora infestans* (Belhaj et al. 2017; Prince et al. 2017). The molecular mechanisms by which *Albugo* strongly suppresses plant immunity are beginning to be revealed (Prince et al. 2017). Further studies are expected to reveal *Albugo* effectors and their plant targets that are likely to be key components of the plant immune system (Kemen et al. 2011;

Table 1.1 Features of cloned *Arabidopsis* resistance genes encoding NLRs mediating detection of the oomycetes *Hyaloperonospora-Hpa* (*RPP*) and *Albugo* (*RAC*, *WRR*) (Herlihy et al. 2019)

Gene	Accession	Col-0 gene/allele	Remark	References	Recognizing	Referencess
RPP1-WsA	Ws	At3g44480* (*rpp1*)	Original *RPP1* locus	Botella et al. (1998)	Unknown	–
RPP1-WsB	Ws	At3g44480* (*rpp1*)	Previous *RPP10* locus	Botella et al. (1998)	ATR1 alleles	Rehmany et al. (2005)
RPP1-WsC	Ws	At3g44480* (*rpp1*)	Previous *RPP14* locus	Botella et al. (1998)	Unknown	–
RPP1-NdA	Nd	At3g44480* (*rpp1*)	Similar but different specificity as *RPP-WsB*	Botella et al. (1998)	ATR1 alleles	Rehmany et al. (2005)
RPP1-EstA	Est	At3g44480* (*rpp1*)	Similar but different specificity as *RPP-WsB*	Goritschnig et al. (2016)	WY motif 2 of ATR1 alleles	Goritschnig et al. (2016)
RPP1-ZdrA	Zdr	At3g44480* (*rpp1*)	Similar but different specificity as *RPP-WsB*	Goritschnig et al. (2016)	WY motif 2 of ATR1 alleles	Goritschnig et al. (2016)
RPP2A	Col-0	At4g19500	Requires *RPP2B*	Sinapidou et al. (2004)	Unknown	–
RPP2B	Col-0	At4g19510	Requires *RPP2A*	Sinapidou et al. (2004)	Unknown	–
RPP4	Col-0	At4g16860	–	Van der Biezen et al. (2002)	ATR4	Asai et al. (2018)
RPP5	*Ler*	At4g16950, paralog of *RPP4*	–	Parker et al. (1997)	ATR5	Bailey et al. (2011)
RPP7	Col-0	At1g58602	–	Eulgem et al. (2007)	Unknown	–
RPP8-Ler	*Ler*	At5g43470 (*rpp8*)	Alleles confer virus resistance	McDowell et al. (1998)	ATR8, not cloned	Gunn et al. (2002)
RPP13-Nd	Nd	At3g46530	Highly polymorphic in *A. thaliana* (Rose et al. 2004)	Bittner-Eddy and Beynon (2001)	ATR13	Allen et al. (2004)
RPP39	Wei-0	At1g61180 and At1g61190	CC-NB-LRR protein signaling through NDR1	Goritschnig et al. (2012)	ATR39	Goritschnig et al. (2012)

RAC1	Ksk-1	At1g31540	Resistance to *Albugo laibachii* Nc14	Borhan et al. (2004)	Unknown	–
WRR4	Col-0	At1g56510	Confers resistance to multiple *A. candida* races	Borhan et al. (2008)	Unknown	–
WRR4B	Ws-2	At1g56540	Resistance to *A. candida* isolate Ac2V	Cevik et al. (2019)	Unknown	–
WRR8	Sf-2	At5g46270	Resistance to *A. candida* isolate Ac2V	Cevik et al. (2019)	Unknown	–
WRR9	Hi-0	At1g63750	Resistance to *A. candida* isolate Ac2V	Cevik et al. (2019)	Unknown	–
WRR12	Ler	At1g17600	Resistance to *A. candida* race AcBoT	Borhan et al. (2008)	Unknown	–

RPP1 orthologs or family members in Col-0 are At3g44000, At3g44480, At3g44630, and At3g44670

* identified as a segregating locus in *H. parasitica* crosses

Links et al. 2011). *Albugo-Arabidopsis* systems are also being developed to better understand how *Albugo* influences plant interactions with other microbes (Ruhe et al. 2016; Jouet et al. 2018), as well as the important phenomenon of nonhost resistance (Cevik et al. 2019).

The experimental advantages and limitations of the *Albugo* species are similar to *H. parasitica*. Like , *Albugo* species cannot be cultured apart from their hosts, but they are relatively straightforward to propagate on susceptible host genotypes (Crute et al. 1993). Long-term storage is possible and genetic crosses between isolates can be performed (Adhikari et al. 2003). Importantly, draft genome sequences are available from two different strains of *A. laibachii* and one of *A. candida* (Kemen et al. 2011; Links et al. 2011). As with *H. parasitica*, the genome data is having a major impact on research with this organism (Kemen and Jones 2012).

1.4 Intracellular Receptors Molecules to Pathogens

Plants have evolved intracellular receptors with nucleotide-binding, leucine-rich repeat domains (NLRs) to detect fungal effectors, or their activity, to trigger immunity (ETI). NLR proteins typically have an N-terminal variable region, either of the coiled-coil type (CC) or Toll/Interleukin-1 receptor-like (TIR), or an RPW8-like domain. The N-terminal domains are followed by a nucleotide-binding domain (NB) and C-terminal leucine-rich repeat region (LRR) (Cui et al. 2015). The genes encoding these intracellular receptors were first identified in *Arabidopsis* through genetic mapping of resistance genes. The *Arabidopsis* genome (Col-0) encodes a total of around 150 NLR proteins, of which only a limited number have an assigned function. The first cloned *NLR* gene was *RPS2* that confers resistance to *Pseudomonas* bacteria expressing the effector AvrRpt2 (Mindrinos et al. 1994). Soon after that also R-genes for resistance to downy mildew and *Albugo* were cloned (Table 1.1), based on natural variability of their functions in different accessions of *Arabidopsis* (Slusarenko and Schlaich 2003; Holub 2008; Coates and Beynon 2010). These included the first R-genes to be cloned against oomycetes (Parker et al. 1997; Botella et al. 1998; McDowell et al. 1998). More than 25 *R* loci conferring isolate-specific resistance to *H. parasitica* have been genetically identified in *Arabidopsis* and are named *RPP*, for recognition of *Peronospora parasitica*, the older name of this pathogen (Slusarenko and Schlaich 2003). So far, eight *RPP* genes have been cloned: *RPP1*, *RPP2*, *RPP4*, *RPP5*, *RPP7*, *RPP8*, *RPP13*, and *RPP39*, and all encode cytoplasmic NLRs. Alleles of *RPP1* confer different isolate-specific resistances (Botella et al. 1998), and also *RPP4* and *RPP5* can be considered allelic variants (Van der Biezen et al. 2002). Many of the other *RPP* loci map to the position of cloned *RPP* genes and could constitute allelic variants (Nemri et al. 2010). Broad resistance to downy mildew, observed in certain *Arabidopsis* accessions, appeared to be mediated by combinations of isolate-specific resistance loci (Lapin et al. 2012). The "helper" NLR proteins ADR1 and NRG1 were identified as signaling components for RPP proteins containing the TIR domain (Castel et al. 2019; Wu et al. 2019a, b). It is becoming clear that ETI can be underpinned by networks of

NLRs, and deconvolution of these networks is an emerging area of interest (Wu et al. 2018). At a broad scale, little is known about whether and how the activity of *RPP* genes is regulated (Lai and Eulgem 2018). In one case, feedback control involving salicylic acid and WRKY proteins has been implicated (Mohr et al. 2010). In another, gene activity is subject to post-transcriptional control involving alternative polyadenylation, which is regulated by histone marks (Tsuchiya and Eulgem 2013; Lai et al. 2019). This research has opened up a new avenue toward understanding the mechanisms and evolution of NLR gene regulation (McDowell and Meyers 2013). Another study identified eQTLs providing resistance against *Hpa*, emphasizing the potential of this pathosystem to understand epigenetic factors that are relevant to plant–pathogen interactions (Furci et al. 2019).

Several avirulence genes from *Hpa* have also been cloned (Table 1.1) (Coates and Beynon 2010; McDowell 2014). In fact, the first oomycete *Avr* gene to be cloned was *ATR13* from *Hpa* (Allen et al. 2004). This gene encodes a small, secreted protein with a signal peptide followed by the conserved protein motif Arg-X-Leu-Arg (Rehmany et al. 2005). Downy mildew pathogens and *Phytophthora* species contain hundreds of putative "RXLR genes" in their genomes, and it is clear that they play an important role in pathogen virulence. RXLR proteins are secreted from pathogen haustoria and enter the interior of plant cells to reprogram plant regulatory networks and thereby promote virulence. However, RXLR proteins can also be recognized inside plant cells by NLR surveillance proteins. Indeed, every *Avr* gene cloned from a downy mildew or *Phytophthora* species to date encodes an RXLR or RXLR-like protein, recognized by the corresponding plant NLR protein. To date, *H. parasitica* Avr proteins recognized by *RPP1*, *RPP13*, *RPP4*, *RPP5*, and *RPP39* have been molecularly identified, and the determinants of their virulence and avirulence functions are under investigation (Allen et al. 2004; Rose et al. 2004; Rehmany et al. 2005; Bailey et al. 2011; Chou et al. 2011; Krasileva et al. 2010; Leonelli et al. 2011; Goritschnig et al. 2012; Steinbrenner et al. 2015). Comparative and molecular studies of *RPP* genes have provided novel insights into the molecular mechanisms and selective forces that drive NLR gene evolution (McDowell and Simon 2006).

The *RPP5*, *RPP1*, and *RPP8* multigene families are physically linked in clusters and are subject to intra- and intergenic recombination, to produce new *NLR* gene variants in host-pathogen arms race (Botella et al. 1998; McDowell et al. 1998; Noel et al. 1999). RPP1 homologs have diversified through repeat duplication and sequence divergence, such that they can detect multiple surfaces of the corresponding effector (Goritschnig et al. 2016). ATR1 is a modular protein that can evolve to escape detection by mutations in any of several surfaces that mediate recognition by RPP1 (Chou et al. 2011). *RPP13* provides an example of a simple locus, in which a single copy gene displays substantial allelic polymorphism, driven by diversifying selection and intra-allelic recombination (Bittner-Eddy and Beynon 2001). Interestingly, the *H. parasitica ATR13* locus, encoding the RXLR effector protein recognized by RPP13, displays similar attributes, suggesting that coevolution has been a major driver for this diversity, along with balancing selection to maintain repertoires of useful alleles in the plant and pathogen populations (Allen et al. 2004). The ATR13 protein displays a novel structure and polymorphisms that mediate recognition specificity by RPP13 map to a single, surface-exposed region (Leonelli et al. 2011).

In the years following *RPP* gene cloning, only the RPP1 protein has been investigated in significant mechanistic detail. This protein binds directly to ATR1 via the LRRs. This interaction displays extensive allelic variability (Krasileva et al. 2010; Steinbrenner et al. 2015). In the absence of ATR1 effector, RPP1 is likely maintained in an inactive state by intramolecular interactions between the N-terminal TIR domain, NB domain, and LRRs. Binding of ATR1 via LRRs disrupts these interactions and permits oligomerization of RPP1, thereby triggering cell death and other immune responses (Krasileva et al. 2010; Steinbrenner et al. 2015). This mode of inactivation/activation appears to be a general aspect of NLR proteins, although details vary between different NLR proteins. As with other NLR proteins, the molecular events between RPP1 activation and deployment of the ultimate cellular immune response remain to be identified.

Several *Arabidopsis* genes for resistance to Albugo have also been cloned (Table 1.1). The *RAC1* gene provides isolate-specific resistance to *Albugo laibachii* Nc14 (Borhan et al. 2004; Cevik et al. 2019). *WRR4* confers broad spectrum resistance to many *Albugo candida* isolates and functions as a transgene in Brassica juncea (Borhan et al. 2008, 2010). These R-genes encode cytoplasmic NLRs that act similar to many of the RPP proteins, and helper NLRs have also been implicated for WRR-mediated resistance (Castel et al. 2019). Interestingly, the phenomenon of nonhost resistance of *Arabidopsis* to *Brassica*-infecting races of *Albugo candida* is mechanistically underpinned by NLR receptors that can be identified by segregation analysis. This led to the molecular cloning of WRR4B, WRR8, WRR9, and WRR12. Some of these genes were shown to provide resistance as transgenes in *Brassica* species (Cevik et al. 2019). Significant achievement into genomics of host resistance is given in Table 1.2 (Herlihy et al. 2019).

1.5 Analysis of Phylogenetic Relationships Among *Brassica* Species for Molecular Mechanisms

The gene pool of elite oilseed rape breeding material has been considerably reduced by emphasis on specific quality traits (Snowdon and Friedt 2004). This narrow genetic base has increased their vulnerability to diseases. Many important fungal pathogens reside in the soil and infect roots, leaves, stems, and fruits of crops, often causing significant yield losses. Fungal diseases such as white rust, black spot, clubroot, Sclerotinia, and blackleg (Ananga et al. 2006); bacterial leaf spot, black rot, and soft rot (Rimmer and Buchwaldt 1995); and insect pests such as aphids, Japanese beetle, and cabbage seedpod weevil (Salisbury et al. 1995) pose a major limitation to the cultivation of this important crop.

Despite being generally susceptible, a growing body of evidence shows that some oilseed rape cultivars are resistant to blackleg disease (Ansan-Melayah et al. 1995, 1997; Ananga et al. 2006). Specifically, species that contain the B-genome (*B. juncea*, *B. nigra*, and *B. carinata*) can display total resistance to blackleg disease

Table 1.2 Significant achievements of genomics of host resistance in crucifers (Herlihy et al. 2019)

Insight	Key References
Cloning of the first R-genes against oomycetes	Parker et al. (1997), Botella et al. (1998), McDowell et al. (1998)
First insights into evolutionary dynamics of R-genes against oomycetes	Botella et al. (1998), McDowell et al. (1998), Bittner-Eddy et al. (2000)
Definition of important immune system regulators	Parker et al. (1996)
Identification and map-based cloning of the first mutants exhibiting gain of resistance to oomycetes	van Damme et al. (2005), van Damme et al. (2008)
Genetic definition of the complexity of the immune signaling network against oomycetes	Aarts et al. (1998), McDowell et al. (2000) Bittner-Eddy and Beynon (2001)
Definition of penetration resistance	Lipka et al. (2005)
Molecular cloning of the first oomycete *Avr* gene	Allen et al. (2004)
First insights into evolutionary dynamics of *Avr* gene loci	Allen et al. (2004), Armstrong et al. (2005), Rehmany et al. (2005)
Identification of the RxLR motif	Rehmany et al. (2005)
Test bed for new technologies to understand effector functions	Fabro et al. (2011), Caillaud et al. (2012)
Discovery that oomycete, fungal, and bacterial effectors can target the same plant signaling hubs	Mukhtar et al. (2011), Wessling et al. (2014)
First insights into genomic basis and evolution of obligate biotrophy	Baxter et al. (2010), Kemen et al. (2011)

throughout the life of the plant (Roy 1984; Ananga et al. 2006). Resistant oilseed rape lines have also been obtained through interspecific crosses between *B. napus* and *B. nigra* (Rimmer et al. 1995; Chevre et al. 1997; Brun et al. 2001). Hence, the use of resistant cultivars has been advocated as the most effective way of limiting yield losses due to blackleg disease (Rimmer et al. 1995).

Identification of resistant cultivars can involve either field trials (disease challenge and plant response data) or predictions based on genetic relationships. Field trials suffer practical drawbacks such as longer time to obtain results and are resource intensive, and the results may be unreliable when the trait is under a strong environmental influence, unlike molecular genetic analyses. Therefore, analysis of the genetic variation and relatedness in the *Brassica* germplasm is needed for genetic resource conservation and plant breeding programs. Resistance to plant pathogens has been shown to be under genetic control (Brun et al. 2001); hence, the genetic relationships among taxa can be used to predict susceptibility or resistance (Ochieng et al. 2007). Understanding the interspecific and intraspecific relationships among cultivated species of *Brassica* will inform better parental selection and to widen the genetic basis of resistance to blackleg through greater variations and the development of new genotypes in oilseed rape cultivars.

Studies on a broader phylogenetic relationships among diploid species including *B. nigra*, *B. rapa*, and *B. oleracea* showed that the B-genome (*B. nigra* B) was supported from a distinct clade separate from the other two (*B. rapa* A, and *B. oleracea* C) (Inaba and Nishio 2002). There is no reported phylogenetic placement of cultivated *B. napus* among these species; however, Inaba and Nishio (2002) suggested that *B. napus*, a tetraploid containing AC-genome, is allied to the *rapa-oleracea* clade. Ananga et al. (2008) assessed the phylogenetic relationships within and among cultivated *B. napus*, and *B. oleracea* using RAPD markers associated with blackleg resistance in *B. nigra* (Ananga et al. 2006). Phylogenetic placement of oilseed rape cultivars from the NWCVT in relation to resistant or susceptible cultivars (Ananga et al. 2006) can help predict their potential response to blackleg. Because these RAPD markers are associated with resistance to a plant pathogen, the resultant phylogenies are predicted to reflect functional genetic variation and affinities. Since the diploid species *B. rapa* and *B. oleracea* each represent a parental genome of the amphidiploid species, the placement of *B. napus*, *B. rapa*, and *B. oleracea* cultivars within the *rapa-oleracea* clade may identify unique variants needed in breeding programs, expansion of the range of useful variation, and for future improvement of heterotic potential (Ananga et al. 2008).

Thirty accessions from USDA germplasm collection representing two diploid *Brassica* species (*Brassica rapa* and *Brassica oleracea* var. *virids*) and 15 tetraploid cultivars (*Brassica napus*) from the national winter canola variety trials (NWCVT) were evaluated using 13 sets of random amplified polymorphic DNA (RAPD) associated with blackleg resistance in *Brassica nigra*. About 126 highly polymorphic bands with an average of 10 per primer were detected. A UPGMA dendrogram showed *B. rapa* as highly diverse and was supported from three different basal branches, while *B. napus* accessions were generally monophyletic. Similarly, all of *B. oleracea* accessions were supported from the same basal node. Generally, the three species were reciprocally paraphylectic, suggesting that the RAPD markers showed both functional relationships and homology, possibly due to selection at the RAPD loci associated with blackleg resistance. Consequently, two potentially susceptible *B. napus* accessions were identified. The high polymorphic information content (PIC) and number of phylogenetically informative bands established RAPD as a useful tool for phylogenetic reconstruction, quantification of genetic diversity for conservation, cultivar classification, and molecular breeding in *Brassica* (Fig. 1.1 Ananga et al. 2008).

1.6 Analysis of NBS-Encoding Genes Between *Brassica* and *Arabidopsis*

The identified 157, 206, and 167 NBS-encoding genes in *A. thaliana*, *B. rapa*, and *B. oleracea* genomes respectively and the total number of NBS-encoding genes in these three species are very close in spite of genome size and WGD/WGT events. Genomic organization and composite phylogenetic analysis facilitate the identification and classification of NBS-encoding genes among *A. thaliana*, *B. rapa* and *B. oleracea*. Expression profiling showing the differential expression pattern of orthologous NBS-encoding genes provides a blueprint for further characterization of these genes in *B. oleracea* and *B. rapa*.

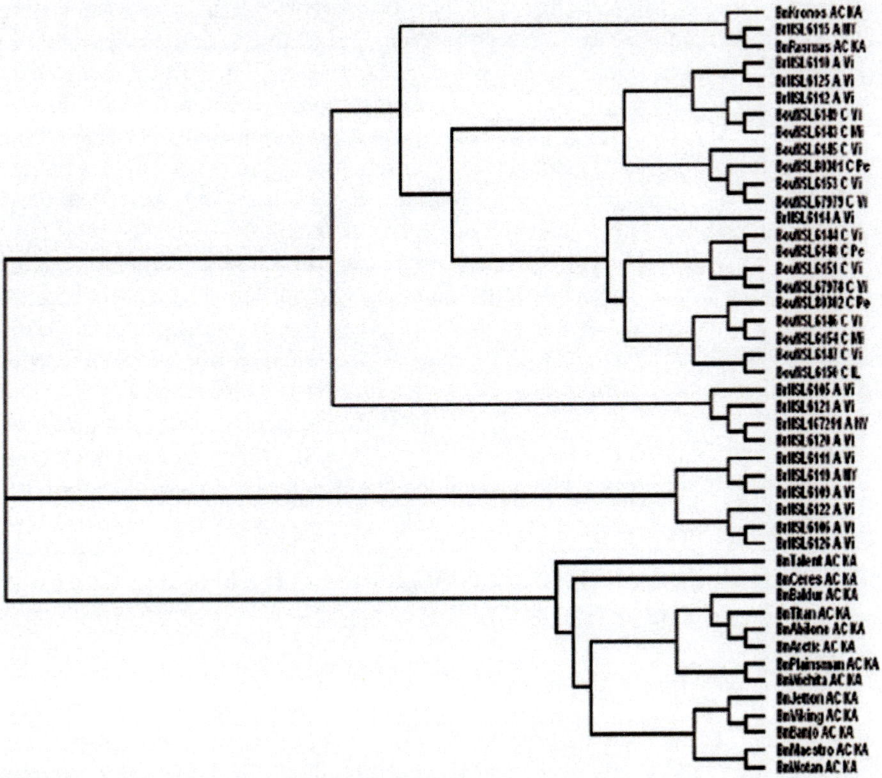

Fig. 1.1 UPGMA cluster showing the genetic relationships among three *Brassica* species (*B. rapa*, *B. oleracea*, *B. napus*) at 13 RAPD loci associated with blackleg resistance. The first two letters in the sample labels refer to the taxonomic abbreviations (Bo—*Brassica oleracea*, Br—*Brassica rapa*, Bn—*Brassica napus*), followed by the Accession number. The genomic grouping (A, C, or AC) precedes the abbreviation for sample (Ananga et al. 2008)

The expression profile of different NBS-coding members can be separated into different groups, indicative of functional divergence. Although orthologous NBS-encoding genes in *B. oleracea* and *B. rapa* are highly divergent, expression pattern divergence among paralogs within a species exceeds the level of divergence among orthologs in each type of NBS-encoding genes. Paralogs might contribute more to functional divergence than orthologs over the evolution of Brassicaceae. Through comparative analysis of tandem duplication and whole genome triplication in NBS-encoding genes among three species, there is fewer paralogous NBS-encoding genes retention after whole genome triplication than those from tandem duplication. So, tandem duplication might play more important influence than whole genome duplication in generation of NBS-encoding genes. Yu et al. (2014a, b) speculated that the quick loss of paralogs from whole genome duplication might be due to the gene dosage imbalance issue. The increase in gene dosage by tandem paralogs might have some advantage to the plant pathogen defense. The evolutionary studies illustrate that CNL-type orthologous genes in *B. rapa* species compared to *A. thaliana* have undergone stronger negative selection than those in *B. oleracea* species compared to *A. thaliana* and opposite to that orthologous genes in *B. oleracea* species experienced stronger evolutionary constraints than those in *B. rapa* species for CNL type R-genes. For TNL-type NBS-encoding genes, Yu et al. (2014a, b) did not observe significant difference between the two species about the orthologous gene pairs using Mann–Whitney U-test. It is indicated that these orthologous NBS-encoding genes in *B. rapa* and *B. oleracea* species may have undergone different selection pressure to resist the same pathogen or some pathogens may act as species-specific pathogens for different *Brassica* species. Through comparative analysis of NBS-encoding genes among *A. thaliana*, *B. rapa*, and *B. oleracea*, it will be easy to explore the evolutionary fate of NBS-encoding genes in *Brassica* lineage after split from *Arabidopsis thaliana* and advance the understanding of disease resistance between *B. oleracea* and *B. rapa* species, which will provide a valuable model for studying functional and evolutionary aspects within the *Brassica* genus and the crucifer lineage (Yu et al. 2014a, b).

1.7 Regulation of Molecular Mechanisms by *Brassica* Genome Complexity

In cruciferous species, *Brassica napus* is a relatively young crop in evolutionary history that results from an accumulated series of polyploidization events (Mason and Snowdon 2016). Prior to hybridization, the whole genome triplication in the diploids (*B. rapa* and *B. oleracea*) was preceded by three events of whole genome duplication in the ancestor genome that have further contributed to complicated gene rearrangement, duplication, and loss (Town et al. 2006; Chen 2007). Despite the genome complexity, *Brassica* species have become models to study evolutionary events and how rich genetic resources can be exploited from the plant (Pires and Gaeta 2011; Liu et al. 2014). Domesticated *B. napus* has a lower genetic diversity compared to its diploid progenitors *B. rapa* and *B. oleracea*, evidenced by similar gene order and content between the three species due to strong genetic bottlenecks during cultivation (Cheung et al. 2009; Mason and Snowdon 2016). By contrast, homoeologous rearrangements/recombination occurs more frequently in resynthesized *B. napus*, thereby contributing greatly toward genetic diversity, which provides a good source of R-genes for breeding resistant *B. napus* cultivars (Gaeta and Chris Pires 2010). The efforts were made to produce introgressed *B. napus* in a novel way, from *B. juncea* (A-genome source) and *B. carinata* (C-genome source) (Chatterjee et al. 2016) (rather than from *B. oleracea* and *B. rapa*) from where more sources of R-genes can then be utilized and identified. While in *B. napus*, the homoeologous genes are often co-expressed (Chen 2007), there could also be bias in homoeolog expression, as found in *B. juncea* (Yang et al. 2016) and *B. rapa* (Cheng et al. 2012). How the different levels of gene expression affect the stability of R-genes remains to be determined. The fate of homoeologous genes, whether they are silenced, expressed, or both copies diverge where the gene function may or may not change (sub- and neo-functionalization) is a result of the genetic and epigenetic interactions between redundant genes (Comai 2005), biased gene expression (Yang et al. 2016), gene fractionation (Subramaniam et al. 2013) or accelerated amino acid sequence evolution and positive selection (Liu and Adams 2010). These polyploidy events could potentially result in a selective advantage, for example, higher seed yield in *B. napus* (Osborn et al. 2003).

Although race-specific R-genes are easier to manipulate for breeding resistant crops compared to polygenic resistance, which involves many genes and the effect of environmental factors, the identification of R-genes in *B. napus* still remains a challenge due to the complexity of its genome. A major QTL for resistance to *S. sclerotiorum* was duplicated through homoeologous non-reciprocal transposition (HNRT) in *B. napus* (Zhao et al. 2006). The quantitative resistance genes against *L. maculans* are affected by gene duplication in homoeologous regions in *B. napus* (Fomeju et al. 2014, 2015). *L. maculans* R-genes, *Lmr1* and *ClmR1* from cultivars Shiralee and Cresor respectively, were mapped on regions that are highly duplicated within and between genomes in *B. napus*, and originated from the common ancestor before divergence of *B. oleracea* and *B. rapa* (Mayerhofer et al. 2005). Evolutionary information regarding resistance genes such as these is useful for breeding purposes, as it highlights specific genera/species likely to be most useful in developing new cultivars with highly effective and durable disease resistances. For *H. parasitica*, the single dominant gene Pp523 was mapped onto two regions on chromosome C8 and another on C5, reflecting triplicated regions in *B. oleracea* (Carlier et al. 2012).

The success stories of cloning *Crr1a*, and *CRa* (Clubroot R-genes encoding TIR-NB-LRR) from *B. rapa* (Hatakeyama et al. 2013) and *LepR3/Rlm2* (Larkan et al. 2013) in *B. napus*, revealed that the R-genes are organized as members of allelic variants or tandem repeats. The clustering effect of the R-genes for these two pathogens provides further evidence of complexity of the genome organization. Clustering of R-genes is also prevalent in association with Clubroot, Blackleg, and Sclerotinia Stem Rot resistance on both *B. rapa* (Kato et al. 2013) and *B. napus* (Delourme et al. 2004; Kato et al. 2013; Li et al. 2015). Such gene clustering effects are common in plant genomes for quick adaptive response to pathogens via recombination events (Hulbert et al. 2001; Meyers et al. 2003). Clustering of R-genes may potentially enable recognition of different pathogen races carrying different avirulence genes, such as in Clubroot R-genes on chromosome A8 (Pang et al. 2014). Epistatic interaction was also observed for Clubroot in *B. napus* (Manzanares-Dauleux et al. 2000), where a single dominant resistance gene does not explain the resistance outcome in its entirety. It was also reported that the same genomic region in *B. napus* can act as both QTL and/or major gene based on different pathotypes of Clubroot (Manzanares-Dauleux et al. 2000; Rocherieux et al. 2004). The clustering and duplication of R-genes can also cause problems in R-gene identification as these genes tend to collapse in genome sequence assemblies. Reliance on the single reference genome sequence can therefore hamper efforts, and future work should incorporate novel technologies such as isolated chromosome sequencing or targeted amplification of genomic regions of interest (Neik et al. 2017).

1.8 Molecular Basis of R-genes Deletion in *Brassica*

The outcome of many plant–pathogen interactions is determined by disease resistance (R)-genes that enable plants to recognize invading pathogens and activate inducible defenses. A typical R-gene allele encodes "race-specific" resistance to only one or a few strains of a single pathogen species. R-gene loci are functionally polymorphic within a plant species and encode alternate alleles that either recognize different strains of the same pathogen or do not recognize any tested pathogen. The molecular studies have revealed that R-genes often reside in complex loci consisting of the R-gene and tightly linked homologs. These complexes can exist in both disease-resistant and disease-susceptible plant genotypes. An R-gene specifies resistance only if the pathogen expresses a corresponding avirulence (*Avr*) gene. If either component is nonfunctional, then the plant is unable to activate resistance responses. This model has successfully described a wide variety of plant–pathogen associations, but little is known about the molecular basis and origin of R-gene polymorphism (Grant et al. 1998).

The most common molecular interpretation of the gene-for-gene hypothesis is that R-genes encode specialized receptors that recognize the direct or indirect products (elicitors) of the corresponding *Avr* genes. One implication of the receptor–elicitor model is that functional polymorphism at *R* loci could arise from gain-of-function mutations that enable the plant to recognize novel pathogen variants. Maintenance of multiple alleles or multiple linked genes with different recognition capabilities at an *R* locus would enable a host population to defend itself against the corresponding pathogens. R-gene polymorphism could also, in principle, arise from loss-of-function mutations in R-genes. Nonfunctional R-gene alleles could be maintained if there is a cost of resistance in the absence of pathogen selection. Nonfunctional alleles could also serve as a repository of divergent sequences that could be contributed to related genes by recombination, thereby, accelerating the evolution of novel R-genes.

The cloning of R-genes against diverse pathogens provides the tools for comparative analysis of R-gene alleles within and between species, which will provide insight into the evolutionary history of R-genes. Grant et al. (1998) described the structure of functional and nonfunctional alleles of the *RPM1* (resistance to *Pseudomonas syringae*, pathovar *maculicola*) bacterial resistance gene in Arabidopsis thaliana and a related crop species, *Brassica napus. RPM1* was initially identified in *A. thaliana* accession Col-0 through its ability to confer resistance to *P. syringae* isolates expressing either *avrRpm1* or *avrB*. The *avrB* and *avrRpm1* avirulence genes are sequence unrelated, thus *RPM1* enables dual-specificity resistance. *RPM1* was isolated by map-based cloning and was shown to encode a protein with a putative amino-terminal leucine zipper, a consensus nucleotide binding site (NBS), and 14 C-terminal leucine-rich repeats (LRRs). The RPM1 protein is thus a member of the largest class of R-proteins functionally characterized to date, the so-called NBS-LRR class. Previous analyses suggested that *RPM1* functional polymorphism in *Arabidopsis* arose from an intraspecific insertion or deletion of the *RPM1* gene. Grant et al. (1998) provided molecular evidence that the evolution of *RPM1* predated the divergence of the Brassicaceae and that independent deletions of *RPM1* have occurred in both *Arabidopsis* and *Brassica* (Grant et al. 1998).

Plant disease resistance (R)-genes confer race-specific resistance to pathogens and are genetically defined on the basis of intraspecific functional polymorphism. Little is known about the evolutionary mechanisms that generate this polymorphism. Most *R* loci examined to date contain alternate alleles and/or linked homologs even in disease-susceptible plant genotypes. In contrast, the resistance to *Pseudomonas syringae* pathovar *maculicola* (*RPM1*) bacterial resistance gene is completely absent (*rpm1-null*) in 5/5 *Arabidopsis thaliana* accessions that lack *RPM1* function. The *rpm1-null* locus contains a 98-bp segment of unknown origin in place of the *RPM1* gene. Grant et al. (1998) undertook comparative mapping of *RPM1* and flanking genes in *Brassica napus* to determine the ancestral state of the *RPM1* locus. They cloned two *B. napus RPM1* homologs encoding hypothetical proteins with 81% amino acid identity to *Arabidopsis RPM1*. Co-linearity of genes flanking *RPM1* is conserved between *B. napus* and *Arabidopsis*. Surprisingly, four additional *B. napus* loci were found in which the flanking marker synteny is maintained but *RPM1* is absent. These *B. napus rpm1-null* loci have no detectable nucleotide similarity to the *Arabidopsis rpm1-null* allele. It was concluded that *RPM1* evolved before the divergence of the Brassicaceae and has been deleted independently in the *Brassica* and *Arabidopsis* lineages. These results suggest that functional polymorphism at (R)-gene loci can arise from gene deletions (Grant et al. 1998).

1.9 Molecular Basis of Pathogen Recognition and Induction of R-Ggenes

Plants have evolved various ways to fend off pathogens. A suite of surface-exposed detectors, named pattern recognition receptors (PRRs, Fig. 1.2), can recognize conserved microbe-associated molecular patterns (MAMPs), such as bacterial flagellin or fungal chitin. Upon detection the PRRs activate a signaling cascade that leads to pattern-triggered immunity (PTI). Most PRRs require a co-receptor to initiate signaling. The receptor FLS2 interacts with the BAK1 co-receptor upon recognition of the MAMP flg22. BAK1 also interacts with other PRRs. PTI signaling via mitogen-activated protein kinase (MAPK) cascades and/or calcium-dependent protein kinases (CDPKs) (Bredow and Monaghan 2019) activates pathogen-nonspecific immune responses such as the production of reactive oxygen species and nitric oxide, cell wall reinforcement, and induction of defense genes (Hein et al. 2009; Nurnberger and Kemmerling 2009). Adapted pathogens are able to suppress PTI with effectors that have evolved to interact directly with host defense-associated proteins, resulting in effector-triggered susceptibility. Bacterial effectors can interfere directly with FLS2 or BAK1 (Toruno et al. 2016). Intracellular nucleotide-binding, leucine-rich repeat (NLR, Fig. 1.2) receptors that recognize these effectors or their activity have evolved in plants as a second layer of pathogen perception. NLR proteins are encoded by plant disease resistance genes and mediate so-called effector-triggered immunity (ETI). This is the molecular basis of the

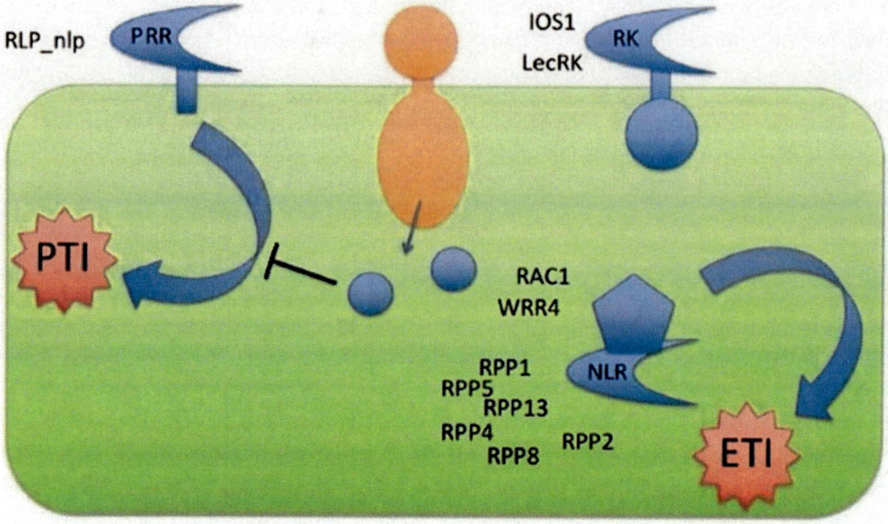

Fig. 1.2 Schematic of PTI and ETI against oomycetes (Herlihy et al. 2019)

genetic model of "gene-for-gene" resistance, in which pathogen "avirulence (Avr) genes" (now known to encode secreted effector proteins) are recognized inside plant cells by plant "resistance (R)-genes" (now known to encode NLR proteins). Resistance to adapted pathogen species is often rapidly broken by loss or mutation of avirulence effector genes or through suppression of ETI by different effectors (Woods-Tor et al. 2018).

The majority of plant PRRs identified so far belong to the family of leucine-rich repeat receptor-like kinases (LRR-RLKs), which has more than 600 members in *Arabidopsis* (Shiu et al. 2004), or receptor-like proteins (LRR-RLPs) that lack a C-terminal kinase domain and interact with RLKs for transducing signals (Gust and Felix 2014). The best-described plant PRR is FLS2 from *Arabidopsis* that recognizes bacterial flagellin, a major structural protein of bacterial flagella. An N-terminal 22-amino acid residue fragment of this protein (flg22) is sufficient to be recognized by FLS2 and to induce PTI (Gomez-Gomez and Boller 2000). Naito et al. (2008) showed that these residues are also essential for flagellum function and motility of bacteria, which makes it difficult for bacteria to circumvent this recognition by evolving different amino acid sequences in this region of the protein. A second well-described plant PRR that seems restricted to Brassicaceae is EFR, which recognizes the bacterial elongation factor Tu (EF-Tu) or its shorter N-terminal fragment elf18 (Kunze et al. 2004). When this PRR is expressed in the solanaceous plants *Nicotiana benthamiana* and tomato (*Solanum lycopersicum*), it confers resistance to a broad spectrum of bacteria (Lacombe et al. 2010), illustrating the potential use of PRRs as new sources for genetic resistance to crop diseases. Fungi can also be recognized by PRRs, for example, through the *Arabidopsis* RLK CERK1 that is required for responses to fungal chitin and bacterial peptidoglycan (Miya et al. 2007;

Willmann et al. 2011), and the RLP30 PRR that mediates recognition of necrotrophic fungi through a currently unknown proteinaceous MAMP (Zhang et al. 2013). However, only a few receptors have been identified so far that mediate the extracellular recognition of oomycete MAMPs (Herlihy et al. 2019).

1.10 Differential Expression of Genes in *Brassica*

Biotic stress is a very devastating stress for the development of cruciferous crops. Several number of genes have been well documented that they are differentially expressed against different biotic stresses. However, when infection with *F. oxysporum* f. sp. *conglutinans*, six *BrWRKY* genes (*BrWRKY4*, *65*, *72*, *97*, *133*, and *141*) showed significantly higher expression and were about eight-, six-, six-, three-, and fivefold higher, respectively (Kayum et al. 2015a). *BrWRKY141* showed about 180-fold higher expression after *P. carotovorum* subsp. *carotovorum* infection (Kayum et al. 2015a). After infection with *F. oxysporum* f. sp. *conglutinans*, *BrAL2*, *3*, *4*, *7*, *9*, *10*, *13*, *14*, and *15* showed severalfold higher expression, and *BrAL2*, *3*, *7*, *9*, *13*, *14*, and *15* were expressed in response to both biotic and abiotic stresses (Kayum et al. 2015b). Ahmed et al. (2012a) reported three chitinase genes' (*BrCLP1*, *BrCLP2*, and *BrCLP3*) responses against *P. carotovorum* subsp. *carotovorum* infection. The defensin-like family protein (BrDLFP) expression significantly increased after infection with *P. carotovorum* subsp. *carotovorum* in Chinese cabbage (Ahmed et al. 2012b). On the other hand, Alfin-like two genes (BoAL8 and BoAL12) were induced after infection with *P. carotovorum* subsp. *carotovorum* in *B. oleracea* (Kayum et al. 2016). Among the 12 thaumatin-like proteins, three expressed differentially after *P. carotovorum* subsp. *carotovorum* infection in cabbage plants (Ahmed et al. 2013).

1.11 Techniques and Approaches Used for Molecular Mechanisms

Technologies that measure some characteristic of a large family of cellular molecules, such as genes, proteins, or small metabolites, have been named by appending the suffix "-omics," as in "genomics." Omics refers to the collective technologies used to explore the roles, relationships, and actions of the various types of molecules that make up the cells of an organism. These technologies include genomics, "the study of genes and their function" (Human Genome Project (HGP) 2003); proteomics, the study of proteins; metabolomics, the study of molecules involved in cellular metabolism; transcriptomics, the study of the mRNA; glycomics, the study of cellular carbohydrates; and lipomics, the study of cellular lipids.

The development of genomic resources like whole genome sequences, genomic survey sequences (GSS), large number of expressed sequence tags (ESTs), larger insert genomic libraries, high-density genomic maps, millions of molecular markers, and high-throughput genome sequences (HTGs) is essential to study the complex defense mechanisms, metabolic pathways, and proteins/genes involved in plant pathogen resistance in *Brassica* crops. The genomic resources have been used for sequencing and a notation, genomic selection, mapping and cloning of genes or QTLs, association mapping, breeding by design, and MAS in crucifer's crops. The advent of next-generation sequencing (NGS) technologies like FLX-454, Illumina, SOLID/ion torrent, pacbio, HIC, bionano oxford nanopore, and helicase has enhanced the possibilities to generate genomic resources and their utilization in *Brassica* improvement programs. The molecular techniques and approaches used for comprehension of molecular mechanisms of crucifers' host resistance are briefly described as follows.

1. Omics: There are "The Technologies" which provide the tools needed to look at the differences in DNA, RNA, proteins, and other cellular molecules between species and among individuals of a species. These types of molecular profiles can vary with cell or tissue exposure to chemicals or drugs and thus have potential use in toxicological assessments. Omics experiments can often be conducted in high-throughput assays that produce tremendous amounts of data on the functional and/or structural alterations within the cell. "These new methods have already facilitated significant advances in understanding the molecular responses to cell and tissue damage, and perturbations in functional cellular systems" (Aardema and Mac Gregor 2003). The -omics technologies will continue to contribute to our understanding of toxicity mechanisms. Regulators are interested in these new technologies but are still sorting out how to incorporate the new information and technologies in regulatory decision-making. For example, the US Food and Drug Administration's Pharmaco-genomic Data submissions guidance document encourages the voluntary submission of genomics data but notes that the field of pharmacogenomics is still in its early developmental stages.
2. Bioinformatics: Bioinformatics is "the science of managing and analyzing biological data using advanced computing techniques" (HGP 2003). Bioinformatics tools include computational tools that mine information from large databases of biological data. These tools are most commonly used to analyze large sets of genomics data. However, bioinformatics tools are also being developed for other types of biological data, such as proteomics. The US National Centre for Biotechnology Information (NCBI) serves as an integrated source of genomics information and bioinformatics tools for researchers. An important bioinformatics tool available at NCBI for proteomics and genomics is the Basic Local Alignment Search Tool (BLAST), which compares gene or protein sequences against databases that contain many archived sequences, in order to find regions of local similarity. The statistical significance of the sequence matches is then calculated, and the results can be used to infer functional and evolutionary relationships.
3. Genomics: The first of the -omics technologies to be developed, genomics has resulted in massive amounts of DNA sequence data requiring great amounts of

computer capacity. Genomics has progressed beyond sequencing of organisms (structural genomics) to identifying the function of the encoded genes (functional genomics). The genome of each species is distinctive, but smaller genomic differences are also observed between each individual of a species. It was originally thought that obtaining the sequence of the human genome would immediately tell us the identity of the human genes. The genome has proved to be much more complex.

4. **Epigenetics:** Epigenetics refers to mechanisms that persistently alter gene expression without actual changes to the gene/DNA sequence. DNA methylation is an example of an epigenetic mechanism. Scientists have shown that DNA methylation is an important component in a variety of chemical-induced toxicities, including carcinogenicity, and is a mechanism that should be assessed in the overall hazard assessment (Watson and Goodman 2002; Mogg et al. 2004).
5. Transcriptomics: Transcriptomics is the study of the transcriptome—the complete set of RNA transcripts that are produced by the genome, under specific circumstances or in a specific cell—using high-throughput methods, such as microarray analysis. Comparison of transcriptomes allows the identification of genes that are differentially expressed in distinct cell populations or in response to different treatments. The term "transcriptome" is now widely understood to mean the complete set of all the ribonucleic acid (RNA) molecules expressed in some given entity, such as a cell, tissue, or organism.
6. Proteomics: Proteins are the primary structural and functional molecules in the cell and are made up of a linear arrangement of amino acids. The linear polypeptide chains are folded into secondary and tertiary structures to form the functional protein. Unlike the static nature of the cell's genes, proteins are constantly changing to meet the needs of the cell. Characterizing the identity, function, regulation, and interaction of all of the cellular proteins of an organism, the proteome, will be a major achievement. Studies of changes in the proteome of cells and tissues exposed to toxic materials, compared to normal cells, are being used to develop an understanding of the mechanisms of toxicity. As proteomics tools become more powerful and widely used, protein and proteome changes in response to exposures to toxic substances (fingerprints or response profiles) will be developed into databases that can be used to classify exposure responses at various levels of organization of the organism, thus providing a predictive in silico toxicology tool.
7. Metabolomics: Metabolomics refers to the comprehensive evaluation of the metabolic state of a cell, organ or organism, in order to identify biochemical changes that are characteristic of specific disease states or toxic insults. Typical metabolomics experiments involve the identification and quantification of large numbers of endogenous molecules in a biological sample (e.g., urine or blood) using chemical techniques such as chromatography and mass spectrometry. The output from these techniques is compared to computerized libraries of mass spectrometry tracings to facilitate identification of the compounds that are present. Environmental stresses such as exposure to chemicals or drugs alter

the metabolic pathways in cells, and metabolite profiling can be used to assess toxic responses/exposures. The name metabolomics was coined in the late 1990s using the word metabolome (Oliver et al. 1998) for systematic functional analysis of the yeast genome. Many of the bio-analytical methods used for metabolomics have been adapted (or in some cases simply adopted) from existing biochemical techniques. Metabolomics is the identification and quantification of all metabolites in a given biological situation.

8. Biometabolomics: Biometabolomics is a term used to describe the destiny of metabolites in biometabolotic reactions (biochemical processes involving metabolites) and measuring of many small molecules in biological systems. These systems can be fluids within our bodies such as urine, blood, saliva, as well as different tissues. Biometabolomics is a discipline of analytic biochemistry in living things. In Biometabolomics, nuclear magnetic resonance (NMR) or mass spectrometery (MS) is mostly used. In the field of obesity and metabolic disease, recent studies employing targeted metabolomics have identified a number of novel metabolites and pathways that may be involved in disease pathogenesis. In biometabolomics, the metabolome represents the collection of all metabolites in a biological cell, tissue, organ, or organism, which are the end products of cellular processes. Metabolites are the intermediates and products of metabolism. The metabolome forms a large network of metabolic reactions, where outputs from one enzymatic chemical reaction are inputs to other chemical reactions.
9. Toxicogenomics: Toxicogenomics compares the genes expressed in organisms that have been exposed to a drug, chemical, or toxin to those of unexposed organisms (negative controls). The up- or downregulation of certain genes or groups of genes may be linked to toxic responses occurring in the organism, and to particular organs or cell types in that organism. The goal of toxicogenomics is to identify patterns of gene expression related to specific chemicals or chemical classes so that these expression patterns can be used as endpoints for assessing toxicity. Thus far, toxicogenomics has been useful in refining animal experiments and identifying mechanisms of toxicity in lab animals where exposures can be controlled. There have also been experiments evaluating gene expression in cell cultures exposed to toxicants, which has been used in limited applications for prediction of in vivo toxicity.
10. Pharmacogenetics: Pharmacogenetics looks at the differences in response to a particular drug that are due to variations in the genetic makeup of individuals. For example, human genetic variation has been implicated in the variability of responses (effectiveness and/or toxicity) seen with some chemotherapeutic drugs (Crews 2006; Hahn et al. 2006).
11. Metagenomics: Metagenomics is the study of a collection of genetic material (genomes) from a mixed community of organisms. Metagenomics usually refers to the study of microbial communities. "Metagenomics" is the two words "meta" and "genomics." So genomics is obtaining the DNA sequence, but meta implies that doing it for many organisms together, and metagenomics is usually used when studying microbial communities where it can't separate one

microbe from another. Like two bacteria grow together, and when take the DNA sequence, that will be together of both bacteria.

12. Pathogenomics**:** Pathogenomics is a field which uses high-throughput screening technology and bioinformatics to study encoded microbe resistance, as well as virulence factors (VFs), which enable a microorganism to infect a host and possibly cause disease. Pathogenomics is a discipline that seeks to mark out virulence factors and their contributions to overall pathogenesis by comparing gene repertoires of pathogenic and nonpathogenic strains/species.
13. **Phylogenomics:** Phylogenomics is the intersection of the fields of evolution and genomics. The term has been used in multiple ways to refer to analysis that involves genome data and evolutionary reconstructions. It is a group of techniques within the larger fields of phylogenetics and genomics. As more complete genomes are sequenced, phylogenetic analysis is entering a new era—that of phylogenomics. One branch of this expanding field aims to reconstruct the evolutionary history of organisms on the basis of the analysis of their genomes. The word "phylogenomics" was first introduced in the context of prediction of gene function for genome-scale data (Eisen 1998), and soon after in the context of phylogenetic inference (O'Brien and Stanyon 1999). The discipline of phylogenomics owes its existence to the advances made in DNA sequencing technology over the past two decades (Metzker 2010). It comprises several areas of research at the interface between molecular and evolutionary biology and has two major goals: (i) to infer phylogenetic relationships between taxa and gain insights into the mechanisms of molecular evolution and (ii) to use multispecies phylogenetic comparisons to infer putative functions for DNA or protein sequences.
14. Pangenomics**:** The pangenome refers to a collection of genomic sequence found in the entire species or population rather than in a single individual; the sequence can be core, present in all individuals, or accessory (variable or dispensable), found in a subset of individuals only. The study of the pangenome is called "pangenomics." The concept of the pangenome was introduced by Tettelin et al. (2005) who described the production of the first ever pangenome, for a bacterial species *Streptococcus agalactiae*. It derives "pan" from the Greek word *παν*, meaning "whole" or "everything," while genome is a commonly used term to describe an organism's complete genetic material. The pangenome can be open or closed (restricted). In the first instance, there appears to be no finite number of genes in the species, and with each new added individual, new genes are incorporated into the pangenome. In the second instance, the gene pool is limited. After a certain number of individuals have been analyzed, the addition of new individuals to the analysis will not result in an expansion of the pangenome.

1.12 Classification of Crucifers' Pathogens on the Basis of Molecular Mechanisms

Cruciferous crops are challenged by 44 pathogens of fungal, bacterial, viral, and nematode origin under natural as well as cultivation environment and cause considerable yield and quality losses depending on disease management practices employed. Sixteen pathogens have been used for dissecting molecular mechanisms of *Brassica* host-pathosystem. These pathogens can be grouped into three different categories, namely, biotrophs, hemibiotrophs, and necrotrophs on the basis of their mode of infection, acquisition of nutrition, and molecular mechanisms of host resistance (Table 1.3). NBS-LRR is the largest class of known plant R-genes, and TIR-NBS-LRR is one of the subclasses of this family. The type of resistance imparted by NBS-LRR is effective against obligate biotrophs or hemibiotrophs but is not effective against necrotrophs.

Fungal plant pathologists or mycologists have attempted to classify pathogens into groups called biotrophs, hemibiotrophs, and necrotrophs. Although these terms are well known and frequently used, disagreements about which pathogens fall into which classes, as well as the precise definition of these terms, have limited their usefulness. Dogmatic properties of these classes have been progressively eroded. However, the genetic analysis of disease resistance, particularly in the model plant *Arabidopsis thaliana*, has provided a biologically meaningful division based on whether defense against fungal pathogens is controlled via the salicylate or jasmonate/ethylene pathways. This mode-of-defense division distinguishes necrotrophs and biotrophs, but it limits the biotroph class to pathogens that possess haustoria (Oliver and Ipcho 2004). The small number and limited range of pathogens that infect *Arabidopsis* means that several interesting questions are still unanswered. Do hemibiotrophs represent a distinct class or a subclass of the necrotrophs? Does this classification help us understand the intricacies of either fungal pathogenicity or plant defense system? However, these three classes of pathogens have been involved in comprehension of molecular mechanisms of host–pathogen interaction and expression of host resistance.

Major characterizations of three groups of pathogens infecting crucifers are as follows:

1. **Biotrophs:** are obligate, possess haustoria, secrete limited amounts of lytic enzymes, cause little damage to the host plant, have a narrow host range, induce hypersensitive cell death in incompatible interactions, are controlled by specific (gene-for-gene) resistance genes, and are controlled by salicylate-dependent defense pathways.
2. Necrotrophs: are non-obligate, have wide host ranges, secrete copious cell-wall-degrading enzymes, produce toxins, are controlled by quantitative resistance genes, and are controlled by jasmonate- and ethylene-dependent defense pathways.
3. Hemibiotrophs: Hemibiotrophs are organisms that are parasitic in living tissue for some time and then continue to live in dead tissue. Biotrophic and hemibiotrophic fungi are successful groups of plant pathogens that require living plant tissue to

Table 1.3 Crucifers' pathogens used for molecular studies

Sr. no.	Pathogen	Disease	GD in number of countries	Host range number	Average/(range) yield loss (%)	Nature of pathogen	Publications on molecular studies
1.	*Albugo candida*	White rust	100	300	34	Biotrophs	122
2.	*Alternaria brassicae*	Alternaria blight	96	4000	45.1	Necrotrophs	183
3.	Colletotrichum *higginsianum*	Anthracnose	100	100	50	Hemibiotrophs	9
4.	*Erysiphe cruciferarum*	Powdery mildew	44	100	27.3	Biotrophs	134
5.	*Fusarium oxysporum*	Wilt	100	100	15.9	Necrotrophs	20
6.	Hyaloperonospora parasitica	Downy mildew	102	99	37.3	Biotrophs	15
7.	*Leptosphaeria maculans*	Blackleg	102	35	42	Hemibiotrophs	142
8.	*Plasmodiophora brassicae*	Clubroot	144	3700	28.5	Biotrophs	273
9.	*Pseudocercosporella capsellae*	White leaf spot	29	22	15	Necrotrophs	4
10.	*Pyrenopeziza brassicae*	Light leaf spot	100	33	22	Hemibiotrophs	9
11.	*Rhizoctonia solani*	Seedling rot	70	200	30	Necrotrophs	5
12.	*Sclerotinia sclerotiorum*	Stem rot	102	600	40	Hemibiotrophs	166
13.	Turnip mosaic virus	TuMV	106	8	–	Biotrophs	77
14.	*Verticillium longisporum*	Wilt	10	9	–	Necrotrophs	27
15.	*Xanthomonas campestris pv. campestris*	Black rot	157	10	–	Hemibiotrophs	73
16.	*Heterodera*	Nematode	91	–	–	Biotrophs	1

– data not available, *GD* geographical distribution

survive and complete their life cycle. Biotrophs have total dependency upon living plant cells, whereas hemibiotrophs have an initial biotrophic lifestyle and a subsequent necrotrophic phase. The necrotrophic lifestyle involves actively killing host plant cells by secreting cell-wall-degrading enzymes and phytotoxins.

Hemibiotrophic fungi represent the most interesting group of pathogens since they use sequential biotrophic and necrotrophic infection strategies to invade and colonize host plants. Transition from the asymptomatic biotrophic phase, characterized by intercellular thick primary hyphae, to the destructive necrotrophic phase, characterized by thin filamentous secondary hyphae, is referred to as the biotrophy-necrotrophy switch (BNS). The BNS and modulation of plant defense in response to BNS is not well understood. One important strategy for studying these dynamic but highly regulated host defenses is to observe gene expression patterns in the host.

The parasite must penetrate preformed and induced physical barriers, such as the waxy cuticle on the leaf surface and the plant cell wall, and withstand preformed and induced antimicrobial chemicals. It must induce release of plant nutrients and acquire them for its own use. Throughout infection, the pathogen must overcome or avoid triggering the host immune system. Biotrophic fungi establish a close association with the host through the development of specialized infection hyphae or haustoria within living plant cells from which nutrients are taken up.

On the other hand, necrotrophic fungi secrete toxins and enzymes that kill host cells and then take up nutrients released from the dead tissue. This latter strategy may limit the capacity of the host plant to mount a defense response including production of antifungal molecules. Hemibiotrophic fungi combine both strategies. An initial biotrophic phase, during which the host's immune system and cell death is actively suppressed, allows invasive hyphae to spread throughout the infected plant tissue. This is followed by a necrotrophic phase during which toxins are secreted by the pathogen to induce host cell death. Until recently, the mechanisms biotrophic and hemibiotrophic fungi use to deal with the host immune system and to manipulate the living host cells were a mystery. However, it has become apparent that, like bacterial pathogens of plants and animals (Hann et al. 2010; Staskawicz et al. 2001), plant pathogenic fungi produce and secrete many so-called effector proteins that interact with the host and play an important role in virulence (Kamoun 2009; Dodds et al. 2009).

In order to facilitate infection, plant pathogens secrete numerous effector proteins into the plant apoplast or cytosol. To understand the plant pathogen effectors, the studies of prokaryotic pathogens, such as the bacterium *Pseudomonas syringae* which uses a type three secretion system (T3SS) to deliver about 30 effectors directly into the plant cytosol, are required (Chang et al. 2005; Schechter et al. 2006). Many of these proteins target PRRs or their signaling partners, thereby interfering with defense signal transduction (Hann et al. 2010; Xiang et al. 2011; Axtell and Staskawicz 2003; Gohre and Robatzek 2008). One of the best characterized is *P. syringae* AvrPto. This small secreted protein inhibits PTI directly by interfering with flg22-induced FLS2 signaling inside the plant cell, allowing the bacteria to grow on (Xiang et al. 2008). Similar to a bacterial T3SS, plant pathogenic nematodes secrete effectors through their stylet, either into the apoplast or via feeding tubes directly into the cytosol of plant cells (Vieira et al. 2011; Vanholme et al. 2004). It is now clear that fungal pathogens also secrete effectors into the plant apoplast or deliver them into host cells where they may act to suppress defense responses or alter host metabolism (Kamoun 2009; Panstruga and Dodds 2009; Ellis et al. 2009). To counteract effector molecules, plants have developed an additional layer of immune recognition based on intracellular NB-LRR (nucleotide-binding–leucine-rich repeats) receptor proteins that can detect individual effectors either directly or indirectly. These receptors are often referred to as resistance (R) proteins and the effectors they recognize as avirulence (Avr) proteins (Dodds and Rathjen 2010; Jones and Dangl 2006). Generally, the plant ETI response to R-protein-mediated recognition is more severe than PTI and frequently results in localized plant cell death, also known as the hypersensitive response (HR). This response is particularly effective against biotrophic pathogens which depend on living host cells for nutrition (Koeck et al. 2011).

1.12.1 Molecular Events During Host–Pathogen Interaction

Biotrophic fungi establish a close association with the host through the development of specialized infection hyphae or haustoria within living plant cells from which nutrients are taken up. On the other hand, necrotrophic fungi secrete toxins and enzymes that kill host cells and then take up nutrients released from the dead tissue. This latter strategy may limit the capacity of the host plant to mount a defense response including production of antifungal molecules. Hemibiotrophic fungi combine both strategies. An initial biotrophic phase, during which the host's immune system and cell death is actively suppressed, allows invasive hyphae to spread throughout the infected plant tissue. This is followed by a necrotrophic phase during which toxins are secreted by the pathogen to induce host cell death. Until recently, the mechanisms biotrophic and hemibiotrophic pathogens use to deal with the host immune system and to manipulate the living host cells were a mystery. However, now it has become apparent that, like bacterial pathogens of plants and animals, plant pathogenic fungi produce and secrete many so-called effector proteins that interact with the host and play an important role in virulence. Plants are able to recognize the presence of pathogens on different levels during infection. Initially, recognition of conserved pathogen-associated molecular patterns (PAMPs) leads to PAMP-triggered immunity (PTI). The second layer of defense is activated when the host recognizes specific pathogen effector proteins, which are produced to suppress PTI and facilitate infection, and is commonly referred to as effector triggered immunity (ETI).

PAMPs recognized by the plant immune system are generally molecules essential to the pathogen and cannot be modified without significant loss of viability. They include the abundant bacterial proteins, elongation factor Tu (EF-Tu) and flagellin, and the fungal cell wall polysaccharide, chitin (Zipfel 2009). During infection,

PAMPs present in the extracellular space (known as the apoplast) are recognized by transmembrane pattern recognition receptors (PRRs) present on the plant plasma membrane, a process that induces PTI (Zipfel 2009). For instance, the Arabidopsis thaliana flagellin receptor, FLS2, recognizes the highly conserved N-terminus of bacterial flagellin (flg22) and activates a MAP kinase pathway to induce expression of defense response genes (Gomez-Gomez and Boller 2002). In addition, antimicrobial compounds and reactive oxygen species are produced and callose is deposited at the site of infection to strengthen the cell wall. Similarly the *Arabidopsis* CERK1 (Petutschnig et al. 2010) and rice CeBip receptors (Kaku et al. 2006) recognize chitin during fungal infection.

1. **Induction of plant pathogen effectors genes:** In order to facilitate infection, plant pathogens secrete numerous effector proteins/genes into the plant apoplast or cytosol. Most of our understanding of plant pathogen effectors comes from studies of prokaryotic pathogens, such as the bacterium *Pseudomonas syringae* which uses a type three secretion system (T3SS) to deliver about 30 effectors directly into the plant cytosol. Many of these proteins target PRRs or their signaling partners, thereby interfering with defense signal transduction (Hann et al. 2010; Xiang et al. 2011). One of the best characterized is *P. syringae* AvrPto. This small secreted protein inhibits PTI directly by interfering with flg22-induced FLS2 signaling inside the plant cell, allowing the bacteria to grow on *A. thaliana* (Xiang et al. 2008). Similar to a bacterial T3SS, plant pathogenic nematodes secrete effectors through their stylet, either into the apoplast or via feeding tubes directly into the cytosol of plant cells (Vieira et al. 2011). It is now clear that fungal pathogens also secrete effectors into the plant apoplast or deliver them into host cells, where they may act to suppress defense responses or alter host metabolism (Kamoun 2009; Panstruga and Dodds 2009; Ellis et al. 2009).
2. **Mechanisms of effectors recognition by plant immune receptors system**: To counteract effector molecules, plants have developed an additional layer of immune recognition based on intracellular NB-LRR (nucleotide-binding–leucine-rich repeats) receptor proteins that can detect individual effectors either directly or indirectly. These receptors are often referred to as resistance (R)-proteins and the effectors they recognize as avirulence (Avr) proteins (Dodds and Rathjen 2010). Generally, the plant ETI response to R-protein-mediated recognition is more severe than PTI and frequently results in localized plant cell death, also known as the hypersensitive response (HR). This response is particularly effective against biotrophic pathogens which depend on living host cells for nutrition.
3. Functions **of effectors genes in biotrophic and hemibiotrophic fungi:** The first fungal effectors were identified in attempts to clone Avr proteins *recognized* by host R-proteins, and it became apparent that many were *recognized* within the host cell cytoplasm. This implies delivery of the effectors into the plant cell during infection, and subsequent studies have directly *visualized* fungal effectors inside host cells (Khang et al. 2010; Rafiqi et al. 2010). How do they get there? No system analogous to the bacterial T3SS has been identified in fungi. The

fungal effectors contain canonical secretion signals and appear to be secreted through the standard endo-membrane pathway. Some insight into the process of movement into the plant cell has come from the finding that oomycete pathogens, which have similar infection strategies to fungi but belong to the Stramenopile kingdom, secrete effector proteins with an N-terminal RxLR (Arg-x-Leu-Arg) host-targeting signal (HTS) (Morgan and Kamoun 2007). *Internalization* appears to be pathogen-independent (Dou et al. 2008). Kale et al. (2010) suggested that the RxLR motif of several oomycete effectors binds to phosphatidyl inositol phosphates (PIPs) on the outer surface of the plant plasma membrane to mediate endocytosis. In contrast, Yaeno et al. (2011) found that PIP binding by these effectors was mediated by the C-terminal domain and not the RxLR region, arguing against a role in effector delivery. Similar pathogen-independent uptake occurs with effectors from the biotrophic flax rust fungus, *Melampsora lini* (Rafiqi et al. 2010), but does not correlate with PIP binding (Gan et al. 2010a). However, unlike oomycete effectors, fungal effector proteins do not share a conserved HTS, and it is not yet known how they are targeted to and enter the plant cell.

In contrast to the relatively small set of effectors produced by bacteria, recent advances through next-generation sequencing and large-scale proteome analysis have now identified hundreds of proteins that are secreted by biotrophic and hemibiotrophic fungi (Dodds 2010; Duplessis et al. 2011). Typically, these secreted proteins have low sequence homology to any known protein, and there is little understanding of their function. Quite often all that is known is that they are recognized by cognate host R-proteins, that is, that they are Avr proteins (Stergiopoulos de Wit 2009; Gan et al. 2010b). Clarification of their role as effectors is hampered by apparent functional redundancy which often masks any obvious phenotype when individual genes are deleted or silenced (Lawrence et al. 2010; Mosquera et al. 2009). Nevertheless, recent research is beginning to reveal the function of increasing numbers of fungal effectors. With next-generation sequencing, we now have the tools to readily obtain whole genome information from most biotrophic and hemibiotrophic fungi, even though many are difficult to study *in vitro*. In many cases, this analysis has revealed that fungal genomes contain a large number of small secreted putative effectors that are rapidly evolving compared to the rest of the genome. This amplification and rapid evolution may be due to the strong selection pressure exerted by the plant's immune recognition system and/or to host species-specific adaptation to different pathogenicity targets. To what extent have effectors in different fungal pathogens evolved to target the same proteins in different host plants? Many of the fungal pathogens that are currently the focus of scientific investigation are pathogens of crop plants, and much of their evolution may have been influenced by domestication of the crop and creation of new pathogenic niches. Nevertheless, the rapid development of genomic tools and advanced microscopic techniques for studies of protein localizations and interactions is providing unprecedented avenues for exploring the roles of pathogen effectors in the interplay between host and their pathogens to reveal molecular mechanisms for better management of crucifers' diseases with durable resistance cultivars.

1.13 Application of Molecular Markers in Molecular Mechanisms of Disease Resistance

1.13.1 Genetic Linkage Map Construction and QTL Identification

One of the most important applications of genetic markers has been the construction of genetic linkage maps. These maps are created by genotyping a large mapping population of segregating individuals and studying the resulting recombination frequencies between genetic markers. This enables establishment of linkage groups of associated markers with an approximate relative position along a chromosome based on their likelihood of being co-inherited. A linkage group will inherently often represent a large proportion of an individual chromosome with imputed recombination points. The abundance of SNPs and their ability to be discovered and genotyped rapidly in a high-throughput manner make them particularly valuable markers for genetic mapping. Importantly, when the same mapping population used to derive a linkage map is phenotyped for segregating traits of interest, such as seed color or flowering time, the association between marker patterns and the phenotypic variation can be quantified. This then enables identification of the genomic regions controlling traits of interest. Where these traits are quantitative, the associated genomic region (s) are known as quantitative trait loci (QTLs). The identification of markers closely linked to genetic loci of interest, including QTL, enables discovery of the underlying, causative gene(s). Prior to the availability of whole genome sequencing technologies, this involved map-based cloning, which used the known sequence of markers directly flanking a locus to amplify and sequence the intervening region for gene candidate identification. Depending on the resolution of the genetic map as defined by marker density and thus distance between flanking markers, this process was often extremely time and resource intensive. Nonetheless, it enabled the first identification of developmentally and agriculturally important genes in many crop and model plant species. In the crop canola, QTL of importance include those for oil yield, oil quality, disease resistance, and pod shatter tolerance, among many others (Kaur et al. 2009; Qiu et al. 2006; Smooker et al. 2011).

1.13.2 Genome Assembly, Physical Mapping, and Synteny Mapping

Genetic linkage maps are highly valuable in helping to assemble contigs of next-generation genome sequencing data into chromosomes. This is achieved by physically mapping genetic marker sequences on these contigs and comparing this to their known relative location on the genetic map. The success of this process depends on the accuracy and robustness of the genetic linkage map, as well as the quality of the original contig sequence assembly. Where markers flanking QTL are physically located on a genome sequence, this enables direct and rapid analysis of the intervening region. With the aid of the plethora of in silico sequence analysis, gene prediction, and annotation tools currently available, candidate genes underlying these loci can be rapidly identified. Polymorphisms in the candidate gene regions between individuals segregating for the trait can further narrow down the causal gene. Moreover, identification and genotyping of additional SNPs in the original mapping population enables fine-mapping, or extremely high density mapping, of the QTL. SNPs found to be causally associated with a trait variation are known as "perfect markers," and these, along with the candidate gene, can be then verified in vitro and applied to molecular-assisted breeding programs (Tollenacre et al. 2012).

In species descended from a common ancestor, the preserved order of at least two homologous genes along chromosomes is known as synteny. Synteny mapping uses the locations of conserved genetic markers on the genetic maps of different species to compare interspecies genome organization. This is useful for analyses of gene and genome evolution and in reconstructing ancestral genomes. During evolution, genome rearrangements, expansion, gene loss, and mutation occur at increasing frequency with genetic distance, reducing synteny between distantly related species. When a region of high synteny between species is identified, this suggests a high level of selection for preserving genome sequence and organization in these regions. Such shared synteny is a basic criterion for establishing the functional orthology of genomic regions in different species and can facilitate rapid identification of conserved, agriculturally important gene regions in related crop species. Furthermore, markers associated with different gene paralogues enable localization and comparison of the specific members of multigene family members. Synteny mapping studies were pioneered in grass species but have been conducted in numerous plant species (Galvao et al. 2012; Zhu et al. 2003).

1.13.3 Association Mapping and Linkage Disequilibrium

Genetic markers that are linked to traits under selection are highly valuable for identifying genetic loci that contribute to phenotypic variation based on linkage disequilibrium (LD). LD refers to the coinheritance of specific genetic markers in ancestrally related individuals at higher frequencies than expected based on recombination distances. Regions that are in high LD may be under high selection pressure for particular allelic combinations, implying a positive relationship between otherwise physically distinct alleles and quantifiable traits. LD mapping, or association mapping, refers to the analysis of statistical associations between genetic markers, usually individual SNPs or SNP haplotypes, and traits (phenotypes) in a collection of individuals. SNP haplotypes, which comprise SNP alleles always found in particular allelic combinations, are found in species with moderate or high levels of LD and may encompass genes or gene clusters. As such, a minimal set of the SNPs normally existing as haplotypes can be used to impute the remainder of the haplotype alleles. This provides the ability to fast-track screening of regions of agronomic interest in breeding programs using a minimal genotyping set. In *Arabidopsis*, identification of linkage disequilibrium based on high-density SNP maps has significantly advanced evolutionary and association genetics studies (Hayward et al. 2012; Cowling and Balázs 2010).

Association studies can either be candidate gene-based or whole genome based. In the candidate gene approach, the aim is to determine correlations between traits of interest and DNA polymorphisms (e.g., SNPs) within candidate genes thought to be involved in those traits. This approach requires prior foresight into the likely biochemistry and genetics of the trait in order to narrow down gene candidates. On the other hand, whole-genome association mapping analyzes association of densely mapped genetic markers across all chromosomes with variation in phenotype to identify potential causal or LD-associated loci. Association mapping has become popular for identifying trait–marker relationships within many species, particularly for mining new alleles in natural populations or germplasm collections, and/or where the creation of large biparental mapping populations may be less feasible. In this approach, genetic markers are screened across natural populations or a diverse collection of individuals in order to associate alleles with phenotypic traits of interest. Since allelic variation in these populations depends on historical recombination and linkage disequilibrium, association studies may produce very high map resolution in species with low levels of LD. LD-based association mapping has been applied in many crop and forage species. In *Brassica napus*, "diversity fixed foundation sets" have been created, comprising a small number of homozygous lines thought to capture a large proportion of the genetic diversity available for the species. Single nucleotide polymorphisms (SNPs) are currently one of the most popular markers for the fine mapping of heritable traits. The availability of large-scale sequencing and SNP genotyping technologies will support genome-wide association studies in important crop species by enabling screening of large sets of polymorphic markers, even in complex polyploid species (Duran et al. 2010).

1.14 Application of Omics Technologies in Molecular Mechanisms of Host Resistance

1.14.1 High-Quality Genome Assemblies

With the introduction of next-generation sequencing (NGS) technology, five out of six *Brassica* crop genomes (*B. rapa*, *B. oleracea*, *B. nigra*, *B. napus* and *B. juncea*) have now been sequenced, with some species having more than one genome from different individuals. The field of *Brassica* genomics has been "revolutionized" by the development of long-read sequencing technologies such as PacBio Single Molecule, Real-Time (PacBio) sequencing, and Oxford Nanopore Technologies (ONT), along with high-throughput physical mapping technologies such as BioNano Optical Mapping and Chromosome Conformation Capture (Hi-C). Valuable genomics resources for interrogating the molecular aspects of *Brassica*–pathogen interactions have been provided via high-quality genome assemblies of *Brassica* species using PacBio sequencing, for example, *B. rapa* cultivar "Chiifu-401-42", *B. oleracea* cultivar "C-8", *B. napus* German winter cultivar "Express 617" and improved *B. napus* "Darmor-bzh", along with the highly contiguous *B. nigra* assembly, both achieved via ONT technology (Table 1.4). These high-quality *Brassica* assemblies resolve the "difficult" genomic regions commonly found in polyploid crops, particularly highly repetitive DNA sequences related to transposable elements (TEs), copy number variation (CNV), presence–absence variation (PAV) and homoeologous exchange, many of which are associated with disease resistance genes.

Although high-quality *Brassica* genome assemblies are available, such reference genomes and other *Brassica* genomes that were previously sequenced represent only a fraction of the *Brassica* morphotypes. The *B. rapa* "Chiifu" is a heading type, while the "Z1" is a sarson type, the cauliflower *B. oleracea* "C-8" is an inflorescence type, and "TO1000" is a leafy type. That genomes have not yet been assembled for other morphotypes in *Brassica* species, such as root or stem tubers in *B. rapa* (turnip), *B. oleracea* (kohlrabi), and *B. napus* (swede), means that may be missing out on much of the extensive genetic diversity present within the various *Brassica* species, but also that a wealth of novel alleles for disease resistance are potentially overlooked if rely on a single reference genome. These issues have driven the development of pangenomes in plants (Neik et al. 2020).

1.14.2 Pangenomics

Pangenomics has been developed to overcome the limitations of relying on a single reference genome and allows more comprehensive genomic variations to be identified from the gene pools represented by many lines within a species. The pangenome of *B. napus*, which was built with eight *B. napus* lines encompassing three ecotypes using a de novo approach, showed the PAV regions were enriched with genes associated with a defense-related response. Further, in the *B. oleracea* and *B. napus* pangenomes, it was found that a large proportion of the disease resistance genes were dispensable, meaning that these genes are not present in all lines. These findings suggest that the R-genes in *Brassica* are highly variable, resulting from the strong selection pressure of arms-race evolution during host–pathogen interaction, superimposed by the frequent homoeologous exchanges between sub-genomes during the domestication process within the *Brassica* lineage. Hence, many candidate R-genes may have been missed from a single reference genome, thus hindering the speed of R-gene cloning in *Brassica* crops (Bayer et al. 2020; Song et al. 2020; Tirnaz et al. 2020).

Pangenomics has identified a large R-gene repertoire, collectively known as resistance gene analogues (RGAs), in *Brassica* species. The RGAs include nucleotide-binding site leucine-rich repeats (NLRs), mainly comprising the TIR-NBS-LRR (TNL) and CC-NBS-LRR (CNL) types, receptor-like protein kinases (RLKs), receptor-like proteins (RLPs), and wall-associated kinases. Using pangenomics, 106 RGAs have been identified within the Blackleg QTL in the

Table 1.4 Summary of the most recent *Brassica* reference genomes useful for omics studies in the *Brassica* pathosystems (Neik et al. 2020)

Reference genome	Approach	Major findings relevant to R-gene study	References
Single genome			
B. napus winter cultivar "Express 617"	PacBio, ONT, Illumina HiSeq, Optical mapping	Resolved break-point sequence at homoeologous exchange regions Cauliflower is the most recent var. to evolve within *Brassica* genus	Lee et al. (2020)
B. oleracea cultivar "C-8"	PacBio, Illumina HiSeq, transcriptomics	It contains more repetitive sequences compared to other *B. oleracea* species	Sun et al. (2019)
B. nigra accession CGN7651	ONT, Hi-C	Hotspot of ALE-type retro element in the centromeric regions showed that these retro elements play an important role in the divergence of	Perumal et al. (2020)
B. rapa cultivar "Chiifu-401-42"	PacBio, Optical mapping, Hi-C	*B. nigra* centromere V3.0 improved repeat reads, defined locations of centromeres, and annotated more genes in these difficult regions. Annotated higher number of TEs	Zhang et al. (2018)
Pangenome			
Eight *B. napus* accessions of three ecotypes	Alignment of de novo assembled genomes against "ZS11"	PAV genes were highly represented by defense response gene	Song et al. (2020)
33 non-synthetic and 20 synthetic *B. napus* accessions	Iterative mapping and assembly using improved "Darmor-bzh" (v8.1) from Bayer et al. (2017) as reference	Homoeologous exchange-related PAV genes highly represented by defense, stress, and auxin pathways	Hurgobin et al. (2018)
Nine *B. oleracea* subspecies and wild type comprising cabbage, kale, Brussels sprouts, kohlrabi, cauliflower, broccoli, and *B. macrocarpa*	Iterative mapping and assembly using Chinese kale rapid cycling line (TO1000) as reference	18.7% of genes showed PAV with annotation of disease resistance genes	Golicz et al. (2016)
Two *B. rapa* subspecies: turnip and rapid cycling	Alignment of de novo assembled genomes against "Chiifu" reference	Peroxidase genes that are involved in phenylpropanoid biosynthesis response	Lin et al. (2014)

(continued)

Table 1.4 (continued)

Reference genome	Approach	Major findings relevant to R-gene study	References
		pathway during biotic stress are unique in turnip, with evidence of copy number variation	

B. napus pangenome while 59 RGAs were detected within the *Sclerotinia*, Fusarium wilt, and Clubroot resistance QTLs in the *B. oleracea* pangenome. These pangenomics studies revealed that different classes of RGAs (RLKs, TNLs, and others) show different percentages of variability across the lines. It was also found in the *B. oleracea* pangenome study, the wild relative *B. macrocarpa* harbors the most RGA candidates, suggesting that a large pool of genetic resources for R-genes can be found in wild *Brassicas*. A super-pangenome was reported that includes the genomes of wild relatives and/or different species within a genus, which adds an additional level of depth for investigating genomic variations within a crop genus. With a super-pangenome developed in *Brassica* crops, identify many more novel candidate disease resistance genes from the wild genotypes, but also develop molecular markers to screen for resistant varieties in the field, therefore not only improving the speed and accuracy of crop breeding but also broadening the *Brassica* gene pool by using novel alleles from wild germplasm (Bayer et al. 2019; Dolatabadian et al. 2019; Neik et al. 2020).

1.14.3 Identification of Candidate QTLs/Genes Using NGS-Based SNP Methods

Many genetic linkage maps of *Brassica* crops have been generated, using bi-parental crossing, selfing, and backcrosses, with molecular markers such as Restriction Fragment Length Polymorphisms (RFLPs), Amplified Fragment Length Polymorphisms (AFLPs), and Randomly Amplified Polymorphic DNA (RAPDs) (Delourme et al. 2018). These molecular markers are often limited by low reproducibility and laborious techniques, thus limiting the quality of marker information in *Brassica* crops. The genomics era, driven by the innovation of high-throughput next-generation sequencing (NGS) technologies, has significantly increased the efficiency of the identification of QTL/candidate genes for disease resistance in *Brassica* crops through the development of genome-wide DNA-based molecular markers. This has brought great improvement in the resolution of genetic maps. Among the DNA-based molecular markers, single nucleotide polymorphism (SNP) markers are most widely used for determining genotypic variation in a given species because they are uniformly distributed and highly abundant in the genome and are amenable in multiple genotyping platforms. With NGS, high-throughput genome-wide SNP marker development can be achieved rapidly and accurately in *Brassica* species through systems such as whole-genome resequencing (WGRS), genotyping-by-sequencing (GBS), and the *Brassica* 60K Illumina Infinium^{TM} 60K SNP array.

WGRS is an omics strategy for obtaining high-quality, high-density SNP markers at a whole-genome level by mapping sequence reads to the *Brassica* reference genome assemblies. In GBS, restriction enzymes are used to digest the genomic DNA and barcodes are used to ligate the fragmented DNA molecules before whole-genome sequencing is performed for SNP discovery. In this way, GBS is less complicated compared to WGRS because the sequencing reads cover only part of the genome instead of the whole genome, thus offering a cost-efficient approach to identifying SNPs yet achieving equally high-quality SNPs with wide applications in crop improvement studies. An extension of GBS called tGBS^{R} using oligonucleotides instead of adaptors has since been developed. The *Brassica* 60K SNP array offers a whole-genome SNP genotyping approach that is highly reproducible for genotyping hundreds of DNA samples in 48 h, making it an attractive option for the routine screening of *Brassica* germplasm. A *Brassica* 60K SNP array data repository called Crop SNPdb has been developed to enable users to access SNP information, which is containing genotypic information of 526 *Brassica* lines, permitting the easy retrieval of whole-genome SNP data for a wide range of downstream data analyses.

These NGS-based SNP genotyping approaches have been applied widely for the QTL mapping of disease resistance traits and identification of candidate genes through genome-wide association studies (GWAS) in *Brassica* crops (Table 1.5). Highlights include the discovery of novel disease resistance QTLs/genes at an unprecedented speed, for example, in Blackleg, *Sclerotinia*, and Clubroot. The other benefit of the application of NGS-based SNP genotyping is the successful breeding of *B. napus* varieties containing multiple improved traits. The breeding of resistant lines was achieved through the introgression of several major QTLs for *Sclerotinia* quantitative resistance from *B. oleracea* into *B. napus* with good seed yield and quality. In addition, bioinformatics pipelines are continuously being improved for whole-genome SNP data analysis. One example is single nucleotide absence polymorphism (SNaP) analysis, which successfully recovered numerous QTLs for *Sclerotinia* and Blackleg resistance in *B. napus* that were lost from the normal filtering of SNP data obtained from the *Brassica* 60K SNP array, with 3.2- and 2.2-fold increases in significant marker-trait associations for *Sclerotinia* and Blackleg resistance, respectively (Neik et al. 2020).

1.14.4 Identification of Candidate R-Gene Using In Silico Methods

The large volume of *Brassica* genomics resources in public databases supports the analysis and interpretation of many complex mechanisms related to *Brassica*–pathogen interaction. Examples include the in silico identification of the Blackleg resistance gene, LepR4, in the C-genome of Korean cabbage (*B. oleracea* var. *capitata*) and the in silico exploration of 641 NBS-LRR-type disease resistance genes in *B. napus*, together highlighting the genomic distribution and structural

Table 1.5 An application of next-generation sequencing (NGS)-based SNP genotyping approaches in resistance studies of *Brassica* diseases (Neik et al. 2020)

Approach	Disease type	*Brassica* sample	Main Findings	References
WGRS, SNP genotyping	–	991 *B. napus* worldwide accessions	Selective-sweep regions enriched with genes related to stress response	Wu et al. (2019a, b)
WGRS, SNP genotyping	–	588 *B. napus* worldwide accessions	A sub-genomic-specific selection contributes toward biotic stress response with several candidate genes identified	Lu et al. (2019)
WGRS, QTL mapping	Black rot	Mapping population of cabbage *B. oleracea* var. *capitata* inbred lines "C1234" (resistant) and "C1184" (susceptible)	21 candidate NBS-LRR genes associated with black rot resistance in *B. oleracea*	Lee et al. (2015)
GBS, GWAS	Blackleg	243 *B. napus* accessions from Canada and China	Significant SNPs were found on chromosome A08 with 25 RGAs identified consisting of NBS, RLK, RLP, and TM-CC type R-genes	Fu et al. (2020)
GBS, GWAS	Sclerotinia	*B. juncea–B. fruticulosa* introgression lines	20 candidate genes mostly located on the A sub-genome of *B. juncea*	Atri et al. (2019)
GBS, GWAS	Sclerotinia	*B. juncea–Erucastrum cardaminoides* introgression lines	QTL region on chromosomes A03 and B03 and candidate genes being LRR-RLK, LRR-PK, and TIR-NBR-LRR	Rana et al. (2019)
tGBS^{R}, GWAS	–	135 *B. oleracea* accessions including var. broccoli Brussels sprout, cabbage cauliflower, Chinese kala, kale, kohlrabi, and savoy cabbage	Resistant phenotype mostly found in kale. Candidate genes encoding pathogenesis-related proteins were mainly found on chromosome C07	Farid et al. (2019)
Brassica 60K SNP array, GWAS	Clubroot	Mapping population of cabbage *B. oleracea* inbred lines "263" and "GZ87"	Significant QTL and novel loci found on C sub-genome	Peng, et al. (2018)
Brassica 60K SNP array,	–	327 *B. napus* worldwide accessions	Selective-sweep regions enriched with	Wei et al. (2017)

(continued)

Table 1.5 (continued)

Approach	Disease type	*Brassica* sample	Main Findings	References
linkage disequilibrium (LD) analysis		comprising three ecotypes	Blackleg and *Sclerotinia* resistance QTLs	
Brassica 60K SNP array, GWAS	Clubroot	472 *B. napus* worldwide accessions	Most candidate genes were found on C sub-genome with novel QTLs and TIR-NBS gene clusters	Li et al. (2016)
Brassica 60K SNP array, GWAS	Sclerotinia	448 worldwide *B. napus* accessions	Two novel loci with 39 candidate genes on C sub-genome	Wu et al. (2016)
Brassica 60K SNP array	Blackleg	Seven *B. napus* seven donor parents for introgression lines	Genomic background of individual varieties and multiple defence-related gene interactions influence the resistance levels	Larkan et al. (2016)

variation of these genes in *B. napus*. Other examples include the in silico evolutionary study of NBS genes in *B. napus*, where comparative genomic analysis highlighted the NBS gene's distribution from its progenitors *B. rapa* and *B. oleracea* in relation to the three main *Brassica* diseases—Blackleg, Clubroot, and *Sclerotinia*. Coupled with modern bioinformatics tools and the integration of multi-omics data sets, in silico methods are powerful tools that rapidly provide accurate and detailed models to answer various research questions ranging from candidate gene identification to evolutionary pathways of resistance mechanisms in both the *Brassica* host and the fungal pathogens. This is a time-, cost-, and manpower-effective means of conducting higher-quality *Brassica* crop improvement research investigations. In addition, database searches for protein motifs associated with disease resistance genes have enabled researchers to identify classes of R-genes in *B. napus* that are associated with Blackleg, *Sclerotinia*, and Clubroot resistance. Stotz et al. (2018) suggest that Clubroot resistance genes are NLR type, Blackleg resistance genes are RLP type, while for *Sclerotinia*, neither NLR nor RLP was involved. Focusing on the NLR genes, comparative genomics and transcriptomics analyses supplemented with a database query on *B. napus* and its progenitors, *B. rapa* and *B. oleracea*, revealed many more NBS (also known as NLR) genes in the C sub-genome of *B. napus*. A number of these genes underly the QTL regions for resistance against Blackleg, *Sclerotinia*, and Clubroot, supporting the concept that the diversification of the R-genes likely happened after interspecific hybridization between *B. rapa* and *B. oleracea* (Ferdous et al. 2020; Fu et al. 2019a, b).

NGS-based bulked segregant analysis (BSA) NGS-based BSA is one of the more recent applications of omics in studying *Brassica*–pathogen interactions. This technique involves bulks/pools of DNA samples with representations of individuals with segregating phenotypes, where the pools are genotyped using NGS, either RNA sequencing (BSR-Seq) or whole-genome resequencing, followed by the detection of QTLs through SNP calling between the bulks (QTL-Seq). The traditional BSA technique allows the screening of many loci. An example is screening for Downy Mildew resistance in lettuce, but it is restricted to the detection of random sequence variation (e.g., RFLPs) and requires intense PCR screening efforts to confirm the molecular markers that are linked with the selected genomic intervals. With the NGS screening of BSA populations, the detection of sequence polymorphisms between the bulks is rapid and effective, as novel variation, such as PAVs or even novel QTLs/genes, can potentially be detected.

Using BSR-Seq, an R-gene for resistance against Blackleg, *Rlm1*, was fine mapped in the *B. napus* cultivar "Quinta," and a candidate gene was identified, BnA07G27460D, that encodes a serine/threonine protein kinase. This gene is homologous to the protein kinase STN7 in *B. rapa*, *B. oleracea*, and *A. thaliana*, which is involved in systemic plant immune responses by regulating reactive oxygen species (ROS)-induced cell signaling at the thylakoid membrane. BSR-Seq has also been applied in characterizing Clubroot resistance in some *Brassica* species, in *B. oleracea*, the first Clubroot major R-gene (*Rcr7*) in the *B. oleracea* cultivar "Tekila" was identified Bo7g108760 was the candidate TNL gene. In Chinese cabbage *B. rapa* var. *pekinensis*, the candidate gene *Rcr2*, which encodes a TIR-NBS-LRR, responsible for Clubroot resistance, has been identified on chromosome A03. In *B. nigra*, a novel Clubroot R-gene, *Rcr6*, was detected (BniBo15819, encoding a TNL gene), which is homologous to chromosome A08 of *B. rapa* and which provides a good source for gene introgression into *B. napus*. Using QTL-Seq, two novel QTL regions associated with Clubroot resistance on chromosomes A07 and A08 were detected in pak choi *B. campestris* var. *chinensis*. A single novel candidate R-gene, also involved in Clubroot resistance, CRd, was identified on chromosome A03 of *B. rapa* (Neik et al. 2020; Chang et al. 2019; Fu et al. 2019a, b).

1.14.5 Resistance Gene Enrichment and Sequencing (RenSeq)

RenSeq is a targeted resequencing method for identifying NLRs. RenSeq in combination with PacBio sequencing was applied in to study the R-gene sequence variants of the White Rust Resistance (WRR) gene against *Albugo candida*. This combinatorial approach was extended to 64 accessions of *A. thaliana* to study the evolution and variability of NLR genes in the model plant *Arabidopsis*, resulting in the construction of a species-wide pan-NLR-ome. Similar to the concept of pangenomics, the pan-NLR-ome is the collection of all the NLR genes and alleles contained within a species, distinguishing between the core and non-core NLRs in terms of the structural variations and diversity. New domain structures of NLRs were identified from the pan-NLR-ome study in *A. thaliana*. This implies that novel R genes can be obtained from the diverse gene pool of NLRs within members of the Brassicaceae family, including *Brassica* crops. The discovery of a repertoire of NLRs is particularly important when pathogen-specific recognition can not only happen in the host species but also in the nonhost species. For example, the nonhost *A. thaliana* displayed ETI-mediated defense against *A. candida* isolates derived from *B. juncea*, *B. rapa*, and *B. oleracea*, implying that R-genes play a conserved role across members within the same family. This implies that screening for novel R-genes for *Brassica* crop improvement should also be applied across nonhost species. Pan-NLR-ome type studies contribute significantly toward the discovery of R-gene diversity in crops (Cevik et al. 2019; Van de Weyer et al. 2019).

1.14.6 Effectoromics

Effectoromics, or effector-based screening, is an omics approach to detecting R-genes in crop plants, although R-gene products may not necessarily interact with effectors directly. In this method, the target effector, as forecast from prediction tools, is transformed using *Agrobacterium* and infiltrated onto the host plants. The host genotypes that give a positive response to the target effector are then subjected to resistance gene mapping using molecular markers. Effectoromics allows screening for potential recognition targets in a particular crop species, for example, immune receptors, and can be applied to screen different crop species with a selection of potential candidates that can be used in inter-crop species. This method was used for the identification of R-genes in the wild potato species *S. pinnatisectum* against the oomycete *Phytophthora infestans* and could be applied to *Brassica* pathosystems for the identification of R-genes (Neik et al. 2020).

1.14.7 Transcriptomics

Plant–pathogen interactions are complex, involving a large number of interconnected molecular mechanisms. Critical for understanding these mechanisms is deciphering which genes are activated in both the host and the pathogen during infection and how these genes affect the expression of others in the pathways. Transcriptome analysis has allowed the monitoring of the molecular cues involved in these interactions. Gene expression analysis has significantly evolved since early techniques and approaches, including RNA sequencing or RNA-seq to overcome the limitations of previous techniques which are considered superior to predecessors in that they can interrogate the whole genome transcriptome of any organism with or without reference genomes and facilitate the discovery of unique genes. They are highly sensitive in detecting lowly expressed genes and have been shown to be highly reproducible. These features make RNA-seq the method of choice in most transcriptome studies, including the analysis of host–pathogen interactions. An advance in this approach, called dual-RNA seq, enables the simultaneous study of gene expression in both the host and pathogen, enabling a real-time and comprehensive analysis of the mechanisms involved in both pathogenesis and the host resistance response (Wang et al. 2009, 2019; Wani and Ashraf 2018).

The integrative approaches such as associative transcriptomics (AT) and bulked RNA sequencing (BSR-Seq) have allowed the incorporation of transcriptome data with genome-wide marker information to increase the power of detection for genomic loci controlling resistance or susceptibility in the host and virulence in the pathogen. These approaches have facilitated a fast-tracked identification of candidate genes underlying these loci and greatly facilitated the dissection of gene expression patterns during pathogen attack in important *Brassica* crops. AT is an RNA-based approach that integrates transcriptome data in GWAS to identify molecular markers associated with a particular trait of interest at marker loci where the levels of gene sequence and gene expression are variable. AT is particularly useful for dissecting trait variation in polyploid species characterized as containing highly duplicated genes displaying variable expression patterns (Fu et al. 2019a, b; Harper et al. 2012).

In *B. napus* resistance against Blackleg, three of the known R-genes have been cloned and were found to encode leucine-rich repeats-receptor like proteins (LRR-RLPs *Rlm2* and *LepR3*) and a wall-associated kinase-like (WAKL) protein (*Rlm9*). The cloning of these genes provided the starting material for using RNA-seq to interrogate the detailed machinery involved in the resistance. With the global transcriptome analysis, it was found that both *LepR3* and *Rlm2* evoked a basal defense response in both compatible and incompatible interactions upon inoculation with *L. maculans* isolates. This suggests that *LepR3* and *Rlm2* may also monitor other molecular patterns produced by *L. maculans* to mount a resistance response in the host plant (Zhou et al. 2019). The *Rlm9* WAKL protein is a type of receptor-like kinase (RLK) localized in the cell wall, which functions to sense environmental and cellular signals. *Rlm9* is only the second WAKL R-gene identified to date; hence, the mechanisms underlying its resistance are yet to be studied in detail. Initially, Larkan et al. (2019) found that *Rlm9* did not seem to have a direct interaction with its counterpart *Avr* gene (*AvrLm5*-9), and it is likely that a mediator molecule, such as damage-associated molecular patterns (DAMPs), is needed to effect resistance. This is supported by the findings of Brutus et al. (2010), which showed that WAKLs can detect DAMPs following pathogen attack. However, this latter mechanism needs to be verified in further studies. Genome-wide transcriptome analysis for this interaction should help uncover how Rlm9 orchestrates race-specific resistance responses against the Blackleg pathogen. Further, as *Rlm9* forms part of the tightly linked R-gene cluster (*Rlm3/4/7/9*) on chromosome A07 of *B. napus*, the cloning of this gene may enable an understanding of the molecular mechanisms of the other genes on this cluster (Neik et al. 2020).

The introgression of the major genes *Rlm2*, *Rlm3*, *LepR1*, and *LepR2* in cultivar "Topas" and *LepR1* and *LepR2* in "Westar" allowed the comparison of the effect of genetic background and R-gene content in host defense expression through genome-wide transcriptome profiling by Haddadi et al. (2019). All introgression lines (ILs) showed an upregulation of genes previously implicated in host defense, including hormone signaling, cell wall thickening, chitin production, and glucosinolate production. Interestingly, these genes have higher levels of expression in LepR1 and Rlm2 compared with LepR2 and Rlm3 lines during the first 3 days of infection. Furthermore, a general trend of delayed defense responses in "Westar" compared with "Topas" ILs was observed, regardless of their R-gene content. This suggests that the genetic background has important effects on resistance. Additionally, there was enhanced expression of the RLK *Brassica napus* (*Bn*) *SOBIR1* (Suppressor of BIR1-1) and salicylic acid-related defense in both LepR1 and Rlm2 lines, consistent with previous investigations.

Transcriptomic studies also showed the involvement of host receptor genes, RLPs, RLKs, TIR-NBS, and WAKLs, in PTI or effector-triggered defense (ETD) in the *B. napus*–*L. maculans* interaction. All these studies suggest that *Avr* genes in *L. maculans* likely play a role in manipulating host resistance in the apoplast environment. The AvrLm1 protein reportedly interacts with the mitogen-activated protein kinase (MAPK) 9 in *B. napus* (BnMPK9). The MAPK9 is implicated as a positive regulator of ROS-mediated abscisic acid (ABA) signaling in the guard cells of the plant, fostering stomatal closure. Larkan et al. (2019) reported the association of *Bn-SOBIR* with *Rlm2*, and it was assumed that the *Rlm2* and *Bn-SOBIR* interaction results in downstream signaling to effect race-specific resistance against the *AvrLm2 L. maculans* pathotype. It is likely that *LepR1* conveys resistance through the same mechanism. Conversely, the expression of *Bn-SOBIR* was low in *Rlm3* introgression lines, which means that Rlm3 not only functions independently of the SOBIR1 interaction but represents another resistance mechanism different from that of other cloned genes. The cloning and transcriptomic analysis of this gene will shed light on the molecular mechanisms governing the operation of this resistance (Neik et al. 2020).

In the *B. napus*–Clubroot pathosystem, several R-genes acting against *P. Brassicae* have been mapped, but only two have been cloned *CRa* and *Crr1a*, which encode TIR-NBS-LRRs. Plant hormones such as ethylene (ET), jasmonic acid (JA), salicylic acid (SA), abscisic acid (ABA), auxin, and cytokinin were implicated in the pyramided lines of *B. napus* containing two Clubroot resistance genes, *PbBa8*.1 and *CRb*, with the candidate genes involved in the hormone signaling pathway identified through comparative RNA-seq. The transcriptome analysis in CRb-containing *B. rapa* lines, at the early stages of *P. Brassicae* infection, confirmed the involvement of pathways typical for ETI-mediated resistance and biotrophic infection. These include the activation of NLR and pathogenesis-related (PR) genes, along with the upregulation of genes for MAPK, WRKY transcription factors, calcium-binding proteins, chitinases, and SA pathway genes (Ji et al. 2021; Shah et al. 2020). However, the transcriptome analysis of Chu et al. (2014) highlighted the induction of JA and ET pathways, implicated in the necrotrophic stage of infection, as important mechanisms of Rcr1-mediated resistance in *B. rapa*, thus highlighting a complex molecular mechanism for *P. Brassicae* resistance in *Brassica* crops. The cloning of the *Rcr1* gene will help to elucidate these hormones' induction dynamics. However, while *Crr1a* and *CRa* have been cloned, the transcriptional control of their resistance has not been widely researched. Genome-wide transcriptome studies should help to elucidate the resistance mechanisms involved in these two key clubroot R-genes. To study the genetic effects of the multiple R-genes in pyramided lines and confirm the role of these introgressed genes in host resistance responses, comparative RNA sequencing could be performed. The two Clubroot-resistant genes, *PbBa8.*1 and *CRb*, were introgressed into a *B. napus*-pyramided line; through comparative RNA sequencing, it was found not only that the pyramided lines displayed a strong multi-gene resistance network during pathogen infection, but also that SA- and ROS-mediated resistance responses played a dominant role in the pyramided lines, supported by Galindo-González et al. (2020) highlighting the importance of SA-mediated resistance in the *B. napus–P. Brassicae* pathosystem.

In case of *Sclerotinia* resistance in *B. napus*, Qasim et al. (2020) detected at least 36 candidate genes representing diverse molecular functions in the resistance response, including TIR-NBS-LRR genes, hormone synthesis, the production of secondary metabolites, and the regulation of transcription factors and several metabolic pathways. One metabolic gene involved in the regulation of the phenylpropanoid pathway that plays a key role in lignin biosynthesis was highly transcribed across time points, and one TIR-NBS-LRR gene, in one of the QTL regions, a widely known R-gene that is associated with a qualitative response, was highly transcribed. As *Sclerotinia* resistance has been known to be quantitatively controlled, the diversity of R-gene host-mediated resistance mechanisms shown in the study mirrors the complexity of quantitative resistance. Some of the described mechanisms are atypical of PTI- and ETI-mediated host defenses, supporting an ongoing discussion challenging the applicability of the conventional two-tier model of plant immunity in explaining quantitative resistance variation. This enigma may be due to the differences in the hosts and the pathogens, as well as the approaches employed for studying genome-wide gene expression. Nevertheless, the increasing availability of transcriptome data generated through various high-throughput platforms results in a better comprehension of the mechanisms underpinning quantitative resistance. Global transcription sequencing also revealed that JA and ET signaling were associated with a resistance response against *S. sclerotiorum* in *B. napus.* A further transcriptome study in the same pathosystem demonstrated the downregulation of *B. napus* NPR1-like gene, *BnaNPR1*, which plays a role in SA and JA signaling, indicating that *S. sclerotiorum* suppresses the expression of *BnaNPR1* during systemic acquired resistance (SAR) for successful invasion into the host cell. It is also noteworthy that *NPR1* genes were activated by *NBS-LRR* genes, as reported in a gene pyramiding study in *B. napus* using two NBS-LRR genes, *BvHs1* $^{pro\text{-}1}$ and *BvcZR3*, obtained from nematode (*Heterodera schachtii*)-resistant sugar beet. The gene interaction network of NPR1 and NBS-LRR should be studied further in relation to other defense-related genes. Comparative transcriptomic analysis of the *B. napus–S. sclerotiorum* pathosystem showed that indolic glucosinolate biosynthesis plays an important role, similar to that found from an overexpression experiment with three glucosinolate genes in *B. napus*, one of which, *BnUGT74B1*, encoding cytochrome P450, enhanced resistance to *S. sclerotiorum*. Transcriptomic analysis in the *B. oleracea–S. sclerotiorum* interaction revealed a total of 45 *B. oleracea* genes involved in Ca^{2+} signaling were upregulated, which is important in ROS generation, and this is consistent with the findings that the resistance mechanism in *B. oleracea* showed ROS generation and increased Ca^{2+} signaling contributing toward resistance outcomes in *B. oleracea* (Wang et al. 2020; Zhong et al. 2019; Ding et al. 2018).

A transcriptome study of downy mildew (*H. Brassicae*)-infected Chinese cabbage lines demonstrated the involvement of protein processing in endoplasmic reticulum and circadian rhythm pathways in resistance mechanisms, in addition, to the downregulation of photosynthetic genes during *H. Brassicae* infection. Xiao et al. (2016) reported a downregulation of energy metabolism genes, particularly those involved in the photosynthetic carbon cycle (PCC). This indicates that the resistance to *H. Brassicae* in *Brassica* crops may be driven by efficient energy metabolism during pathogen invasion.

In an RNA-seq study of *B. oleracea–Fusarium oxysporum* f. sp. *conglutinans* interactions, Ca^{2+}-binding ATPase and aquaporin tonoplast intrinsic protein (TIP), which are involved in Ca^{2+} signaling, were highly expressed in the resistant genotype at 4 h after infection. Similarly, the transcriptome dynamics in *B. oleracea* in response to the black rot pathogen (*X. campestris*), highlighted the role of Ca^{2+} signaling proteins as secondary messengers for several downstream signaling processes, including the activation of several transcription factors responsible for the initiation of SA-mediated host defense (Tortosa et al. 2019). The deep RNA-seq of Liu et al. (2019) found an upregulation of several plant pathogen receptor genes such as chitin elicitor receptor kinase 1, chitin receptor, LRR-RLPK, and WAKL, which are important in the PTI defense response. This led them to conclude that PTI is the primary mechanism for soft rot resistance in Chinese cabbage (*B. rapa* var. *pekinensis*), initiating several downstream signaling pathways for hormone regulation and the production of secondary metabolites and cell wall reinforcement (Neik et al. 2020).

1.14.8 Proteomics

The use of 2D gel electrophoresis (2-DGE) and matrix-assisted laser desorption/ionization time-of-flight mass spectrometry (MALDI-TOF/TOF MS) technology to study the proteomics of plant–pathogen interactions was first reported more than a decade ago. Comparative proteomic analysis of responses to *L. maculans*, between compatible and incompatible interactions in *B. napus* cv. "Surpass 400" with either the virulent isolate UWA 192 or avirulent isolate UWAP11, showed the upregulation of enzymes involved in RuBisCO for CO_2 fixation, H_2O_2 scavenging, and redox metabolism. In *L. maculans*-tolerant *B. carinata*, most of the proteins displayed antioxidant activities. Similarly, in the *B. carinata–L. maculans* interaction, it was found from 2-DGE analysis that proteins related to ROS generation and photosynthetic enzymes were elevated in the resistant genotype 48 h after pathogen infection. A proteomics study in the *B. rapa–P. Brassicae* pathosystem showed *Rcr1* was associated with the ubiquitin-related proteasome system in plant defense reactions, along with the activation of the calcium-independent MAPK signaling pathway, regulation of ROS production via the activity of protein disulfide isomerases, and upregulation of lignin biosynthesis (Marra et al. 2010; Song et al. 2016).

A high-throughput proteomic study using 2-DGE and MALDI-TOF/TOF MS analyses was performed in the *B. oleracea–P. Brassicae* pathosystem to study protein expression during the early stages of host infection, with the highly expressed protein thioredoxin (TRX) enzyme identified, associated with oxidative stress and the pathogen defense response. More than 487 out of 5003 proteins (13.4%) that were identified in *P. Brassicae*-infected Chinese cabbage (*B. rapa* var. *pekinensis*) using isobaric tags for relative and absolute quantitation (iTRAQ)-based proteomic analysis were differentially up- or downregulated, and the proteins that contributed to the defense response included those involved in tryptophan and glutathione biosynthesis and cytokinin signaling (Moon et al. 2020).

In an *H. parasitica* infection study in non-heading Chinese cabbage (*B. campestris* var. *chinensis*), a 2-DGE protein analysis and MALDI-TOF/TOF MS analysis along with transcript mRNA analysis using quantitative RT-PCR suggested a role for a Ca^{2+} signaling pathway as part of the ROS-mediated defense mechanism, with 39% of the genes having no correlation between protein and mRNA levels. Studies have shown that proteomic data may not correlate with transcriptomic data measuring mRNA levels due to post-translational events and protein turnover. To determine post-translational protein modification, an online 2D ion-exchange/reversed-phase HPLC method called Multidimensional Protein Identification Technology (MudPIT) can be used (Lan et al. 2019; Rustagi et al. 2018). Time-course protein profiling in the pathosystem of *B. juncea–Albugo candida* successfully detected proteins that are differentially expressed in the resistant variety such as plant-thaumatin-like protein, superoxide dismutase, glutathione *S*-transferase, cysteine synthase, and red chlorophyll catabolite reductase, suggesting ROS generation plays an important role in *Brassica* host resistance against this pathogen (Kaur et al. 2011).

In a proteomic study applied to a non-pathogenic, arbuscular mycorrhizal fungal species, *Piriformospora indica*, studying the beneficial effect of the fungal endophyte on the *B. napus* host, liquid chromatography-mass spectrometry (LC-MS), coupled with bioinformatics, highlighted significant levels of differentially expressed proteins involved in the stress/defense response during the cell-death colonization phase in the plant roots, elucidating the role of *P. indica* in enhancing *B. napus* resistance against abiotic and biotic stresses. There is opportunity for the further characterization of the genes that encode the stress-response proteins expressed in the *B. napus–P. indica* host symbionts relationship. This could be done through the physical mapping of the genes on the *B. napus* genome assemblies supplemented with transcriptome data (Shrivastava et al. 2018; Neik et al. 2020).

1.15 Application of Omics Approaches Technologies in *Brassica* Host-Pathosystem

1.15.1 Availability of High-Quality Genome Assemblies

High-quality genome assemblies for the major pathogens of *Brassica* are currently available, where long-read sequencing approaches such as ONT MinIon sequencing and PacBio sequencing were applied to assemble the genomes of *L. maculans*, *S. sclerotiorum*, *A. Brassicae*, and *A. alternata*. The genome of *P. Brassicae* was assembled using Illumina Hiseq 2500 technology, while the *A. candida* genome was assembled using Roche/454; both are short-read sequencing technologies. These high-quality genome assemblies of the *Brassica* pathogens revealed that the fungal pathogens contain high genomic variation, including mutations largely induced by transposable elements (TE), large-scale chromosomal re-arrangements, presence–absence variation, and the gain or loss of accessory chromosomes. These genetic events are continuously and actively evolving in the adaptive response to the selection pressure by the host plant resistance mechanism, thus generating high-genome-plasticity regions, which are often found distributed in compartments where most of the virulence genes are housed. The extent of genetic divergence between individuals of the same fungal plant pathogen species is also high, so, reference-based mapping is a challenge, although the diverged regions may not always be associated with virulence. The availability of these high-quality genome assemblies has greatly facilitated the discovery of candidate genes for effectors and virulence factors and significantly advanced our understanding about the pathogen in relation to its evolutionary pattern and species diversity through comparative and population genomics studies. Such deep molecular information will also allow us to uncover the complex *Brassica* host–pathogen interactions, as these omics resources are routinely applied in molecular plant pathology research (Covo 2020; Feurtey et al. 2019).

1.15.2 Transcriptomics of Virulence-Related Genes

Transcriptome analysis has been increasingly applied to study pathogen gene expression during host invasion, allowing the real-time monitoring of the molecular mechanisms involved in pathogenesis. The recent genome-wide transcriptome analysis of Chittem et al. (2020) highlighted the involvement of peroxisome-related pathways, cell wall degradation by various enzymes, and the detoxification of host metabolites as mechanisms of virulence by *S. sclerotiorum* toward *B. napus*. In the *B. napus*–*L. maculans* interaction, Haddadi et al. (2019) reported the upregulation of genes for fungal toxin biosynthesis during the necrotrophic stage of infection. Further, several *L. maculans* effectors have been predicted from RNA-seq data, consistent with the results of Sonah et al. (2016) who detected different sets of genes coding for several effector proteins, where expression was correlated to *L. maculans* lifestyle transition from a biotrophic to necrotrophic stage. This observation provides an additional clue in deciphering the arsenal of virulence mechanisms employed by

L. maculans, one of the most notoriously adaptive disease-causing pathogens of the *Brassica* family. In Clubroot, small secreted proteins (SSPbPs) have been identified that were assumed to play critical functions in primary and secondary infections, leading to hypertrophic tissue development (Chen et al. 2019). For bacterial pathogens, such as *X. campestris* and *P. carotovorum*, a variety of mechanisms are deployed to evade host resistance including the release of extracellular enzymes such as cellulase, mannanase, pectinase, protease, polygalacturonases (PGs), and pectate lyase (Type II secretion system), the injection of effector proteins (Type III secretion system), as well as the production of exopolysaccharides and biofilm formation. These virulence mechanisms can be manipulated by various techniques such as genome editing or developing cultivars that can undermine such mechanisms (Lee et al. 2013). Through the in silico analysis of comparative genomic and transcriptomic data of *S. sclerotiorum*, 80 putative secondary metabolite gene clusters implicated in virulence in *B. napus* were identified in sub-telomeric regions close to transposable elements, with the upregulation of 12 polyketide synthases (PKSs) and enzymes during *S. sclerotiorum* infection of *B. napus*, revealing clues about the virulence pathway in the *B. napus* host. Enzymes associated with secondary metabolites production in *S. sclerotiorum* to suppress host defense mechanisms, such as PKS, non-ribosomal peptide synthase (NRPS), and chalcone synthase (CHS), were upregulated in an RNA-seq experiment studying *B. napus–S. sclerotiorum* interaction to express disease resistance (Graham-Taylor et al. 2020; Seifbarghi et al. 2017).

1.15.3 Secretomics of Pathogenesis

The fungal plant pathogen secretes a whole suite of proteins, collectively known as the secretome, during its interaction with the host. The secretome comprises effector proteins and specific enzymes crucial for host colonization, and the composition of each of these secreted proteins may vary between pathogen types based on the pathogen's mode of nutrition and lifestyle. The identification of these secreted proteins is key to understanding the pathogenicity process of the fungal plant pathogen in the host plant. The availability of rich omics resources and advanced bioinformatics pipelines for diverse fungal plant pathogen species have enabled the quick prediction of effector proteins across kingdom-wide fungal species with different lifestyles and have accelerated the cloning and functional characterization of candidate effectors. An understanding of the structural features of effector proteins, the diversity of the effector genes, and how these genes play a role in the pathogenicity and the evolutionary patterns of the genes are important for uncovering the complexity of the resistance mechanism in *Brassica*–pathogen interactions to support breeding-resistant *Brassica* varieties. Due to sequence diversity of effectors for avoiding recognition by the host immune system, the specific function and mechanism of effectors in inducing pathogenicity in the host are difficult to determine. However, the majority of effector genes can be predicted or identified more accurately and effectively based on the known characteristics of cloned effectors using improved computational, bioinformatics software combined with machine learning. These are including Effector P 2.0, Signal P, and Apoplast P. Bioinformatics tools specific for the identification of transposable elements (TEs) have also been developed. Besides proteinaceous effector molecules, non-proteinaceous effectors in fungal pathogens, such as secondary metabolites, small noncoding RNAs, and their biological roles in pathogenicity, have also been studied in plant–fungus interactions (Van de Wouw and Idnurm 2019; Carreón-Anguiano et al. 2020; Collemare et al. 2019).

The conventional secretion pathway for the proteins during plant–pathogen interaction involves the endoplasmic reticulum–Golgi pathway. However, increasing evidence has shown that some secretomes of plant or fungal proteins are secreted independently of the classical pathway during plant–pathogen interactions. So, it is important that proteins lacking signal peptides within the fungal secretome are not overlooked when identifying candidate effectors. Some signaling molecules produced by phytopathogenic fungal species that play a part in virulence resemble homologous signaling molecules in the host, acting as mimics to evade the plant immune system for successful disease development. For instance, oxylipins, which are important signaling molecules commonly found in animals, plants, and fungi, play a role in growth, development, and the defense response, with one of the examples being jasmonate. In *Brassica*, oxylipins were found to display fungicidal activity against *A. Brassica*e, *L. maculans*, *S. sclerotiorum*, and *Verticillium longisporum*. In phytopathogenic fungi, oxylipins have been found to be involved in disease progression through the modification of the plant host defense mechanisms, an example being *F. oxysporum* hijacking the oxylipin JA signaling pathway in *A. thaliana* (Ruano and Scheuring 2020; Tanveer et al. 2014; Vincent et al. 2020; Deboever et al. 2020; Genva et al. 2019).

The gene expression profile from the *P. brassicae* Pb3 genome assembly revealed that the pathogen contains genes that are associated with the biosynthesis of the plant hormones cytokinin and auxin, suggesting a potential role for these hormones in virulence activity in the host plant, while gene clusters for the synthesis of the ABA hormone were detected in *L. maculans*, suggesting a putative role of ABA production in disease progression in . One of the well-characterized effectors for *P. brassicae* is the benzoic acid (BA)/SA methyltransferase protein (*PbBSMT*), which suppresses host SA signaling during plant defense. The functional role of *PbBSMT* is similar to that of the SABATH methyltransferase gene family in *A. thaliana*, *AtBSMT1*, where the genes play a role in converting SA into methyl salicylate (MeSA), which is the inactive form of SA, thereby compromising the SAR defense response. A transcriptomic study of *P. brassicae* infection in Kohlrabi (*B. oleracea* var. *gongylodes*) showed that *PbBSMT* is one of the highest-expressed pathogen genes in the root gall tissue, playing a role in the local reduction of SA via *PbBMST*-mediated methylation. It was also found from another cloning with *AtBSMT1* vs. *PbBSMT* in *P. brassicae*-infected *Arabidopsis* (host) and *A. candida*-infected *Arabidopsis* (nonhost), comparing the level of SA inactivation in *Arabidopsis*, that *PbBSMT* resulted in higher levels of SA inactivation, meaning *PbBSMT* suppressed the host and nonhost SAR defense mechanisms at a greater level. Multi-omics approaches combining genomics, transcriptomics, proteomics, and metabolomics using computational strategies will allow us to identify suitable mimicking molecules in the fungal and/or host species that trigger stronger plant defense systems during plant–pathogen interactions (Darma et al. 2019; Ciaghi et al. 2019; Pathak et al. 2017).

Beneficial bacterial endophytes found in the apoplast region of *B. napus* have been shown to inhibit the growth of *X. campestris*, *S. sclerotiorum*, and *L. maculans*, which act as natural biological control against *B. napus* diseases. The high-throughput sequencing of the fungal endophytes obtained from healthy roots of tumorous stem mustard (*B. juncea* var. *tumida*) and *P. brassicae*-infected roots of the same plant species showed a more diverse composition of the fungal endophytes in the healthy roots compared to in the diseased roots. This suggests a strong interaction network in the fungal endophyte community that contributes toward the optimum health of the host plant. A combination of secretomic and proteomic analysis of the apoplast fluids will allow us to identify and characterize the diverse apoplast proteins and further elucidate their role in protecting *B. napus* from pathogen invasion (Romero et al. 2018; Tian et al. 2019).

1.15.4 Interactomics of Biological Interaction System

Interactomics is the study of networks of gene and protein interactions in biological systems. Understanding the biological process and pathogenicity mechanisms of the fungal pathogens in *Brassica* crops is crucial for the identification of disease resistance targets. A web-based database called the Pathogen-Host Interactions database (PHI-base) has been set up that stores curated experimental data obtained from host–pathogen studies, encompassing phenotypic data and biological data on pathogenicity, virulence, and effector gene functions from fungal, oomycete, and bacterial pathogens from animal, plant, fungal, and insect host species, with embedded search links including BLAST, PubMed, UniProt Knowledgebase, and others. Complementing PHI-base, PHI-Nets provides information related to networks of molecular and biological protein–protein interactions for the understanding of pathogenicity and virulence mechanisms in *Brassica* host–pathogen relationships (Ahmed et al. 2018; Janowska-Sejda et al. 2019).

1.16 Biometabolomics of *Brassica* Host–Pathogen System

Biometabolomics of molecular plant pathology refers to the study of host plant metabolism changes in response to pathogen infection that provides an understanding of how host–pathogen interaction, through a (de)activation of metabolites and associated signaling pathways, can lead toward a resistant or susceptible outcome for the host. Metabolites associated with black rot infection in *B. oleracea* were identified using liquid chromatography-quadrupole time-of-flight (LC-QTOF)-based metabolite profiling, which revealed that metabolic changes in the host occurred 48 h after infection and implicated photosynthesis, alkaloids, coumarins, and sphingolipids during the infection process. Systems biology was constructed to model the metabolic pathway for JA signaling in the *Brassica–Alternaria* pathophysiology to identify important elements in the regulation of resistance mechanisms and to pinpoint molecular targets for engineering enhanced resistance in *Brassica* crops (Castro-Moretti et al. 2020).

The huge number of data collected from multi-omics technologies will be useful in the construction of network biology, where mathematical models and computational approaches are implemented to predict the pathogenicity and virulence mechanisms in plant–pathogen interactions. Metabolic pathways supported by mathematical modeling can be used to study how cells within a multicellular organism work cooperatively to carry out a particular function. An example of systems biology was carried out for a *A. thaliana–S. sclerotiorum* pathosystem, where a genome-scale metabolic model of *S. sclerotiorum* based on global gene expression was constructed to assess the metabolic activity in different parts of the hyphal cells, supporting the hypothesis that cooperation in *S. sclerotiorum* hyphal cells is necessary for virulence and host colonization. A combination of metabolomics with quantitative genetics was used to discover the potential role of gluconasturtiin in the *B. napus* resistance response against Clubroot and the underlying QTL controlling the trait on chromosome C03 and C09. Gluconasturtiin is a

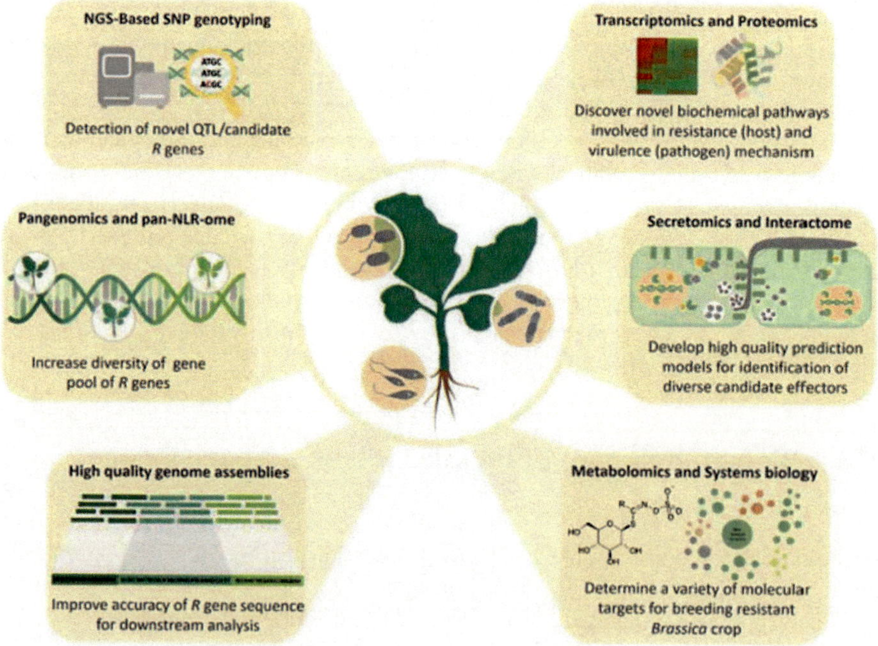

Fig. 1.3 The application of multi-omics technologies in the discovery of novel plant–pathogen interactions in the *Brassica* host pathosystems (Neik et al. 2020)

type of glucosinolate compound associated with the biotic resistance responses of *Brassica* species. The use of multi-omics supplemented with functional studies to discover key resistance pathways involved the soybean–*S. sclerotiorum* pathosystem, where the induction of JA signaling, elevated ROS control, and reprogramming of the phenylpropanoid pathway have been suggested to be important resistance mechanisms. Many more novel plant–pathogen interactions in the *Brassica* pathosystems could be discovered through the application of multi-omics technologies (Fig. 1.3 Botero et al. 2018; Peyraud et al. 2019; Neik et al. 2020).

References

Aardema MJ, Mac Gregor JT (2003) Toxicology and genetic toxicology in the new era of "toxicogenomics": impact of "-omics" technologies. In: Inoue T, Pennie WD (eds) Toxicogenomics. Springer, Tokyo. https://doi.org/10.1007/978-4-431-66999-9_22

Aarts N, Metz M, Holub E, Staskawicz BJ, Daniels MJ, Parker JE (1998) Different requirements for EDS1 and NDR1 by disease resistance genes define at least two R gene-mediated signaling pathways in *Arabidopsis*. Proc Natl Acad Sci U S A 95:10306–10311

Adhikari TB, Liu JQ, Mathur S, Wu CX, Rismmer SR (2003) Genetic and molecular analyses in crosses of Race 2 and Race 7 of *Albugo candida*. Phytopathology 93:959–965

Agrios GN (2005) Plant pathology, 5th edn. Elsevier Academic Press, Burlington, 948p

Ahmed NU, Park JI, Seo MS, Kumar TS, Lee IH, Park BS, Nou IS (2012a) Identification and expression analysis of chitinase genes related to biotic stress resistance in *Brassica*. Mol Biol Rep 39:3649–3657

Ahmed NU, Park JI, Jung HJ, Seo MS, Kumar TS, Lee IH, Nou IS (2012b) Identification and characterization of stress resistance related genes of *Brassica rapa*. Biotechnol Lett 34:979–987

Ahmed NU, Park JI, Jung HJ, Chung MY, Cho YG, Nou IS (2013) Characterization of thaumatin-like gene family and identification of *Pectobacterium carotovorum* subsp.*carotovorum* inducible genes in *Brassica oleracea*. Plant Breed Biotechnol 1:111–121

Ahmed H, Howton TC, Sun Y, Weinberger N, Belkhadir Y, Mukhtar MS (2018) Network biology discovers pathogen contact points in host protein-protein interactomes. Nat Commun 9:2312

Allen RL, Bittner-Eddy PD, Grenville-Briggs LJ, Meitz JC, Rehmany AP, Rose LE, Beynon JL (2004) Host–parasite coevolutionary conflict between *Arabidopsis* and downy mildew. Science 306:1957–1960

Ananga A, Cebert E, Soliman K, Kantety R, Pacumbaba RP, Konan K (2006) RAPD markers associated with resistance to blackleg disease in *Brassica* species. African J Biotechnol 5 (22):2041–2048

Ananga AO, Cebert E, Soliman K, Kantety R, Konan K, Ochieng JW (2008) Phylogenetic relationships within and among *Brassica* species from RAPD loci associated with blackleg resistance. Afr J Biotechnol 7(9):1287–1293

Ansan-Melayah D, Balesdent HM, Buee M, Rouxel T (1995) Genetic characterization of AvrLml, the first avirulence gene of *Leptosphaeria maculans*. Phytopathology 85:1525–1529

Ansan-Melayah D, Balesdent HM, Delourme R, Pilet LM, Tanguy X, Renard M, Rouxel T (1997) Gene-for-gene interactions in the *Leptosphaeria maculans-Brassica napus* pathosystem. Acta Hortic 459:329–334

Aravind L, Koonin EV (1999) Fold prediction and evolutionary analysis of the POZ domain: structural and evolutionary relationship with the potassium channel tetramerization domain. J Mol Biol 285:1353–1361. https://doi.org/10.1006/jmbi.1998.2394

Armstrong MR, Whisson SC, Pritchard L, Bos JI, Venter E, Avrova AO, Rehmany AP, Bohme U, Brooks K, Cherevach I, Hamlin N, White B, Fraser A, Lord A, Quail MA, Churcher C, Hall N, Berriman M, Huang S, Kamoun S, Beynon JL, Birch PRJ (2005) An ancestral oomycete locus contains late blight avirulence gene Avr3a, encoding a protein that is recognized in the host cytoplasm. Proc Natl Acad Sci U S A 102:7766–7771

Asai T, Tena G, Plotnikova J, Willmann MR, Chiu WL, Gomez-Gomez L, Boller T, Ausubel FM, Sheen J (2002) MAP kinase signaling cascade in *Arabidopsis* innate immunity. Nature 415:977–983

Asai S, Rallapalli G, Piquerez SJ, Caillaud MC, Furzer OJ, Ishaque N, Wirthmueller L, Fabro G, Shirasu K, Jones JD (2014) Expression profiling during *Arabidopsis*/downy mildew interaction reveals a highly-expressed effector that attenuates responses to salicylic acid. PLoS Pathog 10: e1004443

Asai S, Furzer OJ, Cevik V, Kim DS, Ishaque N, Goritschnig S, Staskawicz BJ, Shirasu K, Jones JDG (2018) A downy mildew effector evades recognition by polymorphism of expression and subcellular localization. Nat Commun 9:5192

Ascunce MS, Huguet-Tapia JC, Ortiz-Urquiza A, Keyhani NO, Braun EL, Goss EM (2017) Phylogenomic analysis supports multiple instances of polyphyletic in the oomycete Peronosporalean lineage. Mol Phylogenet Evol 114:199–211

Atri C, Akhatar J, Gupta M, Gupta N, Goyal A, Rana K, Kaur R, Mittal M, Sharma A, Singh MP et al (2019) Molecular and genetic analysis of defensive responses of *Brassica juncea—B. fruticulosa* introgression lines to Sclerotinia infection. Sci Rep 9:17089

Axtell MJ, Staskawicz BJ (2003) Initiation of RPS2-specified disease resistance in *Arabidopsis* is coupled to the AvrRpt2-directed elimination of RIN4. Cell 112:369–377

Bailey K, Cevik V, Holton N, Byrne-Richardson J, Sohn KH, Coates M, Woods-Tor A, Aksoy HM, Hughes L, Baxter L, Jones JD, Beynon J, Holub EB, Tor M (2011) Molecular cloning of ATR5 (Emoy2) from *Hyaloperonospora arabidopsidis*, an avirulence determinant that triggers RPP5-mediated defense in *Arabidopsis*. Mol Plant Microbe Interact 24:827–838

Bari R, Jones JD (2009) Role of plant hormones in plant defense responses. Plant Mol Biol 69:473–488. https://doi.org/10.1007/s11103-008-9435-0

Baxter L, Tripathy S, Ishaque N, Boot N, Cabral A, Kemen E, Thines M, Ah-Fong A, Anderson R, Badejoko W, Bittner-Eddy P, Boore JL, Chibucos MC, Coates M, Dehal P, Dele-haunty K, Dong S, Downton P, Dumas B, Fabro G, Fronick C, Fuerstenberg SI, Fulton L, Gaulin E, Govers F, Hughes L, Humphray S, Jiang RHY, Judelson H, Kamoun S, Kyung K, Meijer H, Minx P, Morris P, Nelson J, Phuntumart V, Qutob D, Rehmany A, Rougon-Cardoso A, Ryden P, Torto-Alalibo T, Studholme D, Wang Y, Win J, Wood J, Clifton SW, Rogers J, Van den Ackerveken G, Jones JDG, McDowell JM, Beynon J, Tyler BM (2010) Signatures of adaptation to obligate biotrophy in the *Hyaloperonospora arabidopsidis* genome. Science 330:1549–1551

Bayer PE, Hurgobin B, Golicz AA, Chan CKK, Yuan Y, Lee H, Renton M, Meng J, Li R, Long Y et al (2017) Assembly and comparison of two closely related *Brassica napus* genomes. Plant Biotechnol J 15:1602–1610

Bayer PE, Golicz AA, Tirnaz S, Chan CKK, Edwards D, Batley J (2019) Variation in abundance of predicted resistance genes in the *Brassica oleracea* pangenome. Plant Biotechnol J 17:789–800

Bayer PE, Golicz AA, Scheben A, Batley J, Edwards D (2020) Plant pan-genomes are the new reference. Nat Plants 6:914–920

Belhaj K, Cano LM, Prince DC, Kemen A, Yoshida K, Dagdas YF, Etherington GJ, Schoonbeek HJ, van Esse HP, Jones JD, Kamoun S, Schornack S (2017) *Arabidopsis* late blight: infection of a nonhost plant by *Albugo laibachii* enables full colonization by *Phytophthora infestans*. Cell Microbiol 19:e12628

van der Biezen EA, Freddie CT, Kahn K, Parker JE, Jones JD (2002) *Arabidopsis* RPP4 is a member of the RPP5 multigene family of TIR-NB-LRR genes and confers downy mildew resistance through multiple signaling components. Plant J 29:439–451

Bittner-Eddy PD, Beynon JL (2001) The *Arabidopsis* downy mildew resistance gene, RPP13-Nd, functions independently of NDR1 and EDS1 and does not require the accumulation of salicylic acid. Mol Plant Microbe Interact 14:416–421

Bittner-Eddy PD, Crute IR, Holub EB, Beynon JL (2000) *RPP13* is a simple locus in *Arabidopsis thaliana* for alleles that specify downy mildew resistance to different avirulence determinants in *Peronospora parasitica*. Plant J 21:177–188

Block A, Alfano JR (2011) Plant targets for Pseudomonas syringae type III effectors: virulence targets or guarded decoys? Curr Opin Microbiol 14:39–46

Borhan MH, Brose E, Beynon JL, Holub EB (2001) White rust (*Albugo candida*) resistance loci on three *Arabidopsis* chromosomes are closely linked to downy mildew (*Peronospora parasitica*) resistance loci. Mol Plant Pathol 2:87–95

Borhan MH, Holub EB, Beynon JL, Rozwadowski K, Rimmer SR (2004) The *Arabidopsis* TIR-NB-LRR gene RAC1 confers resistance to *Albugo candida* (white rust) and is dependent on EDS1 but not PAD4. Mol Plant Microbe Interact 17:711–719

Borhan MH, Gunn N, Cooper A, Gulden S, Tor M, Rimmer SR, Holub EB (2008) WRR4 encodes a TIR-NB-LRR protein that confers broad-spectrum white rust resistance in *Arabidopsis thaliana* to four physiological races of *Albugo candida*. Mol Plant Microbe Interact 21:757–768

Borhan MH, Holub EB, Kindrachuk C, Omidi M, Bozorgmanesh Frad G, Rimmer SR (2010) WRR4, a broad-spectrum TIR-NBLRR gene from *Arabidopsis thaliana* that confers white rust resistance in transgenic oilseed *Brassica* crops. Mol Plant Pathol 11:283–291

Botella MA, Parker JE, Frost LN, Bittner-Eddy PD, Beynon JL, Daniels MJ, Holub EB, Jones JD (1998) Three genes of the *Arabidopsis* RPP1 complex resistance locus recognize distinct *Peronospora parasitica* avirulence determinants. Plant Cell 10:1847–1860

Botero D, Alvarado C, Bernal A, Danies G, Restrepo S (2018) Network analyses in plant pathogens. Front Microbiol 9:35. https://doi.org/10.3389/fmicb.2018.00035

Bredow M, Monaghan J (2019) Regulation of plant immune signaling by calcium-dependent protein kinases. Mol Plant Microbe Interact 32:6–19

Brun H, Ruer D, Levivier S, Somda I, Renard M, Chevre MA (2001) Presence in *Leptosphaeria maculans* populations of isolates virulent on resistance introgressed into *Brassica napus* from the *B. nigra* B genome. Plant Pathol 50:69–74

Brutus A, Sicilia F, Macone A, Cervone F, De Lorenzo G (2010) A domain swap approach reveals a role of the plant wall-associated kinase 1 (WAK1) as a receptor of oligogalacturonides. Proc Natl Acad Sci U S A 107:9452

Caillaud MC, Piquerez SJ, Fabro G, Steinbrenner J, Ishaque N, Beynon J, Jones JD (2012) Subcellular localization of the *Hpa* RxLR effector repertoire identifies a tonoplast-associated protein HaRxL17 that confers enhanced plant susceptibility. Plant J 69:252–265

Cao H, Bowling SA, Gordon AS, Dong X (1994) Characterization of an *Arabidopsis* mutant that is nonresponsive to inducers of systemic acquired resistance. Plant Cell 6:1583–1592. https://doi.org/10.1105/tpc.6.11.1583

Cao H, Glazebrook J, Clarke JD, Volko S, Dong X (1997) The *Arabidopsis* NPR1 gene that controls systemic acquired resistance encodes a novel protein containing ankyrin repeats. Cell 88:57–63. https://doi.org/10.1016/S0092-8674(00)81858

Cao H, Li X, Dong XN (1998) Generation of broad–spectrum disease resistance by over-expression of an essential regulatory gene in systemic acquired resistance. Proc Natl Acad Sci U S A 95:6531–6536. https://doi.org/10.1073/pnas.95.11.6531

Carlier JD, Alabaça CA, Coelho PS, Monteiro AA, Leitão JM (2012) The downy mildew resistance locus Pp523 is located on chromosome C8 of *Brassica oleracea* L. Plant Breed 131:170–175. https://doi.org/10.1111/j.1439-0523.2011.01904.x

Carreón-Anguiano KG, Islas-Flores I, Vega-Arreguín J, Sáenz-Carbonell L, Canto-Canché BE (2020) A tool for prediction of effector protein candidates in fungal proteomic databases. Biomolecules 10:712

Castel B, Ngou PM, Cevik V, Redkar A, Kim DS, Yang Y, Ding P, Jones JDG (2019) Diverse NLR immune receptors activate defense via the RPW8-NLR NRG1. New Phytol 222(2):966–980

Castro-Moretti FR, Gentzel IN, Mackey D, Alonso AP (2020) Metabolomics as an emerging tool for the study of plant-pathogen interactions. Metabolites 10:52

Cevik V, Boutrot F, Apel W, Robert-Seilaniantz A, Furzer OJ, Redkar A, Castel B, Kover PX, Prince DC, Holub EB, Jones JDG (2019) Transgressive segregation reveals mechanisms of *Arabidopsis* immunity to *Brassica*-infecting races of white rust *Albugo candida*. Proc Natl Acad Sci U S A 116:2767

Chang JH, Urbach JM, Law TF, Arnold LW, Hu A, Gombar S et al (2005) A high-throughput, near saturating screen for type III effector genes from Pseudomonas syringae. Proc Natl Acad Sci U S A. 102:2549–2554

Chang A, Lamara M, Wei Y, Hu H, Parkin IAP, Gossen BD, Peng G, Yu F (2019) Clubroot resistance gene Rcr6 in *Brassica nigra* resides in a genomic region homologous to chromosome A08 in *B. rapa*. BMC Plant Biol 19:224

Chatterjee D, Banga S, Gupta M, Bharti S, Salisbury PA, Banga SS (2016) Resynthesis of *Brassica napus* through hybridization between *B. juncea* and *B. carinata*. Theor Appl Genet 129:977–990. https://doi.org/10.1007/s00122-016-2677-3

Chen ZJ (2007) Genetic and epigenetic mechanisms for gene expression and phenotypic variation in plant polyploids. Annu Rev Plant Biol 58:377–406. https://doi.org/10.1146/annurev.arplant.58.032806.103835

Chen X, Barnaby JY, Sreedharan A, Huang X, Orbović V, Grosser JW, Wang N, Dong X, Song WY (2013) Over–expression of the citrus gene CtNH1 confers resistance to bacterial canker disease. Physiol Mol Plant Pathol 84:115–122. https://doi.org/10.1016/j.pmpp.2013.07.002

Chen W, Li Y, Yan R, Xu L, Ren L, Liu F, Zeng L, Yang H, Chi P, Wang X et al (2019) Identification and characterization of *Plasmodiophora Brassicae* primary infection effector candidates that suppress or induce cell death in host and nonhost plants. Phytopathology 109:1689–1697

Cheng F, Wu J, Fang L, Sun S, Liu B, Lin K, Bonnema G, Wang X (2012) Biased gene fractionation and dominant gene expression among the subgenomes of *Brassica rapa*. PLoS One 7:e36442. https://doi.org/10.1371/journal.pone.0036442

Cheung F, Trick M, Drou N, Lim YP, Park JY, Kwon SJ, Kim JA, Scott R, Pires JC, Paterson AH, Town C, Bancroft I (2009) Comparative analysis between homoeologous genome segments of *Brassica napus* and its progenitor species reveals extensive sequence-level divergence. Plant Cell 21:1912–1928. https://doi.org/10.1105/tpc.108.060376

Chevre AM, Barret P, Eber F, Dupuy P, Brun H, Tanguy X, Renard M (1997) Selection of stable *Brassica napus-B. juncea* recombinant lines resistant to blackleg (*Leptosphaeria maculans*). 1. Identification of molecular markers, chromosomal and genomic origin of the introgression. Theor Appl Genet 95:1104–1111

Chittem K, Yajima WR, Goswami RS, Del Río Mendoza LE (2020) Transcriptome analysis of the plant pathogen *Sclerotinia sclerotiorum* interaction with resistant and susceptible canola (*Brassica napus*) lines. PLoS One 15:e0229844

Chou S, Krasileva KV, Holton JM, Steinbrenner AD, Alber T, Staskawicz BJ (2011) *Hyaloperonospora arabidopsidis* ATR1 effector is a repeat protein with distributed recognition surfaces. Proc Natl Acad Sci U S A 108:13323–8

Chu M, Song T, Falk KC, Zhang X, Liu X, Chang A, Lahlali R, McGregor L, Gossen BD, Yu F et al (2014) Fine mapping of Rcr1 and analyses of its effct on transcriptome patterns during infection by *Plasmodiophora Brassicae*. BMC Genomics 15:1–20

Ciaghi S, Schwelm A, Neuhauser S (2019) Transcriptomic response in symptomless roots of clubroot infected kohlrabi (*Brassica oleracea* var. *gongylodes*) mirrors resistant plants. BMC Plant Biol 19:288

Coates ME, Beynon JL (2010) *Hyaloperonospora arabidopsidis* as a pathogen model. Annu Rev Phytopathol 48:329–45

Collemare J, O'Connell R, Lebrun MH (2019) Non-proteinaceous effectors: the terra incognita of plant–fungal interactions. New Phytol 223:590–596

Comai L (2005) The advantages and disadvantages of being polyploid. Nat Rev Genet 6:836–846. https://doi.org/10.1038/nrg1711

Conn KL (1986) Leaf epicuticular wax of canola: ultrastructure, chemistry and interaction with *Alternaria Brassicae*. M.Sc. thesis, p 159

Conn KL, Tewari JP (1989) Interactions of *Alternaria Brassicae* conidia with leaf epicuticular wax of canola. Mycol Res 93:240–242. https://doi.org/10.1016/S0953-7562(89)80126-1

Conn KL, Tewari JP, Hadziyev D (1984) The role of epicuticular wax in canola in resistance to *Alternaria Brassicae*. Phytopathology 74:851. (Abstr.)

Conn KL, Tewari JP, Dahiya JS (1988) Resistance to *Alternaria Brassicae* and phytoalexin-elicitation in rapeseed and other crucifers. Plant Sci 55:21–25. https://doi.org/10.1016/0168-9452(88)90037-4

Conn KL, Jejelowa OA, Tewari JP, Bains PS (1991) *Camelina sativa* phytoalexins (Calmaxin and methoxy camalexin) provide resistance against *Alternaria Brassicae*. Can J Plant Pathol 13: Abstract 274

Constantinescu O, Fatehi J (2002) Peronospora-like fungi (Chromista, Peronosporales) parasitic on *Brassica*ceae and related hosts. Nova Hedwigia 74:291–338

Cooper AJ, Latunde-Dada AO, Woods-Tor A, Lynn J, Lucas JA, Crute IR, Holub EB (2008) Basic compatibility of *Albugo candida* in *Arabidopsis thaliana* and *Brassica juncea* causes broad-spectrum suppression of innate immunity. Mol Plant Microbe Interact 21:745–756

Covo S (2020) Genomic instability in fungal plant pathogens. Genes 11:421

Cowling WA, Balázs E (2010) Prospects and challenges for genome-wide association and genomic selection in oilseed *Brassica* species. Genome 53:1024–1028

Crews KR (2006) Individualizing chemotherapeutic treatment of colorectal cancer. Am J Health Syst Pharm 63(suppl 2):S12–S17. https://doi.org/10.2146/ajhp060113

Crute IR, Holub EB, Tor M, Brose E, Beynon JL (1993) The identification and mapping of loci in *Arabidopsis thaliana* for recognition of the fungal pathogens *Peronospora parasitica* (downy mildew) and *Albugo candida* (white blister). In: Nester EW, DPS V (eds) Advances in molecular genetics of plant-microbe interactions. Kluwer Academic, Dordrecht, pp 437–444

Crute IR, Beynon J, Dangl JL, Holub EB, Mauch-Mani B, Slusarenko A, Staskawicz BJ, Ausubel FM (1994) Microbial pathogenesis of *Arabidopsis*. In: Meyerowitz EM, Somerville CR (eds) *Arabidopsis*. Cold Spring Harbor, Cold Spring harbour Laboratory Press, pp 705–748

Cui H, Wang Y, Xue L, Chu J, Yan C, Fu J, Chen M, Innes RW, Zhou JM (2010) *Pseudomonas syringae* effector protein AvrB perturbs *Arabidopsis* hormone signaling by activating MAP kinase 4. Cell Host Microbe 7:164–75

Cui H, Tsuda K, Parker JE (2015) Effector-triggered immunity: from pathogen perception to robust defense. Annu Rev Plant Biol 66:487–511

van Damme M, Huibers RP, Elberse J, Van den Ackerveken G (2008) Arabidopsis DMR6 encodes a putative 2OG-Fe(II) oxygenase that is defense-associated but required for susceptibility to downy mildew. Plant J 54:785–793

Dangl JL, Holub EB, Debener T, Lehnackers H, Ritter C, Crute IR (1992) Genetic definition of loci involved in *Arabidopsis*-pathogen interactions. In: Koncz C, Chua NH, Schell J (eds) Methods in *Arabidopsis* research. World Scientific Publishing, London, pp 393–418

Darma R, Lutz A, Elliott CE, Idnurm A (2019) Identification of a gene cluster for the synthesis of the plant hormone abscisic acid in the plant pathogen *Leptosphaeria maculans*. Fungal Genet Biol 130:62–71

Deboever E, Deleu M, Mongrand S, Lins L, Fauconnier ML (2020) Plant–pathogen interactions: underestimated roles of phyto-oxylipins. Trends Plant Sci 25:22–34

Delaney TP, Friedrich L, Ryals JA (1995) *Arabidopsis* signal transduction mutant defective in chemically and biologically induced disease resistance. Proc Natl Acad Sci U S A 92:6602–6. https://doi.org/10.1073/pnas.92.14.6602

Delourme R, Pilet-Nayel ML, Archipiano M, Horvais R, Tanguy X, Rouxel T, Brun H, Renard M, Balesdent MH (2004) A cluster of major specific resistance genes to *Leptosphaeria maculans* in *Brassica napus*. Phytopathology 94:578–583. https://doi.org/10.1094/PHYTO.2004.94.6.578

Delourme R, Laperche A, Bouchet AS, Jubault M, Paillard S, Manzanares-Dauleux MJ, Nesi N (2018) Genes and quantitative trait loci mapping for major agronomic traits in *Brassica napus*

L. In: Liu S, Snowdon R, Chalhoub B (eds) The *Brassica napus* genome. Springer, Cham, pp 41–85

Deslandes L, Rivas S (2012) Catch me if you can: bacterial effectors and plant targets. Trends Plant Sci 17:644–655

De-Vos M, Van-Oosten VR, van-Poecke RMP, Van-Pelt JA, Pozo MJ, Mueller MJ, Buchala AJ, Métraux JP, Van Loon LC, Dicke M, CMJ P (2005) Signal signature and transcriptome changes of *Arabidopsis* during pathogen and insect attack. Mol Plant Microbe Interact 18:923–937. https://doi.org/10.1094/MPMI-18-0923

Ding Y, Mei J, Chai Y, Yu Y, Shao C, Wu Q, Disi JO, Li Y, Wan H, Qian W (2018) Simultaneous transcriptome analysis of host and pathogen highlights the interaction between *Brassica oleracea* and *Sclerotinia sclerotiorum*. Phytopathology 109:542–550

Dodds PN (2010) Genome evolution in plant pathogens. Science 330:1486–1487

Dodds PN, Rathjen JP (2010) Plant immunity: towards an integrated view of plant-pathogen interactions. Nat Rev Genet 11:539–548

Dodds PN, Rafiqi M, Gan PHP, Hardham AR, Jones DA, Ellis JG (2009) Effectors of biotrophic fungi and oomycetes-pathogenicity factors and triggers of host resistance. New Phytol 183:993–1000

Dolatabadian A, Bayer PE, Tirnaz S, Hurgobin B, Edwards D, Batley J (2019) Characterization of disease resistance genes in the *Brassica napus* pangenome reveals significant structural variation. Plant Biotechnol J 18:969–982

Dou D, Kale SD, Wang X, Jiang RH, Bruce NA, Arredondo FD et al (2008) RXLR-mediated entry of *Phytophthora sojae* effector Avr1b into soybean cells does not require pathogen-encoded machinery. Plant Cell 20:1930–1947

Duplessis S, Cuomo CA, Lin YC, Aerts A, Tisserant E, Veneault-Fourrey C et al (2011) Obligate biotrophy features unraveled by the genomic analysis of rust fungi. Proc Natl Acad Sci US A 108:9166–9171

Duran C, Eales D, Marshall D, Imelfort M, Stiller J, Berkman PJ, Clark T, McKenzie M, Appleby N, Batley J, Basford K, Edwards D (2010) Future tools for association mapping in crop plants. Genome 53:1017–1023

Durrant WE, Dong X (2004) Systemic acquired resistance. Annu Rev Phytopathol 42:185–209. https://doi.org/10.1146/annurev.phyto.42.040803.140421

Dutt M, Barthe G, Irey M, Grosser J (2015) Transgenic citrus expressing an *Arabidopsis* NPR1 gene exhibit enhanced resistance against huanglongbing (HLB citrus greening). PLoS One 10: e0137134. https://doi.org/10.1371/journal.pone.0147657

Dzurilla M, Kutschy P, Tewari JP, Ruzinsky M, Senvicky S, Kovacik V (1998) Synthesis and antifungal activity of new indolypthiazinone derivatives. Coll Czechoslovakia Chem Comm 63:94–102. https://doi.org/10.1135/cccc19980094

Eisen JA (1998) Phylogenomics: improving functional predictions for uncharacterized genes by evolutionary analysis. Genome Res 8:163–167

Ellis JG, Rafiqi M, Gan P, Chakrabarti A, Dodds PN (2009) Recent progress in discovery and functional analysis of effector proteins of fungal and oomycete plant pathogens. Curr Opin Plant Biol 12:399–405

Eulgem T, Tsuchiya T, Wang XJ, Beasley B, Cuzick A, Tör M, Zhu T, McDowell JM, Holub E, Dangl JL (2007) EDM2 is required for RPP7-dependent disease resistance in *Arabidopsis* and affects RPP7 transcript levels. Plant J 49(5):829–839. https://doi.org/10.1111/j.1365-313X.2006.02999.x

Fabro G, Steinbrenner J, Coates M, Ishaque N, Baxter L, Studholme DJ, Korner E, Allen RL, Piquerez SJ, Rougon-Cardoso A, Greenshields D, Lei R, Badel JL, Caillaud MC, Sohn KH, Van den Ackerveken G, Parker JE, Beynon J, Jones JD (2011) Multiple candidate effectors from the oomycete pathogen *Hyaloperonospora arabidopsidis* suppress host plant immunity. PLoS Pathog 7:e1002348

Farid M, Yang RC, Kebede B, Rahman H (2019) Evaluation of *Brassica oleracea* accessions for resistance to *Plasmodiophora brassicae* and identification of genomic regions associated with resistance. Genome 63:91–101

Fatima U, Bhorali P, Borah S, Senthil-Kumar M (2019) Perspectives on the utilization of resistance mechanisms from host and non-host plants for durable protection of *Brassica* crops against Alternaria blight. Peer J 7:e7486. https://doi.org/10.7717/peerj.7486

Felix G, Duran JD, Volko S, Boller T (1999) Plants have a sensitive perception system for the most conserved domain of bacterial flagellin. Plant J 18:265–276

Ferdous MJ, Hossain MR, Park JI, Robin AHK, Natarajan S, Jesse DMI, Jung HJ, Kim HT, Nou IS (2020) In-silico identification and differential expressions of LepR4-syntenic disease resistance related domain containing genes against Blackleg causal fungus *Leptosphaeria maculans* in *Brassica oleracea*. Gene Rep 19:100598

Feurtey A, Stevens DM, Stephan W, Stukenbrock EH (2019) Interspecific gene exchange introduces high genetic variability in crop pathogen. Genome Biol Evol 11:3095–3105

Fisher MC, Henk DA, Briggs CJ, Brownstein JS, Madoff LC, McCraw SL, Gurr SJ (2012) Emerging fungal threats to animal, plant and ecosystem health. Nature 484:186–194

Fomeju BF, Falentin C, Lassalle G, Manzanares-Dauleux MJ, Delourme R (2014) Homoeologous duplicated regions are involved in quantitative resistance of *Brassica napus* to stem canker. BMC Genomics 15:498. https://doi.org/10.1186/1471-2164-15-498

Fomeju BF, Falentin C, Lassalle G, Manzanares-Dauleux MJ, Delourme R (2015) Comparative genomic analysis of duplicated homoeologous regions involved in the resistance of *Brassica napus* to stem canker. Front Plant Sci 6:772. https://doi.org/10.3389/fpls.2015.00772

Friedrich L, Lawton K, Dietrich R, Willits M, Cade R, Ryals J (2001) NIM1 overexpression in *Arabidopsis* potentiates plant disease resistance and results in enhanced effectiveness of fungicides. Mol Plant Microbe Interact 14:1114–1124. https://doi.org/10.1094/MPMI.2001.14.9.1114

Fu ZQ, Dong X (2013) Systemic acquired resistance: turning local infection into Global defense. Annu Rev Plant Biol 64:839–863. https://doi.org/10.1146/annurev-arplant-042811-105606

Fu ZQ, Yan S, Saleh A, Wang W, Ruble J, Oka N, Mohan R, Spoel SH, Tada Y, Zheng N, Dong X (2012) NPR3 and NPR4 are receptors for the immune signal salicylic acid in plants. Nature 486:228–232. https://doi.org/10.1038/nature11162

Fu F, Liu X, Wang R, Zhai C, Peng G, Yu F, Fernando WGD (2019a) Fine mapping of *Brassica napus* Blackleg resistance gene *Rlm1* through bulked segregant RNA sequencing. Sci Rep 9:14600

Fu Y, Zhang Y, Mason AS, Lin B, Zhang D, Yu H, Fu D (2019b) NBS-encoding genes in *Brassica napus* evolved rapidly after allopolyploidization and co-localize with known disease resistance loci. Front Plant Sci 10:26. https://doi.org/10.3389/fpls.2019.00026

Fu F, Zhang X, Liu F, Peng G, Yu F, Fernando D (2020) Identification of resistance loci in Chinese and Canadian canola/rapeseed varieties against *Leptosphaeria maculans* based on Genome-Wide Association Studies. BMC Genomics 21:501

Furci L, Jain R, Stassen J, Berkowitz O, Whelan J, Roquis D, Baillet V, Colot V, Johannes F, Ton J (2019) Identification and characterization of hypo-methylated DNA loci controlling quantitative resistance in *Arabidopsis*. Elife 8:e40655

Gaeta RT, Chris Pires J (2010) Homoeologous recombination in allopolyploids: the polyploid ratchet. New Phytol 186:18–28

Galindo-González L, Manolii V, Hwang SF, Strelkov SE (2020) Response of *Brassica napus* to *Plasmodiophora Brassicae* involves salicylic acid-mediated immunity: an RNA-seq-based study. Front Plant Sci 11:10–25

Galvao VC, Nordstrom KJV, Lanz C, Sulz P, Mathieu J, Pose D, Schmid M, Weigel D, Schneeberger K (2012) Synteny-based mapping- by-sequencing enabled by targeted enrichment. Plant J 71:517–526

Gan PHP, Rafiqi M, Ellis JG, Jones DA, Hardham AR, Dodds PN (2010a) Lipid binding activities of flax rust AvrM and AvrL567 effectors. Plant Signal Behav 5:1272–1775

Gan PHP, Rafiqi M, Hardham AR, Dodds PN (2010b) Effectors of biotrophic fungal plant pathogens. Funct Plant Biol 37:913–918

Gao M, Liu J, Bi D, Zhang Z, Cheng F, Chen S, Zhang Y (2008) MEKK1, MKK1/MKK2 and MPK4 function together in a mitogen-activated protein kinase cascade to regulate innate immunity in plants. Cell Res 18:1190–1198

Genva M, Obounou Akong F, Andersson MX, Deleu M, Lins L, Fauconnier ML (2019) New insights into the biosynthesis of esterified oxylipins and their involvement in plant defense and developmental mechanisms. Phytochem Rev 18:343–358

Glazebrook J (2005) Contrasting mechanisms of defense against biotrophic and necrotrophic pathogens. Annu Rev Phytopathol 43:205–227. https://doi.org/10.1146/annurev.phyto.43.040204.135923

Glazebrook J, Rogers EE, Ausubel FM (1996) Isolation of *Arabidopsis* mutants with enhanced disease susceptibility by direct screening. Genetics 143:973–982

Gohre V, Robatzek S (2008) Breaking the barriers: microbial effector molecules subvert plant immunity. Ann Rev Phytopathol 2008(46):189–215

Goker M, Voglmayr H, García-Blázquez G, Oberwinkler F (2009) Species delimitation in downy mildews: the case of *Hyaloperonospora* in the light of nuclear ribosomal ITS and LSU sequences. Mycol Res 113:308–325

Golicz AA, Bayer PE, Barker GC, Edger PP, Kim H, Martinez PA, Chan CKK, Severn-Ellis A, Mc Combie WR, Parkin IAP et al (2016) The pangenome of an agronomically important crop plant *Brassica oleracea*. Nat Commun 7:13390

Gomez-Gomez L, Boller T (2000) FLS2: an LRR receptor-like kinase involved in the perception of the bacterial elicitor flagellin in *Arabidopsis*. Mol Cell 5:1003–1011

Gomez-Gomez L, Boller T (2002) Flagellin perception: a paradigm for innate immunity. Trends Plant Sci 7:251–256

Goritschnig S, Krasileva KV, Dahlbeck D, Staskawicz BJ (2012) Computational prediction and molecular characterization of an oomycete effector and the cognate *Arabidopsis* resistance gene. PLoS Genet 8:e1002502

Goritschnig S, Steinbrenner AD, Grunwald DJ, Staskawicz BJ (2016) Structurally distinct *Arabidopsis thaliana* NLR immune receptors recognize tandem WY domains of an oomycete effector. New Phytol 210:984–996

Graham-Taylor C, Kamphuis LG, Derbyshire MC (2020) A detailed in silico analysis of secondary metabolite biosynthesis clusters in the genome of the broad host range plant pathogenic fungus *Sclerotinia sclerotiorum*. BMC Genomics 21:7

Grant M, Lamb C (2006) Systemic immunity. Curr Opin Plant Biol 9:414–420. https://doi.org/10.1016/j.pbi.2006.05.013

Grant MR, McDowell JM, Sharpe AG, de Torres ZM, Lydiate DJ, Dangl JL (1998) Independent deletions of a pathogen-resistance gene in *Brassica* and *Arabidopsis*. Proc Natl Acad Sci U S A 95:15843–15848

Gunn ND, Byrne J, Holub EB (2002) Out crossing of two homothallic isolates of *Peronospora parasitica* and segregation of avirulence matching six resistance loci in *Arabidopsis thaliana*. In: Spencer-Phillips P, Gisi U, Lebeda A (eds) Advances in Downy mildew research. Kluwer Academic, Dordrecht, pp 185–188

Gupta SK, Kaushik CD (2002) Metabolic changes in mustard leaf and silique wall due to the infection of Alternaria blight (*Alternaria Brassicae*). Cruciferae Newslett 24:85–86

Gupta SK, Gupta PP, Yadav TP, Kaushik CD (1990) Metabolic changes in mustard due to *Alternaria* leaf blight. Indian Phytopathol 43:64–69

Gupta SK, Gupta PP, Kaushik CD (1995) Changes in leaf peroxidase, polyphenols oxidase, catalase and total phenols due to Alternaria leaf blight in *Brassica* species. Indian J Mycol Plant Pathol 25:175–180

Gust AA, Felix G (2014) Receptor like proteins associates with SOBIR1-type of adaptors to form bimolecular receptor kinases. Curr Opin Plant Biol 21:104–111

Haddadi P, Larkan NJ, Borhan MH (2019) Dissecting R gene and host genetic background effect on the *Brassica napus* defense response to *Leptosphaeria maculans*. Sci Rep 9:6947

Hahn KK, Wolff JJ, Kolesar JM (2006) Pharmacogenetics and irinotecan therapy. Am J Health Syst Pharm 63(22):2211–2217. https://doi.org/10.2146/ajhp060155

Hann DR, Gimenez-Ibanez S, Rathjen JP (2010) Bacterial virulence effectors and their activities. Curr Opin Plant Biol 13:388–393

Harper AL, Trick M, Higgins J, Fraser F, Clissold L, Wells R, Hattori C, Werner P, Bancroft I (2012) Associative transcriptomics of traits in the polyploid crop species *Brassica napus*. Nat Biotechnol 30:798–802

Hatakeyama K, Suwabe K, Tomita RN, Kato T, Nunome T, Fukuoka H, Matsumoto S (2013) Identification and characterization of Crr1a, a gene for resistance to clubroot disease (*Plasmodiophora Brassicae* Woronin) in *Brassica rapa* L. PLoS One 8:e54745. https://doi.org/10.1371/journal.pone.0054745

Hayward A, Dalton-Morgan J, Mason A, Zander M, Edwards D, Batley J (2012) SNP discovery and applications in *Brassica napus*. J Plant Biotechnol 39:49–61

Hein I, Gilroy EM, Armstrong MR, Birch PRJ (2009) The zig-zag-zig in oomycete-plant interactions. Mol Plant Pathol 10:547–562

Hepworth SR, Zhang Y, McKim S, Li X, Haughn GW (2005) BLADE-ON-PETIOLE–dependent signaling controls leaf and floral patterning in Arabidopsis. Plant Cell 17:1434–1448. https://doi.org/10.1105/tpc.104.030536

Herlihy J, Ludwig NR, van den Ackerveken G, McDowell JM (2019) Oomycetes used in *Arabidopsis* research. Arabidopsis Book 17:e0188. https://doi.org/10.1199/tab.0188

Holub EB (2001) The arms race is ancient history in *Arabidopsis*, the wildflower. Nat Rev Genet 2:516–527

Holub EB (2008) Natural history of *Arabidopsis thaliana* and oomycete symbioses. Eur J Plant Pathol 122:91–109

Holub EB, Beynon JL (1996) Symbiology of mouse ear cress (*Arabidopsis thaliana*) and oomycetes. Adv Bot Res 24:227–273

Holub E, Brose E, Tor M, Clay C, Crute IR, Beynon JL (1994a) Phenotypic and genotypic variation in the interaction between *Arabidopsis thaliana* and *Albugo candida*. Mol Plant Microbe Interact 8:916–928

Holub EB, Beynon JL, Crute IR (1994b) Phenotypic and genotypic characterization of interactions between isolates of *Peronospora parasitica* and accessions of *Arabidopsis thaliana*. Mol Plant Microbe Interact 7:223–239

Hulbert SH, Webb CA, Smith SM, Sun Q (2001) Resistance gene complexes: evolution and utilization. Annu Rev Phytopathol 39:285–312. https://doi.org/10.1146/annurev.phyto.39.1.285

Hurgobin B, Golicz AA, Bayer PE, Chan CKK, Tirnaz S, Dolatabadian A, Schiessl SV, Samans B, Montenegro JD, Parkin IAP et al (2018) Homoeologous exchange is a major cause of gene presence/absence variation in the amphidiploid *Brassica napus*. Plant Biotechnol J 16:1265–1274

Ichimura K, Shinozaki K, Tena G, Sheen J, Henry Y, Champion A, Kreis M, Zhang S, Hirt H (2002) Mitogen-activated protein kinase cascades in plants: a new nomenclature. Trends Plant Sci 7:301–308

Inaba R, Nishio T (2002) Phylogenetic analysis of *Brassica*ceae based on the nucleotide sequences of the S-locus related gene, SLRI. Theor Appl Genet 105:1159–1165

Jammes F, Song C, Shin D, Munemasa S, Takeda K, Gu D, Cho D, Lee S, Giordo R, Sritubtim S (2009) MAP kinases MPK9 and MPK12 are preferentially expressed in guard cells and positively regulate ROS-mediated ABA signaling. Proc Natl Acad Sci U S A 106:20520–20525

Jammes F, Yang X, Xiao S, Kwak JM (2011) Two *Arabidopsis* guard cell-preferential MAPK genes, MPK9 and MPK12, function in biotic stress response. Plant Signal Behav 6:1875–1877

Janowska-Sejda EI, Lysenko A, Urban M, Rawlings C, Tsoka S, Hammond-Kosack KE (2019) PHI-Nets: a network resource for Ascomycete fungal pathogens to annotate and identify putative virulence interacting proteins and siRNA targets. Front Microbiol 10:2721

Jejelowo OA, Conn KL, Tewari JP (1991) Relationship between conidial concentration, germling growth, and phytoalexin production by *Camelina sativa* leaves inoculated with *Alternaria Brassicae*. Mycol Res 95:928–934. https://doi.org/10.1016/S0953-7562(09)80089-0

Ji R, Gao S, Bi Q, Wang Y, Lv M, Ge W, Feng H (2021) The salicylic acid signaling pathway plays an important role in the resistant process of *Brassica rapa* L. ssp. *pekinensis* to *Plasmodiophora brassicae* Woronin. J Plant Growth Regul 40:405–422. https://doi.org/10.1007/s00344-020-10105-4

Jones JD, Dangl JL (2006) The plant immune system. Nature 444:323–329

de Jonge R, Peter van Esse H, Kombrink A, Shinya T, Desaki Y, Bours R, van der Krol S, Shibuya N, Joosten MHAJ, Thomma BPHJ (2010) Conserved fungal LysM effector Ecp6 prevents chitin-triggered immunity in plants. Science 329:953–955

Jouet A, Saunders DGO, McMullan M, Ward B, Furzer O, Jupe F, Cevik V, Hein I, Thilliez GJA, Holub E, van Oosterhout C, Jones JDG (2018) *Albugo candida* race diversity, ploidy and host-associated microbes revealed using DNA sequence capture on diseased plants in the field. New Phytol 221:1529–43

Kaku H, Nishizawa Y, Ishii-Minami N, Akimoto-Tomiyama C, Dohmae N, Takio K et al (2006) Plant cells recognize chitin fragments for defense signaling through a plasma membrane receptor. Proc Natl Acad Sci U S A 103:11086–11091

Kale SD, Gu B, Capelluto DGS, Dou D, Feldman E, Rumore A et al (2010) External lipid PI3P mediates entry of eukaryotic pathogen effectors into plant and animal host cells. Cell 142:284–295

Kamoun S (2009) The secretome of plant-associated fungi and oomycetes. In: Deising HB (ed) The mycota. Springer, Berlin, pp 173–180

Kato T, Hatakayema K, Fukino N, Matsumoto S (2013) Fine mapping of the clubroot resistance gene CRb and development of a useful selectable marker in *Brassica rapa*. Breed Sci 63:116–124. https://doi.org/10.1270/jsbbs.63.116

Kaur S, Cogan NO, Ye G, Baillie RC, Hand ML, Ling AE, Mc Gearey AK, Kaur J, Hopkins CJ, Todorovic M, Mountford H, Edwards D, Batley J, Burton W, Salisbury P, Gororo N, Marcroft S, Kearney G, Smith KF, Forster JW, Spangenberg GC (2009) Genetic map construction and QTL mapping of resistance to blackleg (*Leptosphaeria maculans*) disease in Australian canola (*Brassica napus* L.) cultivars. Theor Appl Genet 120:71–83

Kaur P, Jost R, Sivasithamparam K, Barbetti MJ (2011) Proteome analysis of the *Albugo candida-Brassica juncea* pathosystem reveals that the timing of the expression of defense-related genes is a crucial determinant of pathogenesis. J Exp Bot 62:1285–1298

Kayum MA, Jung HJ, Park JI, Ahmed NU, Saha G, Yang TJ, NOU IS (2015a) Identification and expression analysis of WRKY family genes under biotic and abiotic stresses in *Brassica rapa*. Mol Genet Genomics 290:79–95

Kayum MA, Park JI, Ahmed NU, Jung HJ, Saha G, Kang JG, Nou IS (2015b) Characterization and stress-induced expression analysis of Alfin-like transcription factors in *Brassica rapa*. Mol Genet Genomics 290:1299–1311

Kayum MA, Kim HT, Nath UK, Park JI, Kho KH, Cho YG, Nou IS (2016) Research on biotic and abiotic stress related genes exploration and prediction in *Brassica rapa* and *B. oleracea*: a review. Plant Breed Biotechnol 4(2):135–144. https://doi.org/10.9787/PBB.2016.4.2.135

Kemen E, Jones JD (2012) Obligate biotroph parasitism: can we link genomes to lifestyles? Trends Plant Sci 17:448–457

Kemen E, Gardiner A, Schultz-Larsen T, Kemen AC, Balmuth AL, Robert-Seilaniantz A, Bailey K, Holub E, Studholme DJ, Maclean D, Jones JD (2011) Gene gain and loss during evolution of obligate parasitism in the white rust pathogen of *Arabidopsis thaliana*. PLoS Biol 9:e1001094

Khang CH, Park SY, Lee YH, Valent B, Kang S (2008) Genome organization and evolution of the AVR-Pita avirulence gene family in the *Magnaporthe grisea* species complex. Mol Plant-Microbe Interact 21:658–670

Khang CH, Berruyer R, Giraldo MC, Kankanala P, Park SY, Czymmek K, Kang S, Valent B (2010) Translocation of Magnaporthe oryzae effectors into rice cells and their subsequent cell-to cell movement. Plant Cell 22:1388–1403

Khokon MAR, Salam MA, Jammes F, Ye W, Hossain MA, Uraji M, Nakamura Y, Mori IC, Kwak JM, Murata Y (2015) Two guard cell mitogen-activated protein kinases, MPK9 and MPK12, function in methyl jasmonate-induced stomatal closure in *Arabidopsis thaliana*. Plant Biol (Stuttg.) 17:946–952

King SRF, McLellan H, Boevink PC, Armstrong MR, Bukharova T, Sukarta O, Win J, Kamoun S, Birch PRJ, Banfield MJ (2014) *Phytophthora infestans* RXLR effector PexRD2 interacts with host MAPKKKε to suppress plant immune signaling. Plant Cell 26:1345–9.

Klevernic IV, Stafford MJ, Morrice N, Peggie M, Morton S, Cohen P (2006) Characterization of the reversible phosphorylation and activation of ERK8. Biochem J 394:365–373

Koch E, Slusarenko AJ (1900) Fungal pathogens of *Arabidopsis thaliana* (L.) Heyhn. Bot Helvet 100:257–268

Koch E, Slusarenko AJ (1990) *Arabidopsis* is susceptible to infection by a downy mildew fungus. Plant Cell 2:437–5

Koeck M, Hardham AR, Dodds PN (2011) The role of effectors of biotrophic and hemibiotrophic fungi in infection. Cell Microbiol 13(12):1849–1857

Kohmoto K, Otani H, Tsuge T (1995) *Alternaria alternata* pathogens. In: Kohmoto K, Singh US, Singh RP (eds) Vol. 2 Eukaryotes. Oxford Pergamon, Turkey, pp 51–63

Krasileva KV, Dahlbeck D, Staskawicz BJ (2010) Activation of an *Arabidopsis* resistance protein is specified by the in planta association of its leucine-rich repeat domain with the cognate oomycete effector. Plant Cell 22:2444–2458

Krishnia SK, Saharan GS, Singh D (2000) Genetics of Alternaria blight resistance in inter and intraspecific crosses of *Brassica juncea* and *B. carinata*. Ann Biol 16:212–216

Kunze G, Zipfel C, Robatzek S, Niehaus K, Boller T, Felix G (2004) The N terminus of bacterial elongation factor Tu elicits innate immunity in Arabidopsis plants. Plant Cell 16:3496–3507

Lacombe S, Rougon-Cardoso A, Sherwood E, Peeters N, Dahlbeck D, van Esse HP, Smoker M, Rallapalli G, Thomma BPHJ, Staskawicz B, Jones JDG, Zipfel C (2010) Interfamily transfer of a plant pattern-recognition receptor confers broad-spectrum bacterial resistance. Nat Biotechnol 28:365–U394

Lai Y, Eulgem T (2018) Transcript-level expression control of plant NLR genes. Mol Plant Pathol 19:1267–1281

Lai Y, Cuzick A, Lu XM, Wang J, Katiyar N, Tsuchiya T, Le Roch K, McDowell JM, Holub E, Eulgem T (2019) The *Arabidopsis* RRM domain protein EDM3 mediates race-specific disease resistance by controlling H3K9me2-dependent alternative polyadenylation of RPP7 immune receptor transcripts. Plant J 97:646–660

Lan M, Li G, Hu J, Yang H, Zhang L, Xu X, Liu J, He J, Sun R (2019) iTRAQ-based quantitative analysis reveals proteomic changes in Chinese cabbage (*Brassica rapa* L.) in response to *Plasmodiophora Brassicae* infection. Sci Rep 9:12058

Lapin D, Meyer RC, Takahashi H, Bechtold U, Van den Ackerveken G (2012) Broad-spectrum resistance of *Arabidopsis* C24 to downy mildew is mediated by different combinations of isolate-specific loci. New Phytol 196:1171–1181

Larkan NJ, Lydiate DJ, Parkin IA, Nelson MN, Epp DJ, Cowling WA, Rimmer SR, Borhan MH (2013) The *Brassica napus* blackleg resistance gene LepR3 encodes a receptor-like protein triggered by the *Leptosphaeria maculans* effector AVRLM1. New Phytol 197:595–605. https://doi.org/10.1111/nph.12043

Larkan NJ, Yu F, Lydiate DJ, Rimmer SR, Borhan MH (2016) Single R gene introgression lines for accurate dissection of the *Brassica-Leptosphaeria* pathosystem. Front Plant Sci 7:1771

Larkan NJ, Ma L, Haddadi P, Buchwaldt M, Parkin IAP, Djavaheri M, Borhan MH (2019) The *Brassica napus* Wall-Associated Kinase-Like (WAKL) gene *Rlm9* provides race-specific blackleg resistance. Plant J 104(4):892–900. https://doi.org/10.1111/tpj.14966

Lawrence GJ, Dodds PN, Ellis JG (2010) Transformation of the flax rust fungus, *Melampsora lini*: selection via silencing of an avirulence gene. Plant J 61:364–369

Lee DH, Lim JA, Lee J, Roh E, Jung K, Choi M, Oh C, Ryu S, Yun J, Heu S (2013) Characterization of genes required for the pathogenicity of *Pectobacterium carotovorum* subsp. *carotovorum* Pcc21 in Chinese cabbage. Microbiology 159:1487–1496

Lee J, Izzah NK, Jayakodi M, Perumal S, Joh HJ, Lee HJ, Lee SC, Park JY, Yang KW, Nou IS et al (2015) Genome-wide SNP identification and QTL mapping for black rot resistance in cabbage. BMC Plant Biol 15:32

Lee Y, Kim YJ, Kim MH, Kwak JM (2016) MAPK cascades in guard cell signal transduction. Front Plant Sci 7:80

Lee H, Chawla HS, Obermeier C, Dreyer F, Abbadi A, Snowdon R (2020) Chromosome-scale assembly of winter oilseed rape *Brassica napus*. Front Plant Sci 11:496

Leonelli L, Pelton J, Schoeffler A, Dahlbeck D, Berger J, Wemmer DE, Staskawicz B (2011) Structural elucidation and functional characterization of the *Hyaloperonospora arabidopsidis* effector protein ATR13. PLoS Pathog 7:e1002428

Li J, Zhao Z, Hayward A, Cheng H, Fu D (2015) Integration analysis of quantitative trait loci for resistance to *Sclerotinia sclerotiorum* in *Brassica napus*. Euphytica 205:483–489. https://doi.org/10.1007/s10681-015-1417-0

Li L, Luo Y, Chen B, Xu K, Zhang F, Li H, Huang Q, Xiao X, Zhang T, Hu J et al (2016) A genome-wide association study reveals new loci for resistance to clubroot disease in *Brassica napus*. Front Plant Sci 7:1483

Lin K, Zhang N, Severing EI, Nijveen H, Cheng F, Visser RGF, Wang X, de Ridder D, Bonnema G (2014) Beyond genomic variation—comparison and functional annotation of three *Brassica rapa* genomes: a turnip, a rapid cycling and a Chinese cabbage. BMC Genomics 15:250

Links MG, Holub E, Jiang RH, Sharpe AG, Hegedus D, Beynon E, Sillito D, Clarke WE, Uzuhashi S, Borhan MH (2011) De novo sequence assembly of *Albugo candida* reveals a small genome relative to other biotrophic oomycetes. BMC Genomics 12:503

Lipka V, Dittgen J, Bednarek P, Bhat R, Wiermer M, Stein M, Landtag J, Brandt W, Rosahl S, Scheel D, Llorente F, Molina A, Parker J, Somerville S, Lefert PS (2005) Pre- and post-invasion defenses both contribute to non-host resistance in *Arabidopsis*. Science 310:1180–1183

Liu SL, Adams KL (2010) Dramatic change in function and expression pattern of a gene duplicated by polyploidy created a paternal effect gene in the *Brassica*ceae. Genome Biol Evol 27:2817–2828. https://doi.org/10.1093/molbev/msq169

Liu S, Liu Y, Yang X, Tong C, Edwards D, Parkin IA, Zhao M, Ma J, Yu J, Huang S, Wang X, Wang J, Lu K, Fang Z, Bancroft I, Yang TJ, Hu Q, Wang X, Yue Z, Li H, Yang L, Wu J, Zhou Q, Wang W, King GJ, Pires JC, Lu C, Wu Z, Sampath P, Wang Z, Guo H, Pan S, Yang L, Min J, Zhang D, Jin D, Li W, Belcram H, Tu J, Guan M, Qi C, Du D, Li J, Jiang L, Batley J, Sharpe AG, Park BS, Ruperao P, Cheng F, Waminal NE, Huang Y, Dong C, Wang L, Li J, Hu Z, Zhuang M, Huang Y, Huang J, Shi J, Mei D, Liu J, Lee TH, Wang J, Jin H, Li Z, Li X, Zhang J, Xiao L, Zhou Y, Liu Z, Liu X, Qin R, Tang X, Liu W, Wang Y, Zhang Y, Lee J, Kim HH, Denoeud F, Xu X, Liang X, Hua W, Wang X, Wang J, Chalhoub B, Paterson AH (2014) The *Brassica oleracea* genome reveals the asymmetrical evolution of polyploid genomes. Nat Commun 5:3930. https://doi.org/10.1038/ncomms4930

Liu M, Wu F, Wang S, Lu Y, Chen X, Wang Y, Gu A, Zhao J, Shen S (2019) Comparative transcriptome analysis reveals defense responses against soft rot in Chinese cabbage. Hortic Res 6:68

Lu K, Wei L, Li X, Wang Y, Wu J, Liu M, Zhang C, Chen Z, Xiao Z, Jian H et al (2019) Whole-genome resequencing reveals *Brassica napus* origin and genetic loci involved in its improvement. Nat Commun 10:1154

Lucas J, Hayter J, Crute I (1995) The Downy Mildews: host specificity and pathogenesis. In: Singh U, Singh R (eds) Pathogenesis and host specificity in plant diseases. Permagon, Oxford, pp 217–234

Ma L, Djavaheri M, Wang H, Larkan NJ, Haddadi P, Beynon E, Gropp G, Borhan MH (2018) *Leptosphaeria maculans* effector protein *AvrLm1* modulates plant immunity by enhancing MAP Kinase 9 phosphorylation. iScience 3:177–91

Makandar R, Essig JS, Schapaugh MA, Trick HN, Shah J (2006) Genetically engineered resistance to Fusarium head blight in wheat by expression of *Arabidopsis* NPR1. Mol Plant Microbe Interact 19:123–129. https://doi.org/10.1094/MPMI-19-0123

Manzanares-Dauleux MJ, Delourme R, Baron F, Thomas G (2000) Mapping of one major gene and of QTLs involved in resistance to clubroot in *Brassica napus*. Theor Appl Genet 101:885–891. https://doi.org/10.1007/s001220051557

Marra R, Li H, Barbetti MJ, Sivasithamparam K, Vinale F, Cavallo P, Lorito M (2010) Proteomic analysis of the interaction between *Brassica napus* cv. Surpass 400 and virulent or avirulent isolates of *Leptosphaeria maculans*. J Plant Pathol 92:89–101

Mason AS, Snowdon RJ (2016) Oilseed rape: learning about ancient and recent polyploid evolution from a recent crop species. Plant Biol 18:883–892. https://doi.org/10.1111/plb.12462

Mathur B, Chand L (1991) Effect of *Alternaria Brassicae* toxin on photosynthesis and photorespiration in *Brassica* cultivars. Crop Res 4:146–151

Mayerhofer R, Wilde K, Mayerhofer M, Lydiate D, Bansal VK, Good AG, Parkin IA (2005) Complexities of chromosome landing in a highly duplicated genome: toward map-based cloning of a gene controlling blackleg resistance in *Brassica napus*. Genetics 171:1977–1988. https://doi.org/10.1534/genetics.105.049098

McDowell JM (2011) Genomes of obligate plant pathogens reveal adaptations for obligate parasitism. Proc Natl Acad Sci U S A 108:8921–8922

McDowell JM (2014) *Hyaloperonospora arabidopsidis*: a model pathogen of *Arabidopsis*. In: Dean RA, Lichens-Park A, Kole C (eds) Genomics of plant-associated fungi and oomycetes: dicot pathogens. Springer, Berlin, pp 209–234.

McDowell JM, Meyers BC (2013) A transposable element is domesticated for service in the plant immune system. Proc Natl Acad Sci U S A 110:14821–14822

McDowell J, Simon S (2006) Recent insights into R gene evolution. Mol Plant Pathol 7:437–448

McDowell JM, Dhandaydham M, Long TA, Aarts MG, Goff S, Holub EB, Dangl JL (1998) Intragenic recombination and diversifying selection contribute to the evolution of downy mildew resistance at the RPP8 locus of *Arabidopsis*. Plant Cell 10:1861–1874

McDowell JM, Cuzick A, Can C, Beynon J, Dangl JL, Holub EB (2000) Downy mildew (*Peronospora parasitica*) resistance genes in *Arabidopsis* vary in functional requirements for NDR1, EDS1, NPR1, and salicylic acid accumulation. Plant J 22:523–530

McMullan M, Gardiner A, Bailey K, Kemen E, Ward BJ, Cevik V, Robert-Seilaniantz A, Schultz-Larsen T, Balmuth A, Holub E, van Oosterhout C, Jones JD (2015) Evidence for suppression of immunity as a driver for genomic introgressions and host range expansion in races of *Albugo candida*, a generalist parasite. Elife 4:e04550

Medhi BN (1985) Intermating and mutagenesis for release of genetic variability in Indian mustard [*B. juncea* (L) Czern and Coss.]. Ph.D thesis

Meng X, Zhang S (2013) MAPK cascades in plant disease resistance signaling. Annu Rev Phytopathol 51:245–266

Metzker ML (2010) Sequencing technologies — the next generation. Nat Rev Genet 11:31–46

Meyers BC, Kozik A, Griego A, Kuang H, Michelmore RW (2003) Genome-wide analysis of NB-LRR-encoding genes in *Arabidopsis*. Plant Cell 15:809–834

Mindrinos M, Katagiri F, Yu GL, Ausubel FM (1994) The *Arabidopsis thaliana* disease resistance gene RPS2 encodes a protein containing a nucleotide-binding site and leucine-rich repeats. Cell 78:1089–1099

Miya A, Albert P, Shinya T, Desaki Y, Ichimura K, Shirasu K, Narusaka Y, Kawakami N, Kaku H, Shibuya N (2007) CERK1, a LysM receptor kinase, is essential for chitin elicitor signaling in *Arabidopsis*. Proc Nat Acad Sci U S A 104:19613–19618

Mogg K, Philippot P, Bradley BP (2004) Selective attention to angry faces in clinical social phobia. J Abnormal Psychol 113:160–165

Mohr TJ, Mammarella ND, Hoff T, Woffenden BJ, Jelesko JG, McDowell JM (2010) The *Arabidopsis* downy mildew resistance gene RPP8 is induced by pathogens and salicylic acid and is regulated by W box cis elements. Mol Plant Microbe Interact 23:1303–1315

Montillet JL, Leonhardt N, Mondy S, Tranchimand S, Rumeau D, Boudsocq M, Garcia AV, Douki T, Bigeard J, Laurière C (2013) An abscisic acid-independent oxylipin pathway controls stomatal closure and immune defense in *Arabidopsis*. PLoS Biol 11:e1001513

Moon JY, Kim ST, Choi GJ, Kwon SY, Cho HS, Kim HS, Moon JS, Park JM (2020) Comparative proteomic analysis of host responses to *Plasmodiophora Brassicae* infection in susceptible and resistant *Brassica oleracea*. Plant Biotechnol Rep 14:263–274

Morgan W, Kamoun S (2007) RXLR effectors of plant pathogenic oomycetes. Curr Opin Microbiol 10:332–338

Mosquera G, Giraldo MC, Khang CH, Coughlan S, Valent B (2009) Interaction transcriptome analysis identifies *Magnaporthe oryzae* BAS1–4 as biotrophy-associated secreted proteins in rice blast disease. Plant Cell 21:1273–1290

Mou Z, Fan W, Dong X (2003) Inducers of plant systemic acquired resistance regulate NPR function through redox changes. Cell 113:935–944. https://doi.org/10.1016/S0092-8674(03)00429-X

Mukhtar MS, Carvunis AR, Dreze M, Epple P, Steinbrenner J, Moore J, Tasan M, Galli M, Hao T, Nishimura MT, Pevzner SJ, Donovan SE, Ghamsari L, Santhanam B, Romero V, Poulin MM, Gebreab F, Gutierrez BJ, Tam S, Monachello D, Boxem M, Harbort CJ, McDonald N, Gai L, Chen H, He Y, Vandenhaute J, Roth FP, Hill DE, Ecker JR, Vidal M, Beynon J, Braun P, Dangl JL (2011) Independently evolved virulence effectors converge onto hubs in a plant immune system network. Science 333:596–601

Nagy SK, Darula Z, Kállai BM, Bögre L, Bánhegyi G, Medzihradszky KF, Horváth GV, Mészáros T (2015) Activation of AtMPK9 through autophosphorylation that makes it independent of the canonical MAPK cascades. Biochem J 467:167–175

Naito K, Taguchi F, Suzuki T, Inagaki Y, Toyoda K, Shiraishi T, Ichinose Y (2008) Amino acid sequence of bacterial microbe-associated molecular pattern flg22 is required for virulence. Mol Plant-Microbe Interact 21:1165–1174

Neik TX, Barbetti MJ, Batley J (2017) Current status and challenges in identifying disease resistance genes in *Brassica napus*. Front Plant Sci 8:1788. https://doi.org/10.3389/fpls.2017.01788

Neik TX, Amas J, Barbetti M, Edwards D, Batley J (2020) Understanding host-pathogen interactions in *Brassica napus* in the omics era. Plants 9:1336. https://doi.org/10.3390/plants9101336

Nemri A, Atwell S, Tarone AM, Huang YS, Zhao K, Studholme DJ, Nordborg M, Jones JD (2010) Genome-wide survey of *Arabidopsis* natural variation in downy mildew resistance using combined association and linkage mapping. Proc Natl Acad Sci U S A 107:10302–10307

Nishimura S, Kohmoto K (1983) Host-specific toxins and chemical structures from *Alternaria* species. Ann Rev Phytopathol 21:87–116. https://doi.org/10.1146/annurev.phy.21.090183.000511

Noel L, Moores TL, van Der Biezen EA, Parniske M, Daniels MJ, Parker JE, Jones JD (1999) Pronounced intraspecific haplotype divergence at the RPP5 complex disease resistance locus of *Arabidopsis*. Plant Cell 11:2099–2112

Nurnberger T, Kemmerling B (2009) PAMP-triggered basal immunity in plants. Adv Bot Res 51:1–38

O'Brien SJ, Stanyon R (1999) Phylogenomics: ancestral primate viewed. Nature 402:365–366

Ochieng JW, Muigai WT, Ude GN (2007) Phylogenetics in plant biotechnology: principles, obstacles and opportunities for the resource poor. Afr J Biotechnol 6:639–649

Oliver RP, Ipcho SVS (2004) *Arabidopsis* pathology breathes new life into the necrotrophs-vs.-biotrophs classification of fungal pathogens. Mil Plant Pathol 5(4):347–52. https://doi.org/10.1111/j.1364-3703.2004.00228.x

Oliver SG, Winson MK, Kell DB, Baganz F (1998) Systematic functional analysis of the yeast genome. Trends Biotechnol 16:373–378

Osborn TC, Butrulle DV, Sharpe AG, Pickering KJ, Parkin IAP, Parker JS (2003) Detection and effects of a homeologous reciprocal transposition in *Brassica napus*. Genetics 165:1569–1577

Pang W, Liang S, Li X, Li P, Yu S, Lim YP, Piao Z (2014) Genetic detection of clubroot resistance loci in a new population of *Brassica rapa*. Hortic Environ Biotechnol 55:540–547. https://doi.org/10.1007/s13580-014-0079-5

Panstruga R, Dodds PN (2009) Terrific protein traffic: the mystery of effector protein delivery by filamentous plant pathogens. Science 324:748–750

Parker JE, Holub EB, Frost LN, Falk A, Gunn ND, Daniels MJ (1996) Characterization of eds1, a mutation in *Arabidopsis* suppressing resistance to *Peronospora parasitica* specified by several different RPP genes. Plant Cell 8:2033–2046

Parker JE, Coleman MJ, Szabo V, Frost LN, Schmidt R, van der Biezen E, Moores T, Dean C, Daniels MJ, Jones JDG (1997) The *Arabidopsis* downy mildew resistance gene Rpp5 shares similarity to the Toll and interleukin-1 receptors with N and L6. Plant Cell 9:879–894

Parkhi V, Kumar V, Campbell LM, Bell AA, Shah J, Rathore KS (2010) Resistance against various fungal pathogens and reniform nematode in transgenic cotton plants expressing *Arabidopsis* NPR1. Transgenic Res 19:959–975. https://doi.org/10.1007/s11248-010-9374-9

Pathak RK, Baunthiyal M, Shukla R, Pandey D, Taj G, Kumar A (2017) In silico identification of mimicking molecules as defense inducers triggering jasmonic acid mediated immunity against Alternaria Blight disease in *Brassica* species. Front Plant Sci 8:609

Pedras MSC, Smith KC (1997) Sinalexin, a phytoalexin from white mustard elicited by destruxin B and *Alternaria Brassicae*. Phytochemistry 46:833–837. https://doi.org/10.1016/S0031-9422(97)00362-2

Pedras MSC, Zaharia IL, Gai Y, Zhou Y, Ward DE (2001) In planta sequential hydroxylation and glycosylation of a fungal phytotoxin: avoiding cell death and overcoming the fungal invader. Proc Natl Acad Sci U S A 98:747–752. https://doi.org/10.1073/pnas.98.2.747

Peng L, Zhou L, Li Q, Wei D, Ren X, Song H, Mei J, Si J, Qian W (2018) Identification of quantitative trait loci for Clubroot resistance in *Brassica oleracea* with the use of *Brassica* SNP microarray. Front Plant Sci 9:822

Perumal S, Koh CS, Jin L, Buchwaldt M, Higgins EE, Zheng C, Sankoff D, Robinson SJ, Kagale S, Navabi ZK et al (2020) A high-contiguity *Brassica nigra* genome localizes active centromeres and defines the ancestral *Brassica* genome. Nat Plants 6:929–941

Petutschnig EK, Jones AME, Serazetdinova L, Lipka U, Lipka V (2010) The Lysin Motif Receptor-like Kinase (LysM-RLK) CERK1 Is a major chitin-binding protein in *Arabidopsis thaliana* and subject to chitin-induced phosphorylation. J Biol Chem 285:28902–28911

Peyraud R, Mbengue M, Barbacci A, Raffaele S (2019) Intercellular cooperation in a fungal plant pathogen facilitates host colonization. Proc Natl Acad Sci U S A 116:3193

Pieterse CM, Van der Does D, Zamioudis C, Leon-Reyes A, Van Wees SC (2012) Hormonal modulation of plant immunity. Annu Rev Cell Dev Biol 28:489–521. https://doi.org/10.1146/annurev-cellbio-092910-154055

Pires JC, Gaeta RT (2011) Structural and functional evolution of resynthesized polyploids. In: Schmidt R, Bancroft I (eds) Genetics and genomics of the *Brassica*ceae. Springer, New York, pp 195–214

Pitzschke A, Schikora A, Hirt H (2009) MAPK cascade signaling networks in plant defense. Curr Opin Plant Biol 12:421–426

Presti LL, Lanver D, Schweizer G, Tanaka S, Liang L, Tollot M, Zuccaro A, Reissmann S, Kahmann R (2015) Fungal effectors and plant susceptibility. Annu Rev Plant Biol 66:513–545

Prince DC, Rallapalli G, Xu D, Schoonbeek HJ, Cevik V, Asai S, Kemen E, Cruz-Mireles N, Kemen A, Belhaj K, Schornack S, Kamoun S, Holub EB, Halkier BA, Jones JD (2017) *Albugo*-imposed changes to tryptophan-derived antimicrobial metabolite biosynthesis may contribute to suppression of non-host resistance to *Phytophthora infestans* in *Arabidopsis thaliana*. BMC Biol 15:20

Qasim MU, Zhao Q, Shahid M, Samad RA, Ahmar S, Wu J, Fan C, Zhou Y (2020) Identification of QTLs containing resistance genes for Sclerotinia stem rot in *Brassica napus* using comparative transcriptomic studies. Front Plant Sci 11:776

Qiu D, Morgan C, Shi J, Long Y, Liu J, Li R, Zhuang X, Wang Y, Tan X, Dietrich E, Weihmann T, Everett C, Vanstraelen S, Beckett P, Fraser F, Trick M, Barnes S, Wilmer J, Schmidt R, Li J, Li D, Meng J, Bancroft I (2006) A comparative linkage map of oilseed rape and its use for QTL analysis of seed oil and erucic acid content. Theor Appl Genet 114:67–80

Quilis J, Penas G, Messeguer J, Brugidou C, Segundo BS (2008) The *Arabidopsis* AtNPR1 inversely modulates defense responses against fungal, bacterial, or viral pathogens while conferring hypersensitivity to abiotic stresses in transgenic rice. Mol Plant Microbe Interact 21:1215–1231. https://doi.org/10.1094/MPMI-21-9-1215

Rafiqi M, Gan PHP, Ravensdale M, Lawrence GJ, Ellis JG, Jones DA et al (2010) Internalization of flax rust avirulence proteins into flax and tobacco cells can occur in the absence of the pathogen. Plant Cell 22:2017–2032

Rana K, Atri C, Akhatar J, Kaur R, Goyal A, Singh Mohini P, Kumar N, Sharma A, Sandhu Prabhjodh S, Kaur G et al (2019) Detection of first marker trait associations for resistance against *Sclerotinia sclerotiorum* in *Brassica juncea–Erucastrum cardaminoides* introgression lines. Front Plant Sci 10:1015

Rasmussen MW, Roux M, Petersen M, Mundy J (2012) MAP kinase cascades in *Arabidopsis* innate immunity. Front Plant Sci 3:169

Rehmany AP, Gordon A, Rose LE, Allen RL, Armstrong MR, Whisson SC, Kamoun S, Tyler BM, Birch PR, Beynon JL (2005) Differential recognition of highly divergent downy mildew avirulence gene alleles by RPP1 resistance genes from two *Arabidopsis* lines. Plant Cell 17:1839–1850

Rimmer SR, Buchwaldt L (1995) Diseases. In: Kimber DS, Mcgregor DI (eds) *Brassica* Oilseeds. CAB International, Wallingford, pp 65–92

Rimmer SR, Scarth R, McVetty P, Woods D, Gugel R, Séguin-Swartz G (1995) Breeding for resistance to blackleg (stem canker) in Western Canada. Blackleg Newsl 5:11–15

Rocherieux J, Glory P, Giboulot A, Boury S, Barbeyron G, Thomas G, Manzanares-Dauleux MJ (2004) Isolate-specific and broad-spectrum QTLs are involved in the control of clubroot in *Brassica oleracea*. Theor Appl Genet 108:1555–1563. https://doi.org/10.1007/s00122-003-1580-x

Romero FM, Rossi FR, Gárriz A, Carrasco P, Ruíz OA (2018) A bacterial endophyte from apoplast fluids protects canola plants from different phytopathogens via antibiosis and induction of host resistance. Phytopathology 109:375–383

Rose LE, Bittner-Eddy PD, Langley CH, Holub EB, Michelmore RW, Beynon JL (2004) The maintenance of extreme amino acid diversity at the disease resistance gene, RPP13, in *Arabidopsis thaliana*. Genetics 166:1517–1527

Roux F, Voisin D, Badet T, Balague C, Barlet X, Huard-Chauveau C, Roby D, Raffaele S (2014) Resistance to phytopathogens e tutti quanti: placing plant quantitative disease resistance on the map. Mol Plant Pathol 15(5):427–432. https://doi.org/10.1111/mpp.12138

Roy NN (1984) Interspecific transfer of *Brassica juncea* - type high blackleg resistance to *Brassica napus*. Euphytica 33:295–303

Ruano G, Scheuring D (2020) Plant cells under attack: unconventional endo-membrane tracking during plant defense. Plants 9:389

Ruhe J, Agler MT, Placzek A, Kramer K, Finkemeier I, Kemen EM (2016) Obligate biotroph pathogens of the genus *Albugo* are better adapted to active host defense compared to niche competitors. Front Plant Sci 7:820

Rustagi A, Singh G, Agrawal S, Gupta PK (2018) Proteomic studies revealing enigma of plant–pathogen interaction. In: Singh A, Singh IK (eds) Molecular aspects of plant-pathogen interaction. Springer, Singapore, pp 239–264

Saharan GS (1992) Management of rapeseed and mustard diseases. In: Kumar D, Rai M (eds) Advances in oilseed research. Scientific Publisher, Jodhpur

Saharan GS, Verma PR, Meena PD, Kumar A (2014) White rust of crucifers: biology, ecology and management. Springer, Frankfurt, p 244. ISBN 978-81-322-1791-6.

Salisbury PA, Ballinger JD, Wratten N, Plummer MK, Howlett JB (1995) Blackleg disease on oilseed *Brassicas* in Australia - a review. Aust J Exp Agric 3(5):665–674

Sanchez-Vallet A, Saleem-Batcha R, Kombrink A, Hansen G, Valkenburg DJ, Thomma BPHJ, Mesters JR (2013) Fungal effector Ecp6 outcompetes host immune receptor for chitin binding through intrachain LysM dimerization. Elife 2:e00790

Schechter LM, Vencato M, Jordan KL, Schneider SE, Schneider DJ, Collmer A (2006) Multiple approaches to a complete inventory of *Pseudomonas syringae* pv. *tomato* DC3000 type III secretion system effector proteins. Mol Plant-Microbe Int 19:1180–1192

Seifbarghi S, Borhan MH, Wei Y, Coutu C, Robinson SJ, Hegedus DD (2017) Changes in the *Sclerotinia sclerotiorum* transcriptome during infection of *Brassica napus*. BMC Genomics 18:266

Shah J, Zeier J (2013) Long–distance communication and signal amplification in systemic acquired resistance. Front Plant Sci 4:30. https://doi.org/10.3389/fpls.2013.00030

Shah N, Li Q, Xu Q, Liu J, Huang F, Zhan Z, Qin P, Zhou X, Yu W, Zhu L et al (2020) *CRb* and *PbBa8*.1 synergically increases resistant genes expression upon infection of *Plasmodiophora brassicae* in *Brassica napus*. Genes 11:202

Shiu SH, Karlowski WM, Pan RS, Tzeng YH, Mayer KFX, Li WH (2004) Comparative analysis of the receptor-like kinase family in *Arabidopsis* and rice. Plant Cell 16:1220–1234

Shrivastava N, Jiang L, Li P, Sharma AK, Luo X, Wu S, Pandey R, Gao Q, Lou B (2018) Proteomic approach to understand the molecular physiology of symbiotic interaction between *Piriformospora indica* and *Brassica napus*. Sci Rep 8:5773

Sinapidou E, Williams K, Nott L, Bahkt S, Tör M, Crute I, Bittner-Eddy P, Beynon J (2004) Two *TIR:NB: LRR* genes are required to specify resistance to *Peronospora parasitica* isolate Cala2 in *Arabidopsis*. Plant J 38:898–909

Singh DN, Singh NK, Srivastava S (1999) Biochemical and morphological characters in relation to Alternaria blight resistance in rape-seed mustard. Ann Agric Res 20:472–477

Singh V, Kumar K, Bhajan R, Singh PK, Singh RB (2009) Identification of resistance sources against Alternaria blight and white rust in Indian mustard. J Oilseeds Res 26:435–436

Slusarenko AJ, Schlaich NL (2003) Downy mildew of *Arabidopsis thaliana* caused by *Hyaloperonospora parasitica* (formerly *Peronospora parasitica*). Mol Plant Pathol 4:159–170

Smooker AM, Wells R, Morgan C, Beaudoin F, Cho K, Fraser F, Bancroft I (2011) The identification and mapping of candidate genes and QTL involved in the fatty acid desaturation pathway in *Brassica napus*. Theor Appl Genet 122:1075–1090

Snowdon RJ, Friedt W (2004) Molecular markers in *Brassica* oilseed breeding: current status and future possibilities. Plant Breed 123:1–8

Sonah H, Deshmukh RK, Bélanger RR (2016) Computational prediction of effector proteins in fungi: opportunities and challenges. Front Plant Sci 7:126

Song T, Chu M, Lahlali R, Yu F, Peng G (2016) Shotgun label-free proteomic analysis of Clubroot (*Plasmodiophora brassicae*) resistance conferred by the gene Rcr1 in *Brassica rapa*. Front Plant Sci 7:1013

Song JM, Guan Z, Hu J, Guo C, Yang Z, Wang S, Liu D, Wang B, Lu S, Zhou R et al (2020) Eight high-quality genomes reveal pan-genome architecture and ecotype differentiation of *Brassica napus*. Nat Plants 6:34–45

Spoel SH, Koornneef A, Claessens SM, Korzelius JP, Van-Pelt JA, Mueller MJ, Buchala AJ, Métraux JP, Brown R, Kazan K, Van Loon LC, Dong X, Pieterse CMJ (2003) NPR1 modulates cross–talk between salicylate– and jasmonate–dependent defense pathways through a novel function in the cytosol. Plant Cell 15:760–770. https://doi.org/10.1105/tpc.009159

Staskawicz BJ, Mudgett MB, Dangl JL, Galan JE (2001) Common and contrasting themes of plant and animal diseases. Science 292(5525):2285–2289. https://doi.org/10.1126/science.1062013

Steinbrenner AD, Goritschnig S, Staskawicz BJ (2015) Recognition and activation domains contribute to allele-specific responses of an *Arabidopsis* NLR receptor to an oomycete effector protein. PLoS Pathogens 11:e1004665

Stergiopoulos de Wit P (2009) Fungal effector proteins. Ann Rev Phytopathol 47:233–263

Stotz HU, Harvey PJ, Haddadi P, Mashanova A, Kukol A, Larkan NJ, Borhan MH, Fitt BDL (2018) Genomic evidence for genes encoding leucine-rich repeat receptors linked to resistance against the eukaryotic extra- and intracellular *Brassica napus* pathogens *Leptosphaeria maculans* and *Plasmodiophora brassicae*. PLoS One 13:e0198201

Subramaniam S, Wang X, Freeling M, Pires JC (2013) The fate of *Arabidopsis thaliana* homeologous CNSs and their motifs in the paleohexaploid *Brassica rapa*. Genome Biol Evol 5:646–660. https://doi.org/10.1093/gbe/evt035

Subudhi PK, Raut RN (1994) White rust resistance and its association with parental species type and leaf waxiness in *B. juncea* L. Czern and Coss. X *B. napus* L. crosses under the action of EDTA and Gamma ray. Euphytica 74:1–7. https://doi.org/10.1007/BF00033760

Sun D, Wang C, Zhang X, Zhang W, Jiang H, Yao X, Liu L, Wen Z, Niu G, Shan X (2019) Draft genome sequence of cauliflower (*Brassica oleracea* L. var. *botrytis*) provides new insights into the C genome in *Brassica* species. Hortic Res 6:82

Sundaresha S, Rohini S, Appanna VK, Arthikala MK, Shanmugam NB, Shashibhushan NB, Hari Kishore CM, Pannerselvam R, Kirti PB, Udayakumar M (2016) Co-overexpression of *Brassica juncea* NPR1 (BjNPR1) and *Trigonella foenum*-graecum defensin (Tfgd) in transgenic peanut provides comprehensive but varied protection against *Aspergillus flavus* and *Cercospora arachidicola*. Plant Cell Rep 35:1189–1203. https://doi.org/10.1007/s00299-016-1945-7

Tada Y, Spoel SH, Pajerowska-Mukhtar K, Mou Z, Song J, Wang C, Zou J, Dong X (2008) Plant immunity requires conformational charges of NPR1 via S-nitrosylation and thioredoxins. Science 321:952–956. https://doi.org/10.1126/science.1156970

Tanveer T, Shaheen K, Parveen S, Kazi AG, Ahmad P (2014) Plant secretomics. Plant Signal Behav 9:e29426

Tettelin H, Masignani V, Cieslewicz MJ, Donati C, Medini D, Ward NL, Angiuoli SV, Crabtree J, Jones AL, Durkin AS, Deboy RT, Davidsen TM, Mora M, Scarselli M, Ros IM, Peterson JD, Hauser CR, Sundaram JP, Nelson WC, Madupu R, Brinkac LM, Dodson RJ, Rosovitz MJ, Sullivan SA, Daugherty SC, Haft DH, Selengut J, Gwinn ML, Zhou L, Zafar N, Khouri H, Radune D, Dimitrov G, Watkins K, O'Connor KJB, Smith S, Utterback TR, White O, Rubens CE, Grandi G, Madoff LC, Kasper DL, Telford JL, Wessels MR, Rappuoli R, Fraser CM (2005) Genome analysis of multiple pathogenic isolates of *Streptococcus agalactiae*: implications for the microbial "pan-genome". Proc Natl Acad Sci U S A 102:13950–13955

Tewari JP, Conn KL, Dahiya JS (1987) Resistance to *Alternaria Brassicae* in crucifer. In: Proceedings of 7th IRC, Poznan, vol 48, pp 1085–1089

Tewari JP, Conn KL, Dahiya JS (1988) Dynamic response of rapeseed and other crucifers to *Alternaria Brassicae* and *Rhizoctonia solani*. In: Proceedings of 5th international congress on plant pathology, Koyoto, p 242

Thaler JS, Humphrey PT, Whiteman NK (2012) Evolution of jasmonate and salicylate signal crosstalk. Trends Plant Sci 17:260–270. https://doi.org/10.1016/j.tplants.2012.02.010

Thines M (2014) Phylogeny and evolution of plant pathogenic oomycetes, a global over view. Eur J Plant Pathol:1–17

Thines M, Choi YJ, Kemen E, Ploch S, Holub EB, Shin HD, Jones JD (2009) A new species of *Albugo* parasitic to *Arabidopsis thaliana* reveals new evolutionary patterns in white blister rusts (Albuginaceae). Persoonia 22:123–128

Tian X, Wang D, Mao Z, Pan L, Liao J, Cai Z (2019) Infection of *Plasmodiophora Brassicae* changes the fungal endophyte community of tumourous stem mustard roots as revealed by high-throughput sequencing and culture-dependent methods. PLoS One 14:e0214975

Tirnaz S, Edwards D, Batley J (2020) The importance of plant pan-genomes in breeding. In: Kang MS (ed) Quantitative genetics, genomics and plant breeding, 2nd edn. CABI, Boston, pp 27–32

Tollenacre R, Hayward A, Dalton-Morgan J, Campbell E, Lee JRM, Lorenc MT, Manoli S, Stiller J, Raman R, Raman H, Edwards D, Batley J (2012) Identification and characterization of candidate *Rlm4* blackleg resistance genes in *Brassica napus* using next-generation sequencing. Plant Biotechnol J 10:709–715

Tortosa M, Cartea ME, Velasco P, Soengas P, Rodriguez VM (2019) Calcium-signaling proteins mediate the plant transcriptomic response during a well-established *Xanthomonas campestris* pv. *campestris* infection. Hortic Res 6:103

Toruno TY, Stergiopoulos I, Coaker G (2016) Plant-pathogen effectors: cellular probes interfering with plant defenses in spatial and temporal manners. Annu Rev Phytopathol 54:419–441

Town CD, Cheung F, Maiti R, Crabtree J, Haas BJ, Wortman JR, Hine EE, Althoff R, Arbogast TS, Tallon LJ, Vigouroux M, Trick M, Bancroft I (2006) Comparative genomics of *Brassica oleracea* and *Arabidopsis thaliana* reveal gene loss, fragmentation, and dispersal after polyploidy. Plant Cell 18:1348–1359. https://doi.org/10.1105/tpc.106.041665

Tripathi NN, Kaushik CD, Yadav TP, Yadav AK (1978) Studies on the inheritance of Alternaria blight resistance in raya. Indian Phytopathol 31:127

Tripathi NN, Kaushik CD, Yadav TP, Yadav AK (1980) Alternaria leaf spot resistance in raya. Haryana Agric Univ J Res 10:166–168

Tsuchiya T, Eulgem T (2013) An alternative polyadenylation mechanism co-opted to the *Arabidopsis* RPP7 gene through intronic retro-transposon domestication. Proc Natl Acad Sci U S A 110:E3535–E3543

Van Damme M, Andel A, Huibers RP, Panstruga R, Weisbeek PJ, Van den Ackerveken G (2005) Identification of *Arabidopsis* loci required for susceptibility to the downy mildew pathogen *Hyaloperonospora parasitica*. Mol Plant Microbe Interact 18:583–592

Van de Weyer AL, Monteiro F, Furzer OJ, Nishimura MT, Cevik V, Witek K, Jones JDG, Dangl JL, Weigel D, Bemm F (2019) A species-wide inventory of NLR genes and alleles in *Arabidopsis thaliana*. Cell 178:1260–1272.e14

Van de Wouw AP, Idnurm A (2019) Biotechnological potential of engineering pathogen effector proteins for use in plant disease management. Biotechnol Adv 37:107387

Van der Does D, Leon-Reyes A, Koornneef A, Van Verk MC, Rodenburg N, Pauwels L, Goossens A, Korbes AP, Memelink J, Ritsema T, Van Wees SCM, Pieterse CMJ (2013) Salicylic acid suppresses jasmonic acid signaling downstream of SCFCOI1-JAZ by targeting GCC promoter motifs via transcription factor ORA59. Plant Cell 25:744–761. https://doi.org/10.1105/tpc.112.108548

Vanholme B, De Meutter J, Tytgat T, Van Montagu M, Coomans A, Gheysen G (2004) Secretions of plant parasitic nematodes: a molecular update. Gene 332:13–27

Vieira P, Danchin EGJ, Neveu C, Crozat C, Jaubert S, Hussey RS et al (2011) The plant apoplasm is an important recipient compartment for nematode secreted proteins. J Expt Bot 62:1241–1253

Vincent D, Rafiqi M, Job D (2020) The multiple facets of plant-fungal interactions revealed through plant and fungal secretomics. Front Plant Sci 10:1626

Voglmayr H, Riethmüller A, Göker M, Weiß M, Oberwinkler F (2004) Phylogenetic relationships of *Plasmopara, Bremia* and other genera of downy mildews with pyriform haustoria based on Bayesian analysis of partial LSU rDNA sequence data. Mycol Res 108:1011–1024

Wally O, Jayaraj J, Punja ZK (2009) Comparative resistance to foliar fungal pathogens in transgenic carrot plants expressing genes encoding for chitinase, beta–1, 3–glucanase and peroxidase. Eur J Plant Pathol 123:331–342. https://doi.org/10.1007/s10658-008-9370-6

Wang Z, Gerstein M, Snyder M (2009) RNA-Seq: a revolutionary tool for transcriptomics. Nat Rev Genet 10:57

Wang Y, Li J, Hou S, Wang X, Li Y, Ren D, Chen S, Tang X, Zhou JM (2010) A *Pseudomonas syringae* ADP-ribosyl transferase inhibits *Arabidopsis* mitogen-activated protein kinase kinases. Plant Cell 22:2033–2044

Wang B, Kumar V, Olson A, Ware D (2019) Reviving the transcriptome studies: an insight into the emergence of single-molecule transcriptome sequencing. Front Genet 10:384

Wang Z, Ma LY, Li X, Zhao FY, Sarwar R, Cao J, Li YL, Ding LN, Zhu KM, Yang YH et al (2020) Genome-wide identification of the NPR1-like gene family in *Brassica napus* and functional characterization of *BnaNPR1* in resistance to *Sclerotinia sclerotiorum*. Plant Cell Rep 39:709–722

Wani ZA, Ashraf N (2018) Transcriptomic studies revealing enigma of plant-pathogen interaction. In: Singh A, Singh IK (eds) Molecular aspects of plant-pathogen interaction. Springer, Singapore, pp 219–238

Watson RE, Goodman JI (2002) Effects of phenobarbital on DNA methylation in GC-rich regions of hepatic DNA from mice that exhibit different levels of susceptibility to liver tumor genesis. Toxicol Sci 68:51–58

Wei D, Cui Y, He Y, Xiong Q, Qian L, Tong C, Lu G, Ding Y, Li J, Jung C et al (2017) A genome-wide survey with different rapeseed ecotypes uncovers footprints of domestication and breeding. J Exp Bot 68:4791–4801

Wessling R, Epple P, Altmann S, He Y, Yang L, Henz SR, McDonald N, Wiley K, Bader KC, Glasser C, Mukhtar MS, Haigis S, Ghamsari L, Stephens AE, Ecker JR, Vidal M, Jones JD, Mayer KF, Ver Loren van Themaat E, Weigel D, Schulze Lefert P, Dangl JL, Panstruga R, Braun P (2014) Convergent targeting of a common host protein-network by pathogen effectors from three kingdoms of life. Cell Host Microbe 16:364–375

Willmann R, Lajunen HM, Erbs G, Newman MA, Kolb D, Tsuda K, Katagiri F, Fliegmann J, Bono JJ, Cullimore JV, Jehle AK, Gotz F, Kulik A, Molinaro A, Lipka V, Gust AA, Nurnberger T (2011) Arabidopsis lysin-motif proteins LYM1 LYM3 CERK1 mediate bacterial peptidoglycan sensing and immunity to bacterial infection. Proc Nat Acad Sci U S A 108:19824–19829

Win J, Chaparro-Garcia A, Belhaj K, Saunders DGO, Yoshida K, Dong S, Schornack S, Zipfel C, Robatzek S, Hogenhout SA, Kamoun S (2012) Effector biology of plant-associated organisms: concepts and perspectives. Cold Spring Harb Symp Quant Biol 77:235–247

Woods-Tor A, Studholme DJ, Cevik V, Telli O, Holub EB, Tor M (2018) A suppressor/avirulence gene combination in *Hyaloperonospora arabidopsidis* determines race specificity in *Arabidopsis thaliana*. Front Plant Sci 9:265

Wu Y, Zhang D, Chu JY, Boyle P, Wang Y, Brindle ID, Luca VD, Després C (2012) The Arabidopsis NPR1 protein is a receptor for the plant defense hormone salicylic acid. Cell Rep 1:639–647. https://doi.org/10.1016/j.celrep.2012.05.008

Wu J, Zhao Q, Yang Q, Liu H, Li Q, Yi X, Cheng Y, Guo L, Fan C, Zhou Y (2016) Comparative transcriptomic analysis uncovers the complex genetic network for resistance to *Sclerotinia sclerotiorum* in *Brassica napus*. Sci Rep 6:19007

Wu C-H, Derevnina L, Kamoun S (2018) Receptor networks underpin plant immunity. Science 360:1300

Wu D, Liang Z, Yan T, Xu Y, Xuan L, Tang J, Zhou G, Lohwasser U, Hua S, Wang H et al (2019a) Whole-genome resequencing of a worldwide collection of rapeseed accessions reveals the genetic basis of ecotype divergence. Mol Plant 12:30–43

Wu Z, Li M, Dong OX, Xia S, Liang W, Bao Y, Wasteneys G, Li X (2019b) Differential regulation of TNL-mediated immune signaling by redundant helper CNLs. New Phytol 222:938–953

Xiang T, Zong N, Zou Y, Wu Y, Zhang J, Xing W, et al. (2008) *Pseudomonas syringae* effector AvrPto blocks innate immunity by targeting receptor kinases. Curr Biol 18:74–80

Xiang TT, Zong N, Zhang J, Chen JF, Chen MS, Zhou JM (2011) BAK1 is not a target of the *Pseudomonas syringae* effector AvrPto. Mol Plant Microbe Interact 24:100–107

Xiao D, Liu ST, Wei YP, Zhou DY, Hou XL, Li Y, Hu CM (2016) cDNA-AFLP analysis reveals differential gene expression in incompatible interaction between infected non-heading Chinese cabbage and *Hyaloperonospora parasitica*. Hortic Res 3:16034

Yaeno T, Li H, Chaparro-Garcia A, Schornack S, Koshiba S, Watanabe S et al (2011) Phosphatidy linositol monophosphate-binding interface in the oomycete RXLR effector AVR3a is required for its stability in host cells to modulate plant immunity. Proc Natl Acad Sci U S A 108:14682. https://doi.org/10.1073/pnas.1106002108

Yang J, Liu D, Wang X, Ji C, Cheng F, Liu B, Hu Z, Chen S, Pental D, Ju Y, Yao P, Li X, Xie K, Zhang J, Wang J, Liu F, Ma W, Shopan J, Zheng H, Mackenzie SA, Zhang M (2016) The genome sequence of allopolyploid *Brassica juncea* and analysis of differential homoeolog gene expression influencing selection. Nat Genet 48:1225–1232. https://doi.org/10.1038/ng.3657

Yu J, Tehrim S, Zhang F, Tong C, Huang J, Cheng X, Dong C, Zhou Y, Qin R, Hua W, Liu S (2014a) Genome-wide comparative analysis of NBS-encoding genes between *Brassica* species and *Arabidopsis thaliana*. Plant Pathol 56:317–323

Yu J, Tehrim S, Zhang FQ, Tong CB, Huang JY, Cheng XH, Dong C, Zhou Y, Qin R, Hua W, Liu S (2014b) Genome-wide comparative analysis of NBS-encoding genes between *Brassica* species and *Arabidopsis thaliana*. BMC Genomics 15:3. https://doi.org/10.1186/1471-2164-15-3

Zhang FL, Xu JB, Takahata Y (1996) Inheritance of resistance to black leaf spot (*Alternaria Brassicae*) in Chinese cabbage. Cruciferae Newslett 18:134–135

Zhang F, Xu J, Yan H, Li M (1997) Inheritance of resistance to black leaf spot in Chinese cabbage. Acta Agric Boreali-Sin 12:115–119

Zhang Y, Tessaro MJ, Lassner M, Li X (2003) Knockout analysis of *Arabidopsis* transcription factors TGA2, TGA5, and TGA6 reveals their redundant and essential roles in systemic acquired resistance. Plant Cell 15:2647–2653. https://doi.org/10.1105/tpc.014894

Zhang J, Shao F, Li Y, Cui H, Chen L, Li H, Zou Y, Long C, Lan L, Chai J, Chen S, Tang X, Zhou JM (2007) A *Pseudomonas syringae* effector inactivates MAPKs to suppress PAMP-induced immunity in plants. Cell Host Microbe 1:175–185

Zhang W, Fraiture M, Kolb D, Loffelhardt B, Desaki Y, Boutrot FF, Tor M, Zipfel C, Gust A A, Brunner F (2013) *Arabidopsis* receptor-like protein30 and receptor-like kinase suppressor of BIR1-1/EVERSHED mediate innate immunity to necrotrophic fungi. Plant Cell 25:4227–41

Zhang L, Cai X, Wu J, Liu M, Grob S, Cheng F, Liang J, Cai C, Liu Z, Liu B et al (2018) Improved *Brassica rapa* reference genome by single-molecule sequencing and chromosome conformation capture technologies. Hortic Res 5:50

Zhao J, Udall JA, Quijada PA, Grau CR, Meng J, Osborn TC (2006) Quantitative trait loci for resistance to *Sclerotinia sclerotiorum* and its association with a homeologous non-reciprocal transposition in *Brassica napus* L. Theor Appl Genet 112:509–516. https://doi.org/10.1007/s00122-005-0154-5

Zhong X, Xi L, Lian Q, Luo X, Wu Z, Seng S, Yuan X, Yi M (2015) The NPR1 homolog GhNPR1 plays an important role in the defense response of *Gladiolus hybridus*. Plant Cell Rep 34:1063–1074. https://doi.org/10.1007/s00299-015-1765-1

Zhong X, Zhou Q, Cui N, Cai D, Tang G (2019) BvcZR3 and BvHs1pro-1 genes pyramiding enhanced beet cyst nematode (*Heterodera schachtii* Schm.) resistance in oilseed rape (*Brassica napus* L.). Int J Mol Sci 20:1740

Zhou T, Xu W, Hirani AH, Liu Z, Tuan PA, Ayele BT, Daayf F, McVetty PBE, Duncan RW, Li G (2019) Transcriptional insight into *Brassica napus* resistance genes *LepR3* and *Rlm2*-mediated defense response against the *Leptosphaeria maculans* infection. Front Plant Sci 10:823

Zhu HY, Kim DJ, Baek JM, Choi HK, Ellis LC, Kuester H, McCombie WR, Peng HM, Cook DR (2003) Syntenic relationships between *Medicago truncatula* and *Arabidopsis* reveal extensive divergence of genome organization. Plant Physiol 131:1018–1026

Zipfel C (2009) Early molecular events in PAMP-triggered immunity. Curr Opin Plant Biol 12:414–420

Zipfel C, Kunze G, Chinchilla D, Caniard A, Jones JDG, Boller T, Felix G (2006) Perception of the bacterial PAMP EF-Tu by the receptor EFR restricts Agrobacterium-mediated transformation. Cell 125:749–760

Molecular Mechanisms of Host Resistance to Biotrophs

2

Abstract

Molecular mechanisms of crucifers host resistance against biotrophs have been revealed in different *Brassica* species during host–pathogen interactions at different stages of host growth and disease development. The R-genes and QTLs have been identified, functionally characterized, and molecularly mapped on the chromosomes of *Brassica* species expressed during pathogens' interactions. Timing of the expression of defense-related genes plays a crucial role during pathogenesis and host resistance for mounting a successful defense response. A single gene (*Acr*) responsible for conferring resistance to *Albugo* was mapped on a densely populated *Brassica juncea* RFLP map. A single dominant *RAC1* gene in accession Ksk 1 of *Arabidopsis thaliana* confers resistance to *Albugo candida* isolate ACeM1. Two accessions of *A. thaliana* (Ksk1 and Ksk2) were used to identify and map three loci (RAC1, RAC2, and RAC3) of genes that confer resistance to *A. candida*. In *Brassica rapa*, the ACA1 locus was mapped to linkage group 4 and was flanked by RFLP marker loci. *WRR* 4 genes encode a cytoplasmic TIR-NBL-LRR receptor-like protein in Columbia *A. thaliana* and confer broad-spectrum white rust resistance to four races of *A. candida*. Three WRR (*WRR4BCol*-0, *WRR8Sf*-2, and *WRR9Hi*-0) genes against *Ac2V* and a gene *WRR12* (*SOC3*) conferring NHR to AiBoT have been identified. Analysis of allelic variants of BjuA046215 in Indian, east Europe, and Chinese gene pool suggested the presence of three types of alleles. Allele 1 is present in the Chinese Tumida, Allele 2 in Indian, and Allele 3 in east European gene pools. Host resistance in *Brassica* species to powdery mildew is multilayered and multicomponent at both pre- and postpenetration stages. Host resistance is activated either through SA signaling or simultaneous perception of ethylene and jasmonic acid. MLO genes encoding seven transmembrane calmodulin-binding proteins confer broad-spectrum resistance to *Arabidopsis* powdery mildew. The Pmr mutants confer resistance to powdery mildew through altered cell wall composition of

G. S. Saharan et al., *Molecular Mechanism of Crucifer's Host-Resistance*,
https://doi.org/10.1007/978-981-16-1974-8_2

host. Salicylic acid enhances the expression of RPW 8.1 and RPW 8.2, leading to HR or SHL and resistance. *BjNPR1* gene activates SAR to confer broad-spectrum resistance to powdery mildew of *B. juncea*. There is role of WRKY transcription factors and overexpression of R-genes like PMR, MLO, PEN, EDR, MAPK, MAPK 65–3, NPR1, PAD3, PAD4, ED5, SNARE, RLKs, and KDL (AtCEP1) to confer resistance to powdery mildew of crucifers. Powdery mildew resistance genes of *Arabidopsis* have been mapped on chromosomes II (RPW 1), III (RPW 2, 3, 7, 8), IV (RPW 4), and V (RPW 5, 6). The papain-type KDEL Cys EP CEP1 fulfills function in plant defense during late development of *Erysiphe cruciferarum* in close spatial association with the fungal haustorium and haustorial callose encasement.

Keywords

Molecular mechanisms in cruciferous biotrophs · Defense resistant genes · Molecular mapping of R-genes · CNL type R-genes · Non-host resistance to *Albugo candida* · Multiple component resistance mechanisms · Post penetration resistance mechanisms · EDR genes · PMR mutant genes · *Arabidopsis* triple mutant · Induction of R-genes · KDEL function · Identification of seedling and adult plant R-genes · Molecular mapping of Downy mildew (DM) R-genes · Genetics of multiple disease resistance · Age related resistance to DM · Requirement of R-genes · Cloning and mapping of major R-genes · Application of gene silencing in DM · *Brassica* genome sequencing · Mapping of CR genes · Mapping of pathotype specific R-genes · Mapping of QTL · Genome-wide association to identify CR loci · CR QTLs analysis · Linkage markers · Marker assisted selection of CR genes · Genetics of R-genes · Resistance mechanisms to Clubroot · Mapping of TuMV genes · TuMv R-genes · *BjAOX1a* in *Brassica* enhances resistance to TuMV

Resistance to Hyaloperonospora in *Brassica* is controlled by dominant gene at seedling and adult plant stages. White rust R loci on three *Arabidopsis* chromosomes are closely linked to downy mildew R loci. Downy mildew R-genes are differentially expressed. About 129 genes are identified, of which 121 TDFs are upregulated and 8 are downregulated. R-genes of *Arabidopsis* are mapped on five chromosomes in MRC-A, MRC-B, MRC-F, MRC-H, and MRC-J as RPP1 to RPP27. Single transduction pathways used by the different RPP genes are complex and diverse. Stable and transient gene silencing approaches have been applied in downy mildew of crucifers. Strong dependence on genes EDS1 and NDR1 is governed by R-genes' structural type rather than pathogens of *Arabidopsis*. Several clubroot resistance (CR) genes have been identified and mapped on *Brassica rapa, Brassica oleracea, Brassica nigra, Brassica juncea, Brassica napus*, and *Arabidopsis thaliana* chromosomes. *Brassica* species genome sequencing has allowed rapid identification of genome-wide QTLs and GWAS regions for R-genes of clubroot. With the use of transcriptomic and proteomic approaches, more than 1000 *Arabidopsis* genes have been identified which are differentially expressed in Plasmodiophora *brassicae*-infected roots. More than 10 CR genes in *B. rapa* have been identified with pathotypes specificity. The genetic analysis, genetic mapping, and fine mapping on chromosomes of *Brassica* species with molecular markers and pathotypes specificity have been studied in *B. rapa, B. oleracea, B. napus, B. napus* var. *napobrassicae*, and *A. thaliana*. The genes encoding TIR-NBS-LRR class proteins impart resistance against *P. brassicae*. The R-genes are present in clusters on the chromosomes of *Brassica* species. R-genes QTLs using ddRAD-seq have been mapped in *B. rapa* against clubroot pathotypes. The CR QTLs have been identified through GWAS analysis in the natural populations of *B. napus*. Bioinformatic analysis has been used to predict CR QTLs of *Brassica*. For the determination of loss of R-genes, linkage markers of CR in *Brassica* have been identified. Marker-assisted selection of CR genes has reinforced the accumulation of varied R-genes in *Brassica*. The genetic origin of CR genes has been revealed in *Brassica*. Quantitative resistance to clubroot is mediated by transgenerational epigenetic variation in *Arabidopsis*. The use of SNP microarray and proteomic and transcriptome approaches in *Brassica* have revealed multicomponent molecular mechanisms conferred by CR QTLs to *P. brassicae*. In *Brassica,* R-genes and QTLs have been identified with molecular markers and mapped on chromosomes to TuMV isolates. The retr01/retr02/retr03 genes have broad-spectrum resistance to TuMV. Alternative oxidase gene (*BjAOX1a*) in *Brassica* enhances resistance to Turnip mosaic virus. The techniques such as yeast two hybrid (Y2H), biomolecular fluorescence complement (BiFC), and co-immune precipitation (COIP) methods to analyze the resistance interaction between host and TuMV isolates have been employed.

2.1 Introduction

The genome sequencing of a small cruciferous weed, *A. thaliana*, and its establishment as a model for host pathosystem to reveal molecular mechanisms of host–pathogen interactions have transformed traditional sciences like genetics, plant breeding, and plant pathology into molecular genetics, molecular plant breeding, and molecular plant pathology under the umbrella of molecular biology. During the last three decades, significant investigations have been carried out on molecular mechanisms of crucifer's host resistance. Out of 44 pathogens known to infect crucifers, 16 pathogens are of significant important at global level on the basis of geographical distribution, host range, yield losses, and investigations on the molecular studies. On the basis of mode of acquisition of nutrients from the host (life style) and mode of infection, differences in molecular mechanisms of host–pathogen interactions can be divided as biotrophs, hemibiotrophs, and necrotrophs. Biotrophs essentially act as a sink for the host, anabolic assimilates, and keep it alive. Necrotrophs consume the host tissues on invasion. Hemibiotrophs combine both strategies in their life cycle. Therefore, host plants have quite different molecular mechanisms and strategies to deal with these groups of pathogens. Pathogen infection produces a variety of phytohormones depending mainly on the pathogen lifestyle and their mode of infection. Salicylic acid pathway generally provides resistance to biotrophic pathogens whereas JA/ET pathways are commonly associated with resistance to necrotrophic pathogens and to herbivorous pests. Generally, SA and JA signaling pathways operate antagonistically, and thus, elevated resistance against biotrophs is often related with increased susceptibility to necrotrophs and vice versa.

2.2 *Brassica–Albugo*: Molecular Resistance

2.2.1 Identification and Function of Host Defense-Resistant Genes

Plants' defense against colonization by biotrophic pathogens thought to be triggered by either direct or indirect interaction between the pathogen and a corresponding plant-resistant protein. Several resistance genes (*R*-genes) against bacteria, fungi, viruses, and nematodes have been cloned from *A. thaliana* and various crop species (Ellis et al. 1999, 2000; Holub 2001; Hulbert et al. 2001; Takken and Joosten 2000). However, timing of the expression of defense-related genes plays a crucial role during pathogenesis and incompatible interactions, and that the redox balance within the chloroplast may be of crucial importance for mounting a successful defense response. The resistant host to fight off the *A. candida* attack (Kaur et al. 2011) utilizes synergistic and conserved strategies. A single gene (*Acr*) responsible for conferring resistance to this pathogen was mapped on a densely populated *B. juncea* RFLP map. Two closely linked RFLP markers identified (X42 and X83) were 2.3 and 4 cM from the *Acr* locus, respectively (Cheung et al. 1998). These markers may be useful for marker-assisted selection and map-based cloning of these genes.

White rust in natural populations of *A. thaliana* is caused by a distinct subspecies of *A. candida* subsp. *arabidopsis* (Borhan et al. 2008), which offers an attractive model for investigating the molecular basis of broad-spectrum defense suppression. There are numerous examples of receptor-like genes in *A. thaliana* that vary in different modes of defenseregulation for molecular genetic analyses of DM resistance (Eulgem et al. 2004; Holub 2001; Holub et al. 1995; McDowell et al. 1998, 2000; Tor et al. 2002). The majority of plant R-genes encode nucleotide-binding site leucine-rich repeat (NB-LRR)-type protein which can be further grouped into two subclasses based on their N-terminal sequence: those containing a coiled-coil (CC) domain (CC-NB-LRR) or those containing a domain with similarity to *Drosophila* toll and mammalian interleukin-1 receptor (TIR) (TIR-NB-LRR) (Hammond-Kosack and Jones 1997; Jones and Jones 1997; Young 2000). Leucine-rich repeats (LRRs) are involved in protein–protein interactions and occur in a number of proteins with different function (Kobe and Deisenhofer 1994, 1995). Domain exchange between LRR of closely related R-genes supports their role in pathogen recognition (Ellis et al. 1999; Wulff et al. 2001). Variation among R-genes occurs mainly in their LRR domain, typically in the solvent-exposed β-strand/ β-turn structure within the LRR domain. Comparison of this motif among R-gene homologs suggests that the β-strand/ β-turn structure has been under diversifying selection (Bittner-Eddy et al. 2000; Botella et al. 1998; McDowell et al. 1998; Meyers et al. 1998, 2003; Parniske et al. 1997). Evidence for this is based on the ratio of nonsynonymous (*K*a) to synonymous (*K*s) nucleotide substitution at the β-strand/β-turn motif. A *K*a/*K*s ratio > 1 indicates that diversification has occurred under positive selection pressure from the evolving pathogen. This suggests that the β-strand/ β-turn motif may be involved in ligand binding as has been shown for a polygalacturonase-inhibiting protein with its polygalacturonase ligand (using site-directed mutagenesis of the β-strand/ β-turn motif) (Leckie et al. 1999). A chimeric gene constructed from flax rust resistance genes P and P2 showed that amino-acid changes in the β-strand/β-turn motif are sufficient to alter P2 to P specificity (Dodds et al. 2001).

The TIR domains of plant R-proteins are thought to have a similar function to the homolog domains from *Drosophila* toll and human interleukin-1 receptor and act as a signaling domain (Hammond-Kosack and Jones 1997). However, analysis of recombinant alleles of *L* genes from flax indicates that the TIR domain also may play a role in pathogen specificity (Ellis et al. 1999; Luck et al. 2000). R-proteins may interact indirectly with pathogen effect or proteins, which target regulators of plant innate immunity. This idea, developed as the "guard hypothesis" (Dangl and Jones 2001), reasonably explains the interactions shown to be required for resistance mediated by RPM1 and RPS2 in *A. thaliana* (Axtell and Staskawicz 2003; Mackey et al. 2002, 2003), where RIN4 may be targeted by AvrRpm1, AvrB, or AvrRpt2. The R-proteins monitor the state of RIN4 and induce the resistance response when the bacterial Avr proteins interact with RIN4. However, evidence for the guard hypothesis with respect to the TIR-NB-LRR class of *R*-genes has not yet been obtained. Recognition of a pathogen by a plant initiates a rapid response localized to the infection site and manifested by changes in ion flux and production of reactive oxygen species that lead to induction of downstream signals and defense genes (Kombrink and Schmelzer 2001; Morel and Dangl 1997). Initiation of local defense also results in signals that induce systemic-acquired resistance (SAR) in uninfected distal parts of the plant, resulting in broad-spectrum resistance (Dong 2001; Shah and Klessig 1999). The role of salicylic acid (SA) in plant defense and induction of SAR has been shown by treatment of plants with SA or its synthetic analogs such as 2,6-dichloroisonicotinic acid (INA) and benzothiadiazole (Klessig et al. 1994). Furthermore, transgenic plants expressing the bacterial SA-degrading enzyme, NahG, are unable to induce SAR (Delaney et al. 1995).

Several mutants in *A. thaliana* have been identified that affect disease resistance responses associated with defense regulatory genes such as *AtSGT1b* (homolog of the yeast gene SGT1) (Austin et al. 2002; Tor et al. 2002). Similarly, *EDS1* (enhanced disease susceptibility) (Parker et al. 1996), *NDR1* (non-race-specific disease resistance) (Century et al. 1997), *PAD4* (phytoalexin deficient) (Glazebrook et al. 1997), and *RAR1* (homolog of a barley gene required for Mla powdery mildew resistance) (Muskett et al. 2002; Tornero et al. 2002) have been identified. Resistance specified by the *RPS4* gene to the bacterial pathogen *Pseudomonas syringae* expressing avr*Rps4* (Gassmann et al. 1999) and the *Hyaloperonospora parasitica* specified by *RPP1*, *RPP2*, *RPP4,* and *RPP5,* which all encode TIR-NB-LRR proteins, is abolished by *eds1* (Aarts et al. 1998; Parker et al. 1996; Rusterucci et al. 2001). In contrast, many CC-NB-LRR resistance genes are independent of *EDS1* but dependent on *NDR1* (Aarts et al. 1998; Century et al. 1997). Both *EDS1* and *PAD4* encode lipase-like proteins (Falk et al. 1999; Jirage et al. 1999), and function within the same defense pathways that regulate SA accumulation (Feys et al. 2001; Zhou et al. 1998). White rust disease occurs on *A. thaliana* (Holub et al. 1995), and three resistance genes to *A. candida* (*RAC*) isolate *Acem1* were identified (Borhan et al. 2001). Cloning was reported of the first WR resistance gene to isolate *Acem1* of *A. candida* (*RAC1*) from *Ksk-1* accession of *A. thaliana*. They also describe the effect on *RAC*-mediated resistance of standard mutations that were used to characterize defense signaling in downy mildew (DM) resistance.

A single dominant gene, RAC1, in accession Ksk-1 of *A. thaliana*, confers resistance to *A. candida* isolate Acem1. This gene was isolated by positional cloning and is a member of the *Drosophila* toll and mammalian interleukin-1 receptor (TIR) nucleotide-binding site leucine-rich repeat (NB-LRR) class of plant resistance genes. Strong identity of the TIR and NB domains was observed between the predicted proteins encoded by the Ksk-1 allele and the allele from an Acem1-susceptible accession Columbia (Col) (99 and 98%, respectively). However, major differences between the two predicted proteins occur within the LRR domain and mainly are confined to the β-strand/β-turn structure of the LRR. Both proteins contain 14 imperfect repeats. *RAC1*-mediated resistance was analyzed further using mutations in defense regulation, including: *pad4–1*, *eds1–1*, and *NahG*, in the presence of the *RAC1* allele from Ksk-1. White rust resistance was completely abolished by *eds1–1*, but was not affected by either *pad4–1* or *NahG* (Borhan et al. 2004).

Reaction of a genotype to a pathogen depends upon the genetics of both the host plant as well as the pathogen. Hence, based on the genetics of host–pathogen relationship, the response of the host pathogen instruction broadly can be categorized into two classes, that is, race specific and nonspecific. In race specific, the interaction or compatibility between host genotype and pathogen determines the disease reaction to develop or not. Race nonspecific resistance, however, is quantitative in nature and is influenced by environmental conditions largely in different geographical locations over the years. Hence, reaction of genotypes ranges from susceptible to resistance grading(s) under natural conditions termed as horizontal resistance, and a wide range of pathogens come under this category. Understanding whether resistance to a given pathogen is race-specific or nonspecific is, therefore, a prerequisite to selection of breeding strategies and is detected by the presence or absence of crossover interaction between host genotypes and pathogen strains. For this purpose, genotype's main effects and genotype–environment interaction (GGE) biplot is the best method for visualizing important crossover interactions. Thus, understanding of *Brassica* genotypes by *A. candida* interactions is of vital importance in identifying resistant genotypes for specific adaptability. GSL-1, EC 414299, and EC 399299 showed additive gene for horizontal resistance to WR, which can prove good donors in further genetic improvement programs. *Brassica juncea* cvs. Varuna, JMM 07–2, JMM 027–1, and JYM 10 had nonadditive gene action for pathogenicity to WR. PBC 9221, GSL 1, EC 414299, and EC 399299 were very similar in genetic makeup for disease resistance, while Varuna showed maximum divergence in genetic constitution from these strains (AICRPRM 2009). The *B. napus* chromosome segment, carrying the WR resistance gene (Ac2V1), appeared to have recombined with the *B. juncea* DNA since recombinant individuals were identified (Somers et al. 2002).

2.2.1.1 *Albugo*-Specific Primers

Several workers have used markers for selective amplification of *A. candida*. Primer DC6:50-GAG-GGA-CTT-TTGGGT-AATCA-30 (Cooke et al. 2000), and LR-0: 50-GCT-TAA-GTT-CAGCG G-GT-30 reverse complementary to LR-0R (Moncalvo et al. 1995), can be used for further studies.

2.2.2 Molecular Mapping of R-Genes from *Brassica*

A single gene (*Acr*) conferring resistance to *A. candida* was mapped on *B. juncea* RFLP map. A cosegregating RFLP marker (X140a) and two other closely linked RFLP markers (X42 and X83) were identified (Cheung et al. 1998). A locus that accounted for 18.4% of the variation in resistance to WR was mapped to linkage group (LG) 2 near the RAPD marker Z19a. A bacterial resistance gene homologous to *Arabidopsis* RPS2 and six different RGAs were sequenced (Tanhuanppa 2004). A more tightly linked marker for the WR resistance gene, using AFLP in conjunction with bulk segregant analysis, and a PCR-based cleaved amplified polymorphic sequence (CAPS) marker for the closely linked RAPD marker, $OPB06_{1000}$, was developed (Varshney et al. 2004). The partially resistant phenotype appeared to be controlled by a single dominant gene. However, it has variable expression, on 7 and 13 days old young plants, and 34 and 45 days old adult plants, but did not develop hypertrophic growth or stagheads under greenhouse and field conditions (Bansal et al. 1999).

Two accessions of *A. thaliana* (Ksk-1 and Ksk-2) were used to identify and map three loci (RAC1, RAC2, and RAC3) of genes that confer resistance to *A. candida*. The phenotypes associated with these genes were classified as either FN (necrotic flecks on upper surface of cotyledons and no blisters) for RAC2 and RAC3 or FYN (flecks surrounded by yellowing and no blisters) for RAC1. Both phenotypes exhibited rapid death of host cells penetrated by the parasite (hypersensitive response), with callose deposition commonly encasing the haustorium. A fine-scale map interval and cosegregating markers for this locus, which in turn enabled mapping of a previously unnoticed source of resistance in Ksk-1, designated RAC3 that exhibits an FN phenotype hypersensitive to the FYN phenotype of RAC1. RAC3 is closely linked to the RPP8/HRT on chromosome 5, a locus which contains specificities for resistance to DM and turnip crinkle virus. Recombinant inbred also enabled mapping of recessive resistance at RAC2 in Ksk-2 to the bottom arm of chromosome 3, in the 6 cM interval between two DM resistance loci (RPP1 and RPP13) (Borhan et al. 2001) (Fig. 2.1; Tables 2.1 and 2.2).

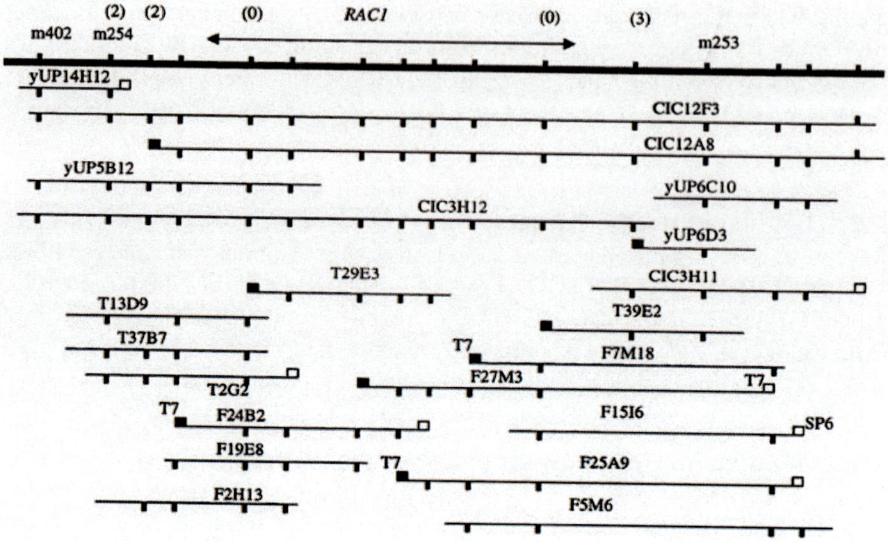

Fig. 2.1 Genetic and physical mapping of the RAC1 locus. Number of recombinants identified in the F6 mapping population is shown in bracket for each molecular marker. Solid bar represents RAC1 interval on chromosome 1. Lines indicate YAC and BAC clones. Boxes at the end of lines are left (solid box) or right (open box) terminal used as anchoring markers. YAC contig was assembled by hybridizing to existing markers and markers developed from their terminal sequences. TAMU (T) and IGF (F) BACs were identified by screening the respective BAC libraries, TAMU (Choi et al. 1995) and IGF (Mozo et al. 1998), with RAC1 flanking markers. BAC clones were ordered by cross-hybridization of their terminal sequences. Based on the physical map for chromosome 1 of *Arabidopsis* (www.arabidopsis.org), the estimated distance for RAC1 interval, delimited by RFLP marker m254 and m253, is about 270 kb (Borhan et al. 2001)

In *B. rapa*, the ACA1 locus was mapped to linkage group 4 and was flanked by RFLP marker loci (Fig. 2.2) (Kole et al. 1996). A quantitative trait locus (QTL) mapping approach using the IP scores detected the same major resistance locus for both races, plus a second minor QTL effect for AC2 on linkage group 2. These results indicate that either a dominant allele at a single locus (Aca1) or two tightly linked loci control seedling resistance to both races of WR in the biennial turnip rape cultivar Per. The map positions of WR resistance genes in *B. rapa* and *B. napus* were compared, and the results indicate that additional loci that have not been mapped may be located. The alignment of these maps to the physical map of the *A. thaliana* genome-identified regions to target for comparative fine mapping using this model organism is presented in Figs. 2.2 and 2.3 (Kole et al. 2002).

The inheritance of avirulence and polymorphic molecular markers were studied in crosses of race 2 (Ac2), isolates MiAc2-B1 or MiAc2-B5 (metalaxyl insensitive and virulent to *B. juncea* cv. Burgonde), and race 7 (Ac7), isolate MsAc7-A1 (metalaxyl-sensitive and virulent to *B. rapa* cv. Torch). Avirulence or virulence of F_2 progeny to *B. rapa* cv. Torch gave 3:1 ratio in each of the three populations, supporting the hypothesis of a single dominant avirulence gene. Amplified fragment

Table 2.1 Genetic position relative to molecular markers in *Arabidopsis thaliana* of three independent loci for resistance to the *Albugo candida* isolate Acem1: RAC1 and RAC3 on chromosomes 1 and 5, respectively, identified in the accession Ksk-1; and RAC2 on chromosome 3 identified in Ksk-2 (Borhan et al. 2001)

		RAC1[c]		RAC2		RAC3	
Marker[a]	Position[b]	*N*	*%R*	*N*	*%R*	*N*	*%R*
m253	I-58	124	6	53	69		
m299	I-62	115	13			57	65
gl-1	III-46			53	12		
p309	III-na			53	3		
pAT389	III-na			53	3		
p3002-2	III-na			45	9		
pm249-2	III-65			54	9		
g2534	III-65			53	13		
m457	III-75	31	55			26	65
m225	V-65					34	72
m331	V-73	43	70			31	36
Cra1	V-na					38	4
nga129	V-na			53	48	36	8
m435	V-80					31	18

[a]RFLP markers include: m253, m299, m331, m457, m225, m435, p3002-2, and p309. PCR-based markers include: g2534, pAT389, pm249-2, Cra1 (CAPS markers), and nga129 (SSLP markers)
[b]Chromosome and relative position (cM) on the unified genetic map of *Arabidopsis* (obtained from TAIR website at www.arabidopsis.org). *na* not available (shown relative to physical order or markers)
[c]N = total number of F6 inbred lines from Wei-1 × Ksk-1 for mapping RAC1 and RAC3, and from Wei × Ksk-2 for mapping RAC2; % R = percentage recombination; data shown for RAC1 was obtained from Holub et al. (1995)

Table 2.2 Key recombinants that define the genetic interval of RAC3, a locus on chromosome 5 of *Arabidopsis thaliana* for resistance to *Albugo candida* (isolate Acem1), relative to molecular markers that were dimorphic between the susceptible accession Wei-1 and the resistant accession Ksk-1 (Borhan et al. 2001)

F_6 Inbred[a]	m331	*RAC3*	Cra1	nga129	m435
2120, 2337	Ksk3[b]	Wei	Wei	Wei	Wei
2377	Ksk	Wei	Wei	Wei	Ksk
2098, 2145, 2273, 2338	Wei	Ksk	Ksk	Ksk	Ksk
2270	Ksk	Ksk	Wei	Wei	Wei
2073	Ksk	Ksk	Ksk	Wei	Wei
2160	Nd	Ksk	Ksk	W/K	Wei
2087	nd	W/K	nd	Ksk	Ksk
2331	Ksk	Ksk	Ksk	Ksk	Wei
2069	nd	Ksk	Ksk	Ksk	W/K

[a]Recombinant inbreds from an outcross between Wei and Ksk-1
[b]Wei = homozygous for Wei-1 dimorphism or uniform susceptibility at *RAC3,* Ksk-1 dimorphism or uniform FN resistance at *RAC3, W/K* heterozygous, *nd* not determined

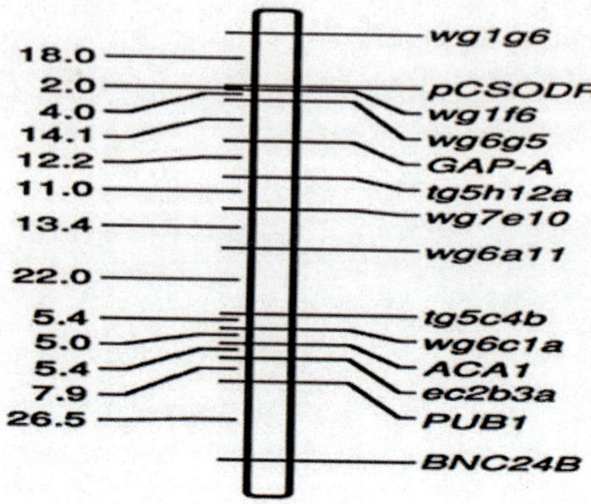

Fig. 2.2 Linkage map of *Brassica rapa* group 4 from analysis of F_3 families derived from cvs. Per × R-500. Locus *ACA1* controls resistance to *A. candida* race 2 and is linked to restriction fragment length polymorphism loci detected by *Brassica* genomic (*wg* and *rg*) and cDNA (*cc*) clones and heterologous probes. Locus *PUB1* controls leaf pubescence. Genetic distance to the left is in centimorgans (Kole et al. 1996)

length polymorphism markers also segregated in regular Mendelian fashion among F_2 progeny derived from two F_1 hybrids (Cr2–5 and Cr2–7) of Cross-2. This first putative avirulence gene in *A. candida* was designated *AvrAc1*. It revealed that a single dominant gene controls avirulence in race Ac2 to *B. rapa* cv. Torch and provides further evidence for the gene-for-gene relationship in the *Albugo–Brassica* pathosystem (Adhikari et al. 2003). *Diplotaxis gomezcampoi* and *Sinapis pubescens* have been identified as WR-resistant wild species (Kumar et al. 2003).

The reaction against WR races 7a and 7v was scored in 20 seedlings from each self-pollinated F_2 individuals. The proportion of resistant plants among these F_3 families varied from zero to 67%. Bulked segregant analysis did not reveal any markers linked with resistance and, therefore, a linkage map with 81 markers was created (Tanhuanppa 2004). A locus that accounted for 18.4% of the variation in resistance to WR was mapped to linkage group (LG) 2 near the RAPD marker Z19a. During the study, a bacterial resistance gene homologous to *Arabidopsis* RPS2 and six different RGAs were sequenced. RPS2 and five of the RGAs were mapped to linkage groups LG1, LG4, and LG9. Unfortunately, none of the RGAs could be shown to be associated with WR resistance (Tanhuanppa 2004).

Various generations, viz., F_1, F_1 (reciprocal), F_2, and DHs, produced from the crosses were inoculated with a zoospore suspension of race 7v of *A. candida*. The partially resistant phenotype appeared to be controlled by a single dominant gene designated as *wpr* with variable expression. The simple inheritance of partial resistance has implications for disease resistance breeding against WR, as this type of resistance can be easily incorporated into elite breeding lines through conventional and DH breeding methods (Bansal et al. 2005). Transfer of resistance to white blister disease between *Brassica* species involving two genotypes each of *B. juncea* and *B. rapa* was studied in hybrids. Hybrids were identified by PCR-based inter-simple sequence repeat (ISSR) markers with both male and female species-specific bands being identified (Gupta et al. 2006).

Two accessions of *A. thaliana* (Ksk-1 and Ksk-2) were used to identify and map three loci (RAC1, RAC2, and RAC3) of genes that confer resistance to *A. candida*

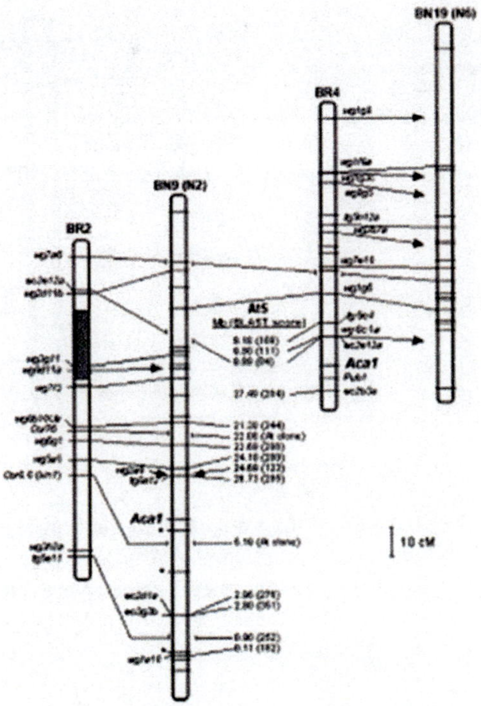

Fig. 2.3 Genetic maps of linkage groups 2 (BR2) and 4 (BR4) from *Brassica rapa* and linkage groups 9 (BN9) (Ferreira et al. 1994), or N2 (Butruille et al. 1999; Osborn et al. 1997; Parkin et al. 1995; Sharpe et al. 1995); and 19 (BN19) (Ferreira et al. 1994), or N6 (Butruille et al. 1999; Parkin et al. 1995; Sharpe et al. 1995) from *B. napus*. *Aca1* on BR4 controls resistance to *A. candida* race 2 (AC2) and race 7 (AC7), and *Aca1* on BN9 controls resistance to *A. candida, Brassica carinata* pathotype. *Pub1* on BR4 controls leaf pubescence. Cross-hatching represents the one-LOD confidence interval for the QTL on BR2 controlling resistance to AC2. RFLP loci detected by *Brassica* genomic (*wg* and *tg*) and cDNA (*ec*) clones on two or more linkage groups are shown and their positions on the different maps are connected by lines. Lines with arrows indicate RFLP loci mapped in *B. rapa* and their approximate positions on N2 and N6 in the maps of Parkin et al. (1995) and Sharpe et al. (1995). The *B. napus* linkage groups are based on "Major" × "Stellar" maps, as described by Osborn et al. 1997, although the relative position of *wg2a6b* (BN9) was estimated based on Butruille et al. (1999). The physical positions in *Arabidopsis* (Mb; chromosome 5) are shown for the DNA sequences that give the strongest matches (BLAST scores in parentheses, except for Cor6.6 and Cor78, which are *Arabidopsis* cDNA clones) to DNA sequences of the RFLP probes surrounding *Aca1* loci. The probe detecting marker locus *wg3h2a* matched a sequence on chromosome 3 of *Arabidopsis* (BLAST score 170); loci marked with asterisks are AFLP markers and were not sequenced (Kole et al. 2002)

(Borhan et al. 2001). Genes for resistance to WR in oilseed *B. rapa* were mapped using a recombinant inbred (RI) population, and a genetic linkage map consisting of 144 RFLP markers, and 3 phenotypic markers. The results indicate that either a dominant allele at a single locus (*Aca1*) or two tightly linked loci control seedling resistance to both races of WR in the biennial turnip rape cultivar Per (Kole et al.

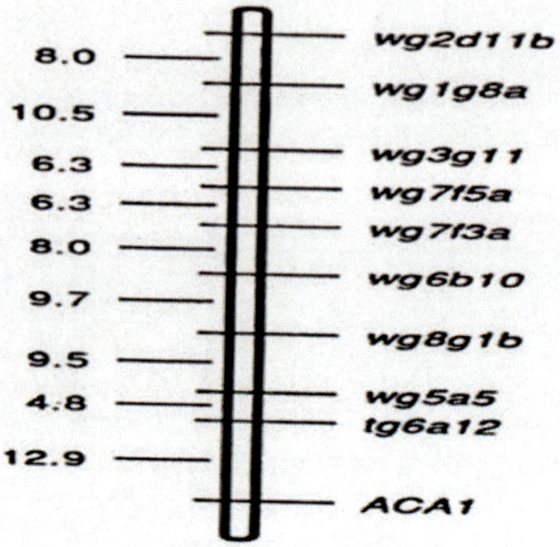

Fig. 2.4 Linkage map of *Brassica napus* group 9 from analysis of doubled haploid lines derived from (Major × Stella) F1. Locus *ACA1* controls resistance to *Albugo candida* isolate ACcar1 and is linked to restriction fragment length polymorphism loci detected by genomic DNA clones and designed *tg* and *wg* on right. Genetic distances (left) in centimorgans (Borhan et al. 2008)

2002). Holub et al. (1995) reported incompatible interactions ranging from reduced blister formation to complete lack of asexual reproduction in the *A. thaliana* and *A. candida* system. The partially resistant phenotype appeared to be controlled by a single recessive gene designated as *wpr* with variable expression. The simple inheritance of partial resistance has implications for disease resistance breeding against WR, as this type of resistance can be easily incorporated into elite breeding lines through conventional and DH breeding methods (Bansal et al. 2005). Information on the genetics and chromosomal location of resistance to these two races of *A. candida* in the *B. rapa* genome would be useful to develop resistant varieties by marker-assisted breeding, and to clone the resistance genes. A single locus controlling resistance to AC2 in *B. rapa* was mapped previously using RFLP markers and a segregating F_2 population from "Per" × "R500" (Kole et al. 1996). However, resistance to AC7 has not been mapped. Resistance to a *B. carinata* pathotype of *A. candida* was mapped as a single locus (*Aca1*) with RFLP markers in *B. napus* (Ferreira et al. 1994). Linkage between the ACA1 locus and nine RFLP loci was observed by Ferreira et al. (1995) on linkage group 9 of *B. napus* RFLP linkage map (Fig. 2.4). Because *B. rapa* is one of the progenitor species of the amphidiploid species *B. napus* and maps of the two species have common RFLP loci (Osborn et al. 1997; Parkin et al. 1995), the potential homology of resistance genes from these species could be investigated by comparative mapping.

Borhan et al. (2008) have characterized a gene designated *WRR4* that encodes a cytoplasmic TIR-NB-LRR receptor-like protein in Columbia *A. thaliana* and confers broad-spectrum WR resistance to four physiological races of *A. candida* described by Pound and Williams (1963) and Liu et al. (1989). Under natural conditions, *A. thaliana* appears to be innately immune as a species to these four *A. candida* races, which instead thrive on their preferred hosts, including different *Brassica* crops and other wild crucifers such as *C. bursa-pastoris.* All four races belong to a predominant group of *A. candida* that is molecularly distinct from the subspecies referred to here as *A. candida* subsp. *arabidopsis,* which is commonly found in the United Kingdom, causing WR in natural *A. thaliana* populations (Fig. 2.5) (Choi et al. 2006;

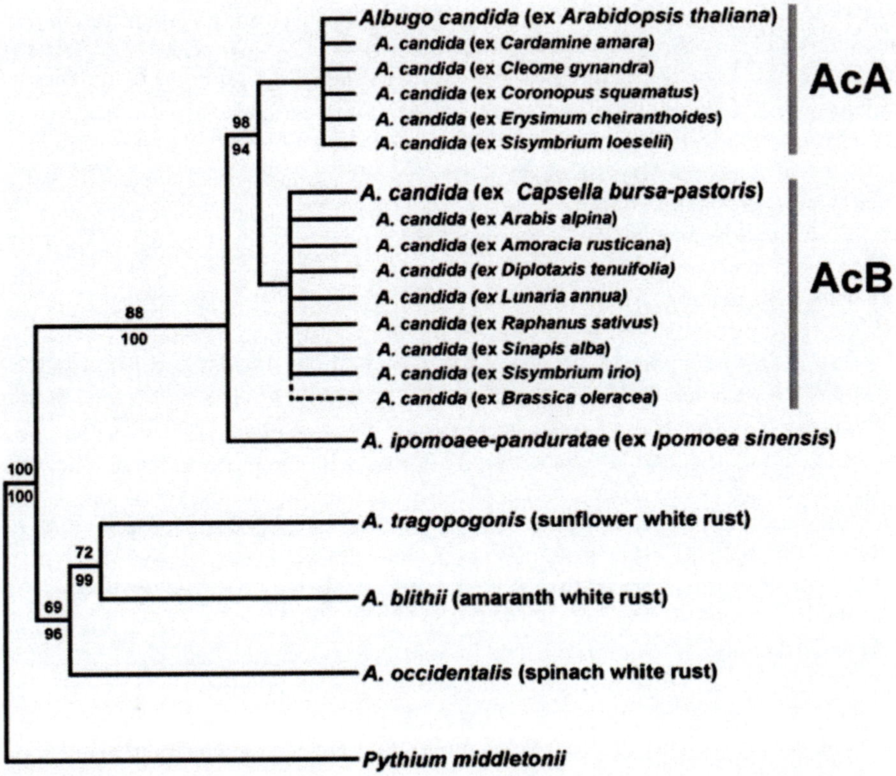

Fig. 2.5 Molecular evidence for two subgroups within *Albugo candida* that are labeled here as AcA (typified by isolates collected from *A. thaliana*) and AcB (typified by *A. candida* race 4 from *Capsella bursa-pastoris*). The phylogenetic relationship between these subgroups was described by Voglmayr and Riethmüller (2006), who compared nuclear large-subunit ribosomal DNA sequences from 60 isolates of *Albugo* spp. in a Bayesian analysis (using Markov chain Monte Carlo analysis over five million generations). An isolate from a *Brassica* host (dashed line) was not included in their study; however, additional data from amplified fragment length polymorphism and internal transcribed spacer 1 sequence comparisons indicate that *Brassica*-derived isolates (including races 2, 7, and 9 from *B. juncea, B. rapa*, and *B. oleracea*, respectively) are grouped with AcB race 4 (Choi et al. 2006; Rehmany et al. 2000)

Rehmany et al. 2000; Voglmayr and Riethmüller 2006). In this context, *WRR4* is similar to other single R-genes that have previously been described including *RB* in the wild potato relative *Solanum bulbocastanum* and *Bs2* in pepper that confer broad-spectrum disease resistance to late blight or bacterial speck, respectively (Tai et al. 1999; Song et al. 2003). *WRR4* (At1g56510) was mapped to a 40-kb region on chromosome 1 containing two additional genes that encode TIRNB-LRR proteins (At1g56520 and At1g56540). Transgenic evidence with subclones from this region and supporting mutant analysis indicated that only one gene (At1g56510) was able to confer WR resistance. A BLAST search of the *A. thaliana* protein database identified the neighboring genes as being the most similar to *WRR4* (74 and 76% similarity, respectively). Meyers et al. (2003) had grouped all three genes in the same clade (TNL-H) of NB-LRR genes in *A. thaliana* Col-0. Interestingly, this clade also includes the stem canker resistance gene *RLM1* (At1g64070; 54% identity and 64% similarity with WRR4) described by Staal et al. (2006), whereas *RPP1-Ws* is the most closely related DM resistance gene (41% identity and 59% similarity). Sequence analysis of the *wrr4–1* EMS mutant allele revealed an amino acid change of C (cysteine) to Y (tyrosine) at position 837. The point mutation lies within the β-strand/β-turn motif of the last LRR of WRR4 (Jones and Jones 1997). Comparison of the solvent-exposed β-strand/β-turn structure within the LRR domain of other known R-genes shows that this motif is hypervariable and under diversifying selection (Bittner-Eddy et al. 2000; Botella et al. 1998; McDowell et al. 1998; Meyers et al. 1998). As with these previous examples, it is plausible that the LRR of *WRR4* is important for recognition specificity of this novel WRR protein.

WRR4 is structurally similar to another TIR-NB-LRR gene called *RAC1* that previously was characterized as a gene conferring *A. candida* subsp. *arabidopsis* resistance in *A. thaliana* (Borhan et al. 2008). As with other examples of TIR-NB-LRR disease resistance genes, *RAC1* and *WRR4*, each require the EDS1 lipase-like protein to confer resistance (Aarts et al. 1998; Feys et al. 2001; Peart et al. 2002). However, unlike the other examples, both of the *WRR* genes appear to function independently from a second lipase-like protein, PAD4. In addition, *WRR4* appears to function independently from FMO1 which is a positive regulator of the EDS1 defense pathway in *A. thaliana* and required for bacterial and DM resistance conferred by other TIR-NB-LRR proteins (Bartsch et al. 2006).

Inducible defense responses involving EDS1 in *A. thaliana* generally have been correlated with SA activity (Wiermer et al. 2005). Enhanced hyphal development of *A. candida* in two mutants (*sid1* and *sid2*) may suggest at least a partial role for constitutive levels of SA in regulating WRR4-mediated resistance because these genes are involved in biosynthesis of SA. In an analogous study, Mellersh and Heath (2003) conducted mutational analyses of nonhost resistance in *A. thaliana* to the basidiomycete rust pathogen from cowpea, *Uromyces vignae,* and concluded that SA plays an important role in restricting compatibility of basidiomycete rusts in *A. thaliana*, as indicated by enhanced hyphal growth and haustorial development of *U. vignae* in the *sid2* mutant, whereas SA-dependent expression of pathogenesis-related proteins (indicative of hypersensitive defense responses) provides no significant contribution to the resistance. A role for SA in *WRR4*-mediated resistance, however, does not appear to involve the elevated expression of the R-protein itself in Columbia *A. thaliana*. Tan et al. (2007) reported that SA treatment of leaf tissue significantly elevated levels of TIR-NB-LRR proteins that are known to confer DM resistance in wild-type Columbia. However, they observed that constitutive expression of WRR4 was among the lowest when compared with other known NB-LRR resistance proteins in nonelicited tissue, with the highest expression of WRR4 in leaf tissue and approximately threefold lower expression in flowers and siliques; the effect of SA treatment on WRR4 expression was negligible. Interestingly, the relationship between SA signals and WRR4 expression may vary markedly in different genetic backgrounds because a sevenfold increase in WRR4 expression was detected in the Libyan accession Mt-0 following treatment of leaf tissue with SA (Tan et al. 2007).

SA is an important molecule for signaling stomatal closure to restrict the entry of motile epiphytic bacteria into the stomatal chamber (Underwood et al. 2007). Guard cells can detect the presence of these bacteria and signal SA-mediated closure via membrane receptors such as FLS2, an LRR receptor-like kinase that detects the nonspecific elicitor protein flagellin (Gomez-Gomez and Boller 2000; Melotto et al. 2006). *Albugo candida* and many basidiomycete rusts such as *U. vignae* also typically attempt to invade a potential host via stomata. Thus, a report by Zipfel et al. (2004) seems particularly fascinating. They used microarray experiments for assessing the response of leaf tissue to flg22 treatment (a fragment of bacterial flagellin protein), and detected an increased expression of WRR4 (approximately sixfold) in wild-type *Landsberg erecta*. A similar flg22-elicited response was observed with the WRR4-paralog At1g56540, whereas most other TIR-NB-LRR proteins were unaffected in expression. Some virulent bacterial pathogens release a phytotoxin coronatine that can suppress the stomatal closure process mediated by FLS2 and SA (Melotto et al. 2006).

A distinction in signaling attributes among members of the TIR-NB-LRR subfamily will be an important factor to investigate for its potential role in determining host specialization among subspecies of *A. candida*. For instance, basic compatibility of *A. candida* subsp. *arabidopsis* in *A. thaliana* involves broad-spectrum suppression of programmed cell death and innate immunity to other biotrophs (Cooper et al. 2008), including suppression of DM resistance conferred by several TIR-NB-LRR genes. However, *RAC1* and *WRR4* represent important exceptions to this phenomenon as receptor-like R-genes in *A. thaliana* that can induce defense in a manner that is nonsuppressible by the corresponding isolates of *A. candida*. This suggests that they differ intrinsically from other TIR-NB-LRR genes, and particularly ones that confer DM resistance, either in timing with an ability to induce a more rapid defense response that preempts defense suppression or perhaps in the biochemistry of the defense-signaling process (e.g., PAD4-dependent defense is suppressible, whereas PAD4-independent defense of *RAC1* and *WRR4* may be nonsuppressible by preinfection with *A. candida* subsp. *arabidopsis*). The latter prediction could be tested by sequential inoculation of transgenic *A. thaliana* Ws-*WRR4*, first with a virulent isolate of *A. candida* subsp. *arabidopsis,* and subsequently with an avirulent isolate of *A. candida* (race 2, 4, or 7), similar to the experiments described by Cooper et al. (2008).

The *WRR4* phenotype in Col-*gl1* does exhibit characteristics of a rapid, full immunity to at least four physiological races of *A. candida*. No symptoms were visible macroscopically, and parasite development invariably was arrested in the first host cell. A rapidly induced host response was evident because an oxidative burst was detected by the accumulation of hydrogen peroxide (shown by DAB staining) following inoculation of Col-*gl1* with the *A. candida* race 4 isolate Acem2. In contrast, the much slower, less-potent attenuation of WR development in Nd-1 did not appear to involve a rapid oxidative burst indicative of host cell death. Interestingly, colonization of Nd-1 by Acem2 was associated with a loss of turgidity, which was not evident with the other three *A. candida* isolates; and Ac9 was strictly unable to reproduce in Nd-1, whereas the other three isolates occasionally were able to produce small white blisters. This subtle phenotypic variation suggests further genetic specificity among interactions of *A. thaliana* with different *A. candida* races, potentially associated with additional receptor-like proteins. Tosa (1992) proposed that interspecific (species-level) variation in innate immunity could be explained in many cases by an accumulation of parasite-recognition genes in natural populations of the host that collectively provide broad-spectrum, species-level disease resistance. In other words, susceptible genotypes (lacking any of the existing R-genes) are a rare occurrence in the host species. Receptor-like genes have been proposed as important determinants of innate immunity in *A. thaliana* to higher taxa (species and subspecies) of parasites from other hosts (Holub 2007). Staal et al. (2006) used transgressive segregation from a recombinant inbred mapping population to identify two TIR-NB-LRR genes that confer resistance to *Leptosphaeria maculans* (stem canker), an important *Brassica* pathogen that is rarely compatible in wild accessions of *A. thaliana. WRR4* provides the first analogous example for WR resistance, which in this case represents a single R-gene that can effectively distinguish between different subspecies of the *A. candida*. The intercross between wild-type Columbia and Ws-3 obviously will provide a more powerful resource for further genetic dissection of non-HR or EDS-independent genes conferring WR resistance in Columbia *A. thaliana*.

Characterization of different defense-signaling attributes among these R-genes will provide an important bridge between the much-needed exploration of *A. thaliana* in its natural habitat and the comparative biochemistry and molecular systems biology research currently being advanced in laboratory research. Investment in *Albugo* genomics, for example, will be essential to determine whether WRR4 confers avirulence recognition of a highly conserved effector protein that is shared among races 2, 4, 7, and 9 of *A. candida*. Given the role of stomata as a preferred site of host entry for bacteria and rust pathogens, it will also be interesting to determine whether *A. candida* produces functional analogs of bacterial flagellin (i.e., a nonspecific elicitor of innate immunity) or coronation (i.e., a hormone-like defense suppressor) and, if so, whether this may relate directly to the effectors protein from *A. candida* that elicits WRR4-mediated recognition or the sustained broad-spectrum defense suppression induced by virulent isolates of *A. candida* subsp. *arabidopsis* (Cooper et al. 2008).

Molecular characterization of *A. candida* effector proteins should be possible because methods have been established for out crossing *A. candida* isolates, bioinformatic identification of effector-like genes from oomycetes, and functional testing of oomycete genes via bacterial delivery (Adhikari et al. 2003; Rentel et al. 2008; Sohn et al. 2007; Whisson et al. 2007; Win et al. 2007). Parasite effectors from *A. candida* will be critical molecular determinants for investigating the relative selective pressures from *A. candida* subsp. *arabidopsis* and races of *A. candida* derived from other hosts on the innate immunity of *A. thaliana* in natural and experimental populations (Fig. 2.6; Borhan et al. 2008).

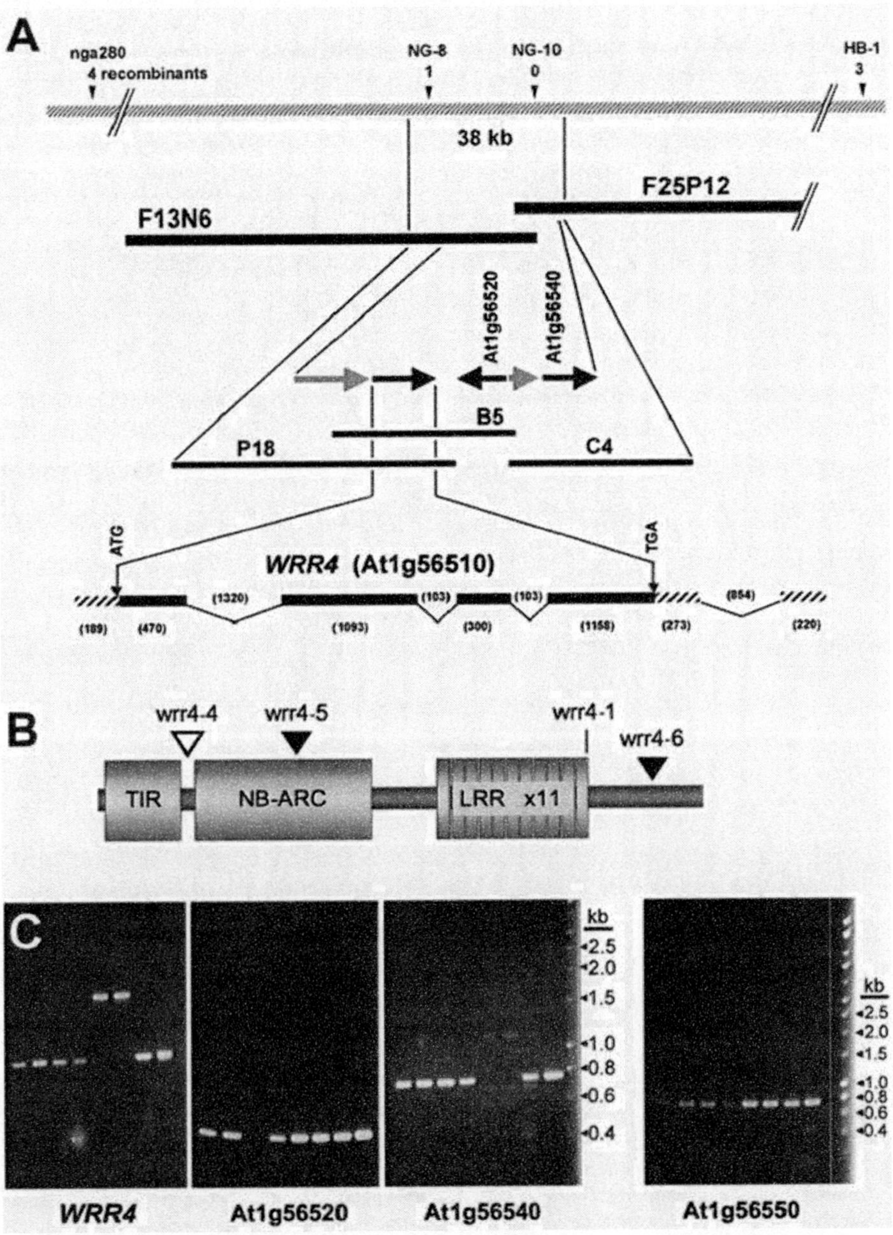

Fig. 2.6 Map-based cloning of *WRR4* from *Arabidopsis thaliana* Columbia and supporting evidence from fast-neutron mutants. (**a**) Position of white rust resistance (*WRR*) gene, *WRR4* (At1g56510), on chromosome 1 (the hatched bar) and position of some of the markers used for mapping *WRR4* indicated by arrow. The bacterial artificial chromosomes spanning across the *WRR4* map contig are shown as solid bars. Cosmid clones spanning across this region are shown by solid lines. Location of exons (solid bars), introns (broken lines), and untranslated regions (hatched bars)

Panjabi-Massand et al. (2010) have tagged two independent loci governing resistance to *A. candida* race 2 V in two east European lines, Heera and Donskaja-IV. Two doubled haploid populations were used; the first population was derived from a cross between Varuna (susceptible Indian type) and Heera (partially resistant east European line), and the second from a cross between TM-4 (susceptible Indian type) and Donskaja-IV (fully resistant east European line). In both the resistant lines, a single major locus was identified to confer resistance to WR. In Heera, the resistance locus AcB1-A4.1 was mapped to linkage group A4, while in Donskaja-IV, the resistant locus AcB1-A5.1 was mapped to linkage group A5. In both the cases, closely linked flanking markers were developed based on synteny between *Arabidopsis* and *B. juncea*. These flanking markers will assist introgression of resistance-conferring loci in the susceptible varieties (Fig. 2.7).

Markers are available for the selection of *B. juncea* plants carrying the resistance gene (AC $2A_1$) to *A. candida* race 2A. Two markers, WR2 and WR3, linked to *A. candida* resistance, flanked the resistance locus AC21 and are very effective in identifying the presence or absence of the resistance gene in the doubled haploid (DH) populations (Prabhu et al. 1998). Eight AFLP markers linked to WR resistance were identified from *B. napus* (Somers et al. 2002) (Fig. 2.8). The *B. napus* chromosome segment, carrying the WR resistance gene (AC2V1), appeared to have recombined with the *B. juncea* DHA since recombinant individuals have been identified (Somers et al. 2002) (Fig. 2.8). Molecular mapping of the locus conferring resistance to *A. candida* in accession BEC-144 of *B. juncea* was accomplished by employing RAPD markers in conjunction with bulked segregant analysis of the F_7 generation RILs. The resistance locus in BEC-144 has been designated as AC2 (+) (Mukherjee et al. 2001).

2.2.3 Molecular Mapping of CNL-Type R-Genes from *Brassica juncea*

Genes conferring resistance to *A. candida* have been mapped in *A. thaliana* (Borhan et al. 2001; Cevik et al. 2019) and in the *Brassica* species—*B. rapa* (AA) (Kole et al.

Fig. 2.6 (continued) are depicted in the lowest bar diagram. (**b**) Diagram of the WRR4 including N-terminal toll-interleukin receptor-like domain (TIR), nucleotide-binding ARC domain (NB-ARC), and 11 leucine-rich repeats (LRRs). The C-terminus is leucine rich but without a repeat structure. Mutations are indicated above, including *wrr4–1* (EMS), which caused a single amino acid change in the last LRR; and TDNA insertion mutants in the first intron (*wrr4–4*, open triangle) in the NB domain (*wrr4–5*, black triangle indicates exon insertion) and non-LRR C-terminal region (*wrr4–6*, exon insertion). (**c**) Gene-specific polymerase chain reaction products were used to detect genetic rearrangements in two fast neutron mutants (*wrr4–2* and *wrr4–3*) that affected *WRR4* and a paralogous gene At1g56540, but not middle paralog (At1g56520) or the next telomeric gene (At1g56550). The bands were generated from the following DNA sources (left to right): Col-*gl1* (*WRR4*), Col-0 (*WRR4*), Ws-0 (*wrr4–0*), Col-*ndr1* (*WRR4*), Col-*ndr1/wrr4–2*, Col-*ndr1/wrr4–3*, Col-*wrr4–1*, and the appropriate BAC clones (F13N6 or F25P12) (Borhan et al. 2008)

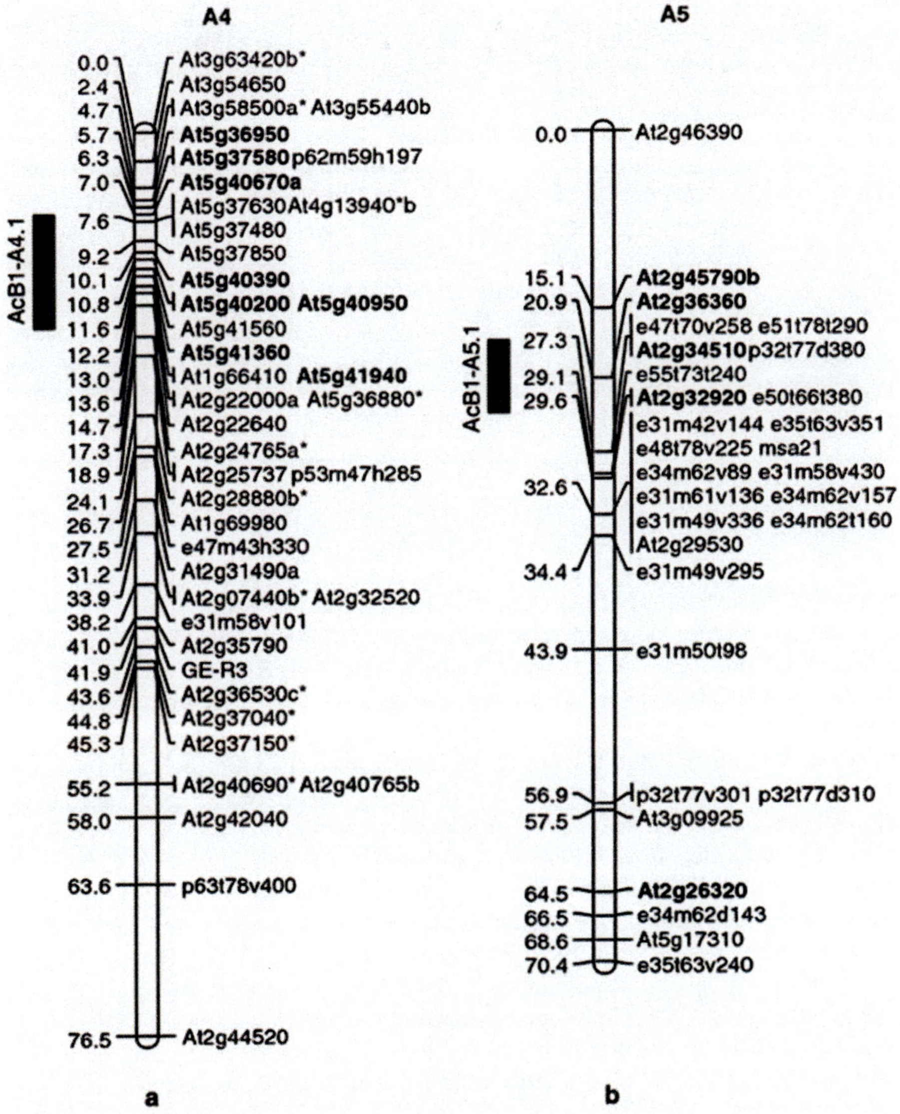

Fig. 2.7 QTL mapping of white rust resistance in two DH populations of *Brassica juncea* (Mukherjee et al. 2001). (**a**) In the VH population, one major QTL (AcB1-A4.1; *black bar*) was mapped in the linkage group A4 at a genetic interval of 7–17 cM, and (**b**) in the TD population, a major QTL (AcB1-A5.1; *black bar*) was detected in the linkage group A5 at a genetic interval of 18–24 cM. The markers highlighted in *bold* represent the new IP markers mapped using syntenic relationship with *Arabidopsis* (Panjabi-Massand et al. 2010)

1996, 2002), *B. juncea* (AABB) (Prabhu et al. 1998; Panjabi-Massand et al. 2010), and *Brassica napus* (AACC) (Ferreira et al. 1995). Arora et al. (2019) reported mapping of resistance-conferring loci in two east European gene pool lines of

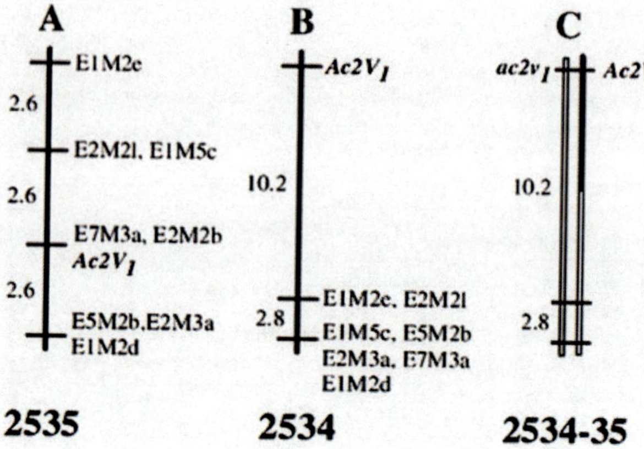

Fig. 2.8 Genetic linkage maps of AFLP markers and the *Ac2V1* locus derived from *Brassica napus* and introgressed into *Brassica juncea*. (**a**) *B. juncea* BC3 F2 population 2535. (**b**) *B. juncea* BC3 F2 population 2534. (**c**) Schematic diagram of the *Ac2V*1 interval in the white rust-resistant *B. juncea* plant 2534–35 showing the *Ac2V1* locus to be heterozygous. *Black and white chromosome segments* represent *B. napus* and *B. juncea*, respectively (Somers et al. 2002)

mustard—Donskaja-IV and Heera. Two different F1 DH populations derived from the crosses—Varuna (susceptible) × Heera (resistant) and TM-4 (susceptible) × Donskaja-IV (resistant)—were used for mapping the resistance conferring loci. Both Varuna and TM4 belong to the Indian gene pool of mustard (Srivastava et al. 2001). Disease assays were carried out using a highly infectious *A. candida* isolate AcB1. In Heera, the resistance-conferring locus AcB1-A4.1 was mapped on LG A4, and in Donskaja-IV, locus AcB1-A5.1 was mapped on LG A5 (Panjabi-Massand et al. 2010). The two loci have been introgressed into four major varieties grown extensively in India. The gene conferring resistance to *A. candida* in the locus AcB1-A5.1 was identified to be aCC-NBS-LRR (CNL)-type R-gene named as BjuA5.WRR.a1(BjuWRR1) that conferred resistance to several isolates of *A. candida* collected from different locations in the mustard growing regions of India (Arora et al. 2019). The vegetable types of mustard, mostly grown in China, constitute the third gene pool of *B. juncea* (Yang et al. 2016). Bhayana et al. (2020) found Tumida, Chinese vegetable type mustard, to be resistant to *A. candida* isolate AcB1. By using an F1 DH population derived from Tumida (resistant) × Varuna (susceptible) cross, they have mapped another major locus on LG BjuA6 of *B. juncea* that is involved with resistance to *A. candida*. Genome assemblies of Tumida (Yang et al. 2016) and Varuna (Paritosh et al. 2019) were used to analyze the genes present in the mapped region. The region was found to contain a CC-NBS-LRR (CNL) type R-gene. The structure and the evolution of the resistance-conferring locus and the candidate R-gene have been reported.

A novel locus in *B. juncea* Tumida that confers resistance to white rust disease caused by *A. candida* has been identified. The resistance-conferring locus AcB1-A6.1 is mapped to LG BjuA6 and is additional to the two loci mapped earlier—AcB1-A4.1 in Heera and AcB1-A5.1 in Donskaja-IV, the two lines belonging to the east European gene pool of mustard (Panjabi-Massand et al. 2010). The locus AcB1-A6.1 contains a CC-NBS-LRR (CNL type R-gene) described as BjuA046215 in the *B. juncea* Tumida genome assembly (Yang et al. 2016). This R-gene is the most likely candidate gene that could be involved with conferring resistance to the isolate AcB1. BjuA046215 has a structure that is typical of CC-NBS-LRR class of R type genes and an inheritance pattern of a single dominant gene usually encountered in effector-triggered immunity (ETI) based on R–Avr interaction (Jones and Dangl 2006; Cevik et al. 2019). BjuA046215 is phylogenetically related to the CNL type R-gene BjuWRR1, identified in the east European gene pool line Donskaja-IV that has been shown to confer resistance to a number of white rust isolates including AcB1 (Arora et al. 2019). BjuA046215 can, therefore, be named as BjuA6.WRR.b1 (BjuWRR2) as it is the second R-gene found to confer resistance to *A. candida* in *B. juncea* (Figs. 2.9 and 2.10).

The most extensive work on the R-genes involved with resistance to *A. candida* has been carried out in *A. thaliana* (Borhan et al. 2004, 2008; Cevik et al. 2019). *A. thaliana* is not a natural host of *A. candida*; therefore, *A. thaliana* mutant line Ws-2-eds1 and two transgressive segregants identified for susceptibility from a MAGIC population have been used for identifying resistance conferring genes (Cevik et al. 2019) in this model plant species. All the identified genes—RAC1, WRR4 (Borhan et al. 2004, 2008), WRR4B, WRR8, WRR9, and WRR12 (Cevik et al. 2019)—belong to the TNL class of NBS-LRR genes. Till now, only two CNL type R-genes, BjuWRR1 (Arora et al. 2019) and BjuWRR2, conferring resistance to *A. candida* have been identified; both the genes belong to the CNLD subgroup. Phylogenetically, the most related gene(s) to the two *B. juncea* genes in *A. thaliana* are AT5G48620, AT5G43470, AT5G35450, and AT1G10920. AT5G43470 has been shown to evolve from AT5G48620 by duplication and transposition of a region containing the latter gene (Meyers et al. 2003). AT5G43470 is a well-characterized gene that confers resistance to some viral pathogens and *Hyaloperonospora parasitica* (Cooley et al. 2000; Takahashi et al. 2002), which causes downy mildew in *A. thaliana* and other species of Brassicaceae; no function has been assigned so far to AT5G48620. The other R-genes belonging to the CNL-D lineage in *A. thaliana* have also been shown to confer resistance to *H. parasitica* (Eulgem et al. 2007). It may be that the related CNL-D lineage R-genes have been recruited in *A. thaliana* ecotypes for resistance to *H. parasitica* and in *Brassica* species for conferring resistance to the more predominant pathogen *A. candida*.

Analysis of allelic variants of BjuA046215 in the Indian (Varuna, Pusa-Jaikisan and Kranti), east European (Donskaja-IV, Cutlass, and Heera), and Chinese (Tumida) gene pool lines suggested the presence of three types of alleles. A clear correlation between the gene pool and the alleles could be observed: Allele 1 present in the Chinese vegetable type Tumida encodes for a complete CNL protein with all the three domains (CC, NB, and LRR), whereas mutations at distinct positions leading to truncation of the LRR domain were observed in the Indian (Allele 2) and the east European gene pools (Allele 3). An analysis of the syntenic regions in the genome assemblies of the *Brassica* species belonging to U's triangle (Nagaharu 1935) showed the presence of one or more orthologs—three paralogs in *B. rapa* (AA) and the A-genome of *B. napus* and one ortholog in *B. oleracea* (CC) and *B. napus* C-genome as well as in the *B. juncea* A-genome. Orthologs identified in other *Brassica* species have not been tested for resistance to *A. candida*. Variability

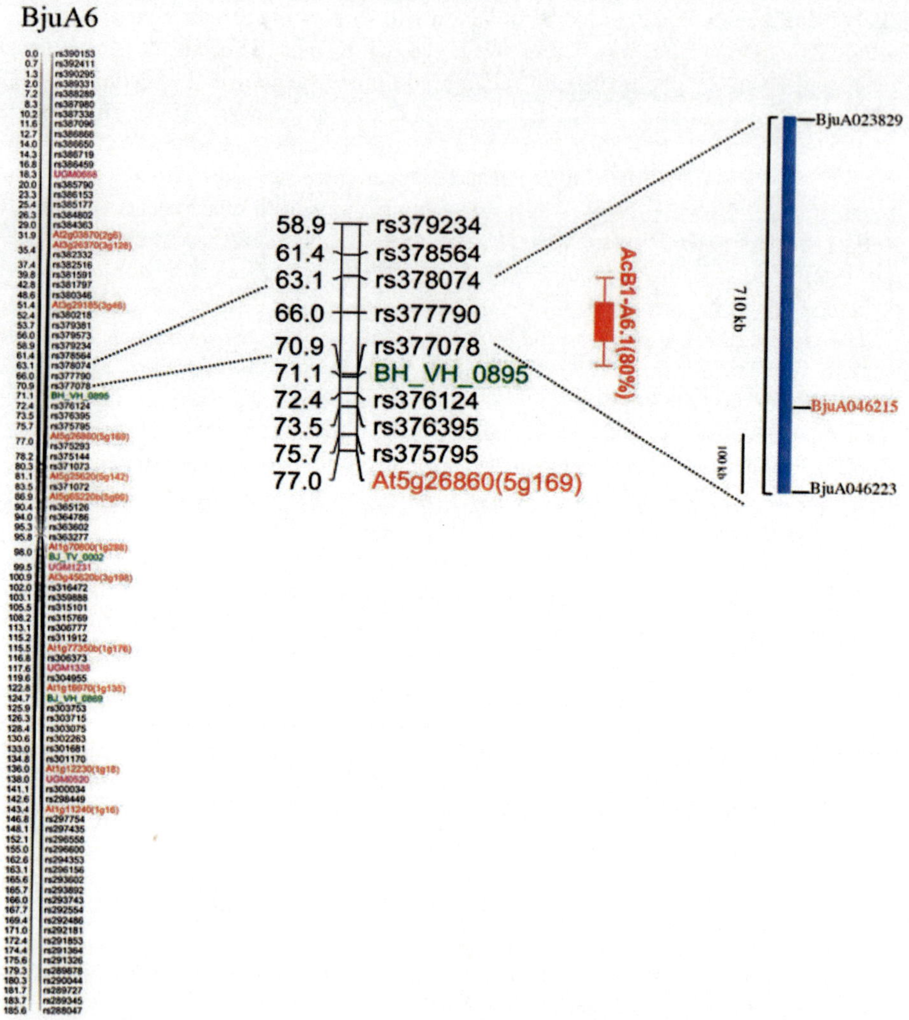

Fig. 2.9 Mapping of white rust resistance in the LG BjuA6. The LG contained 629 polymorphic markers (markers mapping at the same position have been removed). The genic SSR, IP, genic SNPs, and GBS-based SNPs are represented with pink, red, green, and black colors, respectively. A single major QTL (AcB1-A6.1 red bar conferring resistance to white rust isolate AcB1 mapped to an interval of 63.0–70.8 cM. A CNL-type R-gene, Bju046215, was identified in the Tumida genomic sequence spanning the QTL region (Bhayana et al. 2020)

studies in *B. juncea* germplasm have found significantly higher morphological and genetic diversity among Chinese vegetable type mustards in comparison to the germplasm from the other regions (Wu et al. 2009; Chen et al. 2013; Yang et al. 2018).

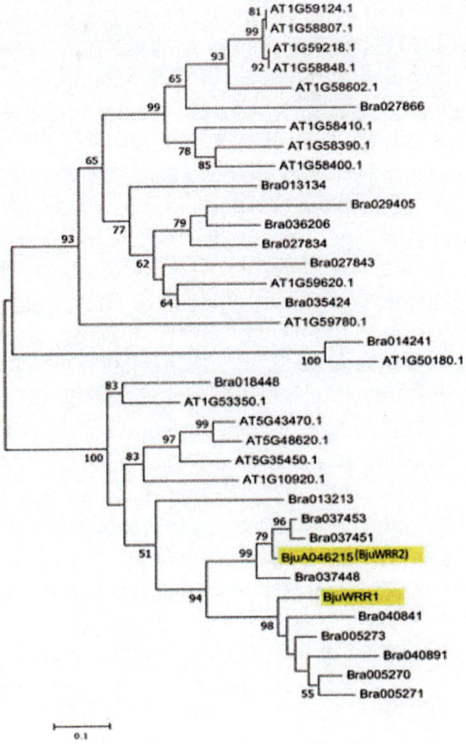

Fig. 2.10 Phylogenetic relationship of NBS-encoding genes belonging to the CNL-D subgroup in *Arabidopsis thaliana*, *Brassica rapa*, and the genes described in *B. juncea*—(BjuWRR1) and BjuA046215 (BjuWRR2) that confer resistance to *Albugo candida* (Bhayana et al. 2020)

It will be useful to look at the allelic diversity for BjuWRR2 in the diverse vegetable and their related oil-yielding types of mustard from the Chinese gene pool of *B. juncea* to identify more R-genes that could be deployed in future to impart durable and broad-spectrum resistance to *A. candida*. Orthologs of BjuWRR2 in the syntenous regions of the related *Brassica* species could also be tested for their potential to confer resistance to white rust in *B. juncea* (Figs. 2.9 and 2.10) (Bhayana et al. 2020).

2.2.4 Mechanisms of *Arabidopsis* Immunity Nonhost Resistance (NHR) to *Albugo candida* Races

Most plant pathogens translocate pathogenicity proteins, called effectors, into host cells; many of these suppress PTI, facilitating colonization. Genetic variation for disease resistance within a plant species is often explained by allelic variation in resistance (R)-genes that encode nucleotide-binding, leucine-rich repeat (NLR) immune receptors. Effector recognition leads to effector-triggered immunity (ETI). Many NLRs carry either toll/interleukin-1 receptor/resistance (TIR-NLRs) or coiled-coil (CC) domains at their N termini (CC-NLRs) and can activate ETI either by directly detecting an effector or indirectly through "guarding" host proteins that are modified by effectors. Unlike CC-NLRs, the function of TIR-NLR proteins requires EDS1 (ENHANCED DISEASE SUSCEPTIBILITY 1), which encodes a lipase-like protein, and forms functional heterodimers in *Arabidopsis* with the related proteins PAD4 (PHYTO ALEXIN DEFICIENT 4) or SAG101 (SENESCENCE-ASSOCIATED GENE 101) (Rietz et al. 2011). Plants are challenged by many potential pathogens, but most plants are resistant to most pathogens and disease is rare. Resistance of a particular plant species against all isolates of a pathogen that can infect other plant species is known as nonhost resistance (NHR). The molecular mechanisms underlying NHR are poorly understood; if all accessions of a species are resistant, genetic analysis of NHR is difficult. Conceivably, NHR or species-level resistance could involve PTI (if effectors cannot suppress PTI), ETI (if effectors do not evade detection), and/or other mechanisms. Fundamental insights into this question are of broad interest. NHR genes that confer complete immunity in a nonhost might confer resistance in susceptible crops and elevate resistance to important crop diseases.

To investigate NHR, Cevik et al. (2019) studied *Albugo candida*, an obligate biotrophic plant pathogen that causes white rust disease in Brassicaceae. In contrast to *A. candida*, *Albugo laibachii* has specialized to cause white rust only on *Arabidopsis*. *A. candida* has many physiological races, each of which specializes on different host species. *Albugo* spp. infection induces a strongly immunocompromised state in host plants, which can enable avirulent races to colonize and reproduce in the same tissue. Sex between different co-colonizing races in the same host could be an important source of new recombinant races. Comparative genomics has revealed extensive genetic exchange between races of *A. candida*, and this genetic exchange could result in races with novel repertoires of effector alleles that, in turn, might enable colonization of new hosts. Therefore, understanding the underlying mechanism of NHR in different *Brassica* species could inform breeding for resistance to *A. candida* (Saharan et al. 2014).

Cevik et al. (2019) investigated adult plant resistance to *A. candida* Race 2 (Ac2V) in diverse *A. thaliana* accessions. While all *Arabidopsis* accessions are resistant to Ac2V, some *A. candida* strains can grow on *Arabidopsis*, but although this pathosystem does not involve NHR to the whole *A. candida* species complex, it is nonetheless instructive. The resistance in *A. thaliana* to Ac2V is due to multiple R-genes, but the R-gene repertoire in different *Arabidopsis* accessions might be distinct, creating the potential for transgressive segregation for susceptibility in recombinant inbreds or other segregating progeny from inter-accession crosses, when screened a population of “MAGIC” inbred lines. These lines result from intercrosses of 19 parents, followed by random intercrossing, and then selfing. These lines have been extensively genotyped. Cevik et al. (2019) inoculated 593 lines and identified two transgressive segregant inbreds (MAGIC.329 and MAGIC.23) that are susceptible in true leaves to Ac2V. However, none of the MAGIC lines tested, nor the 19 parental accessions, are fully susceptible to Race 9 (AcBoT) collected from *Brassica oleracea*. Cevik et al. (2019) defined three loci that contribute resistance to Ac2V, including a known locus, White Rust Resistance 4 (WRR4) on chromosome 1. WRR4 carries two paralogs, WRR4A and WRR4B that can each confer resistance. To investigate AcBoT resistance in *Arabidopsis*, MAGIC.329 with MAGIC.23 were intercrossed. Screening of selfed progeny from these cross-revealed fully susceptible plants at a frequency suggesting that resistance in the two parents is conferred by distinct genes using RenSeq (resistance gene enrichment sequencing). Cevik et al. (2019) identified WRR12 (previously reported as SOC3) as a gene on chromosome 1 that confers AcBoT resistance. These data provide insights into the genetic basis of resistance that restricts pathogen host range and open up a greater subset of the gene pool of crop relatives as a source of genes for crop protection. NHR in one plant species can be defined as complete resistance to pathogens that infect another species. Multiple mechanisms, such as preformed antimicrobial metabolites, and induced defenses, such as PTI and ETI, could contribute to NHR. A better understanding of the mechanisms of NHR could reveal additional genes that confer resistance in crops to plant pathogens.

NHR in *Arabidopsis* against *Brassica*-infecting *A. candida* races has been investigated. All *Arabidopsis* accessions tested are resistant to *B. juncea*-infecting race Ac2V, *B. rapa*-infecting race Ac7V, and *B. oleracea*-infecting race AcBoT. However, Cevik et al. (2019) found that both Col-0-eds1–2 and Ws-2-eds1 are susceptible to all three *A. candida* races, suggesting that NHR to these races might involve TIR-NLR genes. The resistance in different *Arabidopsis* accessions could be mediated by distinct resistance genes. The screened MAGIC lines derived from 19 different *Arabidopsis* parents and identified transgressive segregant lines that are susceptible to Ac2V were used. These susceptible plants enabled to perform genetic analysis to identify resistance genes in multiple *Arabidopsis* accessions.

Three *WRR* (*WRR4BCol*-0, *WRR8Sf*-2, and *WRR9Hi*-0) genes against Ac2V, and a gene, *WRR12* (*SOC3*), conferring NHR to AcBoT were defined, in addition to the previously identified broad-spectrum resistance gene *WRR4ACol*-0. Other investigations have revealed additional *WRR* genes. A point mutation in At1g17610, the neighboring gene of WRR12 encoding a TIR-NB protein, results in chilling sensitive 1 (CHS1), with an auto active defense phenotype. This phenotype could be suppressed by mutations in WRR12, which was therefore named suppressor of chilling sensitive 1–3 (SOC3). SOC3 and CHS1 can associate physically (Zhang et al. 2017).

A phylogenetic analysis using an alignment of the NB-ARC region of TNLs in *Arabidopsis* accession Col-0 reveals that WRR4, WRR4B, and WRR9 are monophyletic, suggesting they shared a more recent common ancestor than with WRR8. This analysis also reveals that WRR12 and CHS1 are located in neighboring expanded clades, many members of which are part of divergently transcribed pairs in the Col-0 genome. This suggests that multiple duplications of an ancestral WRR12/CHS1 pair occurred, similar to the expansion that occurred of RPS4/RRS1-like pairs.

Neither *WRR8* nor *WRR9* confers resistance to *Ac2V* in *B. juncea*, although these genes confer resistance in *Arabidopsis*. *WRR8* also confers resistance to AcBoT in *Arabidopsis*. This could be due to the fact that WRR8- and WRR9-mediated resistance involves a guardee or decoy that is present in *Arabidopsis* but absent or divergent in *Brassica* sp. Indeed, recent publications show that *WRR12*/*SOC3* and

CHS1 form a gene pair and that *WRR12/SOC3*, together with *CHS1*, monitors the homeostasis of E3 ligase SAUL1, a potential guardee that might be targeted by *A. candida* effector(s) (Tong et al. 2017). F2 individuals from crosses between MAGIC.329 and Col-0, Rsch-4, or Ws-2 segregated at a ratio of 13:3, suggesting one dominant and one recessive or haplo-insufficient gene. Identification of a second resistance locus in these F2s will require genotyping fully resistant individuals that lack resistant *WRR4* haplotypes. Crosses between MAGIC.329 and Oy-0 or Sf-2 show 15:1 segregation in the F2, suggesting two independent dominant resistance loci, but genotyping susceptible plants revealed only one locus. All F2 individuals resulting from selfing the F1 between MAGIC.329 × Wu-0 are resistant, suggesting that Wu-0 likely contains >4 resistance loci, so additional loci for resistance to Ac2V and AcBoT likely remain to be discovered. These data suggest that *Arabidopsis* NHR against *Brassica*-infecting *A. candida* races is primarily mediated via ETI, consistent with the expectation that ETI is more likely to contribute to NHR if there is a close evolutionary relationship between the host and nonhost plant species. ETI may contribute to NHR in other plant pathosystems.

NLR-encoding resistance genes recognize pathogen effectors. When *A. candida* races of Ac2V and Ac7V were intercrossed, and F2 individuals obtained and inoculated on *B. rapa* (host for Ac7V but nonhost for Ac2V), a segregation ratio of three avirulent to one virulent was obtained. This supports the hypothesis that resistance to Ac2V in *B. rapa* involves resistance gene-dependent recognition of an Ac2V effector allele that is absent from or different in Ac7V. Specific races of *A. candida*, usually considered a generalist pathogen, colonize a particular host species (McMullan et al. 2015). Conceivably, host and nonhost plants share a common ancestor that was a host for the pathogen. The data suggest that host/race specificity of *A. candida* is determined by the NLR repertoire of the host plant and the recognized effectors of the pathogen race, rather than host compatibility factors. Therefore, some of the NLRs recognizing specific races or multiple races are maintained in different Brassicaceae species. This, in turn, provides an excellent resource to identify WRR genes for different *Brassica* species. By using transgressive segregation to reveal susceptible lines, Cevik et al. (2019) were able to reveal genes that underpin resistance in *Arabidopsis* to *Brassica*-infecting *A. candida* races and show that some of these genes might be useful for elevating crop disease resistance. This strategy could also be applied to identify useful new resistance genes in other crop relatives that show NHR to crop-adapted pathogen races (Cevik et al. 2019).

2.3 *Brassica–Erysiphe*: Molecular Resistance

2.3.1 Multicomponent Mechanisms of Resistance to Powdery Mildew

Host resistance in crucifers to powdery mildews is multilayered and multicomponents both at pre- and postpenetration stages. The intricate immune responses are evolved through accumulation of ROS, H_2O_2, deposition of callose, pectin, cellulose, waxes, silicon, ion fluxes, formation of papilla, cell wall apposition, phenolic compounds, overexpression of R-genes, PR proteins, protein phosphorylation, biosynthesis of phytoalexins, fungal enzymes inhibiters, chito-octamers, triggering of HR, induction of SAR, and nonhost resistance mechanisms. These cytoskeleton components have very important and crucial functional and structural roles in host resistance to powdery mildew pathogens of crucifers. Defenses are activated either through SA signaling or simultaneous perception of ethylene and jasmonic acid (JA). The overexpression of several R-genes in crucifers–powdery mildew host pathosystem induces host resistance. MLO genes encoding seven transmembrane, calmodulin-binding protein confers broad-spectrum resistance to adapted powdery mildews of *Arabidopsis*. *edr* mutants of *Arabidopsis* have a general link between SA-mediated resistance, mitochondrial function, and programmed cell death. The pmr mutants confer resistance to powdery mildew through altered cell wall composition of host. Increased SA enhances the expression of RPW 8.1 and RPW 8.2 leading to HR or SHL and resistance. Bj NPR1 gene activates SAR to confer broad-spectrum resistance to powdery mildew of *B. juncea*. At ROP-regulated At RLCK V1 A3 has a role in basal resistance to powdery mildews. The At MLO2, At MLO6, and At MLO12 triple mutants are resistance to G. *orontii*. The CPR5 gene controls resistance to powdery mildews, and PCD in response to infection by *E. cruciferarum*. There is a role of WRKY transcription factors, and overexpression of R-genes like PMR, MLO, PEN, EDR, MAPK, MAPK 65–3, NPR1, PAD3, PAD4, ED5, SNARE, RLCKs, and KDL (At CEP1) confer R to powdery mildews of crucifers. Higher levels of camalexin contribute to the enhanced R to powdery mildew in Cyp83 a1–3 mutants of *Arabidopsis*. SR1 plays a critical role in powdery mildew resistance by regulating EIN3 and NDR1 expression. There is harmonious coordination between transcriptional regulation and resistance to powdery mildews. The application of *Trichoderma harzianum* and its CF induces (ISR) resistance in crucifers. Mechanisms of nonhost R in crucifers to powdery mildews have been unrevealed which is strong and durable. Nonhost resistance is PEN-gene-mediated at preinvasion and controlled by genes EDS1, PAD4, and SAG (101) at postinvasion of powdery mildew pathogens. In cabbage, a single dominant gene with modifiers controls R to powdery mildew. A single R-gene controls R to powdery mildew in HC-1 and PCC-2 with complete dominance. In *Arabidopsis*, R to PM is polygenic, and based on R-gene *RPW8* or on combination of *RPW* 8 gene complex loci. Powdery mildew resistance genes of *Arabidopsis* have been mapped on chromosomes II (RPW1), III (RPW2, RPW3, RPW7, and RPW8), IV (RPW4), and V (RPW 5, RPW 6). In rapeseed gamma rays, mutagenic plants exhibit R to powdery mildew due to an increase in concentration of unsaturated fatty acids with 18 carbon atoms. Induction of glucosinolates and camalexin plays important roles for resistance to powdery mildew of crucifers. Camalexin biosynthesis and accumulation are affected by WRKY 18, WRKY 40 transcription factor of *Arabidopsis*, and enhances upon *G. orontii* infection to confer resistance. Transfer of powdery mildew resistance to *B. oleracea* from *B. carinata* through embryo rescue has been confirmed. Major gene sources of resistance against powdery mildew of crucifers have been identified (Saharan et al. 2019).

2.3.2 Molecular Mechanisms of Post-penetration Resistance

It is considered as a second line of defense, and RPW8-mediated broad-spectrum resistance, which operates at subcuticular level of host–pathogen interaction. In many plant species that can be colonized by powdery mildew fungi, dedicated dominantly or semi-dominantly inherited resistance (R)-genes provide isolate-specific protection as a second line of defense against the disease (Chelkowski et al. 2003; Bai et al. 2005; Marone et al. 2013). These types of genes typically encode canonical nucleotide-binding site-leucine-rich repeat (NB-LRR/NLR) proteins (Takken and Goverse 2012). R-genes occur typically in multiple allelic forms within plant populations. These polymorphic variants are effective against particular pathogen isolates encoding effectors that are recognized by the respective R-proteins ("gene-for-gene relationship"). It is thought that plant R-proteins either directly or indirectly associate with cognate effector proteins to trigger a boosted defense output that often culminates in a hypersensitive response (HR) associated with local host cell death, thereby restricting pathogen proliferation (Dangl and Jones 2001). However, till date, R-genes that are effective against powdery mildew fungi have not been found in *Arabidopsis*, but in *Brassica* spp., powdery mildew resistance governed by dominant genes have been reported (Abraham 1993; Singh et al. 1997; Saharan and Krishnia 2001). Instead, a polymorphic genetic locus, resistance to powdery mildews *8* (*RPW8*), harboring two unconventional non-NB-LRR-type powdery mildew resistance genes, is a major source of resistance in *Arabidopsis*. The *RPW8* locus has a complex arrangement that differs between *Arabidopsis* accessions. In the resistant ecotype Ms-0, the *RPW8* locus harbors five gene copies that encode small sequence-related basic proteins with a predicted N-terminal transmembrane domain, and one or two C-terminal coiled-coil domains. Of these, two paralogs (tandemly arranged *RPW8.1* and *RPW8.2*) contribute to effective resistance against powdery mildew, while other paralogs (*HR1*, *HR2,* and *HR3*) of the *RPW8* locus are inactive in this respect (Xiao et al. 2001). Phylogenetic analysis on the basis of syntenic loci in the *Arabidopsis* relatives *Arabidopsis lyrata, Brassica rapa,* and *Brassica oleracea* suggests that *RPW8.1* and *RPW8.2* are likely evolved from an *HR3*-like ancestor gene through a series of gene duplication events and subsequent diversification by positive selection (Xiao et al. 2004).

The powdery mildew resistance-conferring *RPW8* locus, which was originally described in the accession Ms-0 (Xiao et al. 1997), shows a widespread distribution in *Arabidopsis* populations. Most powdery mildew-resistant accessions contain a "functional" version of *RPW8.1* and/or *RPW8.2*. The locus is thus a major source of natural powdery mildew resistance in *Arabidopsis* (Orgil et al. 2007; Gollner et al. 2008). Notably, the Col-0 reference accession lacks functional copies of *RPW8.1*/*RPW8.2* and is thus susceptible to all known powdery mildew species that are capable to colonize *Arabidopsis* plants (Xiao et al. 2001). Resistance mediated by

RPW8 occurs after the establishment of haustoria and is typically associated with the accumulation of hydrogen peroxide and localized host cell death, although these responses exhibit some degree of plasticity in different ecotypes (Xiao et al. 2001; Gollner et al. 2008). *RPW8.1* and *RPW8.2* operate through an SA-dependent positive feedback loop, which also promotes transcript accumulation of the two genes (Plate 2.1; Xiao et al. 2003). Consistently, *RPW8*-mediated powdery mildew resistance requires components of SA signaling (*Enhanced Disease Susceptibility (EDS) 1* (At3g48090) and *EDS5* (At4g39030), *Phytoalexin-Deficient (PAD) 4* (At3g52430), *Arabidopsis* nonexpresser of PR genes *(NPR) 1:* At1g64280*1*)) that also play a role in basal defense (Xiao et al. 2005). Overexpression or ectopic expression of RPW8 proteins leads to enhanced resistance against diverse biotrophic pathogens (cauliflower mosaic virus and *Hyaloperonospora arabidopsidis*) but more pronounced susceptibility to necrotrophic pathogens (*Alternaria* and *Botrytis* ssp.; Wang et al. 2007; Ma et al. 2014).

Although RPW8 function in the context of powdery mildew infection is rather evident, information on the protein and its *in planta* activity is limited. In yeast, two-hybrid assays identified 14–3-3λ (AT5g10450) and the phytochrome-associated protein phosphatase type 2C (PAPP2C: At1g22280) as potential RPW8.2 interactors (Yang et al. 2009; Wang et al. 2012). While genetic evidence suggests that the 14–3-3 protein is a positive regulator of RPW8 function, PAPP2C seems to be a negative regulator of cell death and powdery mildew resistance. Notably, following powdery mildew attack, RPW8.2 accumulates at the EHM (Plate 2.1; Wang et al. 2009). In fact, RPW8.2 was the first protein described to localize to this specialized membrane compartment. At the extra-haustorial membrane (EHM), RPW8.2 activates defense signaling via SA and promotes the localized accumulation of hydrogen peroxide and encasement of the haustorial complex (Wang et al. 2009). Targeting of RPW8.2 to the EHM occurs independently of SA accumulation, but requires actin function, and involves transport on secretary vesicles (Wang et al. 2009; Kim et al. 2014). Interestingly, ectopic expression of RPW8.1-YFP or RPW8.2-YFP from the respective native promoters, mutually exchanged promoters, or the constitutive viral 35S promoter results in distinct localization patterns of the proteins and differential resistance phenotypes against powdery mildews. Precise spatiotemporal expression thus appears to be a prerequisite for proper RPW8.2 function (Plate 2.1; Wang et al. 2010; Ma et al. 2014).

To obtain a better understanding of the RPW8.2, protein domains that contribute to its subcellular localization, and defense activity, Wang et al. (2013) functionally analyzed more than one hundred RPW8.2 variants regarding their trafficking and defense properties (Wang et al. 2013). This study revealed single amino acid residues that are critical for the antifungal activity and the induction of cell death. It also uncovered two short stretches rich in basic amino acids that, together with the predicted N-terminal transmembrane domain, define a core targeting signal for the EHM. This region, which comprises 60 amino acids in total, is necessary and sufficient for localization of RPW8.2 to the EHM. Based on the mislocalization of some RPW8.2 mutant variants to the nucleus, and/or plastidic stromules, Kuhn et al.

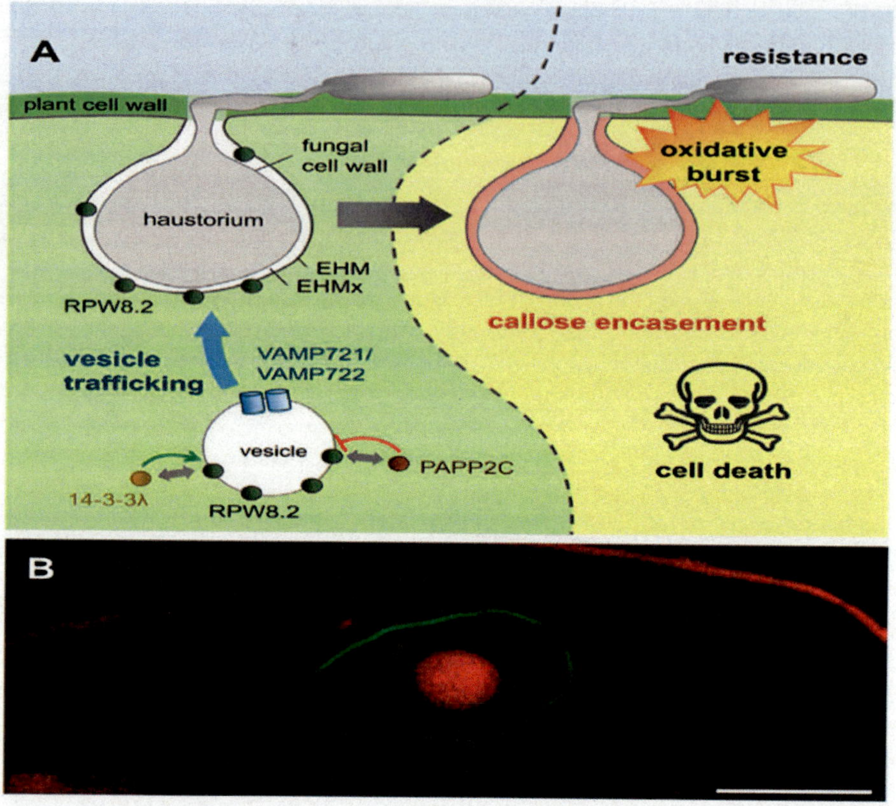

Plate 2.1 RPW8 localizes at the EHM and contributes to cell death upon powdery mildew infection. (**a**) Scheme depicting RPW8.2 function. Left: RPW8.2 interactors and RPW8.2 deposition at the EHM. Right: RPW8.2-triggered oxidative burst and callose encasement of haustoria correlates with subsequent host cell death. (**b**) Confocal laser scanning micrograph of GFP-labeled RPW8.2 (green) in the EHM. Red, propidium iodide-stained plant and fungal structures. Bar = 10 μm (Kuhn et al. 2016)

(2016) proposed the existence of a dedicated membrane trafficking pathway toward the EHM (Wang et al. 2013).

Notably, the two short basic stretches that contribute to EHM localization apparently also play a role in nucleocytoplasmic trafficking of RPW8.2, suggesting that a portion of the RPW8.2 pool might have a function in the nucleus (Huang et al. 2014). Overexpression of nonfunctional, yet EHM-targeted, RPW8.2 versions can exert a dominant-negative effect on functional RPW8.2, thereby compromising RPW8.2-mediated powdery mildew resistance. Such dominant-negative RPW8.2 variants also affect basal defense against powdery mildew and results in an enhanced disease susceptibility (*eds*) phenotype, suggesting the existence of further EHM-localized factors that contribute to basal levels of postpenetration resistance in *Arabidopsis* (Zhang et al. 2015). Widespread presence of a locus that confers broad-spectrum powdery mildew resistance (*RPW8*) might explain why no canonical cytoplasmic NB-LRR-type R-proteins against this disease evolved in *Arabidopsis*. Although such genes are seemingly lacking in natural *Arabidopsis* populations, a heterologously expressed R-protein from monocotyledonous barley can confer isolate-specific powdery mildew resistance in *Arabidopsis*. Transgenic expression of the barley mildew resistance locus A1 (MLA1) coiled-coil NB-LRR-type resistance protein in a partially immune compromised mutant background (*pen2 pad4 senescence-associated gene 101* (*sag101*: At5g14930)) results in isolate-specific resistance against the matching barley powdery mildew (*Bgh*) (Maekawa et al. 2012). This remarkable finding suggests that the signaling machinery acting downstream of MLA1 activation is conserved between monocotyledonous and dicotyledonous plant species, two lineages that diverged ca. 200 million years ago. MLA1 function in resistance responses toward powdery mildew in *Arabidopsis* does not require SA, JA, or ethylene (ET). As in barley, in *Arabidopsis* MLA1 exhibits nucleocytoplasmic partitioning, and its activation upon powdery mildew inoculation results in pronounced and sustained transcriptional reprogramming (Maekawa et al. 2012).

2.3.3 Enhanced Disease Resistance (*EDR*) Genes

Transcriptional activation of *pathogenesis-related* (*PR*) genes is one hallmark of induced defense (Loake and Grant 2007). During a genetic screen aimed to identify novel elements of plant defense, three mutants with enhanced disease resistance (*edr1* (At1g08720), *edr2* (At4g19040), and *edr3* (At3g60190)) that do not express *PR1* (At2g14610) upon inoculation with *Gc* were isolated (Frye and Innes 1998; Frye et al. 2001; Tang and Innes 2002; Tang et al. 2005a, b, 2006). Interestingly, all three mutants show characteristics of "late-acting" resistance (i.e., at 5 to 8 dpi), which is associated with accelerated mesophyll cell death leading to macroscopic patches of lesions and either drastically reduced or absent sporulation. Genetic epistasis analysis revealed that *edr*-mediated resistance is SA-dependent and JA-independent (Frye and Innes 1998; Tang et al. 2005b, 2006).

EDR1 encodes a mitogen-activated protein kinase kinase kinase (MAPKKK) that negatively regulates plant disease resistance (Frye et al. 2001). The *edr1* mutant displays enhanced cell death during infection with the adapted powdery mildew pathogen *Gc*, and in response to drought stress (Frye et al. 2001; Tang et al. 2005a, b). Cell death associated with *edr1* resistance requires the E3 ubiquitin ligases ATL1 (At1g04360) and KEEP ON GOING (KEG: At5g13530). Both E3 proteins are inhibited by interaction with EDR1, and the cell death phenotypes associated with *edr1* are suppressed upon their depletion, indicating that EDR1 acts as a negative regulator of programmed cell death (Serrano et al. 2014). KEG possibly recruits EDR1 to the *trans*-Golgi network (TGN), and in turn EDR1 regulates E3 ligase activity of KEG to further suppress cell death (Gu and Innes 2011; Liu and Stone 2013). Overexpression of *ATL1* causes extensive cell death, which depends on its E3 ligase activity. Strikingly, knockdown of *ATL1* expression does not only interfere with *edr1*-mediated cell death, but causes hypersusceptibility to powdery mildew infection, demonstrating that ATL1 is a positive regulator of pathogen-induced cell death (Serrano et al. 2014). A further link of EDR1 to suppression of cell death is provided by its inhibitory interaction with mitogen-activated protein kinase kinase (MKK) 4 (At1g51660) and MKK5 (At3g21220) that are part of the MAPK cascade fine-tuning plant immunity (Zhao et al. 2014).

EDR2 encodes a mitochondrial protein with a pleckstrin homology domain and a steroidogenic acute regulatory protein-related lipid transfer (START) motif. Both EDR1 and EDR2 function in a common genetic pathway as evidenced by the *edr1 edr2* double mutant, showing resistance phenotypes that are indistinguishable from the respective single mutants (Tang et al. 2005b). In addition, *edr1* and *edr2* both display enhanced senescence in response to ET. Interestingly, mutations in the aminotransferase *AGD2-LIKE* Defense Response Protein1 (*ALD1*: At2g13810) suppress *edr2*-mediated phenotypes including powdery mildew resistance, programmed cell death, and ET-induced senescence, but not the *edr1 edr2* double mutant phenotype (Nie et al. 2011). This raises the question how EDR1 and EDR2 activities are coordinated during the regulation of defense, cell death, and ET-induced senescence. Different from *EDR1* and *EDR2*, *EDR3* seems to function in a separate pathway, since *edr3* does not display an early senescence phenotype. *EDR3* encodes a dynamin-like protein localized partially to mitochondria. Despite the absence of a constitutive cell death phenotype in *Arabidopsis*, the mammalian counterpart of EDR3 plays a role in regulating mitochondrial dynamics associated with programmed cell death (Tang et al. 2006). A fourth *EDR* gene, *EDR4* (At5g05190), with unknown protein function and preferential localization of the gene product at the plasma membrane and endosomal compartments, has been isolated. Like previously identified EDRs, *EDR4* is involved in negative regulation of SA-dependent powdery mildew resistance (Wu et al. 2012). EDR4 functions in the same pathway as EDR1 and EDR2 and interacts with EDR1, recruiting it to fungal penetration sites. The shared phenotypic features of *edr* mutants suggest a general link between SA-mediated resistance, mitochondrial function, and programmed cell death (Ausubel 2005).

2.3.4 Powdery Mildew-Resistant Mutant (*PMR*) Genes

In a genetic screen with the aim to identify susceptibility factors involved in interactions between *Arabidopsis*, and the powdery mildew pathogen *Gc*, six *powdery mildew*-resistant mutants, *pmr1* to *pmr6*, were isolated. Four of the corresponding genes, namely *PMR2* (At1g11310), *PMR4/GSL5*, *PMR5* (At5g58600), and *PMR6* (At3g54920), have been cloned and, to some extent, functionally characterized (Vogel and Somerville 2000; Vogel et al. 2002, 2004; Jacobs et al. 2003; Nishimura et al. 2003; Consonni et al. 2006). The *pmr2* mutant is defective in Mildew Resistance Locus *O* (*MLO*) *2* (At1g11310), which encodes an integral membrane protein of unknown function. *PMR5* belongs to a large plant-specific gene family of unknown function, and *PMR6* encodes a glycosyl-phosphatidyl-inositol (GPI)-anchored pectate lyase-like protein (Vogel et al. 2002, Vogel et al. 2004; Jacobs et al. 2003; Nishimura et al. 2003; Consonni et al. 2006). The latter *pmr* mutants, *pmr5* and *pmr6*, are believed to impact cell wall integrity, further stressing the contribution of the cell wall to powdery mildew resistance. The *Arabidopsis pmr5* mutant exhibits resistance to the adapted powdery mildew fungi *Gc* and *Go*, and enrichment of pectin as well as reduced pectin modification occurs in the cell walls of *pmr5* plants (Vogel et al. 2004). In addition, PMR5 contributes to PEN2-mediated preinvasion resistance to the nonadapted fungus *Magnaporthe oryzae*. The *pen2 pmr5* double mutant shows enhanced penetration success of *M. oryzae* (Maeda et al. 2009), indicating that PMR5 is involved in host and nonhost resistance and emphasizing the importance of cell wall integrity for both types of resistance. PMR6 localizes at the plant cell wall, where it might degrade pectin. In line with this assumption, the *pmr6* mutant displays increased pectin and uronic acid contents. Like *pmr5*, the *pmr6* mutant is resistant to *Gc* and *Go,* which is in both cases independent of SA, ET, and JA signaling (Vogel et al. 2002). The *pmr5 pmr6* double mutant shows increased resistance compared to the respective single mutants, suggesting that the two genes may function separately during plant defense. Furthermore, PMR5 and PMR6 are involved in the regulation of ploidy in mesophyll cells underlying the fungal feeding sites (Chandran et al. 2013).

2.3.4.1 *Arabidopsis* Triple Mutants (mlo2, mlo6, mlo12) Mechanism of Resistance to Powdery Mildew

The major naturally occurring source of resistance effective against powdery mildews in *Arabidopsis* is a gene identified as *resistance to powdery mildew8* (*RPW8*) locus (Xiao et al. 2001; Gollner et al. 2008). This complex locus shows extensive intraspecific genetic variation and confers dominantly inherited resistance against multiple powdery mildew species. The respective genes encode noncanonical resistance proteins that lead to arrest of fungal pathogenesis after host cell penetration (postpenetration resistance). Effective resistance correlates with the encasement of the fungal feeding structures (haustorial complexes) in a callose-containing cell wall matrix (Wang et al. 2009). A different type of powdery mildew resistance is conferred by recessively inherited loss-of-function mutations in specific *mildew resistance locus O* (*MLO*) genes. These genes, which encode integral membrane proteins of unknown biochemical activity, comprise a family of 15 members in *Arabidopsis* (Devoto et al. 2003). Loss-of-function mutations in *MLO2* (At1g11310) results in incomplete resistance against powdery mildew attack that is characterized by a reduction in host cell entry rates by 50%. This coincides with an arrest of hyphal growth prior to the formation of conidiophores in *mlo2* plants, resulting in almost entirely abolished sporulation (Vogel and Somerville 2000; Consonni et al. 2006). Mutations in *MLO6* (At1g61560) and *MLO12* (At2g39200) do not affect powdery mildew interactions on their own. However, it was cooperatively enhanced *mlo2*-conditioned resistance, and in combination with a mutation in *MLO2* causes a complete lack of host cell penetration by fungal sporelings, leading to complete immunity (prepenetration resistance). This type of powdery mildew resistance is best known from barley, where natural and induced *mlo* mutants have been discovered more than 70 years ago and have been successfully employed in agriculture for over 35 years (Jorgensen 1992; Lyng Kjær et al. 2000). The *mlo*-based resistance has been described in several other monocotyledonous and dicotyledonous plant species such as, pea, tomato, and wheat. Hence, *mlo*-based resistance is a seemingly universal phenomenon within angiosperm plant species that are hosts to powdery mildew fungi. At the phenotypical and molecular level, *mlo* resistance resembles the highly effective defense against nonadapted powdery mildews (nonhost resistance). The *mlo2 mlo6 mlo12* mutant shows a spectacular level of resistance against different powdery mildews. Conversely, it slightly enhanced disease symptoms, and in part also pathogen proliferation interactions with some hemibiotrophic/necrotrophic pathogens such as *Alternaria alternata, A. brassicicola,* and *Phytophthora infestans*. These phenotypes might be the indirect consequence of deregulated mesophyll cell death in leaves of the *mlo2 mlo6 mlo12* mutant. Similar to the barley *mlo* mutant, leaves of the triple mutant are subject to spontaneous deposition of callose-containing cell wall appositions, and ultimately premature senescence (Consonni et al. 2006, 2010). The role of *mlo* mutants in providing resistance to powdery mildew fungi is well documented (Kusch and Panstruga 2017).

In order to establish a comprehensive interaction profile of the *Arabidopsis mlo2 mlo6 mlo12* triple mutant, Acevedo-Garcia et al. (2017a) challenged individuals of two independent triple mutant lines with a broad panel of microorganisms along with *Golovinomyces orontii* causing powdery mildew of *Arabidopsis* that are known to be virulent on the *Arabidopsis* Col-0 accession. These two mutant lines represent two entirely independent allele combinations in the genetic background of Col-0. The resulting infection phenotypes were compared with that of Col-0 wild type, which served as control in all experiments. A novel *mlo2 mlo6 mlo12* triple mutant line with near-complete powdery mildew resistance is fully resistant to the adapted powdery mildew pathogen, *G. orontii* (Consonni et al. 2006). Acevedo-Garcia et al. (2017a) generated a second triple mutant line, *mlo2–6 mlo6–4 mlo12–8*, which is based on different T-DNA insertions in the three *AtMLO* genes. As revealed by reverse transcriptase-polymerase chain reaction (RT-PCR) analysis, the T-DNA insertions in these lines result in a lack of full-length *MLO2, MLO6,* and *MLO12* transcripts. The two triple mutant lines grow similarly as Col-0 wild-type plants, but suffer from early leaf senescence. This phenotype is evident by the slightly chlorotic rosette leaves of the two *mlo2 mlo6 mlo12* lines at the age of approximately 6 weeks (Plate 2.2c). The triple mutants upon challenge with *G. orontii* showed that line *mlo2–6 mlo6–4 mlo12–8* fully resembles line *mlo2–5 mlo6–2 mlo12–1* with respect to the macroscopic and microscopic infection phenotypes (Plate 2.2c–e). Unlike the Col-0 control plants, which showed abundant fungal sporulation at 8 days post inoculation (dpi), individuals of both lines lacked visible powdery mildew symptoms (Plate 2.2c). This finding was consistent with analysis at the microscopic level at 48 h post inoculation (hpi), which revealed an early abortion of fungal pathogenesis at the level of host cell entry in the two triple mutants, while Col-0 control plants showed extensive mycelial growth (Plate 2.2d). The line *mlo2–5 mlo6–2 mlo12–1* was entirely resistant, lacking any recognizable host cell

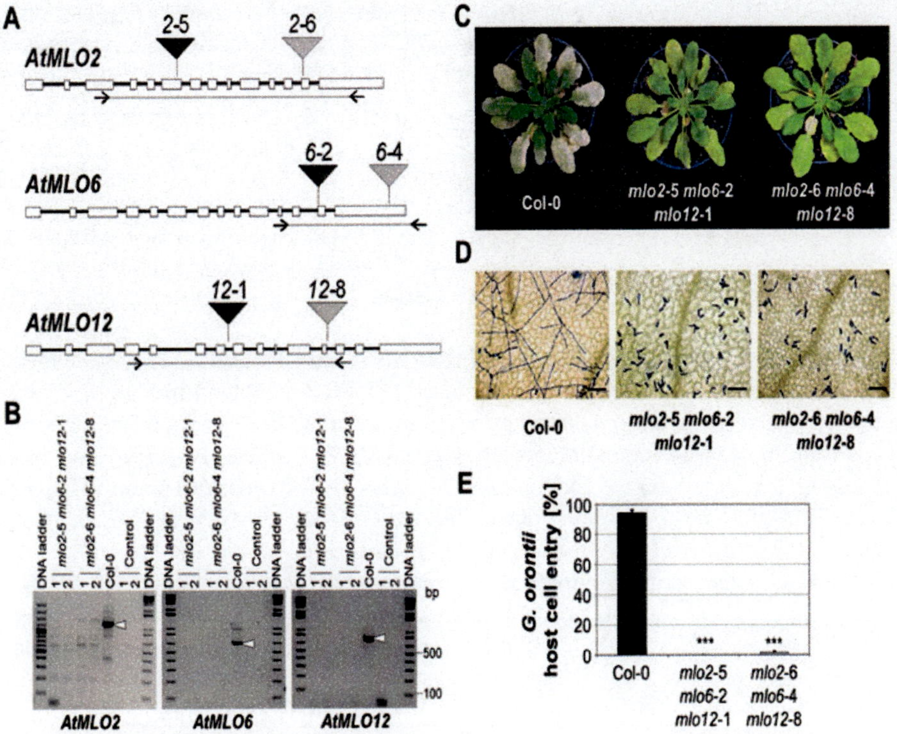

Plate 2.2 The *Golovinomyces orontii* resistance phenotype of the *mlo2–6 mlo6–4 mlo12–8* triple mutant is indistinguishable from the *mlo2–5 mlo6–2 mlo12–1* triple mutant. Six-week-old *Arabidopsis* plants were touch-inoculated with *G. orontii* conidiospores. (**a**) Scheme depicting the T-DNA insertion sites in *MLO2*, *MLO6,* and *MLO12*. Rectangles represent exons, black lines introns. Triangles symbolize the T-DNA insertion sites of the various *mlo* alleles. Lines flanked by inverted arrows (primer binding sites) below the gene models indicate the RTPCR amplicons used to test for *MLO* transcript accumulation in the mutant lines. (**b**) RT-PCR analysis of *MLO2*, *MLO6,* and *MLO12* transcript accumulation. Primer pairs covering the regions indicated in panel A were used to amplify the respective transcript amplicons from cDNA of lines *mlo2–5 mlo6–2 mlo12–1* and *mlo2–6 mlo6–4 mlo12–8* (two individuals each) as well as Col-0 wild-type plants (positive control). RT-PCR reactions without reverse transcription (control 1) and amplification without template (control 2) served as negative controls. White arrowheads indicate RT-PCR products of the expected size in case of Col-0 wild type plants. (**c**) Representative macroscopic infection phenotypes at 8 dpi. (**d**) Light micrographs visualizing fungal pathogenesis at 48 hpi. Leaf samples were cleared in destaining solution, and fungal infection structures subsequently stained with Coomassie Brilliant Blue. Bars = 100 μm. (**e**) Quantitative assessment of host cell entry. Data show the mean ± standard error of the mean (SEM) from three experiments. In each experiment, at least 100 interaction sites from 1–3 leaves of 5 independent plants per genotype were assessed (total of >500 interaction sites per genotype, and experiment). *** Indicates a statistically significant difference from Col-0 ($P < 0.001$) according to a GLM test (binomial distribution) on $n = 3$ independent experimental replicates (Acevedo-Garcia et al. 2017b)

penetration (0% entry rate as judged by the absence of secondary hyphae and discernible haustoria) while line *mlo2–6 mlo6–4 mlo12–8* allowed the occasional formation of fungal microcolonies (Plate 2.2e). The two *mlo2 mlo6 mlo12* triple mutants are essentially equivalent with regard to the level of resistance against the obligate biotrophic powdery mildew pathogen, *G. orontii* (Acevedo-Garcia et al. 2017b).

2.3.5 Powdery Mildew-Resistant Genes

Arabidopsis MLO susceptibility genes were isolated and characterized based on their sequence similarity to barley *MLO* (Consonni et al. 2006) and identified as the *pmr2* mutant in the above-mentioned forward genetic screen (Vogel and Somerville 2000). According to phylogenetic analyses, there are 15 *MLO* genes distributed into five clades in *Arabidopsis*, of which *MLO2*, *MLO6* (At1g61560), and *MLO12* (At2g39200) belong to the same clade (Devoto et al. 2003; Acevedo-Garcia et al. 2014). *mlo2* mutants display reduced penetration success and less sporulation after infection with the adapted powdery mildew fungus *Go* (Consonni et al. 2006). Interestingly, *MLO2* controls penetration success of powdery mildew fungi together with *MLO6* and *MLO12*. While the *mlo6* and *mlo12* single and double mutants do not show any resistance phenotype, they gradually increase resistance of *mlo2* if combined in double and triple mutant combinations, with the *mlo2 mlo6 mlo12* triple mutant being fully resistant (Plate 2.3; Consonni et al. 2006). *MLO* genes encode evolutionary ancient integral membrane proteins with seven transmembrane domains, and unknown biochemical activity (Devoto et al. 2003; Kusch et al. 2016). Besides *Arabidopsis* and barley, mutation of closely related *MLO* genes in tomato, pea, and further plants render these host species resistant to powdery mildew infection, indicating a similar function of the respective proteins (Bai et al. 2008; Humphry et al. 2011). Similar to NHR, *mlo2*-mediated powdery mildew resistance

Plate 2.3 Macroscopic infection phenotypes of Col-0 and the *mlo2 mlo6 mlo12* mutant. Five-week-old wild type (Col-0) and *mlo2 mlo6 mlo12* plants (in Col-0 genetic background) were inoculated with *Golovinomyces orontii,* and photographs were taken 1 week after inoculation (Kuhn et al. 2016)

does not depend on major phytohormone signaling pathways such as those relying on JA, ET, or SA (Consonni et al. 2006). By contrast, all three *PEN* genes are required for *mlo2*-mediated resistance to powdery mildew (Consonni et al. 2006). These findings suggest that *mlo*-mediated resistance and NHR may share overlapping pathways in plant defense (Humphry et al. 2006). Besides the PEN proteins, CYP79B2 (At4g39950) and CYP79B3 (At2g22330), two cytochrome monooxygenases that catalyze the entry step toward the production of diverse indolic metabolites, including the *Arabidopsis*-specific phytoalexin camalexin and indole glucosinolates, are required for *mlo2*-mediated resistance. In contrast to CYP79B2 and CYP79B3, another cytochrome P450 monooxygenase, PAD3 (At3g26830), which catalyzes the final step in camalexin biosynthesis, only plays a minor role in *mlo2*-mediated resistance (Consonni et al. 2010).

2.3.6 Induction and Mechanisms of R-Genes in Pre- and Post-pathogenic Resistance

The pathogen-triggered rearrangement of cellular components correlates with the formation of papillae, and a major radial reorganization of actin filaments underneath attempted powdery mildew entry sites (Kobayashi et al. 1997a, b; Takemoto et al. 2006). Pharmacological treatment of leaves with inhibitors of actin filament polymerization (cytochalasins and latrunculin B), myosin (BDM (2,3-butanedione monoxime)), and NEM (*N*-ethyl-maleimide) results in reduced recruitment of organelles and vesicles toward the site of fungal attack and decreased powdery mildew penetration resistance (Kobayashi et al. 1997a, b; Yun et al. 2003; Yang et al. 2014). Conversely, silencing of genes coding for subclass I actin depolymerization factors (ADFs) increases resistance against *Go* and results in enhanced filament bundling during early *Go* infection (Inada et al. 2016). Together, these findings suggest that intact actin microfilaments and myosin motors are required for successful defense. In fact, single mutants of *MYOSIN XI* genes (*xi-1-1* (At1g17580), *xi-2-1*, *xi-2-2* (At5g43900), *xi-i-1*, *xi-i-2* (At4g33200), *xi-k-1*, *xi-k-2* (At5g20490)), one triple mutant (*xi-1-1*, *xi-2-1*, *xi-k-2*), and one quadruple mutant (*xi-1-1*, *xi-2-1*, *xi-i-1*, *xi-k-2*) exhibit higher penetration frequencies compared to Col-0 wild type upon *Bgh* inoculation. Furthermore, upon challenge with *Gc*, the quadruple mutant shows increased fungal growth, and hyphal branches at 3 dpi, and more conidiophores at 7 dpi compared to Col-0 wild type (Yang et al. 2014). Collectively, these findings indicate that transport activities along the actin cytoskeleton might be crucial for pre- and possibly postinvasive defense against powdery mildews.

SNARE proteins mediate fusion events between vesicular and target membranes. Based on the presence of a critical arginine or glutamine residue in the center of the SNARE domain, this family is divided into R- or Q-SNARE proteins, respectively, where the latter can be further subdivided into Qa-, Qb-, or Qc-SNAREs (Collins et al. 2003; Lipka et al. 2007). PEN1 (Qa-SNARE), SNAP33 (Qb + Qc-SNARE), and VAMP721/722 (R-SNARE) form a ternary SNARE complex that focally accumulates at fungal penetration sites. This complex is required for the timely assembly of papillae, and most likely for the release of pathogen-induced vesicle cargo (Assaad et al. 2004; Kwon et al. 2008; Kwaaitaal et al. 2010). In addition to these SNARE proteins, the TGN-localized Qa-SNAREs of the SYP4 family, which are plant orthologs of the syntaxin 16 in animals, and yeast *Tlg2* (t-SNARE affecting a late Golgi compartment) seem to be required in powdery mildew disease resistance responses. Double mutants, *syp42* (At4g02195) and *syp43* (At3g05710), in plants show increased secondary hyphae formation compared to the Col-0 wild type after inoculation with the nonadapted powdery mildew *E. pisi*, while *Go* infection is unaltered (Uemura et al. 2012). Interestingly, mRFP-VAMP722 partially colocalizes with GFP-tagged SYP43, but not with Venus-SYP61 (At1g28490), another TGN marker. In addition, GFP-SYP43 localizes between the TGN cisternae (labeled with Venus-SYP61), and compartments labeled with mRFP-VAMP722 (Uemura et al. 2012). Furthermore, the TGN-localized KEG ubiquitin ligase, which interacts with EDR1 and regulates transport of membrane-associated proteins to the vacuole, is degraded following the maturation of *Gc* haustoria (Gu and Innes 2012). These observations suggest that KEG might be a plausible virulence target of the powdery mildew fungus. Together, these findings highlight the importance of the TGN during powdery mildew infection.

The ARF-GEF inhibitor brefeldin A (BFA) has been widely used to study the impact of membrane trafficking in powdery mildew interactions. The treatment with BFA hampers penetration resistance to *Bgh* in Col-0 leaves (Nielsen et al. 2012). Additionally, BFA-treated leaves of a *pen1* transgenic line expressing GFP-PEN1 show reduced accumulation of the fusion protein and callose at the sites of attempted fungal penetration. As strong mutants of the well-studied BFA-sensitive ARF-GEF GNOM (At1g13980) are dwarfed, and therefore not suitable for detailed analysis, Nielsen et al. (2012) generated trans-heterozygote plants, carrying two different mutated alleles of GNOM (gnom $^{B409/emb30–1}$). These partially complement the respective nonfunctional domains of the ARF-GEF dimer. *Bgh* infection of genome $^{B409/emb30–1}$ plants reveal an increase in fungal penetration and a delay in callose deposition and papillary GFP-PEN1 accumulation, thus mimicking BFA treatment (Nielsen et al. 2012). Together, these findings suggest that BFA-sensitive GNOM regulates sorting of material to be transported to the papilla, including PEN1 (Nielsen et al. 2012). Notably, BFA treatment of the above-mentioned myosin quadruple knockout mutant (*xi-1-1 xi-2-1 xi-1 xi-k-2*) results in retention of GFP-PEN1 at the plasma membrane, which contrasts its accumulation in BFA bodies in Col-0 epidermal cells. Additionally, accumulation of GFP-PEN1, callose, and autofluorescent material at attempted penetration sites is reduced in the myosin quadruple mutant upon *Bgh* infection (Yang et al. 2014). This experimental outcome implies that members of the myosin XI family are involved in subcellular infection pathways that modulate penetration resistance to powdery mildew.

R-protein *RPW8*.2 localizes to the EHM in cells attacked by the adapted powdery mildew pathogens *Gc* and *Go* (Wang et al. 2009; Micali et al. 2011). Localization studies using *RPW8*.2-YFP under the control of its native promoter in transgenic *Go*-infected Col-0 plants revealed that accumulation of RPW8.2 occurs around mature haustoria that have been partially or completely encased (Fig. 2.11; Micali et al. 2011). Immunogold labeling of RPW8.2-YFP in plants infected with *Gc* supports localization at the EHM, which is reduced after treatment with the actin polymerization inhibitor cytochalasin E (Wang et al. 2009). Overexpression of *ADF6* (At2g31200) in Col-0 plants causes the same response, indicating that intact actin microfilaments are required for successful recruitment of RPW8.2 to the EHM. By

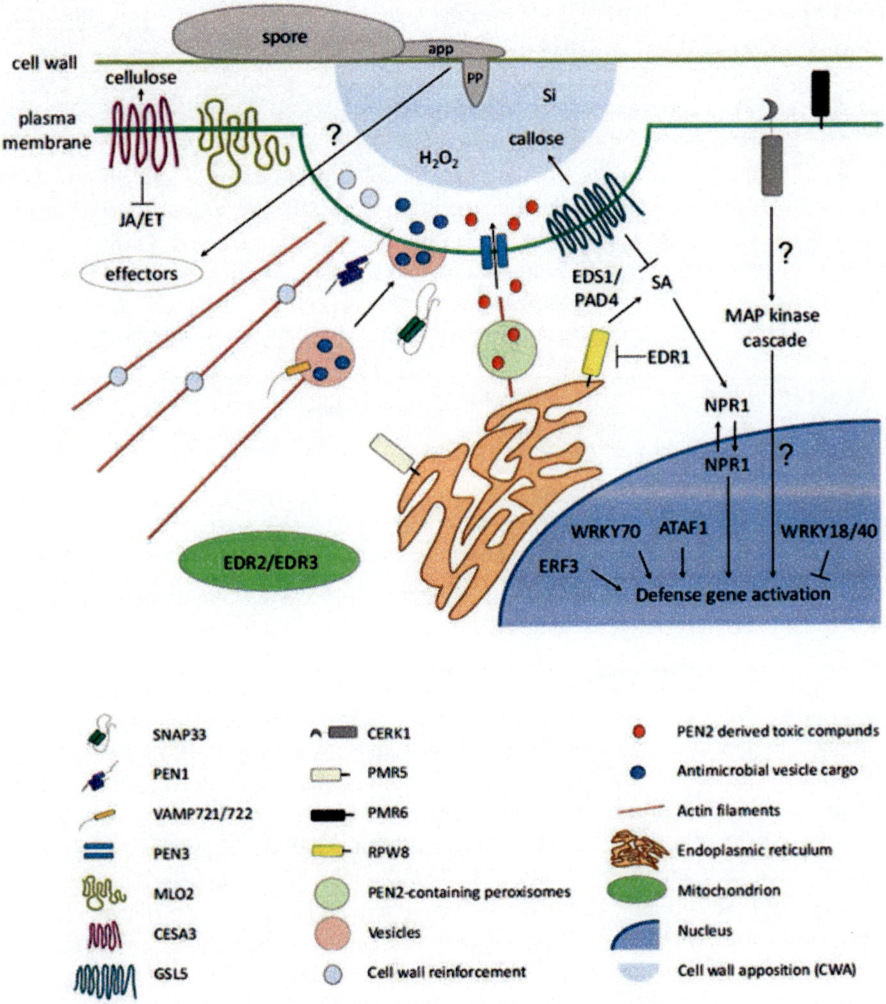

Fig. 2.11 Schematic diagram illustrating genetically-anchored components in *Arabidopsis* powdery mildew susceptibility/resistance. The figure depicts a section of a host cell attacked by a powdery mildew germtube. Components coded by shape and color are explained in the legend below the scheme. *App* appressorium, *pp* penetration peg, *Si* silicon. Question marks indicate presumed links/activities (Micali et al. 2008)

contrast, treatment with oryzalin, a microtubule polymerization inhibitor, does not affect the localization of the resistance protein (Wang et al. 2009). Furthermore, immunogold labeling experiments showed the presence of RPW8.2 in vesicle-like endomembrane compartments on the cytoplasmic side of the callose encasement of the haustorial complex (Wang et al. 2009). A study revealed that the same RPW8.2-containing vesicles colocalize with the R-SNARE proteins VAMP721 and VAMP722. While in the absence of VAMP721 trafficking of RPW8.2 to the EHM is delayed, lack of VAMP722 has a less drastic impact. Reduced EHM targeting efficiency of RPW8.2-YFP in the tested mutants correlates with enhanced *Go* sporulation (Kim et al. 2014). Moreover, delivery of RPW8.2 to the EHM is independent of SA signaling, and PEN1 function, implying that VAMP721/722 vesicles are required for pre- and postinvasive vesicle trafficking pathways in defense against powdery mildews (Wang et al. 2009; Kim et al. 2014).

Host membrane penetration plays a central role during defense against powdery mildew fungi and in other plant–microbe interactions (Dormann et al. 2014; Inada and Ueda 2014; Leborgne-Castel and Bouhidel 2014; Teh and Hofius 2014). Therefore, it is not surprising that pathogens including powdery mildews may attempt to interfere with this pathway. Consistent with this notion, the *Bgh* effecter candidate BEC4 interacts with a member of the ARF-GTPase activating protein (ARF-GAP) family in barley (Schmidt et al. 2014). The *Arabidopsis* ortholog of this protein is AGD5 (At5g54310). Interestingly, *agd5* mutant alleles show considerably elevated *E. pisi*, but unaltered *Go* entry rates. Whether more powdery mildew effectors target the host trafficking machinery should be an object of further investigations (Kuhn et al. 2016).

2.3.7 Function of KDEL (at CEP1) Gene in Powdery Mildew Resistance

Programmed cell death (PCD) is a genetically determined, highly regulated process in all multicellular organisms and a prerequisite for successful development. PCD eliminates tissues and cells serving temporary functions during development such as tapetum cells in anthers and suspensor cells connecting the embryo to the mother plant or nucleus cells of a mature ovule (Zhang et al. 2014; Loapez-Fernaandez and Maldonado 2015; Zhou et al. 2016). Plants furthermore limit the spread of fungal or bacterial pathogens under execution of PCD at the site of infection in a mechanism called the hypersensitive response (HR) (Dickman and Fluhr 2013). Diverse classes of proteases are involved in PCD, including cysteine proteases, serine proteases, aspartic proteases, and metalloproteases. A unique group of papain-type cysteine endopeptidases (CysEPs) is specific for plant PCD and is characterized by a C-terminal KDEL endoplasmic reticulum (ER) retention signal (KDEL CysEPs) with RcCysEP from castor bean (*Ricinus communis*) as the founding member. KDEL CysEPs are not present in mammals or fungi, but are ubiquitous in plants. KDEL CysEPs are synthesized as preproenzymes and are cotranslationally transferred into the ER where the presequence signal peptide is removed. KDEL CysEPs can be stored as enzymatically inactive proenzymes in ER-derived compartments; upon acidification, the KDEL CysEPs are released and the prosequence together with the C-terminal KDEL endoplasmic reticulum retention signal are removed for activation of the enzyme. The mature, enzymatically active KDEL CysEPs exhibit unusual broad substrate specificity. KDEL CysEPs are unique in being able to digest not only cytoplasmic components in tissues that collapse during final stages of PCD; they furthermore are able to digest extensins that form the basic scaffold for cell wall formation. The broad substrate specificity is due to the active site cleft of the KDEL CysEPs that accepts a wide variety of amino acids including proline and glycosylated hydroxyproline of the hydroxyproline-rich glycoproteins of the cell wall.

The respective amino acids, which are decisive for this generally more open appearance of the active site cleft, together with the amino acids defining the catalytic pocket, are highly conserved among all known KDEL CysEPs (Schmid et al. 1998; Beers et al. 2004; Schaller 2004; Than et al. 2004; Helm et al. 2008; Hierl et al. 2012; Howing et al. 2014; Nakano et al. 2014).

In *Arabidopsis*, three KDEL CysEPs—CEP1 (At5g50260), CEP2 (At3g48340), and CEP3 (At3g48350)—have been identified that are expressed in tissues undergoing developmental PCD. Furthermore, *CEP1* was found to be expressed in late response to biotic stress stimuli in the leaf. Two *CEP1* T-DNA insertion lines (SAIL 158 B06 and SALK 01306, both carrying the T-DNA insertion within the third exon) showed enhanced susceptibility to powdery mildew caused by *Erysiphe cruciferarum*. A translational fusion protein of CEP1 with a threefold hemaglutinin tag and the green fluorescent protein under control of the endogenous CEP1 promoter (PCEP1: pre-pro-3xHA-EGFP-AtCEP1-KDEL) rescued the pathogenesis phenotype demonstrating the function of CEP1 in restriction of powdery mildew disease. *atcep1* knockout plants transformed with the nonfunctional reporter including EGFP without the mature CEP1 subunit (PCEP1: prepro-3xHA-EGFP-KDEL) retained susceptibility to *E. cruciferarum*. The spatiotemporal *CEP1*-reporter expression during fungal infection together with microscopic inspection of the interaction phenotype suggested a function of CEP1 in controlling late stages of the compatible interaction including late epidermal cell death (Helm et al. 2008; Hierl et al. 2014; Howing et al. 2014).

Defense responses in *Arabidopsis* are regulated by multiple signal transduction pathways in which salicylic acid (SA), jasmonic acid (JA), and ethylene (ET) function as key signaling molecules. Mutants such as *cpr5* constitutively activate these defense pathways. The *CPR5* gene (*constitutive expression of PR genes 5*, At5g64930) is a regulator of expression of pathogenesis-related genes and participates in signal transduction pathways involved in plant defense and development such as leaf senescence or flowering. It codes for nuclear envelope membrane protein. Loss of *CPR5* leads to spontaneous expression of chlorotic lesions and reduced trichome development. The *cpr5* plants were found to be constitutively resistant to virulent pathogens such as the bacterial pathogen *P. syringae* and *H. arabidopsidis*. Howing et al. (2017) found in public expression data that *CEP1* (At5g50260, Affymetrix ATH1 probe set ID 248545 at; GEO accession GSE5745) is constitutively upregulated in *cpr5* mutants (www.genevestigator.com). They used the *cpr5–2* mutant allele that has a point mutation in the fourth exon leading to a premature stop codon (Trp477stop) in order to analyze a possible contribution of the *CEP1* upregulation to chlorotic leaf lesions in *cpr5*. An increase in *CEP1* expression in *cpr5–2* mutant was measured, which coincided with the appearance of leaf lesions. The expression of *CEP1* was particularly evidenced in leaf cells that surround the chlorotic lesions and presumably underwent cell death. A strong resistance of *cpr5–2* against infection of *E. cruciferarum* and pathogenesis with cell death phenotypes in *cep1 cpr5–2* double mutants as compared to the single mutants was observed. This suggested a contribution of CEP1 to CPR5-controlled cell death (Boch et al. 1998; Clarke et al. 2000; Kirik et al. 2001; Zimmermann et al. 2004; Brininstool et al. 2008; Howing et al. 2017).

Programmed cell death (PCD) is a prerequisite for successful development, and it limits the spread of biotrophic pathogens in a rapid hypersensitive response at the site of infection. KDEL-tailed cysteine endopeptidases (KDEL CysEP) are a subgroup of papain-type cysteine endopeptidases expressed in tissues undergoing PCD. In *Arabidopsis*, three KDEL CysEPs (*AtCEP1*, *AtCEP2*, and *AtCEP3*) are expressed. Howing et al. (2014) have shown that AtCEP1 is a factor of basal resistance to powdery mildew caused by *E. cruciferarum* and is expressed in spatiotemporal association with the late fungal development on *Arabidopsis* leaves. The endoplasmic reticulum-localized proenzyme of AtCEP1 was further visualized at the haustorial complex encased with callose. The *AtCPR5* gene (*constitutive expression of PR genes 5*) is a regulator of expression of pathogenesis-related genes. Loss of *AtCPR5* leads to spontaneous expression of chlorotic lesions, which was associated with enhanced expression of *AtCEP1*. Howing et al. (2017) used the *atcpr5–2* mutant plants and the *atcep1 atcpr5–2* double mutants harboring a nonfunctional reporter (PCEP1: pre-pro-3xHA-EGFP-KDEL) for visualization of *AtCEP1* promoter activity. It is observed that the specific upregulation of *AtCEP1* in direct neighborhood of spreading leaf lesions thus likely represents cells undergoing PCD. A strong resistance of *atcpr5* mutant plants against infection with *E. cruciferarum* is observed. Loss of *AtCEP1* had no obvious influence on the strong resistance of *atcpr5–2* mutant plants against infection with *E. cruciferarum*. However, the area of necrotic leaf lesions associated with *E. cruciferarum* colonies was significantly larger in *atcpr5–2* as compared to *atcep1 atcpr5–2* double mutant plants. The presence of AtCEP1 thus contributes to AtCPR5-controlled PCD at the sites of powdery mildew infection (Howing et al. 2017). Papain-type CysEPs have diverse functions in plant defense to pathogens. The papain-type KDEL CysEP CEP1 fulfills its function in plant defense during late development of *E. cruciferarum* in close spatial association with the fungal haustorium and haustorial callose encasements. Additionally, CPR5 controls resistance to powdery mildew and PCD in response to infection by *E. cruciferarum*. In *cpr5* mutants, *CEP1* is overexpressed in spatiotemporal association with spontaneous cell death. CEP1 contributes to CPR5-controlled PCD triggered by *E. cruciferarum* but is not required to express high-level powdery mildew resistance of *cpr5* mutants (Koh et al. 2005; Misas-Villamil et al. 2016; Howing et al. 2017).

2.4 *Brassica–Hyaloperonospora*: Molecular Resistance

2.4.1 Identification of Seedling and Adult Plant Resistance to Downy Mildew

Resistance to *H. parasitica* has been described in *Brassica oleracea* at the seedling stage (Natti et al. 1967; Monteiro and Williams 1989; Thomas and Jourdain 1992) and in the first true leaves (Thomas and Jourdain 1990). Seedling resistance to *H. parasitica* has been ascribed to a dominantly inherited single gene in Chinese cabbage (Niu et al. 1983), broccoli, and cabbage (Natti et al. 1967). Resistance in cauliflower seedlings has been characterized as a recessive trait based on a single gene (Hoser-Krauze et al. 1984), but within *B. oleracea*, the action of several genes may also control resistance (Hoser-Krauze et al. 1995).

Adult plant resistance to downy mildew has been reported in field trials in cauliflower (Sharma et al. 1991; Mahajan et al. 1995) and has been ascribed to a single gene with dominant effect (Mahajan et al. 1995). From a breeding point of view, assessment of the correspondence between seedling and adult plant resistance may indicate whether resistance at the seedling stage can be used as a reliable indicator of adult plant resistance. Parity between resistance of cotyledons and the first true leaves has been reported (Natti 1958; Monteiro and Williams 1989). Seedlings of six cauliflower cultivars (*Brassica oleracea* convar. *botrytis* var. *botrytis*) were assessed for resistance to a Danish isolate of *H. parasitica* under controlled conditions. Resistance, characterized by restricted sporulation and necrotic dark flecks at the inoculation site on the cotyledons, was expressed in the hybrids 9306 F1, 9311 F1, and the open pollinated cultivar Perfection. Testing of the parent lines and F2 generations of the two resistant hybrids suggested that resistance was a dominantly inherited trait controlled by a single gene. Inoculation of the cultivars with seven isolates, from different geographical origins, showed that the resistance was isolate specific. The two hybrid cultivars expressing cotyledon resistance and two hybrids expressing susceptibility were assessed for adult plant resistance under field conditions. The AUDPC (Area Under Disease Progress Curve), based on disease incidence and severity, revealed significant differences between the cultivars. At harvest, the cultivars exhibited significantly different levels of defoliation and curd attack. The cultivars 9306 F1 and 9311 F1 showed high levels of resistance in all assessments, whereas the two cultivars exhibiting susceptibility at the seedling stage, 9304 F1 and 9305 F1, also exhibited susceptibility through the adult plant stage. Thus, the resistance exhibited under field conditions resembled that identified at the seedling stage under controlled conditions. The results suggest that cotyledon resistance similar to that described could provide resistance throughout the adult plant stage, including curds (Jensen et al. 1999a).

2.4.2 Molecular Mapping of Downy Mildew Resistance Genes

Arabidopsis, a cruciferous weed, has emerged as an important model for investigating the molecular basis of disease resistance in plants. Numerous isolates of *H. parasitica* along with isolates of the bacterium *P. syringae* have been particularly useful for characterizing numerous naturally variable genes that determines the isolates specificity of resistance (so-called R-genes) as well as associated "downstream" defense genes revealed by analyses of artificially derived mutants that exhibit either enhanced susceptibility or enhanced resistance (Holub and Beynon 1997; Shapiro 2000). More than 20 wild-type specificities of resistance to *H. parasitica* (*RPP*) genes have been identified which are distributed among the five chromosomes of *Arabidopsis* (Holub 1997). Genes from four *RPP* loci have so far been published, and they all encode receptor-like proteins containing a conserved nucleotide-binding motif, and a highly variable leucine-rich repeat domain (so-called *NB-LRR* genes) (Bittner-Eddy et al. 2000; Botella et al. 1998; McDowell et al. 1998; Parker et al. 1997). They include examples of functional alleles for different specificities of downy mildew resistance at either a single gene locus in the case of *RPP13* or at a complex locus (multiple genes in a region of <1 cM) in the case of *RPP1*. The simple locus *RPP8/HRT1* is particularly fascinating as an example in which different functional alleles have been described which confer resistance to widely divergent pathogens (Cooley et al. 2000; McDowell et al. 1998). White rust resistance loci on three *Arabidopsis* chromosomes are closely linked to downy mildew (*H. parasitica*) resistance loci. Two accessions of *A. thaliana* (Ksk-1, and Ksk-2) were used to identify and map three loci (RAC1, RAC2, and RAC3) of genes that confer resistance to *Albugo candida* (white rust). The phenotypes associated with these genes were classified as either FN (necrotic flecks on upper surface of cotyledons, and no blisters) for RAC2 and RAC3 or FYN (flecks surrounded by yellowing, and no blisters) for RAC1. Both phenotypes exhibited rapid death of host cells penetrated by the parasite (hypersensitive response), with callose deposition commonly encasing the haustorium. F_6 recombinant inbred lines were produced specifically for the purpose of mapping each RAC locus relative to molecular markers. Dominant resistance at the locus RAC1 in Ksk-1 was previously mapped to chromosome 1 between RFLP markers, m253 and m254, and cosegregating with downy mildew resistance specificity *RPP9* in the accession Wei-0. Borhan et al. (2001) supported a fine-scale map interval and cosegregating markers for this locus, which in turn enabled mapping of a previously unnoticed source of resistance in Ksk-1 designated RAC3 that exhibits an FN phenotype hyperstatic to the FYN phenotype of RAC1. RAC3 is closely linked to the RPP8/HRT on chromosome 5, a locus which contains specificities for resistance to downy mildew, and turnip crinkle virus. Recombinant inbreds also enabled mapping of recessive resistance at RAC2 in Ksk-2 to the bottom arm of chromosome 3, in the 6 cM interval between two downy mildew resistance loci (RPP1 and RPP13).

Phenotypic characterization and molecular mapping of the *A. thaliana* locus RPP5, determining disease resistance to *H. parasitica*, have been observed by Parker et al. (1993). An isolate of *H. parasitica* (denoted NOCO2) was identified that infected *Arabidopsis* plants of the land race Columbia (Col-0) but not plants of land race Landsberg erecta (La-*er*). Segregation analysis of F_2 plants derived from a La-*er* × Col-0 cross established that resistance was inherited as a single locus, denoted *RPP*-5. Macroscopic and microscopic examination of inoculated La-*er* and Col-0 cotyledons showed that restriction of fungal growth in La-*er* was accompanied by massive callose accumulation and death of plant cells in direct contact with points of attempted fungal penetration. La-*er* × Col-0 F1 plants exhibited an intermediate resistance response in all aspects of fungal development, indicating that *RPP*-5 is semidominant in its action. F3 recombinant inbred lines generated between La-*er* and Col-0 were used to map *RPP*-5 to a narrow interval (<1.1 cM) on chromosome 4, utilizing existing restriction fragment length polymorphic (RFLP) markers and newly generated random amplified polymorphic DNA (RAPD) markers. The data provide a basis for the isolation of the *RPP*-5 locus by positional cloning as a first step toward understanding recognition specificity in plant–pathogen interactions at a molecular level (Fig. 2.12).

2.4.3 Genetics of Multiple Disease Resistance in *Brassica*

Multiple disease resistance (MDR) in *B. rapa* to three pathogens showed heritable genetic variation or resistance to each pathogen and a positive genetic correlation between resistance to *H. parasitica* and *Leptosphaeria maculans* (Mitchell-Olds

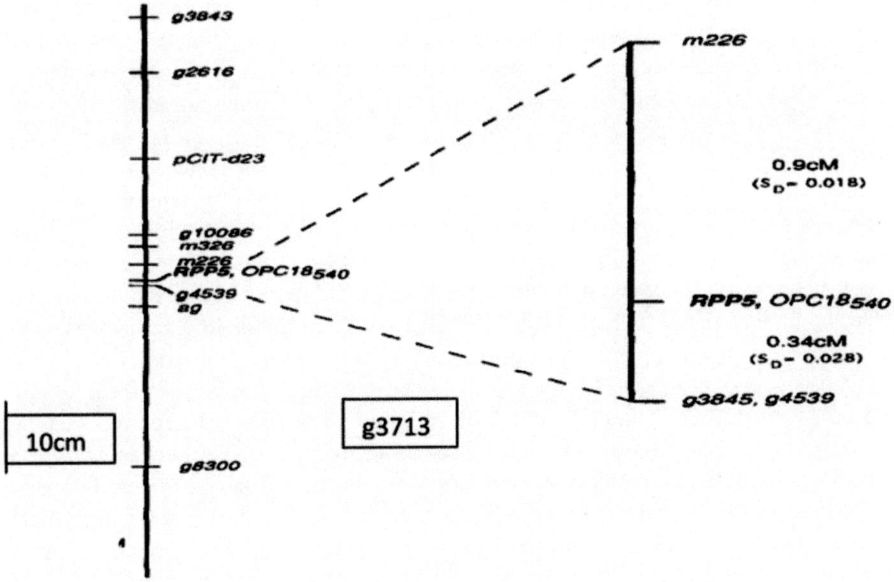

Fig. 2.12 Linkage map of *Arabidopsis* chromosome 4 showing location of RPP-5, relative to cosmid (g) and λ (m) RFLP markers and RAPD (OP) markers, based on the segregation analysis of La-*er* × Col-0 RIs. (**a**) Mapping data derived from segregation analysis of RFLP markers on 100RIs and RAPD markers on 50RIs. (**b**) Map position of RPP-5 relative to closely linked markers, from the analysis of 289 RIs (Parker et al. 1993)

et al. 1995). However, in a later study, Mitchell-Olds and Bradley (1996) found that *Hyaloperonospora*-resistant genotypes grew 6% slower than *Hyaloperonospora*-susceptible genotypes in pathogen-free environments, indicating a significant genetic fitness cost to *Hyaloperonospora* resistance.

Genetic costs of resistance to pathogens may be an important factor maintaining heritable variation for resistance in natural populations. Pleiotropic fitness trade-offs occur when genetic resistance causes reduction in other components of fitness. Although costs of resistance have an important influence on plant–pathogen interactions, few previous studies have detected pleiotropic costs of resistance in the absence of confounding effects of linkage disequilibrium. To avoid this potential problem, Mitchell-Olds and Bradley (1996) performed artificial selection experiments on resistance to two fungal pathogens, *L. maculans* and *H. parasitica*, and compared growth rates of resistant and susceptible genotypes of *Brassica rapa* in the absence of pathogens. *Leptosphaeria* resistance had no effect on growth rate, indicating cost-free defense. In contrast, *Hyaloperonospora*-resistant genotypes grow 6% slower than *Peronospora*-susceptible genotypes in pathogen-free environments, indicating a significant genetic fitness cost to *Hyaloperonospora* resistance. Such genetic trade-offs could maintain genetic variation in the wild. Another factor that might explain heritable variation for resistance is ecological trade-offs, in which genetic resistance to one species causes susceptibility to another. Such ecological trade-offs do not exist for the pathogens studied in this system.

Mitchell-Olds and Bradley (1996) observed that genes conferring resistance could have pleiotropic effects on other aspects of plant performance (Simms 1992). If resistant plants have reduced growth or fecundity in the absence of pathogen attack, then selection may favor intermediate levels of disease resistance. When allocation to constitutive resistance mechanisms varies among genotypes, then resistant genotypes may experience a fitness benefit in the presence of a pathogen, relative to susceptible plants. However, resistant plants would have reduced relative fitness when pathogen attack is reduced or absent, because resources would be expended on unneeded defensive mechanisms. For constitutive defense pathways, a genetic cost to resistance is indicated by a significant negative genetic correlation between growth or fitness in a pathogen-free environment and resistance measured in the presence of pathogens. Few published studies provide clear evidence for pleiotropy causing genetic fitness costs. Berenbaum et al. (1986) showed a negative genetic correlation between insect resistance and fecundity in *Pastinaca sativa*. Han and Lincoln (1994) found a significant negative genetic correlation between growth rate and levels of secondary metabolites in *Diplacus aurantiacus*. In *Escherichia coli*, mutations causing resistance to virus T4 carry pleiotropic costs of resistance, which can be ameliorated by epistatic modifiers (Lenski 1988a, b). Several studies from habitually inbreeding species or near-isogenic lines are consistent with the existence of pleiotropic fitness costs (Bergelson 1994; Brinkman and Frey 1977; Burdon and Muller 1987; Chaplin 1970; Frey and Browning 1971; Simons 1979), but linkage disequilibrium and pleiotropy are confounded in these experiments. Additional evidence from ecological and physiological studies is also suggestive of genetic costs of resistance (Simms 1992).

2.4.4 Expression of Age-Related Resistance (ARR) to Downy Mildew

Disease symptoms in the field start on the lower leaves and progress upward (Natti et al. 1956; Coelho et al. 1998; Jensen et al. 1999a). This disease pattern could be attributed to environmental differences, variation in inoculum availability, and changes in leaf resistance. Downy mildew resistance is affected by plant development in *B. oleracea*. The resistance of mature broccoli plants having eight or more leaves is independent of the resistance of young seedlings (Dickson and Petzoldt 1993), and the plants can be susceptible at the cotyledon stage and resistant at the adult stage (Coelho and Monteiro 2003). In 'Couve Algarvia,' resistance at cotyledon and adult plant stages are under the control of two different genetic systems (Monteiro et al. 2005). Agnola et al. (2003) used leaf discs collected from 17 leaf positions in 10–18-week-old *Brassica* plants to show that incompatible reactions occur regardless of leaf position, but with compatible reactions, and sporulation intensity varies with the position of the leaf on the stem. The increasing resistance to pathogens with plant or leaf aging is a form of resistance referred to as age-related resistance (ARR) (Kim et al. 1989). ARR in *Arabidopsis* is a developmentally regulated and environmentally sensitive defense response to *P. syringae* (Kus et al. 2002). ARR is frequently reported in mature plants in different plant–pathogen interactions: potato–*Phytophthora infestans* (Warren et al. 1971), lettuce–*Bremia lactucae* (Dickinson and Crute 1974), tobacco–*Peronospora tabacina* (Reuveni et al. 1986; Cohen et al. 1987; Wyatt et al. 1991; Hugot et al. 2004), and pepper–*Phytophthora capsici* (Kim et al. 1989). Coelho et al. (2009) conducted experiments under controlled environment conditions, using inoculated leaf discs, to determine the influence of leaf position, plant age, and leaf age on the expression of resistance to downy mildew in various *B. oleracea* genotypes. The upper leaves were more resistant than the lower leaves when 7–19-week-old plants of broccoli and Tronchuda cabbage were tested. Broccoli lines "PCB21.32" and "OL87123–2" were fully susceptible at the cotyledon stage, showed a clear resistance increase from lower to upper leaves at 6 weeks, and "PCB21.32" was fully resistant at 16 weeks after sowing. Immature leaves were more resistant than adjacent fully expanded mature leaves. Susceptibility increased with leaf age when the same leaf was tested at 2- to 4-week intervals. Leaf age and upper-leaf position on the stem had opposite effects on disease score, since younger leaves collected from lower positions and older leaves collected from upper positions tended to score similarly in compatible interactions. The progression of downy mildew from the base of the plant upward on *B. oleracea* in the field could be due to differences in leaf resistance in addition to environmental variation. To maximize the expression of a compatible reaction in adult plants, lower leaves of *Brassica* plants that are at least 12 weeks old should be used. In order to clarify these questions, the use of leaf discs is adequate to assess disease resistance variation within adult plants because it is possible to compare, under the same environmental conditions, the resistance of leaves of different ages or leaves collected from different positions. The use of leaf disc evaluation methods is more informative than field or green house tests, and that can also be used to predict host response to natural epidemics in the field (Coelho et al. 2009).

2.4.5 Different Requirements for Disease Resistance Genes

Mutational analyses in *Arabidopsis* have identified other wild-type genes that are required for R-gene-mediated resistance (Glazebrook et al. 1997). A mutation in *NDR1* (non-race-specific disease resistance) abolished resistance conferred by *RPS2* and *RPS5* to *P. syringae* expressing *avrRpt2* and *avrPph3*, respectively, as well as a dual specificity resistance encoded by *RPM1* to the bacterial genes *avrB* and *avrRpm1* (Century et al. 1995). *ndr1* plants also were compromised in *RPP* gene-mediated resistance to several incompatible *H. parasitica* isolates, suggesting that the wild-type *NDR1* protein may function at a common point downstream from the perception of these prokaryotic and eukaryotic pathogens (Century et al. 1995). Another resistance signaling component is encoded by *EDS1* (enhanced disease susceptibility) as observed by Parker et al. (1996). A mutation in *EDS1* abolished resistance conferred by several *RPP* loci but had no effect on *RPM* 1-specific resistance, suggesting a function before the convergence of downstream pathways (Parker et al. 1996).

The *Arabidopsis* genes *EDS1* and *NDR1* were shown previously by mutational analysis to encode essential components of race-specific disease resistance. Aarts et al. (1998) have examined the relative requirements for *EDS1* and *NDR1* by a broad spectrum of resistance (R)-genes present in three *Arabidopsis* accessions (Columbia, Landsberg-*erecta*, and Wassilewskija). There is a strong requirement for *EDS1* by a subset of *R* loci (*RPP2*, *RPP4*, *RPP5*, *RPP21*, and *RPS4*), conferring resistance to the *H. parasitica* and to *Pseudomonas* expressing the avirulence gene *avrRps4*. The requirement for *NDR1* by these *EDS1*-dependent *R* loci is either weak or not measurable. Conversely, three *NDR1*-dependent *R* loci, *RPS2*, *RPM1*, and *RPS5*, operate independently of *EDS1*. Another *RPP* locus, *RPP8*, exhibits no strong exclusive requirement for *EDS1* or *NDR1* in isolate-specific resistance to *H. parasitica*, although resistance is compromised weakly by *eds1*. Similarly, resistance conditioned by two *EDS1*-dependent *RPP* genes, *RPP4* and *RPP5*, is impaired partially by *ndr1*, implicating a degree of pathway cross-talk. These results provide compelling evidence for the preferential utilization of either signaling component by particular R-genes and thus define at least two disease-resistant pathways. The data also suggest that strong dependence on *EDS1* or *NDR1* is governed by R-protein structural type rather than pathogen class.

2.4.6 Differential Expression of Downy Mildew Resistance Genes

In all identified 129 transcript-derived fragments (TDFs), 121 TDFs were upregulated and 8 were downregulated using cDNA-AFLP (Turck et al. 2004; Xiao et al. 2016). By BLAST searching in the *Brassica* database, these TDFs were classified according to their different functions. The functional categorization showed a complex linkage between proteins encoding by the TDFs. Information obtained from this study may provide a foundation for better understanding of defense mechanisms of the nonheading Chinese cabbage with *H. parasitica* incompatible interaction. Several transcripts encoding the group of PR proteins were differentially expressed in the interaction. TDF3 (β-1, 3-glucanase) was induced within 24 h.p.i., and its expression peaked at 48 h.p.i. The expression levels of TDF8 (*Hapless* 8), TDF16 (a *thaumatin*-like protein), and TDF28 (*PR* 1-like protein) were induced within 24–72 h.p.i. Chen et al. (2008) cloned the full length of β-1, 3-glucanase, *hapless* 8, and *PR* 1 genes and analyzed their expression patterns in response to *H. parasitica* infection in 'Suzhou Qing' cultivar of nonheading Chinese cabbage. The accumulations of these two transcripts were upregulated during the infection period, suggesting that these proteins may participate in the defense reaction for nonheading Chinese cabbage against *H. parasitica*. They also found that expression of TDF39 (*ATMT*-1) peaked at 48 h.p.i. In rice and barley, *MT2A* genes were induced by stresses such as drought, cold treatment, and wounding or in response to pathogen attacks (Ozturk et al. 2002; Degenhardt et al. 2005; Jin et al. 2006). The products of homologous *MT* scavenged the reactive oxygen species (ROS), such as OH to H_2O (Akashi et al. 2004). Evidences suggest that the generation of ROS occurs at early stage in the plant–pathogen interaction. Rapid accumulation of ROS causes oxidative burst that results in hypersensitive cell death and cell wall cross-link (Das et al. 2008). Several signal transduction-related proteins are involved in the plant–fungus interactions. Xiao et al. (2016) identified many TDFs related to the signal transduction, such as TDF42 (*Catalase* 3), TDF43 (*2-Cys PrxB*), TDF45 (*ATP binding*), TDF50 (*ATRER1A*), TDF58 (*Nitrilase*), TDF48 (BcCAM3_A04, calcium ion binding), and TDF68 (*WRKY DNA-binding protein*). Calcium binding-like proteins may have a role in signaling pathways against pathogens and wounding. A number of downstream targets of calmodulin (*CaM*), including *nitric oxide* synthase, barley *MLO protein*, maize Ca^{2+} -CaM, and *transcriptional regulators*, are involved in plant responses to pathogens. Given that *calcium ion-binding* proteins are important modulators of defense response in pathways for pathogen sensing in plants, the *CAM* 3 genes could have a special role as Ca^{2+} sensors during the plant immune response to the fungus *H. parasitica* (Xiao et al. 2016).

WRKY proteins are signal transcriptional factors recognizing the TTGAC (C/T) W-box elements in the promoters of a large number of plant defense-related genes (Turck et al. 2004). Many of WRKY genes are upregulated particularly in pathogen-infected, wounded, or abiotic-treated plants (Pandey and Somssich 2009). In this study, expression of *WRKY DNA*-binding protein peaked at 24 h.p.i., suggesting that the possible role of WRKYs is in the regulation of the genes associated with plant defense responses. However, Xiao et al. (2016) found that expression pattern of TDF60 (*WRKY* gene) determined by *qRT*-PCR was inconsistent with that of cDNA-AFLP. The inconsistence may be caused by different paralogues in the genome. An ethylene response factor (*BcERF1_A01, TDF93*), a regulator of ethylene responses after pathogen attack in *Arabidopsis* (Berrocal-Lobo et al. 2002), may have a key role in the nonheading Chinese cabbage–*H. parasitica* interaction. Studies have been demonstrated that ERFs are involved in regulating the expression of defense-related genes during the disease resistance responses (Singh et al. 2002; Cao et al. 2005). Xiao et al. (2016) found that TDF91 (*BcABCG36_A07, ATP-binding cassette g36*) was inhibited after inoculation by cDNA-AFLP analysis. However, its expression was induced weakly at 24 h.p.i., peaked at 48 h.p.i., and decreased weakly at 72 h.p.i. afterward by qRT-PCR analysis. TDF100 (*BcEDM1_A09*) coding for an enhanced downy mildew 1 homolog was found to be induced during the infection period. Its relationship with the fungi, bacteria, and viruses has been identified to be regulators of R-gene-mediated resistance in other crop species. The studies have revealed that *EDM1* homologue gene *SGT1* is required for pathogen-induced disease-associated cell death during both compatible and incompatible interactions in tobacco (Wang et al. 2010).

Energy metabolism has an important role in plant–pathogen interaction. The photosynthetic carbon cycle (PCC) is part of the dark reactions of photosynthesis and can be roughly divided into three steps: carboxylation, reduction reaction, and regeneration of RuBP. Xiao et al. (2016) found that some TDFs relating to energy metabolism were downregulated, such as TDF76 (*chlorophyll a/b-binding protein*), TDF78 (*SIT4 phosphatase-associated family protein*), and TDF82 (*PP2C-related protein*), whereas some were upregulated, such as TDF79 (*receptor like protein 51*), TDF80 (*NADH-ubiquinone oxidoreductase*), TDF83 (*respiratory burst oxidase protein*), and TDF85 (*rubisco small subunit 1b*). Previous reports have identified that they are involved in PCC cycle, for example, TDF86 (*reduction of transketolase*) inhibited ribulose-1, 5-bisphosphate regeneration and photosynthesis. The expression of energy metabolism-related genes is induced and/or suppressed in photosynthesis during abiotic and biotic stresses. The PCC cycle could provide protection function in energy metabolism during nonheading Chinese cabbage against *H. parasitica* (Xiao et al. 2016).

A number of genes related to protein–protein interaction was induced after inoculation, such as TDF110 (*BcTPR12_A07, tetra-tricopeptide repeat protein*) and TDF111 (*BcZF_A01, zinc-finger family protein*), of which the gene expression of *PAT* is induced in the presence of ozone in *Arabidopsis* (Conklin and Last 1995). The tryptophan biosynthetic enzymes, including anthranilate synthase (*ASA*) and *PAT*, are co-coordinately upregulated at both the messenger RNA and protein level during biotic and abiotic stress (Zhao and Last 1996). Xiao et al. (2016) found that one of *BcZF* orthologous to *A. thaliana* was induced after inoculation. Rizhsky et al. (2004) speculated that a zinc-finger protein is required for the expression of ascorbate peroxidase, which provides some measure of resistance for plant during oxidative stress. The pathogen-induced accumulation of these protein–protein interaction-related genes suggested that these genes may be involved in some defense mechanisms against *H. parasitica* indirectly (Xiao et al. 2016).

Using the cDNA-AFLP method, Xiao et al. (2016) also detected several unknown functional genes. Their biological role is still unclear. Xiao et al. (2016) examined gene expression patterns in an incompatible interaction between nonheading Chinese cabbage 'Suzhou Qing,' and the downy mildew pathogen. They obtained

129 TDFs with different expression patterns and classified functional categories using cDNA-AFLP. Fifteen TDFs were randomly selected for validation of cDNA-AFLP expression patterns using *qRT-PCR*. Results showed that reliability of cDNA-AFLP is suitable for detecting differentially expressed genes. Among the 15 TDFs, four TDFs are related with fungal resistance, namely, TDF14 (*BcLIK1_A01*), TDF42 (*BcCAT3_A07*), TDF75 (*BcAAE3_A06*), and TDF88 (*BcAMT2_A05*). They further compared expression patterns in 'Suzhou Qing', and 'Aijiao Huang' using *qRT-PCR*. Results showed that the four genes displayed similar expression trend in the two lines. Importantly, the expression of genes in the resistant line is higher than that in susceptible line. These genes expression patterns and their putative functions may provide insight in understanding the nonheading Chinese cabbage–downy mildew incompatible interaction. It may also provide a foundation for better understanding molecular mechanisms and can be beneficial in selecting candidate resistance genes for the incompatible interaction between nonheading Chinese cabbage and *H. parasitica*. Further research is needed to study the comparison between compatible and incompatible interactions to identify novel and common genes that regulate nonheading Chinese cabbage–downy mildew pathosystem.

2.4.7 Cloning of Major Resistance Genes

With the emergence and growth of modern molecular biology and cloning techniques, what had long been fantasy became reality. Thus, map-based or positional cloning strategies to isolate "R"-genes could be begun in the early 1990s (Parker et al. 1993). As more "R"-genes were mapped, it became clear that these were not randomly distributed on the chromosomes but that most fell into larger or smaller clusters, often interspersed with major resistance genes, which were active against other pathogens. This led to the coining of the term MRC loci (major recognition gene complexes) with MRC-A, MRC-B, MRC-F, MRC-H, and MRC-J on chromosomes 1, 2, 3, 4, and 5, respectively (Holub 1997). Candidate "R"-genes in *Arabidopsis* were named *RPP* genes for recognition of *H. parasitica* (Crute et al. 1993). Some 19 *RPP* genes are grouped in the three MRCs and four others are scattered at separate loci on chromosomes 1 and 2 (Fig. 2.13; Table 2.3). *RPP3*, which was identified in Oy-0, has not yet been mapped. Of the several *RPP* loci postulated on the basis of interaction patterns with isolates of *H. parasitica*, only a minority have been cloned (Fig. 2.13; Table 2.3). In Zurich, Slusarenko and Schlaich (2003) mapped *RPP11*, which conditions resistance to WELA in RLD, to chromosome 3 at a position 0.4 cM below the marker m249 (Joos et al. 1996). Meanwhile, across the Atlantic at Chapple Hill were chasing *RPP5*/*RPP1*, *RPP13*, and *RPP8*, respectively. Despite all the advantages for molecular genetics which *Arabidopsis* offers, the honor of being the first plant from which an "R"-gene (sensu Flor) was cloned went to tomato, with the isolation of PTO against *P. syringae* pv. *tomato* (Martin et al. 1993). The first *RPP* gene whose cloning was reported was *RPP5* (Parker et al. 1997), followed closely by *RPP1* and *RPP8* (Botella et al. 1998; McDowell et al. 1998). The *Arabidopsis RPP5* gene specifying resistance to the

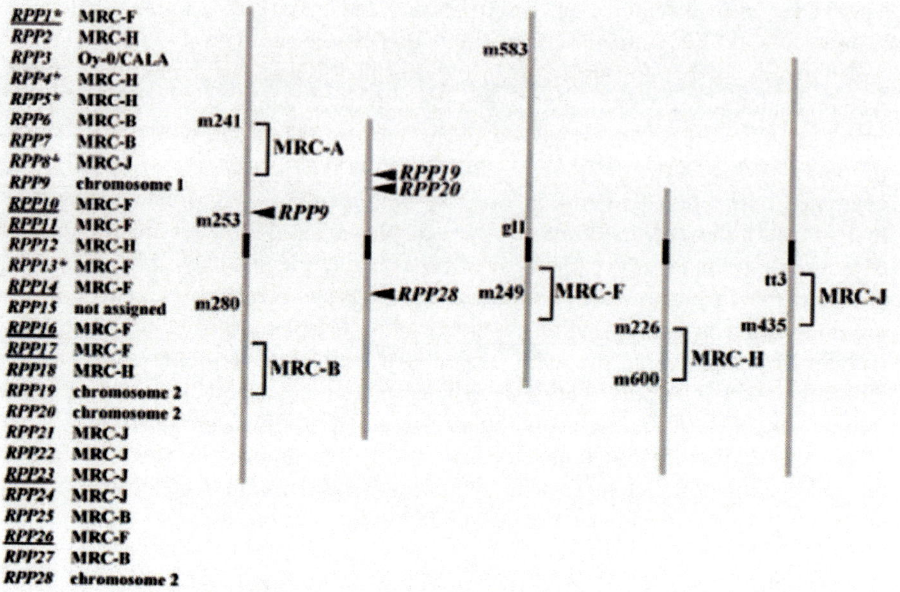

Fig. 2.13 Distribution of the mapped *RPP* genes along the five chromosomes of *A. thaliana*. To the left: a numerical list of the known 27 *RPP* genes with their chromosomal or MRC location given, where known. *RPP3* is not mapped yet, thus the *Arabidopsis* ecotype, and the *Hyaloperonospora parasitica* isolate are given. Underlined RPP genes have been renamed once. RPP genes with an asterisk have been cloned. To the right: graphical representation of the five *Arabidopsis* chromosomes with chromosome 1 to the left and chromosome 5 to the right. The centromeres are shown as black boxes. Mapping markers are given to the left of the chromosome. The region of an MRC is indicated with a black bracket, and the locus of a specific *RPP* gene is shown with a black arrow head to the right of the chromosome (Slusarenko and Schlaich 2003)

downy mildew pathogen *H. parasitica* was positionally cloned by Parker et al. (1997). It encodes a protein that possesses a putative nucleotide-binding site and leucine-rich repeats, and its product exhibits striking structural similarity to the plant resistance gene products N and L6. Like N and L6, the *RPP5* N-terminal domain resembles the cytoplasmic domains of the *Drosophila* toll and mammalian interleukin-1 transmembrane receptors. In contrast to N and L6, which produce predicted truncated products by alternative splicing, *RPP5* appears to express only a single transcript corresponding to the full-length protein. However, a truncated form structurally similar to those of N and L6 is encoded by one or more other members of the *RPP5* gene family that are tightly clustered on chromosome 4. The organization of repeated units within the leucine-rich repeats encoded by the wild-type *RPP5* gene and an *RPP5* mutant allele provides molecular evidence for the heightened capacity of this domain to evolve nove1 configurations and potentially new disease resistance specificities. Interestingly, while *RPP8* conditions resistance against EMCO in Ler-0, alternative alleles condition resistance to turnip crinkle

Table 2.3 Recognition specificities of *RPP* genes. (For each *RPP* gene, the MRC or chromosomal location, the formerly assigned number (where applicable), the ecotype, and the isolate (s) recognized are given (Slusarenko and Schlaich 2003))

Major recognition complexes	RPP gene	*Arabidopsis* ecotype	*Hyaloperonospora parasitica* isolates(s)
MRC-A	No RPP genes	Col-0	
MRC-B	RPP6	Col-0	WELA
	RPP7	Nd-0	HIKS
	RPP25	L*er*-0	AHCO
	RPP27	Ws-0	HIKS
MRC-F	RPP1-WsA formerly RPP-10	Ws-0	NOCO, EMOY, MAKS, CALA
	RPP-WsB formerly RPP-14	Ws-0	NOCO, EMOY, MAKS
	RPP-WsC formerly RPP-1	Nd-1	NOCO
	RPP1-Nd	Nd-1	EMOY, HIKS
	P-1-Nd formerly RPP26	Nd-1	WACO
	RPP13-Nd formerly RPP16	Nd-1	ASWA
	RPP13-Nd formerly RPP17	Nd-1	EMCO
	RPP13-Nd	Nd-1	GOCO, EDCO, MAKS
	RPP13-RLD formerly RPP11	RLD	WELA
MRC-H	RPP2	Col-0	CALA
	RPP4	Col-0	EMOY, EMWA
	RPP5	L*er*-0	NOCO
	RPP12	Ws-0	WELA
	RPP18	Col-0	HIND
MRC-J	RPP8	L*er*-0	EMCO
	RPP8 formerly *RPP23*	L*er*-0	GOWA
	RPP21	L*er*-0	MADI, MAKS
	RPP22	L*er*-0	ASWA
	RPP24	L*er*-0	EDCO
Not in MRCs	RPP3 (not mapped)	Oy-0	CALA
	RPP9	Wei-0	HIKS
	RPP19	Col-0	HIND4
	RPP20	Col-0	WAND
	RPP28	Col-0	HIND2

virus (allele HRT) and cucumber mosaic virus (allele *RCY1*) (Cooley et al. 2000; Takahashi et al. 2002). The *RPP8* locus is a good example of how recombination slippage and domain shuffling lead to new recognition specificities (Cooley et al.

2000; McDowell et al. 1998). The mapping of *RPP13* was published in 1999 (Bittner-Eddy et al. 1999) and was shown to map at the same locus as *RPP11* on chromosome 3 (Joos et al. 1996). Following cloning and characterization, it turned out that *RPP13* and *RPP11* were indeed allelic and recognized different pathogen avirulence determinants (Bittner-Eddy et al. 2000). Thus, *RPP13*, which was originally cloned from the ecotype Niederzenz, was designated *RPP13*-Nd, while *RPP11*, which Slusarenko and Schlaich (2003) had been chasing in the ecotype RLD, was renamed *RPP13*-RLD (Bittner-Eddy et al. 2000). Cloning of the *RPP1* locus from the Ws-0 ecotype revealed that three functional, genetically linked recognition specificities which had been previously designated *RPP1*, *RPP10,* and *RPP14* were present together. These were redesignated *RPP1*-WsC, *RPP1*-WsA, and *RPP1*-WsB, respectively (Botella et al. 1998). Until recently, there has been little progress in characterizing the products of, or cloning avirulence genes from, *H. parasitica*. DNA fingerprinting of the pathogen has been attempted, and Lucas et al. (1988) have established a mating system useful for gene mapping with *H. parasitica*, but this is largely with genetic markers in heterothallic isolates, and the situation is compounded by the homothallism observed for the isolates pathogenic on *Arabidopsis* (Moss et al. 1994; Tham et al. 1994).

Nevertheless, the outcrossing of two homothallic *H. parasitica* isolates revealed subsequent progeny segregation of avirulence loci matching six resistance loci in *A. thaliana* (Gunn et al. 2002). Additionally, the potential for map-based cloning of genes from *H. parasitica* was demonstrated in the report by Rehmany et al. (2003), which described the isolation of a BAC contig of four clones spanning less than 250 kb of DNA over the region containing the ATR1Nd avirulent gene (Slusarenko and Schlaich 2003).

2.4.8 Mapping-Based Cloning of Downy Mildew Resistance Genes

During the last few years, there were some advances in the genetic study of the inheritance of downy mildew resistance, and in the isolation and cloning of resistance genes in *Brassica* species. One locus conferring downy mildew resistance at the cotyledon stage in broccoli (*B. oleracea* convar. *italica*) was genetically mapped by Giovannelli et al. (2002) and located in close linkage to the glucosinolate pathway gene *BoGsl-elong* on a dense map of *B. oleracea* (Gao et al. 2007). A second downy mildew resistance gene at seedling stage was mapped in Chinese cabbage (*Brassica rapa* ssp. *pekinensis*) (Yu et al. 2009). A dominant and monogenically inherited resistance locus expressed at the adult plant stage was identified in broccoli by Coelho et al. (1998) and named *Pp523* (after a pathogen strain). This locus was later located on a new genetic map of RAPD and AFLP markers (Farinho et al. 2004) within a linkage group assigned to the *B. oleracea* chromosome C8 (Carlier et al. 2011). Five DNA markers that defined a genomic region of 8.5 cM encompassing this resistance locus were then cloned, sequenced, and remapped as SCAR and CAPS markers. BLAST queries (www.ncbi.nihl.gov/blast) identified a genomic region syntenic to this *B. oleracea* genome segment at the extremity of the top arm of *A. thaliana* L. chromosome 1 (Farinho et al. 2007).

Map-based, or positional, cloning is a common strategy for isolation of genes responsible for phenotypic differences. This strategy was used for the isolation of most of the >100 reference "R"-genes so far included in the Plant Resistance Genes database (http://prgdb.cbm.fvg.it; Sanseverino et al. 2010). Map-based cloning, with specific variations, was also the central procedure used for the isolation of the *A. thaliana* genes *RPP5* (Parker et al. 1997), *RPP8* (McDowell et al. 1998), *RPP1* (Botella et al. 1998), *RPP4* (Van der Biezen et al. 2002), and *RPP2A/RPP2B* (Sinapidou et al. 2004), the single downy mildew resistance genes so far isolated in the Brassicaceae family.

One of the major steps in map-based cloning is the physical identification of the genomic region where the gene is located. For genomes, still not fully sequenced, this implies the physical mapping of the gene of interest via construction of a contig of large insert DNA clones, usually BACs. Carlier et al. (2011) analyzed the construction of a physical map of a genomic region of 2.9 cM that encompasses the downy mildew resistance locus Pp523 in *B. oleracea*, carried out by exploiting the conserved synteny between *B. oleracea* and *A. thaliana* (Farinho et al. 2007). One major obstacle to overcome was the triplicated nature of *B. oleracea* genome (O'Neill and Bancroft 2000; Lysak et al. 2005; Town et al. 2006).

The selection of BAC clones for construction of the physical map was carried out by screening gridded BAC libraries with DNA over go probes derived from both genetically mapped DNA markers flanking the locus of interest and BAC-end sequences that align to *A. thaliana* sequences within the previously identified syntenic region. The selected BAC clones were consistently mapped to three different genomic regions of *B. oleracea*. Although 83 BAC clones were accurately mapped within a 4.6 cM region surrounding the downy mildew resistance locus *Pp523*, a subset of 33 BAC clones mapped to another region on chromosome C8 that was 60 cM away from the resistance gene, and a subset of 63 BAC clones mapped to chromosome C5. These results reflect the triplication of the *Brassica* genomes since their divergence from a common ancestor shared with *A. thaliana*, and they are consonant with recent analyses of the C-genome of *Brassica napus*. The assembly of a minimal tiling path constituted by 13 (*BoT01*) BAC clones that span the *Pp523* locus sets the stage for map-based cloning of this resistance gene (Carlier et al. 2011).

2.4.8.1 Mapping of R-Genes of *Brassica*

Resistance to downy mildew is thought to be distinct at the *Brassica* seedlings and adult stages. Resistance mapping work has identified several R-genes/loci (Table 2.4). In *B. oleracea*, the first locus conferring resistance in the broccoli seedling stage was placed in a linkage group and was found to be located close to the glucosinolate-related gene *BoGsl elong*. Another single dominant resistance gene expressed at the adult stage was identified in broccoli and was named *Pp523*. The genomic region containing this gene was further analyzed using SCAR and cleaved amplified polymorphic sequence (CAPS) markers, as well as two bacterial artificial

Table 2.4 Resistance genes/QTLs identified and mapped against downy mildew of crucifers (Lv et al. 2020)

Disease	Species	Pathogen race/isolates	Techniques	Results
Downy mildew	*B. oleracea*	–	RAPD, SCAR	A RAPD marker linked to the resistance gene at 3.3 cM
	B. oleracea	–	RAPD, AFLP, ISSR	*Pp523* in a region of 6.7 cM
	B. oleracea	–	RAPD, SCAR, AFLP	*Pp523* in a region of 4.8 cM
	B. oleracea	–	SSR, SRAP	*BoDM1*, close to a glucosinolate pathway gene
	B. rapa	Beijing isolate	AFLP, RAPD, SSR	The major QTL in a region spanning 2.9 cM
	B. rapa	–	RAPD	*BrRHP1* in a 2.2 Mb interval on A01
	B. oleracea	–	RAPD, ISSR, AFLP	*Pp523* on chromosome C8
	B. rapa	–	SNP, SLAF	*sBrDM8* may encode a serine/threonine kinase

chromosome (BAC) libraries. In *B. rapa*, QTLs conferring seedling-stage resistance were discovered. Using a genetic linkage map generated with a DH population, the major effect locus *BraDM* was delimited to a region spanning 2.9 cM in the A08 linkage group. For adult-stage resistance, a monodominant gene, *BrRHP1*, was localized to a 2.2 Mb interval on the A01 linkage group. Over the years, mapping methods based on high-throughput resequencing have greatly promoted the identification of R-genes. Using a high-density SNP-based map, a major locus, *sBrDM8*, was localized to a physical segment of ~228 kb, with one candidate kinase gene, *Bra016457* (Yu et al. 2016; Lv et al. 2020).

The developed markers closely located with these R loci have been adopted for resistance breeding through MAS and have greatly contributed to resistance breeding. Yu et al. (2011) converted the closely linked RAPD marker K14–1030 to a SCAR marker, which greatly improved selection efficiency in the progenies.

2.4.9 Resistance Gene-Mediated Signal Transduction

Signal transduction pathways used by the different RPP genes to mediate resistance are still poorly understood and are complex and diverse. At one extreme, there is *RPP13*-Nd, for which so far no downstream signaling components are known. It functions, independently of EDS1, NDR1, SA or NPR1, which in various combinations are required for the function of several other R-genes (Bittner-Eddy and Benyon 2001). At another extreme is *RPP4*, which seems to depend upon almost all defense-signaling components so far known. Thus, *RPP4*-conditioned resistance against EMOY or EMWA is compromised in Col-0 plants carrying a mutation in EDS1, PAD4; NDR1, PBS3; SGT1b, RAR1; SID1, SID2, or DTH9, respectively (Table 2.5). Additionally, it is dependent on NPR1 and the accumulation of SA and even requires a functional RPS5 gene, which is the R-gene required for the recognition of *P. syringae* pv. *tomato* isolates carrying the *avrPphB* gene. Essentially, all *RPP* genes seem to have individual downstream requirements; being qualitatively or quantitatively dependent on different combinations of, for example, *SGT1b*, *RAR1*, *EDS1*, *PAD4*, or NDR1 (Table 2.5). In mutants, a complete loss of resistance is often not observed, but rather a reduction of resistance which can be additive if two mutations are combined. Thus, resistance in the ecotype Ler against the EMCO isolate (mediated by *RPP8*) was initially reported to be independent of EDS1 and NDR1 (Aarts et al. 1998) but was later found to be significantly compromised in an eds1/ndr1 double mutant (McDowell et al. 2000). Rather than being a straightforward, a linear path from recognition to resistance expression, the picture which is emerging is of a complicated "lattice" or "grid-like" network of interconnecting signaling circuits (Parker 2000). What is known in terms of a rather "linear" summary of signaling components required for establishment of full resistance as mediated by the different R-genes is given in Table 2.5. So far, signaling through *RPP* gene appears to be independent of COI1, JAR1, or EIN2; components required for jasmonate and ethylene signaling, respectively. These genes are required for resistance to necrotrophic pathogens such as *Alternaria* or *Botrytis* (Thomma et al. 1998). A simplified schematic representation of RPP-conditioned signal transduction (Fig. 2.14) has been given by Schlaich and Slusarenko (2009).

2.4.10 Mechanisms and Application of Gene Silencing Techniques to Downy Mildew of Crucifers

Gene silencing is triggered by dsRNA that, under normal cellular conditions, may develop from transposon transcripts, gross overexpression of transcripts (as for viruses), aberrant messenger RNAs, or from nonprotein-coding premicro RNA (miRNA) transcripts (Ruiz et al. 1998; Jensen et al. 1999a, b; Gazzani et al. 2004; Luo and Chen 2007; Ghildiyal et al. 2008). Alternatively, dsRNA molecules can be introduced into cells either as in vitro-synthesized molecules (Timmons and Fire 1998; Clemens et al. 2000) or from antisense or inverted repeat expression vectors (Guo et al. 2003). The basic mechanism of gene silencing follows that dsRNA is bound and cut by the dicer enzyme into shorter dsRNA molecules, typically of 21–24 bp, and termed short-interfering RNAs (siRNAs). Dicer is a multidomain protein that comprises at least two RNase III domains, and additional domains such as dsRNA binding, PAZ, and DEAD box helicase. The PAZ domain may act in concert with the RNase III domains as a molecular measure to define the number of nucleotides of RNA between the two RNase III cut sites on the dsRNA (Zhang et al. 2004). The presence of a helicase domain may assist in unwinding complex secondary structures, processing of long dsRNAs, or unwinding of siRNAs for the next stage of the silencing pathway. Following dicer digestion, the siRNAs strands are

Table 2.5 Contribution of the various signaling components to the resistance reaction mediated by the various RPP resistance proteins (to be read from top to bottom for each *RPP* column) (Slusarenko and Schlaich 2003)

RPP gene	*RPP1-WsA*	*RPP-WsB*	*RPP2*	*RPP4*	*RPP5*	*RPP6*	*RPP7*	*RPP8*	*RPP12*	*RPP13 RLD*	*RPP13 Nd*	*RPP19*	*RPP20*	*RPP21*
Protein structure	TIR-NB-LRR[11]	TIR-NB-LRR[11]	TIR-NB-LRR[22,24]	TIR-NB-LRR[16]	TIR-NB-LRR8		CC-NB-LRR[1,24]	CC-NB-LRR[10]		CC-NB-LRR[12]	CC-NB-LRR[12]			
Ecotype	Ws-0	Ws-0	Col-0	Col-0	L *er*	Col-0	Col-0	Ler	Ws-0	RLD	Nd	Col-0	Col-0	Ler
H. parasitica isolate	CALA, EMOY, HIKS, MAKS, NOCO	EMOY, MAKS, NOCO	CALA	EMOY, EMWA	NOCO	WELA	HIKS	EMCO	WELA	WELA	ASWA, EDCO, EMCO, GOCO, MAKS	HIND4	WAND	MADI, MAKS
RPS5			+[9]	++[9,14]		+[9]	NR[9]				NR[14]	+[9]		
SGT1b	NR[22]		+++[1,22]	+++[1,22]	++[22]	++[1]	+++[1]	NR[22]				+++[1]	++[1]	−/+[22]
RAR1	NR[22]		NR[5,22]	++[4,5,14,16,22]	++[22]	+[5]	NR[4,5] RAR1 + NDR1[4]	NR[22]			NR[14]	+[5]	+++[5]	−/+[22]
EDS1	+++[3,6,20,21]	++ +[3,6,20,21]	+++[3]	+++[3,16]	++ +[3,21]		NR[2,21] EDS1 + NDR1[2,7]	−/+[2]/ NR [3,21]	+++[6]	++[3]	NR[14]			+++[3,21]
NDR1			−/+[3,5,7]	−/+[3,20]/+[5,7,9,14,16]	NR[1]	−/+[5,9]	−/+[2,5,7,9]	NR[3]			NR[14]	−/+[5,9]	−/+[5]	
Oxidative burst[1,20,23]														
PBS3			NR[5]	++[5,16]		NR[5]	NR[5]					+[5]		
PAD1			NR[13]	NR[13]			NR[13]					NR[13]		
PAD2			NR[13]	NR[13]			NR[13]					+[13]		
PAD3			NR[13]	NR[13]			NR[13]					+[13]		
PAD4	+[20,21]	+[20,21]	+++[13]	+++[13,14,16]			NR[13,21]	NR[21]			NR[14]	+++[13]		+[21]
			PAD1 + PAD2[13]											
			PAD2 + PAD3[13]	PAD2 + PAD3[13]										
			PAD1 + PAD3[13]											

SA			−/+[2]	+++[1,2,14,15]	+[21]	++[2,15]	NR[2]	++[2]	+++[15]		NR[14]	−/+[2]	++[2]	
NPR1	++[16,18]	+[16,18]	−/+[2]	+++[1,2,14,16]		−/+[2]	NR[2]	−/+[2]	+[16,18]		NR[14]	++[2]	−/+[2]	
	PHX6 + LSD5[19]			DTH9[16,17]										
	PHX11 + LSD5[19]			SID1[16]										
				SID2[16]										

Quantitative contributions are as follows: *NR* not required, −/+ contributes very little to resistance, + or ++ increasing requirement to resistance, +++ required for resistance. The relative position of the signaling components does not necessarily reflect the correct sequential action of the respective gene product along the signaling pathway. Occurrence of the oxidative burst has been shown only for a few RPPs; however, it is assumed that it is a generally important factor in disease resistance signaling, and hence, it is shown as if it would occur downstream of all RPPs. References given: [1]Tör et al. (2002); [2]McDowell et al. (2000); [3]Aarts et al. (1998); [4]Tornero et al. (2002); [5]Warren et al. (1999); [6]Parker et al. (1996); [7]Century et al. (1995); [8]Parker et al. (1997); [9]Warren et al. (1998); [10]McDowell et al. (1998); [11]Botella et al. (1998); [12]Bittner-Eddy et al. (2000); [13]Glazebrook et al. (1997); [14]Bittner-Eddy and Benyon (2001); [15]Delaney et al. (1994); [16]Van der Biezen et al. 2002; [17]Mayda et al. (2000); [18]Delaney et al. (1995); [19]Morel and Dangl (1999); [20]Rusterucci et al. (2001); [21]Feys et al. (2001); [22]Austin et al. (2002); [23]Aviv et al. (2002); [24]Holub (2001). Components modulating resistance reactions to *H. parasitica* by increasing resistance in compatible interactions are not listed in the table, because they are not attributable to specific RPPs: cpr mutants: Bowling et al. (1994, 1997); Clarke et al. (1998, 2000); Jirage et al. (2001); Yoshioka et al. (2001); lsd mutants: Dietrich et al. (1994); pmr4 mutant: Vogel and Somerville (2000); dnd1 mutant: Yu et al. (1998); ssi mutant: Kachroo et al. (2001); Shah et al. (2001); snc1 mutant: Li et al. (2001); hrl1 mutant: Devadas et al. (2002); cim mutants: Maleck et al. (2002); rin4 mutant: Mackey et al. (2002); son1 mutant: Kim and Delaney (2002)

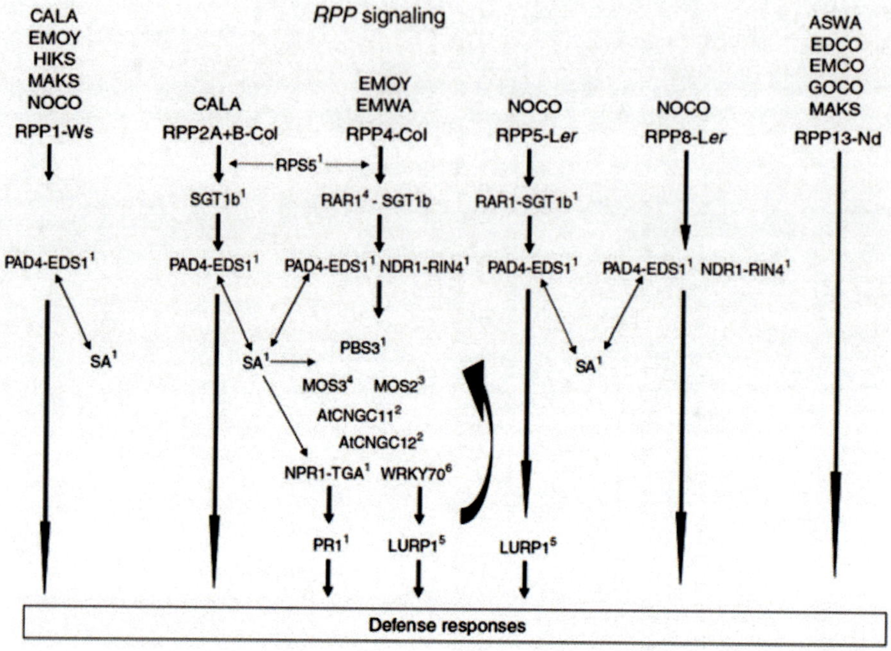

Fig. 2.14 Schematic representation of RPP-conditioned signal transduction. *Hyaloperonospora arabidopsidis* isolates are shown above the corresponding RPPs that recognize them below the isolates (hyphenated with the *Arabidopsis* accession from which they were cloned). RPP1 from Ws (RPP1-Ws) is dependent on PAD4 and EDS1. RPP 2A + B from Col-0 is quantitatively influenced by the resistance protein RPS5 that recognizes the bacterial pathogen *Pseudomonas* and requires SGT1b, PAD4, and EDS1. RPP4 from Col-0 is particularly sensitive to changes in resistance signaling. Thus, mutations in many defense-related genes lower the defense responses to EMOY and EMWA. RPP5 from Ler requires RAR1 and SGT1b as well as PAD4 and EDS1. Furthermore, RPP5 was shown to be dependent on LURP1. RPP8 from Ler is an exception, because it encodes a CCNB-LRR-type resistance protein, yet is dependent not only on NDR1 and RIN4 (like other CC-NB-LRRs) but also on PAD4 and EDS1. No downstream signaling components of RPP13-Nd have so far been reported. Please note: Signaling by the TIR-NBLRR-type RPP genes RPP1-Ws, RPP2A + B-Col-0, RPP4-Col-0, and RPP5-Ler was dependent on PAD4 and EDS1, which have a positive interlocked feedback regulation with salicylic acid (SA) (Slusarenko and Schlaich 2003; Yoshioka et al. 2006; Zhang et al. 2005; Zhang and Li 2005; Knoth and Eulgem 2008; Knoth et al. 2007; Schlaich and Slusarenko 2009)

separated, and the antisense strand is incorporated into the RNA-induced silencing complex (RISC), the core of which is the multidomain argonaute protein. Argonaute proteins contain two major domains, PAZ and PIWI. The antisense strand of the siRNA duplex is bound into the RISC via the PAZ domain and acts as the guide strand, which targets the homologous sequence in the native messenger RNA (mRNA) and degrades it through the slicer activity of the PIWI domain (Collins and Cheng 2005). A related class of small RNA species, called miRNA, also develops through dicer or other RNaseIII enzyme (such as Drosha) digestion of nonprotein-coding pre-miRNA transcripts containing a stem-loop secondary structure. Similar to siRNAs, the mature miRNA may also associate with argonaute proteins to degrade endogenous mRNAs, thus achieving a level of posttranscriptional regulation. Alternatively, the mature miRNA may bind to the homologous messenger RNA (mRNA) to inhibit translation (Brodersen et al. 2008). SiRNAs and fragments released from RISC degradation of mRNAs can act to prime synthesis of more dsRNA from the native mRNA through the action of the RdRP enzyme. RdRP can act both to spread the silencing target signal along the mRNA and to amplify the signal through production of more dsRNA (Alder et al. 2003).

The synthesis of dsRNA by RdRP has been shown to occur in both a primer-dependent and primer-independent manner (Makeyev and Bamford 2002). Secondary siRNAs in *Caenorhabditis elegans* have been shown to exhibit sequence direction bias suggestive of unidirectional priming (Sijen et al. 2001), whereas aberrant mRNAs can be substrates for primer-independent RdRP synthesis of dsRNA (Baulcombe 2004; Luo and Chen 2007).

The major cytoplasmic steps of gene silencing are frequently described as posttranscriptional gene silencing (PTGS) or RNAi. That is, the native gene is transcribed, but the mRNA is degraded before translation can occur. The components of PTGS also provide a link to silencing at the transcriptional level (Verdel et al. 2004; Buhler et al. 2007). In many organisms where PTGS operates, it can also lead to epigenetic changes at the DNA level, principally through de novo methylation of cytosines or through deacetylation and methylation of histone proteins. The outcome of these modifications is silencing at the transcriptional level. Plant proteins involved in these processes are RNA polymerase IV, DNA methyltransferase, histone methyltransferase, histone deacetylase, and chromodomain protein DRD1 (Huettel et al. 2007); RdRP, dicer, and argonaute are also involved in initiating and maintaining TGS. Similar processes also operate in fungi and are best characterized in fission yeast *Schizosaccaromyces pombe*. In *S. pombe*, silencing operates at the transcriptional level through formation of heterochromatin. Similar to plants, this involves dicer, RdRP, and argonaute proteins. *S. pombe* argonaute forms part of the RNA-induced transcriptional silencing (RITS) complex, together with proteins CHP1 and TAS3 (Verdel and Moazed 2005). TAS3 has no discernable domains, but CHP1 is a chromodomain protein. The RITS complex, through CHP1, targets DNA sequences homologous to the siRNA bound in the RITS complex to modulate the deacetylation and methylation of histones, and thus the formation of heterochromatin (Verdel and Moazed 2005). The methylation/acetylation status of histones is thought to then mediate access to the DNA for RNA polymerase II, with any mRNA formed then targeted by RDRP to form dsRNA, then siRNAs; this process maintains the silenced state (Grewal and Jia 2007). Similar to *S. pombe*, TGS that develops from RNAi in the nematode *C. elegans* can persist through several generations, with its basis being the activity of histone deacetylase, histone acetyltransferase, a chromatin remodeling ATPase, and a chromodomain protein (Vastenhouw et al. 2006). PTGS in *C. elegans* and other organisms such as *Neurospora crassa* has been observed as a transient phenomenon that decreases in intensity with time after exposure to the initial silencing stimulus. A component of this "release" from PTGS has been identified from *C. elegans* as a dsRNA-specific RNase termed ERI1, which degrades siRNAs (Kennedy et al. 2004). ERI1 mutants exhibit enhanced and more persistent RNAi and TGS (Kennedy et al. 2004; Iida et al. 2006).

The recent sequencing of the genome of *Arabidopsis* downy mildew pathogen and *Hyaloperonospora arabidopsidis* (*H. parasitica*) (Tyler et al. 2006; http://www.broad.mit.edu/annotation/genome/ *P.infestans*; http://phytophthora.vbi.vt.edu/) has and will continue to assist the discovery of numerous genes and pathways associated with downy mildew biology. Identifying the components of downy mildew gene silencing has also been assisted by the availability of genome sequences. Using the *P. infestans* genome sequence as an example, genes encoding the key components of PTGS are all present: dicer-like, argonaute, and RdRP (Ah Fong et al. 2008). Genes encoding other potential PTGS accessory proteins that contain domains characteristic of proteins that are involved in PTGS in other organisms, such as dsRNA-binding proteins and DEAD-box RNA helicases, are also present (Walker et al. 2008). Using both plant, *S. pombe*, and *C. elegans* as reference systems where TGS has been dissected, genes encoding the major components, such as chromodomain proteins and histone modifying proteins, are all present within oomycete genome sequences in multiple copies as in other organisms. It remains to be demonstrated which of these protein family members contribute to TGS in oomycetes. In support of the observations from laboratory-based studies, genes encoding proteins for de novo methylation of DNA were not identified (Judelson and Tani 2007). Similarly, subunits of RNA polymerase IV, a plant-specific protein apparently involved in de novo DNA methylation were not identified in the *P. infestans* genome.

Model gene-silencing systems *C. elegans*, *S. pombe*, and *A. thaliana* all contain a protein called ERI1, which is an exoribonuclease that digests siRNAs, ultimately leading to a release of PTGS (Kennedy et al. 2004; Iida et al. 2006). The *P. infestans* genome does not seem to contain the gene encoding this protein. The lack of an ERI homolog may in part explain the persistence of gene silencing in the absence of any homologous transgene (van West et al. 1999, 2008; Whisson et al. 2005; Gaulin et al. 2007) and the observation that partial silencing in some *P. infestans* transformants can convert to complete silencing after some time (Ah Fong et al. 2008). The strategies for application of gene silencing in downy mildew of crucifers can be (1) stable gene silencing and (2) transient gene silencing procedures.

2.4.11 Stable Versus Transient Gene Silencing

Each of the two strategies for achieving gene silencing in downy mildew of crucifers has inherent advantages and disadvantages. Stable, transcriptional silencing has the obvious advantages in maintaining undetectable levels of gene expression, such that effectively null genotypes and their associated phenotypes can be studied. This allows very subtle differences in phenotypes to be identified and studied in detail, such as the 50% reduction in cyst germination observed for silencing of NIF genes (Judelson and Tani 2007). However, transcriptional gene silencing is effectively a gene knockout in terms of gene expression, which signifies that genes for which silencing leads to a lethal phenotype are difficult to study by this approach. Transient gene silencing, in which a range of gene expression knockdown levels are observed, may be better suited to the study of these "core biology" genes. The obvious disadvantage of transient silencing is that the genotype and phenotype apparently does not persist for more than 20 days, and thus experiments require repetition on each occasion that the gene under study is to be characterized phenotypically. However, the recent reporting of stable but partial gene silencing may alleviate this difficulty (Ah Fong et al. 2008). Stable gene silencing, whether partial or complete, relies on transformation of downy mildew. Although transformation is routine for model downy mildew species, this is not the situation for many other downy mildews.

2.4.12 Receptor Protein Triggering Downy Mildew Resistance in *Brassica rapa*

Quantitative trait locus (QTL) mapping combined with marker-associated breeding has been one of the most effective methods to find disease resistance genes and create disease-resistant varieties. Genetic analysis of downy mildew resistance in Chinese cabbage firstly initiated several decades before showed that the resistance was determined by a single dominant gene. Yu et al. (2009) demonstrated that the downy mildew resistance is a trait with high heritability in seedling stage of Chinese cabbage and controlled by a major dominant gene as well as some minor modifying genes. A major QTL, effective at every developmental stage, was detected on linkage group A08 by both interval and MQM mapping methods. A single dominant locus designated BrRHP1 was located in a 10.5-cM region on A01 chromosome at the heading stage of Chinese cabbage. In addition, two minor QTLs, located on linkage groups A06 and A04, were effective at the rosette and heading stages (Yu et al. 2016). The resistance of different natural variations was probably conferred mainly by different genes involving in resistant response. The genetic resistance to downy mildew of plants in different growth stages was also somewhat different (Yu et al. 2016). Based on next-generation sequencing technology, specific locus-amplified fragment sequencing (SLAF-seq) was developed for mapping traits of interest, with the advantages of being a technique of high throughput, high accuracy, low cost, and short time (Zhang et al. 2013). An improved SLAF-seq strategy was utilized to construct a high-density bin-map, providing abundant developed single-nucleotide polymorphism (SNP) markers which could be used to locate QTLs for resistance to downy mildew (Yu et al. 2016). Genome-wide resequencing was used to obtain the sequences of different individuals based on reference genomic sequences. On account of the sequence variance analysis of different individuals, plenty of high-density and high-quality SNPs and InDel (insertion/deletion) markers were found which would be efficient tools to assist in QTL fine mapping.

Studies indicate that the R-genes which were cloned not only encoded NBLRR proteins but also encoded receptor-like proteins (RLPs) consisting of an extracellular LRR and transmembrane (TM) domain, as well as receptor-like kinases (RLKs), including an extracellular LRR, TM domain, and a cytoplasmic kinase domain (Verica and He 2002). In the last few decades, an increasing number of RLPs was cloned as immune receptors in plant disease resistance. For example, RPP27 for *Arabidopsis* downy mildew resistance, Ve1 and LeEix1 in tomato, and LepR3 involved in blackleg resistance in *Brassica napus* (Larkan et al. 2013). In particular, extracellular LRR domains in RLPs may play a key role in pathogen recognition. For instance, Cf-4 has two less LRR domains than Cf-9, which leads to differences in resistance responses. In addition, the absence of cytoplasmic kinase domains in RLPs compared with RLKs has enabled RLPs to interact with other RLKs to activate cytoplasmic signaling pathways responsive to pathogen invasion. *Solanum lycopersicum* LRR RLK (SOBIR1) and BRI1-associated receptor kinase 1 (BAK1) were shown to form a complex with Cf-type proteins (Cf-2, Cf-4, Cf-9) and LeEix1 to mediate a disease resistance response (Bar et al. 2010; Liebrand et al. 2013, 2014; Zhang et al. 2013).

Phytohormones play pivotal roles in the defense regulation network, by which plants can translate the pathogen-induced early signaling events into activation of effective defense responses (Pieterse et al. 2009). Among these, jasmonic acid (JA) and salicylic acid (SA) are considered to be the major defense-related phytohormones, in addition to other phytohormones such as ethylene (ET), abscisic acid (ABA), auxin, gibberellins (GAs), cytokinins (CK), and brassinosteroids (BRs) (Shigenaga and Argueso 2016). The studies showed that there is a severe alteration in the level of resistance to specific types of pathogens, depending on the phytohormone involved. SA has been shown to regulate the defense responses triggered by biotrophic pathogens (Thomma et al. 2001). After pathogen infection, increased SA levels both in local and distal parts of plants activates defense responses through nonexpressor of pathogenesis-related genes 1 (NPR1), a component downstream of SA. NPR1 translocates to the nucleus to interact with TGA-type transcription factors (TGA2, TGA5, and TGA6) to initiate the expression of SA-dependent genes including PR genes (Kinkema et al. 2000; Zhang et al. 2003; Dong 2004). Previous studies have demonstrated that WRKY transcription factors and NPR1 exhibit a synergistic effect on the regulation of the expression of PR genes (Mou et al. 2003; Xie et al. 2010).

Zhang et al. (2018) identified candidate downy mildew resistance genes encoding LRR-containing proteins in Br-DM04, within a 160-kb region. Using a transient expression system, overexpression of BrRLP48 could enhance resistance to downy mildew in a susceptible line. Otherwise, the leaves infiltrated with RNAi construct of BrRLP48 could reduce the disease resistance of resistant parent MM. Subsequently, through sequence and expression analysis after downy mildew infection, it was found that BrRLP48 was significantly induced in resistant line probably due to the presence of the putative SA and JA responsive elements in the promoter. Finally, the content of SA in the resistant line MM was dramatically elevated after inoculation, with only the expression of BrRLP48 significantly induced under SA treatment, correlating with an increased expression following pathogen invasion. The results demonstrated the involvement of R-genes in *Brassica* downy mildew resistance and shed light on the molecular mechanism for marker-assisted breeding of disease-resistant varieties in *B. rapa*.

Downy mildew, caused by *H. parasitica*, is a major disease of *Brassica rapa* that causes large economic losses in many *B. rapa*-growing regions of the world (Saharan et al. 2017). The genotype used was based on a double haploid population derived from a cross between the Chinese cabbage line BY and a European turnip line MM, susceptible and resistant to downy mildew, respectively. Zhang et al. (2018) initially located a locus Br-DM04 for downy mildew resistance in a region about 2.7 Mb on chromosome A04, which accounts for 22.3% of the phenotypic variation. Using a large F2 mapping population (1156 individuals), they further mapped Br-DM04 within a 160-kb region, containing 17 genes encoding proteins. Based on sequence annotations for these genes, four candidate genes related to disease resistance, BrLRR1, BrLRR2, BrRLP47, and BrRLP48, were identified. Overexpression of both BrRLP47 and BrRLP48 using a transient expression system significantly enhanced the downy mildew resistance of the susceptible line BY. But only the leaves infiltrated with RNAi construct of BrRLP48 could significantly reduce the disease resistance in resistant line MM. Furthermore, promoter sequence analysis showed that one salicylic acid (SA) and two jasmonic acid responsive transcript elements were found in BrRLP48 from the resistant line, but not in the susceptible one. Real-time PCR analysis showed that the expression level of BrRLP48 was significantly induced by inoculation with downy mildew or SA treatment in the resistant line MM. Based on these findings, it was concluded that BrRLP48 was involved in disease-resistant response and the disease-inducible expression of BrRLP48 contributed to the downy mildew resistance. These findings led to a new understanding of the mechanisms of resistance and lay the foundation for marker-assisted selection to improve downy mildew resistance in *B. rapa* (Zhang et al. 2018).

2.4.13 Different Requirements of *EDS1* and *NDR1* by R-Genes in *Arabidopsis*

The model crucifer *Arabidopsis* has been exploited as a host for bacterial, viral, fungal, and oomycete pathogens, and analysis of natural genetic variation between different plant accessions and pathogen isolates has led to the identification of distinct *R* loci. The cloned *RPS2* and *RPM1* genes, controlling race-specific resistance to the bacterial pathogen *P. syringae*, encode proteins of the LZ-NBSyLRR class. In contrast, the *RPP5* gene mediating isolate-specific resistance to

H. parasitica encodes a protein of the TIR-NBSyLRR class. Mutational analyses in *Arabidopsis* have identified other wild-type genes that are required for R-gene-mediated resistance. A mutation in *NDR1* (non-race-specific disease resistance) abolished resistance conferred by *RPS2* and *RPS5* to *P. syringae* expressing *avrRpt2* and *avrPph3*, respectively, as well as a dual specificity resistance encoded by *RPM1* to the bacterial genes *avrB* and *avrRpm1*. *ndr1* plants also were compromised in *RPP* gene-mediated resistance to several incompatible *H. parasitica* isolates, suggesting that the wild type NDR1 protein may function at a common point downstream from the perception of these prokaryotic and eukaryotic pathogens. Another resistance signaling component is encoded by *EDS1* (enhanced disease susceptibility). A mutation in *EDS1* abolished resistance conferred by several *RPP* loci but had no effect on *RPM1*-specified resistance, suggesting a function before the convergence of downstream pathways (Parker et al. 1996).

It was possible that the *ndr1* and *eds1* mutations may reflect the operation of distinct disease resistance signaling pathways and that particular R-proteins exhibit a preference for one signaling mode. To address this possibility, Aarts et al. (1998) extended analysis of the relative requirements for *EDS1* and *NDR1* to a wide spectrum of *Arabidopsis* *R*-genes in three different genetic backgrounds (accessions). The results show that the requirements for *EDS1* or *NDR1* appear to be mutually exclusive. Furthermore, the data suggest that preferential utilization of either EDS1 or NDR1 is determined more by R-protein structure than pathogen type.

The *Arabidopsis* genes *EDS1* and *NDR1* were shown previously by mutational analysis to encode essential components of race-specific disease resistance. Aarts et al. (1998) examined the relative requirements for *EDS1* and *NDR1* by a broad spectrum of Resistance (R)-genes present in three *Arabidopsis* accessions (Columbia, Landsberg-*erecta*, and Wassilewskija). It was shown that there is a strong requirement for *EDS1* by a subset of *R* loci (*RPP2*, *RPP4*, *RPP5*, *RPP21*, and *RPS4*), conferring resistance to *H. parasitica* and to *Pseudomonas*, expressing the avirulence gene *avrRps4*. The requirement for *NDR1* by these *EDS1*-dependent *R* loci is either weak or not measurable. Conversely, three *NDR1*-dependent *R* loci, *RPS2*, *RPM1*, and *RPS5*, operate independently of *EDS1*. Another *RPP* locus, *RPP8*, exhibits no strong exclusive requirement for *EDS1* or *NDR1* in isolate-specific resistance to *P. parasitica*, although resistance is compromised weakly by *eds1*. Similarly, resistance conditioned by two *EDS1*-dependent *RPP* genes, *RPP4* and *RPP5*, is impaired partially by *ndr1*, implicating a degree of pathway cross-talk. The results provide compelling evidence for the preferential utilization of either signaling component by particular R-genes and thus define at least two disease resistance pathways. The data also suggest that strong dependence on *EDS1* or *NDR1* is governed by R-protein structural type rather than pathogen class (Aarts et al. 1998).

2.5 *Brassica–Plasmodiophora*: Molecular Resistance

2.5.1 Identification and Mapping of R-Genes in the B-Genome of *Brassica* Species

Species containing the B-genome are a useful source of genes when breeding for biotic and abiotic stress tolerance, disease resistance, and oil quality. Transferring disease resistance from *B. nigra* to canola (*B. napus*) has potential to expand the genetic base in canola germplasm. Several CR genes have been identified and mapped in *B. rapa, B. oleracea,* and *B. napus*. More than 10 loci on 5 chromosomes have been identified in the A-genome of *B. rapa*; *Crr1, Crr2, Crr4*, and *CRc* mapped to chromosomes A08, A01, A06, and A02, respectively, and *CRa, CRb, CRk,* CRb^{kato}, *Rcr1,* and *Rcr2* were mapped onto linkage group A03. *CRa*, CRb^{kato}, and *Crr1*, which have been cloned from *B. rapa*, encode toll-interleukin-1 receptor/nucleotide-binding site/leucine-rich repeat (TIR-NBS-LRR or TNL) proteins. In recent studies, six QTLs residing in five CR QTL regions of chromosomes A01, A03, and A08 were identified in *B. rapa*, and three strong QTLs (*Rcr4, Rcr8* and *Rcr9*) were identified in chromosomes A03, A02, and A08, respectively.

In the C-genome of *B. oleracea*, one major resistance gene, *Rcr7*, originating from *B. rapa*, and over 20 QTLs have been identified. So far, at least 20 QTLs involving CR have been identified in the A- and C-genomes of *B. napus*. A major gene, *Pb-Bn1*, which confers resistance to two *P. brassicae* collections, has been mapped to chromosome A03, and a minor QTL for each collection was mapped to linkage groups C2 and C9. Also, 19 QTLs involving race-specific resistance were mapped on 8 chromosomes: A02, A03, A08, A09, C3, C5, C6, and C9. From these 19 QTLs, 4 were closely linked on A03 and 3 were linked on A08. However, mapping and identification of CR genes has not been carried out in *Brassica* species containing the B-genome.

Next-generation sequencing (NGS) has been used to develop the genome sequences of *B. rapa*, *B. oleracea*, *B. nigra*, *B. napus,* and *B. juncea*. The genomes of all *Brassica* species underwent a lineage-specific whole-genome triplication, followed by diploidization that involved substantial genome reshuffling and gene losses. Another application of NGS is gene mapping through mapping by sequencing, which has been used to map important genes in the genomes of organisms such as *A. thaliana* and pok choy. In addition, the CR gene *Rcr7* was successfully mapped in cabbage through analysis of percentage of polymorphic variants (PPV) based on bulked segregant RNA sequencing (BSR-Seq) (Chang et al. 2019).

Brassica species with the B-genome contain a large reservoir of genes conferring resistance to important diseases of *Brassica* such as clubroot and blackleg. Resistance to clubroot in *B. nigra* could be transferred into B-genome species such as *B. juncea* and *B. carinata* by interspecific hybridization or resynthesis of these species. Due to differentiation among the *Brassica* genomes caused by genome rearrangements and gene deletions, the level of homology between B and A/C chromosomes is low, which makes introgression of genes from the B-genome into the A- or C-genomes of canola (*B. napus*) difficult through conventional breeding.

To facilitate the transfer of resistance available in the B-genome into canola, it is necessary to accurately map genes in *B. nigra*, so that molecular markers tightly linked to these resistance genes can be identified for use in marker-assisted selection (MAS). Alternatively, the genes could be isolated from *B. nigra* so that the genes could be delivered into canola through gene transformation.

The gene *Rcr6* was initially mapped against a reference A-genome of *B. rapa*. The genetic location of *Rcr6* was further defined using SNP markers identified though BSR-Seq based on the *B. rapa* reference genome. This demonstrated that mapping by sequencing was possible in a species using the genome sequence from a close relative if its own genome sequence was not available. The gene *Rcr6* was mapped into chromosome A08 using the reference genome from *B. rapa*. A draft genome of *B. nigra* sequenced in Canada subsequently became available. *Rcr6* was then mapped into *B. nigra* chromosome B3 (designated here as B3 Canada), which indicated that the target region of *Rcr6* in B3 Canada of *B. nigra* was homologous to the corresponding region in chromosome A08 of *B. rapa*. *Rcr6* was also mapped using a published *B. nigra* sequence from http://brassicadb.org/brad in China into chromosome B7 China. The Canadian group used a nomenclature published by Lagercrantz and Lydiate (1996), while the Chinese group accepted a proposal for nomenclature described by Panjabi et al. (2008). Therefore, it is not surprising that *Rcr6* was mapped into different chromosomes using these two separately derived sources of the *B. nigra* genome.

RNA-seq and block homoeology analysis had identified synteny between the A and B-genomes in *B. juncea*. *Rcr6* could possibly integrate into A08 of *B. napus* in interspecific crosses between *B. napus* and *B. nigra*, but this speculation has not yet been confirmed. Despite the challenges involved in transferring agronomically important traits from the B-genome species into *B.* napus, a blackleg resistance gene from the B-genome of *B. juncea*, *Rlm6*, has been successfully introgressed into *B. napus*. Substantial effort has been made to identify genes or QTLs for CR in *Brassica* species containing A- or C-genome. In contrast, *Rcr6* is the first CR gene to be finely mapped in a B-genome species. It was located on chromosome B3 Canada, homologous to 14.8–15.4 Mb of chromosome A08. Five genes (*Bra010552*, *Bra010588*, *Bra010589*, *Bra010590,* and *Bra010663*) were identified as encoding the TNL-class of proteins in this interval. The CR gene *Crr1* has previously been mapped into chromosome A08 and cloned. It is highly homologous to gene *Bra020861*, which located in the 10.8 Mb region of chromosome A08. In addition, a strong QTL, *Rcr9*, was also mapped into chromosome A08. The nearest SNP to *Rcr9* was A08_10272562. *Rcr6* was not located on the corresponding genomic region of *Rcr9* or *Crr1* on chromosome A08. The only TNL gene identified in the *Rcr6* interval is FGENESH_B3_72, which is highly homologous to the *B. rapa* TNL gene *Bra010663* and the *B. nigra* TNL gene *BniB015819*. The availability of closely linked SNP markers will facilitate molecular cloning of *Rcr6* from the donor line. Although the cloned CR genes encode the TNL class of proteins, it is possible that clubroot resistance is encoded by other classes of disease resistance genes. Breeding for CR can be severely constrained when the pathotype of interest does not occur near the site of established breeding institutions, so the use of molecular markers in marker-assisted selection can be extremely important. More than 10 robust SNP markers developed by Chang et al. (2019) shown to be tightly linked to *Rcr6* using KASP. These markers could facilitate introgression of *Rcr6* into canola.

Only five pathotypes of *P. brassicae* (2, 3, 5, 6, and 8, based on Williams' differentials) had been identified in Canada, and pathotype 3 was the most prevalent pathotype on canola. Line PI 219576 of *B. nigra*, which was the resistant parent in the cross examined, was highly resistant to all of these pathotypes. Now, 12 new pathotypes of *P. brassicae* virulent on canola were identified using a new Canadian Clubroot Differential set. PI 219576 also showed resistance to all of these new pathotypes. *Rcr6* from PI 219576 was identified through testing for disease reaction to pathotype 3 only, so studies to confirm that *Rcr6* confers resistance to other pathotypes are needed.

RNA-seq produces a large amount of short DNA sequence reads from random places in the transcriptome. The RNA-seq data generated from the R and S bulks in the BC1 population was aligned into the A-genome of *B. rapa* and B-genome of *B. nigra*. The characterized genome-wide variants in the *B. nigra* population carried Rcr6 and demonstrated that the sequence counts assembled into the A-genome of *B. rapa* for each chromosome were correlated to chromosome length from the R and S bulks, which was consistent with a previous study. However, there was no correlation between assembled sequence lengths and the B-genome of *B. nigra*.

A previous study identified that a high proportion of PPV on chromosome A03 of *B. rapa* adjacent to the CR gene *Rcr1* and *Rcr7* was mapped into chromosome C7 of *B. oleracea* through identification of PPV. *Rcr6* was mapped into 10–20 Mb of A08 of *B. rapa* and *B. nigra* B3 Canada. However, the peak region based on PPV on B3 was not closely associated with *Rcr6*. One reason for this could be the low number of variants identified in the region due to a relatively low depth of RNA-seq reads. Greater depth of sequencing for genetic mapping via identification of PPV is recommended for future research. Rcr6 was the first gene for resistance to clubroot identified and mapped in the B-genome of *Brassica* species. The sequencing information from *B. rapa* and *B. nigra* was used for BRS-Seq to map the gene. It resides in the genomic region of *B. nigra* homologous to chromosome A08 of *B. rapa*. It is believed that it could possibly integrate into A08 of canola crop *B. napus*. SNP markers tightly linked to the gene were developed, facilitating canola and mustard breeders for use of MAS in introgression of the CR gene into their cultivars. This gene was mapped into a small interval with one TNL gene (*BniB015819*) identified. This gene can be the candidate of *Rcr6* for gene cloning (Chang et al. 2019).

2.5.2 Identification of R-Genes by *Brassica* Genome Sequencing

The advances in genome sequencing and assembly have allowed us to tap into the rich genetic resources from different crop species and their wild relatives. A reference genome is useful to align and compare all the different QTL mapping experiments. By having a reference genome, the physical location of the markers in QTL maps that are linked with candidate genes can be determined. The genome sequences of the *Brassica* diploids, *B. rapa* (Wang et al. 2011; Cai et al. 2017), *B. oleracea* (Parkin et al. 2014; Liu et al. 2014), and *B. nigra* (Yang et al. 2016) and the amphidiploids *B. napus* (Chalhoub et al. 2014; Bayer et al. 2017) and *B. juncea* (Yang et al. 2016) have been published. This, along with the advanced genotyping tools such as the 60 K *Brassica Infinium* SNP array (Clarke et al. 2016) and genotyping by sequencing (GBS) (Cheng et al. 2016; Scheben et al. 2017), has allowed rapid identification of genome-wide QTL and GWAS regions for R-genes for clubroot (Lee et al. 2016; Li et al. 2016).

Genome sequencing in *Brassicas* has now been expanded to include the *B. oleracea* pan-genome where genomes of nine lines were assembled and compared for structural variation (Golicz et al. 2016). Pan-genome sequencing is an important advance, as the diversity of a species is not represented by a single cultivar, and in *B. oleracea* only 81% genes were present in all lines. Of the 19% variable genes, those associated with disease resistance were found to be among the most prevalent. Therefore, in identification of resistance genes, research should look at the variable genes, not only those in the reference genome sequence. Genome resequencing (Cheng et al. 2016) and pan-genomics (Golicz et al. 2016) approaches are promising tools to make identification, characterization, and cloning of candidate R-genes more achievable and reliable. The pan-genome approach to study structural variation of the genes in *Brassica* genomes, with whole-genome gene expression and methylation studies, has already proved very useful toward uncovering the structure, function, and evolutionary origin of R-genes. Copy number variation (CNV) has already been studied for the AP2/ERF super family involved in plant stress tolerance in *B. napus*. CNV of R-genes in *Brassica* is only now being investigated (Batley et al. 2016) and has been implicated in important crops such as cereals (maize and rice), Solanaceae, and soybean (Springer et al. 2009; McHale et al. 2012; Xu et al. 2012; Saxena et al. 2014; Wei et al. 2016). This gene expansion through structural variation allows the plant to adapt to rapid changes in the environment also (Cheung et al. 2009).

2.5.3 Identification of R-Genes by Genomic Approach

P. brassicae infects all *Brassica* species and the model plant *A. thaliana*. It is possible to apply currently available genomics tools and techniques for the cloning and characterization of CR genes in addition to comparative mapping and identification of candidate genes through transcriptomic analysis. The whole-genome sequence information available for *Arabidopsis* has been used for comparative genome analysis of *Brassica* species. Comparative mapping between *Brassica* and *Arabidopsis* has revealed a conservation of gene order in small chromosomal blocks despite inversions and large-scale deletions (Cavell et al. 1998; O'Neill and Bancroft 2000). This information has been used to align linkage groups containing CR genes in *B. rapa* with those of *Arabidopsis* chromosomes. The genes *Crr1*, *Crr2*, and *CRb* are in synteny with the central region of chromosome 4 of *A. thaliana* (Suwabe et al. 2006; Fig. 2.15). The studies by Jubault et al. (2008) of *A. thaliana* have identified

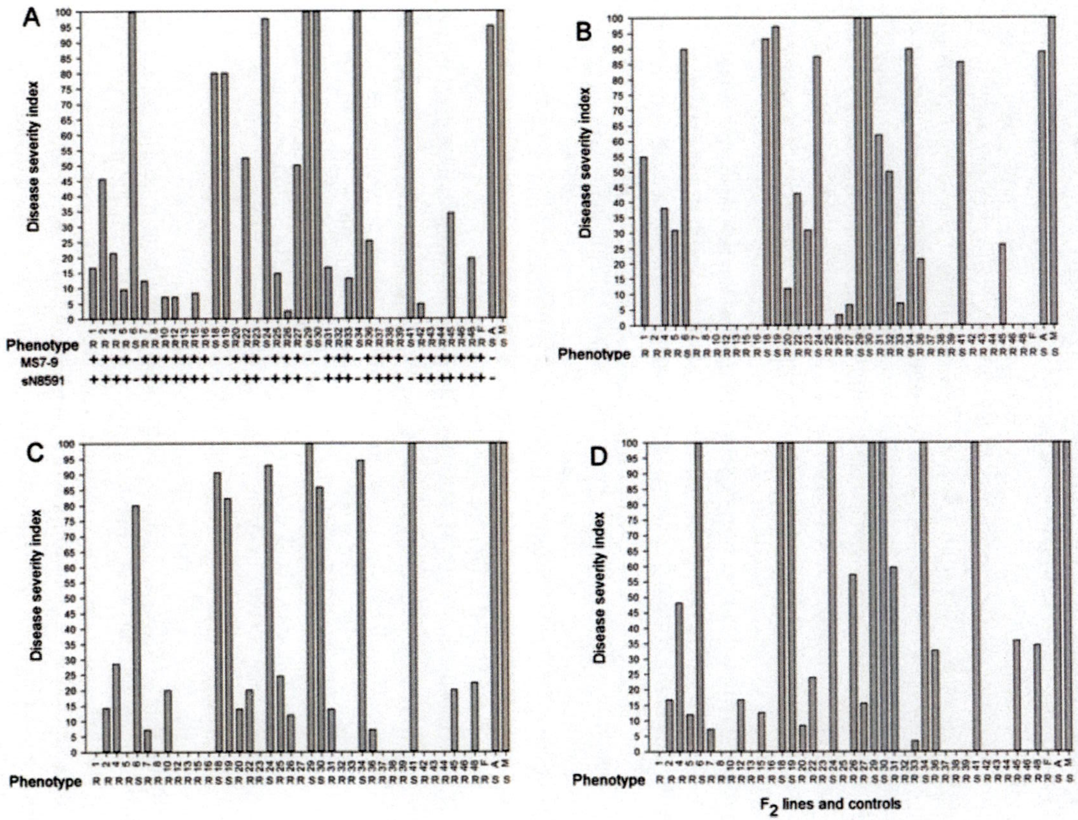

Fig. 2.15 Evaluation of plants from 38 F2 lines for resistance to clubroot. (**a**) Plants were inoculated with pathotype 3 and confirmed with molecular markers MS7–9 and sN8591 flanking Rcr1 described by Chu et al. (2014). "+": heterozygous alleles, one allele from R parent and the other one from S parent, "-": homozygous alleles from S parent; (**b**) inoculated with pathotype 2; (**c**) inoculated with pathotype 5; (**d**) inoculated with pathotype 6. Names of the lines indicate below *X*-axis. Letters F, A, and M represent *B. rapa* cv. or breeding line FN, ACDC, and "Mei Qing Choi," respectively. Disease severity indexes (DSI) were calculated from clubroot severity ratings. R and S are the phenotypes assigned for each line based on DSIs. The experiments were replicated two times with similar results (Yu et al. 2016)

one QTL for clubroot resistance in this region, suggesting the presence of functionally active candidate gene(s). This region of an *Arabidopsis* chromosome has clusters of the CR genes such as leucine-rich repeats (LRRs) and nucleotide-binding sites (NBSs). Similarly, RPP for resistance to *H. parasitica* (downy mildew), RPS for a resistance to *P. syringae* (bacterial blight), and ACD, which accelerates cell death in response to pathogen infection (Suwabe et al. 2006), have been identified. It has been suggested that CR genes may be members of these clusters of resistance genes.

Another novel CR locus, *CRc*, which is independent of all other CR loci, is located on chromosome R2 (Sakamoto et al. 2008). Fuchs and Sacristan (1996) mapped a CR locus (RPB1) in *Arabidopsis* chromosome 1. Their mapping for partial clubroot resistance identified two QTLs in chromosome 5 in F_2 and four QTLs in the RIL population, one each in chromosome 1 and 4 and two in the chromosome 5 of *A. thaliana* (Jubault et al. 2008). The CR QTL region in chromosome 5 colocalized with that containing several resistance gene clusters. These resistance genes could be candidates for clubroot disease resistance.

Fine mapping and detailed analysis of the levels of expression possessed by these genes would help to identify specific ones capable of conferring clubroot resistance. Till the complete genome sequence of only one species of Brassicaceae family, *A. thaliana* is available. It was reported that gene sequence identity of *Brassica* and *Arabidopsis* varies from 75% to 90% (Quiros et al. 2001). However, sequence information from the model species have greatly enhanced cloning and characterization of CR genes in other *Brassica* species because the partial sequences of several accessions of *A. thaliana* showing degrees of response to the pathogen *P. brassicae* collected worldwide were available (Fuchs and Sacristan 1996; Siemens et al. 2002; Nordborg et al. 2005; Alix et al. 2007). Few studies in transcriptomic/expression levels have been completed with respect to disease development. These include investigation of involvement of several metabolic pathways in disease pathogenesis, such as hormonal regulation by auxins (Grsic et al. 1999; Neuhaus et al. 2000), cytokinins (Devos and Prinsen 2006; Siemens et al. 2006), and trehalose synthesis (Brodmann et al. 2002). Grsic et al. (1999) observed that de novo indole-3-acetic acid (IAA) biosynthesis plays a role in symptom development during later stages of disease development; it was suggested that jasmonic acid, which increased during club development, may be involved in the upregulation of enzymes involved in IAA synthesis. Neuhaus et al. (2000) further supported this hypothesis by showing a delay in clubroot development after transforming *A. thaliana* with an antisense construct of Nitrilase 1 and 2, enzymes required for auxin biosynthesis. Siemens et al. (2006) investigated host gene expression during clubroot development in *A. thaliana* using an ATH1 microarray at two points in time, an early initial stage of infection and a later stage at which 60% of the host root cells were colonized.

More than 1000 genes were observed as being associated with the growth and cell cycle, sugar phosphate metabolism, and defense genes that were differentially expressed between infected versus control plants. Upregulation of auxin biosynthesis genes such as nitrilases and members of the GH3 family and downregulation of cytokinin homeostasis were observed. It was further observed that lines overexpressing cytokinin oxidase/hydrolases were resistant to clubroot, thereby strongly suggesting cytokinin is a key factor in clubroot development. Of the 312 genes identified as defense- and disease resistance-related genes, only 5 and 7% were upregulated at the first and the second points in time. This functional analysis of gene expression in clubroot-resistant and susceptible lines provided initial preliminary information.

2.5.4 Identification of R-Genes by Transcriptome and Proteomic Approaches

With the use of transcriptomic and proteomic approaches, the mechanisms of *P. brassicae* infection, pathogenesis, host specificity, and expression of host resistance has been elucidated. To date, the complete genome sequences of all species of Brassicaceae is available. The gene sequence identity of *Brassica* and *Arabidopsis* varies from 75% to 90% (Quiros et al. 2001). Using a full genome Affymetrix chip, Siemens et al. (2006) found that more than 1000 *Arabidopsis* genes are differentially expressed in *P. brassicae*-infected roots at either 10 or 23 days after inoculation versus noninoculated roots. These included genes associated with growth and cell cycle, sugar phosphate metabolism, and defense. The involvement of plant hormones in club development further supported, genes involved in auxin homeostasis, such as nitrilases and members of the GH3 family, are upregulated, whereas genes involved in cytokinin homeostasis are strongly downregulated at both time points. Cytokinin oxidase/dehydrogenase overexpressing lines are disease resistant and clearly indicate the importance of cytokinin as a key factor in clubroot disease development. Most known defense- or resistance-related genes are either not differentially expressed or downregulated at 23 versus 10 days after inoculation. Feng et al. (2012a, b) observed that the expression of several canola genes homologous to *Arabidopsis* genes found to be downregulated at 35 days after inoculation, are highly expressed at early stages (7 days after inoculation), and downregulated at later stages (42 days after inoculation) of infection. At the later stages of infection, the galls were forming and the resting spores were mature. Thus, the host genes involved in resistance/susceptibility may no longer be as active as during the early infection stages. Another explanation is that during the later stages, the pathogen already has successfully suppressed the expression of most resistance-related genes.

Proteome-level analysis, a powerful tool for high-throughput global protein expression analysis using two-dimensional gel electrophoresis (2-DE) coupled with mass spectrometry (MS) and bioinformatics, has been extensively used in the identification of proteins that are differentially expressed in plants in response to various abiotic or biotic stresses including those caused by plant pathogens. In order to investigate whether the pattern of specific host proteins is changed during clubroot formation, Hansen et al. (1994) employed 2-DE to compare the root protein profiles of *B. oleracea* L. with *B. oleracea* infected with *P. brassicae* 4 weeks after inoculation. They found that several proteins present in noninfected roots were absent or strongly reduced in the infected roots. In a proteomic study of a Chinese cabbage cultivar and two *P. brassicae* isolates, Ito et al. (1996) were able to resolve 133 differentially abundant protein spots, relative to noninfected controls, in infected roots 3 to 4 weeks after inoculation. The 133 protein spots were further classified into 6 groups, including spots that were enhanced in the susceptible response (59 spots), unique to the susceptible response (6 spots), repressed in the susceptible response (9 spots), enhanced in the resistant response (35 spots), unique to the resistant response (5 spots), and repressed in the resistant response (19 spots). N-terminal amino acid sequencing revealed that one protein (25 kDa, pI 7.0) that was enhanced in the susceptible response group showed high homology with pathogenesis-related protein group 5. These results suggest that the patterns of gene expression are different in the susceptible and the resistant responses.

In a differential protein analysis of infected versus noninfected roots and hypocotyls of *Arabidopsis* with 2-DE and MS, Devos et al. (2006) reported that 12% (46/390) of the visualized *Arabidopsis* proteins showed an altered abundance 4 days after inoculation with *P. brassicae* compared with the noninoculated plants, including proteins involved in metabolism, cell defense, cell differentiation, and detoxification. The metabolism-related enzymes adenosine kinase 2 (ADK2) and fructose bisphosphatealdolase (FBA) were found to be downregulated at 4 days after inoculation. Downregulation of ADK2 during the early stages of *P. brassicae* infection may keep the elevated levels of active cytokinins high, which is consistent with the increased endogenous isopentenyl adenine content. Downregulation of FBA upon *P. brassicae* infection suggests a flow toward the production of glucose, which correlates with the fact that galls of *P. brassicae* act as a carbon sink that may be induced by the high cytokine level present at the infection site (Devos et al. 2006). Myrosinase, an enzyme that can catalyze indole-3-methyl glucosinolate degradation to form indole-3-acetic acid, was also found to be six times more abundant in infected *Arabidopsis* at 4 days after inoculation. During *P. brassicae* infection, a number of detoxification-related proteins are found to be upregulated, including catalase, glutathione S-transferase, thioredoxin, and ferredoxin-nitrate reductase. Out of the 46 differentially regulated proteins, 11 proteins, including ADK2, FBA, peroxiredoxin, α-tubulin, and heat shock protein 70, are downregulated and 35 are upregulated, including myrosinase, glutathione S-transferase, ferredoxin-nitrite reductase, and pectin methyl-esterase (Devos et al. 2006).

The changes in the root protein profile were examined by 2-DE at 12, 24, 48, and 72 h after inoculation of a susceptible canola (*B. napus*) genotype with *P. brassicae* (Cao et al. 2008). A total of 20 protein spots were identified as either upregulated (13 spots) or downregulated (7 spots). Decreased abundance of adenosine kinase, which is involved in cytokine homeostasis, supported the previous reports that cytokinins play a key role in the early phases of *P. brassicae* infection (Devos et al. 2006). An approximately sixfold reduction in caffeoyl-CoA O-methyltransferase abundance suggested a reduction in host lignin biosynthesis after inoculation and is consistent with the compatible nature of the *B. napus*–*P. brassicae* interaction. Levels of enzymes involved in the metabolism of reactive oxygen species, such as copper/zinc superoxide dismutase and cytochrome c-oxidase, declined sharply at 12 h after inoculation, but increased at 24–72 h.

Protein identification by MS analysis revealed that the 20 differentially regulated proteins can be classified into proteins involved in lignin biosynthesis, cytokinin metabolism, glycolysis, intracellular calcium homeostasis, and the detoxification of reactive oxygen species.

2.5.5 Identification of Pathotype-Specific R-Genes

More than 10 clubroot resistance genes including *CRa*, *CRb*, *CRc*, *CRk*, *Crr1*, *Crr2*, *Crr3*, and *Crr4* were identified in vegetable cultivars of *B. rapa*. These genes were identified based on the reaction to the *P. brassicae* pathotypes prevalent in Japan, Korea, and China, which would likely be classed into pathotypes 2 or 4 based on Williams' differential hosts. Earlier five pathotypes of *P. brassicae* (2, 3, 5, 6, and 8, again based on Williams' differentials) occurred in Canada. Pathotype 3 was the most prevalent on canola. *Rcr1* was mapped on chromosome A03 based on resistance to pathotype 3, but the donor cv. FN was resistant to each of the pathotypes identified in Canada. To determine if *Rcr1* confers resistance against the range of Canadian pathotypes, an F2 population consisting of 38 lines was developed. The F2 lines were equivalent to self-pollinated BC1 populations (BC1S1). There were two genotypes "Rr," heterozygous at Rcr1 locus, which should be resistant and "rr," homozygous susceptible alleles, which should be susceptible phenotype in F1. Theoretically, segregation for resistance and susceptibility from "Rr" plants in the F2 lines is 3:1 and that from "rr" should be 0:1. Ideally, X2 analysis would have been performed. The small number of plants available for evaluation of plant resistance in this study made X2 analysis impractical. Yu et al. (2016) speculated that the eight F2 lines with high severity were from "rr" F1 plants and the 30 F2 lines with low to moderate severity were from "Rr" F1 plants. Similar results were obtained by repeating the experiments. In addition, the results were confirmed by analysis of the molecular markers (Fig. 2.16; Table 2.6). Thus, the genotypes of the F1 plants determined by the phenotypes in F2 lines in this study were reliable, even though 7 to 14 plants each line for each pathotype were tested per replicate. In some of the F2 lines that were derived from "Rr" F1 plants, their DSIs were as low as zero. This can be caused by segregating distortion so no "rr" plants appeared in the small populations. In addition, disease escapes from "rr" plants could contribute to the zero DSIs. Resistance to pathotypes 2, 3, 5, and 6 was completely linked in each of the 38 lines. However, variation was usually found among DSIs of the same F2 lines responding to different pathotypes.

The reason for the variation can be that the proportions of the genotypes (RR, Rr, and rr) for those "Rr"-derived F2 lines against each pathotype were not exactly same. It also can be caused by some escapes from "rr" plants especially for those "rr"--derived F2 lines. The original intent was to assess 200 F1 plants in this study, but only these 38 plants produced sufficient seed in the F2 generation for assessment. Resistance to pathotype 8 was also associated with resistance to pathotype 3 in one interaction of the assessment, but there was not sufficient seed to repeat that portion of the study. Resistance to this range of pathotypes indicates that Rcr1 could confer

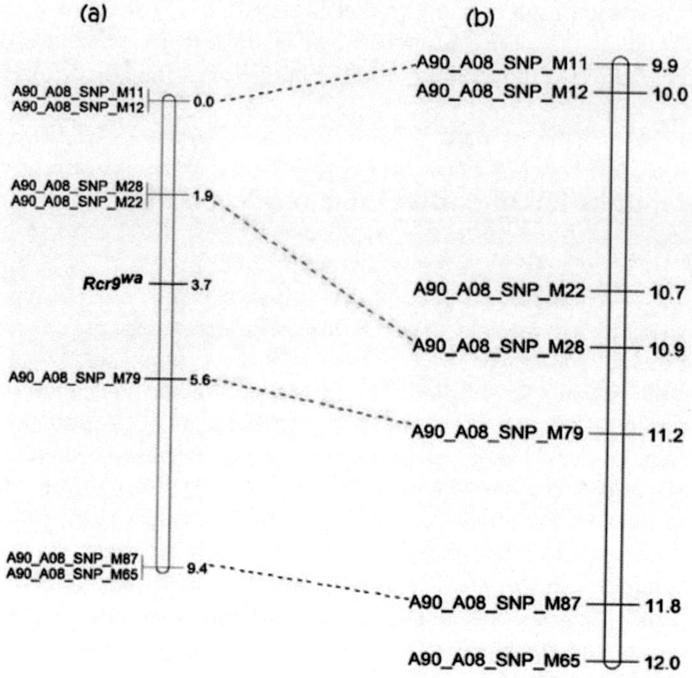

Fig. 2.16 Genetic linkage map based on BC1 populations of *Brassica rapa* line 96–6990-2; (**a**) genetic map of the region in which the Rcr9wa gene is located (genetic distance on right), and (**b**) physical location of the Rcr9wa region (in bases, on right), with SNP markers connected with a broken line between the maps (Karim et al. 2020)

broad-spectrum resistance to many of the pathotypes of *P. brassicae* that are present in Canada. Another possibility is that linked genes could control the resistance to this range of pathotypes.

RNA-seq generates a wealth of short DNA sequence reads from random places in the transcriptome. Transcriptome analysis and identification of SNPs based on RNA-seq have been carried out in many organisms, including *Brassicas* such as *B. napus* and *B. rapa*. The characterization of variants in the *B. rapa* population carried *Rcr1* and identified the most probable candidates for *Rcr1* based on DNA variants and development of SNP markers tightly linked to the gene for MAS. In addition, SNP variants were found to be the most common DNA sequence variation, accounting for over 80% of the *B. rapa* transcriptome. Similar results were reported in rice, with an average of 99,955 putative SNPs and 14,617 putative InDels and the human genome, where SNPs made up about three million of the four million DNA sequence. RNA-seq of human airway epithelial cells revealed that over 90% of variants were SNPs.

A wealth of SNPs and InDels were identified in the *B. rapa* genome using both the SSA and PSA approaches, which generally produced similar results. The SSA method, in which three biological replicates of R and S bulks were assessed,

Table 2.6 Association between disease reaction phenotype and SNP markers in the F1 segregating populations (Yu et al. 2016)

SNP marker	Gene	A03 location[a]	Phenotype[b]	Number of plants with marker[c]	
				+	−
SNP_A03_11	Bra019413	24,352,616	R	30	0
			S	0	8
SNP_A03_104	Bra019413	24,352,908	R	30	0
			S	0	8
SNP_A03_08	Bra019413	24,353,289	R	30	0
			S	0	8
SNP_A03_09	Bra019412	24,371,044	R	30	0
			S	0	8
SNP_A03_102	Bra019410	24,375,018	R	30	0
			S	0	8
SNP_A03_12	Bra019410	24,375,572	R	30	0
			S	0	8
SNP_A03_13	Bra019410	24,375,979	R	30	0
			S	0	8
SNP_A03_14	Bra019410	24,376,507	R	30	0
			S	0	8
SNP_A03_103	Bra019410	24,378,671	R	30	0
			S	0	8
SNP_A03_16	Bra019409	24,382,012	R	30	0
			S	0	8
SNP_A03_101	Bra019409	24,382,587	R	30	0
			S	0	8
SNP_A03_18	Bra019406	24,393,429	R	30	0
			S	0	8
SNP_A03_19	Bra019406	24,393,825	R	30	0
			S	0	8
SNP_A03_32	Bra019406	24,394,435	R	30	0
			S	0	8

[a]Physical location on chromosome A03 based on *Brassica rapa* reference genome v1.5
[b]Phenotypes in 38 F1 plants were determined in F2
[c]Genotyping SNP markers in the Rcr1 target region using KASP method: "+" allele from the resistant parent FN; "-" allele from the susceptible parent ACDC

provided estimates of variance that can be used to assess the statistical significance of variation of SNPs and InDels. This approach is very useful for identifying differentially expressed genes through RNA-seq. In contrast, the PSA approach, which merged the data across biological replicates, is much less amenable to statistical analysis. However, it produced a greater depth of sequencing data, providing more reliable results in discovery of variants.

In both R and S samples, variants from the reference genome DNA sequences in *B. rapa* Chinese cabbage "Chiifu" v1.5 were very frequent. One factor that likely

Table 2.7 Number of polymorphic variants identified in each gene located in the Rcr1 interval in the simple sequence assembly (SSA) and pooled sequence assembly (PSA) analysis (Yu et al. 2016)

Gene	No. of variants (SSA)[a]	No of variants (PSA)
Bra019409	81.7 ± 1.2 a	115
Bra019410	75.0 ± 0.9 b	85
Bra019413	34.7 ± 0.4 c	52
Bra019406	33.3 ± 0.1 cd	37
Bra038754	31.3 ± 0.5 cd	39
Bra019390	29.0 ± 0.8 de	38
Bra038757	24.0 ± 0.3 e	31
Bra019403	16.7 ± 0.5 f	18
Bra038753	14.3 ± 0.3 fg	20
Bra038755	13.0 ± 0.6 fgh	17
Bra038756	11.3 ± 0.2 gh	9
Bra019395	9.7 ± 0.3 ghi	12
Bra038748	8.0 ± 0.1 hij	11
Bra019391	5.7 ± 0.8 ijk	10
Bra038749	5.3 ± 0.2 ijk	5
Bra019400	5.3 ± 0.9 ijk	4
Bra019412	4.0 ± 0.4 jk	4
Bra019398	4.0 ± 0.5 jk	9
Bra038758	2.0 ± 0.1 k	3
Bra019407	1.7 ± 0.1 k	3
Bra019405	1.3 ± 0.1 k	2
		≤0.05

[a]Mean (± SE) followed by the same letter do not differ at *P*

contributed to this high frequency of variants was that the parental lines were pak choy and canola, rather than Chinese cabbage. Deep sequencing, with coverage of 120-fold of the reference genome in R and 109-fold in S samples, could also have contributed to the wealth of variants. More than 70% of the variants between R and S bulks were monomorphic in all chromosomes except A03, where a significantly higher percentage of polymorphic variants were present. The high frequency of polyvariants could indicate that there was a higher level of difference between the R and S bulks in *Rcr1* and its surrounding region than in other parts of genome.

Rcr1 was fine mapped in an interval of 0.28 centi-Morgan (cM), flanking by microsatellite markers MS7–9 and sN8591 on *B. rapa* chromosome A03. Yu et al. (2016) identified more than 400 variants (Table 2.7) in the target region. Fourteen SNP markers that reside in five genes (*Bra019406*, *Bra019409*, *Bra019410*, *Bra019412*, and *Bra019413*) in the target region were genotyped using KASP method. Each of the markers cosegregated with *Rcr1*, indicating that they were completely linked to Rcr1 and closer to *Rcr1* than the markers previously described by Chu et al. (2014). However, it was not possible to define the *Rcr1* location further with the recombinants in the mapping population, because no recombination events could be identified in the set of recombinants. Therefore, it was essential to obtain accurate SNP and InDel profile on the genes of interest through analysis of RNA-seq data. Four of the five genes identified encode proteins related to disease resistance that belong to TNL families. Nonsynonymous, nonsense, frame shift, and synonymous variants were identified in the TNL genes. It is unlikely that the genes *Bra019412* and *Bra019413* are the candidates for *Rcr1*; although they showed differential expression between R and S bulks, few polyvariants, especially variants that cause changes in amino acid sequence, were identified in these genes. *Bra019409* and *Bra019410* carried higher number of variants that could cause changes in amino acid sequence and higher number of synonymous variants, previously considered as "silent," but which can cause changes in protein expression, conformation, and function (Sauna and Kimchi-Sarfaty 2011). Therefore, it is more likely that either of these genes is a candidate for *Rcr1*. It is also possible that a combination of the TNL genes is required for the expression of *Rcr1* resistance. *Rcr1* was mapped in the same interval as one of the cloned clubroot resistance gene *CRa*. However, it was not illustrated how *CRa* was related to the TNL genes in the target region. The complete *CRa* coding sequence is 4223 bp (GenBank: AB751517.1). Yu et al. (2016) searched the *B. rapa* genome using the 4223 bp sequence at http://brassicadb.org/brad/blastPage.php and found that *CRa* was homologous to the four TNL genes *Bra019410, Bra019412, Bra019409,* and *Bra019413* with score and E value at 1905, 0.0; 823, 0.0; 739, 0.0; and 504, e-141, respectively. This indicates that *CRa* is a gene homologous to the TNL genes especially *Bra019410*, but obviously different from any of the TNLs. Identifying the TNL gene that corresponds with Rcr1 and relationship of *Rcr1* with *CRa* will be addressed after Rcr1 has been cloned.

Breeding for clubroot resistance in canola at many locations in Canada is severely constrained because the pathogen, or the pathotype of interest, does not occur at or near the site of established breeding institutions. Use of highly specific markers in MAS could be to accurately assess the reaction of lines under controlled (i.e., indoor) conditions. A number of techniques are available for genotyping SNPs. NGS is an emerging method of SNP genotyping that is being increasingly adopted for discovery applications. However, this method can be expensive and time-consuming in terms of informatics needs and currently generates datasets with a large proportion of missing data. KASP offers cost-effective and scalable flexibility in applications. In total, 14 robust SNP markers were confirmed that were completely associated with *Rcr1*, based on testing the recombinants identified from 1587 plants. Most of the SNPs proved to be very reliable markers in use for introgression of *Rcr1* into elite canola breeding lines. Furthermore, MAS with the polymorphic SNP markers was confirmed to have detected *Rcr1* in the BC1 population described by Chu et al. (2014) with 100%. Therefore, it is concluded that these SNP markers can provide an effective and robust basis for introgression of *Rcr1* into canola using MAS.

2.5.6 Mapping of Clubroot Resistance Genes in *Brassica* Species

Details of clubroot resistance QTLs, their number, and location on specific chromosomes in each species are as under.

2.5.6.1 *Brassica rapa*

Genetic analysis and genetic mapping of CR genes are well studied in *B. rapa*. All eight possible CR genes present in *B. rapa* have been identified through QTL mapping using a range of resistant sources and marker systems (Table 2.8). Kuginuki et al. (1997) identified three random amplified polymorphic DNA (RAPD) markers linked to a CR locus using turnip cv. Siloga as a source of resistance. Suwabe et al. (2003) identified *Crr1* and *Crr2*, two loci that originated from cv Siloga, using F_2 populations and simple sequence repeat (SSR) marker systems. It was concluded that these two loci are complementary for clubroot resistance. Inoculation of test plants with mild and virulent isolates of *P. brassicae* showed that plants with a homozygous resistance locus had greater resistance to clubroot than those having a heterozygous resistance locus. In addition to *Crr1* and *Crr2*, *Crr4*, a weak QTL, was also detected using a similar F_2 population (Suwabe et al. 2006). Five genes involved in glucosinolates biosynthesis were mapped, and GSL-BoELONG and GSL-BoPro were found to be linked to the Pb-Bo (Anju) 1 and Bo (GC) 1 loci (Nagaoka et al. 2010). Hirai et al. (2004) identified and mapped a novel locus named *Crr3*, which originated from the clubroot resistant turnip cv. Milan White. Another CR locus, *CRb*, derived from turnip cv. Gelria R, was mapped in an interval of 3 cM using sequence characterized amplified region (SCAR) markers converted from amplified fragment length polymorphism (AFLP) markers (Piao et al. 2004). Homology analysis of SCAR markers with that of *Arabidopsis* sequence database identified that the region containing the *CRb* locus is homologous to the central part of the *A. thaliana* chromosome 4 (Fig. 2.16). These are also homologs with the region of the *B. rapa* chromosome containing *Crr1* and *Crr2* CR QTL. *Brassica rapa* BAC clones containing these SCAR markers have been identified and located on a specific chromosome in clubroot resistance cv. CR Shinki through fluorescent in situ hybridization (FISH) (Plate 2.4). Colocalization of these BAC clones was found in *B. rapa* chromosome 1. This locus is independent of *Crr1*, *Crr2*, *Crr3*, and *Crr4* (Saito et al. 2006). Sakamoto et al. (2008) identified *CRk* and *CRc* loci derived from turnip cv. Debra using QTL analysis of two F_2 populations and two isolates of *P. brassicae*. In total, eight CR loci were mapped and allocated to five different chromosomes of *B. rapa*. The genes *Crr1, Crr2, Crr3, Crr4*, and *CRc* are mapped to R8, R1, R3, R6, and R2, respectively, of the internationally agreed upon *B. rapa* reference genetic map. It is noteworthy that *CRa, CRb*, and *CRk* with *Crr3* are mapped on the same linkage group of R3, but they are not located in the same chromosome region, except for *CRk* and *Crr3* (Piao et al. 2009). A clubroot resistance gene (*Rcr2*) was identified and fine mapped in Chinese cabbage cv. "Jazz" using single-nucleotide polymorphism (SNP) markers identified from bulked segregant RNA sequencing (BSR-Seq), and molecular markers were developed for use in marker-assisted selection. In total, 203.9 million raw reads were generated from one pooled resistant (R) and one pooled susceptible (S) sample, and > 173,000 polymorphic SNP sites were identified between the R and S samples.

One significant peak was observed between 22 and 26 Mb of chromosome A03, which had been predicted by BSR-Seq to contain the causal gene *Rcr2*. There were 490 polymorphic SNP sites identified in the region. A segregating population

Table 2.8 Genetic mapping of clubroot resistance loci in *Brassica rapa* (Piao et al. 2009)

Resistant source	Population	Isolate[a]	Locus	Types of DNA marker	Flanking markers	Chromosome with interval (cM)	References
ECD02	F2	Race 2	*CRa*	RFLP	HC352b, HC181	R3 (15 cM)	Matsumoto et al. (1998)
				STS	HC352b-SCAR		Sakamoto et al. (2008)
Gelria R	F_2	Race 4	*CRb*	SCAR	TCR05, TCR09	R3 (3. 0 cM)	Piao et al. (2004)
Siloga	F_2	Ano-o1 (race 2)	*Crr1*	SSR	BRMS-297	R8 (1.6 cM)	Suwabe et al. (2006)
		Wakayama-01	*Crr2*	SSR	BRMS-088	R1 (2.2 cM)	
			Crr4	RFLP	BRMS-100, BRMS-096, BN288D, WE24–1	R6 (2.7 cM)	
Milan White	F_3	Ano-01	*Crr3*	STS	OPC11–1S, OPC11–2S	R3 (10 cM)	Hirai et al. (2004)
Debra	F_2	M85 (race 2), K04	*CRk*	STS	HC688, OPC11–2S	R3 (9.1 cM)	Sakamoto et al. (2008)
	F_2		*CRc*	AFLP	E14M3–02	R2 (5.1 cM)	
					E15M4–006		

[a]Isolate characterization based on the Williams' classification (Williams 1966)

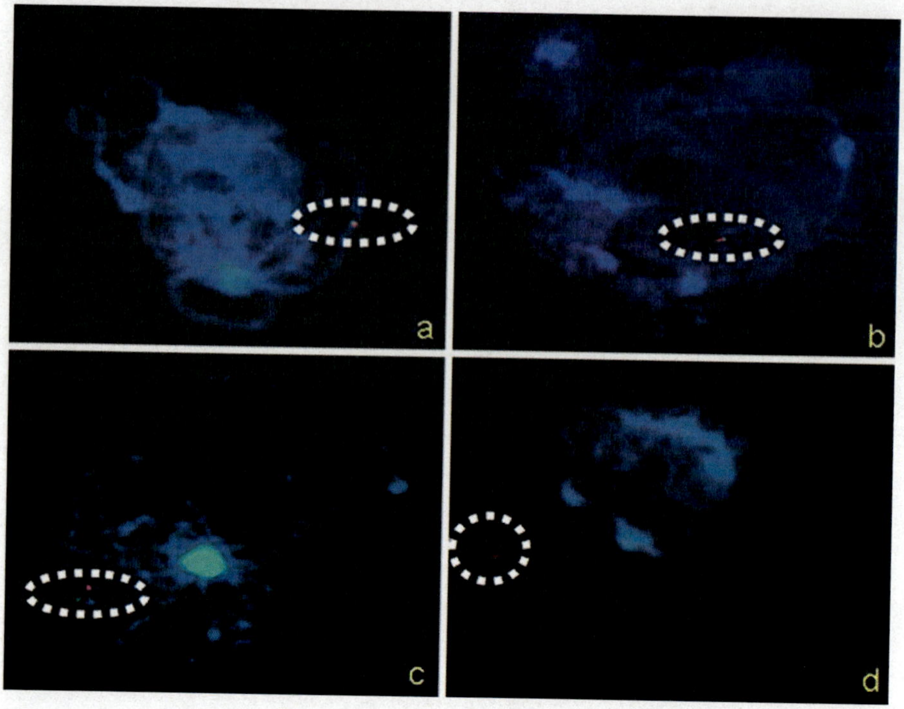

Plate 2.4 FISH localization of BAC clones containing SCAR markers (TCR05 and TCR09) linked to clubroot resistance locus in *B. rapa* CR Shinki genome. (**a**) BAC clones KBrH060E03 (green) and KBrH097J16 (red) from same contig harboring TCR05 marker showing localization in the same position in *B. rapa* CR Shinki chromosome 1. (**b**) BAC clones KBrH083J12 (green) and KBrH103M15 (red) from same contig harboring TCR09 marker showing localization in same region. (**c**) BAC clones KBrH97J16 (green, harboring TCR05) and KBrH88B11 (red, harboring TCR09) from different contig linked to TCR05 and TCR09 markers showing localization in different regions of the same chromosome. (**d**) BAC clones KBrH115F22 (green, harboring TCR05) and KBrH144K19 (red, harboring TCR09) from different contig linked to TCR05 and TCR09 markers showing localization in different regions of the same chromosome (Piao et al. 2009)

consisting of 675 plants was analyzed with 15 SNP sites in the region using the Kompetitive allele-specific PCR method, and *Rcr2* was fine mapped between two SNP markers, SNP_A03_32 and SNP_A03_67, with 0.1 and 0.3 cM from *Rcr2*, respectively. Five SNP markers cosegregated with *Rcr2* in this region. Variants were identified in 14 of 36 genes annotated in the *Rcr2* target region. The numbers of poly variants differed among the genes. Four genes encode TIR-NBS-LRR proteins and two of them, *Bra019410* and *Bra019413*, had high numbers of polymorphic variants and so are the most likely candidates of *Rcr2* (Huang et al. 2017).

Eight Chinese cabbage germplasms were screened using published clubroot-resistant (CR) loci–/gene-linked markers. A CR gene *Crr3* potential carrier "85–74" was detected which linked to marker BRSTS61; however, "85–74" shows different responses to local pathogens "LAB-19," "LNND-2," and "LAB-

10" from "CR-73" which harbors *Crr3*. Peng et al. (2018) used a next-generation sequencing-based bulked segregant analysis approach combined with genetic mapping to detect CR genes in an F_2 segregant population generated from a cross between the Chinese cabbage inbred lines "85–74" (CR) and "BJN3–1" (clubroot susceptible). The "85–74" line showed resistance to a local pathogen "LAB-19" which was identified as race 4; a genetic analysis revealed that a single dominant gene conferred the resistance. The CR gene named *CRd* was mapped to a 60 kb (1 cM) region between markers yau389 and yau376 on chromosome A03. *CRd* is located upstream of *Crr3* which was confirmed based on the physical positions of *Crr3* linked markers. The identification of *CRd*-linked markers can be applied to marker-assisted selection in the breeding of new CR cultivars of Chinese cabbage and other *Brassica* crops.

2.5.6.2 *Brassica oleracea*

A number of DNA markers linked to CR loci in *B. oleracea* have been developed. Two to nine QTLs have been identified (Table 2.9) (Landry et al. 1992; Figdore et al. 1993; Grandclément and Thomas 1996; Voorrips et al. 1997; Moriguchi et al. 1999; Rocherieux et al. 2004; Nomura et al. 2005). Two QTLs, *CR2a* and *CR2b*, showing resistance to race 2 of *P. brassicae* and contributing 58 and 15%, respectively, of the phenotypic variation, were identified using swede cv. Wilhelmsburger as a resistance source (Landry et al. 1992). Three QTLs showing resistance to race 7 were identified using broccoli (Figdore et al. 1993). Voorrips et al. (1997) identified two QTLs, *pb-3* and *pb-4*, and a minor QTL contained in cv. Bindsachsener using a multiple QTL mapping approach, which analyzed the fresh weight of galls. The additive effects of two major loci were responsible for 68% of the difference between the parents and for 60% of the genetic variance among the means of DH lines. One QTL for resistance to clubroot disease was identified on linkage group 3 using resistant kale (K269) under conditions of natural infection (Moriguchi et al. 1999). Based on the quantitative analysis of F3 family using controlled environments with four single-spore isolates and one field isolate from four *P. brassicae* isolates, Rocherieux et al. (2004) found two to five QTLs depending on the pathotype used. Of the nine QTLs fully identified, *Pb-Bol* is common to all isolates and accounts for 20.7–80.7% of the phenotypic variation, whereas the rest were specific to one, two, or three isolates. Nomura et al. (2005) identified three QTLs, QTL1, QTL3, and QTL9, using a kale line (K-269) as the resistant parent, which was similar to that used by Moriguchi et al. (1999). Therefore, one of these QTLs is probably similar to the one QTL that is located at the end of LG3. The SCAR markers converted from RAPD and RFLP markers linked to these QTLs were evaluated for F2 and F3 plants. It was observed that F2 individuals with three QTLs expressed very high clubroot resistance, similar to that of the kale parent, whereas the F2 and F3 plants carrying a single QTL expressed only intermediate resistance (Nomura et al. 2005).

At least 22 QTLs have been found in *B. oleracea* so far. The discovery of several CR QTLs indicates that the clubroot resistance in *B. oleracea* is controlled at several QTLs, further confirming the complex genetic basis of clubroot resistance in *B. oleracea*. Because these mapping studies used different CR sources and isolates,

Table 2.9 Genetic mapping of clubroot resistance loci in *Brassica oleracea* (Piao et al. 2009)

Resistant source	Population	Isolate	Loci	Types of DNA marker	Flanking markers	LG with interval (cM)	Reference
Wilhelmsburger	F2	Race[b] 2	CR2a	RFLP	2NF11, 2ND3	LG6 (22)	Landry et al. (1992)
			CR2b		3NE4a, 3ND3	LG1 (12)	
Broccoli (OSU[a] CR-7)	F2	Race 7	3	RFLP	14a	LG1	Figdore et al. (1993)
					48	LG4	
					177b	LG9	
Kale (C10)	F2	16/31/31[c]	At least 2	RAPD	OPL6–780, OPB11–740, OPA18–14900, OPA4-700, OPE20–1250, OPA1-1880, OPA16–510	–	Grandclement et al. (1996)
Indsachsener	DH	Field isolate	*pb-3*	RFLP	4NE11a	LG3	Voorrips et al. (1997)
			pb-4	AFLP	2NA8c	LG1	
Kale (K269)	F2	Race 1 and 3	1	RAPD, RFLP	WG6A1, WG1G5	LG3 (2.6)	Moriguchi et al. (1999)
Kale (K269)	F2	Three field isolates	QTL1	SCAR	SCA02a 2, SCB50b,SCB74c SOPT15a, SCA25	LG1	Nomura et al. (2005)
			QTL3			LG3	
			QTL9			LG9	
Kale (C10)	F2	P1 (Ms6, eH), P2 (K92), P4 (K92–16), P7 (Pb137–522)[d]	*Pb-Bo1*	RAPD, RFLP, ACGM	Ae05.8800, T2	LG1 (18.4)	Rocherieux et al. (2004)
			Pb-Bo2		PBB38a, r10.1200	LG2 (19.3)	

			Pb-Bo3		ae15.100, RGA8.450	LG3 (13.9)	
			Pb-Bo4		ELI3.983, aa9.983,	LG4 (3.1)	
			Pb-Bo5a		PBB7b, ae05.135	LG5 (32.5)	
			Pb-Bo5b		ELI3.115, a18.1400	LG5 (12.0)	
			Pb-Bo8		c01.980, t16.500	LG8 (10.2)	
			Pb-Bo9a		aj16.570, W22B.400	LG9 (24.1)	
			Pb-Bo9b		a04.1900, ae03.136	LG9 (1.4)	

[a]*OSU* Oregon State University
[b]Isolate characterization based on Williams' classification (Williams 1966)
[c]Isolate characterization based on ECD set (Buczacki et al. 1975)
[d]Isolate characterization based on Somé et al. (1996)

although the primer and marker sequence are disclosed, the comparison of these QTLs is impossible. To understand the genetics and genomics of CR loci in *B. oleracea* in detail, development of common PCR-based markers is required. In addition, comparative studies suggested that the genomes of *Brassica* species have evolved from a common ancestor. It is worthwhile investigating whether CR genes in *Brassica* species might have common origins and how their mechanisms of evolution are maintained. Comparative studies of CR genes or their linked markers should provide new insights into these processes (Piao et al. 2009).

2.5.6.3 *Brassica napus*

Twenty-two QTLs involving clubroot resistance have been identified in *B. napus* (Table 2.10). Manzanares-Dauleux et al. (2000) mapped the major gene *Pb-Bn1*, which confers resistance to two single-spore isolates (SSI) of *P. brassicae*, onto linkage group DY4. Based on the quantitative resistance expressed against each SSI, they also found at least two additive QTLs on chromosomes DY4 and DY15, respectively. In addition, epistatic interactions between nine regions with or without additive effects have been located. The total phenotypic variation accounted for by additive and epistatic QTLs ranged from 62% to 81.4% depending on the *P. brassicae* isolate used. Analyses of double haploid (DH) populations, in which CR genes are derived from ECD04, identified one major gene and two recessive genes (Diederichsen et al. 2006). A resynthesized *B. napus* was developed by crossing cv. Bohmerwaldkohl (*B. oleracea*) and ECD-04 (*B. rapa*). From this, the CR DH line "263/11" was obtained (Diederichsen and Sacristan 1996). Werner et al. (2008) analyzed the DH population derived from a cross of 263/11, and the susceptible cv. Express using 7 *P. brassicae* isolates. Nineteen QTLs expressing resistance to seven isolates were detected on eight chromosomes, N02, N03, N08, N09, N13, N15, N16, and N19. All QTLs were found to be race-specific. The total phenotypic variation accounted for ranged from 20.8% to 79.6% depending on the pathogen isolate used. Among the 19 QTLs detected, 4 were closely linked to each other on chromosome N03; 3 were linked also to chromosome N08. In *B. rapa*, genes *CRa, CRb, CRk*, and *Crr3* are located on chromosome R3, which corresponds with N03 in *B. napus*. Genes *CRk* and *Crr3* are located in the similar region of *PbBn-k-2, PbBn-1-1, PbBn-01:60–1* on N03. However, genes *CRa* and *CRb* are independent from them. *PbBn-01.07–2, PbBn-l-2*, and *PbBn-a-1* are linked to BRMS088 on chromosome N08 in *B. napus*, which is also linked with *Crr1* on R8 in *B. rapa*. Further, studies of these loci, using common markers, might explain whether they are identical. The QTLs located on N03 and N19 contribute strong effects and confer broad-spectrum resistance. The clubroot resistance QTLs/genes mapped/fine mapped in different *Brassica* species are given in Table 2.11. It has been suggested that CR genes may be members of these clusters of resistance genes. The genes *Crr1*, *Crr2*, and *CRb,* however, are distributed on three different chromosomes in *B. rapa*, R8, R1, and R3, respectively (Saito et al. 2006; Suwabe et al. 2006). Based on this observation, it has been suggested that the evolution of CR genes occurs by one of two routes. First, clubroot resistance was originally controlled by a single major gene in the ancestral genome, which later differentiated and diverged as functionally

Table 2.10 Genetic mapping of clubroot resistance loci in *Brassica napus* (Piao et al. 2009)

Resistant source	Population	Isolate	Loci	Types of DNA marker	Flanking markers	LG with interval (cM)	Reference
Winter dwarf line	DH	P4 (K92–16)[a]	Pb-Bn1	RAPD	OPG03.960, OPV09.2100	DY4 (8.3)	Manzanares-Dauleux et al. (2000)
		P7 (Pb137–522)	Two QTL		OPC18.1250, OPD20.760 OPQ01.930, OPG13.950	DY2 (11.8) DY15 (3. 7)	
ECD04	DH	1	One dominant and at least two recessive	SSR	HMR337, HMR388, HMR307	MS06 (4.0)	Diederichsen et al. (2006)
Bohmer waldkohl	DH	01:60[b]	PbBn-01:60–1	AFLP	158_241, 128_330	N03 (13.6)	Werner et al. (2008)
ECD-04		0.1:07	PbBn-01:60–3		84_174, 1_103	N19 (6.4)	
		e4x04a	PbBn-01:60–4		107_370, 166_215	N19 (4.3)	
			PbBn-01:07–1		107_106, 128_330	N03 (7.7)	
			PbBn-01:07–2		84_258, 79_168	N08 (2.0)	
			PbBn-01:07–3		160_186, 160_193	N13 (2.0)	
			PbBn-e4x04–1		163_448, 159_296	N19 (4.4)	
			PbBn-a-1		84_258, 153_370	N08 (5.4)	
Bohmer waldkohl	DH	1	PbBn-l-1	AFLP	107_106, 128_330	N03 (7.7)	Werner et al. (2008)
ECD-04		k Korporal	PbBn-l-2		84_258, 79_168	N08 (2.0)	
			PbBn-k-1		154_103, 152_373	N02 (7.5)	
			PbBn-k-2		107_106, 128_330	N03 (7.7)	
			PbBn-k-3		79_75, HMR1382a	N15 (4.8)	
			PbBn-Korp-1		6_450, 165_156	N09 (3.1)	
			PbBn-Korp-2		107_366, 146_363	N09 (9.0)	

(continued)

Table 2.10 (continued)

Resistant source	Population	Isolate	Loci	Types of DNA marker	Flanking markers	LG with interval (cM)	Reference
			PbBn-Korp-3		158_154, 148_158	N09 (1.3)	
			PbBn-Korp-4		19_155, 166_146	N16 (12.6)	
			PbBn-Korp-5			N16 (31.7)	

[a]Isolate characterization based on Somé et al. (1996)
[b]Isolate characterization based on ECD set (Buczacki et al. 1975)

Table 2.11 Clubroot resistances QTL/gene mapped/fine mapped in different *Brassica* species (Hirani and Li 2015)

Brassica species (R sources)	Populations	LG	QTL/genes
B. rapa (Chinese cabbage)	BC1	A03	*Rcr1* fine mapped
B. rapa (Chinese cabbage)	F2	A03	*CRb* fine mapped
B. rapa (G004 line)	F2	A08	*Crr1a* fine mapped
B. rapa (Chinese cabbage)		A03	*CRa* fine mapped
B. oleracea (Anju)	DH	O2, O5,	pb-Bo(Anju)1, pb-Bo(GC)1
B. napus (synthetic line)	DH	N02, N03, N08, N13, N15, N16, and N19	Nineteen QTL identified on different LGs
B. oleracea (kale)	F2:3	LG1, 2, 5	Nine QTL (Pb-Bo1 to Pb-Bo9) with phenotypic variance 20–88%
B. rapa (Shinki)	F2	A03	*CRb*
Brassica oleracea (Bindsachsener)	DH	–	Two QTL (pb-3 and pb-4)
B. rapa (Chinese cabbage)	F2, BC1	A03	CR gene fined mapped
B. rapa (turnip line)	F2	A03, A08	Two major QTL (Pb-Br3, Pb-Br8 and
B. rapa (European turnip)	F2:3	A03	*Crr3*
B. rapa (G004)	F2	A06	*Crr4*
B. rapa (Chinese cabbage)	F2	A03 and A02	*CRk and CRc*

duplicate genes during the course of evolution in the *Brassica* genome (Suwabe et al. 2006). Second, route might be that the resistance genes for clubroot were originally clustered in that region in the ancestral genome which was later distributed into different genomic regions following chromosomal rearrangement in *Brassica*. Current *Brassica* species, which diverged 17–18 million years ago from *Arabidopsis*, are evolutionarily believed to be derived from whole-genome triplication and rearrangement of one ancestral genome (Lagercrantz 1998; O'Neill and Bancroft 2000; Yang et al. 2006).

This hypothesis explains why these CR genes are dispersed and located on different chromosomes in *B. rapa*. Saito et al. (2006) suggested that the genomic region around *Crr3* exhibits homology to the top of the long arm of *Arabidopsis* chromosome 3, and possibly also to *CRk* (Sakamoto et al. 2008). It was concluded that *Crr3* has a different origin from that of *Crr1, Crr2*, and *CRb*. Gene *CRk* is independent of the CR genes *Crr1, Crr2, CRa*, and *CRb* but has a similar QTL region with *Crr3* (Sakamoto et al. 2008). Another novel CR locus, *CRc*, which is independent of all other CR loci, is located on chromosome R2 (Sakamoto et al. 2008). Fuchs and Sacristan (1996) mapped a CR locus (RPB1) in *Arabidopsis* chromosome 1. Their mapping for partial clubroot resistance identified two QTLs in chromosome 5 in F_2 and four QTLs in the RIL population, one each in chromosome 1 and 4 and two in the chromosome 5 of *A. thaliana* (Jubault et al. 2008). The CR QTL region in chromosome 5 colocalized with that containing several resistance gene clusters. These resistance genes could be candidates for clubroot disease resistance. Fine mapping and detailed analysis of the levels of expression possessed by these genes would help to identify specific ones capable of conferring clubroot resistance. The complete genome sequence of only one species of Brassicaceae family, *A. thaliana*, is available. It has been reported that gene sequence identity of *Brassica* and *Arabidopsis* varies from 75% to 90% (Quiros et al. 2001). Therefore, sequence information from the model species will greatly enhance cloning and characterization of CR genes in other *Brassica* species because the partial sequences of several accessions of *A. thaliana* showing degrees of response to the pathogen *P. brassicae* collected worldwide are available now (Fuchs and Sacristan 1996; Siemens et al. 2002; Nordborg et al. 2005; Alix et al. 2007). Only a few studies in transcriptomic/expression levels have been completed with respect to disease development. These include investigation of involvement of several metabolic pathways in disease pathogenesis, such as hormonal regulation by auxins (Grsic et al. 1999; Neuhaus et al. 2000), cytokinins (Devos and Prinsen 2006; Siemens et al. 2006), and trehalose synthesis (Brodmann et al. 2002). Grsic et al. (1999) observed that de novo indole-3-acetic acid (IAA) biosynthesis plays a role in symptom development during later stages of disease development; it was suggested that jasmonic acid, which increased during club development, may be involved in the upregulation of enzymes involved in IAA synthesis. Neuhaus et al. (2000) further supported this hypothesis by showing a delay in clubroot development after transforming *A. thaliana* with an antisense construct of Nitrilase 1 and 2, enzymes required for auxin biosynthesis. Siemens et al. (2006) investigated host gene expression during clubroot development in *A. thaliana* using an ATH1 microarray at two points in time, an early initial stage of infection and a later stage at which 60% of the host root cells were colonized.

More than 1000 genes were observed as being associated with the growth and cell cycle, sugar phosphate metabolism, and defense genes that were differentially expressed between infected versus control plants were observed. Upregulation of auxin biosynthesis genes such as nitrilases and members of the GH3 family and downregulation of cytokinin homeostasis were observed. It was further observed that lines overexpressing cytokinin oxidase/ hydrolases were resistant to clubroot, thereby strongly suggesting cytokinin is a key factor in clubroot development. Of the 312 genes identified as defense- and disease resistance-related genes, only 5 and 7% were upregulated at the first and the second points in time. This functional analysis of gene expression in clubroot-resistant and susceptible lines provided initial preliminary information. To strengthen further the understanding of genes and genetic networks involved in the mechanisms of clubroot disease and host interaction, more detail analyses at the transcriptomic level are needed (Piao et al. 2009).

2.5.6.4 *Brassica napus* var. *napobrassicae* (Rutabaga)

The resistance to *P. brassicae* pathotype 3 in Rutabaga-BF is controlled by a major gene located on chromosome A8, where the resistant phenotype exerts dominance effect over susceptibility. The genomic region carrying this gene also confers resistance to four other *P. brassicae* pathotypes (Hasan and Rahman 2016). Nine clubroot resistance loci have been identified and mapped on five chromosomes of the *Brassica* A-genome (Piao et al. 2004; Sakamoto et al. 2008; Chu et al. 2014). Five of these loci, namely *CRa*, *CRb*, *Crr3*, *CRk*, and *Rcr1*, are mapped on chromosome A3 (Sakamoto et al. 2008; Chu et al. 2014), while the loci *Crr2*, *CRc*, and *Crr4* are mapped on chromosome A1 (Suwabe et al. 2006), A2 (Sakamoto et al. 2008), and A6 (Suwabe et al. 2006), respectively. Suwabe et al. (2006) reported the locus *Crr1* and mapped this on chromosome A8 of *B. rapa*. This locus was found to be flanked by the SSR markers BRMS-088 and BRMS-173 and accounted for 26.8% of the total phenotypic variation for resistance to *P. brassicae* isolate Wakayama-01 (Suwabe et al. 2006). According to Suwabe et al. (2012), the genetic distance between these two flanked SSR markers was 3.9 cM; however, neither of these markers showed polymorphism between the resistant and susceptible parents (Rutabaga-BF and A07-29NI). Suwabe et al. (2012) and Hatakeyama et al. (2013) revealed that the genomic region carrying *Crr1* is comprised of two gene loci; the locus with major effect was named as *Crr1a* and the minor locus was named as *Crr1b*. They also reported that the gene *Crr1a* is composed of four open reading frames (ORFs) and showed similarity with a TIR-NBS-LRR-type R-gene.

NBS-LRR is the largest class of known plant disease resistance genes (R-genes), and TIR-NBS-LRR is one of the subclasses of this gene family (Akira and Hemmi 2003). Yu et al. (2014a, b) reported 157 and 206 NBS-LRR protein encoding genes in *B. oleracea* and *B. rapa*, respectively, and 167 in *A. thaliana*. NBS-LRR imparts plant immunity through hypersensitive response by monitoring the status of plant proteins targeted by pathogen virulence effector (Nandety et al. 2013). This type of resistance, imparted by NBS-LRR, is effective against obligate biotrophic or hemibiotrophic pathogens, but it is not effective against necrotrophs (Glazebrook 2005). Genes encoding TIR-NBS-LRR class proteins have been described to confer clubroot resistance in *B. rapa* (Ueno et al. 2012; Hatakeyama et al. 2013). Swiderski et al. (2009) reported that the TIR domain of the TIR-NBSLRR gene plays a signaling role in the induction of the defense response, and the LRR domain plays a role in recognition specificity (Collier and Moffett 2009). Hatakeyama et al. (2013) observed a complete susceptibility to clubroot disease in *B. rapa* owing to the lack of more than half of the TIR domain. Swiderski et al. (2009) also observed a complete loss of function of the TIR-NBS-LRR gene due to a mutation at the TIR domain (Hasan and Rahman 2016). Hasan and Rahman (2016) performed a BLASTn search of the cDNA of the *Crr1a* reported by Hatakeyama et al. (2013). This identified the two genomic regions 11,105,764 to 11,635,861 nt and 15,371,893 to 15,723,317 nt of the A8 chromosome of *B. rapa* cultivar Chiifu-401 sequence assembly version 1.5 (Cheng et al. 2011; www.brassicadb.org). Based on this, they designed a total of 78 SSR markers from the genomic region of 10,353,228 to 12,646,253 nt and 13 markers from the region of 13,604,267 to 16,005,042 nt, and screened these markers for polymorphism between the resistant and susceptible parents. However, only 12 markers from the region of 10,353,228 to 12,646,253 nt and 4 markers from the region of 13,604,267 to 16,005,042 nt were polymorphic between the parents and mapped on the linkage group A8. Composite interval mapping analysis indicated that the resistance locus in Rutabaga-BF is flanked by the SSR markers sS1702 and A08_5024. These two markers were found to be located at the genomic region of 10,692,602 to 11,617,968 nt on chromosome A8 of *B. rapa* cultivar Chiifu-401 assembly version 1.5 (Cheng et al. 2011). It was anticipated that the resistant parent Rutabaga-BF carries *Crr1a* and confers resistance to *P. brassicae* pathotypes 2, 3, 5, 6, and 8, or a cluster of genes conferring resistance to different pathotypes is present in this genomic region. Based on fine mapping of a 1.6 cM genetic region of A8, located between the markers BRMS-088 and BRMS-173, and comparison of this region with *A. thaliana* chromosome 4, Suwabe et al. (2012) found evidence that this region carries two clubroot resistance genes. The evaluation of the DH population for resistance to multiple pathotypes also provides evidence that a cluster of clubroot resistance genes might be present in chromosome A8. Clusters of other resistance genes, such as resistance to blackleg disease caused by *Leptosphaeria maculans*, can also be found in the *Brassica* genome as reported by Delourme et al. (2004) for four *Rlm* genes (*Rlm3*, *Rlm4*, *Rlm7*, and *Rlm9*) on chromosome A10. Clusters of disease resistance genes has also been reported in other plant species, such as lettuce (Meyers et al. 1998) and potato (Kuang et al. 2005; Hasan and Rahman 2016).

2.5.6.5 Mapping of Pathotype Specific R-Genes of *Brassica*

In *B. rapa*, several important CR genes conferring complete resistance in accessions against specific pathogen isolates have been identified. The mapping and cloning of the *CRb*/*CRa* loci took over 20 years. *CRa* was mapped and the candidate gene encodes a TIR-NBSLRR. Another locus, *CRb*, from the Chinese cabbage cultivar CR Shinki, was extensively mapped to a final 84 kb region. Kato et al. (2012) identified another CR resistance locus, CRb^{Kato}, in Akiriso Chinese cabbage. Hatakeyama et al. (2017) further determined that CRb^{Kato} and *CRa* was the same TIR-NB-LRR gene, whereas *CRb* might be a different locus. Another example is the *Crr1*–4 genes from turnip, which were initially primarily mapped using different molecular markers and populations. Through fine mapping, Hatakeyama et al. (2013) discovered that *Crr1* consists of two genes: *Crr1a* and *Crr1b*. The former encodes a TIR-NB-LRR and was functionally confirmed. With the development of genomic and molecular genetics, several loci were further identified using newly developed marker techniques. Yu et al. (2016) applied BSA-seq and identified a novel resistance gene, *Rcr1*, and two candidates encoding TIR-NB-LRRs. Huang et al. (2017) adopted KASP markers and BSR-seq strategies to finely map *Rcr2* to a 0.4 cM interval, identifying two TIRNBS-LRRs as candidates. Using BSA-seq, Peng et al. (2018) identified the new locus *CRd* in a 60 kb region on chromosome A03, which is located upstream of *Crr3*.

In *B. oleracea*, CR resistance appears to be determined by quantitative genes. Figdore et al. (1993) first identified three QTLs conferring resistance to *Pb* race 7 in broccoli. In the resistant kale line C10, Grandclément and Thomas (1996) performed QTL detection with RAPD markers and the results indicated at least two types of genetic mechanisms. Rocherieux et al. (2004) further found two to five QTLs depending on which of five pathotypes were used and *Pb-Bo1* was uncovered for all *Pb* isolates, accounting for 20.7–80.7% of the phenotypic variation. In another resistant kale line, K269, Moriguchi et al. (1999) and Nomura et al. (2005) detected two and three loci, respectively, conferring resistance to different isolates. In cabbage, Voorrips et al. (1997) first reported two major QTLs, *pb*-3 and *pb*-4. Nagaoka et al. (2010) identified a major QTL, *pbBo (Anju) 1*, from the cabbage accession Anju. Lee et al. (2016) employed the genotyping by sequencing (GBS) technique and a QTL survey to reveal two and one major loci for races 2 and 9, respectively. These loci showed positions close to the previously identified resistance loci in *B. oleracea* but in distinct locations from those discovered in *B. rapa*, indicating divergence of R loci between the *Brassica* A- and C-genomes.

For *B. napus*, a few loci conferring resistance to various isolates have been characterized. Landry et al. (1992) identified two QTLs controlling CR resistance to race 2, which contributed 58% and 15% of the observed phenotypic variation. Manzanares-Dauleux et al. (2000) reported the mapping of R loci in Darmor-bzh and identified one major gene, *Pb-Bn1*. Using a DH population, Werner et al. (2008) detected 19 QTLs that conferred resistance to 7 different isolates, but none of them could confer resistance to all these isolates. Fredua-Agyeman and Rahman (2016) mapped canola CR resistance to a DNA segment that comprised 12 markers linked to the *CRa* locus, indicating its possible A-genome origin. Hasan and Rahman (2016) used rutabaga-derived populations for resistance mapping and characterized a genomic segment on chromosome A8 conferring resistance to all five tested pathotypes. GWAS enables rapid detection of recombinants and variations using natural populations based on whole-genome SNP data. Li et al. (2016) first applied GWAS to 472 accessions to identify CR resistance with the 60 K *Brassica infinium* SNP. A total of nine loci were characterized through integrative analysis, with seven of them being novel and six of them being in the C-genome. The closely linked markers and resistance genes have been widely used in *Brassica* CR resistance breeding, generating a series of resistant cultivars that successfully control CR in many areas. For example, considering that high resistance is found mostly in *B. rapa*, researchers have frequently applied interspecies crossing to facilitate R-gene transfer combined with MAS and phenotype evaluation (Hirani et al. 2016; Liu et al. 2018; Lv et al. 2020).

2.5.6.6 Mapping of a QTL Using Ddrad-Seq in *Brassica Rapa* Against Clubroot

Both qualitative and quantitative patterns of clubroot disease resistance have been identified in *B. rapa*. Over the last two decades, a total of 14 different loci (*Crr1, Crr2, Crr3, Crr4, CRa, CRb, CRc, CRk, PbBa3.1, PbBa3.3, Rcr2, Rcr4, Rcr8,* and *Rcr9*) have been identified that are believed to govern clubroot resistance (CR) in *B. rapa*. The major gene-based CR is due to pathotype-specific reactions with *P. brassicae*, indicating that one or a few resistance genes control resistance to each isolate. Laila et al. (2019) used double digest restriction site-associated DNA sequencing (ddRAD-seq) in a comparatively robust F2 mapping population to identify genetic variations in nucleotide positions between the resistant and susceptible genotypes. Further, a high-resolution melting (HRM) analysis was conducted to validate the SNPs that had high LOD (logl0 of the likelihood ratio) scores according to QTL mapping. Yu et al. (2017) identified another clubroot resistance QTL, *Rcr9*, on chromosome A08 of *B. rapa* that provides resistance against a newly identified pathotype 5x. A PCR-based assay was conducted using a *Crr1* (Bra020861), also a candidate of *Rcr9*-specific marker to confirm that the presence of the *Crr1* locus was not associated with Seosan resistance and therefore that the newly identified source of resistance (*CRs*) must represent a novel locus. Finally, a separate PCR-based assay was conducted to show that another candidate of *Rcr9* gene (*Bra020936*)-specific marker was unable to differentiate between Seosan-isolate resistant and susceptible F2 lines.

Laila et al. (2019) investigated a quantitative trait locus (QTL) in *B. rapa* conferring resistance to a Korean *P. brassicae* pathotype isolate, Seosan. Crossed resistant and susceptible parental were analyzed for the segregation pattern in an F2 population of 348 lines. They identified and mapped a novel clubroot resistance QTL using the same mapping population that included susceptible Chinese cabbage and resistant turnip lines. Forty-five resistant and 45 susceptible F2 lines along with their parental lines were used for double digest restriction site-associated DNA sequencing (ddRAD-seq). High resolution melting (HRM)-based validation of SNP positions was conducted to confirm the novel locus. A 3:1 ratio was observed for resistant: susceptible genotypes, which is in accordance with Mendelian segregation. ddRAD-seq identified a new locus, *CRs*, on chromosome A08 that was different from the clubroot resistance (CR) locus, *Crr1*. HRM analysis validated SNP positions and constricted CRs region. Four out of 17 single-nucleotide polymorphism (SNP) positions were within a 0.8-Mb region that included three NBS-LRR candidate genes but not *Crr1* (Fig. 2.17). The newly identified CRs locus is a novel clubroot resistance locus, as the cultivar Akimeki bears the previously known *Crr1* locus but remains susceptible to the Seosan isolate. These results could be exploited to develop molecular markers to detect Seosan-resistant genotypes and develop resistant Chinese cabbage cultivars (Laila et al. 2019).

2.5.6.7 Genome-Wide Association to Identify CR Loci in *Brassica napus*

In order to select suitable model for association mapping, GWAS was performed using four general liner models (GLM, Q, K, and P) and two mixed liner models (QCK and PCK) for IF-DI (DI in IF), IF-IR (IR in IF), GH-DI (DI in GH), and GH-IR (IR in GH), respectively. As the Q-Q plot showed, the distribution of observed -log10 (p) from P + K model was the closest to the expected distribution for all four traits, which led to a low level of false-positive signals. Therefore, the association signals were identified with PCK model subsequently. To avoid ignoring the micro effective locus, significant associated SNP (sSNP, $P < 5.01 \times 10^{5}$) and potential association SNP (pSNP, $5.01 \times 10^{5} < P < 1.00 \times 10^{4}$) were introduced. If a confidence interval of a locus included one or more sSNPs, it signifies association

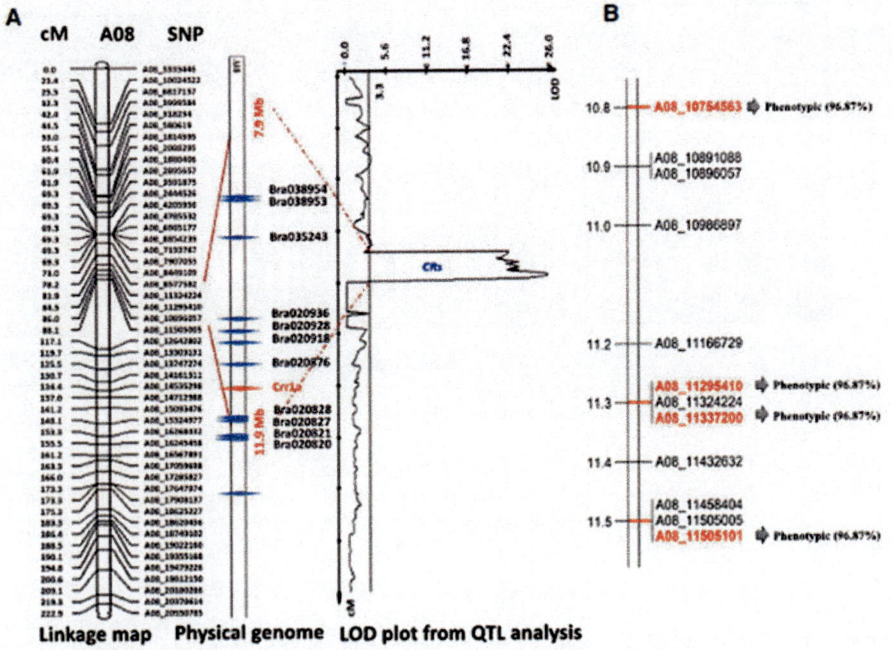

Fig. 2.17 The clubroot disease resistance linkage map identified in chromosome A08 from Chinese cabbage (*Brasssica rapa* var. *pekenensis*) (**a**). The mapped locus is resistant to the pathotype 4 isolate Seosan. The SNP haplotyping was done by ddRAD-seq using a mapping population containing 45 resistant and 45 susceptible F2 genotypes and 3 resistant and 3 susceptible parents. ddRAD-seq was accomplished using 2152 markers. QTL mapping was done using Windows QTL cartographer with 1000 permutations and a significance level of 0.05. Four HRM probes separated resistant and susceptible members of the F2 population (**b**) (Laila et al. 2019)

locus. Similarly, if a confidence interval of a locus did not have sSNP, and only included two or more pSNPs, is potential association locus. In total, seven significant association loci and two potential association loci were identified with these two indicators (Fig. 2.18),

For IF-DI, one significant association locus on BnC03 (named *MCR-C3*) was identified, with a peak SNP (highest significant) Bn-scaff_17521_1-p419499 which contributed to 4.81% of phenotypic variance. One potential association locus on BnC09 (named *MCR-C9*) was detected, with a peak SNP Bn-scaff_15576_1-p660538 which explained 4.21% of phenotypic variance (Table 2.12). For IF-IR, one significant association locus on BnA04 (named *MCR-A04*) was mapped, with a peak SNP Bn-A04-p16156157 that contributed 4.72% to phenotypic variance. It was noteworthy that *MCR-C3* in IF-DI was also detected in IF-IR. For GH-DI, one significant association locus *SCR-C6* was detected on BnC06 with a peak SNP Bn-scaff_16064_1-p26703, which explained 5.26% of phenotypic variance. Another significant association locus *SCR-A10a* was identified on BnA10 with a

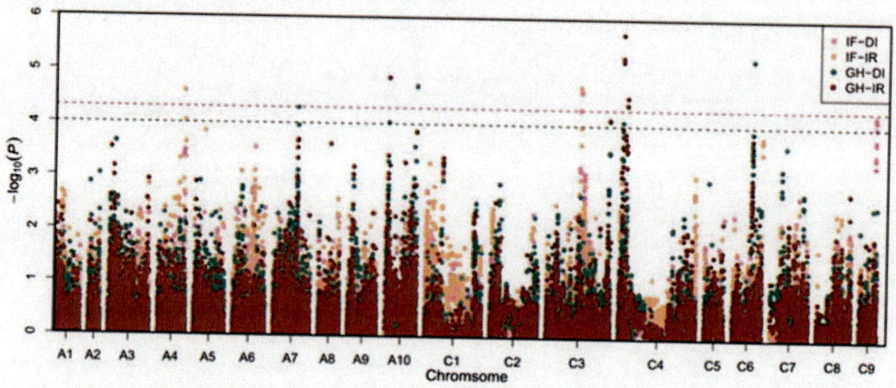

Fig. 2.18 Manhattan plots of association analysis using the mixed liner model (MLM) model P + K. The pink plots, yellow plots, green plots, and brown plots represent the associated signals for infected field-disease index (IF-DI), infected field-incidence rate (IF-IR), greenhouse-disease index (GH-DI), and greenhouse-incidence rate (GH-IR), respectively. The black- and red-dashed horizontal lines depict the two significant thresholds that are ($\log_{10}1/19{,}945 = 4.30$) and ($\log_{10}2/19{,}945 = 4.00$) (Li et al. 2016)

Table 2.12 The identified QTL information through GWAS analysis in the natural population of *Brassica napus* (Li et al. 2016)

Trait	Locus[a]	Chr	SNP No.[b]	MSS-R^2 (%)[c]	Position range (bp)
IF-DI	MCR-C3	C03	4	4.81	21,372,298–21,810,339
	MCR-C9	C09	0 (8)	4.21	41,753,926–41,967,651
IF-IR	MCR-A4	A04	1 (1)	4.96	16,356,482–16,733,484
	MCR-C3	C03	1	4.88	21,372,298
GH-DI	SCR-A10a	A10	1	4.96	15,476,654
	SCR-C6	C06	1	5.9	25,596,721
GH-IR	SCR-A10b	A10	1	5.19	885,083
	SCR-C3	C03	0 (6)	4.34	58,088,063–58,097,249
	SCR-C4a	C04	4	6.51	2,498,886–2,511,207
	SCR-C4b	C04	8	4.79	8,065,329–8,102,210

[a]QTL name identified in this study; MCR: QTL identified in infected field; SCR: QTL identified in green house

[b]The number out of bracket depict the no. of significant associated SNP and the number in bracket represent the no. of potential associated SNP

[c]Percentage of phenotypic variance explained by that of peak SNP marker

peak SNPBn-A10-p16087066 that accounted for 4.71% of phenotypic variance. For GH-IR, three significant association loci were identified in total.

Two loci were both on BnC04 (named *SCR-C4a* and *SCR-C4b*), which explained 5.72 and 4.54% of phenotypic variance, respectively. Their stone named *SCR-A10b* was on BnA10, with a peak SNP Bn-A10-p3966740 that explained 4.87% of phenotypic variance. In addition, one potential association locus *SCR-C3* was mapped on BnC03 with a peak SNPBn-scaff_18559_1-p166394 which accounted for 4.12% of phenotypic variance (Table 2.1).

2.5.7 Environ Effects on CR Genes

To understand the effects of allelic variations on CR in rapeseed, the combined effects of DI of different QTLs were analyzed by Li et al. (2016) in IF and GH, respectively (Fig. 2.19). The genotype and phenotypic variance (*R*2) of each QTL was substituted by that of each peak marker of the corresponding QTL. On this basis, 472 accessions were grouped into three classes which contained 1, 2, and 3 favorable alleles in IF, respectively (Fig. 2.19a). Because of too less accession contained in one group if all accessions were grouped by containing 1–6 favorable alleles in GH, all these six loci were merged and divided into three groups which contained 1–2, 3–4, and 5–6 favorable alleles to statistical analyze the DI in GH (Fig. 2.19b). The results indicated that the more favorable alleles (resistant alleles) had, the lower DI (more

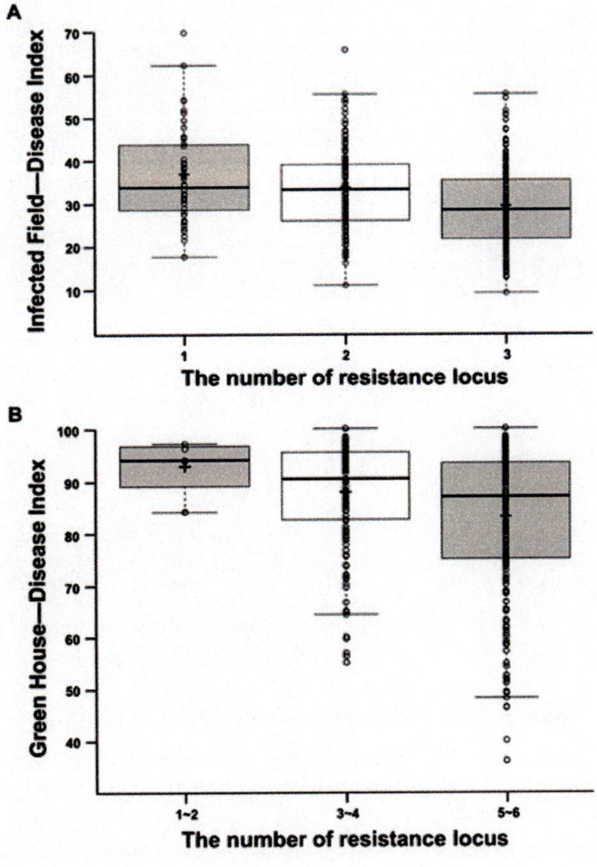

Fig. 2.19 (Box plot for DI in two environments). (**a**) Box plots for DI in IF. (**b**) Box plots for DI in GH. The middle line indicates the median, the plus sign shows the mean, the box represents the range of the 25th to 75th percentiles of the total data, the whiskers show the interquartile range, the two-line segments outside of the box indicates the boundary of normal value, and the out dots means the outliers (Li et al. 2016)

resistance to *P. brassicae*) was demonstrated (Fig. 2.19). In IF, the DI of the accession shield three resistant alleles can be reduced by 20.62% compared with that of holding one resistant allele. Similarly, the DI of the accessions held 5–6 resistant alleles can be reduced by 10.56% compared with that of only holding 1–2 resistant alleles in GH. The above results revealed that the CR was mainly controlled by additive effect. Therefore, the CR for *B. napus* can be improved by polymerization of all identified resistant alleles.

2.5.7.1 *Brassica rapa*

A resistance gene designated as Rcr3 was mapped initially based on the percentage of polymorphic variants using bulked segregant RNA sequencing (BSR-Seq) and further mapped using Kompetitive allele-specific PCR. DNA variants were identified by assembling short reads against a reference genome of *B. rapa*. *Rcr3* was mapped into chromosome A08. It was flanked by single-nucleotide polymorphism (SNP) markers (A90_A08_SNP_M12 and M16) between 10.00 and 10.23 Mb, in an interval of 231.6 Kb. There were 32 genes in the *Rcr3* interval. Three genes (*Bra020951, Bra020974*, and *Bra020979*) were annotated with disease resistance mechanisms, which are potential candidates for Rcr3. Another resistance gene, designated as $Rcr9^{wa}$, for resistance to pathotype 5X was mapped, with the flanking markers (A90_A08_SNP_M28 and M79) between 10.85 and 11.17 Mb using the SNP sites identified through BSR-Seq for Rcr3.

There were 44 genes in the $Rcr9^{wa}$ interval, three of which (*Bra020827, Bra020828, Bra020814*) were annotated as immune system process-related genes, which are potential candidates for $Rcr9^{wa}$ (Karim et al. 2020).

2.5.8 The Novel Loci Detected by GWAS

To compare the GWAS results with the previous reported CRQTLs/genes, these quinces of CR genes and markers of QTLs related to CR were collected in *B. napus*, *B. rapa*, and *B. oleracea*, and then the blastn or e-PCR with a threshold of 1E–10 was performed to *B. napus* reference genome (*Darmor-bzh*) to search their homoeologous regions in *B. napus* (Fig. 2.20). In *B. rapa*, two homoeologous regions of *CRa* were identified on BnA03 (22,864,716–22,877,171 bp) and BnC07 (38,863,628–38,883,622 bp), respectively (Fig. 2.20). Three homoeologous regions of *Crr1* were found on BnA08 (9,456,084–9,467,947 bp), BnC03 (54,084,701–54,109,095), and BnC07 (38,863,628–38,869,983 bp), respectively (Fig. 2.20). One homoeologous region of *CRb* was identified on BnA03 (21,682,961–22,238,961 bp). Two homoeologous regions of *Rcr1* were found on BnA03 (22,797,049–23,009,129 bp) and BnC07 (36,229,328–39,036,510 bp) in *B. napus*. Homoeologous regions of other QTLs in *B. rapa* were not identified, because of their limited marker information or the physical positions of forward and reverse markers linked with the QTLs were not on the same chromosome in *B. napus* genome. Because of the limited reports on CR studies in *B. oleracea*, only three QTLs, *CRQTL-YC*, *CRQTL-GN_1*, and *CRQTL-GN_2*, were searched. Two

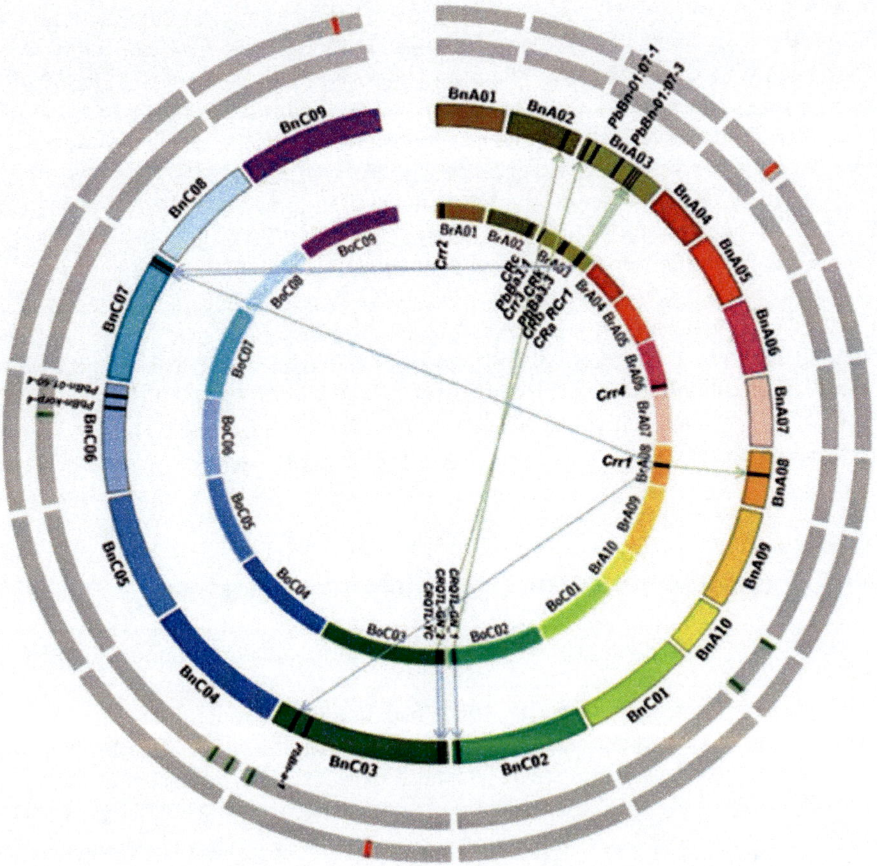

Fig. 2.20 Integrative physical map of clubroot resistance (CR) sites based on the mapping results of the *Brassica rapa*, *Brassica oleracea,* and *Brassica napus***.** The right-hand side of the innermost circle depicts the 10 chromosomes of *B. rapa*; the black squares on that represent the CR loci identified in *B. rapa*. The left-hand side of the innermost circle depicts the nine chromosomes of *B. oleracea*; the black squares on that represent the CR loci identified in *B. oleracea*. The second circle (from the innermost one) depicts the 19 chromosomes of *B. napus*; the black squares on that represent the CR loci which were identified in previous reports in *B. napus* (without arrow) or the homoeologous regions (with arrow) of the black squares on the innermost circle. The green squares on the third circle represent the CR loci identified in greenhouse in this study. The red squares on the outermost circle represent the CR loci identified in infected field (Li et al. 2016)

homoeologous regions of *CRQTL-YC* were identified on BnA03 and BnC03, respectively. Also, two homoeologous regions of *CRQTL-GN_1* were detected on BnA02 (22,576,577–22,970,418 bp) and BnC02 (42,737,365–43,546,906 bp), and one homoeologous region of *CRQTL-GN_2* was mapped on BnC03 (1,185,066–2,835,468 bp; Fig. 2.20). Moreover, it was attempted to ensure the QTL physical position in *B. napus* according to the providing markers information in previous reports. Marker sequences of five QTLs reported on previous study in *B. napus* were searched and the physical position of *PbBn-01:07–1* and *PbBn-korp-4* were confirmed successfully. The physical positions of these two QTLs, BnA03 (5,386,730–5,386,845 bp) and BnC06 (27,067,129–27, 067, 246 bp), were proved to be in accordance with the previous reports. However, the physical positions of other three QTLs were not in accordance with the previous studies. The QTL *PbBn-a-1* located on BnA08 was mapped on BnC03, and the *PbBn-01:07–3* located on BnC03 was mapped on BnA03 by Li et al. 2016 (Fig. 2.20). It was found that the three QTLs identified in IF study were all novel, in which regions there were no homoeologous regions of CR QTLs reported in previous studies (Fig. 2.20). The QTL *SCR-C3* identified in GH was not located on the confidence intervals of one homoeolog of *Crr1*, but was close to it. It was as an old CR locus. The same situation was on *SCR-C6*, which closed to the physical position of *PbBn-korp-4*. The other four QTLs identified were novel. Therefore, the GWAS had stronger ability to detect new loci on a given trait.

2.5.9 Prediction of CR QTLs by Bioinformatic Analyses

To predict the candidate genes of CR QTLs, the candidate region (confidence interval) for each QTL was confirmed. There were 542 predicted genes in all 9-candidate regions, 471 genes of which had functional annotations in total. To gain insights into the functionality of above genes, Li et al. (2016) performed GO enrichment analysis using Blast2GO (Conesa et al. 2005). The result showed that eight pathways of the molecular function related to ADP binding and methylthiopropyl-desulfoglucosinolate sulfotransferase activity were enriched, the pathways of which were highly significant with TIR-NBS-LRR genes. It was interesting that two TIR-NBS gene clusters held 28 TIR-NBS genes which located in *SCR-C6* (10 TIR-NBS genes) and *MCR-C9* (18 TIR-NBS genes; Table 2.13) were participated in the function of ADP binding. The results indicated that the TIR-NBS gene family may be associated with CR. In order to get more evidence to predict candidate genes, the homoeologous genes of DEGs from the transcriptome data after inoculation *P. brassicae* in *B. rapa* were identified in *napus* genome (Fig. 2.21). There were five SNPs involved in the *MCR-C3* LD block; the candidate gene region was 21.72–21.92 Mb (194.6 Kb) in BnC03, where 44 genes were included, and four of which were the homoeologous genes of four DEGs identified in *B. rapa* (Fig. 2.21a). Seven SNPs were also involved in the LD block for *SCR-C3*, in which the candidate region was 57.86–58.10 Mb (339.3 Kb) in BnC03, and 16 genes were predicted in this region. Only one homoeologous gene *BnaC03g68270D* of DEGs corresponding in *B. rapa* was searched (Fig. 2.21a). Similarly, 9, 1, 2, 1, 6, 2, and 4 candidate genes obtained from the *B. rapa* homoeologous genes were identified in the candidate regions of *MCR-A4*, *SCR-A10a*, *SCR-A10b*, *SCR-C4a*, *SCR-C4b*, *SCR-C6*, and *MCR-C9*, respectively (Fig. 2.21). Overall, the 30 homoeologous genes of DEGs from *B. rapa* were likely

Table 2.13 The details of the TIR-NBS gene clusters in the candidate gene regions of *SCR-C6* and *MCR-C9* (Li et al. 2016)

Locus	Candidate gene region (Mb)	TIR-NBS gene cluster	Description
SCR-C6	25.09–26.22	BnaC06g23900D	Disease resistance protein (TIR-NBS class)
		BnaC06g23910D	Disease resistance protein (TIR-NBS class)
		BnaC06g23920D	Disease resistance protein (TIR-NBS class)
		BnaC06g23930D	Disease resistance protein (TIR-NBS class)
		BnaC06g23940D	Disease resistance protein (TIR-NBS-LRR class)
		BnaC06g23950D	Disease resistance protein (TIR-NBS-LRR class) family
		BnaC06g23970D	Disease resistance protein (TIR-NBS class)
		BnaC06g23980D	Disease resistance protein (TIR-NBS-LRR class) family
		BnaC06g24000D	Disease resistance protein (TIR-NBS class)
		BnaC06g24010D	Disease resistance protein (TIR-NBS-LRR class)
MCR-C9	41.72–42.80	BnaC09g39420D	Disease resistance protein (TIR-NBS-LRR class), putative
		BnaC09g39430D	Disease resistance protein (TIR-NBS-LRR class) family
		BnaC09g39440D	Disease resistance protein (TIR-NBS-LRR class) family
		BnaC09g39450D	Disease resistance protein (TIR-NBS-LRR class) family
		BnaC09g39460D	Disease resistance protein (TIR-NBS-LRR class) family
		BnaC09g39470D	Disease resistance protein (TIR-NBS-LRR class) family
		BnaC09g39490D	Disease resistance protein (TIR-NBS-LRR class) family
		BnaC09g39500D	Disease resistance protein (TIR-NBS-LRR class) family
		BnaC09g39520D	Disease resistance protein (TIR-NBS-LRR class) family
		BnaC09g39560D	Disease resistance protein (TIR-NBS-LRR class) family
		BnaC09g39570D	Disease resistance protein (TIR-NBS-LRR class) family
		BnaC09g39590D	Disease resistance protein (TIR-NBS-LRR class) family

(continued)

Table 2.13 (continued)

Locus	Candidate gene region (Mb)	TIR-NBS gene cluster	Description
		BnaC09g39630D	Disease resistance protein (TIR-NBS-LRR class) family
		BnaC09g39890D	Disease resistance protein (TIR-NBS-LRR class) family
		BnaC09g39900D	Disease resistance protein (TIR-NBS-LRR class) family
		BnaC09g40030D	Disease resistance protein (TIR-NBS-LRR class)
		BnaC09g40060D	Disease resistance protein (TIR-NBS-LRR class)
		BnaC09g40250D	Disease resistance protein (TIR-NBS-LRR class), putative

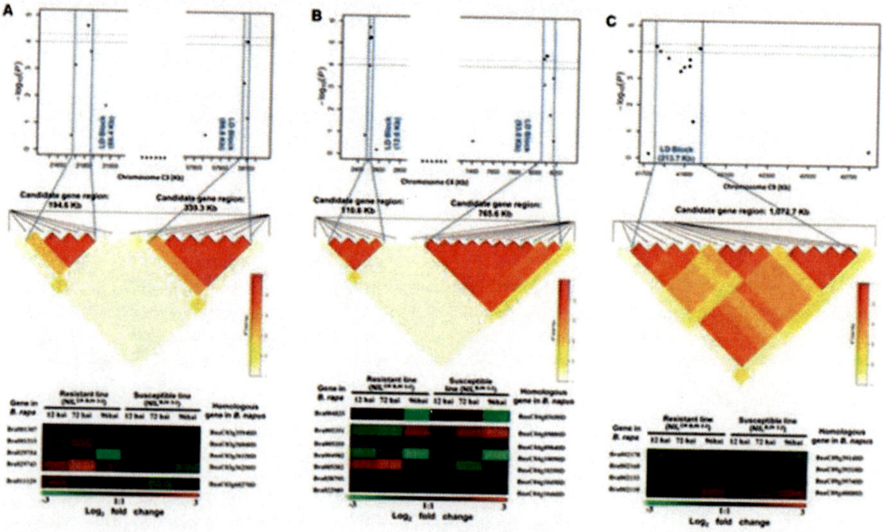

Fig. 2.21 The candidate regions and predicting candidate genes for part of quantitative trait loci (QTLs) identified in this study. (**a**) The candidate regions and predicting candidate genes for MCR-C3 and SCR-C3. (**b**) The candidate regions and predicting candidate genes for SCR-C4a and SCR-C4b. (**c**) The candidate regions and predicting candidate genes for MCR-C9. Haplotype block in strong LD ($R^2 > 0.4$) with the most significant associated SNPs are shown between the blue dashed line. The chromosome region between the two flanking markers of the LD block is defined as candidate gene region for each QTL. Heat maps of the differentially expressed genes (DEGs) from transcriptome data of *Brassica rapa*, which were performed by Chen et al. (2015); the shade of color represents the $\log_2$ fold changes (inoculated/mock-inoculated) of the DEGs from *B. rapa*. The genes at the left of heat maps are the DEGs in *B. rapa*, and the genes at the right of heat maps are the homoeologous genes of DEGs from *B. rapa* in *B. napus* (Li et al. 2016)

candidate genes as these 9 QTLs. However, the more evidence needs to be obtained by functional analysis of these genes.

2.5.10 Linkage Markers of Clubroot Resistance in *Brassica*

The loss of resistance genes during breeding process is a very common phenomenon. For the determination of such genes, Kuginuki et al. (1997) observed linkage markers for a CR locus in *Brassica rapa*. Two CR loci, *Crr1* and *Crr2*, were identified in a doubled-haploid (DH) line, G004 (Suwabe et al. 2003). The former locus corresponds to Kuginuki's CR locus because the linkage marker RA12–75 is tightly linked to BRMS-088, a marker for *Crr1*. These are based on CR traits derived from a European fodder turnip, Siloga. The study conducted by Yoshikawa (1993) on the CR turnips, Milan White, was found to have polygenic resistance. To elucidate the genetics of CR of this turnip, Hirai et al. (2004) analyzed an inbred line having the resistant trait from Milan White. An inbred turnip (*Brassica rapa* syn. *campestris*) line, N-WMR-3, which carries the trait of clubroot resistance (CR) from a European turnip, Milan White, was crossed with a clubroot-susceptible doubled haploid line, A9709. A segregating F3 population was obtained by single-seed descent of F2 plants and used for a genetic analysis. Segregation of CR in the F3 population suggested that CR is controlled by a major gene. Two RAPD markers, OPC11–1 and OPC11–2, were obtained as candidates of linkage markers by bulked segregant analysis.

These were converted to sequence-tagged site markers, by cloning and sequencing of the polymorphic bands, and named OPC11–1S and OPC11–2S, respectively. The specific primer pairs for OPC11–1S amplified a clear dominant band, while the primer pairs for OPC11–2S resulted in codominant bands. Frequency distributions and statistical analyses indicate the presence of a major dominant CR gene linked to these two markers. The present marker for CR was independent of the previously found CR loci, *Crr1* and *Crr2*. Genotypic distribution and statistical analyses did not show any evidence of CR alleles on *Crr1* and *Crr2* loci in N-WMR-3. *B. rapa* has at least three CR loci. Therefore, the new CR locus was named *Crr3*. The present locus may be useful in breeding CR Chinese cabbage cultivars to overcome the decay of present CR cultivars (Hirai et al. 2004).

2.5.11 Marker-Assisted Selection of Clubroot Resistance Genes

To ensure stable production, more than one resistance genes need to be accumulated into a single CR cultivar. Early studies on clubroot resistance found that at least three major resistance genes existed in turnips, and that cultivars with such qualitative CR genes showed highly stable resistance (Crute et al. 1980; Toxopeus and Janssen 1975; Wit and van de Weg 1964). These genes, which are probably contained in source materials of CR Chinese cabbage, differ in their protective action. Genetic accumulation can bring elevated resistance corresponding to the wide pathogenicity of clubroot fungus. Marker-assisted selection (MAS) is indispensable in CR breeding because it allows for the acquisition of higher resistance by combining different resistance genes (Hirai 2006). DNA markers for clubroot resistance have been developed in major *Brassica* crops. Seven major CR loci, *CRa, CRb, CRk, CRc, Crr1, Crr2*, and *Crr3*, have been identified for Chinese cabbage using various types of markers (Hirai et al. 2004; Matsumoto et al. 1998; Piao et al. 2004; Sakamoto et al. 2008; Suwabe et al. 2003), though mutual relations among them are somewhat ambiguous, because different resistant materials and pathogenic isolates are employed for detecting the loci.

The complexity of the clubroot pathogen in particular makes an exact evaluation of resistance genes difficult. Matsumoto et al. (2012) attempted to accumulate three CR genes, *CRa, CRk*, and *CRc*, through marker-assisted selection. Five doubled haploid CR lines with an individual CR locus were used as breeding materials. The CR lines were crossed with each other. A subsequent selection for resistance was performed using sequence characterized amplified region markers in segregating generations. As a result, four homozygous lines for three resistance genes and the F_1 hybrids between them were developed. CR pyramiding lines were inoculated with six field isolates of *P. brassicae*. The homozygous lines for three CR genes, whether selfed or crossed, exhibited exceedingly high resistance against all of the isolates. Morphological characters of F_1 hybrids were comparable to those of a control cultivar, but the degree of heterosis was less than expected, which is probably because of genetic similarity of the parents. The results prove that clubroot resistance can be reinforced through the accumulation of varied resistance genes in *B. rapa*.

2.5.12 R-Gene Hot-spots in *Brassica*

The CR QTL mapping results in *B. rapa* showed that 6 of 11 CR genes/QTLs were on BrA03, which indicated that there were CR hotspots existed in *B. rapa*. Now, the genome sequences of *B. rapa*, *B. oleracea*, and *B. napus* have also been sequenced which will accelerate their genetic improvements. All reported QTLs/loci information (linkage markers, candidate genes, and positions information) related to CR were collected in these three species. Most of the reported CR QTLs/genes were mapped on *B. napus* genome through bioinformatic analyses. The integration results illustrated the CR hotspots also existed in *B. napus*, which were the regions of BnA03 (21–23 Mb), BnC03 (1–2 Mb), and BnC07 (36–39 Mb). It was interesting that the top of BnA03 was homoeologous with the top of BnC03, and the bottom of BnA03 was homoeologous with the bottom of BnC07 (Chalhoub et al. 2014). Therefore, the evolution and origin of BnA03, and the genes related to disease-resistance on BnA03, are worth to deep study. These hotspots may belong to rare variation sites, which were difficult to detect in natural population, or they were peculiar in *B. rapa* (Li et al. 2016).

2.5.13 The Molecular Regulation of R-Genes

There were more than 70 R-genes which have been cloned from plants (Liu et al. 2007). Most of the cloned R-genes were NBS-LRR family (Belkhadir et al. 2004; McHale et al. 2006). It was interesting that two R-genes-controlled CR in *B. rapa* were all TIR-NBS-LRR genes, which encoded large and abundant proteins involved in the detection of diverse pathogens. Similarly, the enrichment analysis also found eight pathways which were highly significant with TIR-NBS-LRR genes, and there were two TIR-NBS gene clusters in candidate regions (Table 2.13). NBS-LRR proteins could recognize the specialized pathogen effectors of ETI (also called avirulence proteins). Therefore, TIR-NBS-LRR gene family is most likely to play an important role in the process of the clubroot disease, which needs a deep study in *Brassica*. Besides that, plant hormones, mainly salicylic acid (SA), jasmonic acid (JA), auxins, and cytokinins, also played a role in compatible interactions between *Arabidopsis* and *P. brassicae* (Siemens et al. 2006; Lemarié et al. 2015). It also enlightens on predicting candidate genes of above pathways and understanding the resistant molecular mechanism. The 30 candidate genes identified through the data from RNA-seq in *B. rapa* just only provided clues for candidate genes prediction and understanding the molecular regulation of CR in *B. napus*. However, more experiments need to be carried out to obtain more evidence to identify or evidence these candidate genes. For example, the RNA-seq analysis can be carried out by using of some CR and susceptible lines from the natural population (Li et al. 2016).

2.5.14 Genetics of R-Genes

The genetic basis of clubroot resistance (CR) has been determined in economically important cruciferous species: *Brassica rapa, B. oleracea*, *B. napus*, *Raphanus sativus*, and *A. thaliana*. In *B. rapa*, mainly in Chinese cabbage, one minor and seven major CR genes introduced from European fodder turnips have been identified. Mapping of these CR genes localized *Crr1* on R8, *Crr2* on R1, *CRc* on R2, and *Crr4* on R6 linkage groups of Chinese cabbage. Genes *Crr3, CRa, CRb*, and *CRk* mapped to R3, but at two separate loci, *CRa* and *CRb*, are independent of *Crr3* and *CRk*, which are closely linked. Further, analysis suggested that *Crr1, Crr2*, and *CRb* have similar origins in the ancestral genome as in chromosome 4 of *A. thaliana*. Genetic analysis of clubroot resistance genes in *B. oleracea* suggests that these are quantitative traits. Twenty-two quantitative trait loci (QTLs) have been mapped in different linkage groups of *B. oleracea*. In *B. napus*, genetic analysis of clubroot resistance was found to be governed by one or two dominant genes. The quantitative analysis approach, however, proved that these are polygenic. In total, at least 19 QTLs have been detected on 8 chromosomes of *B. napus*, N02, N03, N08, N09, N13, N15, N16, and N19. The chromosomal location of the other six QTLs is not clear. Cloning of these QTLs or resistance loci is possible. Progress in genomics, particularly the techniques of comparative mapping and genome sequencing, supplements with cloning and allows improved characterization of CR genes.

Further, the development of DNA markers linked to CR genes will in turn hasten the breeding of clubroot-resistant crucifer's cultivars.

2.5.15 The Genetic Origin of Clubroot Resistance

In the past decades, genetic maps of *Brassica* have been constructed by means of RFLPs, AFLPs, and RAPDs to better understand their genetic makeup. These studies have contributed to analyses of complicated quantitative traits and to comparisons of the organization of the chromosomes of *Brassica*. Since the DNA sequences of homologous genes in the related taxa are quite similar, clones have been extensively used as RFLP markers to elucidate the genomic colinearity between species (Osborn and Lukens 2003). The degree of genomic conservation has been elucidated by comparative linkage mapping, and conserved genomic segments have been identified by micro- and macrosynteny studies in various *Brassica* species (Cavell et al. 1998; Lagercrantz 1998; O'Neill and Bancroft 2000; Ryder et al. 2001). However, because of the high rate of genomic segment replication, RFLP has been shown at times to detect more than one locus in *Brassica* genomes (Lagercrantz and Lydiate 1996). Although duplication/triplication of genomic segments permits the evolution of genes, it also leads to considerable complexity in evaluating genomic colinearity within species. For these reasons, a linkage map based on more precise molecular markers that would allow discrimination between homologous and homeologous regions is required for an accurate comparative analysis in *Brassica* (Suwabe et al. 2006).

The single sequence repeat (SSR)-based linkage map was constructed in *B. rapa* by Suwabe et al. (2006). It includes 113 SSR, 87 RFLP, and 62 RAPD markers. It consists of 10 linkage groups with a total distance of 1005.5 cM and an average distance of 3.7 cM. SSRs are distributed throughout the linkage groups at an average of 8.7 cM. Synteny between *B. rapa* and a model plant, *A. thaliana*, was analyzed. A number of small genomic segments of *A. thaliana* were scattered throughout an entire *B. rapa* linkage map. This points out the complex genomic rearrangements during the course of evolution in cruciferae. A 282.5-cM region in the *B. rapa* map was in synteny with *A. thaliana*. Of the three QTL (*Crr1, Crr2*, and *Crr4*) for clubroot resistance identified, synteny analysis revealed that two major QTL regions, *Crr1* and *Crr2*, overlapped in a small region of *Arabidopsis* chromosome 4. This region belongs to one of the disease-resistance gene clusters (MRCs) in the *A. thaliana* genome. These results suggest that the resistance genes for clubroot originated from a member of the MRCs in a common ancestral genome and subsequently were distributed to the different regions they now inhabit in the process of evolution (Fig. 2.23; Suwabe et al. 2006).

2.5.16 Quantitative Resistance to Clubroot Mediated by Transgenerational Epigenetic Variation in *Arabidopsis*

Numerous studies (Aoun et al. 2017; Zheng et al. 2017) have reported that plant responses to abiotic (temperature, drought) and biotic stresses could be associated with epigenetic variation in addition to nucleotidic variation. It has been shown that *Arabidopsis* mutants, altered in the maintenance of DNA methylation in the CG, CHG, and CHH contexts, showed strong resistance to *P. syringae* pv. tomato strain DC3000. The role of epigenetics in the expression of adaptive plant traits thus suggests that epigenetic variability could be used for generating stress-tolerant or resistant plants. Furthermore, the occurrence of natural DNA methylation variants (epialleles) in plants and their implication in evolution suggest that epialleles could be considered as a source of variability in plant breeding. In this "epigenetic breeding" approach (Gallusci et al. 2017), two conditions are needed: transgenerational inheritance of epialleles and a clear connection between epigenotype and observed phenotype. Previous studies demonstrated that epialleles could be stably transmitted across at least eight generations (Johannes et al. 2009; Teixeira et al. 2009), and that such heritable differences in DNA methylation could be associated with heritable phenotypic variation for several complex traits (Johannes et al. 2009; Reinders et al. 2009). However, linking heritable phenotypic variation to *epigenetic* variation remains challenging because of the difficulty in teasing apart its effects to that of DNA sequence variation in natural settings (Johannes et al. 2008; Quadrana and Colot 2016).

This problem can, however, be greatly alleviated in *Arabidopsis* by using experimental populations of so-called epigenetic recombinant inbred lines (epiRIL), which show extensive epigenetic variation but limited DNA sequence variation. One such population was indeed used to build a genetic map based solely on heritable differences in DNA methylation (differentially methylated regions, DMR) (Colome-Tatche et al. 2012) and to identify epigenetic QTL (QTLepi) for several complex traits (Cortijo et al. 2014; Kooke et al. 2015; Aller et al. 2018).

Liégard et al. (2019) used this same epiRIL population to determine the impact of heritable differences in DNA methylation on the response of *Arabidopsis* to clubroot. It was found that *ddm1* mutants were less susceptible to clubroot than the wild-type Col-0 and that the assessed subset of 123 epiRIL displayed a wide range of continuous phenotypic responses to clubroot. Twenty QTLepi were detected across the five chromosomes, with a bona fide epigenetic origin for 16 of them. It was demonstrated that heritable differences in DNA methylation also could contribute to quantitative resistance to clubroot. Six QTLepi colocalized with previously identified clubroot resistance genes and QTL in *Arabidopsis*, revealing that quantitative resistance to clubroot in natural accessions could be controlled by both nucleotidic and epigenetic variations (Liégard et al. 2019). Quantitative disease resistance, often influenced by environmental factors, is thought to be the result of DNA sequence variants segregating at multiple loci. However, heritable differences in DNA methylation, so-called transgenerational epigenetic variants, also could contribute to quantitative traits. Liégard et al. (2019) used the epigenetic

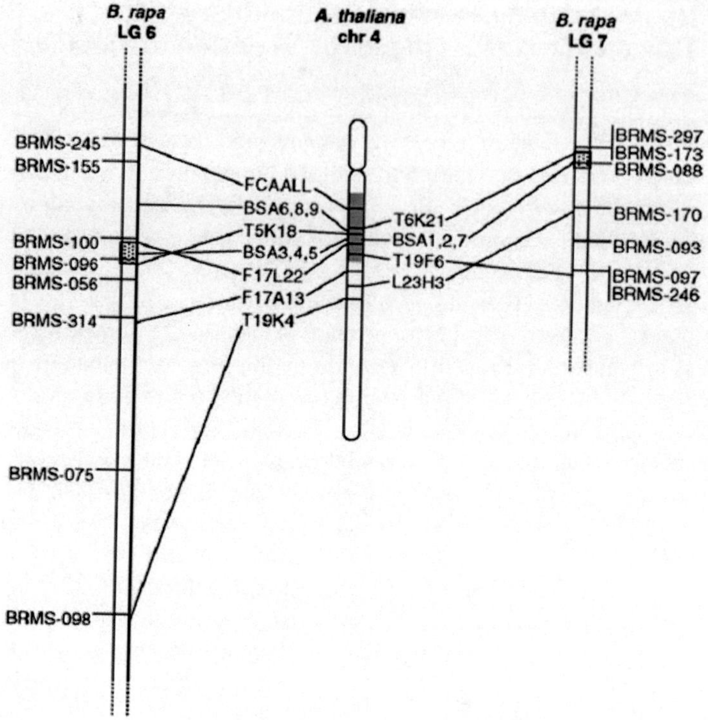

Fig. 2.22 Synteny map of the *Crr1* and *Crr2* regions (stippled area) with Brassica *rapa* and *A. thaliana*. The linkage groups (LG) of *B. rapa* and the chromosome (chr) of *A. thaliana* are presented as vertical bars. The *Arabidopsis* BAC clones that correspond to the SSR loci of *B. rapa* are connected by lines. The shaded region indicates the cluster of disease resistance genes (MRC-H) in the *Arabidopsis* chromosome (Suwabe et al. 2006)

recombinant inbred lines (epiRIL) derived from the cross ddm1–2 9 Col-0, which show extensive epigenetic variation but limited DNA sequence variation (Fig. 2.22).

Quantitative loci under epigenetic control (QTLepi) mapping were carried out on 123 epiRIL infected with *P. brassicae* and using various disease-related traits. EpiRIL displayed a wide range of continuous phenotypic responses. Twenty QTLepi were detected across the five chromosomes, with a bona fide epigenetic origin for 16 of them. The effect of five QTLepi was dependent on temperature conditions. Six QTLepi colocalized with previously identified clubroot resistance genes and QTL in *Arabidopsis*. Colocalization of clubroot resistance QTLepi with previously detected DNA-based QTL reveals a complex model in which a combination of allelic and epiallelic variations interacts with the environment to lead to variation in clubroot quantitative resistance (Liégard et al. 2019).

2.5.17 Identification of QTLs for Clubroot Resistance with the Use of *Brassica* SNP Microarray

Increasing clubroot resistance (CR) of *B. oleracea* by ascertaining the molecular mechanisms has been the key focus in modern *B. oleracea* breeding. In order to identify the quantitative trait loci (QTLs) associated with CR in *B. oleracea*, 94 F_2 vegetative lines which were developed by tissue culture of selfed seeds from the F_1 generation between a clubroot-resistant *B. oleracea* inbred line and a susceptible line were identified for disease incidence and six CR-associated traits under a lab inoculation by Plasmodiophora *brassicae* were genotyped with the 60 K *Brassica* SNP array. Significant correlations were detected for numbers of fibrous roots and *P. brassicae* content in roots with disease incidence. Nine linkage groups were constructed from 565 bins, which covered around 3000 SNPs, spanning 1028 cM of the *B. oleracea* genome with an average distance of 1.82 cM between adjacent bins. A total of 23 QTLs were identified for disease incidence and the other two correlated traits, individually explaining 6.1–17.8% of the phenotypic variation. Several overlaps were detected among traits, including one three-traits-overlapped locus on linkage group C08 and two important overlapped regions between the two CR-associated traits on C06. The QTLs were compared with known CR loci/genes (Peng et al. 2018).

2.5.18 Resistance Mechanisms in *Brassica* to Clubroot

Understanding the defense response and molecular mechanism in a host against a pathogen is of great significance in the breeding of resistant varieties. Proteomic and transcriptome studies were conducted in *Arabidopsis*, *B. rapa*, *B. oleracea*, and *B. juncea*, revealing many factors associated with mechanisms of resistance to *P. brassicae* in these host species. Among these are cell wall modifications, cell cycle alterations, the production of secondary metabolites, calcium and hormone signaling and homeostasis, the activation of pathogen-associated molecular patterns, the activity of effector receptors (resistance genes) and pathogenesis-related (PR) genes, the induction of transcriptional factors and mitogen-activated protein kinases, or the formation of reactive oxygen species (ROS) (Chen et al. 2016; Irani et al. 2018; Luo et al. 2018; Siemens et al. 2006; Zhang et al. 2016; Zhao et al. 2017). However, these provided limited reference values for understanding the resistance mechanism of rapeseed because no transcriptome study has been reported in rapeseed, mainly due to the lack of resistant resources in *B. napus*. Among the six *Brassica* spp. in U's triangle (Nagaharu 1935), innate clubroot resistance (CR) is found frequently in *B. rapa* (particular turnip), *B. oleracea*, and *B. nigra* (Carlsson et al. 2004; Crisp et al. 1989; Hasan et al. 2012; Manzanares-Dauleux et al. 2000) but rarely in *B. napus*. On the other hand, genetic studies have revealed different modulation patterns of resistance genes or loci on CR in *B. napus* (possibly in quantitative manner) (Neik et al. 2017; Piao et al. 2009; Rahman et al. 2014) and in *B. rapa* (qualitative plus quantitative manner) (Piao et al. 2002; Sakamoto et al.

2008; Suwabe et al. 2006). Therefore, studies in rapeseed are necessary to understand its resistance mechanism against *P. brassicae*.

Chinese *B. napus* variety, Huashuang 5R, exhibited high resistance to the most prevalent pathotype of *P. brassicae* in China (pathotype 4) (Chai et al. 2014; Zhan 2017). In order to understand the response in clubroot-resistant *B. napus* to *P. brassicae*, transcriptomic, physiological, biochemical, and histochemical assays were conducted in the between a susceptible *B. napus* genotype and a resistant genotype ZHE-226 that carries the same CR locus (*PbBa8.1*) with Huashuang 5R (Zhan 2017). The data revealed great differences between the two rapeseed genotypes in response to *P. brassicae*, particularly on activation of signaling networks and modulation of ROS and programmed cell death (PCD) (Mei et al. 2019).

Exploring the mechanism of plant resistance has become the basis for selection of resistance varieties, but reports on revealing resistant mechanism in *Brassica napus* against *P. brassicae* are rare. RNA-seq was conducted in the clubroot-resistant *B. napus* breeding line ZHE-226 and in the clubroot-susceptible rapeseed cultivar Zhongshuang 11 at 0, 3, 6, 9, and 12 days after inoculation. Strong alteration was detected specifically in ZHE-226 as soon as the root hair infection happened, and significant promotion was found in ZHE-226 on cell division or cell cycle, DNA repair and synthesis, protein synthesis, signaling, antioxidation, and secondary metabolites. Combining results from physiological, biochemical, and histochemical assays highlighted an effective signaling in ZHE-226 in response to *P. brassicae*. This response consists of a fast initiation of receptor kinases by *P. brassicae*; the possible activation of host intercellular G proteins which might, together with an enhanced Ca^{2+} signaling, promote the production of reactive oxygen species; and programmed cell death in the host. Meanwhile, a strong ability to maintain homeostasis of auxin and cytokinin in ZHE-226 might effectively limit the formation of clubs on host roots. The study provides initial insights into resistance mechanism in rapeseed to *P. brassicae* (Mei et al. 2019).

2.5.19 Proteomic Approach to Identify Clubroot R-Genes

Proteomics has been used in studying plant responses to biotic stresses based on differentially accumulated proteins (DAPs) as well as their functional annotation (Quirino et al. 2010). Two-dimensional electrophoresis (2-DE) used commonly to quantify and compare proteins in different samples (Gao et al. 2012; Wu et al. 2014a, b), but technical challenges have limited its application in the postgenomics era, especially for large-scale global profiling of proteome. Quantitative proteomics, especially the use of label-free shotgun techniques, has become a popular approach to replace the 2-DE in studying plant–pathogen interactions (Novo et al. 2014). Critical metabolic or signaling pathways may be identified via functional annotation of DAPs, complementing the findings from other studies such as transcriptomic analysis. A label-free shotgun proteomic approach was used to profile and compare the proteomes of *Brassica rapa* carrying and not carrying the CR gene *Rcr1* in

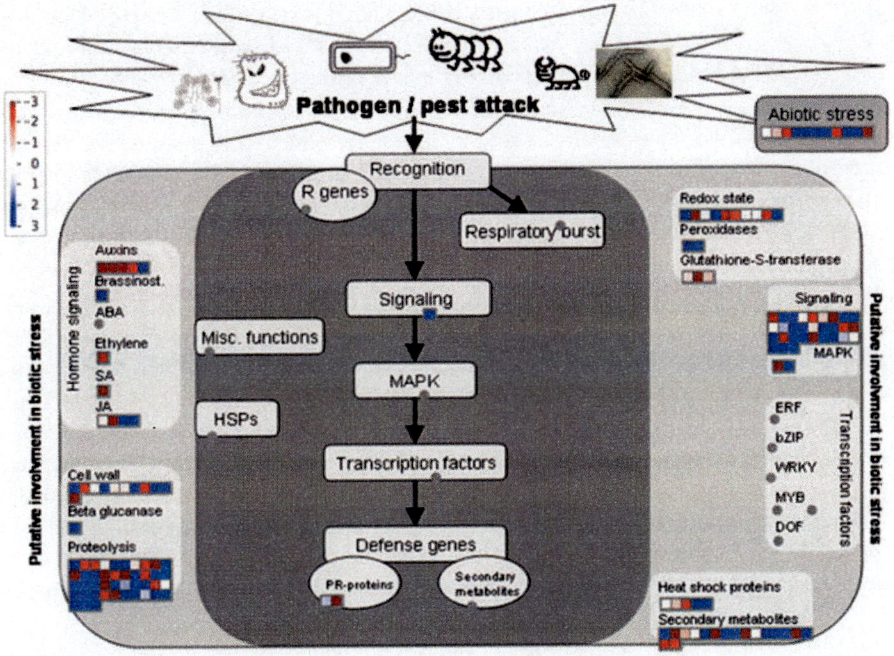

Fig. 2.23 Annotation of 145 DAPs in relation to signaling pathways involved in biotic/abiotic stress responses based on the analysis using the software Map man. The blue and red colors indicate up- and downregulation, respectively, and gray circles indicate no DAPs identified in these categories (Song et al. 2016)

response to *P. brassicae* infection. A total of 527 differentially accumulated proteins (DAPs) have been identified between the resistant (with *Rcr1*) and susceptible (without *Rcr1*) samples, and functional annotation of these DAPs indicates that the perception of *P. brassicae* and activation of defense responses are triggered via a unique signaling pathway distinct from common modes of recognition receptors reported with many other plant–pathogen interactions. This pathway appears to act in a calcium-independent manner through a not-well-defined cascade of mitogen-activated protein kinases and may require the ubiquitin-26S proteasome found to be related to abiotic stresses, especially the cold-stress tolerance in other studies (Fig. 2.23). Both upregulation of defense-related and downregulation of pathogenicity-related metabolism were observed in plants carrying *Rcr1*, and these functions may all contribute to the CR mediated by *Rcr1*. These results, combined with those of transcriptomic analysis reported earlier, improved the understanding of molecular mechanisms associated with *Rcr1* and CR at large and identified candidate metabolites or pathways related to specific resistance mechanisms. Deploying CR genes with different modes of action may help to improve the durability of CR (Song et al. 2016).

2.6 *Brassica–Turnip Mosaic Virus*: Molecular Resistance

2.6.1 Mapping of R-Genes in *Brassica rapa* to *TuMV*

Based on the inheritance analysis, several resistance genes were mapped and cloned. Rusholme et al. (2007) established a genetic map of *B. rapa* based on 213 RFLP markers in 120 BC1 individuals and mapped a dominant gene (ConTR01) to TuMV isolate CDN1 on the upper portion of chromosome R8. Zhang et al. (2008) used a population of 100 double haploid (DH) lines, derived from the F1 generation through microspore culture, to construct a linked map using 376 molecular markers (235 amplified fragment length polymorphism (AFLP), 129 random amplified polymorphic DNA (RAPD), 10 simple sequence repeat (SSR), 1 sequence characterized amplified region (SCAR), and 1 morphological marker; they identified four quantitative trait loci (QTLs) for the TuMV isolates Tu1, Tu2, Tu3, and Tu4. Subsequently, Zhang et al. (2009) used 183 DH lines derived from a cross of two Chinese cabbage lines to map QTLs related to TuMV C4 resistance and found three QTLs localizing on linkage groups R3, R4, and R6. Ma et al. (2010) cloned a cDNA for TuMV-induced gene in nonheading Chinese cabbage that was involved in plant resistance against TuMV infection and was named *BcTuR3* (TIRNB-LRR type). Another gene from *B. rapa*, *TuRB01b*, was identified by Lydiate et al. (2014) on a 2.9 Mb region of A6; comparative analysis indicated that TuRB01 is the first dominant resistance gene to be mapped (Walsh et al. 1999), and *TuRB01b* represent similar or identical alleles at the same A-genome resistance locus. Qian et al. (2013) used bulk segregant analysis (BSA) to identify SSR and insertion-deletion polymorphism (indel) markers linked to the gene retr02 for resistance to TuMV isolate C4 and located the gene on chromosome A4 of the *B. rapa* genome, encoding the eIF (iso) 4E protein. Kim et al. (2013) mapped a gene, trs, located on A4 that conferred resistance to TuMV isolates CHN2/CHN3/CHN4/CHN5 and which may be another recessive gene tightly linked to *retr02* or another allele. In general, recessive genes can impart resistance to multiple TuMV isolates. In particular, the *retr01/retr02/retr03* genes have broad-spectrum resistance to the TuMV potyvirus, which may facilitate their application for breeding (Li et al. 2019).

2.6.2 Mapping of R-Genes in *Brassica napus* to *TuMV*

Few resistance genes have been mapped in *B. napus*, most of which are dominant (Table 2.14). Walsh et al. (1999) employed a set of DH lines extracted from the oilseed rape B_1S_1 population to develop an RFLP map of the *B. napus* genome; they mapped the first resistance gene *TuRB01* on linkage group N6 of the *B. napus* A-genome, indicating that the gene may have originated from *B. rapa*. The dominant gene *TuRB01* could impart resistance to a wide range of TuMV isolates, including the European, American, and Taiwanese (China) pathotyping systems. Moreover, a second resistance locus, *TuRB02*, controlling the degree of susceptibility to isolate CHN1 was identified on the C-genome linkage group N14 (Walsh et al. 1999).

Table 2.14 Genes for resistance to *TuMV*, mapped or cloned from *Brassica* crops (Li et al. 2019)

Resistance gene	Species (genome)	Line	Mapping markers	Inheritance and mapping	Pathotype/isolate	Reference
TuRB01	*B. napus* (A)	N-o-9	RFLP	Single dominant, mapped to a 7.2 cM interval on N6	1	Walsh et al. (1999)
TuRB02	*B. napus* (C)	N-o-9	RFLP	Dominant, mapped to N14	CHN1, JPN1	Walsh et al. (1999)
TuRB03	*B. napus* (A)	22S	AFLP, SSR	Single dominant, mapped to a 7.9 cM interval on N6	CDN1	Hughes et al. (2003)
TuRB04	*B. napus* (A)	165	–	Single dominant, unmapped	1, 3	Jenner et al. (2002, 2003)
TuRB05	*B. napus* (A)	165	–	Single dominant, unmapped	1, 3	Jenner et al. (2002, 2003)
retr01	*B. rapa* (A)	RLR22	RFLP, SSR	Recessive, mapped to R4, and the candidate encoding eIF(iso)4E	1, 3, 4, 7, 8, 9, 12	Rusholme et al. (2007)
ConTR01	*B. rapa* (A)	RLR22	RFLP, SSR	Dominant, mapped to R8, and the candidate is an eIF(iso)4E-encoding gene	1, 3, 4, 7, 8, 9, 12	Rusholme et al. (2007)
–	*B. rapa* (A)	91–112	AFLP, RAPD, SSR, SCAR	Four QTLs	Tu1, Tu2, Tu3, Tu4	Zhang et al. (2008)
–	*B. rapa* (A)	Y195–93	AFLP, RAPD, SSR, SCAR	Three QTLs on R3, R4, and R6	C4	Zhang et al. (2009)
BcTuR3	*B. rapa* (A)	Duanbaigeng	–	Possibly related gene, TIR-NB-LRR type	–	Ma et al. (2010)
rnt1	*B. rapa* (A)	AS9	SSR	Dominant, mapped on R6	UK1	Fujiwara et al. (2011)
TuRBCH01	*B. rapa* (A)	Q048	AFLP, SSR	A single dominant gene on R6	C5	Wang et al. (2011)
retr02	*B. rapa* (A)	BP8407	SSR, indel	Recessive, mapped to A4, candidate is Bra035393, encoding eIF(iso)4E	C4	Qian et al. (2013)
Trs		SB18/SB22		Recessive, mapped to A4		Kim et al. (2013)

(continued)

Table 2.14 (continued)

Resistance gene	Species (genome)	Line	Mapping markers	Inheritance and mapping	Pathotype/isolate	Reference
	B. rapa (A)		CAPS, SCAR, SNP		CHN2, CHN3, CHN4, CHN5	
TuMV-R	*B. rapa* (A)	VC40	SSR, indel, SNP	Mapped to a 0.34 Mb region on A6, with six candidates	–	Chung et al. (2014)
TuRB01b	*B. rapa* (A)	TD34-S1	RFLP	Single dominant, mapped to a 2.9 Mb region on A6	1	Lydiate et al. (2014)
TuRB07	*B. rapa* (A)	VC1/VC40	SSR	Single dominant, mapped to A6, candidate gene Bra018863 (CC-NB-LRR)	C4	Jin et al. (2014)
TuRBCS01	*B. rapa* (A)	8407	EST, SSR, indel	1.98 Mb region on A4	C4	Li et al. (2015)
TuRBJU01	*B. juncea* (unknown)	Oasis CI	–	Dominant, unmapped	8 (WA-Ap1)	Nyalugwe et al. (2015, 2016)
retr03	*B. juncea* (A)	VC029	SNP	Recessive, encoding eIF2Bb	ZJ	Shopan et al. (2017)

AFLP amplified fragment length polymorphism, *CAPS* cleaved amplified polymorphic sequence, *EST* expressed sequence tag, *Indel* insertion deletion, *QTL* quantitative trait locus, *RAPD* random amplified polymorphic DNA, *RFLP* restriction fragment length polymorphism, *SCAR* sequence characterized amplified region, *SNP* single-nucleotide polymorphism, *SSR* simple sequence repeat

Hughes et al. (2003) constructed a backcross generation from a cross between resistant and susceptible *B. napus* to identify the dominant and monogenic resistance to TuMV pathotype 4. They then used the AFLP and SSR markers to map the resistance gene *TuRB03* on chromosome N6. Jenner et al. (2002) used the segregating populations derived from the *B. napus* 165 line, which was resistant to infection by TuMV isolates belonging to pathotypes 1 and 3, to demonstrate that the P3 and CI proteins from TuMV were pathogenicity determinants for two different dominant resistance genes (*TuRB04* and *TuRB05*); *TuRB04* was found to be epistatic to the *TuRB05* gene.

2.6.3 Mapping of R-Genes in *Brassica juncea* to *TuMV*

To date only two genes have been mapped in *B. juncea*, *TuRBJU01* and *retr03*. Nyalugwe et al. (2015) employed an isolate of TuMV pathotype 8 to inoculate 69 *B. juncea* lines, establishing a standard for distinguishing TuMV resistance phenotypes and identifying a single incompletely dominant TuMV resistance gene, *TuRBJU01*. This gene was located on the A-genome and was, in some ways, related to the *TuRB01* and *TuRB03* genes in *B. napus*. Shopan et al. (2017) developed segregating populations with the TuMV-resistant and TuMV-susceptible lines of *B. juncea* through bulked segregant analysis and sequencing of resistant/susceptible plant lines. Their results indicated that a single recessive gene, retr03, controls resistance to the TuMV isolate ZJ that is an allele of the eukaryotic translation initiation factor 2B-beta (*eIF2Bb*).

2.6.4 Molecular Mechanisms of R-Genes in *Brassica*

With the development of genomics and molecular biology techniques, molecular mechanisms are becoming well characterized once resistance genes are cloned. The protein–protein interaction between host plants and pathogens is an active area of research that can not only deepen our understanding of virus–plant interaction, but also facilitate resistance breeding. A variety of plant–virus interaction proteins have been characterized using techniques such as yeast two-hybrid (Y2H), bimolecular fluorescence complement (BiFC), and coimmunoprecipitation (CoIP) methods to analyze the resistance interaction between plants and various TuMV isolates.

2.6.4.1 Dominant Genes of *Brassica*

Dominant resistance genes are usually found to code for NLR proteins; however, there have been no reports of NLR genes for TuMV resistance in *Brassica* crops. In the *Brassica*–TuMV interaction, researchers have identified the cytoplasmic inclusion (CI) protein as the viral avirulence determinant and interacted with the host resistance genes *TuRB01*, *TuRB01b,* and *TuRB04*, while P3 is the viral avirulence determinant for TuRB03 and TuRB05 (Jenner et al. 2000, 2002, 2003; Walsh and Jenner 2002). Jenner et al. (2000) made sequence comparisons of an infectious cDNA clone of the UK1 isolate of TuMV (avirulent on *TuRB01*) and a spontaneous mutant capable of infecting *B. napus* plants possessing *TuRB01* and suggested that a single-nucleotide change in the cylindrical inclusion (CI) protein-coding region (gene) of the virus was responsible for the altered phenotype. This is the first example of a potyvirus CI gene acting as a determinant for a genotype-specific resistance interaction (Jenner et al. 2002). In *B. rapa*, Rusholme et al. (2007) confirmed the location of the ConTR01 candidate *BraA*.eIF (*iso*) 4E.c from line RLR22 on chromosome A8. Sequencing *BraA.eIF* (*iso*) 4E.c from line RLR22 and the susceptible *B. rapa* line R-o-18 showed four amino acid differences (L36F, V52A, T80I, and Q150P). Their attempts to identify an interaction between the R-o-18 or RLR22 alleles of *BraA.eIF*(*iso*)4E.c and the viral protein VPg (viral protein genome-linked) of TuMV isolate C4 in yeast two-hybrid experiments were unsuccessful, indicating that the TuMV isolates could interact with different copies of *BraA.eIF* (*iso*)4E.

2.6.4.2 Recessive Genes of *Brassica*

The passive mechanisms of plant virus resistance indicate that loss, deletion, or mutation of a required host factor may cause recessive resistance to the virus (Dinkova et al. 2016). Recessive resistance genes primarily belong to the plant eukaryotic initiation factor 4E (eIF4E) family, which are well-known host factors that play a critical role in the infection of several potyviruses. The interaction between VPg of potyviruses and eIF4E (eukaryotic initiation factor 4E) or eIF (iso) 4E of the host determines the virulence (Wittmann et al. 1997; Robaglia and Caranta 2006; Beauchemin et al. 2007). This eIF4E-mediated resistance often confers strong and broad-spectrum resistance (Yeam et al. 2007; Mazier et al. 2011; Rodríguez-Hernández et al. 2012).

In *Brassica*, the *eIF* (*iso*) 4E gene has been shown to be strongly linked to the *Brassica* recessive resistance genes *retr01, retr02,* and *trs* (Rusholme et al. 2007; Kim et al. 2013; Qian et al. 2013). The recessive gene *retr02* was identified and assumed to be an *eIF* (*iso*) 4E-encoding gene (Qian et al. 2013). Moreover, it has been shown that *retr02* and *retr01* are the same (Nellist et al. 2014). Nellist et al. (2014) reported that *retr01/retr02* broad-spectrum resistance to TuMV occurs because of a natural mechanism based on the mis-splicing of the *eIF* (*iso*) 4E allele in *B. rapa*. The existence of multiple copies of *eIF* (*iso*) 4E has enabled redundancy in the host plant's translational machinery, resulting in diversification and emergence of the resistance. Kim et al. (2014) determined the key amino acid residues in the *B. rapa eIF* (*iso*) 4E protein for interaction with TuMV VPg using the Y2H technique. In addition, in Y2H and BiFC assays, Kim et al. (2014) showed that TuMV VPg could not interact with eIF4E, but only with eIF (iso) 4E of *B. rapa*. Finally, some SNPs have been identified, which may affect the interaction between eIF (iso) 4E and VPg; these include SNP $T_{106}C$ in *BraA.eIF (iso) 4E.c* and SNP $A_{154}C$ in VPg (Li et al. 2018). After *retr01* and *retr02* genes, the *retr03* gene was cloned, which was an allele of the eukaryotic translation initiation factor 2B-beta (*eIF2Bb*). eIF2B acts as a guanine nucleotide exchange factor (GEF) for its GTP-binding protein partner eIF2 via interaction with eIF2-GTP at an early step in translation initiation and therefore represents a new class of virus resistance gene conferring resistance against a pathogen (Shopan et al. 2017; Li et al. 2019).

2.6.5 Alternative Oxidase Gene (*BjAOX1a*) in *Brassica* Enhances Resistance to *Turnip Mosaic Virus* (*TuMV*)

The alternative oxidase (AOX) is the unique terminal oxidase of cyanide-insensitive pathway in the mitochondrial inner membranes. AOX participates in the non-protontranslocating transfer of electrons and reduces oxygen to water without ATP production (Feng et al. 2010). AOX are encoded by a small family of nuclear genes in a wide variety of monocotyledonous and eudicotyledonous plants and can be classified into two subfamilies: *AOX1* and *AOX2* (Moore and Albury 2008). *AOX1* can be stimulated by different stresses, for example, cold treatment (Li et al. 2011), light treatment (Yoshida et al. 2006), and cucumber mosaic virus (CMV) (Murphy et al. 2004). Moreover, AOX1 can be provoked by salicylic acid (SA), hydrogen peroxide (H_2O_2), nitric oxide (NO), ethylene (ET), and other signaling molecules in many different plants (Fu et al. 2010). However, AOX2 are usually constitutive or developmentally expressed in plants (Schmid et al. 2005).

AOX may help to improve the adaptation of plants to stresses by controlling the cellular metabolism (Arnholdt-Schmitt et al. 2006). It was reported that AOX has been a necessary component in maintaining the balance of carbon and nitrogen metabolism under low temperature (Watanabe et al. 2008). In *A. thaliana*, cauliflower mosaic virus (CaMV) resistance was enhanced by antimycin A (AA) (Love et al. 2005), but the resistance to Tobacco mosaic virus (TMV) was not increased in tobacco mutants with overexpression of AOX (Ordog et al. 2002; Gilliland et al. 2003). The systemic movement of CMV in *A. thaliana* was delayed by AA treatment, which led to an increased CMV resistance, but CMV resistance was not induced by AA in squash (Meyers et al. 2005).

Reactive oxygen species (ROS) production is a continuous process inside the cell. Its equilibrium is maintained between productions and scavenging under normal metabolic conditions (Song et al. 2009). ROS play various roles in plants and are suggested to be involved in the defense against pathogens (Pei et al. 2011). ROS production is associated with hypersensitive response (HR) at sites of infection by incompatible pathogens (Mur et al. 2008). It was reported that AOX could alleviate ROS production in the mitochondria as well as in the whole cell, by which it can influence plants' resistance to virus and other stresses such as chilling, ozone, etc. (Watanabe et al. 2008). The AOX-mediated changes in ROS production may regulate the activity of downstream elements in the salicylhydroxamic acid (SHAM)-sensitive pathway (Wang et al. 2010). This evidence indicates that the levels of mitochondrial ROS production affected by AOX proteins could influence plant defense against viruses (Kiraly et al. 2008). On the other hand, the high level of the alternative oxidase protein accelerated the systemic movement of TMV and accumulation of ROS in *N. benthamiana* plants (Murphy et al. 2004). Hence, the role of AOX and ROS in plant defense signaling may be very complicated, and its efficacy may depend on the host–virus combination.

The roles of *Brassica juncea AOX1a* gene (*BjAOX1a*) and ROS in TuMV resistance was investigated with the relationship between AOX and ROS. However, the complicated genetic background of mustard made it challenging to study the function of AOX in TuMV resistance. Furthermore, most of the past investigations focused on the inoculation of detached leaves. Zhu et al. (2012) mainly studied the response of systemic leaves in mustard under TuMV infection. An alternative oxidase (AOX) gene, designated as *BjAOX1a*, of mustard (*Brassica juncea*) was cloned by reverse transcription-polymerase chain reaction (RT-PCR). The full-length cDNA of *BjAOX1a* was 1346 bp in size and containing an open reading frame (ORF) of 1083 bp in size. The predicted amino acid sequence exhibited 82.17% homology to the alternative oxidase of *A. thaliana*. Analysis of functional regions revealed that the *BjAOX1a* protein contained several metal-binding regions, α-helical regions, and cysteine residues, similar to other AOX1 proteins. Pretreatment of mustard plants with salicylhydroxamic acid (SHAM), an inhibitor of AOX protein, inhibited AOX activity and promoted systemic movement of the coat protein of turnip mosaic virus (TuMV), whereas pretreatment of plants with antimycin A (AA), a cytochrome pathway inhibitor, inhibited the systemic movement of the coat protein of TuMV. Furthermore, H_2O_2 and O^{2-} contents were enhanced following SHAM pretreatment. Thus, AOX1 might alleviate reactive oxygen species (ROS) and enhance resistance of mustard plants to TuMV and could serve as a critical component in TuMV resistance (Zhu et al. 2012).

2.6.6 Expression of Resistance-Modulating Genes in *TuMV-Arabidopsis* Pathosystem

Turnip mosaic virus (*TuMV*) is one of the most important plant viruses worldwide. It has a very wide host range infecting at least 318 species in over 43 families, such as Brassicaceae, Fabaceae, Asteraceae, or Chenopodiaceae from dicotyledons. Plant NADPH oxidases, the respiratory burst oxidase homologues (RBOHs), are a major source of reactive oxygen species (ROS) during host–pathogen interactions. The functions of RBOHs in different plant–pathogen interactions have been analyzed using knockout mutants, but little focus has been given to plant–virus responses. To reveal the response after mechanical inoculation with *TuMV* in *Arabidopsis rbohD* and *rbohF* transposon knockout mutants, Otulak-Koziel et al. (2020) analyzed ultrastructural changes after *TuMV* inoculation. The development of the *TuMV* infection cycle was promoted in rbohD plants, suggesting that RbohD plays a role in the *Arabidopsis* resistance response to *TuMV*. The *rbohF* and *rbohD*/F mutants display less *TuMV* accumulation, and a lack of virus cytoplasmic inclusions were observed; these observations suggest that RbohF promotes viral replication and increases susceptibility to TuMV. The *rbohD/F* displayed a reduction in H_2O_2 but enhanced resistance similarly to *rbohF*. This dominant effect of the *rbohF* mutation could indicate that RbohF acts as a susceptibility factor. Induction of hydrogen peroxide by *TuMV* was partially compromised in *rbohD* mutants, whereas it was almost completely abolished in *rbohD*/F, indicating that these oxidases are responsible for most of the ROS produced in this interaction. The pattern of in situ H_2O_2 deposition after infection of the more resistant rbohF and *rbohD*/F genotypes suggests a putative role of these species on systemic signal transport.

The ultrastructural localization and quantification of pathogenesis-related protein 1 (PR1) indicate that ROS produced by these oxidases also influence PR1 distribution in the *TuMV-A. thaliana* pathosystem. There was highest activation of PR1 in *rbohD* and Col-0, indicating a correlation between PR1 accumulation and susceptibility to *TuMV*. The specific localization of PR1 in the most resistant genotypes after *TuMV* inoculation may indicate a connection of PR1 induction with susceptibility, which may be characteristic for this pathosystem. There is importance of NADPH oxidases RbohD and RbohF in the regulation of the *TuMV* infection cycle in *Arabidopsis*. This may help provide a better understanding of the molecular mechanisms modulating *A. thaliana–TuMV* interactions to express host resistance.

2.6.6.1 *RbohD* Limits Accumulation of *TuMV* in *Arabidopsis*

Plant NADPH oxidases have emerged as important regulators of many responses to the environment, and many studies document the role of plant NADPH oxidases in different plant–pathogen interactions. However, the function of these enzymes in plant–virus interactions remains poorly investigated. The two *Arabidopsis* NADPH oxidase genes, *RbohD* and *RbohF*, are the highest expressed homologues that are pleiotropic and act often redundantly in many immune responses to a wide range of pathogens. Using *Arabidopsis rbohD* and *rbohF* transposon mutants, Otulak-Koziel et al. (2020) demonstrated through ultrastructural studies that *rbohD* supports more *TuMV* proliferation compared to Col-0 (and *rbohF*), especially at later times (Fig. 2.24). *TuMV* infection effects were more severe in *rbohD* tissues, where *TuMV* induced typical potyviral cytoplasmic inclusions not only in mesophyll cells but also in phloem and xylem cells, providing evidence for systemic virus spread. Moreover, virus particles and inclusions were observed at 3 dpi and were more intense than in the other genotypes. Consequently, in situ quantification of virus particles in these tissues showed that rbohD supports the highest accumulation of *TuMV* (Fig. 2.25). It suggests that *RbohD* contributes to stop *TuMV* proliferation and spread. The *RbohD* mutant was found to be more susceptible than wild-type *Arabidopsis* to infection with different biotrophic and necrotrophic fungi and bacteria (Chen et al. 2017; Kadota et al. 2014) and other potyviruses.

The potato *RbohD* orthologue limits the systemic spread of a tagged Potato Virus Y construct (PVY^{N}-GFP) to upper (noninoculated) leaves. It seems that *RbohD* has an important contribution to the resistance response to potyvirus. The high levels of *RbohD* transcript were associated with resistance and hypersensitive response (HR) to PVY. However, *RbohD* contribution to resistance is not general to all pathosystems— for example, *rbohD* supports less *Alternaria brassicola* biomass growth than in wild type or *rbohF*, suggesting that *RbohD* does not develop a specific antifungal activity and does not act as a resistance factor, but rather could play a role as a cell death regulator. In other pathosystems with avirulent and virulent

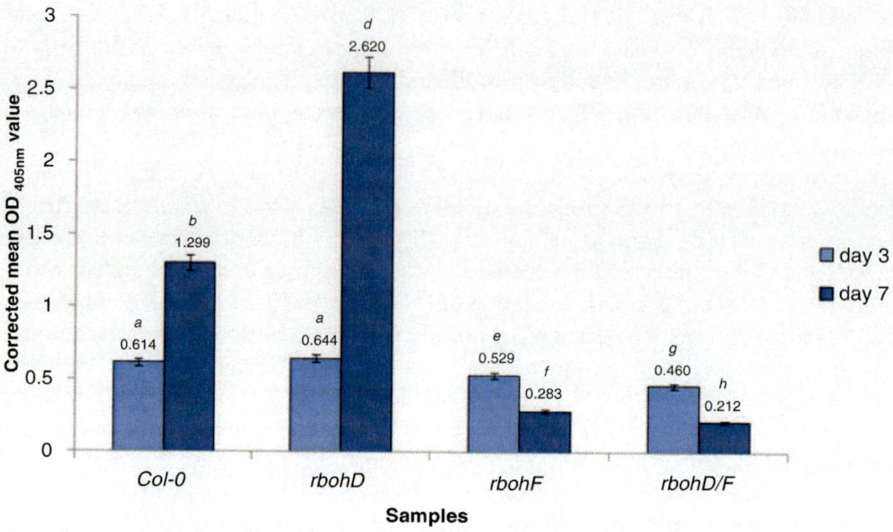

Fig. 2.24 *TuMV* detection and relative virus concentration assessment in *A. thaliana* Col-0 and *rbohD, rbohF,* and *rbohD/F* mutants at 3 and 7 days after inoculation. Values represented are mean of corrected OD405nm values. Significant differences between classes at p 0.05 level of significance were assessed by analysis of variance (ANOVA) with post hoc Tukey HSD. The statistically significant values are marked by letters above chart bars (Otulak-Koziel et al. 2020)

bacteria or with the necrotrophic fungus *Botrytis cinerea*, the responses in *rbohD* (or *rbohF*) were not accompanied by an enhanced growth of the pathogen. Therefore, resistance is not a general trait associated with *RbohD* (Torres et al. 2002, 2005).

2.6.6.2 *RbohF* Promotes *TuMV* Virulence in *Arabidopsis*

Intriguingly, *rbohF* and *rbohD*/F display less *TuMV* accumulation at 3 dpi than Col-0 (Fig. 2.24), also confirmed by the detection of *TuMV* in situ at the ultrastructural level. After virus inoculation, *rbohF* and *rbohD*/F mutants displayed changes in their chloroplasts, induction of multivesicular structures, or the appearance of single membranes with tubular structures as an effect of *TuMV* inoculation. These changes have been documented in other infections with viruses and may be an effect of cell wall rebuilding. The cell wall rebuilding was especially noticeable in the *TuMV–rbohD*/F interaction and was often accompanied with phenolic-like compounds located inside xylem tracheary elements. Additionally, *rbohF* and especially *rbohD*/F mutants exhibited local necrosis at the 7-day time point, typical of the hypersensitive response (HR). Importantly, these ultrastructural changes in *rbohF* or *rbohD*/F were associated with fewer virus particles, as compared to Col-0 and *rbohD* interactions. Moreover, *rbohF* or *rbohD*/F particles were located in the vacuole and occasionally in small vacuole-like structures or vesicles in the mesophyll cell of the inoculated leaf, and no cytoplasmic inclusions were observed. This less severe *TuMV* infection in these mutants and the fact that virus was not able to create the

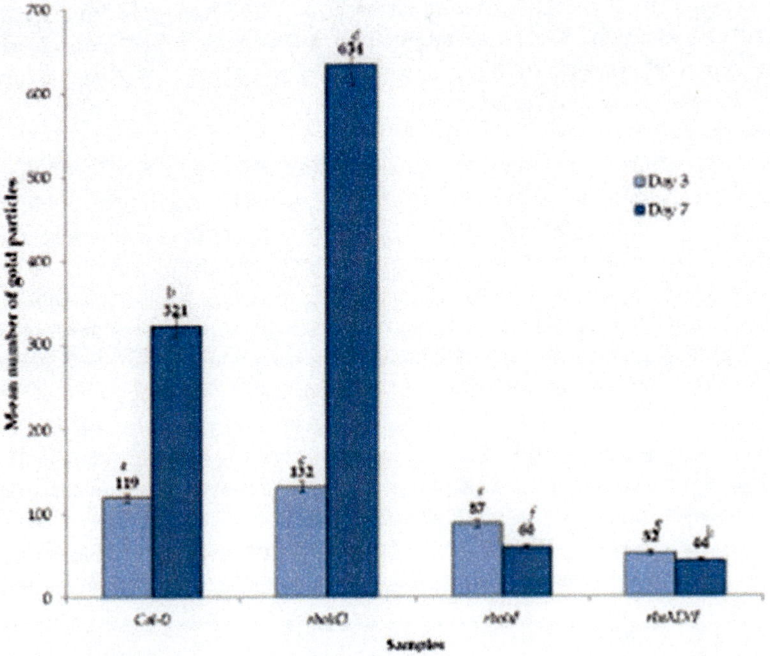

Fig. 2.25 Quantification gold particles associated with *TuMV* in *A. thaliana* Col-0, *rbohD, rbohF,* and double *rbohD/F* mutants at 3 and 7 days after inoculation. Significant differences between classes at $P < 0.05$ level of significance by ANOVA with post hoc Tukey HSD were assessed. The statistically significant values are marked by letters above chart bars (Otulak-Koziel et al. 2020)

inclusions needed for further spread through host cells suggest that *RbohF* promotes viral replication and increases susceptibility to *TuMV*. Although *RbohF* has mostly been associated with the establishment of immunity, some studies document its role in susceptibility, as *rbohF* displays increased resistance in response to several pathogens compared to the wild type. Thus, again, resistance is not always associated with *RbohF*, and its function may vary depending on the pathosystem. However, it is noticeable that *rbohD* and *rbohF* mutants display opposed effects in the *Arabidopsis–TuMV* interaction, suggesting that *RbohD* and *RbohF* play different and opposing roles in response to this virus. Usually, studies on the double mutant *rbohD*/F document an additive effect in response to pathogens as well as abiotic stress. It indicates that ROS produced by these two oxidases may have a qualitative (spatial or temporal) difference to signal these opposite functions (Marino et al. 2012; Otulak-Koziel et al. 2020; Kwak et al. 2003).

2.6.6.3 *RbohD* and *RbohF* Are Responsible for Most ROS Produced during *TuMV* Infection

The quantification in situ, the location of H_2O_2 by histochemical densitometry on the electron micrographs of infected tissues was performed by Otulak-Koziel et al.

(2020) (Fig.2.26). This H_2O_2 accumulation after *TuMV* was transitory, since Col-0 showed an induction of H_2O_2 accumulation at 3 dpi that was reduced at 7 dpi. Production of ROS was partially compromised in the NADPH oxidase mutants, with *rbohD* showing a greater reduction in levels of H_2O_2 than in *rbohF*. These data suggest that *RbohD* is responsible for most ROS produced during *TuMV* infection, as has been shown in the response to many other pathogens. The *rbohD*/F showed an additive reduction in ROS, suggesting than these two oxidases contribute to ROS production. A similar "gradation" in H_2O_2 levels in these mutants has also been documented in response to other pathogens. Interestingly, it revealed a different pattern of H_2O_2 localization between susceptible/resistant genotypes. In Col-0 and *rbohD*, the most susceptible genotypes, H_2O_2 decorated the vesicular and membranous structures and was also detected inside vacuoles and in the area between the cell wall and plasmalemma—areas where a profusion of virus particles were present. By contrast, in *rbohF* and *rbohD*/F, the most resistant genotypes, the deposits of hydrogen peroxide were linked to the cell wall, near plasmodesmata and in the area around necrotizing cells. This pattern of H_2O_2 deposition evidences a possible longer distance transport of these reactive species in these genotypes and/or a putative role in plasmodesmatal permeability, a factor that could contribute to the resistance response to the virus.

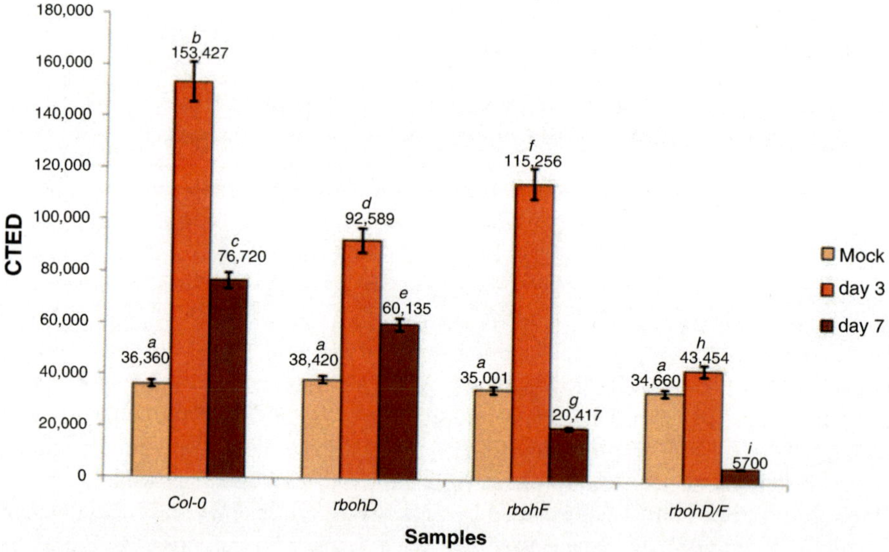

Fig. 2.26 Corrected total electron density (CTED) of cerium (IV) perhydroxide precipitate in mock- and *TuMV*-inoculated *A. thaliana* Col-0, *rbohD, rbohF*, and double mutant *rbohD/F* at 3 and 7 dpi. Significant differences between classes at $P < 0.05$ level of significance by ANOVA with post hoc Tukey HSD were assessed. The statistically significant values are marked by letters above chart bars (Otulak-Koziel et al. 2020)

In addition, these resistant genotypes displayed apparent necrosis, typical of the HR-like reaction, with characteristic rebuilding of cell walls and deposition of phenolic-like compounds inside xylem elements. The commonly accepted hypothesis predicts that the suppression of virus spread and multiplication during HR is due to the necrosis, which leads to virus resistance. H_2O_2 has been identified as a key signaling molecule promoting HR-cell death, with the contribution of NADPH oxidases (in *Arabidopsis RbohD* and *RbohF*) as the usual source of ROS defense-related ROS generation (Torres and Dangl 2005). However, these ROS can also inhibit necrosis in some pathosystems. Even in some plant–virus interactions, downregulation of ROS and activation of antioxidant enzymes do associate with the suppression of virus-induced necrotization and susceptibility. However, Otulak-Koziel et al. (2020) observed that the greater reduction in H_2O_2 in *rbohD* correlate with the lower level of necrosis and the greatest susceptibility to *TuMV*, suggesting that *RbohD*-dependent ROS contributes to restrict *TuMV* infection. Similar findings, where downregulation of RBOH-dependent ROS drove the suppression of the HR, have been documented in other plant–pathogen interactions such as in response to the hemibiotrophic bacterium *P. syringae* or the hemibiotrophic oomycete *P. infestans*. Although *rbohF* also displayed lower ROS production compared to Col-0, this genotype exhibited extensive necrosis and was more resistant, which indicates that *RbohF*-dependent ROS contributes to negatively regulate the HR and promotes susceptibility to this virus. This reduction in Rboh-dependent ROS with an increase in the HR has also been observed in response to some pathosystems such as to an avirulent *P. syringae* strain or to the biotrophic oomycete *Hyaloperonospora arabidopsidis*. Thus, ROS might serve different signaling functions in different types of disease resistance and in the HR. Even the dual role for the same oxidase has been suggested, with ROS produced by *RbohD* acting as positive or negative regulator of cell death in different cellular contexts during the same *Arabidopsis–A. brassicicola* interaction. The different localization of *TuMV*-induced H_2O_2 in *rbohD* and *rbohF* mutants suggests that RbohD and RbohF drive ROS generation at different locations, and that these different pools of ROS could be responsible for these opposed functions. This may be related to their differential pattern of expression. Although both oxidases are transcriptionally upregulated by pathogens, they present a differential spatiotemporal expression pattern that contributes to fine-tune *Rboh*-dependent ROS production. Intriguingly, *rbohD*/F displayed additive reduction in H_2O_2 but presented the same enhanced resistance than *rbohF*. This dominant effect of the *rbohF* mutation could indicate that RbohF acts as a susceptibility factor. It has also been documented that *red clover necrotic mosaic virus* (RCNMV), a *Dianthus* virus, hijacks the host generation of reactive oxygen species during infections while RCNMV replication proteins were associated with the ROS-generating system and triggered the ROS-burst, similarly to *Brome mosaic virus* replication which also depends on ROS. Alternatively, since *rbohD*/F (and *rbohF*) plants displayed HR-like necrosis, the enhanced dominant resistance in *rbohD*/F could be associated with the compensatory activation of other defense pathways leading to cell death/necrosis (Kadota et al. 2014; Hyodo et al. 2017).

2.6.6.4 PR1 Accumulation Correlates with *TuMV* Virulence

Pathogenesis-related proteins (PRs) are usually induced during pathogen infection and biotic stresses and are classically associated with resistance. It has been hypothesized that H_2O_2 is able to upregulate the defense gene PR1. However, regarding localization and quantification of *Arabidopsis*, PR1 protein does not support these ideas. PR1 protein quantification indicated that PR1 deposition was induced as an effect of *TuMV* inoculation compared to mock-inoculated plants in all genotypes analyzed (Fig. 2.27). However, similarly to CTED analyses of H_2O_2 precipitates, PR1 depositions significantly decreased between 3 and 7 dpi. Notably, the analyses revealed that the highest buildup of PR1 was observed in the most susceptible lines, *rbohD* and Col-0, whereas the weakest accumulation of PR1 was observed in *rbohF* and *rbohD*/F, the mutants that restricted virus concentration.

It suggests a correlation between PR1 accumulation and susceptibility. Indeed, it has been suggested in other plant–virus interactions that PR1 is rather related to susceptibility than to resistance. The expression of PR1 appears to largely differ among experimental systems and seems to depend on various factors including plant species, plant organs, cell type, and the tissue colonization by different pathogens. During the infection of *Capsicum chinense* Jacq., with virulent isolates of Pep MMoV, the peak of PR1 expression and virus accumulation occurred at 4 dpi. Noticeably, in the *Arabidopsis–A. brassicola* pathosystem, *rbohD* plants also displayed increased PR1 protein accumulation and enhanced susceptibility to this

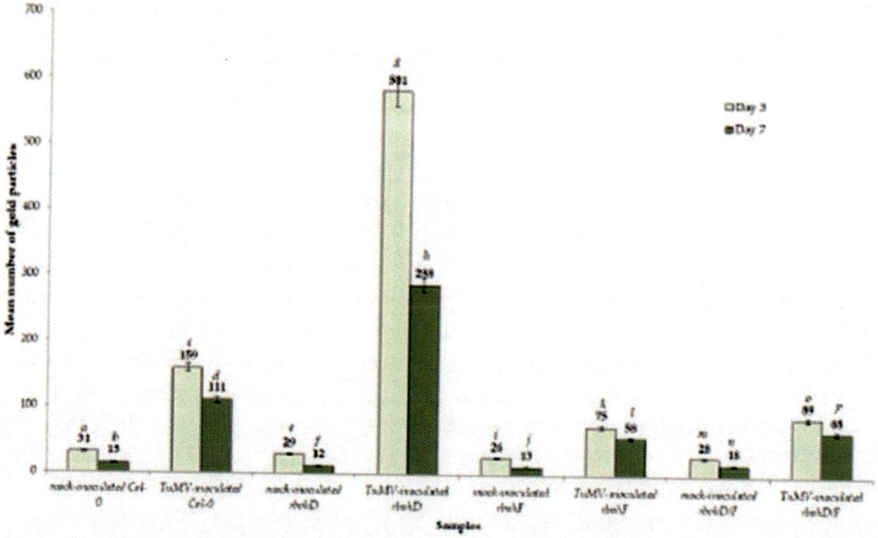

Fig. 2.27 Quantification of PR1 antigen localization in mock- and *TuMV*-inoculated *A. thaliana* Col-0, *rbohD*, *rbohF*, and double *rbohD/F* mutant at 3 and 7 days after inoculation. Significant differences between classes at $P < 0.05$ level of significance by ANOVA with post hoc Tukey HSD were assessed. The statistically significant values are marked by letters above chart bars (Otulak-Koziel et al. 2020)

pathogen compared to Col-0. In addition, ultrastructural studies revealed a different location of PR1 accumulation between susceptible and resistant interactions. The most susceptible interactions (Col-0-*TuMV* and *rbohD-TuMV*) presented strong PR1 depositions in the cell wall at the sites where virus particles and inclusions were present. However, in *atrbohF* or *atrbohD*/F, the lines where the virus infection cycle was limited, PR1 appeared around multivesicular bodies and rebuilt cell wall areas. These data suggest that more than the amount of PR1 protein itself, the areas where PR1 deposition occur are more important for the limitation of the virus infection cycle. Immunofluorescence localization of PR1 indicates the preferential accumulation of PR1 in the leaf epidermis, within the basis of trichomes, and in vascular tissues. This observation was identical to that of Hoegen et al. (2002) reported PR1b in epidermal cells, glandular trichomes, and leaf vascular bundles. The higher accumulation of PR1 protein in the more susceptible lines was probably because of the larger amount of plant cells responding to the multiplying virulent pathogen. However, PR1 is not believed to possess antiviral activity, and SA (salicylic acid)-mediated responses to incompatible virus infection are almost activated by NPR1-independent pathways. Nevertheless, the specific localization of PR1 in the most resistant genotypes after *TuMV* inoculation may indicate a connection of PR1 induction with susceptibility, which may be characteristic for this pathosystem (Kachroo et al. 2000; Wong et al. 2002; Otulak-Koziel et al. 2020).

References

Aarts N, Metz M, Holub E, Staskawicz BJ, Daniels MJ, Parker JE (1998) Different requirements for EDS1 and NDR1 by disease resistance genes define at least two R gene-mediated signaling pathways in *Arabidopsis*. Proc Natl Acad Sci U S A 95:10306–10311

Abraham V (1993) Transfer of resistance to powdery mildew from *Brassica carinata* to Indian mustard (*Brassica juncea*). In: National Symposium on Oilseed Research and Development in India. Status and Strategies Abstr. August 2–5, 1993, Hyderabad, p 86

Acevedo-Garcia J, Kusch S, Panstruga R (2014) Magical mystery tour: MLO proteins in plant immunity and beyond. New Phytol 204:273–281

Acevedo-Garcia J, Gruner K, Reinstadler A, Kemen A, Kemen E, Cao L, Takken FLW, Reitz MU, Schafer P, O'Connell RJ, Kusch S, Kuhn H, Panstruga R (2017a) The powdery mildew-resistant *Arabidopsis mlo2 mlo6 mlo12* triple mutant displays altered infection phenotypes with diverse types of phytopathogens. Sci Rep 7:9319

Acevedo-Garcia J, Spencer D, Thieron H, Reinstadler A, Hammond-Kosack K, Phillips AL, Panstruga R (2017b) Mlo-based powdery mildew resistance in hexaploid bread wheat generated by a non-transgenic TILLING approach. Plant Biotechnol J 15:367–378

Adhikari TB, Liu JQ, Mathur S, Wu CX, Rismmer SR (2003) Genetic and molecular analyses in crosses of race 2 and race 7 of *Albugo candida*. Phytopathology 93:959–965

Agnola B, Boury S, Monot C, Quillevere A, Herve Y, Silue D (2003) Evidence that a leaf-disk test allows assessment of isolate-specific resistance in *Brassica oleracea* crops against downy mildew (*Peronospora parasitica*). Eur J Plant Pathol 109:471–478. https://doi.org/10.1023/A:1024217223829

Ah Fong AMV, Bormann-Chung CA, Judelson HS (2008) Optimization of transgene-mediated is silencing in *Phytophthora infestans* and its association with small-interfering RNAs. Fungal Genet Biol 45:1197–1205

AICRPRM-AR (2009) Annual progress report. All India co-ordinated research project on rapeseed-mustard. ICARDirectorate of Rapeseed-Mustard Research, Bharatpur

Akashi K, Nishimura N, Ishida Y, Yokota A (2004) Potent hydroxyl radical-scavenging activity of drought-induced type-2 metallothionein in wild watermelon. Biochem Biophys Res Commun 323:72–78

Akira S, Hemmi H (2003) Recognition of pathogen-associated molecular patterns by TLR family. Immunol Lett 85(2):85–95

Alder MN, Dames S, Gaudet J, Mango SE (2003) Gene silencing in *Caenorhabditis elegans* by transitive RNA interference. RNA 9:25–32

Alix K, Lariagon C, Delourme R, Manzanares-Dauleux MJ (2007) Exploiting natural genetic diversity and mutant resources of *Arabidopsis thaliana* to study the *A. thaliana-Plasmodiophora brassicae* interaction. Plant Breed 126:218–221

Aller EST, Jagd LM, Kliebenstein DJ, Burow M (2018) Comparison of the relative potential for epigenetic and genetic variation to contribute to trait stability. G3 8:1733–1746

Aoun N, Tauleigne L, Lonjon F, Deslandes L, Vailleau F, Roux F, Berthomé R (2017) Quantitative disease resistance under elevated temperature: genetic basis of new resistance mechanisms to *Ralstonia solanacearum*. Front Plant Sci 8:1387

Arnholdt-Schmitt B, Costa JH, de Melo DF (2006) AOX—a functional marker for efficient cell reprogramming under stress? Trends Plant Sci 11:281–287

Arora H, Padmaja KL, Paritosh K, Mukhi N, Tewari A, Mukhopadhyay A, Gupta V, Pradhan AK, Pental D (2019) BjuWRR1, a CC-NB-LRR gene identified in *Brassica juncea*, confers resistance to white rust caused by *Albugo candida*. Theor Appl Genet 132:2223–2236. https://doi.org/10.1007/s00122-019-03350-z

Assaad FF, Qiu JL, Youngs H, Ehrhardt D, Zimmerli L, Kalde M, Wanner G, Peck SC, Edwards H, Ramonell K, Somerville CR, Thordal-Christensen H (2004) The PEN1 syntaxin defines a novel cellular compartment upon fungal attack and is required for the timely assembly of papillae. Mol Biol Cell 156(1):5118–5129

Austin MJ, Muskett P, Kahn K, Feys BJ, Jones JD, Parker JE (2002) Regulatory role of SGT1 in early *R*. gene-mediated plant defenses. Science 295:2077–2080

Ausubel FM (2005) Are innate immune signaling pathways in plants and animals conserved? Nat Immunol 6:973–979

Aviv DH, Rusterucci C, Iii BF, Dietrich RA, Parker JE, Dangl JL (2002) Runaway cell death, but not basal disease resistance, in lsd1 is SA- and NIM1/NPR1-dependent. Plant J 29:381–391

Axtell MJ, Staskawicz BJ (2003) Initiation of *RPS2*-specified disease resistance in *Arabidopsis* is coupled to the AvrRpt2-directed elimination of RIN4. Cell 112:369–377

Bai YL, van der Hulst R, Bonnema G, Marcel BC, Meijer-Dekens F, Niks RE, Lindhout P (2005) Tomato defense to *Oidium neolycopersici*: dominant *Ol* genes confer isolate-dependent resistance via a different mechanism than recessive *ol-2*. Mol Plant Microbe Interact 186(1):354–362

Bai YL, Pavan S, Zheng Z, Zappel NF, Reinstadler A, Lotti C, De Giovanni C, Ricciardi L, Lindhout P, Visser R, Theres K, Panstruga R (2008) Naturally occurring broad-spectrum powdery mildew resistance in a central American tomato accession is caused by loss of *Mlo* function. Mol Plant Microbe Interact 216(1):30–39

Bansal VK, Thiagarajah MR, Stringam GR, Tewari JP (1999) Inheritance of partial resistance to race 2 of *Albugo candida* in canola-quality mustard (*Brassica juncea*) and its role in resistance breeding. Plant Pathol 48:817–822

Bansal VK, Tewari JP, Stringam GR, Thiagarajah MR (2005) Histological and inheritance studies of partial resistance in the *Brassica napus–Albugo candida* host– pathogen interaction. Plant Breed 124:27–32

Bar M, Sharfman M, Ron M, Avni A (2010) BAK1 is required for the attenuation of ethylene-inducing xylanase (Eix)-induced defense responses by the decoy receptor LeEix1. Plant J 63:791–800. https://doi.org/10.1111/j.1365-313X.2010.04282.x

Bartsch M, Gobbato E, Bednarek P, Debey S, Schultze JL, Bautor J, Parker JE (2006) Salicylic acid-independent ENHANCED DISEASE SUSCEPTIBILITY1 signaling in *Arabidopsis* immunity and cell death is regulated by the monooxygenase FMO1 and the Nudix hydrolase NUDT7. Plant Cell 18:1038–1051. https://doi.org/10.1105/tpc.105.039982

Batley J, Dolatabadian A, Yang H, Severn-Ellis A, Alamery S, Tollenaere R, Bayer P, Hurgobin B, Golicz AA, Edwards D (2016) The more the merrier? Investigating copy number variation in *Brassica* disease resistance. In: Plant and animal genome conference (PAG) Asia (Singapore)

Baulcombe D (2004) RNA silencing in plants. Nature 431:356–363

Bayer PE, Hurgobin B, Golicz AA, Chan CK, Yuan Y, Lee H, Renton M, Meng J, Li R, Long Y, Zou J, Bancroft I, Chalhoub B, King GJ, Batley J, Edwards D (2017) Assembly and comparison of two closely related *Brassica napus* genomes. Plant Biotechnol J 15:1602–1610

Beauchemin C, Boutet N, Laliberte JF (2007) Visualization of the interaction between the precursors of Vpg, the viral protein linked to the genome of turnip mosaic virus, and the translation eukaryotic initiation factor iso4e in planta. J Virol 81:775–782

Beers EP, Jones AM, Dickerman AW (2004) The S8 serine, C1A cysteine and A1 aspartic protease families in *Arabidopsis*. Phytochemistry 65:43–58

Belkhadir Y, Subramaniam R, Dangl JL (2004) Plant disease resistance protein signaling: NBS-LRR proteins and their partners. Curr Opin Plant Biol 7:391–399. https://doi.org/10.1016/j.pbi.2004.05.009

Berenbaum MR, Zangerl AR, Nitao JK (1986) Constraints on chemical evolution: wild parsnip and the parsnip webworm. Evolution 40:1215–1228

Bergelson J (1994) The effects of genotype and the environment on costs of resistance in lettuce. Am Nat 143:349–359

Berrocal-Lobo M, Molina A, Solano R (2002) Constitutive expression of *ETHYLENE-RESPONSE-FACTOR1* in *Arabidopsis* confers resistance to several necrotrophic fungi. Plant J 29:23–32

Bhayana L, Paritosh K, Arora H, Yadava SK, Singh P, Nadan D, Mukhopadhyay A, Gupta V, Pradhan AK, Pental D (2020) A mapped locus on LG A6 of *Brassica juncea* line tumida conferring resistance to white rust contains a CNL type R gene. Front Plant Sci 10:1690. https://doi.org/10.3389/fpls.2019.01690

Bittner-Eddy PD, Beynon JL (2001) The *Arabidopsis* downy mildew resistance gene, RPP13-Nd, functions independently of NDR1 and EDS1 and does not require the accumulation of salicylic acid. Mol Plant Microbe Interact 14:416–421

Bittner-Eddy PD, Can C, Gunn N, Pinel M, Tor M, Crute I, Holub EB, Beynon J (1999) Genetic and physical mapping of the *RPP13* locus, in *Arabidopsis*, responsible for specific recognition of several *Peronospora parasitica* (downy mildew) isolates. Mol Plant Microbe Interact 12:792–802

Bittner-Eddy PD, Crute IR, Holub EB, Beynon JL (2000) *RPP13* is a simple locus in *Arabidopsis thaliana* for alleles that specify downy mildew resistance to different avirulence determinants in *Peronospora parasitica*. Plant J 21:177–188

Boch J, Verbsky ML, Robertson TL, Larkin JC, Kunkel BN (1998) Analysis of resistance gene-mediated defense responses in *Arabidopsis thaliana* plants carrying a mutation in CPR5. Mol Plant Microbe Interact 11:1196–1206

Borhan MH, Brose E, Beynon JL, Holub EB (2001) White rust (*Albugo candida*) resistance loci on three *Arabidopsis* chromosomes are closely linked to downy mildew (*Peronospora parasitica*) resistance loci. Mol Plant Pathol 2:87–95

Borhan MH, Holub EB, Beynon JL, Rozwadowski K, Rimmer SR (2004) The *Arabidopsis* TIR-NB-LRR gene RAC1 confers resistance to *Albugo candida* (white rust) and is dependent on EDS1 but not PAD4. Mol Plant Microbe Interact 17:711–719

Borhan MH, Gunn N, Cooper A, Gulden S, Tor M, Rimmer SR, Holub EB (2008) WRR4 encodes a TIR-NB-LRR protein that confers broad-spectrum white rust resistance in *Arabidopsis thaliana* to four physiological races of *Albugo candida*. Mol Plant Microbe Interact 21:757–768

Botella MA, Parker JE, Frost LN, Bittner-Eddy PD, Beynon JL, Daniels MJ, Holub EB, Jones JD (1998) Three genes of the *Arabidopsis* RPP1 complex resistance locus recognize distinct *Peronospora parasitica* avirulence determinants. Plant Cell 10:1847–1860

Bowling SA, Guo A, Gordon AS, Klessig DF, Dong X (1994) A mutation in *Arabidopsis* that leads to constitutive expression of systemic acquired resistance. Plant Cell 6:1845–1857

Bowling SA, Clarke JD, Liu Y, Klessig DF, Dong X (1997) The *cpr5* mutant of *Arabidopsis* expresses both NPR1-dependent and NPR1-independent resistance. Plant Cell 9:1573–1584

Brininstool G, Kasili R, Simmons LA, Kirik V, Hulskamp M, Larkin JC (2008) Constitutive expresser of pathogenesis-related genes 5 affects cell wall biogenesis and trichome development. BMC Plant Biol 8:58

Brinkman A, Frey KJ (1977) Growth analysis of isoline-recurrent parent grain yield differences in oats. Crop Sci 17:426–430

Brodersen P, Sakvarelidze-Achard L, Bruun-Rasmussen M, Dunoyer P, Yamamoto YY, Sieburth L, Voinnet O (2008) Widespread translational inhibition by plant miRNAs and siRNAs. Science 320:1185–1190

Brodmann A, Schuller A, Ludwig-Muller J, Aeschbacher RA, Wiemken A, Boller T, Wingler A (2002) Induction of trehalase in *Arabidopsis* plants infected with the trehalose producing pathogen *Plasmodiophora brassicae*. Mol Plant Microbe Interact 15:693–700

Buczacki ST, Toxopeus H, Mattusch P, Johnston TD, Dixon GR, Hobolth LA (1975) Study of physiologic specialization in *Plasmodiophora brassicae*: proposals for attempted rationalization through an international approach. Trans Br Mycol Soc 65:295–303. https://doi.org/10.1016/s0007-1536(75)80013-1

Buhler M, Haas W, Gygi SP, Moazed D (2007) RNAi-dependent and –independent RNA turnover mechanisms contribute to heterochromatic gene silencing. Cell 129:707–721

Burdon JJ, Muller WJ (1987) Measuring the cost of resistance to *Puccinia coronata* in *Avena fatua*. J Appl Ecol 24:191–200

Butruille DV, Guries RP, Osborn TC (1999) Linkage analysis of molecular markers and quantitative trait loci in populations of inbred backcross lines of *Brassica napus*. Genetics 133:950–964

Cai C, Wang X, Liu B, Wu J, Liang J, Cui Y, Cheng F, Wang X (2017) *Brassica rapa* genome 2.0: a reference upgrade through sequence re-assembly and gene re-annotation. Mol Plant 10:649–651

Cao Y, Wu Y, Zheng Z, Song F (2005) Overexpression of the rice EREBP-like gene *OsBIERF3* enhances disease resistance and salt tolerance in transgenic tobacco. Physiol Mol Plant Pathol 67:202–211

Cao T, Srivastava S, Rahman MH, Kav NNV, Hotte N, Deyholos MK, Strelkov SE (2008) Proteome-level changes in the roots of *Brassica napus* as a result of *Plasmodiophora brassicae* infection. Plant Sci 174:97–115

Carlier JD, Alabaça CS, Sousa NH, Coelho PS, Monteiro AA, Paterson AH, Leitao JM (2011) Physical mapping in a triplicated genome: mapping the downy mildew resistance locus Pp523 in *Brassica oleracea* L. G3 (Bethesda) 1:593–601. https://doi.org/10.1534/g3.111.001099

Carlsson M, Bothmer RV, Merker A (2004) Screening and evaluation of resistance to downy mildew (*Peronospora parasitica*) and clubroot (*Plasmodiophora brassicae*) in genetic resources of *Brassica oleracea*. Hereditas 141:293–300. https://doi.org/10.1111/j.1601-5223.2004.01818.x

Cavell AC, Lydiate DC, Parkin IAP, Dean C, Trick M (1998) Collinearity between a 30-centimorgan segment of *Arabidopsis thaliana* chromosome 4 and duplicated regions within the *Brassica napus* genome. Genome 41:62–69

Century KS, Holub EB, Staskawicz BJ (1995) NDR1, a locus of *Arabidopsis thaliana* that is required for disease resistance to both bacterial and fungal pathogen. Proc Natl Acad Sci U S A 92:6597–6601

Century KS, Shapiro AD, Repetti PP, Dahlbeck D, Holub E, Staskawicz BJ (1997) *NDR1*, a pathogen-induced component required for *Arabidopsis* disease resistance. Science 278:1963–1965

Cevik V, Boutrot F, Apel W, Robert-Seilaniantz A, Furzer OJ, Redkar A, Castel B, Kover PX, Prince DC, Holub EB, Jones JDG (2019) Transgressive segregation reveals mechanisms of

Arabidopsis immunity to *Brassica*-infecting races of white rust *Albugo candida*. Proc Natl Acad Sci 116:2767–2773

Chai AL, Xie XW, Shi YX, Li BJ (2014) Research status of clubroot (*Plasmodiophora brassicae*) on cruciferous crops in China. Can J Plant Pathol 36:142–153. https://doi.org/10.1080/07060661.2013.868829

Chalhoub B, Denoeud F, Liu S, Parkin IA, Tang H, Wang X, Chiquet J, Belcram H, Tong C, Samans B, Corréa M, Da Silva C, Just J, Falentin C, Koh CS, Le Clainche I, Bernard M, Bento P, Noel B, Labadie K, Alberti A, Charles M, Arnaud D, Guo H, Daviaud C, Alamery S, Jabbari K, Zhao M, Edger PP, Chelaifa H, Tack D, Lassalle G, Mestiri I, Schnel N, Le Paslier MC, Fan G, Renault V, Bayer PE, Golicz AA, Manoli S, Lee TH, Thi VH, Chalabi S, Hu Q, Fan C, Tollenaere R, Lu Y, Battail C, Shen J, Sidebottom CH, Wang X, Canaguier A, Chauveau A, Bérard A, Deniot G, Guan M, Liu Z, Sun F, Lim YP, Lyons E, Town CD, Bancroft I, Wang X, Meng J, Ma J, Pires JC, King GJ, Brunel D, Delourme R, Renard M, Aury JM, Adams KL, Batley J, Snowdon RJ, Tost J, Edwards D, Zhou Y, Hua W, Sharpe AG, Paterson AH, Guan C, Wincker P (2014) Early allopolyploid evolution in the post-neolithic *Brassica napus* oilseed genome. Science 345(6199):950–953

Chandran D, Rickert J, Cherk C, Dotson BR, Wildermuth MC (2013) Host cell ploidy underlying the fungal feeding site is a determinant of powdery mildew growth and reproduction. Mol Plant Microbe Interact 26:537–545

Chang A, Lamara M, Wei Y, Hu H, Parkin IAP, Gossen BD, Peng G, Yu F (2019) Clubroot resistance gene *Rcr6* in *Brassica nigra* resides in a genomic region homologous to chromosome A08 in *B. rapa*. BMC Plant Biol 19:224. https://doi.org/10.1186/s12870-019-1844-5

Chaplin JF (1970) Associations among disease resistance, agronomic characteristics, and chemical constituents in flue-cured tobacco. Agron J 62:87–91

Chelkowski J, Tyrka M, Sobkiewicz A (2003) Resistance genes in barley (*Hordeum vulgare* L.) and their identification with molecular markers. J Appl Genet 44:291–309

Chen XF, Hou XL, Zhang JY, Zheng JQ (2008) Molecular characterization of two important antifungal proteins isolated by downy mildew infection in non-heading Chinese cabbage. Mol Biol Rep 35(4):621–629

Chen J, Jing J, Zhan Z, Zhang T, Zhang C, Piao Z (2013) Identification of novel QTLs for isolate-specific partial resistance to *Plasmodiophora brassicae* in *Brassica rapa*. PLoS One 8:e85307. https://doi.org/10.1371/journal.pone.0085307

Chen J, Pang W, Chen B, Zhang C, Piao Z (2015) Transcriptome analysis of *Brassica rapa* near-isogenic lines carrying clubroot-resistant and -susceptible alleles in response to *Plasmodiophora brassicae* during early infection. Front Plant Sci 6:1183. https://doi.org/10.3389/fpls.2015.01183

Chen L, Zhang X, Xu H, Song B, Fan X (2016) Introgression of clubroot resistance into an elite pak choi inbred line through marker-assisted introgression breeding. Plant Breed 135:471–475. https://doi.org/10.1111/pbr.12379

Chen D, Cao Y, Li H, Kim D, Ahsan N, Thelen J, Stacey G (2017) Extracellular ATP elicits DORN1-mediated *RbohD* phosphorylation to regulate stomatal aperture. Nat Commun 8:1–13

Cheng F, Liu S, Wu J, Fang L, Sun S, Liu B, Li P, Hua W, Wang X (2011) BRAD, the genetics and genomics database for *Brassica* plants. BMC Plant Biol 11:136. https://doi.org/10.1186/1471-2229-11-136

Cheng F, Wu J, Cai C, Fu F, Liang J, Borm T, Zhuang M, Zhang Y, Zhang F, Bonnema G, Wang X (2016) Genome resequencing and comparative varicose analysis in a *Brassica rapa* and *Brassica oleracea* collection. Sci Data 3:160119

Cheung WY, Gugel RK, Landry BS (1998) Identification of RFLP markers linked to the white rust resistance gene (Acr) in mustard [*Brassica juncea* (L.) Czern & Coss.]. Genome 41:626–628. https://doi.org/10.1139/g98-043

Cheung F, Trick M, Drou N, Lim YP, Park JY, Kwon SJ, Kim JA, Scott R, Pires JC, Paterson AH, Town C, Bancroft I (2009) Comparative analysis between homoeologous genome segments of *Brassica napus* and its progenitor species reveals extensive sequence-level divergence. Plant Cell 21:1912–1928. https://doi.org/10.1105/tpc.108.060376

Choi S, Creelman RA, Mullet JE, Wing RA (1995) Construction and characterisation of a bacterial artificial chromosome library of *Arabidopsis thaliana*. Weeds World 2:17–20

Choi YJ, Hong SB, Shin HD (2006) Genetic diversity within the *Albugo candida* complex (Peronosporales, Oomycota) inferred from phylogenetic analysis of ITS rDNA and COX2 mtDNA sequences. Mol Phylogenet Evol 40:400–409

Chu M, Song T, Falk KC, Zhang X, Liu X, Chang A, Lahlali R, McGregor L, Gossen BD, Peng G, Yu F (2014) Fine mapping of Rcr1 and analyses of its effect on transcriptome patterns during infection by *Plasmodiophora brassicae*. BMC Genomics 15:1166. https://doi.org/10.1186/1471-2164-15-1166

Chung H, Jeong YM, Mun JH, Lee SS, Chung WH, Yu HJ (2014) Construction of a genetic map based on high-throughput SNP genotyping and genetic mapping of a *TuMV* resistance locus in *Brassica rapa*. Mol Gen Genomics 289:149–160

Clarke JD, Liu Y, Klessig DF, Dong X (1998) Uncoupling *PR* gene expression from NPR1 and bacterial resistance, characterization of the dominant *Arabidopsis cpr 6-1* mutant. Plant Cell 10:557–569

Clarke JD, Volko SM, Ledford H, Ausubel FM, Dong X (2000) Roles of salicylic acid, jasmonic acid, and ethylene in *cpr*-induced resistance in *Arabidopsis*. Plant Cell 12:2175–2190

Clarke WE, Higgins EE, Plieske J, Wieseke R, Sidebottom C, Khedikar Y, Batley J, Edwards D, Meng J, Li R, Lawley CT, Pauquet J, Laga B, Cheung W, Iniguez-Luy F, Dyrszka E, Rae S, Stich B, Snowdon RJ, Sharpe AG, Ganal MW, Parkin I (2016) A high-density SNP genotyping array for *Brassica napus* and its ancestral diploid species based on optimized selection of single-locus markers in the allotetraploid genome. Theor Appl Genet 129:1887–1899

Clemens JC, Worby CA, Simonson-Leff N, Muda M, Maehama T, Hemmings BA, Dixon JE (2000) Use of double-stranded RNA interference in *Drosophila* cell lines to dissect signal transduction pathways. Proc Natl Acad Sci U S A 97:6499–6503

Coelho PS, Monteiro AA (2003) Expression of resistance to downy mildew at cotyledon and adult plant stages in *Brassica oleracea* L. Euphytica 133:279–284. https://doi.org/10.1023/A:1025787608919

Coelho P, Bahcevandziev K, Valério L, Monteiro A, Leckie D, Astley D, Crute IR, Boukema I (1998) The relationship between cotyledon and adult plant resistance to downy mildew (*Peronospora parasitica*) in *Brassica oleracea*. Acta Hortic 459:335–342. https://doi.org/10.17660/ActaHortic.1998.459.39

Coelho PS, Valério L, Monteiro AA (2009) Leaf position, leaf age and plant age affect the expression of downy mildew resistance in *Brassica oleracea*. Eur J Plant Pathol 125:179–188

Cohen Y, Peer S, Balass O, Coffey MD (1987) A fluorescent technique for studying growth of *Peronospora tabacina* on leaf surfaces. Phytopathology 77(2):201–204. https://doi.org/10.1094/Phyto-77-201

Collier SM, Moffett P (2009) NB-LRRs work a "bait and switch" on pathogens. Trends Plant Sci 14 (10):521–529

Collins RE, Cheng X (2005) Structural domains in RNAi. FEBS Lett 579:5841–5849

Collins NC, Thordal-Christensen H, Lipka V, Bau S, Kombrink E, Qiu JL, Huckelhoven R, Stein M, Freialdenhoven A, Somerville SC, Schulze-Lefert P (2003) SNARE-protein-mediated disease resistance at the plant cell wall. Nature 4256(1):973–977

Colome-Tatche M, Cortijo S, Wardenaar R, Morgado L, Lahouze B, Sarazin A, Etcheverry M, Martin A, Feng S, Duvernois-Berthet E, Labadie K, Wincker P, Jacobsen SE, Jansen RC, Colot V, Johannes F (2012) Features of the *Arabidopsis* recombination landscape resulting from the combined loss of sequence variation and DNA methylation. Proc Natl Acad Sci U S A 109:16240–16245

Conesa A, Gotz S, Garcia-Gomez JM, Terol J, Talon M, Robles M (2005) Blast2GO: a universal tool for annotation, visualization and analysis in functional genomics research. Bioinformatics 21:3674–3676. https://doi.org/10.1093/bioinformatics/bti610

Conklin PL, Last RL (1995) Differential accumulation of antioxidant mRNAs in *Arabidopsis thaliana* exposed to ozone. Plant Physiol 109:203–212

Consonni C, Humphry ME, Hartmann HA, Livaja M, Durner J, Westphal L, Vogel J, Lipka V, Kemmerling B, Schulze-Lefert P, Somerville SC, Panstruga R (2006) Conserved requirement for a plant host cell protein in powdery mildew pathogenesis. Nat Genet 386(1):716–720

Consonni C, Bednarek P, Humphry M, Francocci F, Ferrari S, Harzen A (2010) Tryptophan-derived metabolites are required for antifungal defense in the *Arabidopsis mlo2* mutant. Plant Physiol 152:1544–1561. https://doi.org/10.1104/pp.109.147660

Cooke DEL, Drenth A, Duncan JM, Wagels G, Brasier CM (2000) A molecular phylogeny of *Phytophthora* and related oomycetes. Fungal Genet Biol 30:17–32

Cooley MB, Pathirana S, Wu HJ, Kachroo P, Klessig DF (2000) Members of the *Arabidopsis* HRT/RPP8 family of resistance genes confer resistance to both viral and oomycete pathogens. Plant Cell 12:663–676. https://doi.org/10.1105/tpc.12.5.663

Cooper AJ, Latunde-Dada AO, Woods-Tor A, Lynn J, Lucas JA, Crute IR, Holub EB (2008) Basic compatibility of *Albugo candida* in *Arabidopsis thaliana* and *Brassica juncea* causes broad-spectrum suppression of innate immunity. Mol Plant Microbe Interact 21:745–756

Cortijo S, Wardenaar R, Colome-Tatche M, Gilly A, Etcheverry M, Labadie K, Caillieux E, Hospital F, Aury JM, Wincker P, Roudier F, Jansen RC, Colot V, Johannes F (2014) Mapping the epigenetic basis of complex traits. Science 343:1145–1148

Crisp P, Crute IR, Sutherland RA, Angell SM, Bloor K, Burgess H, Gordon PL (1989) The exploitation of genetic resources of *Brassica oleracea* in breeding for resistance to clubroot (*Plasmodiophora brassicae*). Euphytica 42:215–226

Crute IR, Gray AR, Crisp P, Buczacki ST (1980) Variation in *Plasmodiophora brassicae* and resistance to clubroot disease in brassicas and allied crops—a critical review. Plant Breed Abstr 50:91–104

Crute IR, Holub EB, Tor M, Brose E, Beynon JL (1993) The identification and mapping of loci in *Arabidopsis thaliana* for recognition of the fungal pathogens *Peronospora parasitica* (downy mildew) and *Albugo candida* (white blister). In: Nester EW, Verma DPS (eds) Advances in molecular genetics of plant-microbe interactions. Kluwer Academic, Dordrecht, pp 437–444

Dangl JL, Jones JD (2001) Plant pathogens and integrated defense responses to infection. Nature 411:826–833. https://doi.org/10.1038/35081161

Das R, Roy A, Dutta N, Majumder H (2008) Reactive oxygen species and imbalance of calcium homeostasis contributes to curcumin induced programmed cell death in *Leishmania donovani*. Apoptosis 13:867–882

Degenhardt J, Al-Masri AN, Kurkcuoglu S, Szankowski I, Gau AE (2005) Characterization by suppression subtractive hybridization of transcripts that are differentially expressed in leaves of apple scab-resistant and susceptible cultivars of *Malus domestica*. Mol Genet Genome 273:326–335

Delaney TP, Uknes S, Vernooij B, Friedrich L, Weinmann K, Negrotto D, Gaffney T, Gut-Rella M, Kessmann H, Ward E, Ryals J (1994) A central role of salicylic acid in plant disease resistance. Science 266:1247–1250

Delaney TP, Friedrich L, Ryals JA (1995) *Arabidopsis* signal transduction mutant defective in chemically and biologically induced disease resistance. Proc Natl Acad Sci U S A 92:6602–6606. https://doi.org/10.1073/pnas.92.14.6602

Delourme R, Pilet-Nayel ML, Archipiano M, Horvais R, Tanguy X, Rouxel T, Brun H, Renard M, Balesdent MH (2004) A cluster of major specific resistance genes to *Leptosphaeria maculans* in *Brassica napus*. Phytopathology 94:578–583. https://doi.org/10.1094/PHYTO.2004.94.6.578

Devadas SK, Enyedi A, Raina R (2002) The *Arabidopsis hrl1* mutation reveals novel overlapping roles for salicylic acid, jasmonic acid and ethylene signaling in cell death and defense against pathogens. Plant J 30:467–480

Devos S, Prinsen E (2006) Plant hormones: a key in clubroot development. Commun Agric Appl Biol Sci 71(3 Pt B):869–872

Devos S, Laukens K, Deckers P, Van Der Straeten D, Beeckman T, Inze D, van Onckelen H, Witters E, Prinsen E (2006) A hormone and proteome approach to picturing the initial metabolic events during *Plasmodiophora brassicae* infection on *Arabidopsis*. Mol Plant Microbiol Inter 19:1431–1433

Devoto A, Hartmann HA, Piffanelli P, Elliott C, Simmons C, Taramino G, Goh CS, Cohen FE, Emerson BC, Schulze-Lefert P, Panstruga R (2003) Molecular phylogeny and evolution of the plant-specific seven-transmembrane MLO family. J Mol Evol 56:77–88

Dickinson CH, Crute IR (1974) The influence of seedling age and development on the infection of lettuce by Bremia lactucae. Ann Appl Biol 76:49–56

Dickman MB, Fluhr R (2013) Centrality of host cell death in plant-microbe interactions. Annu Rev Phytopathol 51:543–570. https://doi.org/10.1146/annurev-phyto-081211-173027

Dickson MH, Petzoldt R (1993) Plant age and isolate source affect expression of downy mildew resistance in broccoli. Hort Sci 28:730–731

Diederichsen E, Sacristan MD (1996) Disease response of resynthesized *Brassica napus* L. lines carrying different combinations of resistance to *Plasmodiophora brassicae* worn. Plant Breed 115:5–10. https://doi.org/10.1111/j.1439-0523.1996.tb00862.x

Diederichsen E, Beckmann J, Schondelmeier J, Dreyer F (2006) Genetics of clubroot resistance in *Brassica napus* 'Mendel'. Acta Hortic 706:307–312. https://doi.org/10.17660/ActaHortic.2006.706.35

Dietrich RA, Delaney TP, Uknes SJ, Ward ER, Ryals JA, Dangl JL (1994) *Arabidopsis* mutants simulating disease resistance response. Cell 77:565–577

Dinkova TD, Martinez-Castilla L, Cruz-Espíndola MA (2016) The diversification of eif4e family members in plants and their role in the plant-virus interaction. In: Evolution of the protein synthesis machinery and its regulation. Springer International, Cham, pp 187–205

Dodds P, Lawrence G, Ellis J (2001) Six amino acid changes confined to the leucine-rich repeat beta-strand/beta-turn motif determine the difference between the P and P2 rust resistance specificities in flax. Plant Cell 13:163–178

Dong X (2001) Genetic dissection of systemic acquired resistance. Curr Opin Plant Biol 4:309–314

Dong XN (2004) NPR1, all things considered. Curr Opin Plant Biol 7:547–552. https://doi.org/10.1016/j.pbi.2004.07.005

Dormann P, Kim H, Ott T, Schulze-Lefert P, Trujillo M, Wewer V, Huckelhoven R (2014) Cell-autonomous defense, re-organization and trafficking of membranes in plant-microbe interactions. New Phytol 204:815–822

Ellis JG, Lawrence GJ, Luck JE, Dodds PN (1999) Identification of regions in alleles of the flax rust resistance gene L that determines differences in gene-for-gene specificity. Plant Cell 11:495–506

Ellis JG, Dodds P, Pryor T (2000) Structure, function and evolution of plant disease resistance genes. Curr Opin Plant Biol 3:278–284

Eulgem T, Weigman VJ, Chang HS, McDowell JM, Holub EB, Glazebrook J, Zhu T, Dangl JL (2004) Three genetically separable R gene-signaling pathways converge to regulate a largely overlapping transcriptome. Plant Physiol 135:1129–1144

Eulgem T, Tsuchiya T, Wang XJ, Beasley B, Cuzick A, Tör M, Zhu T, McDowell JM, Holub E, Dangl JL (2007) EDM2 is required for RPP7-dependent disease resistance in *Arabidopsis* and affects RPP7 transcript levels. Plant J 49(5):829–839. https://doi.org/10.1111/j.1365-313X.2006.02999.x

Falk A, Feys BJ, Frost LN, Jones JDG, Daniels MJ, Parker JE (1999) EDS1, an essential component of R gene-mediated disease resistance in Arabidopsis has homology to eukaryotic lipases. Proc Natl Acad Sci U S A 96:3292–3297

Farinhó M, Coelho P, Carlier J, Svetleva D, Monteiro A, Leitao J (2004) Mapping of a locus for adult plant resistance to downy mildew in broccoli (*Brassica oleracea* convar. *Italica*). Theor Appl Genet 109:1392–1398. https://doi.org/10.1007/s00122-004-1747-0

Farinho M, Coelho P, Monteiro A, Leitao J (2007) SCAR and CAPS markers flanking the *Brassica oleracea* L. Pp523 downy mildew resistance locus demarcate a genomic region syntenic to the top arm end of *Arabidopsis thaliana* L. chromosome 1. Euphytica 157:215–221. https://doi.org/10.1007/s10681-007-9414-6

Feng HQ, Sun K, Li MQ, Li HY, Li X, Li Y, Wang YF (2010) The expression, function and regulation of mitochondrial alternative oxidase under biotic stresses. Mol Plant Pathol 11:429–440

Feng J, Hwang SF, Strelkov SE (2012a) Studies into primary and secondary infection processes by *Plasmodiophora brassicae* on canola. Plant Pathol 62(1):177–183. https://doi.org/10.1111/j.1365-3059.2012.02612.x

Feng J, Xiao Q, Hwang SF, Strelkov SE, Gossen BD (2012b) Infection of canola by secondary zoospores of *Plasmodiophora brassicae* produced on a nonhost. Eur J Plant Pathol 132:309–315

Ferreira ME, Williams PH, Osborn TC (1994) RFLP mapping of *Brassica napus* using doubled haploid lines. Theor Appl Genet 89:615–621

Ferreira ME, Williams PH, Osborn TC (1995) Mapping of locus controlling resistance to *Albugo candida* in *Brassica napus* using molecular markers. Phytopathology 85:218–220

Feys BJ, Moisan LJ, Newman MA, Parker JE (2001) Direct interaction between the *Arabidopsis* disease resistance signaling proteins, EDS1 and PAD4. EMBO J 20:5400–5411

Figdore SS, Ferreira ME, Slocum MK, Williams PH (1993) Association of RFLP markers with trait loci affecting clubroot resistance and morphological characters in *Brassica oleracea* L. Euphytica 69(1/2):33–44

Fredua-Agyeman R, Rahman H (2016) Mapping of the clubroot disease resistance in spring *Brassica napus* canola introgressed from European winter canola cv. 'Mendel'. Euphytica 211:201–213. https://doi.org/10.1007/s10681-016-1730-2

Frey KJ, Browning JA (1971) Association between genetic factors for crown rust resistance and yield in oats. Crop Sci 11:757–760

Frye CA, Innes RW (1998) An *Arabidopsis* mutant with enhanced resistance to powdery mildew. Plant Cell 10:947–956

Frye CA, Tang D, Innes RW (2001) Negative regulation of defense responses in plants by a conserved MAPKK kinase. Proc Natl Acad Sci U S A 98:373–378

Fu LJ, Shi K, Gu M, Zhou YH, Dong DK, Liang WS, Song FM, Yu JQ (2010) Systemic induction and role of mitochondrial alternative oxidase and nitric oxide in a compatible tomato–*tobacco mosaic virus* interaction. Mol Plant Microbe Interact 23:39–48

Fuchs H, Sacristan MD (1996) Identification of a gene in *Arabidopsis thaliana* controlling resistance to clubroot and characterization of the resistance response. Mol Plant Microbe Interact 9:91–97

Fujiwara A, Inukai T, Kim BM, Masuta CI (2011) Combinations of a host resistance gene and the CI gene of *turnip mosaic virus* differentially regulate symptom expression in *Brassica rapa* cultivars. Arch Virol 156:1575–1581

Gallusci P, Dai Z, Génard M, Gauffretau A, Leblanc-Fournier N, Richard-Molard C, Vile D, Brunel-Muguet S (2017) Epigenetics for plant improvement: current knowledge and modeling avenues. Trends Plant Sci 22:610–623

Gao M, Li G, Yang B, Qiu D, Farnham M, Quiros C (2007) High-density *Brassica oleracea* linkage map: identification of useful new linkages. Theor Appl Genet 115:277–287

Gao S, Liu T, Li Y, Wu Q, Fu K, Chen J (2012) Understanding resistant germplasm-induced virulence variation through analysis of proteomics and suppression subtractive hybridization in a maize pathogen *Curvularia lunata*. Proteomics 12:3524–3535. https://doi.org/10.1002/pmic.201200105

Gassmann W, Hinsch ME, Staskawicz BJ (1999) The *Arabidopsis* RPS4 bacterial-resistance gene is a member of the TIR-NBS-LRR family of disease-resistance genes. Plant J 20(3):265–277. https://doi.org/10.1046/j.1365-313X.1999.t01-1-00600.x

Gaulin E, Haget N, Khatib M, Herbert C, Rickauer M, Bottin A (2007) Transgenic sequences are frequently lost in *Phytophthora parasitica* transformants without reversion of the transgene-induced silenced state. Can J Microbiol 53:152–157

Gazzani S, Lawrenson T, Woodward C, Headon D, Sablowski R (2004) A link between mRNA turnover and RNA interference in *Arabidopsis*. Science 306:1046–1048

Ghildiyal M, Seitz H, Horwich MD, Li C, Du T, Lee S, Xu J, Kittler EL, Zapp ML, Weng Z, Zamore PD (2008) Endogenous siRNAs derived from transposons and mRNAs in *Drosophila* somatic cells. Science 320:1077–1081

Gilliland A, Singh DP, Hayward JM, Moore CA, Murphy AM, YorkCJ SJ, Carr JP (2003) Genetic modification of alternative respiration has differential effects on antimycin a induced versus salicylic acid-induced resistance to tobacco mosaic virus. Plant Physiol 132:1518–1528

Giovannelli JL, Farnham MW, Wang M, Strand AE (2002) Development of sequence characterized amplified region markers linked to downy mildew resistance in broccoli. J Am Soc Hortic 127 (4):597–601

Glazebrook J (2005) Contrasting mechanisms of defense against biotrophic and necrotrophic pathogens. Annu Rev Phytopathol 43:205–227. https://doi.org/10.1146/annurev.phyto.43.040204.135923

Glazebrook J, Zook M, Mert F, Kagan I, Rogers EE, Crute IR, Holub EB, Hammerschmidt R, Ausubel FM (1997) Phytoalexin-deficient mutants of *Arabidopsis* reveal that PAD4 encodes a regulatory factor and that four PAD genes contribute to downy mildew resistance. Genetics 146:381–392

Golicz AA, Bayer PE, Barker GC, Edger PP, Kim H, Martinez PA, Chan CK, Severn-Ellis A, McCombie WR, Parkin IA, Paterson AH, Pires JC, Sharpe AG, Tang H, Teakle GR, Town CD, Batley J, Edwards D (2016) The pangenome of an agronomically important crop plant *Brassica oleracea*. Nat Commun 7:13390. https://doi.org/10.1038/ncomms13390

Gollner K, Schweizer P, Bai Y, Panstruga R (2008) Natural genetic resources of *Arabidopsis thaliana* reveals a high prevalence and unexpected phenotypic plasticity of RPW8-mediated powdery mildew resistance. New Phytol 1776(1):725–742

Gomez-Gomez L, Boller T (2000) FLS2: an LRR receptor-like kinase involved in the perception of the bacterial elicitor flagellin in *Arabidopsis*. Mol Cell 5:1003–1011

Grandclément C, Thomas G (1996) Detection and analysis of QTL based on RAPD markers for polygenic resistance to *Plasmodiophora brassicae* worn. In *Brassica oleracea* L. Theor Appl Genet 93:86–90

Grandclement C, Laurens F, Thomas G (1996) Genetic analysis of resistance to clubroot (*Plasmodiophora brassicae* Woron.) in two *Brassica oleracea* groups (subsp. *acephala* and subsp. *botrytis*) through diallel analysis. Plant Breed 115:152156

Grewal SI, Jia S (2007) Heterochromatin revisited. Nat Rev Genet 8:35–46

Grsic S, Kirchheim B, Pieper K, Fritsch M, Hilgenberg W, Ludwig-Muller J (1999) Induction of auxin biosynthetic enzymes by jasmonic acid and in clubroot diseased Chinese cabbage plants. Physiol Plant 105:521–531. https://doi.org/10.1034/j.1399-3054.1999.105318.x

Gu Y, Innes RW (2011) The KEEP ON GOING protein of *Arabidopsis* recruits the ENHANCED DISEASE RESISTANCE1 protein to trans-golgi network/early endosome vesicles. Plant Physiol 155:1827–1838

Gu Y, Innes RW (2012) The KEEP ON GOING protein of *Arabidopsis* regulates intracellular protein trafficking and is degraded during fungal infection. Plant Cell 24:4717–4730

Gunn ND, Byrne J, Holub EB (2002) Out crossing of two homo-thallic isolates of *Peronospora parasitica* and segregation of avirulence matching six resistance loci in *Arabidopsis thaliana*. In: Spencer-Phillips P, Gisi U, Lebeda A (eds) Advances in downy mildew research. Kluwer Academic, Dordrecht, pp 185–188

Guo HS, Fei JF, Xie Q, Chua NH (2003) A chemical-regulated inducible RNAi system in plants. Plant J 34:383–392

Gupta K, Prem D, Nashaat NI, Agnihotri A (2006) Response of interspecific *Brassica juncea*/ *Brassica rapa* hybrids and their advanced progenies to *Albugo candida* (white blister). Plant Pathol 55(5):679–689. https://doi.org/10.1111/j.1365-3059.2006.01433.x

Hammond-Kosack KE, Jones JD (1997) Plant disease resistance genes. Annu Rev Plant Physiol Plant Mol Biol 48:575–607

Han K, Lincoln DE (1994) The evolution of carbon allocation to plant secondary metabolites: a genetic analysis of cost in *Diplacus aurantiacus*. Evolution 48:1550–1563

Hansen AK, Andersen HV, Svendsen O (1994) Studies on the diagnosis of Tyzzer's disease in laboratory rat colonies with antibodies against *Bacillus pilifornis* (*Clostridiurn piliforrne*). Lab Anirn Sci 44:424–429

Hasan MJ, Rahman H (2016) Genetics and molecular mapping of resistance to *Plasmodiophora brassicae* pathotypes 2, 3, 5, 6, and 8 in rutabaga (*Brassica napus* var. *napobrassica*). Genome 59:805–815. https://doi.org/10.1139/gen-2016-0034

Hasan M, Strelkov SE, Howard RJ, Rahman H (2012) Screening of *Brassica* germplasm for resistance to *Plasmodiophora brassicae* pathotypes prevalent in Canada for broadening diversity in clubroot resistance. Can J Plant Sci 92:501–515

Hatakeyama K, Suwabe K, Tomita RN, Kato T, Nunome T, Fukuoka H, Matsumoto S (2013) Identification and characterization of Crr1a, a gene for resistance to clubroot disease (*Plasmodiophora brassicae* Woron) in *Brassica rapa* L. PLoS One 8:e54745. https://doi.org/10.1371/journal.Pone.0054745

Hatakeyama K, Niwa T, Kato T, Ohara T, Kakizaki T, Matsumoto S (2017) The tandem repeated organization of NB-LRR genes in the clubroot- resistant CRb locus in *Brassica rapa* L. Mol Genet Genomics 292:397–405. https://doi.org/10.1007/s00438-016-1281-1

Helm M, Schmid M, Hierl G, Terneus K, Tan L, Lottspeich F, Kieliszewski MJ, Gietl C (2008) KDEL-tailed cysteine end peptidases involved in programmed cell death, inter-calation of new cells and dismantling of extension scaffolds. Am J Bot 95:1049–1062

Hierl G, Vothknecht U, Gietl C (2012) Programmed cell death in *Ricinus* and *Arabidopsis*: the function of KDEL cysteine peptidases in development. Physiol Plant 145:103–113

Hierl G, Howing T, Isono E, Lottspeich F, Gietl C (2014) Ex vivo processing for maturation of Arabidopsis KDELtailed cysteine endopeptidase2 (AtCEP2) pro-enzyme and its storage in endoplasmic reticulum derived organelles. Plant Mol Biol 84:605–620. https://doi.org/10.1007/s11103-013-0157-6

Hirai M (2006) Genetic analysis of clubroot resistance in *Brassica* crops. Breed Sci 56:223–229

Hirai M, Harada T, Kubo N, Tsukada M, Suwabe K, Matsumoto S (2004) A novel locus for clubroot resistance in *Brassica rapa* and its linkage markers. Theor Appl Genet 108:639–643

Hirani AH, Li G (2015) Understanding the genetics of clubroot resistance for effectively controlling this disease in *Brassica* species. In: Plants for the future. Intech Open, London. https://doi.org/10.5772/60936

Hirani AH, Gao F, Liu J, Fu G, Li G (2016) Transferring clubroot resistance from Chinese cabbage (*Brassica rapa*) to canola (*B. napus*). Can J Plant Pathol 38(1):82–90. https://doi.org/10.1080/07060661.2016.1141799

Hoegen E, Strömberg A, Pihlgren U, Kombrik E (2002) Primary structure and tissue-specific expression of the pathogenesis-related protein PR-1b in potato. Mol Plant Pathol 3:329–345

Holub EB (1997) Organization of resistance genes in *Arabidopsis*. In: Crute IR, Holub EB, Burdon JJ (eds) The gene-for-gene relationship in plant-parasite interaction. CAB International, Wallingford, pp 5–26

Holub EB (2001) The arms race is ancient history in *Arabidopsis*, the wildflower. Nat Rev Genet 2:516–527

Holub EB (2007) Natural variation in innate immunity of a pioneer species. Curr Opin Plant Biol 10:415–424

Holub EB, Beynon JL (1997) Symbiology of mouse ear cress (*Arabidopsis thaliana*) and oomycetes. Adv Bot Res 24:228–273

Holub EB, Brose E, Tor M, Clay C, Crute IR, Beynon JL (1995) Phenotypic and genotypic variation in the interaction between *Arabidopsis thaliana* and *Albugo candida*. Mol Plant Microbe Interact 8:916–928

Hoser-Krauze J, Lakowska-Ryk E, Antosik J (1984) Resistance of cauliflower and broccoli (*B. oleracea* var. *botrytis* L.) seedlings to downy mildew (*Peronospora parasitica*). Cruciferae. NewsLett 9:92–93

Hoser-Krauze J, Lakowska-Ryk E, Antosik J (1995) The inheritance of broccoli (*Brassica oleracea* var *botrytis*) leaf resistance to downy mildew—*Peronospora parasitica* (Pers. ex. Fr.). J Appl Genet 36:27–33

Howing T, Huesmann C, Hoefle C, Nagel M-K, Isono E, Hueckelhoven R, Gietl C (2014) Endoplasmic reticulum KDEL-tailed cysteine endo-peptidases 1of *Arabidopsis* (AtCEP1) is involved in pathogen defense. Front Plant Sci 5:58. https://doi.org/10.3389/fpls.2014.00058

Howing T, Dann M, Hoefle C, Huckelhoven R, Gietl C (2017) Involvement of *Arabidopsis thaliana* endoplasmic reticulum KDEL- tailed cysteine endopeptidase 1 (AtCEP1) in powdery mildew-induced and AtCPR5-controlled cell death. PLoS One 12(8):e0183870

Huang YJ, Qi A, King GJ, Fitt BDL (2014) Assessing quantitative resistance against *Leptosphaeria maculans* (phoma stem canker) in *Brassica napus* (oilseed rape) in young plants. PLoS One 9: e84924. https://doi.org/10.1371/journal.pone.0084924

Huang Z, Peng G, Liu X, Deora A, Falk KC, Gossen BD, McDonald MR, Yu F (2017) Fine mapping of a clubroot resistance gene in Chinese cabbage using SNP markers identified from bulked segregant RNA sequencing. Front Plant Sci 8:1448. https://doi.org/10.3389/fpls.2017.01448

Huettel B, Kanno T, Daxinger L, Bucher E, van der Winden J, Matzke AJ, Matzke M (2007) RNA-directed DNA methylation mediated by *DRD1* and *pol IVb:* a versatile pathway for transcriptional gene silencing in plants. Biochim Biophys Acta 1769:358–374

Hughes SL, Hunter PJ, Sharpe AG, Kearsey MJ, Lydiate DJ, Walsh JA (2003) Genetic mapping of the novel turnip mosaic virus resistance gene TuRB03 in *Brassica napus*. Theor Appl Genet 107:1169–1173

Hugot K, Riviere MP, Moreilhon C, Dayem MA, Cozzitorto J, Arbiol G, Barbry P, Weiss C, Galiana E (2004) Coordinated regulation of genes for secretion in tobacco at late developmental stages:association with resistance against oomycetes. Plant Physiol 134:858–870. https://doi.org/10.1104/pp.103.034173

Hulbert SH, Webb CA, Smith SM, Sun Q (2001) Resistance gene complexes: evolution and utilization. Annu Rev Phytopathol 39:285–312. https://doi.org/10.1146/annurev.phyto.39.1.285

Humphry M, Consonni C, Panstruga R (2006) *Mlo*-based powdery-mildew immunity: silver bullet or simply non-host resistance. Mol Plant Pathol 76(1):605–610

Humphry M, Reinstadler A, Ivanov S, Bisseling T, Panstruga R (2011) Durable broad-spectrum powdery mildew resistance in pea *er1* plants is conferred by natural loss-of-function mutations in *PsMLO1*. Mol Plant Pathol 12:866–878. https://doi.org/10.1111/j.1364-3703.2011.00718.x

Hyodo K, Hashimoto K, Kuchitsu K, Suzuki N, Okuno T (2017) Harnessing host ROS-generating machinery for the robust genome replication of a plant RNA virus. Proc Natl Acad Sci U S A 114:1282–1290

Iida T, Kawaguchi R, Nakayama JI (2006) Conserved ribonuclease, Eri1, negatively regulates heterochromatin assembly in fission yeast. Curr Biol 16(14):1459–1464. https://doi.org/10.1016/j.cub.2006.05.061

Inada N, Ueda T (2014) Membrane trafficking pathways and their roles in plant-microbe interactions. Plant Cell Physiol 55:672–686

Inada N, Higaki T, Hasezawa S (2016) Nuclear function of subclass I actin-depolymerizing factor contributes to susceptibility in *Arabidopsis* to an adapted powdery mildew fungus. Plant Physiol 170:1420–1434

Irani S, Trost B, Waldner M, Nayidu N, Tu J, Kusalik AJ, Todd CD, Wei Y, Bonham-Smith PC (2018) Transcriptome analysis of response to *Plasmodiophora brassicae* infection in the *Arabidopsis* shoot and root. BMC Genomics 19:23

Ito S, Tanaka S, Miyanaga C, Takabayashi S, Yano S, Kameya-Iwaki M (1996) Constitutive and inducible proteins in the root of Chinese cabbage infected with *Plasmodiophora brassicae*. J Phytopathol 144:89–95

Jacobs AK, Lipka V, Burton RA, Panstruga R, Strizhov N, Schulze-Lefert P, Fincher GB (2003) An *Arabidopsis* Callose synthase, GSL5, is required for wound and papillary Callose formation. Plant Cell 15:2503–2513

Jenner CE, Sanchez F, Nettleship SB, Foster GD, Ponz F, Walsh JA (2000) The cylindrical inclusion gene of turnip mosaic virus encodes a pathogenic determinant to the Brassica resistance gene TuRB01. Mol Plant-Microbe Interact 13:1102–1108

Jenner CE, Tomimura K, Ohshima K, Hughes SL, Walsh JA (2002) Mutations in turnip mosaic virus P3 and cylindrical inclusion proteins are separately required to overcome two *Brassica napus* resistance genes. Virology 300:50–59

Jenner CE, Wang X, Tomimura K, Ohshima K, Ponz F, Walsh JA (2003) The dual role of the potyvirus P3 protein of turnip mosaic virus as a symptom and a virulence determinant in brassicas. Mol Plant Microbe Interact 16:777–784

Jensen BD, Hockenhull J, Munk L (1999a) Seedling and adult plant resistance to downy mildew (*Peronospora parasitica*) in cauliflower (*Brassica oleracea* convar. *botrytis* var. *botrytis*). Plant Pathol 48:604–612. https://doi.org/10.1046/j.1365-3059.1999.00388.x

Jensen BD, Værbak S, Munk L, Andersen SB (1999b) Characterization and inheritance of partial resistance to downy mildew, *Peronospora parasitica*, in breeding material of broccoli, *Brassica oleracea* con var. *botrytis* var. *italica*. Plant Breed 118:549–554

Jin S, Cheng Y, Guan Q, Liu D, Takano T, Liu S (2006) A metallothionein-like protein of rice (rgMT) functions in *E. coli* and its gene expression is induced by abiotic stresses. Biotechnol Lett 28:1749–1753

Jin M, Lee SS, Ke L, Kim JS, Seo MS, Sohn SH, Park BS, Bonnema G (2014) Identification and mapping of a novel dominant resistance gene, TuRB07 to *turnip mosaic virus* in *Brassica rapa*. Theor Appl Genet 127:509–519

Jirage D, Tootle TL, Reuber TL, Frost LN, Feys BJ, Parker JE, Ausubel FM, Glaze Brook J (1999) *Arabidopsis thaliana* PAD4 encodes a lipase-like gene that is important for salicylic acid signaling. Proc Natl Acad Sci U S A 96:13583–13588

Jirage D, Zhou N, Cooper B, Clarke JD, Dong X, Glaze Brook J (2001) Constitutive salicylic acid-dependent signaling in *cpr1* and *cpr6* mutants requires *PAD4*. Plant J 26:395–407

Johannes F, Colot V, Jansen RC (2008) Epigenome dynamics: a quantitative genetics perspective. Nat Rev Genet 9:883–890

Johannes F, Porcher E, Teixeira FK, Saliba-Colombani V, Simon M, Agier N, Bulski A, Albuisson J, Heredia F, Audigier P, Bouchez D, Dillmann C, Guerche P, Hospital F, Colot V (2009) Assessing the impact of transgenerational epigenetic variation on complex traits. PLoS Genet 5:e1000530

Jones JD, Dangl JL (2006) The plant immune system. Nature 444:323–329

Jones DA, Jones JDG (1997) The role of leucine-rich repeat proteins in plant defenses. Adv Bot Res 24:89–167

Joos HJ, Mauch-Mani B, Slusarenko AJ (1996) Molecular mapping of the *Arabidopsis* locus *RPP11* which conditions isolate-specific hypersensitive resistance against downy mildew in ecotype RLD. Theor Appl Genet 92:281–284

Jorgensen JH (1992) Discovery, characterization and exploitation of *Mlo* powdery mildew resistance in barley. Euphytica 63:141–152

Jubault M, Hamon C, Gravot A, Lariagon C, Delourme R, Bouchereau A, Manzanares-Dauleux MJ (2008) Differential regulation of root arginine catabolism and polyamine metabolism in clubroot-susceptible and partially resistant *Arabidopsis* genotypes. Plant Physiol 146:2008–2019

Judelson HS, Tani S (2007) Transgene-induced silencing of the zoosporogenesis-specific *NIFC* gene cluster of *Phytophthora infestans* involves chromatin alterations. Eukaryot Cell 6:1200–1209

Kachroo P, Yoshioka K, Shah J, Dooner HK, Klessig DF (2000) Resistance to turnip crinkle virus in *Arabidopsis* is regulated by two host genes and is salicylic acid dependent but npr1, ethylene, and jasmonate independent. Plant Cell 12:677–690

Kachroo P, Shanklin J, Shah J, Whittle EJ, Klessig DF (2001) A fatty acid desaturase modulates the activation of defense signaling pathways in plants. Proc Natl Acad Sci U S A 98:9448–9453

Kadota Y, Sklenar J, Dherbyshire P, Stransfeld L, Asai S, Notukakis V, Jones JDG, Shirasu K, Jones A, Zipfel C (2014) Direct regulation of the NADPH oxidase RBOHD by the PRR-associated kinase BIK1 during plant immunity. Mol Cell 54:43–55

Karim MM, Dakouri A, Zhang Y, Chen Q, Peng G, Strelkov SE, Gossen BD, Yu F (2020) Two clubroot-resistance genes, Rcr3 and Rcr9wa, mapped in *Brassica rapa* using bulk segregant RNA sequencing. Int J Mol Sci 21:5033. https://doi.org/10.3390/ijms21145033

Kato T, Hatakayema K, Fukino N, Matsumoto S (2012) Identification of a clubroot resistance locus conferring resistance to a *Plasmodiophora brassicae* classified into pathotype group 3 in

Chinese cabbage (*Brassica rapa* L.). Breed Sci 62:282–287. https://doi.org/10.1270/jsbbs.62.282

Kaur P, Jost R, Sivasithamparam K, Barbetti MJ (2011) Proteome analysis of the *A. candida–B. juncea* pathosystem reveals that the timing of the expression of defence-related genes is a crucial determinant of pathogenesis. J Exp Bot 62:1285–1298

Kennedy S, Wang D, Ruvkun G (2004) A conserved siRNA-degrading RNase negatively regulates RNA interference in *Colletotrichum elegans*. Nature 427:645–649

Kim HS, Delaney TP (2002) *Arabidopsis SON1* is an F-box protein that regulates a novel induced defense response independent of both salicylic acid and systemic acquired resistance. Plant Cell 14:1469–1482

Kim YJ, Hwang BK, Park KW (1989) Expression of age-related resistance in pepper plants infected with *Phytophthora capsici*. Plant Dis 73:745–747. https://doi.org/10.1094/PD-73-0745

Kim J, Kang WH, Yang HB, Park S, Jang C, Yu HJ, Kang BC (2013) Identification of a broad spectrum recessive gene in *Brassica rapa* and molecular analysis of the eIF4E gene family to develop molecular markers. Mol Breed 32:385–398

Kim H, O'Connell R, Maekawa-Yoshikawa M, Uemura T, Neumann U, Schulze-Lefert P (2014) The powdery mildew resistance protein RPW8.2 is carried on VAMP721/722 vesicles to the extrahaustorial membrane of haustorial complexes. Plant J 79:835–847

Kinkema M, Fan W, Dong X (2000) Nuclear localization of NPR1 is required for activation of PR gene expression. Plant Cell 12:2339–2350. https://doi.org/10.1105/tpc.12.12.2339

Kiraly L, Hafez YM, Fodor J, Kiraly Z (2008) Suppression of tobacco mosaic virus-induced hypersensitive-type necrotization in tobacco at high temperature is associated with down -regulation of NADPH oxidase and superoxide and stimulation of dehydroas-corbate reductase. J Gen Virol 89:799–808

Kirik V, Bouyer D, Schobinger U, Bechtold N, Herzog M, Bonneville JM, Hulskamp M (2001) CPR5 is involved in cell proliferation and cell death control and encodes a novel transmembrane protein. Curr Biol 11:1891–1895

Klessig DF, Malamy J, Hennig J, Sanchez-Casas P, Indulski J, Grynkiewicz G, Chen Z (1994) Induction, modification, and perception of the salicylic acid signal in plant defense. Biochem Soc Symp 60:219–229

Knoth C, Eulgem T (2008) The oomycete response gene *LURP1* is required for defense against *Hyaloperonospora parasitica* in *Arabidopsis thaliana*. Plant J 55(1):53–64

Knoth C, Ringler J, Dangl JL, Eulgem T (2007) *Arabidopsis* WRKY70 is required for full RPP4—mediated disease resistance and basal defense against *Hyaloperonospora parasitica*. Mol Plant Microbe Interact 20:120–128

Kobayashi Y, Kobayashi I, Funaki Y, Fujimoto S, Takemoto T, Kunoh H (1997a) Dynamic reorganization of microfilaments and microtubules is necessary for the expression of non-host resistance in barley coleoptile cells. Plant J 116(1):525–537

Kobayashi Y, Yamada M, Kobayashi I, Kunoh H (1997b) Actin microfilaments are required for the expression of nonhost resistance in higher plants. Plant Cell Physiol 386(1):725–733

Kobe B, Deisenhofer J (1994) The leucine-rich repeat: a versatile binding motif. Trends Biochem Sci 19:415–421

Kobe B, Deisenhofer J (1995) Proteins with leucine-rich repeats. Curr Opin Struct Biol 5:409–416

Koh S, Andre A, Edwards H, Ehrhardt D, Somerville S (2005) *Arabidopsis thaliana* subcellular responses to compatible *Erysiphe cichoracearum* infections. Plant J 446(1):516–529

Kole C, Teutonico R, Mengistu A, Williams PH, Osborn TC (1996) Molecular mapping of a locus controlling resistance to *Albugo candida* in *Brassica rapa*. Phytopathology 86:367–369

Kole C, Williams PH, Rimmer SR, Osborn TC (2002) Linkage mapping of genes controlling resistance to white rust (*Albugo candida*) in *Brassica rapa* (syn. *Campestris*) and comparative mapping to *Brassica napus* and *Arabidopsis thaliana*. Genome 45:22–27

Kombrink E, Schmelzer E (2001) The hypersensitive response and its role in local and systemic disease resistance. Eur J Plant Pathol 107:69–78

Kooke R, Johannes F, Wardenaar R, Becker F, Etcheverry M, Colot V, Vreugdenhil D, Keurentjes JJB (2015) Epigenetic basis of morphological variation and phenotypic plasticity in *Arabidopsis thaliana*. Plant Cell 27:337–348

Kuang H, Wei F, Marano MR, Wirtz U, Wang X, Liu J, Shum WP, Zaborsky J, Tallon LJ, Rensink W, Lobst S, Zhang P, Tornqvist CE, Tek A, Bamberg J, Helgeson J, Fry W, You F, Luo MC, Jiang J, Buell CR, Baker B (2005) The R1 resistance gene cluster contains three groups of independently evolving, type I R1 homologues and shows substantial structural variation among haplotypes of *Solanum demissum*. Plant J 44(1):37–51. https://doi.org/10.1111/j.1365-313X.2005.02506.x

Kuginuki Y, Ajisaka H, Yui M, Yoshikawa H, Hida K, Hirai M (1997) RAPD markers linked to a clubroot-resistance locus in *Brassica rapa* L. Euphytica 98:149–154

Kuhn H, Kwaaitaal M, Kusch S, Acevedo-Garcia J, Wu H, Panstruga R (2016) Biotrophy at its best: novel findings and unsolved mysteries of the *Arabidopsis*-powdery mildew pathosystem. Arabidopsis Book 14:e0184. https://doi.org/10.1199/tab.0184

Kumar GR, Bhat SR, Prakash S, Chopra VL (2003) Development of novel white rust resistant genetic stocks in crop Brassica by somatic hybridization. Plant Biotechnology 2002 and beyond. In: Proceedings of the 10th IAPTC & B Congress, Orlando, Florida, USA, 23–28 June, 2002. Kluwer Academic, Dordrecht, pp 555–558

Kus JV, Zaton K, Sarkar R, Cameron RK (2002) Age-related resistance in *Arabidopsis* is a developmentally regulated defense response to *Pseudomonas syringae*. Plant Cell 14:479–490

Kusch S, Panstruga R (2017) Mlo-based resistance: an apparently universal "weapon" to defeat powdery mildew disease. Mol Plant Microbe Interact 30:179–189. https://doi.org/10.1094/mpmi-12-16-0255-cr

Kusch S, Pesch L, Panstruga R (2016) Comprehensive phylogenetic analysis sheds light on the diversity and origin of the MLO family of integral membrane proteins. Genome Biol Evol 8:878–895

Kwaaitaal M, Keinath NF, Pajonk S, Biskup C, Panstruga R (2010) Combined bimolecular fluorescence complementation and forster resonance energy transfer reveals ternary SNARE complex formation in living plant cells. Plant Physiol 152:1135–1147. https://doi.org/10.1104/pp.109.151142

Kwak JM, Mori IC, Pei ZM, Leonhardt N, Torres MA, Dangl JL, Bloom RE, Bodde B, Jones JDG, Schroader JI (2003) NADPH oxidase *AtrbohD* and *AtrbohF* genes function in ROS-dependent ABA signaling in *Arabidopsis*. EMBO J 22:2623–2633

Kwon C, Panstruga R, Schulze-Lefert P (2008) Les liaisons dangereuses: immunological synapse formation in animals and plants. Trends Immunol 296(1):159–166

Lagercrantz U (1998) Comparative mapping between *Arabidopsis thaliana* and *Brassica nigra* indicates that *Brassica* genomes have evolved through extensive genome replication accompanied by chromosome fusions and frequent rearrangements. Genetics 150:1217–1228

Lagercrantz U, Lydiate DJ (1996) Comparative genome mapping in *Brassica*. Genetics 144 (4):1903–1910

Laila R, Park JI, Robin AHK, Natarajan S, Vijayakumar H, Shirasawa K, Isobe S, Kim HT, Nou IS (2019) Mapping of a novel clubroot resistance QTL using ddRAD-seq in Chinese cabbage (*Brassica rapa* L.). BMC Plant Biol 19:13. https://doi.org/10.1186/s12870-018-1615-8

Landry BS, Hubert N, Crete R, Chiang MS, Lincoln SE, Etoh T (1992) A genetic map for *Brassica oleracea* based on RFLP markers detected with expressed DNA sequences and mapping of resistance gene to race 2 of *Plasmodiophora brassicae*. Genome 35:409–420

Larkan NJ, Lydiate DJ, Parkin IA, Nelson MN, Epp DJ, Cowling WA, Rimmer SR, Borhan MH (2013) The *Brassica napus* blackleg resistance gene LepR3 encodes a receptor-like protein triggered by the *Leptosphaeria maculans* effector AVRLM1. New Phytol 197:595–605. https://doi.org/10.1111/nph.12043

Leborgne-Castel N, Bouhidel K (2014) Plasma membrane protein trafficking in plant-microbe interactions: a plant cell point of view. Front Plant Sci 5:735. https://doi.org/10.3389/fpls.2014.00735

Leckie F, Mattei B, Capodicasa C, Hemmings A, Nuss L, Aracri B, De Lorenzo G, Cervone F (1999) The specificity of polygalacturonase-inhibiting protein (PGIP): a single amino acid

substitution in the solvent-exposed beta-strand/beta-turn region of the leucine-rich repeats (LRRs) confers a new recognition capability. EMBO J 18:2352–2363
Lee Y, Kim YJ, Kim MH, Kwak JM (2016) MAPK cascades in guard cell signal transduction. Front Plant Sci 7:80
Lemarié S, Robert-Seilaniantz A, Lariagon C, Lemoine J, Marnet N, Jubault M, Manzanares-Dauleux MJ, Gravot A (2015) Both the jasmonic acid and the salicylic acid pathways contribute to resistance to the biotrophic clubroot agent *Plasmodiophora brassicae* in *Arabidopsis*. Plant Cell Physiol 56:2158–2168
Lenski RE (1988a) Experimental studies of pleiotropy and epistasis in *Escherichia coli*. I. Variation in competitive fitness among mutants resistant to virus T4. Evolution 42:425–432
Lenski RE (1988b) Experimental studies of pleiotropy and epistasis in *Escherichia coli*. II. Maladaptive effects associated with resistance to virus T4. Evolution 42:433–440
Li Y, Mei D, Huang Y, Xu Y, Roger W, Guy K, Richard B (2001) Biochemical changes of Zhongyou 821 after inoculation with mycelia of *Sclerotinia sclerotiorum*. Chin J Oil Crop Sci 23:63–65
Li QY, Tong HS, Zhang ZG, Zhao ZZ, Song XY (2011) Inheritance and development of EST-SSR marker associated with turnip mosaic virus resistance in Chinese cabbage. Can J Plant Sci 91:707–715
Li Q, Zhang X, Zeng Q et al (2015) Identification and mapping of a novel *turnip mosaic virus* resistance gene TuRBCS01 in Chinese cabbage (*Brassica rapa* L.). Plant Breed 134:221–225
Li L, Luo Y, Chen B, Xu K, Zhang F, Li H, Huang Q, Xiao X, Zhang T, Hu J, Li F, Wu X (2016) A genome-wide association study reveals new loci for resistance to clubroot disease in *Brassica napus*. Front Plant Sci 7:1483. https://doi.org/10.3389/fpls.2016.01483
Li G, Qian W, Zhang S, Zhang S, Li F, Zhang H, Fang Z, Wu J, Wang X, Sun R (2018) Variability in eukaryotic initiation factors iso4E in *Brassica rapa* influences interactions with the viral protein linked to the genome of turnip mosaic virus. Sci Rep 8:13588
Li G, Lv H, Zhang S, Zhang S, Li F, Zhang H, Qian W, Fang Z, Sun R (2019) *TuMV* management for *Brassica* crops through host resistance: retrospect and prospects. Plant Pathol 68:1035–1044. https://doi.org/10.1111/ppa.13016
Liebrand TWH, van den Berg GCM, Zhang Z, Smit P, Cordewener JHG, America AH, Sklenar J, Am J, Tameling WI, Robatzek S, Thmma BP, Joosten MH (2013) Receptor-like kinase SOBIR1/EVR interacts with receptor-like proteins in plant immunity against fungal infection (vol 110, pg 10010, 2013). Proc Natl Acad Sci U S A 110:13228–13228. https://doi.org/10.1073/pnas.1312631110
Liebrand TW, van den Burg HA, Joosten MH (2014) Two for all: receptor-associated kinases SOBIR1 and BAK1. Trends Plant Sci 19:123–132. https://doi.org/10.1016/j.tplants.2013.10.003
Liégard B, Baillet V, Etcheverry M, Joseph E, Lariagon C, Lemoine J, Evrard A, Colot V, Gravot A, Manzanares-Dauleux MJ, Jubault M (2019) Quantitative resistance to clubroot infection mediated by transgenerational epigenetic variation in *Arabidopsis*. New Phytol 22 (1):468–479. https://doi.org/10.1111/nph.15579
Lipka V, Kwon C, Panstruga R (2007) SNARE-ware: the role of SNARE domain proteins in plant biology. Annu Rev Cell Dev Biol 23:147–174
Liu H, Stone SL (2013) Cytoplasmic degradation of the *Arabidopsis* transcription factor ABSCISIC ACID INSENSITIVE 5 is mediated by the RING-type E3 ligase KEEP ON GOING. J Biol Chem 288:20267–20279
Liu Q, Rimmer SR, Scarth R (1989) Histopathology of compatibility and incompatibility between oilseed rape and *Albugo candida*. Plant Pathol 38:176–182
Liu J, Liu X, Dai L, Wang G (2007) Recent progress in elucidating the structure, function and evolution of disease resistance genes in plants. J Genet Genomics 34:765–776
Liu S, Liu Y, Yang X, Tong C, Edwards D, Parkin IA, Zhao M, Ma J, Yu J, Huang S, Wang X, Wang J, Lu K, Fang Z, Bancroft I, Yang TJ, Hu Q, Wang X, Yue Z, Li H, Yang L, Wu J, Zhou Q, Wang W, King GJ, Pires JC, Lu C, Wu Z, Sampath P, Wang Z, Guo H, Pan S, Yang L,

Min J, Zhang D, Jin D, Li W, Belcram H, Tu J, Guan M, Qi C, Du D, Li J, Jiang L, Batley J, Sharpe AG, Park BS, Ruperao P, Cheng F, Waminal NE, Huang Y, Dong C, Wang L, Li J, Hu Z, Zhuang M, Huang Y, Huang J, Shi J, Mei D, Liu J, Lee TH, Wang J, Jin H, Li Z, Li X, Zhang J, Xiao L, Zhou Y, Liu Z, Liu X, Qin R, Tang X, Liu W, Wang Y, Zhang Y, Lee J, Kim HH, Denoeud F, Xu X, Liang X, Hua W, Wang X, Wang J, Chalhoub B, Paterson AH (2014) The *Brassica oleracea* genome reveals the asymmetrical evolution of polyploid genomes. Nat Commun 5:3930. https://doi.org/10.1038/ncomms4930

Liu C, Yang Z, He P, Munir S, Wu Y, Ho H, He Y (2018) Deciphering the bacterial and fungal communities in clubroot-affected cabbage rhizosphere treated with *Bacillus subtilis* XF-1. Agric Ecosyst Environ 256:12–22

Loake G, Grant M (2007) Salicylic acid in plant defense—the players and protagonists. Curr Opin Plant Biol 10(5):466–472

Loapez-Fernaandez MP, Maldonado S (2015) Programmed cell death in seeds of angiosperms. J Integr Plant Biol 57:996–1002

Love AJ, Yun BW, Laval V, Loake GJ, Milner JJ (2005) Cauliflower mosaic virus, a compatible pathogen of Arabidopsis, engages three distinct defense-signaling pathways and activates rapid systemic generation of reactive oxygen species. Plant Physiol 139:935–948

Lucas JA, Crute IR, Sherriff C, Gordon PL (1988) The identification of a gene for race-specific resistance to *Peronospora parasitica* (downy mildew) in *Brassica napus* var. *oleifera* (oilseed rape). Plant Pathol 37:538–545

Luck JE, Lawrence GJ, Dodds PN, Shepherd KW, Ellis JG (2000) Regions outside of the leucine-rich repeats of flax rust resistance proteins play a role in specificity determination. Plant Cell 12:1367–1377

Luo Z, Chen Z (2007) Improperly terminated, unpolyadenylated mRNA of sense transgenes is targeted by RDR6-mediated RNA silencing in *Arabidopsis*. Plant Cell 19:943–958

Luo Y, Dong D, Su Y, Wang X, Peng Y, Peng J, Zhou C (2018) Transcriptome analysis of *Brassica juncea* var. *tumida* Tsen responses to *Plasmodiophora brassicae* primed by the biocontrol strain *Zhihengliuella aestuarii*. Funct Integr Genomics 18:301–314. https://doi.org/10.1007/s10142-018-0593-0

Lv H, Fang Z, Yang L, Zhang Y, Wang Y (2020) An update on the arsenal: mining resistance genes for disease management of *Brassica* crops in the genomic era. Hort Res 7:34. https://doi.org/10.1038/s41438-020-0257-9

Lydiate DJ, Pilcher RL, Higgins EE, Walsh JA (2014) Genetic control of immunity to *turnip mosaic virus* (*TuMV*) pathotype 1 in *Brassica rapa* (Chinese cabbage). Genome 57:419–425

Lyng Kjær MF, Newton AC, Atzema JL, Baker SJ (2000) The barleymlo-gene: an important powdery mildew resistance source. Agronomie 20:745–756

Lysak MA, Koch MA, Pecinka A, Schubert I (2005) Chromosome triplication found across the tribe Brassiceae. Genome Res 15(4):516–525

Ma JF, Hou XL, Xiao D, Qi L, Wang F, Sun F, Wang Q (2010) Cloning and characterization of the BcTUR3, gene related to resistance to *turnip mosaic virus* (*TuMV*) from non-heading Chinese cabbage. Plant Mol Biol Rep 28:588–596

Ma XF, Li Y, Sun JL, Wang TT, Fan J, Lei Y, Huang YY, Xu YJ, Zhao JQ, Xiao S, Wang WM (2014) Ectopic expression of *RESISTANCE TO POWDERY MILDEW 8.1* confers resistance to fungal and oomycete pathogens in *Arabidopsis*. Plant Cell Physiol 55:1484–1496

Mackey D, Holt BF III, Wiig A, Dangl JL (2002) RIN4 interacts with *Pseudomonas syringae* type III effector molecules and is required for RPM1-mediated resistance in *Arabidopsis*. Cell 108:743–754

Mackey D, Belkhadir Y, Alonso JM, Ecker JR, Dangl JL (2003) *Arabidopsis* RIN4 is a target of the type III virulence effector AvrRpt2 and modulates RPS2-mediated resistance. Cell 112:379–389

Maeda K, Houjyou Y, Komatsu T, Hori H, Kodaira T, Ishikawa A (2009) AGB1 and PMR5 contribute to PEN2-mediated pre-invasion resistance to *Magnaporthe oryzae* in *Arabidopsis thaliana*. Mol Plant Microbe Interact 22:1331–1340

Maekawa T, Kracher B, Vernaldi S, Ver Loren van Themaat E, Schulze-Lefert P (2012) Conservation of NLR-triggered immunity across plant lineages. Proc Natl Acad Sci U S A 109:20119–20123

Mahajan V, Gill HS, More TA (1995) Inheritance of downy mildew resistance in Indian cauliflower (group III). Euphytica 86:1–3. https://doi.org/10.1007/BF00035932

Makeyev EV, Bamford DH (2002) Cellular RNA-dependent RNA polymerase involved in post transcriptional gene silencing has two distinct activity modes. Mol Cell 10:1417–1427

Maleck K, Neuenschwander U, Cade RM, Dietrich RA, Dangl JL, Ryals JA (2002) Isolation and characterization of broad-spectrum disease-resistant *Arabidopsis* mutants. Genetics 160:1661–1671

Manzanares-Dauleux MJ, Delourme R, Baron F, Thomas G (2000) Mapping of one major gene and of QTLs involved in resistance to clubroot in *Brassica napus*. Theor Appl Genet 101:885–891. https://doi.org/10.1007/s001220051557

Marino D, Dunand C, Puppo A, Pauly N (2012) A burst of plant NADPH oxidases. Trends Plant Sci 17:9–15

Marone D, Russo MA, Laido G, De Leonardis AM, Mastrangelo AM (2013) Plant nucleotide binding site-leucine rich repeat (NBS-LRR) genes: active guardians in host defense response. Int J Mol Sci 14:7302–7326

Martin GB, Brommonschenkel SH, Chunwongse J, Frary A, Ganal MW, Spivy R, Wu T, Earle ED, Tanksley SD (1993) Map-based cloning of a protein kinase gene conferring disease resistance in tomato. Science 262:1432–1436

Matsumoto E, Yasui C, Ohi M, Tsukada M (1998) Linkage analysis of RFLP markers for clubroot resistance and pigmentation in Chinese cabbage (*Brassica rapa* ssp. *pekinensis*). Euphytica 104:79–86

Matsumoto E, Ueno H, Aruga D, Sakamoto K, Hayashida N (2012) Accumulation of three clubroot resistance genes through marker-assisted selection in Chinese cabbage (*Brassica rapa* ssp. *pekinensis*). Engei Gakkai Zasshi 81:184–190. https://doi.org/10.2503/jjshs1.81.184

Mayda E, Mauch-Mani B, Vera P (2000) *Arabidopsis dth9* mutation identifies a gene involved in regulating disease susceptibility without affecting salicylic acid-dependent responses. Plant Cell 12:2119–2128

Mazier M, Flamain F, Nicolaï M, Sarnette V, Caranta C (2011) Knock-down of both eIF4E1 and eIF4E2 genes confers broad-spectrum resistance against poty viruses in tomato. PLoS One 6: e29595

McDowell JM, Dhandaydham M, Long TA, Aarts MG, Goff S, Holub EB, Dangl JL (1998) Intragenic recombination and diversifying selection contribute to the evolution of downy mildew resistance at the RPP8 locus of *Arabidopsis*. Plant Cell 10:1861–1874

McDowell JM, Cuzick A, Can C, Beynon J, Dangl JL, Holub EB (2000) Downy mildew (*Peronospora parasitica)* resistance genes in *Arabidopsis* vary in functional requirements for *NDR1*, *EDS1*, *NPR1*, and salicylic acid accumulation. Plant J 22:523–530

McHale L, Tan X, Koehl P, Michelmore RW (2006) Plant NBS-LRR proteins: adaptable guards. Genome Biol 7:212. https://doi.org/10.1186/gb-2006-7-4-212

McHale LK, Haun WJ, Xu WW, Bhaskar PB, Anderson JE, Hyten DL, Gerhardt DJ, Jeddeloh JA, Stupar RM (2012) Structural variants in the soybean genome localize to clusters of biotic stress response genes. Plant Physiol 159:1295–1308. https://doi.org/10.1104/pp.112.194605

McMullan M, Gardiner A, Bailey K, Kemen E, Ward BJ, Cevik V, Robert-Seilaniantz A, Schultz-Larsen T, Balmuth A, Holub E, van Oosterhout C, Jones JDG (2015) Evidence for suppression of immunity as a driver for genomic introgressions and host range expansion in races of *Albugo candida*, a generalist parasite. Elife 4:e04550. https://doi.org/10.7554/eLife.04550.001

Mei J, Guo Z, Wang J, Feng Y, Ma G, Zhang C, Qian W, Chen G (2019) Understanding the resistance mechanism in *Brassica napus* to clubroot caused by *Plasmodiophora brassicae*. Phytopathology 109:810–818. https://doi.org/10.1094/PHYTO-06-18-0213-R

Mellersh DG, Heath MC (2003) An investigation into the involvement of defense signaling pathways in components of the nonhost resistance of *Arabidopsis thaliana* to rust fungi also reveals a model system for studying rust fungal compatibility. Mol Plant Microbiol Inter 16:398–404

Melotto M, Underwood W, Koczan J, Nomura K, He SY (2006) Plant stomata function in innate immunity against bacterial invasion. Cell 126:969–980

Meyers BC, Chin DB, Shen KA, Sivaramakrishnan S, Lavelle DO, Zhang Z, Michaelmore RW (1998) The major resistance gene cluster in lettuce is highly duplicated and spans several mega bases. Plant Cell 10(11):1817–1832

Meyers BC, Kozik A, Griego A, Kuang H, Michelmore RW (2003) Genome-wide analysis of NB-LRR-encoding genes in *Arabidopsis*. Plant Cell 15:809–834

Meyers BC, Kaushik S, Nandety RS (2005) Evolving disease resistance genes. Curr Opin Plant Biol 8:129–134

Micali C, Gollner K, Humphry M, Consonni C, Panstruga R (2008) The powdery mildew disease of *Arabidopsis*: a paradigm for the interaction between plants and biotrophic fungi. Arabidopsis Book 6:e0115. https://doi.org/10.1199/tab.0115

Micali CO, Neumann U, Grunewald D, Panstruga R, O'Connell R (2011) Biogenesis of a specialized plant–fungal interface during host cell internalization of Golovinomyces orontii haustoria. Cell Microbiol 13(2):210–226. https://doi.org/10.1111/j.1462-5822.2010.01530.x

Misas-Villamil JC, van der Hoorn RA, Doehlemann G (2016) Papain-like cysteine proteases as hubs in plant immunity. New Phytol 212:902–907

Mitchell-Olds T, James RV, Palmer MJ, Williams PH (1995) Genetics of *Brassica rapa* (syn. *Campestris*). 2. Multiple disease resistance to three fungal pathogens: *Peronospora parasitica*, *Albugo candida* and *Leptosphaeria maculans*. Heredity 75:362–369

Mitehell-Olds T, Bradley RD (1996) Genetics of *Brassica rapa*. 3. Cost of disease resistance to three fungal pathogens. Evolution 50(5):1859–1865

Moncalvo JM, Wang HH, Hseu RS (1995) Phylogenetic relationships in *Ganoderma* inferred from the internal transcribed spacers and 25S ribosomal DNA sequences. Mycologia 87:223–238

Monteiro AA, Williams PH (1989) The exploration of genetic resources of Portuguese cabbage and kale for resistance of several Brassica diseases. Euphytica 41:215–225. https://doi.org/10.1007/BF00021588

Monteiro AA, Coelho PS, Bahcevandziev K, Valério L (2005) Inheritance of downy mildew resistance at cotyledon and adult-plant stages in 'Couve Algarvia' (*Brassica oleracea* var. *tronchuda*). Euphytica 141:85–92. https://doi.org/10.1007/s10681-005-5696-8

Moore AL, Albury MS (2008) Further insights into the structure of the alternative oxidase: from plants to parasites. Biochem Soc T36:1022–1026

Morel J, Dangl JL (1997) The hypersensitive response and the induction of cell death in plants. Cell Death Differ 4:671–683

Morel JB, Dangl JL (1999) Suppressors of the *Arabidopsis lsd5* cell death mutation identify genes involved in regulating disease resistance responses. Genetics 151:305–319

Moriguchi K, Kimizuka-Takagi C, Ishii K, Nomura K (1999) A genetic map based on RAPD, RFLP, isozyme, morphological markers and QTL analysis for clubroot resistance in *Brassica oleracea*. Breed Sci 49:257–265. https://doi.org/10.1270/jsbbs.49.257

Moss NA, Crute IR, Lucas JA (1994) Laboratory production of oospores of *Peronospora parasitica* (crucifer downy mildew) and the recovery and characterization of sexual progeny from crosses between isolates with different host specificity. Plant Pathol 43:713–725

Mou Z, Fan W, Dong X (2003) Inducers of plant systemic acquired resistance regulate NPR function through redox changes. Cell 113:935–944. https://doi.org/10.1016/S0092-8674(03)00429-X

Mozo T, Fischer S, Shizuya H, Altmann T (1998) Construction and characterization of the IGF Arabidopsis BAC library. Mol Gen Genet 258:562–570

Mukherjee AK, Mohapatra T, Varshney A, Sharma R, Sharma RP (2001) Molecular mapping of a locus controlling resistance to *Albugo candida* in Indian mustard. Plant Breed 120:483–487. https://doi.org/10.1046/j.1439-0523.2001.00658.x

Mur LAJ, Kenton P, Lloyd AJ, Ougham H, Prats E (2008) The hypersensitive response; the centenary is upon us but how much do we know? J Exp Bot 59:501–520

Murphy AM, Gilliland A, York CJ, Hyman B, Carr JP (2004) High-level expression of alternative oxidase protein sequences enhances the spread of viral vectors in resistant and susceptible plants. J Gen Virol 85:3777–3786

Muskett PR, Kahn K, Austin MJ, Moisan LJ, Sadanandom A, Shirasu K, Jones JDG, Parker JE (2002) *Arabidopsis* RAR1 exerts rate-limiting control of R gene-mediated defenses against multiple pathogens. Plant Cell 14:979–992

Nagaharu U (1935) Genome analysis in *Brassica* with special reference to the experimental formation of *B. napus* and peculiar mode of fertilization. Japan J Bot 7:389–452

Nagaoka T, Doullah MA, Matsumoto S, Kawasaki S, Ishikawa T, Hori H, Okazaki K (2010) Identification of QTLs that control clubroot resistance in *Brassica oleracea* and comparative analysis of clubroot resistance genes between *B. rapa* and *B. oleracea*. Theor Appl Genet 120:1335–1346. https://doi.org/10.1007/s00122-010-1259-z

Nakano RT, Yamada K, Bednarek P, Nishimura M, Hara-Nishimura I (2014) ER bodies in plants of the Brassicales order: biogenesis and association with innate immunity. Front Plant Sci 5:73

Nandety RS, Caplan JL, Cavanaugh K, Perroud B, Wroblewski T, Michelmore RW, Meyers BC (2013) The role of TIR-NBS and TIR-X proteins in plant basal defense responses. Plant Physiol 162(3):1459–1472

Natti JJ (1958) Resistance of broccoli and other crucifers to downy mildew. Plant Dis Rep 42:656–662

Natti JJ, Harvey GER, Sayre CB (1956) Factors contributing to the increase of downy mildew of broccoli in New York state and its control with fungicides and agrimycin. Plant Dis Rep 40:118–124

Natti JJ, Dickson MH, Aktin JD (1967) Resistance of *Brasssica oleracea* cultivars to downy mildew. Phytopathology 57:144–147

Neik TX, Barbetti MJ, Batley J (2017) Current status and challenges in identifying disease resistance genes in *Brassica napus*. Front Plant Sci 8:1788. https://doi.org/10.3389/fpls.2017.01788

Nellist CF, Qian W, Jenner CE, Moore JD, Zhang S, Wang X, Briggs WH, Barker GC, Sun R, Walsh JA (2014) Multiple copies of eukaryotic translation initiation factors in *Brassica rapa* facilitate redundancy, enabling diversification through variation in splicing and broad-spectrum virus resistance. Plant J 77:261–268

Neuhaus K, Grsic-Rausch S, Sauerteig S, Ludwig-Muller J (2000) *Arabidopsis* plants transformed with nitrilase 1 or 2 in antisense direction are delayed in clubroot development. J Plant Physiol 156:756–761

Nie H, Wu Y, Yao C, Tang D (2011) Suppression of *edr2*-mediated powdery mildew resistance, cell death and ethylene-induced senescence by mutations in *ALD1* in *Arabidopsis*. J Genet Genomics 38:137–148

Nielsen ME, Feechan A, Bohlenius H, Ueda T, Thordal-Christensen H (2012) *Arabidopsis* ARF-GTP exchange factor, GNOM, mediates transport required for innate immunity and focal accumulation of syntaxin PEN1. Proc Natl Acad Sci U S A 109:11443–11448

Nishimura MT, Stein M, Hou BH, Vogel JP, Edwards H, Somerville SC (2003) Loss of a callose synthase results in salicylic acid-dependent disease resistance. Science 3016(1):969–972

Niu XK, Leung H, Williams PH (1983) Sources and nature of resistance to downy mildew and turnip mosaic in Chinese cabbage. J Am Soc Hort Sci 5:775–778

Nomura K, Minegishi Y, Kimizuka-Takagi C, Fujioka T, Moriguchi K, Shishido R, Ikehashi H (2005) Evaluation of F2 and F3 plants introgressed with QTL for clubroot resistance in cabbage developed by using SCAR markers. Plant Breed 124:371–375

Nordborg M, Hu TT, Ishino Y, Jhaveri J, Toomajian C, Zheng HG, Bakker E, Calabrese P, Gladstone J, Goyal R, Jakobsson M, Kim S, Morozov Y, Padhukasahasram B, Plagnol V, Rosenberg NA, Shah C, Wall JD, Wang J, Zhao KY, Kalbfleisch T, Schulz V, Kreitman M, Bergelson J (2005) The pattern of polymorphism in *Arabidopsis thaliana*. PLoS Biol 3:e196

Novo JVJ, Pascual J, Lucas RS, Romero-Rodriguez C, Ortega MR, Lenz C, Valledor L (2014) Fourteen years of plant proteomics reflected in "proteomics": moving from model species and

2-DE based approaches to orphan species and gel-free platforms. Proteomics 15:5–6. https://doi.org/10.1002/pmic.2014000349

Nyalugwe EP, Barbetti MJ, Jones RAC (2015) Studies on resistance phenotypes to turnip mosaic virus, in five species of Brassicaceae, and identification of a virus resistance gene in *Brassica juncea*. Eur J Plant Pathol 141:647–666

Nyalugwe EP, Barbetti MJ, Clode PL, Jones RAC (2016) Programmed cell death pathways induced by early plant-virus infection are determined by isolate virulence and stage of infection. Plant Pathol 65:1518–1528

O'Neill CM, Bancroft I (2000) Comparative physical mapping of segments of the genome of *Brassica oleracea* var. *alboglabra* that are homoeologous to sequenced regions of chromosomes 4 and 5 of *Arabidopsis thaliana*. Plant J 33(2):233–243

Ordog SH, Higgins VJ, Vanlerberghe GC (2002) Mitochondrial alternative oxidase is not a critical component of plant viral resistance but may play a role in the hypersensitive response. Plant Physiol 129:1858–1865

Orgil U, Araki H, Tangchaiburana S, Berkey R, Xiao S (2007) Intraspecific genetic variations, fitness cost and benefit of RPW8, a disease resistance locus in *Arabidopsis thaliana*. Genetics 1766(1):2317–2333

Osborn TC, Lukens L (2003) The molecular genetic basis of flowering time variation in Brassica species. Biotechnol Agric For 52:69–86

Osborn TC, Kole C, Parkin IAP, Sharpe AG, Kuiper M, Lydiate DJ, Trick M (1997) Comparison of flowering time genes in *Brassica rapa*, *B napus* and *Arabidopsis thaliana*. Genetics 146:1123–1129

Otulak-Koziel K, Koziel E, Bujarski JJ, Frankowska-Lukawska J, Torres MA (2020) Respiratory burst oxidase homologs RBOHD and RBOHF as key modulating componenets of response in *turnip mosaic virus- Arabidopsis thaliana* (L.) heyhn system. Int J Mol Sci 21:8510. https://doi.org/10.3390/ijms21228510

Ozturk ZN, Talame V, Deyholos M, Michalowski CB, Galbraith DW, Gozukirmizi N, Tuberosa R, Bohnert H (2002) Monitoring large-scale changes in transcript abundance in drought- and salt-stressed barley. Plant Mol Biol 48:551–573

Pandey SP, Somssich IE (2009) The role of WRKY transcription factors in plant immunity. Plant Physiol 150:1648–1655

Panjabi P, Jagannath A, Bisht NC, Padmaja KL, Sharma S, Gupta V, Pradhan AK, Pental D (2008) Comparative mapping of *Brassica juncea* and *Arabidopsis thaliana* using intron polymorphism (IP) markers: homoeologous relationships, diversification and evolution of the a, B and C *Brassica* genomes. BMC Genomics 9:113. https://doi.org/10.1186/1471-2164-9-113

Panjabi-Massand P, Yadava SK, Sharma P, Kaur A, Kumar A, Arumugam N, Sodhi YS, Mukhopadhyay A, Gupta V, Pradhan AK, Pental D (2010) Molecular mapping reveals two independent loci conferring resistance to *Albugo candida* in the east European germplasm of oilseed mustard *Brassica juncea*. Theor Appl Genet 121:137–145. https://doi.org/10.1007/s00122-010-1297-6

Paritosh K, Yadava SK, Singh P, Bhayana L, Mukhopadhyay A, Gupta V, Bisht NC, Zhang J, Kudrna D, Copetti D, Wing RA, Lachagari VBR, Pradhan AK, Pental D (2019) A chromosome-scale assembly of allotetraploid *Brassica juncea* (AABB) elucidates comparative architecture of the a and B genomes. bioR xiv. https://doi.org/10.1101/681080

Parker JE (2000) Signaling in plant disease resistance. In: Dickinson M, Beynon J (eds) Molecular plant pathology. Annual plant reviews, vol 4. Sheffield Academic Press, Sheffield, pp 144–174

Parker JE, Szabo V, Staskawicz BJ, Lister C, Dean C, Daniels MJ, Jones JDG (1993) Phenotypic characterization and molecular mapping of the *Arabidopsis thaliana* locus *RPP5*, determining disease resistance to *Peronospora parasitica*. Plant J 4:821–831

Parker JE, Holub EB, Frost LN, Falk A, Gunn ND, Daniels MJ (1996) Characterization of eds1, a mutation in *Arabidopsis* suppressing resistance to *Peronospora parasitica* specified by several different RPP genes. Plant Cell 8:2033–2046

Parker JE, Coleman MJ, Szabo V, Frost LN, Schmidt R, van der Biezen E, Moores T, Dean C, Daniels MJ, Jones JDG (1997) The *Arabidopsis* downy mildew resistance gene Rpp5 shares similarity to the toll and interleukin-1 receptors with N and L6. Plant Cell 9:879–894

Parkin I, Sharpe AG, Keith DJ, Lydiate DJ (1995) Identification of the A and C genomes in the amphidiploid *Brassica napus* (oilseed rape). Genome 38:1122–1131

Parkin IA, Koh C, Tang H, Robinson SJ, Kagale S, Clarke WE, Town CD, Nixon J, Krishnakumar V, Bidwell SL, Denoeud F, Belcram H, Links MG, Just J, Clarke C, Bender T, Huebert T, Mason AS, Pires JC, Barker G, Moore J, Walley PG, Manoli S, Batley J, Edwards D, Nelson MN, Wang X, Paterson AH, King G, Bancroft I, Chalhoub B, Sharpe AG (2014) Transcriptome and methylome profiling reveals relics of genome dominance in the mesopolyploid *Brassica oleracea*. Genome Biol 15(6):R77. https://doi.org/10.1186/gb-2014-15-6-r77

Parniske M, Hammond-Kosack KE, Golstein C, Thomas CM, Jones DA, Harrison K, Wulff BB, Jones JD (1997) Novel disease resistance specificities result from sequence exchange between tandemly repeated genes at the Cf-4/9 locus of tomato. Cell 91:821–832

Peart JR, Cook G, Feys BJ, Parker JE, Baulcombe DC (2002) An EDS1 orthologue is required for N-mediated resistance against tobacco mosaic virus. Plant J 29:569–579

Pei D, Ma H, Zhang Y, Ma Y, Wang W, Geng H, Wu J, Li C (2011) Virus-induced gene silencing of a putative glutathione s-transferase gene compromised ol-1-mediated resistance against powdery mildew in tomato. Plant Mol Biol Rep 29(4):972–978. https://doi.org/10.1007/s11105-011-0331-4

Peng L, Zhou L, Li Q, Wei D, Ren X, Song H, Mei J, Si J, Qian W (2018) Identification of quantitative trait loci for clubroot resistance in *Brassica oleracea* with the use of *Brassica* SNP microarray. Front Plant Sci 9:822. https://doi.org/10.3389/fpls.2018.00822

Piao ZY, Park YJ, Choi SR, Hong CP, Park JY, Choi YS, Lim YP (2002) Comversion of an AFLP marker linked to clubroot resistance gene in Chinese cabbage into a SCAR marker. J Korean Soc Hortic Sci 43:653–659

Piao ZY, Deng YQ, Choi SR, Park YJ, Lim YP (2004) SCAR and CAPS mapping of *CRb*, a gene conferring resistance to *Plasmodiophora brassicae* in Chinese cabbage (*Brassica rapa* ssp. *pekinensis)*. Theor Appl Genet 108:1458–1465. https://doi.org/10.1007/s00122-003-1577-5

Piao Z, Ramchiary N, Lim YP (2009) Genetics of clubroot resistance in *Brassica* species. J Plant Growth Regul 28:252–264. https://doi.org/10.1007/s00344-009-9093-8

Pieterse CM, Leon-Reyes A, Van der Ent S, Van Wees SC (2009) Networking by small-molecule hormones in plant immunity. Nat Chem Biol 5:308–316. https://doi.org/10.1038/nchembio.164

Pound GS, Williams PH (1963) Biological races of *Albugo candida*. Phytopathology 53:1146–1149

Prabhu KV, Somers DJ, Rakow G, Gugel RK (1998) Molecular markers linked to white rust resistance in mustard *Brassica juncea*. Theor Appl Genet 97:865–870

Qian W, Zhang SJ, Zhang SF, Li F, Zhang H, Wu J, Wang X, Walsh JA, Sun R (2013) Mapping and candidate-gene screening of the novel turnip mosaic virus resistance gene retr02 in Chinese cabbage (*Brassica rapa* L.). Theor Appl Genet 126:179–188

Quadrana L, Colot V (2016) Plant transgenerational epigenetics. Annu Rev Genet 50:467–491

Quirino BF, Candido ES, Campos PF, Franco OL, Kruger RH (2010) Proteomic approaches to study plant–pathogen interactions. Phytochemistry 71:351–362. https://doi.org/10.1016/j.phytochem.2009.11.005

Quiros CF, Grellet F, Sadowski J, Suzuki T, Li G, Wroblewski T (2001) *Arabidopsis* and *Brassica* comparative genomics: sequence, structure and gene content in the ABI1-Rps2-Ck1 chromosomal segment and related regions. Genetics 157:1321–1330

Rahman H, Peng G, Yu F, Falk KC, Kulkarni M, Selvaraj G (2014) Genetics and breeding for clubroot resistance in Canadian spring canola (*Brassica napus* L.). Can J Plant Pathol 36:122–134. https://doi.org/10.1080/07060661.2013.862571

Rehmany AP, Lynn JR, Tor M, Holub EB, Beynon JL (2000) A comparison of *Peronospora parasitica* (downy mildew) isolates from *Arabidopsis thaliana* and *Brassica oleracea* using amplified fragment length polymorphism and internal transcribed spacer 1 sequence analyses. Fungal Genet Biol 30:95–103. https://doi.org/10.1006/fgbi.2000.1216

Rehmany AP, Grenville LJ, Gunn ND, Allen RL, Paniwnyk Z, Byrne J, Whisson SC, Birch PR, Beynon JL (2003) A genetic interval and physical contig spanning the *Peronospora parasitica* (at) avirulence gene locus ATR1Nd. Fungal Genet Biol 38:33–42

Reinders J, Wulff BBH, Mirouze M, Mari Ordonez A, Dapp M, Rozhon W, Bucher E, Theiler G, Paszkowski J (2009) Compromised stability of DNA methylation and transposon immobilization in mosaic *Arabidopsis* epigenomes. Genes Dev 23:939–950

Rentel MC, Leonelli L, Dahlbeck D, Zhao B, Staskawicz BJ (2008) Recognition of the *Hyaloperonospora parasitica* effector ATR13 triggers resistance against oomycete, bacterial, and viral pathogens. Proc Natl Acad Sci U S A 105:1091–1096

Reuveni M, Tuzun S, Cole JS, Siegel MR, Kuc J (1986) The effects of plant age and leaf position on the susceptibility of tobacco to blue mold caused by *Peronospora tabacina*. Phytopathology 76 (4):455–458

Rietz S, Stamm A, Malonek S, Wagner S, Becker D, Medina-Escobar N, Vlot AC, Feys BJ, Niefind K, Parker JE (2011) Different roles of enhanced disease susceptibility1 (EDS1) bound to and dissociated from phytoalexin Deficient4 (PAD4) in *Arabidopsis* immunity. New Phytol 191:107–119. https://doi.org/10.1111/j.1469-8137.2011.03675.x

Rizhsky L, Davletova S, Liang H, Mittler R (2004) The zinc-finger protein Zat12 is required for cytosolic ascorbate peroxidase 1 expression during oxidative stress in *Arabidopsis*. J Biol Chem 279:11736–11743

Robaglia C, Caranta C (2006) Translation initiation factors: a weak link in plant RNA virus infection. Trends Plant Sci 11:40–45

Rocherieux J, Glory P, Giboulot A, Boury S, Barbeyron G, Thomas G, Manzanares-Dauleux MJ (2004) Isolate-specific and broad-spectrum QTLs are involved in the control of clubroot in *Brassica oleracea*. Theor Appl Genet 108:1555–1563. https://doi.org/10.1007/s00122-003-1580-x

Rodríguez-Hernández AM, Gosalvez B, Sempere RN, Burgos L, Aranda MA, Truniger V (2012) Melon RNA interference (RNAi) lines silenced for cm-eIF4E show broad virus resistance. Mol Plant Pathol 13:755–763

Ruiz MT, Voinnet O, Baulcombe DC (1998) Initiation and maintenance of virus-induced gene silencing. Plant Cell 10:937–946

Rusholme RE, Higgins E, Walsh JA, Lydiate DJ (2007) Genetic control of broad-spectrum resistance to *turnip mosaic virus* in *Brassica rapa* (Chinese cabbage). J Gen Virol 88:3177–3186

Rusterucci C, Aviv DH, Holt BF, Dangl JL, Parker JE (2001) The disease resistance signaling components EDS1 and PAD4 are essential regulators of the cell death pathway controlled by LSD1 in *Arabidopsis*. Plant Cell 13:2211–2224

Ryder CD, Smith LB, Teakle GR, King GJ (2001) Contrasting genome organization: two regions of the *Brassica oleracea* genome compared with collinear regions of the *Arabidopsis thaliana* genome. Genome 44:808–817

Saharan GS, Krishnia SK (2001) Multiple disease resistance in rapeseed and mustard. In: Nagarajan S, Singh DP (eds) Role of resistance in intensive agriculture. Kalyani Pub, New Delhi, pp 98–108

Saharan GS, Verma PR, Meena PD, Kumar A (2014) White rust of crucifers: biology, ecology and disease management. Springer Verlag, Frankfurt, p 244. ISBN 978-81-322-1791-6

Saharan GS, Mehta N, Meena PD (2017) Downy mildew disease of crucifers: biology, ecology and disease management, vol 56. Springer Verlag, Singapore, p 357. ISBN 978-981-10-7499-8

Saharan GS, Naresh M, Meena PD (2019) Powdery mildew disease of crucifers: biology, ecology and disease management. Springer Verlag, Singapore, p 362. ISBN 978-981-13-9852-0

Saito M, Kubo N, Matsumoto S, Suwabe K, Tsukada M, Hirai M (2006) Fine mapping of the clubroot resistance gene, Crr3, in *Brassica rapa*. Theor Appl Genet 114:81–91. https://doi.org/10.1007/s00122-006-0412-1

Sakamoto K, Saito A, Hayashida N, Taguchi G, Matsumoto E (2008) Mapping of isolate-specific QTLs for clubroot resistance in Chinese cabbage (*Brassica rapa* L. ssp. *pekinensis*). Theor Appl Genet 117:759–767. https://doi.org/10.1007/s00122-008-0817-0

Sanseverino W, Roma G, De Simone M, Faino L, Melito S, Stupka E, Frusciante L, Ercolano MR (2010) PRGdb: a bioinformatics platform for plant resistance gene analysis. Nucleic Acids Res 38(Database issue):D814–D821

Sauna ZE, Kimchi-Sarfaty C (2011) Understanding the contribution of synonymous mutations to human disease. Nat Rev Genet 12:683–691. https://doi.org/10.1038/nrg3051

Saxena RK, von Wettberg E, Upadhyaya HD, Sanchez V, Songok S, Saxena K, Kimurto P, Varshney RK (2014) Genetic diversity and demographic history of *Cajanus* spp. illustrated from genome-wide SNPs. PLoS One 9:e88568. https://doi.org/10.1371/journal.pone.0088568

Schaller A (2004) A cut above the rest: the regulatory function of plant proteases. Planta 220:183–197

Scheben A, Batley J, Edwards D (2017) Genotyping-by-sequencing approaches to characterize crop genomes: choosing the right tool for the right application. Plant Biotechnol J 15:149–161. https://doi.org/10.1111/pbi.12645

Schlaich NL, Slusarenko A (2009) Downy mildew of *Arabidopsis* caused by *Hyaloperonospora arabidopsidis* (formerly *Hyaloperonospora parasitica*). In: Lamour K, Kamoun S (eds) Oomycete genetics and genomics: diversity, interaction, and research tools. https://doi.org/10.1002/9780470475898.ch13

Schmid M, Simpson D, Kalousek F, Gietl C (1998) A cysteine endopeptidase with a C-terminal KDEL motif isolated from castor bean endosperm is a marker enzyme for the ricinosome, a putative lytic compartment. Planta 206:466–475

Schmid M, Davison TS, Henz SR, Pape UJ, Demar M, Vingron M, Scholkopf B, Weigel D, Lohmann JU (2005) A gene expression map of *Arabidopsis thaliana* development. Nat Genet 37:501–506

Schmidt SM, Kuhn H, Micali C, Liller C, Kwaaitaal M, Panstruga R (2014) Interaction of a *Blumeria graminis* f. sp. *hordei* effector candidate with a barley ARF-GAP suggests that host vesicle trafficking is a fungal pathogenicity target. Mol Plant Pathol 15:535–549

Serrano I, Gu Y, Qi D, Dubiella U, Innes RW (2014) The *Arabidopsis* EDR1 protein kinase negatively regulates the ATL1 E3 ubiquitin ligase to suppress cell death. Plant Cell 26:4532–4546

Shah J, Klessig DF (1999) Salicylic acid: signal perception and transduction. In: Hooykaas MA, Libbenga KR (eds) Biochemistry and molecular biology of plant hormones. Elsevier Science B. V, New York, pp 513–541

Shah J, Kachroo P, Nandi A, Klessig DF (2001) A recessive mutation in the *Arabidopsis SSI2* gene confers SA- and NPR1-independent expression of *PR* genes and resistance against bacterial and oomycete pathogens. Plant J 25:563–574

Shapiro A (2000) Using *Arabidopsis* mutants to delineate disease resistance signaling pathways. Can J Plant Pathol 22:199–216

Sharma BR, Dhiman JS, Thakur JC, Singh A, Bajaj KL (1991) Multiple disease resistance in cauliflower. Adv Hort Sci 5:30–34

Sharpe AG, Parkin IA, Keith DJ, Lydiate DJ (1995) Frequent nonreciprocal translocations in the amphidiploid genome of oilseed rape (*Brassica napus*). Genome 28:1112–1121

Shigenaga AM, Argueso CT (2016) No hormone to rule them all: interactions of plant hormones during the responses of plants to pathogens. Semin Cell Dev Biol 56:174–189. https://doi.org/10.1016/j.semcdb.2016.06.005

Shopan J, Mou H, Zhang L, Zhang C, Ma W, Walsh JA, Hu Z, Yang J, Zhang M (2017) Eukaryotic translation initiation factor 2B-beta (eIF2Bb), a new class of plant virus resistance gene. Plant J 90:929–940

Siemens J, Nagel W, Ludwig-Muller J, Sarcristán MD (2002) The interaction of *Plasmodiophora brassicae* and *Arabidopsis thaliana*: parameters for disease quantification and screening mutant lines. J Phytopathol 150:592–605

Siemens J, Keller I, Sarx J, Kunz S, Schuller A, Nagel W, Schmulling T, Parniske M, Ludwig-Muller J (2006) Transcriptome analysis of *Arabidopsis* clubroots indicate a key role for cytokinins in disease development. Mol Plant Microbe Interact 19:480–494. https://doi.org/10.1094/MPMI-19-0480

Sijen T, Fleenor J, Simmer F, Thijssen KL, Parrish S, Timmons L, Plasterk RH, Fire A (2001) On the role of RNA amplification in dsRNA-triggered gene silencing. Cell 107:465–476

Simms EL (1992) Costs of plant resistance to herbivory. In: Fritz RS, Simms EL (eds) Plant resistance to herbivores and pathogens: ecology, evolution, and genetics. University of Chicago Press, Chicago, pp 392–425

Simons MD (1979) Influence of genes for resistance to *Puccinia coronata* from *Arena sterilis* on yield and rust reaction of cultivated oats. Phytopathology 69:450–452

Sinapidou E, Williams K, Nott L, Bahkt S, Tor M, Crute I, Bittner-Eddy P, Beynon J (2004) Two *TIR:NB: LRR* genes are required to specify resistance to *Peronospora parasitica* isolate Cala2 in *Arabidopsis*. Plant J 38:898–909

Singh D, Chandra N, Gupta PP (1997) Inheritance of powdery mildew resistance in interspecific crosses of Indian and Ethiopian mustard. Ann Biol 13:73–77

Singh KB, Foley RC, Onate-Sanchez L (2002) Transcription factors in plant defense and stress responses. Curr Opin Plant Biol 5:430–436

Slusarenko AJ, Schlaich NL (2003) Downy mildew of *Arabidopsis thaliana* caused by *Hyaloperonosora parasitica* (formerly *Peronospora parasitica*). Mol Plant Pathol 4:159–170

Sohn KH, Lei R, Nemri A, Jones JD (2007) The downy mildew effector proteins ATR1 and ATR13 promote disease susceptibility in *Arabidopsis thaliana*. Plant Cell 19:4077–4090

Somé A, Manzanares MJ, Laurens F, Baron F, Thomas G, Rouxel F (1996) Variation for virulence on *Brassica napus* L. amongst *Plasmodiophora brassicae* collections from France and derived single-spore isolates. Plant Pathol 45:432–439. https://doi.org/10.1046/j.1365-3059.1996.d01-155.x

Somers DJ, Rakow G, Rimmer SR (2002) *Brassica napus* DNA markers linked to white rust resistance in *Brassica juncea*. Theor Appl Genet 104:1121–1124

Song J, Bradeen JM, Naess SK, Raasch JA, Wielgus SM, Haberlach GT, Liu J, Kuang H, Austin-Phillips S, Buell CR, Helgeson JP, Jiang J (2003) Gene RB cloned from *Solanum bulbocastanum* confers broad spectrum resistance to potato late blight. Proc Natl Acad Sci U S A 100:9128–9133

Song XS, Wang YJ, Mao WH, Shi K, Zhou YH, Nogues S, Yu JQ (2009) Effects of cucumber mosaic virus infection on electron transport and antioxidant system in chloroplasts and mitochon-dria of cucumber and tomato leaves. Physiol Plant 135:246–257

Song T, Chu M, Lahlali R, Yu F, Peng G (2016) Shotgun label-free proteomic analysis of clubroot (*Plasmodiophora brassicae*) resistance conferred by the gene *Rcr1* in *Brassica rapa*. Front Plant Sci 7:1013

Springer NM, Ying K, Fu Y, Ji T, Yeh CT, Jia Y, Wu W, Richmond T, Kitzman J, Rosenbaum H, Iniguez AL, Barbazuk WB, Jeddeloh JA, Nettleton D, Schnable PS (2009) Maize inbreds exhibit high levels of copy number variation (CNV) and presence/absence variation (PAV) in genome content. PLoS Genet 5:e1000734. https://doi.org/10.1371/journal.pgen.1000734

Srivastava A, Gupta V, Pental D, Pradhan A (2001) AFLP-based genetic diversity assessment amongst agronomically important natural and some newly synthesized lines of *Brassica juncea*. Theor Appl Genet 102:193–199. https://doi.org/10.1007/s001220051635

Staal J, Kaliff M, Bohman S, Dixelius C (2006) Transgressive segregation reveals two *Arabidopsis* TIR-NB-LRR resistance genes effective against *Leptosphaeria maculans*, causal agent of blackleg disease. Plant J 46:218–230

Suwabe K, Tsukazaki H, Iketani H, Hatakeyama K, Fujimura M, Nunome T, Fukuoka H, Matsumoto S, Hirai M (2003) Identification of two loci for resistance to clubroot (*Plasmodiophora brassicae* Woronin) in *Brassica rapa* L. Theor Appl Genet 107:997–1002

Suwabe K, Tsukazaki H, Iketani H, Hatakeyama K, Kondo M, Fujimura M, Nunome T, Fukuoka H, Hirai M, Matsumoto S (2006) Simple sequence repeat-based comparative genomics between *Brassica rapa* and *Arabidopsis thaliana*: the genetic origin of clubroot resistance. Genetics 173:309–319

Suwabe K, Suzuki G, Nunome T, Hatakeyama K, Mukai Y, Fukuoka H, Matsumoto S (2012) Microstructure of a *Brassica rapa* genome segment homoeologous to the resistance gene cluster on *Arabidopsis* chromosome 4. Breed Sci 62:170–177. https://doi.org/10.1270/jsbbs.62.170

Swiderski MR, Birker D, Jones JDG (2009) The TIR domain of TIR-NB-LRR resistance proteins is a signaling domain involved in cell death induction. Mol Plant Microbe Interact 22(2):157–165

Tai TH, Dahlbeck D, Clark ET, Gajiwala P, Pasion R, Whalen MC, Stall RE, Staskawicz BJ (1999) Expression of the Bs2 pepper gene confers resistance to bacterial spot disease in tomato. Proc Natl Acad Sci U S A 96:14153–14158. https://doi.org/10.1073/pnas.96.24.14153

Takahashi H, Miller J, Nozaki Y, Takeda M, Shah J, Hase S, Ikegami M, Ehara Y, Dinesh-Kumar SP, Sukamto (2002) RCY1, an *Arabidopsis thaliana* RPP8/HRT family resistance gene, conferring resistance to cucumber mosaic virus requires salicylic acid, ethylene and a novel signal transduction mechanism. Plant J 32:655–667. https://doi.org/10.1046/j.1365-313X.2002.01453.x

Takemoto D, Jones D, Hardham A (2006) Re-organisation of the cytoskeleton and endoplasmic reticulum in the *Arabidopsis pen1-1* mutant inoculated with the non-adapted powdery mildew *Blumeria graminis* f. sp. *hordei*. Mol Plant Pathol 76(1):553–563

Takken FL, Goverse A (2012) How to build a pathogen detector: structural basis of NB-LRR function. Curr Opin Plant Biol 15:375–384

Takken FL, Joosten MH (2000) Plant resistance genes: their structure, function and evolution. Eur J Plant Pathol 106:699–713

Tan X, Meyers BC, Kozik A, West MAL, Morgante M, St Clair DA, Bent AF, Michelmore RW (2007) Global expression analysis of nucleotide binding site-leucine rich repeat-encoding and related genes in Arabidopsis. BMC Plant Biol 7:56. Published online

Tang D, Innes RW (2002) Overexpression of a kinase-deficient form of the EDR1 gene enhances powdery mildew resistance and ethylene-induced senescence in *Arabidopsis*. Plant J 32:975–983

Tang D, Ade J, Frye CA, Innes RW (2005a) Regulation of plant defense responses in *Arabidopsis* by EDR2, a PH and START domain-containing protein. Plant J 446(1):245–257

Tang D, Christiansen KM, Innes RW (2005b) Regulation of plant disease resistance, stress responses, cell death and ethylene signaling in *Arabidopsis* by the EDR1 protein kinase. Plant Physiol 1386(1):1018–1026

Tang D, Ade J, Frye CA, Innes RW (2006) A mutation in the GTP hydrolysis site of *Arabidopsis* dynamin-related protein 1E confers enhanced cell death in response to powdery mildew infection. Plant J 476(1):75–84

Tanhuanppa P (2004) Identification and mapping of resistance gene analogs and a white rust resistance locus in *B. rapa* ssp *oleifera*. Theor Appl Genet 108:1039–1046

Teh OK, Hofius D (2014) Membrane trafficking and autophagy in pathogen-triggered cell death and immunity. J Exp Bot 65:1297–1312

Teixeira FK, Heredia F, Sarazin A, Roudier F, Boccara M, Ciaudo C, Cruaud C, Poulain J, Berdasco M, Fraga MF, Voinnet O, Wincker P, Esteller M, Colot V (2009) A role for RNAi in the selective correction of DNA methylation defects. Science 323:1600–1604

Tham FY, Lucas JA, Wilson ZA (1994) DNA fingerprinting of *Peronospora parasitica*, a biotrophic fungal pathogen of crucifers. Theor Appl Genet 88:490–496

Than ME, Helm M, Simpson DJ, Lottspeich F, Huber R, Gietl C (2004) The 2.0-a crystal structure of the KDEL-tailed cysteine endopeptidase from germinating endosperm of *Ricinus communis* confirms its function in the final stage of programmed cell death. J Mol Biol 336:1103–1116

Thomas CE, Jourdain EL (1990) Evaluation of broccoli and cauliflower germplasm for resistance to race 2 of *Peronospora parasitica*. Hort Sci 25:1429–1431

Thomas CE, Jourdain EL (1992) Resistance to race 2 of *Peronospora parasitica* in U.S. plant introductions of *Brassica oleracea* var. *capitata*. Hort Sci 27:1120–1122

Thomma BP, Eggermont K, Penninckx IA, Mauch-Mani B, Vogelsang R, Cammue BP, Broekaert WF (1998) Separate jasmonate-dependent and salicylate-dependent defense-response pathways in *Arabidopsis* are essential for resistance to distinct microbial pathogens. Proc Natl Acad Sci U S A 95(25):15107–15111. https://doi.org/10.1073/pnas.95.25.15107

Thomma BP, Penninckx IA, Broekaert WF, Cammue BP (2001) The complexity of disease signaling in *Arabidopsis*. Curr Opin Immunol 13:63–68. https://doi.org/10.1016/S0952-7915(00)00183-7

Timmons L, Fire A (1998) Specific interference by ingested dsRNA. Nature 395:854

Tong M, Kotur T, Liang W, Vogelmann K, Kleine T, Leister D, Brieske C, Yang S, Ludke D, Wiermer M, Zhang Y, Li X, Hoth S (2017) E3 ligase SAUL1 serves as a positive regulator of PAMP-triggered immunity and its homeostasis is monitored by immune receptor SOC3. New Phytol 215:1516–1532

Tor M, Gordon P, Cuzick A, Eulgem T, Sinapidou E, Mert-Turk F, Can C, Dangl JL, Holub EB (2002) *Arabidopsis* SGT1b is required for defense signaling conferred by several downy mildew resistance genes. Plant Cell 14:993–1003

Tornero P, Chao RA, Luthin WN, Goff SA, Dangl JL (2002) Large-scale structure-function analysis of the Arabidopsis RPM1 disease resistance protein. Plant Cell 14(2):435–450. https://doi.org/10.1105/tpc.010393

Torres MA, Dangl JL (2005) Functions of the respiratory burst oxidase in biotic interactions, abiotic stress and development. Curr Opin Plant Biol 8:397–403

Torres MA, Dangl JL, Jones JDG (2002) *Arabidopsis* gp91(phox) homologues *AtrbohD* and *AtrbohF* are required for accumulation of reactive oxygen intermediates in the plant defense response. Proc Natl Acad Sci U S A 99:517–522

Torres MA, Jones DGJ, Dangl JL (2005) Pathogen-induced, NADPH oxidase-derived reactive oxygen intermediates suppress spread of cell death in *Arabidopsis thaliana*. Nat Genet 37:1130–1134

Tosa Y (1992) A model for the evolution of formae speciales and races. Phytopathology 82:728–729

Town CD, Cheung F, Maiti R, Crabtree J, Haas BJ, Wortman JR, Hine EE, Althoff R, Arbogast TS, Tallon LJ, Vigouroux M, Trick M, Bancroft I (2006) Comparative genomics of *Brassica oleracea* and *Arabidopsis thaliana* reveal gene loss, fragmentation, and dispersal after polyploidy. Plant Cell 18:1348–1359. https://doi.org/10.1105/tpc.106.041665

Toxopeus H, Janssen A (1975) Clubroot resistance in turnip II. The "slurry" screening method and clubroot races in the Netherlands. Euphytica 24(3):751–755

Turck F, Zhou A, Somssich IE (2004) Stimulus-dependent, promoter-specific binding of transcription factor WRKY1 to its native promoter and the defense-related gene *PcPR1-1* in parsley. Plant Cell 16:2573–2585

Tyler BM, Tripathy S, Zhang X, Dehal P, Jiang RH, Aerts A, Arredondo FD, Baxter L, Bensasson D, Beynon JL, Chapman J, Damasceno CMB, Dorrance AE, Dou D, Dickerman AW, Dubchak IL, Garbelotto M, Gijzen M, Gordon SG, Govers F, Grunwald NJ, Huang W, Ivors KL, Jones RW, Kamoun S, Krampis K, Lamour KH, Lee MK, McDonald WH, Medina M, Meijer HJG, Nordberg EK, Maclean DJ, Ospina-Giraldo MD, Morris PF, Phuntumart V, Putnam NH, Rash S, Rose JKC, Sakihama Y, Salamov AA, Savidor A, Scheuring CF, Smith BM, Sobral BWS, Terry A, Torto-Alalibo TA, Win J, Xu Z, Zhang H, Grigoriev IV, Rokhsar DS, Boore JL (2006) Phytophthora genome sequences uncover evolutionary origins and mechanisms of pathogenesis. Science 313:1261–1266

Uemura T, Kim H, Saito C, Ebine K, Ueda T, Schulze-Lefert P, Nakano A (2012) Qa-SNAREs localized to the frans-Golgi network regulate multiple transport pathways and extracellular disease resistance in plants. Proc Natl Acad Sci U S A 109:1784–1789

Ueno H, Matsumoto E, Aruga D, Kitagawa S, Matsumura H, Hayashida N (2012) Molecular characterization of the *CRa* gene conferring clubroot resistance in *Brassica rapa*. Plant Mol Biol 80:621–629. https://doi.org/10.1007/s11103-012-9971-5

Underwood W, Melotto M, He SY (2007) Role of plant stomata in bacterial invasion. Cell Microbiol 9(7):1621–1629. https://doi.org/10.1111/j.1462-5822.2007.00938.x

Van der Biezen EA, Freddie CT, Kahn K, Parker JE, Jones JD (2002) *Arabidopsis* RPP4 is a member of the RPP5 multigene family of TIR-NB-LRR genes and confers downy mildew resistance through multiple signalling components. Plant J 29:439–451

van West P, Kamoun S, van't Klooster JW, Govers F (1999) Internuclear gene silencing in *Phytophthora infestans*. Mol Cell 3:339–348

van West P, Shepherd SJ, Walker CA, Li S, Appiah AA, Grenville-Briggs LJ, Govers F, Gow NAR (2008) Internuclear gene silencing in *Phytophthora infestans* is established through chromatin remodelling. Microbiology 154:1482–1490

Varshney A, Mohapatra T, Sharma RP (2004) Development and validation of CAPS and AFLP markers for white rust resistance gene in *Brassica juncea*. Theor Appl Genet 109:153–159

Vastenhouw NL, Brunschwig K, Okihara KL, Muller F, Tijsterman M, Plasterk RHA (2006) Gene expression: long-term gene silencing by RNAi. Nature 442:882

Verdel A, Moazed D (2005) RNAi-directed assembly of heterochromatin in fission yeast. FEBS Lett 579(26):5872–5878

Verdel A, Jia S, Gerber S, Sugiyama T, Gygi S, Grewal SIS, Moazed D (2004) RNAi-mediated targeting of heterochromatin by the RITS complex. Science 303:672–676

Verica JA, He ZH (2002) The cell wall-associated kinase (WAK) and WAK-like kinase gene family. Plant Physiol 129:455–459. https://doi.org/10.1104/pp.011028

Vogel J, Somerville S (2000) Isolation and characterization of powdery mildew resistant *Arabidopsis* mutants. Proc Natl Acad USA 976(1):1897–1902

Vogel JP, Raab TK, Schiff C, Somerville SC (2002) PMR6, a pectate lyase-like gene required for powdery mildew susceptibility in Arabidopsis. Plant Cell 14(9):2095–2106

Vogel JP, Raab TK, Somerville CR, Somerville SC (2004) Mutations in PMR5 result in powdery mildew resistance and altered cell wall composition. Plant J 40(6):968–978

Voglmayr M, Riethmüller A (2006) Phylogenetic relationships of *Albugo candida* (white blister rusts) based on LSU rDNA sequence and oospore data. Mycol Res 110:75–85

Voorrips RE, Jongerius MC, Kanne HJ (1997) Mapping of two genes for resistance to clubroot (*Plasmodiophora brassicae*) in a population of doubled haploid lines of *Brassica oleracea* by means of RFLP and AFLP markers. Theor Appl Genet 94:75–82. https://doi.org/10.1007/s001220050384

Walker CA, Koppe M, Grenville-Briggs LJ, Avrova AO, Horner NR, McKinnon AD, Whisson SC, Birch PRJ, van West P (2008) A putative DEAD-box RNA-helicase is required for normal zoospore development in the late blight pathogen *Phytophthora infestans*. Fungal Genet Biol 45:954–962

Walsh JA, Jenner CE (2002) *Turnip mosaic virus* and the quest for durable resistance. Mol Plant Pathol 3:289–300

Walsh JA, Sharpe AG, Jenner EC, Lydiate DJ (1999) Characterization of resistance of turnip mosaic virus in oilseed rape (*Brassica napus*) and genetic mapping of TuRB01. Theor Appl Genet 99:1149–1154

Wang H, Ngwenyama N, Liu Y, Walker JC, Zhang S (2007) Stomatal development and patterning are regulated by environmentally responsive mitogen-activated protein kinases in *Arabidopsis*. Plant Cell 19:63–73

Wang Z, Mao H, Dong C, Ji R, Cai L, Fu H, Liu S (2009) Over expression of *Brassica napus MPK4* enhances resistance to *Sclerotinia sclerotiorum* in oilseed rape. Mol Plant Microbiol Inter 22:235–244. https://doi.org/10.1094/MPMI-22-3-0235

Wang Y, Li J, Hou S, Wang X, Li Y, Ren D, Chen S, Tang X, Zhou JM (2010) A *Pseudomonas syringae* ADP-ribosyl transferase inhibits *Arabidopsis* mitogen-activated protein kinase kinases. Plant Cell 22:2033–2044

Wang X, Wang H, Wang J, Sun R, Wu J, Liu S, Bai Y, Mun JH, Bancroft I, Cheng F, Huang S, Li X, Hua W, Wang J, Wang X, Freeling M, Pires JC, Paterson AH, Chalhoub B, Wang B, Hayward A, Sharpe AG, Park BS, Weisshaar B, Liu B, Li B, Liu B, Tong C, Song C, Duran C, Peng C, Geng C, Koh C, Lin C, Edwards D, Mu D, Shen D, Soumpourou E, Li F, Fraser F, Conant G, Lassalle G, King GJ, Bonnema G, Tang H, Wang H, Belcram H, Zhou H,

Hirakawa H, Abe H, Guo H, Wang H, Jin H, Parkin IA, Batley J, Kim JS, Just J, Li J, Xu J, Deng J, Kim JA, Li J, Yu J, Meng J, Wang J, Min J, Poulain J, Wang J, Hatakeyama K, Wu K, Wang L, Fang L, Trick M, Links MG, Zhao M, Jin M, Ramchiary N, Drou N, Berkman PJ, Cai Q, Huang Q, Li R, Tabata S, Cheng S, Zhang S, Zhang S, Huang S, Sato S, Sun S, Kwon SJ, Choi SR, Lee TH, Fan W, Zhao X, Tan X, Xu X, Wang Y, Qiu Y, Yin Y, Li Y, Du Y, Liao Y, Lim Y, Narusaka Y, Wang Y, Wang Z, Li Z, Wang Z, Xiong Z, Zhang Z (2011) *Brassica rapa* genome sequencing project consortium. The genome of the mesopolyploid crop species *Brassica rapa*. Nat Genet 43:1035–1039. https://doi.org/10.1038/ng.919

Wang WM, Ma XF, Zhang Y, Luo MC, Wang GL, Bellizzi M, Xiong XY, Xiao SY (2012) papp2c interacts with the atypical disease resistance protein RPW8.2 and negatively regulates salicylic acid-dependent defense responses in *Arabidopsis*. Mol Plant 5:1125–1137

Wang Q, Li J, Hu L, Zhang T, Zhang G, Lou Y (2013) OsMPK3 positively regulates the JA signaling pathway and plant; resistance to a chewing herbivore in rice. Plant Cell Rep 32:1075–1084. https://doi.org/10.1007/s00299-013-1389-2

Warren RC, King JE, Colhoun J (1971) Reaction of potato leaves to infection by *Phytophthora infestans* in relation to position on the plant. Trans Br Mycol Soc 57:501–514

Warren RF, Henk A, Mowery P, Holub E, Innes RW (1998) A mutation within the leucine-rich repeat domain of the Arabidopsis disease resistance gene RPS5 partially suppresses multiple bacterial and downy mildew resistance genes. Plant Cell 10(9):1439–1452

Warren RF, Merritt PM, Holub E, Innes RW (1999) Identification of three putative signal transduction genes involved in *R*-gene-specified disease resistance in *Arabidopsis*. Genetics 152:401–412

Watanabe CK, Hachiya T, Terashima I, Noguchi K (2008) The lack of alternative oxidase at low temperature leads to a disruption of the balance in carbon and nitrogen metabolism, and to an up-regulation of antioxidant defense systems in *Arabidopsis thaliana* leaves. Plant Cell Environ 31:1190–1202

Wei C, Chen J, Kuang H (2016) Dramatic number variation of R genes in solanaceae species accounted for by a few R gene subfamilies. PLoS One 11(2):e0148708. https://doi.org/10.1371/journal.pone.0148708

Werner S, Diederichsen E, Frauen M, Schondelmaier J, Jung C (2008) Genetic mapping of clubroot resistance genes in oilseed rape. Theor Appl Genet 116:363–372. https://doi.org/10.1007/s00122-007-0674-2

Whisson SC, Avrova AO, Van West P, Jones JT (2005) A method for double-stranded RNA-mediated transient gene silencing in *Phytophthora infestans*. Mol Plant Pathol 6:153–163

Whisson SC, Boevink PC, Moleleki L, Avrova AO, Morales JG, Gilroy EM, Armstrong MR, Grouffaud S, van West P, Chapman S, Hein I, Toth IK, Pritchard L, Birch PRJ (2007) A translocation signal for delivery of oomycete effector proteins into host plant cells. Nature 450:115–118

Wiermer M, Feys BJ, Parker JE (2005) Plant immunity: the EDS1 regulatory node. Curr Opin Plant Biol 8:383–389

Williams PH (1966) A system for the determination of races of *Plasmodiophora brassicae* that infect cabbage and rutabaga. Phytopathology 56:624–626

Win J, Morgan W, Bos J, Krasileva KV, Cano LM, Chaparro Garcia A, Ammar R, Staskawicz BJ, Kamoun S (2007) Adaptive evolution has targeted the C-terminal domain of the RxLR effectors of plant pathogenic oomycetes. Plant Cell 19:2349–2369

Wit F, van de Weg M (1964) Clubroot-resistance in turnips (*Brassica campestris* L.). Euphytica 13:9–18. https://doi.org/10.1007/BF00037512

Wittmann S, Chatel H, Fortin MG, Laliberte JF (1997) Interaction of the viral protein genome linked of turnip mosaic potyvirus with the translational eukaryotic initiation factor (iso) 4e of *Arabidopsis thaliana* using the yeast two-hybrid system. Virology 234:84–92

Wong CE, Carson RAJ, Carr JP (2002) Chemically induced virus resistance in *Arabidopsis thaliana* is independent of pathogenesis-related protein expression and the NPR1 gene. Mol Plant Microbe Interact 15:75–81

Wu XM, Chen BY, Lu G, Wang HZ, Xu K, Guizhan G, Song Y (2009) Genetic diversity in oil and vegetable mustard (*Brassica juncea*) landraces revealed by SRAP markers. Genet Resour Crop Evol 56:1011–1022. https://doi.org/10.1007/s10722-009-9420-8

Wu Y, Zhang D, Chu JY, Boyle P, Wang Y, Brindle ID, Luca VD, Després C (2012) The *Arabidopsis* NPR1 protein is a receptor for the plant defense hormone salicylic acid. Cell Rep 1:639–647. https://doi.org/10.1016/j.celrep.2012.05.008

Wu J, Lee DY, Wang Y, Kim ST, Baek SB, Kim SG, Kang KY (2014a) Protein profiles secreted from phylloplane of rice leaves free from cytosolic proteins: application to study rice-*Magnaporthe oryzae* interactions. Physiol Mol Plant Pathol 88:28–35. https://doi.org/10.1016/j.pmpp.2014.08.003

Wu S, Shan L, He P (2014b) Microbial signature-triggered plant defense responses and early signaling mechanisms. Plant Sci 228C:118–126. https://doi.org/10.1016/j.plantsci.2014.03.001

Wulff BB, Thomas CM, Smoker M, Grant M, Jones JD (2001) Domain swapping and gene shuffling identify sequences required for induction of an Avr-dependent hypersensitive response by the tomato Cf- 4 and Cf-9 proteins. Plant Cell 13:255–272

Wyatt SE, Pan SQ, Kuc J (1991) β-1, 3-glucanase, chitinase, and peroxidase activities in tobacco tissues resistant and susceptible to blue mould as related to flowering, age and sucker development. Physiol Mol Plant Pathol 39:433–440. https://doi.org/10.1016/0885-5765(91)90009-7.188

Xiao SY, Ellwood S, Findlay K, Oliver RP, Turner JG (1997) Characterization of three loci controlling resistance of *Arabidopsis thaliana* accession Ms-0 to two powdery mildew diseases. Plant J 126(1):757–768

Xiao SY, Ellwood S, Calis O, Patrick E, Li TX, Coleman M, Turner JG (2001) Broad-spectrum mildew resistance in *Arabidopsis thaliana* mediated by *RPW8*. Science 2916(1):118–120

Xiao S, Brown S, Patrick E, Brearley C, Turner JG (2003) Enhanced transcription of the *Arabidopsis* disease resistance genes RPW8.1 and RPW8.2 via a salicylic acid–dependent amplification circuit is required for hypersensitive cell death. Plant Cell 15:33–45

Xiao S, Emerson B, Ratanasut K, Patrick E, O'Neill C, Bancroft I, Turner JG (2004) Origin and maintenance of a broad-spectrum disease resistance locus in *Arabidopsis*. Mol Biol EvoI 21:1661–1672

Xiao S, Calis O, Patrick E, Zhang G, Charoenwattana P, Muskett P, Parker JE, Turner JG (2005) The atypical resistance gene, RPW8, recruits components of basal defence for powdery mildew resistance in *Arabidopsis*. Plant J 42:95–110

Xiao D, Liu ST, Wei YP, Zhou DY, Hou XL, Li Y, Hu CM (2016) cDNA-AFLP analysis reveals differential gene expression in incompatible interaction between infected non-heading Chinese cabbage and *Hyaloperonospora parasitica*. Hort Res 3:16034

Xie C, Zhou X, Deng X, Guo Y (2010) PKS5, a SNF1-related kinase, interacts with and phosphorylates NPR1, and modulates expression of WRKY38 and WRKY62. J Genet Genomics 37:359–369. https://doi.org/10.1016/S1673-8527(09)60054-0

Xu G, Guo C, Shan H, Kong H (2012) Divergence of duplicate genes in exon–intron structure. Proc Natl Acad Sci 109(4):1187–1192

Yang TJ, Kim JS, Kwon SJ, Lim KB, Choi BS, Kim JA, Jin M, Park JY, Lim MH, Kim HI, Lim YP, Kang JJ, Hong JH, Kim CB, Bhak J, Bancroft I, Park BS (2006) Sequence-level analysis of the diploidization process in the triplicated FLOWERING LOCUS C region of *Brassica rapa*. Plant Cell 18(6):1339–1347

Yang Y, Zhang H, Li G, Li W, Wang X, Song F (2009) Ectopic expression of MgSM1, a Ceratoplatan in family protein from *Magnaporthe grisea*, confers broad-spectrum disease resistance in *Arabidopsis*. Plant Biotechnol J 7(8):763–777. https://doi.org/10.1111/j.1467-7652.2009.00442.x

Yang L, Qin L, Liu G, Peremyslov VV, Dolja VV, Wei Y (2014) Myosins XI modulate host cellular responses and penetration resistance to fungal pathogens. Proc Natl Acad Sci U S A 111:13996–14001

Yang J, Liu D, Wang X, Ji C, Cheng F, Liu B, Hu Z, Chen S, Pental D, Ju Y, Yao P, Li X, Xie K, Zhang J, Wang J, Liu F, Ma W, Shopan J, Zheng H, Mackenzie SA, Zhang M (2016) The genome sequence of allopolyploid *Brassica juncea* and analysis of differential homoeolog gene expression influencing selection. Nat Genet 48:1225–1232. https://doi.org/10.1038/ng.3657

Yang K, Nath UK, Biswas MK, Kayum MA, Yi GE, Lee J, Yang TJ, Nou IS (2018) Whole-genome sequencing of Brassica oleracea var. capitata reveals new diversity of the mitogenome. PLoS One 13(3):e0194356. https://doi.org/10.1371/journal.pone.0194356

Yeam I, Cavatorta JR, Ripoll DR, Kang BC, Jahn MM (2007) Functional dissection of naturally occurring amino acid substitutions in eIF4E that confers recessive potyvirus resistance in plants. Plant Cell 19:2913–2928

Yoshida K, Terashima I, Noguchi K (2006) Distinct roles of the cytochrome pathway and alternative oxidase in leaf photosynthesis. Plant Cell Physiol 47:22–31

Yoshikawa H (1993) Studies on breeding of clubroot resistance in cole crops. Bull Natl Res Inst Veg Ornam Plants Tea Jpn Ser A 7:1–165

Yoshioka K, Kachroo P, Tsui F, Sharma SB, Shah J, Klessig DF (2001) Environmentally sensitive, SA-dependent defense responses in the *cpr22* mutant of *Arabidopsis*. Plant J 26:447–459

Yoshioka K, Moeder W, Kang HG, Kachroo P, Masmoudi K, Berkowitz G, Klessig DF (2006) The chimeric *Arabidopsis* CYCLIC NUCLEOTIDE-GATED ION CHANNEL11/12 activates multiple pathogen resistance responses. Plant Cell 18:747–763

Young ND (2000) The genetic architecture of resistance. Curr Opin Plant Biol 3:285–290

Yu IC, Parker JE, Bent AF (1998) Gene-for-gene disease resistance without the hypersensitive response in the *Arabidopsis dnd1* mutant. Proc Natl Acad Sci U S A 95:7819–7824

Yu S, Zhang F, Yu R, Zou Y, Qi J, Zhao X, Yu Y, Zhang D, Li L (2009) Genetic mapping and localization of a major QTL for seedling resistance to downy mildew in Chinese cabbage (*Brassica rapa* ssp. *pekinensis*). Mol Breed 23:573–590. https://doi.org/10.1007/s11032-009-9257-z

Yu S, Zhang F, Zhao X, Yu Y, Zhang D (2011) Sequence-characterized amplified region and simple sequence repeat markers for identifying the major quantitative trait locus responsible for seedling resistance to downy mildew in Chinese cabbage (*Brassica rapa* ssp. *pekinensis*). Plant Breed 130:580–583. https://doi.org/10.1111/j.1439-0523.2011.01874.x

Yu J, Tehrim S, Zhang F, Tong C, Huang J, Cheng X, Dong C, Zhou Y, Qin R, Hua W, Liu S (2014a) Genome-wide comparative analysis of NBS-encoding genes between *Brassica* species and *Arabidopsis thaliana*. Plant Pathol 56:317–323

Yu J, Tehrim S, Zhang F, Tong C, Huang J, Cheng X, Dong C, Zhou Y, Qin R, Hua W, Liu S (2014b) Genome-wide comparative analysis of NBS-encoding genes between *Brassica* species and *Arabidopsis thaliana*. BMC Genomics 15:3. https://doi.org/10.1186/1471-2164-15-3

Yu F, Zhang X, Huang Z, Chu M, Song T, Falk KC, Deora A, Chen Q, Zhang Y, McGregor L, Gossen BD, McDonald MR, Peng G (2016) Identification of genome-wide variants and discovery of variants associated with *Brassica rapa* clubroot resistance gene Rcr1 through bulked segregant RNA sequencing. PLoS One 11:e0153218. https://doi.org/10.1371/journal.pone.0153218

Yu F, Zhang X, Peng G, Falk KC, Strelkov SE, Gossen BD (2017) Genotyping-by-sequencing reveals three QTL for clubroot resistance to six pathotypes of *Plasmodiophora brassicae* in *Brassica rapa*. Sci Rep 7:4516. https://doi.org/10.1038/s41598-017-04903-2

Yun BW, Atkinson HA, Gaborit C, Greenland A, Read ND, Pallas JA, Loake GJ (2003) Loss of actin cytoskeletal function and EDS1 activity, in combination, severely compromises non-host resistance in Arabidopsis against wheat powdery mildew. Plant J 346(1):768–777

Zhan Z (2017) Clubroot resistant germplasm innovation and transferring of resistant locus PbBa8.1 from turnip into canola elite variety. Huazhong agricultural university, Wuhan, China (in Chinese with English abstract)

Zhang Y, Li X (2005) A putative nucleoporin 96 is required for both basal defense and constitutive resistance responses mediated by suppressor of npr 1-1 constitutive l. Plant Cell 17:1306–1316

Zhang Y, Tessaro MJ, Lassner M, Li X (2003) Knockout analysis of *Arabidopsis* transcription factors TGA2, TGA5, and TGA6 reveals their redundant and essential roles in systemic acquired resistance. Plant Cell 15:2647–2653. https://doi.org/10.1105/tpc.014894

Zhang H, Kolb FA, Jaskiewicz L, Westhof E, Filipowicz W (2004) Single processing center models for human dicer and bacterial RNase III. Cell 118:57–68

Zhang Y, Cheng YT, Bi D, Palma K, Li X (2005) MOS2, a protein containing G-patch and KOW motif, is essential for innate immunity in *Arabidopsis thaliana*. Curr Biol 15:1936–1942

Zhang FL, Wang M, Liu XC, Zhao XY, Yang JP (2008) Quantitative trait loci analysis for resistance against turnip mosaic virus based on a doubled-haploid population in Chinese cabbage. Plant Breed 127:82–86

Zhang XW, Yuan YX, Wang XW (2009) QTL mapping for TuMV resistance in Chinese cabbage [*Brassica campestris* L. ssp. *pekinensis* (Lour.)Olssom]. Acta Hortic Sin 36:731–736

Zhang W, Fraiture M, Kolb D, Loffelhardt B, Desaki Y, Boutrot FF, Tor M, Zipfel C, Gust AA, Brunner F (2013) *Arabidopsis* receptor-like protein30 and receptor-like kinase suppressor of BIR1-1/EVERSHED mediate innate immunity to necrotrophic fungi. Plant Cell 25:4227–4241

Zhang T, Zhao Z, Zhang C, Pang W, Choi SR, Lim YP, Piao Z (2014) Fine genetic and physical mapping of the *CRb* gene conferring resistance to clubroot disease in *Brassica rapa*. Mol Breed 34:1173–1183. https://doi.org/10.1007/s11032-014-0108-1

Zhang Q, Berkey R, Pan Z, Wang W, Zhang Y, Ma X, King H, Xiao S (2015) Dominant negative RPW8.2 fusion proteins reveal the importance of haustorium-oriented protein trafficking for resistance against powdery mildew in *Arabidopsis*. Plant Signal Behav 10:e989766. https://doi.org/10.4161/15592324.2014.989766

Zhang X, Liu Y, Fang Z, Li Z, Yang L, Zhuang M, Zhang Y, Lv H (2016) Comparative transcriptome analysis between broccoli (*Brassica olercea* var. *italica*) and wild cabbage (*Brassica macrocarpa* Guss) in response to *Plasmodioiphora brassicae* during different infection stages. Front Plant Sci 7:1929

Zhang L, Zhang F, Melotto M, Yao J, He SY (2017) Jasmonate signaling and manipulation by pathogens and insects. J Exp Bot 68:1371–1385

Zhang B, Li P, Su T, Li P, Xin X, Wang W, Zhao X, Yu Y, Zhang D, Yu S, Zhang F (2018) BrRLP48, encoding a receptor-like protein, involved in downy mildew resistance in *Brassica rapa*. Front Plant Sci 9:1708. https://doi.org/10.3389/fpls.2018.01708

Zhao J, Last RL (1996) Coordinate regulation of the tryptophan biosynthetic pathway and indolic phytoalexin accumulation in *Arabidopsis*. Plant Cell 8:2235–2244

Zhao C, Nie H, Shen Q, Zhang S, Lukowitz W, Tang D (2014) EDR1 physically interacts with MKK4/MKK5 and negatively regulates a MAP kinase cascade to modulate plant innate immunity. PLoS Genet 10(5):e1004389. https://doi.org/10.1371/journal.pgen.1004389

Zhao Y, Bi K, Gao Z, Chen T, Liu H, Xie J, Cheng J, Fu Y, Jiang D (2017) Transcriptome analysis of *Arabidopsis thaliana* in response to *Plasmodiophora brassicae* during early infection. Front Microbiol 8:673

Zheng X, Chen L, Xia H, Wei H, Lou Q, Li M, Li T, Luo L (2017) Transgenerational epimutations induced by multigeneration drought imposition mediate rice plant's adaptation to drought condition. Sci Rep 7:39843

Zhou N, Tootle TL, Tsui F, Klessig DF, Glaze Brook J (1998) PAD4 functions upstream from salicylic acid to control defense responses in *Arabidopsis*. Plant Cell 10:1021–1030

Zhou LZ, Howing T, Muller B, Hammes UZ, Getl C, Dresselhaus T (2016) Expression analysis of KDEL CysEPs programmed cell death markers during reproduction in *Arabidopsis*. Plant Reprod 29:265–272

Zhu L, Li Y, Ara N, Yang J, Zhang M (2012) Role of a newly cloned alternative oxidase gene (BjAOX1a) in turnip mosaic virus (TuMV) resistance in mustard. Plant Mol Biol Rep 30:309–318

Zimmermann P, Hirsch-Hoffmann M, Hennig L, Gruissem W (2004) Gene vestigator. *Arabidopsis* microarray data base and analysis toolbox. Plant Physiol 136:2621–2632

Zipfel C, Robatzek S, Navarro L, Oakeley EJ, Jones JDG, Felix G, Boller T (2004) Bacterial disease resistance in *Arabidopsis* through flagellin perception. Nat Intl J Sci 428:764–767

Molecular Mechanisms of Host Resistance to Hemibiotrophs and Necrotrophs

3

Abstract

The molecular mechanisms of nonhost resistance to *Alternaria* in *Brassica* indicated activation of defense-related genes earlier in nonhost plants than the host plants. Six QTLs governing resistance to *Alternaria brassicae* have been identified, five of them are population-specific while one QTL is common between all the three mapping populations. In *Brassica*, defense-related genes (PR proteins) including defensins (β-glucanases and chitinases) have been identified against *Alternaria*. Different defensin genes in *B. juncea* have been characterized for their structures, evolution, cellular location, and regulation in response to SA, JA, and *Alternaria* infection. Chitinase genes in *Brassica* have been identified to breed *Alternaria*-resistant cultivars. A single R-locus RCH-1 has been identified at the tip of chromosomes 4 in *Arabidopsis thaliana* ecotype Eil-O against *Colletotrichum higginsianum.* The locus RCHz maps to an extensive cluster of R-loci known as MRC-J in the *Arabidopsis* ecotype Ws-O. In *B. oleracea*, two types of R-genes have been identified. A type with single dominant genes is effective against *Fusarium* race 1 and stable under high or low temperatures. B type is polygenic resistance, and it is unstable under high temperature above 24 °C. Type A resistance to *FOC* 1 race 1 conferred by a dominant single gene, *FOC1*, has been mapped and molecular markers have been developed. R-genes and QTLs have been identified from *B. rapa*, *B. juncea*, *B. nigra*, and *B. napus* to provide resistance to *Leptosphaeria*. The LepR1 and lepR2 genes provide cotyledon and adult plant resistance in *B. napus* to *Leptosphaeria.* The genetic background is useful markers to select the most suitable genotype as a recipient parent while developing *Brassica* cvs. with *Lm* resistance. The combined effect of R-genes and host genetic background generates stronger defense responses through induction of genes related to callose deposition, upregulation of ROS, and SA and JA pathways.

G. S. Saharan et al., *Molecular Mechanism of Crucifer's Host-Resistance*,
https://doi.org/10.1007/978-981-16-1974-8_3

Keywords

Molecular mechanisms · Hemibiotrophs · NHR to *Alternaria* · Quantitative resistance · Cloning of R-genes · Defensin genes · Chitinase genes · *Leptosphaeria* AVR genes · Expression of *Cf9* and *Avr* 9 genes · Mapping of R-genes · Breakdown of R-genes · Gene expression · Cysteine-rich protein kinase genes · Stable QTL's R-genes · Mapping of loci · *Lm-AvrLm9* R-gene · Race specific R-genes · Introgression of R-genes · Mechanism of R-genes · Phytoalexins · Over expression of R-genes · *OsPGIP2* gene · *BnWRKY33* gene · *MPK4* gene · *AtWRKY28* · *AtWRKY75* · Microsatellite markers · NBS-LRR encoding genes · Expression of microRNAs · Breakdown of R-genes

Expression of *Cf9* and *Avr* 9 genes in *Brassica* induces resistance to *Leptosphaeria maculans* (*Lm*). R-genes and QTLs from *B. rapa*, *B. juncea*, and *B. napus* have been molecularly mapped with markers to provide resistance to *Lm*. Two R-genes, *LepR3* and *Rlm2*, have been cloned along with seven corresponding avirulence (*Avr*) genes (*Avr lm1*, *AvrLm6*, *AvrLm3*, *AvrLm5/AvrLmJ1*, *AvrLm4-7*, *Avr Lm6*, and *AvrLm11*). Molecular basis of assessment of R-genes breakdown in *Brassica* cvs can be predicted by frequencies of avirulence/virulence alleles in populations of *Lm*. In resistant and susceptible lines of *B. napus*, gene expression profile differs to confer resistance to *Lm*. Cystein-rich protein kinase genes and multienvironment stable six QTLs suitable for quantitative blackleg resistance in *B. napus* have been identified. Marker-associated mapping of QTLs in *Brassica* will help identification of desirable alleles and facilitate QTL introgression. For stable resistance to *Lm*, eight QTL regions for resistance on AO2, AO9, AO10, CO1, and CO9 have been identified in *B. napus* under diverse growing environments. Nine of the 18 blackleg R-genes from *Brassica* species correspond to genetically defined *Avr* genes from *Lm*. Mechanisms of resistance to light leaf spot (*Pyrenopeziza*) in *Brassica* is controlled by two R-genes. QTLs controlling partial resistance in *Brassica* at leaf and stem infection stages by *Sclerotinia* have been identified. Resistance to *Sclerotinia* functions at host tissue breakdown, preformed antifungal compounds, and production of phytoalexins. Resistance in *Brassica* to *Sclerotinia* is governed by a complex interplay of minor effect genes designated as quantitative disease resistance (QDR). The candidates defense-related genes have been mapped by genome-wide association combined with other omics approaches. Overexpression of genes like *BnNPR1*, *OsPGIP2*, *BnWRKY33*, and *BnMPK4* enhances resistance in *B. napus* to *Sclerotinia*, whereas overexpression of *AtWRKY28* and AtWRKY75 genes in *Arabidopsis* enhanced resistance to oxalic acid and *Sclerotinia*. The microsatellite marker for genome-wide associations in *B. napus* has been identified three loci DSRC4, DSRC6, and DSRC8 for *Sclerotinia* rot resistance. SSR-based association mapping identified six markers' loci associated to *Sclerotinia* resistance in A- and B-genomes of *Brassica*. Genome-wide association analysis of 84ILS recognized a large number of SNPs associated to *Sclerotinia* resistance in chromosomes AO3, AO6, and BO3.

In *B. napus*, one major RQTL contributed by R53 on chromosome C5 and minor QTL contributed by Express 617 on chromosome C1 confer resistance to *Verticillium longisporum*. Black rot (*XCC*) resistance genes and QTLs in different *Brassica* species have been identified on different chromosomes to *Xanthomonas* races 1 and 4. Seven NBS-LRR-encoding genes linked to black rot resistance in *B. oleracea* have been identified on four QTL regions and chromosomes CO1, CO3, CO6, CO7, and CO8, respectively. Micro RNAs (miR156, miR167, miR169, and miR90) play a role in *B. oleracea* resistance enhancement to *XCC*. R-gene1 conferring resistance to race 1 has been identified in *B. carinata*, *B. juncea*, and *B. nigra*.

3.1 Introduction

The development in the science and technology through modern approaches like omics and genomic have facilitated to investigate molecular mechanisms of crucifer's host resistance against different biotic and abiotic stresses. The molecular mechanisms of host resistance have been investigated in *Brassica* crops using three different categories of host pathosystem, viz., biotrophs, hemibiotrophs, and necrotrophs. The molecular mechanisms of R-genes resistance imparted by NBS-LRR are effective against biotrophs or hemibiotrophs but are not effective against necrotrophs. Salicylic acid (SA), jasmonic acid (JA), and abscisic acid (ABA) are the key modulators involved in signaling pathways during host defense. Jasmonic acid and ethylene-mediated signaling is predominantly involved in general defense against necrotrophs. NBS-LRR genes are intracellular receptors that can recognize the pathogen invasion by binding to pathogen effector proteins in host plant and trigger various defense signal transduction to yield hypersensitive response and inhibit pathogen development. The NBS domain binds and hydrolyzes ATP and GTP while LRR domain is liable for protein–protein interaction. NBS domains are of various motifs like P-loop and kinase 2 motif. NBS-LRR genes are grouped into TIR-BSS-LRR (TNL) and CC-NBS-LRR (ONL). Many NBS-LRR genes have been identified in *Brassica* species for the defense against different pathogens infecting *Brassica* crops. The primary plant defense (pre-penetration) against fungal pathogen relies on cell surface-mediated defense signaling initiated by recognition of PAMPs, secondary (post-penetration) defense responses are often induced by the Salicylic acid (SA) or Jasmonic acid/Ethylene (JA/ET) phytohormone signaling pathways.

3.2 *Brassica–Alternaria*: Molecular Resistance

3.2.1 The Molecular Mechanisms of Nonhost Resistance (NHR) to *Alternaria* Species

Defense responses are evoked in both host and nonhost infections; however, the NHR involves not only earlier induction but also a robust defense response compared to the host plant. Both *A. brassicicola* and *A. brassicae* affect almost the same cruciferous crops and generally exhibit similar symptoms (Saharan et al. 2016; Meena et al. 2016). However, the literature indicates that *Arabidopsis* and *S. alba*, which belong to the same family (Brassicaceae), show NHR to *A. brassicicola* (Thomma et al. 1999; Pedras et al. 2001) but act as hosts for *A. brassicae* (Hansen and Earle 1997; Sharma et al. 2002). A study showed *Arabidopsis* accession Gre-0 to be highly susceptible to *A. brassicae* (Mandal et al. 2018). Therefore, understanding the NHR mechanisms in these plants against *A. brassicicola* will indirectly illustrate the resistance mechanisms operating during the interaction of *A. brassicae* with nonhost plants or absent during the interaction with these susceptible host plants. In *A. thaliana* and *S. alba*, early and high-level induction of defense-related genes, namely, pathogenesis-related-1 (PR1), β-1,3 glucanase (PR2), and chitinase (PR3), occurs compared to *B. juncea* after infection with *A. brassicae* or *A. brassicicola* (Narusaka et al. 2003; Van Wees et al. 2003; Ghose et al. 2008; Nayanakantha et al. 2016; Mandal et al. 2018). Moreover, the expression of PR3, encoding chitinase, remained undetected in *B. juncea* after pathogen infection, but its expression was highly induced in *S. alba* (Nayanakantha et al. 2016). However, both *A. thaliana* and *S. alba* actively secrete chitinase enzymes, which hydrolyze the fungal cell wall and release chitin fragments (Narusaka et al. 2003; Chatterjee et al. 2013). These chitin fragments are recognized by some receptors that activate effective defense responses against the pathogens (Wan et al. 2008). This suggests that the defense-related genes are activated earlier in nonhost plants than in host plants as a part of the NHR strategy, and the chitinase capable of degrading fungal cell walls plays an important role in restricting pathogen growth at early stages of infection. Furthermore, the receptors involved in chitin recognition in *Arabidopsis*, CERK 1 or LysM receptor-like kinase (LysM-RLK) and phytosulfokine receptor kinase (PSK), were found to be induced after *A. brassicicola* infection (Wan et al. 2008; Loivamaki et al. 2010). Wan et al. (2008) also showed that a mutation in the LysM-RLK gene inhibits the induction of all chitin-responsive defense genes and compromises the NHR to *A. brassicicola*. Similarly, *Arabidopsis* homologs for RLK receptors were identified in *S. alba*, and their expression was highly induced after infection with *A. brassicicola*; the genes were downregulated in *B. juncea* (Ghose et al. 2008). This indicates that chitin digestion from the fungal cell wall and its perception by the plant may play an essential role during defense signaling in nonhost plants. The NHR response involves the stimulation of a signal transduction cascade after the perception of a pathogen by the plant cell, which initiates the activation of protein kinases and members of MAP kinases and subsequently leads to defense gene activation in nonhost plants (Lawrence et al. 2008). MAPK6 expression was found to be highly induced for a longer duration in *S. alba* plants infected with *A. brassicicola*, but it was downregulated in *B. juncea* (Taj et al. 2010). These findings lead to speculate that MAPK6 might be involved in imparting NHR and targeting many downstream components that are involved in effective defense response.

Salicylic acid (SA), jasmonic acid (JA), and abscisic acid (ABA) are the key modulators involved in phytohormone signaling during plant defense. JA-mediated signaling is predominantly involved in general defense against necrotrophic pathogens (Glazebrook 2005). In *Arabidopsis*, JA-mediated activation of the defense response occurs against *A. brassicicola* (Thomma et al. 1998; Van Wees et al. 2003), while *S. alba* challenged with the same pathogen induces ABA- and JA-mediated defense responses (Mazumder et al. 2013). In contrast, susceptible *B. juncea* plants induce SA-mediated defense signaling pathways (Thomma et al. 1998; Van Wees et al. 2003; Mazumder et al. 2013). It is likely that there is a significant amount of crosstalk between phytohormones and a convergence of two or more signaling pathways, which play a role in deciding whether disease progression will occur or the defense pathways will overcome the pathogen. Considering these studies together, it can be summarized that early signaling events in nonhost plants have a major role in imparting NHR to *A. brassicicola* (Fatima et al. 2019).

3.2.2 Identification and Mapping of Quantitative Disease Resistance to *Alternaria brassicae* in *Arabidopsis*

Broad host-range necrotrophs (BHNs), such as *Sclerotinia sclerotiorum* and *Alternaria brassicae*, can infect different plant species. BHNs typically deploy a diverse arsenal of effectors, including cell-wall degrading enzymes (CWDEs), phytotoxic compounds, and reactive oxygen species (ROS), to induce necrosis. The diversity of virulence strategies thus warrants a multifaceted defense by the host to successfully ward off the attack. In contrast, narrow host-range necrotrophs (NHNs) tend to rely on host-specific toxins (HSTs) that are directed at specific targets present only in some species or subtypes of a particular species. The recognition of these HSTs by the host machinery thus leads to susceptibility. Plant disease resistance can be either qualitative, which is conferred by single resistance (R)-genes, or quantitative, which is mostly mediated by multiple genes. Host resistance against BHNs is known to be usually quantitative. Some of the quantitative resistance loci (QRLs) for the paradigmatic BHNs, such as *S. sclerotiorum*, have been identified, but the underlying mechanisms of most of these QRLs are unknown. Few genes, which recognize the HSTs of NHNs, have been identified. Unlike NHNs, necrotrophs such as *A. brassicae* and *A. brassicicola* infect only the members of the Brassicaceae family, including the wild and cultivated species (Sharma et al. 2002). These necrotrophs thus represent an intermediate class between BHNs and NHNs. The genetic architecture of resistance to these necrotrophs is relatively unexplored when compared with that of BHNs and NHNs. There are currently no known resistance loci identified in any of the natural hosts (*Brassica* crops) for resistance to *A. brassicae*.

Arabidopsis has been used as a model host to study the host–pathogen interactions of many plant pathogens. Most of the *Arabidopsis* accessions tested with *A. brassicicola* have shown complete resistance to the pathogen (Kagan and Hammerschmidt 2002; Mukherjee et al. 2009). In the crosses between highly susceptible and resistant accessions, both major and minor loci conferring resistance were mapped. Further, it was hypothesized that different loci/mechanisms conferring resistance might exist in different accessions of *Arabidopsis* (Rajarammohan et al. 2018).

Quantitative disease resistance (QDR) is the predominant form of resistance against necrotrophic pathogens. The genes and mechanisms underlying QDR are not well known. The *Arabidopsis–Alternaria brassicae* pathosystem was used to uncover the genetic architecture underlying resistance to *A. brassicae* in a set of geographically diverse *Arabidopsis* accessions. *Arabidopsis* accessions revealed a rich variation in the host responses to the pathogen, varying from complete resistance to high susceptibility. Genome-wide association (GWA) mapping revealed multiple regions to be associated with disease resistance. A subset of genes prioritized based on gene annotations and evidence of transcriptional regulation in other biotic stresses was analyzed using a reverse genetics approach employing T-DNA insertion mutants. The mutants of three genes, namely At1g06990 (GDSL-motif lipase), At3g25180 (CYP82G1), and At5g37500 (GORK), displayed an enhanced susceptibility relative to the wild type. These genes are involved in the development of morphological phenotypes (stomatal aperture) and secondary metabolite synthesis, thus defining some of the diverse facets of quantitative resistance against *A. brassicae* (Rajarammohan et al. 2018). Three biparental mapping populations were developed from three resistant accessions, viz. CIBC-5, Ei-2, and Cvi-0, and two susceptible accessions—Gre-0 and Zdr-1 (commonly crossed to CIBC-5 and Ei-2). A total of six quantitative trait loci (QTLs) governing resistance to *A. brassicae* were identified, five of which were population-specific while one QTL was common between all the three mapping populations. Interestingly, the common QTL had varying phenotypic contributions in different populations, which can be attributed to the genetic background of the parental accessions. The presence of both common and population-specific QTLs indicates that resistance to *A. brassicae* is quantitative, and that different genes may mediate resistance to the pathogen in different accessions. Two of the QTLs had moderate-to-large effects, one of which explained nearly 50% of the variation. The large-effect QTLs may therefore contain genes that could play a significant role in conferring resistance even in heterologous hosts (Rajarammohan et al. 2017).

3.2.3 Identification, Cloning, and Sequencing of Resistant Genes

The antifungal effect of hevein, the chitin binding lectin from rubber plant (*Hevea brasiliensis*), has been analyzed in transgenic plants for potential control of *Alternaria* blight in Indian mustard. A cDNA encoding hevein was transferred into Indian mustard (*B. juncea* cv. RLM-198). Southern analysis of the putative transgenic has shown the integration of the transgene. Northern and Western analyses proved that the integrated transgene is expressed in the transgenics. In whole plant bioassay under glasshouse conditions, transgenics possess parameters that are associated with resistance such as longer incubation and latent period, smaller necrotic lesion size, lower disease intensity, and delayed senescence (Kanrar et al. 2002).

β-Amino-butyric acid (BABA) pretreatment of *Brassica* plants provides protection against the necrotrophic pathogen *A. brassicae*. The achieved resistance level is much higher than that seen after salicylic acid (SA) and jasmonic acid (JA) pretreatments. BABA pretreatment to leaves, 1 day before inoculation, although leads to an inhibition of the oxidative burst and a decrease in SA levels, but neither influences lipoxygenase activity nor causes callose deposition at the site of inoculation. Expression of two marker genes of the SA and JA pathways, namely *PR1* and *PDF1.2*, enhances in response to BABA pretreatment. BABA-induced resistance is mediated through an enhanced expression of pathogenesis-related protein genes, independent of SA and JA accumulation (Kamble and Bhargava 2007). The identification and cloning of *hsr203j* homologues from tolerant and susceptible genotypes of *B. juncea* through RT-PCR analysis have been accomplished. In silico analysis of the sequences isolated from susceptible and tolerant genotypes of *B. juncea* shows the presence of conserved abhydrolase domain having role in cell death. Motif analysis indicates that motif 19 that functions in prenylation is found exclusively in tolerant genotypes and motif 12 having myristoylation site is found in susceptible genotypes. Various defense-related important *cis*- and *trans*-acting factors are also found in these homologues. This suggests that these *hsr203j* like homologues of *Brassica* play an important role in differential defense against *Alternaria* blight—a recalcitrant disease caused by *A. brassicae* (Mishra et al. 2010).

Cysteine-rich antimicrobial peptides isolated from plants have emerged as potential resource for protection of plants against phytopathogens. The significance of an antimicrobial peptide, *Pm AMPI*, isolated from western white pine (*Pinus monticola*), in providing canola resistance against several phytopathogenic fungi, has been reported by Verma et al. (2012). The cDNA encoding *Pm AMPI* was successfully incorporated into the genome of *B. napus*, and its in planta expression conferred greater protection against *A. brassicae*, *Leptosphaeria maculans*, and *Sclerotinia sclerotiorum*. In vitro experiments with proteins extracted from transgenic canola expressing *Pm AMPI* demonstrated its inhibitory activity by reducing growth of fungal hyphae. In addition, the in vitro-synthesized peptide also inhibited the growth of fungi. Generating transgenic crops expressing *Pm AMPI* may be an effective and versatile method to protect susceptible crops against multiple phytopathogens.

To develop resistance against *A. brassicae*, the barley antifungal genes class II chitinase (*AAA56786*) and type I ribosome-inactivating protein (RIP; *AAA32951*) were coexpressed in Indian mustard via *Agrobacterium*-mediated transformation. The stable integration and expression of transgenes in T_0 plants were confirmed by Southern blot and Western analysis. The transgenic lines showing inheritance in Mendalian fashion (3:1) were further evaluated by in vitro studies and under greenhouse conditions for resistance against *A. brassicae*. The transgenic plants showed up to 44% reduction in *A. brassicae* hyphal growth in in vitro antifungal assays. In greenhouse screening, the transgenic plants sprayed with *A. brassicae* spores showed resistance through delayed onset of the disease and restricted number, size, and expansion of lesions as compared to wild-type plants. The expression of chitinase and RIP from a heterologous source in *B. juncea* provides subsequent protection against *Alternaria* leaf spot and can be helpful in increasing the production of Indian mustard (Chhikara et al. 2012).

Mutation in the gene *PAD3*, encoding a cytochrome *P450*, abolishes the biosynthesis of camalexin (Glazebrook and Ausubel 1994) and results in enhanced susceptibility to necrotrophic fungi (Thomma et al. 1999). The phytoanticipins glucosinolates (GS) are sulphonated thioglycosides comprising a common glycone moiety with a variable a glycone side chain and are considered the major secondary metabolites of Brassicaceae (Fahey et al. 2001). Upon tissue damage, GS come into contact with myrosinases, a specific class of β-thioglucosidases, which are stored separately in the cell. Hydrolysis of GS by myrosinases yields isothiocyanates (ITC), nitrile, and epithionitriles. The most common breakdown products, ITC, exhibit toxicity toward several plant pathogens including bacteria, fungi, insects, and nematodes (Fahey et al. 2001). It has been demonstrated that ITC produced by *Arabidopsis thaliana* significantly inhibit growth of some fungal pathogens in plants (Tierens et al. 2001), although recent results suggest that apart from their direct toxic effects, GS breakdown products may also act by modulation of plant defense signaling (Brader et al. 2006). During host infection, *A. brassicicola* is exposed to high levels of antimicrobial defense compounds, such as indolic phytoalexins, and glucosinolate breakdown products. To investigate the transcriptomic response of *A. brassicicola* when challenged with brassicaceous defense metabolites, suppression subtractive hybridization (SSH) was performed to generate two cDNA libraries from germinated conidia treated either with allyl isothiocyanate (AI-ITC) or with camalexin. Following exposure to AI-ITC, *A. brassicicola* displays a response similar to that experienced during oxidative stress. A substantial subset of differentially expressed genes is related to cell protection against oxidative damage. Treatment of *A. brassicicola* conidia with the phytoalexin camalexin appears to activate a compensatory mechanism to preserve cell membrane integrity, and among the camalexin-elicited genes, several are involved in sterol and sphingolipid biosynthesis. The transcriptomic analysis suggests that protection against the two tested compounds also involves mechanisms aimed at limiting their intracellular accumulation, such as melanin biosynthesis (in case of camalexin exposure only) and drug efflux. From the AI-ITC and the camalexin differentially expressed genes identified, 25 are selected to perform time course studies during interactions with brassicaceous hosts. In Planta, upregulation of all the selected genes is observed during infection of *Raphanus sativus*, whereas only a subset is overexpressed during the incompatible interaction with *A. thaliana* ecotype Columbia (Sellam et al. 2007; Saharan et al. 2016; Fig. 3.1).

3.2.4 Identification and Characterization of Defensin Genes in *Brassica juncea* Against *Alternaria brassicae*

The molecular mechanisms underlying activation of plant defense responses are exceedingly complex, and understanding how plants defend themselves against a range of pathogens is of crucial importance for successful managing of crops to get

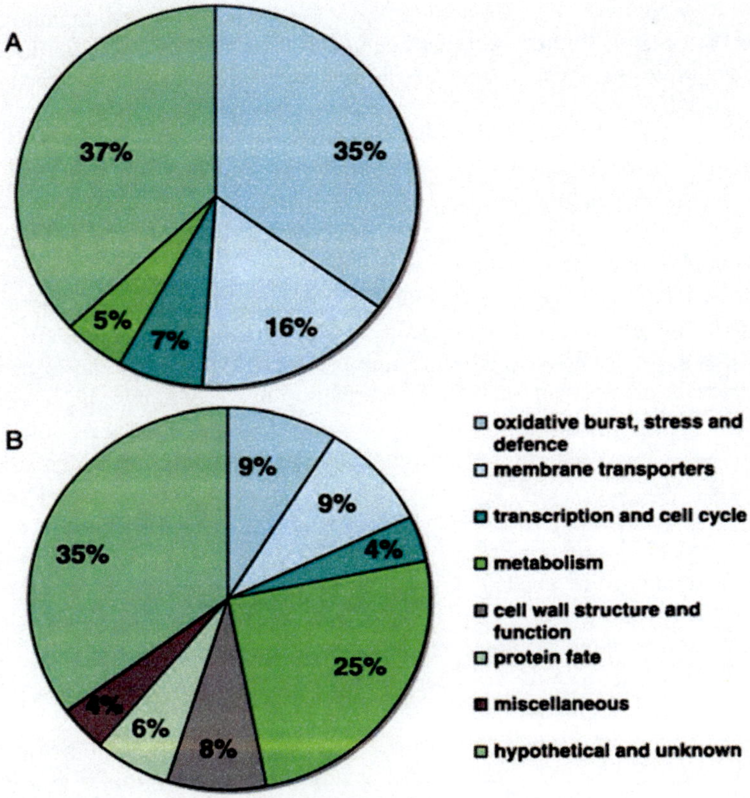

Fig. 3.1 Functional classification of up-regulated genes in *Alternaria brassicicola*, (**A**) AI-ITC- and (**B**) camalexin-treated conidia according to their putative biological function (Sellam et al. 2007)

higher returns. Once plant defense responses are activated at the site of infection, a systemic defense is often triggered in distal plant parts to protect these undamaged tissues against subsequent invasion by the pathogen. Systemic defense is controlled by a complex signal transduction network that coordinates the overall defense responses of the plant. Resistance to biotrophic pathogens is mediated through phytohormones such as salicylic acid (SA) and necrotrophic pathogens by jasmonic acid (JA) and ethylene (ET) signaling pathways (Loake and Grant 2007). Various novel PR proteins are induced collectively as a group of the most important inducible defense-related antifungal proteins. They are associated with the development of systemic acquired resistance in wide range from cell wall rigidification to signal transduction and antimicrobial activity. They are the key modulators of plant defense, which cause activation of a different set of defense-related genes, including plant defensin genes. Plant defensins, which belong to PR-12 family of PR proteins, are cysteine-rich, highly basic prominent cationic peptides in plants with different roles in defense. They are one of the largest families of antifungal peptide molecules.

Since the beginning of 1990s, many cationic plant cysteine-rich antimicrobial peptides have been studied, and plant defensins were first described in the seeds of wheat and barley. There are two major hypotheses, which explain the mechanism of action of antimicrobial defensin: the carpet model and the pore model. In both models, defensins are described to interact with the negatively charged molecules present at the cell membrane of pathogens, causing an increase of its permeabilization, leading to cell leakage and death by necrosis. In addition to being antimicrobial, plant defensins are also involved in the biotic stress response, as well as plant growth and development.

In *Brassica*, only a few defense-related genes have been reported. These include PR proteins including few plant defensins, β-glucanases, and chitinases. Tiwari et al. (2017) studied the effect of methyl jasmonate on disease severity and expression of plant defensin gene during *A. brassicae* infection in *Arabidopsis* and reported that transcripts of *PDF1.2* accumulated at a greater level upon challenge inoculation with *A. brassicae* along with jasmonic acid compared to treatment containing pathogen as well as jasmonic acid alone. However, no attempts have been undertaken to find the number of defensin genes present in *B. juncea* or in *Alternaria*-resistant *C. sativa*, and the structural and functional characteristics of defensins from above two genotypes remain to be unclear. Rawat et al. (2017) isolated the full-length pathogen-inducible plant-defensin gene from *B. juncea* and have reported its similarity with gamma thionin and knottin families of plant antimicrobial peptides. Efforts are being made in order to have sufficient knowledge about the genes induced during infection and their regulation measures. Chaturani et al. (2019) characterized different defensin genes in *B. juncea*, *C. sativa*, and related species to find their structure, evolution, and cellular localization and to study their regulation in response to SA, JA, and *Alternaria* infection.

Plants are attacked by various biotic and abiotic factors throughout their lives, and how efficient and effective the defense mechanisms are determine the survival. Defensins belong to pathogenesis-related (PR) protein class-12. They are one of the key modulators in plant defense mechanisms, which have diverse functions including inhibitory effect on a broad range of phytopathogenic fungi, and are involved in biotic stress response. In depth, study on defensins gives an insight into their role in defense, which further can be used in conferring disease resistance in crop plants. In silico analysis identified 56 putative defensins in *C. sativa* (resistant to *Alternaria* blight), whereas only six were identified in *B. juncea* (susceptible to *Alternaria* blight). Multiple sequence alignment and consensus analysis confirmed the conserved eight cysteine residues, and structural analysis revealed the presence of an α-helix and triple strand antiparallel β-sheets. Defensin genes identified were having a single intron, and identification of methyl Jasmonate (MeJA) responsive elements in promoter analysis confirms their regulation under jasmonic acid (JA) signaling pathway. These defensins were phylogenetically related to other *Brassica* species as well as to *Arabidopsis* indicating their close evolutionary relationship (Fig. 3.2). Expression analysis indicated that defensin is responsive to JA, salicylic acid (SA), *Alternaria* infection, and wounding. The antagonistic effect between SA and JA was observed in defensin gene expression in response to their

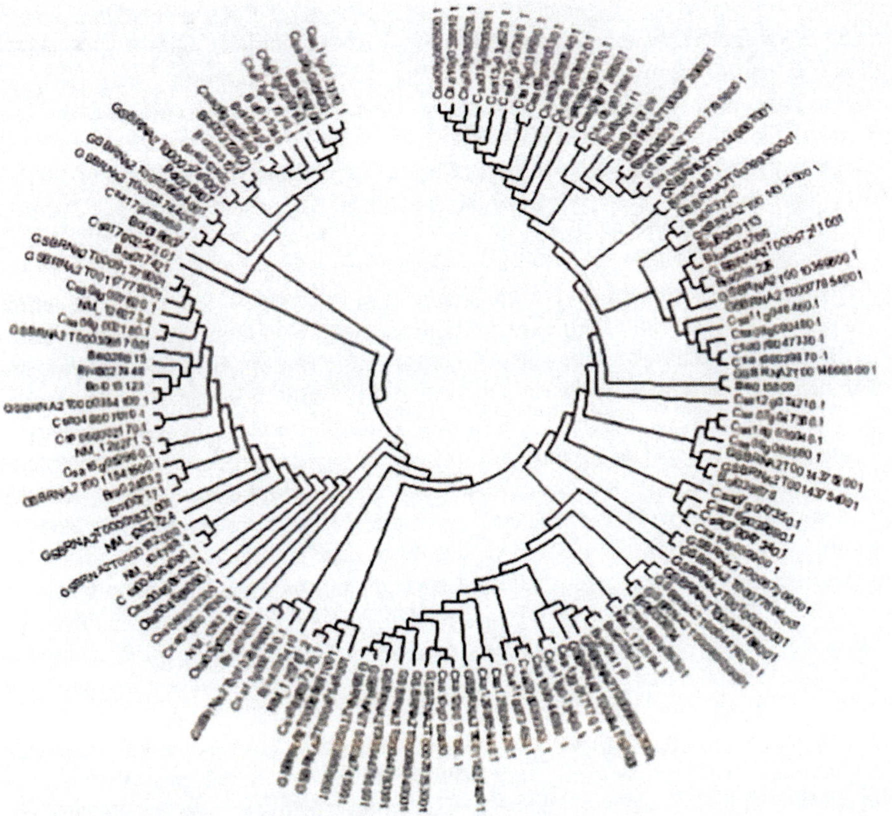

Fig. 3.2 The phylogenetic analysis of *Brassica juncea*, *Brassica rapa*, *Brassica oleracea*, *Brassica napus*, *Camelina sativa*, and *Arabidopsis thaliana* defensin proteins by neighbor-joining method (Chaturani et al. 2019)

exogenous application. It is an important foundation for further investigation on defensin as a possible source for creating disease resistance in transgenics (Chaturani et al. 2019).

3.2.5 Identification and Characterization of Chitinase Genes in *Brassica juncea* in Response to *Alternaria brassicae*

Chitinases belong to the group of pathogenesis-related (PR) proteins that provides protection against fungal pathogens. The genome-wide identification and characterization of chitinase gene family in two important oilseed crops *B. juncea* and *C. sativa*, belonging to family Brassicaceae, identified 47 and 79 chitinase genes in the genomes of *B. juncea* and *C. sativa*, respectively. Phylogenetic analysis of chitinases in both the species revealed four distinct subgroups, representing different classes of chitinases (I–V). Microscopic and biochemical study revealed the role of reactive oxygen species (ROS) scavenging enzymes in disease resistance of *B. juncea* and *C. sativa*.

Furthermore, qRT-PCR analysis showed that expression of chitinases in both *B. juncea* and *C. sativa* was significantly induced after *Alternaria brassicae* infection. However, the fold change in chitinase gene expression was considerably higher in *C. sativa* compared to *B. juncea*, which further proves their role in *C. sativa* disease resistance to *A. brassicae*.

The identified 47 and 79 chitinase genes in *B. juncea* and *C. sativa*, respectively, were unevenly distributed on chromosomes in both crop species. Based on amino acid sequence similarities, chitinases of both species were classified into five classes (I–V), and these classes are clustered into four major groups of phylogenetic trees. Members of the same group or subgroup reveal highly conserved exon/intron composition, conserved domain, and motifs. Furthermore, the synteny analysis revealed that chitinase gene family in *C. sativa*, *B. rapa*, and *B. oleracea* crop plants have expanded through WGT, segmental and tandem duplication events, and this gene family is under positive selection pressure. Enzyme assays also revealed the possible role in defense in response to *A. brassicae* infection. The qRT-PCR expression analysis revealed that chitinase genes are significantly induced by *A. brassicae* infection, and thus provides insights about their potential role in disease resistance. Moreover, chitinase genes of *C. sativa* showed greater fold change in expression compared to *B. juncea* following *A. brassicae* inoculation and provides further evidence for increased disease resistance of *C. sativa*. This fact has been further proved, as more number of defense-related *cis*-elements was present in upstream regions of chitinase genes of *C. sativa* compared to *B. juncea*. Hence, this leads to the conclusion that *C. sativa* chitinase genes can be effectively used for the development of disease resistant varieties of *Brassica* crop species (Mir et al. 2020).

3.2.6 Bioassay of Molecular Tolerance Mechanisms in *Brassica* to *Alternaria*

Alternaria fungi are known to cause damage to photosynthetic machinery of plants. Chlorophyll and carotenoid pigments are necessary for plant photosynthesis, which are present in photosystems I and II and impart their role in light harvesting. Carotenoid inhibits oxidative stress by quenching singlet oxygen (1O2) and triplet chlorophyll (3Chl) and thus protects photosynthetic machinery. *Alternaria* negatively affects photosynthetic activities via necrosis development in leaves ultimately causing reduction in chlorophyll and carotenoid content. Reduction in chlorophyll content on pathogen invasion indicates cell damage in canola tissues. The photosynthetic pigments decreased with time after inoculation. It was observed that pathogen-treated plants had lost their photosynthetic pigments. Minimum chlorophyll (chl "*a*," chl "*b*," and total chlorophyll) and carotenoid contents were observed in infected leaves of susceptible genotypes showing that tolerant genotypes have potential to retain their photosynthetic pigment under stress (Munir et al. 2020; Martinez et al. 2018).

Borah et al. (1978) proposed that reduction in chlorophyll contents (chlorophyll "*a*" and "*b*") might be attributed to inhibition of synthesis instead of degradation of existing pigments. Such changes in photosynthetic attributes are general signs of stress. Therefore, sustaining chlorophyll content in plants upon pathogen invasion is vital, as it will permit plant cell to continue photosynthesis.

Munir et al. (2020) showed that TSP contents were increased with disease progression in mustard and rapeseed plants. On contrary to photosynthetic pigments, TSP contents increased in susceptible genotypes as compared to tolerant ones. Parallel trend was also observed by Onifade and Agboola (2003), who suggested that proliferation of pathogen synthesizes numerous enzymatic proteins and causes, occasionally, rearrangement of nutritive composition of substrate due to formation of many degraded byproducts thereby enhancing its protein content. Amino acids act as a substrate for causal pathogen during host–pathogen interplay. They might also be involved in metabolic phenomenon interlinked with disease tolerance and exerting fungistatic effects via synthesis of infection-specific proteins, e.g., glyceollin accumulation in soybean tissues upon pathogen invasion (Mathpal et al. 2011).

The oxidative burst or overproduction of ROS belongs to the earliest defense responses against pathogen invasion in plants. To counterbalance the effect of oxidative stress, plants have developed an arsenal of defense mechanisms against pathogen outbreak. High SOD activity was observed in all genotypes due to high ROS accumulation as SOD acts as first line of defense against oxidative burst and dismutates O_2- to H_2O_2 and O_2. The SOD activity was higher in tolerant genotypes. Ehsani-Moghaddam et al. (2006) also observed high SOD activity in resistant strawberry genotypes against *Mycosphaerella fragariae* and concluded that resistant genotype possesses higher SOD activity and contributes in efficient antioxidant mechanism. Higher SOD activity can be a selection tool for plant tolerance against diseases. Likewise, CAT and POD counter overproduction of H_2O_2 and play central role in plant defense response. Several evidences support defensive role of POD activity in disease tolerance mechanism against *Alternaria* via production of phenolics, phytoalexins, and glycoproteins. Increase in CAT activity was observed from 4 to 48 hpi and decreased after 48 hpi as shown in Fig. 3.3. The increase in CAT activity during 0–48 hpi showed scavenging of excessive H_2O_2 produced in plants. Reduction in CAT activity occurs after 48 hpi due to overproduction/accretion of H_2O_2 and might be due to enhanced proteolysis induced by oxidative burst. Durner and Klessig (1996) reported that decrease in CAT activity was a part of plant defense response to protect it against pathogens, and plants can tolerate excessive H_2O_2 concentration when CAT activity was least (Munir et al. 2020).

Genetic control of rapeseed and mustard genotypes tolerant against *A. brassicicola* plays a critical role in determining the pathosystem interaction outcome. The assessment of *A. brassicicola* tolerance among rapeseed and mustard revealed that dark leaf spot disease induced an extreme alteration in plant biochemistry that cause reduction in photosynthetic area, defoliation, accelerated senescence, and ultimately poor yield in susceptible genotypes as compared to tolerant ones. Physiobiochemical defense response, as evidenced by tolerant genotypes via augmented activities of defense enzymes, is a vital sign of their role in *Brassica–*

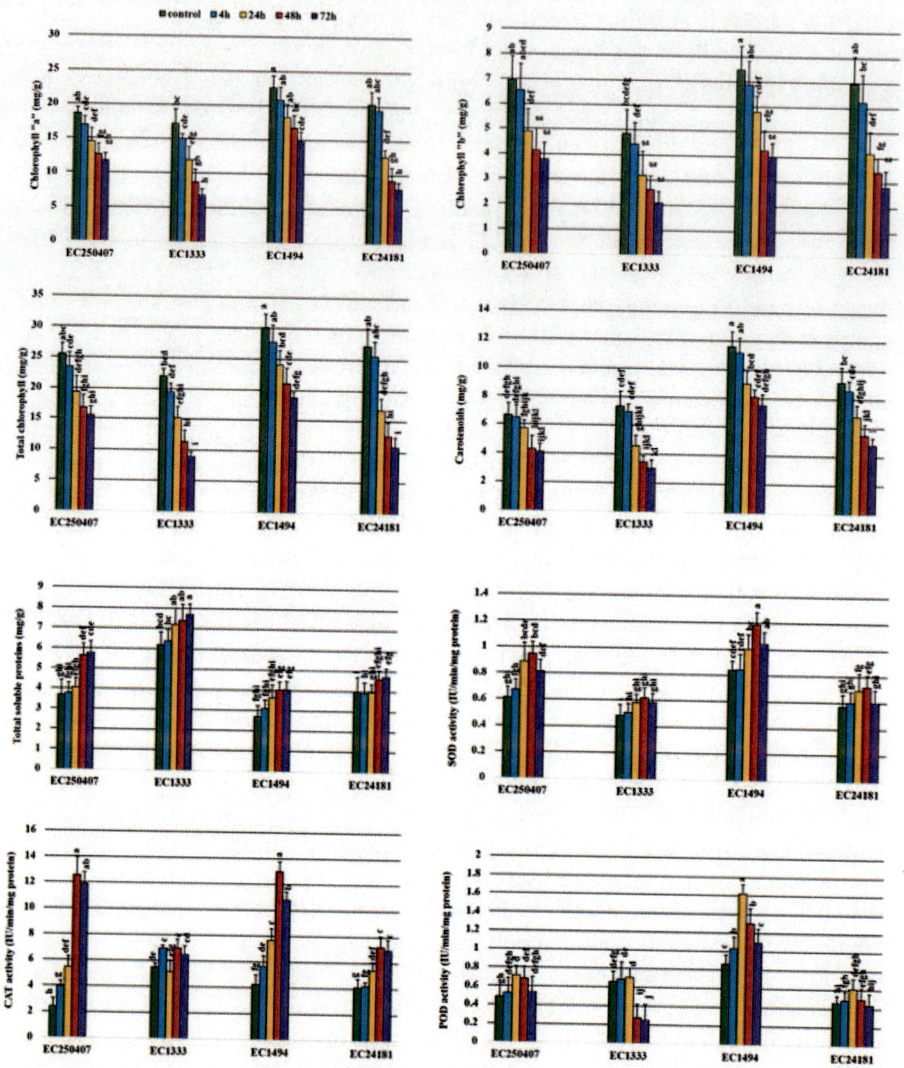

Fig. 3.3 Molecular and biochemical assays of the *Alternaria brassicicola–Brassicaceae* pathosystem. Genotypes (40-day old seedlings) were inoculated with spore suspension. Samples were taken from 4 to 72 hpi. Processed samples were used for several bioassays comprising of photosynthetic pigments, three redox enzymatic activities, and total soluble proteins, with at least three replication per sample. Bars represent means with standard deviation. Datasets for each parameter were analyzed with two-way ANOVA ($\alpha = 0.05$), with post hoc Tukey's HSD whenever significant ($P < 0.05$). Interactions between genotypes and time (hpi) were recorded. Data points with same letters do not differ significantly (Munir et al. 2020)

Alternaria interactions and its tolerance. Currently, these physiological, morphobiochemical marker indexes can be used as probes for rapid screening of germplasm. However, moderate tolerant germplasm EC250407 (mustard) and

EC1494 (rapeseed) can be utilized for future experiments and can serve as tolerant material for black spot disease.

3.3 *Brassica–Colletotrichum*: Molecular Resistance

3.3.1 *Arabidopsis–Colletotrichum* Pathosystem as a Model for Molecular Research

Colletotrichum is a large ascomycete genus comprising more than 190 species, many of which cause devastating diseases on a large range of agricultural and horticultural crops worldwide. Among species of Colletotrichum, *C. higginsianum* is classified in a main phylogenetic clade within the *C. destructum* complex and causes anthracnose disease on a wide range of cruciferous plants, such as species of *Brassica* and *Raphanus* as well as the model plant *Arabidopsis thaliana*. Since most *A. thaliana* ecotypes are susceptible to *C. higginsianum*, the pathogen can be regarded as adapted for *A. thaliana*. As a typical hemibiotrophic fungus, *C. higginsianum* develops a series of specialized infection structures including germ tubes, appressoria, primary biotrophic hyphae (BH), and secondary necrotrophic hyphae (NH). Thus, *C. higginsianum* is one of the best-studied species within the genus Colletotrichum because of its interesting infection strategy and the ease with which it can be cultured axenically and transformed with high efficiency by T-DNA transfer mediated by *Agrobacterium tumefaciens*. Furthermore, complete genome sequences and transcriptome data are available. For these reasons, the *C. higginsianum–Arabidopsis* pathosystem has become an attractive model for research on the molecular basis of fungal pathogenicity and plant–fungal interactions (Yan et al. 2018).

3.3.2 Identification and Characterization of Resistance Genes

Molecular and biochemical bases of cultivar resistance to Colletotrichum spp. have been investigated using genetically diverse materials. Innate disease resistance responses in plants are triggered by a dual surveillance system composed of nucleotide-binding-leucine rich repeat (NB-LRR) proteins encoded by resistance genes and pattern recognition receptors (PRRs). The two layers are often called MAMPs-triggered immunity (MTI) and effector-triggered immunity (ETI). By a combination of quantitative trait locus (QTL) and Mendelian mapping, a single putative R locus RCH1 was identified, at the tip of chromosome 4, in the resistant *A. thaliana* ecotype Eil-0 against *C. higginsianum*. By using map-based cloning and natural variation analysis of 19 *Arabidopsis* ecotypes, another dominant resistance locus RCH2 was identified against *C. higginsianum*. The locus RCH2 maps to an extensive cluster of disease-resistance loci known as MRC-J in the *Arabidopsis* ecotype Ws-0. These indicate that *Arabidopsis* resistance to *C. higginsianum* is controlled by a gene-for-gene interaction. In *A. thaliana*, NB-LRR-type resistance (R)-genes to *Pseudomonas syringae* 4 (RPS4) and to *Ralstonia solanacearum*

1 (RRS1-R) was reported to also confer resistance to *C. higginsianum*. RRS1-R and RPS4 were also found as a complex that could help detect effectors, which target WRKY proteins. Therefore, effectors in *C. higginsianum* that target WRKY proteins may be more likely to act as *Avr* genes in *Brassica* (Birker et al. 2009; Sarris et al. 2015; Yan et al. 2018).

3.4 *Brassica–Fusarium*: Molecular Resistance

3.4.1 Mapping of R-Genes of *Brassica*

Most resistance resources have been identified in *B. oleracea* (Table 3.1). Specifically, two types of resistance have been characterized, i.e., A and B. Type A resistance is stable under high or low temperature and follows a single dominant inheritance pattern; type B polygenic resistance is unstable under high temperatures above 24 °C. The type A single dominant resistance gene for Foc race 1 has been explored extensively in the last several years. The Fusarium wilt R-gene *FocBo1* was first mapped to linkage group seven using both BSA and QTL analysis by Pu et al. (2012). Lv et al. (2013, 2014) generated a genetic linkage map based on a cabbage DH population and mapped the R-gene *FOC1* to a 1.8 cM interval between two adjacent InDel markers. The mapped candidate gene *FOC1* is repredicted as *Bol037156*, which encodes a TIR-NBS-LRR, using an enlarged population. Shimizu et al. (2015) also mapped the resistance locus *FocBo1* by using 139 recombinant F2 plants and identified a candidate gene, *Bra012688*. The two mapped candidates are homologous with high identity. However, the functions of these genes remain to be identified. Type A resistance to Foc race 1 conferred by a dominant single gene, *FOC1*, has been successfully mapped and molecular markers have been developed and applied to generate various resistant cultivars. In addition, MAS using these

Table 3.1 Resistance genes/QTLs identified and mapped against *Fusarium* wilt of crucifers (Lv et al. 2020)

Disease	Species	Pathogen race/ isolates	Techniques	Results
Fusarium wilt	*B. oleracea*	Cong: 1-1 strain	SSR	A linked marker at 1.2 cM
	B. oleracea	FGL3-6, race 1	InDel	*FOC1* in an interval of 1.8 cM
	B. oleracea	FGL3-6, race 1	InDel	The candidate is a repredicted *Bol037156*
	B. rapa	Cong: 1-1 strain	RNA-seq	Two candidate R-genes identified: *Bra012688* and *Bra012689*
	B. oleracea	Cong: 1-1 strain	SSR	The candidate is *Bra012688*
	B. oleracea	FGL3-6, race 1	SRR	A high-efficiency marker located 75 kb from the resistance gene

markers has been combined with other breeding methods to promote the breeding process. Lv et al. (2014) reported the use of isolated microspore cultures with MAS to rapidly obtain target DH lines with FW resistance, which could be used directly in resistance breeding, thereby shortening the breeding period by 2–3 years (Lv et al. 2020).

3.5 *Brassica–Leptosphaeria*: Molecular Resistance

3.5.1 Molecular Characterization of AVR Genes

Compared with other *Brassica* pathogens, the molecular characterization of avirulence genes in *L. maculans* is more widely studied. The pathogen adapts to its new host via strict interaction at the molecular level. For example, the interaction between *Rlm3* and *AvrLm3* is suppressed by *AvrLm4-7* when attacked by the *L. maculans* isolate that carries this allele, although this isolate can be uncommon in the Canadian rapeseed fields (Zhang et al. 2016). There is positive influence of the frequency of the virulent allele *avrLm1* on *avrLm6* due to genetic linkage between the two alleles in the pathogen genome (Van de Wouw et al. 2010a). This observation is important when decision-making is required for which variety to rotate in the field. For example, selecting a variety with *Rlm1* directly after growing a variety containing *Rlm6* would not be advised, as there would be higher infection due to the presence of *avrLm1*.

The sources of Blackleg R-genes were more frequently found and more diversified in the winter-type *B. napus* compared to the spring-type cultivars (Rouxel et al. 2003b), where *Rlm4* is the most common R-gene in the latter (Rouxel et al. 2003b; Marcroft et al. 2012). Both *B. rapa* and *B. napus* have proven to be good sources of Blackleg resistance genes on the A-genome. Some genes from *B. rapa* have been introgressed into *B. napus*, such as the *LepR*-series of R-genes from *B. rapa* ssp. *sylvestris* (Yu et al. 2005, 2008, 2013). Researchers have also attempted to exploit the B-genome for resistance, and R-genes/QTL that are highly resistant to *L. maculans* have been introgressed from the B-genome (Plieske et al. 1998), particularly into *B. napus* (Fredua-Agyeman et al. 2014). The resistance level in the introgressed *B. napus* was the same as that in the donor B-genome source, although expression of phenotype was shown to be influenced by temperature (Plieske et al. 1998). These novel sources of R-genes introgressed into *B. napus* will enable continued resistance against the pathogen, through expanding the number of resistance alleles available for breeding, particularly in times of climatic uncertainty.

Both *LepR1* and *LepR2* are involved in cotyledon resistance and are also associated with resistance at the adult stage in *B. napus* (Yu et al. 2005). However, the mechanism of resistance at both stages is different because Yu et al. (2005) showed that presence of *LepR1* provided complete resistance, while *LepR2* only offered partial resistance at the cotyledon stage, yet both genes ensured complete resistance at the adult stage. A very different pattern of resistance exists for the Chinese *B. napus* lines derived from winter-type and spring-type parents, where

R-gene-mediated resistance in these lines conferred by both *Rlm3* and *Rlm4* was effective at both seedling and adult plant stages against *L. maculans* (Zhang et al. 2016). Similarly, the Australian winter *B. napus* lines harboring either *Rlm1*, *Rlm3*, or a combination of both display stable resistance against *L. maculans* where resistance is effective at both seedling and adult plant stages (Light et al. 2011). Such examples, taken together, imply that while R-genes can be effective at both cotyledon and adult stages, they can also act interdependently or dependently at different plant developmental stages and are likely affected by the genotype of host and pathogen as well as by environmental conditions. The R-gene effective at seedling stage can also lie in the same QTL region as that of adult plant resistance (Raman et al. 2012b). Whether this adult plant resistance is mediated by the same R-gene or by quantitative nonspecific resistance genes is yet to be confirmed. An increased understanding of this molecular basis for resistance, for example, through further genome sequencing and gene cloning, will enable the genes to be deployed more effectively. Further, expression of R-gene-mediated resistance can also be expressed differently across other host plant components; for example, *Rlm1* and *Rlm4* exhibit major gene resistance on *B. napus* pods, not only on cotyledons (Elliott et al. 2016).

Besides qualitative resistance, quantitative resistance has also been assessed and reported in *B. napus* against *L. maculans* (Huang et al. 2009). Methods to determine quantitative resistance more efficiently such as inoculation at different parts of the leaves under controlled conditions have been evaluated (Huang et al. 2014). Several QTL responsible for quantitative resistance against *L. maculans* have been identified in French, European, and Australian *B. napus* cultivars (Huang et al. 2016; Larkan et al. 2016a). However, the genetic control of quantitative resistance against Blackleg is poorly understood as it is very difficult to compare QTL, even with the same molecular markers and lines being utilized, due to environmental differences and a lack of knowledge of the pathogen presence in the field (Neik et al. 2017).

3.5.2 Molecular Mechanisms of Host Gene Background on R-Genes Effects to Defense Responses

Plants are exposed to a myriad of microorganism in their environment. However, natural physical barriers and chemical deterrents guard the plants from the majority of these microbes. The small number of microorganisms that overcome this passive defense still have to evade recognition by the plant cell surface and cytoplasmic receptors that have evolved to perceive conserved structural molecules and virulence factors termed pathogen-associated molecular patterns (PAMP) and effectors, respectively. Effector-triggered immunity (ETI) is often manifested as a rapid and strong defense response leading to induction of cell death, also known as hypersensitive response (HR), at the site of infection to arrest pathogen growth beyond the point of entry. HR provides early and robust resistance against *Leptosphaeria maculans* (*Lm*), the cause of blackleg disease in *Brassica* crops. The gene expression profile of *B. napus* (*Bn*) cotyledons infected with *Lm* reveals the transition from biotrophy to necrotrophy as infection progresses in a compatible interaction. Genes related to salicylic acid (SA) pathway are induced at the earlier stages of infection (3 days after inoculation; dai) while expression of jasmonic acid (JA) pathway genes are linked to plant response to necrotrophic pathogens. Gene necrosis and ethylene-inducing peptide-1 (*Nep-1*)-like protein are well-known markers of transition from biotrophy to necrotrophy in fungi. Becker et al. (2017) reported the importance of SA and JA pathways in a *Bn* line carrying the *LepR1* resistance gene in response to a *Lm* isolate carrying the corresponding *AvrLep1* effector. Upon perception of *Lm*, an array of host genes, functionally defined as proteases and protease inhibitors, chitinases, peroxidases, transcription factors (WRKY, AP2/EREBP, MYB), genes related to the production of the secondary metabolites, and genes involved in plant cell wall reinforcement, were differentially expressed (Haddadi et al. 2016, 2019).

Despite being a qualitative trait, the immunity response triggered by R-gene *Avr* gene recognition varies in phenotype; from a highly localized response seen as minute necrosis, trailing necrosis, no visual symptoms or contained pathogen growth and sporulation. Both genotypes of the host and functional variation of R-genes cause variation in the interaction phenotype. R-protein activation and recognition of pathogen effector proteins often requires inter- and/or intramolecular interaction and complex formation with other host proteins. These receptor complexes affect R-protein function and consequently plant–pathogen interaction phenotype. Natural variation in *Arabidopsis* has served as a tool to dissect the genetic basis of polymorphism in plant to interaction with pathogens (Holub 2001, 2008).

Race-specific R-genes have been widely used in breeding for *Brassica napus* resistance to blackleg disease caused by *Leptosphaeria maculans*. Out of 19 R-genes reported from *Brassica* species, 11 of them (*Rlm1*, *Rlm2*, *Rlm3*, *Rlm4*, *Rlm7*, *Rlm9*, *Rlm*11, *LepR1*, *LepR2*, *LepR3*, and *LepR4*) originated from the A-genome of *Bn* and *Brassica rapa* (*Br*). The genetics of the *Bn-Ln* pathosystem has been greatly advanced by the cloning of *Lm Avr* genes; *AvrLm1*, *2*, *3*, *4–7*, *5–9*, *6*, and *11*; and the characterization of the *Bn* R-genes *LepR3* and *Rlm2*. Host differential lines are indispensable for genotyping plant pathogen races. In the studies, *Rlm* and *LepR* genes were introgressed into common susceptible *Bn* doubled-haploid lines Topas (DH16516) or Westar (N-o-1). This led to generation of seven Topas introgression lines (T-*Rlm1*, T-*Rlm2*, T-*Rlm3*, T-*Rlm4*, T-*LepR1*, T-*LepR2*, T-*LepR3*), which share 92.9–98.9% of their genomic background with the susceptible parental lines, and two Westar introgression lines (1065; W-*LepR1*, 1135; W-*LepR2*). These introgression lines provide a unique tool to compare the function of different R-genes in a common genotype background and also for the dissection of the effect of host genetic background on the defense responses triggered by the same R-gene. To gain insight into the molecular mechanism of host genetic background and R-gene effect, Haddadi et al. (2019) conducted a comprehensive transcriptome analysis by performing RNA-seq (1.5 billion raw reads—168 samples) on *Lm*-infected cotyledons at 0, 3, 6, and 9 dai and described the differences in gene expression profiles and defense pathways among these treatments and their correlation with the variation in visual interaction phenotypes.

Plant response to pathogen infection leads to significant changes in the plant's transcript profile. Despite overlaps and commonality between responses of different plant species to various pathogens, variation in plant phenotypic and molecular interactions to pathogens also occurs due to differences in R-;genes and host genetic backgrounds. Natural variation in *Arabidopsis* has been exploited to capture genetics for defense polymorphism against many fungal and bacterial pathogens. For some pathogens such as *Lm*, genetics and genomics of the defense response needs to be investigated using its natural host, as *Arabidopsis* does not provide an ideal model system. Haddadi et al. (2019) took the advantage of several well-defined introgression lines each harboring individual R-genes against blackleg pathogen in the *Bn* cv. Topas, a common susceptible genotype. They monitored phenotypic interaction and changes in gene expression profile due to immunity response triggered by the race-specific resistance genes *Rlm2*, *Rlm3*, *LepR1*, and *LepR2* in two susceptible *Bn* genetic backgrounds, Topas and Westar.

Comparing the transcript profiles of Topas and Westar and also W-*LepR1* with T-*LepR1* and W-*LepR2* with T-*LepR2* revealed a delay in defense response in Westar compared to Topas. While at the earliest time point, i.e., 3 dai, changes in a limited number of defense-related genes were noticed in Topas, the transcript profile of infected Westar plants was the same as Westar mock-inoculated controls. Similarly, judging by the defense-related GO terms, immunity triggered by *LepR1* and *LepR2* was less intense in the Westar background compared to the immunity response induced by these same genes in Topas. Early induction of genes related to cell wall strengthening and production of antifungal compounds such as chitinases and lipases occurred in Topas but were not observed in Westar. The GO enrichment of camalexin in Topas early in infection was observed as compared to Westar. An *Arabidopsis* Phytoalexin Deficient 3 (*pad3*) mutant that is defective in camalexin production has been reported to be more susceptible to *Lm* infection. The combined effect of R-gene and host genetic background in generating a stronger defense response was also evident by expansion of DEG for SA, JA, and Et pathways in T-*LepR1* and T-*LepR2* compared to W-*LepR1* and W-*LepR2*, respectively. Clustering of R-gene ILs confirmed the boosting effect of the Topas genetic background on the level of gene expression at 3 dai for *LepR1* in Topas compared to *LepR1* in Westar. Differentially expressed genes in Topas versus Westar background could prove to be useful as markers to select the most suitable genotype as a recipient parent when developing *Bn* cultivars with resistance to *Lm*. The effect of R-genes on spontaneity and robustness of defense response was measured based on the number and scope of defense-related genes by comparing differentially expressed genes in T-*Rlm2*, T-*Rlm3*, T-*LepR1*, and T-*LepR2*. Based on the microscopy observation of the strength of the interaction phenotype and the genes' ability to limit the pathogen growth over the course of infection, these R-genes could be ordered as T-*LepR1*/T-*Rlm2*; T-*LepR2*; and finally T-*Rlm3* from the most robust to a weaker defense response. As shown in the heat map of DEG at 3 dai, in T-*LepR1*, T-*Rlm2*, and to certain extent in T-*LepR2*, a strong induction of chitin-responsive genes, a known PAMP, and induction of genes related to callose deposition, upregulation of ROS related genes, induction of SA and JA and to a lesser extent glucosinolate pathways, regulation of hypersensitive response, and induction of downstream MAP kinases were the most prominent and well-documented indicators of plant immunity responses. It was only at 6 dai that these pathways were upregulated to the same extent in T-*LepR2*; however, in the case of T-*Rlm3*, expression of defense-related pathways was significantly less than in the other Topas introgression lines.

PAMP-triggered immunity (PTI) provides basal defense upon detection of conserved pathogen molecules while effector-triggered immunity (ETI) provides a rapid and strong defense in response to pathogen virulence (effector) genes. Haddadi et al. (2019) investigated the PTI and ETI defense against *Lm* by comparing the DEG with the *Arabidopsis* genes associated with PTI and ETI as reported by Dong et al. (2015). This comparison revealed strong induction of PTI- and ETI-related genes in T-*LepR1* and T-*Rlm2*. It has also been reported that Nonrace-Specific Disease Resistance 1 (NDR1) is a conserved downstream regulator of *R* signaling. Interaction of NDR1 and RIN4 results in transduction of extracellular pathogen-derived signals. The difference in expression of *Bn* genes with homology to *NDR1* related to ETI in *Lm*-inoculated compared with mock-treated cotyledons of Topas, Westar, and the ILs revealed activation of these genes in incompatible hosts at 3 dai, with comparatively higher activity in T-*Rlm2*, T-*LepR1*, and T-*LepR2*.

Among several WRKY transcription factors, WRKY11 and WRKY17 have both been reported to be involved in JA-dependent defense response. The results showed upregulation of both in the ILs as compared to Topas and Westar. The WRKY33 transcription factor is reported to be important for plant resistance to hemibiotrophic and necrotrophic pathogens and to be involved in response to chitin, production of secondary metabolic and the phytoalexin biosynthetic pathway (Fig. 3.4a, c). Bimolecular fluorescence complementation previously revealed that WRKY33 interacts with nuclear-encoded SIGMA FACTOR BINDING PROTEIN 1 (SIB1) and SIB2. Both SIB1 (VQ16) and SIB2 (VQ23) contain a short VQ motif that is important for interaction with WRKY33. Transcripts associated with *WRKY33*, *VQ16*, and *VQ23* homologues were enriched at 3 dai in the ILs (Fig. 3.5). Comparative transcriptomic analysis identified all three copies of *VQ16* were suppressed in the *Lm*-infected Topas and Westar plants. The data presented in Fig. 3.5 presents a model describing the possible role of VQ proteins and WRKY genes in the induction and suppression of defense against *Lm*. Quantification of expression of *VQ16*, *WRKY33*, and *PDF1.2* was conducted by Droplet Digital PCR (ddPCR) which confirmed the RNA-seq expression data.

While monitoring the receptor complex-associated *SOBIR1* homologues, it was noted that these were strongly expressed in T-*LepR1* and T-*Rlm2*, with expression peaking early in the time course (3 dai). It was reported that the *Arabidopsis thaliana* LRR-receptor-like kinase (LRR-RLK) suppressor of Bir-1 (AtSOBIR1) interacts with LRR-RLPs resistance genes (Liebrand et al. 2013). It has been demonstrated that SOBIR1 binds with both Rlm2 and its allelic variant LepR3, which are membrane-bound receptor-like proteins (RLPs). This result suggests that SOBIR1 is also required for successful LepR1 defense response, and that *LepR1* may also encode an RLP. Conversely, very low expression of the *SOBIR1* homologues was observed during the *Rlm3*-induced defense response, which could indicate that Rlm3

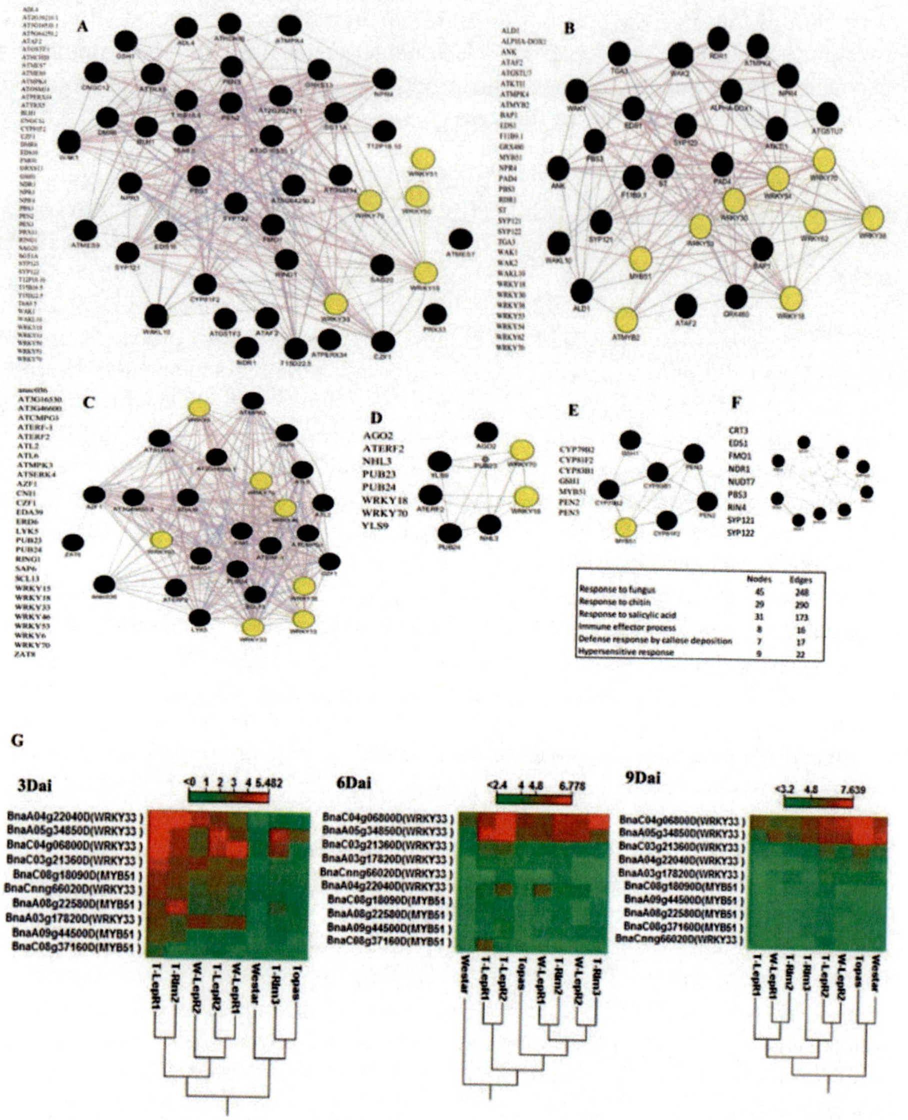

Fig. 3.4 Predicted networks are involved in early response (3 dai) to *Leptosphaeria maculans* (*Lm*) in Introgression Lines (ILs). A predicted network that is involved in (**a**) response to fungus, (**b**) response to salicylic acid, (**c**) chitin response, (**d**) immune effector process, (**e**) defense response by cell wall thickening or callose deposition, and (**f**) hypersensitive response at 3 dai. TFs are highlighted in yellow. (**g**) Heat maps present the difference in expression of *Brassica napus* genes with homology to WRKY33 and MYB51 in *Lm*-inoculated compared with mock-treated cotyledons of Topas, Westar, and ILs at 3.6 and 9 dai. The colors correspond to log 2 RPKM (Infected–Uninfected) ranging from red (high) to green (low). Euclidean distance for the distances measure and complete linkage for clusters linkage criteria were selected (Haddadi et al. 2019)

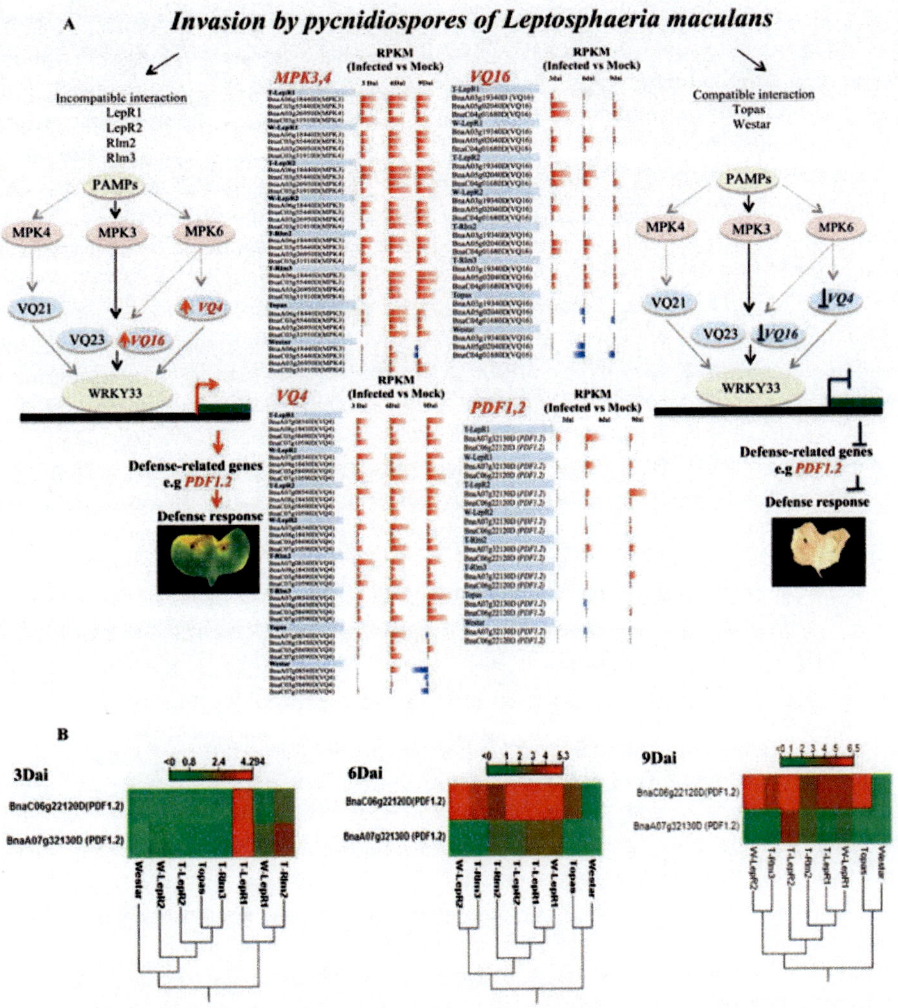

Fig. 3.5 (**a**) An overview of MPK-VQ-JA signaling in *Brassica napus* and *Leptosphaeria maculans* (*Lm*) interaction. Identification of networks involved in earlier response to blackleg in ILs provided the evidence that WRKY33 is involved in response to *Lm*. Accumulation of transcripts associated with *WRKY33*, *VQ16*, and*VQ23* homologues were enriched by 3 dai in resistance lines. RNA-seq analysis revealed all three copies of *VQ16* is suppressed after invasion by pycnidiospores of *Leptosphaeria maculans* in Topas and Westar; consequently one copy of *PDF1.2* is highly suppressed in susceptible lines while it is highly accumulated in resistance lines. (**b**) Heat maps present the difference in expression of *B. napus* genes with homology to JA (*PDF1.2*) in *Lm*-inoculated compared with mock-treated cotyledons of Topas, Westar, and resistance lines at 3, 6, 9 dai. The colors correspond to log 2 RPKM (Infected–Uninfected) ranging from red (high) to green (low). Euclidean distance for the distances measure and complete linkage for clusters linkage criteria were selected (Haddadi et al. 2019), and plant R-proteins in triggering the initial immunity response, there are clear differences in the dynamics of defense-related gene expression, and this is influenced by the host genetic background. Robust immune response and arrest of the pathogen at the site of penetration is highly desirable when developing resistant cultivars. Furthering, information with regards to how R-genes interact with host genotype background will help ensure selection of the best germplasm for robust expression of resistance (Haddadi et al. 2019)

may function independent of SOBIR1. The data showed a repression of cytokinin (CK) responsive genes during early infection (3 dai) of *Lm* in Topas and resistance lines while the induction of CK was observed in Westar. CK level in *Bn* cotyledons increases upon *Lm* infection. The role of CK has been reported in various pathosystems. A search for the genes involved in the CK pathway identified isopentenyl transferase (IPT) and adenosine kinase (AK) in the genome of *Lm*. The observations that the level of CK was elevated in the *Lm*-infected tissues need to be further explored to determine the origin of CK (pathogen or host) and its importance in defense against *Lm* (Trda et al. 2017).

The Quantitative trait locus (QTL) studies suggested a role for cysteine-rich protein kinase genes in quantitative resistance to blackleg disease in *Brassica napus*. Association of CRK11 within a functional network of plant immunity-related genes (based on DEG at 3 dai) further supports its importance in defense against *Lm*. In the case of quantitative resistance to pathogens, additive effect of genes and the combined interaction of host genotype background and environment account for variation in phenotypes. However, immunity response triggered by single R-genes is generally thought to be less variable. By taking advantage of R-gene introgression lines described, it was observed that, despite the involvement of the same pathogen effector interaction with host genotype, background will help ensure selection of the best germplasm for robust expression of resistance (Haddadi et al. 2019).

3.5.3 Expression of *Cf9* and *Avr* 9 Genes in *Brassica* Induces Resistance to *Leptosphaeria*

Among plant defense reactions to pathogen infection, particular interest has been given to the hypersensitive response (HR). The HR is a rapid and strictly localized cell death at the infection site in the host plant, limiting the spread of the pathogen and preventing its propagation. On the basis of the gene-for-gene concept of Flor (1971), it is widely accepted that a dominant plant resistance (R)-gene and the corresponding dominant pathogen avirulence (*Avr*) gene are the basic components required for a HR. The HR can be seen as an early defense response, involving a cascade of plant defense reactions such as the production of reactive oxygen species, and followed by cell wall fortification, and the production of many defense-related proteins and compounds (Greenberg 1997; Hammond-Kosack and Jones 1996; Hutcheson 1998). It is expected that artificial induction of the HR could be an important strategy to engineer a broad disease resistance in plants. The HR must be strictly regulated, however, in order to avoid the uncontrolled spread of cell death throughout the whole plant (De Wit 1992). Several approaches have been followed to artificially induce a HR in plants. Strittmatter et al. (1995) proposed a strategy on the basis of the pathogen-inducible expression of a bacterial ribonuclease. Keller et al. (1999) induced a HR in transgenic tobacco plants by pathogen-induced expression of the elicitor cryptogein. Another possibility to artificially induce a HR in plants is the use of the two-component system (De Wit 1992). The induction of cell death and endogenous defense responses caused by placing an *Avr* gene under the control of a pathogen-inducible promoter in a genetic background carrying the corresponding R-gene could provide a general mechanism for engineering broad disease resistance (De Wit 1992).

The tomato *Cf9* resistance gene and the corresponding *Avr9* avirulence gene are candidate genes that could be used in the two-component system. The *Cf9* gene confers resistance to particular races of *Cladosporium fulvum* that express the corresponding avirulence gene *Avr9* (De Wit 1992; Hammond-Kosack et al. 1994). *Cf9* and *Avr9* genes have been cloned and characterized (Jones et al. 1994; Van den Ackerveken et al. 1992, 1993; Van Kan et al. 1991). Injection of the Avr9 peptide into leaves of *Cf9* tomato plants induces an oxidative burst, electrolyte leakage, production of ethylene, salicylic acid (SA), pathogenesis-related (PR) proteins, and hypersensitive cell death at the injection site (Hammond-Kosack et al. 1996; May et al. 1996; Wubben et al. 1996). Moreover, the *Cf9–Avr9* system has been transferred successfully to other *Solanaceae* spp. such as tobacco and potato (Hammond-Kosack et al. 1994, 1998; Honee et al. 1995, 1998).

Artificial induction of an Avr9-mediated HR in non-*Solanaceous Cf9* transgenic plants might be difficult to achieve. There are several indications that the functional transfer of genes between distantly related plants families is limited (Tai et al. 1999). To date there has been only one particular case where a tomato *Cf4–Avr4* gene cassette was expressed transiently in a distantly related plant species (*Asteraceae*) (Van der Hoorn et al. 2000). Hennin et al. (2001) investigated whether the specificity of the *Cf9–Avr9* interaction was retained when introduced into a distantly related plant species. To address this, they transformed the *Cf9* and *Avr9* genes into oilseed rape, *Brassica napus* spp. *oleifera* L. Oilseed rape is a suitable species because of its well-established transformation protocol (De Block et al. 1989) and susceptibility to different types of pathogens. It was investigated whether the *Cf9* and *Avr9* genes could be functionally expressed in oilseed rape and whether the presence of their gene products induces defense responses that are effective in disease control. Studies reveal that whether the *Cf9* gene product responded functionally to the corresponding *Avr9* gene product when introduced in a heterologous plant species could successfully express the *Cf9* gene under control of its own promoter and the *Avr9* or *Avr9R8K* genes under control of the p35S1 promoter in transgenic oilseed rape. It was demonstrated that the transgenic oilseed rape plants produced the Avr9 elicitor with the same specific necrosis-inducing activity as reported for *Cladosporium fulvum*. An Avr9-dependent HR was induced in *Cf9* oilseed rape upon injection of intercellular fluid containing Avr9. It showed Avr9-specific induction of *PR1*, *PR2*, and *Cxc750* defense genes in oilseed rape expressing *Cf9*. *Cf9* × *Avr9* oilseed rape did not result in seedling death of the F1 progeny, independent of the promoters used to express the genes.

The F1 (*Cf9* × Avr9) plants, however, were quantitatively more resistant to *Leptosphaeria maculans*. Phytopathological analyses revealed that disease development of *L. maculans* was delayed when the pathogen was applied on an Avr9-mediated HR site. This demonstrated that the *Cf9* and *Avr9* gene can be functionally expressed in a heterologous plant species and that the two components confer an increase in disease resistance (Hennin et al. 2001).

3.5.4 Molecular Mapping and Cloning of R-Genes of *Brassica*

Most blackleg resistance genes/QTLs originated from the *B. napus* A-genome (Table 3.2). Ferreira et al. (1995) first applied a double haploid (DH) population from *B. napus* to localize the major locus LEM1 on N7. Using a similar method, Dion et al. (1995) identified another major gene, *LmFr1*. Mayerhofer et al. (1997) detected a major locus, *LmR1*, and cosegregating markers were developed. Delourme et al. (2004) reported the mapping of resistance loci in two genomic regions, and a cluster consisting of five R-genes was proposed as the candidate. Fine mapping work was conducted extensively after 2010. Long et al. (2011) identified two resistance genes, BLMR1 and BLMR2, and fine mapping of BLMR1 resulted in the closest marker distance of 0.13 cM. Jestin et al. (2011) used an association mapping method to characterize the molecular diversity using 128 oilseed rape accessions and identified five novel alleles. Raman et al. (2012a) positioned a new major locus, *Rlm4*, and the deposited region was further analyzed, with several candidates being characterized. In addition, blackleg resistance loci have been transferred from wild relatives of *B. rapa* and *B. oleracea* to *B. napus*. Yu et al. (2005, 2008) mapped blackleg resistance derived from the wild relative and *LepR1-LepR3* was identified. Larkan et al. (2013, 2014) employed map-based cloning to isolate *LepR3*, which encoded an RLP, representing the first cloned blackleg disease resistance gene. Lv et al. (2020) further isolated the *Rlm2* gene, which is an allelic variant of *LepR3*. They cloned another blackleg resistance gene, *Rlm9*, which encodes a wall-associated kinase-like protein, a newly discovered class of race-specific plant RLK resistance genes. In addition to the major locus, some QTLs have also been characterized, including six and four that are stable under different environmental conditions (Huang et al. 2016).

Brassica cultivars with improved resistance to blackleg are frequently cultivated due to extensive R-gene mapping work. In addition, MAS is often integrated with other breeding methods to shorten the breeding period. For instance, Yu et al. (2013) described the successful introgression of blackleg resistance from wild *B. rapa* subsp. sylvestris to *B. napus* via interspecific hybridization and MAS, which generates a series of resistant cultivars. In addition, based on both the major genes and QTLs identified, the next breeding effort could involve a combination of qualitative and quantitative loci to provide more durable resistance (Brun et al. 2010).

3.5.4.1 Mapping of Marker-Linked R-Genes in *Brassica napus*

To identify new alleles, especially for quantitative resistance (QR), Raman et al. (2020) analyzed 177 doubled haploid (DH) lines derived from an RP04/Ag-Outback cross. DH lines were evaluated for QR under field conditions under shade house conditions using the "ascospore shower" test. DH lines were also characterized for qualitative R-gene-mediated resistance via cotyledon tests with two differential single spore isolates, IBCN17 and IBCN76, under glasshouse conditions. Based on 18,851 DArTseq markers, a linkage map representing 2019 unique marker bins was constructed and then utilized for QTL detection. Marker regression analysis

Table 3.2 Resistance genes/QTLs identified and mapped against blackleg of crucifers (Lv et al. 2020)

Disease	Species	Pathogen race/ isolates	Techniques	Results
Blackleg	*B. napus*	PG2 isolate PHW1245	RFLP	LEM1 on A-genome linkage group N7
	B. napus	Leroy	RFLP	A major gene, *LmFr1*, and a minor locus
	B. nigra	Four isolates	RAPD	Resistance gene on LG B4
	B. napus	Four isolates	RAPD, RFLP	*LmR1* in A-genome linkage group N7
	B. juncea	Isolate 314	RAPD	Resistance gene in LG B8
	B. napus	Field experiment	RAPD, RFLP	Four major genomic regions
	B. napus	Five isolates	RAPD	*Rlm1*, *Rlm3*, *Rlm4*, *Rlm7*, and *Rlm9* in LG10
	B. napus	–	RFLP, SCAR	*LmR1* and *ClmR1* mapped to the same genetic interval in N7
	B. juncea	PG2 isolate	RFLP	*LMJR1* on LG J13 and *LMJR2* in J18
	B. rapa	PG2 and PG3	RFLP	*LepR1* in N2 and *LepR2* and *LepR3*
	B. rapa	31 Isolates	Microsatellite	*LepR3* at an interval of 2.9 cM in LG N10
	B. napus	Isolate 87-41	SRAP	*BLMR1*, with the closet marker of 0.13 cM, and *BLMR2*
	B. napus	Field experiment	SSR	Seven alleles located close to the previous QTLs and five novel alleles
	B. napus	11 isolates	SSR, SRAP	14 QTLs, with the locus *Rlm4* on major qualitative chromosome A7
	B. napus	–	NGS, BIA	Several candidates for *Rlm4* on A7
	B. napus	S005, P042 and others	Function analysis	*LepR3* encodes an RLP
	B. napus	Isolate 165 and others	Function analysis	*Rlm2*, an allelic variant of *LepR3*
	B. napus		Function analysis	*Rlm9* encodes an RLK
	B. napus	Field experiment	SSR	17 QTLs, with six stable ones
	B. napus	WA30 or v23.1.3	DArT	Four QTLs, with a 49 gene QTL interval on chromosome A01

identified 22 significant marker associations for resistance, allowing identification of two race-specific resistance R-genes, *Rlm3* and *Rlm4*, and 21 marker associations for QR loci. At least three SNP associations for QR were repeatedly detected on chromosomes A03, A07, and C04 across phenotyping environments. Physical mapping of markers linked with these consistent QR loci on the *B. napus* genome assembly revealed their localization in close proximity of the candidate genes of *B. napus BnaA03g26760D* (A03), *BnaA07g20240D* (A07), and *BnaC04g02040D* (C04). Annotation of these candidate genes revealed their association with protein kinase and jumonji proteins implicated in defense resistance.

Both *Rlm3* and *Rlm4* genes identified in this DH population did not show any association with resistance loci detected under either field or shade house conditions (ascospore shower) suggesting that both genes are ineffective in conferring resistance to *L. maculans* in Australian field conditions. The identified sequence-based molecular markers for dissecting *R* and QR loci to *L. maculans* in a canola DH population from the RP04/Ag-Outback cross may be very useful for breeding resistant cultivars (Raman et al. 2020).

3.5.5 Effective and Durable Resistance in *Brassica* Cultivars

Brassica crops are one of the major oilseed crops of the world. Blackleg, caused by the fungal pathogen *Leptosphaeria maculans*, is a serious disease of canola (*B. napus*) in North America, Australia, Europe, and many other regions around the world (Fitt et al. 2006). As an environmentally friendly strategy, genetic resistance is generally very effective in disease control. Both seedling resistance controlled by major or seedling R-genes and adult plant resistance (APR) mediated by quantitative resistance (minor) genes to *L. maculans* have been identified in *B. napus* varieties (Jestin et al. 2011, 2015). R-genes confer race-specific resistance and follow the gene-for-gene concept proposed by Flor (1971). To date, at least 18 major R-genes against *L. maculans* have been identified in *Brassica* species: *Rlm1*, *Rlm2*, *Rlm3*, *Rlm4*, *Rlm7*, and *Rlm9* from *B. napus*, which have been mapped to two *B. napus* linkage groups, N7 and N10 (Ferreira et al. 1995; Mayerhofer et al. 1997; Ansan-Melayah et al. 1998; Zhu and Rimmer 2003; Rimmer 2006; Delourme et al. 2006); *Rlm8* and *Rlm11* from *B. rapa* (Balesdent et al. 2002, 2013); *Rlm5* and *Rlm6* from *B. juncea* (Chevre et al. 1997; Balesdent et al. 2002); *Rlm10* from *B. nigra* (Chevre et al. 1996; Eber et al. 2011); *LepR1*, *LepR2*, *LepR3*, *LepR4*, and *RlmS* from resynthesized *B. rapa* subsp. *sylvestris* (Yu et al. 2005, 2007, 2008; Van De Wouw et al. 2009); and *BLMR1* and *BLMR2* from Surpass 400 (Long et al. 2011). To date, two R-genes, *LepR3* (that interacts with *AvrLm1*) and *Rlm2*, have been cloned (Larkan et al. 2013, 2015). By contrast, at least seven of the corresponding avirulence (*Avr*) genes have been cloned: *AvrLm1* (Gout et al. 2006), *AvrLm2* (Ghanbarnia et al. 2015), *AvrLm3* (Plissonneau et al. 2016), *AvrLm5/AvrLmJ1* (Van de Wouw et al. 2014), *AvrLm4*-7 (Parlange et al. 2009), *AvrLm6* (Fudal et al. 2007), and *AvrLm11* (Balesdent et al. 2013). The cotyledon inoculation assay has been used to identify resistance to *L. maculans* (Williams and Delwiche 1979; Rimmer and Van den Berg 1992; Rouxel et al. 2003b; Marcroft et al. 2012). The characterization of R-genes in a given canola variety can be achieved by analyzing its interactions with a set of *L. maculans* isolates carrying known avirulence genes. Based on reactions to isolates with known avirulence alleles, Rouxel et al. (2003b) deduced race-specific resistance genes to blackleg in accessions of *B. napus* mainly originating from Europe. Marcroft et al. (2012) identified seedling resistance genes in Australian *B. napus* varieties using *L. maculans* isolates harboring known avirulence genes. In Canada, blackleg resistance breeding programs have successfully developed resistant varieties for commercial release. However, R-genes for blackleg resistance in Canadian *B. napus* varieties are unknown (Rimmer 2006). Both seedling and adult plant resistance play important roles in blackleg control. It has been shown that a combination of major gene resistance and adult plant resistance can provide effective and durable resistance against blackleg (Kiyosawa 1982; Brun et al. 2010). Selection of blackleg-resistant breeding materials is usually based on field evaluations without genetic characterization of R-genes (Rouxel et al. 2003b). Moreover, the interaction between specific R-genes and their corresponding avirulence genes in the seedling stage occurs.

3.5.6 Molecular Basis for Assessment of Breakdown of Resistance in *Brassica* Cultivars

The *AvrLm6* gene, which was adjacent to a single-copy noncoding sequence at the 3′ end, had six different RIP alleles conferring virulence. The degree (intensity) of RIP mutation in these single-copy sequences was proportional to the proximity of flanking repetitive DNA that had undergone RIP mutation. The potential leakage of RIP mutations into closely linked genes highlights the power of RIP to lead to major evolutionary changes to genes such as effectors that play an important role in fungal living strategies. The breakdown of "sylvestris" resistance was associated with an eightfold increase in frequencies of isolates lacking *AvrLm1*. Although this gene was embedded in repetitive DNA, no RIP alleles were identified. The frequencies of virulence alleles (both deletion and RIP mutation) of *AvrLm6* increased sixfold, even though cultivars with the complementary resistance gene, *Rlm6*, had not been sown on Eyre Peninsula (Van de Wouw et al. 2010a). The close linkage and genomic location of *AvrLm1* and *AvrLm6* might have led to a selective sweep, whereby selection at *AvrLm1* affected the frequency of alleles of *AvrLm6* through hitchhiking (Barton et al. 2013). Thus, strong selection imposed by widespread deployment of one resistance gene also may lead to breakdown of resistance conferred by another gene, if the two complementary avirulence genes are closely linked. Clearly the genomic environment, as well as extent of exposure to resistance genes in *B. napus* cultivars, affects evolution of avirulence effectors in this fungal pathogen.

Monitoring changes in virulence frequencies in Australian populations of the blackleg fungus is armed with knowledge about the field biology and genomic analysis of *L. maculans*. The monitoring of changes in frequency of virulence of fungal populations across Australia is with the aim of preventing another breakdown of disease resistance. Since 2009, cultivars with different complements of resistance genes sown in 32 trial sites covering *Brassica* growing regions of Australia have been assessed for disease severity. Resistance genes in canola cultivars are identified by the use of 12 differential *L. maculans* isolates. On this basis of responses (susceptible or resistant) to these isolates, seven resistance groups (A–G) that include a range of resistance genes have been defined (Marcroft et al. 2012). Stubble from each site represents local fungal populations, which can be analyzed in high-throughput laboratory assays to quantify regional frequencies of alleles of avirulence genes. In these assays, blackleg-infested stubble is wetted and placed in a wind tunnel. This triggers release of ascospores, which are deposited onto tape from which ascospore DNA is extracted and analyzed. The quantitative polymerase chain reaction (qPCR) detects the frequency of alleles of particular avirulence genes (*AvrLm1*, *AvrLm6*), where virulence is due to deletion, while PCR and pyrosequencing detects single-nucleotide base-pair changes where virulence is due to a particular point mutation (*AvrLm4*) (Van de Wouw and Howlett 2012; Van de Wouw et al. 2010b). The total number of ascospores is estimated by qPCR of the internal transcribed spacer region of the ribosomal DNA, a genomic region present in all isolates. This is the first example of a high-throughput molecular assay that can distinguish genotypes of airborne spores. This type of assay can be applied to other diseases that involve airborne inocula and where the genetic basis of virulence in the pathogen has been identified. Analysis of data on changes over time in disease severity of cultivars, and in frequencies of avirulence/virulence alleles in populations released from canola stubble, allows predictions of risk of disease outbreaks in different geographic regions. If an epidemic is predicted, farmers are advised to plant a different canola cultivar with a different complement of resistance genes (Howlett et al. 2015).

3.5.7 Difference in Gene Expression Profile of Resistant and Susceptible *Brassica napus* Lines

The differences in the pathogen's gene activity during infection of susceptible and resistant lines have been assessed by Becker et al. (2019). Principle component analysis (PCA) suggests similar mRNA profiles at 1 dpi regardless of host genotype, but divergent responses at 3 dpi as indicated by independent clusters of samples (Fig. 3.6a). To elucidate the genes contributing to these responses, differential gene expression was investigated between *L. maculans* colonizing resistant and susceptible plants at both time points (Fig. 3.6b) and between time points (Fig. 3.6c). At 1 dpi, only 13 *L. maculans* genes, including hypothetical proteins and two putative pectate lyases (*LEMA_T033320; LEMA_T001630*; File S1), were differentially expressed between the two interactions (Fig. 3.6b). At 3 dpi, this increased to 927 *L. maculans* DEGs (~8% of the *L. maculans* genome) that were influenced by host genotype. Differential gene expression analyses between time points identified 633 and 928 *L. maculans* DEGs that respond during infection of susceptible and resistant lines, respectively, and an additional 554 DEGs that were shared between interactions (Fig. 3.6c). Together, data revealed global shifts in *L. maculans* gene expression during infection—that is influenced by host genotype.

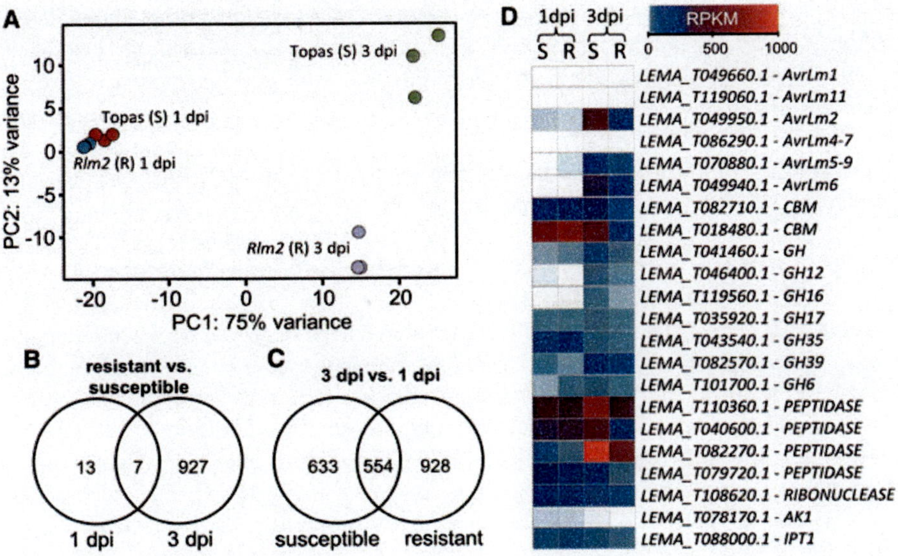

Fig. 3.6 *Leptosphaeria maculans* gene expression during its infection of susceptible and resistant hosts. (**a**) PCA showing relationships between pathogen transcriptomes in infected treatments. Transcriptomes cluster together at 1 dpi and diverge by 3 dpi. (**b**) Significantly ($P < 0.05$) differentially expressed genes between *L. maculans* infection on resistant hosts versus susceptible hosts. (**c**) Significantly ($P < 0.05$) differentially expressed genes between treatments at 3 and 1 dpi. (**d**) Selected gene set of virulence factors and known effectors, and their expression in reads per kilobase per million mapped (RPKM). A more intense red color denotes higher expression 115 × 73 mm (300 × 300 dpi) (Becker et al. 2019)

Expression of known *L. maculans Avr* genes during infection showed that *AvrLm2* and *AvrLm5-9* are the first *Avr* genes expressed by the pathogen with detectable expression at 1 dpi and ~100-fold higher expression at 3 dpi (Fig. 3.6d). Expression of *AvrLm6* was first detected at 3 dpi, and *AvrLm1*, *4-7*, and *11* were undetected (FPKM = 0) in this data set are likely to be absent in strain 00-100.

To determine putative *L. maculans* effectors involved in pathogen establishment, Becker et al. (2019) identified secretary signals within the N-terminal domain of translated DEGs. The top five most upregulated genes (3 dpi vs. 1 dpi) within the predicted *L. maculans* secretome contained *AvrLm2* and *AvrLm6*, along with three hypothetical proteins of unknown function (*LEMA_T082980.1*, *LEMA_T054900.1*, *LEMA_T086540.1*), highlighting the ability of analysis to identify a combination of putative and characterized pathogen effectors. Further analysis of the secretome identified *L. maculans* degradative enzymes and *Avr* genes with higher gene expression during colonization of susceptible plants at 3 dpi (Fig. 3.6d). This included *AvrLm2*, *5-9*, *6*, putative carbohydrate binding molecules (CBMs) with homology to starch hydrolyzing amylases (*LEMA_T082710.1*, *LEMA_T018480.1*), glucoside hydrolases (*GH6,12*, *16*, *17*, *35*, *39*), peptidases (*LEMA_T110360.1*;

LEMA_T040600.1, *LEMA_T082270.1*, *LEMA T079720.1*), and a ribonuclease (*LEMA_T108620.1*). The production of cytokinins by *L. maculans* has been shown in vitro, and cytokinin biosynthetic enzyme AK1 activity increases disease severity (Trda et al. 2017). Both *ADENOSINE KINASE 1* (*AK1*) and *ISOPENTENYLTRANSFERASE 1* (*IPT1*), involved in cytokinin production, were expressed at 1 dpi with decreasing expression by 3 dpi in both lines. In resistant plants, there was no detectable expression of *AK1* at 3 dpi.

3.5.7.1 Early Differences in Gene Expression Between Resistant and Susceptible Plants

Notable differences in pathogen gene activity have been observed between compatible and incompatible interactions, when examined host responses in plant cells directly adjacent to the inoculation site. At 1 dpi, 2572 (47%) upregulated DEGs (uDEGs) were shared between both genotypes, with an additional 541 (9.9%) uDEGs specific to susceptible plants and 778 (14.2%) specific to resistant plants. At 3 dpi, the number of shared uDEGs decreased to 2484 (31.0%), with 1193 (14.9%) and 2105 (26.3%) uDEGs specific to susceptible and resistant plants, respectively. Similar trends in the number and distribution of downregulated DEGs were observed in response to *L. maculans* inoculation (Fig. 3.7a, b). Together, data indicated a more conserved response between lines at 1 dpi and an overall stronger response to *L. maculans* in resistant plants at the mRNA level (Becker et al. 2019).

3.5.8 Identification of Cysteine-Rich Protein Kinase Genes in Quantitative Resistance to Blackleg in *Brassica napus*

The HR response of race-specific R-genes often provides a visual phenotype, indicating an incompatible interaction and allowing for the determination of pathogen virulence. This distinction is used to separate specific R-gene interactions from quantitative resistance which can provide effective "adult plant resistance" (APR) within a crop variety through the cumulative action of multiple resistance loci. APR is usually measured at the end of the growing season in field trials. APR is particularly important for combating diseases of *Brassica* crops in which R-gene-mediated resistance is lacking, such as *Sclerotinia* stem rot (*Sclerotinia sclerotiorum*) and *Verticillium* wilt (*Verticillium longisporum*) or for diseases where pathogen populations often display a rapid adaptation toward R-gene-mediated resistance, such as in the case of blackleg disease, caused by the hemibiotrophic fungal pathogen *Leptosphaeria maculans*.

Avoidance of R-gene-mediated resistance by *L. maculans* can occur both rapidly and in a geographically localized fashion when a pathogen population is under heavy selection pressure. A rapid decline in the efficiency of the blackleg R-gene *Rlm1* in controlling the disease in Europe highlighted the evolutionary potential of the pathogen. A high frequency of mutation and deletion of the *L. maculans* avirulence gene *AvrLm4-7* was reported to occur within a small plot area sown continually to

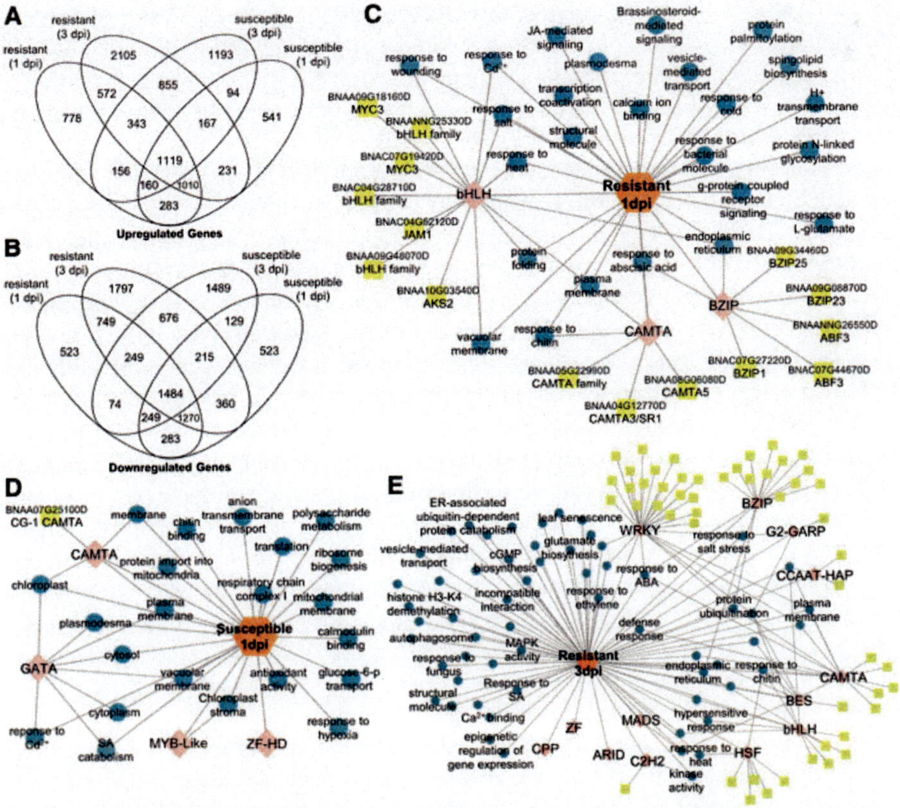

Fig. 3.7 Differential gene expression analyses and predictive transcription factor networks. (**a**) Significantly ($P < 0.05$) upregulated genes in each treatment as compared to their mock-inoculated controls. Four-way VENN diagram shows number of unique or shared genes between treatments. (**b**) Significantly ($P < 0.05$) downregulated genes in each treatment as compared to their mock-inoculated controls. Four-way VENN diagram shows number of unique or shared genes between treatments. (**c**) Predicted network showing transcription factors (yellow squares), DNA binding motifs (pink diamonds), and gene ontology terms (teal circles) enriched in resistant-specific differentially expressed genes at 1 day post inoculation (orange hexagon). Gray lines show predicted connections between transcription factors, motifs, gene ontology terms, or the resistant-specific gene set. (**d**) Predicted network showing genes specifically upregulated in susceptible hosts at 1 day post inoculation. (**e**) Predicted network showing genes specifically upregulated in resistant hosts at 3 days post inoculation. 190 × 180 mm (300 × 300 dpi) (Becker et al. 2019)

B. napus harboring *Rlm7*, while virulent pathotypes remained undetectable in samples from the surrounding local pathogen population. High rates of infection were observed in some areas of Australia in canola varieties carrying the R-gene *LepR3* only 3 years after first commercial release of the material, though this rapid loss of effective resistance may have been aided by preexposure to *Rlm1* varieties, as avirulence toward *LepR3* and *Rlm1* is conferred by the same *L. maculans* avirulence gene; *AvrLm1* (Larkan et al. 2013).

B. napus cultivars containing only APR usually show no difference in the development of leaf lesions when compared with susceptible cultivars, yet they restrict the development of internal stem infection by the pathogen, resulting in lower levels of crown canker formation. This is in contrast to R-gene-mediated resistance which leads to arrest of *L. maculans* growth at the site of infection on cotyledons and leaves. When major R-gene-mediated resistance is avoided by virulent strains within the mixed pathogen population, APR reduces the selection and proliferation of virulent pathotypes in crop residues and the potential for catastrophic crop loss in following seasons (Delourme et al. 2014). While R-gene-mediated resistance can often be detected efficiently and rapidly by observing hypersensitive response after inoculation of *B. napus* cotyledons with well-characterized *L. maculans* isolates, assessment of APR is much more difficult. Resistance needs to be measured either through field-based studies or under controlled conditions through infection with single spore-derived *L. maculans* isolates and assessment of stem infection in plants grown for several months. Assessment of APR in field-based studies can be difficult considering the complexity of plant–pathogen environment interactions. Populations of *L. maculans* in most disease nurseries are genetically heterogeneous mixtures arising from sexual recombination, and variation of pathotypes should be expected both within a trial site and between trial years. Also, variation of host response due to heterozygosity of *B. napus* lines may be confused for polygenic control of resistance. There has also been a widely-held view that blackleg APR is race nonspecific, based largely on experience of the French variety Jet Neuf, which provided durable resistance to blackleg disease over many years in Europe and was also utilized in early efforts to improve blackleg resistance in Australian germplasm. However, more recent studies utilizing single *L. maculans* isolates have questioned the "race-nonspecific" nature of blackleg APR.

Maintenance of strong APR in canola varieties can most efficiently be achieved through marker-assisted breeding based on the molecular characterization of quantitative trait loci (QTLs) associated with resistance (Rimmer 2006). The French variety Darmor, derived from Jet Neuf, is the most extensively studied *B. napus* variety harboring quantitative resistance to *L. maculans*. A doubled-haploid (DH) population produced from a cross between Darmor-bzh and the susceptible Korean cultivar Yudal (DY) was utilized to map 10 QTL contributing to blackleg resistance, with four of the QTL detected stably across 2 years of field testing. The resistance was further analyzed in Darmor × Samouraï (DS) DH and F2 populations, revealing four QTLs that were common to both the DY and DS populations. Near-isogenic lines (NILs) were also produced for four Darmor QTL; *LmA2*, *LmA9*, *LmC2*, and *LmC4*, though only *LmA2* was fully validated as having a significant effect on reducing disease severity. Blackleg APR has also been assessed in several Australian varieties, revealing multiple QTL that are potentially common to several Australian and French cultivars. Little is known about the molecular basis of APR to *L. maculans* infection in *Brassica* species. While two race-specific genes responsible for ETI-mediated blackleg resistance, *LepR3* and *Rlm2*, have been cloned from *B. napus* and shown to encode extracellular leucine-rich repeat (eLRR) receptor-like proteins recognizing the *L. maculans* effectors *AvrLm1* and *AvrLm2*, respectively. No genes underpinning blackleg resistance QTL have been identified. Infection of *B. napus* by *L. maculans* results in attempted physical restriction of the pathogen by the host, via callose deposition, while an increased lignification responses has also been reported for APR varieties. *L. maculans* infection triggers induction of the salicylic acid (SA) signaling pathway, which plays a critical role in plant defense. SA signaling can be triggered in *B. napus* by purified *L. maculans* cell wall components and is greatly induced during ETI, along with the ethylene signaling pathway and H_2O_2 accumulation. However, these studies have all focused on early infection events in the cotyledons of *B. napus* seedlings; nothing is known about which defense mechanisms may be active against the invading hyphae as they grow asymptomatically through the petiole and stem.

Several stable blackleg resistance QTLs, with resistance alleles derived from AG-Castle and AV-Sapphire have been identified. Two blackleg-resistant Australian *B. napus* varieties were released in 2002 and 2003, respectively. Larkan et al. (2016a) used Topas/AG-Castle (TC) and Topas/AV-Sapphire (TS) DH populations to assess the APR of the varieties over multiple years at two locations, performed both single- and multienvironment QTL mapping, and defined the physical locations of the QTL relative to the recently released *B. napus* Darmor-bzh reference genome (Chalhoub et al. 2014), allowing for the identification of candidate defense-related genes. Resistance to the blackleg disease of *Brassica* crops caused by *Leptosphaeria maculans* is determined by both race-specific resistance (R)-genes and quantitative resistance loci (QTL), or adult-plant resistance (APR). While the introgression of R-genes into breeding material is relatively simple, QTL are often detected sporadically, making them harder to capture in breeding programs. For the effective deployment of APR in crop varieties, resistance QTL need to have a reliable influence on phenotype in multiple environments and be well defined genetically to enable marker-assisted selection (MAS). Doubled-haploid populations produced from the susceptible *B. napus* variety Topas and APR varieties AG Castle and AV-Sapphire were analyzed for resistance to blackleg in two locations over 3 years and 4 years, respectively. Three stable QTL were detected in each population, with two loci appearing to be common to both APR varieties. Physical delineation of three QTL regions was sufficient to identify candidate defense-related genes, including a cluster of cysteine-rich receptor-like kinases contained within a 49 gene QTL interval on chromosome A01. Individual *L. maculans* isolates were used to define the physical intervals for the race-specific R-genes Rlm3 and *Rlm4* and to identify QTL common to both field studies and the cotyledon resistance response. Through multienvironment QTL analysis have been identified and delineated four significant and stable QTL suitable for MAS of quantitative blackleg resistance in *B. napus*, with identified candidate genes, which potentially play a role in quantitative defense responses to *L. maculans* (Larkan et al. 2016a).

3.5.9 Identification of Environmentally Stable QTLs for Resistance to Blackleg of *Brassica*

Generally, quantitative resistance is considered to be more durable and less likely to be rendered ineffective by new virulent pathogen races than resistance mediated by single R-genes that is frequently associated with "boom and bust" cycles (Delourme et al. 2014). The use of such quantitative resistance is vital for nonintensive cropping systems where farmers cannot afford fungicides and their crops are threatened by widespread epidemics of diseases. The severity of epidemics is predicted to increase with global warming (Butterworth et al. 2010). The use of host resistance to control this disease is becoming ever more important. Resistance against *L. maculans* relies on two types of resistance that operate in *B. napus* (Rimmer 2006).

Firstly, major resistance (R)-gene-mediated resistance operates at the leaf infection stage, after airborne ascospores produced on crop debris have landed on leaves of the new crop, germinated and begun to penetrate leaves through stomata (Huang et al. 2003). R-gene-mediated resistance protects the plant from development of phoma leaf spots at the young plant stage (i.e., no phoma leaf spots in autumn) and subsequently prevents the development of phoma stem canker at the adult plant stage (e.g., no phoma stem canker in the following spring/summer). Therefore, R-gene-mediated resistance against *L. maculans* is also referred as complete resistance. Secondly, quantitative resistance, mediated by QTL, operates as the pathogen is spreading symptomless along the leaf petiole toward the stem or growing in stem tissues (Huang et al. 2009, 2014). Quantitative resistance does not prevent the development of phoma leaf spots at the young plant stage but decreases the severity of phoma stem canker at the adult plant stage. Therefore, quantitative resistance against *L. maculans* is also referred as partial resistance (Delourme et al. 2006). Successful breeding of oilseed rape cultivars for control of phoma stem canker in Australia and France has led to an improvement in quantitative resistance with time (Jestin et al. 2011). Since R-gene-mediated resistance is race-specific and is often rapidly rendered ineffective by changes in *L. maculans* populations, if it is deployed commercially in a large area over more than 3 years (Sprague et al. 2006), combining R-gene resistance with quantitative resistance provides a more robust crop protection strategy (Delourme et al. 2014). However, effective detection of quantitative resistance in field conditions is only possible in the absence of effective R-genes. Therefore, QTL for resistance against *L. maculans* can be identified only in mapping populations or in germplasm collections that do not segregate for effective R-genes. Few studies have identified QTL for resistance against *L. maculans*. One French winter oilseed rape cultivar Darmor was used as a source of resistance in two genetic backgrounds (Jestin et al. 2012). Pilet et al. (2001) showed that both the genetic background and the environment influenced detection of QTL. However, four QTL for resistance to *L. maculans* were detected consistently. QTL analyses done in Australia in different populations also showed that environmental conditions influenced detection of resistance QTL (Kaur et al. 2009a, b; Raman et al. 2012b). One of the limitations in the use of resistance QTL in breeding is their inconsistency due to genotype × environment interactions (McDonald 2010). It is essential for breeders to develop oilseed rape cultivars with resistance that is effective in different environmental conditions. Although four QTL for resistance to *L. maculans* were consistently detected in two genetic backgrounds over two growing seasons in France (Pilet et al. 1998, 2001), it was not clear whether these QTL could be considered as stable QTL. There is evidence that environmental factors, especially temperature, affect the effectiveness of both R-gene-mediated resistance and quantitative resistance against *L. maculans* (Huang et al. 2006, 2009). Identification of stable resistance is important both for breeding and for understanding mechanisms of resistance. Identification of *B. napus* QTL for resistance against *L. maculans* have been described that are less sensitive to environmental factors, thus facilitating selection of stable QTL in breeding for effective, stable resistance (Huang et al. 2016).

Six stable QTL for resistance against *L. maculans* have been identified by QTL × environment interaction analysis using data from five winter oilseed rape field experiments. Quantitative trait locus (QTL)-mediated resistance against *L. maculans* in *B. napus* is considered to be race nonspecific and potentially durable. Identification and evaluation of QTL for resistance to *L. maculans* is important for breeding oilseed rape cultivars with durable resistance. An oilseed rape mapping population was used to detect QTL for resistance against *L. maculans* in five winter oilseed rape field experiments under different environments. A total of 17 QTL involved in "field" quantitative resistance against *L. maculans* were detected and collectively explained 51% of the phenotypic variation. The number of QTL detected in each experiment ranged from two to nine and individual QTL explained 2–25% of the phenotypic variation. QTL × environment interaction analysis suggested that six of these QTL were less sensitive to environmental factors, so they were considered to be stable QTL. Markers linked to these stable QTL will be valuable for selection to breed for effective resistance against *L. maculans* in different environments, which will contribute to sustainable management of the disease (Huang et al. 2016).

3.5.10 Molecular Characterization of Near Isogenic Lines at QTL for Quantitative Resistance to *Leptosphaeria* in *Brassica*

The development of near isogenic lines (NILs) for quantitative trait loci (QTLs) using marker-assisted selection is a reliable method for validating the additive effect of QTL, because the effect of individual or combined QTL alleles is more accurately estimated in a homogeneous genetic background. This method was successfully used to validate QTL when NILs were generated by backcrosses using either one of the QTL mapping population parents as the recurrent parent (Maeda et al. 2006) or new genetic backgrounds (Richardson et al. 2006). With the same objective, NILs were produced from heterogeneous inbred families derived from recombinant inbred line mapping populations (Loudet et al. 2007) or from breeding populations (Pumphrey et al. 2007). NILs are valuable material for studying QTL × environment (Steele et al. 2007) or QTL × genetic background (Lecomte et al. 2004) interactions.

NILs can also be used for an improved phenotypic characterization of QTL effect (Ioannidou et al. 2003; Wissuwa and Ae 2001) or for fine mapping and identifying candidate genes underlying QTL (Wan et al. 2006; Loudet et al. 2007).

The genetic basis of quantitative resistance in the French winter oilseed rape "Darmor," derived from "Jet Neuf," was studied in two genetic backgrounds. In the "Darmor *bzh*" × "Yudal" cross, Pilet et al. (1998) identified a total of ten resistance QTL, four of which were associated with decreased stem canker severity and decreased plant death in two seasons of Weld experiments. Analysis of progeny derived from a "Darmor" × "Samourai" cross, consisting of one double haploid (DH) population and one $F_{2:3}$ family population, identified six QTL in the DH population and four QTL in the $F_{2:3}$ families (Pilet et al. 2001). Out of 16 loci detected in all, only four QTL were common to both the "Darmor-*bzh*" × "Yudal" and "Darmor" × "Samourai" crosses (Delourme et al. 2008).

The aim was to develop NILs for four stem canker resistance QTL in order to validate the effect of individual QTL in the susceptible background "Yudal." These four QTL were chosen because they were the most stable. Three of these QTL were stably detected in the two DH populations (on linkage groups A2, C2, and C4). The fourth QTL (on linkage group A9) was specific to the "Darmor-*bzh*" × "Yudal" cross but was detected throughout the 2-year trial. For each QTL, tall and dwarf lines were generated since it was shown that the *Bzh* dwarf gene might have an effect on resistance to *L. maculans* (Pilet et al. 1998). In order to characterize the introgressed segment, new molecular markers were designed within the targeted QTL regions either by deriving sequenced characterized amplified regions (SCARs) (Paran and Michelmore 1993) from random amplified polymorphic DNA (RAPD) or amplified fragment length polymorphism (AFLP) markers or by taking advantage of the synteny with the *Arabidopsis thaliana* genome (Parkin et al. 2005). The derived NILs were then evaluated for their level of stem canker resistance in the Weld, which allowed to clearly validating the effect of QTL *LmA2* and to a lesser extent the effect of QTL *LmA9* in four different Weld sites.

The genetic basis of quantitative resistance in the French winter oilseed rape "Darmor," which was derived from "Jet Neuf," was examined in two genetic backgrounds. Stable QTL involved in blackleg resistance across year and genetic backgrounds were identified. The near isogenic lines (NILs) were produced in the susceptible background "Yudal" for four of these QTL using marker-assisted selection. Various strategies were used to develop new molecular markers, which were mapped in these QTL regions. These were used to characterize the length and homozygosity of the "Darmor-*bzh*" introgressed segment in the NILs. Individuals from each NIL were evaluated in blackleg disease Weld trials and assessed for their level of stem canker in comparison to the recurrent line "Yudal." The effect of QTL *LmA2* was clearly validated, and to a lesser extent, QTL *LmA9* also showed an effect on the disease level. This work provides valuable material that can be used to study the mode of action of genetic factors involved in *L. maculans* quantitative resistance (Delourme et al. 2008).

3.5.11 Molecular Mapping of Qualitative and Quantitative Loci for Resistance to *Leptosphaeria*

Host resistance genes have been catalogued using differential sets of *L. maculans* isolates and/or using molecular markers. These studies evaluated blackleg resistance on the basis of pathogen infection on cotyledons, stem (canker), and survival under field conditions. Up to now, 14 major loci (*Rlm1-10* and *LepR1 to LepR4*) conferring resistance to specific races of *L. maculans* have been identified (Yu et al. 2005, 2008). The linkage mapping studies revealed that at least five resistance genes (*Rlm1*, *Rlm3*, *Rlm4*, *Rlm 7*, and *Rlm9*) are localized in a cluster within a 40-cM genomic region on chromosome A7 (Delourme et al. 2004, 2006). This genomic region showed extensive inter- and intragenomic duplications, as well as intrachromosomal tandem duplications (Mayerhofer et al. 2005), whether some of these R-genes are allelic remains unknown (Raman et al. 2012b).

The effectiveness of resistance for some of the major genes has decreased in some cultivars within a few years of their release, limiting their usefulness in managing blackleg disease (Rouxel et al. 2003a). Brun et al. (2010) demonstrated that a major resistance gene (*Rlm6*) is more durable when expressed in a genetic background that has quantitative resistance, indicating the need to identify, and combine, both existing and new qualitative and quantitative genes for blackleg resistance. Raman et al. (2012a) described (1) the construction of a linkage map of the *B. napus* doubled-haploid (DH) population derived from Skipton/Ag-Spectrum designated as "SASDH," (2) the determination of the inheritance and location of blackleg resistance genes, and (3) the identification of molecular markers linked with resistance loci, applying a whole-genome mapping approach, with the aim of providing canola breeders with tools for routine marker-assisted selection.

Raman et al. (2012b) described the construction of a genetic linkage map, comprising 255 markers, based upon simple sequence repeats (SSR), sequence-related amplified polymorphism, sequence tagged sites, and EST-SSRs and the localization of qualitative (race-specific) and quantitative (race-nonspecific) trait loci controlling blackleg resistance in a doubled-haploid population derived from the Australian canola (*Brassica napus* L.) cultivars Skipton and Ag-Spectrum using the whole-genome average interval mapping approach. Marker regression analyses revealed that at least 14 genomic regions with LOD C 2.0 were associated with qualitative and quantitative blackleg resistance, explaining 4.6–88.9% of genotypic variation. A major qualitative locus, designated *RlmSkipton* (*Rlm4*), was mapped on chromosome A7, within 0.8 cM of the SSR marker *Xbrms075*. Alignment of the molecular markers underlying this QTL region with the genome sequence data of *B. rapa* L. suggests that *RlmSkipton* is located approximately 80 kb from the *Xbrms075* locus. Molecular marker-*RlmSkipton* linkage was further validated in an F2 population from Skipton/Ag-Spectrum. The results show that SSR markers linked to consistent genomic regions are suitable for enrichment of favorable alleles for blackleg resistance in canola breeding program.

3.5.12 Association Mapping of QTLs in *Brassica* for *Leptosphaeria*

In the *B. napus*–*L. maculans* pathosystem, the search for quantitative resistance factors has focused on one source of resistance, the variety Darmor, for which resistance quantitative trait loci (QTLs) were detected in two genetic backgrounds (Pilet et al. 1998, 2001). These QTLs, as in the majority of studies on quantitative traits, were identified by linkage analysis using populations derived from a biparental cross. However, the use of this population type has some limitations. Firstly, the limited number of recombination events occurring during the construction of mapping populations results in poor resolution for quantitative traits, depending on the size of the population. Secondly, only two alleles at any given locus can be studied simultaneously (Flint-Garcia et al. 2003), at least in diploid species. Thirdly, the parental genotypes are often not representatives of the germplasm pool that is actively used in breeding programs. Thus, the markers linked to QTL are not always transferable to other genetic backgrounds, limiting their usefulness in marker-assisted selection (MAS) (Snowdon and Friedt 2004). Hence, in these studies, there is a substantial time-lag between QTL discovery and marker-assisted crop improvement practices, due to the need to confirm the presence and stability of the QTL (Ersoz et al. 2008).

To overcome these limitations, other methods have recently emerged, such as association mapping, which can be a powerful tool to exhaustively identify QTL in plants. In a well-designed association study, the results can be more immediately applied to MAS (Ersoz et al. 2008). Association mapping (AM) or linkage disequilibrium mapping (LD-mapping) or association analysis is a method that exploits the variation in a collection of genetically diverse materials (composed of unrelated individuals or unknown pedigrees) to uncover a significant association between a trait and a gene or a molecular marker on the basis of linkage disequilibrium. AM offers the advantage that historical and evolutionary recombination can be exploited at the population level, and all natural genetic diversities (larger number of alleles studied) can be used in order to obtain a high-resolution map. Moreover, no pedigree or cross is required, making it easier to produce the data (Aranzana et al. 2005; Jestin et al. 2011).

Since a genome-wide approach is not always possible, e.g., for species with a large genome and/or with limited genomic resources, AM can also be used as a validation method based on candidate genes (Thornsberry et al. 2001) or markers linked to QTL previously detected by classical methods. Nevertheless, one important point of AM studies in plants (Zhao et al. 2007) as well as in humans (Marchini et al. 2004) is that the population structure and/or family relatedness can result in spurious correlations, leading to an elevated false-positive rate for marker–trait association. Different statistical methods (Price et al. 2006; Yu et al. 2006) that account for population structure and/or family relatedness can then be used to reduce the false positives and thus increase the power of AM (Zhao et al. 2007). To validate the QTL previously identified in Darmor for quantitative resistance to *L. maculans* in diverse genetic backgrounds and establish the interest for MAS of the markers located within the QTL, Jestin et al. (2011) analyzed marker polymorphisms within an oilseed rape breeding line collection and used different AM models taking or not the population structure and/or family relatedness into consideration.

Quantitative resistance factors appear to be important components for effective and durable control of this pathogen. Quantitative trait loci (QTLs) for stem canker resistance have been identified in the Darmor variety. However, before these QTLs can be used in marker-assisted selection (MAS) to breed resistant varieties, they must be validated in a wide range of genetic backgrounds. Jestin et al. (2011) used an association mapping approach to confirm the markers located within the QTL previously identified in Darmor and established their usefulness in MAS. Characterization of 128 oilseed rape lines for the molecular diversity showing a large spectrum of responses to *L. maculans* infection, using 72 pairs of primers for simple sequence repeat and other markers. The different association mapping models were used which either do or do not take into account the population structure and/or family relatedness. In all, 61 marker alleles were found to be associated with resistance to stem canker. Some of these markers were associated with previously identified QTL, which confirms their usefulness in MAS. Markers located in regions not harboring previously identified QTL were also associated with resistance, suggesting that new QTL or allelic variants are present in the collection. All of these markers associated with stem canker resistance will help identify accessions carrying desirable alleles and facilitate QTL introgression (Fig. 3.8; Jestin et al. 2011).

3.5.13 Identification of Stable QTLs in *Brassica napus* to Blackleg Resistance

There is a wide range of genetic variation in resistance to blackleg among canola cultivars. Two types of resistance, qualitative resistance conferred by race-specific R-genes and quantitative resistance (QR) conferred by non-race-specific genes (QTL), have been deployed to combat this disease in commercial cultivars. To date, at least 15 R-genes have been identified in *B. napus* and its related species, *B. rapa* and *B. juncea* (Raman et al. 2013). Although these R-genes provide "complete" protection from *L. maculans* infection, this resistance often becomes ineffective, particularly when the frequency of *avr* alleles is high in nature (Van de Wouw et al. 2010a, b, 2014). Even deployment of R-gene pyramids in cultivars such as Surpass400 and Hyola50 has not provided long-term durability against highly diverse *L. maculans* races in Australia. Research has shown that QR is more "durable" and provides long-term protection to host plants (Delourme et al. 2014). However, identification of such resistance is very difficult to achieve particularly when R-gene(s) are present in the same cultivar/accession. Several canola-breeding programs rely on screening germplasm under blackleg nurseries across multiple locations and selecting the best standing plants or surviving lines at the end of the growing season. This practice is quite effective but is not highly efficient for making genetic gains by incorporating QR in breeding lines.

Genetic mapping of QR via traditional QTL and association (genome-wide association studies) mapping approaches has enabled the identification of

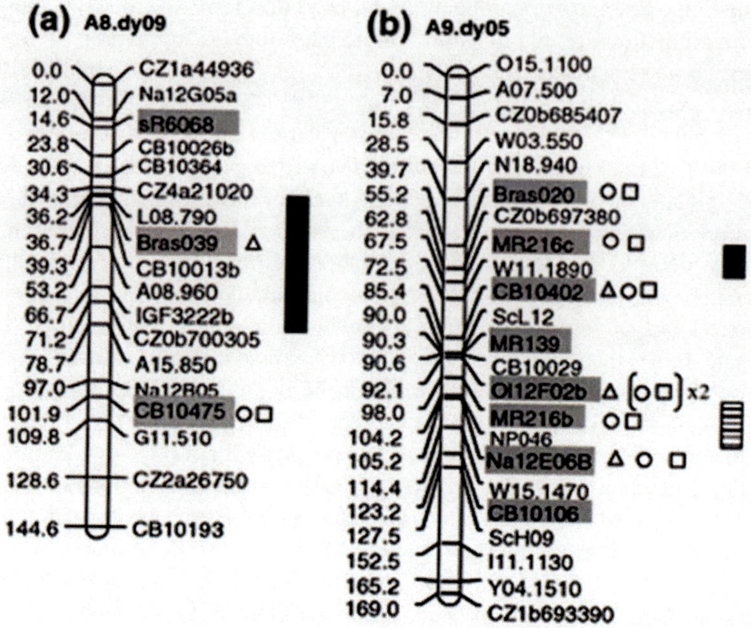

Fig. 3.8 Comparison between the association mapping study and linkage analysis for quantitative stem canker resistance. Linkage groups A8.dy09 (**a**) and A9.dy05 (**b**) of doubled haploid (DH) Darmor-bzh Yudal (DY) population. The marker alleles tested in association mapping, with an allelic frequency higher than 5%, are highlighted in gray. The circle, square, and triangle are the marker (Darmor allele or Yudal allele or both Darmor and Yudal alleles (92)) declared as significantly ($P < 0.05$) associated with stem canker resistance in the GLM, K MLM and K+ Q MLM, respectively. The hatched bars correspond to QTL detected by Pilet et al. (1998) in a 152 DH DY population and the black bars represent the QTL detected on the 279 DH DY populations by Jestin et al. (2011)

quantitative loci for resistance to *L. maculans* on the genetic/physical maps of *B. napus* (Kumar et al. 2018). One of the major limitations of QTL mapping studies in genetic improvement programs has been the stability of QTL across environments and the magnitude of QTL effects in different genetic background. Of the various genetic analyses/QTL identification studies conducted to date, Darmorbzh/Yudal (DY) DH population of *B. napus* has been extensively evaluated for resistance to blackleg in diverse environments across Europe (Jestin et al. 2012; Kumar et al. 2018). However, the effectiveness of QR and those QTL identified in Europe have not been tested against highly diverse pathogenic Australian *L. maculans* populations. Raman et al. (2018) evaluated the DYDH population under Australian field, glasshouse, and shade house conditions using an ascospore shower test and identified QR loci. They further compared the physical locations of QTL that are identified in Australia, with those identified in France and United Kingdom to identify consistent genomic regions involved in resistance across continents. It has revealed functional genes underlying such regions as well as to provide target regions (molecular markers) for marker-assisted and genomic selection in canola breeding programs (Raman et al. 2018).

To study the genomic regions involved in quantitative resistance (QR), 259–276 DH lines from Darmor-bzh/Yudal (DYDH) population were assessed for resistance to blackleg under shade house and field conditions across 3 years. In different experiments, the broad-sense heritability varied from 43% to 95%. A total of 27 significant quantitative trait loci (QTLs) for QR were detected on 12 chromosomes and explained between 2.14% and 10.13% of the genotypic variance. Of the significant QTL, at least seven were repeatedly detected across different experiments on chromosomes A02, A07, A09, A10, C01, and C09. Resistance alleles were mainly contributed by 'Darmor-bzh' but 'Yudal' also contributed few of them. The results suggest that plant maturity and plant height may have a pleiotropic effect on QR in conditions tested. It was confirmed that *Rlm9*, which is present in 'Darmor-bzh,' is not effective to confer resistance in Australian field conditions. Comparative mapping showed that several R-genes coding for nucleotide-binding leucine-rich repeat (LRR) receptors map in close proximity (within 200 kb) of the significant trait-marker associations on the reference 'Darmor-bzh' genome assembly. More importantly, eight significant QTL regions were detected across diverse growing environments: Australia, France, and United Kingdom. These stable QTL identified herein can be utilized for enhancing QR in elite canola germplasm via marker-assisted or genomic selection strategies (Raman et al. 2018).

Putative candidate genes for resistance loci were identified in *B. napus*. Physical localization of R-genes from *A. thaliana* and *B. napus* in the vicinity of significant trait-marker association suggested that these genomic regions may be associated with defense-related genes. For example, *B. napus* R-genes such as *BnaA02g25160D* with CC–NBS–LRR domain, *BnaA02g36900D* with TIR domain, *BnaA09g14320D*, *BnaC09g06020D*, and *BnaC09g44520D* with TIR–NBS–LRR domains and *A. thaliana* R-genes such as *AT2G05940*.1 (RPM1-INDUCED PROTEIN KINASE) which encodes a receptor-like cytoplasmic kinase that phosphorylates the host target RIN4, leading to the activation of a plant innate immune receptor RPM1, were detected within 200 kb from significant SNP associations. AT1G56510.1 (*BnaC09g06020D*) is reported to confer resistance to races of *Albugo candida*. The TIR, NBS, LRR, protein kinase proteins are well known to confer disease resistance in plants (Césari et al. 2014). Several transcripts involved in *L. maculans* infection in resistant and susceptible lines were identified in *B. napus* lines upon infection with *L. maculans*, at the cotyledonary stage (Becker et al. 2017). Raman et al. (2018) compared the physical position of those genes with the genomic regions associated with resistance in DY population. They found two genes, *BnaA07g32130D* and *BnaC01g36910D*, within 200 kb from the transcripts, which are shown to be accumulated upon infection with *L. maculans* (Becker et al. 2017). The *BnaA07g32130D* (PDF1.2) gene model relates to the *A. thaliana* gene (*AT5G44420*.1) and belongs to the plant defensin (PDF) family and encodes an ethylene- and jasmonate-responsive plant defensin, while the *BnaC01g36910D* was mapped within 100 kb from a significant QTL for field canker and survival on C01.

Strategies to incorporate into breeding programs were suggested by Raman et al. (2018). Various breeding programs prefer to introgress qualitative and quantitative resistance genes with major allelic effects to *L. maculans*. It was found that eight significant QTL regions for QR on A02, A07, A09, A10, C01, and C09, which were detected across diverse growing environments: Australia, France, and United Kingdom. Almost all QTL had small allelic effects and accounted for less than 10% of the genotypic variance individually. Some of the allelic effects have been validated in DH populations derived from 'Darmor-bzh/Bristol' and 'Darmor/Samourai' and diversity panels. Therefore, these validated and stable loci can be introgressed for enhancing QR in elite canola germplasm via molecular marker assays such as KASP and sequence capture probes. Given that QR is difficult to select and confounded with environmental factors, genomic selection could be performed while also targeting loci associated with high grain yield, flowering time, and tolerance to water and nutrient stress in the canola improvement programs (Raman et al. 2018).

3.5.14 Cloning and Characterization of *Leptosphaeria Maculans* R-Genes (*Lm*) *AvrLm9* Gene

Nine of the 18 blackleg R-genes reported from *Brassica* species correspond to genetically defined *Avr* genes from *Lm* (Larkan et al. 2016b; Plissonneau et al. 2017; Raman et al. 2013). Two *B. napus* R-genes, *LepR3* and *Rlm2*, effective against *Lm* isolates with corresponding *Avr* genes, *AvrLm1* and *AvrLm2*, respectively, have been cloned (Larkan et al. 2013, 2015). *LepR3* and *Rlm2* are allelic and encode membrane-bound receptor-like proteins (RLPs). The matching *Avr* genes, *AvrLm1* and *AvrLm2*, and several other *Lm Avr* genes, including *AvrLm3*, *AvrLm4–7*, *AvrLm6*, *AvrLm11*, and *AvrLmJ1*, have been cloned (Balesdent et al. 2013; Fudal et al. 2007; Ghanbarnia et al. 2015; Gout et al. 2006; Parlange et al. 2009; Plissonneau et al. 2016; Van de Wouw et al. 2014). All of the *Avr* proteins are predicted to be secreted into the host apoplast based on the presence of an N-terminal secretion signal peptide. They all encode small cysteine-rich proteins of unknown function, with the exception of *AvrLm1*, which contains only one C residue (Gout et al. 2006). Although most of the R–*Avr* interactions between *Lm* and *B. napus* fit the classic gene-for-gene model, there are several examples of deviation from this model (Petit-Houdenot and Fudal 2017). The first example to be reported was *AvrLm4–7*, a single *Avr* gene that is perceived by both *Rlm4* and *Rlm7*. A point mutation abolishes *Rlm4* recognition without affecting *Rlm7* recognition (Parlange et al. 2009). Another example is the dual specificity of the single *Avr* gene *AvrLm1*, which is recognized by both *LepR3* and *Rlm1*. *LepR3* and *Rlm1* are two independent R loci located on chromosomes A10 and A07, respectively (Larkan et al. 2013, 2016b).

Point mutation, deletion, and transposon insertion are frequently reported as gain-of-virulence mechanisms largely driven by the plasticity of the repeat-rich region of the genome, where it is enriched for predicted small secreted proteins and harbors all known *Lm* effectors (Rouxel et al. 2011). Gain of virulence as a result of the suppressive effect of other *Avr* genes has been reported for *Lm*. Plissonneau et al. (2016) cloned *AvrLm3* and showed that AvrLm3 recognition by the *B. napus* R-protein *Rlm3* is masked in the presence of *AvrLm4–7*. *AvrLm3* and *AvrLm4–7* are physically linked; both are located on super contig 12 (SC12) of the Lm genome (Plissonneau et al. 2016). Two other *Avr* genes, *AvrLm9* and *AvrLep1**, have been reported to be closely linked to *AvrLm4–7*, forming a group of tightly linked effector genes on SC12 (Balesdent et al. 2005; Ghanbarnia et al. 2012). Several other effectors, *AvrLm1*, *AvrLm2*, *AvrLm6*, and the predicted effector *LmCys2*, form a second group of tightly linked effectors on SC6 of the *Lm* genome (Fudal et al. 2007; Ghanbarnia et al. 2015; Gout et al. 2006; Van de Wouw et al. 2014). Singleton *Lm Avr* genes have also been reported: *AvrLm11* on SC22 and *AvrLmJ1* on SC7 (Balesdent et al. 2013; Van de Wouw et al. 2014). *AvrLmJ1* was characterized as being avirulent on *B. juncea* carrying an unknown R-gene. However, a report by Plissonneau et al. (2017) has shown that *AvrLmJ1* is in fact *AvrLm5* based on colocalized mapping and the demonstration that Lm isolates carrying *AvrLmJ1* are also avirulent on *B. juncea* differential lines carrying *Rlm5*. Therefore, Ghanbarnia et al. (2018) refer to *AvrLmJ1* as *AvrLm5*. The cloning of *AvrLm9* and its characterization, as the second *Avr* gene from *Lm* for which recognition by the host is suppressed by AvrLm4–7, has been made when mapped the *AvrLm9* phenotype to SC7 and discovered that it is an allele of *AvrLm5*.

The cloning of *Lm* effector *AvrLm9* which is recognized by the resistance gene *Rlm9* in *B. napus* cultivar Goeland has been done by Ghanbarnia et al. (2018). *AvrLm9* was mapped to scaffold 7 of the *Lm* genome, cosegregating with the previously reported *AvrLm5* (previously known as *AvrLmJ1*). Comparison of *AvrLm5* alleles among the 37 resequenced *Lm* isolates and transgenic complementation identified a single point mutation correlating with the *AvrLm9* phenotype. Therefore, this gene was renamed as *AvrLm5*-9 to reflect the dual specificity of this locus. *Avrlm5*-9 transgenic isolates were avirulent when inoculated on the *B. napus* cultivar Goeland. The expression of *AvrLm5*-9 during infection was monitored by RNA sequencing. The recognition of *AvrLm5*-9 by *Rlm9* is masked in the presence of *AvrLm4*-7, another *Lm* effector. *AvrLm5*-9 and *AvrLm4*-7 do not interact, and *AvrLm5*-9 is expressed in the presence of *AvrLm4*-7. *AvrLm5*-9 is the second *Lm* effector for which host recognition is masked by *AvrLm4*-7. An understanding of this complex interaction will provide new opportunities for the engineering of broad-spectrum recognition (Ghanbarnia et al. 2018).

3.5.15 Mechanisms of Quantitative Resistance in *Brassica* Cotyledons to *Leptosphaeria*

Genetic resistance is a cornerstone of blackleg management and is usually classified as either qualitative or quantitative. The former is controlled by single resistance (R)-genes, while the latter is often, though not always, polygenic. While many of the R-genes have been identified, quantitative resistance (QR) is not well understood.

QR can be attributed to multiple genomic regions in *B. napus*, with many of the same loci found in multiple canola cultivars. QR to blackleg in canola is believed to be expressed primarily in adult plants. However, 74-44 BL, a Canadian canola cultivar, has consistently shown QR to stem canker in adult canola, as well as to infection in cotyledons by *Lm*. Poland et al. (2009) postulated that plant QR might be due to weaker versions of R-genes, alterations in plant morphology or development, phytoalexin production, variants of innate immunity, or signal transduction associated genes. QR in canola to *Lm* might also be attributed to uncharacterized R-genes.

RNA sequencing (RNA-seq) has provided valuable insights into the interactions between canola and blackleg in the initial stages of cotyledon infection in the absence of genetic resistance, in canola with and without major resistance genes, as well as the genes that are potentially involved in other plant–pathogen interactions. Hao et al. (2016) used RNA-seq to explore QR to rust in wheat. In addition, Joshi et al. (2016) used RNA-seq to identify genes involved in resistance to *Sclerotinia* in *B. napus*. Haddadi et al. (2016) found that, in the absence of any known resistance, genes related to initial lignin biosynthesis, as well as those involved in biosynthesis, breakdown of glucosinolates and cell surface receptors (PAMP and effector recognition) were upregulated. In contrast, transcription factors, proteases, and protease inhibitors, peroxidases, and chitinases were expressed to a lesser extent within blackleg lesions. However, it is not known if any of these host responses can be induced in seedlings. Larkan et al. (2016a) found evidence that a cluster of receptor-like kinases could be involved in the QR of adult canola plants against blackleg. Consistent with this finding, Haddadi et al. (2016) reported that one of the cell surface receptors expressed differentially in blackleg-infected seedlings was also a receptor-like kinase. Additional studies have identified regions in the *B. napus* genome that are potentially involved in QR to *Leptosphaeria*. It is therefore useful to explore the modes of action for QR against *Leptosphaeria*.

Next-generation sequencing approaches might help relate phenotypic observations, such as those obtained from microscopy, to molecular mechanisms. Fluorescent microscopy of proteins tagged with fluorophores, such as green fluorescent protein (GFP), provides valuable information about plant colonization by microbes, including the canola-blackleg pathosystem. Similarly, imaging of the ROS in canola cotyledons can complement RNA-seq data. This study presents data on *Lm* colonization in cotyledons of Westar (susceptible) and 74-44 BL (expressing QR and carrying two specific R-genes in a hybrid background) inoculated with a GFP-expressing *Leptosphaeria* isolate. To explore the genes that are differentially expressed between canola cultivars in seedlings is an attempt to gain insights into the potential mechanisms of QR in the resistant cultivar 74-44 BL.

Using resistant cultivars is a common approach to managing blackleg of canola/rapeseed caused by *Leptosphaeria maculans*. Quantitative resistance (QR), as opposed to major-gene resistance, is of interest because it is generally more durable, due to its multigenetic basis. However, the mechanisms and genes underlying QR are mostly unknown. Potential QR modes of action in 74-44 BL were explored. This Canadian canola cultivar showed moderate but consistent race-nonspecific resistance at the cotyledon and adult plant stages. A susceptible cultivar, Westar, was

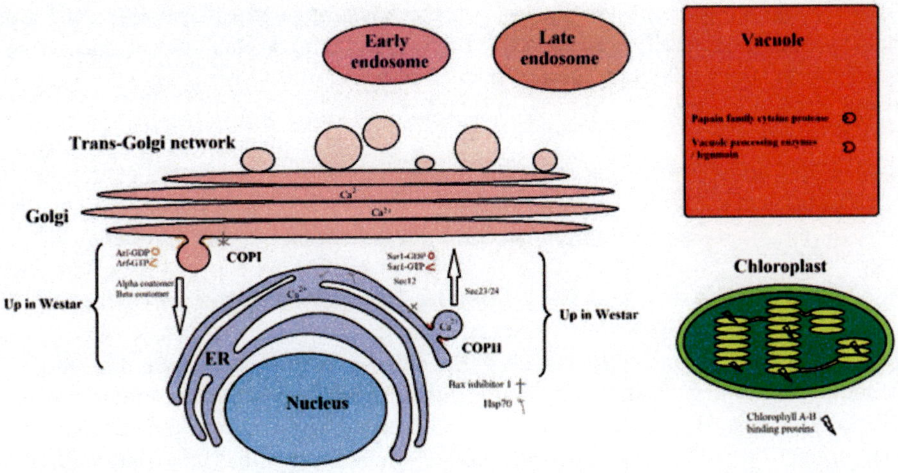

Fig. 3.9 Proposed model on how some of the most highly-expressed DEGs (differentially expressed genes) might interact, potentially resulting in programed cell death. *ER* Endoplasmic reticulum (Hubbard et al. 2020)

used as a control. After inoculation, the lesions developed more slowly on the cotyledons of 74-44 BL than those of Westar. The RNA sequencing (RNA-seq) was used to identify genes and their functions, putatively related to this resistance, and found that genes involved in programmed cell death (PCD), reactive oxygen species (ROS), signal transduction, or intracellular endomembrane transport were most differentially expressed. ROS production was assessed in relation to *Leptosphaeria* hyphal growth and lesion size; it occurred beyond the tissue colonized by *Leptosphaeria* in 74-44 BL and appeared to trigger rapid cell death, limiting cotyledon colonization by *Leptosphaeria*. In contrast, *Leptosphaeria* grew more rapidly in Westar, often catching up with the ring of ROS and surpassing lesion boundaries. It appears that QR in 74-44 BL cotyledons is associated with limited colonization by *Leptosphaeria* possibly mediated via ROS. The RNA-seq data also showed a link between ROS, signal transduction, and endomembrane vesicle trafficking, as well as PCD in the resistance. These results provide a starting point for a better understanding of the mechanisms behind QR against *Leptosphaeria* in canola (Fig. 3.9; Hubbard et al. 2020).

3.5.16 Status of Major Gene and Polygenic Resistance to *Leptosphaeria* in *Brassica*

Different sources of resistance to *L. maculans* have been identified and introduced into *B. napus* breeding lines and cultivars. Many studies on the inheritance of resistance have been done at both seedling and adult plant growth stages. Two types of resistance are usually distinguished. The first type is a qualitative resistance, which is expressed from the seedling to the adult plant stage in cotyledons and leaves, and is generally considered as single-gene race-specific resistance. The second type is a quantitative adult plant resistance, which is a partial resistance usually thought to be race nonspecific and mediated by many genes. In Europe, Canada, and Australia, many resistant cultivars have been registered, but there is evidence of breakdown of race-specific resistance in response to rapid evolution of *L. maculans* populations. Therefore, understanding the genetic basis of resistance in oilseed rape is strategically important for management of resistant cultivars.

3.5.16.1 Identification of Race-Specific Resistance Genes in Different *Brassica* Species to *Leptosphaeria*

Differential interactions in the *Brassica–L. maculans* pathosystem were first studied at the seedling stage using a cotyledon inoculation test (Williams and Delwiche 1979). The first *B. napus* differential set consisted of three cultivars, 'Westar' (susceptible control, spring oilseed type), 'Quinta,' and 'Glacier' (winter oilseed types) (Mengistu et al. 1991). Using this differential set, *L. maculans* isolates were classified into three Pathogenicity Groups (PG), i.e., PG2 (avirulent on 'Quinta' and 'Glacier'), PG3 (avirulent on 'Quinta' but virulent on 'Glacier'), and PG4 (virulent on all three cultivars). Badawy et al. (1991) replaced 'Westar' with winter *B. napus* cultivar 'Lirabon' and added 'Jet Neuf,' leading to the description of six PG, termed A1-A6, resulting from a subdivision of each of the previous groups into two PG (virulent or avirulent on 'Jet Neuf,' respectively). Other race-specific interactions were described using other differential sets including other *Brassica* species (Cargeeg and Thurling 1980; Ballinger et al. 1991; Kutcher et al. 1993; Kuswinanti et al. 1999). Genetic studies demonstrated a number of gene-for-gene interactions between *B. napus* and *L. maculans* and both avirulence genes (*AvrLm*) in the pathogen and their corresponding resistance genes (*Rlm*) in the host have been identified. Race-specific resistance to isolates of *L. maculans* with the corresponding avirulence allele results in an incompatible interaction that inhibits infection from germinated ascospores or conidia and subsequent development of leaf lesions (Delourme et al. 2006).

3.5.16.2 Identification and Mapping of R-Genes Identified in *Brassica napus* to *Leptosphaeria*

The first race-specific resistance genes were identified in 'Quinta' and 'Glacier' cultivars in the original differential set (Rimmer and van den Berg 1992). Gene-for-gene *B. napus/L. maculans* interactions (*Rlm1/AvrLm1* in 'Quinta'-PG3; *Rlm2/AvrLm2* in 'Glacier'-PG2 interactions) were demonstrated through the use of segregating populations of both plant and pathogen (Ansan-Melayah et al. 1995, 1998). Other dominant race-specific resistance genes have been described through genetic studies involving different oilseed rape cultivars/lines and different *L. maculans* isolates (Table 3.3). Some of these genes have been positioned on *B. napus* linkage maps (Ferreira et al. 1995; Mayerhofer et al. 1997; Delourme et al. 2004; Rimmer 2006). Mapping studies showed that some of the resistance genes are organized in clusters. Zhu and Rimmer (2003) found two closely linked but distinct

Table 3.3 Identification and mapping of R-genes in *Brassica napus* genotypes to *Leptosphaeria* (Delourme et al. 2006)

B. napus cultivar/ genotypes	*L. maculans* isolate: Name and information[a]	*L. maculans* isolate: Genotype[b]	Resistance gene	Location in *B. napus* LG[c]	Reference
Quinta	11.26.11 (PG3; A2)	***AvrLm1*** *avrLm2*	Rlm1	LG10 = N7	Ansan-Melayah et al. (1998)
		avrLm3 avrLm4			
		AvrLm7 avrLm9			Delourme et al. (2004)
Maxol	v11.1.2 (PG3; A2)	***AvrLm1*** *avrLm2*	Rlm1	LG10 = N7	Balesdent et al. (2002)
		avrLm3 avrLm4			
		AvrLm7 avrLm9			Delourme et al. (2004)
Glacier	14.3.01 (PG2; A4)	*AvrLm1* ***AvrLm2***	Rlm2	LG16 = N10	Ansan-Melayah et al. (1998)
		avrLm3 avrLm4			Delourme et al. (2004)
		AvrLm7 avrLm9			
Major	PHW1245 (PG2; A3)	*AvrLm1 AvrLm2*	LEM1 = Rlm4	LG6 = N7	Ferreira et al. (1995)
		avrLm3 ***AvrLm4***			
		AvrLm7 avrLm9			
Shiralee	Canadian isolates	–	LmR1 = Rlm4?	LG6 = N7	Mayerhofer et al. (1997)
Cresor	Field population (Saskatchewan)	–	LmFr1	–	Dion et al. (1995)
Cresor	Field population (Saskatchewan)	–	aRLMc	LG6 = N7	Rimmer (2006)
Maluka	Pl86.12 (PG2; –)	–	CRLMm = Rlm4?	LG6 = N7	Rimmer (2006)
RB87–62	Pl86.12 (PG2; –)	–	CRLMrb = Rlm4?	LG6 = N7	Rimmer (2006)
			aRLMrb		Zhu and Rimmer (2003)
DH88–752	Pl86.12 (PG2; –)	–	cRLMj	LG6 = N7	Rimmer (2006)
			aRLMj		Zhu and Rimmer (2003)
Quinta	v23.2.1 (PG4; A5)	*avrLm1 avrLm2*	Rlm4	LG10 = N7	Balesdent et al. (2001)

(continued)

Table 3.3 (continued)

B. napus cultivar/ genotypes	*L. maculans* isolate		Resistance gene	Location in *B. napus* LG[c]	Reference
	Name and information[a]	Genotype[b]			
		avrLm3 ***AvrLm4***			Delourme et al. (2004)
		AvrLm7 avrLm9			
Maxol	19.2.01 (PG4; A1)	*avrLm1 avrLm2*	Rlm3	LG10 = N7	Balesdent et al. (2002)
		AvrLm3 *avrLm4*			Delourme et al. (2004)
		avrLm7 avrLm9			
23.1.1	A290 (PG4; A1)	*avrLm1 avrLm2*	Rlm7	LG10 = N7	Balesdent et al. (2002)
		avrLm3 avrLm4			Delourme et al. (2004)
		AvrLm7 *avrLm9*			
Darmor	IBCN56	*AvrLm1? AvrLm2?*	Rlm9	LG10 = N7	Balesdent et al. (2002)
		AvrLm3? avrLm4			Delourme et al. (2004)
		avrLm7 ***AvrLm9***			

[a]Pathogenicity groups are indicated as PG2-PG4 (Mengistu et al. 1991) and as A1-A6 (Badawy et al. 1991)
[b]AvrLm genes matching the Rlm genes studied are indicated in boldface
[c]Linkage groups LG10 and LG16 are from the Lombard and Delourme (2001) genetic map; LG6 is from the Ferreira et al. (1994) genetic map; LG N7 and N10 are from the Parkin et al. (1995) genetic map

loci mediating resistance at the seedling and adult plant stage, respectively, in two *B. napus* breeding lines ('RB87-62' and 'DH88-752'). These genes all mapped to Linkage Group 6 (LG6) of the genetic map published by Ferreira et al. (1994). The two resistance loci in line 'RB87-62' mapped more than 40 cM away from those in line 'DH88-752', but only 5–10 cM separated the seedling and adult plant resistance loci of each line (Rimmer 2006). Three other resistance genes (*LEM1*, *LmR1*, and *cRLMm*, present in 'Major', 'Shiralee,' and 'Maluka', respectively) have also been mapped onto this linkage group (Ferreira et al. 1995; Mayerhofer et al. 1997; Rimmer 2006). From comparing their locations on the LG6 linkage group, it seems that LmR1 is different from LEM1 but LmR1 could be identical to cRLMm since 'Shiralee' and 'Maluka' share a similar pedigree and produce similar interactions with *L. maculans* (Mayerhofer et al. 1997). Based on differential interactions with a series of *L. maculans* isolates, it seems likely that the seedling resistance genes in 'Maluka' (cRLMm) and 'RB87-62' (cRLMrb) are equivalent (Rimmer 2006). Delourme et al. (2004) have mapped five race-specific resistance genes (*Rlm1*, *Rlm3*, *Rlm4*, *Rlm7*, and *Rlm9*) on LG10 and one gene (*Rlm2*) on LG16 of the genetic map published by Lombard and Delourme (2001). Rlm1 is clearly distinct from Rlm3 and Rlm4 because they both occur in one cultivar and they map to different positions. *Rlm3* and *Rlm4* are found in many cultivars but rarely seem to be present together in a single cultivar. Similarly, Rlm3 and Rlm7 have not been found in the same cultivar. Thus, *Rlm3*, *Rlm4*, *Rlm7*, and *Rlm9* could be a cluster of tightly linked genes, or a single gene with different alleles, or a combination of both. Both LG6 of the genetic map published by Ferreira et al. (1994) and LG10 published by Lombard and Delourme (2001) seem to correspond to LG N7 of the genetic map described by Parkin et al. (1995). Thus, some of the genes described on these LG might be the same. The genes LEM1 and cRLMm are almost certainly identical to Rlm4, present in 'Major' and 'Maluka' (Rouxel et al. 2003a). Additionally, LmR1 in 'Shiralee' and cRLMrb in 'RB87-62' might also correspond to Rlm4. The different locations of LmR1 and LEM1 (Mayerhofer et al. 1997) might be due to the homeologous reciprocal translocation that can occur between LG N16 and LG N7 close to the position of LEM1 (Osborn et al. 2003). Such homeologous reciprocal translocation can affect recombination and precise mapping in this region using parents with or without the translocation. Definite conclusions on identity of or distinctness between these Rlm genes will be possible only through a precise characterization of *B. napus/L. maculans* interactions using differential *L. maculans* isolates selected or genetically bred to carry single (or as few as possible) identified avirulence (Avr) genes (Balesdent et al. 2002), through allelism tests or, in the longer term, by cloning and sequence comparison of the resistance genes. An improved host differential set comprising fixed cultivars or lines possessing a minimum number of *Rlm* genes has been developed (Balesdent et al. 2005). It consists of 'Westar' (no R-genes, susceptible control), 'Columbus' (*Rlm1-Rlm3*), 'Bristol' (*Rlm2-Rlm9*), '22-1-1' (*Rlm3*), 'Jet Neuf' (*Rlm4*), '150-2-1' (*B. juncea* line, *Rlm5*, not yet characterized at the *Rlm9* locus), 'Darmor-MX' (*Rlm6*, not yet characterized at the *Rlm9* locus), '23-1-1' (*Rlm7*), '156-2-1' (*B. rapa* line, *Rlm8*, not yet characterized at the *Rlm9* locus), and 'Goeland'

(*Rlm9*). The host genotypes carrying genes originating from *B. napus* are freely available, so that a common nomenclature can be used to simplify the identification of genes for resistance to *L. maculans* in different genotypes (Delourme et al. 2006).

3.5.16.3 R-Genes Identified in *Brassica* Species to *Leptosphaeria*

Resistance to *L. maculans* in germplasm of other Brassicaceae species related to *B. napus* has also been studied. Few resistance genes were found by screening different accessions of the two diploid progenitors of oilseed rape, *B. oleracea* (CC, $2n = 18$) and *B. rapa* (AA, $2n = 20$). An extensive screening of *B. oleracea* germplasm in the main European Gene Banks was done at "Instituto Superior de Agronomia" (ISA Lisbon). The differential isolates were BBA62908, harboring AvrLm1, AvrLm2, and AvrLm4 alleles (Rouxel et al. 2003a), and three "PG4" European isolates harboring none of these avirulence alleles. Of the 392 accessions tested, a few occasionally reacted to one of the "PG4" isolates, but none was resistant to the isolate BBA62908, suggesting the absence of *Rlm1*, *Rlm2*, or *Rlm4* in *B. oleracea* genotypes. These data, which are consistent with the data of Mithen et al. (1987) and Rimmer and van den Berg (1992), confirm that no major resistance genes to *L. maculans* originate from *B. oleracea*. However, in one closely related species, *B. insularis* ($2n = 18$), two dominant resistance genes were detected in a segregating population obtained from a *B. oleracea* × *B. insularis* hybrid (Mithen and Lewis 1988).

This screening of genetic resources also encompassed 555 accessions of *B. rapa*, including accessions of vars *chinensis*, *japonica*, *parachinensis*, *pekinensis*, *perviridis*, *rapifera*, and *trilocularis* and a few wild accessions. Most (95.5%) of these accessions were fully susceptible to all four *L. maculans* isolates. However, 12 (2%) accessions were resistant to all four isolates and ten (1.8%) accessions were resistant to isolate BBA62908 and susceptible to the three "PG4" isolates. These data suggest that the resistant accessions of *B. rapa* could harbor genes previously identified in *B. napus* such as *Rlm1*, *Rlm2*, or *Rlm4*. To test this hypothesis, limited screening was done through collaboration between IPK Gatersleben and INRA-PMDV. Sixty-two *B. rapa* var. *oleifera* accessions, a few *B. rapa* var. sylvestris accessions, and wild accessions were inoculated with differential isolates BBA62908 [race Av1-2-4-5-6-7-(8)], v11.1.1 [Av5-6-7-8], v11.1.2 [Av1-5-6-7-8], and v23.2.1 [Av4-5-6-7-8]. Twenty-two percent of the accessions were susceptible to all isolates and 48.3% of the accessions showed either a heterogeneous or a homogeneous resistance to all four isolates. Of these, four accessions have *Rlm1*, three accessions have *Rlm4*, and two accessions have both genes. The resistant accessions were investigated using a wide range of differential isolates. In at least one accession (CR1478), self-pollination of one fully resistant plant generated a line expressing the *Rlm7* resistance. Screening of progeny of another resistant accession (156.1.1) showed monogenic control by *Rlm8* interacting with the novel single-gene avirulence *AvrLm8* (Balesdent et al. 2002). In a few accessions, resistance was observed against all or most isolates tested, suggesting occurrence of undescribed major resistance genes. Dominant resistance genes were also identified in two *B. rapa* cultivars (Crouch et al. 1994; Chevre et al. 2003) and a cluster of race-specific genes, effective at the cotyledon stage, was identified in one source (Chevre et al. 2003).

3.5.16.4 Introgression of R-Genes to *Leptosphaeria* from *Brassica* Species to *Brassica napus*

The *B. rapa* genes were introduced into the *B. napus* genome either through production of a synthetic oilseed rape crossed to *B. napus* (Crouch et al. 1994) or by direct crosses between *B. napus* and *B. rapa* (Chevre et al. 2003) (Table 3.4). Genetic studies with lines obtained from the synthetic *B. napus* indicated the presence of three genes introgressed from *B. rapa* var. *sylvestris* on different *B. napus* linkage groups; *LepR1* and *LepR2* were mapped, respectively, onto *B. napus* A-genome LG N2 and LG N10 of the Parkin et al. (1995) genetic map (Yu et al. 2005), and *LepR3* was identified from new commercial cultivars Surpass 400 (Li and Cowling 2003) and Hyola 60. *LepR3* was mapped onto *B. napus* LG N10 about 15 cM below *LepR2* (Yu et al. 2004). The LG N10 is the LG where *Rlm2* mapped (Delourme et al. 2004). In seedling assays, *LepR1* behaved as a dominant allele and was resistant to all except one *L. maculans* isolates. The *LepR3* gene was described as a dominant gene (Li and Cowling 2003), whereas *LepR2* was incompletely dominant to most isolates, with the phenotype of the heterozygotes more similar to that of the susceptible parent than to that of the homozygous resistant lines. Isolates virulent on *LepR2* have been identified. Resistance conferred by *LepR3* has been overcome in some parts of Australia (Li et al. 2003; Sprague et al. 2006). Thus, these three genes are race specific. A recessive gene has also been identified in *B. napus* lines derived from *B. rapa* var. *sylvestris*. The cluster of race-specific dominant *B. rapa* resistance genes (Chevre et al. 2003) has been transferred into *B. napus* genetic backgrounds with or without polygenic resistance. This cluster was introgressed into a different *B. napus* linkage group.

The *Brassica* species with the B-genome, *B. nigra* (BB, $2n = 16$), *B. juncea* (AABB, $2n = 36$), and *B. carinata* (BBCC, $2n = 34$) have been described as highly resistant to *L. maculans* under field conditions (Rimmer and van den Berg 1992). Based on cotyledon and stem resistance ratings, Keri et al. (1997) suggested that resistance in *B. juncea* is mediated by two genes. This is consistent with genetic data obtained with *L. maculans*, which showed that the interaction was governed by two avirulence genes termed *AvrLm5* and *AvrLm6* (Balesdent et al. 2002). The corresponding resistance genes were fixed, respectively, in a *B. juncea* line originating from 'Aurea' (*Rlm5*) and in the series of introgressed *B. napus* MX lines developed at INRA Rennes (*Rlm6*) (Chevre et al. 1997; Balesdent et al. 2002, 2005). The resistance genes were introgressed into *B. napus* either by hand pollination between the donor species and *B. napus* cultivars/lines or by symmetric or asymmetric protoplast fusion (Table 3.4) and the resulting hybrids were backcrossed to *B. napus*. Whatever the screening methods used, all genes detected were dominant, except for one recessive gene introgressed from *B. juncea* (Saal et al. 2004) and three genes acting in a complex interaction (Pang and Halloran 1996a). Evaluation of different *B. napus*-*B. nigra* addition lines carrying resistance has suggested that a number of different resistance genes occur in the B-genome (Zhu et al. 1993; Chevre

Table 3.4 Introgression of R-genes from *Brassica* species into *Brassica napus* (Delourme et al. 2006)

Donor species		Resistance tests	Genetic control	References
Diploid species	*B. rapa* (AA, 2*n* = 20)	Cotyledon, leaf, field	Dominant gene	Crouch et al. (1994)
		Field	Dominant gene	Li and Cowling (2003)
		Cotyledon	Dominant genes	Chevre et al. (2003)
		Cotyledon, field	Two genes, *LepR1* and *LepR2*	Yu et al. (2005)
	B. nigra (BB, 2*n* = 16)	Leaf, stem	–	Sjodin and Glimelius (1989)[a]
		Petiole	Three additional chromosomes	Zhu et al. (1993)
		Cotyledon, field	One additional chromosome	Chevre et al. (1996)
		Petiole	Dominant gene, *PhR1*	Plieske et al. (1998)
		Leaf	Two independent dominant genes,	Dixelius (1999)[a]
			LmBR2 and *LmBR3*	
		Cotyledon, leaf	Three independent genes	Dixelius and Wahlberg (1999)[a]
		Stem, field	–	Ogbonnaya et al. (2003)
	Sinapis arvensis (SarSar, 2*n* = 18)	Cotyledon, stem	One additional chromosome	Snowdon et al. (2000)
	Arabidopsis thaliana (AtAt, 2*n* = 10)	Leaf	–	Bohman et al. (2002)
Allotetraploid species	*B. juncea* (AABB, 2*n* = 36)	Cotyledon, field	Dominant gene (s)	Roy (1978, 1984)
		Seed, cotyledon, leaf	Dominant character	Sacristan and Gerdemann (1986)
		Leaf, stem	–	Sjodin and Glimelius (1989)[a]
		Stem	Three genes with interaction	Pang and Halloran (1996a)
		Cotyledon, field	Dominant gene, JLm1	Chevre et al. (1997), Barret et al. (1998)
		Petiole	Dominant gene, PhR2	Plieske et al. (1998)

(continued)

Table 3.4 (continued)

Donor species		Resistance tests	Genetic control	References
		Cotyledon, stem	–	Winter et al. (1999)
		Leaf	Dominant gene, LmBR1	Dixelius (1999)[a]
		Cotyledon, leaf	Three independent genes	Dixelius and Wahlberg (1999)[a]
		Cotyledon	Recessive gene, $r_j$lm2	Saal et al. (2004)
	B. carinata (BBCC, 2*n* = 34)	Leaf, stem	–	Sjodin and Glimelius (1989)[a]
		Petiole	Dominant gene, PhR3	Plieske et al. (1998)
		Cotyledon, leaf	Three independent genes	Dixelius and Wahlberg (1999)[a]

– no fusion
[a]Material produced from symmetric and asymmetric protoplast fusion

et al. 1996). Resistance genes, introgressed from *B. nigra*, *B. juncea*, or *B. carinata* into the *B. napus* genome are all on the same B-genome region (Dixelius 1999). Furthermore, Plieske et al. (1998) found that resistance genes from these three species all introgressed into the same *B. napus* linkage group. In all the introgression lines obtained by sexual crosses, resistance genes from the B-genome were introgressed into A-genome linkage groups of *B. napus* (Roy 1978; Barret et al. 1998; Plieske et al. 1998). However, from their location on *B. napus* genetic maps, it seems that different genes were introgressed. This result was confirmed by the different interactions from different introgressed lines (Saal et al. 2004).

Other sources of resistance are available in less closely related species such as *Arabidopsis thaliana*, *Sinapis arvensis*, *Coincya monensis*, *Diplotaxis muralis*, *Diplotaxis tenuifolia*, or *Raphanus raphanistrum* (Chen and Seguin-Swartz 1999; Winter et al. 1999; Snowdon et al. 2000; Bohman et al. 2002). Some gene introgressions have been attempted by crosses to *B. napus* or by asymmetric protoplast fusion for *Arabidopsis* (Table 3.4). Resistant addition lines have been obtained from *B. napus*–*S. arvensis* hybrids (Snowdon et al. 2000). Bohman et al. (2002) showed that an introgression of genes carried by chromosome 3 of *A. thaliana* confers adult leaf resistance in *B. napus*.

3.5.16.5 Correlation Between Race-Specific Resistance at Seedling and Adult Stages

Comparisons between seedling (cotyledon test) and adult (petiole or stem inoculation in glasshouse or field tests) resistance screening tests have produced either significant positive (McNabb et al. 1993; Bansal et al. 1994) or nonsignificant

(Ballinger and Salisbury 1996; Pang and Halloran 1996b) correlations. These differences may be explained by differences between sources of resistance studied (conferring either race-nonspecific quantitative resistance versus race-specific resistance or a combination of both resistance types) and differences in combinations of avirulence genes between *L. maculans* isolates used in controlled environment tests and *L. maculans* populations in field tests. Another explanation is that isolates may interact with each other. Mahuku et al. (1996) reported that the weakly virulent *L. biglobosa* can induce resistance in *B. napus* to the highly virulent *L. maculans*. The effectiveness of race-specific resistance genes at growth stages later than seedlings has been clearly demonstrated. The effect of LEM1 in 'Major' was detected using a stem inoculation test (Ferreira et al. 1995). The LmFr1 gene from 'Cresor' accounted for 57–84% of the variation in resistance in a segregating doubled haploid (DH) population in field trials, depending on the year/location of the trial (Dion et al. 1995). The Rlm1 gene in 'Maxol' explained 70% of the phenotypic variation for resistance in a field trial (Delourme et al. 2004). The *Rlm7* resistance is 100% effective in France because nearly 100% of field isolates of *L. maculans* harbor *AvrLm7* (Balesdent et al. 2005). Similarly, cultivars/lines with race-specific resistance genes introgressed from *B. rapa* var. *sylvestris* (*LepR1*, *LepR2*, and *LepR3*) were highly resistant to *L. maculans* in field trials (Li and Cowling 2003; Yu et al. 2005), except for those with LepR3, which has been overcome, so that large yield losses have occurred in regions of south eastern Australia (Sprague et al. 2006). Furthermore, resistance genes introgressed into *B. napus* from *B. nigra* (Chevre et al. 1996) or *B. juncea* (Roy 1984; Chevre et al. 1997) are generally effective in field trials. Conversely, Zhu and Rimmer (2003), comparing the results of cotyledon and stem inoculation tests on lines 'RB87-62' and 'DH88-762,' concluded that distinct but linked genes were effective in each line at the seedling and adult stages. The effect of LEM1 was not detected in field trials where *L. maculans* isolates were predominantly of the same pathogenicity group (PG2) as the isolate used to identify this gene at the seedling stage. However, some isolates in the *L. maculans* field population were highly virulent on 'Major.' A difference in avirulence allele composition between *L. maculans* isolates could explain the contrasting response of 'Major' in controlled environment and field experiments (Ferreira et al. 1995) since PG2 isolates can be either virulent or avirulent on lines with resistance conferred by Rlm4 (Badawy et al. 1991). Consequently, in the field, the effect of a race-specific resistance gene will depend on the *L. maculans* population structure, i.e., on the frequency of the corresponding avirulence allele. However, the threshold frequency of the virulence allele at which the corresponding resistance gene is no longer effective in protecting the crop is not known (Brun et al. 2004).

3.5.16.6 Quantitative Resistance in *Brassica napus*

A high level of field resistance to *L. maculans* in the absence of effective race-specific resistance genes has been observed in winter European *B. napus* cultivars such as 'Jet Neuf,' which is one of the best-known sources of quantitative resistance to *L. maculans*. Cultivar Jet Neuf was widely grown all over Europe during the

1970s and 1980s and is still very resistant to *L. maculans*. The major sources of resistance used in the Australian *B. napus* breeding programs have been Japanese spring types and French winter types (Roy et al. 1983). Although Japanese lines such as 'Chikuzen' and 'Chisaya' are only moderately resistant to *L. maculans*, resistant selections from crosses between these and other lines were obtained. Two other Japanese cultivars ('Norin 20' and 'Mutu') also showed resistance and have been widely used in breeding programs (Salisbury and Wratten 1999). There is usually no difference in the development of phoma leaf spot symptoms on young plants between cultivars with quantitative resistance to *L. maculans* and cultivars without it, but later in the season stem cankers do not develop or are less severe on the cultivars with quantitative resistance than those without it. *L. maculans* can survive and reproduce on even the most resistant lines (Marcroft et al. 2004). As quantitative resistance is partial, when *L. maculans* inoculum concentrations are high, it may not prevent large yield losses (Salisbury et al. 1995; Khangura and Barbetti 2001; Marcroft et al. 2003).

Screening for quantitative resistance is primarily done by assessment of stem cankers on mature plants in field nurseries where plants have been exposed to the locally prevalent mixture of *L. maculans* races. Phoma stem canker severity is assessed using a disease index based on the extent of external and internal necrosis at the crown (stem base) of plants sampled just before harvest. Controlled environment tests for quantitative resistance using inoculation of leaves, petioles or stems with *L. maculans* have also been proposed (Newman and Bailey 1987; Kutcher et al. 1993; McNabb et al. 1993; Bansal et al. 1994; Ballinger and Salisbury 1996; Pang and Halloran 1996b). With these tests, the correct evaluation of the quantitative resistance of a *B. napus* genotype depends on the *L. maculans* isolate used. Since the effect of a race-specific resistance gene is detectable at later growth stages, *L. maculans* isolates that are virulent against any race-specific resistance gene (s) present in the genotype to be tested must be used. Similarly, a cultivar carrying a new race-specific resistance gene that is effective against all or most of the *L. maculans* isolates in a field population cannot be evaluated for quantitative resistance in that field. A controlled environment test can be done, provided a *L. maculans* isolate virulent against that particular race-specific resistance gene is used. Little information is available on the genetic control of quantitative resistance to *L. maculans*. Ferreira et al. (1995) detected two QTL, which were associated with field resistance in Manitoba, on LG12 and LG21. The genetic basis of quantitative resistance in the French winter oilseed rape 'Darmor,' derived from 'Jet Neuf,' has been studied. In the 'Darmor-bzh' × 'Yudal' cross, Pilet et al. (1998) identified a total of ten QTL for resistance, of which four were associated with decreased stem canker severity and decreased plant death in both seasons of field experiments. Analysis of progeny derived from a 'Darmor' × 'Samourai' cross, consisting of one DH population and a number of $F_{2:3}$ families identified six QTL in the DH population and four QTL in the $F_{2:3}$ families (Pilet et al. 2001). Out of a total of 16 loci detected in the four cultivars, only four QTL were common to the 'Darmor-bzh' × 'Yudal' and 'Darmor' × 'Samourai' crosses. Pilet et al. (2001) concluded that the genetic background contributes greatly to the observed QTL and that the concentration of *L. maculans* inoculum at each location is probably important in revealing QTL with small contributions to overall field resistance to *L. maculans*.

The genomic regions carrying the most consistent resistance QTL in 'Darmor' do not correspond to the two regions on LG10 and LG16 identified as carrying race-specific resistance genes to *L. maculans* (Delourme et al. 2004). The position of Rlm2 on LG16 corresponds to a QTL identified for adult plant resistance in the 'Darmor' × 'Samourai' DH population (Pilet et al. 2001). The cultivar Samourai carries both the resistance allele at this QTL and *Rlm2*. Since no French isolates of *L. maculans* carry *AvrLm2* (Rouxel et al. 2003a), two hypotheses can be proposed to explain this colocation; either the *Rlm2* gene has a residual effect at the adult plant stage, similar to that suggested in other pathosystems, or genes linked to *Rlm2* are responsible for part of the variation for resistance at this QTL (Delourme et al. 2006).

3.6 *Brassica–Pyrenopeziza*: Molecular Resistance

3.6.1 Mechanism of Resistance to Light Leaf Spot (*Pyrenopeziza brassicae*) in *Brassica*

The resistance for hemibiotrophic pathogen *Pyrenopeziza brassicae* has been introgressed into oilseed rape (*Brassica napus*). *P. brassicae* causes light leaf spot, a polycyclic disease initiated in the UK in the autumn by airborne ascospores produced following sexual reproduction of the pathogen on senescent plant debris (Gilles et al. 2000a). After infection by airborne *P. brassicae* ascospores, the fungus grows biotrophically and asymptomatically between the cuticle and the epidermal cells of the leaf, until the first symptoms of asexual sporulation are observed (white *P. brassicae* acervuli breaking through the leaf surface) (Gilles et al. 2000b). The conidia contained in the acervuli are splash dispersed to cause secondary infections of leaves, stems, meristems, and other tissues (Gilles et al. 2001). No clear hypersensitive response following infection of *B. napus* by *P. brassicae* has been reported (Boys et al. 2007). The only report of R-gene-mediated resistance was by Bradburne et al. (1999), who suggested that there were two resistance genes segregating in two mapping populations of doubled haploid (DH) lines produced following introgression of genetic material from *B. rapa* (A-genome) and *B. oleracea* var. *atlantica* (C-genome) into *B. napus* (amphidiploid AC-genome) via synthetic lines. These resistance genes were not finely mapped and the resistant phenotypes were not investigated further (Boys et al. 2011).

In controlled environment and field experiments, the sources of genetic resistance against *P. brassicae*, R-gene-mediated resistance introduced into *B. napus* slowed growth of *P. brassicae*, prevented asexual sporulation on living tissue, but did not prevent sexual sporulation on senescent tissue. The resistance did not operate in the manner typical of R-gene-mediated resistance against hemibiotrophs. *P. brassicae* infected the resistant lines but did not elicit an immediate hypersensitive response preventing further fungal growth. Instead, it grew sparsely as subcuticular hyphae within green leaves, until a "dark flecking" phenotype associated with collapse of epidermal cells was observed approximately 10 days post inoculation. This resistance may be more durable than that of a typical R-gene because it reduces secondary infection by splash-dispersed conidia but does not apply selection by preventing the pathogen from completing its life cycle (Boys et al. 2011).

3.7 *Brassica–Sclerotinia*: Molecular Resistance

3.7.1 Mapping of R-Genes of *Brassica*

Almost all the mapping work in this context has focused on *B. napus*; however, only partial resistance has been characterized in both the A- and C-genomes (Table 3.5). Zhao and Meng (2003) first identified three QTLs for leaf resistance and three other QTLs for stem resistance in the seedling and adult stages, respectively, but no common QTLs. Zhao et al. (2006) identified eight and one QTL involved in two segregating DH populations, with each explaining 6–22% of the observed variance, still with no common QTLs. Yin et al. (2010) detected ten, one, and ten QTLs in one DH population using three inoculation procedures, and only two common QTLs were detected. Wu et al. (2013) identified three QTLs at the seedling stage for leaf resistance and ten QTLs for stem resistance at the adult stage.

Two major QTLs could be detected repeatedly and a candidate resistance gene, *BnaC.IGMT5*, was first identified. These studies revealed abundant QTLs but seldom common ones, indicating the complicated genetic structure of these plants. Recently, the release of the *B. napus* genome sequence has strongly facilitated mapping work. Fomeju et al. (2014) first adopted a genome-wide association study (GWAS) using 116 materials genotyped with 3228 SNPs and the results indicated that 64 genomic regions are involved in SR resistance. Wei et al. (2015) combined GWAS and SNP array analyses using 347 *B. napus* accessions and 17 significant regions were located on the A8 and C6 chromosomes. These SNPs on Chr. A8 were placed in a 409 kb segment, with candidate genes being suggested. Using a similar method, Wu et al. (2016) genotyped 448 accessions and 26 SNPs corresponding to three loci were associated with SR resistance. In total, 39 candidates were proposed. Gyawali et al. (2016) performed a GWAS using microsatellite markers in a global collection of 152 accessions and found that 34 loci were significantly associated. To date, many loci opposing SR have been characterized but none has been functionally characterized. Considering that high resistance to SR in *B. napus* is not available, researchers tend to investigate wild *Brassica* relatives for novel germplasm, such as *Berteroa incana* and *Brassica cretica*. MAS combined with distant hybridization plays significant roles in resistance transfer. Mei et al. (2015) successfully introgressed resistance from wild *B. incana* into *B. napus* through hexaploidy hybridization and MAS using newly developed simple sequence repeat (SSR) markers and phenotype evaluation.

Table 3.5 List of candidate genes identified through GWAS (genome-wide association study) and transcriptome sequencing (Wu et al. 2016)

Loci	Candidate gene	*Arabidopsis thaliana* locus	Description
DSRC4	BnaC04g39510D	AT2G28110.1	Exostosin family protein
	BnaC04g39650D	AT2G28315.1	Nucleotide/sugar transporter family protein
	BnaC04g39750D	AT2G28400.1	Unknown protein
	BnaC04g39840D	AT2G28570.1	Unknown protein
	BnaC04g40010D	AT2G28780.1	Unknown protein
	BnaC04g40020D	AT2G28790.1	Pathogenesis-related thaumatin super family protein
	BnaC04g40170D	AT2G28950.1	Expansin A6
	BnaC04g40340D	AT2G29260.1	NAD(P)-binding Rossmann-fold super family protein
	BnaC04g40460D	AT2G29350.1	Senescence-associated gene 13
	BnaC04g40550D	AT2G29460.1	Glutathione *S*-transferase tau 4 (GSTU4)
	BnaC04g40560D	AT2G29470.1	Glutathione *S*-transferase tau 3 (GSTU3)
	BnaC04g40580D	AT2G29500.1	HSP20-like chaperones super family protein
	BnaC04g40700D	AT2G29660.1	Zinc finger (C2H2 type) family protein
	BnaC04g40820D	AT2G30040.1	MAPKKK14
	BnaC04g40850D	AT2G30050.1	Transducin family protein/WD-40 repeat family protein
	BnaC04g40890D	AT2G30110.1	Ubiquitin-activating enzyme 1
	BnaC04g40960D	AT2G30160.1	Mitochondrial substrate carrier family protein
	BnaC04g41120D	AT2G30490.1	Cinnamate-4-hydroxylase (C4H)
	BnaC04g41130D	AT2G30490.1	Cinnamate-4-hydroxylase (C4H)
DSRC6	BnaC06g24000D	AT4G09420.1	Disease resistance protein (TIR-NBS class)
	BnaC06g24010D	AT1G72840.2	Disease resistance protein (TIR-NBS-LRR class)
	BnaC06g24200D	AT1G72180.1	Leucine-rich receptor-like protein kinase family protein
	BnaC06g24220D	AT1G72170.1	Domain of unknown function (DUF543)
	BnaC06g24320D	AT1G72470.1	Exocyst subunit exo70 family protein D1 (EXO70D1)
	BnaC06g24360D	AT1G72360.3	Ethylene-responsive transcription factor (ERF73)
	BnaC06g24500D	AT1G70780.1	Unknown protein
	BnaC06g24560D	AT1G70700.1	TIFY domain/Divergent CCT motif family protein
	BnaC06g24620D	AT1G70600.1	Ribosomal protein L18e/L15 super family protein
	BnaC06g24700D	AT1G70370.2	Polygalacturonase 2 (PG2)
	BnaC06g24750D	AT1G70260.1	Nodulin MtN21/Eam A-like transporter family protein
	BnaC06g24770D	AT1G70230.1	TRICHOME BIREFRINGENCE-LIKE 27 (TBL27)
DSRC8	BnaC08g16800D	AT1G15860.2	Domain of unknown function (DUF298)

(continued)

Table 3.5 (continued)

Loci	Candidate gene	*Arabidopsis thaliana* locus	Description
	BnaC08g16820D	AT2G32060.2	Ribosomal protein L7Ae/L30e/S12e/Gadd45 family protein
	BnaC08g16850D	AT1G16030.1	Heat shock protein 70B (Hsp70b)
	BnaC08g16900D	AT1G16260.2	Wall-associated kinase family protein
	BnaC08g16960D	AT1G16610.1	Arginine/serine-rich 45 (SR45)
	BnaC08g16970D	AT1G16670.1	Protein kinase super family protein
	BnaC08g17030D	AT1G16740.1	Ribosomal protein L20
	BnaC08g17050D	AT1G16760.1	Protein kinase protein with adenine nucleotide alpha hydrolases-like domain

3.7.2 Mechanisms of Host Resistance

It is probably not reasonable to expect pronounced resistance against a fungus with such a wide host range within one of its host species or even a genus. In addition, strain specificity in regard to pathogenicity to various hosts has not been reported. The dearth of reports before 1968 indicates that many researchers formerly accepted the idea that resistance to *S. sclerotiorum* does not exist. In earlier studies, field resistance to *S. minor* has been observed in red and white clover (Aldrich 1974) and alfalfa (Elgin and Beyer 1968). Escape from *S. sclerotiorum* infection due to type of growth habit has been reported in lettuce (Newton and Sequeira 1972), sunflower (Laclerca 1973), and beans. Differences in susceptibility of cvs. breeding lines and plant introductions are noted in soybean (Grau and Bissonnette 1974), peanut (Porter et al. 1975), and sunflower (Orellana 1975). Orellana (1975) attributed tolerance of sunflower to enhanced growth and lignification of host tissues in response to long-day treatment. Kanbe et al. (1997) suggested that the growth of *S. sclerotiorum* hyphae invading resistant strains of alfalfa is inhibited due to browning of the host cells.

The physiology of *Sclerotinia* disease resistance has not been studied adequately and in fact, disease resistance among many susceptible genera of dicotyledonous plants has not been found. Monocots generally are immune or very resistant (Lumsden 1979). Three general types of resistance to *Sclerotinia* spp. have been observed. First resistance of tissue to breakdown is possibly associated with nutrition of the fungus, second presence of preformed antifungal materials and third formation of phytoalexins. In cruciferous crops, preformed materials have been found to be associated with resistance of onion and potato to *S. sclerotiorum* infection (Echandi and Walker 1957). Unidentified substances from resistant potato tissue inhibit maceration of susceptible radish, cucumber, and carrot tissues by extracts from *S. sclerotiorum* cultures. Immune onion tissue extracts completely prevent maceration. In resistant cvs. of *B. juncea*, phenolics accumulate in diseased stems accumulate at the infection site, and there is a relatively low level of enzyme activity compared to that in the susceptible cv. (Rai et al. 1979). According to Tewari and Conn (1992), the pathogenesis of *S. sclerotiorum* in rapeseed reduces due to sequestration of oxalic acid by calcium. If enough calcium is applied to chemically tie up all the oxalic acid produced by *S. sclerotiorum*, infection is not likely to take place. Constitutive overexpression of a protein involved in plant defense mechanisms to disease is one of the strategies proposed to increase plant tolerance to fungal pathogens. A hybrid endochitinase gene under a constitutive promotor has been introduced by *Agrobacterium*-mediated transformation into a winter-type oil-seed rape (*B. napus* var. *oleifera*) inbred line. When progeny from transformed plants are challenged by pathogens, plants exhibit an increased tolerance to disease as compared with the nontransgenic parental plants (Grison et al. 1996).

3.7.3 Production of Phytoalexins

S. sclerotiorum produces a somewhat selective phytotoxin "sclerin" which is phytotoxic to three cruciferous species (*Brassica napus*, *B. juncea*, and *Sinapsis alba*) susceptible to *Sclerotinia* stem rot disease, causing severe necrosis and chlorosis, but not to a resistant species (*Erucastrum gallicum*). Oleic acid, the major fatty acid isolated from sclerotia of *S. sclerotiorum*, is responsible for the toxic activity of extracts of sclerotia to brine shrimp larvae (*Artemia salina*). Phytoalexin elicitation in the leaves of *E. gallicum* results in the isolation of three known phytoalexins: indole-3-acetonitrile, arvelexin and 1-methoxyspirobrassinin. Considering that the resistance of *E. gallicum* to *S. sclerotiorum* is potentially transferable to *B. rapa*, a susceptible rape species and that arvelexin and 1-methoxyspirobrassinin are not produced by *B. rapa*. These phytoalexins may become useful markers for resistance against *S. sclerotiorum* (Pedras and Ahiahonu 2004). In soybean, Glyphosate resistant lines S20-B9 and P 93 B01 produces more phytoalexins than glyphosate susceptible S19-20 and P 9281 (Nelson et al. 2002).

3.7.4 Polygenic Architecture of Quantitative Resistance to *Sclerotinia sclerotiorum*

Plant immune response to necrotrophs is governed by a complex interplay of minor-effect genes, which results in a full continuum of resistance phenotypes in natural plant populations, designated as quantitative disease resistance (QDR) (Roux et al. 2014). The genetic determinants of QDR are complex, and the underlying genetic components can be common with, but are generally not limited to, PTI and ETI response genes (Iakovidis et al. 2016).

In the case of *S. sclerotiorum*, its host plants, such as the model plant *A. thaliana*, *B. napus* and soybean, show symptoms ranging from high susceptibility to relative tolerance to the pathogen, corresponding to a typical QDR response (Chen and Wang 2005; Liu et al. 2005; Perchepied et al. 2010), which involves allelic variation at different quantitative trait loci (QTLs) since the continuous distribution of heritable phenotypes must result from combinations of genetic loci (Corwin and Kliebenstein 2017). However, although a large body of mapping information on

QTLs is available for the QDR to the pathogen (Zhao and Meng 2003; Bert et al. 2004; Micic et al. 2004; Ronicke et al. 2005; Zhao et al. 2006; Yin et al. 2010; Behla et al. 2017; Wu et al. 2013; Wei et al. 2014), relatively little is known about the molecular basis of QTLs.

Recent research technologies have developed efficient omic tools to better understand the genetic and molecular mechanisms regarding plant QDR to *S. sclerotiorum*. Chalhoub et al. (2014) predicted a total of 181 and 245 putative NBS-LRR resistance genes on the A and C subgenomes of the *B. napus* genome through examining large-scale genomic data. Interestingly, from these genes, Li et al. (2015) found a total of 26 candidate NBS-LRR genes associated with resistance to *S. sclerotiorum* through integrating and comparing QTLs for resistance to this pathogen from previous mapping efforts. Correspondingly, a bioinformatic study revealed that the *S. sclerotiorum* genome encodes a large set of candidate effector proteins (Guyon et al. 2014). R-gene-mediated resistance commonly results in rapid cellular desiccation and death at the site of attempted infection that constitutes a hypersensitive response (HR) (Wright and Beattie 2004; Dodds and Rathjen 2010). Consistently, some researchers claimed that they observed HR-like lesions on *S. sclerotiorum*-inoculated stems (Uloth et al. 2013, 2015; Ming et al. 2016) or cotyledons (Garg et al. 2008, 2010b; Uloth et al. 2014; Ge et al. 2015). These data suggested that *R*-mediated resistance seems to exist in the interaction of *B. napus* with *S. sclerotiorum*. Considering the continuous distribution of disease resistance phenotype in host populations, a typical characterization of QDR, roles of these candidate R-genes in this resistance need to be confirmed further, because R-mediated resistance can be seen as an extreme of the phenotypic spectrum, in which the switch from susceptibility to resistance in plant populations is reduced to a minimum of detectable transition states (Roux et al. 2014).

With the development of high-throughput sequencing technology, genome-wide association study (GWAS) based on linkage disequilibrium (LD) has emerged as an important tool for identifying small-to-moderate effect loci associated with resistance to *S. sclerotiorum*. For example, based on GWAS, studies identified two QDR genes, coding for the POQR prolyl oligo peptidase and the actin-related protein complex isoform 4, respectively, for *S. sclerotiorum* in the model plant *A. thaliana* (Badet et al. 2017, 2019). However, in most cases of crops, GWAS cannot lead directly to the gene(s) at a given locus because of insufficient marker density and linkage disequilibrium. Thus, GWAS data are usually combined with other omic experiments, such as microarray study, RNA sequencing (RNA-seq), to interpret the results, which can increase the confidence in identifying candidate defense-associated (CDA) genes (Corwin and Kliebenstein 2017). To date, this approach has been employed to interpret GWAS results associated with resistance to *S. sclerotiorum* in two crops, *B. napus* and soybean. On the basis of data from these reports in the two crops, the candidate genes related with defense mechanisms (Table 3.6) and the following features in resistance to *S. sclerotiorum*. CDA genes identified by using a QTL meta-analysis also are considered (Table 3.6) as important.

Table 3.6 Candidate defense-related genes mapped by genome-wide association study (GWAS) combined with other omics experiments (Wang et al. 2019)

Tag	Group	Related role	Plant	Gene	Protein	Annotation	References
I	Recognition	Recognition of MAMPs	*Brassica napus*	*BnaC06g24200D*	A Leucine-rich receptor-like protein kinase family protein	An RLP-like kinase	Wu et al. (2016)
I	Recognition	Recognition of MAMPs	*Glycine max*	*Glyma13 g03360*	A PR5-like receptor kinase	A serine/threonine receptor kinase	Zhao et al. (2015)
I	Recognition	Recognition of MAMPs	*Phaseolus vulgaris*	*Phvul.008G173600*	A receptor-like protein	An RLP	Vasconcellos et al. (2017)
I	Recognition	Recognition of DAMPs	*Brassica napus*	*BnaC08g16900D*	A wall-associated kinase family protein	The cell wall-associated protein	Wu et al. (2016)
I	Recognition	Recognition of DAMPs	*Brassica napus*	*BnaC06g24700D*	The polygalac-turonase2 (PG2)	Cell wall modification	Wu et al. (2016)
I	Recognition	Recognition of DAMPs	*Glycine max*	*Glyma.18G116400*	A probable PG	Cell wall modification	Wei et al. (2017)
I	Recognition	Recognition of DAMPs	*Glycine max*	*Glyma.18G116401*	A cellulose synthase	Cell wall modification	Wei et al. (2017)
I	Recognition	Recognition of DAMPs	*Glycine max*	*Glyma.18G116402*	A pectate lyase	Cell wall modification	Wen et al. (2018)
I	Recognition	Recognition of DAMPs	*Phaseolus vulgaris*	*Phvul.001G236600*	A wall-associated receptor kinase protein	Recognizing cell wall changes	Vasconcellos et al. (2017)
I	Recognition	Recognition of the pathogen effectors	*Brassica napus*	*BnaC06g30610D*	A leucine-rich repeat (LRR) family protein	The R-protein	Wei et al. (2016)
I	Recognition	Recognition of the pathogen effectors	*Brassica napus*	*BnaC06g24000D*	A TIR-NBS class protein	The R-protein	Wei et al. (2016)
I	Recognition	Recognition of the pathogen effectors	*Brassica napus*	*BnaC06g24010D*	A TIR-NBS-LRR class	The R-protein	Wei et al. (2016)

I	Recognition	Recognition of the pathogen effectors	*Glycine max*	*Glyma.09G062100*	A LRR family protein	The R-protein	Wen et al. (2018)
I	Recognition	Recognition of the pathogen effectors	*Glycine max*	*Glyma.09G062100*	NB-ARC domain protein	Regulating R-protein	Wen et al. (2018)
I	Recognition	Recognition of the pathogen effectors	*Glycine max*	*Glyma.16G135200*	A NB-ARC domain protein	Regulating R-protein	Wen et al. (2018)
I	Recognition	Recognition of the pathogen effectors	*Glycine max*	*Glyma.16G135200*	A NB-ARC domain protein	Regulating R-protein	Wen et al. (2018)
I	Recognition	Recognition of the pathogen effectors	*Glycine max*	*Glyma.16G159200*	A NB-ARC domain protein	Regulating R-protein	Wen et al. (2018)
I or II	Recognition or signal transduction	?	*Brassica napus*	*BnaC08g16970D*	A Protein kinase superfamily protein	?	Wu et al. (2016)
I or II?	Recognition or signal transduction	?	*Glycine max*	*Glyma.14G049600*	A phosphatase	?	Wen et al. (2018)
II	Signal transduction	Receives the signals from PRRs	*Brassica napus*	*BnaC04g40820D*	The MAPKKK14	The MAPK cascade	Wu et al. (2016)
II	Signal transduction	Signaling	*Brassica napus*	*BnaC04g40340D*	A NAD(P)-binding Rossmann-fold superfamily protein	Systemic acquired resistance (SAR)	Wu et al. (2016)
II	Signal transduction	SA signaling	*Glycine max*	*Glyma.01G104100*	The isochorismate synthase	Synthesis of salicylic acid	Wei et al. (2017)
II	Signal transduction	JA signaling	*Glycine max*	*Glyma.16 g134400*	The carboxyl methyltransferase	JA signaling	Wen et al. (2018)

(continued)

Table 3.6 (continued)

Tag	Group	Related role	Plant	Gene	Protein	Annotation	References
II	Signal transduction	JA signaling	*Phaseolus vulgaris*	*Phvul.001G240400*	The coronatine-insensitive protein 1 (COI1)	The jasmonate receptor	Vasconcellos et al. (2017)
II	Signal transduction	ET signaling	*Brassica napus*	*BnaC06g24360D*	An ethylene-responsive transcription factor (ERF73)	ET signaling	Wu et al. (2016)
II	Signal transduction	ET signaling	*Phaseolus vulgaris*	*Phvul.002G055700*	An ERF	ET signaling	Vasconcellos et al. (2017)
II	Signal transduction	ET signaling	*Phaseolus vulgaris*	*Phvul.002G055800*	An ERF	ET signaling	Vasconcellos et al. (2017)
II	Signal transduction	ET signaling	*Phaseolus vulgaris*	*Phvul.006G183100*	An ERF	ET signaling	Vasconcellos et al. (2017)
II	Signal transduction	ET signaling	*Phaseolus vulgaris*	*Phvul.006G183200*	An ERF	ET signaling	Vasconcellos et al. (2017)
III	Defense response	The PR proteins	*Brassica napus*	*BnaC06g30470D*	The β-1,3-glucanase	The PR-2 family protein	Wei et al. (2016)
III	Defense response	The PR proteins	*Brassica napus*	*BnaC04g40020D*	A PR thaumatin superfamily protein	The PR protein	Wu et al. (2016)
III	Defense response	Secondary metabolite	*Glycine max*	*Glyma13 g04031*	The MYB domain protein 33	Controlling secondary metabolism	Zhao et al. (2015)
III	Defense response	Secondary metabolite	*Glycine max*	*Glyma.18G113400*	A putative MYB transcription factor	Controlling secondary metabolism	Wei et al. (2017)
III	Defense response	Secondary metabolite	*Phaseolus vulgaris*	*Phvul.005G115500*	A MYB domain protein	Controlling secondary metabolism	Vasconcellos et al. (2017)
III	Defense response	Secondary metabolite enzyme	*Brassica napus*	*BnaA08g19770D*	A glucosidase	The cleavage of the glucoside	Wei et al. (2016)
III	Defense response	The secondary metabolite enzyme	*Brassica napus*	*BnaC04g41120D*	A cinnamate-4-hydroxylase (C4H)	The biosynthesis of monolignols and anthocyanins	Wu et al. (2016)

III	Defense response	The secondary metabolite enzyme	*Brassica napus*	*BnaC04g41130D*	A C4H	The biosynthesis of monolignols and anthocyanins	Wu et al. (2016)
III	Defense response	The secondary metabolite enzyme	*Glycine max*	*Glyma.04G198000*	An acyltransferase	Secondary metabolism biosynthesis	Wen et al. (2018)
III	Defense response	The secondary metabolite enzyme	*Glycine max*	*Glyma.16G158100*	A UDP-glucosyl-transferase	Secondary metabolism biosynthesis	Wen et al. (2018)
III	Defense response	The secondary metabolite enzyme	*Brassica napus*	*BnaC06g37610D (BnaC.IGMT5.a)*	An indole glucosinolate methyltransferase	Secondary metabolite	Wu et al. (2013), Wei et al. (2016)
III	Defense response	Detoxification	*Glycine max*	*Glyma.16G158100*	A glucuronosyl-transferases	Detoxification mechanism	Wen et al. (2018)
III	Defense response	Detoxification	*Glycine max*	*Glyma.06G106100*	An oxalate exchanger-related (OER) protein	Detoxification of oxalic acid	Wen et al. (2018)
III	Defense response	Detoxification	*Glycine max*	*Glyma.07G218800*	A protein encoded by an OER gene that do not overlap	Detoxification of oxalic acid	Wen et al. (2018)
III	Defense response	Detoxification	*Glycine max*	*Glyma.13G087200*	A protein encoded by an OER gene that do not overlap with GWAS-identified loci	Detoxification of oxalic acid	Wen et al. (2018)
III	Defense response	Detoxification	*Glycine max*	*Glyma.19G159000*	A protein encoded by an OER gene that do not overlap with GWAS-identified loci	Detoxification of oxalic acid	Wen et al. (2018)
III	Defense response	Detoxification and oxidative protection	*Glycine max*	*Glyma.01G106000*	A tau class glutathione *S*-transferase (GST) protein	Xenobiotic detoxification, reduction, or oxidative protection	Wei et al. (2017)
III	Defense response	Oxidative protection	*Brassica napus*	*BnaC04g40550D*	The GST tau4 (GSTU4)	An antioxidant defense	Wu et al. (2016)

(continued)

Table 3.6 (continued)

Tag	Group	Related role	Plant	Gene	Protein	Annotation	References
III	Defense response	Oxidative protection	*Brassica napus*	*BnaC04g40560D*	The GSTU3	An antioxidant defense	Wu et al. (2016)
III	Defense response	Antioxidant	*Brassica napus*	*BnaC06g31020D*	A GSTU protein	An antioxidant defense	Wei et al. (2016)
III	Defense response	Antioxidant	*Brassica napus*	*BnaC06g31030D*	A GSTU protein	An antioxidant defense	Wei et al. (2016)
III	Defense response	Antioxidant	*Brassica napus*	*BnaC06g31040D*	A GSTU protein	An antioxidant defense	Wei et al. (2016)
III	Defense response	ROS production	*Phaseolus vulgaris*	*Phvul.003G164600*	The peroxidase	ROS accumulation	Vasconcellos et al. (2017)
III	Defense response	Controlling HR	*Glycine max*	*Glyma.06G107800*	The serine hydroxyl methyltransferase	Controlling HR	Wen et al. (2018)
III	Defense response	Cell cycle, cell autophagy	*Glycine max*	*Glyma13 g04020*	A member of the RINT-1/TIP-1 family	Radiation-induced checkpoint control, Golgi transport	Zhao et al. (2015)
III	Defense response	Unknown	*Brassica napus*	*BnaC04g40700D*	The Zinc finger (C2H2 type) family protein	The transcription factor	Wu et al. (2016)
III	Defense response	Unknown	*Brassica napus*	*BnaC06g30580D*	The DHHC-type zinc finger protein (ZFP)	The transcription factor	Wei et al. (2016)
–a	Unknown	Defense-associated proteins	*Brassica napus*	*BnaC06g30160D*	The β-xylosidase	Hydrolysis reaction of xyloglucan oligosaccharides	Wei et al. (2016)
–a	Unknown	Defense-associated proteins	*Glycine max*	*Glyma.11G084200*	A GRIP-like protein	Targeting the Golgi	Wei et al. (2017)
–a	Unknown	Defense-associated proteins	*Glycine max*	*Glyma.09 g281900*	A *O*-methyl-transferase	*O*-methyltransferase	Wen et al. (2018)

–a Unknown

1. Resistance to *S. sclerotiorum* is determined by minor QTLs. These phenotypic contributions of the GWAS-identified loci were low, with each locus explaining less than 10% of the observed phenotypic variance in resistance to *S. sclerotiorum*. This is supported by QTL mapping in which resistance to *S. sclerotiorum* is a trait with very complex genetic underpinnings determined by multiple minor QTLs.
2. The predominant group of genes linked to GWAS-identified loci as potential causal genes were those involved in downstream defense responses including pathogenesis-related proteins, ROS production, detoxification, oxidative protection, and secondary metabolite enzymes. This implies that QDR to *S. sclerotiorum* is a function of a variety of cellular processes and not simply pathogen detection and signal transduction.
3. Interestingly, many potential resistance (R)-genes are identified as potential causal genes by GWAS. It has been suggested that upstream signaling components, such as R-protein, in the plant pathogen response are typically encoded by medium-to-large-effect loci, which is supported by the large number of studies that investigated the quantitative genetics of wheat resistance to wheat stripe rust (Fu et al. 2009), *Arabidopsis* resistance to *Xanthomonas campestris* (Huard-Chauveau et al. 2013) and *Fusarium oxysporum* (Diener and Ausubel 2005; Shen and Diener 2013), and rice resistance to *Magnaporthe oryzae* (Ballini et al. 2008; Miah et al. 2013; Kang et al. 2016; Raboin et al. 2016). In the case of resistance to *S. sclerotiorum*, all of these potential candidate R-genes are located in small-effect loci. Is ETI response to *S. sclerotiorum*, if it exists, a quantitative trait? A recent report has shown that an ETI response in *Arabidopsis* resistance to *Pseudomonas syringae*, a hemibiotroph, is a quantitative trait, in which a single effector, HopAM1, was used to identify quantitative natural variation in the response to this effector (Iakovidis et al. 2016). Considering the identified large set of candidate effector protein from *S. sclerotiorum* genome (Guyon et al. 2014) and a potential double feeding lifestyle of *S. sclerotiorum*, it is possible to use these candidate effectors to test if there is quantitative variation in ETI response to *S. sclerotiorum* signals.
4. In the case of *B. napus*, an amphidiploid formed by interspecific hybridization of the two diploid species *B. rapa* (AA, $n = 10$) and *B. oleracea* (CC, $n = 9$), the GWAS data showed that the majority of potential causal genes for the pathogen were identified in the C-genome (C9 and C6), but not in A-genome (Table 3.6), although it has been known that putative resistance-related genes in C-genome also are observed in the syntenic region on A-genome (Mei et al. 2013). These observations suggested that *B. oleracea*, not *B. rapa*, may be a good source of QDR genes for *S. sclerotiorum*. Further, it has been reported that a few Chinese *B. oleracea* var. capitata genotypes exhibit high-level stem and leaf resistances to *S. sclerotiorum* (Mei et al. 2011; Ming et al. 2016). In contrast, there is no genome specificity of QDR genes for this pathogen in soybean.
5. Comparing with existing biparental populations, the GWAS populations use more lines and also utilize the increased number of meiotic generations to provide increased recombination and potentially increased mapping resolution (Nordborg

and Weigel 2008; Atwell et al. 2010; Alonso-Blanco et al. 2016). However, so far, the GWAS-identified loci/genes for resistance to *S sclerotiorum* in each analysis collectively explained a small portion of the phenotypic variation, and few loci/genes could be detected repeatedly in different populations for the same species (Table 3.6). This suggests that these GWASs is still largely underpowered given the number of accessions and the effect of residual population structure or adaptive genetic variation unaccounted for in these studies (Platt et al. 2010; Brachi et al. 2015). Thus, future studies with even more powerful populations are required (Wang et al. 2019).

3.7.5 Overexpression of R-Genes in *Brassica* Enhances Resistance to *Sclerotinia*

Plants are able to protect themselves from the pathogen infection through the deployment of various induced defense responses. These defense responses are dependent on a complex network of transduction pathways and are mediated by a number of signaling molecules, including salicylic acid (SA) and jasmonic acid (JA). Based on studies on the interaction of the model plant *Arabidopsis thaliana* with pathogens, a general defense model was proposed, in which SA-mediated defense response provides protection from biotrophic pathogens, whereas JA-mediated defense response is against necrotrophs. In the case of the necrotrophic fungus *S. sclerotiorum*, the results of studies on the host gene expression profiling appear to conform to the general defense model, in which *S. sclerotiorum* infection induces expression of genes associated with JA defense response in *B. napus*, while expression of genes associated with SA defense response are not induced. Further, study with *Arabidopsis* mutants showed that the impairment in SA signaling did not affect susceptibility to the pathogen. However, roles of SA signaling in defense against *S. sclerotiorum* are challenging, since a study showed that the application of a biologically active analog of SA enhanced resistance to the necrotrophic *S. sclerotiorum* in *B. napus*, suggesting a possible positive role of SA signaling in this resistance. This result from the *B. napus-S. sclerotiorum* pathosystem appears to conflict with the general defense model. However, its molecular evidences are still lacking. SA signaling is important not only in plant defense against pathogens but also in mediating a type of broad-spectrum, systemic disease resistance known as systemic acquired resistance (SAR). This signaling is dependent on the transcription co activator "nonexpresser of *PR1* genes 1" (*NPR1*), an important regulator of plant immunity (Dong 2004).

NPR1 was first cloned from *A. thaliana*. The typical characteristic of *AtNPR1* is that its protein sequence contains an N-terminal BTB/POZ domain, a central ankyrin-repeat domain and a C-terminal trans activation domain. *AtNPR1* is the receptor of SA and assist TGA transcription factors to activate the expression of *PR1*, the marker gene of SA-mediated defense response. The mutation occurred in *AtNPR1* block induction of genes related to SA defense response in *Arabidopsis* plants and, consequently, resulted in enhanced susceptibility to pathogens. Hence, *AtNPR1* is the key positive regulator of SA defense response.

In fact, researchers have used the *Arabidopsis* mutant *npr1* to investigate the role of *AtNPR1* in defense to *S. sclerotiorum*. It was reported that *npr1* mutant showed enhanced susceptibility to this pathogen, whereas another study reported that *npr1* mutant did not show increased susceptibility. These results from the *npr1* mutant are contradictory. Overexpression researches may give a new hint. However, the resistance to the pathogen is not yet assessed in *Arabidopsis* plants overexpressing *AtNPR1*. To date, some *NPR1* homologs have been cloned from various crop species, and overexpression researches of these *NPR1* homologs have been performed on many important pathosystems. For example, in *Arabidopsis* plants, overexpression of *AtNPR1* was found to be able to enhance resistance to pathogens including *P. syringae*, *H. parasitica*, and *E. cichoracearum*. The *B. juncea NPR1* homolog was cloned from this crop species and found that its overexpression in this crop confers resistance to *Alternaria brassicae* and *E. cruciferarum* (Ali et al. 2017). In *B. napus*, however, there has not been a specific report examining whether overexpression of *B. napus NPR1* (*BnaNPR1*) affects resistance against *S. sclerotiorum*, the most important pathogen of this crop.

3.7.5.1 Overexpression of NPR1 Genes in *B. napus* Confers Resistance to *Sclerotinia*

A new *NPR1* homolog (*BnaNPR1*) is cloned from *B. napus*, and its role in the crop for regulating defense response and improving disease resistance against *S. sclerotiorum* is evaluated by using the overexpression approach. These analyses allowed to (1) identify the positive role of *BnaNPR1* in resistance against *S. sclerotiorum*, and (2) provide further molecular evidence about the positive role of SA defense response in this resistance. The breeding of *S. sclerotiorum*-resistant oilseed rape cultivars using traditional methods are difficult. Engineering resistance by genetic transformation is pursued as an important strategy to control diseases caused by this devastating pathogen. Interestingly, in the induced expression experiment, the downexpression of *BnaNPR1* was observed in *B. napus* infected with *S. sclerotiorum*, which will raise the exploitation value of *BnaNPR1* overexpression. Thus, Wang et al. (2020) proposed that the strategies for utilization of *BnaNPR1* to improve resistance to *S. sclerotiorum* would be overexpression (Wang et al. 2020).

It has been established that *Arabidopsis thaliana Non-expresser of Pathogenesis-Related Genes 1* (*NPR1*), a key regulator of salicylic acid (SA) signaling, plays an important role in plant defense against pathogens. However, little is known about the *B. napus* (*Bna*) *NPR1* gene and its role in defense to *S. sclerotiorum*. Wang et al. (2020) cloned a new *NPR1* homolog (*BnaNPR1*) from *B. napus*. The new cloned *BnaNPR1* exhibits 68.35% identity with *AtNPR1* in protein level, and its expression is strongly activated by the SA treatment that, in turn, can enhance resistance to *S. sclerotiorum*. Further, transgenic *Nicotiana benthamiana* and *B. napus* overexpressing *BnaNPR1* showed significantly enhanced resistance to *S. sclerotiorum*. Further experiments showed that after *S. sclerotiorum* infection, transgenic plants activated the expression of genes associated with SA defense response but suppressed genes associated with JA signaling. These results indicated that *BnaNPR1* plays a positive role in resistance of *B. napus* against *S. sclerotiorum*, which provides molecular evidence about the positive role of SA signaling in this resistance. Interestingly, it was revealed that the induced expression of *BnaNPR1* is suppressed during the *S. sclerotiorum* infection. Therefore, Wang et al. (2020) proposed that the strategies for utilization of *BnaNPR1* to improve resistance to *S. sclerotiorum* would be over expression.

3.7.5.2 Overexpression of *OsPGIP2* Gene in *Brassica napus* Confers Resistance to *Sclerotinia sclerotiorum*

Polygalacturonase inhibitor proteins (PGIPs) seem not only to recognize different surface motifs of functionally related structurally variable PGs but also to shift the breakdown process toward generating larger peptide fragments, which are damage-associated molecular pattern (DAMP)-active oligogalacturonides (OGs) (Federici et al. 2006; Maulik et al. 2012; Kalunke et al. 2015). Transgenic *Arabidopsis thaliana* plants expressing a particular PGIP–PG chimera exhibit accumulated OGs and enhanced resistance to a variety of pathogens, providing direct evidence for the function of OGsas in vivo elicitors of plant defense responses (Benedetti et al. 2015). Furthermore, treatment of plants with exogenous OGs promotes the production of ROS, up-regulates the expression of β-1,3-glucanase, increases the transcripts of phenylalanine ammonia-lyase (PAL) to accumulate phytoalexins and lignin (D'Ovido et al. 2004; Ridley et al. 2001; Ferrari et al. 2013), and promotes changes in gene expression in the salicylic acid (SA), ethylene (ET), and jasmonic acid (JA) pathways (Dupont et al. 2006; Aziz et al. 2007; Osorio et al. 2008). Because the OG receptor WALL-ASSOCIATED KINASE 1 (WAK1) is capable of activating the kinase domain of the elongation factor Tu (EF-TU) receptor (EFR), OGs seem to be involved in PTI responses to necrotrophic pathogens (Mengiste 2012). The "PG–PGIP" interaction mechanism provides a theoretical basis for plant disease resistance (Benedetti et al. 2011; Gutierrez 2012). Studies have demonstrated that overexpressing exogenous *PGIPs* is associated with a reduction of symptoms caused by bacteria, fungal pathogens, and insects in several crops including Chinese cabbage (Wang et al. 2015; Chotechung et al. 2016). In addition to *PvPGIP2* (*Phaseolus vulgaris*) and *BnPGIP2* (Akhgari et al. 2012; Huangfu et al. 2014), only a few other *PGIP* genes cause *S. sclerotiorum* resistance when overexpressed in *B. napus*. Although there are at least 16 *PGIP* genes that are highly induced by *S. sclerotiorum* infection in *B. napus* (Hegedus et al. 2008), few commercial rapeseed varieties exhibit partial resistance (Wu et al. 2016). *Arabidopsis thaliana* lines over expressing *BnPGIP2* exhibit smaller necrotic lesions than wild type plants, but no long-term effect on *S. sclerotiorum* disease progression is observed (Bashi et al. 2013).

Wang et al. (2018) transformed the exogenous *OsPGIP2* gene and introduced it into two different rapeseed backgrounds, and then assessed the enhancement of *S. sclerotiorum* resistance in the transgenic rapeseed lines. The results indicated that the ectopic expression of *OsPGIP2* in rapeseed not only conferred resistance to *S. sclerotiorum* at both the seedling and adult stages, but also improved seed quality traits in the transgenic lines. Moreover, transient expression experiments showed that OsPGIP2 interacts with *S. sclerotiorum* SsPG3 and SsPG6 and, based on RNA sequencing analysis, it is suggested that *OsPGIP2* played an important role in the defense mechanisms in the transgenic lines. The results demonstrated that *OsPGIP2* might be an effective gene for developing *S. sclerotiorum*-resistant rapeseed varieties (Wang et al. 2018). *Oryza sativa* polygalacturonase-inhibiting protein 2 (*OsPGIP2*) was found to effectively enhanced rapeseed immunity against *S. sclerotiorum* infection. Leaf extracts of *B. napus* plants overexpressing *OsPGIP2* showed enhanced *S. sclerotiorum* resistance by delaying pathogen infection. The constitutive expression of *OsPGIP2* in rapeseed plants provided a rapid and effective defense response, which included the production of reactive oxygen species, interactions with *S. sclerotiorum* polygalacturonases (SsPG3 and SsPG6), and effects on the expression of defense genes. RNA sequencing analysis revealed that the pathogen induced many differentially expressed genes associated with pathogen recognition, redox homeostasis, mitogen-activated protein kinase signaling cascades, hormone signaling pathways, pathogen-/defense-related genes, and cell wall-related genes. The overexpression of *OsPGIP2* also led to constitutively increased cell wall cellulose and hemicellulose contents in stems without compromising seed quality (Fig. 3.10). The results demonstrate that *OsPGIP2* plays a major role in rapeseed defense mechanisms, and proposed a model for *OsPGIP2*-conferred resistance to *S. sclerotiorum* in these plants (Wang et al. 2018).

3.7.5.3 Overexpression of *Brassica napus BnWRKY33* Gene Enhances Resistance to *Sclerotinia*

Studies on various host–pathogen interactions have shown that plant WRKY transcription factors are essential for defense. For the *B. napus–S. sclerotiorum* interaction, little direct evidence has been found with regard to the biological roles of specific *WRKY* genes in host resistance. Wang et al. (2014) isolated a *B. napus WRKY* gene, *BnWRKY33*, and found that the gene is highly responsive to *S. sclerotiorum* infection. Transgenic *B. napus* plants overexpressing *BnWRKY33* showed markedly enhanced resistance to *S. sclerotiorum*, constitutive activation of the expression of *BnPR1* and *BnPDF1.2*, and inhibition of H_2O_2 accumulation in response to pathogen infection. Further, they isolated a mitogen-activated protein (MAP) kinase substrate gene, *BnMKS1*, and found that not only can BnWRKY33 interact with BnMKS1, which can also interact with BnMPK4, using the yeast two-hybrid assay, consistent with their collective nuclear localization, but also BnWRKY33, BnMKS1, and BnMPK4 are substantially and synergistically expressed in response to *S. sclerotiorum* infection. In contrast, the three genes showed differential expression in response to phytohormone treatments. Together, these results suggest that *BnWRKY33* plays an important role in *B. napus* defense to *S. sclerotiorum*, which is most probably associated with the activation of the salicylic acid (SA)- and jasmonic acid (JA)-mediated defense response and inhibition of H_2O_2 accumulation, and have a potential mechanism in which BnMPK4–BnMKS1–BnWRKY33 exist in a nuclear localized complex to regulate resistance to *S. sclerotiorum* in oilseed rape (Wang et al. 2014).

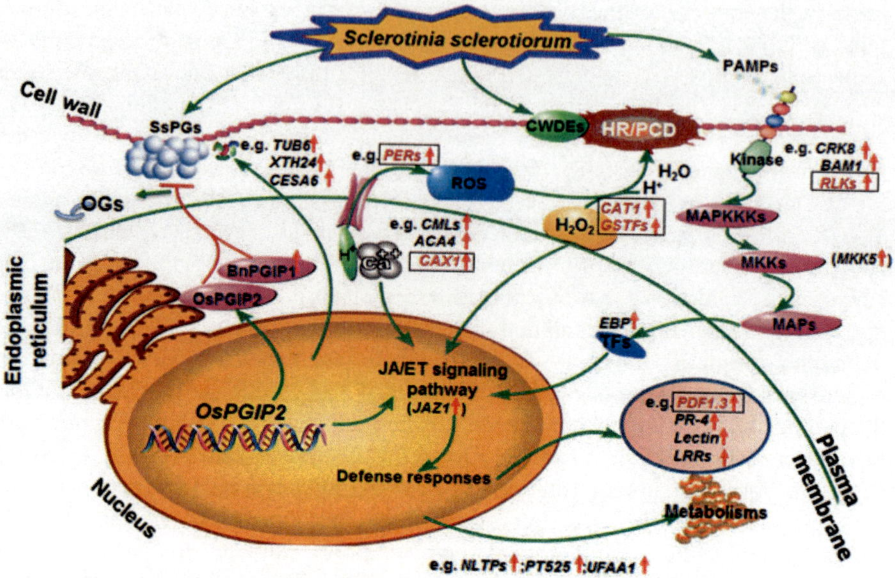

Fig. 3.10 A model for the molecular mechanism by which *OsPGIP2* confers *Sclerotinia sclerotiorum* resistance through increased activation of defense mechanisms. Upward arrows next to gene names highlight the upregulation of differentially expressed genes, as revealed in RNA-seq analysis. Genes highlighted in boxes are the common upregulated DEGs in the 7-5 and T45 backgrounds. *CWDEs* cell wall-degrading enzymes, *HR/PCD*, hypersensitive response/ programmed cell death, *MAPs* mitogen-activated proteins, *MAPKKKs* MAP-kinase-kinase kinases, *MKKs* MAP-kinase kinases, *OGs* oligogalacturonides, *PAMPs* pathogen-associated molecular patterns, *PGIP* polygalacturonase inhibitor protein, *ROS* reactive oxygen species, *SsPGs S. sclerotiorum* polygalacturonases, *TFs* transcription factors (Wang et al. 2018)

3.7.5.4 Overexpression of *Brassica napus MPK4* Gene Enhances Resistance to *Sclerotinia*

Studies on the *Arabidopsis thaliana* MPK4 loss-of-function mutant have implicated that *AtMPK4* is involved in plant defense regulation, and its effect on disease resistance varies in different plant–pathogen interactions. Wang et al. (2009) isolated a *B. napus* mitogen-activated protein kinase, *BnMPK4*, and found that *BnMPK4* along with *PDF1.2* are inducible in resistant line Zhongshuang9 but both are consistently suppressed in susceptible line 84039 after inoculation with *S. sclerotiorum*. Transgenic oilseed rape overexpressing *BnMPK4* markedly enhances resistance to *S. sclerotiorum* and *Botrytis cinerea*. Further experiments showed that transgenic plants inhibited growth of *S. sclerotiorum* and constitutively activated *PDF1.2* but decreased H_2O_2 production and constitutively suppressed *PR-1* expression. Treatment of roots of the transgenic plants with H_2O_2 solution resulted in enhanced susceptibility to the two pathogens. The results support the idea that *MPK4* positively regulates jasmonic acid-mediated defense response, which might play an important role in resistance to *S. sclerotiorum* in oilseed rape.

3.7.5.5 Overexpression of *AtWRKY28* and *AtWRKY75* Genes in *Arabidopsis* Enhances Resistance to Oxalic Acid and *Sclerotinia*

Based on *Arabidopsis* microarray, Chen et al. (2013) found eight WRKY genes were upregulated with oxalic acid (OA) challenge; *AtWRKY28* and *AtWRKY75* overexpression lines showed enhanced resistance to OA and *Sclerotinia sclerotiorum*. The WRKY transcription factors are involved in various plant physiological processes and most remarkably in coping with diverse biotic and abiotic stresses. Oxalic acid (OA) is an important pathogenicity-determinant of necrotrophic phytopathogenic fungi, such as *Sclerotinia sclerotiorum* and *Botrytis cinerea*. The identification of differentially expressed genes under OA stress should facilitate our understanding of the pathogenesis mechanism of OA-producing fungi in host plants, and the mechanism of how plants respond to OA and pathogen infection. *Arabidopsis* oligo microarray studies revealed that eight WRKY genes were upregulated upon OA challenge. The *Arabidopsis* plants overexpressing *AtWRKY28* and *AtWRK75* showed enhanced resistance to OA and *S. sclerotiorum* simultaneously. Furthermore, results showed that overexpression of *AtWRKY28* and *AtWRK75* induced oxidative burst in host plants, which suppressed the hyphal growth of *S. sclerotiorum*, and consequently inhibited fungal infection. Gene expression profiling indicates that both *AtWRKY28* and *AtWRKY75* are transcriptional regulators of salicylic acid (SA)- and jasmonic acid/ethylene (JA/ET)-dependent defense signaling pathways, AtWRKY28 and *AtWRKY75* mainly active JA/ET pathway to defend *Arabidopsis* against *S. sclerotiorum* and oxalic acid stress (Chen et al. 2013).

3.7.6 Microsatellite Markers for Genome-Wide Association Mapping of Partial Resistance in *Brassica napus* to *Sclerotinia*

Plant breeders in China were the first to develop partially resistant varieties, such as Zhongyou 821 and Zhongshuang 9 (Li et al. 1999; Wang et al. 2004). Quantitative resistance loci (QTL) were subsequently mapped in several Chinese cultivars, and breeding lines (Zhao and Meng 2003; Zhao et al. 2006; Yin et al. 2010; Wu et al. 2013), one European cultivar (Wei et al. 2014), and a wild *Brassica* species *B. incana* (Mei et al. 2013) all demonstrating that *Sclerotinia* resistance is a complex trait. However, the level of resistance was relatively low and prompted researchers to screen more *B. napus* varieties, breeding lines and land races (Zhao et al. 2004; Bradley et al. 2006), other *Brassicas* (Li et al. 2007; Mei et al. 2011), and crucifer species (Uloth et al. 2013). Introgression of resistance from distantly related species into oilseed rape has been undertaken (Garg et al. 2010a; Navabi et al. 2010; Ding et al. 2013), but is hampered by linkage drag of undesirable traits, varying levels of sterility, and occasionally insurmountable crossing barriers.

Mapping of *Sclerotinia* resistance loci has been carried out in populations derived from a cross between susceptible and resistant parents. However, marker–trait associations can also be identified in a collection of unrelated lines using the genome-wide association mapping (GWAM) approach. This method exploits the nonrandom association between phenotypic traits and molecular markers in genetically diverse germplasm in which random recombination events have accumulated over millennia (Flint-Garcia et al. 2003; Rafalski 2010). In *B. napus*, GWAM has so far been used to map glucosinolate content (Hasan et al. 2008), oil content (Zou et al. 2010), phenolic compounds (Rezaeizad et al. 2011), and several seed traits (Honsdorf et al. 2010). Jestin et al. (2011) used association mapping to identify loci conferring quantitative resistance to the fungal pathogen *Leptosphaeria maculans* causing blackleg in oilseed rape. GWAM was employed to map resistance to *Sclerotinia* in *B. napus*, sunflower, and soybean (Wei et al. 2015; Fusari et al. 2012; Bastien et al. 2014). Wei et al. (2015) were able to identify three loci associated with resistance in primarily Chinese *B. napus* lines using a disease screening method that involved wounding of the plant tissue. Gyawali et al. (2016) set out to screen more diverse collection of *B. napus* germplasm from various parts of the world using an inoculation method that resembles natural infection. The objectives were to identify new sources of *Sclerotinia* resistance and to utilize the GWAM approach at various stringencies to demarcate a larger set of molecular markers contributing to resistance.

In preparation for genome-wide association mapping (GWAM) of *Sclerotinia* resistance in *B. napus*, 152 accessions from diverse geographical regions were screened with a single Canadian isolate, #321. Plants were inoculated by attaching mycelium plugs to the main stem at full flower. Lesion lengths measured 7, 14, and 21 days after inoculation were used to calculate the area under the disease progress curve (AUDPC). Depth of penetration was noted and used to calculate percent soft and collapsed lesions ($\% \ s + c$). The two disease traits were highly correlated ($R = 0.93$). Partially resistant accessions (AUDPC < 7 and $\% \ s + c < 2$) were identified primarily from South Korea and Japan with a few from Pakistan, China, and Europe. Genotyping of accessions with 84 simple sequence repeat markers provided 690 polymorphic loci for GWAM. The general linear model in TASSEL best fitted the data when adjusted for population structure (STRUCTURE), GLM + Q. After correction for positive false discovery rate, 34 loci were significantly associated with both disease traits of which 21 alleles contributed to resistance, while the remaining enhanced susceptibility. The phenotypic variation explained by the loci ranged from 6% to 25%. Five loci mapped to published quantitative trait loci conferring *Sclerotinia* resistance in Chinese lines (Gyawali et al. 2016).

3.7.7 Marker–Trait Association for Resistance to *Sclerotinia* in *Brassica juncea–Erucastrum cardaminoides* Introgression Lines

A set of 96 *Brassica juncea–Erucastrum cardaminoides* introgression lines (ILs) were developed with genomic regions associated with *Sclerotinia* stem rot (*Sclerotinia sclerotiorum*) resistance from a wild Brassicaceous species *E. cardaminoides*. ILs was assessed for their resistance responses to stem inoculation with *S. sclerotiorum*, over three crop seasons. Initially, ILs were genotyped with transferable SSR markers and subsequently through genotyping by sequencing. SSR-based association mapping identified six marker loci associated to resistance in both A- and B-genomes. Subsequent genome-wide association analysis (GWAS) of 84 ILs recognized a large number of SNPs associated to resistance, in chromosomes A03, A06, and B03. Chromosomes A03 and A06 harbored the maximum number of resistance-related SNPs. Annotation of linked genomic regions highlighted an array of resistance mechanisms in terms of signal transduction pathways, hypersensitive responses, and production of antifungal proteins and metabolites. Of major importance was the clustering of SNPs, encoding multiple resistance genes on small regions spanning approximately 885 kb region on chromosome A03 and 74 kb on B03. Five SNPs on chromosome A03 (6,390,210-381) were associated with LRR-RLK (receptor-like kinases) genes that encode LRR protein kinase family proteins. Genetic factors associated with pathogen-associated molecular patterns (PAMPs) and effector triggered immunity (ETI) were predicted on chromosome A03, exhibiting 11 SNPs (6,274,763-994). These belonged to three R-genes encoding TIR-NBS-LRR proteins. Marker–trait associations (MTAs) identified will facilitate marker-assisted introgression of these critical resistances, into new cultivars of *B. juncea* initially and, subsequently, into other crop *Brassica* species (Rana et al. 2019).

3.7.8 Genome-Wide Association Identifies New Loci for *Sclerotinia* Rot Resistance in *Brassica napus*

The breeding for SSR resistance is challenging, as no immune germplasm has been identified in *B. napus* or its close relatives. Nevertheless, breeding practices and genetic studies have repeatedly demonstrated that resistance performance between various *B. napus* genotypes differs dramatically and that quantitative resistance is the most important form of SSR resistance breeding in *B. napus* (Ge et al. 2012; Taylor et al. 2015; Wei et al. 2015). To understand the genetic basis of quantitative resistance, several studies have performed quantitative trait locus (QTL) mapping using biparental populations, derived normally from crosses between a partially resistant parent and a susceptible parent (Wu et al. 2013; Wei et al. 2014). A number of QTLs for SSR resistance have been mapped based on these studies, and conserved QTLs have been identified on chromosomes A9 and C6 through integration analyses of these QTLs based on *B. napus* genome sequences (Li et al. 2015). Despite these successes, biparental QTL mapping suffers from two fundamental limitations: first, only allelic diversity that segregates between the parents can be assayed, and second, the limited number of recombination events in the biparental population places a limit on the mapping resolution (Korte and Farlow 2013). As a complement to the detection of QTLs, genome-wide association studies (GWAS) have been implemented to overcome the two main limitations of biparental QTL mapping approaches described above. Moreover, their center lease of *B. napus* genome sequences (Chalhoub et al. 2014), together with *Brassica* single-nucleotide polymorphism (SNP) array technologies, provide an unprecedented opportunity to conduct GWAS in *B. napus*. Wei et al. (2015) conducted GWAS on SSR resistance in 347 *B. napus* accessions and identified 17 significant associations on A8 and C6, five of which were located on A8 and 12 on C6. The broad-sense heritability of stem resistance on the population was 84%, but the two loci on A8 and C6 explained only a small part of the observed phenotypic variation (Wei et al. 2015). Considering the complexity of the genetic underpinnings of SSR resistance, continuing efforts are required to identify more significant loci/genes through GWAS.

GWAS for SSR resistance was performed in *B. napus* using an association panel with 448 accessions, which were genotyped using *Brassica* 60K *Infinium* R SNP arrays (Liu et al. 2016). The resistance performance of the panel was investigated in two consecutive years using detached stem inoculation assays. A total of 26 SNPs corresponding to three loci were identified through GWAS. Candidate genes for three loci were predicted based on the differentially expressed genes identified through recent transcriptomics analyses (Wei et al. 2015; Wu et al. 2016).

Breeding for SSR resistance in *B. napus*, as in other crops, relies only on germplasms with quantitative resistance genes. A better understanding of the genetic basis for SSR resistance in *B. napus* thus holds promise for the genetic improvement of disease resistance. A genome-wide association study (GWAS) for SSR resistance in *B. napus* was performed using an association panel of 448 accessions genotyped with the *Brassica* 60K *Infinium* R single-nucleotide polymorphism (SNP) array. A total of 26 SNPs corresponding to three loci, *DSRC4*, *DSRC6*, and *DSRC8*, were associated with SSR resistance (Table 3.5). Haplotype analysis showed that the three favorable alleles for SSR resistance exhibited cumulative effects. After aligning SSR resistance, quantitative trait loci (QTLs) were identified. In studies to the *B. napus* reference genome, one locus (*DSRC6*) was found to be located within the confidence interval of a QTL identified in previous QTL mapping studies and another two loci (*DSRC4* and *DSRC8*) were considered novel loci for SSR resistance. A total of 39 candidate genes were predicted for the three loci based on the GWAS combining with the differentially expressed genes identified in previous transcriptomics analyses (Wu et al. 2016).

3.8 *Brassica–Verticillium*: Molecular Resistance

3.8.1 Molecular Mechanisms of Resistance Against *Verticillium longisporum* in *Brassica napus*

Brassica napus has a comparatively narrow gene pool which originated in a limited geographic region of southern Europe through spontaneous hybridizations between a restricted number of turnip rape (*B. rapa*) and cabbage (*B. oleracea*) genotypes (Allender and King 2010). Genetic diversity in elite breeding material has been further eroded by continuous selection for seed quality traits since the 1970s, in particular through introgression of double-low (00) seed quality (zero erucic acid, low glucosinolate content) essentially from two varieties as donors of superior quality (Allender and King 2010). As a consequence, 00-quality winter oilseed rape has a relatively low genetic diversity and lacks a broad spectrum of disease resistances. Extensive screening of diverse *B. napus* germplasm for resistance to *V. longisporum* under controlled greenhouse conditions revealed no resistance sources (Happstadius et al. 2003), and current European winter oilseed rape cultivars exhibit only low levels of tolerance against *V. longisporum*. On the other hand, *V. longisporum* resistance has been successfully transferred from the two progenitor species into resynthesized *B. napus* lines via the embryo rescue technique (Happstadius et al. 2003; Rygulla et al. 2007a, b). The quantitative resistance in these lines is predominantly derived from the *B. oleracea* C-genome donor. The genetic basis of resistance to the systemic spread of *V. longisporum* in *B. napus* and other *Brassica* species is unknown. A proteome expression study identified proteins upregulated upon *V. longisporum* infection in roots, hypocotyls, and leaves, including the typical pathogen defense-related enzymes b-1, 3-glucanase, peroxidase, PR4, and endochitinase (Floerl et al. 2008). Histochemical analysis of *B. napus* and *B. oleracea* genotypes with different levels of resistance revealed that the resistance response to systemic *V. longisporum* spread is localized internally within the hypocotyls, and that the buildup of physical barriers, deposition of cell wall bound phenolic compounds, and lignin modification within the vascular system play a crucial role in the defense reaction (Eynck et al. 2009). Similarly, various structural barriers that restrict invading hyphae of *V. dahliae* and other *Verticillium* species have been described in roots and stems of cotton, potato, and tomato (Klosterman et al. 2009; Xu et al. 2011).

A number of studies characterizing the interaction of *Verticillium* species have been performed in the model crucifer *Arabidopsis thaliana*, a close relative of *Brassica* crops. In contrast, there are only limited reports on the genetic basis of resistance to *Verticillium* ssp. in crop species (Klosterman et al. 2009). For example, the gene Ve1 conferring resistance to *V. dahliae* and *V. albo-atrum* race 1 has been identified in tomato (*Solanum lycopersicum*). Ve1 encodes a cell surface receptor that belongs to the extracellular leucine-rich repeat (LRR) class of receptor-like proteins. *Arabidopsis* plants transformed with tomato Ve1 showed resistance to *V. dahliae* and *V. albo-atrum* race 1, but not to *V. longisporum* (Fradin et al. 2011). In *A. thaliana* a single dominant locus on chromosome IV, VET1, was found to confer resistance to *Verticillium* infection (Veronese et al. 2003). Haffner et al. (2010) found that genes involved in flowering control on *A. thaliana* chromosome IV are localized within quantitative trait loci (QTLs) for resistance to systemic spread, although further studies are required to determine if these regions coincide with VET1. Also nearby are nucleotide-binding LRR protein genes of the RPP5 superfamily, with toll/interleukin-1 receptor domains controlling resistance to the fungal pathogen *H. parasitica* (Haffner et al. 2010).

In different *A. thaliana* ecotypes, high infection rates are not always correlated with severe disease symptoms such as stunting which can be observed in *B. napus* after greenhouse infections (Floerl et al. 2010; Veronese et al. 2003; Haffner et al. 2010). Analyses of mutants deficient in hormone signaling suggest that the jasmonic acid and ethylene signaling pathways, but not salicylic acid signaling, are involved in resistance activation in *A. thaliana* (Fradin et al. 2011; Pantelides et al. 2010). In contrast, however, salicylic acid and its glucoside were detected in xylem sap from *B. napus* roots and hypocotyls, and increased in the shoots above the hypocotyls after infection by *V. longisporum*, whereas jasmonic acid and abscisic acid levels remained unchanged (Ratzinger et al. 2009). Since *A. thaliana* is not naturally infected by *V. longisporum*, however, the *A. thaliana–V. longisporum* pathosystem has limitations for simplistic transfer of information on putative resistance mechanisms to *B. napus*. It is not known whether different races or pathotypes of *V. longisporum* exist. Nevertheless, genetic mapping in a doubled haploid (DH) population, with resistance derived from a resynthesized *B. napus* line originating from a cross between white cabbage (*B. oleracea* ssp. *oleracea* convar. *capitata*) and a winter turnip rape (*B. rapa* ssp. *oleifera*) as genome donors, revealed two major QTL for *V. longisporum* resistance on chromosomes C4 and C5. After infection with either a mixture of five Swedish *V. longisporum* isolates or a single-spore isolate, respectively, these two QTL consistently exhibited significant effects on resistance in multiple greenhouse environments (Rygulla et al. 2008). Obermeier et al. (2013) identified genomic regions involved in *V. longispourm* disease development and resistance in different *B. napus* genetic backgrounds. To achieve this, a QTL mapping approach was applied to identify QTL that are stable across environments and breeding populations, and to develop markers to pyramid complementary resistance loci from resynthesized *B. napus* and commercial oilseed rape lines. The QTL analyses of phenolic compounds in hypocotyl tissues were performed to determine their relevance in the *V. longisporum* disease reaction and resistance in *B. napus* (Obermeier et al. 2013).

A genetic approach was used to identify quantitative trait loci (QTLs) for *V. longisporum* resistance and metabolic traits potentially influencing resistance in a *B. napus* mapping population. Resistance to *V. longisporum* was mapped in a doubled haploid (DH) population from a cross between the partially resistant winter oilseed rape variety Express 617 and a resistant resynthesized *B. napus* line, R53. One major resistance QTL contributed by R53 was identified on chromosome C5, while a further, minor QTL contributed by Express 617 was detected on chromosome C1. Markers flanking the QTL also significantly correlated with *V. longisporum* resistance in four further DH populations derived from crosses between elite oilseed rape cultivars and other resynthesized *B. napus* lines originating from genetically and geographically diverse *Brassica* A- and C-genome donors. The tightly-linked markers developed enable the combination of favorable alleles for novel resistance loci from resynthesized *B. napus* materials with existing resistance loci from commercial breeding lines. HPLC analysis of hypocotyls from infected DH lines revealed that concentrations of a number of phenylpropanoids were correlated with *V. longisporum* resistance. QTL for some of these phenylpropanoids were also found to co-localize with the QTL for *V. longisporum* resistance (Figs. 3.11 and 3.12). Genes from the phenylpropanoid pathway are suggested as candidates for *V. longisporum* resistance (Obermeier et al. 2013).

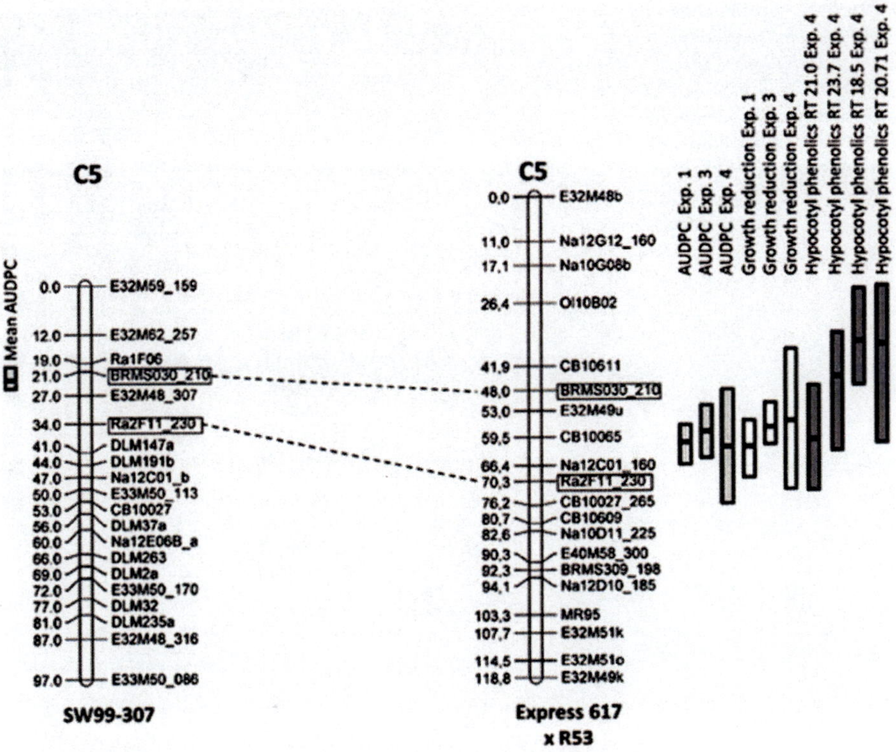

Fig. 3.11 Comparison of quantitative trait loci for *Verticillium longisporum* resistance-related traits and soluble phenolic metabolites in the hypocotyl localized on chromosome C5. Blocks indicate confidence intervals of the QTL. Data for the population SW99-307 are from Rygulla et al. (2008). Marker alleles BRMS030_210 and Ra2F11_230 are designated BRMS030 and Ra2F11_b, respectively, in Rygulla et al. (2008). *Exp* experiment (Obermeier et al. 2013)

3.9 *Brassica–Xanthomonas*: Molecular Resistance

3.9.1 Molecular Markers and Mapping of R-Genes of *Brassica*

Durable resistance to black rot can be achieved using the genes present on the A- and B-genomes of *Brassica* species, by accumulating strong race-specific genes in the genetic background of cultivars containing race-nonspecific genes through advanced molecular breeding methods. The advent of molecular marker approaches has accelerated crop-breeding research and is the most popular technique among plant breeders in the twenty-first century. DNA markers are widely exploited for assessing the genetic diversity in collections of germplasm resources, allelic variation in the genes governing target traits, molecular mapping of genes associated with target traits, identification of QTL, and further exploiting the linked markers in marker-

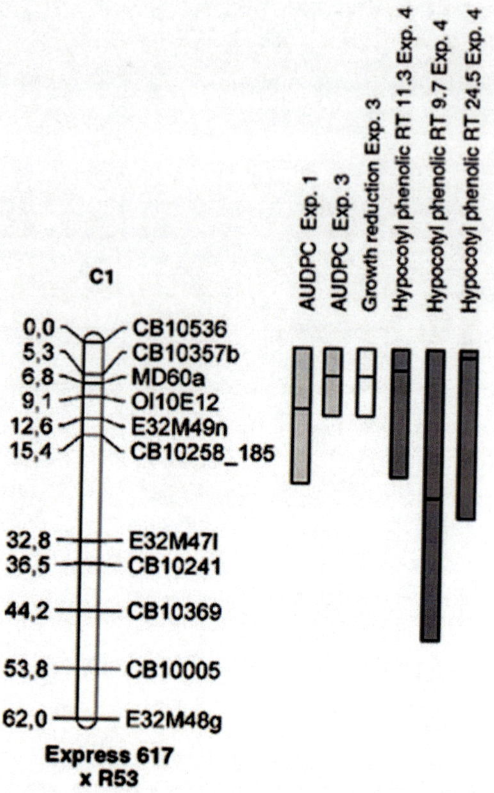

Fig. 3.12 Comparison of quantitative trait loci for *Verticillium longisporum* resistance and soluble phenolic metabolites in the hypocotyl localized on chromosome C1. Blocks indicate confidence intervals of the QTL. *Exp* experiment (Obermeier et al. 2013)

assisted selection (MAS) of desirables genes in breeding programs (Collard and Mackill 2008; Jiang 2015). The identification of closely linked DNA markers to the target traits/genes/QTLs has accelerated crop breeding programs via different molecular marker approaches, viz., marker-assisted selection (MAS), marker-assisted backcrossing (MABC), marker-assisted recurrent selection (MARS), marker-assisted gene pyramiding, marker-assisted pedigree selection (MAPS), and combined marker-assisted selection (Ye and Smith 2008; Ribaut et al. 2010; Ragimekula et al. 2013).

The identification of closely linked markers to resistance could be useful for early selection of plants in breeding programs, thereby enhancing black rot resistance breeding. In this regard, the use of BSA (bulk segregant analysis) in conjunction with PCR-based DNA markers and next-generation sequencing techniques (NGS) has proven to be a very efficient technique for identification of closely linked markers cosegregating with genes underlying monogenic traits (Song et al. 2017; Dong et al. 2018). Furthermore, the identification of closely linked markers to the disease resistance QTL can permit analysis of the consistency of QTLs effects across a wide range of environments and genetic background, and enhance the frequency of favorable alleles during selection (Camargo et al. 1995). Several R-genes/QTL and cosegregating DNA markers for resistance to Xcc have been reported in cruciferous crops on different linkage groups (Table 3.7) (Soengas et al. 2007; Kaur et al. 2009b; Doullah et al. 2011; Kifuji et al. 2013; Tonu et al. 2013; Saha et al. 2014; Lee et al. 2015a, b; Sharma et al. 2016; Singh et al. 2018).

For black rot resistance, there are reports on identification of RAPD markers linked to Xcc resistance locus. Ignatov et al. (2000) found one RAPD marker "WE (22)980" linked to locus resistance to race 4 of black rot in *B. rapa*. Using the F2 mapping populations from a cross between the susceptible cauliflower cultivar 'Snow Ball' and resistant line 11B-1-12, eight amplified polymorphic RAPD linked to disease resistant plants were reported. The segregation of these linked markers with resistance to Xcc suggested that resistance to black rot race 4 is governed by a single dominant gene (Tonguc et al. 2003). Similarly, in cabbage (*B. oleracea* var. *capitata* L.) one RAPD marker 'C-111000' which flanked the resistance gene at a distance of 3.1 cM was found using the mapping population of 200 F2 plants (January King 9 Golden Acre), where January King was the resistant parent (Kaur et al. 2009a). Owing to the dominant nature of RAPD markers and to increase the reproducibility, they are being converted to sequence characterized amplified region (SCAR) markers. Saha et al. (2014) in bulk segregant analysis in resistant and susceptible bulk of F2 progeny of cauliflower genetic sources revealed seven differentiating polymorphic markers (three RAPD, two ISSR, and two SSR) out of 102 markers screened via BSA in resistant and susceptible bulks of F2 progeny. Subsequently, the genotyping of an entire F2 population was carried out using these polymorphic markers, and a genetic linkage map covering 74.7 cM distance was constructed. The major locus *Xca1bo* was mapped in a 1.6 cM interval flanked by the markers RAPD 04_{833} and ISSR 11_{635}. The Xca1bo locus was located on chromosome 3. The linked markers can be successfully utilized for marker-assisted resistance breeding to black rot disease in cauliflower (Saha et al. 2014).

Polygenic resistance to black rot was reported in Chinese cabbage accession B162 and was confirmed with observed broad spectrum resistance to isolates of Xcc in B162 (Taylor et al. 2002). Soengas et al. (2007) studied the genetics of broad-spectrum resistance in *B. rapa* accession 'B162,' using the analysis of QTL resistance to race 1 and 4 of the black rot bacteria. A linkage map of *B. rapa* comprising ten linkage groups with a total map distance of 664 cM was developed, utilizing an F2 mapping population of 114 plants resulting from a cross between the *B. rapa* susceptible inbred line R-o-18 and B162. Correlation was found for resistance to both the races 1 and 4 and a highly significant QTL cluster was located on linkage group A06. On the linkage groups A02 and A09, two additional QTL for race 4 resistance were found. Markers closely linked to these QTL could assist in the transfer of the resistance into different *B. rapa* or *B. oleracea* cultivars.

The third-generation molecular markers, single-nucleotide polymorphism (SNP) markers, have been widely used. The advent of next-generation sequencing (NGS) technology and low-cost genome sequencing have led to the identification of a large number of SNPs to design arrays for major crops (Kifuji et al. 2013; Chen et al. 2013; Lee et al. 2015a, b). SNPs have proved to be most abundant and universal markers depicting genetic variation even among the individuals of same species.

Table 3.7 The black rot resistance genes and QTLs in different *Brassica* crops (Singh et al. 2018)

Gene locus/QTL	Chr/ LG	Source/ accession	Xcc races	Marker name	Sequence (5′–3′)	Mapping distance (cM)	Reference (s)
Xca1bo	III	BR161 (*B. oleracea*).	1	RAPD04	ACGTAGCGTC	1.6	Saha et al. (2014)
				ISSR11	GTGAGAGAGAGAGAGAYT	1.6	
QTL-1	C2	*B. oleracea*	1	BoCL5989 s	TCGGTGAGTACCATCTCTTGGA		Kifuji et al. (2013)
				BoCL5545 s	TACGCGGTTCAAGTGATGAACT		
XccR1d-1, XccR1i-1	A06	B162 (*B. rapa*)	1	E11M50_280b (AFLP)	EcoR1 (E? AA), Mse1 (M? CAT)	20.5	Soengas et al. (2007)
				E12M48_171r (AFLP)	EcoR1 (E? AC), Mse1 (M? CAC)	15.7	
XccR4d-1, XccR4i-1	A06	B162 (*B. rapa*)	4	E12M61_215b (AFLP)	EcoR1 (E? AC), Mse1 (M? CTG)	33.5	Soengas et al. (2007)
				E12M61_215b (AFLP)	EcoR1 (E? AC), Mse1 (M? CTG)	33.7	
XccR4i-2	A02	B162 (*B. rapa*)	4	E11M59_178r (AFLP)	EcoR1 (E? AA), Mse1 (M? CTA)	27.6	Soengas et al. (2007)
XccR4i-3	A09	*B. rapa*	4	E12M48_1 >330b (AFLP)	EcoR1 (E? AC), Mse1 (M? CAC)	27.2	Soengas et al. (2007)
BRQTL-C1_1	C1	*B. oleracea*		BnGMS301	AATATGCAGCATTCTAGACAAA	2.8	Lee et al. (2015a, b)
BRQTL-C1_2	C1	L. var. capitata		BoESSR089	ATGATCAGCGAAACCACTCC	28.1	
BRQTL-C3	C3			B041F06-2	–	7.6	
BRQTL-C6	C6			Ol10-G06	–	9.5	

XccBo (Reiho)1	C5	*B. oleracea*	1	BoGMS1330 (SSR)	AGGAGAAGAAGGAAGATACCA	90.7	Tonu et al. (2013)
XccBo (Reiho)2	C8			BoGMS0971 (SSR)	TAATCCGAACAACACGAA	58.0	
XccBo (GC)1	C9			CB10459 (SSR)	CCTGCTTTTGCTCTGTTC	81.0	
Xca1bc	B7	*B. carinata*	1	At1g70610 (ILP)	TGGGTTATCTTCGCTGCGTT	6.2	Sharma et al. (2016)
		A. braun		At1g71865 (ILP)	TTGCGTCTCCAGATCTCAA	12.8	
				Na14-G02 (SSR)	TTCCCTTTATTGAGCAAGCTG	30.1	
QTLs between CAM1–GSA1, F12-R12e–BORED	LG2 and LG9	*B. oleracea*	–	CAPS, SRAP	–		Doullah et al. (2011)
Xcc resistant gene		*B. oleracea* (January King)	–	C-111000 RAPD	AAAGCTGCGG	3.1	Kaur et al. (2009a, b)

Kifuji et al. (2013) used newly developed EST-SNP markers for analysis of QTLs for black rot resistance in a resistant cabbage line 'Early Fuji' by determining EST sequences. They detected three QTLs, i.e., QTL-1, QTL-2, and QTL-3, on linkage groups C2, C4, and C5, respectively, and among these, QTL-1 was found to be a major QTL. The identified SNP markers in QTL-1 are considered useful for MAS for *Xcc* resistance in *B. oleracea* cultivars. In another study, SNPs were validated and derived cleaved amplified polymorphic sequence (dCAPS) markers for black rot resistance in cabbage were developed. After performing QTL analysis, three significant QTL, viz., BRQTL-C1_1 and BRQTL-C1_2 on chromosome C01 and BRQTL-C3 on chromosome C03, were detected. Among these, BRQTL-C1_2 had the highest LOD score, additive effect, and variance explained. Thus, the identified markers and QTL are useful for marker-assisted breeding of *Brassica* crops (Lee et al. 2015a, b). Sharma et al. (2016) reported the identification of ILP (Intron length polymorphism) and microsatellites markers linked to the Xca1bc locus in *B. carinata* (BBCC) on linkage group B-7 against resistance to black rot race 1. Thus, the information regarding genetics of race-specific resistance, location of resistant genes, and identification of linked molecular markers provides vital assistance in the development of resistant varieties and DNA markers for marker-assisted selection (MAS) and introgression of resistance genes to the cultivars of Brassica crops via marker-assisted back cross breeding and other advanced molecular tools (Singh et al. 2018).

3.9.2 NBS-LRR-Encoding Genes for Black Rot Resistance in *Brassica*

To resist the pathogen infection, plants have self-developed innate immune system comprising multilayered network of defense proteins against pathogen invasion. Among these layered networks, the effector-triggered immunity (ETI) is mediated inside the cell encoded by plants disease resistance genes (R-genes). Various classes of R-genes have been recognized according to domain organization (Sagi et al. 2017), and the most abundant class of R-genes is those belonging to NBS-LRR (nucleotide-binding site and carboxyl/C-terminal leucine-rich repeat domain) family (Mun et al. 2009; Lee et al. 2015a, b; Sagi et al. 2017). These NBS-LRR genes are widely distributed in plants. NBS-LRR genes are intracellular receptors that can recognize the pathogen invasion by binding to pathogen effector proteins in host plant and trigger various defense signal transductions to yield hypersensitive response and inhibit pathogen development. The NBS domain binds and hydrolyzes ATP and GTP, while LRR domain is liable for protein–protein interactions (Afrin et al. 2018). The NBS domains are indulged in signal transduction and consist of various motifs, like P-loop, kinase-2 motifs (Marone et al. 2013). NBS-LRR genes are further grouped into two subclasses on the basis of presence or absence of amino/N-terminal domains, and the first subfamily is TIR-NBS-LRR (TNL), which consists of proteins carrying *Drosophila* toll and interleukin1-like receptor (TIR) domain at N-terminal. The second subfamily is CC-NBS-LRR (CNL), which encompasses coiled-coil domain proteins (Sagi et al. 2017; Afrin et al. 2018). Many NBS-LRR

Table 3.8 NBS-LRR-encoding genes linked to black rot resistance in different *Brassica* species (Singh et al. 2018)

QTL region	Crop	Genes	Chr/ LG	Reference(s)
BRQTL-C1_1	Cabbage	Bo1g056920	C01	Parkin et al. (2014), Lee et al. (2015a, b)
BRQTL-C1_2	Cabbage	Bo1g057060/070	C01	Parkin et al. (2014), Lee et al. (2015a, b)
		Bo1g086130		
		Bo1g087610		
		Bo1g091560		
		Bo1g094680/710[a]		
		Bo1g103860		
BRQTL-C3	Cabbage	Bo3g060060/070/080/100/110/130/140[a]	C03	Parkin et al. (2014), Lee et al. (2015a, b)
BRQTL-C6	Cabbage	Bo6g025490	C06	Parkin et al. (2014), Lee et al. (2015a, b)
		Bo6g031330/350/360/380		
	Cabbage	Bol003711	C01	Afrin et al. (2018)
		Bol010559	C08	
		Bol029866	C03	
		Bol031422	C08	
		Bol040045	C06	
		Bol042095	C07	
		Bol042121	C07	

genes have been identified in *Brassica* species for the defense against different pathogens affecting *Brassica* crops (Lee et al. 2015a, b; Table 3.8). The eight NBS-LRR-encoding genes have been reported displaying BRQTL-C1_1 and BRQTL-C1_2 QTLs and seven and five NBS-LRR-encoding genes within 1 Mb region of marker closely associated near the BRQTL-C3 region and BRQTL-C6 region, respectively. Afrin et al. (2018) identified seven NBS-encoding genes linked to black rot resistance in *B. oleracea* var. *capitata* L (Table 3.8). Further analysis of these NBS-encoding genes will pave the way for new insights into mechanisms of resistance to black rot in *Brassica* crops (Singh et al. 2018).

3.9.3 Expression of MicroRNAs in *Brassica* Enhances Resistance to *Xanthomonas*

The molecular processes involved in the ability of Xcc to infect the plant have been explored by different "omic" techniques, which have contributed to a better understanding of the plant–pathogen interaction. Although plant–*X. campestris* interaction has been studied by transcriptomic and proteomic approaches, the analysis of small RNAs, such as microRNAs (miRNAs) approach is also useful. miRNAs are a class of endogenous sRNAs of 21–23 nucleotides, which have been associated to various biological processes including development, hormone signaling, flowering, and responses to abiotic and biotic stresses. The abundance of miRNAs is tissue specific and associated with the developmental stage of the organism. As well-conserved molecules, miRNAs are involved in many molecular interaction networks, including mechanisms necessary to establish plant–pathogen interaction. Many studies about the role of miRNAs in plant–pathogen interaction suggest that miRNAs, such as miR167 and miR390, finely tune and actively participate in plant self-defense responses against various pathogens. Studies in *Arabidopsis thaliana* suggest an miRNA-mediated modulation of multiple plant hormone pathways, regulating stress-responsive genes in response to the bacterium *Pseudomonas syringae*. Both miR156 and miR169, characterized as modulators of transcription factors involved in many physiological processes, have also been described as differentially expressed in some pathosystems. In transgenic assays, miR156 expression seems to be upregulated in *A. thaliana* infected with *TYMV p69* virus while repressed in rust fungus-infected *Pinus taeda*.

Different methods for miRNA expression quantification are still being improved, especially when dealing with nonmodel species and regarding the choice of the most appropriate normalizing RNA, which is crucial in RT-qPCR experiments. A recent study of *Citrus–Xanthomonas* interaction revealed that stem-loop RT-qPCR is an efficient method in the detection of conserved miRNAs in nonmodel plants (Alizadeh et al. 2017). Perez-Quintero et al. (2012) analyzed the interaction between *Manihot esculenta* and *Xanthomonas axonopodis* pv. *manihotis* (*Xam*) and identified 68 miRNA families, most of which were differentially expressed in infection conditions suggesting that they may play an important role in plant defense against *Xam*. Santos et al. (2019) chose miR156, miR167, miR169, miR390, miR771, and miR823 that were upregulated in response to *Xam* to verify their expression profile in the interaction between susceptible and resistant *B. oleracea* plants infected with *X. campestris* pv. *campestris*. MicroRNAs (miRNAs) play an important role in plant defense modulation, and therefore, the analysis of these molecules can help better understand plant–pathogen interactions. Santos et al. (2019) observed the differential expression of four miRNAs that seem to participate in the plant response to Xcc infection. Northern Blot and RT-qPCR techniques were used to measure miRNA expression in resistant (Uniao) and susceptible (Kenzan) cultivars. From six miRNAs analyzed, four were detected and differentially expressed, showing a down- and upregulated expression profile in susceptible and resistant cultivars, respectively. These results suggest that miR156, miR167, miR169, and miR390 could play a role in *B. oleracea* resistance enhancement against Xcc and could be explored as potential resistance markers in *B. oleracea*-Xcc interaction (Santos et al. 2019).

3.9.4 Identification of Quantitative Resistance in *Brassica oleracea* to *Xanthomonas*

The search of sources of resistance is complicated due to the existence of nine races of the pathogen. Initially, identified six races (1–6), and later added three additional ones (7–9). Race identification was based on avirulence/virulence patterns to six differential host genotypes, involving five resistance genes in the plant and five avirulence genes in the pathogen. Worldwide, race 1 is one of the most virulent and widespread races in *B. oleracea* crops, accounting for more than 90% of black rot disease around the world together with race 4. Following extensive screening of *B. oleracea* accessions concluded that resistance to races 1 and 4 was either nonexistent or very rare, whereas resistance to less important races (2, 3 and 6) was frequently found. Dominant R-gene-mediated effector-triggered immunity (ETI) is considered to be the most efficient form of resistance in plants. R-genes encode proteins by recognizing directly or indirectly corresponding effectors, mostly followed by hypersensitive response. R-gene conferring resistance to race 1 (R1) is present in the B-genomes of *B. carinata* (BBCC), *B. juncea* (AABB), and *B. nigra* (BB) (Taylor et al. 2002; Vicente et al. 2001). However, this gene is absent in *B. oleracea* (CC). Although, resistance of cauliflower line SN455 has been reported to be determined by a recessive allele of a single gene. Black rot resistance of most *B. oleracea* lines is considered to be under quantitative control, because ETI fails to provide durable and broad-spectrum resistance. Increasing attention has been focused to quantitative resistance. Few R-genes are responsible for quantitative disease resistance. On the contrary, identified genes accounting for quantitative resistance represent a broad range of molecular functions, including, for example, kinases and ABC transporters (Debieu et al. 2016).

Secondary metabolites can also be involved in quantitative resistance. *B. oleracea* is known for its characteristic and high content of glucosinolates (GSLs) and phenolic compounds. Upon cellular disruption, GSLs are hydrolyzed to various bioactive breakdown products, by the endogenous enzymes myrosinases, which have been proved to be toxic to pathogens. The accumulation of GSLs in transgenic *Arabidopsis* plants enhanced resistance to bacteria. Besides, GSLs may play a role in plant defense against *Xcc* in *B. oleracea*. There is a potent role of phenolic compounds in disease resistance. Methanolic extracts from *B. rapa*, containing GSLs and phenolic compounds, inhibited the growth of *Xcc*. The genetic and metabolic basis of phenolics and GSLs accumulation was dissected through analysis of total phenolics concentration, and its individual components in the Bol TBDH doubled haploid mapping population of *B. oleracea*. QTLs that had an effect on phenolics and GSLs concentration were integrated, resulting in 33 consensus QTLs controlling phenolics traits and 18 QTLs controlling GSLs (Francisco et al. 2016; Sotelo et al. 2014).

Race 1 of *Xanthomonas campestris* pv. *campestris* is the most virulent and widespread in *B. oleracea*, and resistance is under quantitative control. Knowledge about the genetics of this resistance would help in designing strategies to control initial stages of invasion and development of the disease. QTL analysis of the resistance in the Bol TBDH mapping population was performed. Resistance was measured with five traits related to initial stages of the invasion, success of infection, and spread of the pathogen. Four single-trait QTLs of resistance were found, from which one represent novel variation. After performing multitrait QTL, Iglesias-Bernabe et al. (2019) concluded that spread of *Xcc* is related to the size of the leaf. Individuals from the mapping population follow two different strategies to cope with the spread of the disease: reducing lesion size or maintain more area of the leaf photosynthetically active, being more tolerant to *Xcc* invasion (Table 3.9). Mechanisms underlying variation for resistance may be related to different aspects of plant immunity, including the synthesis of glucosinolates and phenolics (Iglesias-Bernabe et al. 2019).

3.10 Molecular Bases for Assessment of Breakdown of R-Genes in *Brassica*

The *AvrLm6* gene, which was adjacent to a single-copy noncoding sequence at the 3′ end, had six different RIP alleles conferring virulence. The degree (intensity) of RIP mutation in these single-copy sequences was proportional to the proximity of flanking repetitive DNA that had undergone RIP mutation. The potential leakage of RIP mutations into closely linked genes highlights the power of RIP to lead to major evolutionary changes to genes such as effectors that play an important role in fungal living strategies. The breakdown of "sylvestris" resistance was associated with an eightfold increase in frequencies of isolates lacking *AvrLm1*. Although this gene was embedded in repetitive DNA, no RIP alleles were identified. The frequencies of virulence alleles (both deletion and RIP mutation) of *AvrLm6* increased sixfold, even though cultivars with the complementary resistance gene, *Rlm6*, had not been sown on Eyre Peninsula (Van de Wouw et al. 2010a). The close linkage and genomic location of *AvrLm1* and *AvrLm6* might have led to a selective sweep, whereby selection at *AvrLm1* affected the frequency of alleles of *AvrLm6* through hitchhiking (Barton et al. 2013). Thus, strong selection imposed by widespread deployment of one resistance gene also may lead to breakdown of resistance conferred by another gene, if the two complementary avirulence genes are closely linked. Clearly the genomic environment, as well as extent of exposure to resistance genes in *B. napus* cultivars, affects evolution of avirulence effectors in this fungal pathogen.

Monitoring changes in virulence frequencies in Australian populations of the blackleg fungus is armed with knowledge about the field biology and genomic analysis of *L. maculans*. The monitoring of changes in frequency of virulence of fungal populations across Australia is with the aim of preventing another breakdown of disease resistance. Since 2009, cultivars with different complements of resistance genes sown in 32 trial sites covering *Brassica* growing regions of Australia have been assessed for disease severity. Resistance genes in canola cultivars are identified by the use of 12 differential *L. maculans* isolates. On this basis of responses (susceptible or resistant) to these isolates, seven resistance groups (A–G) that include

Table 3.9 Characteristics of single-trait QTLs for resistance traits and leaf size found in the BolTBDH mapping population inoculated with race 1 of *Xanthomonas campestris* pv. *campestris* (*Xcc*) in two consecutive inoculations (Iglesias-Bernabe et al. 2019)

QTL	Linkage group	Position cM	Trait	Profile LOD	LOD threshold	Additive effect	partR2	QTLs of resistance in other populations
QTLs related to resistance								
Inoculation 1								
Xcc1.1	1	76 (46–81)	INDEX 0	3.91	2.34	0.898	12.66	BRQTL-CL-1, BRQTL-CL-2 (1)
Xcc6.1	6	23 (9–48)	DIS AREA	2.70	2.16	0.389	7.89	BRQTL-C6 (1)
Xcc8.1	8	44 (37–53)	DIS AREA	3.73	2.16	0.423	9.69	XccBo (Reiho)2 (2)
Inoculation 2								
Xcc9.1	9	2 (0–10)	%DIS AREA	3.87	2.28	−3.310	12.76	
QTLs related to leaf size								
Area6.1	6	38 (32–47)	TOT AREA	7.66	2.23	7.363	25.74	
Area8.1	8	41 (28–59)	TOT AREA	2.23	2.23	3.109	5.68	
Area9.1	9	9 (3–17)	TOT AREA	4.72	2.23	4.796	14.28	
Inoculation 2								
a6.2	6	63 (53–73)	TOT AREA	3.14	2.29	3.751	9.17	
Area9.2	9	5 (0–15)	TOT AREA	6.01	2.29	6.528	21.57	

Shown are those overlapping QTLs of resistance to *Xcc* found in other mapping populations. (1), (2) INDEX0 (disease index based on a visual 1–9 rating scale taken 6 days after inoculation), DIS AREA (diseased leaf area measured in cm2), % DISAREA (percentage of diseased leaf area), TOT AREA (total leaf area measured in cm^2)

Table 3.10 The performance and heritability of clubroot resistance of the natural population in *Brassica napus* (Li et al. 2016)

Environment	Trait	Minimum	Maximum	Mean	SE	h^2 (%)
Infected field	DI	9.3	75.0	31.39	5.92	64.8
	IR	18.2	100.0	48.11	8.70	55.0
Greenhouse	DI	36.1	100.0	84.77	9.67	78.2
	IR	58.3	100.0	96.49	7.64	69.4

DI disease index, *IR* incidence rate (%)
h^2, broad-sense heritability

a range of resistance genes have been defined (Marcroft et al. 2012). Stubble from each site represents local fungal populations, which can be analyzed in high-throughput laboratory assays to quantify regional frequencies of alleles of avirulence genes. In these assays blackleg-infested stubble is wetted and placed in a wind tunnel. This triggers release of ascospores, which are deposited onto tape from which ascospore DNA is extracted and analyzed. The quantitative polymerase chain reaction (qPCR) detects the frequency of alleles of particular avirulence genes (*AvrLm1*, *AvrLm6*), where virulence is due to deletion, while PCR and pyrosequencing detects single-nucleotide base-pair changes where virulence is due to a particular point mutation (*AvrLm4*) (Van de Wouw and Howlett 2012; Van de Wouw et al. 2010b). The total number of ascospores is estimated by qPCR of the internal transcribed spacer region of the ribosomal DNA, a genomic region present in all isolates. This is the first example of a high-throughput molecular assay that can distinguish genotypes of airborne spores. This type of assay can be applied to other diseases that involve airborne inocula and where the genetic basis of virulence in the pathogen has been identified. Analysis of data on changes over time in disease severity of cultivars, and in frequencies of avirulence/virulence alleles in populations released from canola stubble, allows predictions of risk of disease outbreaks in different geographic regions. If an epidemic is predicted, farmers are advised to plant a different canola cultivar with a different complement of resistance genes (Table 3.10) (Howlett et al. 2015).

References

Afrin KS, Rahim MA, Park JI, Natarajan S, Kim H-T, Nou IS (2018) Identification of NBS-encoding genes linked to black rot resistance in cabbage (*Brassica oleracea* var. *capitata*). Mol Biol Rep 45:773–785

Akhgari AB, Motallebi M, Zamani MR (2012) Bean polygalacturonase-inhibiting protein expressed in transgenic *Brassica napus* inhibits polygalacturonase from its fungal pathogen *Rhizoctonia solani*. Plant Prot Sci 48:1–9

Aldrich DTA (1974) The Legume-a reappraisal of its place in today's farming. J Br Grassland Soc 24:249

Ali S, Mir ZA, Tyagi A, Mehari H, Meena RP, Bhat JA, Yadav P, Papalou P, Rawat S, Grover A (2017) Overexpression of *NPR1* in *Brassica juncea* confers broad spectrum resistance to fungal pathogens. Front Plant Sci 8:1693. https://doi.org/10.3389/fpls.2017.01693

Alizadeh M, Askari H, Najafabadi MS (2017) Temporal expression of three conserved putative microRNAs in response of Citrus × Limon to *Xanthomonas citri* subsp. *citri* and *Xanthomonas fuscans* subsp. *aurantifolii*. Biotechnologia 98(3):257–264

Allender CJ, King GJ (2010) Origins of the amphiploid species *Brassica napus* L. investigated by chloroplast and nuclear molecular markers. BMC Plant Biol 10:54

Alonso-Blanco C, Andrade J, Becker C, Bemm F, Bergelson J, Borgwardt K, Cao J, Chae E, Dezwaan TM, Ding W, Ecker JR, Exposito-Alonso M, Farlow A, Fitz J, Gan X, Grimm DG, Hancock AM, Henz SR, Holm S, Horton M, Jarsulic M, Kerstetter RA, Korte A, Korte P, Lanz C, Lee CR, Meng D, Michael TP, Mott R, Muliyati NW, Nagele T, Nagler M, Nizhynska V, Nordborg M, Novikova PY, Pico FX, Platzer A, Rabanal FA, Rodriguez A, Rowan BA, Salome PA, Schmid KJ, Schmitz RJ, Seren U, Sperone FG, Sudkamp M, Svardal H, Tanzer MM, Todd D, Volchenboum SL, Wang C, Wang G, Wang X, Weckwerth W, Weigel D, Zhou X (2016) 1,135 genomes reveal the global pattern of polymorphism in *Arabidopsis thaliana*. Cell 166(2):481–491. https://doi.org/10.1016/j.cell.2016.05.063

Ansan-Melayah D, Balesdent MH, Buee M, Rouxel T (1995) Genetic characterization of *AvrLm1*, the first avirulence gene of Leptosphaeria maculans. Phytopathology 85:1525–1529

Ansan-Melayah D, Balesdent MH, Delourme R, Pilet ML, Tanguy X, Renard M, Roulx T (1998) Genes for race-specific resistance against blackleg disease in *Brassica napus* L. Plant Breed 117:373–378. https://doi.org/10.1111/j.1439-05231998.tb01956.x

Aranzana MJ, Kim S, Zhao KY, Bakker E, Horton M, Jakob K, Lister C, Molitor J, Shindo C, Tang CL, Toomajian C, Traw B, Zheng HG, Bergelson J, Dean C, Marjoram P, Nordborg M (2005) Genome-wide association mapping in *Arabidopsis* identifies previously known flowering time and pathogen resistance genes. PLoS Genet 1:531–539

Atwell S, Huang YS, Vilhjálmsson BJ, Willems G, Horton M, Li Y, Meng D, Platt A, Tarone AM, Hu TT, Jiang R, Muliyati NW, Zhang X, Amer MA, Baxter I, Brachi B, Chory J, Dean C, Debieu M, de Meaux J, Ecker JR, Faure N, Kniskern JM, Jones JD, Michael T, Nemri A, Roux F, Salt DE, Tang C, Todesco M, Traw MB, Weigel D, Marjoram P, Borevitz JO, Bergelson J, Nordborg M (2010) Genome-wide association study of 107 phenotypes in *Arabidopsis thaliana* inbred lines. Nature 465(7298):627–631

Aziz A, Gauthier A, Bézier A, Poinssot B, Joubert JM, Pugin A, Heyraud A, Baillieul F (2007) Elicitor and resistance-inducing activities of beta-1,4 cellodextrins in grapevine, comparison with beta-1,3 glucans and alpha-1,4 oligogalacturonides. J Exp Bot 58:1463–1472

Badawy HMA, Hoppe HH, Koch E (1991) Differential reactions between the genus *Brassica* and aggressive single spore isolates of *Leptosphaeria maculans*. J Phytopathol 131:109–119

Badet T, Voisin D, Mbengue M, Barascud M, Sucher J, Sadon P, Balague C, Roby D, Raffaele S (2017) Parallel evolution of the POQR prolyl oligo peptidase gene conferring plant quantitative disease resistance. PLoS Genet 13(12):e1007143. https://doi.org/10.1371/journal.pgen.1007143

Badet T, Léger O, Barascud M, Voisin D, Sadon P, Vincent R, Le Ru A, Balagué C, Roby D, Raffaele S (2019) Expression polymorphism at the ARPC4 locus links the actin cytoskeleton with quantitative disease resistance to *Sclerotinia sclerotiorum* in *Arabidopsis thaliana*. New Phytol 222(1):480–496. https://doi.org/10.1111/nph.15580

Balesdent MH, Attard A, Ansan-Melayah D, Delourme R, Renard M, Rouxel T (2001) Genetic control and host range of avirulence toward *Brassica napus* cultivars Quinta and Jet Neuf in *Leptosphaeria maculans*. Phytopathology 91:70–76

Balesdent MH, Attard A, Kuhn ML, Rouxel T (2002) New avirulence genes in the phytopathogenic fungus *Leptosphaeria maculans*. Phytopathology 92:1122–1133. https://doi.org/10.1094/PHYTO.2002.92.10.1122

Balesdent MH, Barbetti M, Hua L, Sivasithamparam K, Gout L, Rouxel T (2005) Analysis of Leptosphaeria maculans race structure in a worldwide collection of isolates. Phytopathology 95:1061–1071

Balesdent MH, Fudal I, Ollivier B, Bally P, Grandaubert J, Eber F, Chèvre AM, Leflon M, Rouxel T (2013) The dispensable chromosome of *Leptosphaeria maculans* shelters an effector gene conferring avirulence towards *Brassica rapa*. New Phytol 198:887–898. https://doi.org/10.1111/nph.12178

Ballinger DJ, Salisbury PA (1996) Seedling and adult plant evaluation of race variability in *Leptosphaeria maculans* on *Brassica* species in Australia. Aust J Exp Agric 36:485–488

Ballinger DJ, Salisbury PA, Kadkol GP (1991) Race variability in *Leptosphaeria maculans* and the implication for resistance breeding in Australia. In: Proc. 8th Intl Rapeseed Congr, Saskatoon, Canada, vol 1, pp 226–231

Ballini E, Morel JB, Droc G, Price A, Courtois B, Notteghem JL, Tharreau D (2008) A genome-wide meta-analysis of rice blast resistance genes and quantitative trait loci provides new insights into partial and complete resistance. Mol Plant-Microbe Interact 21:859–868

Bansal VK, Kharbanda PD, Stringam GR, Thiagarajah MR, Tewari JP (1994) A comparison of greenhouse and field screening methods for blackleg resistance in doubled haploid lines of *Brassica napus*. Plant Dis 78:276–281

Barret P, Guerif J, Reynoird JP, Delourme R, Eber F, Renard M, Chevre AM (1998) Selection of stable *Brassica napus–Brassica juncea* recombinant lines resistant to blackleg (*Leptosphaeria maculans*). 2. A 'to and fro' strategy to localise and characterise interspecific introgressions on the *B. napus* genome. Theor Appl Genet 96:1097–1103

Barton NH, Etheridge AM, Kelleher J, Veber A (2013) Genetic hitchhiking in spatially extended populations. Theor Pop Biol 87:757–789. https://doi.org/10.1016/j.tpb.2012.12.001

Bashi ZD, Rimmer SR, Khachatourians GG, Hegedus DD (2013) *Brassica napus* polygalacturonase inhibitor proteins inhibit *Sclerotinia sclerotiorum* polygalacturonase enzymatic and necrotizing activities and delay symptoms in transgenic plants. Can J Microbiol 59:79–86

Bastien M, Sonah H, Belzile F (2014) Genome wide association mapping of *Sclerotinia sclerotiorum* resistance in soybean with genotyping-by-sequencing approach. Plant Genome 7:1. https://doi.org/10.3835/plantgenome2013.10.0030

Becker MG, Zhang X, Walker PL, Wan JC, Millar JL, Khan D, Granger MJ, Cavers JD, Chan AC, Fernando DW, Belmonte MF (2017) Transcriptome analysis of the Brassica napus-Leptosphaeria maculans pathosystem identifies receptor, signaling and structural genes underlying plant resistance. Plant J 90:573–586

Becker MG, Haddadi P, Wan J, Adam L, Walker P, Larkan NJ, Daayf F, Borhan MH, Belmonte MF (2019) Transcriptome analysis of *Rlm2*-mediated host immunity in the *Brassica napus–Leptosphaeria maculans* pathosystem. Mol Plant-Microbe Interact 32(8):1001–1012. https://doi.org/10.1094/MPMI-01-19-0028-R

Behla R, Hirani AH, Zelmer CD, Yu F, Fernando WGD, McVetty P, Li G (2017) Identification of common QTL for resistance to *Sclerotinia sclerotiorum* in three doubled haploid populations of *Brassica napus (L.)*. Euphytica 213(11):260. https://doi.org/10.1007/s10681-017-2047-5

Benedetti M, Bastianelli E, Salvi G, Lorenzo GD, Caprari C (2011) Artificial evolution corrects a repulsive amino acid in polygalacturonase inhibiting proteins (PGIPs). J Plant Pathol 93:89–95

Benedetti M, Pontiggia D, Raggi S, Cheng Z, Scaloni F, Ferrari S, Ausubel FM, Cervone F, De Lorenzo G (2015) Plant immunity triggered by engineered *in vivo* release of oligogalacturonides, damage-associated molecular patterns. Proc Natl Acad Sci U S A 112:5533–5538

Bert PF, Dechamp-Guillaume G, Serre F, Jouan I, de Labrouhe DT, Nicolas P, Vear F (2004) Comparative genetic analysis of quantitative traits in sunflower (*Helianthus annuus L.*) 3. Characterization of QTL involved in resistance to *Sclerotinia sclerotiorum* and *Phoma macdonaldi*. Theor Appl Genet 109(4):865–874. https://doi.org/10.1007/s00122-004-1701-1

Birker D, Heidrich K, Takahara H, Narusaka M, Deslandes L, Narusaka Y, Reymond M, Parker JE, O'Connell R (2009) A locus conferring resistance to *Colletotrichum higginsianum* is shared by four geographically distinct *Arabidopsis* accessions. Plant J 60:602–613

Bohman S, Wang M, Dixelius C (2002) Arabidopsis thaliana-derived resistance against *Leptosphaeria maculans* in a *Brassica napus* genomic background. Theor Appl Genet 105:498–504

Borah RC, Bajaj KL, Bhatia IS (1978) Biochemical changes in tea leaves after infection with red rust (*Cephaleurosparasiticus*). J Phytopathol 93:208–217

Boys EF, Roques SE, Ashby AM, Evans N, Latunde-Dada AO, Thomas JE, West JS, Fitt BDL (2007) Resistance to infection by stealth: *Brassica napus* (winter oilseed rape) and *Pyrenopeziza brassicae* (light leaf spot). Eur J Plant Pathol 118:307–321

Boys EF, Roques SE, West JS, Werner CP, King GJ, Dyer PS, Fitt BDL (2011) Effects of R gene-mediated resistance in *Brassica napus* (oilseed rape) on asexual and sexual sporulation of *Pyrenopeziza brassicae* (light leaf spot). Plant Pathol 61(3):543–554

Brachi B, Meyer CG, Villoutreix R, Platt A, Morton TC, Roux F, Bergelson J (2015) Co-selected genes determine adaptive variation in herbivore resistance throughout the native range of *Arabidopsis thaliana*. Proc Natl Acad Sci U S A 112(13):4032–4037. https://doi.org/10.1073/pnas.1421416112

Bradburne R, Majer D, Magrath R, Werner CP, Lewis B, Mithen R (1999) Winter oilseed rape with high levels of resistance to *Pyrenopeziza brassicae* derived from wild *Brassica* species. Plant Pathol 48:550–558

Brader G, Mikkelsen MD, Halkier BA, Palva ET (2006) Altering glucosinolate profiles modulates disease resistance in plants. Plant J 46(5):758–767. https://doi.org/10.1111/j.1365-313X.2006.02743.x

Bradley CA, Henson RA, Porter PM, LeGare DG, Del Río LE, Khot SD (2006) Response of canola cultivars to *Sclerotinia sclerotiorum* in controlled and field environments. Plant Dis 90:215–219

Brun H, Huteau V, Ermel M, Eber F, Chevre AM, Renard M (2004) Field behaviour of oilseed rape genotypes carrying major resistance genes exposed to different *Leptosphaeria maculans* populations. Int Organiz Biol Contr Bull 27:95–100

Brun H, Chèvre AM, Fitt BD, Powers S, Besnard AL, Ermel M, Huteau V, Marquer B, Eber F, Renard M, Andrivon D (2010) Quantitative resistance increases the durability of qualitative resistance to *Leptosphaeria maculans* in *Brassica napus*. New Phytol 185:285–299. https://doi.org/10.1111/j.1469-8137.2009.03049.x

Butterworth MH, Semenov MA, Barnes A, Moran D, West JS, Fitt BDL (2010) North-south divide; contrasting impacts of climate change on crop yields in Scotland and England. J R Soc Interface 7:123–130. https://doi.org/10.1098/rsif.2009.0111

Camargo LEA, Williams PH, Osborn TC (1995) Mapping of quantitative loci controlling resistance to *Brassica oleracea* to *Xanthomonas campestris* pv. *campestris* in the field and green house. Phytopathology 85:1296–1300

Cargeeg LA, Thurling N (1980) Contribution of hostpathogen interactions to the expression of the blackleg disease of spring rape (*Brassica napus* L.) caused by *Leptosphaeria maculans* (Desm.) Ces.et de Not. Euphytica 29:465–476

Césari S, Kanzaki H, Fujiwara T, Bernoux M, Chalvon V, Kawano Y, Shimamoto K, Dodds P, Terauchi R, Kroj T (2014) The NB-LRR proteins RGA4 and RGA5 interact functionally and physically to confer disease resistance. EMBO J 33:1941–1959

Chalhoub B, Denoeud F, Liu S, Parkin IA, Tang H, Wang X, Chiquet J, Belcram H, Tong C, Samans B, Corréa M, Da Silva C, Just J, Falentin C, Koh CS, Le Clainche I, Bernard M, Bento P, Noel B, Labadie K, Alberti A, Charles M, Arnaud D, Guo H, Daviaud C, Alamery S, Jabbari K, Zhao M, Edger PP, Chelaifa H, Tack D, Lassalle G, Mestiri I, Schnel N, Le Paslier MC, Fan G, Renault V, Bayer PE, Golicz AA, Manoli S, Lee TH, Thi VH, Chalabi S, Hu Q, Fan C, Tollenaere R, Lu Y, Battail C, Shen J, Sidebottom CH, Wang X, Canaguier A, Chauveau A, Bérard A, Deniot G, Guan M, Liu Z, Sun F, Lim YP, Lyons E, Town CD, Bancroft I, Wang X, Meng J, Ma J, Pires JC, King GJ, Brunel D, Delourme R, Renard M, Aury JM, Adams KL, Batley J, Snowdon RJ, Tost J, Edwards D, Zhou Y, Hua W, Sharpe AG, Paterson AH, Guan C, Wincker P (2014) Plant genetics. Early allopolyploid evolution in the post-neolithic *Brassica napus* oilseed genome. Science 345(6199):950–953. https://doi.org/10.1126/science.1253435

Chatterjee M, Mazumder M, Basu D (2013) Functional analysis of the promoter of a glycosyl hydrolase gene induced in resistant *Sinapis alba* by *Alternaria brassicicola*. Phytopathology 103:841–850. https://doi.org/10.1094/PHYTO-11-12-0303-R

Chaturani GDG, Mir ZA, Mishra P, Grover A (2019) Identification and characterization of defensin genes in *Brassica juncea* and *Camelina sativa* and analysis of their expression in response to biotic and abiotic stress. Acta Sci Agric 3(12):96–112

Chen CY, Seguin-Swartz G (1999) Reaction of wild crucifers to *Leptosphaeria maculans*, the causal agent of blackleg of crucifers. Can J Plant Pathol 21:361–367

Chen Y, Wang D (2005) Two convenient methods to evaluate soybean for resistance to *Sclerotinia sclerotiorum*. Plant Dis 89(12):1268–1272. https://doi.org/10.1094/PD-89-1268

Chen X, Liu J, Lin G, Wang A, Wang Z, Lu G (2013) Overexpression of AtWRKY28 and AtWRKY75 in Arabidopsis enhances resistance to oxalic acid and *Sclerotinia sclerotiorum*. Plant Cell Rep 32(10):1589–1599. https://doi.org/10.1007/s00299-013-1469-3

Chen L, Hao L, Parry MAJ, Phillips AL, Hu YG (2013) Progress in TILLING as a tool for functional genomics and improvement of crops. J Integr Plant Biol 56:425–443

Chevre AM, Eber F, This P, Barret P, Tanguy X, Brun H, Dlseny M, Renard M (1996) Characterization of *Brassica nigra* chromosomes and of blackleg resistance in *B. napus–B. nigra* addition lines. Plant Breed 115:113–118. https://doi.org/10.1111/j.1439-05231996.tb00884.x

Chevre AM, Barret P, Eber F, Dupuy P, Brun H, Tanguy X, Renard M (1997) Selection of stable *Brassica napus-B. juncea* recombinant lines resistant to blackleg (*Leptosphaeria maculans*). I. Identification of molecular markers, chromosomal and genomic origin of the introgression. Theor Appl Genet 95:1104–1111

Chevre AM, de Leon AP, Jenczewski E, Eber F, Delourme R, Renard M, Brun H (2003) Introduction of blackleg resistance from *Brassica rapa* into *Brassica napus*. In: Proc 11th Intl Rapeseed Congr, Copenhagen, Denmark, vol 1, pp 32–35

Chhikara S, Chaudhury D, Dhankher OP, Jaiwal PK (2012) Combined expression of a barley class II chitinase and type I ribosome inactivating protein in transgenic *Brassica juncea* provides protection against *Alternaria brassicae*. Plant Cell Tissue Organ Cult 108:83–89

Chotechung S, Somta P, Chen J, Yimram T, Chen X, Srinives P (2016) A gene encoding a polygalacturonase-inhibiting protein (PGIP) is a candidate gene for bruchid (Coleoptera: bruchidae) resistance in mungbean (*Vigna radiata*). Theor Appl Genet 129:1673–1683

Collard BCY, Mackill DJ (2008) Marker-assisted selection: an approach for precision plant breeding in the twenty-first century. Philos Trans R Soc Lond Ser B Biol Sci 363:557–572

Corwin JA, Kliebenstein DJ (2017) Quantitative resistance: more than just perception of a pathogen. Plant Cell 29:655–665. https://doi.org/10.1105/tpc.16.00915

Crouch JH, Lewis BG, Mithen RF (1994) The effect of A-genome substitution on the resistance of *Brassica napus* to infection by *Leptosphaeria maculans*. Plant Breed 112:265–278

D'Ovido R, Mattei B, Robert S, Bellincampi D (2004) Polygalacturonases, polygalacturonase-inhibiting proteins and pectic oligomers in plant–pathogen interactions. Biochim Biophys Acta Prot Proteom 1696(2):237–244. https://doi.org/10.1016/j.bbapap.2003.08.012

De Block M, Tenning P, de Brouwer D (1989) Transformation of *Brassica napus* and *Brassica oleracea* using *Agrobacterium tumefaciens* and the expression of the bar and neo genes in the transgenic plants. Plant Physiol 91(2):694–701. https://doi.org/10.1104/pp.91.2.694

De Wit PJGM (1992) Molecular mechanisms of gene-for-gene systems in plant-fungus interactions and the application of avirulence genes in control of plant pathogens. Annu Rev Phytopathol 30:391–418

Debieu M, Huard-Chauveau C, Genissel A, Roux F, Roby D (2016) Quantitative disease resistance to the bacterial pathogen *Xanthomonas campestris* involves an *Arabidopsis* immune receptor pair and a gene of unknown function. Mol Plant Pathol 17(4):510–520. https://doi.org/10.1111/mpp.12298

Delourme R, Pilet-Nayel ML, Archipiano M, Horvais R, Tanguy X, Rouxel T, Brun H, Renard M, Balesdent MH (2004) A cluster of major specific resistance genes to *Leptosphaeria maculans* in *Brassica napus*. Phytopathology 94:578–583

Delourme R, Chèvre AM, Brun H, Rouxel T, Balesdent MH, Dias JS, Salisbury P, Renard M, Rimmer SR (2006) Major gene and polygenic resistance to *Leptosphaeria maculans* in oilseed rape (*Brassica napus*). Eur J Plant Pathol 114:41–52. https://doi.org/10.1007/s10658-005-2108-9

Delourme R, Brun H, Ermel M, Lucas MO, Vallee P, Domin C, Walton G, Li H, Sivasithamparam K, Barbetti MJ (2008) Expression of resistance to Leptosphaeria maculans in Brassica napus double haploid lines in France and Australia is influenced by location. Ann Appl Biol 153:259–269

Delourme R, Bousset L, Ermel M, Duffe P, Besnard AL, Marquer B, Fudal I, Linglin J, Chadœuf J, Brun H (2014) Quantitative resistance affects the speed of frequency increase but not the diversity of the virulence alleles overcoming a major resistance gene to *Leptosphaeria maculans* in oilseed rape. Infect Genet Evol 27:490–499. https://doi.org/10.1016/j.meegid.2013.12.019

Diener AC, Ausubel FM (2005) Resistance to *Fusarium oxysporum* 1, a dominant *Arabidopsis* disease-resistance gene, is not race specific. Genetics 171:305–321

Ding Y, Mei J, Li Q, Liu Y, Wan H, Wang L, Becker HC, Qian W (2013) Improvement of *Sclerotinia sclerotiorum* resistance in *Brassica napus* by using *B. oleracea*. Genet Resour Crop Evol 60:1615–1619. https://doi.org/10.1007/s10722-013-9978-z

Dion Y, Gugel RK, Rakow GFW, Seguin-Swartz G, Landry BS (1995) RFLP mapping of resistance to the blackleg disease [causal agent, *Leptosphaeria maculans* (Desm.) Ces.et de Not.] in canola (*Brassica napus* L.). Theor Appl Genet 91:1190–1194

Dixelius C (1999) Inheritance of the resistance to *L. maculans* of *B. nigra* and *B. juncea* in near isogenic lines of *B. napus*. Plant Breed 118:151–156

Dixelius C, Wahlberg S (1999) Resistance to *L. maculans* is conserved in a specific region of *Brassica* B genome. Theor Appl Genet 99:368–372

Dodds PN, Rathjen JP (2010) Plant immunity: towards an integrated view of plant-pathogen interactions. Nat Rev 11:539–548. https://doi.org/10.1038/nrg2812

Dong X (2004) NPR1, all things considered. Curr Opin Plant Biol 7:547–552. https://doi.org/10.1016/j.pbi.2004.07.005

Dong X, Jiang Z, Peng YL, Zhang Z (2015) Revealing shared and distinct gene network organization in *Arabidopsis* immune responses by integrative analysis. Plant Physiol 167:1186–1203

Dong W, Wu D, Li G, Wu D, Wang Z (2018) Next-generation sequencing from bulk segregant analysis identifies a dwarfism gene in watermelon. Sci Rep 8:2908. https://doi.org/10.1038/s41598-018-21293-1

Doullah MAU, Mohsin GM, Ishikawa K, Hori H, Okazaki K (2011) Construction of a linkage map and QTL analysis for black rot resistance in *Brassica oleracea* L. Int J Nat Sci 1:1–6

Dupont J, Reverchon M, Cloix L, Froment P, Ramé C (2006) Transcriptional analysis of calcium-dependent and calcium-independent signalling pathways induced by oligogalacturonides. J Exp Bot 57:2847–2865

Durner J, Klessig DF (1996) Salicylic acid is a modulator of tobacco and mammalian catalases. J Biol Chem 271:28492–28501. https://doi.org/10.1074/jbc.271.45.28492. PMID: 8910477

Eber F, Lourgant K, Brun H, Lodé M, Huteau V, Coriton O et al (2011) Analysis of *Brassica nigra chromosomes allows identification* of a new effective *Leptosphaeria maculans* resistance gene introgressed in *Brassica napus*. In: 13th International Rapeseed Congress, Prague

Echandi E, Walker JC (1957) Pectolytic enzymes produced by *Sclerotinia sclerotiorum*. Phytopathology 47:303–306

Ehsani-Moghaddam B, Charles MT, Carisse O, Khanizadeh S (2006) Superoxide dismutase responses of strawberry cultivars to infection by *Mycosphaerella fragariae*. J Plant Physiol 163:147–153. https://doi.org/10.1016/j.jplph.2005.04.025. PMID: 16399005

Elgin JH, Beyer EH (1968) Evaluation of selected alfalfa clover for resistance to *Sclerotinia trifoliorum* Erikss. Crop Sci 8:265–266

Elliott VL, Marcroft SJ, Howlett BJ, Van De Wouw AP (2016) Gene-for-gene resistance is expressed in cotyledons, leaves and pods, but not during late stages of stem colonization in the *Leptosphaeria maculans–Brassica napus* pathosystem. Plant Breed 135:200–207. https://doi.org/10.1111/pbr.12343/

Ersoz ES, Yu J, Buckler ES (2008) Applications of linkage disequilibrium and association mapping in crop plants. In: Varshney R, Tuberosa R (eds) Genomic assisted crop improvement. Vol. I: Genomics approaches and platforms. Springer, Berlin, pp 97–120

Eynck C, Koopmann B, Karlovsky P, von Tiedemann A (2009) Internal resistance in winter oilseed rape inhibits systemic spread of the vascular pathogen *Verticillium longisporum*. Phytopathology 99:802–811

Fahey JW, Zalemann AT, Talahay P (2001) The chemical diversity and distribution of glucosinolates and isothiocyanates among plants. Phytochemistry 56:5–51

Fatima U, Bhorali P, Borah S, Senthil-Kumar M (2019) Perspectives on the utilization of resistance mechanisms from host and nonhost plants for durable protection of *Brassica* crops against Alternaria blight. Peer J 7:e7486. https://doi.org/10.7717/peerj.7486

Federici L, Di Matteo A, Fernandez-Recio J, Tsernoglou D, Cervone F (2006) Polygalacturonase inhibiting proteins: players in plant innate immunity? Trends Plant Sci 11:65–70

Ferrari S, Savatin DV, Sicilia F, Gramegna G, Cervone F, Lorenzo GD (2013) Oligogalacturonides: plant damage-associated molecular patterns and regulators of growth and development. Front Plant Sci 4:49

Ferreira ME, Williams PH, Osborn TC (1994) RFLP mapping of Brassica napus using doubled haploid lines. Theor Appl Genet 89:615–621

Ferreira ME, Rimmer SR, Williams PH, Osborn TC (1995) Mapping loci controlling *Brassica napus* resistance to *Leptosphaeria maculans* under different screening conditions. Phytopathology 85:213–217

Fitt BDL, Brun H, Barbetti MJ, Rimmer SR (2006) World-wide importance of phoma stem canker (Leptosphaeria maculans and L. biglobosa) on oilseed rape (Brassica napus). Eur J Plant Pathol 114:3–15

Flint-Garcia SA, Thornsberry JM, Buckler ES (2003) Structure of linkage disequilibrium in plants. Annu Rev Plant Biol 54:357–374

Floerl S, Druebert C, Majcherczyk A, Karlovsky P, Kues U, Polle A (2008) Defense reactions in the apoplastic proteome of oilseed rape (*Brassica napus* var. *napus*) attenuate *Verticillium longisporum* growth but not disease symptoms. BMC Plant Biol 8:129

Floerl S, Druebert C, Aroud HI, Karlovsky P, Polle A (2010) Disease symptoms and mineral nutrition in *Arabidopsis thaliana* in response to *Verticillium longisporum* V143 infection. J Plant Pathol 92:693–700

Flor HH (1971) Current status of the gene-for-gene concept. Annu Rev Phytopathol 9(1):275–296. https://doi.org/10.1146/annurev.py.09.090171.001423

Fomeju BF, Falentin C, Lassalle G, Manzanares-Dauleux MJ, Delourme R (2014) Homoeologous duplicated regions are involved in quantitative resistance of *Brassica napus* to stem canker. BMC Genomics 15:498. https://doi.org/10.1186/1471-2164-15-498

Fradin EF, Abd-El-Haliem A, Masini L, van den Berg GCM, Joosten MHAJ, Thomma BPHJ (2011) Interfamily transfer of tomato Ve1 mediates *Verticillium* resistance in *Arabidopsis*. Plant Physiol 156:2255–2265

Francisco M, Joseph B, Caligagan H, Li B, Corwin JA, Lin C, Kerwin RE, Burow M, Kliebenstein DJ (2016) Genome wide association mapping in *Arabidopsis thaliana* identifies novel genes involved in linking allyl glucosinolate to altered biomass and defense. Front Plant Sci 7:1010

Fredua-Agyeman R, Coriton O, Huteau V, Parkin IA, Chèvre AM, Rahman H (2014) Molecular cytogenetic identification of B genome chromosomes linked to blackleg disease resistance in *Brassica napus* × *B. carinata* interspecific hybrids. Theor Appl Genet 127:1305–1318. https://doi.org/10.1007/s00122-014-2298-7

Fu D, Uauy C, Distelfeld A, Blechi A, Epstein L, Che X, Sela H, Fahima T, Dubcovsky J (2009) A kinase-START gene confers temperature-dependent resistance to wheat stripe rust. Science 323:1357–1360

Fudal I, Ross S, Gout L, Blaise F, Kuhn ML, Eckert MR, Cattolico L, Bernard-Samain S, Balesdent MH, Rouxel T (2007) Heterochromatin-like regions as ecological niches for avirulence genes in the Leptosphaeria maculans genome: map-based cloning of AvrLm6. Mol Plant-Microbe Interact 20:459–470

Fusari CM, Di Rienzo JA, Troglia C, Nishinakamasu V, Moreno MV, Maringolo C, Quiroz F, Alvarez D, Escande A, Hopp E, Heinz R, Lia VV, Paniego NB (2012) Association mapping in sunflower for *Sclerotinia* head rot resistance. BMC Plant Biol 12:93. https://doi.org/10.1186/1471-2229-12-93

Garg H, Sivasithamparam K, Banga SS, Barbetti MJ (2008) Cotyledon assay as a rapid and reliable method of screening for resistance against *Sclerotinia sclerotiorum* in *Brassica napus* genotypes. Aust Plant Pathol 37:106–111

Garg H, Atri C, Sandhu PS, Kaur B, Renton M, Banga SK, Singh H, Singh C, Barbetti MJ, Banga SS (2010a) High level of resistance to *Sclerotinia sclerotiorum* in introgression lines derived from hybridization between wild crucifers and the crop *Brassica* species *B. napus and B. juncea*. Field Crop Res 117:51–58. https://doi.org/10.1016/j.fcr.2010.01.013

Garg H, Li H, Sivasithamparam K, Kuo J, Barbetti MJ (2010b) The infection processes of *Sclerotinia sclerotiorum* in cotyledon tissue of a resistant and a susceptible genotype of *Brassica napus*. Ann Bot 106:897–908. https://doi.org/10.1093/aob/mcq196

Ge XT, Li YP, Wan ZJ, You MP, Finnegan PM, Banga SS, Sandhu PS, Garg H, Salisbury PA, Barbetti MJ (2012) Delineation of *Sclerotinia sclerotiorum* pathotypes using differential resistance responses on *Brassica napus* and *B. juncea* genotypes enables identification of resistance to prevailing pathotypes. Field Crop Res 127:248–258

Ge XT, You MP, Barbetti MJ (2015) Virulence differences among *Sclerotinia sclerotiorum* isolates determines host cotyledon resistance responses in Brassicaceae genotypes. Eur J Plant Pathol 143:527–541. https://doi.org/10.1007/s10658-015-0696-6

Ghanbarnia K, Lydiate DJ, Rimmer SR, Li G, Kutcher HR, Larkan NJ, McVetty PBE, Fernando WGD (2012) Genetic mapping of the Leptosphaeria maculans avirulence gene corresponding to the LepR1 resistance gene of Brassica napus. Theor Appl Genet 124:505–513

Ghanbarnia K, Fudal I, Larkan NJ, Links MG, Balesdent MH, Profotova B, Fernando DWG, Rouxel T, Borhan H (2015) Rapid identification of the Leptosphaeria maculans avirulence gene AvrLm2 using an intraspecific comparative genomics approach. Mol Plant Pathol 16:699–709

Ghanbarnia K, Ma L, Larkan NJ, Haddadi P, Fernando WGD, Borhan MH (2018) *Leptosphaeria maculans* AvrLm9: a new player in the game of hide and seek with AvrLm4-7. Mol Plant Pathol 19(7):1754–1764

Ghose K, Dey S, Barton H, Loake GJ, Basu D (2008) Differential profiling of selected defense-related genes induced on challenge with *Alternaria brassicicola* in resistant white mustard and their comparative expression pattern in susceptible India mustard. Mol Plant Pathol 9 (6):763–775. https://doi.org/10.1111/j.1364-3703.2008.00497.x

Gilles T, Evans N, Fitt BDL, Jeger MJ (2000a) Epidemiology in relation to methods for forecasting light leaf spot (Pyrenopeziza brassicae) severity on winter oilseed rape (Brassica napus) in the UK. Eur J Plant Pathol 106:593–605

Gilles T, Fitt BDL, Kennedy R, Welham SJ, Jeger MJ (2000b) Effects of temperature and wetness duration on conidial infection, latent period and asexual sporulation of Pyrenopeziza brassicae on leaves of oilseed rape. Plant Pathol 49:498–508

Gilles T, Ashby AM, Fitt BDL, Cole T (2001) Development of Pyrenopeziza brassicae apothecia on oilseed rape debris and agar. Mycol Res 105:705–714

Glazebrook J (2005) Contrasting mechanisms of defense against biotrophic and necrotrophic pathogens. Annu Rev Phytopathol 43:205–227. https://doi.org/10.1146/annurev.phyto.43.040204.135923

Glazebrook J, Ausubel FM (1994) Isolation of phytoalexin-deficient mutants of *Arabidopsis thaliana* and characterization of their interactions with bacterial pathogens. Proc Natl Acad Sci U S A 91:8955–8959

Gout L, Fudal I, Kuhn ML, Blaise F, Eckert M, Cattolico L, Balesdent MH, Rouxel T (2006) Lost in the middle of nowhere: the AvrLm1 avirulence gene of the Dothideomycete *Leptosphaeria maculans*. Mol Microbiol 60:67–80

Grau CF, Bissonnette HL (1974) Whetzelinia stem rot of soybean in Minnesota. Plant Dis Rep 58:693–695

Greenberg JT (1997) Programmed cell death in plant-pathogen interactions. Annu Rev Plant Physiol Plant Mol Biol 48:525–545

Grison R, Grezes BB, Schneider M, Lucante N, Olsen L, Leguay JJ, Toppan A (1996) Field tolerance to fungal pathogens of *Brassica napus* constitutively expressing a chimeric chitinase gene. Nat Biotechol 14:643–646

Gutierrez G (2012) SPR and differential proteolysis/MS provide further insight into the interaction between PGIP2 and EPGs. Fungal Biol 116:737–746

Guyon K, Balague C, Roby D, Raffaele S (2014) Secretome analysis reveals effector candidates associated with broad host range necrotrophy in the fungal plant pathogen *Sclerotinia sclerotiorum*. BMC Genomics 15:336. https://doi.org/10.1186/1471-2164-15-336

Gyawali S, Harrington M, Durkin J, Horner K, Parkin IA, Hegedus DD, Bekkaoui D, Buchwaldt L (2016) Microsatellite markers used for genome-wide association mapping of partial resistance to *Sclerotinia sclerotiorum* in a world collection of *Brassica napus*. Mol Breed 36:72. https://doi.org/10.1007/s11032-016-0496-5

Haddadi P, Ma L, Wang H, Borhan MH (2016) Genome-wide transcriptome analyses provides insights into the lifestyle transition and effector repertoire of Leptosphaeria maculans during colonization of Brassica napus seedlings. Mol Plant Pathol 17(8):1196–1210. https://doi.org/10.1111/mpp.12356

Haddadi P, Larkan NJ, Borhan MH (2019) Dissecting R gene and host genetic background effect on the *Brassica napus* defense response to *Leptosphaeria maculans*. Sci Rep 9:6947. https://doi.org/10.1038/s41598-019-43419-9

Haffner E, Karlovsky P, Diederichsen E (2010) Genetic and environmental control of the *Verticillium* syndrome in *Arabidopsis thaliana*. BMC Plant Biol 10:235

Hammond-Kosack KE, Jones JDG (1996) Resistance gene dependent plant defense responses. Plant Cell 8:1773–1791

Hammond-Kosack KE, Harrison K, Jones JDG (1994) Developmentally regulated cell death on expression of the fungal avirulence gene *Avr9* in tomato seedlings carrying the disease-resistance gene Cf9. Proc Natl Acad Sci U S A 91:10445–10449

Hammond-Kosack KE, Silverman P, Raskin I, Jones JDG (1996) Race-specific elicitors of *Cladosporiumfulvum* induce changes in cell morphology and the synthesis of ethylene and SA in tomato cells carrying the corresponding Cf resistance gene. Plant Physiol 110:1381–1394

Hammond-Kosack KE, Tang S, Harrison K, Jones JDG (1998) The tomato Cf9 disease resistance gene functions in tobacco and potato to confer responsiveness to the fungal avirulence gene product Avr9. Plant Cell 10:1–17

Hansen LH, Earle ED (1997) Somatic hybrids between *Brassica oleracea* (L.) and *Sinapis alba* (L.) with resistance to *Alternaria brassicae* (Berk) Sacc. Theor Appl Genet 94:1078–1085

Hao Y, Wang T, Wang K, Wang X, Fu Y, Huang L, Kang Z (2016) Transcriptome analysis provides insights into the mechanisms underlying wheat plant resistance to stripe rust at the adult plant stage. PLoS One 11:e0150717

Happstadius I, Ljungberg A, Kristiansson B, Dixelius C (2003) Identification of *Brassica oleracea* germplasm with improved resistance to *Verticillium* wilt. Plant Breed 122:30–34

Hasan M, Friedt W, Pons-Kuhnemann J, Freitag NM, Link K, Snowdon RJ (2008) Association of gene-linked SSR markers to seed glucosinolate content in oilseed rape (*Brassica napus* ssp. *napus*). Theor Appl Genet 116:1035–1049

Hegedus DD, Li R, Buchwaldt L, Parkin I, Whitwill S, Coutu C, Bekkaoui D, Rimmer SR (2008) *Brassica napus* possesses an expanded set of polygalacturonase inhibitor protein genes that are differentially regulated in response to *Sclerotinia sclerotiorum* infection, wounding and defense hormone treatment. Planta 228:241–253

Hennin C, Hofte M, Diederichsen E (2001) Functional expression of Cf9 and Avr9 genes in *Brassica napus* induces enhanced resistance to *Leptosphaeria maculans*. Mol Plant-Microbe Interact 14(9):1075–1085

Holub EB (2001) The arms race is ancient history in *Arabidopsis*, the wildflower. Nat Rev Genet 2:516–527

Holub EB (2008) Natural history of *Arabidopsis thaliana* and oomycete symbioses. Eur J Plant Pathol 122:91–109

Honee G, Melchers LS, Vleeshouwers VGAA, Van Roekel JSC, De Wit PJGM (1995) Production of the AVR9 elicitor from the fungal pathogen *Cladosporium fulvum* in transgenic tobacco and tomato plants. Plant Mol Biol 29:909–920

Honee G, Buitink J, Jabs T, De Kloe J, Sijbolts F, Apotheker M, Weide R, Sijen T, Stuiver M, De Wit PJGM (1998) Induction of defense-related response in Cf9 tomato cells by the *Avr9* elicitor peptide of *Cladosporium fulvum* is developmentally regulated. Plant Physiol 117:809–820

Honsdorf N, Becker HC, Ecke W (2010) Association mapping for phenological, morphological, and quality traits in canola quality winter rapeseed (*Brassica napus* L.). Genome 53:899–907

van der Hoorn RA, Laurent F, Roth R, De Wit PJ (2000) Agroinfiltration is a versatile tool that facilitates comparative analyses of Avr9/Cf-9-induced and Avr4/Cf-4-induced necrosis. Mol Plant-Microbe Interact 13:439–446. https://doi.org/10.1094/MPMI.2000.13.4.439

Howlett BJ, Lowe RGT, Marcroft SJ, Van de Wouw AP (2015) Evolution of virulence in fungal plant pathogens: exploiting fungal genomics to control plant disease. Mycologia 107 (3):441–451. https://doi.org/10.3852/14-317

Huang YJ, Toscano-Underwood C, Fitt BDL, Hu XJ, Hall AM (2003) Effects of temperature on ascospore germination and penetration of oilseed rape (*Brassica napus*) leaves by A-group or B-group *Leptosphaeria maculans* (phoma stem canker). Plant Pathol 52:245–255. https://doi.org/10.1046/j.1365-3059.2003.00813.x

Huang YJ, Evans N, Li ZQ, Eckert M, Chevre AM, Renard M, Fitt BDL (2006) Temperature and leaf wetness duration affect phenotypic expression of Rlm6-mediated resistance to Leptosphaeria maculans in Brassica napus. New Phytol 170:129–141

Huang YJ, Pirie EJ, Evans N, Delourme R, King GJ, Fitt BDL (2009) Quantitative resistance to symptomless growth of *Leptosphaeria maculans* (phoma stem canker) in *Brassica napus* (oilseed rape). Plant Pathol 58:314–323. https://doi.org/10.1111/j.1365-3059.2008.01957.x

Huang YJ, Qi A, King GJ, Fitt BDL (2014) Assessing quantitative resistance against *Leptosphaeria maculans* (phoma stem canker) in *Brassica napus* (oilseed rape) in young plants. PLoS One 9: e84924. https://doi.org/10.1371/journal.pone.0084924

Huang YJ, Jestin C, Welham SJ, King GJ, Manzanares-Dauleux MJ, Fitt BD, Delourme R (2016) Identification of environmentally stable QTL for resistance against *Leptosphaeria maculans* in oilseed rape (*Brassica napus*). Theor Appl Genet 129:169–180. https://doi.org/10.1007/s00122-015-2620-z

Huangfu H, Guan C, Jin F, Yin C (2014) Prokaryotic expression and protein function of *Brassica napus* PGIP2 and its genetic transformation. Plant Biotechnol Rep 8:171–181

Huard-Chauveau C, Perchepied L, Debieu M, Rivas S, Kroj T, Kars I, Bergelson J, Roux F, Roby D (2013) An atypical kinase under balancing selection confers broad-spectrum disease resistance in *Arabidopsis*. PLoS Genet 9:e1003766

Hubbard M, Zhai C, Peng G (2020) Exploring mechanisms of quantitative resistance to *Leptosphaeria maculans* (Blackleg) in the cotyledons of canola (*Brassica napus*) based on transcriptomic and microscopic analyses. Plants 9:864

Hutcheson SW (1998) Current concepts of active defense in plants. Annu Rev Phytopathol 36:59–90

Iakovidis M, Teixeira PJ, Exposito-Alonso M, Cowper MG, Law TF, Liu Q, Vu MC, Dang TM, Corwin JA, Weigel D, Dangl JL, Grant SR (2016) Effector-triggered immune response in *Arabidopsis thaliana* is a quantitative trait. Genetics 204(1):337–353. https://doi.org/10.1534/genetics.116.190678

Iglesias-Bernabe L, Madloo P, Rodriguez VM, Francisco M, Soengas P (2019) Dissecting quantitative resistance to *Xanthomonas campestris* pv. *campestris* in leaves of *Brassica oleracea* by QTL analysis. Sci Rep 9:2015. https://doi.org/10.1038/s41598-019-38527-5

Ignatov A, Kuginuki Y, Hidam K (2000) Distribution and inheritance of race-specific resistance to *Xanthomonas campestris* pv. *campestris* in *Brassica rapa* and *B. napus*. J Russ Phytopathol Soc 1:89–94

Ioannidou D, Pinel A, Brugidou C, Albar L, Ahmadi N, Ghesquiere A, Nicole M, Fargette D (2003) Characterization of the effects of a major QTL of the partial resistance to rice yellow mottle virus using a near-isogenic line approach. Physiol Mol Plant Pathol 63:213–221

Jestin C, Lode M, Vallée P, Domin C, Falentin C, Horvais R, Coedel S, Manzanares-Dauleux MJ, Delourme R (2011) Association mapping of quantitative resistance for *Leptosphaeria maculans* in oilseed rape (*Brassica napus* L.). Mol Breed 27:271–287. https://doi.org/10.1007/s11032-010-9429-x

Jestin C, Vallee P, Domin C, Manzanares-Dauleux MJ, Delourme R (2012) Assessment of a new strategy for selective phenotyping applied to complex traits in *Brassica napus*. Open J Genet 2 (4):190

Jestin C, Bardol N, Lode M, Duffe P, Domin C, Vallee P, Mangin B, Manzanares-Dauleux MJ, Delourme R (2015) Connected populations for detecting quantitative resistance factors to Phoma stem canker in oilseed rape (Brassica napus L.). Mol Breed 35:167

Jiang G-L (2015) Molecular marker-assisted breeding: a plant breeder's review. In: Advances in plant breeding strategies: breeding, biotechnology and molecular tools. Springer, New York, NY, pp 431–472

Jones DA, Thomas CM, Hammond-Kosack KE, Balint-Kurti PJ, Jones JD (1994) Isolation of the tomato Cf-9 gene for resistance to *Cladosporium fulvum* by transposon tagging. Science 266:789–793. https://doi.org/10.1126/science.7973631

Joshi RK, Megha S, Rahman MH, Basu U, Kav NNA (2016) Global study of transcriptome dynamics in canola (Brassica napus L.) responsive to *Sclerotinia sclerotiorum* infection using RNA-Seq. Gene 590:57–67

Kagan IA, Hammerschmidt R (2002) Arabidopsis ecotype variability in camalexin production and reaction to infection by *Alternaria brassicicola*. J Chem Ecol 28:2121–2140

Kalunke RM, Tundo S, Benedetti M, Cervone F, De Lorenzo G, D'Ovidio R (2015) An update on polygalacturonase-inhibiting protein (PGIP), a leucine-rich repeat protein that protects crop plants against pathogens. Front Plant Sci 6:146

Kamble A, Bhargava S (2007) β-amino-butyric acid-induced resistance in *Brassica juncea* against the necrotrophic pathogen *Alternaria brassicae*. J Phytopathol 155(3):152–158. https://doi.org/10.1111/j.1439-0434.2007.01209.x

Kanbe M, Fujimoto F, Mizukami Y, Inami S, Fukaya K (1997) Increase of resistance of alfalfa to *Sclertotinia* crown and stem rot through recurrent selection. Breed Sci 47:347–351

Kang H, Wang Y, Peng S, Zhang Y, Xiao Y, Wang D, Qu S, Li Z, Yan S, Wang Z, Liu W, Ning Y, Korniliev P, Leung H, Mezey J, McCouch SR, Wang GL (2016) Dissection of the genetic architecture of rice resistance to the blast fungus *Magnaporthe oryzae*. Mol Plant Pathol 17 (6):959–972. https://doi.org/10.1111/mpp.12340

Kanrar S, Venkateswari JC, Kirti PB, Chopra VL (2002) Transgenic expression of hevein, the rubber tree lectin, in Indian mustard confers protection against *Alternaria brassicae*. Plant Sci 162:441–448

Kaur R, Shivani Saxena B, Kanwar HS, Dohroo NP, Majeed S, Sharma DR (2009a) Detecting RAPD markers associated with black rot resistance in cabbage (*Brassica oleracea* var. *capitata*). Fruit Veg Cereal Sci Biotechnol 13:12–15

Kaur S, Cogan NO, Ye G, Baillie RC, Hand ML, Ling AE, McGearey AK, Kaur J, Hopkins CJ, Todorovic M, Mountford H, Edwards D, Batley J, Burton W, Salisbury P, Gororo N, Marcroft S, Kearney G, Smith KF, Forster JW, Spangenberg GC (2009b) Genetic map construction and QTL mapping of resistance to blackleg (*Leptosphaeria maculans*) disease in Australian canola (*Brassica napus* L.) cultivars. Theor Appl Genet 120:71–83. https://doi.org/10.1007/s00122-009-1160-9

Keller H, Pamboukdjian N, Ponchet M, Poupet A, Delon R, Verrier JL, Roby D, Ricci P (1999) Pathogen induced elicitin production in transgenic tobacco generates a hypersensitive response and nonspecific disease resistance. Plant Cell 11:223–235

Keri M, vanden Berg CGJ, McVetty PBE, Rimmer SR (1997) Inheritance of resistance to Leptosphaeria maculans in *Brassica juncea*. Phytopathology 87:594–598

Khangura RK, Barbetti MJ (2001) Prevalence of blackleg (Leptosphaeria maculans) on canola (Brassica napus) in Western Australia. Aust J Exp Agric 41:71–40

Kifuji Y, Hanzaea H, Terasawa Y, Nishio T (2013) QTL analysis of black rot resistance in cabbage using newly developed EST-SNP markers. Euphytica 190:289–295

Kiyosawa S (1982) Genetics and epidemiological modeling of breakdown of plant disease resistance. Annu Rev Phytopathol 20:93–117

Klosterman SJ, Atallah ZK, Vallad GE, Subbarao KV (2009) Diversity, pathogenicity, and management of *Verticillium* species. Annu Rev Phytopathol 47:39–62

Korte A, Farlow A (2013) The advantages and limitations of trait analysis with GWAS: a review. Plant Methods 9:29. https://doi.org/10.1186/1746-4811-9-29

Kumar V, Paillard S, Fopa-Fomeju B, Falentin C, Deniot G, Baron C, Vallee P, Manzanares-Dauleux MJ, Delourme R (2018) Multi-year linkage and association mapping confirm the high number of genomic regions involved in oilseed rape quantitative resistance to blackleg. Theor Appl Genet 131:1627–1643

Kuswinanti T, Sock J, Hoppe HH (1999) Virulence pattern of aggressive isolates of *Leptosphaeria maculans* on an extended set of *Brassica* differentials. Zeitschrift für Pflanzenkrankheiten und Pflanzenschutz 106:12–20

Kutcher HR, van den Berg CGJ, Rimmer SR (1993) Variation in pathogenicity of *Leptosphaeria maculans* on *Brassica* spp. based on cotyledon and stem reactions. Can J Plant Pathol 15:253–258

Laclerca P (1973) Influence de facteurs hereditairs sur la resistance apparente du edible beans. Crop Sci 18:275–279

Larkan NJ, Lydiate DJ, Parkin IA, Nelson MN, Epp DJ, Cowling WA, Rimmer SR, Borhan MH (2013) The *Brassica napus* blackleg resistance gene LepR3 encodes a receptor-like protein triggered by the *Leptosphaeria maculans* effector AVRLM1. New Phytol 197:595–605. https://doi.org/10.1111/nph.12043

Larkan NJ, Lydiate DJ, Yu F, Rimmer SR, Borhan MH (2014) Co-localization of the blackleg resistance genes *Rlm2* and *LepR3* on *Brassica napus* chromosome A10. BMC Plant Biol 14:387. https://doi.org/10.1186/s12870-014-0387-z

Larkan NJ, Ma L, Borhan H (2015) The Brassica napus receptor-like protein RLM2 is encoded by a second allele of the LepR3/Rlm2 blackleg resistance locus. Plant Biotechnol J 13:983–992

Larkan NJ, Raman H, Lydiate DJ, Robinson SJ, Yu F, Barbulescu DM, Raman R, Luckett DJ, Burton W, Wratten N, Salisbury PA, Rimmer SR, Borhan MH (2016a) Multi-environment QTL studies suggest a role for cysteine-rich protein kinase genes in quantitative resistance to blackleg disease in *Brassica napus*. BMC Plant Biol 16:183. https://doi.org/10.1186/s12870-016-0877-2

Larkan NJ, Yu F, Lydiate DJ, Rimmer SR, Borhan MH (2016b) Single R gene introgression lines for accurate dissection of the *Brassica-Leptosphaeria* pathosystem. Front Plant Sci 7:1771. https://doi.org/10.3389/fpls.2016.01771

Lawrence CB, Mitchell TK, Craven KD, Cho YR, Cramer RA, Kim KH (2008) At death's door: alternaria pathogenicity mechanisms. Plant Pathol J 24:101–111

Lecomte L, Saliba-Colombani V, Gautier A, Gomez-Jimenez MC, Duffé P, Buret M, Causse M (2004) Fine mapping of QTL of chromosome 2 affecting the fruit architecture and composition of tomato. Mol Breed 13:1–14

Lee J, Izzah NK, Jayakodi M et al (2015a) Genome-wide SNP identification and QTL mapping for black rot resistance in cabbage. BMC Plant Biol 15:32. https://doi.org/10.1186/s12870-015-0424-6

Lee Y-G, Jeong N, Kim JH, Lee K, Kim KH, Pirani A, Ha BK, Kang ST, Park BS, Moon JK, Kim N, Jeong SC (2015b) Development, validation and genetic analysis of a large soybean SNP genotyping array. Plant J 81:625–636

Li CX, Cowling WA (2003) Identification of a single dominant allele for resistance to blackleg in *Brassica napus* 'Surpass 400'. Plant Breed 122:485–488

Li Y, Chen J, Bennett R, Kiddle G, Wallsgrove R, Huang Y, He Y (1999) Breeding, inheritance, and biochemical studies on *Brassica napus* cv. Zhongyou 821: tolerance to *Sclerotinia sclerotiorum* (stem rot). In: Wratten N, Salisbury PA (eds) Proc 10th. Intl Rapeseed Cong Canberra, Australia, September 26-29, p 61

Li H, Sivasithamparam K, Barbetti MJ (2003) Breakdown of a *Brassica rapa* subsp. *sylvestris* single dominant blackleg resistance gene in *Brassica napus* rapeseed by *Leptosphaeria maculans* field isolates in Australia. Plant Dis 87:752

Li CX, Li H, Siddique AB, Sivasithamparam K, Salisbury P, Banga SS, Banga S, Chattopadhyay C, Kumar A, Singh R, Singh D, Agnihotri A, Liu SY, Li YC, Tu J, Fu TD, Wang YF, Barbetti MJ (2007) The importance of the type and time of inoculation and assessment in the determination of resistance in *Brassica napus* and *B. juncea* to *Sclerotinia sclerotiorum*. Aust J Agric Res 58:1198–1203. https://doi.org/10.1071/AR07094

Li J, Zhao Z, Hayward A, Cheng H, Fu D (2015) Integration analysis of quantitative trait loci for resistance to *Sclerotinia sclerotiorum* in *Brassica napus*. Euphytica 205:483–489. https://doi.org/10.1007/s10681-015-1417-0

Li L, Luo Y, Chen B, Xu K, Zhang F, Li H, Huang Q, Xiao X, Zhang T, Hu J, Li F, Wu X (2016) A genome-wide association study reveals new loci for resistance to clubroot disease in Brassica napus. Front Plant Sci 7:1483. https://doi.org/10.3389/fpls.2016.01483

Liebrand TWH, van den Berg GCM, Zhang Z, Smit P, Cordewener JHG, America AH, Sklenar J, Jones AM, Tameling WI, Robatzek S, Thmma BP, Joosten MH (2013) Receptor-like kinase SOBIR1/EVR interacts with receptor-like proteins in plant immunity against fungal infection. Proc Natl Acad Sci U S A 110:10010–10015. https://doi.org/10.1073/pnas.1312631110

Light KA, Gororo NN, Salisbury PA (2011) Usefulness of winter canola (*Brassica napus*) race-specific resistance genes against blackleg (causal agent *Leptosphaeria maculans*) in southern Australian growing conditions. Crop Pasture Sci 62:162–168. https://doi.org/10.1071/CP10187

Liu S, Wang H, Zhang J, Fitt BD, Xu Z, Evans N, Liu Y, Yang W, Guo X (2005) *In vitro* mutation and selection of doubled-haploid *Brassica napus* lines with improved resistance to *Sclerotinia sclerotiorum*. Plant Cell Rep 24:133–144. https://doi.org/10.1007/s00299-005-0925-0

Liu Y, Huang X, Li M, He P, Zhang Y (2016) Loss-of-function of *Arabidopsis* receptor-like kinase BIR1 activates cell death and defense responses mediated by BAK1 and SOBIR1. New Phytol 212:637–645

Loake G, Grant M (2007) Salicylic acid in plant defense--the players and protagonists. Curr Opin Plant Biol 10(5):466–472

Loivamaki M, Stuhrwohldt N, Deeken R, Steffens B, Roitsch T, Hedrich R, Sauter M (2010) A role for PSK signaling in wounding and microbial interactions in *Arabidopsis*. Physiol Plant 139 (4):348–357. https://doi.org/10.1111/j.1399-3054.2010.01371.x

Lombard V, Delourme R (2001) A consensus linkage map for rapeseed (*Brassica napus* L.): construction and integration of three individual maps from DH populations. Theor Appl Genet 103:491–507

Long Y, Wang Z, Sun Z, Fernando DW, Mcvetty PB, Li G (2011) Identification of two blackleg resistance genes and fine mapping of one of these two genes in a *Brassica napus* canola cultivar 'Surpass 400'. Theor Appl Genet 122:1223–1231. https://doi.org/10.1007/s00122-010-1526-z

Loudet O, Saliba-Colombani V, Camilleri C, Calange F, Gaudon V, Koprovova A, North KA, Kopriva S, Daniel-Vedele F (2007) Natural variation for sulphate content in *Arabidopsis thaliana* is highly controlled by APR2. Nat Genet 39:896–900

Lumsden RD (1979) Histology and physiology of pathogenesis in plant disease caused by *Sclerotinia* species. Phytopathology 69:890–896

Lv H, Yang L, Kang J, Wang Q, Wang X, Fang Z, Liu Y, Zhuang M, Zhang Y, Lin Y, Yang Y, Xie B, Liu B, Liu JS (2013) Development of InDel markers linked to Fusarium wilt resistance in cabbage. Mol Breed 32:961–967. https://doi.org/10.1007/s11032-013-9925-x

Lv H, Fang Z, Yang L, Zhang Y, Wang Q, Liu Y, Zhuang M, Yang Y, Xie B, Liu B, Liu J, Kang J, Wang X (2014) Mapping and analysis of a novel candidate Fusarium wilt resistance gene FOC1 in *Brassica oleracea*. BMC Genomics 15:1094. https://doi.org/10.1186/1471-2164-15-1094

Lv H, Fang Z, Yang L, Zhang Y, Wang Y (2020) An update on the arsenal: mining resistance genes for disease management of Brassica crops in the genomic era. Hortic Res 7:34. https://doi.org/10.1038/s41438-020-0257-9

Maeda H, Matshushita K, Iida S, Sunohara Y (2006) Characterization of two QTLs controlling resistance to rice stripe virus detected in a Japanese upland rice line, Kanto 72. Breed Sci 56:359–364

Mahuku GS, Hall R, Goodwin PH (1996) Co-infection and induction of systemic acquired resistance by weakly and highly virulent isolates of *Leptosphaeria maculans* in oilseed rape. Physiol Mol Plant Pathol 49:61–72

Mandal S, Rajarammohan S, Kaur J (2018) *Alternaria brassicae* interactions with the model *Brassicaceae* member *Arabidopsis thaliana* closely resembles those with mustard (*Brassica juncea*). Physiol Mol Biol Plants 24:51–59. https://doi.org/10.1007/s12298-017-0486-z

Marchini J, Cardon LR, Phillips MS, Donnelly P (2004) The effects of human population structure on large genetic association studies. Nat Genet 36:512–517

Marcroft SJ, Sprague SJ, Pymer SJ, Salisbury PA, Howlett BJ (2003) Factors affecting production of inoculum of the blackleg fungus Leptosphaeria maculans in south-eastern Australia. Aust J Exp Agric 43:1231–1236

Marcroft SJ, Sprague SJ, Salisbury PA, Howlett BJ (2004) Potential for using host-resistance to reduce production of pseudothecia and ascospores of *Leptosphaeria maculans*, the blackleg pathogen of *Brassica napus*. Plant Pathol 53:468–474

Marcroft SJ, Elliott VL, Cozijnsen AJ, Salisbury PA, Howlett BJ, Van De Wouw AP (2012) Identifying resistance genes to *Leptosphaeria maculans* in Australian *Brassica napus* cultivars based on reactions to isolates with known avirulence genotypes. Crop Pasture Sci 63:338–350. https://doi.org/10.1071/CP11341

Marone D, Russo MA, Laido G, De Leonardis AM, Mastrangelo AM (2013) Plant nucleotide binding site-leucine rich repeat (NBS-LRR) genes: active guardians in host defense response. Int J Mol Sci 14:7302–7326

Martinez V, Nieves-Cordones M, Lopez-Delacalle M, Rodenas R, Mestre TC, Garcia-Sanchez F et al (2018) Tolerance to stress combination in tomato plants: new insights in the protective role of melatonin. Molecules 23:1–20. https://doi.org/10.3390/molecules23030535. PMID: 29495548

Mathpal P, Punetha H, Tewari AK, Agrawal S (2011) Biochemical defense mechanism in rapeseed-mustard genotypes against alternaria blight disease. J Oilseed Brassica 2:2011

Maulik A, Sarkar AI, Devi S, Basu S (2012) Polygalacturonase-inhibiting proteins–leucine-rich repeat proteins in plant defense. Plant Biol 14:22–30

May M, Hammond-Kosack KE, Jones JDG (1996) Involvement of reactive oxygen species, glutathione metabolism, and lipid peroxidation in the Cf-gene-dependent defense responses of tomato cotyledons induced by race-specific elicitors of *Cladosporium fulvum*. Plant Physiol 110:1367–1379

Mayerhofer R, Good AG, Bansal VK, Thiagarajah MR, Stringam GR (1997) Molecular mapping of resistance to *Leptosphaeria maculans* in Australian cultivars of *Brassica napus*. Genome 40:294–301. https://doi.org/10.1139/g97-041

Mayerhofer R, Wilde K, Mayerhofer M, Lydiate D, Bansal VK, Good AG, Parkin IA (2005) Complexities of chromosome landing in a highly duplicated genome: toward map-based cloning of a gene controlling blackleg resistance in *Brassica napus*. Genetics 171:1977–1988. https://doi.org/10.1534/genetics.105.049098

Mazumder M, Das S, Saha U, Chatterjee M, Bannerjee K, Basu D (2013) Salicylic acid mediated establishment of the compatibility between *Alternaria brassicicola* and *Brassica juncea* is mitigated by abscisic acid in *Sinapis alba*. Plant Physiol Biochem 70:43–51. https://doi.org/10.1016/j.plaphy.2013.04.025

McDonald B (2010) How can we achieve durable disease resistance in agricultural ecosystems? New Phytol 185:3–5. https://doi.org/10.1111/j.1469-8137.2009.03108.x

McNabb WM, vanden Berg CGJ, Rimmer SR (1993) Comparison of inoculation methods for selection of plant resistant to *Leptosphaeria maculans* in *Brassica napus*. Can J Plant Sci 73:1199–1207

Meena PD, Jambhulkar SJ, Gupta R, Meena HS, Singh D (2016) Rapid screening technique for Alternaria blight resistance in Indian mustard (*Brassica juncea* L.) using cotyledonary leaf method. J Plant Pathol 98(3):463–469

Mei J, Qian L, Disi JO, Yang X, Li Q, Li J, Frauen M, Cai D, Qian W (2011) Identification of resistant sources against *Sclerotinia sclerotiorum* in *Brassica* species with emphasis on *B. oleracea*. Euphytica 177:393–399

Mei J, Ding Y, Lu K, Wei D, Liu Y, Disi JO, Li J, Liu L, Liu S, McKay J, Qian W (2013) Identification of genomic regions involved in resistance against *Sclerotinia sclerotiorum* from wild *Brassica oleracea*. Theor Appl Genet 126:549–556. https://doi.org/10.1007/s00122-012-2000-x

Mei J, Liu Y, Wei D, Wittkop B, Ding Y, Li Q, Li J, Wan H, Li Z, Ge X, Frauen M, Snowdon RJ, Qian W, Friedt W (2015) Transfer of *Sclerotinia* resistance from wild relative of *Brassica oleracea* into *Brassica napus* using a hexaploidy step. Theor Appl Genet 128:639–644

Mengiste T (2012) Plant immunity to necrotrophs. Annu Rev Phytopathol 50:267–294

Mengistu A, Rimmer SR, Koch E, Williams PH (1991) Pathogenicity grouping of isolates of *Leptosphaeria maculans* on *Brassica napus* cultivars and their disease reaction profiles on rapid-cycling Brassicas. Plant Dis 75:1279–1282

Miah G, Rafii MY, Ismail MR, Puteh AB, Rahim HA, Asfaliza R, Latif MA (2013) Blast resistance in rice: a review of conventional breeding to molecular approaches. Mol Biol Rep 40 (3):2369–2388. https://doi.org/10.1007/s11033-012-2318-0

Micic Z, Hahn V, Bauer E, Schon CC, Knapp SJ, Tang S, Melchinger AE (2004) QTL mapping of *Sclerotinia* midstalk-rot resistance in sunflower. Theor Appl Genet 109(7):1474–1484. https://doi.org/10.1007/s00122-004-1764-z

Ming PY, Uloth MB, Xi XL, Banga SS, Banga SK, Barbetti MJ (2016) Valuable new resistances ensure improved management of *Sclerotinia* stem rot (*Sclerotinia sclerotiorum*) in horticultural and oilseed brassica species. J Phytopathol 164(5):291–299. https://doi.org/10.1111/jph.12456

Mir ZA, Ali S, Shivaraj SM, Bhat JA, Singh A, Yadav P, Rawat S, Paplao PK, Grover A (2020) Genome-wide identification and characterization of Chitinase gene family in *Brassica juncea* and *Camelina sativa* in response to *Alternaria brassicae*. Genomics 112(1):749–763

Mishra A, Pandey D, Goel A, Kumar A (2010) Molecular cloning and in-silico analysis of functional homologous of hypersensitive response genes induced during pathogenesis of Alternaria Blight in two genotypes of *Brassica*. J Proteom Bioinformatics 3:244–248

Mithen RF, Lewis BG (1988) Resistance to *Leptosphaeria maculans* in hybrids of *Brassica oleracea* and *Brassica insularis*. J Phytopathol 123:253–258

Mithen RF, Lewis BG, Heaney RK, Fenwick GR (1987) Resistance of leaves of *Brassica* species to *Leptosphaeria maculans*. Trans Br Mycol Soc 88:525–531

Mukherjee AK, Lev S, Gepstein S, Horwitz BA (2009) A compatible interaction of *Alternaria brassicicola* with *Arabidopsis thaliana* ecotype DiG: evidence for a specific transcriptional signature. BMC Plant Biol 9(1):31

Mun JH, Yu HJ, Park S, Park B-S (2009) Genome-wide identification of NBS-encoding resistance genes in *Brassica rapa*. Mol Gen Genomics 282:617–631

Munir S, Shahzad SN, Qureshi MK (2020) Acuities tolerance mechanisms via different bioassay during *Brassicaceae-Alternaria brassicicola* interaction and its impact on yield. PLoS One 15 (12):e0242545. https://doi.org/10.1371/journal.pone.0242545

Narusaka Y, Narusaka M, Seki M, Ishida J, Nakashima M, Kamiya A, Enju A, Sakurai T, Satoh M, Kobayashi M (2003) The cDNA microarray analysis using an *Arabidopsis pad3* mutant reveals the expression profiles and classification of genes induced by *Alternaria brassicicola* attack. Plant Cell Physiol 44:377–387. https://doi.org/10.1093/pcp/pcg050

Navabi ZK, Strelkov SE, Good AG, Thiagarajah MR, Rahman MH (2010) *Brassica* B-genome resistance to stem rot (*Sclerotinia sclerotiorum*) in a doubled haploid population of *Brassica napus* × *Brassica carinata*. Can J Plant Pathol 32:237–246. https://doi.org/10.1080/07060661.2010.484229

Nayanakantha NMC, Rawat S, Ali S, Grover A (2016) Differential expression of defense related genes in *Sinapis alba* and *Brassica juncea* upon the infection of *Alternaria brassicae*. Trop Agric Res 27(3):123–136

Neik TX, Barbetti MJ, Batley J (2017) Current status and challenges in identifying disease resistance genes in *Brassica napus*. Front Plant Sci 8:1788. https://doi.org/10.3389/fpls.2017.01788

Nelson KA, Renner KA, Hammerschmidt R (2002) Effects of protoporphyrinogen oxidase inhibitors on soybean (*Glycine max* L.) response, *Sclerotinia sclerotiorum* disease development and phytoalexin production by soybean. Weed Technol 16:353–359

Newman PL, Bailey DJ (1987) Screening for resistance to canker (*Leptosphaeria maculans*) in winter oilseed rape (*Brassica napus* ssp. *oleifera*). Plant Pathol 36:346–354

Newton HC, Sequeira L (1972) Possible sources of resistance in lettuce to *Sclerotinia sclerotiorum*. Plant Dis Rep 56:875–878

Nordborg M, Weigel D (2008) Next-generation genetics in plants. Nature 456(7223):720–723. https://doi.org/10.1038/nature07629

Obermeier C, Hossain MA, Snowdon R, Knufer J, von Tiedemann A, Friedt W (2013) Genetic analysis of phenylpropanoid metabolites associated with resistance against *Verticillium longisporum* in *Brassica napus*. Mol Breed 31:347–361

Ogbonnaya FC, Halloran GM, Marcroft SJ, Pang ECK, Gororo N (2003) Progress in the utilisation of *Brassica nigra* in breeding for resistance to blackleg (*Leptosphaeria maculans*). In: Proc 11th Int Rapeseed Congress, Copenhagen, Denmark, vol 1, pp 39–41

Onifade AK, Agboola YAJ (2003) Effect of fungal infection on proximate nutrient composition of coconut (*Cocos nucifera* L.) fruit. J Food Agric Environ 1:141–142

Orellana RG (1975) Photoperiod influence on the susceptibility of sunflower to Sclerotinia stalk rot. Phytopathology 65:1293–1298

Osborn TC, Butruille DV, Sharpe AG, Pickering KJ, Parkin IAP, Parker JS, Lydiate DJ (2003) Detection and effects of a homeologous reciprocal transposition in *Brassica napus*. Genetics 165:1569–1577

Osorio S, Castillejo C, Quesada MA, Medina-Escobar N, Brownsey GJ, Suau R, Heredia A, Botella MA, Valpuesta V (2008) Partial demethylation of oligogalacturonides by pectin methyl esterase 1 is required for eliciting defense responses in wild strawberry (*Fragaria vesca*). Plant J 54:43–55

Pang ECK, Halloran GM (1996a) The genetics of adult plant blackleg (*Leptosphaeria maculans*) resistance from *Brassica juncea* in *B. napus*. Theor Appl Genet 92:382–387

Pang ECK, Halloran GM (1996b) The genetics of blackleg [Leptosphaeria maculans (Desm.) Ces.et De Not.] resistance in rapeseed (*Brassica napus* L.). II. Seedling and adult-plant resistance as quantitative traits. Theor Appl Genet 93:941–949

Pantelides IS, Tjamos SE, Paplomatas EJ (2010) Ethylene perception via ETR1 is required in *Arabidopsis* infection by *Verticillium dahliae*. Mol Plant Pathol 11:191–202. https://doi.org/10.1111/j.1364-3703.2009.00592.x

Paran I, Michelmore RW (1993) Development of reliable PCR-based markers linked to downy mildew resistance genes in lettuce. Theor Appl Genet 85:985–993

Parkin IAP, Sharpe AG, Keith DJ, Lydiate DJ (1995) Identification of the A and C genomes of amphidiploid *Brassica napus* (oilseed rape). Genome 38:1122–1131

Parkin IA, Gulden SM, Sharpe AG, Lukens L, Trick M, Osborn TC, Lydiate DJ (2005) Segmental structure of the *Brassica napus* genome based on comparative analysis with *Arabidopsis thaliana*. Genetics 171:765–781

Parkin IA, Koh C, Tang H, Robinson SJ, Kagale S, Clarke WE, Town CD, Nixon J, Krishnakumar V, Bidwell SL, Denoeud F, Belcram H, Links MG, Just J, Clarke C, Bender T, Huebert T, Mason AS, Pires JC, Barker G, Moore J, Walley PG, Manoli S, Batley J, Edwards D, Nelson MN, Wang X, Paterson AH, King G, Bancroft I, Chalhoub B, Sharpe AG (2014) Transcriptome and methylome profiling reveals relics of genome dominance in the mesopolyploid *Brassica oleracea*. Genome Biol 15(6):R77

Parlange F, Daverdin G, Fudal I, Kuhn ML, Balesdent MH, Blaise F, Grezes-Besset B, Rouxel T (2009) *Leptosphaeria maculans* avirulence gene AvrLm4-7 confers a dual recognition specificity by the *Rlm4* and *Rlm7* resistance genes of oilseed rape, and circumvents Rlm4-mediated recognition through a single amino acid change. Mol Microbiol 71:851–863. https://doi.org/10.1111/j.1365-2958.2008.06547.x

Pedras MSC, Ahiahonu PWK (2004) Phytotoxin production and phytoalexin elicitation by the phytopathogenic fungus *Sclerotinia sclerotiorum*. J Chem Ecol 30:2163–2179

Pedras MSC, Zaharia IL, Gai Y, Zhou Y, Ward DE (2001) In planta sequential hydroxylation and glycosylation of a fungal phytotoxin: avoiding cell death and overcoming the fungal invader. Proc Natl Acad Sci U S A 98:747–752. https://doi.org/10.1073/pnas.98.2.747

Perchepied L, Balague C, Riou C, Claudel-Renard C, Riviere N, Grezes-Besset B, Roby D (2010) Nitric oxide participates in the complex interplay of defense-related signaling pathways controlling disease resistance to *Sclerotinia sclerotiorum* in *Arabidopsis thaliana*. Mol Plant-Microbe Interact 23:846–860. https://doi.org/10.1094/MPMI-23-7-0846

Perez-Quintero AL, Quintero A, Urrego O, Vanegas P, Lopez C (2012) Bioinformatic identification of cassava miRNAs differentially expressed in response to infection by *Xanthomonas axonopodis* pv. *manihotis*. BMC Plant Biol 12:29. http://www.biomedcentral.com/1471-2229/12/29

Petit-Houdenot Y, Fudal I (2017) Complex interactions between fungal avirulence genes and their corresponding plant resistance genes and consequences for disease resistance management. Front Plant Sci 8:1072. https://doi.org/10.3389/fpls.2017.01072

Pilet ML, Delourme R, Foisset N, Renard M (1998) Identification of loci contributing to quantitative field resistance to blackleg disease, causal agent *Leptosphaeria maculans* (Desm.) Ces.Et de Not., in winter rapeseed (*Brassica napus* L.). Theor Appl Genet 96:23–30

Pilet ML, Duplan G, Archipiano M, Barret P, Baron C, Horvais R, Tanguy X, Lucas MO, Renard M, Delourme R (2001) Stability of QTL for field resistance to blackleg across two genetic backgrounds in oilseed rape. Crop Sci 41:197–205

Platt A, Vilhjalmsson BJ, Nordborg M (2010) Conditions under which genome-wide association studies will be positively misleading. Genetics 186(3):1045–1052. https://doi.org/10.1534/genetics.110.121665

Plieske J, Struss D, Robbelen G (1998) Inheritance of resistance derived from the B-genome of *Brassica* against *Phoma lingam* in rapeseed and the development of molecular markers. Theor Appl Genet 97:929–936. https://doi.org/10.1007/s001220050973

Plissonneau C, Daverdin G, Ollivier B, Blaise F, Degrave A, Fudal I, Rouxel T, Balesdent MH (2016) A game of hide and seek between avirulence genes *AvrLm4-7* and *AvrLm3* in *Leptosphaeria maculans*. New Phytol 209(4):1613–1624

Plissonneau C, Rouxel T, Chèvre AM, Van De Wouw AP, Balesdent MH (2017) One gene-one name: the *AvrLmJ1* avirulence gene of *Leptosphaeria maculans* is AvrLm5. Mol Plant Pathol 19:1012. https://doi.org/10.1111/mpp.12574

Poland J, Balint-Kurti P, Wisser RJ, Pratt RC, Nelson RJ (2009) Shades of gray: the world of quantitative disease resistance. Trends Plant Sci 14:21–29. https://doi.org/10.1016/j.tplants.2008.10.006

Porter DM, Beute MK, Wynne JC (1975) Resistance of peanut germplasm to *Sclerotinia sclerotiorum*. Peanut Sci 2:78–80

Price AL, Patterson NJ, Plenge RM, Weinblatt ME, Shadick NA, Reich D (2006) Principal components analysis corrects for stratification in genome-wide association studies. Nat Genet 38:904–909

Pu Z, Shimizu M, Zhang Y, Nagaoka T, Hayashi T, Hori H, Matsumoto S, Fujimoto R, Okazaki K (2012) Genetic mapping of a Fusarium wilt resistance gene in *Brassica oleracea*. Mol Breed 30:809–818. https://doi.org/10.1007/s11032-011-9665-8

Pumphrey MO, Bernardo R, Anderson JA (2007) Validating the Fhb1 QTL for Fusarium head blight resistance in near-isogenic wheat lines developed from breeding populations. Crop Sci 47:200–206

Raboin LM, Ballini E, Tharreau D, Ramanantsoanirina A, Frouin J, Courtois B, Ahmadi N (2016) Association mapping of resistance to rice blast in upland field conditions. Rice (NY) 9(1):59. https://doi.org/10.1186/s12284-016-0131-4

Rafalski JA (2010) Association genetics in crop improvement. Curr Opin Plant Biol 13:174–180

Ragimekula N, Varadarajula NN, Mallapuram SP, Gangimeni G, Reddy RK, Kondreddy HR (2013) Marker assisted selection in disease resistance breeding. J Plant Breed Genet 01:90–109

Rai JN, Singh RP, Saxena VC (1979) Phenolics and cell wall degrading enzymes in *Sclerotinia* infection of resistant and susceptible plants of *Brassica juncea*. Indian J Mycol Plant Pathol 9:160–165

Rajarammohan S, Kumar A, Gupta V, Pental D, Pradhan AK, Kaur J (2017) Genetic architecture of resistance to *Alternaria brassicae* in *Arabidopsis thaliana*: QTL mapping reveals two major resistance-conferring loci. Front Plant Sci 8:260

Rajarammohan S, Pradhan AK, Pental D, Kaur J (2018) Genome-wide association mapping in *Arabidopsis* identifies novel genes underlying quantitative disease resistance to *Alternaria brassicae*. Mol Plant Pathol 19:1719–1732

Raman R, Taylor B, Lindbeck K, Coombes N, Barbulescu D, Salisbury P, Raman H (2012a) Molecular mapping and validation of *Rlm1* genes for resistance to *Leptosphaeria maculans* in canola (*Brassica napus* L). Crop Pasture Sci 63:1007–1017. https://doi.org/10.1071/CP12255

Raman R, Taylor B, Marcroft S, Stiller J, Eckermann P, Coombes N, Rehman A, Lindbeck K, Luckett D, Wratten N, Batley J, Edwards D, Wang X, Raman H (2012b) Molecular mapping of qualitative and quantitative loci for resistance to *Leptosphaeria maculans* causing blackleg disease in canola (*Brassica napus* L.). Theor Appl Genet 125:405–418. https://doi.org/10.1007/s00122-012-1842-6

Raman H, Raman R, Larkan N (2013) Genetic dissection of blackleg resistance loci in rapeseed (*Brassica napus* L.). In: Andersen SB (ed) Plant breeding from laboratories to fields. InTech, Rijeka, pp 85–120

Raman H, Raman R, Diffey S, Qiu Y, McVittie B, Barbulescu DM, Salisbury PA, Marcroft S, Delourme R (2018) Stable quantitative resistance loci to blackleg disease in canola (Brassica napus L.) over continents. Front Plant Sci 9:9

Raman R, Diffey S, Barbulescu DM, Coombes N, Luckett D, Salisbury P, Cowley R, Marcroft S, Raman H (2020) Genetic and physical mapping of loci for resistance to blackleg disease in canola (*Brassica napus* L.). Sci Rep 10:4416. https://doi.org/10.1038/s41598-020-61211-y

Rana K, Atri C, Akhatar J, Kaur R, Goyal A, Singh MP, Kumar N, Sharma A, Sandhu PS, Kaur G, Barbetti MJ, Banga SS (2019) Detection of first marker trait associations for resistance against *Sclerotinia sclerotiorum* in *Brassica juncea–Erucastrum cardaminoides* introgression lines. Front Plant Sci 10:1015. https://doi.org/10.3389/fpls.2019.01015

Ratzinger A, Riediger N, Tiedemann AV, Karlovsky P (2009) Salicylic acid and salicylic acid glucoside in xylem sap of *Brassica napus* infected with *Verticillium longisporum*. J Plant Res 122:571–579

Rawat S, Ali S, Mittra B, Grover A (2017) Expression analysis of chitinase upon challenge inoculation to Alternaria wounding and defense inducers in *Brassica juncea*. Biotechnol Rep 13:72–79

Rezaeizad A, Wittkop B, Snowdon R, Hasan M, Mohammadi V, Zali A, Friedt W (2011) Identification of QTLs for phenolic compounds in oilseed rape (*Brassica napus* L.) by association mapping using SSR markers. Euphytica 177:335–342

Ribaut JM, de Vicente MC, Delannay X (2010) Molecular breeding in developing countries: challenges and perspectives. Curr Opin Plant Biol 13:213–218

Richardson KL, Vales MI, Kling JG, Mundt CC, Hayes PM (2006) Pyramiding and dissecting disease resistance QTL to barley stripe rust. Theor Appl Genet 113:485–495

Ridley BL, O'Neill MA, Mohnen D (2001) Pectins: structure, biosynthesis, and oligogalacturonide-related signaling. Phytochemistry 57:929–967

Rimmer SR (2006) Resistance genes to *Leptosphaeria maculans* in *Brassica napus*. Can J Plant Pathol 28:S288–S297. https://doi.org/10.1080/07060660609507386

Rimmer SR, Van den Berg CGJ (1992) Resistance of oilseed *Brassica* spp. to blackleg caused by *Leptosphaeria maculans*. Can J Plant Pathol 14:56–66. https://doi.org/10.1080/07060669209500906

Ronicke S, Hahn V, Vogler A, Friedt W (2005) Quantitative trait loci analysis of resistance to *Sclerotinia sclerotiorum* in sunflower. Phytopathology 95(7):834–839. https://doi.org/10.1094/PHYTO-95-0834

Roux F, Voisin D, Badet T, Balague C, Barlet X, Huard-Chauveau C, Roby D, Raffaele S (2014) Resistance to phytopathogens e tutti quanti: placing plant quantitative disease resistance on the map. Mol Plant Pathol 15(5):427–432. https://doi.org/10.1111/mpp.12138

Rouxel T, Penaud A, Pinochet X, Brun H, Gout L, Delourme R, Schmit J, Balesdent MH (2003a) A 10-year survey of populations of *Leptosphaeria maculans* in France indicates a rapid adaptation towards the *Rlm1* resistance gene of oilseed rape. Eur J Plant Pathol 109:871–881. https://doi.org/10.1023/A:1026189225466

Rouxel T, Willner E, Coudard L, Balesdent MH (2003b) Screening and identification of resistance to *Leptosphaeria maculans* (stem canker) in *Brassica napus* accessions. Euphytica 133:219–231. https://doi.org/10.1023/A:1025597622490

Rouxel T, Grandaubert J, Hane JK, Hoede C, Van de Wouw AP, Couloux A, Dominguez V, Anthouard V, Bally P, Bourras S, Cozijnsen AJ, Ciuffetti LM, Degrave A, Dilmaghani A, Duret L, Fudal I, Goodwin SB, Gout L, Glaser N, Linglin J, Kema GH, Lapalu N, Lawrence CB, May K, Meyer M, Ollivier B, Poulain J, Schoch CL, Simon A, Spatafora JW, Stachowiak A, Turgeon BG, Tyler BM, Vincent D, Weissenbach J, Amselem J, Quesneville H, Oliver RP, Wincker P, Balesdent MH, Howlett BJ (2011) Effector diversification within compartments of the Leptosphaeria maculans genome affected by repeat-induced point mutations. Nat Commun 2:202

Roy NN (1978) A study on disease variation in the populations of an interspecific cross of *Brassica juncea* L. x *Brassica napus* L. Euphytica 27:145–149

Roy NN (1984) Interspecific transfer of *Brassica juncea*-type high blackleg resistance to *Brassica napus*. Euphytica 33:295–303

Roy NN, Fisher HM, Tarr A (1983) Wesbrook – a new prime variety of rapeseed. In: Proc Fourth Australian Rapeseed Agronomists and Breeders Workshop, Lyndoch. 4 p

Rygulla W, Friedt W, Seyis F, Luhs W, Eynck C, von Tiedemann A, Snowdon RJ (2007a) Combination of resistance to *Verticillium longisporum* from zero erucic acid *Brassica oleracea* and oilseed *Brassica rapa* genotypes in resynthesized rapeseed (*Brassica napus*) lines. Plant Breed 126:596–602

Rygulla W, Snowdon RJ, Eynck C, Koopmann B, von Tiedemann A, Luhs W, Friedt W (2007b) Broadening the genetic basis of *Verticillium longisporum* resistance in *Brassica napus* by interspecific hybridisation. Phytopathology 97:1391–1396

Rygulla W, Snowdon RJ, Friedt W, Happstadius I, Cheung WY, Chen D (2008) Identification of quantitative trait loci for resistance against *Verticillium longisporum* in oilseed rape (*Brassica napus*). Phytopathology 98:215–221

Saal B, Brun H, Glais I, Struss D (2004) Identification of a *Brassica juncea*-derived recessive gene conferring resistance to *Leptosphaeria maculans* in oilseed rape. Plant Breed 123:505–511

Sacristan MD, Gerdemann M (1986) Different behavior of *Brassica juncea* and *Brassica carinata* as sources of *Phoma lingam* resistance in experiments of interspecific transfer to Brassica napus. Plant Breed 97:304–314

Sagi MS, Deokar AA, Taran B (2017) Genetic analysis of NBS-LRR gene family in chickpea and their expression profiles in response to ascochyta blight infection. Front Plant Sci 8:838

Saha P, Kalia P, Sharma P, Sharma TR (2014) Race-specific genetics of resistance to black rot disease [*Xanthomonas campestris* pv. *campestris* (Xcc) (Pammel) Dowson] and the development of three random amplified polymorphic DNA markers in cauliflower. J Hortic Sci Biotechnol 89:480–486

Saharan GS, Mehta N, Meena PD (2016) Alternaria blight of crucifers: biology, ecology and disease management, vol 326. Springer, Singapore. ISBN 978-981-10-0019-5

Salisbury PA, Wratten N (1999) *Brassica napus* breeding. In: Salisbury PA, Potter T, McDonald G, Green AG (eds) Canola in Australia: the first thirty years. Organising Committee of the 10th International Rapeseed Congress, Canberra, ACT, pp 29–35

Salisbury PA, Ballinger DJ, Wratten N, Plummer KM, Howlett BJ (1995) Blackleg disease on oilseed Brassica in Australia: a review. Aust J Exp Agric 35:665–672

Santos C, Nogueira FCS, Domont GB, Fontes W, Prado GS, Habibi P, Santos VO, Oliveira-Neto OB, Grossi-de-Sá MF, Jorrín-Novo JV, Franco OL, Mehta A (2019) Proteomic analysis and functional validation of a *Brassica oleracea* endochitinase involved in resistance to *Xanthomonas campestris*. Front Plant Sci 10:414. https://doi.org/10.3389/fpls.2019.00414

Sarris PF, Duxbury Z, Huh SU, Ma Y, Segonzac C, Sklenar J, Derbyshire P, Cevik V, Rallapalli G, Saucet SB, Wirthmueller L, Menke FLH, Sohn KH, Jones JDG (2015) A plant immune receptor detects pathogen effectors that target WRKY transcription factors. Cell 161:1089–1100

Sellam A, Dongo A, Guillemette T, Hudhomme P, Simoneau P (2007) Transcriptional responses to exposure to the brassicaceous defense metabolites camalexin and allyl-isothiocyanate in the necrotrophic fungus *Alternaria brassicicola*. Mol Plant Pathol 8:195–208

Sharma G, Kumar VD, Haque A, Bhat SR, Prakash S, Chopra VL (2002) *Brassica* coenospecies: a rich reservoir for genetic resistance to leaf spot caused by *Alternaria brassicae*. Euphytica 125:411–417

Sharma BB, Kalia P, Yadava DK, Singh D, Sharma TR (2016) Genetics and molecular mapping of black rot resistance locus Xca1bc on chromosome B-7 in Ethiopian mustard (*Brassica carinata* A. Braun). PLoS One 11:e0152290. https://doi.org/10.1371/journal.pone.0152290

Shen Y, Diener AC (2013) *Arabidopsis thaliana* resistance to *Fusarium oxysporum* 2 implicates tyrosine-sulfated peptide signaling in susceptibility and resistance to root infection. PLoS Genet 9(5):e1003525. https://doi.org/10.1371/journal.pgen.1003525

Shimizu M, Pu ZJ, Kawanabe T, Kitashiba H, Matsumoto S, Ebe Y, Sano M, Funaki T, Fukai E, Fujimoto R, Okazaki K (2015) Map–based cloning of a candidate gene conferring *Fusarium* yellows resistance in *Brassica oleracea*. Theor Appl Genet 128:119–130. https://doi.org/10.1007/s00122-014-2416-6

Singh S, Dey SS, Bhatia R, Batley J, Kumar R (2018) Molecular breeding for resistance to black rot [*Xanthomonas campestris* pv. *campestris* (Pammel) Dowson] in *Brassicas*: recent advances. Euphytica 214:196

Sjodin C, Glimelius K (1989) Transfer of resistance against *Phoma lingam* to *B. napus* by asymetric somatic hybridization combined with toxin selection. Theor Appl Genet 78:513–520

Snowdon RJ, Friedt W (2004) Molecular markers in *Brassica* oilseed breeding: current status and future possibilities. Plant Breed 123:1–8

Snowdon RJ, Winter H, Diestel A, Sacristan MD (2000) Development and characterization of *Brassica napus-Sinapis arvensis* addition lines exhibiting resistance to *Leptosphaeria maculans*. Theor Appl Genet 101:1008–1014

Soengas P, Hand P, Vicente JG, Pole JM, Pink DAC (2007) Identification of quantitative trait loci for resistance to *Xanthomonas campestris* pv. *campestris* in *Brassica rapa*. Theor Appl Genet 114:637–645

Song J, Li Z, Liu Z, Guo Y, Qiu L-J (2017) Next-generation sequencing from bulk-segregant analysis accelerates the simultaneous identification of two quantitative genes in soybean. Front Plant Sci 8:919. https://doi.org/10.3389/fpls.2017.00919

Sotelo T, Soengas P, Velasco P, Rodriguez VM, Cartea ME (2014) Identification of metabolic QTLc and candidate genes for glucosinolate synthesis in *Brassica oleracea* leaves, seeds and flower buds. PLoS One 9:e91428

Sprague SJ, Balesdent MH, Brun H, Hayden HL, Marcroft S, Pinochet X, Rouxel T, Howlett BJ (2006) Major gene resistance of *Brassica napus* (oilseed rape) is overcome by changes in virulence of populations of *Leptosphaeria maculans* in France and Australia. Eur J Plant Pathol 114:33–40

Steele KA, Virk DS, Kumar R, Prasad SC, Witcombe JR (2007) Field evaluation of upland rice lines selected for QTLs controlling root traits. Field Crop Res 101:180–186

Strittmatter G, Janssens J, Opsomer C, Botterman J (1995) Inhibition of fungal disease development in plants by engineering controlled cell death. Bio/Technology 13:1085–1089

Tai TH, Dahlbeck D, Clark ET, Gajiwala P, Pasion R, Whalen M, Stall RE, Staskawicz BJ (1999) Expression of the Bs2 pepper gene confers resistance to bacterial spot disease of tomato. Proc Natl Acad Sci U S A 96:14153–14158

Taj G, Agarwal P, Grant M, Kumar A (2010) MAPK machinery in plants: recognition and response to different stresses through multiple signal transduction pathways. Plant Signal Behav 5:1370–1378. https://doi.org/10.4161/psb.5.11.13020

Taylor JD, Conway J, Roberts SJ, Astley D, Vicente JG (2002) Source and origin of resistance to *Xanthomonas campestris* pv. *campestris* in *Brassica* genomes. Phytopathology 92:105–111

Taylor A, Coventry E, Jones JE, Clarkson JP (2015) Resistance to a highly aggressive isolate of *Sclerotinia sclerotiorum* in *Brassica napus* diversity set. Plant Pathol 64:932. https://doi.org/10.1111/ppa.12327

Tewari JP, Conn KH (1992) Control of Sclerotinia on canola by calcium application. Comm Proj 9101:17

Thomma BP, Eggermont K, Penninckx IA, Mauch-Mani B, Vogelsang R, Cammue BP, Broekaert WF (1998) Separate jasmonate-dependent and salicylate-dependent defense-response pathways in *Arabidopsis* are essential for resistance to distinct microbial pathogens. Proc Natl Acad Sci U S A 95(25):15107–15111. https://doi.org/10.1073/pnas.95.25.15107

Thomma BP, Eggermont K, Tierens KF, Broekaert WF (1999) Requirement of functional ethylene-insensitive 2 gene for efficient resistance of *Arabidopsis* to infection by *Botrytis cinerea*. Plant Physiol 121:1093–1102

Thornsberry JM, Goodman MM, Doebley J, Kresovich S, Nielsen D, Buckler ES (2001) Dwarf polymorphisms associate with variation in flowering time. Nat Genet 28:286–289

Tierens KF, Thomma BP, Brouwer M, Schmidt J, Kistner K, Porzel A, Mauch-Mani B, Cammue BP, Broekaert WF (2001) Study of the role of antimicrobial glucosinolate-derived isothiocyanates in resistance of *Arabidopsis* to microbial pathogens. Plant Physiol 125:1688–1699

Tiwari S, Pandey D, Gaur M, Kumar A (2017) Effect of methyl jasmonate on disease severity and expression of plant defensin gene during *Alternaria brassicae* infection in *Arabidopsis*. Int J Curr Microbiol App Sci 6(7):857–865

Tonguc M, Earle ED, Griffiths PD (2003) Segregation distortion of *Brassica carinata* derived black rot resistance in *Brassica oleracea*. Euphytica 134:269–276

Tonu NN, Doullah MA, Shimizu M, Karim MM, Kawanabe T, Fujimoto R, Okazaki K (2013) Comparison of positions of QTLs conferring resistance to *Xanthomonas campestris* pv. *campestris* in *Brassica oleracea*. Am J Plant Sci 4:11–20

Trda L, Baresova M, Sasek V, Novakova M, Zahajska L, Dobrev PI, Motyka V, Burketova L (2017) Cytokinin metabolism of pathogenic fungus *Leptosphaeria maculans* involves isopentenyl-transferase, adenosine kinase and cytokinin oxidase/dehydrogenase. Front Microbiol 8:1374

Uloth MB, You MP, Finnegan PM, Banga SS, Banga SK, Sandhu PS, Phillip HY, Salisbury A, Brbetti MJ (2013) New sources of resistance to *Sclerotinia sclerotiorum* for crucifer crops. Field Crop Res 154:40–52. https://doi.org/10.1016/j.fcr.2013.07.013

Uloth M, You MP, Finnegan PM, Banga SS, Yi H, Barbetti MJ (2014) Seedling resistance to *Sclerotinia sclerotiorum* as expressed across diverse cruciferous species. Plant Dis 98:184–190. https://doi.org/10.1094/PDIS-06-13-0612-RE

Uloth MB, You MP, Barbetti MJ (2015) Host resistance to Sclerotinia stem rot in historic and current *Brassica napus* and *Brassica juncea* varieties: critical management implications. Crop Pasture Sci 66:841–848. https://doi.org/10.1071/CP15064

Van de Wouw AP, Howlett BJ (2012) Estimating frequencies of virulent isolates in field populations of a plant pathogenic fungus, *Leptosphaeria maculans*, using high-throughput pyro sequencing. J Appl Microbiol 113:1145–1153. https://doi.org/10.1111/jam.2012.113.issue-5

Van De Wouw AP, Marcroft SJ, Barbetti MJ, Hua L, Salisbury PA, Gout L, Rouxel T, Howlett BJ, Balesdent MH (2009) Dual control of avirulence in *Leptosphaeria maculans* towards a *Brassica napus* cultivar with 'sylvestris -derived' resistance suggests involvement of two resistance genes. Plant Pathol 58:305–313

Van de Wouw AP, Cozijnsen AJ, Hane JK, Brunner PC, McDonald BA, Oliver RP, Howlett BJ (2010a) Evolution of linked avirulence effectors in *Leptosphaeria maculans* is affected by genomic environment and exposure to resistance genes in host plants. PLoS Pathog 6: e1001180. https://doi.org/10.1071/CP16411

Van de Wouw AP, Stonard JF, Howlett BJ, West JS, Fitt BDL, Atkins SD (2010b) Determining frequencies of avirulent alleles in airborne *Leptosphaeria maculans* inoculum using quantitative PCR. Plant Pathol 59:809–818. https://doi.org/10.1111/j.1365-3059.2010.02311.x

Van de Wouw AP, Lowe RGT, Elliott CE, Dubois DJ, Howlett BJ (2014) An avirulence gene, AvrLmJ1, from the blackleg fungus, Leptosphaeria maculans, confers avirulence to Brassica juncea cultivars. Mol Plant Pathol 15:523–530

Van den Ackerveken GFJM, Van Kan JAL, De Wit PJGM (1992) Molecular analysis of the avirulence gene avr9 of *Cladosporium fulvum* fully supports the gene-for-gene hypothesis. Plant J 2:359–366

Van den Ackerveken GFJM, Vossen P, De Wit PJGM (1993) The AVR9 race-specific elicitor of *Cladosporium fulvum* is processed by endogenous and plant proteases. Plant Physiol 103:91–96

Van Kan JAL, Van den Ackerveken GFJM, De Wit PJGM (1991) Cloning and characterization of the cDNA encoding the AVR9 race-specific elicitor of *Cladosporium fulvum*. Mol Plant-Microbe Interact 4:52–59

Van Wees SC, Chang HS, Zhu T, Glazebrook J (2003) Characterization of the early response of *Arabidopsis* to *Alternaria brassicicola* infection using expression profiling. Plant Physiol 132:606–617. https://doi.org/10.1104/pp.103.022186

Vasconcellos RC, Oraguzie OB, Soler A, Arkwazee H, Myers JR, Ferreira JJ, Song Q, McClean P, Miklas PN (2017) Meta-QTL for resistance to white mold in common bean. PLoS One 12(2): e0171685. https://doi.org/10.1371/journal.pone.0171685

Verma S, Yajima WR, Rahman MH, Shah S, Liu JJ, Ekramoddoullah AKM, Kav NNV (2012) A cysteine-rich antimicrobial peptide from *Pinus monticola* (PmAMP1) confers resistance to multiple fungal pathogens in canola (*Brassica napus*). Plant Mol Biol 79(1–2):61–74

Veronese P, Narasimhan ML, Stevenson RA, Zhu JK, Weller SC, Subbarao KV, Bressan RA (2003) Identification of a locus controlling *Verticillium* disease symptom response in *Arabidopsis thaliana*. Plant J 35:574–587

Vicente JG, Conway J, Roberts SJ, Taylor JD (2001) Identification and origin of *Xanthomonas campestris* pv. *campestris* races and related pathovars. Phytopathology 91:492–499

Wan XY, Wan JM, Jiang L, Wang JK, Zhai HQ, Weng JF, Wang HL, Lei CL, Wang JL, Zhang X, Cheng ZJ, Guo XP (2006) QTL analysis for rice grain length and fine mapping of an identified QTL with stable and major effects. Theor Appl Genet 112:1258–1270

Wan JR, Zhang XC, Neece D, Ramonell KM, Clough S, Kim SY, Stacey MG, Stacey G (2008) A LysM receptor-like kinase plays a critical role in chitin signaling and fungal resistance in *Arabidopsis*. Plant Cell 206(1):471–481

Wang HZ, Liu GH, Zheng YB, Wang XF, Yang Q (2004) Breeding of the *Brassica napus* cultivar Zhongshuang 9 with high-resistance to *Sclerotinia sclerotiorum* and dynamics of its important defense enzyme activity. Sci Agric Sin 37:23–28

Wang Z, Mao H, Dong C, Ji R, Cai L, Fu H, Liu S (2009) Over expression of *Brassica napus MPK4* enhances resistance to *Sclerotinia sclerotiorum* in oilseed rape. Mol Plant-Microbe Interact 22:235–244. https://doi.org/10.1094/MPMI-22-3-0235

Wang Y, Cheng X, Shan Q, Zhang Y, Liu J, Gao C, Qiu JL (2014) Simultaneous editing of three homoalleles in hexaploid bread wheat confers heritable resistance to powdery mildew. Nat Biotechnol 32:947–951

Wang X, Wu J, Liang J, Cheng F, Wang X (2015) *Brassica* database (BRAD) version 2.0: integrating and mining Brassicaceae species genomic resources. Database 2015:bav093. https://doi.org/10.1093/database/bav093

Wang Z, Yang C, Chen H, Wang P, Wang P, Song C, Zhang X, Wang D (2018) Multi-gene co-expression can improve comprehensive resistance to multiple abiotic stresses in *Brassica napus* L. Plant Sci 274:410–419. https://doi.org/10.1016/j.plantsci.2018.06.014

Wang Z, Bao L-L, Zhao F-Y, Tang M-Q, Chen T, Li Y, Wang B-X, Fu B, Fang H, Li G-Y, Cao J, Ding L-N, Zhu K-M, Liu S-Y, Tan X-L (2019) BnaMPK3 Is a key regulator of defense responses to the devastating plant pathogen *Sclerotinia sclerotiorum* in Oilseed rape. Front Plant Sci 10:91. https://doi.org/10.3389/fpls.2019.00091

Wang Z, Zhao FY, Tang MQ, Chen T, Bao LL, Cao J, Li YL, Yang YH, Zhu KM, Liu S, Tan XL (2020) *BnaMPK6* is a determinant of quantitative disease resistance against *Sclerotinia sclerotiorum* in oilseed rape. Plant Sci 291:110362

Wei D, Mei J, Fu Y, Disi JO, Li J, Qian W (2014) Quantitative trait loci analyses for resistance to *Sclerotinia sclerotiorum* and flowering time in *Brassica napus*. Mol Breed 34:1797–1804

Wei L, Jian H, Lu K, Filardo F, Yin N, Liu L, Qu C, Li W, Du H, Li J (2015) Genome-wide association analysis and differential expression analysis of resistance to Sclerotinia stem rot in *Brassica napus*. Plant Biotechnol J 14:1368. https://doi.org/10.1111/pbi.12501

Wei L, Jian H, Lu K, Filardo F, Yin N, Liu L, Qu C, Li W, Du H, Li J (2016) Genome-wide association analysis and differential expression analysis of resistance to Sclerotinia stem rot in *Brassica napus*. Plant Biotechnol J 14:1368–1380. https://doi.org/10.1111/pbi.12501

Wei W, Mesquita ACO, Figueiro AA, Wu X, Manjunatha S, Wickland DP, Hudson ME, Juliatti FC, Clough SJ (2017) Genome-wide association mapping of resistance to a Brazilian isolate of *Sclerotinia sclerotiorum* in soybean genotypes mostly from Brazil. BMC Genomics 18:849. https://doi.org/10.1186/s12864-017-4160-1

Wen Z, Tan R, Zhang S, Collins PJ, Yuan J, Du W, Gu C, Ou S, Song Q, An YQC, Boyse JF, Chilvers MI, Wang D (2018) Integrating GWAS and gene expression data for functional characterization of resistance to white mold in soya bean. Plant Biotechnol J 16:1825–1835. https://doi.org/10.1111/pbi.12918

Williams PH, Delwiche PA (1979) Screening for resistance to blackleg of crucifers in the seedling stage. In: Proceedings Eucarpia Conference on the Breeding of Cruciferous Crops, Wageningen, The Netherlands, pp 164–170

Winter H, Gaertig S, Diestel A, Sacristan MD (1999) Blackleg resistance of different origin transferred into *Brassica napus*. In: Proc 10th Intl Rapeseed Congr, Canberra, Australia. http://www.regional.org.au/papers/index.htm

Wissuwa M, Ae N (2001) Further characterization of two QTLs that increase phosphorus uptake of rice (*Oryza sativa* L.) under phosphorus deficiency. Plant Soil 237:275–286

Wright CA, Beattie GA (2004) *Pseudomonas syringae* pv. *tomato* cells encounter inhibitory levels of water stress during the hypersensitive response of *Arabidopsis thaliana*. Proc Natl Acad Sci U S A 101(9):3269–3274. https://doi.org/10.1073/pnas.0400461101

Wu J, Cai G, Tu J, Li L, Liu S, Luo X, Zhou L, Fan C, Zhou Y (2013) Identification of QTLs for resistance to Sclerotinia stem rot and BnaC.IGMT5.a as a candidate gene of the major resistant QTL SRC6 in *Brassica napus*. PLoS One 8:e67740. https://doi.org/10.1371/journal.pone.0067740

Wu J, Zhao Q, Yang Q, Liu H, Li Q, Yi X, Cheng Y, Guo L, Fan C, Zhou Y (2016) Comparative transcriptomic analysis uncovers the complex genetic network for resistance to *Sclerotinia sclerotiorum* in *Brassica napus*. Sci Rep 6:19007. https://doi.org/10.1038/srep19007

Wubben JP, Lawrence CB, De Wit PJGM (1996) Differential induction of chitinases and 1,3-β-glucanase gene expression in tomato by *Cladosporium fulvum* and its race-specific elicitors. Physiol Mol Plant Pathol 48:105–116

Xu YH, Liu R, Yan L, Liu ZQ, Jiang SC, Shen YY, Wang XF, Zhang DP (2011) Light-harvesting chlorophyll a/b-binding proteins are required for stomatal response to abscisic acid in *Arabidopsis*. J Exp Bot 63:1095–1106

Yan D, Scott RL, Biederman JA, Barron-Gafford G, Yang X, Moore DJP, Dannenberg MP, Wang X, Knowles JF, Yang J, Smith WK (2018) Quantifying plant physiological response to water stress with high-frequency, near-surface observations of chlorophyll fluorescence and photochemical reflectivity. In: Presented at 2018 AGU Fall Meeting, Washington, DC, 10–14 Dec 2018, Abstr B31N-2694

Ye G, Smith KF (2008) Marker assisted gene pyramiding for inbred line development: basic principles and practical guidelines. Int J Plant Breed 2:1–10

Yin X, Yi B, Chen W, Zhang W, Tu J, Fernando WGD, Fu T (2010) Mapping of QTLs detected in a *Brassica napus* DH population for resistance to *Sclerotinia sclerotiorum* in multiple environments. Euphytica 173:25–35. https://doi.org/10.1007/s10681-009-0095-1

Yu F, Lydiate DJ, Rimmer SR (2004) Identification and mapping of a third blackleg resistance locus in Brassica napus derived from B. rapa subsp. sylvestris. In: Plant, Animal & Microbe Genomes XII Conference, San Diego, California, USA

Yu F, Lydiate DJ, Rimmer SR (2005) Identification of two novel genes for blackleg resistance in *Brassica napus*. Theor Appl Genet 110:969–979. https://doi.org/10.1007/s00122-004-1919-y

Yu J, Pressoir G, Briggs WH, Bi IV, Yamasaki M, Doebley JF, McMullen MD, Gaut BS, Nielsen DM, Holland JB, Kresovich S, Buckler ES (2006) A unified mixed-model method for association mapping that accounts for multiple levels of relatedness. Nat Genet 38:203–208

Yu F, Lydiate DJ, Hahn K, Kuzmicz S, Hammond C, Rimmer SR (2007) Identification and mapping of a novel blackleg resistance locus *LepR4* in the progenies from *Brassica napus* x *B. rapa* subsp. Sylvestris. In: 12th International Rapeseed Congress, Wuhan

Yu F, Lydiate DJ, Rimmer SR (2008) Identification and mapping of a third blackleg resistance locus in *Brassica napus* derived from *B. rapa* subsp. *sylvestris*. Genome 51:64–72. https://doi.org/10.1139/G07-103

Yu F, Gugel RK, Kutcher HR, Peng G, Rimmer SR (2013) Identification and mapping of a novel blackleg resistance locus LepR4 in the progenies from *Brassica napus* × *B. rapa* subsp. *sylvestris*. Theor Appl Genet 126:307–315. https://doi.org/10.1007/s00122-012-1919-2

Zhang X, Peng G, Parks P, Hu B, Li Q, Jiang L, Niu Y, Huang S, Fernando WGD (2016) Identifying seedling and adult plant resistance of Chinese *Brassica napus* germplasm to *Leptosphaeria maculans*. Plant Pathol 66:752–762. https://doi.org/10.1111/ppa.12626

Zhao J, Meng J (2003) Genetic analysis of loci associated with partial resistance to *Sclerotinia sclerotiorum* in rapeseed (*Brassica napus* L.). Theor Appl Genet 106(4):759–764. https://doi.org/10.1007/s00122-002-1171-2

Zhao J, Peltier AJ, Meng J, Osborn TC, Grau CR (2004) Evaluation of Scleroinia stem rot resistance in oilseed *Brassica napus* using an inoculation technique under green house conditions. Plant Dis 88:1033–1039

Zhao J, Udall JA, Quijada PA, Grau CR, Meng J, Osborn TC (2006) Quantitative trait loci for resistance to *Sclerotinia sclerotiorum* and its association with a homeologous non-reciprocal transposition in *Brassica napus* L. Theor Appl Genet 112:509–516

Zhao J, Wang J, An L, Doerge RW, Chen ZJ, Grau CR, Meng J, Osborn TC (2007) Analysis of gene expression profiles in response to *Sclerotinia sclerotiorum* in *Brassica napus*. Planta 227:13–24. https://doi.org/10.1007/s00425-007-0586-z

Zhao M, Cai C, Zhai J, Lin F, Li L, Shreve J, Thimmapuram J, Hughes TJ, Meyers BC, Ma J (2015) Coordination of microRNAs, phasiRNAs, and NB-LRR Genes in response to a plant pathogen: insights from analyses of a set of soybean Rps gene near-isogenic lines. Plant Genome 8:1–13. https://doi.org/10.3835/plantgenome2014.09.0044

Zhu B, Rimmer SR (2003) Inheritance of resistance to *Leptosphaeria maculans* in two accessions of *B. napus*. Can J Plant Pathol 3:98–103

Zhu JS, Struss D, Robbelen G (1993) Studies on resistance to Phoma lingam in *Brassica napus-Brassica nigra* addition lines. Plant Breed 111:192–197

Zou J, Zhu JL, Huang SM, Tian ET, Xiao Y, Fu DH, Tu JX, Fu TD, Meng J (2010) Broadening the avenue of intersubgenomic heterosis in oilseed *Brassica*. Theor Appl Genet 120:283–290

4 Biometabolomics of Disease Resistance to Biotrophs

Abstract

Biometabolomics of disease resistance plays a significant role in induction of immunity signaling pathways and defense responses. With the perception of pathogens, signaling is initiated that results in the execution of a broad range of defense responses to stop the infection and pathogenesis of invading pathogens. Polar metabolites of *Brassica* plants including glucosinolates, polar indole metabolites, soluble compounds, and wall-bound phenolics are induced to confer resistance to *Albugo*. The phytoalexins cyclobrassinin, rapalexin, and brassilexin are inhibitory to *Albugo* zoospores release.

Keywords

Biometabolomics to biotrophs · Accumulation of phytoalexins · Components of host defense pathways · Induction of defense-related genes · Accumulation of JA/ET-responsive genes · Tryptophan-derived antimicrobials · Biochemical basis of resistance · Genes encoding camalexin · Extracellular deposition of proteins · Silicon-mediated resistance · Mechanisms of R-genes · Induction of salicylate · NPR1, PAD4, and EDS5 · Chitin gene · Mechanisms of *NPR1* gene · *MAP65*-3 gene · Receptor-like cytoplasmic kinases (RLCK) genes · Molecular mechanisms of camalexin synthesis · Salicylic acid-mediated signaling resistance · Jasmonic acid (JA), and ethylene (ET) mediated resistance · WRKY transcription · Hormonal signaling-induced transcription · Gene regulation · Non-host resistance · Mutagenic resistance · Induction of biomolecules · Synthesis of phytoalexins · Antipathogenic molecules · Resistance gene-signal transduction · R-genes pyramiding · Defense signaling transduction · Lignin biosynthesis · Metabolic pathway · Hormones related genes · Biometabolomic resistance

G. S. Saharan et al., *Molecular Mechanism of Crucifer's Host-Resistance*,
https://doi.org/10.1007/978-981-16-1974-8_4

Strong and accelerated activation of defense-related transcriptional programs in *mlo2 mlo6 mlo12* triple mutants of *Arabidopsis* has been observed against powdery mildew with jasmonic acid/ethylene (JA/ET) signaling and biosynthesis of indole-derived antimicrobial compounds. Defense compounds like camalexin, indole glucosinolates, and Indole cyanogenic metabolites provide full immunity in *mlo2 mlo6 mlo12* mutant to powdery mildew. Salicylic acid (SA) plays an important role in resistance to powdery mildew of *Arabidopsis* required for signaling of mutant's edr1, edr2, and edr4. Genes encoding camalexin biosynthesis for powdery mildew resistance have been determined. Pre-penetration resistance to powdery mildew of *Arabidopsis* operates through cuticle, and epicuticular wax when pathogen exerts pressure to penetrate cell wall, host activates, mechanosensors to activate immune signaling gene *CERK1* to contribute basal resistance. Resistance is expressed through papilla formation which contains callose, silicon, ROS, and phenolic compounds. Induction of salicylate, NPR1, PAD4, and EDS5 in *Arabidopsis* provides resistance to powdery mildew. Chito-oligomers in *Brassica* induce defense responses to powdery mildew. Salicylic acid dependent transcription of *RPW* 8.1 and *RPW* 8.2 genes leads to accumulation of SA and formation of SHL or HR in *Brassica* to confer powdery mildew resistance. *BjNPR1* transgenic *B. juncea* plants exhibit partial resistance to powdery mildew through activation of SAR. The expression of gene-encoding AtMAP 65-3 in *A. thaliana* downregulates SA signaling for powdery mildew infection. Loss of At MAP 65-3 function in the map65-3 mutant reduces infection at the stage of leaf penetration. There is role of AtROP-regulated AtRLCK VI A3 in basal resistance to *Erysiphe cruciferarum.* The PAD 3 overexpressing plants accumulate higher camalexin and MiMick resistant eyp 8301-3 mutant to powdery mildew. Powdery mildew resistance of atg 2 plants depends on SA signaling. Autophagy contributes to suppression of cell death and defense response to powdery mildew. JA/Et-induced defense responses are effective against virulent powdery mildew fungi if stimulated constitutively, artificially, or systemically despite biotrophic nature of the interaction. Expression of WRKY TFs in *Brassica* enhances resistance to powdery mildew in edr1 plants. SR 1 plays a critical role in powdery mildew resistance by regulating EIN 3 and NDR1 expression. Identification of defense-related TFs in the regulation indicates recognition of pathogen, response initiation and defense execute an efficient immune output. In *Arabidopsis,* WRKY 70 is transcriptionally induced by SA signaling and mediates basal resistance to powdery mildew. Transcriptional regulation and expression of several genes in response to powdery mildew interaction with host and nonhosts is regulated by signaling molecules like SA, JA, and ET pathways differentially. The application of *Trichoderma harzianum* and cultural filtrate provides systemic resistance to powdery mildew in *Brassica.* Phytohormones abscisic acid (ABA) is involved in NHR against powdery mildew of *Brassica.* Two distinct pre-haustorial nonhost defense mechanisms mediated by genes *PEN1, SNAP 33, VAMP 721, VAMP 722, PEN 2*, and *PEN* 3 are subject to pathogen-induced cell polarization and mediate the induction of antimicrobial compounds, which is common in both. *Arabidopsis* NHR to powdery mildew is based on multicomponent and independently effective defense layers. *PEN* gene-mediated pre-invasion resistance and post-invasion immunity is controlled by *EDS1, PAD4*, and *SAG* 101 genes. Mutagenic resistance to powdery mildew is expressed by biochemical changes in the form of fatty acids, enzymes, and biosynthetic pathways.

Induction of several biomolecules in *Brassica* confers resistance to downy mildew. The synthesis of phytoalexins in *A. thaliana* is mediated by genes *PAD1* to *PAD4* to provide resistance against downy mildew. In resistant genotypes of *Brassica*, cell wall lignification provides resistance to downy mildew. Salicylic acid is important for defense against most of *Hyaloperonospora parasitica* races as signal amplifier to induce antipathogenic molecules to provide downy mildew resistance. Downy mildew resistance signal pathways for the expression of R-gens are very complex.

R-genes pyramiding in *Brassica* for clubroot resistance strongly trigger multiple resistance pathways by hormonal signal transduction, secondary metabolites, transcription factors, and other process via multi-gene network controlled by SA, ROS, PCD, and R line *ZHE-226* gene. The pyramided line 618 R showed higher number of DEGs than the parental lines. PbBa 8.1 shared higher number of DEGs than CRb. SA and ROS signaling pathways are significant in pyramided and parental lines. There is significant indication of PR proteins, thaumatin, and lipid transfer proteins in the pyramided line as compared to parental lines. Lignin biosysnthesis enhances resistance to *Plasmodiophora brassicae* at early stage of infection. Flavonoids act as defense compounds or antioxidants for clubroot. MET serves as the major energy source required for defense responses mediated by Rcr1 against clubroot. To activate defense response in *Brassica* to clubroot, there is induction of signal molecules like SA, BA, SABATH enzymes, JA, IAA, GA, and MeSA. At the later stages of *P. brassicae* infection, the gene expression is different in *B. napus* cvs. differing in resistance for induction of hormones like CK, IAA, SA, JA, ABA, and ET. In general gene expression was low in resistant cultivar and higher in susceptible ones. There is differential expression of hormones in leaves verses roots and formation of galls in resistant and susceptible cultivars indicating organ and host resistance specificity in production of hormones. However, SA signal transduction is upregulated mainly in the R-genotypes as comparison to susceptible ones. The biometabolomic resistance to clubroot in *Brassica* is expressed by NBS-LRR proteins of PbBa 1 and CRb which recognize the elicitors released by *P. brassicae* and trigger defense signal transduction. The key pathways are SA, JA, MAPK, ET, Ca^{++}, and ROS. RBOH and phenylpropanoid biosynthesis activate ROS, SA signaling, PCD, HR, and antioxidant enzymes to enhance defense responses in *Brassica* to clubroot. Numerous genes for immune-related functional categories are associated with SA-mediated responses. JA-mediated responses are mostly inhibited especially in clubroot resistant genotypes.

4.1 Introduction

The biometabolomics of disease resistance plays a significant role in induction of immunity signaling pathways and defense responses. Following the perception of pathogens, signaling is initiated that ultimately results in the execution of a broad range of defense responses that stop the invading pathogens. In *Arabidopsis*, many genetic screens have been performed to identify genes that are important for defense signaling. Some of these screens were focused on signaling downstream of specific R-proteins, while others were aimed to identify more general signal transduction components. The important plant proteins, EDS1 and NPR1, play crucial roles in the interaction with pathogens and are well studied in *Arabidopsis*. EDS1 and NPR1 are closely linked to the immunity-related hormone salicylic acid (SA) that is also key for resistance to several pathogens. Multi-omics approaches combining genomics, transcriptomics, proteomics, and biometabolomics using computational strategies will allow us to identify suitable mimicking molecules in the pathogen and/or host species that trigger stronger plant defense system during plant–pathogen interaction

EDS1 (enhanced disease susceptibility 1) is a lipase-like protein that is required for the function of a subgroup of NLRs, those with a Toll–interleukin 1 receptor domain called *TIR-NLR*. In addition, EDS1 has a function in oxidative stress responses and as a positive regulator in basal resistance to virulent pathogens making the *eds1* mutant enhanced disease-susceptible (Wiermer et al. 2005). Two sequence-related signaling partners, PAD4 (phytoalexin-deficient 4) and SAG101 (senescence-associated gene 101), form complexes with EDS1 and have signaling functions in the cytoplasm and nucleus. The NLR helper protein NRG1 also resides in this complex and plays a role in cell death signaling. Interestingly, EDS1 was found to associate with AvrRps4, a *Pseudomonas* effector, and to interact with TNLs, like RPS4 that mediate recognition of AvrRps4. It was hypothesized that EDS1, being important for basal resistance, could be a target for pathogen effectors and thus serve as a molecular bridge to mediate recognition by R-proteins that guard the EDS1 protein in addition to its signaling functions. This hypothesis is currently debated and under further investigation. The *eds1* mutant was originally identified in a screen for loss of resistance to downy mildew (Parker et al. 1996). EDS1 was shown to be required for the function of *TIR-NLR* genes such as *RPP1*, *RPP4*, *RPP5*, and *WRR4*. Virulent strains of *H. parasitica* and *Albugo* are also able to cause more severe infections on *eds1* mutant plants because of the reduced level of basal resistance. *Arabidopsis eds1* mutants also enhance susceptibility to a number of pathogens including *H. parasitica*. The lowered basal immunity is associated with reduced production of the defense hormone SA that is essential for defense against biotrophic pathogens. EDS1 positively regulates SA accumulation, acting upstream of SA, but the *EDS1* gene is also activated at the transcriptional level by SA. EDS1 thus seems to be a part of an SA-associated positive feedback loop of plant defense (Lapin et al. 2019).

SA is not only crucial for resistance to downy mildew, but also affects basal and R-gene-mediated resistance to many other pathogens of crucifers. Exogenous application of the hormone SA is sufficient to trigger efficient defense responses to a broad range of pathogens, and in particular to downy mildew and hemi-biotrophic species. SA (2-hydroxybenzoic acid) is synthesized via two main routes, the isochorismate (IC) and phenylalanine ammonia lyase (PAL) route. The *Arabidopsis* IC route goes via two IC synthases (ICS) and downstream steps that differ from the canonical pathway defined in bacteria (Torrens-Spence et al. 2019). The *ics1* (*sid2*) mutant has received the majority of experimental attention to date. *Ics1* mutants exhibit 90% lower levels of SA, and levels are ~95% lower in the *ics1 ics2* double mutant following powdery mildew infection. The *ICS1* gene is induced by MAMPs and by NLR-mediated pathogen recognition leading to high local levels of SA that activate defense. Besides local defenses, SA plays an important role, together with pipecolic acid, in systemic acquired resistance (SAR) that enhances immunity of distant plant tissues to protect against future infections. The export protein EDS5 has been implicated in synthesis of pipecolic acid, in addition to its well-known role in SA biosynthesis, revealing a surprising instance of immune convergence (Rekhter et al. 2019).

Downstream of SA, the central NPR1 (NONEXPRESSOR of PR GENES 1) protein plays a crucial role in the transcriptional activation of SA-induced defense in *Arabidopsis*. Mutation of *NPR1* leads to loss of basal resistance to (hemi-) biotrophic pathogens and of SAR. NPR1 proteins form a multimeric complex in the plant cell cytoplasm. Upon triggering of the plant immune system SA accumulates and subsequently elicits the thioredoxin-mediated reduction of a cysteine residue in NPR1. This results in the release of monomeric NPR1 that can travel into the nucleus where it interacts with TGA transcription factors to activate transcription of defense genes, like pathogenesis-related (*PR*) genes. NPR1 and the two paralogous *Arabidopsis* proteins NPR3 and NPR4 have been shown to bind SA and could thus constitute the SA immune receptors (Ding et al. 2018; Innes 2018).

The RPP proteins vary significantly in their requirements for functionality of these signaling components. For example, resistance mediated by the TIR-NLR protein RPP5 is strongly compromised by mutations that nullify EDS1, NPR1, or SA biosynthesis. On the other hand, the CC-NLR proteins RPP7 and RPP13 retain almost full functionality in these backgrounds. These differences underscore the complexity of ETI regulation. It is well known that many proteins are specifically produced during plant defense. In particular, many secreted and vacuolar proteins have been classified as pathogenesis-related proteins. Also in *Arabidopsis* many PR proteins have been identified, for example, PR1, PR2, and PR5 that were found in apoplastic fluid of *Arabidopsis* treated with 2,6-dichloroisonicotinic acid (INA), which is a mimic of SA. Several of these PR proteins have anti-pathogenic activities, while others contribute to resistance. One of the most used marker genes of SA-induced plant defense is *PR1*. Plants have multiple *PR1* genes encoding different protein variants with different activities. Overexpression of several *PR1* genes in different plant species also increased their resistance to pathogens. PR1 has been shown to bind sterols, indicative of defensive mode of action that is based on restriction of this important nutrient for pathogens (Gamir et al. 2017). The PR2 proteins are ß-1,3-beta-glucanases that are thought to attack the cell wall of invading

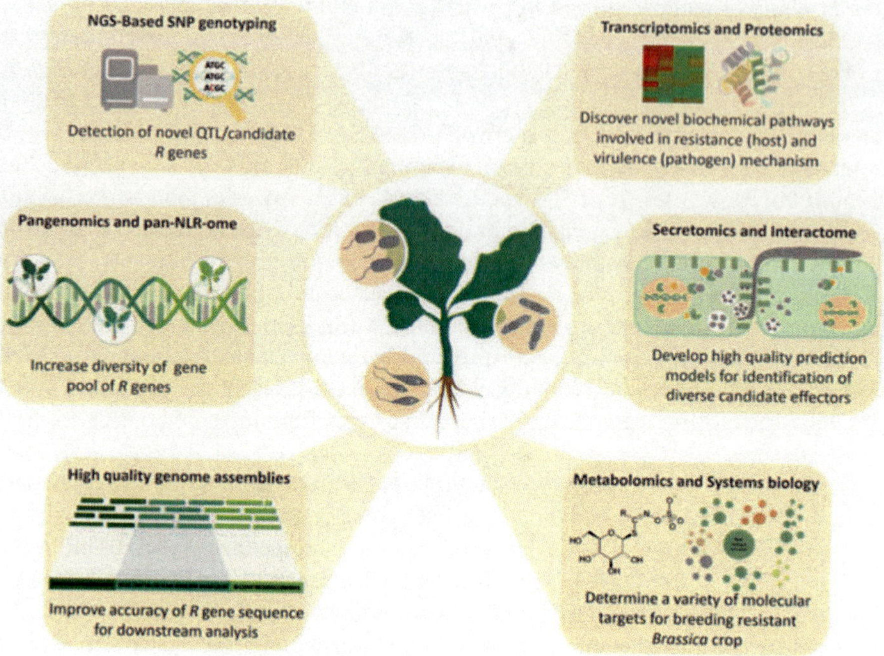

Fig. 4.1 The application of multi-omics technologies in the discovery of novel plant–pathogen interactions in the *Brassica* pathosystems (Neik et al. 2020)

fungal pathogens. Several cases of enhanced resistance of plants overexpressing *PR2* to pathogens are well known.

Other non-proteinaceous compounds that are important for resistance to pathogens in *Arabidopsis* are small secondary metabolites. Important anti-pathogenic compounds produced by *Arabidopsis* are for instance indole-glucosinolates (iGS) and the phytoalexin camalexin. Mutant plants that no longer produce either iGS or camalexin are only slightly more susceptible to *P. brassicae* and other pathogens. A very strong gain in susceptibility was observed in the double mutant *cyp79b2 cyp79b3* that is blocked in the production of indole-3-aldoxime, which is a common precursor for iGS and camalexin. Camalexin appears to contribute to resistance to *Arabidopsis* downy mildew, as several *phytoalexin-deficient* (*pad*) mutants are slightly more susceptible to *H. parasitica*. A combination of metabolomics with quantitative genetics was used to discover the potential role of gluconasturtiin in the *B. napus* resistance response against Clubroot and the underlying QTL controlling the trait on chromosome C03 and C09. Gluconasturtiin is a type of glucosinolate compound associated with the biotic resistance responses of *Brassica* species. With the use of multi-omics supplemented with functional studies key resistance pathways involved in the *Brassica* host pathosystem have been revealed (Fig. 4.1; Singh 2017; Wagner et al. 2019).

4.2 *Brassica–Albugo*: Biometabolomic Resistance

4.2.1 Accumulation of Phytoalexins and Polar Metabolites

In general, the metabolic responses of plants vary according to the type of stress. These responses can be rather specific since the metabolic pool of plant defenses is composed of a variety of constitutive and induced metabolites. Phytoalexins are induced antimicrobial metabolites produced de novo by plants in response to biotic or abiotic stress, whereas phytoanticipins are constitutive metabolites with a defensive role (Pedras et al. 2007). *Brassica rapa* canola is reported to produce several phytoalexins after $CuCl_2$ sprays (Rouxel et al. 1991). The metabolite profiles of phenylpropanoids in turnip leaves after insect infestation (Widarto et al. 2006), and methyl jasmonate sprays (Liang et al. 2006a, b) are reported.

Obligate biotrophs are capable of penetrating and colonizing plant host tissues in ways that prevent immediate recognition by the host. It has been suggested that a biotroph may induce production or suppression of different proteins in the host, and suppress induced resistance mechanisms such as phytoalexin production and hypersensitive host cell death (Mendgen and Hahn 2002; O'Connell and Panstruga 2006). Recently, a mutant of *A. thaliana* that accumulated significantly higher levels of the phytoalexin camalexin was reported to be more resistant than wild-type plants to virulent strains of the biotrophic pathogen *H. parasitica* (Veronese et al. 2004). Nonetheless, little is known about the chemical defense pathways of plants, including phytoalexin elicitation and accumulation during biotrophic infections (Mendgen and Hahn 2002; O'Connell and Panstruga 2006).

Phytoalexins, phytoanticipins, and polar metabolites from leaves of *B. rapa* cvs. Torch and Reward with four races of *A. candida* provide a model to analyze metabolic responses in compatible and incompatible interactions between a crucifer and a biotroph. This work established consistent phytoalexin production in response to inoculation with a biotroph (*A. candida* races 2V, 2A, 7V, and 7A). Although the accumulation of spirobrassinin, cyclobrassinin, and rutalexin in leaves inoculated with *A. candida* races 7V, 7A, 2A, or 2V was similar during the first 4 days, their concentrations decreased, and eventually undetectable, in incompatible interactions; in the compatible interactions the concentrations of all three phytoalexins increased, with the highest concentration at 7–10 days post-inoculation. These results suggest that during the initial stage of the interaction, leaves of *B. rapa* have a similar response to avirulent and virulent races of *A. candida* with respect to the accumulation of chemical defenses. After this stage, despite the higher phytoalexin concentration, the "compatible" races could overcome the plant defense system for further infection, but growth of the "incompatible" races was inhibited. The induction of invertase activity and defense proteins occurred very rapidly in leaves of *A. thaliana* after the initial challenge with *A. candida* in the incompatible interaction. By contrast, in compatible interactions (*A. candida–A. thaliana*) invertase activity, accumulation of sugars, and the repression of photosynthetic gene expression occurred several days after infection (Chou et al. 2000; Pedras et al. 2008).

Lower amounts of cyclobrassinin and higher amounts of rutalexin in infected leaves, and absence of rutalexin in $CuCl_2$-sprayed plants, suggest that these metabolites (and/or their biosynthetic precursors) play an important role in the plant–pathogen interaction. Results of bioassays showed that cyclobrassinin and brassilexin were stronger inhibitors of *A. candida* than rutalexin; this apparent redirection of the phytoalexin pathway toward increased amounts of rutalexin, and decreased level of cyclobrassinin and brassilexin accumulation might be caused by the pathogen. This exchange would, therefore, favor the pathogen and which is consistent with the lack of brassilexin production in infected plants, since brassilexin is produced only in $CuCl_2$-sprayed leaves. This hypothesis appears reasonable since cyclobrassinin is a biosynthetic precursor of both rutalexin and brassilexin, and the biosynthesis of rutalexin from cyclobrassinin is unlikely to involve more than three enzymatic steps (oxidation, hydrolysis, and methylation) (Pedras et al. 2007). Alternatively, *A. candida* might be able to detoxify both cyclobrassinin and brassilexin, similar to necrotrophic fungi (Pedras and Ahiahonu 2005). Although indolyl-3-acetonitrile, arvelexin, and caulilexin C were previously reported as phytoalexins of a few plant species (Pedras et al. 2007), these metabolites are phytoanticipins but not phytoalexins in cv. Torch and Reward because they were detected in control leaves also, and their production increased upon elicitation. In a study using 43 accessions of *Brassica* species, it was found that a *B. rapa* line 29 produced the phytoalexins cyclobrassinin, brassilexin, and cyclobrassinin sulfoxide 48 h after spraying with $CuCl_2$ (Rouxel et al. 1991).

Different types of polar metabolites, including esters and glucosides of hydroxycinnamic acid and their malic derivatives, flavonoids, ionone glucosides, amino acids, and nucleotides, were isolated from leaves of cv. Torch infected with *A. candida* (Pedras et al. 2008). Among this large pool, quantitative differences were detected in leaf extracts by HPLC for indole glucosinolates and tryptophan. In fact the production of 4-methoxyglucobrassicin (13b) was substantially higher in sprayed than in control leaves (Pedras et al. 2008). These results are consistent with the accumulation of higher amounts of arvelexin (10, $tR = 14.5$ min) in sprayed-leaves, as glucosinolates are known to give nitriles, isothiocyanates, etc., upon enzymatic degradation (Bones and Rossiter 2006). As well, since crucifer phytoalexins are biosynthesized from tryptophan, an increase in its concentration was expected; the highest concentration of tryptophan was detected 8 days after elicitation. Nonetheless, since tryptophan is a biosynthetic precursor of many different metabolites, its concentration increase could be related with other biosynthetic pathways. Earlier, it was reported that polar metabolites from cruciferous plants, including glucosinolates (Fahey et al. 2001), polar indole metabolites (Hahlbrock et al. 2003), soluble compounds (Hagemeier et al. 2001), and wall-bound phenolics (Tan et al. 2004), were induced and might play an important role in crucifer defense. Glucosyl indole-3-carboxylic acids were identified from *A. thaliana* roots infected with *Pythium sylvaticum* (Bednarek et al. 2005). As well, monolignol glucosides, for example, coniferin, were characterized from *A. thaliana* roots infected with *P. sylvaticum,* but their biological functions were not established (Whetten and Sederoff 1995).

Naturally occurring phenylpropanoids such as hydroxycinnamic acid, caffeic acid, ferulic acid, and sinapic acid can be present in plants as glucose esters and glucosides, or as esters of organic acids, such as quinic acid, malic acid, and tartaric acid. The 6 new methyl malates 14e–j were isolated from canola leaves and their structures were established using spectroscopic data and comparison with the corresponding free acid derivatives. The four hydroxycinnamoyl malic acids 14a–d as well as 5-hydroxyferuloyl malic acid were detected and characterized in extracts of turnip leaves by NMR analysis (Liang et al. 2006a, b). The hydroxycinnamoyl malic acids 14a–d is also isolated from radish (Brandl et al. 1984). These phenolic derivatives appeared to be related with UV resistance, but not with pathogen defense (Li et al. 1993; Hagemeier et al. 2001). No substantial quantitative differences in the accumulation of these metabolites were detected in biotically or abiotically stressed canola leaves (Pedras et al. 2008). The correlations observed between phytoalexin production in infected leaves of canola and rapeseed (biotrophic elicitation), and the outcome of the plant–pathogen interaction, suggest that *A. candida* is able to elude the plant defense mechanisms by redirecting the phytoalexin biosynthetic pathway. Considering both the nonpolar and polar metabolite profiles, it is clear that canola metabolic responses to the biotroph *A. candida* are distinct from responses to abiotic stress.

4.2.2 Phytoalexins and Metabolites from Zoosporangia

The main chemical components of zoosporangia of *A. candida* races 2V and 7V were determined to ensure that induced metabolites (detected or isolated) were produced by canola leaves and not by sporangia. Zoosporangia contained the phytoalexin spirobrassinin (1, 7 nmol/g of zoosporangia), rapalexins A and B (5, <2 nmol/g of zoosporangia), and trehalose (37 mg/g of zoosporangia) (Pedras et al. 2008). The presence of spirobrassinin was not surprising since this is a phytoalexin produced in leaves of many Brassicas including brown mustard (*B. juncea*), canola and rapeseed (*B. rapa*) from where zoosporangia of races 2V and 7V were isolated (spirobrassinin may adhere to zoosporangia surface). On the other hand, rapalexins A and B are much less abundant than spirobrassinin, and therefore, their presence was less expected. Nonetheless, because these compounds are induced by $CuCl_2$ as well, there is no doubt that they are produced by the plant. Trehalose is a storage carbohydrate in microorganisms accumulated mainly during starvation, and has been connected with stress protection (Rolland et al. 2006), as well as thermo- and cryo-protector and protein stabilizer (Voit 2003). Trehalose is one of the most important storage carbohydrates as it is present in almost all living organisms except mammals (Benaroudj et al. 2001).

It is worthy to note that the phytoalexins cyclobrassinin, rapalexin A, and brassilexin displayed substantially more potent inhibitory activity against zoospore release at much lower concentrations (ca. 2 mg/L) than the commercial fungicides Benomyl and Ridomil MZ (ca. 50 mg/L), because Ridomil MZ was found to be one of the most active fungicides against *A. candida* (Godika and Pathak 2005), whereas rapalexin A was observed a good lead structure to develop natural protection agents against pathogen. It would be of great interest to analyze lines of *Brassica* species to determine if the production of cyclobrassinin, rapalexin A, and brassilexin correlates positively with resistance to *A. candida* (Pedras et al. 2008).

4.3 *Brassica-Erysiphe*: Biometabolomic Resistance

4.3.1 Components of Host Defense Pathways for Powdery Mildew Resistance of *Arabidopsis* mlo2 mlo6 mlo12 Triple Mutant

The powdery mildew disease control is mediated by application of fungicides and/or genetically encoded types of resistance. Besides typically short-lived isolate-specific immunity, loss-of-function mutations in specific members of the Mildew Resistance Locus O (MLO) gene family provide stable protection (Jorgensen 1992; Brown 2015; Kusch et al. 2016). This durable, broad-spectrum type of immunity is conserved among monocots and dicots, including several important crop plants (Acevedo-Garcia et al. 2016; Kusch et al. 2016; Kusch and Panstruga 2017). Employment of mlo mutant plants in agriculture was pioneered by mutations in a single Mlo gene in diploid barley (Jorgensen 1992). Recent developments in targeted breeding and gene editing enable the transfer of this trait to further important crop plants with more complex genetic features that are less likely to acquire natural mutations in MLO genes (Wang et al. 2014; Acevedo-Garcia et al. 2016). However, despite the great potential that interference with MLO protein function offers to agriculture, the biochemical basis of the resistance is still incompletely understood. In *Arabidopsis*, three MLO proteins, MLO2 (At2g11310), MLO6 (At1g61560), and MLO12 (At2g39200), cooperatively mediate susceptibility to powdery mildew fungi (Consonni et al. 2006). The strongest susceptibility defect is caused by mutation of MLO2, while MLO6 and MLO12 have no detectable effect when mutated individually. Double mutant combinations of mlo2 with mlo6 or mlo12 gradually increase the resistance phenotype of the mlo2 single mutant, and lack of all three proteins results in complete penetration resistance (Consonni et al. 2006). Although MLO2, MLO6, and MLO12 show functional overlap in terms of mediating powdery mildew susceptibility, they seem to be only partially redundant as indicated by their differential quantitative impact on penetration resistance and distinct expression patterns (*Arabidopsis* eFP Browser is http://bar.utoronto.ca/efp_*Arabidopsis*/cgi-bin/efpWeb.cgi) (Winter et al. 2007; Bhardwaj et al. 2011). Consequently, components that have been recognized as contributors to mlo2-mediated resistance have to be validated for their impact on complete penetration resistance of the mlo2 mlo6 mlo12 triple mutant (Kuhn et al. 2017).

In *Arabidopsis*, mutation of components of nonhost resistance (NHR) abrogates penetration resistance of mlo2 mutant plants. The two defense pathways associated with the PENETRATION 1 (PEN1; At3g11820) and PEN2/PEN3 (At2g44490/At1g59870) proteins are major factors of NHR and critical for mlo2-mediated resistance (Stein et al. 2006; Bednarek et al. 2009). PEN1/Syntaxin of Plants

(SYP) 121 encodes a Soluble *N*-Ethylmalemide-Sensitive Factor Attachment Protein Receptor (t-SNARE) potentially involved in the exocytosis of defense compounds (Kwon et al. 2008; Kwaaitaal et al. 2010). Notably, the situation in *Arabidopsis* is reminiscent of barley, where the t-SNARE Ror2, the ortholog of PEN1, is required for full powdery mildew resistance of mlo mutants (Collins et al. 2003). The b-thioglucoside glucohydrolase PEN2 (Lipka et al. 2005) and the ABC transporter PEN3 (Stein et al. 2006) contribute to hydrolysis of tryptophan (Trp)-derived indole glucosinolates and putative secretion of downstream products with potential antimicrobial activity (Bednarek et al. 2009; Clay et al. 2009; Lu et al. 2015). Besides PEN2 and PEN3, the cytochrome P450 monooxygenases CYP79B2 (At4g39950) and CYP79B3 (At2g22330) are required for mlo2-mediated resistance (Consonni et al. 2010). The two functionally redundant enzymes catalyze the conversion of Trp to indole-3-acetaldoxime (IAOx), a precursor of several indolic metabolites, including the phytoalexin camalexin and indole glucosinolates (Hull et al. 2000; Mikkelsen et al. 2000; Zhao et al. 2002).

A further link of the PEN genes to mlo2-mediated resistance is provided by their co-expression with MLO2 and a set of genes whose transcripts accumulate after biotic stress and stimulation with microbe-associated molecular patterns (MAMPs) (Humphry et al. 2010). This finding and several commonalities shared between mlo phenotypes and NHR in *Arabidopsis* and barley (Humphry et al. 2006) led to the conclusion that both types of penetration resistance have a mechanistic overlap, with mlo mutations potentially mediating an accelerated NHR response effective even against adapted pathogens (Trujillo et al. 2004; Humphry et al. 2006).

The resistance mechanism(s) employed by mlo2 mlo6 mlo12 mutants was examined by Kuhn et al. (2017) with tested known genetic suppressors of mlo2 resistance for their impact on mlo triple mutant plants. Surprisingly, all known contributors to mlo2-mediated resistance were found to be fully dispensable for the mlo2 mlo6 mlo12 resistance phenotype. To identify defense pathways differentially activated in the mlo2 mlo6 mlo12 mutant, which might help to identify novel components critical for its resistance phenotype, they performed comparative transcript profiling of triple mutant and wild-type plants at early time points upon powdery mildew challenge. The results point to a strong and accelerated activation of defense-related transcriptional programs in mlo2 mlo6 mlo12 triple mutant plants, which are consequently able to tolerate the lack of individual or even several key defense components (Kuhn et al. 2017).

4.3.1.1 MLO2, MLO6, and MLO12 Retards Induction of Defense-Related Genes During Powdery Mildew Infection

Performing comparative analysis of microarray data, Genevestigator studies, and investigation of GO term over representation, Kuhn et al. (2017) found that the mlo2 mlo6 mlo12 mutant shows an accelerated/increased accumulation of transcripts associated with defense, JA/ET signaling and biosynthesis of indole-derived antimicrobials at early time points after *G. orontii* inoculation. The finding is reminiscent of the results of a previous study, which used a cDNA array to probe differences in transcript accumulation between powdery mildew-challenged barley wild type (Mlo genotype) and mlo mutant plants (Zierold et al. 2005). It was from epidermal cells for hybridization to the array when RNA is used and it was found that ca. 10% of the genes (311 out of 3128) were reproducibly upregulated in response to attack by Bgh, with the mlo genotype typically showing a more pronounced differential regulation (Zierold et al. 2005).

Experiments based on laser capture micro-dissection have revealed distinctive transcript accumulation at powdery mildew infection sites in *Arabidopsis* (Chandran et al. 2010). The lack of a constitutively altered transcript profile and boosted expression of many genes in the mlo2 mlo6 mlo12 genotype hint at a strong impact of quantitative or kinetic rather than qualitative alterations which contributing to the powdery mildew phenotype. A fast and robust induction of multiple plant immune pathways could explain why mutations of PEN genes in various combinations are insufficient to break resistance in mlo2 mlo6 mlo12 plants. This contrasts the situation for mlo2 mutants, in which qualitative changes in single defense pathways, such as PEN1-mediated secretion or PEN2/PEN3-mediated biosynthesis and delivery of glucosinolate-derived antimicrobials, strongly contribute to the resistance phenotype (Consonni et al. 2006). Interestingly, according to microarray data, all three PEN genes are induced in mlo2 mlo6 mlo12 and wild-type Col-0 plants at 8 and 12 h after inoculation with *G. orontii*. In conclusion, while the two pathways represented by PEN1 and PEN2/PEN3 are generally involved in powdery mildew resistance, their necessity is overcome in mlo2 mlo6 mlo12 mutant plants.

4.3.1.2 Transcript Accumulation of JA/ET-Responsive Genes in the *mlo2, mlo6, mlo12* Mutant Upon Powdery Mildew Challenge

JA/ET-related defense has preferentially been linked to resistance against necrotrophic parasites (Thomma et al. 2001; Glazebrook 2005). Still, it is effective against adapted powdery mildews if stimulated constitutively, artificially or systemically (Kuhn et al. 2016), but usually either suppressed or failed to be elicited by virulent powdery mildews (Zimmerli et al. 2004; Antico et al. 2012). By contrast, non-adapted powdery mildews trigger JA/ET-dependent plant defense (Zimmerli et al. 2004). Remarkably, GO term overrepresentation, microarray analysis, and results of qRT-PCR assays indicated enhanced induction of mRNAs associated with JA/ET signaling in mlo2 mlo6 mlo12 vs. Col-0 plants. Kuhn et al. (2017) detected a statistically significant over accumulation of the JA-responsive *PDF1.2*, *VSP2*, and *LOX2* transcripts in the *mlo2 mlo6 mlo12* genotype upon *G. orontii* challenge already at a very early time point after inoculation. This finding resembles the situation of the *Arabidopsis*-Bgh nonhost interaction, where PDF gene expression also was more strongly induced at early time points after inoculation than in a compatible interaction (Zimmerli et al. 2004). This commonality further corroborates potential mechanistic overlap between NHR and mlo resistance. The JA-induced genes *PDF1.2* and *VSP2*, *LOX2* are responsive to two different branches of JA signaling that depend on the transcription factors (TFs) ORA59/ERF1 and MYC2, respectively (Pieterse et al. 2012; Vos et al. 2013; Wasternack and Hause 2013). This fact may explain the differential expression kinetics of *PDF1.2* on the one hand and *VSP2* and *LOX2* on the other hand observed following challenge with

G. orontii. Together with JA/ET-responsive genes, also *ORA59*, *ERF1*, and *ERF2* mRNAs accumulated to higher levels in *mlo2 mlo6 mlo12* plants vs. Col-0. All three TFs are positive regulators of JA and/or ET-responsive defense gene expression and in turn induced by JA/ET treatment; *Arabidopsis* eFP Browser (Winter et al. 2007). However, in contrast to *PDF1*.2 transcripts, *ORA59* mRNA levels were significantly higher in the mlo triple mutant than in the wild type, where it increased only at later stages. Moreover, *MYC2* was not induced to higher levels in the triple mutant, which together might suggest that early *PDF1*.2, *VSP2* and *LOX2* expression during infection is regulated at least partially independently from an induced expression of these two TFs. Consequently, despite the accumulation of *ORA59*, *ERF1*, and *ERF2* mRNAs, distinct factors might contribute to regulation of JA/ET-responsive genes in the context of powdery mildew challenge of *mlo2 mlo6 mlo12* plants. Alternatively, the increase in *PDF1*.2, *VSP2*, and *LOX2* transcript levels in the *mlo2 mlo6 mlo12* triple mutant might be due to constitutively elevated protein levels of these TFs.

Three findings suggest that the expression of JA/ET responsive genes in the *mlo2 mlo6 mlo12* mutant infected with *G. orontii* occurs independently of canonical JA/ET signaling: First, ET and JA levels are unaltered in mlo triple mutant plants after *G. orontii* inoculation. Second, the *mlo2 mlo6 mlo12 ora59* mutant does not show altered resistance to *G. orontii*. Third, prototypical ET/JA signaling components are dispensable for mlo2-mediated resistance (Consonni et al. 2006). Developmentally controlled enhanced SA levels in *mlo2* single and *mlo2 mlo6 mlo12* triple mutant plants have been reported previously (Consonni et al. 2006). Kuhn et al. (2017) could not detect altered SA levels after inoculation with *G. orontii*. This finding is consistent with previous analysis of a barley mlo mutant, which also lacked alterations in SA levels following powdery mildew (*Bgh*) challenge (Huckelhoven et al. 1999). It is therefore unlikely that antagonistic (Zander et al. 2010, 2012, 2014; Van der Does et al. 2013) or synergistic/reduced antagonistic (Mur et al. 2006; Liu et al. 2016) effects of SA might impact JA/ET-related gene expression in mlo triple mutant plants during fungal infection. In fact, interference with SA abundance and Non-expresser of Pr Genes 1 (NPR)1-mediated SA signaling, responsible for SA/JA antagonism (Spoel et al. 2003), merely suppresses premature senescence of *mlo2* plants, uncoupling this phenotype from penetration resistance (Consonni et al. 2006). So, *mlo*-mediated resistance has so far not been shown to depend on canonical defense hormone signaling (Consonni et al. 2006). Whether or not, a noncanonical induction of JA/ET-related defense gene expression is required for *mlo2 mlo6 mlo12*-mediated resistance remains to be investigated. This might involve transcriptional activation of JA/ET-related genes via non-COI1-dependent de-repression of JA signaling, possibly including components of SA signaling such as NPR3 and NPR4 and degradation of JAZ (Jasmonate Zim Domain) proteins, which act as repressors of JA-signaling (Liu et al. 2016; Kuhn et al. 2017).

4.3.1.3 Tryptophan-Derived Antimicrobials Are Dispensable for *mlo2, mlo6, mlo12*-Mediated Resistance

Tryptophan-derived antimicrobials such as camalexin and indole glucosinolates contribute to mlo2-mediated resistance (Consonni et al. 2010). In line with this finding, Kuhn et al. (2017) discovered transcripts of biosynthetic enzymes for tryptophan-derived antimicrobials as well as levels of camalexin and the indole glucosinolate 4MI3G (hyper-)accumulated in *G. orontii*-inoculated leaves of *mlo2mlo6 mlo12* and *mlo2 mlo6 mlo12 pen2* plants, respectively. Remarkably, elevated accumulation of 4MI3G in response to *G. orontii*, but not upon *Bgh* challenge, required in addition to the *pen2* the *mlo2 mlo6 mlo12* mutations. Consequently, in contrast to non-adapted *Bgh*, the adapted fungus either fails to elicit or actively suppresses 4MI3G biosynthesis in wild-type plants, for transcript accumulation of many defense-related genes. A key role for MLO proteins in defense suppression by adapted powdery mildew pathogens has been previously suggested (Panstruga 2005). It is still not unclear whether the fungus exerts the proposed defense suppression directly via the MLO proteins (prior to host cell entry through manipulation of MLO activity by apoplastic effector proteins), or whether the presence of the MLO proteins is primarily a prerequisite for successful host cell entry and the observed defense suppression merely a secondary effect (via effector proteins secreted by the fungal haustorium).

Surprisingly, loss of function of CYP79B2/CYP97B3, completely depleting the plant for tryptophan-derived antimicrobials (Zhao et al. 2002; Glawischnig et al. 2004), did not interfere with penetration resistance of the mlo triple mutant. This is noteworthy since similar to the PEN proteins, CYP79B2/CYP97B3 contribute to resistance against adapted and non-adapted powdery mildews and are required for *mlo2*-mediated resistance against *G. orontii* (Consonni et al. 2010). However, abrogation of CYP79B2/CYP79B3 function likely has a broader effect on the plant defense arsenal than loss of PEN activity. In *Arabidopsis*, biosynthesis of several defense compounds, such as camalexin (Ahuja et al. 2012), indole glucosinolates (Zhao et al. 2002), and indolic cyanogenic metabolites (4-hydroxyindole-3-carbonyl nitrile, 4-OHICN) (Rajniak et al. 2015), relies on CYP79B2/CYP79B3 activity. The fact that all these antimicrobial metabolites are dispensable for full immunity in the *mlo2 mlo6 mlo12* mutant suggests the involvement of additional highly potent and possibly yet undiscovered defense pathways in *Arabidopsis* (Kuhn et al. 2017).

4.3.2 Biochemical Basis of Powdery Mildew Resistance

An adapted pathogen can recognize specific host-derived signals that trigger pathogen differentiation and expression of virulence, or both. Therefore, pathogens must adapt to the chemical composition of their hosts. This leads to the evolution of an enzymatic toolbox, which the pathogen uses to overcome structural and chemical barriers of its hosts or to metabolize host-derived substrates. In addition, plants possess robust innate immunity that involves defense responses triggered after the recognition of pathogen-derived elicitors (pathogen-associated molecular patterns or pathogen effectors) or host-derived elicitors (damage-associated molecular patterns) (Boller and Felix 2009; Maekawa et al. 2011). Consequently, pathogens require host-specific effector molecules that match host targets for the suppression of immunity, and reprogramming of the host for the demands of the pathogen. Loss of susceptibility may therefore result from altered host immunity (gain of resistance functions) or, in a stricter sense, from changes in host components that are required by the pathogen for pathogenesis, but do not directly operate in the regulation of defense (de Almeida Engler et al. 2005; Huckelhoven 2005; Pavan et al. 2010; Huckelhoven et al. 2013). In contrast with wild-type *Arabidopsis* plants, loss-of-function mutants of the cytochrome P450 monooxygenase CYP83A1 gene are barely susceptible to the biotrophic ascomycete *Erysiphe cruciferarum* (Weis et al. 2013), which causes powdery mildew on many Brassicaceae (Adam et al. 1999; Saharan et al. 2005, 2019). The cytochrome P450 gene family in *Arabidopsis* comprises 244 genes (and 28 pseudogenes), and constitutes one of the largest gene families in plants. P450 enzymes function as monooxygenases in the biosynthesis of diverse metabolites, including pigments, phytohormones, and lignin, or defense compounds, such as flavonoids, alkaloids, or glucosinolates (Bak et al. 2011). The synthesis of aliphatic glucosinolates is divided into three stages: the chain elongation of the amino acid, formation of the glucosinolate core structure, and finally, secondary modifications (Wittstock and Halkier 2002). The two CYP83 proteins, CYP83A1 and CYP83B1, phylogenetically belong to the CYP71 clade (Hansen et al. 2001; Bak et al. 2011). They non-redundantly function in the core structure synthesis of glucosinolates by catalyzing the initial conversion of aldoximes to thiohydroximates (Bak and Feyereisen 2001; Naur et al. 2003). CYP83A1 has higher substrate specificity for methionine-derived aldoximes in the synthesis of aliphatic glucosinolates, whereas CYP83B1 preferentially converts tryptophan-derived aldoximes in the synthesis of indole-glucosinolates (Bak and Feyereisen 2001; Naur et al. 2003).

In plant–herbivore interactions, glucosinolates, and their hydrolysis products, such as isothiocyanate, nitriles, or epithionitriles, function as deterrents against generalist herbivores, but also as attractants for specialized insects. The knowledge about the function of glucosinolates in plant–fungus interactions are more limited (Bednarek and Osbourn 2009). However, in *Arabidopsis*, the peroxisome-associated myrosinase penetration 2 (PEN2) hydrolyzes 4-methoxyindol-3-ylmethylglucosinolate to bioactive products involved in resistance to non-adapted powdery mildew fungi (Lipka et al. 2005; Bednarek et al. 2009). Indole-glucosinolates are further important to balance the mutualistic interaction of *Arabidopsis* with the beneficial root endophyte *Piriformospora indica* (Jacobs et al. 2011; Nongbri et al. 2012). Aliphatic glucosinolates are important in resistance to lepidopteran larvae, to non-adapted bacterial pathogens, and to the necrotrophic fungus Sclerotinia sclerotiorum (Beekwilder et al. 2008; Fan et al. 2011; Stotz et al. 2011). However, little information is available about the role of aliphatic glucosinolates in interactions with haustorium-forming fungi. Powdery mildew fungi are obligate biotrophic ascomycete pathogens with a high degree of specialization to a limited range of hosts. They form haustoria from appressoria that sense, and directly penetrate the host cuticle, and cell wall (Green et al. 2002; Huckelhoven and Panstruga 2011). *Erysiphe cruciferarum* is a typical powdery mildew fungus adapted to Brassicaceae (Adam et al. 1999). Hence, it can normally cope with glucosinolates as a result of an unknown mechanism. Weis et al. (2014) have observed a loss of *CYP83A1* which changes the metabolic composition of *Arabidopsis* in a manner that greatly influences the interaction with powdery mildew pathogen.

Aliphatic glucosinolates function in the chemical defense of Capparales. The cytochrome P450 83A1 monooxygenase (CYP83A1) catalyzes the initial conversion of methionine-derived aldoximes to thiohydroximates in the biosynthesis of glucosinolates, and thus cyp83a1 mutants have reduced levels of aliphatic glucosinolates. Loss of CYP83A1 function leads to dramatically reduced parasitic growth of powdery mildew fungus *Erysiphe cruciferarum* on *Arabidopsis thaliana*. The *cyp83a1* mutants support less conidial germination, and appressorium formation of *E. cruciferarum* on the leaf surface, and post-penetration conidiophore formation by the fungus. By contrast, a myb28-1 myb29-1 double mutant, which totally lacks aliphatic glucosinolates, shows a high level of susceptibility to *E. cruciferarum*. The *cyp83a1* mutants also lack very-long-chain aldehydes on their leaf surface. Such aldehydes support appressorium formation by *E. cruciferarum* in vitro. In addition, when chemically complemented with the C26 aldehyde *n*-hexacosanal, *cyp83a1* mutants can again support appressorium formation. The mutants further accumulate 5-methylthiopentanaldoxime, the potentially toxic substrate of CYP83A1. Loss of powdery mildew susceptibility by *cyp83a1* may be explained by a reduced supply of the fungus with inductive signals from the host, and an accumulation of potentially fungitoxic metabolites (Weis et al. 2014).

4.3.3 Induction of Biochemical Metabolites

Powdery mildew fungi are biotrophic pathogens that form a complex interface, the haustorium, between the host plant and the pathogens. The pathogen acts as an additional sink, competing with host sinks, resulting in considerable modification of photo assimilate production, and partitioning within the host tissue. The biotrophic interaction of *Arabidopsis*/*Erysiphe cichoracearum* induces number of responses: Glc uptake in host tissues is enhanced after fungal infection; this coincides with the induction of expression of the monosaccharide transporter gene, *Arabidopsis sugar transport protein 4* (*AtSTP4*), in infected leaves; invertase activity, and transcript levels for a cell wall invertase, *Atβfruct1*, increase substantially in *Arabidopsis* during infection of pathogen. Before infection, *Arabidopsis* plants transformed with an *AtSTP4* promoter-β-glucuronidase construct show expression mainly in sink tissues such as roots; after infection, *AtSTP4* expression is induced in the mature leaves, and increases over the 6-day time period. Sections of infected leaves stained for β-glucuronidase shows that *AtSTP4* expression is not confined to infected epidermal cells but is also evident in a wider range of cells, including those of the vascular tissue (Fotopoulos et al. 2003).

The finding that infection by powdery mildew induces *AtSTP4* expression in a wide range of host cells despite the fungus being confined to the epidermis implicates the role of signaling mechanisms in this response. There is evidence that defense response genes are induced by elevated sugar levels (Ehness et al. 1997; Herbers et al. 2000), and hexose transport into the host cells may be enhanced to cope with this increased energy demand, and/or to reduce availability of sugars to the pathogen. Enhanced cell wall invertase reduces Suc loading (Stitt et al. 1990; von Schaewen et al. 1990), and an increase in hexose availability as a consequence could increase the concentration gradient of hexoses to the fungus. The transfer of carbon to the fungus, primarily as Glc, is thought to occur down a concentration gradient created by fungal solute uptake, and use. The nature of the sugar transporters at the haustorial membrane in the powdery mildew fungi is unknown, but by analogy with rust, carriers may exist at this membrane to facilitate transport into the mycelium. A major goal will be to dissect the signaling pathway in a compatible interaction that results in the diversion of host resources to the fungus (Fotopoulos et al. 2003)

4.3.4 Genes Encoding Camalexin Biosynthesis for Powdery Mildew Resistance

To protect themselves against pathogens, plants have evolved intricate immune responses that include accumulation of reactive oxygen species, deposition of callose, enhanced expression of *pathogenesis-related* (*PR*) genes, and biosynthesis of phytoalexins. Phytoalexins are low molecular mass secondary metabolites that are induced by both biotic and abiotic stress. During pathogen infection, plants synthesize a wide variety of structurally different phytoalexins to defend against pathogen invasion (Hammerschmidt 1999; Pedras et al. 2011). Camalexin, 3-thiazol-20-yl-indole, is one of the major phytoalexins of *Arabidopsis thaliana,* and was long considered to be the only phytoalexin of *Arabidopsis*, until the discovery of rapalexin A (Pedras and Adio 2008). The biosynthesis and regulation of camalexin in *Arabidopsis* remain only partially understood, and the full scope of camalexin functions also remains to be defined. Camalexin is derived from tryptophan and requires many cytochrome P450s, including CYP79B2, CYP71A13, and CYP71B15 [which corresponds to the camalexin-deficient mutant *phytoalexin deficient 3* (*PAD3*); Nafisi et al. 2007; Schuhegger et al. 2007a, b]. The *Arabidopsis* transcription factors WRKY33, WRKY18, and WRKY40 appear to be involved in the regulation of camalexin biosynthesis (Qiu et al. 2008; Pandey et al. 2010; Mao et al. 2011). Camalexin plays an important role in the response to necrotrophic pathogens *Alternaria brassicicola,* and *Botrytis cinerea* (Thomma et al. 1999; Ferrari et al. 2003; Kliebenstein et al. 2005; Nafisi et al. 2007), and the oomycete *Phytophthora brassicae* (Schlaeppi et al. 2010), as well as the biotrophic fungus *Golovinomyces orontii* (Consonni et al. 2010; Pandey et al. 2010). Although camalexin produces broad-spectrum resistance to many species of plant pathogens, how it functions remains unclear (Liu et al. 2016).

In addition to camalexin, plants synthesize other related secondary metabolites, such as glucosinolates, that participate in the defense response. Plant cells usually store glucosinolates in stable forms; during insect and/or pathogen attack, myrosinases hydrolyze these stable forms in to active compounds (Halkier and Gershenzon 2006). According to their side-chain radical, glucosinolates can be divided in to aliphatic glucosinolates, indole glucosinolates, and aromatic glucosinolates. Many cytochrome P450s function in glucosinolate synthesis, including CYP83A1 and CYP83B1.The *Arabidopsis* cytochrome P450 monooxygenase CYP83A1 participates in the biosynthesis of aliphatic glucosinolates from aliphatic oximes, whereas CYP83B1, the *Arabidopsis* protein most similar to CYP83A1, functions in the biosynthesis of indole glucosinolates. The biosynthetic pathways of alkyl glucosinolates and indole glucosinolates affect each other; for instance, the *cyp83a1-2* (also called *ref2-1*) mutant produces slower levels of aliphatic glucosinolates, but accumulates higher levels of indole-derived glucosinolates compared with wild type (Hemm et al. 2003; Naur et al. 2003; Sonderby et al. 2010). The biosynthesis of glucosinolates, especially indole glucosinolates, shares the intermediate product indole-3-acetaldoxime (IAOx) with biosynthetic pathways that produce many other secondary metabolites or hormones like camalexin, and indole-3-acetic acid (IAA), respectively (Hemm et al. 2003; Grubb and Abel 2006; Nafisi et al. 2007). Although the regulation of callose biosynthesis, in response to bacterial elicitors of *Arabidopsis* immunity, requires 4-methoxy-indol-3-yl methylglucosinolate (4MI3G; Bednarek et al. 2009; Clay et al. 2009), how 4MI3G and related metabolites participate in the immune response is not well-understood. Powdery mildew fungi, as biotrophic pathogens, infect many plant species, and cause huge agricultural losses worldwide. The plant hormone salicylic acid (SA) plays an important role in resistance to powdery mildew in *Arabidopsis*, and many mutants showing enhanced resistance to powdery mildew require SA signaling for their resistance phenol type; these mutants include *edr1* (*enhanced disease resistance1*), *edr2*, and *edr4* (Frye and Innes 1998; Frye et al. 2001; Tang et al. 2005a, b; Zhao et al. 2014; Wu et al. 2015). Liu et al. (2016) characterized an *Arabidopsis* mutant that exhibits enhanced resistance to a variety of powdery mildew species. A mutation in the gene-encoding cytochrome P450 monooxygenase CYP83A1, a component of the glucosinolate pathway, leads to higher accumulation of camalexin, and enhanced resistance to the powdery mildew fungus *Golovinomyces cichoracearum*, which is consistent with the previous finding that *cyp83a1* exhibits increased resistance to powdery mildew fungus *Erysiphe cruciferarum* (Weis et al. 2013). The *cyp83a1-3* accumulates higher levels of camalexin. The mutations in genes affecting camalexin production suppress the resistance of *cyp83a1-3*, indicating that higher accumulation of camalexin in *cyp83a1-3* mutants contributes to their enhanced powdery mildew resistance. Small secondary metabolites, including glucosinolates, and the major phytoalexin camalexin, play important roles for immunity in *Arabidopsis thaliana*. Liu et al. (2016) isolated an *Arabidopsis* mutant with increased resistance to the powdery mildew fungus *Golovinomyces cichoracearum,* and identified a mutation in the gene-encoding cytochrome P450 83A1 monooxygenase (CYP83A1), which functions in glucosinolate biosynthesis. The *cyp83a1-3* mutant exhibited enhanced defense responses to *G. cichoracearum,* and double mutant analysis showed that this enhanced resistance requires NPR1, EDS1, and PAD4, but not SID2 or EDS5. In *cyp83a1-3* mutants, the expression of genes related to camalexin synthesis increased upon *G. cichoracearum* infection. Significantly, the *cyp83a1-3* mutant also accumulated higher levels of camalexin. Decreasing camalexin levels by mutation of the camalexin synthetase gene *PAD3* or the camalexin synthesis regulator *AtWRKY33* compromised the powdery mildew resistance in these mutants. Consistent with these observations, overexpression of *PAD3* increased camalexin levels, and enhanced resistance to *G. cichoracearum.* It indicates that accumulation of higher levels of camalexin contributes to increased resistance to powdery mildew of *A. thaliana*.

Camalexin biosynthesis and accumulation are affected by wrky18 wrky40 transcription factor of *Arabidopsis*. Camalexin accumulation enforces defenses to penetrating pathogens, and is the main *Arabidopsis* phytoalexin induced at infection sites (Rauhut and Glawischnig 2009). Genes encoding all three enzymes of camalexin biosynthesis were upregulated in microarrays. Conversion of indole-3-acetaldoxime (IAOx) to indole-3-acetonitrile (IAN), catalyzed by CYP71A13, marks a committed step (Nafisi et al. 2007), whereas PAD3 catalyzes the final step in camalexin biosynthesis (Schuhegger et al. 2006). Pandey et al. (2010) examined expression changes for both of these genes over a time course in WT, and wrky18 wrky40 plants by qPCR. Compared with uninfected plants, *G. orontii*-infected wrky18 wrky40 plants had >15-fold elevated CYP71A13 transcript levels, whereas infected WT plants showed only a fourfold increase. Similarly, the wrky18 wrky40 mutant accumulated overall more PAD3 transcripts than WT plants (Fig. 4.2b). Camalexin levels was observed prior to 24 hpi with *G. orontii* and compared with WT plants, the wrky18 wrky40 mutant already had elevated levels of camalexin in uninfected tissue (Fig. 4.2a). Nevertheless, these levels increased significantly in both genotypes upon infection, with the mutant accumulating 18-fold higher concentrations of camalexin than WT. These findings substantiated microarray studies, and revealed that loss-of-WRKY18 and WRKY40 functions resulted in increased biosynthesis and accumulation of camalexin, which is further strongly enhanced upon *G. orontii* infection. The preexisting higher camalexin levels found in uninfected wrky18 wrky40 plants may in part be due to the nearly twofold elevated transcript levels observed for CYP79B2, CYP71A13, and CYP71B15/PAD3 (Pandey et al. 2010).

4.3.5 Synthesis of Structural and Functional Biochemical Components of Host Resistance

In *Arabidopsis* and other crucifers species infected cells react to penetration attempts through the formation of cell wall apposition (CWAs) (Plate 4.1), which (despite

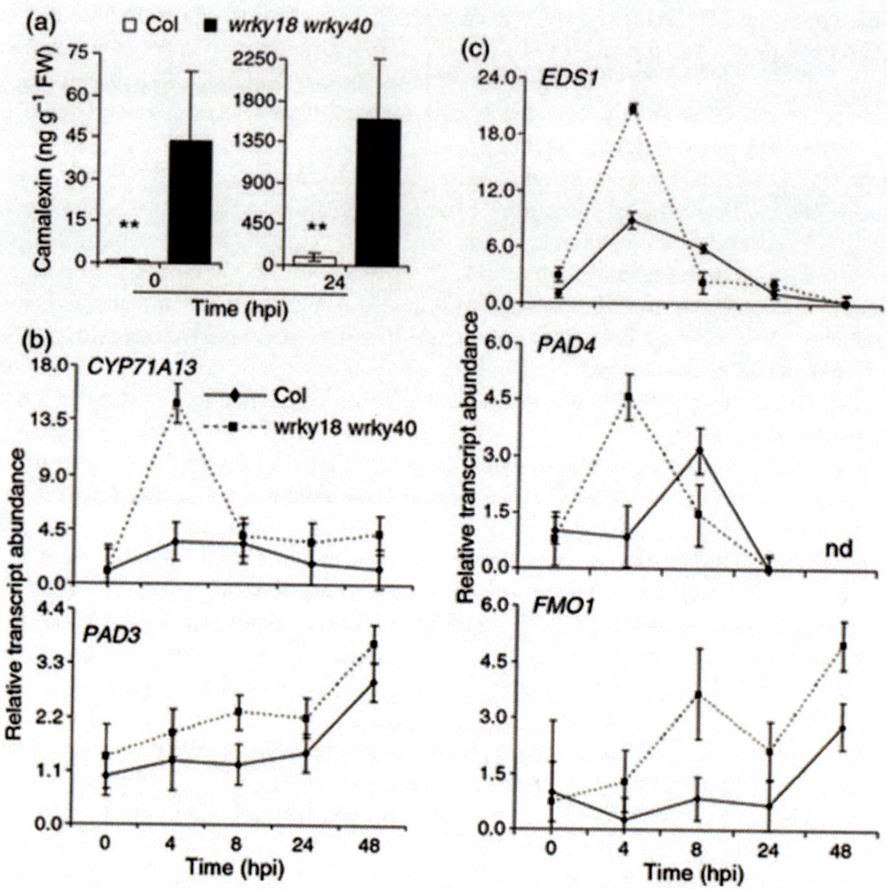

Fig 4.2 Loss-of-WRKY18/40 functions upregulate the accumulation and biosynthesis of camalexin, and the EDS1 signaling pathway upon *Golovinomyces orontii* infection. (**a**) Camalexin levels were determined in WT (open bars), and wrky18 wrky40 (solid bars) plants before (0 hpi), and at 24 hpi with *G. orontii*. (**b**) Temporal expression of *G. orontii*-induced host genes CYP71A13 and PAD3 essential for camalexin biosynthesis, and (**c**) of EDS1, PAD4, and FMO1 in WT (solid lines), and wrky18 wrky40 (broken lines) plants as determined by qPCR at the indicated time-points. Samples were collected, and gene expression levels were calculated with respect to time 0. **Student's *t*-test, $n = 10$, $P < 0.05$ (Pandey et al. 2010)

sparse experimental evidence supporting this view) are commonly thought to provide both physical cell wall reinforcements and a chemical antimicrobial blockade against the invading pathogen (Hardham et al. 2007; Huckelhoven 2007). Papillae are formed within a few hours after fungal penetration attempts in response to adapted and non-adapted pathogens. Remarkably, physical damage (needle puncture) can also induce the formation of papilla-like structures, which suggests that it is the pathogen-triggered wounding of plant cells that prompts the formation of CWAs (Aist 1976). However, detailed analysis of the structures formed in response to

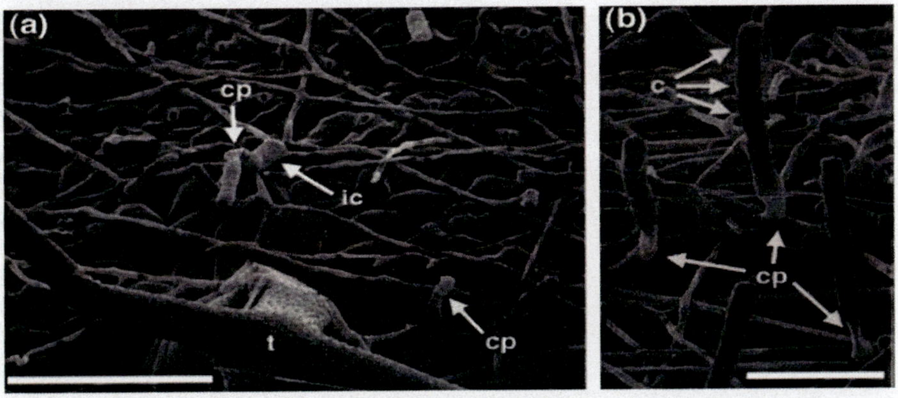

Plate 4.1 Scanning electron micrographs of *Erysiphe orontii*. (**a**) A 5-day-old colony of *E. orontii* on a Columbia leaf. *i c* initial conidium, *cp* conidiophore, *t* trichome. Scale bar = 100 μm. (**b**) Mature conidiophores on a pad4 stem. *cp* conidiophore, *c* conidium. Scale = 50 μm (Reuber et al. 1998)

non-biotic, mechanical wounds revealed that they are distinct in their composition, and likely in their function from fungus-induced CWAs (Russo and Bushnell 1989). In addition, no papilla formation was detected in response to mechanical (needle) stimulation of parsley suspension culture cells, despite detection of other responses associated with pathogen recognition such as reactive oxygen intermediates (ROI) production, and induction of several elicitor-response genes (Gus-Meyer et al. 1998). Although there is no absolute association between papilla formation and resistance to fungal penetration, delayed formation of CWAs is correlated with enhanced fungal infection (Assaad et al. 2004). CWAs are composed of an apparently amorphous mixture of cellulose, pectin, callose, lignins, phenolics, silicon, H_2O_2, and derivatives as well as dedicated de novo-synthesized antimicrobial metabolites (phytoalexins), and fungal enzyme inhibitors (Plate 4.1). Some of these compounds are delivered within secretory vesicles while others are synthesized on the spot by cell wall-resident enzymes (Huckelhoven 2007).

Each component of the CWA likely plays a role in defense. Structural components such as pectin, callose, and cellulose are thought to provide physical reinforcement to the plant cell wall at the site of attack (Zeyen et al. 2002). In addition, released moieties of these carbohydrate polymers as well as enzymes involved in their synthesis have potent signaling capabilities, and are thought to be involved in the modulation of downstream defense responses. Callose synthase GSL5/PMR4 protein (and/or callose itself) appears to be a negative regulator of SA-dependent defense responses, and mutations in this gene render *Arabidopsis* resistant to *G. orontii* and *G. cichoracearum* (Jacobs et al. 2003; Nishimura et al. 2003). Silicon (Si) is another component found in papillae (Plate 4.1) that was long thought to provide structural reinforcement to the cell wall (Zeyen et al. 2002). More recently, a comparative transcriptome analysis of *Arabidopsis* plants either supplied with or lacking Si revealed that silicon modulates gene expression in a pathogen-dependent manner (Fauteux et al. 2006). Specifically, it appears to dampen general stress responses associated with pathogen infection, and partially restores expression of genes encoding components of primary metabolism to normal conditions without affecting the expression of defense-related genes. Reduction of biotic stress responses results in reduced powdery mildew density, and conidiation on leaves, effects that had been historically reported for a variety of other plant species treated with Si (Ghanmi et al. 2004). To complete defense at the cell wall, pathogen-induced antimicrobial compounds (phytoalexins) are actively delivered to the site of fungal contact. Components of the basal defense machinery such as the PEN2 glycosyl hydrolase (associated with peroxisomes; Plate 4.1) and the plasma membrane-resident PEN3 ABC transporter (Plate 4.1) are thought to cooperate and generate toxic antimicrobial compounds that are pumped at the site of attack (Lipka et al. 2005; Stein et al. 2006). The delivery of cell wall components and antimicrobial compounds at CWAs is thought to occur at least in part in membrane-bound vesicles that move along tracks of actin microfilaments (Plate 4.1). This view is supported by the observation of massive cytoskeletal reorganization of plant cells upon pathogen infection. Reminiscent of findings in other plant-powdery mildew interactions (Kobayashi et al. 1997a; Opalski et al. 2005) fluorescence microscopy of GFP-tagged *Arabidopsis* cytoskeletal proteins revealed a rapid concentration of actin arrays toward the site of attack by the non-adapted fungus *B. graminis* f. sp. *hordei*. This rearrangement was accompanied by a mobilization of the nucleus, and endoplasmic reticulum to the site of the emergent penetration peg, and later toward the developing haustorium (Takemoto et al. 2006).

Golgi-derived bodies circulate and make frequent stops below the incipient penetration peg, consistent with the hypothesis that the plant protein synthesis and secretory machinery is actively recruited to the emerging infection structures (Hardham et al. 2007). Mutations in PEN1, originally identified as a component of powdery mildew nonhost resistance, did not affect actin reorganization upon challenge with the non-adapted barley powdery mildew fungus (Takemoto et al. 2006), suggesting that this protein acts downstream of actin reorganization. However, in barley cells, chemical and genetic interference with actin polymerization significantly affects basal defense, and blocks at least partially nonhost resistance against inappropriate powdery mildews (Kobayashi et al. 1997a, b; Miklis et al. 2007). The scenario is similar in *Arabidopsis* (Yun et al. 2003), and suggests an important functional as well as structural role for cytoskeletal components in plant defense responses (Schmidt and Panstruga 2008).

It is interesting to note that plant cytoskeleton and organelle responses documented for powdery mildew infections are strikingly similar to those observed during infection of several plant species with beneficial arbuscular mycorrhizal (AM) fungi (Strack et al. 2003).

In this case, the plant cell guides the formation of intracellular arbuscules (analogous in structure, and function to haustoria) through the formation of microtubule tracks, and by reorganization of the actin network. In tomato, inhibition of microfilament formation interferes with the proper development of the fungus on root tissue (Timonen et al. 2006). In pathogenic powdery mildew–plant interactions,

Plate 4.2 *Arabidopsis* plants infected with *Erysiphe orontii*, and epifluorescence micrographs of infected *Arabidopsis* leaves stained for callose. (**a–f**) Plants were infected by settling tower. Photographs were taken 12 days post-inoculation. Arrows indicate representative areas of infection. (**a**) Col-0; (**b**) pad4–1; (**c**) npr1–1; (**d**) eds 5–1; (**e**) La-*er*; (**f**) nah G. (**g–j**) Leaves stained with aniline blue to detect callose 24 h after infection with *E. orontii* (magnification = 180×). (**g**) Col-0. *i* initial conidium, *a* appressorial germ tube, *c* cell wall apposition. (**h**) pad4. (**i**) npr1–1. (**j**) eds 5 (Reuber et al. 1998)

components of both fungal and plant origin are implicated in the formation of the haustorium. During post-penetration, the plant plasma membrane invaginates, and the haustorium starts to develop, initially as a bud, then as a relatively smooth globular body, and finally matures into an intricate, highly convoluted, and branched structure (Koh et al. 2005; Plate 4.2). The haustorium is a unicellular structure shielded from the plant environment by its own plasma membrane, cell wall, extra-haustorial matrix (an amorphous environment separating plant, and fungal structures), and extra-haustorial membrane (EHM), a unique interface between the host and the pathogen, thought to be composed of a modified plant plasma membrane (Bracker 1968).

In the studies of *G. cichoracearum* haustoria on *Arabidopsis*, fluorescently labeled host plasma membrane markers spread as far as the two haustorial neckbands that separate fungal from plant structures. Below the neckbands, the EHM has a distinct architecture, and function from the plant and fungal membranes (Koh et al. 2005). ATPase and other enzymatic activities normally found at the plant plasma membrane are lacking in the EHM (Spencer-Phillips and Gay 1981). The EHM enveloping haustoria of the pea powdery mildew, *Erysiphe pisi*, is twice as thick as normal host plasma membrane, is highly convoluted in healthy haustorial complexes, and is resistant to detergent treatments (Gil and Gay 1977). The biogenesis and molecular composition of EHMs covering the haustoria of biotrophic pathogens represents indeed one of the major mysteries of plant–powdery mildew interactions. Under optimal conditions, the life-span and transport capacity of a single haustorium appears to be sufficient to allow the fungus to complete its life cycle (Shirasu et al. 1999). However, defense responses deployed by the host can interfere with optimal haustorium development and function. In *Arabidopsis,* cells infected with *G. orontii*, mature haustoria are frequently encased (partially or fully) in what appears to be an extension of the papillary structure, characterized by presence of callose (Micali et al. 2008). These encasements, when complete, appear to crush haustorial bodies, and may restrict passage of nutrients to the fungus, and pathogen effectors to the host. As such they may represent one of the several layers of resistance to powdery mildew infection (Micali et al. 2008).

4.3.6 Induction of Biomolecules for Pre-penetration Resistance Mechanisms

Most of the information on crucifers host resistance to powdery mildew has been generated using *Arabidopsis*–powdery mildew host pathosystem. The first barriers of powdery mildew pathogens encountered during infection are the cuticle and epicuticular waxes overlying the plant cell wall (Malinovsky et al. 2014). Powdery mildew presumably employs mainly hydrostatic pressure to penetrate this preformed perimeter of epidermal cells. Accordingly, the plant can sense the pathogen in several ways. Firstly, the pressure exerted on the plant cell might activate plant mechanosensors (Bhat et al. 2005; Ellinger and Voigt 2014). Secondly, damage-associated molecular patterns (DAMPs) released by the breakdown of the plant cell wall or pathogen-associated molecular patterns (PAMPs) released by the fungus can be detected by pattern recognition receptors (PRRs) and activate immune signaling (Boller and Felix 2009).

The carbohydrate polymer chitin is a major constituent of fungal cell walls, and when exogenously applied to *Arabidopsis* activates PAMP-triggered immune responses. Chitin is perceived by the membrane-localized PRRs chitin elicitor receptor kinase 1 (CERK1: At3g21630) (Miya et al. 2007) and the lysin motif receptor-like kinases 4/5 (LYK4/5: At2g23770/At2g33580) (Cao et al. 2014). Application of PAMPs, including chitin, leads to the accumulation of defense-related proteins, and the deposition of callose at seemingly random locations in treated tissues (Gomez-Gomez et al. 1999; Luna et al. 2011; Underwood and Somerville 2013). This phenomenon is similar to the localized formation of papillae, and suggests that PAMP-induced PRR activation alone can trigger the establishment of papilla-like structures. The adapted powdery mildew *Galovinomyces cichoracearum (GC)* shows increased sporulation on *cerk1* mutants in comparison to wild-type plants, which suggests that signaling through CERK1 contributes to basal resistance to the powdery mildew disease (Wan et al. 2008). It is presently unknown whether *lyk4* or *lyk5* mutants are more susceptible to powdery mildew as well. Presumably, plants can perceive further powdery mildew-derived PAMPs. However, the only other known molecule from a powdery mildew pathogen that activates defense gene expression and decreases fungal growth in various cereals after application is a soluble carbohydrate elicitor isolated from conidia of the wheat powdery mildew pathogen, *Blumeria graminis* f. sp. *tritici* (*Bgt*) (Schweizer et al. 2000).

4.3.6.1 Papilla Formation

Plant cell responses to the early detection events include polarization of cellular organelles, and rearrangement of cytoskeletal elements (microtubules and actin filaments) below the attack site (Schmelzer 2002; Huckelhoven and Panstruga 2011). Underneath the attempted penetration site defense-related proteins focally accumulate (Assaad et al. 2004; Bhat et al. 2005; Kwon et al. 2008; Meyer et al. 2009; Kwaaitaal et al. 2010). Furthermore, the plasma membrane is altered locally and gains lipid raft-like properties (Bhat et al. 2005). Both virulent and non-virulent powdery mildew fungi induce the formation of a small, dome-like structure called papilla below the incipient fungal appressorium (Collins et al. 2003; Assaad et al. 2004; Koh et al. 2005). Among other components such as membranous vesicles, the papilla contains callose, silicon, reactive oxygen species (ROS), and phenolic compounds. The resulting structure is believed to reinforce the cell wall to prevent fungal invasion (Zeyen et al. 2002). This hypothesis is supported by a correlation between the timing of papilla formation and powdery mildew resistance. Mutation of the target membrane soluble *N*-ethylmalemide-sensitive factor attachment protein receptor (t-SNARE) penetration (PEN) 1/syntaxin of plants (SYP) 121 (At3g11820) and the vesicle-associated SNAREs vesicle-associated membrane proteins (VAMPs) 721/722 (At1g04750/At2g33120) delays papilla formation and increases *Bgh* penetration successfully (Assaad et al. 2004; Kwon et al. 2008; Bohlenius et al. 2010). The ubiquitin ligase *Arabidopsis* toxicos en lavadura 31 (ATL31: At5g27420), which is a regulator of responses to changes in the cellular carbon/nitrogen ratio, interacts with PEN1 in co-immuno-precipitation experiments. Transient expression assays in *Nicotiana benthamiana* leaves inoculated with *Blumeria graminis* f. sp. *hordei* (*Bgh*) show accumulation of ubiquitination activity deficient ATL31C143S-GFP around papillae, and in vesicle-like structures near papillae, while fluorescence is undetectable for ATL31-GFP. Furthermore, overexpression of ATL31 in *Arabidopsis pen1*-1 mutants results in enhanced penetration resistance to *Bgh,* and faster formation of papillae (Maekawa et al. 2014). The same observation was made for *RAB GTPASE homolog 4c* (*RABA4c:* At5g47960) overexpression lines (Ellinger et al. 2014), suggesting that only the timely formation of papillae can restrict invasion by powdery mildew fungi.

4.3.6.2 Callose Deposition

Accumulation of the β-1,3 polyglucan callose, as the major constituent of papillae, is a generic response to pathogen challenge (Ellinger and Voigt 2014). Together with the β-1,4 polyglucan cellulose, callose generates a three-dimensional network of ~250 nm fibrils, which can provide protection against cell wall hydrolysis by fungal enzymes (Plate 4.3; Eggert et al. 2014). Callose could not be unambiguously linked to resistance for a long time, and work on a mutant of the *GSL5/PMR4* gene initially even suggested that loss of papillary callose reduces sporulation of adapted powdery mildews (Jacobs et al. 2003; Nishimura et al. 2003). However, the inhibition of post-invasive fungal growth in *pmr4* mutants relies on hyper-induced salicylic acid (SA) responses upon powdery mildew attack. When the *pmr4* knockout mutation is combined with a further mutation that leads to a loss of SA biosynthesis or signaling the increased resistance is compromised (Nishimura et al. 2003). Except for the loss of callose, papillae of *pmr4* mutant plants have a similar appearance as papillae of wild-type plants (Nishimura et al. 2003). While loss of callose in the *pmr4* mutant has only limited impact on penetration resistance (Jacobs et al. 2003; Ellinger et al. 2013), increased callose deposition after powdery mildew attack caused by *PMR4* overexpression results in full penetration resistance to both *Gc* and *Bgh*. The latter effect seems to correlate with structural differences of papillae in

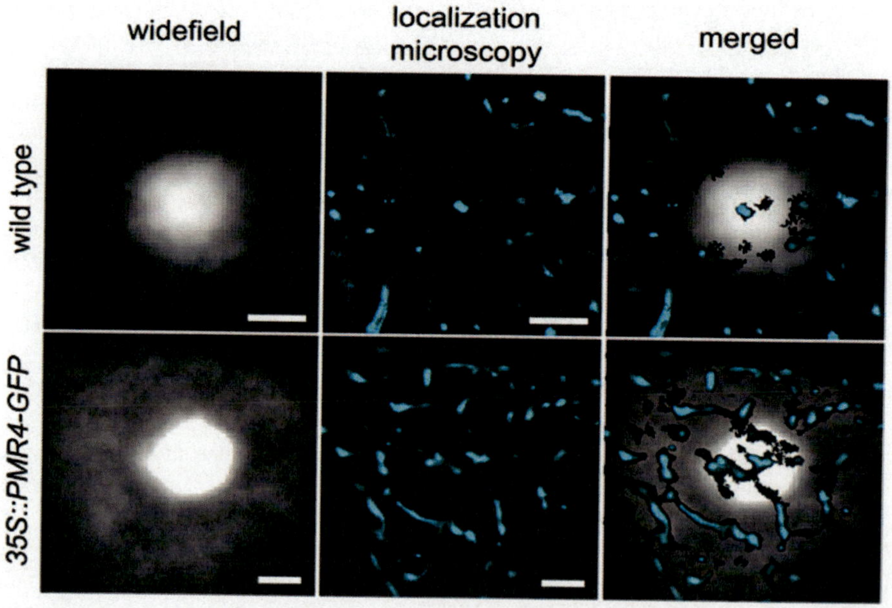

Plate 4.3 Nanoscale resolution of callose polymer fibrils in pathogen-induced cell wall papillae. Three-week-old *Arabidopsis* wild type, and pathogen-resistant *PMR4-GFP* overexpressing lines (*P35S:: PMR4-GFP*) were inoculated with the adapted powdery mildew *Golovinomyces orontii*. Localization microscopy (dSTORM: direct stochastical optical reconstruction microscopy) of aniline blue-stained callose polymer fibrils in pathogen-induced papillae at sites of attempted fungal penetration at 12 hpi in rosette leaves. Scale bars = 2 μm (Kuhn et al. 2016)

PMR4 overexpression lines compared to the wild type, as the transgenic lines show larger cores of callose-dense deposits, whereas wild-type papillae display a more diffuse structure (Naumann et al. 2013). Together these findings indicate that additional papillary components support the contribution of callose to prevent fungal penetration (Ellinger et al. 2013). The PMR4-GFP fusion protein focally accumulates at the powdery mildew attack site, and its presence coincides with the occurrence of callose deposits. The callose accumulations in the *PMR4-GFP* overexpression line are not only enlarged, but also deposited in a layer facing the fungus on top of the cellulose microfibrilar network (Eggert et al. 2014). The increase in the proportion of callose presumably protects the cellulose component of papillae from enzymatic digestion (Eggert et al. 2014). In contrast to the increased post-penetration resistance in *pmr4* mutants, the increase in resistance caused by *PMR4-GFP* overexpression is independent from SA- or jasmonic acid (JA)-mediated defense (Ellinger et al. 2013). Similar to what was reported for barley (Bohlenius et al. 2010), ADP ribosylation factor-GTP exchange factor (ARF-GEF)-mediated vesicle trafficking is essential for callose accumulation in papillae in *Arabidopsis* (Nielsen et al. 2012). This ARF-GEF-dependence indicates that either PMR4 accumulation at fungal attack sites, the delivery of callose precursors, or the callose deposition process itself involves vesicle-mediated transport processes.

PMR4 interacts with and acts as an effector of the small GTPase of the Ras (rat sarcoma) superfamily, RABA4c (Ellinger et al. 2014). *RABA4c* expression is transiently upregulated prior to callose deposition in response to biotic stress. Knockouts of *RABA4c* exhibit a delayed increase of callose synthase activity, slightly reduced numbers of callose deposits, and slightly increased *Gc* penetration rates. By contrast, overexpression of *RABA4c* results in full penetration resistance to *Gc,* and hastens and increases callose deposition. Both effects depend on the presence of PMR4 and RABA4c GTPase activity. The RABA4c localization to membranes is independent on the prenylation of its C-terminal CaaX motif ("C" cysteine, "a" aliphatic amino acid, "X" variable amino acid). A C-terminal RABA4c-mCitrine fusion lacking this lipid modification still localizes to membranes, though solely when PMR4 is present, which supports the finding that both proteins physically interact in planta (Ellinger et al. 2014). Rab (Ras-related in brain) GTPases play major roles in virtually all vesicle trafficking processes in eukaryotic cells. To what extent PMR4 localization to the plasma membrane or to focal accumulation sites depends on RABA4c-dependent vesicle trafficking pathways is unknown (Kuhn et al. 2016).

4.3.7 Host Resistance by Extracellular Deposition of Proteins into Papillae

The discovery that components of a SNARE protein complex are involved in penetration resistance suggests that these proteins directly control vesicle fusion at the powdery mildew attack site (Collins et al. 2003; Kwon et al. 2008). After vesicle fusion and cargo release, SNARE proteins are usually recycled, and stay on the cytosolic side of the plasma membrane (Kwon et al. 2008). Surprisingly, in case of the focal accumulation of SNARE proteins at attempted fungal entry sites this is not the case. Instead, fluorescent fusions of PEN1, soluble *N*-ethylmaleimide-sensitive factor adaptor protein (SNAP) 33 (At5g61210; a t-SNARE), and the ATP-binding cassette (ABC) transporter PEN3 (At1g59870) accumulate within papillae and haustorial encasements, and therefore end up in the extracellular (apoplastic) space. Within cell wall appositions, GFP-PEN1 co-localize with the lipophilic fluorescent tracer of endosomes, FM4-64, indicating that membrane material co-accumulates with these proteins in papillae and haustorial encasements (Meyer et al. 2009; Nielsen et al. 2012). As demonstrated by electron microscopy and co-localization with the Rab-like GTPase MVB marker ARA6/RABF1-GFP (At3g54840), MVBs focally accumulate at pathogen attack sites. It is therefore conceivable that MVBs contribute to extracellular deposition of otherwise intracellularly localized proteins (An et al. 2006; Meyer et al. 2009; Nielsen et al. 2012). According to this hypothesis, vesicles containing PEN1, SNAP33, and PEN3 may sort and incorporate into the lumen of MVBs after endocytosis. These MVBs might subsequently fuse with the plasma membrane, which could explain extracellular protein delivery (Meyer et al. 2009). The extent and significance of the extracellular deposition of otherwise intracellular proteins for powdery mildew resistance is currently unknown (Kuhn et al. 2016).

4.3.8 Silicon-Mediated Resistance

Silicon (Si) contributes to powdery mildew resistance of cereals and various other plant species (Fauteux et al. 2005). Accordantly, watering *Arabidopsis* plants with silicon results in a lower powdery mildew disease incidence, although *Arabidopsis* lacks dedicated Si transporters (Ghanmi et al. 2004). Accumulation of insoluble Si at powdery mildew attack sites led to the hypothesis of Si acting as a simple physical barrier (Belanger et al. 2002). However, not in every case presence of insoluble Si correlates with increased resistance to fungal penetration. Consequently, a physiological or biochemical role in mediating cellular resistance has been postulated (Belanger et al. 2003). While Si fertilization alone has a minor effect on transcript abundance, *Gc* inoculation of Si fertilized plants versus *Gc* inoculation of non-Si supplemented plants attenuated the magnitude of powdery mildew-induced downregulation of genes by more than 25% (Fauteux et al. 2006). As many of these powdery mildew-repressed genes are related to primary metabolism, the Si-mediated reduced downregulation might indicate stress alleviation. Consequently, Si feeding potentially facilitates a more efficient response to powdery mildew infection (Fauteux et al. 2006). This hypothesis is further corroborated by transgenic *Arabidopsis* plants stably expressing the wheat Si transporter *Ta*Lsi1. These plants have increased Si levels, and concomitantly further enhanced powdery mildew resistance in the presence of Si compared to wild-type plants (Vivancos et al. 2015).

4.3.9 Mechanisms of R-Genes Regulation for Altered Cell Wall Composition of Host Resistance

As in the pmr6 mutants, three lines of evidence indicate that the resistance mechanism operating in the pmr5 mutant does not require the activation of either the SA or JA/ethylene defense pathways. First, pmr5 plants did not constitutively express high levels of either PR1 or PDF1.2 mRNA indicating that resistance is not mediated by constitutive activation of the SA or JA/ethylene pathways. Secondly, neither PR1 nor PDF1.2 was induced to high levels after inoculation indicating that the resistance is not mediated by a hyper-activation of either of these signal transduction pathways. Thirdly, and most convincingly, mutants or transgenes that block signaling through the SA or JA/ethylene pathways did not abolish powdery mildew resistance in pmr5 plants. Thus, pmr5-mediated resistance is independent of the activation of known defense pathways. Therefore, PMR5 is either required for fungal growth or the pmr5 mutation activates a novel defense pathway. Unlike resistance attributed to the majority of resistance genes and disease-resistant mutants, the attenuation of powdery mildew growth on pmr5 and pmr6 did not require cell death as shown by the lack of cell death below fungal colonies. In the course of looking for resistance-associated cell death, it was noticed that both pmr5 and pmr6 had micro lesions along veins on a subset of the oldest leaves. This phenotype was not correlated with resistance because only a small subset of the oldest leaves had lesions, yet all leaves were highly resistant. This phenotype can be phenocopied by heat treatment of wild-type plants (Vogel et al. 2004). Importantly, heat-treated plants were still susceptible to powdery mildew. It was the micro lesions as a pleiotropic effect of the *pmr5* and *pmr6* mutations unrelated to disease resistance. These micro lesions may be responsible for the slightly elevated basal level of PR1 observed in pmr5 and pmr6. That both pmr5 and pmr6 have these micro lesions underscores the similarity of these mutants.

As both *pmr5* and *pmr6* were fully susceptible to *P. syringae* pv. *tomato* and *H. parasitica*, the resistance is not due to the activation of a broad-spectrum defense pathway, like systemic acquired resistance. Both *pmr5* and *pmr6* were resistant to *E. orontii* indicating that the resistance is effective against isolates from two powdery mildew species. Thus, *pmr5* and *pmr6* resistance is qualitatively different than resistance conferred by either gene-for-gene resistance genes or previously described disease-resistant mutants. The FTIR spectra from pmr5 epidermal cell walls were similar to the spectra from pmr6 suggesting that both mutants have increased pectin, and the pectin had lower methyl esterification or O-acetylation relative to wild type. Moreover, like *pmr6*, the major FTIR spectral features associated with cellulose and xyloglucan shifted in energy in *pmr5* cell walls, suggesting an alteration in the hydrogen-bonding environment. Interestingly, the *pmr5 pmr6-3* double mutant had higher levels of uronic acid than either pmr5 or pmr6 indicating that these two mutants interact synergistically to increase uronic acid content. The synergistic effect on uronic acid content along with the similarity of pmr5 and pmr6 phenotypes (e.g., powdery mildew resistance, morphology, micro lesions, cell wall composition) suggests that these two mutations affect parallel pathways that regulate some aspects of pectin biosynthesis either directly or indirectly. Furthermore, PMR5 is predicted to be associated with the endoplasmic reticulum by a hydrophobic N-terminal signal sequence, and PMR6 is predicted to locate to the exterior side of the plasma membrane via a glycosylphosphatidylinositol anchor. Thus, it is unlikely that these two proteins interact directly. To address the possibility that the increase in uronic acid observed in the pmr5 pmr6-3 double mutant was due to an indirect effect of cell size, Vogel et al. (2004) determined the relationship between cell size, and pectin content in three dwarf mutants that were not directly related to disease resistance or pectin metabolism. Results indicated that, while the wild type, the three dwarfs, and pmr5 did show a weak correlation between cell size and pectin content, the large increase in uronic acid observed for the pmr5 pmr6-3 double mutant could not be attributed solely to decreased cell size. Thus, it is possible that the increase in pectin in pmr5 pmr6-3 cell walls restricts cell expansion, and this in turn limits cell size.

The cell wall is very dynamic, and responds to physiological stresses and altered substrate availability with compensating changes in organization (Gillmore et al. 2002). To assess whether the changes in pectin content inferred from the FTIR spectra were associated with any compensating changes in other components, the cell wall neutral sugar content was measured. Aside from the approximately 50% reduction in fucose in the double mutant, all other statistically significant changes in neutral sugars were modest. As approximately two-thirds of the fucose in the *Arabidopsis* leaf cell wall is found in xyloglucan, the decreased fucose in the double mutant may suggest decreased xyloglucan fucosylation (Perrin et al. 2003; Zablackis et al. 1995). The presence of relatively normal amounts of xylose in the double mutant suggests that the amount of xyloglucan is not strongly altered. Pmr5 pmr6-3 cell walls had small but significant increases in arabinose and galactose, suggesting increased abundance of the galactose and arabinose-containing side chains of the pectin and rhamnogalacturonan I. The characterization of pmr5 revealed that pmr5-mediated resistance does not require the activation of the SA or JA/ethylene defense pathways, does not require cell death, and is not broad-spectrum. In addition, the phenotype of pmr5 plants is very similar to pmr6 plants. Taken together, these data suggest that pmr5 and pmr6 employ similar mechanisms to limit fungal growth and that this mechanism is unrelated to known defense signaling pathways. There are several possible explanations for the disease resistance of the pmr5 mutant. This mutant may be a less hospitable host for powdery mildews. It is evident that, the pmr5 extra-haustorial matrix may have altered composition, especially of modified pectins, decreasing nutrient transport to the fungus or the powdery mildew pathogen may have limited ability to digest the pmr5 outer epidermal cell wall. Alternatively, the pmr5 cell wall may carry latent signaling molecules that are released upon powdery mildew infections to activate novel defenses (Vorwerk et al. 2004). Whatever the basis for disease resistance, the pmr5 and pmr6 mutants highlight the importance of cell wall composition in plant–pathogen interactions (Vogel et al. 2004).

4.3.10 Induction of Salicylate, NPR1, PAD4, and EDS5 in Powdery Mildew Resistance to *Arabidopsis*:

The powdery mildew fungus, *E. orontii* infection of *Arabidopsis* elicits the strong accumulation of PR1, BGL2, and PR5 mRNAs (Fig. 4.3). In several plants, salicylate has been shown to act as a signal molecule in the activation of PR gene expression (Yang et al. 1997), suggesting that a salicylate-dependent signaling pathway is important for limiting *E. orontii* growth in *Arabidopsis*. It is consistent with observations that npr1 and pad4 mutants as well as transgenic nahG plants exhibit enhanced susceptibility to *E. orontii*. *Arabidopsis* npr1/nim1/sai1 mutants fail to activate PR1, BGL2, and PR5 gene expression in response to exogenous salicylate, demonstrating that the NPR1 protein acts in a signal transduction cascade which responds to a salicylate signal (Cao et al. 1994; Delaney et al. 1995).

However, the npr1 mutant accumulates essentially wild-type levels of BGL2 and PR5 mRNAs, and about 10% of the wild-type levels of PR1 mRNA in response to *P. syringae* (Glazebrook et al. 1996). Similarly, Reuber et al. (1998) observed only a modest reduction in the levels of BGL2 and PR5 mRNAs, and about 20% of the wild-type levels of PR1 mRNA following *E. orontii* infection of npr1–1. Assuming that the npr1 mutants are not leaky, the activation of PR gene expression in the npr1 mutants by *P. syringae* and *E. orontii* suggests that there are NPR1-independent pathogen-activated pathways that can lead to PR gene expression. There is a possibility that the npr1 mutants could retain some activity, since a transgenic line that is suppressed for NPR1 expression exhibits very low PR1 activity after *P. syringae* infection (Cao et al. 1998). It has been observed that nahG plants exhibit very low levels of PR1 induction following *E. orontii* infection suggests that salicylate-dependent pathways are required for almost all of the PR1 expression in response to *E. orontii*. These results are consistent with those of Zhao and Last (1996), who found that PR1 expression in response to virulent *P. syringae* is completely abolished in nahG plants. Moreover, the fact that nahG plants are more susceptible to *E. orontii* than npr1 plants and exhibit lower levels of PR1 induction following *E. orontii* infection suggests the existence of an NPR1-independent but

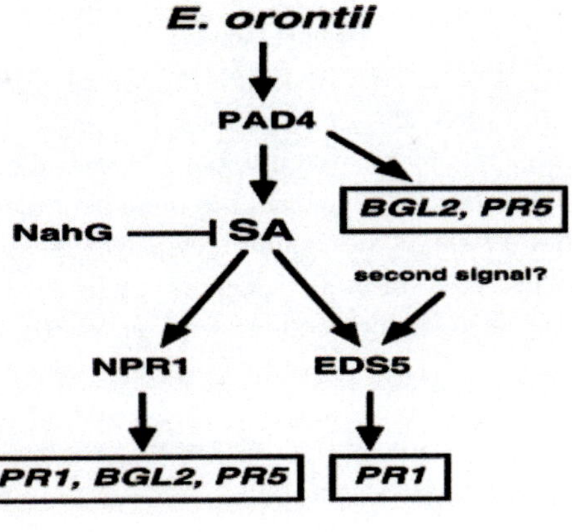

Fig. 4.3 Model of gene induction in *Arabidopsis* by *Erysiphe orontii* (Reuber et al. 1998)

salicylate-dependent defense gene activation pathway. There is precedent for such pathways as camalexin induction in response to *P. syringae* is salicylate-dependent but NPR1-independent (Zhao and Last 1996). Because the nahG plants retain significant BGL2 and PR5 expression, there must also be pathogen-activated salicylate-independent pathways leading to BGL2 and PR5 expression. Further, evidence that both salicylate-independent pathways and salicylate-dependent but NPR1-independent pathways are involved in limiting the growth of *E. orontii* is provided by the pad4 mutant. The pad4 mutant is the most susceptible to *E. orontii* of all the mutants tested. The pad4–1 mutant was recently reported to be deficient in the accumulation of salicylic acid in response to infection by virulent *P. syringae* strains (Zhou et al. 1998), indicating that PAD4 functions upstream of salicylic acid, and NPR1. However, Reuber et al. (1998) found less than 10% accumulation of PR1, BGL2, and PR5 mRNAs following infection of pad4 with *E. orontii*, compared to 20% accumulation of PR1, and greater than 50% accumulation of BGL2, and PR5 in npr1 mutants. The small amount of PR gene expression in pad4 could be due to leakiness of the pad4 allele that have been used for the production of a small amount of salicylate through a PAD4-independent pathway, or induction through an *E. orontii*-dependent but salicylate independent pathway. Evidence for the existence of salicylate-independent defense gene activation pathways is provided by the observation that pad4 is significantly more deficient in BGL2 and PR5 expression than the nahG transgenics. If this interpretation is correct, PAD4 must clearly play a pivotal role in the presumptive salicylate independent pathway(s). The eds5–1 also showed reduced PR1 accumulation in response to *E. orontii*. Almost all of the PR1 expression in response to *E. orontii* seems to be salicylate dependent. Therefore, it is most likely that EDS5 also functions in a salicylate-dependent pathway. The additive effects of the npr1–1 and eds5–1mutations on PR1 expression in the eds5–1 npr1–1 double mutant suggest that the EDS5, and NPR1 gene products may act in parallel signal transduction pathways, both of which are necessary for full expression of PR1 in response to pathogen infection. If this is the case, the EDS5-dependent pathway must also require a second, salicylate independent pathogen-generated signal to induce PR gene expression, since the npr1–1 mutation completely blocks expression of PR genes in response to exogenously applied salicylate. However, since it is not clear whether the npr1–1 mutation is a complete loss of function mutation, and because the nature of the eds5–1 mutation is unknown, it is also possible that neither mutation results in a null phenotype and that the EDS5 and NPR1 gene products function in the same pathway. In any case, the fact that the double npr1–1 eds5–1 mutant is more susceptible than either the eds5–1 or npr1–1 mutant that EDS5 plays a significant role in limiting the extent of *E. orontii* infection. There is no accumulation of salicylate in uninfected leaves since there is no PR gene induction. This is expected for a non-necrotizing pathogen such as *E. orontii* which does not typically induce systemic acquired resistance (Reuber et al. 1998).

4.3.11 Induction of Chitin Gene for Powdery Mildew Resistance

Chitin is a major component of fungal walls and insect exoskeletons. Plants produce chitinases upon pathogen attack, and chito-oligomers induce defense responses in plants, though the exact mechanism behind this response is unknown. Using the ATH1 Affy metrix microarrays consisting of about 23,000 genes, Ramonell et al. (2002) examined the response of *Arabidopsis* (*Arabidopsis thaliana*) seedlings to chito-octamers, and hydrolyzed chitin after 30 min of treatment. The expression patterns elicited by the chito-octamer and hydrolyzed chitin were similar. Microarray expression profiles for several genes were verified via northern analysis or quantitative reverse transcription-PCR. T-DNA insertion mutants for nine chito-oligomer responsive genes have been characterized. Three of the mutants were more susceptible to the fungal pathogen, powdery mildew, than wild type as measured by conidiophore production. These three mutants included mutants of genes for two disease resistance-like proteins and a putative E3 ligase. The isolation of loss-of-function mutants with enhanced disease susceptibility provides direct evidence that the chito-octamer is an important oligosaccharide elicitor of plant defenses. Also, this study demonstrates the value of microarray data for identifying new components of uncharacterized signaling pathways. The location of chitin present in different structures of powdery mildew pathogen (*Ec*) has been revealed through TEM (Plate 4.4).

Using genomics techniques, Ramonell et al. (2005) identified several genes that may play a role in the chitin-mediated defense pathway in *Arabidopsis*. The results showed that T-DNA insertions in three genes, At2g35000 (a putative RING zinc-finger protein), At2g34930, and At5g25910 (putative disease resistance-like proteins), resulted in plants that were more susceptible to powdery mildew than wild type, indicating that these genes may play a role in the defense response of *Arabidopsis* to powdery mildew. Thus, the general elicitor, chitin, has been linked to a defense response in plants indicating that it plays a role in plant responses to fungal pathogens. In addition, this study demonstrates the power of microarray data for identifying potential targets for mutation in uncharacterized signaling pathways. Additional experiments will now be necessary to elucidate the precise functions of these genes in defense responses and chito-oligomer recognition in *Arabidopsis*.

4.3.12 R-Genes-Mediated Expression of SHL/HR and Resistance

Arabidopsis plants with multiple copies of genomic fragments containing *RPW8.1* and *RPW8.2*, transcriptionally regulated by their native promoters, developed apparently SHL that were associated with enhanced expression of *RPW8.1*, *RPW8.2*, and *PR* genes. In the transgenic line S24, SHL appeared as isolated necrotic spots that enlarged to form necroses that resembled the HR induced by powdery mildew pathogens on plants containing *RPW8.1* and *RPW8.2*, such as resistant accession Ms-0 and Col-0 transgenic line S5. SHL developed in S24 and other lines in which the transgenes *RPW8.1* and *RPW8.2* were under the control of their native promoters

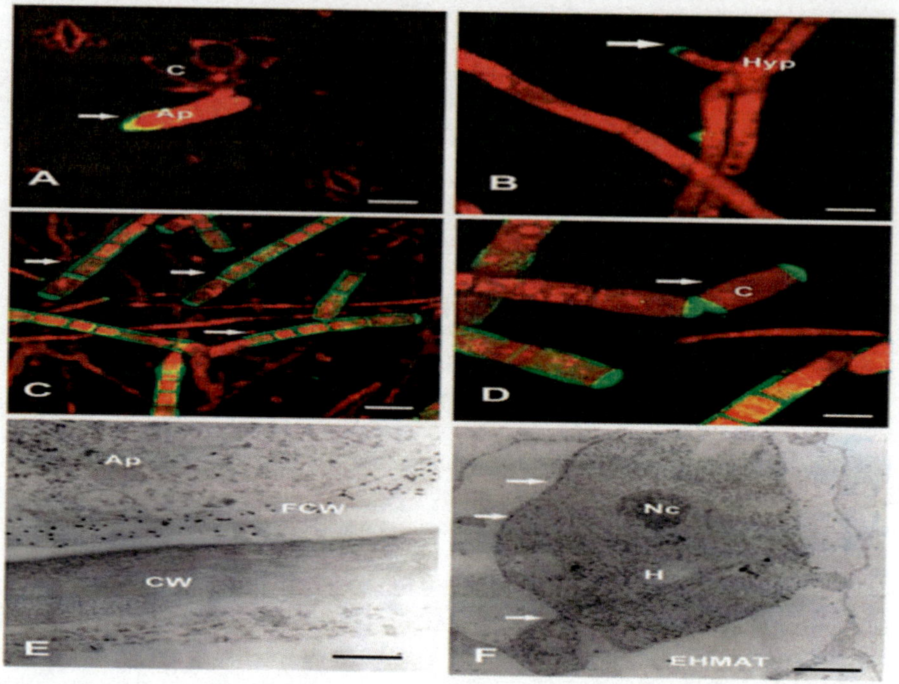

Plate 4.4 Chitin localization in the powdery mildew, *Erysiphe cichoracearum*. Confocal images are presented in (**a–d**), and transmission electron micrographs in (**e**, **f**). In (**a–d**), samples were stained with PI (red channel) to highlight fungal structures, although plant structures can be stained with this nonspecific stain, and chitin was localized with the lectin, WGA-Alexa Fluor 488 (green channel). In (**e**, **f**), WGA colloidal gold conjugates were used to localize chitin. (**a**) A merged confocal micrograph showing chitin localized to the tip of the appressorium (arrow; 1 dpi). Plant guard cells are also partially stained with PI in this image. Bar = 12 μm. *C* Conidium, *Ap* appressorium. (**b**) Chitin localization at the growing tip of hypha (arrow) but not on the elongated hyphal region (3 dpi). *Hyp* Hyphae. Bar = 11 μm. (**c**) Chitin labeling occurs in the developing conidia still attached to conidiophores (arrows; 7 dpi). Bar = 31 μm. (**d**) A mature conidium (arrow) detaching from a conidiophore shows strong chitin localization at the both ends. Bar = 16 μm. (**e**) Chitin occurs in the fungal appressorial cell wall but not on the plant cell wall. *CW* plant outer epidermal cell wall, *FCW* fungal cell wall. Bar = 2.15 μm. (**f**) Chitin is found in the haustorial cell wall (arrows). *EHMAT* extrahaustorial matrix, *H* haustorium, *Nc* haustorial nucleus. Bar = 9.21 μm (Ramonell et al. 2002)

but not in lines in which these transgenes were under the control of the 35S promoter. However, the level of the transcripts for *RPW8.1* and *RPW8.2* was not greatly different between these different lines. To investigate this discrepancy, Xiao et al. (2003) used a reporter for the *RPW8.1* promoter, and observed that in progeny of line S24, this reporter was activated only in cells at the margin of spreading lesions, which represented <10% of the leaf tissue. Apparently, there would have been at least ten times more transcripts of *RPW8.1* and *RPW8.2* in these S24 cells destined to become necrotic than in cells of lines in which the genes were expressed from the 35S promoter. These results indicate that the native promoters of *RPW8.1* and*RPW8.2* probably are required for SHL. It is possible that any lines with similarly high expression of the transgenes under the control of the 35S promoter would not survive. By contrast, overexpression of the R-gene *Pto* controlled by the 35S promoter in tomato results in the non-propagative spontaneous death of isolated cells (Tang et al. 1999). However, these lesions were barely visible to the naked eye and much smaller than the HR lesions induced by avirulent strains of the pathogen.

There was a general positive correlation between SHL development and level of *RPW8.1* and *RPW8.2*. It indicates that transcripts *RPW8.1* and *RPW8.2* were induced by *E. cichoracearum* pathogen (Xiao et al. 2001), which led to speculate that *RPW8*-mediated HR and SHL involve a self-amplification mechanism. By introducing a *GUS* reporter for the putative *RPW8.1* promoter into a background containing multiple copies of *RPW8.1* and *RPW8.2*, Xiao et al. (2003) clearly demonstrated that transcription of *RPW8.1* was further enhanced in the localized area where SHL developed, most likely as a consequence of the expression of *RPW8.1* and *RPW8.2* above a threshold level that has yet to be determined. Thus, the promoters of *RPW8.1* and probably *RPW8.2* seem to be critical for the transcriptional self-amplification. Analysis of the 1-kb sequence upstream of the *RPW8.1* and *RPW8.2* translational starts revealed three W-box elements in the promoter region of *RPW8.1* (TTGACC at −162 bp, TTGACT at −282 bp, and AGTCAA at −135 bp in front of the translational start), and 2 in the promoter region of *RPW8.2* (TTGACT at −281 and −526 bp). W-boxes are *cis*-acting elements often found in promoters of many SA and pathogen-responsive genes, such as *NPR1* and *PR1* (Lebel et al. 1998; Yu et al. 2001). They are binding sites for the WRKY family of transcription factors for the transcriptional regulation of defense-related genes (Rushton and Somssich 1998; Eulgem et al. 2000). Therefore, it was anticipated that *RPW8.1* and *RPW8.2* would be induced by SA.

SA is required for HR and for the expression of disease resistance in many plants (Malamy and Klessig 1992; Shirasu et al. 1997; Dempsey et al. 1999; Alvarez 2000). In *Arabidopsis*, SA also may function in a feedback amplification circuit involving the defense signaling genes *EDS1*, *PAD4*, and *EDS5* that amplifies the resistance response (Falk et al. 1999; Jirage et al. 1999; Feys et al. 2001; Rusterucci et al. 2001). These responses have similarity to the cell death in the *lsd5* mutant, which is SA dependent and SA induced (Weymann et al. 1995). SA also is essential to *RPW8.1* and *RPW8.2*-dependent HR, and plants containing the *nahG* transgene which did not express the HR (Xiao et al. 2001). SA is critical for *RPW8*-mediated SHL, because the *nahG* transgene suppressed SHL in line S24. Moreover, S24 plants in which SHL was initiated also had high levels of endogenous SA, and this finding correlated positively with the enhanced expression of *RPW8.1* and *RPW8.2*. Significantly, environmental conditions that suppressed SHL also suppressed both the accumulation of SA and the enhanced transcription of *RPW8.1* and *RPW8.2*. These data suggest that the expression of *RPW8.1* and *RPW8.2* above a threshold level leads to SA accumulation, because SA did not accumulate nor did SHL develop, in the single-copy line S5. Paradoxically, perhaps, by SA application to S24 plants that SA regulates *RPW8.1* and *RPW8.2* expression directly, enhancing SHL in otherwise suppressive conditions has been shown by Xiao et al. (2003).

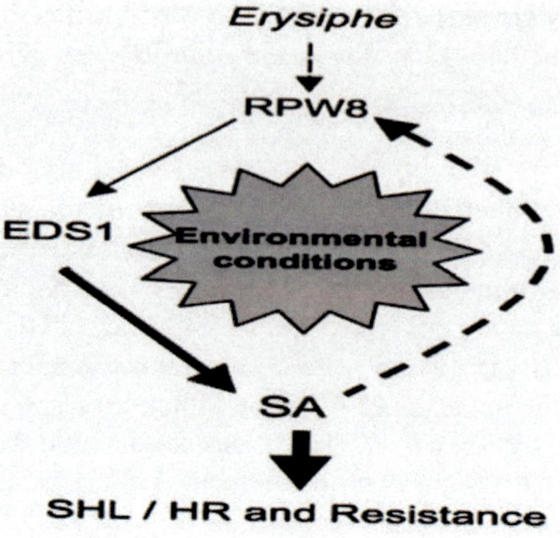

Fig. 4.4 Model for *RPW8*-mediated SHL, HR, and resistance. In this model, the recognition of *Erysiphe* pathogens by *RPW8* triggers defense responses via an SA-dependent pathway (Xiao et al. 2001), leading to SA accumulation. Increased SA further enhances the expression of *RPW8.1* and *RPW8.2* via a feedback amplification circuit, leading to HR, and resistance or SHL. Environmental conditions that suppress this amplification circuit suppress HR, and disease resistance or SHL (Xiao et al. 2003)

Significantly, Col-0 plants containing the *GUS* reporter for the putative *RPW8.1* promoter did not develop lesions when treated with SA, but GUS activity increased after SA application.

Xiao et al. (2003) studies support a model for the *RPW8*-dependent SHL, HR, and resistance in which SA mediates the transcriptional self-amplification of *RPW8* (Fig. 4.4). This feedback circuit amplifies the defense response stimulated by *RPW8* independently of the pathogen. Xiao et al. (2003) proposed that unknown local stimuli trigger this amplification circuit in line S24, leading to local necroses. *RPW8.1* promoter–*GUS* reporter experiments provide compelling independent evidence that SA enhances *RPW8.1* transcription through the promoter element, even in lines lacking the *RPW8* genes. Further, evidence for this model showed that the *RPW8.1* promoter is activated only in incipient lesions, and at the leading edges of developing lesions. This finding indicates that *RPW8.1* expression, and *RPW8.2* expression as well, is programmed to occur just ahead of the advancing lesion but not throughout the leaf. A corollary to this is that local signaling molecules, which include SA, must define the advance of lesions into tissue in which the transcription of *RPW8* genes is activated. This model also could explain why SHL did not develop in *35S*::*RPW8* lines: this promoter is not activated by SA. SA-dependent transcription of *RPW8.1* and *RPW8.2* forms part of an amplification circuit that leads to the accumulation of SA and SHL or HR. Presumably, therefore, SA formed during the expression of resistance to different pathogens also would induce the accumulation of transcripts of *RPW8.1* and *RPW8.2*. This raises the possibility that *RPW8.1* and *RPW8.2*, and possibly other members of this gene family as well (Xiao et al. 2001), contribute to the expression of HR against other pathogens. However, *RPW8.1* and

RPW8.2 couple the specific recognition of powdery mildew pathogens to the induction of this defense response, and a future challenge is to understand the basis of this specificity (Xiao et al. 2003).

4.3.13 Mechanisms of *NPR1* Gene in Powdery Mildew Resistance

Plants are very often exposed to a variety of biotic stresses, and thus have evolved multidimensional defense approaches to survive or retain their fitness (Roux et al. 2014). The plants display both preformed and inducible defense mechanisms to overcome pathogen challenges. However, much stronger and long lasting is inducible defense response such as systemic acquired resistance (SAR). Most of the PR proteins such as glucanases, chitinases, thaumatins, and defensins possess antifungal activities, and are known to play important role in disease resistance. Exogenous application of SA or its analogs have been also revealed to activate SAR pathway in plants (Durrant and Dong 2004; Makandar et al. 2006). Conversely, *Arabidopsis thaliana* plants expressing NahG transgene which codes for salicylate hydroxylase (SA-degrading enzyme) were deficient in accumulating SA, and hence failed to activate SAR (Delaney et al. 1995). In addition to SA, a group of heterogeneous proteins are crucial for the activation of SAR. Among them are the *NPR1* proteins, a key regulator in the SA-mediated SAR signal transduction pathway. The quest to discover the SA receptor led to the discovery of a regulatory or transcription co-factor protein *NPR1* (Cao et al. 1994). However, many studies have revealed that NPR1 is linked to SA signaling; however, its role as SA receptor remains largely unknown. In this context, Wu et al. (2012) have reported that NPR1 is the receptor for SA pathway in *Arabidopsis*. In addition, two NPR1 paralogs, namely, NPR3 and NPR4, bind SA and control the proteasome-mediated degradation of NPR1 protein through their interaction with NPR1 (Fu et al. 2012).

After pathogen infection, plants produce a variety of phytohormones, their composition, quantity, and timing significantly varies among plant species, and depends mainly on the pathogens lifestyle, and their mode of infection (Devos et al. 2005). SA pathway generally provides resistance to biotrophic pathogens, whereas jasmonic acid/ethylene (JA/ET) pathways are commonly associated with resistance to necrotrophic pathogens, and to herbivorous pests (Glazebrook 2005; Bari and Jones 2009). Generally, SA and JA signaling pathways operate antagonistically, and thus, elevated resistance against biotrophs is often related with increased susceptibility to necrotrophs, and vice versa (Grant and Lamb 2006). Many regulatory components involved in SA/JA crosstalk have been identified; among them is NPR1 which plays a crucial role in regulating SA-mediated suppression of the JA pathway (Spoel et al. 2003; Pieterse et al. 2012; Thaler et al. 2012; van der Does et al. 2013). Furthermore, SA/JA antagonism is commonly found in many plant species under various taxonomic groups and therefore seems to be evolutionaril conserved (Thaler et al. 2012).

As first discovered in *Arabidopsis*, various AtNPR1 homologs have been isolated thereafter in many agriculturally important crops (Chen et al. 2013; Zhong et al.

2015). NPR1 is a multigene family in *Arabidopsis* with multifaceted functions. The genes AtNPR1 and AtNPR2 are notably considered as a key regulator of SAR (Cao et al. 1997, 1998; Zhang et al. 2003) while AtNPR3 and AtNPR4 are known as negative regulator of SAR (Fu et al. 2012). Moreover, another group of AtNPR1 homologs are AtBOP1 and AtBOP2, which are related with lateral organ development (Hepworth et al. 2005). However, most of the studies were carried out on *Arabidopsis* NPR1 (AtNPR1). Structurally, AtNPR1 and its homologs contain an ankyrin repeat, N-terminal BTB/POZ broad-complex, Tramtrack, Bric a brac/poxvirus, and zinc finger domains, respectively (Cao et al. 1997; Aravind and Koonin 1999). In non-induced state, NPR1 exists as an inactive oligomer form into cytosolic region. However, after SA accumulation, the redox status of the cell changes which leads to dissociation of the inactive oligomer NPR1 to active monomers, and their translocation to the nucleus where they bind to TGA factors there by inducing PR genes (Mou et al. 2003; Tada et al. 2008). Transcriptional studies have shown that NPR1 is expressed at low levels in mock plants but is induced significantly after microbial attack or treatment with SA or its biologically active analogs. Many studies have revealed NPR1 mutant (NPR1) plants are more prone to diseases, and also show altered expression of defense marker PR genes compared to NPR1 expressing plants (Glazebrook et al. 1996; Cao et al. 1997). Furthermore, NPR1 also plays a role in cross talk of SA/JA signaling pathways and in antagonistic effect of SA on JA signaling (Spoel et al. 2003). Hence, NPR1 is considered as the positive regulator of SA-mediated plant immune responses. To explore the defense role of NPR1 against bacterial, viral, and fungal pathogens, various overexpression studies have been carried out in both model and crop plant systems. NPR1 mediates the SA-induced expression of pathogenesis-related (PR) genes and SAR. Overexpression of NPR1 results in the increase of transcript levels of antifungal genes like *PR1*, *PR2* (glucanase), and *PR5* (thaumatin), which are universally known to have antifungal activity. Many studies have revealed the potential antifungal activity of these PR genes against wide range of fungal pathogens. PR gene activity is regulated at the level of redox-dependent nuclear transport of NPR1. For example, overexpression of NPR1 in *Arabidopsis* plants confers enhanced disease resistance to bacterial and fungal infections (Cao et al. 1998; Friedrich et al. 2001). Transgenic carrot plants overexpressing AtNPR1 exhibit high disease resistance not only to biotrophs (*Erysiphe heraclei*) but also to necrotrophic fungal pathogen (*Botrytis cinerea*, *Sclerotinia sclerotiorum*, and *Alternaria radicina*), respectively (Wally et al. 2009). Parkhi et al. (2010) also reported that cotton transgenic plants expressing AtNPR1 exhibited broad spectrum of disease resistance not only to fungal pathogens but also to nematodes. Additionally, tobacco plants overexpressing *Malus hupehensis* NPR1 confers resistance to *Botrytis cinerea* as well as induces battery of pathogen-related genes. Furthermore, studies have revealed that rice and wheat plants overexpressing NPR1 gene confers broad spectrum of disease resistance against most disastrous pathogens *Magnaporthe oryzae*, *Fusarium verticillioides*, and *Fusarium oxysporum*, respectively (Makandar et al. 2006; Quilis et al. 2008). NPR1 overexpression in *A. thaliana* was reported to result in an increase in the transcript levels of PR genes, hence proves that NPR1-dependent PR gene-mediated disease resistance (Cao et al. 1997; Friedrich et al. 2001). Most recently, overexpression of NPR1 was also revealed to confer disease resistance against broad range of pathogens in different crops (Dutt et al. 2015; Sundaresha et al. 2016). These results revealed that NPR1 is a potential candidate gene for developing disease resistant transgenic crops against multiple pathogens. Therefore, genetic transformation of defense regulatory genes (like *NPR1*) that controls the function of multiple defense genes are potential sources for developing broad spectrum, and long-lasting disease resistance against pathogens in *B. juncea*. *NPR1* (non-expresser of pathogen-related gene 1) is a bon afide receptor of salicylic acid (SA) which modulates multiple immune responses in plants especially activation of induced, and systemic acquired resistance (SAR). Ali et al. (2017) have isolated and characterized a new NPR1 homolog (BjNPR1) from *B. juncea*. The phylogenetic tree constructed based on the deduced sequence of BjNPR1 with homologs from other species revealed that BjNPR1 grouped together with other known NPR1 proteins of Cruciferae family, and was nearest to *B. napus* (Fig. 4.5). Furthermore, expression analysis showed that BjNPR1 was upregulated after SA treatment and fungal infection but not by jasmonic acid or abscisic acid. To understand the defensive role of this gene, Ali et al. (2017) generated *B. juncea* transgenic lines overexpressing BjNPR1, and further confirmed by PCR and Southern blotting. The transgenic lines showed no phenotypic abnormalities, and constitutive expression of BjNPR1 activates defense signaling pathways by priming the expression of antifungal PR genes. Moreover, BjNPR1 transgenic lines showed enhanced resistance to *Alternaria brassicae* and *Erysiphe cruciferarum* as there was delay in symptoms and reduced disease severity than non-transgenic plants. In addition, the rate of disease spreading to uninfected or distal parts was also delayed in transgenic plants thus suggesting the activation of SAR. It is suggested that BjNPR1 is involved in broad spectrum of disease resistance against fungal pathogens including powdery mildew of *B. juncea* (Ali et al. 2017).

BjNPR1 plants showed improved resistance against powdery mildew. To assess the resistance level of BjNPR1 transgenic lines against powdery mildew, plants were infected and disease scoring was done for 1–3 weeks. In non-transgenic plants, higher number of *E. cruciferarum* colonies was observed than transgenic lines on 7th, 12th, and 17th day after infection (Plate 4.5a). Based on colony count, there was approximately 50% reduction in newly formed colonies between transgenic lines and non-transgenic plants (Plate 4.5b). At 17th day of infection, transgenic plants showed powdery mildew infection with a disease scale of 3–4 (30–40%), while non-transformed leaves (wild type) revealed 7–8 (70–80%) of disease incidence, respectively (Plate 4.5c). In addition, *E. cruciferarum*-mediated cell death was examined in BjNPR1 transgenic, and wild-type plants at different time points using trypan blue staining and light microscopy. Based on microscopic observations, more cell death was observed in control than that of transgenic plants (Plate 4.6a). To further investigate the role of BjNPR1 in improving powdery mildew disease resistance, the growth or fungal biomass of *E. cruciferarum* in BjNPR1 transgenic lines with wild-type plants was compared using light microscopy. The picture in Plate 4.6b reveals overexpression lines (L2 and L5) showed significant reduction of

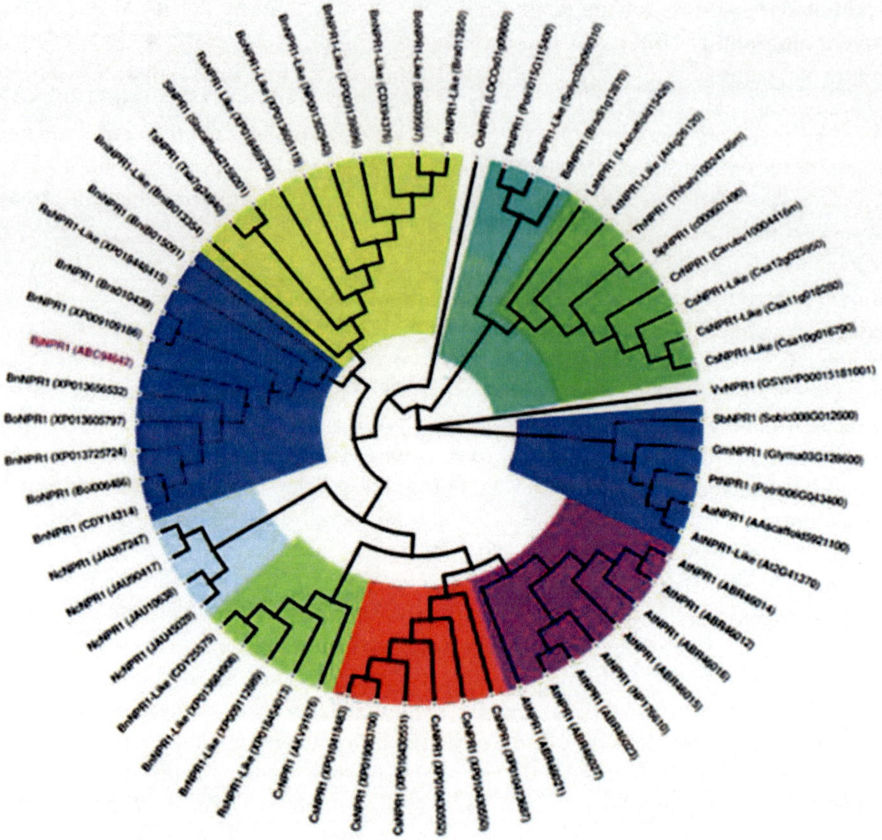

Fig. 4.5 Phylogenetic analysis of BjNPR1 with other NPR1 proteins from different plant species. The deduced amino acid sequences of BjNPR1 was retrieved from NCBI Gen Bank, and were further aligned with Clustal W using MEGA7.1 bioinformatic tool. The tree was generated using Maximum-Likelihood (ML) method with 1000 bootstrap replicates. Gen Bank IDs of each NPR1 protein sequence are given in the brackets behind the species names (Ali et al. 2017)

fungal biomass of *E. cruciferarum* than wild-type plants at 7th, 12th, and 17th dpi. However, overexpression of BjNPR1 could not inhibit the growth of *E. cruciferarum* completely, thereby providing only partial resistance to powdery mildew disease. Also, more number of leaves and pods were infected in non-transgenic plants as compared to transgenic lines. These results indicate that BjNPR1 transgenic plants exhibited partial resistance to powdery mildew infection, which was sufficient to delay the spread of infection in noninfected leaves or other parts (Ali et al. 2017).

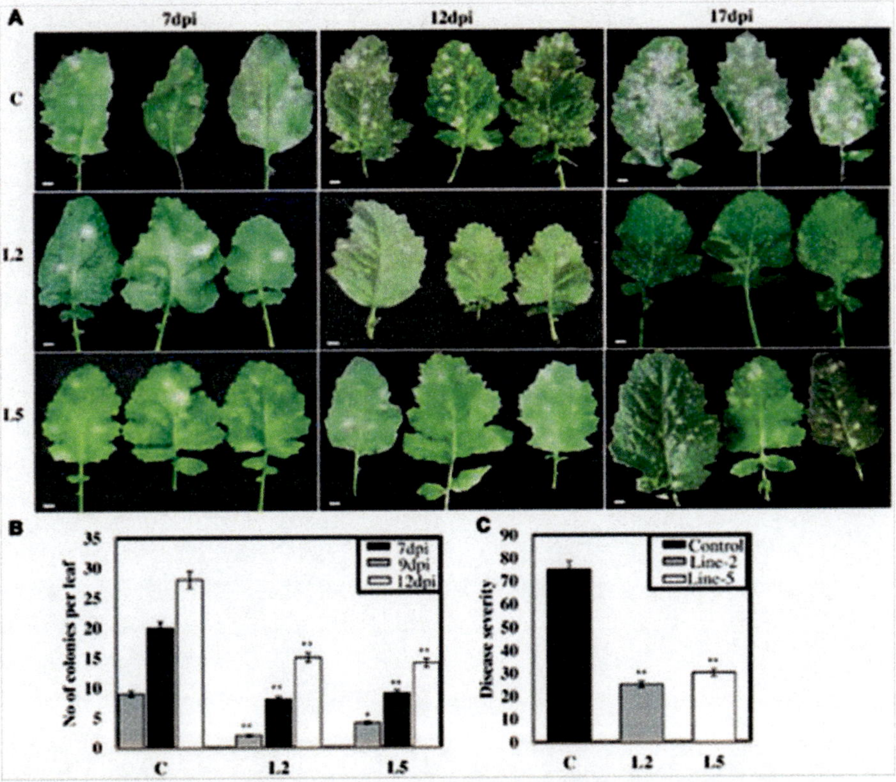

Plate 4.5 Screening of BjNPR1 transgenic lines for powdery mildew disease resistance. Forty-days-old wild-type plants and Bj NPR1 transgenic plants were infected with *E. cruciferarum,* and disease scoring was done at different time intervals. (**a**, **b**) BjNPR1 transgenic lines (L2 and L5) showed reduced number of *E. cruciferarum* colonies than wild-type plants at 7th, 12th, and 18th dpi. (**c**) *E. cruciferarum* disease severity in Bj NPR1 transgenic lines and wild-type plants. Bar = 35 μm. The asterisks indicate statistically significant differences between the BjNPR1 transgenic and control (non-transgenic) plants after powdery mildew infection ($P < 0.05$; $P < 0.01$) (Ali et al. 2017)

4.3.14 Role of *MAP65*-3 Gene in Powdery Mildew Resistance

The microtubule (MT) cytoskeleton is a highly flexible and dynamic polar structure of the plant cell, assembled from tubulin heterodimers. It is involved in nuclear and cell division, in cell morphogenesis and expansion, and in intracellular transport (Wasteneys and Galway 2003; Wasteneys 2004; Hamada 2014). MTs also play a role in plant in responses to biotic and biotic stress exposure, and their rearrangements accompany both defense and successful infection by symbiotic and pathogenic microbes (Schmidt and Panstruga 2008; Hardham 2013). MT rearrangements occur during arbuscular mycorrhizal (Genre et al. 2005) and rhizobial symbiosis (Vassileva et al. 2005), during the formation of plant parasitic

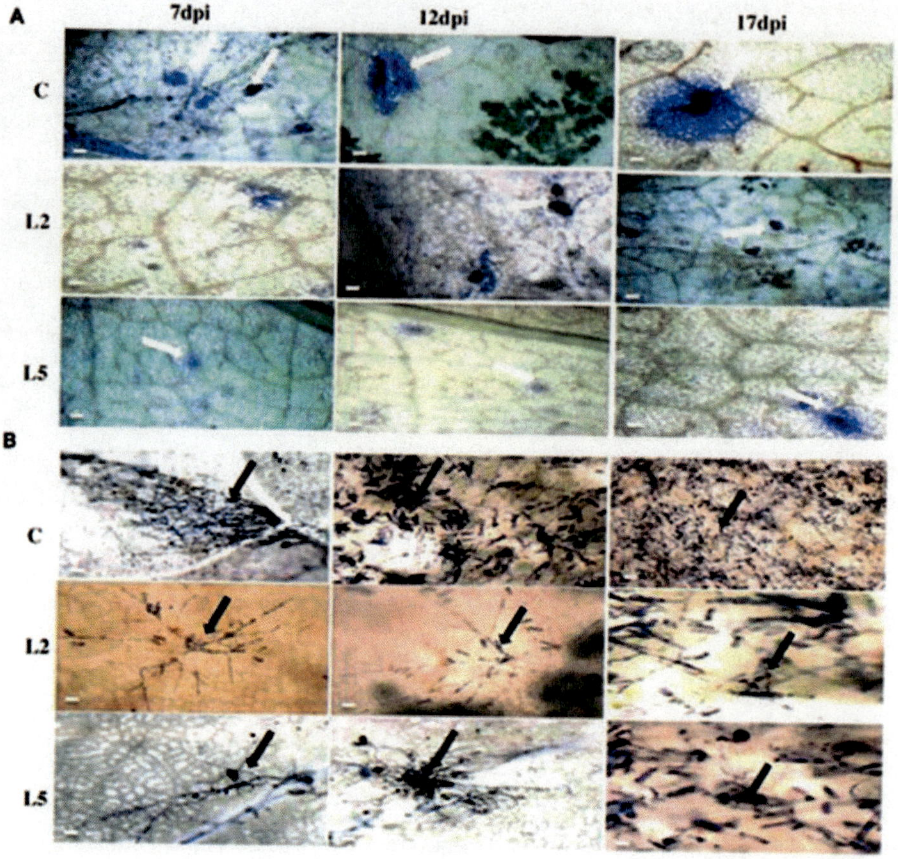

Plate 4.6 Microscopic examination of cell death, and fungal biomass in BjNPR1 transgenic and wild-type plants using trypan blue staining. (**a**) Microscopic examination of *Erysiphe cruciferarum*-mediated cell death in BjNPR1 lines and wild-type plants are shown with bold white arrows. (**b**) *E. cruciferarum* spore load or biomass in BjNPR1 transgenic lines and wild-type plants at various dpi after trypan blue staining are highlighted with bold black arrows. Bar = 30 mm (Ali et al. 2017)

nematode feeding sites (Caillaud et al. 2008a; de Almeida and Favery 2011), following virus attack (Martiniere et al. 2009) or following infection by filamentous oomycetes or fungi (Kobayashi et al. 1994; Baluska et al. 1995; Cahill et al. 2002; Takemoto et al. 2003; Hardham et al. 2008; Hoefle et al. 2011).

Microtubule-associated proteins (MAPs) and their regulatory kinases and phosphatases are instrumental for microtubule dynamics (Wasteneys 2004; Gardiner 2013; Hamada 2014). They, and the small Rho of Plants (ROP) GTPases that regulate the MT cytoskeleton (Mucha et al. 2011), have been shown to determine plant susceptibility to viruses and fungi (Kragler et al. 2003; Ouko et al. 2010; Hoefle et al. 2011; Poraty-Gavra et al. 2013). The *Arabidopsis thaliana* MAP65-3 (AtMAP65-3) is a critical module giant cell ontogenesis, and for successful pathogen development (Caillaud et al. 2008b), plant MAP65s are involved in the spatially and temporally regulated binding and bundling of MTs (Chan et al. 1999; Hamada 2014). In *A. thaliana*, nine members of this family were identified (Hussey et al. 2002), and individual members have particular functions with respect to different MT arrays. AtMAP65-3 is only associated with mitotic MT arrays (Muller et al. 2004; Caillaud et al. 2008b; Ho et al. 2011). The protein organizes both spindle morphogenesis and phragmoplast expansion (Muller et al. 2004; Caillaud et al. 2008b; Ho et al. 2011). Consequently, AtMAP65-3 loss-of-function mutants are dwarf, with both shoots and roots being stunted, and polynucleate, hypertrophied cells with aberrant cell wall stubs occur frequently (Muller et al. 2004; Caillaud et al. 2008b; Ho et al. 2011).

Plants protect themselves against pathogenic microorganisms by combining constitutive and induced defense mechanisms. The induction of plant defenses involves the recognition of compounds derived from the pathogen, called pathogen-associated molecular patterns (PAMPs). Pattern-triggered immunity (PTI) results from PAMP perception, which leads to the activation of signaling cascades, and the subsequent induction of defense-related genes (Zipfel et al. 2004; Jones and Dangl 2006). Pathogens are able to suppress these defenses by secreting effector proteins that manipulate host cell functions. In turn, plants evolved resistance proteins, which allow recognition of these effectors or their activities. This leads to effector-triggered immunity (ETI) and activation of the hypersensitive response (HR). The HR involves local programmed cell death that prevents pathogen spreading within the plant (Zipfel et al. 2004; Jones and Dangl 2006). Both PTI and ETI/HR involve mitogen-activated protein kinase (MAPK) cascades, the production of reactive oxygen species (ROS), and the transcriptional activation of genes, which, among others, encode antimicrobial pathogenesis-related (PR) proteins. The signaling pathways of PTI or ETI are fine-tuned by plant signaling molecules such as salicylic acid (SA), jasmonic acid (JA), and ethylene (ET) (Glazebrook 2005; Pieterse et al. 2012). The hormone SA plays a major role in plant resistance to (hemi-) biotrophic pathogens (Pieterse et al. 2012). In *A. thaliana*, SA synthesis occurs in plastids via isochorismate synthase 1 (ICS1 or SID2), and is triggered by pathogens. SA can be exported to the cytosol by the transporter enhanced disease susceptibility 5 (EDS5).

SA accumulated in the cytoplasm can be converted to SA glucoside (SAG), which is stored in the vacuole, and hydrolyzed back to SA when needed. Elevated levels of total SA (free SA plus SAG) have been correlated with the induction of defense gene expression and enhanced plant resistance (Pieterse et al. 2012). In addition, SA is a key regulator of plant immunity through its antagonistic interaction with ET and JA pathways (Glazebrook 2005). Different from SA, JA, and ET accumulate mainly in response to necrotrophic pathogens. An increasing number of mutants with reduced susceptibility to plant pathogens are described, and breeding for loss of susceptibility becomes a new strategy to achieve disease resistance (de Almeida-Engler et al. 2005; Dangl et al. 2013; Huckelhoven et al. 2013). Loss of disease susceptibility is frequently caused by the deregulation of SA- or JA-dependent plant defense signaling, by the impairment of cellular rearrangements, or by the limitation of nutrient supply for the pathogen (Dangl et al. 2013;

Huckelhoven et al. 2013; Lapin and Van den Ackerveken 2013; Van Schie and Takken 2014). The expression of the gene-encoding AtMAP65-3 is strongly induced in *A. thaliana* upon infection by two biotrophic filamentous pathogens, the oomycete *Hyaloperonospora arabidopsidis* (*Hpa*) and the powdery mildew fungus *Erysiphe cruciferarum* (*Ec*). Both pathogens develop haustoria inside host cells that constitute the feeding structures for nutrient supply (O'Connell and Panstruga 2006). The plants mutated in *AtMAP65-3* are impaired in their susceptibility to both filamentous pathogens. Mutants accumulate increased levels of SA and constitutively express genes encoding PR proteins in the leaves. Increased SA accumulation is not responsible for the mutant dwarfism, indicating that AtMAP65-3 exerts a dual role in positively regulating plant growth and development, and in negatively regulating plant defense responses. After infection, powdery mildew pathogen (*E. cruciferarum*) induces the formation of intracellular bulbous structures called haustoria, which are required for the biotrophic lifestyle. The microtubule-associated protein AtMAP65-3 plays a critical role in organizing cytoskeleton microtubule arrays during mitosis and cytokinesis. This renders the protein essential for the development of giant cells, which are the feeding sites induced by root knot nematodes. The *At- MAP65-3* expression is also induced in leaves upon infection by the downy mildew oomycete, and the powdery mildew fungus. Loss of *AtMAP65-3* function in the *map65-3* mutant dramatically reduced infection by both pathogens, predominantly at the stages of leaf penetration. Whole transcriptome analysis showed an over-represented, constitutive activation of genes involved in salicylic acid (SA) biosynthesis, signaling, and defense execution in *map65-3*, whereas jasmonic acid (JA)-mediated signaling was downregulated. Preventing SA synthesis and accumulation in *map65-3* rescued plant susceptibility to pathogens, but not the developmental phenotype caused by cytoskeleton defaults. AtMAP65-3 thus has a dual role. It positively regulates cytokinesis, thus plant growth, and development, and negatively interferes with plant defense against filamentous biotrophs. Quentin et al. (2016) suggested that downy mildew, and powdery mildew stimulates *AtMAP65-3* expression to downregulate SA signaling for infection.

4.3.15 Role of Receptor-Like Cytoplasmic Kinases (RLCK) Genes in Powdery Mildew Resistance

Receptor-like cytoplasmic kinases (RLCKs) belong to the receptor-like kinase (RLK) super family that is involved in a variety of biological processes like plant growth, development, and immune responses (Afzal et al. 2008). RLCKs share a conserved serine/threonine (Ser/Thr) kinase domain (Afzal et al. 2008; Gish and Clark 2011) together with the large and diverse transmembrane RLK protein family. Together over 600 RLKs and RLCKs exist in *Arabidopsis,* and over 1000 in rice (Shiu et al. 2004). In contrast to RLKs, RLCKs do not possess an extracellular and transmembrane domain resulting in their cytoplasmic localization. However, some RLCKs are anchored to the plasma membrane through myristoylation motifs

(Murase et al. 2004; Tang et al. 2008; Veronese et al. 2006). As protein kinases, RLCKs transmit intracellular signals through phosphorylation of target proteins or through RLK complex-mediated transphosphorylation events. Although little is known about their precise biological functions, several RLCKs were reported to be, either alone or in concert with RLKs, involved in plant development and immunity (Lin et al. 2013). In this regard, RLCKs play a role in the embryonic patterning process (Bayer et al. 2009), self-incompatibility (Murase et al. 2004), organ separation (Burr et al. 2011), ethylene (Laluk et al. 2011,) and brassinosteroid signaling (Sreeramulu et al. 2013; Tang et al. 2008). In plant immunity, RLCKs are involved in both pathogen-associated molecular pattern (PAMP)-triggered immunity (PTI) and effector-triggered immunity (ETI) (Lu et al. 2010; Swiderski and Innes 2001; Zhang et al. 2010). RLCKs are divided into 13 subfamilies (RLCK I-XIII) (Shiu et al. 2004). *Arabidopsis* RLCKs of the subfamily VI A interact with the plant-specific Rho family of small monomeric G proteins called "Rho of plants" (RAC/ROPs) (Jurca et al. 2008; Molendijk et al. 2008). A function of the VI subfamily of RLCKs as RAC/ROP downstream signaling effectors in plants was supported by RAC/ROP GTPase-dependent activation of RLCKs in *Arabidopsis, Medicago truncualata* (*M. truncualata*), and barley (*Hordeum vulgare* L.) (Dorjgotov et al. 2009; Huesmann et al. 2012). RAC/ROP proteins regulate processes like cell development, hormone signaling, cytoskeleton rearrangement, and plant disease resistance or susceptibility via various downstream effector proteins including RLCKs (Berken 2006; Nibau et al. 2006). RAC/ROPs act as molecular switches, transducing extracellular signals into intracellular responses by shuttling between an inactive GDP-bound, and an activated GTP-bound state. Several regulatory molecules including guanine nucleotide exchange factors (ROPGEFs), GTPase activating proteins (ROPGAPs), and guanine nucleotide dissociation inhibitors (GDIs) adjust the balance between these two forms (Nibau et al. 2006). Furthermore, RAC/ROPs are divided into two phylogenetic subgroups (type I, and type II) depending on their posttranslational lipid modifications, which anchor the active proteins in the plasma membrane (Winge et al. 2000). Besides the 11 characterized RAC/ROP proteins in *Arabidopsis* (Li et al. 2001), 6 and 7 RAC/ROPs are described in barley (Schultheiss et al. 2003) and rice (Chen et al. 2010), respectively. In *Arabidopsis*, AtROP4 and AtROP6 are involved in the auxin binding protein 1 (ABP1)-mediated formation of lobed pavement cells through organization of the actin, and microtubule cytoskeleton as well as in pathogen response (Fu et al. 2005, 2009; Poraty-Gavra et al. 2013; Xu et al. 2010). In addition, AtROP4 and AtROP6 are described as activators of AtRLCKs. Nevertheless, the functional knowledge about RAC/ROP-regulated RLCK signaling in plants is limited. Recently, the barley RLCK ROP binding kinase1 (HvRBK1), which is closely related to AtRLCK VI A3, was shown to serve as RAC/ROP effector in the barley-barley powdery mildew interaction (Huesmann et al. 2012). HvRBK1 interacted with the susceptibility factor HvRACB, which is required for successful invasion of intact barley epidermal cells by the biotrophic fungus *Blumeria graminis* f. sp. *hordei*, the causal agent of powdery mildew disease (Hoefle et al. 2011; Schultheiss et al. 2002). All RLCKs described as RAC/ROP interactors are members of the RLCK VI subfamily that is divided into group A and B based on their domain structure (Jurca et al. 2008; Molendijk et al. 2008). In *Arabidopsis*, the RLCK VI subfamily shows upregulation of gene expression under abiotic stress or hormone treatments as well as in response to the pathogens *Botrytis cinerea* and *Phytophthora infestans* (Jurca et al. 2008; Molendijk et al. 2008). Reiner et al. (2014) have described the HvRBK1-related *Arabidopsis* AtRLCK VI A3 as direct molecular interactor of *Arabidopsis* RAC/ROPs. AtRLCK VI A3 interacts with AtROPs in yeast, and shows increased kinase activity in the presence of constitutively activated (CA) AtROP6 in vitro. Furthermore, *AtRLCK VIA3* mutant lines show a reduced growth phenotype and an increased number in trichome branching. Finally, a slight increase in susceptibility toward the powdery mildew fungus *Erysiphe cruciferarum* was observed in *AtRLCK VIA3* mutant lines suggesting a function of AtRLCK VIA3 in the *Arabidopsis*-powdery mildew pathosystem. RLCKs of the VIA subfamily are possible downstream effectors of the small monomeric G proteins of the plant-specific Rho family, called "Rho of plants" (RAC/ROPs). *Arabidopsis thalina* AtRLCK VI A3 acts as a molecular interactor of AtROPs. In *Arabidopsis* epidermal cells, transient co-expression of plasma membrane located constitutively activated (CA) AtROP4 or CA AtROP6 resulted in the recruitment of green fluorescent protein (GFP)-tagged AtRLCK VI A3 to the cell periphery. Intrinsic kinase activity of AtRLCK VI A3 was enhanced in the presence of CA AtROP6 in vitro, and further suggested a functional interaction between the proteins. In the interaction of the biotrophic powdery mildew fungus *Erysiphe cruciferarum,* and its host plant *Arabidopsis*, *Atrlck VI A3* mutant lines supported enhanced fungal reproduction. Furthermore *Atrlck VI A3* mutant lines showed slightly reduced size, and an increase in trichome branch number compared to wild-type plants. There is a role of the AtROP-regulated AtRLCK VI A3 in basal resistance to *E. cruciferarum* as well as in plant growth, and cellular differentiation during trichome morphogenesis (Reiner et al. 2014).

4.3.16 Molecular Mechanisms of Camalexin Synthesis for Powdery Mildew Resistance

The *cyp83a1-3* mutant displays enhanced resistance to powdery mildews, including *G. cichoracearum*. The levels of SA accumulation and *PR* gene expression in *cyp83a1-3* mutants were similar to the wild-type, and double mutant analysis showed that resistance in *cyp83a1-3* was dependent on *PAD4*, *EDS1*, and *NPR1* but independent of *SID2* and *EDS5*. Mutations in *SID2* or *EDS5* suppressed SA accumulation induced by powdery mildew but did not suppress the disease resistance, indicating that resistance in *cyp83a1-3* is not caused by enhanced SA signaling. These observations indicate that *cyp83a1-3*-mediated resistance differs from that in the *edr1* and *edr2* mutants (Frye et al. 2001; Tang et al. 2005b). Moreover, the *cyp83a1-3* mutant accumulates more camalexin upon powdery mildew infection, and mutations in *PAD3* or *WRKY33* suppressed both the disease resistance and the high accumulation of camalexin, indicating a link between camalexin levels and responses to powdery mildew (Plate 4.7a). Consistent with a role of camalexin in

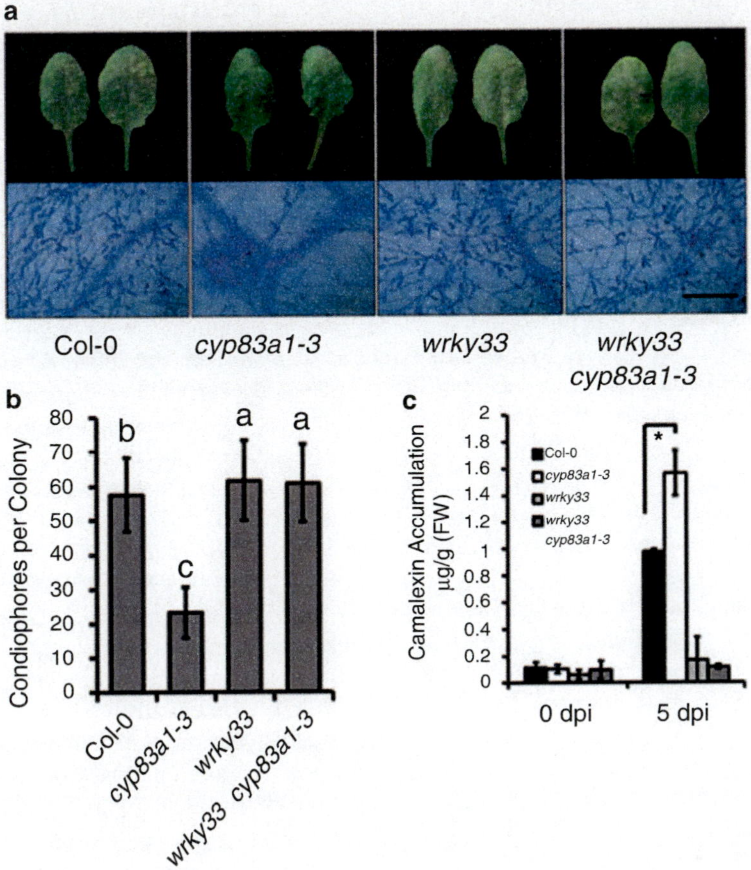

Plate 4.7 The resistance phenotype and high levels of camalexin in *cyp83a1-3* are suppressed by mutation of *WRKY33*. (**a**) Four-weeks-old wild-type, *cyp83a1-3*, *wrky33*, and *wrky33 cyp83a1-3* double-mutant plants were infected with *Golovinomyces cichoracearum*. Representative leaves were removed, and stained with trypan blue at 8 dpi, bar = 200 μm. (**b**) Quantification of fungal growth of the plants in (**a**) at 5 dpi by counting the number of conidiophores per colony. Results represent the mean and standard deviation in three independent experiments ($n = 30$; $P < 0.01$, nested ANOVA). (**c**) Camalexin accumulation of the plants in (**a**) was determined at 0, and 5 dpi. Results represent the mean, and standard deviation in three independent experiments ($n = 3$). Asterisk represents statistically significant difference from wild-type ($P < 0.01$, nested ANOVA) (Liu et al. 2016)

powdery mildew resistance, the *PAD3*-overexpressing plants accumulated more camalexin and mimicked the resistance phenotype of the *cyp83a1-3* mutant. Taken together, these findings indicate that the higher level of camalexin contributes to the enhanced resistance to *G. cichoracearum* observed in the *cyp83a1-3* mutant (Plate 4.7b). Loss-of-function mutations in two transcription factors, WRKY18 and WRKY40, result in the accumulation of higher levels of camalexin as well as both pre-invasive and post-invasive resistance against the powdery mildew fungus *G. orontii* (Pandey et al. 2010; Schon et al. 2013). The PAD3 (a key enzyme in camalexin biosynthesis) is required for the pre-invasive resistance (but surprisingly not for the post-invasive resistance) against *G. orontii* in a *wrky18 wrky40* background (Schon et al. 2013). Moreover, Schon et al. (2013) also report that *wrky18 wrky40* plants do not show increased resistance against two other powdery mildews, *G. cichoracearum* and *E. cruciferarum.* It may be related to differences in camalexin levels in the two mutants or to differences in susceptibility of different *G. cichoracearum* isolates to camalexin. The function of CYP83A1 has been studied previously. An earlier study showed that overexpression of *CYP83A1* could rescue the auxin-excess phenotype of *cyp83b1/rnt1/sur2* mutants (Bak and Feyereisen 2001). Further studies showed that levels of many phenylpropanoid pathway-derived products were reduced in the *cyp83a1-2/ref2-1* mutant, indicating cross talk between the pathways producing aliphatic-glucosinolates and indole-glucosinolates (Hemm et al. 2003; Naur et al. 2003). The studies showed that CYP83A1 interacts with BAX INHIBITOR-1, a cell death suppressor in plants and animals. The loss-of-function mutants *cyp83a1-1* and *cyp83a1-2* displayed enhanced resistance to the powdery mildew fungus *E. cruciferarum* (Weis et al. 2013).

A more study measured the levels of several glucosinolates in *cyp83a1-1* mutants, but found only marginally increased amounts of indole-derived glucosinolates. The *cyp83a1* mutants lack very-long-chain aldehydes, and accumulate more 5-methylthiopentanaldoxime (5-MPTO), a potentially toxic substrate of CYP83A1 (Weis et al. 2014). Very-long-chain aldehydes promote germination, and appressorium formation of *E. cruciferarum*, it was proposed that lack of very-long-chain aldehydes cause the resistance phenotypes in *cyp83a1* mutants (Weis et al. 2014).

Liu et al. (2016) observed that the high level of camalexin contributes to resistance to the powdery mildew fungus *G. cichoracearum*. It is worth noting that different species of powdery mildew, *G. cichoracearum,* and *E. cruciferarum* were used by Liu et al. (2016) and earlier by Weis et al. (2014). Although *cyp83a1* mutants displayed enhanced resistance to both powdery mildew strains, the mechanisms could differ. Consistent with this notion, several studies observed differences in infection phenotypes between different powdery mildew species in *Arabidopsis*, *G. cichoracearum,* and *G. orontii* have different host ranges/responses (Plotnikova et al. 1998) and many *Arabidopsis* accessions have different responses to the powdery mildew species *E. cruciferarum* UEA1 and *G. cichoracearum* UCSC1 (Adam et al. 1999). Liu et al. (2016) showed that *pad3* and *wrky33* suppressed the accumulation of camalexin and the enhanced resistance in *cyp83a1-3* mutants, indicating a role of camalexin in *cyp83a1-3*-mediated resistance (Plate 4.8a). It would be interesting to examine the responses of *pad3 cyp83a1-3* and *wrky33 cyp83a1-3* mutants to *E. cruciferarum*, and to measure the levels of very-long- chain aldehydes and 5-MPTO in those mutants. It is also possible that both very-long-chain aldehydes and camalexin contribute to resistance against *G. cichoracearum* and *E. cruciferarum.*

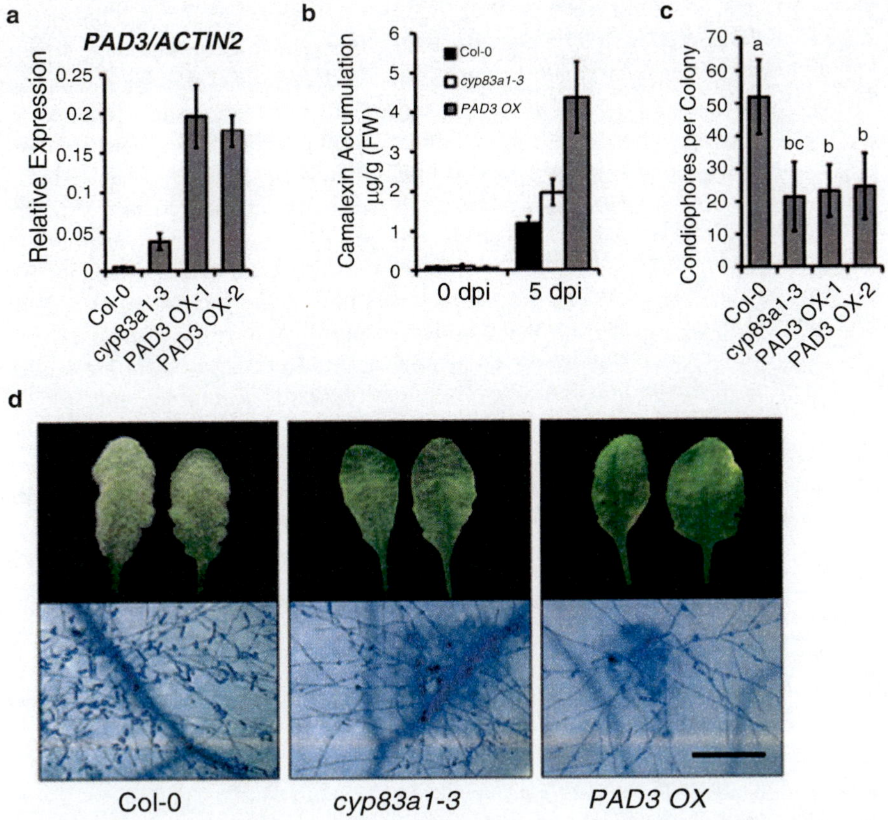

Plate 4.8 *PAD3* overexpressing plants accumulate higher levels of camalexin, and display a *cyp83a1-3*-like resistance phenotype. (**a**) The accumulation of *PAD3* transcript in 4-weeks-old plants was examined by quantitative real-time PCR. Results represent the mean and standard deviation in three independent experiments ($n = 4$). (**b**) Four-weeks-old plants were infected with *G. cichoracearum,* and camalexin accumulation was determined at 0 and 5 dpi. Three *PAD3*-OX independent lines were tested, and similar results were obtained in these lines. One representative *PAD3*-OX line (*PAD3*-OX-*1*) is shown. Results represent the mean and standard deviation in three experiments ($n = 3$). (**c**) Quantification of fungal growth of wild-type, *cyp83a1-3*, and *PAD3*-OX plants at 5 dpi by counting the number of conidiophores per colony. Results represent the mean and standard deviation in three independent experiments ($n = 3$). Different letters represent statistically significant differences ($P < 0.01$, nested ANOVA). (**d**) Four-weeks-old wild-type, *cyp83a1-3*, and *PAD3*-OX-1 plants were infected with *G. cichoracearum*. Three *PAD3*-OX independent lines were examined, and similar results were obtained in these three lines. Representative leaves were removed and stained with trypan blue at 8 dpi, bar = 200 µm (Liu et al. 2016)

CYP83A1 functions in the biosynthesis of aliphatic glucosinolates, and it is not clear that how the mutation of the glucosinolate synthetase gene *CYP83A1* affects the accumulation of camalexin. The possibility may be that it may cause cross talk with in the complicated metabolic network. In this scenario, the biosynthetic pathway of aliphatic-glucosinolates, which involves CYP83A1 and indole-glucosinolates, which share the IAOx intermediate with camalexin (Nafisi et al. 2007; Schuhegger et al. 2007a; Bottcher et al. 2009) can affect each other. So when the aliphatic-glucosinolates pathway is blocked in the *cyp83a1* mutant, the pathway for indole-derived products, including indole-glucosinolates and camalexin, is enhanced. Indole-glucosinolates are known to contribute to plant immunity (Bednarek 2012). *Arabidopsis* penetration2 (PEN2), which initiates indole-glucosinolate metabolism, plays an important role in penetration resistance (Bednarek et al. 2009; Clay et al. 2009).

In addition to increased levels of camalexin, the *cyp83a1* mutant has several other aberrant phenotypes including decreased levels of very-long-chain aldehydes and alterations in many metabolites that could also contribute to resistance, so the exact mechanism of powdery mildew resistance in *cyp83a1* needs to be further studied. Consistent with this notion, *PAD3*-overexpression plants accumulate much higher levels of camalexin than the *cyp83a1-3* mutant, but no further increase in powdery mildew resistance was observed (Plate 4.8b, c). It would be interesting to measure very-long-chain aldehydes and indole-derived products in *pad3 cyp83a1* double mutants to further examine whether altered camalexin levels or differences in other metabolites are responsible for the powdery mildew resistance phenotype in *cyp83a1* (Plate 4.8d; Liu et al. 2016).

4.3.17 Molecules and Phytohormones Triggering Defense Signaling Pathways

The first line of plant defense (pre-penetration) against fungal pathogens largely relies on cell surface-mediated defense signaling initiated by recognition of PAMPs, and secondary (post-penetration) defense responses are often induced by the salicylic acid (SA) or Jasmonic acid/ ethylene (JA/ET) phytohormone signaling pathways.

4.3.17.1 Salicylic Acid-Mediated Signaling to Powdery Mildew Resistance

The fact that npr 1, pad 4, and eds 5 mutants are susceptible to *G. orontii* indicates that NPR1, PAD 4, and EDS 5 play key roles in defense signaling pathways. EDS 5 plays a role in salicylate dependent signal transduction. A model of gene induction in *Arabidopsis* by *G. orontii* has been suggested by Reuber et al. (1998) shown in Fig. 4.3. As for other biotrophic interactions, SA-mediated defense signaling plays a pivotal role in *Arabidopsis* defense against adapted powdery mildew fungi: SA-dependent gene expression and immune responses increase in *Arabidopsis* leaves upon infection with powdery mildews, and contribute to restriction of colony expansion and reproduction of the fungi (Zimmerli et al. 2004; Chandran et al. 2009). Consequently, many mutants with defects in SA biosynthesis, accumulation, and signaling exhibit enhanced susceptibility (hyper susceptibility) to *Go* and *Gc* (Dewdney et al. 2000; Chandran et al. 2009; Zhang et al. 2015).

Likewise, interference with SA accumulation by transgenic expression of NahG, a *Pseudomonas* salicylate hydroxylase that degrades SA, increases susceptibility against *Gc* isolates (Ederli et al. 2015; Zhang et al. 2015). Despite these genetic indications for an involvement of SA in defense against powdery mildew infection, there is currently only limited direct evidence for increased SA levels during powdery mildew infection (Fabro et al. 2008).

As the final steps of SA biosynthesis in the leaf take place in chloroplasts (Strawn et al. 2007), export of the hormone is required for elevated cytosolic and nuclear SA levels. Consequently, loss-of-function mutation of DP-E2F-like 1 (DEL1: At3g48160), a transcriptional repressor of the gene encoding the plastidic SA exporter EDS5/salicylic acid induction deficient (SID) 1 (At4g39030), results in enhanced SA-dependent resistance against *Go* (Chandran et al. 2014). This phenotype correlates with elevated basal SA levels and increased transcript abundance of SA-responsive genes in the *del1* mutant (Chandran et al. 2014). Strikingly, DEL1 also promotes cell proliferation by repressing genes involved in the induction of endo-reduplication (Vlieghe et al. 2005; Lammens et al. 2008). Together these findings suggest that DEL1-mediated control of SA levels regulates the balance between growth and immunity in developing leaves. The translation of distinct SA levels into specific defense responses occurs by the action of NPR proteins in *Arabidopsis* (Pajerowska-Mukhtar et al. 2013; Seyfferth and Tsuda 2014; Yan and Dong 2014). The outcome of SA-mediated signaling depends on subcellular SA levels and the abundance of active NPR1 in the nucleus. Its paralogs, the SA receptors NPR3 (At5g45110) and NPR4 (At4g19660), cooperatively fine-tune NPR1 degradation by their competitive SA concentration-dependent interaction with nuclear NPR1. Moderately elevated SA levels, as present during systemic acquired resistance (SAR), allow the accumulation of active NPR1 protein in the nucleus. Subsequent interaction of NPR1 with TGA transcription factors (binding to a TGACG nucleotide motif) promotes gene expression of SA-responsive genes and induces SA-mediated defense. The relevance of NPR1 for the *Arabidopsis* powdery mildew interaction is substantiated by identification of NPR1 as a protein–protein interaction network hub during *Go* infection (Jiang et al. 2016). This NPR1 interaction network includes the TGA-interacting GLUTAREDOXIN 480 (GRX480/ROXY19: At1g28480) involved in regulating SA/JA antagonism (Zander et al. 2012), and several TGA transcription factors (TGA1: At5g65210, TGA2: At5g06950, TGA3: At1g22070, TGA7: At1g77920) that can further regulate the expression of defense-related genes. Genes that show SA-dependent transcript accumulation during powdery mildew infection encode proteins involved in redox regulation, vacuolar transport, secretion, and signaling-relevant processes such as Ca^{2+} homeostasis and SA/JA cross talk (Chandran et al. 2009).

ROP GTPases (Rho (RAS homolog) of plants) are molecular switches and key regulators of immunity (Kawano et al. 2014). Mutation of the *Arabidopsis* ROP-GAPs, *ROPGAP1* (At5g22400) and *ROPGAP4* (At3g11490), trapping their (yet unidentified) target ROPs in the active state, results in enhanced susceptibility to *E. cruciferarum* (Hoefle et al. 2011; Huesmann et al. 2011). Accordantly, expression of an inactive (dominant negative) ROP6 (rop6dn; At4g35020) variant, unable to interact with downstream effectors, results in reduced penetration by *Go*. This correlates with an increased transcript abundance of SA-responsive genes, such as *PR1*, and elevated SA-mediated defense responses. However, *Go* resistance of the rop6dn transgenic plants is uncoupled from SA signaling (Poraty-Gavra et al. 2013). Nevertheless, these results, together with previous findings in barley (Hoefle et al. 2011; Scheler et al. 2016), suggest a positive role of active ROPs in mediating susceptibility to adapted powdery mildews.

A number of proteins whose deficiency leads to enhanced powdery mildew resistance, such as EDR1 to 4, lesion initiation 2 (LIN2: At1g03475), and GSL5/PMR4 are associated with the repression of SA-mediated defense. This is indicated by requirement of SA for the increased disease resistance phenotypes of the respective mutants (Vorwerk et al. 2007; Zhang et al. 2007; Wawrzynska et al. 2010; Guo et al. 2013; Wu et al. 2015). The MAPKKK EDR1 gene negatively regulates SA-dependent defense responses and cell death. In consequence, *edr1* mutants are constitutively primed for SA-inducible defense which might occur *via* the regulation of the MAPKs MPK3 (At3g45640) and MPK6 (At2g43790) (Beckers et al. 2009). The role of EDR1 in the control of SA signaling probably relies on its interaction with MKK4 and MKK5, the upstream MAPKKs activating MPK3 and MPK6 (Zhao et al. 2014). EDR1 negatively affects MKK4 and MKK5 levels, presumably resulting in repression of the MKK4/MKK5-MPK3/MPK6 cascade involved in the induction of SA signaling (Zhao et al. 2014; Wu et al. 2015). Mutations in *MKK4, MKK5,* or *MPK3* (but not in *MPK6*) suppress the *edr1* phenotype, indicating a requirement of these kinases for *edr1*-mediated powdery mildew resistance. The same holds true for *edr4,* which is in line with the need of EDR4 for the relocation of EDR1 to the powdery mildew penetration site (Wu et al. 2015). Strikingly, overexpression of MKK4 or MKK5 causes *edr1*-like resistance and powdery mildew-induced cell death, pointing to an additional role for the MKK4/MKK5-MPK3/MPK6 kinase cascade in SA-induced cell death, parallel to its contribution to SA-regulated gene expression (Zhao et al. 2014).

NPR3-mediated NPR1 degradation at high SA levels promotes the onset of cell death. However, while no powdery mildew phenotype has been reported for *npr3* (Liu et al. 2005), *npr1,* and *npr4* mutant plants are more susceptible to *Gc* (Reuber et al. 1998; Liu et al. 2005; Humphry et al. 2010). Thus, it remains elusive to which extent NPR3-mediated cell death contributes to defense against powdery mildews. In addition to NPR3, also NPR4 adds to pathogen-triggered cell death (Pajerowska-Mukhtar et al. 2013; Kumar 2014; Yan and Dong 2014). A role for SA-mediated cell death responses in powdery mildew defense is supported by correlation of enhanced SA signaling with increased powdery mildew-induced cell death in several of the above mentioned resistant mutants (Guo et al. 2013). Mutants impaired in autophagy further corroborate a role of cell death in powdery mildew resistance, as some of them display early leaf senescence and spontaneous cell death, which in several cases coincides with increased powdery mildew resistance (Yoshimoto et al. 2009; Wang et al. 2011a, b). Autophagy targets organelles and cytosolic proteins for vacuolar/lysosome-mediated degradation (Liu and Bassham 2012). The mutant of *autophagy-related 2 (ATG2:* At3g19190), impaired in the early steps of autophagosome biogenesis, exhibits severe powdery mildew-induced cell death, and increased resistance when challenged with *Gc,* while susceptibility toward *Go* is unaltered (Yoshimoto et al. 2009; Wang et al. 2011a, b). Powdery mildew resistance of *atg2* plants depends on SA signaling, while cell death is partially independent of SA. In conclusion, autophagy contributes to suppression of cell death and defense response to powdery mildew fungi; however, the mechanisms by which autophagy controls these processes are yet unknown (Wang et al. 2011a, b).

4.3.17.2 Jasmonic Acid (JA) and Ethylene (ET)-Mediated Signaling to Powdery Mildew Resistance

Salicylic acid (SA)-mediated defense appears to act mainly against biotrophic pathogens, while the JA and ET pathways are preferentially linked to resistance against necrotrophic parasites (Thomma et al. 2001; Glazebrook 2005). Induced systemic resistance (ISR) confers JA/ET-mediated protection of shoot tissues *via* root-to-shoot signaling. ISR is initiated by interactions with beneficial fungi or bacteria such as arbuscular mycorrhiza or plant growth-promoting rhizobacteria in the root, and has proven effective against necrotrophs and herbivores (Pieterse et al. 2012). Despite the fact that powdery mildew fungi are obligate biotrophs, root colonization with the putative plant growth-promoting basidiomycetes *Piriformospora indica* reduces *Go* conidiation in a JA signaling-dependent manner (Stein et al. 2008). This finding suggests that besides SA, JA also contributes for resistance against powdery mildew fungi (Fig. 4.6). Accordingly, *Bgh* inoculation of *Arabidopsis* induces expression of genes that are controlled by the JA/ET signaling pathways (Zimmerli et al. 2004). By contrast, although endogenous JA levels are enhanced during the formation of haustoria by *Gc*, this does not result in transcript accumulation of JA/ET-responsive genes (Reuber et al. 1998; Nishimura et al. 2003; Zimmerli et al. 2004; Glazebrook 2005; Fabro et al. 2008). Nevertheless, constitutive or ectopic activation of the JA/ET pathway due to elevated JA/ET levels in the mutant of *cellulose synthase 3/constitutive expression of VSP 1* (*CESA3/CEV1*: At5g05170) or treatment of Col-0 with methyl-JA enhances resistance against *Gc*. This effect depends on the JA receptor component coronatine-insensitive protein 1 (COI1: At2g39940; Ellis et al. 2002a, b; Zimmerli et al. 2004). These findings suggest that although elicitation of JA/ET-mediated defense signaling seems to be restricted to incompatible powdery mildew host interactions, JA/ET-induced defense responses are effective against virulent powdery mildew fungi if stimulated constitutively, artificially, or systemically, despite the biotrophic nature of the interaction. Consequently, JA/ET signaling must either be suppressed or failed to be elicited by adapted powdery mildews during a successful infection (Fig. 4.6; Antico et al. 2012).

4.3.18 WRKY Transcription Factors for Powdery Mildew Resistance

Expression of resistance genes or defense-related genes is also known as host transcriptional programming. Transcriptional changes in response to powdery

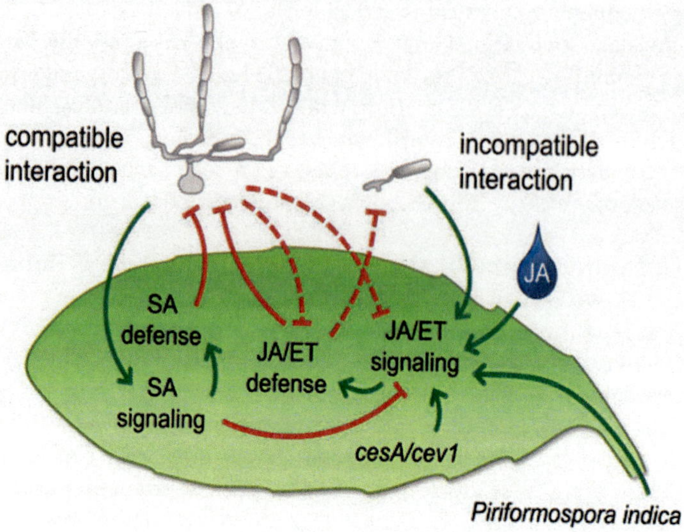

Fig. 4.6 Integration of phytohormone signaling in defense against powdery mildew fungi. Although only incompatible powdery mildew–host interactions elicit JA/ET-mediated defense, JA/ET-induced defense responses are effective against virulent powdery mildew fungi if stimulated constitutively (*cesA/cev1*), artificially (JA treatment), or systemically (*Piriformospora indica* root colonization). These findings suggest that virulent fungi suppress JA/ET signaling during compatible interactions. This suppression might involve the antagonistic action of SA signaling. Solid lines indicate experimentally supported impacts, while dashed lines indicate speculative connections (Kuhn et al. 2016)

mildew inoculation reflect a combination of both activation of defense after recognition of the pathogen and host cell manipulation by the fungal invader. *Arabidopsis* responds to powdery mildew attack with the differential regulation of defense-related genes. While, many genes are induced in host, and nonhost interactions, changes in gene expression interactions, indicating that virulent fungi might suppress gene expression related to basal occur more rapidly, and are often more pronounced in nonhost interactions than in host defense (Zimmerli et al. 2004). A large subset of the powdery mildew-responsive genes is transcriptional factors (TFs). These induced or repressed TFs further transcriptionally regulate secondary up- or downregulated genes and thus enable the coordinated expression of genes in fine-tuned expression networks (Zimmerli et al. 2004; Fabro et al. 2008; Chandran et al. 2009; Christiansen et al. 2011). Among the genes that show altered transcript accumulation in response to *Go,* members of the plant-specific WRKY (single amino acid letter code for tryptophan-arginine-lysine-tyrosine) TFs represent the most prominent TF family (Chandran et al. 2009). WRKY TFs are key regulators of pathogen-triggered changes in gene expression that act as transcriptional activators or repressors in various homo- and heterodimer combinations.

They function up- and downstream of hormone signaling pathways, are involved in the antagonistic control of SA, and JA/ET signaling pathways, and can be regulated by mitogen activated protein kinases (MAPKs) (Bakshi and Oelmüller 2014; Buscaill and Rivas 2014; Caarls et al. 2015, 2016). The involvement of WRKYs in defense of *Arabidopsis* against powdery mildews is indicated by transcriptional changes of WRKY-encoding genes in response to *Go,* and an enrichment of WRKY-targeted W-box *cis*-regulatory elements in promoters of genes differentially transcribed upon powdery mildew challenge (Chandran et al. 2009). Furthermore, expression of WRKY TFs is enhanced in powdery mildew resistant *edr1* plants relative to the wild type in response to *Gc*, and genes whose promoters contain W-boxes are likewise enriched in this dataset. As powdery mildew resistance in *edr1* plants depends on the MPKK4/5-MPK3/6 cascade, the expression of WRKY TFs might be regulated *via* this pathway (Christiansen et al. 2011).

In barley, the MLA10 NB-LRR interacts with the *Bgh* AVR_{A10} (avirulence A10) effector, and induces transcriptional changes by inhibition of *Hv*WRKY1 and *Hv*WRKY2. Both TFs supposedly act as transcriptional repressors of genes involved in basal defense and effector-triggered immunity (ETI; Shen et al. 2007). Similarly, the closely related *Arabidopsis* TFs *WRKY18* (At4g31800) and *WRKY40* (At1g80840), whose transcription is rapidly induced during powdery mildew infection, negatively regulate defense against *Go*. Together these results indicate functional conservation of the defense-repressive role of this WRKY subfamily (Shen et al. 2007). Altered pathogen-induced transcriptional reprogramming in the *Go*-resistant *wrky18 wrky40* double mutant corroborates the negative impact of WRKY18/40 on defense-related gene expression (Pandey et al. 2010; Schon et al. 2013). Chromatin immune-precipitation (ChIP) experiments revealed binding of WRKY40 to W-box containing promoter regions of *EDS1*, the AP2 (apetala 2)-type TF gene *redox responsive transcription factor 1* (*RRTF1*: At4g34410), and to *jasmonate-zim-domain protein 8* (*JAZ8*: At1g30135), a member of the JA-signaling repressor gene family (Pandey et al. 2010). Thus, WRKY18/40 TFs seem to repress the transcription of positive defense regulators such as EDS1 and positively modulate JA-signaling (Fig. 4.7; Pandey et al. 2010; Schon et al. 2013). Although the regulatory role of *Hv*WRKY1/2 and WRKY18/40 is conserved between barley, and *Arabidopsis*, an *Arabidopsis* R-protein interfering with WRKY18/40 function, remains to be identified (Kuhn et al. 2016). The conservation of MLA1 functionality and induction of MLA-dependent defense gene expression in response to *Bgh* might indicate the existence of a respective MLA analog in *Arabidopsis* (Maekawa et al. 2012). In contrast to WRKY18 and WRKY40, WRKY70 (At3g56400) contributes to resistance of *Arabidopsis* to *Gc,* and inactivation of the respective gene results in increased susceptibility to this pathogen (Li et al. 2006). *WRKY70* overexpression coincides with a partially NPR1-dependent suppression of JA responsive genes, indicating a role of this TF in the control of SA/JA cross talk (Li et al. 2006; Caarls et al. 2015).

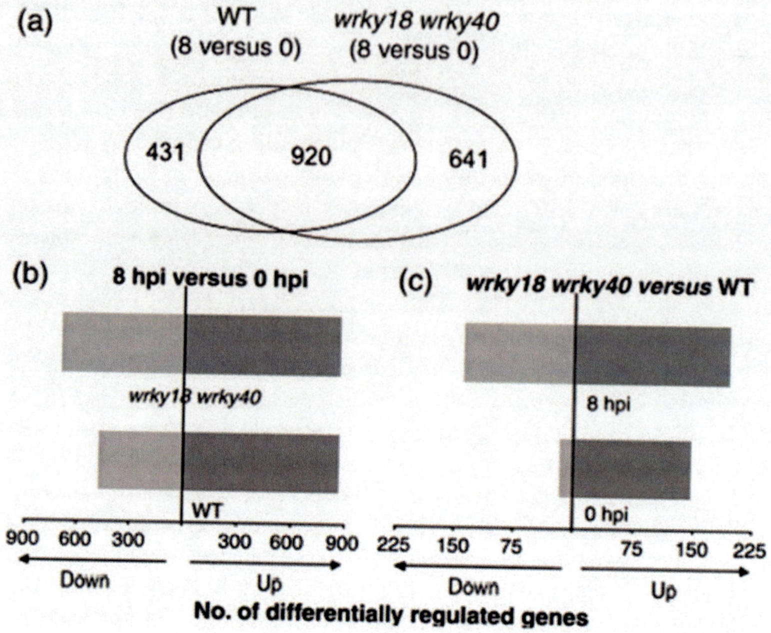

Fig. 4.7 Early transcriptional reprogramming in WT, and wrky18 wrky40 plants upon *Golovinomyces orontii* infection. (**a**) Venn diagram illustrating total number and overlap of genes affected in WT, and wrky18 wrky40 8 h post *G. orontii* infection. (**b**) Changes in the transcript profiles of wrky18 wrky40 (upper panel), and WT (lower panel) plants 8 hpi compared with their respective profiles in the noninfected (0 hpi) states. (**c**) Differentially regulated genes between wrky18 wrky40, and WT plants at 0 h (lower panel), and 8 hpi (upper panel). Numbers along solid lines (*x*-axis) indicate genes upregulated (dark gray) and downregulated (light gray) (Pandey et al. 2010)

4.3.19 Molecule (Hormone) Signaling-Induced Transcriptional Reprogramming During R to Powdery Mildew

SA signaling contributes to gene expression during the *Arabidopsis*–powdery mildew interaction. This involves *Go*-induced expression of genes related to Ca^{2+} signaling, and genes coding for redox regulators that contribute to NPR1 activation (Chandran et al. 2009). Transcript accumulation of SA-responsive TFs, SA biosynthesis genes (*isochorismate synthase 1* (*ICS1*)/*SID2*: At1g74710), and SA-responsive pathogenesis-related genes such as *PR1* emphasizes the predominant contribution of SA signaling to powdery mildew-induced gene expression. In line with this notion, SA signaling-dependent resistance of the *wrky18 wrky40* double mutant correlates with massive *Go*-induced transcriptional reprogramming (Pandey et al. 2010; Schon et al. 2013). The *edr1* mutation, which enhances SA-dependent powdery mildew resistance, affects accumulation of defense-related transcripts in response to *Gc* including transcripts encoding WRKY and AP2/ET-response element binding factor (ERF) TFs (Christiansen et al. 2011). Furthermore, genes encoding proteins associated with ROS production and the endomembrane system are induced in infected *edr1* plants. Powdery mildew-induced enrichment of the latter together with transcripts associated with secretion suggests that the secretory pathway may play an important role in *edr1*-mediated immunity (Christiansen et al. 2011). This assumption is in agreement with the relocalization of EDR1 from the ER to the plant-fungal interface during *Gc* infection (Christiansen et al. 2011; Wu et al. 2015).

TFs of the AP2/ERF family particularly regulate genes related to JA/ET signaling. Besides WRKY TFs, AP2/ERF TFs, and transcripts associated with AP2/ERF response elements (GCC-boxes) are overrepresented among genes upregulated during powdery mildew infection in the *edr1* mutant (Christiansen et al. 2011). *Arabidopsis* ERF6 (At4g17490) and ERF104 (At5g61600) are phosphorylated by MPK6 and/or MPK3, indicating a regulation of these ERFs by defense-related MAPK cascades (Bethke et al. 2009; Meng and Zhang 2013; Tsuda and Somssich 2015).

Furthermore, *ERF1* (At3g23240) and *ERF2* (At5g47220) transcripts accumulate upon *Go* infection (Chandran et al. 2009). A role of ERF1 in defense against powdery mildews is in line with the finding that *ERF1* overexpression results in enhanced resistance to *Go* (Gu et al. 2002; Chandran et al. 2009). *Octadecanoid-responsive Arabidopsis AP2/ERF 59* (*ORA59*: At1g06160), a master regulator of ERF-controlled JA/ET signaling, has been identified, together with other ET/JA-responsive genes, as differentially regulated in SA-deficient mutants *versus* wild-type plants upon *Go* infection. This suggests that ORA59 modulates the cross talk between SA, and JA/ET signaling during powdery mildew-induced defense responses (Chandran et al. 2009). In agreement with this finding, ORA59 was identified as a major target for SA antagonism (Zander et al. 2012, 2014; Van der Does et al. 2013; Caarls et al. 2015). A dominant allele of *signal responsive1* (*SR1*: At2g22300), encoding a calmodulin-binding TF that regulates ET-induced senescence by binding to the *ethylene insensitive 3* (*EIN3*: At3g20770) promoter, has been identified as a gain-of-function suppressor of *edr2*-mediated resistance to *Gc* (Nie et al. 2012). SR1 binds to the promoter of *non-race-specific disease resistance1* (*NDR1*: At5g06320). NDR1, a membrane-associated protein, contributes to plant immunity mediated by several coiled-coil NB-LRRs and SAR (Century et al. 1997; Shapiro and Zhang 2001; Zhang and Shapiro 2002). An *sr1* null mutant conditions resistance to *Gc*, while an additional mutation in *ndr1* suppresses this phenotype. Together these data suggest that SR1 plays a critical role in powdery mildew resistance, possibly by regulating *EIN3* and *NDR1* expression (Nie et al. 2012).

Integrative analysis of protein interaction networks and transcriptomics during *Go* infection revealed a negative correlation between development-related and defense-related genes/proteins (Jiang et al. 2016). While many defense-related genes are induced after *Go* infection, the majority of genes linked to development are downregulated. Interestingly, auxin-related genes are overrepresented among nodes connecting the defense, and development subnetworks. Together these findings emphasize that defense is triggered at the expense of developmental programs, and that regulation of this trade-off involves auxin (Jiang et al. 2016).

4.3.20 Harmonious Coordination Between Transcriptional Regulation and R to Powdery Mildew

Despite its compatible nature, the *Arabidopsis*–*Gc* interaction elicits expression of genes related to nonhost resistance (NHR) (Chandran et al. 2010). Remarkably, in compatible interactions, neither activation of NHR genes nor of SA-induced defense is sufficient to confer resistance (Chandran et al. 2009, 2010). The PEN1/SNAP33/VAMP722 and the PEN2/PEN3 pathways are important determinants of NHR, and are additionally required for *mlo2*-mediated immunity. In line with their pivotal role in defense, *PEN1, PEN2, PEN3, SNAP33,* and *MLO2* share a substantial amount of co-expressed genes, with the majority of transcripts accumulating in response to biotic stresses and PAMP treatment (Humphry et al. 2010). Notably, many transcripts of this regulon accumulate in *Bgh*-inoculated *Arabidopsis* plants ectopically expressing the barley MLA1 R-protein. This implies a substantial overlap of MLA1-dependent transcriptional regulation and basal resistance against powdery mildew fungi (Humphry et al. 2010; Maekawa et al. 2012).

Consistent with the role of PEN2, and PEN3 in indole glucosinolate biosynthesis and secretion, many of the co-expressed genes are associated with the glucosinolate pathway. The gene-encoding myeloblastosis (MYB) 51 TF (At1g18570) is a major regulator of defense-related expression of glucosinolate biosynthesis genes (Humphry et al. 2010). Transcriptomic evaluation of the powdery mildew-resistant *wrky18 wrky40* double mutant revealed that WRKY18 and WRKY40 suppress crucial biosynthesis genes of the indolic phytoalexin camalexin (Pandey et al. 2010). Accordingly, increased expression of camalexin and indole glucosinolate biosynthesis genes after pathogen challenge in *wrky18 wrky40* plants correlates with the enhanced accumulation of the phytoalexin camalexin and 4MI3G (4-methoxyindol-3-ylmethyl-glucosinolate), an indole glucosinolate intermediate relevant for powdery mildew resistance, in this mutant (Pandey et al. 2010; Schon et al. 2013). Loss of function of the crucial 4MI3G biosynthesis gene *CYP81F2* suppresses *wrky18 wrky40*-mediated inhibition of host cell entry, indicating that the indolic metabolite is required for penetration resistance of the double mutant against *Go* (Schon et al. 2013).

The group of genes co-expressed with *PEN3* further showed a significant over-representation of components involved in Ca^{2+} signaling such as calcium/calmodulin-dependent protein kinases (CCaMKs; Humphry et al. 2010; Campe et al. 2015). Consistently, components of the Ca^{2+} homeostasis and signaling machinery such as *CAM9/CALMODULIN-LIKE* (*CML*) *9* (At3g51920) and calreticulin-encoding genes are rapidly induced after *Go* inoculation, and *CML38* (At1g76650) is constitutively upregulated in the highly resistant *wrky18 wrky40* mutant (Chandran et al. 2009; Pandey et al. 2010). Co-expression of genes encoding proteins involved in secondary metabolite biosynthesis (such as *PEN2* or cytochrome P450s like

CYP83B1 (At4g31500)) together with genes whose products mediate exocytosis/extrusion (SNAREs, exocyst subunits, and ABC transporters like *PEN3*) suggests that production and secretion of antimicrobial compounds is transcriptionally attuned. Additional co-regulation of receptor-like kinase genes, transcripts of Ca^{2+} signaling components, and the heterotrimeric G-protein β and g subunits *AGB1* (At4g34460) and *AGG1* (At3g63420) indicates that also recognition, and signaling components are co-expressed with key components of antifungal defense (Humphry et al. 2010; Lorek et al. 2013). Enhanced host cell entry by *Go* and *E. pisi* of Gβ-deficient mutants emphasizes the importance of second messenger signaling via heterotrimeric G-protein components for powdery mildew resistance (Lorek et al. 2013). Finally, identification of defense-related TFs in the regulon indicates that recognition of pathogens, response initiation, and defense execution are transcriptionally coordinated to enable an efficient immune output.

4.3.21 Transcription Factors and Gene Regulation for Powdery Mildew Resistance

In plants, recognition of pathogen attack and/or its entry and initiation of defense responses results in a dramatic reprogramming of gene expression (Schenk et al. 2000, Zimmerli et al. 2004). Whether mediated by direct interaction between a pathogen receptor and transcription factors (Shen et al. 2007) or, as more often observed in animal systems, by a complex signal transduction cascade (mediated by MAP kinases; Asai et al. 2002), the complete activation of defense responses relies in part on the activity of so-called WRKY transcriptional regulators (Eulgem and Somssich 2007). In *Arabidopsis*, the WRKY family is composed of 72 expressed members displaying various degrees of functional redundancy with respect to the diverse biological roles they are involved in, including plant defense. WRKY70 is transcriptionally induced by SA signaling, and mediates basal resistance to a variety of pathogens including *G. cichoracearum* (Li et al. 2006; Wang et al. 2006; Ulker et al. 2007; Plate 4.1). In contrast, WRKY18 and WRKY40 act as negative regulators of basal defense against *G. orontii* (Plate 4.1), and double mutants in these genes are fully resistant to fungal attack (Shen et al. 2007). In the analogous interaction of the monocot barley with its cognate powdery mildew pathogen, Bgh, WRKY18, and 40 counterparts WRKY1, and WRKY2 were shown to interact with the CC domain of MLA immune receptors (Shen et al. 2007). Under normal (pathogen-free) conditions, WRKY1 and WRKY2 are proposed to suppress the transcription of genes involved in basal defense. Upon pathogen challenge and activation by a matching fungal avirulence protein, the MLA receptor is re-localized to the nucleus where it recruits WRKY1 and WRKY2 away from the promoters of defense genes. As expected, silencing of these WRKY transcription factors results in enhanced basal defense against Bgh (Shen et al. 2007; Eckey et al. 2004).

In addition to WRKY transcription factors, the NAC transcription factors, barley NAC6, and its *Arabidopsis* homolog ATAF1, were identified as positive regulators of transcription involved in early defense against powdery mildews. Silencing (barley) or knock out (*Arabidopsis*) of the respective gene resulted in decreased penetration resistance upon inoculation with Bgh while overexpression (in barley) resulted in increased resistance (Plate 4.1). These findings suggest a conserved contribution of NAC transcription factors in pre-haustorial defense (Jensen et al. 2007). A member of another group of transcription factors, the wheat ethylene response factor ERF3, has also been suggested to act as a positive regulator of early defense against powdery mildews (Zhang et al. 2007; Plate 4.1). ERF3 expression is induced by pathogen challenge (Bgh), and the gene product operates as a GCC box-binding transcriptional activator of SA signaling genes (Zhang et al. 2007). Similarly, heterologous overexpression of other members of this family from tomato in *Arabidopsis* resulted in enhanced resistance to the adapted powdery mildew, *G. orontii*, further corroborating their role as positive regulators of transcription in pathogen defense (Gu et al. 2002).

4.3.22 Transcriptional (Genes) Regulation, and Expression in Response to Powdery Mildew Infection

The broad range of transcriptional changes that occur in *Arabidopsis* during powdery mildew infection (Zimmerli et al. 2004; Fabro et al. 2008) reflect both defense responses mounted by the plant as well as possible manipulation of the host by the fungal pathogen. With the sequencing of the *Arabidopsis* genome and availability of novel technologies for transcript profiling (oligonucleotide-based microarray chips), it has been possible to peek into the gene expression changes that occur during the infection process. Several dedicated websites such as Genevestigator (https://www.genevestigator.ethz.ch/) and ATGenExpress, (http://www.*Arabidopsis*.org/info/expression/ATGenExpress.jsp) provide publicly available collections of microarray data on *Arabidopsis* challenged with a variety of pathogens including powdery mildews, which represent a largely unexplored treasure of information. However, the spatial resolution of microarray data is limiting, since most microarray experiments are performed on whole leaf samples, and as such only provide a gross overview about changes in transcript abundance in pathogen-challenged leaf tissue. Since, powdery mildews exclusively colonize the epidermal layer of the leaf, many tissue-specific transcriptional changes are likely to be masked by proportionately larger amounts of mesophyll transcripts. Even within the epidermis, attacked cells, and successfully colonized cells each differ in their transcript profile from non-attacked cells (Gjetting et al. 2004, 2007). Owing to these limitations, expression levels of numerous genes in infected host cells might be considerably underestimated. Employment of techniques that allow transcript profiling of powdery mildew-challenged *Arabidopsis* leaves at cellular resolution, for example, based on RNA extraction via micro capillaries or by laser micro-dissection (Kehr 2003), is thus urgently required to refine the currently existing data sets (Micali et al. 2008).

Earlier transcriptional changes in *Arabidopsis* compared the response observed in host (*G. cichoracearum*) and nonhost interactions (Bgh; Zimmerli et al. 2004) at

8, 18, and 24 h post-inoculation. Many of the differentially regulated genes displayed a similar response in both host and nonhost combinations. However, in the nonhost interaction, the response was more rapid, and of greater amplitude than in the compatible interaction, supporting the hypothesis that common defense machinery is activated upon attack by adapted and non-adapted pathogens, and reinforcing the notion that adapted pathogens are able to partially suppress basal defense in their host species. Several genes that exhibit lower transcript abundance after pathogen challenge, a majority encode proteins that are involved in photosynthesis, metabolism, transport, and transcription/translation. Consistent with these functions, half of the gene products are either known or predicted to be chloroplast residents. This observation corroborates further studies indicating that *Arabidopsis* suffers up to 30% reduction in the gain of dry weight following Bgh infections (Zimmerli et al. 2004). This decrease is more modest in compatible infections with *G. cichoracearum*. An inverse relationship between defense responses and plant growth has been also observed in interactions between cereals and powdery mildews, and underscores the metabolic costs associated with activation of defense pathways (Swarbrick et al. 2006; Wright et al. 1995a, b). Intriguingly, a recent microarray analysis of the compatible interaction between *Arabidopsis* and *G. cichoracearum* revealed that host photosynthesis is likely stimulated by the fungus, through the induction of genes involved in chlorophyll binding, and chlorophyll *a* oxygenase (Fabro et al. 2008). Though it is conceivable that an increase in photosynthetic rate is required, as the metabolic demands of the infected epidermal tissue increase, these findings are somewhat contradictory to those obtained in the study of Zimmerli et al. (2004). Manipulation of the host metabolism by the fungal parasite thus likely involves up- and downregulation of photosynthetic genes.

However, many gene transcripts related to defense responses accumulate upon both host and nonhost powdery mildew attack (transcripts encoding glycosyl hydrolases and β-1,3-glucanase). Consistently, many of the respective genes contain in their upstream promoter region the cis-regulatory element OCS, which is known to be responsive to auxin, SA, and hydrogen peroxide. Transcripts specifically accumulating during nonhost interactions encode defensins and derive from genes that are regulated by JA and ethylene. SA signaling therefore does not appear to play an important role in nonhost defense against Bgh. Alternatively, SA-controlled genes may only become activated once the penetration defenses have failed. However, analysis of the transcriptional response of *Arabidopsis* leaves to chitin (the main fungal cell wall polysaccharide), and chitin derivatives identified several genes involved in defense responses (Ramonell et al. 2002). Genes encoding plant-derived chitinases exhibit higher transcript accumulation upon pathogen challenge compared to control conditions, and chitin oligomers obtained as fungal cell wall degradation products constitute potent signaling molecules responsible for induced resistance to *G. cichoracearum*. This resistance response appears to be activated through signaling routes that are distinct from the SA, JA, and ET pathways (Ramonell et al. 2002). Genes encoding PR5, two WRKY transcription factors, and several disease resistance proteins belonging to the TIR-NB-LRR family were induced upon chitin treatment (Ramonell et al. 2002). Genes with reduced transcript abundance included

ETR2, encoding a transmembrane protein kinase with an LRR domain, the defensin PDF2.5, RAR1 (encoding a protein with two zinc-finger-like domains that is required for the accumulation of many R-proteins), and several other genes coding for zinc-finger proteins. To functionally validate the microarray experiments, T-DNA insertion mutants within seven of the differentially expressed genes were tested for susceptibility to *G. cichoracearum*. Three of the seven mutants, encoding a putative RING zinc-finger, and two putative disease resistance proteins, displayed enhanced susceptibility to the pathogen, further reinforcing the role of chitin sensing and signaling in plant defense responses (Ramonell et al. 2002). The critical role played by chitin perception in plant defense responses was recently strengthened by the cloning of the presumptive *Arabidopsis* chitin (co-)receptor LysM Receptor-Like Kinase1 gene (LysM RLK1 or CERK1) by two independent groups (Miya et al. 2007, Wan et al. 2008). The gene codes for a plasma membrane-resident protein with a single transmembrane domain, three extracellular LysM domains and a cytoplasmic serine/threonine kinase domain (Plate 4.1). Mutations in this gene abolish virtually all changes in gene transcription normally observed upon chitin perception by the plant, without affecting other elicitor perception such as flagellin sensing, suggesting that the protein is part of the chitin receptor complex (Wan et al. 2008). It is conceivable that in analogy to the FLS2 receptor kinase, which is responsible for flagellin-triggered signaling (Asai et al. 2002), LysM RLK1/CERK feeds into MAP kinase signaling and defense gene activation (Plate 4.1). Interestingly, although loss of LysM RLK1 caused enhanced susceptibility to the host fungus *G. cichoracearum* at the post-penetration stage (Wan et al. 2008), it did not affect the plant response to the adapted hemibiotrophic fungus *Colletotrichum higginsianum,* and resulted in only a marginal increase in susceptibility to the non-adapted necrotrophic fungus *Alternaria brassicicola*. These observations suggest that pathogen perception by the plant is microbe-specific and implies the existence of more elicitor-specific receptors yet to be discovered (Micali et al. 2008).

In addition to genes involved in defense responses, genes that are induced because they are likely required for pathogen growth have also been identified (Fotopoulos et al. 2003). The observation that *Arabidopsis* leaves infected with *G. cichoracearum* have an enhanced sugar uptake compared to noninfected leaves prompted an investigation of sugar transporter expression upon powdery mildew challenge. *G. cichoracearum* induces expression of STP4, encoding a glucose transporter as well as βfruct1, coding for a cell wall invertase, in several cell types of the leaf (most notably the epidermis and vascular bundle) irrespective of whether they are penetrated by the fungus or not. This transcriptional induction is sustained for up to 6 days post-inoculation, and is not reproducible in a nonhost interaction with Bgh. A similar role for sugar transporters in fungal development was also found in the interaction between the rust fungus *Uromyces fabae* and broad bean (*Vicia faba*). However, in this case, the hexose transporter is of fungal origin and resides in the haustorial membrane (Voegele et al. 2001), suggesting that exchange of nutrients at the plant/biotroph interface requires transport facilitators on both sides. It is not surprising that sugars, and therefore sugar biosynthesis and transport, would be important for fungal growth and development, and enhanced by successful fungal pathogens. The mechanisms leading to this genetic and physiological manipulation remain an enigma. However, new efforts into powdery mildew studies promise to lift the veil on the mystery (Micali et al. 2008).

4.3.23 Role of *Trichoderma* in Systemic Resistance to Powdery Mildew

Induced resistance (IR) varies according to different signals. Systemic acquired resistance (SAR) and induced systemic resistance (ISR) are two forms of systemic resistance. In both SAR and ISR, plant defenses are released by a previous infection, thus, biotic and abiotic factors play a role in resistance; most agents reduce disease in the infected plants by 20–85% (Walters et al. 2013). ISR is mediated by the NPR1 gene, which is a key gene involved in disease resistance, and phenotypically similar to SAR. ISR can be defined by induction of defenses in plants against many pathogens via application of plant growth-promoting microorganisms in the soil, as well as direct spreading on plants (Hossain et al. 2008; Maksimov et al. 2011), whereas SAR is usually caused by a pathogen attack locally. However, the molecular pathways of each systemic resistance are different; ISR depends on two pathways that respond to ethylene, and jasmonic acid (Verhagen et al. 2004), whereas SAR depends on the salicylic acid (SA) responsiveness (Spoel et al. 2003). The SA pathway controls the expression of pathogenesis-related (PR) proteins, such as PR1, PR2, and PR5, whereas the JA–ET pathway regulates the expression of a different group of defense genes, that is, PR3 and PDF1.2 in *Arabidopsis*; these PR proteins have been defined as PR2, PR5, PDF1.2, and basic chitinase (PR3) (Thomma et al. 1998).

Trichoderma spp. is soil-borne, produces green spores, and is among the ascomycetes that are widespread throughout the world (Schuster and Schmoll 2010). The fungi of the genus *Trichoderma*, which comprises a group of plant growth-promoting fungi, can colonize the intercellular parts of plant roots, and stimulate systemic resistance in all parts of the plant, the actions of these fungi suppress some plant diseases by direct mycoparasitism or antibiosis, as well as indirect induced resistance (IR) (Yedidia et al. 1999; Sharma and Sain 2004). The interaction between *Trichoderma* spp. and plant is correlated with transcriptome and systemic modulations of the plant proteome (Sharma and Sain 2004; Hermosa et al. 2012). *Trichoderma* spp. stimulates ISR through hormonal and molecular pathways in a JA/ET-dependent manner (Shoresh et al. 2010; Korolev et al. 2008). A study has demonstrated that fungi mainly affect the pathogenesis-related genes by increasing their expression levels, thereby resulting in resistance of the treated plants to the leaf pathogens (Harel et al. 2014).

T. harzianum is the most effective biocontrol agent against a wide range of plant pathogens and has been recommended for the integrated management of crucifer's diseases (Saharan et al. 2005, 2014, 2016, 2017, 2019). *T. harzianum* TH12 isolates are used to produce a cell-free culture filtrate (CF), which has been initially characterized as an elicitor agent of induced systemic resistance to *Sclerotinia* stem rot (SSR) on *B. napus* AACC; CF is proven to effectively control pathogen when sprayed onto rapeseed crops (Alkooranee et al. 2015). The effects of *E. cruciferarum* in *R. alboglabra* and *B. napus* treated or non-treated with *T. harzianum* and its CF have been examined. Alkooranee et al. (2015) investigated the gene expression of common genes involved in plant disease resistance. The effects of powdery mildew on *B. napus* and *R. alboglabra* were also investigated in terms of three aspects: first, the effects of *E. cruciferarum* on *B. napus* and *R. alboglabra* in comparison with the noninfected samples at six different time points; second, the effects of *T. harzianum* and its CF on *B. napus* and *R. alboglabra* in comparison with non-treated samples at six different time points; and third, the effects of *E. cruciferarum* on *B. napus* and *R. alboglabra* treated with *T. harzianum* and its CF at six different time points of infection, namely, 1, 2, 4, 6, 8, and 10 days post-inoculation (dpi).

Trichoderma harzianum TH12 is a microbial pesticide for certain rapeseed diseases. The mechanism of systemic resistance induced by TH12 or its cell-free culture filtrate (CF) in *Brassica napus* (AACC) and *Raphanus alboglabra* (RRCC) to powdery mildew disease caused by *Erysiphe cruciferarum* was investigated by Alkooranee et al. (2015). The investigation was on the cellular and molecular aspects of *B. napus* and *R. alboglabra* infected with *E. cruciferarum*. The histological study showed the resistance of *R. alboglabra* to powdery mildew disease. The growth of fungal colonies was not observed on *R. alboglabra* leaves at 1, 2, 4, 6, 8, and 10 days post-inoculation (dpi); whereas this was clearly observed on *B. napus* leaves after 6 dpi. In addition, the gene expression of six plant defense-related genes, namely, PR-1, PR-2 (a marker for SA signaling), PR-3, PDF 1.2 (a marker for JA/ET signaling), CHI620, and CHI570, for both genotypes were analyzed in the leaves of *B. napus* and *R. alboglabra* after treatment with TH12 or CF, and compared with the non-treated ones. The qRT-PCR results showed that the PR-1 and PR-2 expression levels increased in *E. cruciferarum*-infected leaves, but decreased in the TH12-treated leaves compared with leaves treated with CF. The expression levels of PR-3 and PDF1.2 decreased in plants infected by *E. cruciferarum*. However, expression levels increased when the leaves were treated with TH12. It was disclosed that the nature of gene expression in *B. napus* and *R. alboglabra* to explore the resistance pathways in the leaves of both genotypes infected and noninfected by powdery mildew, and inoculated or non-inoculated with elicitor factors. Results suggested that *R. alboglabra* exhibited resistance to powdery mildew disease, and the application of *T. harzianum* and its CF are a useful tool to facilitate new protection methods for resistant or susceptible plants (Fig. 4.8 A-F).

4.3.24 Mechanisms of Nonhost Resistance in Crucifers to Powdery Mildew

In nature, plants are constantly challenged by above- and belowground attack by a vast array of potential pathogens that belong to distinct phyla, and exhibit different infection strategies, and lifestyles. Thus, it is surprising that natural plant

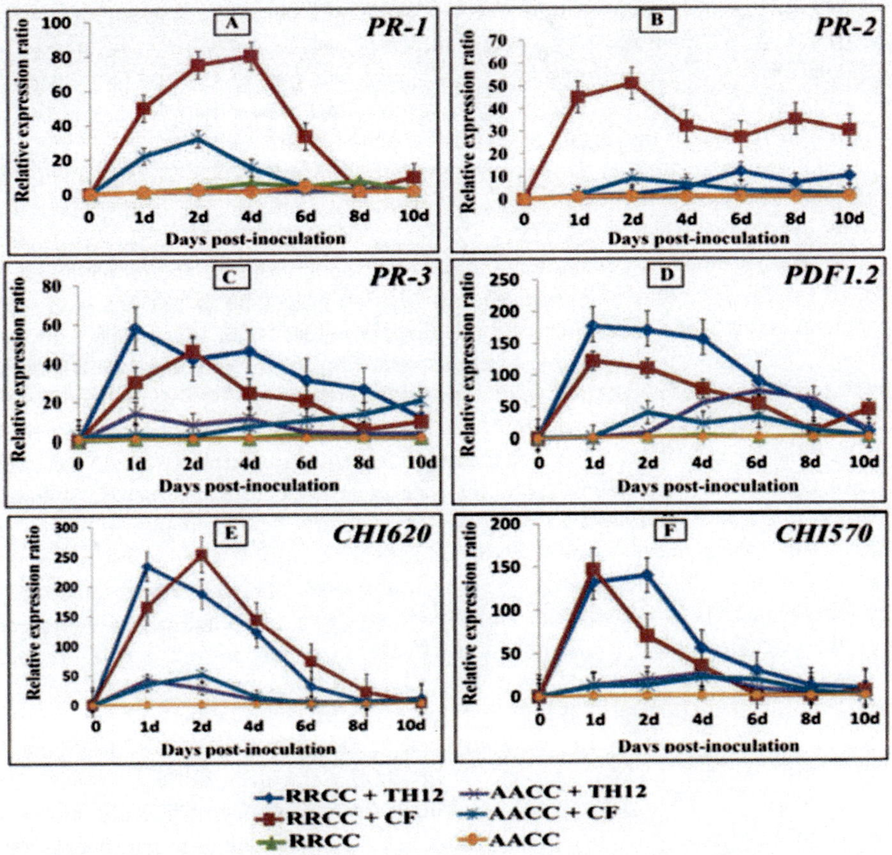

Fig. 4.8 Expression of defense-related genes of *Brassica napus* and *Raphanus alboglabra* genotypes of 6-weeks-old inoculated with suspensions 10 mL of TH12, and its cell-free culture filtrate (CF) separately treated by soil drenching. Three potted for each time non-inoculated served as control plants. Leaves were collected 1, 2, 4, 6, 8, and 10 days post-inoculation. Total RNA was extracted, and cDNA was synthesized. Expression levels of the PR-1, PR-2, PDF1.2 (glucanase; BGL2), PR-3 (basic chitinase), CHI620, and CHI570 (chitinase) genes were monitored by RT q-PCR. The expression levels of genes were compared with the expression level of GAPDH (Alkooranee et al. 2015).

communities normally look rather healthy. The reason is that most plants are immune to the majority of would-be pathogens, and susceptible to only a relatively small number of adapted microbes. Consequently, disease is the exception, and not the rule. Disease resistance shown by an entire plant species to all genetic variants of a non-adapted pathogen species (or bacterial pathovar [pv] or fungal forma specialis [f.sp.]) is the most common form of plant immunity, and termed nonhost resistance (NHR) (Mysore and Ryu 2004; Numberger and Lipka 2005). Nonhost resistance has been defined as the capacity of a particular plant species to resist infection by all genetic variants of a pathogen that normally infect other host species (Thordal-

Christensen 2003). Nonhost resistance to pathogens can be as strong, and broad resistance of all members (ecotypes, landraces, cultivars) of a given plant species against all known isolates or races of a given pathogen species or host species-associated forms (*formae specialis*). Regarding attacks by phytopathogenic fungi, nonhost resistance is the default case because plants in natural plant communities are frequently encountered by propagules (spores) of inappropriate (sub) species that have not co-evolved with the respective plant host and, therefore, possess blunt weapons of attack.

The unlikelihood of a plant being attacked by an appropriate fungal pathogen is based on the fact that its spores cannot actively target hosts but are passively transported by water splashing or wind dispersal, and often touch down on the "wrong" plant species, a so-called nonhost. An important exception is, of course, susceptible crop monocultures where spore dispersal even over short to medium distances leads to successful new infection, and finally to epidemics that endanger or destroy the harvest. Two models of nonhost resistance are currently in the focus of interest. The first one postulates the absence of adapted fungal effectors, thereby leading to a non-compromised PAMP-triggered default defense response, which is durable in nature and also known as basal resistance in host systems. The second model postulates the presence of stacks of multiple resistance (R)-genes encoding proteins of the NBS-LRR type that simultaneously recognize a number of pathogen-derived avirulence (Avr) genes encoding effector proteins and that lead to durable resistance by functional redundancy. Several of these nonhost R-genes may very well be also mediating race-specific resistance against appropriate host pathogens, because dual functionality of one and the same R-gene against two non-related pathogens has been shown (Jones and Dangl 2006; Schweizer 2007).

In other words, NHR delimits the host range of phytopathogenic microorganisms, and from an evolutionary perspective, impinges on their radiation and speciation potential. Overcoming NHR and the subsequent lifecycle completion represents the hallmarks of basic compatibility. Thus, adapted pathogens need to evade or suppress the plant's basal defense machinery, and to manipulate plant cell functions to their own advantage. To accomplish this mission, pathogens utilize a repertoire of effector molecules that target distinct plant mechanisms (Speth et al. 2007). Infrequent historical host range shifts are indicative of the generally robust and durable nature of NHR (Heath 2000). Conceptually, the stability of NHR has been proposed to be the consequence of several successive layers of protective mechanisms that comprise both constitutive barriers and inducible reactions (Numberger and Lipka 2005). Preformed physical and chemical barriers such as a rigid cell wall and toxic phytoanticipins are frequently cited as controlling invasion success of some non-adapted parasites. It is now generally accepted that inducible components of NHR and basal resistance (which defines the plant's ability to reduce the disease severity of adapted pathogens) both involve, as a first step, non-self recognition of the intruder. This is mediated by transmembrane pattern recognition receptors (PRRs) that perceive slowly evolving microbial- or pathogen-associated molecular patterns (MAMPS or PAMPs) (Schweizer 2007). Downstream cell-autonomous responses of PAMP-triggered immunity include MAP-kinase signaling, production of reactive oxygen species (ROS), ethylene, ion fluxes, transcriptional induction of pathogenesis-related (PR) genes, protein phosphorylation, and callose deposits (Bittel and Robatzek 2007). Unlike cultivar-specific effector-triggered host plant immunity, which is mediated by resistance (R) proteins, NHR often (but not always) permits plants to terminate pathogenesis of invaders without a hypersensitive cell death response (HR) (Mysore and Ryu 2004). Chisholm et al. (2006) and Jones and Dangl (2006) proposed a conceptual framework for plant immunity evolution that suggests a sequential interplay of PAMP-triggered immunity (PTI), effector-triggered susceptibility (ETS), and immunity (ETI). With reference to this concept, NHR can be considered to be the consequence of ineffective microbial effectors, resulting in no suppression of PTI (and/or ETI). Alternatively, (individual) effectors might not have been selected to evade recognition and could thus be recognized in nonhost plants, resulting in ETI.

Arabidopsis is a host plant for at least four distinct powdery mildews, the two closely related species *Golovinomyces cichoracearum* (*Gc*) and *G. orontii* (*Go*), as well as *Oidium neolycopersici* (*On*) and *Erysiphe cruciferarum* (*Ec*). Interestingly, *Ec* appears to be closely related to the pea pathogen *Erysiphe pisi* (*Ep*) (Saenz and Taylor 1999). Comparative cytological analyses of *Arabidopsis* inoculations with non-adapted and adapted powdery mildew species provided some conceptual clues: unlike compatible powdery mildews such as *Go*, which readily invades *Arabidopsis* (ca. 70% entry rate within 24 h post inoculation [hpi]), the grass and legume powdery mildews *Bgh* and *Ep* fail to penetrate approximately 95 and 75% of attacked epidermal pavement cells, respectively (Lipka et al. 2005). Entry failure correlates with timely and localized defense responses, which is reflected in dynamic cytoskeleton rearrangements, organelle transport, protein translocation, secretion processes, and focal cell wall remodeling (formation of multilayered, callose containing papilla) at sites of attempted fungal ingress (Hardham et al. 2007; Huckelhoven 2007). These pre-invasion defense mechanisms are backed up by post-invasion resistance: at the remaining 5 (25%) of successful *Bgh* (*Ep*) entry sites haustoria become encased in callose, and the attacked epidermal cells undergo an HR-like cell death response, which is accompanied by microscopically detectable whole-cell auto fluorescence. This ultimately terminates any further fungal development and prohibits colonization. By contrast, efficient invasion and cell death suppression appear to represent hallmarks of compatible and co-evolved biotrophic interactions. It is conceivable that adapted powdery mildews employ effector molecule transfer to interfere with both pre- and post-invasion defense mechanisms, and to establish basic compatibility. Non-adapted isolates might entirely lack the required effector repertoire or host-specific transport machinery. Alternatively, lack of co-evolution might result in production of effector variants that are either nonfunctional or subject to recognition. It is plausible that the extent of taxonomic, evolutionary, and ecological separation between the interaction partners has a major impact on these alternative scenarios (O'Connell and Panstruga 2006; Lipka et al. 2008).

4.3.25 Components of Nonhost Resistance

Powdery mildew species either can have a wide or narrow host range or might even be specialized to a single host plant species. *Golovinomyces orontii* (*Go*) is virulent on *Arabidopsis*, on other Brassicaceae species, as well as on Solanaceae and Cucurbitaceae species, but does not infect Rosaceae or Asteraceae (Plotnikova et al. 1998). In contrast to *Go*, the pea powdery mildew pathogen *E. pisi* and the barley powdery mildew *Bgh* are not able to cause disease on *Arabidopsis*, as they show little penetration success and no completion of their asexual life cycle. Consequently, they do not give rise to any visible epiphytic colonization and symptom formation (Lipka et al. 2005). This is essentially due to NHR of plants against pathogens, which by definition is resistance of an entire plant species against all genetic variants of a microbial species (Lipka et al. 2005; Nurnberger and Lipka 2005). Mechanistically, NHR seems to be equivalent to basal defense or innate immunity, and supposedly relies mainly on preformed defenses and MAMP-triggered immune responses (Thordal-Christensen 2003; Nurnberger and Lipka 2005). Accordingly, components involved in NHR often not only contribute to defense against non-adapted but also against adapted pathogens. In the case of filamentous phytopathogens, NHR can be subdivided in pre- and post-invasive resistance. Pre-invasive NHR restricts the penetration of fungal and oomycete pathogens, including powdery mildews, whereas post-invasive NHR eventuates if the non-adapted pathogen succeeds in host cell entry, and frequently results in an HR associated with local host cell death. This response is mainly effective against biotrophic pathogens as it deprives the invader of nutrients (Glazebrook 2005).

Several main components from two distinct pathways of pre-invasive NHR have been identified so far. The first pathway relies on PEN1, which is believed to mediate exocytosis of potentially harmful cargo upon pathogen attack by forming ternary SNARE complexes with SNAP33 and VAMP721/722 (Kwon et al. 2008; Kwaaitaal et al. 2010). The second pathway includes PEN2 (At2g44490) and PEN3. PEN2 is a tail-anchored β-thioglucoside glucohydrolase that synthesizes indole glucosinolates from a tryptophan-derived precursor (Bednarek et al. 2009; Clay et al. 2009; Fuchs et al. 2016). It contains a carboxy terminal tail anchor that targets the protein to peroxisomal and outer mitochondrial membranes. Requirement of the cytochrome P450 monooxygenase CYP81F2 (At5g57220) for PEN2-mediated resistance to powdery mildew penetration indicates its involvement in pathogen-induced production of 4-substituted indol-3-ylmethyl-glucosinolate (I3G) substrates of PEN2 (Bednarek et al. 2009; Clay et al. 2009). Consistently, CYP81F2-RFP localizes to the ER membrane, focally accumulating at sites of *Bgh* attack. This suggests that PEN2 substrate production occurs in close proximity to PEN2-decorated mitochondrial subpopulations that are recruited to sites of attempted fungal invasion (Fuchs et al. 2016). The products of PEN2 are thought to be exported by the PEN3 ABC transporter (Stein et al. 2006). Mutants of *PEN3* result in pathogen-inducible, PEN2-dependent overaccumulation of an indole compound (4-*O*-β-D-glucosyl-indol-3-yl formamide) in leaves. This suggests that PEN3 is involved in the transport of this indole or a precursor during pathogen defense (Lu et al. 2015). PEN3 interacts with calmodulin (CAM)7 (At3g43810), and *cam7* mutant plants are more susceptible to the non-adapted fungal pathogens *Bgh* and the Asian soybean rust fungus *Phakopsora pachyrhizi*, demonstrating that CAM7 and therefore Ca^{2+} sensing/ transmission of Ca^{2+} signals is an important factor of NHR in *Arabidopsis* (Campe et al. 2015). Fluorophore-tagged PEN1, PEN2, and PEN3 focally accumulate at powdery mildew penetration sites (Assaad et al. 2004; Lipka et al. 2005; Stein et al. 2006). Recruitment of PEN1 and PEN3 fusions with GFP to infection sites can be triggered by MAMPs, but distinct mechanisms contribute to the transport of the two proteins (Underwood and Somerville 2013). Interestingly, PEN1 and PEN3 accumulate in the apoplast at sites of papilla formation (Meyer et al. 2009; Underwood and Somerville 2013). Importantly, although the *pen* and *cam7* mutants allow increased host cell entry, subsequent host cell death due to post-penetration NHR restricts infection success. Key components of post-invasive NHR are EDS1, PAD4, and SAG101, all of which are required, to different degrees, for full resistance against various pathogens (Feys et al. 2005; Wiermer et al. 2005; Rietz et al. 2011; Wagner et al. 2013). Single mutations in *EDS1*, *SAG101,* and *PAD4* and the respective double mutants in combination with *pen2* are insufficient to allow sporulation of *Bgh* and *E. pisi* on *Arabidopsis*. However, the pre- and post-invasive NHR-deficient mutant *pen2 pad4 sag101* enables these non-adapted pathogens to form secondary hyphae. *E. pisi* even causes macroscopically visible powdery mildew symptoms resulting from moderate conidiation, and *Bgh* occasionally forms conidiospores (Lipka et al. 2005). Besides extracellular papilla formation, a prominent aspect of pre-invasive NHR is the focal accumulation of various cellular components, and organelles toward sites of attempted pathogen invasion. These structures include secretory vesicles, peroxisomes, mitochondria, and the ER (Koh et al. 2005; Bohlenius et al. 2010). In *Arabidopsis* these rearrangements further comprise the accumulation of proteins with defense functions, for example, the PEN proteins (Assaad et al. 2004; Lipka et al. 2005; Stein et al. 2006) and the callose synthase PMR4/GSL5 (Ellinger et al. 2013), at *Bgh* attack sites. This focal aggregation requires reorganization of the actin cytoskeleton toward sites of fungal ingress, emphasizing the central role of actin-based transport processes in pre-invasive NHR, and plant antifungal immunity in general (Takemoto et al. 2006; Underwood and Somerville 2008, 2013; Feechan et al. 2011; Yang et al. 2014).

Another component of the NHR to powdery mildew is the *Arabidopsis* phospholipase Dδ (PLDδ: At4g35790), which is involved in the biosynthesis of phosphatidic acid (Wang 2004). Phosphatidic acid can serve as a precursor for membrane phospholipids or as a signaling molecule, and may play a role in plant defense, as its levels increase after MAMP perception or recognition of various pathogen effectors (Van der Luit et al. 2000; de Jong et al. 2004; Andersson et al. 2006; Kirik and Mudgett 2009). A PLDδ fusion with GFP accumulates around papillae at sites of attempted *Bgh* penetration. Additionally, the *pldδ* mutant allows increased cell entry by *Bgh* and *E. pisi*, and shows delayed upregulation of early MAMP-responsive genes after chitin treatment (Pinosa et al. 2013). Finally, the phytohormone abscisic acid (ABA) seems to be involved in NHR against powdery mildews. The NAC transcription factor (TF) *Arabidopsis thaliana* activating factor1 (ATAF1:

At1g01720) contributes to defense against *Bgh*. Loss of ATAF1 partially compromises *Bgh* penetration resistance, which correlates with the induction of ABA biosynthesis and transcript accumulation of ABA-responsive genes (Jensen et al. 2007, 2008). By contrast, endogenous ABA levels are decreased after inoculation of wild-type plants with *Bgh*. ATAF1-dependent suppression of ABA levels after pathogen challenge suggests that ATAF1 acts as attenuator of ABA signaling in order to mediate efficient penetration resistance against *Bgh* (Jensen et al. 2008). A better understanding of the genetics and molecular mechanism understanding plant nonhost resistance which is strong and durable, and bears the potential for targeted employment of the valuable trait to control host pathogens. The use of mutants and gene-silencing approaches in *A. thaliana* opens up new possibilities to discover genes involved in nonhost resistance and to unravel the underlying mechanisms.

4.3.26 Mechanisms of Powdery Mildew Penetration Control

In order to quit descriptive phenomenology, genetic screens for chemically induced *Arabidopsis* mutants with altered nonhost interactions upon *Bgh* inoculation were recently conducted. These efforts allowed the identification of three genes penetration1 (PEN1), PEN2, and PEN3 (Lipka et al. 2005; Collins et al. 2003; Stein et al. 2006), whose products limit the entry success of non-adapted powdery mildews. Single mutants of these genes exhibit several-fold enhanced invasion frequencies of *Bgh* and *Ep*, but no increase in overall susceptibility owing to concomitant HR-like cell death of invaded epidermal cells. Systematic gene interaction analyses suggested that PEN1 and PEN2 act in two distinct penetration control mechanisms, and that PEN2 cooperates with PEN3 (Lipka et al. 2005; Stein et al. 2006). These findings were further substantiated by the fact that *pen2* and *pen3* mutants, but not *pen1* mutants, also show enhanced entry of the non-adapted hemibiotrophic oomycete *Phytophthora infestans,* and impaired basal resistance to the necrotrophic ascomycete *Plectosphaerella cucumerina.* PEN1 encodes a plasma membrane-anchored syntaxin with a SNARE (soluble *N*-ethylmaleimide-sensitive factor attachment protein receptor) domain. SNARE domain-containing proteins are key players in vesicle-associated membrane fusion, and secretion processes, including exocytosis and endocytosis. The presence of a functional homolog (required for MLO resistance 2 [ROR2]) in the distantly related monocot species barley supports the idea that PEN1-mediated defense represents an evolutionarily ancient mechanism. Cell biological analyses with functional GFP-PEN1 fusions showed a pathogen-induced accumulation in lipid raft like plasma membrane micro domains at sites of attempted ingress by powdery mildews.

PEN1 gene product plays a role for timely secretion, and cell wall remodeling processes (Plate 4.9). However, genetic evidence for PEN1 repressor activity in multiple defense responses recently challenged simple mechanistic interpretations. Kwon et al. (2008) could show that PEN1 forms a pathogen-induced ternary complex with the adaptor SNARE synaptosomal-associated protein 33 (SNAP33), and two endomembrane compartment-associated SNARE proteins, vesicle-

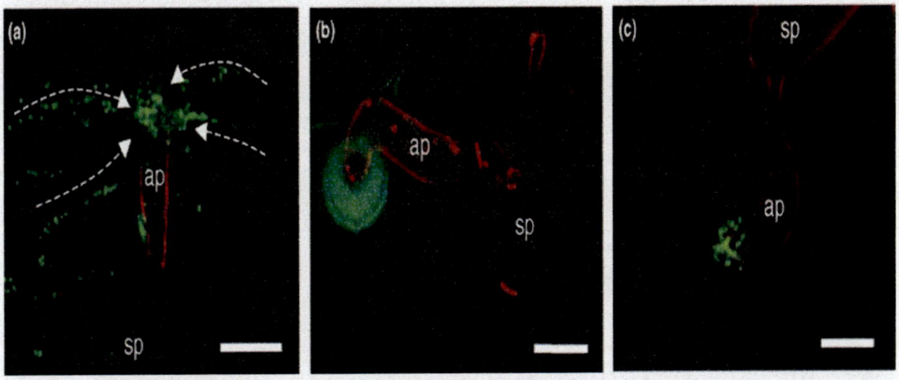

Plate 4.9 Pre-invasion resistance manifests at the cell periphery. Confocal imaging of GFP-fusion proteins in transgenic *Arabidopsis* plants inoculated with barley powdery mildew spores reveals dynamic changes in subcellular protein localization, secretory transport processes, and organelle delocalization at incipient fungal entry sites. (**a**) Endomembrane compartments tagged with GFP-labeled *Arabidopsis* R-SNARE VAMP722 move to (indicated by white dashed arrows), and accumulate at a fungal interaction site. (**b**) Focal accumulation of GFP-labeled ABC transporter PEN3 in a lipid raft-like plasma membrane microdomain. (**c**) Peroxisome-associated PEN2-GFP fusions concentrate at a contact site between plant, and fungal invader. Fungal structures were stained with FM 4–64. *sp* spore, *ap* appressorial germ tube. Bars = 10 μm (Lipka et al. 2008)

associated membrane protein (VAMP) 721, and VAMP722. Moreover, transgenic expression of functional GFPVAMP722 fusions driven by endogenous promoter sequences revealed an inducible cell autonomous expression in epidermal pavement cells that are under powdery mildew attack. The fluorescent fusion proteins tag endomembrane compartments that move along defined routes to sites of attempted fungal penetration (Plate 4.9).

These experiments provide the first evidence for SNARE complex formation in plants, and in analogy to yeast and mammal model systems corroborate a likely role of PEN1, SNAP33, and VAMP721/722 in pathogen-induced exocytotic secretion (Fig. 4.9). Interestingly, VAMP721 and VAMP722 have a second functionally redundant role in a default secretory pathway, suggesting that an ancient transport mechanism involving these VAMP proteins was co-opted to form part of a secretory plant immune response. Homozygous/heterozygous knockout mutant combinations of vamp721 and vamp722 result in haplo insufficient nonhost, and basal resistance to *Ep*, *Go,* and the adapted oomycete *Hyaloperonospora parasitica* (*Hp*). The experimentally challenging purification of VAMP721/722-tagged compartments, and subsequent comparative proteomic/metabolomic cargo analyses might provide further insights into the PEN1-mediated defense mechanism (Lipka et al. 2005, 2007, 2008; Collins et al. 2003; Stein et al. 2006; Assaad et al. 2004; Bhat et al. 2005; Kwon et al. 2008).

PEN2 encodes one out of 48 predicted *Arabidopsis* family 1 glycoside hydrolases (F1GHs). F1GHs are required for the enzymatic activation of small molecules from inactive glycosidic precursor metabolites, and play diverse roles in plant biology.

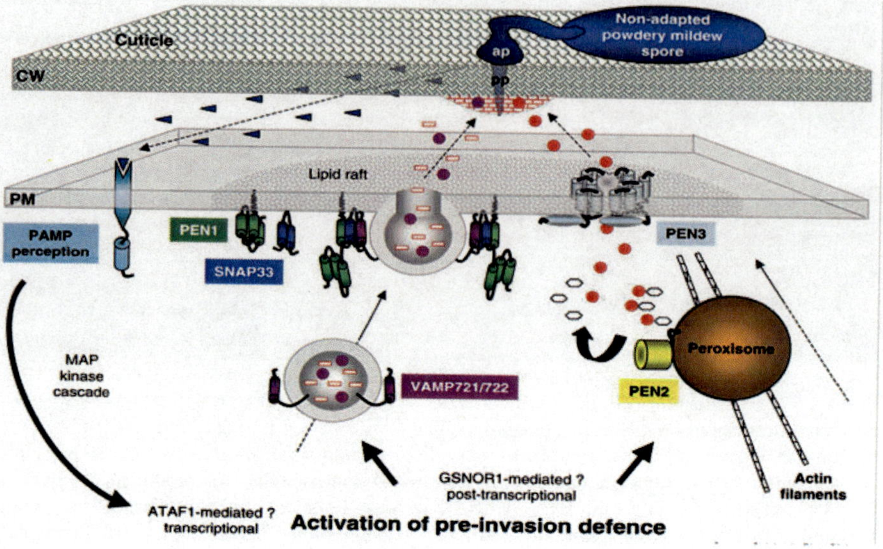

Fig. 4.9 Schematic overview of pre-invasion resistance mechanisms. Barley powdery mildew spores try to penetrate the cuticle and cell wall (CW) of host and nonhost plant by means of appressorium (ap) and penetration peg formation (pp). On the nonhost plant *Arabidopsis*, PRR-mediated recognition of fungal PAMPs (blue triangles) is likely to induce MAP-kinase signaling and ATAF1-mediated (?) transcriptional activation of the pre-invasion defense machinery. Post-translational control (e.g., via GSNOR1-mediated S-nitrosylation) represents another regulatory layer. The plasma membrane (PM)-localized syntaxin PEN1 and ABC-transporter PEN3 accumulate in a lipid raft-like microdomain. PEN1 forms a SNARE complex with the membrane-anchored adaptor SNARE SNAP33 and endomembrane-compartment-associated R-SNAREs VAMP721/722. SNARE complex formation drives secretion of cell wall precursors (red rectangles) and/or antimicrobial compounds (purple dots) at sites of attempted fungal invasion. PEN3 discharges potentially toxic a glycons (red dots) that were catalytically released from nontoxic glycosidic (black hexagons) precursors by PEN2 enzyme activity. PEN2 is associated with the periphery of peroxisomes. These are known to shuttle along a focally reorganized actin cytoskeleton. Together, PEN1/SNAP33/VAMP721/722 and PEN2/PEN3-mediated defense mechanisms contain the majority of fungal invasion attempts (Lipka et al. 2008)

Catalytic activity was shown to be required for PEN2 function in powdery mildew entry control, suggesting that the PEN2 product(s) may have direct or indirect antifungal activity. Metabolite profiling experiments with *pen2* knockout lines suggest tryptophan-derived metabolites as potential substrates and point to a myrosinase-like activity of PEN2. Fluorescence labeling revealed an association of the PEN2 protein with the periphery of peroxisomes, which are subject to pathogen-induced cell polarization (Plate 4.9). Intracellular transport of plant peroxisomes is known to occur along actin filaments, and genetic interference with actin dynamics has recently confirmed that pathogen-induced actin cytoskeleton reorganization plays a major role for efficient pre-haustorial NHR to powdery mildews. Thus, peroxisomal production of PEN2 substrate(s) and cytoskeleton-mediated peroxisome delivery to fungal entry sites represents a potential underlying mechanism

(Fig. 4.9). Phylogenetic analyses suggest that the PEN2 gene represents an evolutionarily recent acquisition of the *Arabidopsis* genome. This finding demonstrates that NHR is not an exclusively ancient, fixed, and nonadaptive trait, and might also explain why *pen2* mutations affect a wide range of pathogen interactions (Lipka et al. 2005, 2008; Xu et al. 2004; Miklis et al. 2007; Consonni et al. 2006).

PEN3 encodes one of the 15 pleiotropic drug resistance (PDR) ATP-binding cassette (ABC) transporters present in the *Arabidopsis* genome. These are ubiquitous transmembrane proteins that function in the ATP-dependent transport of a wide variety of substrates across extracellular and intracellular membranes. Like PEN1, PEN3 resides in the plasma membrane and accumulates at sites of attempted fungal penetration (Plate 4.9). Interestingly, lack of PEN3 results in hyper-activation of SA-dependent defense signaling pathways, which explains the initially surprising enhanced basal resistance of pen3 mutants to adapted powdery mildews such as *Gc* and *Go*. Intriguingly, this phenotype is significantly attenuated in pen2pen3 double mutants. One possible explanation is that concerted PEN2 and PEN3 activity drives enzymatic release and energy-dependent apoplast secretion of toxic compounds at interaction sites with invading fungi (Fig. 4.9). In the absence of PEN2 its toxic product would no longer accumulate inside pen3 mutant cells, and thus not unintentionally activate SA-dependent hypersensitive responses (Stein et al. 2006; Crouzet et al. 2006).

The two distinct pre-haustorial nonhost defense mechanisms (defined by PEN1/SNAP33/VAMP721/VAMP722, and PEN2/PEN3, respectively) have in common that they are both subject to pathogen-induced cell polarization (Plate 4.9) and that they mediate the discharge of antimicrobial compounds (Fig. 4.9). This highly organized battle array and the timely commitment of forces are strikingly reminiscent of the execution of immune responses at the immunological synapse in vertebrate T cells. Thus, plant innate immunity and vertebrate adaptive immunity feature mechanistic parallels that are worth further exploration (Kwon et al. 2008; Lipka et al. 2008).

Broad-spectrum resistance to adapted powdery mildews is conferred by loss-of-function mutant alleles of mildew resistance locus O (MLO) genes in both barley and *Arabidopsis*. Powdery mildew fungi are believed to require a subset of functional MLO proteins as compatibility factors ("entry portals") for successful invasion of host epidermis cells. Lack of these potential powdery mildew effector targets results in efficient pre-invasion resistance that shares many characteristics with NHR to powdery mildews. Intriguingly, mlo resistance in *Arabidopsis* requires all PEN genes described so far. Moreover, experiments showed that transient overexpression of constitutively active variants of a calcium-dependent protein kinase (CDPK) affects both mlo and nonhost resistance in barley. Together, these findings imply that powdery mildew fungi utilize a common and evolutionarily old host cell entry mechanism (dependent on wild-type MLO) and that mlo resistance represents a reestablishment of pre-invasion NHR (Consonni et al. 2006; Humphry et al. 2006; Freymark et al. 2007). Other components that were recently claimed to contribute to *Arabidopsis* pre-invasion NHR against non-adapted powdery mildews include the *S*-nitrosoglutathione reductase GSNOR1 and the plant transcription factor ATAF1, respectively. Although the entry success of *Bgh* and *Bgt* on *ataf1-1* and *atgsnor1-3* mutants was only moderately enhanced (from 9% to 14%, and from 1% to 3%, respectively), these results suggest both transcriptional and post-transcriptional regulation of penetration resistance (Feechan et al. 2005; Jensen et al. 2007).

4.3.27 Mechanism of Post-penetration Defense

Arabidopsis mutants showing enhanced entry are still nonhost plants for non-adapted powdery mildews owing to effective post-entry cell death. Systematic analyses with multiple mutant combinations revealed that post-haustorial NHR requires enhanced susceptibility 1 (EDS1), phytoalexin-deficient 4 (PAD4), and senescence-associated gene 101 (SAG101). These lipase-like proteins constitute a regulatory node that is essential for basal defense against *Hp*, SA-mediated signaling, and a subset of R-gene-mediated resistance pathways. Single mutants for *EDS1*, *PAD4*, and *SAG101* have no or only a minor effect on pre-invasion resistance to powdery mildews. However, at successful invasion sites HR is less frequent and allows ecto-parasitic secondary hyphal growth and microcolony formation of *Bgh* and *Ep*. Intriguingly, microscopic inspection revealed that NHR to *Ep*, but not to *Bgh*, was fully compromised on *pen2pad4* and *pen3eds1* double mutants, indicated by occasional conidiophore formation. On *pen2pad4sag101* triple mutants *Bgh* was capable of completing its lifecycle, while *Ep* growth and reproduction was even macroscopically detectable. Thus, removal of only three genes is sufficient to make *Arabidopsis* a fully susceptible host plant for the non-adapted pea powdery mildew and allows the monocot pathogen *Bgh* to establish basic compatibility. This is remarkable, as it suggests that both pathogens do not lack any principal adaptations to establish (presumably) complex biotrophic interactions and to complete their lifecycles. The quantitative differences in virulence between *Ep* and *Bgh* possibly reflect the closer evolutionary species, and effector complement relatedness between *Ep* and the adapted powdery mildews *Ec*. Comparative powdery mildew effector studies are urgently required to unequivocally resolve this issue. In summary, *Arabidopsis* NHR to non-adapted biotrophic powdery mildews is based upon two successive, multicomponent, and independently effective defense layers: PEN-gene-mediated pre-invasion resistance and post-invasion immunity, controlled by EDS1, PAD4, and SAG101 (Lipka et al. 2005, 2008; Stein et al. 2006; Wiermer et al. 2005).

HR-like cell death execution and its control by the EDS1/PAD4/SAG101 node suggested a possible contribution of R-gene-mediated ETI to post-invasion resistance against *Bgh* and *Ep*. The proteins required for MLA12 resistance (RAR1) and suppressor of *G2* allele of SKP1 (SGT1) function as cofactors in heat shock protein 90 (HSP90)-mediated stabilization of R-protein complexes, and represent a genetic convergence point for EDS1/PAD4/SAG101, and other R-gene signaling pathways. Double and triple mutant combinations of either *pen2* or *pen3* with *rar1* and *sgt1b* had a significantly weaker effect on colonization success of *Bgt* or *Ep* if compared with combinations with *eds1*, *pad4*, and *sag101*, and did not result in lifecycle completion. Hence, R-genes appear to play a minor if any role in post-haustorial

Arabidopsis NHR to the tested non-adapted powdery mildews. Perhaps such a scenario is more likely in interactions where would-be pathogen, and nonhost plants share a more recent compatible ancestry for example, in barley/wheat interactions with the wheat/barley, powdery mildew, or where a brisk intraspecific or interspecific exchange of effector molecules determines a pathogen's host range (Stein et al. 2006; Azevedo et al. 2006; Noel et al. 2007; Schweizer 2007).

4.3.28 Mutagenic Resistance to Powdery Mildew

The method of induced mutagenesis is a powerful tool for creating new genotypes with an increased yield, quality, and resistance to diseases and pests (Ahloowalia et al. 2004). Upon irradiation with gamma rays, the variation of the resulting genetic alterations has high frequency and wider range of morphological and biochemical changes (Skoric et al. 2008). There is evidence that the mutations may also change the expression of certain enzymes in the biosynthetic pathway of fatty acids (Auld et al. 1992; Pleines and Friendt 1989). The irradiation of rapeseed with low doses of gamma rays of 100 and 150 Gy has a stimulating effect on the synthesis of linoleic and linolenic acids. Conversely, at higher doses of the mutagen factor, the concentration of these fatty acids decreases while the amount of oleic acid increases (Rahimi and Bahrani 2011). The fatty acids synthesized in plants play an important role by participating in the composition of tissue hormones and in cell and nuclear membrane structures (Kolattukudy 1974). It is assumed that unsaturated fatty acids in plant cells are involved in cutin and suberin synthesis (Pinot et al. 1999). Cutin, which acts as first barrier that protects plant organs (leaves and fruits) from dehydration providing protection against pathogens and pests, is composed of C16 and C18 fatty acids. Suberin, which is predominantly composed of hydroxyl and epoxy unsaturated fatty acids, is a structural component of the plant underground parts, roots and tubers (Kolattukudy 1980). C18 fatty acids in plants play a role in building the defense mechanism against diseases and pests (Kachroo and Kachroo 2009). It has been found that hydroxyl fatty acids also inhibit the development of *Erysiphe graminis, Puccinia recondita,* and *Phytophthora infestans* (Hou and Forman 2000). The powdery mildew development is favored by meteorological factors, such as temperature and high humidity (Penaud 1999; Enright and Cipollini 2007). Fungicide application is often not efficient enough due to the pathogen's rapid spread in favorable conditions. The effects of gamma radiation on the fatty acid composition of rapeseed oil and the increased resistance to *E. cruciferarum* have been determined by Petkova et al. (2014). Biochemical changes in the oil after gamma-ray irradiation are also associated with changes in the susceptibility to powdery mildew.

Petkova et al. (2014) assessed through the gas chromatography the difference in the content of saturated and unsaturated fatty acids in the seeds of the nonirradiated variants, and the ones irradiated with two doses of gamma radiation. A summary of the data for all 3 years of the fatty acid composition of the Abacus hybrid oil, and its M0, M1, and M2 was decomposing the populations. The chemical analysis showed generally that the irradiated plants showed a lower content of oleic acid, and a higher content of linoleic acid and linolenic acid compared with the nonirradiated variants. The data in Table 4.1 shows a tendency toward increase in the total content of raw fats in mutant populations of M1 and M2 of Abacus as compared with nonirradiated plants. This, in turn, is associated with a reduction in the total content of saturated fatty acids, and an increase in the content of unsaturated fatty acids.

There is an inverse relationship between the values of the total fat content and the content of saturated and unsaturated fatty acids (Table 4.1). The rapeseed oil of the Abacus hybrid showed higher content of saturated fatty acids as compared with the M2 and M3 generations of the irradiated variants. In contrast, the percentage content of unsaturated fatty acids is influenced positively by irradiation with gamma rays and is comparatively high in M2 A1-100 and M2 A-150 92.7% and 93.7%, compared with the baseline 89.7%. The data in Table 4.2 shows the extent of damage by the powdery mildew agent on plants of the Abacus variety, and its gamma-ray irradiation-treated variants. It also shows the arithmetic mean values and their errors, the accuracy indicators, and the variation coefficients. The data show that the highest extent of damage on the stems and branches in the range of 98.9–99% ($p = 0.1\%$) is observed with the M0 Abacus generation, irradiated with absorbed dose 100 and 150 Gy gamma rays, compared with the nonirradiated variant (Table 4.3).

The irradiation did not have a positive impact on the sustainability of pods to *E. cruciferarum.* The differences found are insignificant compared with the corresponding baseline group. In contrast, the M1 and M2 generations were shown to have a proven lower extent of damage by the phytopathogen as compared with the baseline. It is assumed that the increased resistance demonstrated is due to the higher content of unsaturated fatty acids in the seeds of the irradiated plants. Unsaturated fatty acids can be used as a means of biological control of plant diseases on global scales. Increasing the dose of the mutagen leads to an increase in the concentration of unsaturated fatty acids with 18 carbon atoms and to a decrease in the extent of damage by *Erysiphe cruciferarum* (Petkova et al. 2014).

4.4 *Brassica–Hyaloperonospora*: Biometabolomic Resistance

4.4.1 Induction of Biomolecules

The presence and absence of natural biochemical compounds like glucosinolates and other phenolic compounds play a significant role in providing resistance to the host plant. There is a correlation between high levels of flavor volatiles (e.g., allylisothiocyanate) released by tissue damage and the limitation of fungal growth in both wild and cultivated *Brassica* lines. In cultivated *Brassicas*, breeding has resulted in reduced levels of flavor volatiles with a consequent reduction in their general resistance to *H. parasitica*. The resistance to *H. parasitica* in the cabbage cultivar "January King" may be attributed to the high concentration of allylisothiocyanate (Greenhalgh and Mitchell 1976). In France, cauliflower cvs. C300 and Maudez resistant to downy mildew contained highest sinigrin content which can act as biochemical marker for identification of downy mildew resistance (Menard

Table 4.1 Assessing the extent of powdery mildew damage on stems, branches, and pods, as a percentage of the infected plant area on rapeseed plants from Abacus, and M0, M1, and M2 generation of gamma irradiated plants (Petkova et al. 2014)

	Extent of powdery mildew damage (%)											
	Stems				Branches				Pods			
Variants	Mean ± SD	*D*	*t*	Sign.	Mean ± SD	*D*	*t*	Sign.	Mean ± SD	*D*	*t*	Sign.
Abacus control	79.3 ± 1.31				90.5 ± 1.07				97.6 ± 0.49			
M0-Abascus-100 Gy	99.3 ± 0.18	20.09	17.62	+++	98.9 ± 0.22	8.4	16.43	+++	99.12 ± 0.2	1.52	1.36	ns
M0-Abascus-150 Gy	99 ± 0.21	19.69	17.27	+++	99 ± 0.24	8.5	15.71	+++	98.9 ± 0.31	1.3	1.04	ns
M1-Abascus-100 Gy	28 ± 1.52	−51.31	19.23	− − −	40.5 ± 1.34	−50	32.53	− − −	49.9 ± 1.17	−47.7	26.2	− − −
M1-Abascus-150 Gy	26.7 ± 1.26	−52.61	34.1	− − −	34.8 ± 1.23	−55.7	29.46	− − −	47.4 ± 1.02	−50.2	29.01	− − −
M2-Abascus-100 Gy	10.3 ± 0.58	−79.31	74.6	− − −	21.1 ± 0.77	−69	42.37	− − −	27.1 ± 1.04	−70.5	40.5	− − −
M2-Abascus-150 Gy	11.5 ± 0.56	−67.81	39.8	− − −	22.7 ± 1.31	67.8	35.12	− − −	27.5 ± 1.11	−70.1	39.8	− − −

Mean average value in % from 30 repletion; *D*—difference against control; *t*—Student's *t* distribution
ns nonsignificant
$P > 5\%$+, $P > 1\%$ ++, $P > 0.1\%$ +++, $P < 5\%$ −, $P < 1\%$ − −, $P < 0.1\%$ − − −

Table 4.2 Total fat, saturated, and unsaturated fatty acids in the seeds of *Abacus*, and M0, M1, and M2 segregating generations of Abacus, treated with absorbed dose of 100 and 150 Gy gamma rays (Petkova et al. 2014)

Fats and fatty acids (%)	*Abacus* control	M0 A-100	M0 A-150	M1 A-100	M1 A-150	M2 A-100	M2 A-150
Content of crude fat	35.9	40.7	39.5	43.2	45.1	42.5	47.1
Saturated fatty acids	10.1	9.3	7.5	6.3	6.8	6.3	6.9
Unsaturated fatty acids	89.7	82.3	83.5	85.2	86.2	92.7	93.7

Table 4.3 Fatty acid composition of rapeseed oil from the seeds of Abacus, and M0, M1, and M2 generation of irradiated plants with 100 and 150 Gy gamma rays (Petkova et al. 2014)

Fats and fatty acids (%)	Abacus control	M0 A-100	M0 A-150	M1 A-100	M1 A-150	M2 A-100	M2 A-150
C12:0 Lauric acid	–	0.1	0.1	0.1	0.1	0.1	0.2
C14:0 Myristic acid	0.1	0.1	0.2	0.2	0.2	0.1	0.1
C16:0 Palmic acid	7.5	6.7	6.3	6.5	6.7	5.9	5.7
C18:0 Stearic acid	2.3	1.8	2.0	1.9	1.2	0.3	0.3
C18:1 Oleic acid	66.3	69.3	67.0	67.8	68.5	67.5	73.0
C18:2 Linoleic acid	17.2	15.5	17.4	17.2	18.3	19.1	15.9
C18:3 Linoleic acid	4.4	3.1	3.5	4.2	5.9	6.9	4.6

et al. 1999). The incidence and severity of downy mildew is positively correlated with glucosinolate concentration in seeds of oilseed rape (Rawlinson et al. 1989). In oilseed rape, downy mildew severity is lower on cultivars with high concentration of glucosinolate (>100/μmol g^{-1}), and greater on those with a lower concentrations (<15 μmol g^{-1}). In all cultivars grown in the UK, the incidence is lower in mid-February when glucosinolate products in the leaves reach a maximum level (Anonymous 1985).

A large number of *B. napus* species with different glucosinolate and erucic acid contents have been screened for resistance to four isolates of *H. parasitica* at the cotyledon stage (Nashaat and Rawlinson 1994). Two groups of accessions with different resistance factors were identified. The first group was different from that of the cultivar "Cresar" which has an isolate-specific gene for resistance to *H. parasitica*, and the second group was identical to that of "Cresar." There was moderate to full susceptibility at the cotyledon stage but no clear differential response to any of the isolates. Those with high glucosinolate and high erucic acid

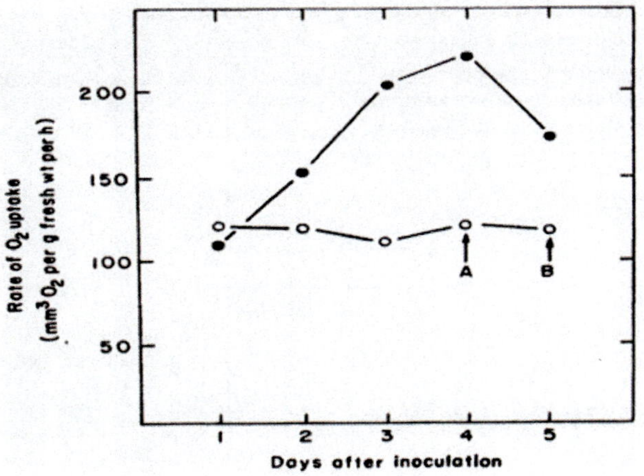

Fig. 4.10 Rates of O_2 uptake of infected and uninfected cotyledons at various times after inoculation: (●-●), infected; (○-○), uninfected; A, visible signs of sporulation (Thornton and Cooke 1974)

content were significantly less susceptible than those with high glucosinolate and low erucic acid, or low glucosinolate and low erucic acid content.

In preliminary experiments on the effect of treatment with abiotic elicitors on disease reaction in oilseed rape seedlings, salicylic acid, and methyl jasmonate reduced the severity of infection when cotyledons were subsequently inoculated with *H. parasitica* (Doughty et al. 1995). Many marked shifts were observed in the metabolic processes of plant tissues following infection by biotrophic parasites. These included changes in respiration, photosynthesis, nucleic acid, and protein synthesis, and phenol metabolism. There could also be changes in the translocation and accumulation of nutrients and in the levels of endogenous growth substances.

The respiration rate was raised sharply soon after *H. parasitica* infection of cabbage cotyledons, and reached a maximum, almost twice that of uninfected controls, at the time of the initiation of sporulation (Fig. 4.10). The chlorophyll content of infected and noninfected cotyledons did not differ significantly at any time (Fig. 4.11). The increased respiration rate of the infected tissues did not reflect any significant changes to the pentose phosphate pathway in this infection since no acyclic polyhydric alcohols were detected in soluble extracts of either infected leaves or fungal conidia (Thornton and Cooke 1974) (Figs. 4.12 and 4.13).

Pathogenesis in *Brassica-Hyaloperonospora* combinations was observed to be accompanied by large increases in electrolyte leakage (Figs. 4.14, 4.15, 4.16, and 4.17), and increased activity of glucosidase (Fig. 4.15), ribonuclease (Fig. 4.16), and peroxidase (Fig. 4.17) (Kluczewski and Lucas 1982). The large increase in glucosidase was of pathogen origin while enhanced ribonuclease activity was due to a new post-inflectional form of the enzyme. In vivo infected leaves of *B. juncea* produced cellulase, endo-PMG, and endo-PG (Singh et al. 1980). There are a number of natural biochemical compounds present in host tissues which may influence the defense mechanism of crucifers against downy mildew infection. The role of phenolic compounds, glucosinolates, phytoalexins, and flavor volatile compounds in providing resistance to crucifers against downy mildew infection has been

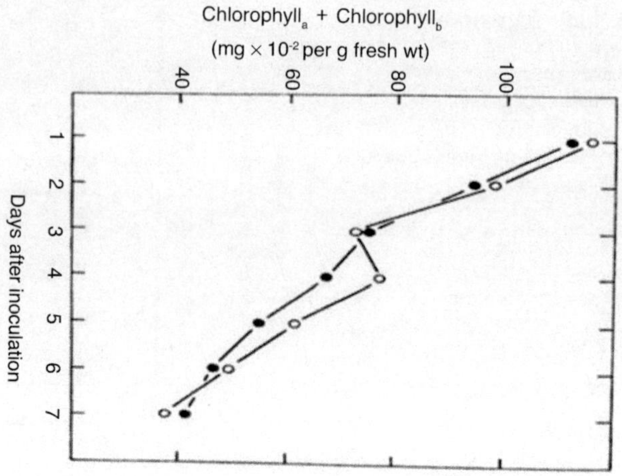

Fig. 4.11 Chlorophyll-*a* plus chlorophyll-*b*, content of infected, and uninfected cotyledons at various times after inoculation: (●-●), infected; (○-○), uninfected (Thornton and Cooke 1974)

explained in our book "Downy mildew disease of crucifers: biology, ecology and disease management," chapter 12, "Host resistance," section 12.6 "Biochemical basis of resistance" (Saharan et al. 2017).

4.4.2 Synthesis of Phytoalexins

A phytoalexin–deficient (pad) mutant in *Arabidopsis thaliana* Columbia (Col-0) was used by Glazebrook et al. (1997) to investigate the role of phytoalexin camelexin (3-thiazol-2yl-indole) in disease resistance. Mutations in PAD1, PAD2, and PAD4 caused enhanced susceptibility to *Pseudomonas syringae* pv. *maculicola* strain ES4326 (Psm ES4326), while mutations in PAD3 or PAD 5 did not. Camelexin was not detected in the double mutants pad 1-1, pad 2-1, pad 1-1, pad 3-1 or pad 2-1, pad 3-1. Growth of Psm ES4326 in pad 1-1, pad 2-1 was greater than that in pad1-1 or pad2-1 plants, while bacterial growth in pad 1-1, pad 3-1, and pad 2-1, pad 3-1 plants was similar to that in pad1-1 and pad 2-1 plants, respectively. The pad 4-1 mutation caused reduced camelexin synthesis in response to Psm4326 infection but not in response to infection by *Cochliobolus carbonum*, indicating that PAD 4 has a regulatory function. PAD 1, PAD 2, PAD 3, and PAD 4 are all required for resistance to downy mildew (*H. parasitica*). The pad 4-1 mutation caused the most dramatic change, exhibiting full susceptibility to four out of six Col-incompatible *H. parasitica* isolates. Each combination of double mutants between pad1-1, pad2-1, and pad3-1 exhibited additive shifts to moderate or full susceptibility to most of the isolates.

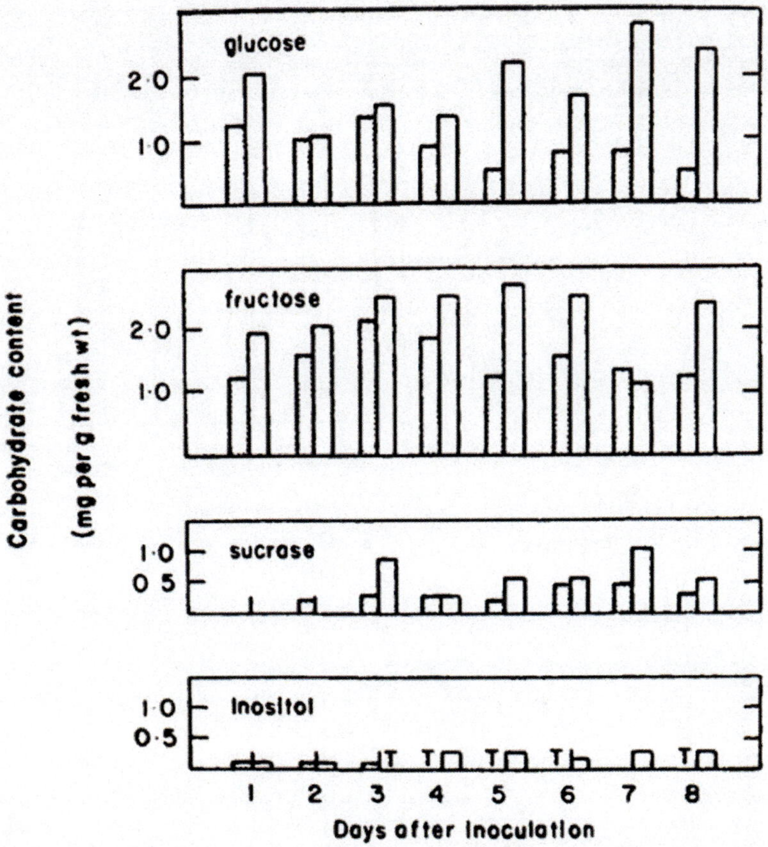

Fig. 4.12 Carbohydrate content of the alcohol soluble fraction of infected, and uninfected cotyledons at various times after inoculation with *Hyaloperonospora parasitica*: ■ infected, □ uninfected, *T* trace (indicating that the peak height of the TMS derivative was indeterminable at an attenuation of 20 × 10^3) (Thornton and Cooke 1974)

4.4.3 Induction of Lignification of Host Cells

Lignin formation was observed in cell walls of parenchyma of Japanese radish root infected with downy mildew fungus. The observed lignin was mainly composed of guaincylpropane units, and apparently differed from syringyl lignin which was present in vessels of the healthy tissues. Different pathways for lignin biosynthesis in the healthy and the diseased tissues were proposed (Tables 4.4 and 4.5; Figs. 4.18, 4.19, and 4.20) by Asada and Matsumoto (1972). Specific isoperoxidases synthesized de novo in diseased tissues were presumed to play an important role in the formation of the guaincyl lignin (Ohguchi et al. 1974; Ohguchi and Asada 1975). Lignin was also formed in cell walls of parenchyma of Japanese radish root which was infiltrated with 700 × *g* supernatant of homogenate of downy mildew

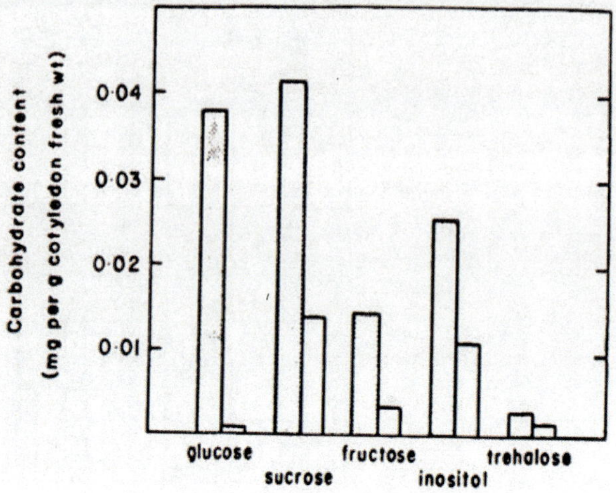

Fig. 4.13 Principal carbohydrates of the alcohol soluble fraction of sporangia from infected cotyledons and control washings, 7 days after inoculation with *Hyaloperonospora parasitica*. ■ infected, □ uninfected (Thornton and Cooke 1974)

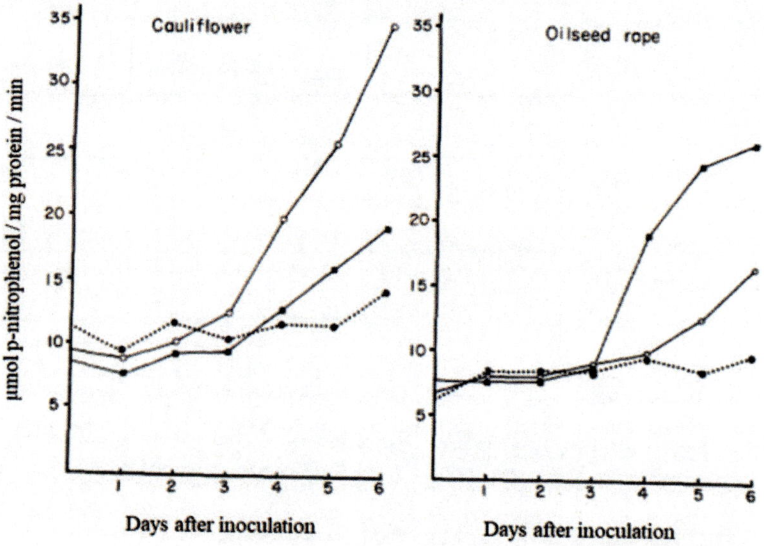

Fig. 4.14 Conductivity changes of deionized glass-distilled water containing samples of uninfected cotyledons (...) and cotyledons infected (-) by *Hyaloperonospora parasitica* isolate from cauliflower (O), and oilseed rape (–) (Kluczewski and Lucas 1982). Each point represents the mean of four replicates

infected tissues. It began to form about 12 h after infiltration of the homogenate (Matsumoto et al. 1978). The effective component in the homogenate for the induction of lignification is dialyzable, which is highly water soluble, and seems to resemble monilicolin A (Asada et al. 1975). The lignification inducing factor (LIF) plays a significant role in the induction of systemic resistance. Resistance to

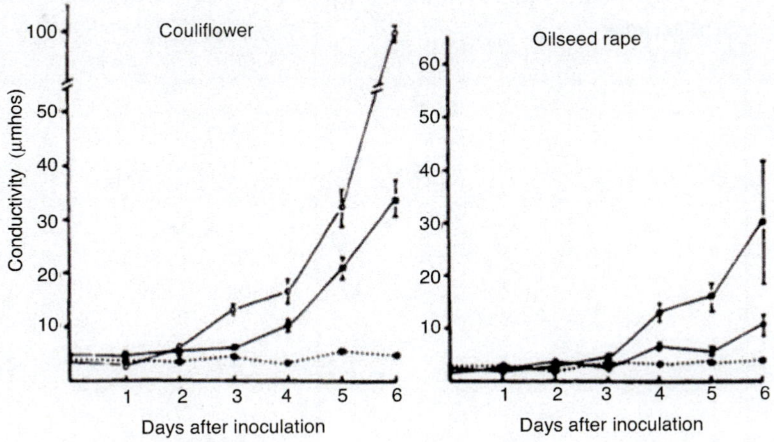

Fig. 4.15 β-glucosidase activity in extracts of control cotyledons (...) and cotyledons infected (-) by either CI (O) or R1 (–) isolate of *Hyaloperonospora parasitica* (Kluczewski and Lucas 1982)

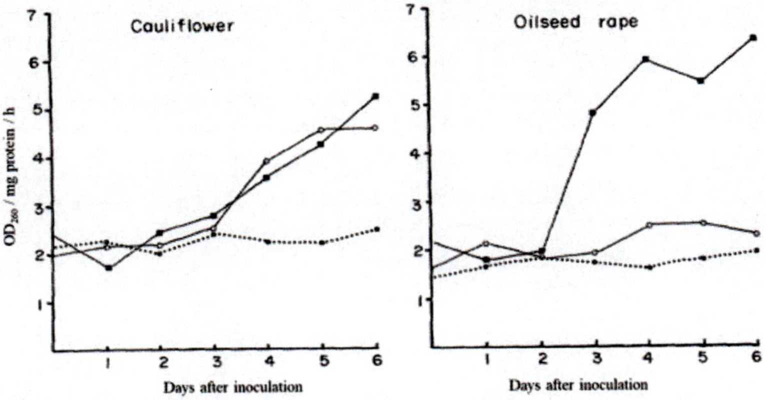

Fig. 4.16 Acid ribonuclease activity in extracts of control cotyledons (...) and cotyledons infected (-) by *Hyaloperonospora parasitica* isolates CI (O) and R1 (–) (Kluczewski and Lucas 1982)

H. parasitica was induced in roots of susceptible radish cultivars when they were preliminarily inoculated with the pathogens or wounded. An increase in L-phenylalanine ammonia lyase (PAL) activity and lignification of cell walls occurred in these tissues after challenge inoculation with downy mildew. Histochemical observations indicated that the cell walls were lignified in the tissues beyond the site of fungal attack (Matsumoto and Asada 1984). High peroxidase activity was located around the lignified cell walls (Asada and Matsumoto 1969; Ohguchi et al. 1974). The

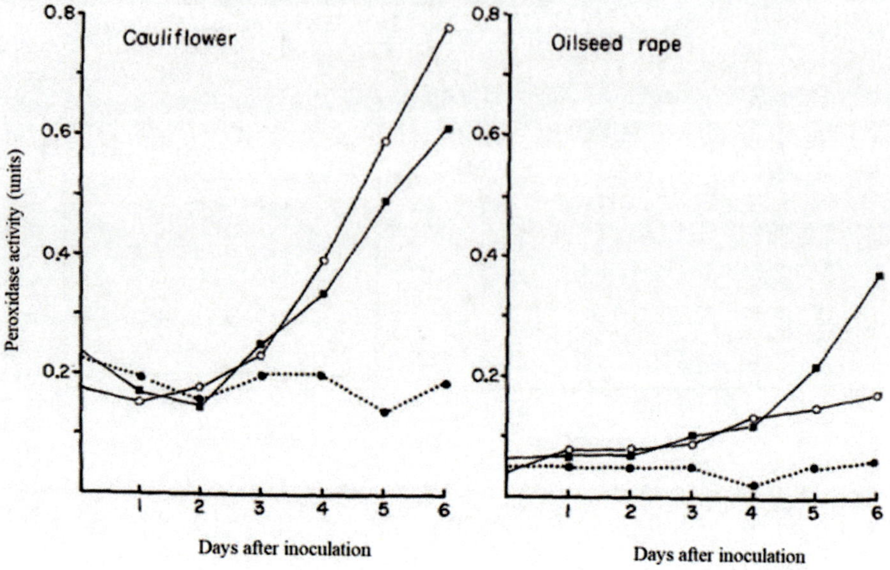

Fig. 4.17 Peroxidase activity in extracts of control cotyledons (...) and cotyledons infected by either isolate CI (○) or R1 (–) of *Hyaloperonospora parasitica* (-) (Kluczewski and Lucas, 1982)

Table 4.4 Amounts of degradation products by alkaline nitrobenzene oxidation of the isolated lignin (Asada and Matsumoto 1972)

Lignin	Product	Amount (%)	Ratio
Healthy	*p*-Hydroxybenzaldehyde (H)	0.39	V/H 6.05
	Vanillin (V)	2.36	S/H 3.90
	Syringaldehyde (S)	1.52	S/V 0.64
	Total	4.27	
Diseased	*p*-Hydroxybenzaldehyde (H)	0.99	V/H 2.75
	Vanillin (V)	2.72	S/H 0.0
	Syringaldehyde (S)	0.00	S/V 0.0
	Total	3.71	

higher amount of lignin accumulation was present in the middle portion of the cell wall (Asada and Matsumoto 1971). The most resistant cvs of Japanese radish have lignified cell walls prior to attack by *H. parasitica*, and the hyphae could not penetrate cells (Matsumoto 1994). Following infection of radish by *H. parasitica,* the deposition of lignin in host cell walls may have a role to play in nonspecific limitations of the growth of biotrophic fungi. Ohguchi and Asada (1975) have defined the pathways and enzymes involved in lignin biosynthesis in radish following infection by *H. parasitica*.

The formulae were obtained from the following equations.

Table 4.5 Elemental compositions and empirical formulae of the isolated lignins and the related compounds (Asada and Matsumoto 1972)

Lignin	C (%)	H (%)	OCH_3 (%)	Formula
Healthy root	64.38	7.27	18.93	$C_9H_{10.3}O_{2.20}(OCH_3)_{1.16}$
Diseased root	63.47	5.74	12.58	$C_9H_{8.33}\ O_{2.80}(OCH_3)_{0.75}$
Birch[a]	58.82	6.49	21.51	$C_9H_{9.03}O_{2.77}(OCH_3)_{1.58}$
Spruce[a]	63.48	6.35	14.84	$C_9H_{8.83}\ O_{2.37}(OCH_3)_{0.96}$
DHP[b]	64.00	6.00	16.90	$C_9H_{8.21}O_{2.50}(OCH_3)_{1.02}$
Coniferyl alc.	66.67	6.67	17.22	$C_9H_9O_2(OCH_3)_1$

[a]From Bjorkman and Person (1957)
[b]Dehydrogenation polymerization product, from Nozu (1967)

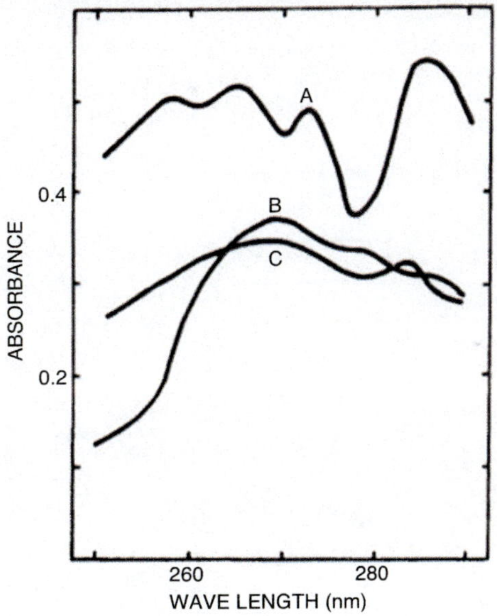

Fig. 4.18 The UV absorption spectra of the diseased parenchyma cell wall (A), the vessel wall (B), and the healthy parenchyma cell wall (C) of the Japanese radish root (Asada and Matsumoto 1972)

$$X = \frac{108.09[\mathrm{H}(\%) - 31.035 \times \mathrm{OCH_3}(\%)]}{1.008D} = \frac{108.09\,[\mathrm{O}(\%) - 31.035 \times \mathrm{OCH_3}(\%)]}{16D}$$

$$Z = \frac{108.9 \times \mathrm{OCH_3}(\%)}{31.035D}$$

$$D = \mathrm{C}(\%) - \frac{12.01}{31.035 \times \mathrm{OCH_3}(\%)}$$

4.4.4 Induction of Antipathogenic Molecules for Disease Resistance

Plants can employ chemical defense either as preformed antimicrobial substances (phytoanticipins) or as induced antimicrobial substances which accumulate after contact with the pathogen (phytoalexins) (VanEtten et al. 1994). In *Arabidopsis*, as in other *Brassicas*, glucosinolates come into question as potential phytoanticipins.

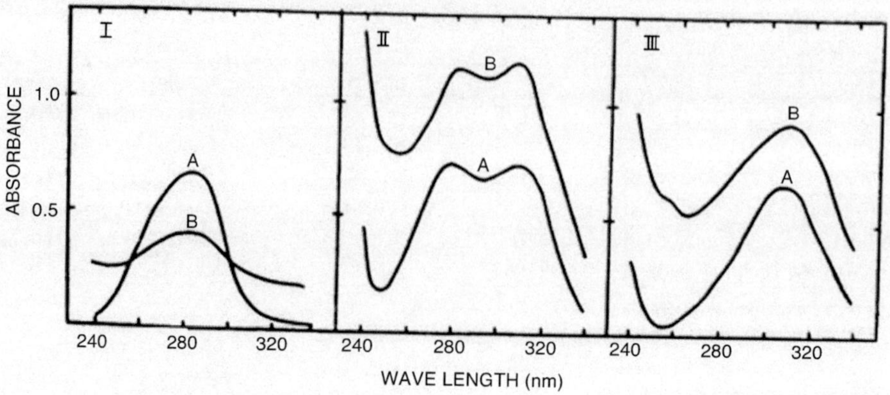

Fig. 4.19 UV absorption spectra of the authentic compounds (A) and the degradation products (B) obtained from the extraction of paper chromatograms. I = *p*-hydroxybenzaldehyde; II = vanillin; III = syringaldehyde (Asada and Matsumoto 1972)

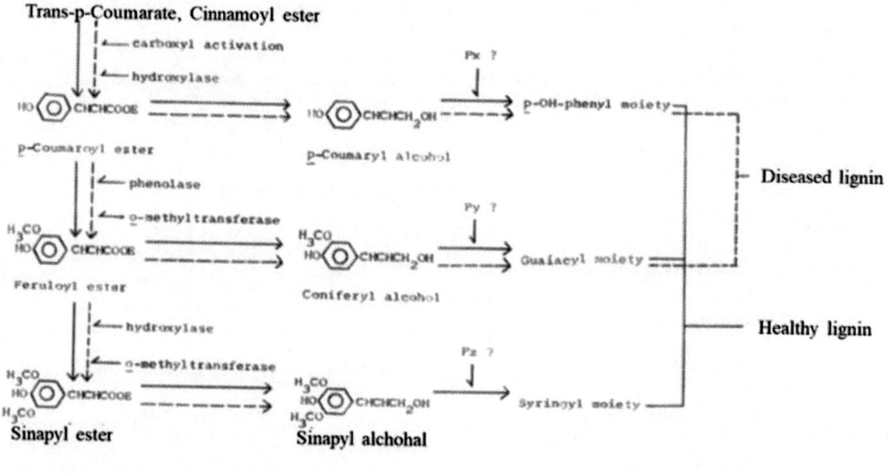

Fig. 4.20 Suggested pathway of lignin biosynthesis in healthy (full lines) and diseased (broken lines) plants. Px, Py, Pz: Peroxidase isoenzymes x, y, z (Asada and Matsumoto 1972)

However, phytoanticipins are generally correlated with nonhost resistance (Mansfield 2000), and there is no evidence that glucosinolates play a role in the interaction of *Arabidopsis* with *H. parasitica*. The accumulation of defense gene transcripts such as *PAL* and *CHS* and the increase in peroxidase activity in cell suspension cultures treated with elicitor was reported by Davis and Ausubel (1989) indicated that stimulation of phenylpropanoid metabolism in *Arabidopsis* could be expected to be associated with resistance, as observed in several other plant species (Dixon and Paiva 1995). In the *Arabidopsis*/*H. parasitica* interaction an early oxidative burst of H_2O_2 production is observed on penetration of the epidermis by an avirulent race of the pathogen. Following the oxidative burst is the genetically programmed hypersensitive cell death response (HR). In an HR against an avirulent *Pseudomonad* a shift from house-keeping to defense metabolism which affected an estimated 10% of the transcriptome was demonstrated (Scheideler et al. 2002), and this is presumably similar in the HR against *H. parasitica*. Since *H. parasitica* is an obligate biotroph, the rapid hypersensitive response (HR) which occurs in the epidermal cells adjacent to the penetration hyphae, and sometimes an additional few cells in the mesophyll (Koch and Slusarenko 1990), would be expected to be sufficient to effectively condition resistance by preventing the establishment of the highly co-evolved nutritional relationship between host cell and pathogen, which depends on haustoria. Nevertheless, the HR to *H. parasitica* is associated with the accumulation of at least one antimicrobial phytoalexin, camalexin (Slusarenko and Mauch-Mani 1991; Tsuji et al. 1992), and it was reported that the phytoalexin deficient mutants (*pad1*-1, *pad2*-1, and *pad3*-1, and the double mutants *pad1*-1/*pad2*-1, *pad1*-1/*pad3*-1, and *pad2*-1/*pad3*-1) showed increased susceptibility in incompatible combinations with five races of *H. parasitica*, namely, EMOY, EMWA, CALA, HIKS, and HIND (Glazebrook et al. 1997).

Camalexin, the only *Arabidopsis* phytoalexin so far described, is an indole thiazole derivative, as are other phytoalexins so far reported from the Brassicaceae. Thus, camalexin is synthesized from tryptophan, and not from phenylalanine, which is a precursor for phenolics in plants. This metabolic distinction was made use of to assess the relative contributions of phenolics and phenolic polymers to the resistance of *Arabidopsis* to *H. parasitica*, separately from phytoalexins (Mauch-Mani and Slusarenko 1996). Thus, in *Arabidopsis,* it is possible to inhibit phenolic metabolism without directly interfering with phytoalexin synthesis (Fig. 4.21). Using specific inhibitors of *PAL* (2-aminoindan-2-phosphonic acid, AIP) and cinnamyl alcohol dehydrogenase (2-hydroxyphenyl-aminosulphinyl acetic acid 1,1-dimethyl ester, OH-PAS) to inhibit phenolic metabolism in general or lignifications in particular, it was shown that loss of lignification caused a mild shift toward susceptibility, but that a more general inhibition of the phenolic metabolism resulted in complete susceptibility (Mauch-Mani and Slusarenko 1996). Feeding salicylic acid (SA) back into the system restored resistance in the presence of AIP, and was interpreted as showing the dependence of resistance expression on SA. These results conform to the observations of Delaney et al. (1994) for transgenic *Arabidopsis* carrying the bacterial *nahG* gene, and are therefore unable to accumulate SA. However, despite the overwhelming effect of SA, the contribution of lignification to resistance against *H. parasitica* was shown, and this highlights the multifactorial nature of the resistant response. It has been suggested that the SA that is required for *PR1* gene expression and SAR might be predominantly synthesized via isochorismate whereas the SA which modulates cell death in the HR might arise predominantly from phenylalanine (Fig. 4.21; Wildermuth et al. 2001).

The idea that plants might lignify not only their own walls to strengthen them as barriers to pathogen spread (Hijwegen 1963) but that the pathogens themselves might be inactivated by polymerization of monilignols to lignin in their walls as part of the peroxidase catalyzed intercellular free radical condensation reaction was proposed by Hammerschmidt and Kuc (1982) and Ride (1983). However,

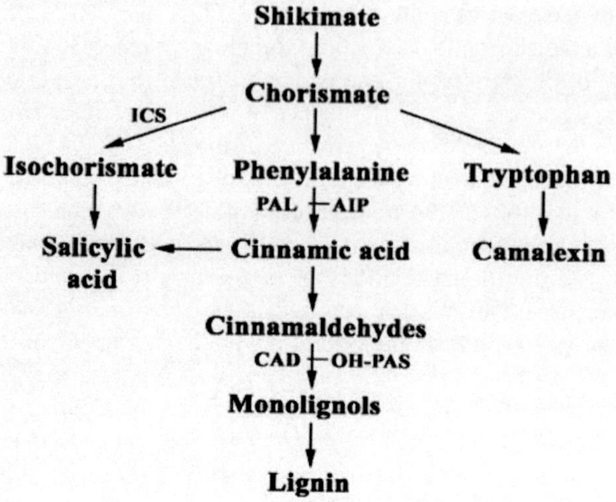

Fig. 4.21 Simplified scheme of the biosynthesis of the defense-related compounds camalexin, salicylic acid, and lignin in *Arabidopsis*. Chorismate is the first branch point, since camalexin arises via tryptophan, while salicylic acid is synthesized via isochorismate and phenylalanine, and lignins arise via phenylalanine. Chorismate is converted to isochorismate by isochorismate synthase (*ICS*). Phenylalanine ammonia-lyase (*PAL*) convert phenylalanine to cinnamic acid, and is specifically inhibited by aminoindan phosphonic acid (*AIP*). From cinnamic acid, the pathway branch to produce salicylic acid or, via cinnamaldehydes, and monolignols, lignin. The conversion from cinnamaldehydes to monolignols by cinnamyl alcohol dehydrogenase (*CAD*) is inhibited by hydroxyphenyl-aminosulphinyl acetic acid dimethyl ester (OH-PAS) (Slusarenko and Schlaich 2003)

Slusarenko and Schlaich (2003) have found evidence for the active lignification of hyphae in intercellular spaces (Mauch-Mani and Slusarenko 1996). *Arabidopsis* host-pathosystem is not an easy subject for biochemical study, and comparative studies are complicated by leaf-age-and ecotype-specific differences in basal enzyme activity levels (Mauch-Mani et al. 1993). After the inoculation of plants with virulent or avirulent isolates of *H. parasitica*, Slusarenko and Schlaich (2003) did not observe a significant pattern of change in the activities of superoxide dismutase, catalase, ascorbate peroxidase, lipolytic acyl-hydrolase, lipoxygenase, or linolenic acid 13-hydroperoxide decomposing activity. All these enzymes have been reported to be important in one or more pathosystems. It is possible that in *Arabidopsis* these enzymes have no role in resistance against *H. parasitica* or, because only relatively few cells show a HR, any changes are diluted out in comparison with the bulk of nonstimulated cells in the leaf. In contrast, transcripts for lipoxygenase were reported to be induced by treatment with *Pseudomonas* (Melan et al. 1993), which leads to a quantitatively greater leaf response than in the *H. parasitica* pathosystem where Slusarenko and Schlaich (2003) found no increased transcript levels using the same probes (Mauch-Mani et al. 1993; Slusarenko 1996). Thus, SA seems to be important for defense against most *H. parasitica* races, presumably because of its role as a signal amplifier, and it seems that lignification and camalexin might be needed for full resistance, since some resistance is lost when lignifications is blocked or in *pad4* or *pad1* + 2, 1 + 3 or 2 + 3 mutants.

4.4.5 Resistance Gene-Mediated Signal Transduction

Signal transduction pathways used by the different *RPP* genes to mediate resistance are still poorly understood, and are complex, and diverse. At one extreme, there is *RPP13*-Nd, for which so far no downstream signaling components are known. It functions independently of EDS1, NDR1, SA or NPR1, which in various combinations are required for the function of several other "R"-genes (Bittner-Eddy and Beynon 2001). At another extreme is *RPP4*, which seems to depend upon almost all defense signaling components so far known. Thus, *RPP4*-conditioned resistance against EMOY or EMWA is compromised in Col-0 plants carrying a mutation in EDS1, PAD4; NDR1, PBS3; SGT1b, RAR1; SID1, SID2, or DTH9, respectively (Table 2.5, Chap. 2). Additionally, it is dependent on NPR1 and the accumulation of SA and even requires a functional *RPS5* gene, which is the "R"-gene required for the recognition of *Pseudomonas syringae* pv. *tomato* isolates carrying the *avrPphB* gene. Essentially, all *RPP* genes seem to have individual downstream requirements; being qualitatively or quantitatively dependent on different combinations of, for example, *SGT1b*, *RAR1*, *EDS1*, *PAD4,* or NDR1 (Table 2.5, Chap. 2). In mutants, a complete loss of resistance is often not observed, but rather a reduction of resistance which can be additive if two mutations are combined. Thus, resistance in the ecotype Ler against the EMCO isolate (mediated by *RPP8*) was initially reported to be independent of *EDS1* and *NDR1* (Aarts et al. 1998) but was later found to be significantly compromised in an eds1/ndr1 double mutant (McDowell et al. 2000). Rather than being a straight forward, a linear path from recognition to resistance expression, the picture which is emerging is of a complicated "lattice" or "grid-like" network of interconnecting signaling circuits (Parker 2000). What is known in terms of a rather "linear" summary of signaling components required for establishment of full resistance as mediated by the different "R"-genes is given in Table 2.5, Chap. 2. Thus so far, signaling through *RPP* gene appears to be independent of COI1, JAR1, or EIN2, components required for jasmonate and ethylene signaling, respectively. These genes are required for resistance to necrotrophic pathogens such as *Alternaria* or *Botrytis* (Thomma et al. 1998). A simplified schematic representation of RPP-conditioned signal transduction (Fig. 2.14, Chap. 2) has been given by Slusarenko and Schlaich (2003).

4.5 *Brassica-Plasmodiophora*: Biometabolomic Resistance

4.5.1 R-Genes Pyramiding and Signaling Pathways

The success of gene pyramiding in a cultivar depends on several factors, such as the number of genes to be transferred and the ease of genotyping and phenotyping the plants. Validation of the number of resistance genes pyramided in a single plant or a cultivar is a challenging task, as a plant carrying one or multiple resistance genes can show similar resistance phenotype unless the different resistance genes confer resistance to different pathotypes. Also, testing a plant for resistance to multiple pathotypes is not feasible in a backcross breeding program. In this regard, development of gene-specific markers for use in marker-assisted selection would facilitate the gene-pyramiding task greatly. Among the cultivated *Brassicas*, resistance to *P. brassicae* in *B. rapa* often shows race-specific reaction controlled by major genes, while resistance in *B. oleracea* is often controlled by multiple QTL. In the case of *B. napus*, some of the swede or rutabaga (*B. napus* var. *napobrassica*) genotypes have been reported to possess resistance to different *P. brassicae* pathotypes.

RNA-seq is technique that allows fast and extensive study of transcriptional levels of gene expression. A number of published transcriptome analysis results of the defense responses of Brassica crops against *P. brassicae* at early infection stages at different time points, and resistant and susceptible lines were compared (Chen et al. 2016a, b; Wang et al. 2019). The results have shown that despite the presence of more robust effector-triggered immunity, the salicylic acid (SA) signaling pathway plays a pivotal role in clubroot resistance in *Brassica rapa*, which contains the *CRb* resistance gene (Chen et al. 2016a), and strongly and rapidly accelerates receptor kinase and G proteins in *B. napus* (ZHE-226) resistance lines that contain the *PbBa8*.1 resistance gene, which may trigger reactive oxygen species (ROS), programmed cell death (PCD), and the hormonal signal transduction signaling pathway (Mei et al. 2019). Previous studies were based on lines that were resistant or susceptible to clubroot isolates, and the resistant lines consisted of single resistance genes, in either a homozygous or heterozygous state at the locus. Shah et al. (2020) pyramided two resistant genes (*CRb* and *PbBa8*.1) and develop a homozygous line. The lines show strong resistance against a number of *P. brassicae* field isolates compared with heterozygous pyramided *B. napus* and single resistant homozygous and heterozygous lines (Shah et al. 2019). Historically, the molecular mechanism of the strong resistance of pyramided lines has been poorly understood. Therefore, a comparative transcriptomic analysis between 618R and its parental lines (305R and 409R) was performed in order to understand the superior resistance mechanisms of the pyramided line and to elucidate the genetic compatibility of the two resistant genes in a single line.

The pyramiding of *PbBa8*.1 and *CRb* resistant genes strongly induces resistance against a number of *P. brassicae* field isolates (Shah et al. 2019). To understand the molecular mechanism of how these two resistance genes elevate resistance in a single line, Shah et al. (2020) selected only one time point to comparatively investigate the genetic coordination of the pyramided line (618R) with its parental lines

(305R and 409R), rather than focus on the change in the transcriptome at different time points after clubroot pathogen inoculation. It was shown that *PbBa8*.1 shares a greater number of DEGs than *CRb* and plays a dominant role, with a higher number of nonadditive genes than the pyramided line. This insight revealed that the pyramided line strongly triggers multiple resistance pathways, including the plant–pathogen interaction, plant hormone signal transduction, secondary metabolites, transcription factors, and other processes, indicating that the plant has responded via a multi-gene network to pathogen attack. Most interestingly, it was shown that the SA signaling pathway was higher in 409R than 305R, and the ROS response was greater than signaling in 305R. However, higher expressions were evaluated in 618R controlled by SA and ROS in the pyramided line. It will provide comprehensive transcriptomic insight into uncovering the interaction of pyramiding resistance genes in a single line, and this architecture further facilitates breeding programs with improved resistance efficiency in *B. napus*.

To cope with *P. brassicae*, a total of 9797, 13,323, and 8807 DEGs were identified in 305R, 618R, and 409R, respectively. qRT-PCR results validated the DEGs selected from transcriptome profiling. The pyramided line, 618R, showed a higher number of DEGs than the parental lines, indicating that 618R may trigger different genes or gene complexes that control the resistance response. However, *PbBa8.1* shared a higher number of DEGs than *CRb*, indicating that it plays a dominant role in the pyramided line, on the challenge of pyramided and parental *B. napus* lines to the clubroot pathogen, *P. brassicae*, pathotype 4. Significant changes in plant–pathogen interaction, cell wall modification, plant hormone signaling, metabolic process, and oxidative and plant defense response were observed. Plants have evolved a series of defense mechanisms in the presence of pathogens during their co-evolution; these mechanisms are instigated by pathogen effectors and host R-proteins. PTI is the first defense mechanism, in which plants detect the pathogen through PAMPs regulated by PRR proteins (Bernoux et al. 2011). PTI is involved in the induction of plant hormone signal transduction and plant–pathogen interaction at the late stage of *B. rapa* resistant lines containing the *CRd* gene when exposed to *P. brassicae* (Fu et al. 2019). In contrast, in the *B. rapa* near-isogenic line (CR BJN3-2) and *B. napus* resistance line (ZHE-226), no significant changes were detected when compared with the susceptible lines after inoculation with *P. brassicae* (Chen et al. 2016a; Mei et al. 2019). The second layer of immunity is the recognition of effectors, which leads to the initiation of ETI. Plants have evolved specific resistance proteins (R-proteins) that recognize pathogenic virulence molecule effectors, leading to ETI (Jones and Dangl 2006). Shah et al. (2020) identified 13 NBS-LRR R-proteins in 618R, which was twofold higher than in any of the parent lines. RPS2, RPS3, and RPS5 were strongly induced in 618R as compared to 305R and 409R. It was reported that RPS2 confers resistance against downy mildew in grapevines (Li et al. 2015). RPM1 (RPS3) and RPS2 also confer resistance against Fusarium wilt in bananas (Sun et al. 2019). The strong upregulation of R-genes in 618R suggests that the pyramiding of resistance genes enhances the ETI response, which plays a significant role in maintaining host resistance.

Calcium is considered to be an imperative secondary messenger in the signal transduction pathway, which plays a vital role in plants in the production of HR in response to biotic stress (Aldon et al. 2018; Wang et al. 2019). The downregulation of CMLs, CIPKs, and CDPKs was highly suppressed in 618R as compared to 305R and 409R. Ca^{2+} also triggers RBOH proteins, which participate in ROS production. A significant upregulation of RBOH factors was found in 618R, followed by 305R, indicating higher ROS activity in the pyramided and 305R lines, inhibiting the invasion and colonization of the clubroot pathogen.

4.5.1.1 Defense Signaling Transduction in Pyramided Lines and Parental Lines

Signaling transduction was activated soon after recognition of the pathogen for the establishment of the defense system. The phytohormones SA, JA, and ET are mainly involved in defense processes (Sun et al. 2019). *P. brassicae* is a biotrophic pathogen, and resistance is mainly controlled by the SA-mediated signaling pathway in this regard (Glazebrook 2005). It was revealed that SA was significantly accumulated in the pyramided and parental lines, reflecting its crucial role in clubroot defense. It is in agreement with previous works, which showed that the SA pathway mediates clubroot resistance in *Brassica* crops (Chen et al. 2016a). The PR protein had a superior impact in the response to *P. brassicae*, serving as a guard to the plants for further infection, the hypersensitive response (HR), or systemic-acquired resistance (SAR) (Jain and Khurana 2018). Shah et al. (2020) identified a significant induction of PR proteins, thaumatin, and lipid transfer proteins in the pyramided line as compared to the parental lines. Additionally, the induction of defensin (PDF) after inoculation with the clubroot pathogen supplemented the PR role in the plant disease resistance response. These findings support the reports showing the upregulation of PR proteins and PDF in plant defense in response to *P. brassicae* in *Brassica oleracea* and *Arabidopsis thaliana* resistant lines (Glazebrook 2005; Jain and Khurana 2018; Irani et al. 2018). PR proteins are involved in the resistance response against the clubroot pathogen and that the pyramiding of resistance genes enhances PR activities. However, SA is a pivotal factor in the SAR response, and the SA signaling pathway is mainly effective for biotrophic pathogens. Nonexpressor of pathogenesis-related (NPR) is a master regulator of plant SAR, conferring immunity, and SAR induction leads to the accumulation of SA signaling molecules, resulting in the induction of defense genes through NPR1 activation (Mou et al. 2003). NPR1 further interacts with TGA, and these TGA factors bind with PR1 gene promoters, which is necessary for the SA response (Mou et al. 2003; Backer et al. 2019). It was observed that *NIM1*, *NIM2*, and *PR1* were significantly upregulated in 618R, followed by 409R and 305R. The suppression of the SA-mediated pathway in susceptible *A. thaliana* and the upregulation of NPR1 and PR1 genes in CR-BJN3-2 were both reported earlier after inoculation with *P. brassicae*. The upregulation of SA signaling transduction was reported in resistant genotypes. It is suggested that enhanced SA levels in host cells suppress *P. brassicae* colonization.

The interaction between SA and JA is antagonistic, and it was observed that genes related to JA signaling and biosynthesis (*LOX3*, *LOX4*) were extremely downregulated in the pyramided and parental lines. However, differential regulations of *LOX2* were detected between 618R, 409R, and 305R. The TIFY protein and *MYC2* are related to JA signaling and are activators, which were significantly downregulated in the pyramided line as compared to the parental lines. Besides this, three genes of MAPK related to JA biosynthesis were profoundly downregulated in the pyramided line as compared to the parental lines. This is in agreement with the claim that JA biosynthesis is expressed in disease resistance to necrotrophic pathogens and is suppressed in biotrophic pathogens (Sun et al. 2019; Bai et al. 2013). The stress-regulating gene involved in the negative regulation of cell death, the BRI1 protein, was strongly downregulated in the pyramided and 305R lines as compared 409R, indicating that *PbBa8*.1 plays a dominant role in the pyramided line in this regard.

ROS is a well-known secondary messenger in cellular processes, and RBOHs are critical enzymes that produce ROS in response to hormonal and environmental signals and are activated by Ca^{2+} (Sagi and Fluhr 2001; Chapman et al. 2019). ROS-associated signaling that suppresses pathogen growth determines PCD and HR (Mei et al. 2019). It also works with other signaling molecules and induces SA and its regulators, the NPR1 and TGA transcription factors (Fu and Dong 2013), and ROS-mediated defense plays a central key role in cross talk with other molecular mechanisms, having great potential for resistant genotype selection (Ali et al. 2018). The six genes related to RBOH, regarded as key initiators of ROS, were strongly induced in 618R and 305R as compared to 409R, indicating a higher ROS ability in *PbBa8*.1 than in *CRb*. An elevation of RBOH in resistant lines is in agreement with previous studies in *B. napus* and *B. oleracea* (Wang et al. 2019; You and Chan 2015). PERs maintain ROS homeostasis in plant cells and restrict ROS-dependent damage or finely tune ROS-dependent signal transduction with regard to downregulated scavenging systems (Torres et al. 2006). Five genes related to PER were strongly downregulated in 618R as compared with the parental lines; however, *PERP7* was only repressed in 618R. The antioxidant enzymes GST, SOD, APX, and CAT, which play a significant role in the scavenging of ROS, were strongly regulated after the inoculation of *P. brassicae*. SOD activity was greatly induced in 618R and 305R as compared to 409R. APX, which maintains ROS homeostasis in plant cells and restricts ROS-dependent signaling transduction, was downregulated in the pyramided and 305R lines, while CAT3, which protects cells from ROS, was profoundly expressed in 618R. However, higher ROS activity in the resistance NIL line CR-BJN3-2 (Chen et al. 2013) was observed, and that *PbBa8*.1 played a stronger role than *CRb*.

4.5.1.2 Transcription Factor (TF) Response to *Plasmodiophora Brassicae*

TFs respond soon after the pathogen attack in defense signaling transduction. The *ERF*, *WRKY*, *AUX/IAA*, *MYC2*, and *MYB* were identified during the plant–pathogen interaction. ERF TFs bind to a GCC-motif in the promoter of JA and ET responsive genes, thereby positively or negatively regulating gene expression (Caarls et al.

2016). Elevation of *ERF11* was observed in 618R-T and 409R-T, and negative regulation was ascertained in 305R-T, indicating that there may be negative regulators in resistance lines containing *PbBa8*.1 for the purpose of the downregulation of the JA signaling pathway. Negative regulations of ERF11 JA-responsive defense expressions were reported in *B. napus* when exposed to *Fusarium oxysporum* (Hejna et al. 2019). WRKY TFs bind with downstream elements and play an important role in biotic and abiotic responses (Hussain et al. 2019; Xiong et al. 2019), whereas the *WRKY70* TF represents an antagonistic interaction between SA and JA (Li et al. 2004). It positively regulates the SA-mediated signaling pathway and negatively regulates the JA-mediated pathway, and together with *WRKY TF*-54, 53, and 46, it promotes the defense response in biotrophs by inducing SA-mediated PR gene expression (Li et al. 2004, 2006). The *WRKY70* TF was greatly induced in 618R-T and 409R-T as compared to 305R-T after inoculation with *P. brassicae*, indicating the induction of SA signaling and the repression of JA signaling genes induced by PR gene expression, also enhancing resistance in *B. napus* lines. An inconsistency was observed for auxin-related TFs, such as *SAUR* genes, where downregulation was observed in *IAA18*. *MYC2* genes were strongly downregulated in the pyramided and 305R lines as compared to 409R, and related genes were also inconsistent in their regulations. However, these factors are also involved in plant growth and development, and the regulation of these genes might be dependent on the pathogenic response over different time periods. Overall, these results indicate that TFs play an important role in plant defense and that the pyramiding of *PbBa8*.1 and *CRb* enhances both the expressions of genes and their number.

PbBa8.1 shares a greater number of DEGs than *CRb* and plays a dominant role in the pyramided line. Comparative correlation analysis of different log-fold changes demonstrated a higher number of nonadditive genes in the pyramided line. This insight revealed that the pyramided line strongly triggers multiple resistance pathways, including the host–pathogen interaction, plant hormone signal transduction, secondary metabolites, transcription factors, and other processes, indicating that the plant responds via a multi-gene network to pathogen attack. Comparative analysis revealed that, among the DEGs, some DEGs were differentially regulated and expressed in the pyramided line, perhaps due to cross talk or overexpression. The SA signaling pathway was stronger in 409R, and ROS response was greater than signaling in 305R. However, higher expressions were evaluated in 618R, indicating that the resistance to *P. brassicae* was mainly controlled by SA and ROS in the pyramided line. These results provide a comprehensive transcriptomic insight into uncovering the interaction of the pyramiding of resistance genes in a single line, and this architecture further facilitates breeding programs with the aim of improved resistance efficiency in *B. napus* (Shah et al. 2020).

4.5.2 Lignin Biosynthesis

Lignification is a mechanism that involves the enhancement of the cell wall in order to resist infection by foreign pathogens (Vance et al. 1980). Siemens et al. (2006) reported that some *A. thaliana* genes involved in cell division and expansion were upregulated 10 and 23 days after *P. brassicae* inoculation. CCoAOMT was suppressed at 48 hpi (Cao et al. 2008). The expression of *4CL*, *CCR1*, and *CAD5*, which encode enzymes involved in the lignin biosynthesis pathway, was suppressed 7 days after *P. brassicae* inoculation (Agarwal et al. 2011; Jubault et al. 2013). The early stage of infection by *P. brassicae*, *CCR1* and *CCoAOMT* were upregulated and *CAD5* was downregulated at 24 hpi, while *CAD6* and *CAD9* were upregulated at 48 hpi. Following the activation of the lignin (phenyl propanoids) biosynthesis pathway, lignin precursors (coumaryl alcohol, coniferyl alcohol, and sinapyl alcohol) accumulated. At the same time, the expression of many cell wall-related genes was also enhanced (Fig. 4.22). Lignification has been proved to induce systemic resistance in cucumber (Hammerschmidt and Kuc 1982); cell wall hydroxyproline enhancement and lignin deposition were reported to participate the resistance of cucumber to *Cladosporium cucumerinum* (Hammerschmidt et al. 1984); cell wall-bound phenolic acid and lignin contents were related to the resistance of date palm to *Fusarium oxysporum* (El Modafar and El Boustani 2001). The lignin begins to accumulate in order to enhance resistance to *P. brassicae* at the early stage of infection.

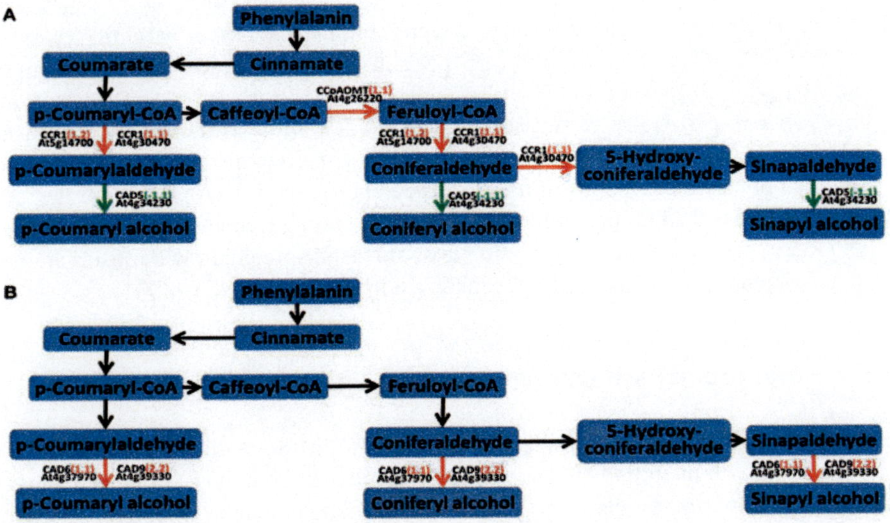

Fig. 4.22 Lignin pathway analyses of DEGs in *Arabidopsis thaliana* during early infection by *Plasmodiophora brassicae*. Analysis of the lignin (phenylpropanoid biosynthesis) pathway of DEGs was performed using MAPMAN software. Red arrows mean upregulated genes and green mean downregulated. Black ones mean no change. (**a**) 24 h after inoculation; (**b**) 48 h after inoculation. The pathway frames are from the MAPMAN software database (Zhao et al. 2017)

4.5.3 Flavonoids as Defense Compounds or Antioxidants

Flavonoids are known to be stress protectors, signaling molecules for the establishment of symbiosis, and defense compounds (Dixon 2001; Ryder et al. 1987; Wasson et al. 2009). Flavonoid accumulation might also be associated with defense reactions in the case of clubroot development. Careful monitoring of the disease parameters revealed differences at low infection pressure, and several of the mutant lines were slightly more tolerant. This could be interpreted as a role of flavonoids as pathogenicity factors. When simultaneous abiotic stress factors were applied, the flavonoid-deficient mutant *tt4* was more affected than L*er* in shoot growth. The Buer and Djordjevic (2009) described aberrant phenotypes on various *tt* mutants when grown under stress conditions. On the other hand, the root-deficient mutant *myb12-KO* was not affected in shoot growth when compared with the wild-type. Flavonoids might therefore be necessary to protect against simultaneous aboveground stress factors. In addition, flavonoids could act as antioxidants that scavenge reactive oxygen species, which could be a result of biotic, but also abiotic, stresses associated with various cellular dysfunctions.

Protection against oxidative stress or oxidative burst may be likely, because no indications for a hypersensitive response were found in *Arabidopsis* galls (Siemens et al. 2002, 2006). Moreover, their role as antioxidants in root galls might be connected with the degradation of IAA. In white clover, flavonoids were postulated to be involved in the regulation of auxin breakdown by peroxidase (Mathesius 2001).

1. Many flavonoids have antioxidative activity, but naringenin, in particular, which accumulates at high concentrations in root galls, does not show high antioxidative activity with DPPH.
2. Horseradish peroxidase catalyzes IAA degradation in vitro and horseradish peroxidase can be inhibited by flavonoids in vitro. However, when IAA was simultaneously incubated with flavonoids and peroxidase, IAA was not protected from degradation. On the contrary, flavonoids seem to accelerate the degradation of IAA by horseradish peroxidase in vitro. This is in agreement with observations that formonetin accelerates IAA breakdown in vitro (Mathesius 2001).

4.5.4 Metabolic Pathway Mediated by Clubroot Resistance (CR)

Several processes of primary metabolism were also differentially regulated in CR mediated by *Rcr1*, including enhanced MET and reduced glycolysis. Plant defense is an energy-consuming process resulting from enhanced demand on certain metabolism and cytological activities. Glycolysis and MET are two major pathways for plant respiration which can be stimulated during defense responses to generate ATP and carbon skeletons. It has been demonstrated that glycolysis is contributing to disease resistance in several plant–pathogen interactions. However, the expression of proteins related to glycolysis, especially the major regulator phosphofructokinase

(PFK), was reduced while those related to MET were upregulated in resistant samples (Song et al. 2016). This result suggests that MET serve as the major energy source required for defense responses mediated by *Rcr1* against clubroot. It is also possible that the downregulation of glycolysis serves as another defense response by inhibiting certain pathogen-induced metabolism (carbohydrate or auxin) favoring excessive cell growth and division. During a compatible interaction between *Arabidopsis* and *P. brassicae*, an increased carbon flow toward glycolysis was observed (Jubault et al. 2013; Schuller et al. 2013).The pathogen-induced arginine catabolism appeared to be inhibited in resistant samples. In *Arabidopsis*, proline deriving from the arginine catabolism accumulated substantially in a susceptible ecotype upon *P. brassicae* infection while such an accumulation was not observed in the partially resistant ecotype Bur-0 (Jubault et al. 2008). The abundance in 11-pyyroline-5-carboxylate synthase (Bra005012) was decreased in resistant samples. This enzyme catalyzes the rate-limiting step of proline biosynthesis responding to environmental stresses (Zheng et al. 2014). Event woof the enzymes involved in arginine biosynthesis were upregulated, and the downregulation of this key enzyme in proline biosynthesis may indicate that the *P. brassicae*-induced arginine catabolism and proline biosynthesis are both disrupted in resistant samples carrying *Rcr1* (Song et al. 2016).

4.5.5 Induction of Signal Molecules for Defense Response

Salicylic acid (SA) is a crucial signal molecule for the activation of plant defense responses. The regulation of SA in the plant cell is achieved, in part, by its maintenance in a range of different forms. SA is rendered largely inactive in localized defense induction by SA glucosyltransferase-mediated conjugation to glucose (forming SA β-glucoside) or methylation to form methyl salicylate (MeSA). Methylation of SA from the *S*-adenosine-L-methionine (SAM) methyl group donor is achieved by members of the distinct SABATH group of methyltransferases. These enzymes have been shown to have similar and/or overlapping methylation activity against both SA and benzoic acid (BA). In addition to SA and BA, other SABATH enzymes act on diverse secondary metabolites, including jasmonic acid (JA), indole-3-acetic acid (IAA), and gibberellic acid.

The role of MeSA in plant defense and its interrelationship with benzoic acid/salicylic acid methyltransferase (BSMT) enzyme activity are not precisely understood. Because of its greater hydropho-debicity than SA, MeSA is more volatile and membrane permeable and hence can move more easily between cells. MeSA can be converted back to free, active SA via the activity of a MeSA esterase. Evidence that MeSA is a long-distance systemic acquired resistance (SAR) signal has been presented in tobacco. MeSA has been shown to be fully or partially required for the development of SAR, depending on the amount of light following infection. Both biotic and abiotic stresses induced the transcription of *Arabidopsis AtBSMT1* in leaves. Studies of *Arabidopsis* containing knockouts of *AtBSMT1* showed a near complete reduction in MeSA production following pathogen infection.

Overexpression of the rice *OsBSMT1* gene in *Arabidopsis* led to reduced SA accumulation, constitutive MeSA production, and greater susceptibility to attack by both *Pseudomonas syringae* and *Golovinomyces orontii* (Ludwig-Muller et al. 2015).

Much research into pathogen infection strategies has focused on protein molecules that are secreted into cells to shut down plant defenses, redirect nutrient flows, promote tissue colonization and influence signaling. Bacterial proteins are the best understood effectors, and roles for a number have been confirmed, including proteases and E3 ligases. By contrast, the molecular mechanisms of effectors from eukaryotes are less well understood; avirulence proteins from oomycetes and fungi have been shown to translocate into cells, but evidence for specific biochemical roles for these molecules is accumulating slowly. It is presumed that *P. brassicae* must secrete effectors into host cells to cause the wholesale cellular and tissue rearrangements seen in clubroot disease. However, the understanding of the molecular biology of *P. brassicae* is rudimentary. Of the 76 genes examined in the principal study investigating full-length *P. brassicae* cDNAs to date, 17 had predicted signal sequences, but few, if any, of these presented a clear role in pathogenesis based on sequence similarity. A single *P. brassicae* gene from a subsequent collection has been characterized in some detail at the biochemical level; a putatively secreted protein, Pro1, was shown to have in vitro serine protease activity, although the actual substrate for this protease in planta remains unidentified. Furthermore, the addition of Pro1 to canola (*Brassica napus*) root exudates increased the positive effect of these exudates on the germination rate of resting spores, indicating a positive role for Pro1 during spore germination. The first report of *P. brassicae* transformation has been published, but it has not been possible so far to demonstrate the expression of the transgenes in *P. brassicae*.

In contrast with most biotrophic eukaryotes, in which feeding structures are bordered by the plant cell membrane, the intracellular growth of *P. brassicae* means that secreted proteins are likely to enter the plant cytoplasm directly. The comprehensive data on SA metabolism and transport during clubroot disease, and SA methyltransferase from *P. brassicae* by which this pathogen may manipulate SA in host cells has been elucidated by Ludwig-Muller et al. (2015). Plant infection by phytomyxids requires a sophisticated subversion of host cells. In *P. brassicae*, much attention has been paid to the regulation of the plant hormones auxin and cytokinin during clubroot establishment. Less is known about SA, ethylene, and JA signaling, which are important for plant defense in other plant–pathogen interactions. SA is a pivotal signal in local and systemic defense. However, our understanding of SA signaling during interactions with plant root pathogens lags behind that for foliar pathogens. Although the SA level in control plants decreased with time, in *P. brassicae*-infected plants it increased and this was particularly obvious in the leaves. At 28 dai, both roots and leaves of infected plants contained more SA than those from control plants; such large-scale SA accumulation occurs in a number of plant–pathogen interactions, although it is unclear whether this is a plant defense signaling process or a secondary effect of cellular stress (Loake and Grant 2007). MeSA was emitted from leaves and moved from roots to the upper plant tissue.

Furthermore, it was possible to inhibit MeSAD4 to SAD4 conversion in infected plants using the artificial analogue TFA, which competitively inhibits the activity of the MeSA esteraseSABP2 (Park et al. 2007). The levels of labeled SA found in infected plants could derive from a plant esterase activity. Indeed, the addition of the esterase inhibitor resulted in lower SAD4 and higher MeSAD4 amounts (Fig. 4.23). This finding suggests that MeSA found in leaves partially originates from the site of *P. brassicae* infection, raising the question of whether this MeSA is produced by *Arabidopsis*, by *P. brassicae* itself or by the action of both organisms (Fig. 4.24).

Similarity searches showed that proteins from a wide range of organisms contain the pfam03492 domain. At a primary sequence level, the *P. brassicae* BSMT protein is only distantly related to the plant SABATH enzymes, and phylogenies constructed from full-length protein sequences gave few clues as to the functions of these genes; the relationships between clusters of sequences were unstable and the plant enzymes formed a discrete cluster from other organisms. The *PbBSMT* sequence did not group robustly with any other sequences, consistent with the phylogenetic separation of Rhizaria from other studied organisms, and a long period of vertical rather than horizontal evolution.

Despite these phylogenetic distances to known methyltransferases, *PbBSMT* heterologously expressed in *E. coli* catalyzed the methylation of SA, BA, and AA, confirming that a protein from an organism outside of Plantae has SABATH-like methyltransferase activity. The activity of *PbBSMT* against these particular substrates, but not against a range of other molecules, was consistent with protein alignments and modeling, which showed that *PbBSMT* had higher structural similarity to the crystal structures of CbSAMT than to the closely related AtIAMT (Zhao et al. 2008). Most SA-methylating enzymes have preference for the SA substrate, whereas the conversion of BA and/or AA is less pronounced (Huang et al. 2012). The AtBSMT enzyme differs from PbBSMT in showing only moderate methylation of BA, AA, and 4-hydroxybenzoate (Chen et al. 2003).

Quantitative PCR assays, as well as evidence from an expressed sequence tag (EST) study (Bulman et al. 2006) and callus culture transcriptome work (Bulman et al. 2011; Burki et al. 2010), indicated that *PbBSMT* transcripts were at low levels during early infection and highest when galls were present. At 4 dai, the *PbBSMT* levels were near the detection limit of the quantitative PCR method. By 7 dai, an increase in *PbBSMT* transcription was observed, which is not inconsistent with a role in the modification of SA, as initial *P. brassicae* penetration of the root cortex, signaling the onset of secondary infection, can be detected as early as 5 dai (Siemens et al. 2009b). Although it can be argued that the expression of a *P. brassicae* gene throughout the infection process is consistent with a role in the modification of plant defenses (Feng et al. 2010), increased expression of *PbBSMT* later in infection is congruent with its postulated role; if PbBSMT hinders SA-mediated plant defense, it would indeed be most required during secondary infection when that response is strongest. A variety of studies have suggested that initial primary infection by plasmodiophorids meets a limited response from the plant (Siemens et al. 2009a). That PbBSMT methylation of SA would be a plausible mechanism for rendering SA inactive in triggering defense mechanisms (Seskar et al. 1998) is supported by a

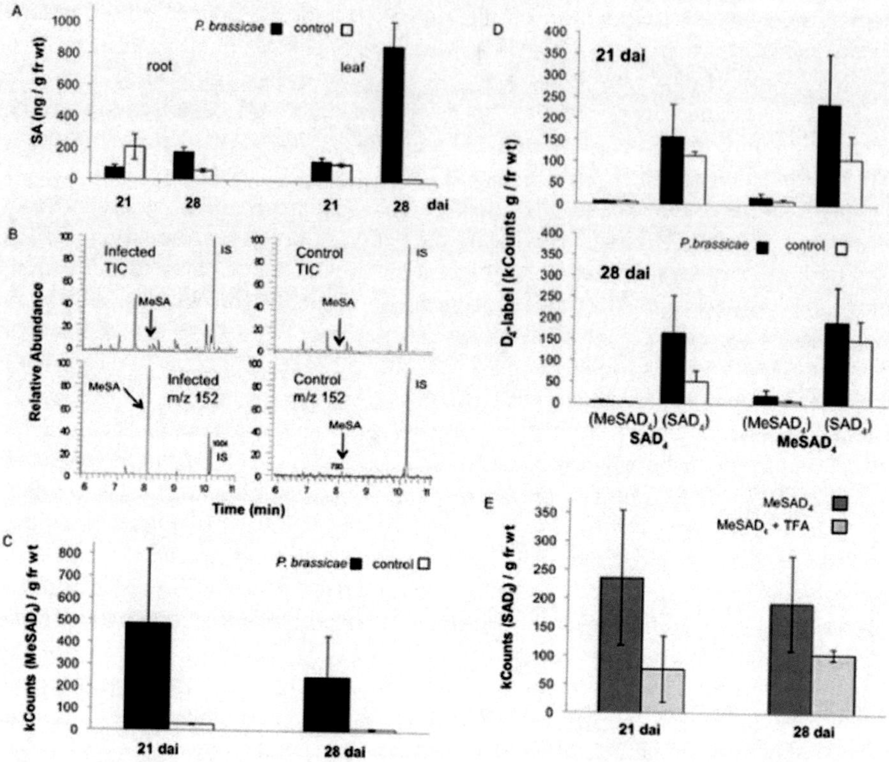

Fig. 4.23 Analysis of salicylic acid (SA) and methyl salicylate (MeSA) transport in *Plasmodiophora brassicae*-infected and control *Arabidopsis*. For disease progression, see also photographs in Plate 4.10. (**a**) Gas chromatography-mass spectrometry (GC-MS) measurement of SA, for two time points, in roots and leaves of infected and control plants. The results are means of three experiments. (**b**) Emission of volatiles determined by GC-MS from leaves of infected and control plants at 28 days after infection (dai). The two right panels show a control sample; the two left panels show a sample from infected roots. The measurement was repeated twice and similar results were obtained. The two top panels show the total ion chromatogram; the two bottom panels represent the selected ion current of *m*/*z* 151–152 which is typical for MeSA. MeSA was identified by its mass spectrum; it was found in infected plants; in controls, no MeSA could be determined. The complete spectra are shown in (**c**) The amount of deuterium-labeled MeSA (MeSAD4) in the roots of infected and control plants after incubation (for 5 h) with deuterium-labeled SA (SAD4). MeSAD4 = [2H4] MeSA; SAD4 = [2H4] SA. The measurements were performed at 21 and 28 dai. The experiment was repeated three times. (**d**) The amount of D4-labeled compounds (SAD4 and MeSAD4) in the leaves of infected and control plants at 21 and 28 dai. SAD4, roots were incubated in [2H4] SA; MeSAD4, roots were incubated in [2H4] MeSA. Data are means ± SE of three independent experiments. (**e**) The amount of SAD4 in leaves of infected *Arabidopsis* after the incubation of roots with either MeSAD4 or MeSAD4 and esterase inhibitor (TFA). The differences were significant between treatments in (**e**) at $P \leq 0.05$. The experiments in (**c**–**e**) were repeated three times with independently cultivated plants over different seasons (Ludwig-Muller et al. 2015)

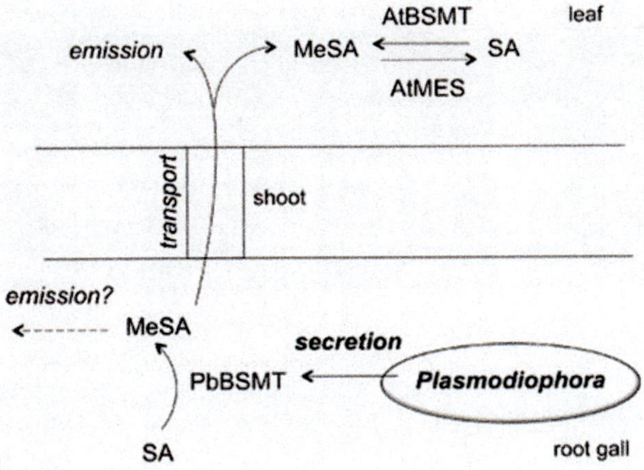

Fig. 4.24 A model showing the possible role of PbBSMT in salicylic acid (SA)-dependent plant defense based on our findings. *Plasmodiophora brassicae* secretes PbBSMT into the host cell, where it methylates the defense signal SA. The resulting methyl salicylate (MeSA) fails to upregulate plant defense reactions and is transmitted to leaves, where it is emitted or converted back to SA (Ludwig-Muller et al. 2015)

number of other studies. Following the overexpression of *OsBSMT* in *Arabidopsis*, MeSA levels were elevated, SA levels were reduced and susceptibility to infection by a biotrophic pathogen was increased (Koo et al. 2007). Similarly, overexpression of *AtBSMT1* in *Arabidopsis* resulted in an increase in MeSA and compromised SAR (Liu et al. 2010). Knockdown of SAMT in tomato plants led to reduced MeSA production and increased resistance to the root pathogen *Fusarium oxysporum* (Ament et al. 2010).

In accordance with a function for SA in defense against *P. brassicae*, treatment of *Arabidopsis* roots with 0.5 mM SA before inoculation completely reduced clubroot symptoms (Agarwal et al. 2011). The possible role for PbBSMT in the methylation of SA is complemented by the additional observation that the At*bsmt1* mutant did not show altered disease symptoms, suggesting that AtBSMT is not responsible for the SA to MeSA conversion seen during clubroot disease. Measurement of SA metabolism in the *Atbsmt* mutant awaits confirmation, as does the possible role in SA methylation of other *Arabidopsis* proteins, which must also be considered. The proposed model for the function of PbBSMT in the light of MeSA transport and emission from *Arabidopsis* is shown in Fig. 4.24. It summarizes the idea that the plant's defense route via SA can be altered by an SA methyltransferase from the protist, so that MeSA is the major transport form and finally emitted from leaves. However, future experiments, including more mutants and complementation of plants with the *PbBSMT* gene, could lead to an understanding of this mode of action.

With PbBSMT, the second protein with possible function in pathogenicity from *P. brassicae* was characterized. Although the protease Pro1 has been proposed to play a role in a completely different process, that is, germination of resting spores via its proteolytic activity (Feng et al. 2010), PbBSMT is active during later stages of the

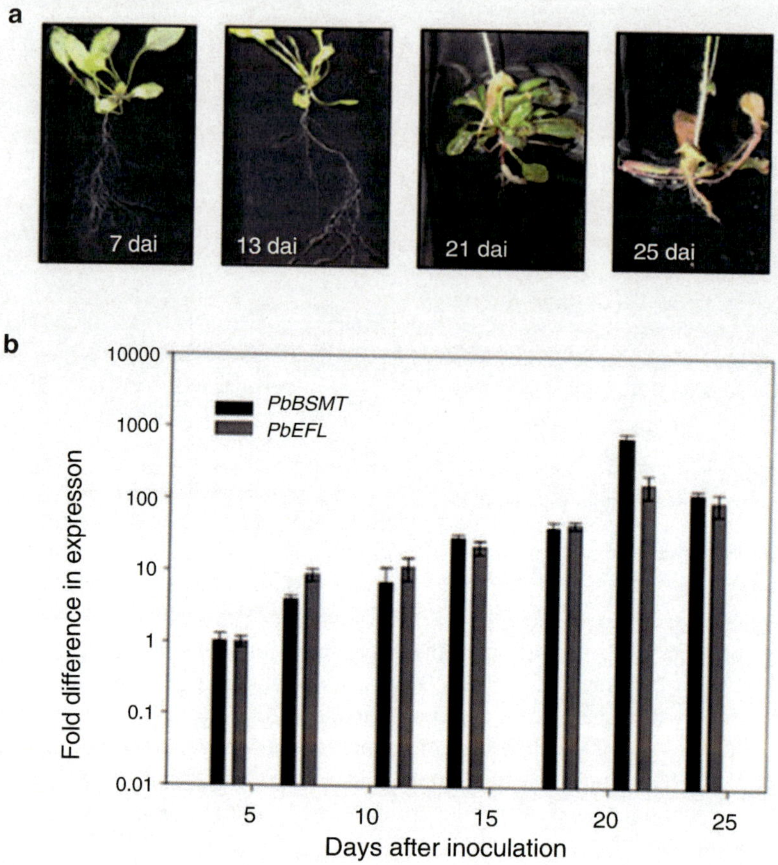

Plate 4.10 Transcript analysis of *PbBSMT* during infection in roots of *Arabidopsis* plants at 4–24 days after infection (dai). (**a**) Time course of disease development. Examples of roots of infected plants at 7, 13, 21, and 25 dai are shown because of their distinctive stages in clubroot symptom progression. (**b**) Samples were analyzed for *PbBSMT* and *PbEFL* expression by quantitative polymerase chain reaction (PCR) at different time points after inoculation (dai). Fold difference in expression compared with day 4 normalized to *Arabidopsis glyceraldehyde-3-phosphate dehydrogenase* (*GAPDH*) is shown. Normalization of data on *Arabidopsis Actin8* (data not shown) gave similar results (Ludwig-Muller et al. 2015)

life cycle, as presumed from the expression analysis (Plate 4.10). Although methylation of SA via PbBSMT would be a novel pathogenicity approach, studies of other microorganisms have pointed to related infection strategies. The *Pseudomonas syringae* virulence factor coronatine functions in part to reduce SA-mediated plant defenses; it has been shown to repress isochorismate synthase gene 1 and to activate AtBSMT1, thus increasing the production of the volatile MeSA and decreasing SA at infection sites (Attaran et al. 2009; Zheng et al. 2012). The smut fungus *Ustilago maydis* secretes an effector encoding a chorismate mutase into maize (Djamei et al. 2011).

This effector converts chorismate to prephenate, making chorismate less available as a precursor for plant SA synthesis. Hence, although *U. maydis* has been proposed to indirectly modulate SA levels, *P. brassicae* may directly metabolize SA, as the protein has a putative secretion signal and could therefore act in the host cell. A pfam03492 motif was found in a gene from the phytopathogenic bacterium *Pectobacterium wasabiae*, and Nykyri et al. (2012) postulated that this may be a horizontally acquired plant gene involved in pathogenesis. Although this represents an excellent candidate for further characterization, as phylogenies indicated that this gene is not horizontally acquired from plants.

As PbBSMT is active against more than one substrate, it has another target in planta. Nevertheless, it is presumed that AA is not this target as MeAA was not emitted from infected plants. Similarly, it is possible that the signal sequence does not target the PbBSMT enzyme to the plant cytoplasm, but instead to a location within the *P. brassicae* plasmodia. As plasmodia are in close contact with the host cytoplasm, a role for detoxification of SA can still be made. Unlike many other candidate effectors, the primary DNA sequence provides a tantalizing clue as to the function of the PbBSMT protein. It was confirmed that PbBSMT has in vitro SA methylation capability and has provided supporting evidence that such activity by PbBSMT may be important during infection. Future studies to confirm the role of this protein during clubroot progression will include the localization of PbBSMT in infected plant cells (Catanzariti et al. 2006; Kemen et al. 2005) and the transformation and analysis of *Arabidopsis* mutants (*Atbsmt1*) with the *PbBSMT* gene for a more in-depth characterization of the in planta function of this protein (Ludwig-Muller et al. 2015).

4.5.5.1 Salicylic Acid-Mediated Immunity in *Brassica* Cultivars to *Plasmodiophora*

Genetic and molecular studies to increase resistance to *P. brassicae* have traditionally focused on finding resistance genes or resistance loci that can be introgressed into cultivars by classical breeding approaches using diverse marker techniques. Two such genes (*Crr1* and *Crr2*) were first identified in *Brassica rapa* L. in an F2 segregating population, showing that they acted together to confer resistance (Suwabe et al. 2003). *Crr1* was subsequently characterized as consisting of two genes, *Crr1a* corresponding to a TIR-NBS-LRR protein and a second locus (*Crr1b*) that was not characterized, but which seems to have a minor effect on resistance (Hatakeyama et al. 2013). Another resistance gene (*CRa*) was originally introgressed from turnip (*B. rapa* ssp. *rapifera*) and mapped by linkage analysis in Chinese cabbage (*B. rapa* ssp. *pekinensis*) (Matsumoto et al. 1998), and then fully characterized as a TIR-NBS-LRR protein (Ueno et al. 2012). The *Rcr1* locus, which was linked to clubroot resistance in *B. rapa* ssp. *chinensis*, was associated with differential expression of defense mechanisms in genotypes with or without the allele (Chu et al. 2014). The region encompassing this marker includes at least 35 genes, of which 2 (R-genes) have the larger number of polymorphisms associated with differences in resistance (Yu et al. 2016).

Mapping in *B. napus* has shown different sources of resistance located on diverse chromosome locations depending on the specific genotype/cultivar–isolate interaction. Such resistance can come from a major gene (Manzanares-Dauleux et al. 2000; Hasan and Rahman 2016; Zhang et al. 2016), or from several QTLs (Jung 2008; Li et al. 2016) that do not necessarily match resistance genes found in progenitor genomes or in other genomes from the family. This fact is probably related to genetic rearrangements which are common during polyploidization events (Szadkowski et al. 2011; Xu et al. 2012). For example, in *Brassica oleracea* L., some QTLs were found to co-localize with previously identified CR loci for that species, but are not analogous to CR loci from *B. rapa* (Lee et al. 2016). In *A. thaliana* (L.) Heynh, several QTLs related to resistance do not map to previously characterized monogenic resistance in a different genotype (Jubault et al. 2008), and QTL regions encompass R-genes, as well as genes related to the cell wall, antimicrobial compounds, the oxidative burst, auxin metabolism, cell expansion, and root architecture. These data support a mix of quantitative and qualitative resistance that is highly dependent on the host–pathogen interaction, and highlights the need for exploring key molecular regulatory mechanisms beyond R-genes.

Transcriptome studies in the clubroot pathosystem have allowed the generation of a molecular landscape in susceptible and resistant interactions, which can serve as a scaffold for understanding the genomic basis of resistance. Studies analyzing host transcriptomic changes first used microarrays, but later switched to RNA-seq analysis, which allows discovery of novel transcripts and has a larger dynamic range for transcript level detection. Microarray analysis of the clubroot susceptible *A. thaliana* ecotype Col-0 challenged with *P. brassicae* showed transcriptional changes in cell growth, carbohydrate and defense mechanisms, and modulation of genes related to auxin, cytokinin, and brassinosteroid metabolism (Siemens et al. 2006; Agarwal et al. 2011; Schuller et al. 2014). An alternate approach using microarrays was undertaken by using *A. thaliana* Bur-0, which showed partially resistant and susceptible reactions, respectively, to two different *P. brassicae* isolates (Jubault et al. 2013). The study indicated that the partially resistant interaction delays pathogen control of host metabolism and induces stronger or earlier defense responses.

The incremental shift toward use of RNA-seq has validated some of the results from microarrays, but also increased the level of information in genome-scale studies. One of the early studies with RNA-seq used *B. rapa* ssp. *chinensis* lines segregating for the clubroot resistance gene *Rcr1* (Chu et al. 2014). Resistant lines carrying the R-gene upregulated jasmonate and ethylene metabolism, reinforced cell walls through callose deposition, and used indole-containing metabolites as defense mechanisms. The studies with *B. rapa* ssp. *pekinensis* highlighted how resistant interactions involve receptors of pathogen recognition, pathogenesis-related proteins, salicylic acid signaling, and secondary metabolism as key elements of defense (Chen et al. 2016a; Jia et al. 2017). In *Arabidopsis*, responses at 24 and 48 h after inoculation (period of primary infection by *P. brassicae*) of the susceptible ecotype Col-0 included secondary metabolite synthesis (flavonoids), lignification, hormone signaling (auxins and cytokinins), and receptor-like kinase activation (Zhao et al. 2017). Surveys of the same interaction at 17, 20, and 24 days post inoculation (dpi) in shoots and roots showed downregulation of primary metabolism (photosynthesis, carbohydrate metabolism) and cell wall modification, while an upregulation of secondary metabolism-related genes (glucosinolates, camalexin, phytoalexins) was detected mainly in shoots (Irani et al. 2018). Roots showed features of disease development through upregulation of cell wall modification enzymes and auxin response genes, resulting in cell growth, division, and expansion.

As RNA-seq becomes an important tool in understanding the molecular mechanisms of defense against *P. brassicae*, specific interactions between particular pathotypes and host genotypes can now be studied using this technology. Addressing specific pathotype–host genotype interactions is necessary to dissect constitutive and induced defenses and how these mechanisms may be altered when new interactions arise. Furthermore, a focus on understanding the responses of *B. napus* to specific pathotypes would enable a direct assessment of important genes in this host, rather than looking for homologous genes in close relatives. Selecting candidate genes in B. napus for functional validation, or alteration of resistance through gene editing, can shorten the time required for the development of new genotypes/cultivars with increased host resistance.

Galindo-Gonzalez et al. (2020) followed disease progression through a time-course in two *B. napus* genotypes having a contrasting interaction with pathotype 5X of *P. brassicae*, as classified on the Canadian Clubroot Differential set (Strelkov et al. 2018). This pathotype can overcome the resistance in most CR canola cultivars and is one of numerous novel pathotypes of the clubroot pathogen identified in western Canada in recent years (Strelkov et al. 2016). The study shows that the host (especially the moderately resistant genotype) relies heavily on salicylic acid-mediated immunity and that several defense mechanisms mimic defense mechanisms of plants when faced with infection by nematodes. The research presents complete description of the molecular responses in this specific interaction, outlining previously reported key defense genes, and shedding light on candidate genes that behave as defense and susceptibility factors.

The clubroot is managed most effectively by planting resistant host varieties, new pathotypes of *P. brassicae* have emerged that can overcome the resistance. Whole genome analyses provide both a toolbox and a systemic view of molecular mechanisms in host–pathogen interactions, which can be used to design new breeding strategies to increase *P. brassicae* resistance. Galindo-Gonzalez et al. (2020) used RNA-seq to evaluate differential gene expression at 7, 14, and 21 days after inoculation (dai) of two *B. napus* genotypes with differential responses to *P. brassicae* pathotype 5X. Gall development was evident at 14 dai in the susceptible genotype (the oilseed rape "Brutor"), while gall development in the resistant genotype (the rutabaga (*B. napus*) "Laurentian") was limited and not visible until 21 dai. Immune responses were better sustained through the time-course in "Laurentian," and numerous genes from immune-related functional categories were associated with salicylic acid (SA)-mediated responses. Jasmonic acid (JA)-mediated responses seemed to be mostly inhibited, especially in the resistant genotype. The upregulation of standard defense-related proteins, like chitinases and thaumatins, was evident in "Laurentian." The enrichment, in both host genotypes, of functional categories for syncytium

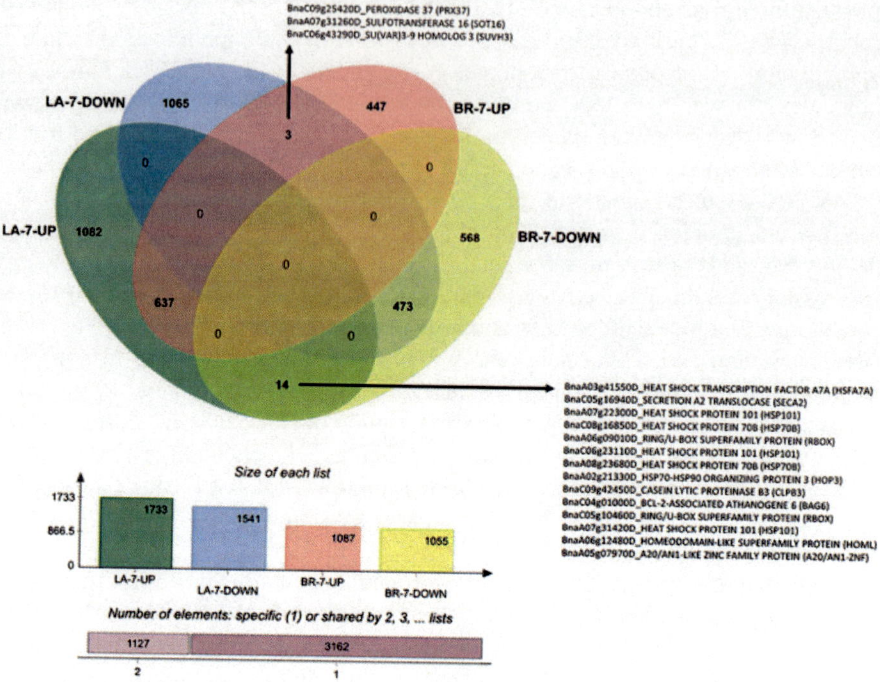

Fig. 4.25 Venn diagram of DEGs 21 dai. The diagram depicts the number of genes with common, opposite, and distinct expression patterns in *Brassica napus* "Laurentian" and "Brutor" in response to *Plasmodiophora brassicae* pathotype 5X. The annotation of genes that demonstrated opposite expression between the genotypes is shown. *LA* Laurentian, *BR* Brutor (Galindo-Gonzalez et al. 2020)

formation and response to nematodes indicated that cell enlargement during *P. brassicae* infection, and the metabolic processes therein share similarities with the response to infection by nematodes that produce similar anatomical symptoms. An analysis of shared genes between the two genotypes at different time-points, confirmed that the nematode-like responses occurred earlier for "Brutor," along with cell metabolism and growth changes. Additionally, the susceptible cultivar turned off defense mechanisms earlier than "Laurentian" (Fig. 4.25). Collectively, the study showed the importance of SA in triggering immune responses and suggested some key resistance and susceptibility factors that can be used in future studies for resistance breeding through gene-editing approaches (Galindo-Gonzalez et al. 2020).

4.5.6 Expression of Hormones-Related Genes During Biometabolomic Resistance

4.5.6.1 Cytokinin-Related Genes

Gene expression was followed in the later stages of *P. brassicae* infection on *B. napus* cvs differing in resistance between 15 and 49 dai (Fig. 4.26). The

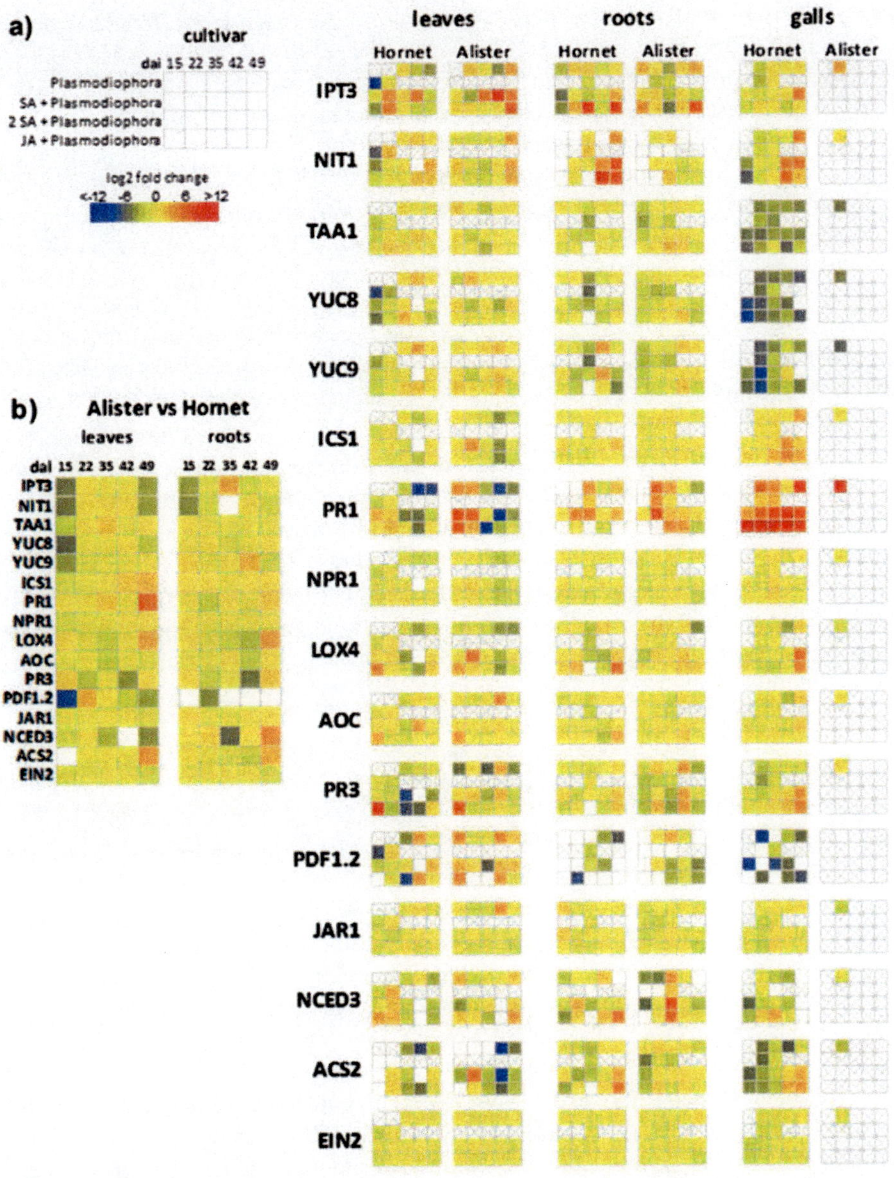

Fig. 4.26 Expression patterns of hormone-related genes. Mean qRT-PCR values of the experimental variants are related to (**a**) control at each time point (leaves are related to leaves of the respective cultivar, roots, and galls are related to roots of the respective cultivar). Results are organized as in the top scheme. Non-analyzed variants are crossed. White fields correspond to non-detected expression. (**b**) Control Alister plants (leaves or roots) are compared with control Hornet plants (Prerostova et al. 2018)

expression of the CK biosynthetic gene *isopentenyl transferase IPT3* was diminished in infected leaves in Hornet between 42 and 49 dai, and in Alister at only 42 dai. Downregulation of the CK plastid biosynthetic pathway might reflect a promoted senescence of infected leaves in the later stages of infection. This assumption is in accordance with enhanced *IPT3* expression after double SA application, which promoted stress tolerance. After JA application, *IPT3* expression had generally similar levels as in the *P. brassicae*-infected variant. In roots of both cultivars, *IPT3* expression was generally low, with exception of JA-treated plants (in Hornet, 35 dai and 49 dai, in Alister, 49 dai). The *IPT3* expression was slightly enhanced in galls.

4.5.6.2 Auxin-Related Genes

The expression of four IAA biosynthetic genes was investigated—*nitrilase 1* (*NIT1*; indole-3-acetonitrile pathway), *tryptophan aminotransferase* (*TAA1;* indole-3-pyruvic pathway), and *indole-3-pyruvate monooxygenase 8* and *9* (*YUC8* and *YUC9* via indole-3-pyruvic acid). *P. brassicae* infection led to their mild transient suppression in Alister leaves, and their upregulation was found in Hornet leaves from 35 dai on. In Hornet and Alister roots, a contrasting behavior was observed in the expression levels of *NIT1,* the auxin biosynthetic gene that is most commonly involved in the response to infection. In the sensitive cultivar, the gene expression was higher than in the resistant one, further increasing especially after JA application. A similar response was found in Hornet galls.

4.5.6.3 Salicylic Acid-Related Genes

The expression of the SA biosynthetic gene *isochorismate synthase 1* (*ICS1*) was higher in leaves of Alister than in those of Hornet already at control conditions. This was associated with the elevated expression of the SA-inducible gene *PR1*. During later stages of *P. brassicae* infection, a suppression of *ICS1* expression was detected especially in Alister. However, the *PR1* expression in Alister was still enhanced within 15 and 22 dai. Double SA application resulted in the stimulation of PR1 expression in Alister leaves 15 and 22 dai, but only a moderate elevation was observed in Hornet ones. JA application was generally associated with the downregulation of *ICS1*. The expression of the central component of the SA signaling pathway, *NPR1*, exhibited only minor changes during infection, and after hormone treatment in leaves (with a tendency to decrease in Alister). Alister exhibited higher *ICS1* expression under control conditions, and also in roots compared to Hornet. *P. brassicae* infection was associated with a slight elevation of *ICS1* expression in Hornet, and somewhat more in Alister. Also *PR1* expression was enhanced in both cultivars (till 35 dai). Double SA treatment resulted in the upregulation of *PR1* expression in both cultivars. After JA treatment *ICS1* profiles resembled those of infected plants, *PR1* expression was often under the detection limit. In roots, the *NPR1* expression did not show any significant change. In galls, a mild elevation of *ICS1* expression was detected in Hornet, a moderate increase being found after double SA and JA treatments. *P. brassicae* infection was associated with a profound stimulation of *PR1* expression over time. Double SA treatment gradually upregulated *PR1* expression in Hornet. Also JA treatment coincided with high *PR1* expression.

4.5.6.4 Jasmonic Acid-Related Genes

The expression of the JA biosynthetic gene lipoxygenase 4 (*LOX4*) decreased upon *P. brassicae* infection in leaves of both cultivars, more in Alister. In contrast, a mild elevation of the other gene encoding a JA biosynthetic enzyme—allene oxide cyclase (*AOC*), was observed in the later stages. JA application resulted in the transient upregulation of both genes in both cultivars 15 dai, possibly via a positive feedback loop. The JA-inducible gene PR3 was suppressed by *P. brassicae* infection especially in the resistant Alister. Double SA treatment resulted in a substantial decrease of *PR3* expression in Hornet; in Alister, it abolished the effects of infection. JA application led to a strong transient increase of *PR3* expression in both cultivars 15 dai, corresponding well to the transient elevation of JA levels after foliar application. *P. brassicae* infection stimulated *PDF1.2* expression in leaves of both cultivars at the later stages (42–49 dai). The expression of JA amido synthetase (*JAR1*), which conjugates JA to isoleucine, was generally not affected by *P. brassicae* infection. JA treatment had in the long-term a negative effect on *JAR1* expression in Alister leaves. The expression patterns of both *LOX4* and *AOC* followed similar profiles in roots as in leaves after *P. brassicae* infection. Double SA treatment decreased *LOX4* transcript levels in Alister roots 15 dai. JA had generally a positive effect on *LOX4* expression in roots. *PR3* expression showed a slightly decreasing trend in roots after *P. brassicae* infection. Both SA treatments led to the downregulation of *PR3* in Alister roots over a prolonged period. *P. brassicae* infection was associated with the decrease of *PDF1.2* expression in Hornet roots. The expression of *JAR1* was upregulated in Hornet and downregulated in Alister 15 dai after infection in all infected variants except for JA treatment. *LOX4* and *AOC* expression were upregulated in galls of Hornet 49 dai, especially after double SA and JA treatments, but the expression of the former gene was reduced in Alister galls. *PR3* and *PDF1.2* expression was low in Hornet galls; however, the expression of *PR3* increased after double SA and JA applications during later stages. *JAR1* expression was lower in Alister than in Hornet after *P. brassicae* infection.

4.5.6.5 Abscisic Acid-Related Genes

The expression of the gene encoding the rate-limiting step in ABA biosynthesis, *9-cis-epoxycarotenoid dioxygenase* (*NCED3*), showed no distinct trend in the expression under control conditions in leaves of both cultivars. Similarly, no consistent trend was observed after *P. brassicae* infection with or without SA treatment. A mild elevation was found after JA application in leaves 15 dai, which was more profound in Hornet than in Alister. In roots, also a similar variable situation concerning the *NCED3* expression was found. After JA application a small elevation was found between 35 and 49 dai. In addition, the expression of *NCED3* was elevated in galls of JA-treated plants of the cultivar Hornet.

4.5.6.6 Ethylene-Related Genes

The expression of the ethylene biosynthetic gene *1-amino cyclopropane-1-carboxylic acid synthetase2* (*ACS2*) was decreased by *P. brassicae* infection in leaves and roots of both cultivars and the effect was partially diminished by both SA and JA treatments. *ACS2* expression was higher in roots of the cultivar Hornet compared to Alister. In addition, *ACS2* expression was upregulated 42 and 49 dai in Hornet galls treated with JA. A slightly lower effect was found in the case of the double SA application. The expression of a crucial component of ethylene signaling pathway EIN2 was not significantly affected by any treatment either in leaves or in roots (Prerostova et al. 2018).

4.5.7 Expression of Hormones in Different Cultivars

The expression of hormones in the *P. brassicae* infected resistant and susceptible cvs. have been characterized. In the early stages of infection, there is a change in growth promoting hormones in response to level of host resistance. During later phase of infection for gall formation, the amount and rate of hormonal production is high in susceptible cvs. than the resistant ones. There is a differential expression of hormones in leaves verses roots and formation of galls in R and S cvs. indicating organ and host resistance specificity in production of hormones.

4.5.7.1 The Early Response Is Characterized by Changes in Growth-Promoting Hormones

Colonized cells were demonstrated to exhibit a stimulation of cell division and elongation with the elevation of CKs in roots during the early response to *P. brassicae* infection. The impact of *P. brassicae* infection on hormone and gene expression data are shown in Fig. 4.27. Active CKs are partially produced by the pathogen, as indicated by the conversion of a precursor to an active CK, and by the occurrence of two putative IPT genes in the genome. In *Brassica rapa* and *Arabidopsis thaliana*, Prerostova et al. (2018) detected the increase of active CKs, predominantly caused by cis-zeatin, already 2 dai in roots and leaves of both resistant and susceptible cultivars. In the disease-sensitive cultivar Hornet, the elevation of the CK content was maintained for 10 days in both organs, while in the resistant cultivar Alister, it is dropped to the control level 5 dai, and further decreased below the controls in leaves 7 dai and in roots 10 dai. Transient CK elevation is in accordance with the upregulation of several IPT genes during the first phase of infection in inoculated roots reported for *Brassica rapa* (Chinese cabbage) (Ludwig-Muller 2009a, b). The CK downregulation seems to be part of plant defense, as plants overexpressing CK degrading enzyme cytokinin oxidase/dehydrogenase are more resistant to the clubroot disease (Siemens et al. 2006). This effect can be interpreted as the reduction of the signals involved in cell division, because the elevated CKs trigger the local re-initiation of cell cycle progression in the root cortex, as indicated by upregulation of cyclin CYCB1;1 expression described by Devos et al. (2006).

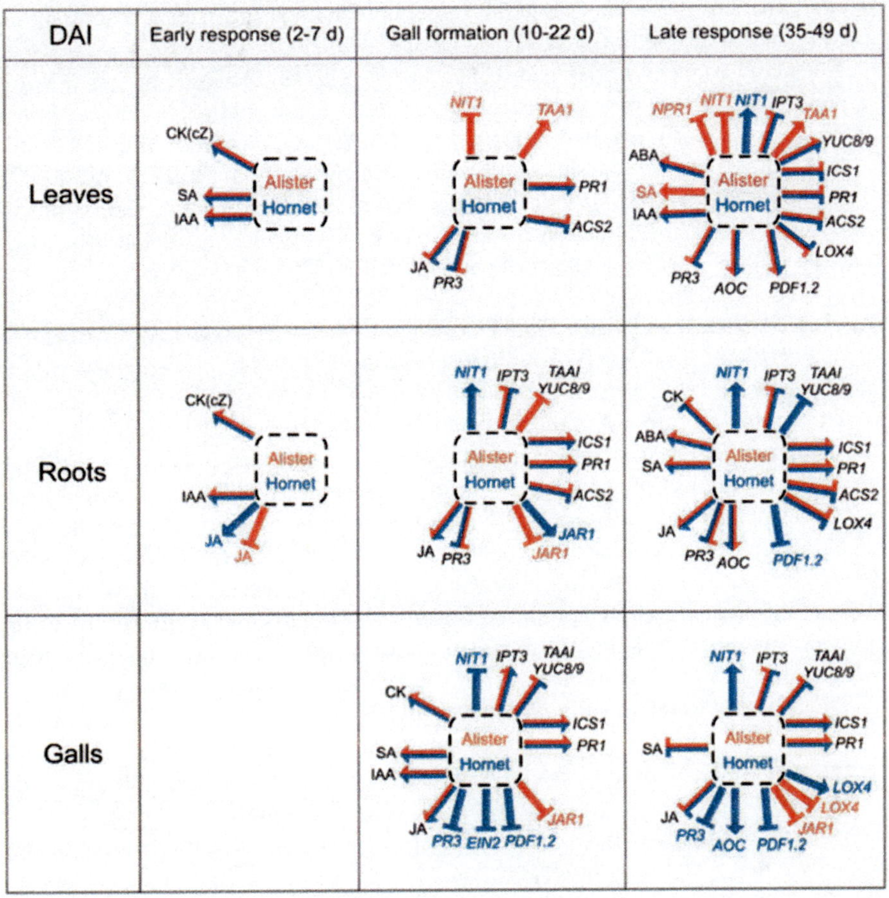

Fig. 4.27 Summary of the *Plasmodiophora brassicae* impact on *Brassica napus* plants during infection progression. Up- or downregulation of phytohormone contents and gene expression levels in Alister (specific response—red letters and arrows), and Hornet (specific response—blue letters and arrows) leaves, roots, and galls detected during early phase (2–7 dai), gall formation (10–22 dai), and late phase (35–49 dai). Response common for both cultivars—black letters (Prerostova et al. 2018)

Stimulation of cell division also requires the presence of auxin. The elevated content of IAA was reported in Brassica campestris 10 dai. In *Brassica napus*, upregulation of IAA was detected after 3 dai, to a higher extent in roots. In *Arabidopsis*, an enhanced response of auxin-related genes was observed from 6 dai. Due to the importance of active polar auxin transport, it seems that the host auxin needs to be moved to the site of infection. In addition, the signaling pathways responsible for auxin also need to be functioning. A transient elevation of IAA contents in leaves 2 dai in Alister, and 2–5 dai in Hornet was observed. In roots, IAA content was promoted between 2 and 7 dai in Alister, and from 2 dai to 15 dai in

Hornet. The stronger induction in the susceptible cultivar indicates a prominent role for IAA in disease progression.

Not many changes in defense-related hormones were detected during the early response. The content of SA in roots was similar as in the controls, while the content of JA was transiently elevated in Hornet roots between 2 and 5 dai, and in Alister at 7 dai. Irani et al. (2018) showed that *P. brassicae* stimulate the expression of JA-related genes in *A. thaliana*, but not genes related to SA. It seems that the onset of *P. brassicae* infection is associated with colonization of the root tissue, but still without visible symptoms (Ludwig-Muller 2009a, b).

4.5.7.2 Differences in the Gall Formation of the Susceptible Cultivar Hornet and Resistant Cultivar Alister

The later phase of *P. brassicae* infection can be characterized by gall formation, which was more profound in the susceptible cultivar Hornet. During this stage, the CK content in leaves was higher in Alister (at 15–42 dai) than in Hornet which may reflect the better physiological state of the more tolerant cultivar. CK downregulation was found in the roots of both cultivars during the later stage. Malinowski et al. (2016) reported that CKs are diminished in *A. thaliana* after infection (16 dai). In galls of both cultivars, however, high CK levels were detected 22 dai.

In addition, auxin might also be responsible for enlarged cells during the development of large sporulating plasmodia and resting spores. In roots, the susceptible cultivar Hornet exhibited increased IAA content until 15 dai. In both cultivars, the IAA level was elevated in galls 22 dai (to a higher extent in Hornet), which might be related to cell expansion governed by auxin. In Hornet galls, NIT1 transcription was upregulated 35 dai, and especially at 49 dai. Several reports describe the upregulation of transcription of *nitrilase* isoforms in clubroot tissues compared to controls in different host plants such as *Brassica rapa* (turnip, Chinese cabbage) or *Brassica juncea*, and specific expression of *nitrilase 1* and *2* genes in developing galls of *Arabidopsis thaliana*, as well as with the upregulation of myrosinases and glucosinolate metabolism. The difference between IAA levels and the expression of IAA-biosynthetic genes may be given by the upregulation of the expression of GH3 genes (encoding auxin conjugating enzymes), as was shown in the case of *Arabidopsis thaliana*. It may indicate that plant defense includes IAA conjugation, that is, the inactivation of the excess of IAA involved in gall formation (Jahn et al. 2013).

SA is the key hormone in plant combat with biotrophic pathogens. SA increase in roots was more profound in Alister (35–49 dai) than in Hornet, where it occurred later (42–49 dai). In Alister galls, SA level was enhanced at 22 and 42 dai, while its enhancement was delayed in Hornet (only between 42 and 49 dai). The expression of SA biosynthetic gene *ICS1* was higher in Alister roots between 35 and 42 dai, whereas its expression was lower in Hornet root galls. The induction of defense-related responses in susceptible infected plants is of limited extent and duration. In *A. thaliana*, an upregulation of the corresponding *ICS1* transcript was reported at 21 but not 28 dai. The responses to *P. brassicae* infection were followed in the resistant genotype *Brassica macrocarpa* Guss. and in the susceptible genotype

Brassica oleracea var. *italica*, the SA signal transduction was upregulated mainly in the resistant genotype, in comparison with the susceptible one. Only minor differences were observed in the expression of the central component of the SA signaling pathway *NPR1*. Regulation of its activity seems to be realized predominantly by oligomerization of the resulting protein, and subsequent differential cellular localization. PR1 expression was upregulated in Hornet roots 35–49 dai, and in Alister already 22–42 dai. It has a role as a marker for SA-dependent defense. Constitutive upregulation of PR1 expression in the cpr1 mutant of *A. thaliana* resulted in more clubroot resistant plants. The experiments with SA deficient/insensitive and constitutively active mutants indicate that SA is very important, but not sufficient in defense to *P. brassicae* (Lovelock et al. 2016).

During the later stages, the JA content was elevated in the roots of both cultivars, which indicates that JA might not play a role in the defense of the resistant cultivar to the clubroot disease. Jasmonic acid content was especially high during the gall initiation, occurring around 22 dai. As JA is able to induce nitrilase activity, there could be a link between jasmonate and auxin metabolism, and not so much of a signal for plant defense. The gene for the rate-limiting enzyme of ethylene biosynthesis, ACS2, exhibited lower expression in roots and galls of the susceptible cultivar Hornet compared to the resistant Alister. Knaust and Ludwig-Muller (2013) reported for *A. thaliana* that the expression of another gene for the ethylene biosynthetic enzyme ACO2 was upregulated throughout the whole experimental period of 10 to 28 dai. However, the expression of a crucial component of the ethylene signaling pathway EIN2 was not significantly affected in the two *Brassica* cultivars. Ethylene signaling mutants of *A. thaliana* were more susceptible to *P. brassicae* infection, thereby making larger galls at a lower infection pressure. The ethylene might be needed to restrict gall size under normal conditions (Prerostova et al. 2018).

4.5.7.3 The Impacts of Salicylic Acid and Jasmonic Acid Applications

Taking into account that SA is a crucial hormone in the response to biotrophic pathogens, Prerostova et al. (2018) tested the effect of exogenous SA applied before infection or twice before infection and 15 dai, when the gall formation starts. The impact of exogenous SA in the fight with the clubroot pathogen is dependent on the time point and method of administration as well as on the plant material. Agarwal et al. (2011) reported that a very early incubation with SA significantly reduced the clubroot symptoms of *A. thaliana*, which was confirmed for *Brassica oleracea* (broccoli). However, a later treatment of *A. thaliana* plants with SA during infection had no effect on disease severity. The exogenous SA was rather quickly metabolized, as already 2 days after its application of only a moderate amount of the hormone was left. At the time point of the second foliar spraying 15 dai, a very high peak of SA was detected 2 h after application. SA metabolism may be partly due to the host plant tendency to re-establish hormone homeostasis disturbed by the application of the hormone, or it may be a part of the strategy of the pathogen to downregulate SA levels, and thus suppress plant defense. Indeed, *P. brassicae* was reported to have high SA-methyltransferase activity, creating an SA transport form, which can be translocated from the galls to the upper plant parts where it is emitted. Consequently, *A. thaliana* plants overexpressing the methyltransferase gene were more susceptible to clubroot. A similar mode of action was observed in the case of *Pseudomonas syringae*, which produces coronatine, a compound structurally similar to JA-isoleucine. Coronatine induces the JA signaling pathway, which leads to the suppression of the SA pathway. Also, SA glucosylation might be considered in the SA deactivation. SA application before infection had a negligible effect on gall formation. However, SA application at the onset of gall formation prevented disease development in the more resistant cultivar Alister, indicating that in cv. Alister, the defense is already partially working, and therefore, only small amounts of SA increase resistance. The fact that the JA-metabolic mutant jar1, which is unable to produce the active JA metabolite JA-isoleucine, is more sensitive to *P. brassicae* infection indicates that JA signaling is also involved in the response to *P. brassicae* infection. Nevertheless, exogenous application of JA at 15 dai promoted gall formation in Hornet, but also in the more resistant cultivar Alister. Similarly, the partially resistant *Arabidopsis* ecotype Bur was reported to activate after infection of the SA pathway, while the sensitive *Arabidopsis* ecotype Col-0 activated the JA pathway. When more virulent pathotype of *P. brassicae* (isolate e2 instead of eH isolate) was used, even Bur did not activate the SA pathway. These data indicate a complex cross talk among these two pathways which depends on the mutual ratio of their activities (Lemarie et al. 2015; Jubault et al. 2013).

4.5.7.4 Organ Specificity: Leaves Versus Roots and Galls

Shoot/root communication is fast and intensive within the plant. This can be demonstrated by changes of leaf hormone levels upon attack by soil pathogens as well as by elevation of SA or JA in roots after their foliar application. SA has been proven to be the key hormone in defense against biotrophs. In some cases, when exogenous SA did not decrease clubroot growth, it diminished at least the stress-induced inhibition of shoot growth. Thus, the (partial) preservation of active CK and auxin content in leaves as shown for both cultivars might be a measure of the shoot fitness, and not the consequence of the pathogen impact, as is seen in roots. In the latter, the regulation of CK and auxin levels (including the degradation with cytokinin oxidase/dehydrogenase CKX or conjugation with GH3) seems to reflect the combat between the pathogen and plant. Hormone homeostasis in leaves might either be maintained through transport, degradation, or conjugation, or through de novo synthesis. Upregulation of transcripts for auxin biosynthesis genes in both cultivars, such as *NIT1* and the two *YUC* genes, indicates the stimulation of IAA synthesis (Lovelock et al. 2016).

The modulation of SA and JA content in leaves reflects the intensive activation of the defense, also in organs that are not directly exposed to the pathogen. The SA level was increased in Alister leaves between 35 and 49 dai, while in Hornet leaves, the increase occurred earlier, being maintained until the end of the experimental period (between 15 and 49 dai). In the roots, the SA level increased earlier, and to a higher extent in the resistant than in the susceptible cultivar. Methylation of SA by the protist's methyltransferase can result in a form that is more readily transported into the leaves of *A. thaliana*. The methylester of SA can be converted back to free SA in the leaves using specific esterases. The role of such esterases and methyltransferases for SA metabolism of *B. oleracea* during clubroot infection was demonstrated. Not only in the roots but also in the leaves differentially expressed genes were found that could be involved in SA metabolism (Manoharan et al. 2016).

In the case of JA, the situation might be a bit different. The JA levels were upregulated in leaves of the cultivar Alister only (15–35 dai), the expression of the JA biosynthetic gene *LOX4* was downregulated in leaves of both cultivars. Finally, the fitness of the plant is also dependent on the water status. In roots, the content of abscisic acid (ABA) was enhanced in both cultivars during later stages of the infection, a bit higher in the susceptible Hornet than in the resistant Alister cultivar, which confirmed earlier data where ABA was also increased in *Brassica* galls at a later stage of disease development. Ludwig-Muller (2009a) proposed, based on the upregulation of many transcripts related to drought stress during later stages of the clubroot disease, a strong influence of ABA. The drought stress could be attributed to the destroyed root tissues that include the vasculature. The content of ABA was enhanced in leaves and roots of both cultivars at 42 and 49 dai. ABA has obviously a dual role in the regulation of stomata aperture and stimulation of the expression of stress-related genes. However, the ABA elevation in leaves seems to be rather a consequence of impaired water transport (leaf wilting) than modulation of the transcriptome (Prerostova et al. 2018).

4.5.8 Schematic Model of Host–Pathogen Interaction to Express Biometabolomic Resistance

Based on the comparative transcriptome analysis results, the superior molecular resistance to clubroot is reviewed in Fig. 4.28. In this model, the NBS-LRR protein of *PbBa*.1 and *CRb* recognizes the elicitors released by *P. brassicae* and triggers defense signal transduction. The key pathways associated here are SA, JA, MAPK, ET, Ca^{2+}, and ROS. The activation of SA signaling transduction and downregulation via JA signaling and biosynthesis affects the defense-related WRKY and ERF transcription factor families, which regulate downstream defense mechanisms. On the other hand, RBOH and phenylpropanoid biosynthesis activate ROS activities that also enhance the SA signaling and biosynthesis, PCD, and the HR pathway. Antioxidant enzymatic regulations may also contribute to the defense response.

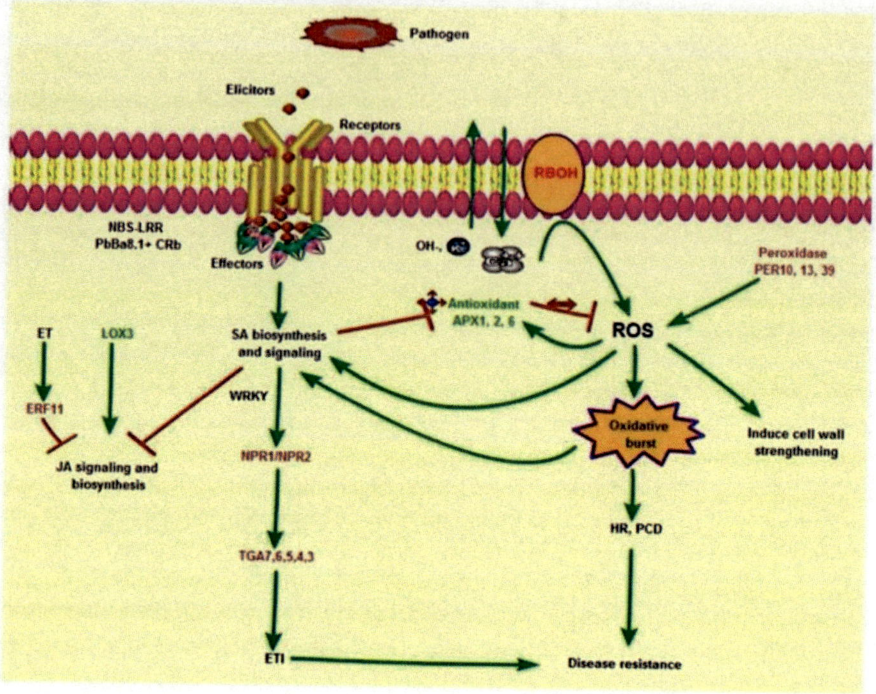

Fig. 4.28 Schematic diagram showing the plant defense signaling network in *Brassica napus* resistance lines in response to *Plasmodiophora brassicae* (Shah et al. 2020)

References

Aarts N, Metz M, Holub E, Staskawicz BJ, Daniels MJ, Parker JE (1998) Different requirements for EDS1 and NDR1 by disease resistance genes define at least two R gene-mediated signaling pathways in *Arabidopsis*. Proc Natl Acad Sci U S A 95:10306–10311

Acevedo-Garcia J, Spencer D, Thieron H, Reinstadler A, Hammond-Kosack K, Phillips AL (2016) mlo-based powdery mildew resistance in hexaploid bread wheat generated by a non-transgenic TILLING approach. Plant Biotechnol J 15:367–378. https://doi.org/10.1111/pbi.12631

Adam L, Ellwood S, Wilson I, Xiao S, Saenz G, Oliver R, Turner JG, Somerville S (1999) Comparison of *Erysiphe cichoracearum* and *E. cruciferarum* and a survey of 360 *Arabidopsis thaliana* accessions for their resistance to these two powdery mildew pathogens. Mol Plant Microbe Interact 12:1031–1043

Afzal AJ, Wood AJ, Lightfoot DA (2008) Plant receptor-like serinethreonine kinases: roles in signaling and plant defense. Mol Plant Microbe Interact 21:507–517

Agarwal A, Kaul V, Faggian R, Rookes JE, Ludwig-Muller J, Cahill DM (2011) Analysis of global host gene expression during the primary phase of the *Arabidopsis thaliana–Plasmodiophora brassicae* interaction. Funct Plant Biol 38:462–478

Ahloowalia BS, Maluszynski M, Nichterlein K (2004) Global impact of mutation-derived varieties. Euphytica 135:187–201

Ahuja I, Kissen R, Bones AM (2012) Phytoalexins in defense against pathogens. Trends Plant Sci 17:73–90. https://doi.org/10.1016/j.tplants.2011.11.002

Aist JR (1976) Papillae and related wound plugs of plant cells. Annu Rev Phytopathol 14:145–163

Aldon D, Mbengue M, Mazars C, Galaud JP (2018) Calcium signaling in plant biotic interactions. Int J Mol Sci 19(3):665. https://doi.org/10.3390/ijms19030665

Ali S, Mir ZA, Tyagi A, Mehari H, Meena RP, Bhat JA, Yadav P, Papalou P, Rawat S, Grover A (2017) Overexpression of *NPR1* in *Brassica juncea* confers broad spectrum resistance to fungal pathogens. Front Plant Sci 8:1693. https://doi.org/10.3389/fpls.2017.01693

Ali M, Cheng Z, Ahmad H, Hayat S (2018) Reactive oxygen species (ROS) as defences against a broad range of plant fungal infections and case study on ROS employed by crops against *Verticillium dahliae* wilts. J Plant Interact 13:353–363. https://doi.org/10.1080/17429145.2018.1484188

Alkooranee JT, Liu S, Aledan TR, Yin Y, Li M (2015) First report of powdery mildew caused by *Erysiphe cruciferarum* on *Brassica napus* in China. Plant Dis 99(11):1651

de Almeida EJ, Favery B (2011) The plant cytoskeleton remodelling in nematode induced feeding sites. In: Jones J, Gheysen G, Fenoll C (eds) Genomics and molecular genetics of plant–nematode interactions. Springer, Heidelberg, pp 369–393

de Almeida EJ, Favery B, Engler G, Abad P (2005) Loss of susceptibility as an alternative for nematode resistance. Curr Opin Biotechnol 16:112–117

Alvarez ME (2000) Salicylic acid in the machinery of hypersensitive cell death and disease resistance. Plant Mol Biol 44:429–442

Ament K, Krasikov V, Allmann S, Rep M, Takken FLW, Schuurink RC (2010) Methyl salicylate production in tomato affects biotic interactions. Plant J 62(1):124–134. https://doi.org/10.1111/j.1365-313X.2010.04132.x

An Q, Ehlers K, Kogel KH, van Bel AJE, Huckelhoven R (2006) Multivesicular compartments proliferate in susceptible and resistant *MLA12*-barley leaves in response to infection by the biotrophic powdery mildew fungus. New Phytol 172:563–576

Andersson MX, Kourtchenko O, Dangl JL, Mackey D, Ellerström M (2006) Phospholipase-dependent signalling during the AvrRpm1-and AvrRpt2-induced disease resistance responses in *Arabidopsis thaliana*. Plant J 47:947–959

Anonymous (1985) Foliar diseases of *Brassicas*. Research report 1984, horticulture. Dublin, Ireland; An Foras Taluntais, pp 45–46

Antico CJ, Colon C, Banks T, Ramonell KM (2012) Insights into the role of jasmonic acid-mediated defenses against necrotrophic and biotrophic fungal pathogens. Front Biol 7:48–56. https://doi.org/10.1007/s11515-011-1171-1

Aravind L, Koonin EV (1999) Fold prediction and evolutionary analysis of the POZ domain: structural and evolutionary relationship with the potassium channel tetramerization domain. J Mol Biol 285:1353–1361. https://doi.org/10.1006/jmbi.1998.2394

Asada Y, Matsumoto I (1969) Formation of lignin-like substance in the root tissue of Japanese radish plant infected by downy mildew fungus. Ann Phytopathol Soc Japan 35:160–167

Asada Y, Matsumoto I (1971) Microspectrophotometric observations on the cell walls of Japanese radish (*Raphanus sativus*) root infected by *Peronospora parasitica*. Physiol Plant Pathol 1:377–383

Asada Y, Matsumoto I (1972) The nature of lignin obtained from downy mildew-infected Japanese radish root. Phytopathol Z 73:208–214

Asada Y, Ohguchi T, Matsumoto I (1975) Lignin formation in fungus infected plants. Rev Plant Protect Res 8:104–113

Asai S, Furzer OJ, Cevik V, Kim DS, Ishaque N, Goritschnig S, Staskawicz BJ, Shirasu K, Asai T, Tena G, Plotnikova J, Willmann MR, Chiu WL, Gomez-Gomez L, Boller T, Ausubel FM, Sheen J (2002) MAP kinase signalling cascade in *Arabidopsis* innate immunity. Nature 415:977–983

Assaad FF, Qiu JL, Youngs H, Ehrhardt D, Zimmerli L, Kalde M, Wanner G, Peck SC, Edwards H, Ramonell K, Somerville CR, Thordal-Christensen H (2004) The PEN1 syntaxin defines a novel cellular compartment upon fungal attack and is required for the timely assembly of papillae. Mol Biol Cell 156(1):5118–5129

Attaran E, Zeier TE, Griebel T, Zeier J (2009) Methyl salicylate production and jasmonate signaling are not essential for systemic acquired resistance in Arabidopsis. Plant Cell 21:954–971

Auld D, Heikkinen M, Erickson D, Sernyk J, Romero J (1992) Rapeseed mutants with reduced levels of polyunsaturated fatty acids and increased levels of oleic acid. Crop Sci 32:657–662

Azevedo C, Betsuyaku S, Peart J, Takahashi A, Noel L, Sadanandom A, Casais C, Parker J, Shirasu K (2006) Role of SGT1 in resistance protein accumulation in plant immunity. EMBO J 25:2007–2016

Backer R, Naidoo S, van den Berg N (2019) The non-expresser of pathogenesis-related genes 1 (NPR1) and related family: mechanistic insights in plant disease resistance. Front Plant Sci 10:102. https://doi.org/10.3389/fpls.2019.00102

Bai TT, Xie WB, Zhou PP, Wu ZL, Xiao WC, Zhou L, Sun J, Ruan XL, Li HP (2013) Transcriptome and expression profile analysis of highly resistant and susceptible banana roots challenged with *Fusarium oxysporum* f. sp. *cubense* tropical race 4. PLoS One 8:e73945. https://doi.org/10.1371/journal.pone.0073945

Bak S, Feyereisen R (2001) The involvement of two p450 enzymes, CYP83B1 and CYP83A1, in auxin homeostasis and glucosinolate biosynthesis. Plant Physiol 127:108–118. https://doi.org/10.1104/pp.127.1.108

Bak S, Beisson F, Bishop G, Hamberger B, Hofer R, Paquette S, Werck-Reichhart D (2011) Cytochromes p450. Arabidopsis Book 9:e0144

Bakshi M, Oelmüller R (2014) WRKY transcription factors: Jack of many trades in plants. Plant Signal Behav 9:e27700. https://doi.org/10.4161/psb.27700

Baluska F, Bacigalova K, Oud JL, Hauskrecht M, Kubica S (1995) Rapid reorganization of microtubular cytoskeleton accompanies early changes in nuclear ploidy and chromatin structure in post-mitotic cells of barley leaves infected with powdery mildew. Protoplasma 185:140–151

Bari R, Jones JD (2009) Role of plant hormones in plant defence responses. Plant Mol Biol 69:473–488. https://doi.org/10.1007/s11103-008-9435-0

Bayer M, Nawy T, Giglione C, Galli M, Meinnel T, Lukowitz W (2009) Paternal control of embryonic patterning in *Arabidopsis thaliana*. Science 323:1485–1488

Beckers GJ, Jaskiewicz M, Liu Y, Underwood WR, He SY, Zhang S, Conrath U (2009) Mitogen-activated protein kinases 3 and 6 are required for full priming of stress responses in *Arabidopsis thaliana*. Plant Cell 21:944–953

Bednarek P (2012) Chemical warfare or modulators of defence responses—the function of secondary metabolites in plant immunity. Curr Opin Plant Biol 15:407–414. https://doi.org/10.1016/j.pbi.2012.03.002

Bednarek P, Osbourn A (2009) Plant–microbe interactions: chemical diversity in plant defense. Science 324:746–748

Bednarek P, Schneider B, Svatos A, Oldham NJ, Hahlbrock K (2005) Structural complexity, differential response to infection, and tissue specificity of indolic and phenylpropanoid secondary metabolism in *Arabidopsis* roots. Plant Physiol 138:1058–1107

Bednarek P, Pislewska-Bednarek M, Svatos A, Schneider B, Doubsky J, Mansurova M, Humphry M, Consonni C, Panstruga R, Sanchez-Vallet A, Molina A, Schulze-Lefert P (2009) A glucosinolate metabolism pathway in living plant cells mediates broad-spectrum antifungal defense. Science 323:101–106. https://doi.org/10.1126/science.1163732

Beekwilder J, van Leeuwen W, van Dam NM, Bertossi M, Grandi V, Mizzi L, Soloviev M, Szabados L, Molthoff JW, Schipper B, Verbocht H, de Vos RCH, Morandini P, Arts MGM, Bovy A (2008) The impact of the absence of aliphatic glucosinolates on insect herbivory in *Arabidopsis*. PLoS One 3:e2068

Belanger RR, Bushnell WR, Dik AJ, Carver TLW (2002) The powdery mildews: a comprehensive treatise. American Phytopathology Society (APS Press), St. Paul

Belanger RR, Benhamou N, Menzies JG (2003) Cytological evidence of an active role of silicon in wheat resistance to powdery mildew (*Blumeria graminis* f. sp. *tritici*). Phytopathology 93:402–412

Benaroudj N, Lee DH, Goldberg AL (2001) Trehalose accumulation during cellular stress protects cells and cellular proteins from damage by oxygen radicals. J Biol Chem 276:24261–24267. https://doi.org/10.1074/jbc.M101487200

Berken A (2006) ROPs in the spotlight of plant signal transduction. Cell Mol Life Sci 63:2446–2459

Bernoux M, Ve T, Williams S, Warren C, Hatters D, Valkov E, Zhang X, Ellis JG, Kobe B, Dodds PN (2011) Structural and functional analysis of a plant resistance protein TIR domain reveals interfaces for selfassociation, signaling, and autoregulation. Cell Host Microbe 9:200–211

Bethke G, Unthan T, Uhrig JF, Poschl Y, Gust AA, Scheel D, Lee J (2009) Flg22 regulates the release of an ethylene response factor substrate from MAP kinase 6 in *Arabidopsis thaliana* via ethylene signaling. Proc Natl Acad Sci U S A 106:8067–8072

Bhardwaj V, Meier S, Petersen LN, Ingle RA, Roden LC (2011) Defense responses of *Arabidopsis thaliana* to infection by *Pseudomonas syringae* are regulated by the circadian clock. PLoS One 6:e26968. https://doi.org/10.1371/journal.pone.0026968

Bhat RA, Miklis M, Schmelzer E, Schulze-Lefert P, Panstruga R (2005) Recruitment and interaction dynamics of plant penetration resistance components in a plasma membrane microdomain. Proc Natl Acad Sci U S A 102:3135–3140

Bittel P, Robatzek S (2007) Microbe-associated molecular patterns (MAMPs) probe plant immunity. Curr Opin Plant Biol 10:335–341

Bittner-Eddy PD, Beynon JL (2001) The *Arabidopsis* downy mildew resistance gene, RPP13-Nd, functions independently of NDR1 and EDS1 and does not require the accumulation of salicylic acid. Mol Plant Microbe Interact 14:416–421

Bjorkman A, Person B (1957) Studies on finely divided wood. Part 2. The properties of lignins extracted with neutral solvents from softwoods and hardwoods. Svensk Papperstidning 5:158–169

Bohlenius H, Mørch SM, Godfrey D, Nielsen ME, Thordal-Christensen H (2010) The multivesicular body-localized GTPase ARFA1b/1c is important for callose deposition and ROR2 syntaxin-dependent preinvasive basal defense in barley. Plant Cell 22:3831–3844

Boller T, Felix G (2009) A renaissance of elicitors: perception of microbe-associated molecular patterns and danger signals by pattern-recognition receptors. Annu Rev Plant Biol 60:379–406. https://doi.org/10.1146/annurev.arplant.57.032905.105346

Bones AM, Rossiter JT (2006) The enzymic and chemically induced decomposition of glucosinolates. Phytochemistry 67:1053–1067

Bottcher C, Westphal L, Schmotz C, Prade E, Scheel D, Glawischnig E (2009) The multifunctional enzyme CYP71B15 (PHYTOALEXIN DEFICIENT3) converts cysteine-indole-3-acetonitrile to camalexin in the indole-3-acetonitrile metabolic network of *Arabidopsis thaliana*. Plant Cell 21:1830–1845

Bracker CE (1968) Ultra-structure of the haustorial apparatus of *Erysiphe graminis* and its relationship to the epidermal cell of barley. Phytopathology 58:12–30

Brandl W, Herrmann K, Grotjahn LZ (1984) Hydroxycinnamoyl esters of malic acid in small radish (*Raphanus sativus* L. var. *sativus*). Zeitschrift fur Naturforschung 39c:515–520

Brown JKM (2015) Durable resistance of crops to disease: a Darwinian perspective. Annu Rev Phytopathol 53:513–539. https://doi.org/10.1146/annurev-phyto-102313-045914

Buer CS, Djordjevic MA (2009) Architectural phenotypes in the transparent testa mutants of *Arabidopsis thaliana*. J Exp Bot 60:751–763

Bulman SR, Siemens J, Ridgway HJ, Eady C, Conner AJ (2006) Identification of genes from the obligate intracellular plant pathogen, *Plasmodiophora brassicae*. FEMS Microbiol Lett 264:198–204

Bulman S, Candy JM, Fiers M, Lister R, Conner AJ, Eady CC (2011) Genomics of biotrophic, plant-infecting plasmodiophorids using *in vitro* dual cultures. Protist 162:449–461

Burki F, Kudryavtsev A, Matz MV, Aglyamova GV, Bulman S, Fiers M, Keeling PJ, Pawlowski J (2010) Evolution of Rhizaria: new insights from phylogenomic analysis of uncultivated protists. BMC Evol Biol 10:377–395

Burr CA, Leslie ME, Orlowski SK, Chen I, Wright CE, Daniels MJ, Liljegren SJ (2011) CAST AWAY, a membrane-associated receptor-like kinase, inhibits organ abscission in *Arabidopsis*. Plant Physiol 156:1837–1850

Buscaill P, Rivas S (2014) Transcriptional control of plant defence responses. Curr Opin Plant Biol 20:35–46

Caarls L, Pieterse CMJ, Van Wees SCM (2015) How salicylic acid takes transcriptional control over jasmonic acid signaling. Front Plant Sci 6:170. https://doi.org/10.3389/fpls.2015.00170

Caarls L, Van der Does D, Hickman R, Jansen W, Verk MCV, Proietti S, Lorenzo O, Solano R, Pieterse CMJ, Van Wees SCM (2016) Assessing the role of ETHYLENE RESPONSE FACTOR transcriptional repressors in salicylic acid-mediated suppression of jasmonic acid-responsive genes. Plant Cell Physiol 58:266–278. https://doi.org/10.1093/pcp/pcw187

Cahill D, Rookes J, Michalczyk A, McDonald K, Drake A (2002) Microtubule dynamics in compatible and incompatible interactions of soybean hypocotyl cells with *Phytophthora sojae*. Plant Pathol 51:629–640

Caillaud MC, Abad P, Favery B (2008a) Cytoskeleton reorganization, a key process in root-knot nematode-induced giant cell ontogenesis. Plant Signal Behav 3:816–818

Caillaud MC, Lecomte P, Jammes F, Quentin M, Pagnotta S, Andrio E, de Almeida EJ, Marfaing N, Gounon P, Abad P, Favery B (2008b) MAP65-3 microtubule-associated protein is essential for nematode-induced giant cell ontogenesis in *Arabidopsis*. Plant Cell 20(2):423–437

Campe R, Langenbach C, Leissing F, Popescu GV, Popescu SC, Goellner K, Beckers GJM, Conrath U (2015) ABC transporter PEN3/PDR8/ABCG36 interacts with calmodulin that, like PEN3, is required for *Arabidopsis* nonhost resistance. New Phytol 209:294–306

Cao H, Bowling SA, Gordon AS, Dong X (1994) Characterization of an *Arabidopsis* mutant that is nonresponsive to inducers of systemic acquired resistance. Plant Cell 6:1583–1592. https://doi.org/10.1105/tpc.6.11.1583

Cao H, Glazebrook J, Clarke JD, Volko S, Dong X (1997) The *Arabidopsis* NPR1 gene that controls systemic acquired resistance encodes a novel protein containing ankyrin repeats. Cell 88:57–63. https://doi.org/10.1016/S0092-8674(00)81858

Cao H, Li X, Dong XN (1998) Generation of broad-spectrum disease resistance by overexpression of an essential regulatory gene in systemic acquired resistance. Proc Natl Acad Sci U S A 95:6531–6536. https://doi.org/10.1073/pnas.95.11.6531

Cao T, Srivastava S, Rahman MH, Kav NNV, Hotte N, Deyholos MK, Strelkov SE (2008) Proteome-level changes in the roots of *Brassica napus* as a result of *Plasmodiophora brassicae* infection. Plant Sci 174(1):97–115

Cao Y, Liang Y, Tanaka K, Nguyen CT, Jedrzejczak RP, Joachimiak A, Stacey G (2014) The kinase LYK5 is a major chitin receptor in *Arabidopsis* and forms a chitin-induced complex with related kinase CERK1. eLife 3:e03766. https://doi.org/10.7554/eLife.03766

Catanzariti AM, Dodds PN, Lawrence GJ, Ayliffe MA, Ellis JG (2006) Haustorially expressed secreted proteins from flax rust are highly enriched for avirulence elicitors. Plant Cell 18:243–256

Century KS, Shapiro AD, Repetti PP, Dahlbeck D, Holub E, Staskawicz BJ (1997) *NDR1*, a pathogen-induced component required for *Arabidopsis* disease resistance. Science 278:1963–1965

Chan J, Jensen CG, Jensen LC, Bush M, Lloyd CW (1999) The 65-kDa carrot microtubule-associated protein forms regularly arranged filamentous cross-bridges between microtubules. Proc Natl Acad Sci U S A 96:14931–14936

Chandran D, Tai YC, Hather G, Dewdney J, Denoux C, Burgess DG, Ausubel FM, Speed TP, Wildermuth MC (2009) Temporal global expression data reveal known and novel salicylate-impacted processes and regulators mediating powdery mildew growth and reproduction on *Arabidopsis*. Plant Physiol 149:1435–1451

Chandran D, Inada N, Hather G, Kleindt CK, Wildermuth MC (2010) Laser micro dissection of *Arabidopsis* cells at the powdery mildew infection site reveals site-specific processes and regulators. Proc Natl Acad Sci U S A 107:460–465. https://doi.org/10.1073/pnas.0912492107

Chandran D, Rickert J, Huang Y, Steinwand MA, Marr SK, Wildermuth MC (2014) Atypical E2F transcriptional repressor DEL1 acts at the intersection of plant growth and immunity by controlling the hormone salicylic acid. Cell Host Microbe 15:506–513

Chapman JM, Muhlemann JK, Gayomba SR, Muday GK (2019) RBOH-dependent ROS synthesis and ROS scavenging by plant specialized metabolites to modulate plant development and stress responses. Chem Res Toxicol 32:370–396. https://doi.org/10.1021/acs.chemrestox.9b00028

Chen F, D'Auria JC, Tholl D, Ross JR, Gershenzon J, Noel JP, Pichersky E (2003) An *Arabidopsis thaliana* gene for methylsalicylate biosynthesis, identified by a biochemical genomics approach, has a role in defense. Plant J 36:577–588

Chen S, Zou J, Cowling WA, Meng J (2010) Allelic diversity in a novel gene pool of canola-quality *Brassica napus* enriched with alleles from *B. rapa* and *B. carinata*. Crop Pasture Sci 61:483–492. https://doi.org/10.1071/CP09327

Chen X, Barnaby JY, Sreedharan A, Huang X, Orbović V, Grosser JW, Wang N, Dong X, Song WY (2013) Over-expression of the citrus gene CtNH1 confers resistance to bacterial canker disease. Physiol Mol Plant Pathol 84:115–122. https://doi.org/10.1016/j.pmpp.2013.07.002

Chen J, Pang W, Chen B, Zhang C, Piao Z (2016a) Transcriptome analysis of *Brassica rapa* near-isogenic lines carrying clubroot-resistant and—susceptible alleles in response to *Plasmodiophora brassicae* during early infection. Front Plant Sci 6:1–14. https://doi.org/10.3389/fpls.2015.01183

Chen L, Zhang X, Xu H, Song B, Fan X (2016b) Introgression of clubroot resistance into an elite pak choi inbred line through marker-assisted introgression breeding. Plant Breed 135:471–475. https://doi.org/10.1111/pbr.12379

Chisholm ST, Coaker G, Day B, Staskawicz BJ (2006) Host-microbe interactions: shaping the evolution of the plant immune response. Cell 124:803–814

Chou HM, Bundock N, Rolfe SA, Scholes JD (2000) Infection of *Arabidopsis thaliana* leaves with *Albugo candida* (white blister rust) causes a reprogramming of host metabolism. Mol Plant Pathol 1:99–113

Christiansen KM, Gu Y, Rodibaugh N, Innes RW (2011) Negative regulation of defence signalling pathways by the EDR1 protein kinase. Mol Plant Pathol 12:746–758

Chu M, Song T, Falk KC, Zhang X, Liu X, Chang A, Lahlali R, McGregor L, Gossen BD, Peng G, Yu F (2014) Fine mapping of Rcr1 and analyses of its effect on transcriptome patterns during infection by *Plasmodiophora brassicae*. BMC Genomics 15:1166. https://doi.org/10.1186/1471-2164-15-1166

Clay NK, Adio AM, Denoux C, Jander G, Ausubel FM (2009) Glucosinolate metabolites required for an *Arabidopsis* innate immune response. Science 323:95–101. https://doi.org/10.1126/science.1164627

Collins NC, Thordal-Christensen H, Lipka V, Bau S, Kombrink E, Qiu JL, Hückelhoven R, Stein M, Freialdenhoven A, Somerville SC, Schulze-Lefert P (2003) SNARE-protein-mediated disease resistance at the plant cell wall. Nature 256(1):973–977

Consonni C, Humphry ME, Hartmann HA, Livaja M, Durner J, Westphal L, Vogel J, Lipka V, Kemmerling B, Schulze-Lefert P, Somerville SC, Panstruga R (2006) Conserved requirement for a plant host cell protein in powdery mildew pathogenesis. Nat Genet 386(1):716–720

Consonni C, Bednarek P, Humphry M, Francocci F, Ferrari S, Harzen A (2010) Tryptophan-derived metabolites are required for antifungal defense in the *Arabidopsis mlo2* mutant. Plant Physiol 152:1544–1561. https://doi.org/10.1104/pp.109.147660

Crouzet J, Trombik T, Fraysse AS, Boutry M (2006) Organization and function of the plant pleiotropic drug resistance ABC transporter family. FEBS Lett 580:1123–1130

Dangl JL, Horvath DM, Staskawicz BJ (2013) Pivoting the plant immune system from dissection to deployment. Science 341:746–751

Davis KR, Ausubel F (1989) Characterization of elicitor-induced defense responses in suspension-cultured cells of *Arabidopsis*. Mol Plant Microbe Interact 2:363–368

Delaney TP, Uknes S, Vernooij B, Friedrich L, Weymann K, Negrotto D, Gaffney T, Gut-Rella M, Kessmann H, Ward E, Ryals J (1994) A central role for salicylic acid in plant disease resistance. Science 266:1247–1250

Delaney TP, Friedrich L, Ryals JA (1995) *Arabidopsis* signal transduction mutant defective in chemically and biologically induced disease resistance. Proc Natl Acad Sci U S A 92:6602–6606. https://doi.org/10.1073/pnas.92.14.6602

Dempsey DA, Shah J, Klessig DF (1999) Salicylic acid and disease resistance in plants. Crit Rev Plant Sci 18:547–575

Devos S, Vissenberg K, Verbelen JP, Prinsen E (2005) Infection of Chinese cabbage by Plasmodiophora brassicae leads to a stimulation of plant growth: impacts on cell wall metabolism and hormone balance. New Phytol 166(1):241–250

Devos S, Laukens K, Deckers P, Van Der Straeten D, Beeckman T, Inze D, van Onckelen H, Witters E, Prinsen E (2006) A hormone and proteome approach to picturing the initial metabolic events during *Plasmodiophora brassicae* infection on Arabidopsis. Mol Plant-Microbe Interact 19:1431–1433

Dewdney J, Reuber TL, Wildermuth MC, Devoto A, Cui J, Stutius LM, Drummond EP, Ausubel FM (2000) Three unique mutants of *Arabidopsis* identify *eds* loci required for limiting growth of a biotrophic fungal pathogen. Plant J 24:205–218

Ding Y, Sun T, Ao K, Peng Y, Zhang Y, Li X, Zhang Y (2018) Opposite roles of salicylic acid receptors NPR1 and NPR3/NPR4 in transcriptional regulation of plant immunity. Cell 173:1454–1467.e1415

Dixon RA (2001) Natural products and plant disease resistance. Nature 411:843–847

Dixon RA, Paiva NL (1995) Stress induced phenyl-propanoid metabolism. Plant Cell 7:1085–1097

Djamei A, Schipper K, Rabe F, Ghosh A, Vincon V, Kahnt J, Osorio S, Tohge T, Fernie AR, Feussner I, Feussner K, Meinicke P, Stierhof YD, Schwarz H, Macek B, Mann M, Kahmann R (2011) Metabolic priming by a secreted fungal effector. Nature 478:395–398

Dorjgotov D, Jurca ME, Fodor-Dunai C, Szucs A, Otvos K, Klement E, Biro J, Feher A (2009) Plant Rho-type (Rop) GTPase-dependent activation of receptor-like cytoplasmic kinases in vitro. FEBS Lett 583:1175–1182

Doughty KJ, Bennett RN, Nashaat NI, Schrijvers S, Kiddle G, Pye BJ, Mitchell SE, Wallsgrove RM (1995) The response of oilseed rape (*Brassica napus* L.) seedlings to *Peronospora parasitica* and *Alternaria brassicae* following treatment with salicylic acid or methyl jasmonate. In: Proceedings of the 9th international rapeseed congress, Cambridge, UK, July 4-7, 1995, vol 3, pp 971–973

Durrant WE, Dong X (2004) Systemic acquired resistance. Annu Rev Phytopathol 42:185–209. https://doi.org/10.1146/annurev.phyto.42.040803.140421

Dutt M, Barthe G, Irey M, Grosser J (2015) Transgenic citrus expressing an *Arabidopsis* NPR1 gene exhibit enhanced resistance against huanglongbing (HLB; citrus greening). PLoS One 10: e0137134. https://doi.org/10.1371/journal.pone.0147657

Eckey C, Korell M, Leib K, Biedenkopf D, Jansen C, Langen G, Kogel KH (2004) Identification of powdery mildew-induced barley genes by cDNA-AFLP: functional assessment of an early expressed MAP kinase. Plant Mol Biol 55:1–15

Ederli L, Dawe A, Pasqualini S, Quaglia M, Xiong L, Gehring C (2015) *Arabidopsis* flower specific defense gene expression patterns affect resistance to pathogens. Front Plant Sci 6:79. https://doi.org/10.3389/fpls.2015.00079

Eggert D, Naumann M, Reimer R, Voigt CA (2014) Nanoscale glucan polymer network causes pathogen resistance. Sci Rep 4:4159. https://doi.org/10.1038/srep04159

Ehness R, Ecker M, Godt DE, Roitsch T (1997) Glucose and stress independently regulate source and sink metabolism and defence mechanisms via signal transduction pathways involving protein phosphorylation. Plant Cell 9:1825–1841

El Modafar C, El Boustani E (2001) Cell wall-bound phenolic acid and lignin contents in date palm as related to its resistance to *Fusarium oxysporum*. Biol Plant 44:125–130. https://doi.org/10.1023/A:1017942927058

Ellinger D, Voigt CA (2014) Callose biosynthesis in *Arabidopsis* with a focus on pathogen response: what we have learned within the last decade. Ann Bot 114:1349–1358

Ellinger D, Naumann M, Falter C, Zwikowics C, Jamrow T, Manisseri C, Somerville SC, Voigt CA (2013) Elevated early callose deposition results in complete penetration resistance to powdery mildew in *Arabidopsis*. Plant Physiol 161:1433–1444

Ellinger D, Glockner A, Koch J, Naumann M, Sturtz V, Schütt K, Manisseri C, Somerville SC, Voigt CA (2014) Interaction of the *Arabidopsis* GTPase RabA4c with its effector PMR4 results in complete penetration resistance to powdery mildew. Plant Cell 26:3185–3200

Ellis C, Karafyllidis I, Turner JG (2002a) Constitutive activation of jasmonate signaling in an *Arabidopsis* mutant correlates with enhanced resistance to *Erysiphe cichoracearum*, *Pseudomonas syringae*, and *Myzus persicae*. Mol Plant Microbe Interact 15:1025–1030. https://doi.org/10.1094/MPMI.2002.15.10.1025

Ellis C, Karafyllidis I, Wasternack C, Turner JG (2002b) The *Arabidopsis* mutant cev1 links cell wall signaling to jasmonate and ethylene responses. Plant Cell 14:1557–1566. https://doi.org/10.1105/tpc.002022

Enright S, Cipollini D (2007) Infection by powdery mildew *Erysiphe cruciferarum* (Erysiphaceae) strongly affects growth and fitness of *Alliaria petiolata* (Brassicaceae). Am J Bot 94:1813–1820

Eulgem T, Somssich IE (2007) Networks of WRKY transcription factors in defense signaling. Curr Opin Plant Biol 10:366–371

Eulgem T, Rushton PJ, Robatzek S, Somssich IE (2000) The WRKY super family of plant transcription factors. Trends Plant Sci 5:199–206. https://doi.org/10.1016/S1360-1385(00)01600-9

Fabro G, Di Rienzo JA, Voigt CA, Savchenko T, Dehesh K, Somerville S, Alvarez ME (2008) Genome-wide expression profiling *Arabidopsis* at the stage of *Golovinomyces cichoracearum* haustorium formation. Plant Physiol 146:1421–1439

Fahey JW, Zalcmann AT, Talalay P (2001) The chemical diversity and distribution of glucosinolates and isothiocyanates among plants. Phytochemistry 56:5–51

Falk A, Feys BJ, Frost LN, Jones JD, Daniels MJ, Parker JE (1999) EDS1, an essential component of R gene-mediated disease resistance in *Arabidopsis* has homology to eukaryotic lipases. Proc Natl Acad Sci U S A 96:3292–3297

Fan J, Crooks C, Creissen G, Hill L, Fairhurst S, Doerner P, Lamb C (2011) *Pseudomonas* sax genes overcome aliphatic isothiocyanate-mediated non-host resistance in *Arabidopsis*. Science 331:1185–1188

Fauteux F, Remus-Borel W, Menzies JG, Belanger RR (2005) Silicon and plant disease resistance against pathogenic fungi. FEMS Microbiol Lett 249:1–6. https://doi.org/10.1016/j.femsle.2005.06.034

Fauteux F, Chain F, Belzile F, Menzies JG, Belanger RR (2006) The protective role of silicon in the *Arabidopsis*-powdery mildew pathosystem. Proc Natl Acad Sci U S A 103:17554–17559

Feechan A, Kwon E, Yun BW, Wang Y, Pallas JA, Loake GJ (2005) A central role for S-nitrosothiols in plant disease resistance. Proc Natl Acad Sci U S A 102:8054–8059

Feechan A, Kabbara S, Dry IB (2011) Mechanisms of powdery mildew resistance in the Vitaceae family. Mol Plant Pathol 12:263–274

Feng J, Hwang R, Hwang SF, Strelkov SE, Gossen BD, Zhou QX, Peng G (2010) Molecular characterization of a serine protease Pro1 from *Plasmodiophora brassicae* that stimulates resting spore germination. Mol Plant Pathol 11(4):503–512. https://doi.org/10.1111/J.1364-3703.2010.00623.X

Ferrari S, Plotnikova JM, De Lorenzo G, Ausubel FM (2003) *Arabidopsis* local resistance to *Botrytis cinerea* involves salicylic acid and camalexin and requires EDS4 and PAD2, but not SID2, EDS5 or PAD4. Plant J 35:193–205. https://doi.org/10.1046/j.1365-313X.2003.01794.x

Feys BJ, Moisan LJ, Newman MA, Parker JE (2001) Direct interaction between the *Arabidopsis* disease resistance signaling proteins, EDS1 and PAD4. EMBO J 20:5400–5411

Feys BJ, Wiermer M, Bhat RA, Moisan LJ, Medina-Escobar N, Neu C, Cabral A, Parker JE (2005) *Arabidopsis* SENESCENCE-ASSOCIATED GENE101 stabilizes and signals within an ENHANCED DISEASE SUSCEPTIBILITY1 complex in plant innate immunity. Plant Cell 17:2601–2613

Fotopoulos V, Gilbert MJ, Pittman JK, Marvier AC, Buchanan AJ, Sauer N, Hall JL, Williams LE (2003) The monosaccharide transporter gene, *AtSTP4,* and the cell-wall invertase, *Atfruct1*, are induced in *Arabidopsis* during infection with the fungal biotroph *Erysiphe cichoracearum*. Plant Physiol 132(2):821–829

Freymark G, Diehl T, Miklis M, Romeis T, Panstruga R (2007) Antagonistic control of powdery mildew host cell entry by barley calcium-dependent protein kinases (CDPKs). Mol Plant-Microbe Interact 20:1213–1221

Friedrich L, Lawton K, Dietrich R, Willits M, Cade R, Ryals J (2001) NIM1 overexpression in *Arabidopsis* potentiates plant disease resistance and results in enhanced effectiveness of fungicides. Mol Plant Microbe Interact 14:1114–1124. https://doi.org/10.1094/MPMI.2001.14.9.1114

Frye CA, Innes RW (1998) An *Arabidopsis* mutant with enhanced resistance to powdery mildew. Plant Cell 10:947–956

Frye CA, Tang D, Innes RW (2001) Negative regulation of defense responses in plants by a conserved MAPKK kinase. Proc Natl Acad Sci U S A 98:373–378

Fu ZQ, Dong X (2013) Systemic acquired resistance: turning local infection into global defense. Annu Rev Plant Biol 64:839–863. https://doi.org/10.1146/annurev-arplant-042811-105606

Fu Y, Gu Y, Zheng Z, Wasteneys G, Yang Z (2005) *Arabidopsis* interdigitating cell growth requires two antagonistic pathways with opposing action on cell morphogenesis. Cell 120:687–700

Fu D, Uauy C, Distelfeld A, Blechl A, Epstein L, Chen X, Sela H, Fahima T, Dubcovsky J (2009) A kinase-START gene confers temperature-dependent resistance to wheat stripe rust. Science 323 (5919):1357–1360. https://doi.org/10.1126/science.1166289

Fu ZQ, Yan S, Saleh A, Wang W, Ruble J, Oka N et al (2012) NPR3 and NPR4 are receptors for the immune signal salicylic acid in plants. Nature 486:228–232. https://doi.org/10.1038/nature11162

Fu F, Liu X, Wang R, Zhai C, Peng G, Yu F, Fernando WGD (2019) Fine mapping of *Brassica napus* blackleg resistance gene Rlm1 through bulked segregant RNA sequencing. Sci Rep 9 (1):14600

Fuchs R, Kopischke M, Klapprodt C, Hause G, Meyer AJ, Schwarzlander M, Fricker MD, Lipka V (2016) Immobilized subpopulations of leaf epidermal mitochondria mediate PEN2-dependent pathogen entry control in *Arabidopsis*. Plant Cell 28:130–145

Galindo-Gonzalez L, Manolii V, Hwang SF, Strelkov SE (2020) Response of *Brassica napus* to *Plasmodiophora brassicae* involves salicylic acid-mediated immunity: an RNA-Seq-based study. Front Plant Sci 11:1025. https://doi.org/10.3389/fpls.2020.01025

Gamir J, Darwiche R, Van't Hof P, Choudhary V, Stumpe M, Schneiter R, Mauch F (2017) The sterol-binding activity of PATHOGENESIS-RELATED PROTEIN 1 reveals the mode of action of an antimicrobial protein. Plant J 89:502–509

Gardiner J (2013) The evolution and diversification of plant microtubule associated proteins. Plant J 75:219–229

Genre A, Chabaud M, Timmers T (2005) Arbuscular mycorrhizal fungi elicit a novel intracellular apparatus in *Medicago truncatula* root epidermal cells before infection. Plant Cell 17:3489–3499

Ghanmi D, McNally DJ, Benhamou N, Menzies JG, Belanger RR (2004) Powdery mildew of *Arabidopsis thaliana*: a pathosystem for exploring the role of silicon in plant-microbe interactions. Physiol Mol Plant Pathol 64:189–199

Gil F, Gay JL (1977) Ultra-structural and physiological properties of the host interfacial components of haustoria of *Erysiphe pisi in vivo* and *in vitro*. Physiol Plant Pathol 10:1–4

Gilmore B, Myers JR, Kean D (2002) Completion of testing of *Phaseolus coccineus* plant introductions (Pls) for white mold, *Sclerotinia sclerotiorum*, resistance. Ann Rep Bean Improv Coop, p. 45

Gish LA, Clark SE (2011) The RLK/Pelle family of kinases. Plant J66:117–127

Gjetting T, Carver TL, Skot L, Lyngkjaer MF (2004) Differential gene expression in individual papilla-resistant and powdery mildew-infected barley epidermal cells. Mol Plant Microbe Interact 176(1):729–738

Gjetting T, Hagedorn PH, Schweizer P, Thordal-Christensen H, Carver TL, Lyngkjaer MF (2007) Single-cell transcript profiling of barley attacked by the powdery mildew fungus. Mol Plant Microbe Interact 206(1):235–246

Glawischnig E, Hansen BG, Olsen CE, Halkier BA (2004) Camalexin is synthesized from indole-3-acetaldoxime, a key branching point between primary and secondary metabolism in *Arabidopsis*. Proc Natl Acad Sci U S A 101:8245–8250. https://doi.org/10.1073/pnas.0305876101

Glazebrook J (2005) Contrasting mechanisms of defense against biotrophic and necrotrophic pathogens. Annu Rev Phytopathol 43:205–227. https://doi.org/10.1146/annurev.phyto.43.040204.135923

Glazebrook J, Rogers EE, Ausubel FM (1996) Isolation of *Arabidopsis* mutants with enhanced disease susceptibility by direct screening. Genetics 143:973–982

Glazebrook J, Zook M, Mert F, Kagan I, Rogers EE, Crute IR, Holub EB, Hammerschmidt R, Ausubel FM (1997) Phytoalexin-deficient mutants of *Arabidopsis* reveal that PAD4 encodes a regulatory factor and that four PAD genes contribute to downy mildew resistance. Genetics 146:381–392

Godika S, Pathak AK (2005) Control of white rust and Alternaria blight diseases of mustard by foliar sprays of Ridomil. Pestol 29:9–10

Gomez-Gomez L, Felix G, Boller T (1999) A single locus determines sensitivity to bacterial flagellin in *Arabidopsis thaliana*. Plant J 18:277–284

Grant M, Lamb C (2006) Systemic immunity. Curr Opin Plant Biol 9:414–420. https://doi.org/10.1016/j.pbi.2006.05.013

Green JR, Carver TL, Gurr SJ (2002) The formation and function of infection and feeding structures. In: Belanger RR, Bushnell WR, Dik AJ, Carver TL (eds) The powdery mildews: a comprehensive treatise. APS Press, St Paul, pp 66–82

Greenhalgh JR, Mitchell ND (1976) The involvement of flavour volatiles in the resistance to downy mildew of wild and cultivated forms of *Brassica oleracea*. New Phytol 77:391–398

Grubb CD, Abel S (2006) Glucosinolate metabolism and its control. Trends Plant Sci 11:89–100. https://doi.org/10.1016/j.tplants.2005.12.006

Gu YQ, Wildermuth MC, Chakravarthy S, Loh YT, Yang C, He X, Han Y, Martin GB (2002) Tomato transcription factors Pti4, Pti5, and Pti6 activate defense responses when expressed in *Arabidopsis*. Plant Cell 14:817–831

Guo CY, Wu GH, Xing J, Li WQ, Tang DZ, Cui BM (2013) A mutation in a coproporphyrinogen III oxidase gene confers growth inhibition, enhanced powdery mildew resistance and powdery mildew-induced cell death in *Arabidopsis*. Plant Cell Rep 32:687–702

Gus-Meyer S, Naton B, Hahlbrock K, Schmelzer E (1998) Local mechanical stimulation induces components of the pathogen defense response in parsley. Proc Natl Acad Sci U S A 146 (1):8398–8403

Hagemeier J, Schneider B, Oldham NJ, Hahlbrock K (2001) Accumulationof soluble and wall-bound indolic metabolites in *Arabidopsis thaliana* leaves infected with virulent or avirulent *Pseudomonas syringae* pathovar *tomato* strains. Proc Natl Acad Sci U S A 98:753–758

Hahlbrock K, Bednarek P, Ciolkowski I, Hamberger B, Heise A, Liedgens H, Logemann E, Nürnberger T, Schmelzer E, Somssich IE et al (2003) Non-self recognition, transcriptional reprogramming, and secondarymetabolite accumulation during plant/pathogen interactions. Proc Natl Acad Sci U S A 100:14569–14576

Halkier BA, Gershenzon J (2006) Biology and biochemistry of glucosinolates. Annu Rev Plant Biol 57:303–333. https://doi.org/10.1146/annurev.arplant.57.032905.105228
Hamada T (2014) Microtubule organization and microtubule-associated proteins in plant cells. Int Rev Cell Mol Biol 312:1–52
Hammerschmidt R (1999) Phytoalexins: what have we learned after 60 year? Annu Rev Phytopathol 37:285–306
Hammerschmidt R, Kuc J (1982) Lignification as a mechanism for induced systemic resistance in cucumber. Physiol Plant Pathol 20:61–71. https://doi.org/10.1016/0048-4059(82)90024-8
Hammerschmidt R, Lamport DTA, Muldoon EP (1984) Cell wall hydroxyproline enhancement and lignin deposition as an early event in the resistance of cucumber to *Cladosporium cucumerinum*. Physiol Plant Pathol 24:43–47
Hansen CH, Du L, Naur P, Olsen CE, Axelsen KB, Hick AJ, Pickett JA, Halkier BA (2001) CYP83b1 is the oxime-metabolizing enzyme in the glucosinolate pathway in *Arabidopsis*. J Biol Chem 276:24790–24796
Hardham AR (2013) Microtubules and biotic interactions. Plant J 75:278–289
Hardham AR, Jones DA, Takemoto D (2007) Cytoskeleton and cell wall function in penetration resistance. Curr Opin Plant Biol 10:342–348. https://doi.org/10.1016/j.pbi.2007.05.001
Hardham AR, Takemoto D, White RG (2008) Rapid and dynamic sub-cellular reorganization following mechanical stimulation of *Arabidopsis* epidermal cells mimics responses to fungal and oomycete attack. BMC Plant Biol 8:63
Harel TM, Mehar ZHI, Rav-David D, Elad Y (2014) Systemic resistance to gray mold induced in tomato by Benzothiadiazole and *Trichoderma harzianum* T39. Phytopathology 104:150–157
Hasan MJ, Rahman H (2016) Genetics and molecular mapping of resistance to *Plasmodiophora brassicae* pathotypes 2, 3, 5, 6, and 8 in rutabaga (*Brassica napus* var. *napobrassica)*. Genome 59:805–815. https://doi.org/10.1139/gen-2016-0034
Hatakeyama K, Suwabe K, Tomita RN, Kato T, Nunome T, Fukuoka H, Matsumoto S (2013) Identification and characterization of Crr1a, a gene for resistance to clubroot disease (*Plasmodiophora brassicae* Worn.) in *Brassica rapa* L. PLoS One 8:e54745. https://doi.org/10.1371/journal.pone.0054745
Heath MC (2000) Non-host resistance and nonspecific plant defences. Curr Opin Plant Biol 3:315–319
Hejna O, Havlickova L, He Z, Bancroft I, Curn V (2019) Analysing the genetic architecture of clubroot resistance variation in Brassica napus by associative transcriptomics. Mol Breed 39:112. https://doi.org/10.1007/s11032-019-1021-4
Hemm MR, Ruegger MO, Chapple C (2003) The *Arabidopsis* ref2 mutant is defective in the gene encoding CYP83A1 and shows both phenylpropanoid and glucosinolate phenotypes. Plant Cell 15:179–194. https://doi.org/10.1105/tpc.006544
Hepworth SR, Zhang Y, McKim S, Li X, Haughn GW (2005) BLADE-ON-PETIOLE-dependent signaling controls leaf and floral patterning in *Arabidopsis*. Plant Cell 17:1434–1448. https://doi.org/10.1105/tpc.104.030536
Herbers K, Takahata Y, Melzer M, Mock H-P, Hajirezaei M, Sonnewald U (2000) Regulation of carbohydrate partitioning during the interaction of potato virus Y with tobacco. Mol Plant Pathol 1:51–59
Hermosa R, Viterbo A, Chet I, Monte E (2012) Plant-beneficial effects of *Trichoderma* and of its genes. Microbiology 158:17–25
Hijwegen T (1963) Lignification, a possible mechanism of active response against pathogens. Neth J Plant Pathol 69:314–317
Ho CMK, Hotta T, Guo F, Roberson RW, Lee YRJ, Liu B (2011) Interaction of antiparallel microtubules in the phragmoplast is mediated by the microtubule-associated protein MAP65-3 in *Arabidopsis*. Plant Cell 23:2909–2923
Hoefle C, Huesmann C, Schultheiss H, Bornke F, Hensel G, Kumlehn J, Huckelhoven R (2011) A barley ROP GTPase ACTIVATING PROTEIN associates with microtubules and regulates entry of the barley powdery mildew fungus into leaf epidermal cells. Plant Cell 23:2422–2439

Hossain MY, Jasmine S, Ibrahim AHM, Ahmed ZF, Rahman MM, Ohtomi J (2008) Length-weight and length-length relationships of ten small fish species from the Ganges (Bangladesh). J Appl Ichthyol 25:117–119

Hou CT, Forman RJ (2000) Growth inhibition of plant pathogenic fungi by hydroxy fatty acids. J Ind Microbial Biochem 24:275–276

Huang R, Hippauf F, Rohrbeck D, Haustein M, Wenke K, Feike J, Sorelle N, Piechulla B, Barkman TJ (2012) Enzyme functional evolution through improved catalysis of ancestrally nonpreferred substrates. Proc Natl Acad Sci U S A 109:2966–2971

Huckelhoven R (2005) Powdery mildew susceptibility and biotrophic infection strategies. FEMS Microbiol Lett 245:9–17

Huckelhoven R (2007) Cell wall-associated mechanisms of disease resistance and susceptibility. Annu Rev Phytopathol 45:101–127

Hückelhoven R, Panstruga R (2011) Cell biology of the plant powdery mildew interaction. Curr Opin Plant Biol 14:738–746

Huckelhoven R, Fodor J, Preis C, Kogel KH (1999) Hypersensitive cell death and papilla formation in barley attacked by the powdery mildew fungus are associated with hydrogen peroxide but not with salicylic acid accumulation. Plant Physiol 119:1251–1260. https://doi.org/10.1104/pp.119.4.1251

Huckelhoven R, Eichmann R, Weis C, Hoefle C, Proels RK (2013) Genetic loss of susceptibility: a costly route to disease resistance? Plant Pathol 62(Suppl 1):56–62

Huesmann C, Hoefle C, Huckelhoven R (2011) ROPGAPs of *Arabidopsis* limit susceptibility to powdery mildew. Plant Signal Behav 6:1691–1694

Huesmann C, Reiner T, Hoefle C, Preuss J, Jurca ME, Domoki M, Feher A, Huckelhoven R (2012) Barley ROP binding kinase1 is involved in microtubule organization and in basal penetration resistance to the barley powdery mildew fungus. Plant Physiol 159:311–320

Hull AK, Vij R, Celenza JL (2000) *Arabidopsis* cytochrome P450s that catalyze the first step of tryptophan-dependent indole-3-acetic acid biosynthesis. Proc Natl Acad Sci U S A 97:2379–2384. https://doi.org/10.1073/pnas.040569997

Humphry M, Consonni C, Panstruga R (2006) mlo-based powdery mildew immunity: silver bullet or simply non-host resistance? Mol Plant Pathol 7:605–610. https://doi.org/10.1111/j.1364-3703.2006.00362.x

Humphry M, Bednarek P, Kemmerling B, Koh S, Stein M, Gobel U, Stuber K, Pislewska-Bednarek M, Loraine A, Schulze-Lefert P, Somerville S, Panstruga R (2010) A regulon conserved in monocot and dicot plants defines a functional module in antifungal plant immunity. Proc Natl Acad Sci U S A 107:21896–21901. https://doi.org/10.1073/pnas.1003619107

Hussain A, Noman A, Khan MI, Zaynab M, Aqeel M, Anwar M, Ashraf MF, Liu Z, Raza A, Mahpara S (2019) Molecular regulation of pepper innate immunity and stress tolerance: an overview of WRKY TFs. Microb Pathog 135:103610. https://doi.org/10.1016/j.micpath.2019.103610

Hussey PJ, Hawkins TJ, Igarashi H, Kaloriti D, Smertenko A (2002) The plant cytoskeleton: recent advances in the study of the plant microtubule-associated proteins MAP-65, MAP-190 and the Xenopus MAP215-like protein, MOR1. Plant Mol Biol 50:915–924

Innes R (2018) The positives and negatives of NPR: a unifying model for salicylic acid signaling in plants. Cell 173:1314–1315

Irani S, Trost B, Waldner M, Nayidu N, Tu J, Kusalik AJ, Todd CD, Wei Y, Bonham-Smith PC (2018) Transcriptome analysis of response to *Plasmodiophora brassicae* infection in the Arabidopsis shoot and root. BMC Genomics 19:23. https://doi.org/10.1186/s12864-017-4426-7

Jacobs AK, Lipka V, Burton RA, Panstruga R, Strizhov N, Schulze-Lefert P, Fincher GB (2003) An *Arabidopsis* callose synthase, GSL5, is required for wound and papillary callose formation. Plant Cell 15:2503–2513

Jacobs S, Zechmann B, Molitor A, Trujillo M, Petutschnig E, Lipka V, Kogel KH, Schcafer P (2011) Broad-spectrum suppression of innate immunity is required for colonization of *Arabidopsis* roots by the fungus *Piriformospora indica*. Plant Physiol 156:726–740

Jahn L, Mucha S, Bergmann S, Horn C, Staswick P, Steffens B, Siemens J, Ludwig-Muller J (2013) The clubroot pathogen (*Plasmodiophora brassicae*) influences auxin signaling to regulate auxin homeostasis in *Arabidopsis*. Plants 1:726–749. https://doi.org/10.3390/plants2040726

Jain D, Khurana JP (2018) Molecular aspects of plant–pathogen interaction. In: Role of pathogenesis-related (PR) proteins in plant defense mechanism. Springer, Singapore, pp 265–281

Jensen MK, Rung JH, Gregersen PL, Gjetting T, Fuglsang AT, Hansen M, Joehnk N, Lyngkjaer MF, Collinge DB (2007) The *Hv*NAC6 transcription factor: a positive regulator of penetration resistance in barley and *Arabidopsis*. Plant Mol Biol 65:137–150

Jensen MK, Hagedorn PH, de Torres-Zabala M, Grant MR, Rung JH, Collinge DB, Lyngkjaer MF (2008) Transcriptional regulation by an NAC (NAM-ATAF1, 2-CUC2) transcription factor attenuates ABA signalling for efficient basal defence towards *Blumeria graminis* f. sp. *hordei* in *Arabidopsis*. Plant J 56:867–880

Jia H, Wei X, Yang Y, Yuan Y, Wei Y, Zhao F, Yang Y, Yao S, Wang Q, Tian Z, Zhang B, Wei X (2017) Root RNA-seq analysis reveals a distinct transcriptome landscape between clubroot susceptible and clubroot-resistant Chinese cabbage lines after *Plasmodiophora brassicae* infection. Plant Soil 421:93–105. https://doi.org/10.1007/s11104-017-3432-5

Jiang Z, Dong X, Zhang Z (2016) Network-based comparative analysis of *Arabidopsis* immune responses to *Golovinomyces orontii* and *Botrytis cinerea* infections. Sci Rep 6:19149. https://doi.org/10.1038/srep19149

Jirage D, Tootle TL, Reuber TL, Frost LN, Feys BJ, Parker JE, Ausubel FM, Glazebrook J (1999) *Arabidopsis thaliana* PAD4 encodes a lipase-like gene that is important for salicylic acid signaling. Proc Natl Acad Sci U S A 96:13583–13588

Jones JD, Dangl JL (2006) The plant immune system. Nature 444:323–329

de Jong CF, Laxalt AM, Bargmann BO, de Wit PJ, Joosten MH, Munnik T (2004) Phosphatidic acid accumulation is an early response in the *Cf-4/Avr4* interaction. Plant J 39:1–12

Jorgensen JH (1992) Discovery, characterization and exploitation of Mlo powdery mildew resistance in barley. Euphytica 63:141–152

Jubault M, Hamon C, Gravot A, Lariagon C, Delourme R, Bouchereau A, Manzanares-Dauleux MJ (2008) Differential regulation of root arginine catabolism and polyamine metabolism in clubroot-susceptible and partially resistant *Arabidopsis* genotypes. Plant Physiol 146:2008–2019

Jubault M, Lariagon C, Gravot A, Delourme R, Manzanares-dauleux MJ (2013) Partial resistance to clubroot in *Arabidopsis* is based on changes in the host primary metabolism and targeted cell division and expansion capacity. Funct Integr Genomics 13:191–205. https://doi.org/10.1007/s10142-013-0312-9

Jung JSEC (2008) Genetic mapping of clubroot resistance genes in oilseed rape. Theor Appl Genet 116:363–372. https://doi.org/10.1007/s00122-007-0674-2

Jurca ME, Bottka S, Feher A (2008) Characterization of a family of*Arabidopsis* receptor-like cytoplasmic kinases (RLCK class VI). Plant Cell Rep 27:739–748

Kachroo A, Kachroo P (2009) Fatty acid-derived signals in plant defense. Annu Rev Phytopathol 47:153–176

Kawano Y, Kaneko-Kawano T, Shimamoto K (2014) Rho family GTPase-dependent immunity in plants and animals. Front Plant Sci 5:522

Kehr J (2003) Single cell technology. Curr Opin Plant Biol 66(1):617–621

Kemen E, Kemen AC, Rafiqi M, Hempel U, Mendgen K, Hahn M, Voegele RT (2005) Identification of a protein from rust fungi transferred from haustoria into infected plant cells. Mol Plant Microbe Interact 18:1130–1139

Kirik A, Mudgett MB (2009) SOBER1 phospholipase activity suppresses phosphatidic acid accumulation and plant immunity in response to bacterial effector AvrBsT. Proc Natl Acad Sci U S A 106:20532–20537

Kliebenstein DJ, Rowe HC, Denby KJ (2005) Secondary metabolites influence *Arabidopsis*/*Botrytis* interactions: variation in host production and pathogen sensitivity. Plant J 44:25–36. https://doi.org/10.1111/j.1365-313X.2005.02508.x

Kluczewski SM, Lucas JA (1982) Development and physiology of infection by the downy mildew fungus *Peronospora parasitica* (Pers. ex Fr.) Fr. in susceptible and resistant *Brassica* species. Plant Pathol 31:373–389

Knaust A, Ludwig-Muller J (2013) The ethylene signalling pathway is needed to restrict root gall growth in *Arabidopsis* after infection with the obligate biotrophic protist *Plasmodiophora brassicae*. J Plant Growth Regul 32:9–21

Kobayashi I, Kobayashi Y, Hardham AR (1994) Dynamic reorganization of microtubules and microfilaments in flax cells during the resistance response to flax rust infection. Planta 61:237–247

Kobayashi Y, Kobayashi I, Funaki Y, Fujimoto S, Takemoto T, Kunoh H (1997a) Dynamic reorganization of microfilaments and microtubules is necessary for the expression of non-host resistance in barley coleoptile cells. Plant J 116(1):525–537

Kobayashi Y, Yamada M, Kobayashi I, Kunoh H (1997b) Actin microfilaments are required for the expression of nonhost resistance in higher plants. Plant Cell Physiol 386(1):725–733

Koch E, Slusarenko AJ (1990) Fungal pathogens of *Arabidopsis thaliana* (L.) Heyhn. Bot Helvet 100:257–268

Koh S, Andre A, Edwards H, Ehrhardt D, Somerville S (2005) *Arabidopsis thaliana* subcellular responses to compatible *Erysiphe cichoracearum* infections. Plant J 446(1):516–529

Kolattukudy PE (1974) Biosynthesis of a hydroxy fatty acid polymer, cutin. Identification and biosynthesis of 16-oxo-9- or 10-hydroxypalmitic acid, a novel compound in *Vicia faba*. Biochemistry 13:1354–1363

Kolattukudy PE (1980) Biopolyester membranes of plants: cutin and suberin. Science 208:990–1000

Koo YJ, Kim MA, Kim EH, Song JT, Jung C, Moon JK, Kim JH, Seo HS, Song SI, Kim JK, Lee JS, Cheong JJ, Choi YD (2007) Overexpression of salicylic acid carboxyl methyltransferase reduces salicylic acid-mediated pathogen resistance in *Arabidopsis thaliana*. Plant Mol Biol 64:1–15

Korolev N, David DR, Elad Y (2008) The role of phytohormones in basal resistance and *Trichoderma*-induced systemic resistance to *Botrytis cinerea* in *Arabidopsis thaliana*. BioControl 53:667–683

Kragler F, Curin M, Trutnyeva K, Gansch A, Waigmann E (2003) MPB2C, a microtubule-associated plant protein binds to and interferes with cell-to-cell transport of tobacco mosaic virus movement protein. Plant Physiol 132:1870–1883

Kuhn H, Kwaaitaal M, Kusch S, Acevedo-Garcia J, Wu H, Panstruga R (2016) Biotrophy at its best: novel findings and unsolved mysteries of the *Arabidopsis*-powdery mildew pathosystem. Arabidopsis Book 14:e0184. https://doi.org/10.1199/tab.0184

Kuhn H, Lorek J, Kwaaitaal M, Consonni C, Becker K, Micali C, Ver Loren van Themaat E, Bednarek P, Raaymakers TM, Appiano M, Bai Y, Meldau D, Baum S, Conrath U, Feussner I, Panstruga R (2017) Key components of different plant defense pathways are dispensable for powdery mildew resistance of the *Arabidopsis* mlo2 mlo6 mlo12 triple mutant. Front Plant Sci 8:1006. https://doi.org/10.3389/fpls.2017.01006

Kumar D (2014) Salicylic acid signaling in disease resistance. Plant Sci 228:127–134

Kusch S, Panstruga R (2017) mlo-based resistance: an apparently universal "weapon" to defeat powdery mildew disease. Mol Plant Microbe Interact 30:179–189. https://doi.org/10.1094/mpmi-12-16-0255-cr

Kusch S, Pesch L, Panstruga R (2016) Comprehensive phylogenetic analysis sheds light on the diversity and origin of the MLO family of integral membrane proteins. Genome Biol Evol 8:878–895

Kwaaitaal M, Keinath NF, Pajonk S, Biskup C, Panstruga R (2010) Combined bimolecular fluorescence complementation and förster resonance energy transfer reveals ternary SNARE complex formation in living plant cells. Plant Physiol 152:1135–1147. https://doi.org/10.1104/pp.109.151142

Kwon C, Panstruga R, Schulze-Lefert P (2008) Les liaisons dangereuses: immunological synapse formation in animals and plants. Trends Immunol 29(1):159–166

Laluk K, Luo H, Chai M, Dhawan R, Lai Z, Mengiste T (2011) Biochemical and genetic requirements for function of the immune response regulator BOTRYTIS-INDUCED KINASE1 in plant growth, ethylene signaling, and PAMP-triggered immunity in *Arabidopsis*. Plant Cell 23:2831–2849

Lammens T, Boudolf V, Kheibarshekan L, Zalmas LP, Gaamouche T, Maes S, Vanstraelen M, Kondorosi E, La Thangue NB, Govaerts W, Inze D, De Veylder L (2008) Atypical E2F activity restrains APC/C^{CCS52A2} function obligatory for endocycle onset. Proc Natl Acad Sci U S A 105:14721–14726

Lapin D, Van den Ackerveken G (2013) Susceptibility to plant disease: more than a failure of host immunity. Trends Plant Sci 18:546–554

Lapin D, Kovacova V, Sun X, Dongus J, Bhandari DD, Born PV, Bautor J, Guarneri N, Stuttmann J, Beyer A, Parker JE (2019) A coevolved EDS1-SAG101-NRG1 module mediates cell death signaling by TIR-domain immune receptors. bioRxiv: 572826.

Lebel E, Heifetz P, Thorne L, Uknes S, Ryals J, Ward E (1998) Functional analysis of regulatory sequences controlling PR-1 gene expression in *Arabidopsis*. Plant J 16:223–233

Lee J, Izzah NK, Choi BS, Joh HJ, Lee SC, Perumal S, Seo J, Ahn K, Jo EJ, Choi GJ, Nou IS, Yu Y, Yang TJ (2016) Genotyping-by-sequencing map permits identification of clubroot resistance QTLs and revision of the reference genome assembly in cabbage (*Brassica oleracea* L.). DNA Res 23:29–41. https://doi.org/10.1093/dnares/dsv034

Lemarie S, Robert-Seilaniantz A, Lariagon C, Lemoine J, Marnet N, Jubault M, Manzanares-Dauleux MJ, Gravot A (2015) Both the jasmonic acid and the salicylic acid pathways contribute to resistance to the biotrophic clubroot agent *Plasmodiophora brassicae* in *Arabidopsis*. Plant Cell Physiol 56:2158–2168

Li J, Ou-Lee TM, Raba R, Amundson RG, Last RL(1993) *Arabidopsis* flavonoid mutants are hypersensitive to UV-B irradiation. Plant Cell 5:171–179

Li H, Shen JJ, Zheng ZL, Lin Y, Yang Z (2001) The Rop GTPase switch controls multiple developmental processes in Arabidopsis. Plant Physiol 126:670–684

Li R, Rimmer R, Buchwaldt L, Sharpe AG, Seguinswartz G, Coutu C, Hegedus DD (2004) Interaction of *Sclerotinia sclerotiorum* with a resistant *Brassica napus* cultivar: expressed Sequence Tag Analysis identifies genes associated with fungal pathogenesis. Fungal Genet Biol 41:735–753

Li J, Brader G, Kariola T, Palva ET (2006) WRKY70 modulates the selection of signaling pathways in plant defense. Plant J 46(3):477–491

Li J, Zhao Z, Hayward A, Cheng H, Fu D (2015) Integration analysis of quantitative trait loci for resistance to *Sclerotinia sclerotiorum* in *Brassica napus*. Euphytica 205:483–489. https://doi.org/10.1007/s10681-015-1417-0

Li GL, Qian W, Zhang SJ, Zhang SF, Li F, Zhang H, Wu J, Wang ZW, Sun RF (2016) Development of gene-basedmarkers for the Turnip mosaic virus resistance gene retr02 in *Brassica rapa*. Plant Breed 135:466–470

Liang YS, Choi YH, Kim HK, Linthorst HJM, Verpoorte R (2006a) Metabolomic analysis of methyl jasmonate treated *Brassica rapa* leaves by 2-dimensional NMR spectroscopy. Phytochemistry 67(22):2503–2511

Liang YS, Kim HK, Lefeber AWM, Erkelens C, Choi YH, Verpoorte R (2006b) Identification of phenylpropanoids in methyl jasmonate treated *Brassica rapa* leaves using two dimensional nuclear magnetic resonance spectroscopy. J Chromatogr A 1112:148–155

Lin W, Ma X, Shan L, He P (2013) Big roles of small kinases: thecomplex functions of receptor-like cytoplasmic kinases in plant immunity and development. J Integr Plant Biol 55:1188–1197

Lipka V, Dittgen J, Bednarek P, Bhat R, Wiermer M, Stein M, Landtag J, Brandt W, Rosahl S, Scheel D, Llorente F, Molina A, Parker J, Somerville S, Lefert PS (2005) Pre- and post-invasion defenses both contribute to non-host resistance in *Arabidopsis*. Science 310:1180–1183

Lipka V, Kwon C, Panstruga R (2007) SNARE-ware: the role of SNARE-domain proteins in plant biology. Annu Rev Cell Dev Biol 23:147–174

Lipka U, Fuchs R, Lipka V (2008) *Arabidopsis* non-host resistance to powdery mildews. Curr Opin Plant Biol 11:404–411

Liu Y, Bassham DC (2012) Autophagy: pathways for self-eating in plant cells. Annu Rev Plant Biol 63:215–237

Liu G, Holub EB, Alonso JM, Ecker JR, Fobert PR (2005) An *Arabidopsis* NPR1-like gene, NPR4, is required for disease resistance. Plant J 41:304–318. https://doi.org/10.1111/j.1365-313X.2004.02296x

Liu LJ, Zhang YY, Tang SY, Zhao QZ, Zhang ZH, Zhang HW, Dong L, Guo H, Xie Q (2010) An efficient system to detect protein ubiquitination by agroinfiltration in *Nicotiana benthamiana*. Plant J 61:893–903. https://doi.org/10.1111/j.1365-313X.2009.04109.x

Liu Y, Huang X, Li M, He P, Zhang Y (2016) Loss-of-function of *Arabidopsis* receptor-like kinase BIR1 activates cell death and defense responses mediated by BAK1 and SOBIR1. New Phytol 212:637–645

Loake G, Grant M (2007) Salicylic acid in plant defence—the players and protagonists. Curr Opin Plant Biol 10(5):466–472

Lorek J, Griebel T, Jones AM, Kuhn H, Panstruga R (2013) The role of *Arabidopsis* heterotrimeric G-protein subunits in MLO2 function and MAMP-triggered immunity. Mol Plant Microbe Interact 26:991–1003

Lovelock DA, Sola I, Marschollek S, Donald CE, Rusak G, van Pée KH, Ludwig-Müller J, Cahill DM (2016) Analysis of salicylic acid-dependent pathways in Arabidopsis thaliana following infection with Plasmodiophora brassicae and the influence of salicylic acid on disease. Mol Plant Pathol 17:1237–1251. https://doi.org/10.1111/mpp.12361

Lu D, Wu S, Gao X, Zhang Y, Shan L, He P (2010) A receptor-like cytoplasmic kinase, BIK1, associates with a flagellin receptor complex to initiate plant innate immunity. Proc Natl Acad Sci U S A 107:496–501

Lu X, Dittgen J, Pislewska-Bednarek M, Molina A, Schneider B, Svatos A, Doubsky J, Schneeberger K, Weigel D, Bednarek P, Schulze-Lefert P (2015) Mutant allele-specific uncoupling of PENETRATION3 functions reveals engagement of the ATP-binding cassette transporter in distinct tryptophan metabolic pathways. Plant Physiol 168:814–827. https://doi.org/10.1104/pp.15.00182

Ludwig-Muller J (2009a) Glucosinolates and the clubroot disease: defense compounds or auxin precursors? Phytochem Rev 8:135–148

Ludwig-Muller J (2009b) Plant defence—what can we learn from clubroots? Austral Plant Pathol 38:318–324

Ludwig-Muller J, Julke S, Geiß K, Richter F, Mithofer A, Sola I, Rusak G, Keenan S, Bulman S (2015) A novel methyltransferase from the intracellular pathogen *Plasmodiophora brassicae* methylates salicylic acid. Mol Plant Pathol 16(4):349–364

Luna E, Pastor V, Robert J, Flors V, Mauch-Mani B, Ton J (2011) Callose deposition: a multifaceted plant defense response. Mol Plant Microbe Interact 24:183–193

Maekawa T, Kufer TA, Schulze-Lefert P (2011) NLR functions in plant and animal immune systems: so far and yet so close. Nat Immun 12:817–826

Maekawa T, Kracher B, Vernaldi S, Ver Loren van Themaat E, Schulze-Lefert P (2012) Conservation of NLR-triggered immunity across plant lineages. Proc Natl Acad Sci U S A 109:20119–20123

Maekawa S, Inada N, Yasuda S, Fukao Y, Fujiwara M, Sato T, Yamaguchi J (2014) The carbon/nitrogen regulator *ARABIDOPSIS* TOXICOS EN LEVADURA31 controls papilla formation in response to powdery mildew fungi penetration by interacting with SYNTAXIN OF PLANTS121 in *Arabidopsis*. Plant Physiol 164:879–887

Makandar R, Essig JS, Schapaugh MA, Trick HN, Shah J (2006) Genetically engineered resistance to Fusarium head blight in wheat by expression of *Arabidopsis* NPR1. Mol Plant Microbe Interact 19:123–129. https://doi.org/10.1094/MPMI-19-0123

Maksimov V, Abizgil'dina RR, Pusenkova LI (2011) Plant growth promoting rhizobacteria as alternative to chemical crop protectors from pathogens (review). Appl Biochem Microbiol 47 (4):373–385

Malamy J, Klessig DF (1992) Salicylic acid and plant disease resistance. Plant J 2:643–654. https://doi.org/10.1111/j.1365-313X.1992.tb00133.x

Malinovsky FG, Fangel JU, Willats WGT (2014) The role of the cell wall in plant immunity. Front Plant Sci 5:178. https://doi.org/10.3389/fpls.2014.00178

Malinowski R, Novak O, Borhan MH, Spichal L, Strnad M, Rolfe SA (2016) The role of cytokinins in clubroot disease. Eur J Plant Pathol 145:543–557. https://doi.org/10.1007/s10658-015-0845-y

Manoharan RK, Shanmugam A, Hwang I, Park JI, Nou IS (2016) Expression of salicylic acid-related genes in *Brassica oleracea* var. *capitata* during *Plasmodiophora brassicae* infection. Genome 59:379–391. https://doi.org/10.1139/gen-2016-0018

Mansfield JW (2000) Antimicrobial compounds and resistance. The role of phytoalexins and phytoanticipins. In: Slusarenko A, Fraser RSS, van Loon LC (eds) Mechanisms of resistance to plant diseases. Kluwer Academic Publishers, Dordrecht, pp 325–370

Manzanares-Dauleux MJ, Delourne R, Baron F, Thomas G (2000) Mapping of one major gene and of QTLs involved in resistance to clubroot in *Brassica napus*. Theor Appl Genet 101:885–891. https://doi.org/10.1007/s001220051557

Mao G, Meng X, Liu Y, Zheng Z, Chen Z, Zhang S (2011) Phosphorylation of a WRKY transcription factor by two pathogen-responsive MAPKs drives phytoalexin biosynthesis in *Arabidopsis*. Plant Cell 23:1639–1653. https://doi.org/10.1105/tpc.111.084996

Martiniere A, Gargani D, Uzest M, Lautredou N, Blanc S, Drucker M (2009) A role for plant microtubules in the formation of transmission specific inclusion bodies of cauliflower mosaic virus. Plant J 58:135–146

Mathesius U (2001) Flavonoids induced in cells undergoing nodule organogenesis in white clover are regulators of auxin breakdown by peroxidase. J Exp Bot 52:419–426

Matsumoto I (1994) Lignin induction in the root tissues of Japanese radish infected with downy mildew fungus. Mem Coll Agric Ehime University, Ehime

Matsumoto I, Asada Y (1984) A role of lignification-inducing factor in resistance induction demonstrated in Japanese radish downy mildew and cucumber anthracnose. Ann Phytopathol Soc Japan 50:483–490

Matsumoto I, Ohguchi T, Inoue M, Asada Y (1978) Lignin induction in roots of Japanese radish by an homogenate of downy mildew-infected root tissue. Ann Phytopathol Soc Japan 44:22–27

Matsumoto E, Yasui C, Ohi M, Tsukada M (1998) Linkage analysis of RFLP markers for clubroot resistance and pigmentation in Chinese cabbage (*Brassica rapa* ssp. *pekinensis*). Euphytica 104:79–86

Mauch-Mani B, Slusarenko AJ (1996) Production of salicylic acid precursors is a major function of phenylalanine ammonia-lyase in the resistance of *Arabidopsis* to *Peronospora parasitica*. Plant Cell 8:203–212

Mauch-Mani B, Croft KPC, Slusarenko AJ (1993) The genetic basis of resistance of *Arabidopsis thaliana* L. Heyhn. to *Peronospora parasitica*. In: Davis KR, Hammerschmidt R (eds) *Arabidopsis thaliana* as a model for plant–pathogen interactions. APS Press, St Paul, pp 5–20

McDowell JM, Cuzick A, Can C, Beynon J, Dangl JL, Holub EB (2000) Downy mildew (*Peronospora parasitica*) resistance genes in *Arabidopsis* vary in functional requirements for NDR1, EDS1, NPR1, and salicylic acid accumulation. Plant J 22:523–530

Mei J, Guo Z, Wang J, Feng Y, Ma G, Zhang C, Qian W, Chen G (2019) Understanding the resistance mechanism in *Brassica napus* to clubroot caused by *Plasmodiophora brassicae*. Phytopathology 109:810–818. https://doi.org/10.1094/PHYTO-06-18-0213-R

Melan MA, Dong X, Endara ME, Davis KR, Ausubel FM, Peterman TK (1993) An *Arabidopsis thaliana* lipoxygenase gene can be induced by pathogens, abscisic acid, and methyl jasmonate. Plant Physiol 101:441–450

Menard R, Larue JP, Silue D, Thouvenot D (1999) Glucosinolates in cauliflower as biochemical markers for resistance against downy mildew. Phytochemistry 52:29–35

Mendgen K, Hahn M (2002) Plant infection and the establishment of fungal biotrophy. Trends Plant Sci 7:352–356

Meng X, Zhang S (2013) MAPK cascades in plant disease resistance signalling. Annu Rev Phytopathol 51:245–266

Meyer D, Pajonk S, Micali C, O'Connell R, Schulze-Lefert P (2009) Extracellular transport and integration of plant secretory proteins into pathogen-induced cell wall compartments. Plant J 57:986–999

Micali C, Gollner K, Humphry M, Consonni C, Panstruga R (2008) The powdery mildew disease of *Arabidopsis*: a paradigm for the interaction between plants and biotrophic fungi. Arabidopsis Book 6:e0115. https://doi.org/10.1199/tab.0115

Mikkelsen MD, Hansen CH, Wittstock U, Halkier BA (2000) Cytochrome P450 CYP79B2 from *Arabidopsis* catalyzes the conversion of tryptophan to Indole-3-acetaldoxime, a precursor of indole glucosinolates and Indole-3-acetic acid. J Biol Chem 275:33712–33717. https://doi.org/10.1074/jbc.M001667200

Miklis M, Consonni C, Bhat RA, Lipka V, Schulze-Lefert P, Panstruga R (2007) Barley mlo modulates actin-dependent and actin-independent antifungal defense pathways at the cell periphery. Plant Physiol 144(2):1132–1143

Miya A, Albert P, Shinya T, Desaki Y, Ichimura K, Shirasu K, Narusaka Y, Kawakami N, Kaku H, Shibuya N (2007) CERK1, a LysM receptor kinase, is essential for chitin elicitor signaling in *Arabidopsis*. Proc Natl Acad Sci U S A 104:19613–19618

Molendijk AJ, Ruperti B, Singh MK, Dovzhenko A, Ditengou FA, Milia M, Westphal L, Rosahl S, Soellick T-R, Uhrig J, Weingarten L, Huber M, Palme K (2008) A cysteine-richreceptor-like kinase NCRK and a pathogen-induced proteinkinase RBK1 are Rop GTPase interactors. Plant J 53:909–923

Mou Z, Fan W, Dong X (2003) Inducers of plant systemic acquired resistance regulates NPR function through redox changes. Cell 113:935–944. https://doi.org/10.1016/S0092-8674(03)00429-X

Mucha E, Fricke I, Schaefer A, Wittinghofer A, Berken A (2011) Rho proteins of plants—functional cycle and regulation of cytoskeletal dynamics. Eur J Cell Biol 90:934–943

Muller S, Smertenko A, Wagner V, Heinrich M, Hussey PJ, Hauser MT (2004) The plant microtubule-associated protein AtMAP65-3/PLE is essential for cytokinetic phragmoplast function. Curr Biol 14:412–417

Mur LAJ, Kenton P, Atzorn R, Miersch O, Wasternack C (2006) The outcomes of concentration-specific interactions between salicylate and jasmonate signaling include synergy, antagonism, and oxidative stress leading to cell death. Plant Physiol 140:249–262. https://doi.org/10.1104/pp.105.072348

Murase K, Shiba H, Iwano M, Che FS, Watanabe M, Isogai A, Takayama S (2004) A membrane-anchored protein kinase involved in Brassica self-incompatibility signaling. Science 303:1516–1519

Mysore KS, Ryu CM (2004) Non-host resistance how much do we know? Trends Plant Sci 9:97–104

Nafisi M, Goregaoker S, Botanga CJ, Glawischnig E, Olsen CE, Halkier BA, Glazebrook J (2007) *Arabidopsis* cytochrome P450 mono-oxygenase 71A13 catalyzes the conversion of indole-3-acetaldoxime in camalexin synthesis. Plant Cell 19:2039–2052

Nashaat NI, Rawlinson CJ (1994) The response of oilseed rape (*Brassica napus* ssp. *oleifera*) accessions with different gluconsinolate and erucic acid contents to four isolates of *Peronospora parasitica* (downy mildew) and the identification of new sources of resistance. Plant Pathol 43:278–285

Naumann M, Somerville S, Voigt C (2013) Differences in early callose deposition during adapted and non-adapted powdery mildew infection of resistant *Arabidopsis* lines. Plant Signal Behav 8: e24408. https://doi.org/10.4161/psb.24408

Naur P, Petersen BL, Mikkelsen MD, Bak S, Rasmussen H, Olsen CE, Halkier BA (2003) CYP83A1 and CYP83B1, two non-redundant cytochrome P450 enzymes metabolizing oximes in the biosynthesis of glucosinolates in *Arabidopsis*. Plant Physiol 133:63–72. https://doi.org/10.1104/pp.102.019240

Neik TX, Amas J, Barbetti M, Edwards D, Batley J (2020) Understanding host-pathogen interactions in *Brassica napus* in the omics era. Plants 9:1336. https://doi.org/10.3390/plants9101336

Nibau C, Wu HM, Cheung AY (2006) RAC/ROP GTPases: 'hubs' forsignal integration and diversification in plants. Trends Plant Sci 11:309–315

Nie H, Zhao C, Wu G, Wu Y, Chen Y, Tang D (2012) SR1, a calmodulin-binding transcription factor, modulates plant defense and ethylene-induced senescence by directly regulating NDR1 and EIN3. Plant Physiol 158:1847–1859. https://doi.org/10.1104/pp.111.192310

Nielsen ME, Feechan A, Bohlenius H, Ueda T, Thordal-Christensen H (2012) *Arabidopsis* ARF-GTP exchange factor, GNOM, mediates transport required for innate immunity and focal accumulation of syntaxin PEN1. Proc Natl Acad Sci U S A 109:11443– 11448

Nishimura MT, Stein M, Hou BH, Vogel JP, Edwards H, Somerville SC (2003) Loss of a callose synthase results in salicylic acid-dependent disease resistance. Science 3016(1):969–972

Noel LD, Cagna G, Stuttmann J, Wirthmuller L, Betsuyaku S, Witte CP, Bhat R, Pochon N, Colby T, Parker JE (2007) Interaction between SGT1 and cytosolic/nuclear HSC70 chaperones regulates *Arabidopsis* immune responses. Plant Cell 19:4061–4076

Nongbri PL, Johnson JM, Sherameti I, Glawischnig E, Halkier BA, Oelmculler R (2012) Indole-3-acetaldoxime-derived compounds restrict root colonization in the beneficial interaction between *Arabidopsis* roots and the endophyte *Piriformospora indica*. Mol Plant-Microbe Interact 25:1186–1197

Nozu Y (1967) Studies on the biosynthesis of lignin. IIL Dehydrogenative polymerization of coniferyl alcohol by peroxidase. J Biodiem 62:519–530

Nurnberger T, Lipka V (2005) Non-host resistance in plants: new insights into an old phenomenon. Mol Plant Pathol 6:335–345

Nykyri J, Niemi O, Koskinen P, Nokso-Koivisto J, Pasanen M, Broberg M, Plyusnin I, Toronen P, Holm L, Pirhonen M, Palva ET (2012) Revised phylogeny and novel horizontally acquired virulence determinants of the model soft rot phytopathogen *Pectobacterium wasabiae* SCC3193. PLoS Pathog 8:e1003013

O'Connell RJ, Panstruga R (2006) Tete a tete inside a plant cell: establishing compatibility between plants and biotrophic fungi and oomycetes. New Phytol 171:699–718

Ohguchi T, Asada Y (1975) Dehydrogenation polymerization products of phydroxycinnamyl alcohols by isoperoxidases obtained from downy mildew infected roots of Japanese radish (*Raphanus sativus*). Physiol Plant Pathol 5:183–192

Ohguchi T, Yamashita Y, Asada Y (1974) Isoperoxidases of Japanese radish root infected by downy mildew fungus. Ann Phytopathol Soc Japan 40:419–426

Opalski KS, Schultheiss H, Kogel KH, Huckelhoven R (2005) The receptor-like MLO protein and the RAC/ROP family G-protein RACB modulate actin reorganization in barley attacked by the biotrophic powdery mildew fungus *Blumeria graminis* f.sp. *hordei*. Plant J 41:291–303

Ouko MO, Sambade A, Brandner K, Niehl A, Pena E, Ahad A, Heinlein M, Nick P (2010) Tobacco mutants with reduced microtubule dynamics are less susceptible to TMV. Plant J 62:829–839

Pajerowska-Mukhtar KM, Emerine DK, Mukhtar MS (2013) Tell me more: roles of NPRs in plant immunity. Trends Plant Sci 18:402–411

Pandey SP, Roccaro M, Schon M, Logemann E, Somssich IE (2010) Transcriptional reprogramming regulated by WRKY18 and WRKY40 facilitates powdery mildew infection of *Arabidopsis*. Plant J 64:912–923. https://doi.org/10.1111/j.1365-313X.2010.04387.x

Panstruga R (2005) Serpentine plant MLO proteins as entry portals for powdery mildew fungi. Biochem Soc Trans 33:389–392. https://doi.org/10.1042/BST0330389

Park SW, Kaimoyo E, Kumar D, Mosher S, Klessig DF (2007) Methyl salicylate is a critical mobile signal for plant systemic acquired resistance. Science 318:113–116

Parker JE (2000) Signalling in plant disease resistance. In: Dickinson M, Beynon J (eds) Molecular plant pathology. Annual plant reviews, vol 4. Sheffield Academic Press, Sheffield, pp 144–174

Parker JE, Holub EB, Frost LN, Falk A, Gunn ND, Daniels MJ (1996) Characterization of eds1, a mutation in *Arabidopsis* suppressing resistance to *Peronospora parasitica* specified by several different RPP genes. Plant Cell 8:2033–2046

Parkhi V, Kumar V, Campbell LM, Bell AA, Shah J, Rathore KS (2010) Resistance against various fungal pathogens and reniform nematode in transgenic cotton plants expressing *Arabidopsis* NPR1. Transgenic Res 19:959–975. https://doi.org/10.1007/s11248-010-9374-9

Pavan S, Jacobsen E, Visser RG, Bai Y (2010) Loss of susceptibility as a novel breeding strategy for durable and broad-spectrum resistance. Mol Breed 25:1–12

Pedras MSC, Adio AM (2008) Phytoalexins and phytoanticipins from the wild crucifers *Thellungiella halophila* and *Arabidopsis thaliana*: rapalexin A, wasalexins and camalexin. Phytochemistry 69:889–893. https://doi.org/10.1016/j.phytochem.2007.10.032

Pedras MSC, Ahiahonu WK (2005) Metabolism and detoxification of phytoalexins and analogs by phytopathogenic fungi. Phytochemistry 66(4):391–411

Pedras MSC, Gadagi RS, Jha M, Srma-Mamillapalle VK (2007) Detoxification of the phytoalexin brassinin by isolates of *Leptosphaeria maculans* pathogenic on brown mustard involves an inducible hydrolase. Phytochemistry 68:1572–1578

Pedras MSC, Zheng QA, Gadagi RS, Rimmer SR (2008) Phytoalexins and polar metabolites from the oilseeds canola and rapeseed: differential metabolic responses to the biotroph *Albugocandida* and to abiotic stress. Phytochemistry 69:894–910

Pedras MSC, Yaya EE, Glawischnig E (2011) The phytoalexins from cultivated and wild crucifers: chemistry and biology. Nat Prod Rep 28:1381–1405. https://doi.org/10.1039/C1np00020a

Penaud A (1999) Chemical control and yield losses caused by *Erysiphe cruciferarum* on oilseed rape in France. In: Wratten N, Salisbury PA (eds) Proceedings of the 10th international rapeseed congress. The Regional Institute Ltd, Canberra, pp 1–8. CD-ROM, Doc. No. 327

Perrin RM, Jia X, Wagner TA, O'Neill MA, Sarria R, York WS, Raikhel NV, Keegstra K (2003) Analysis of xyloglucan fucosylation in *Arabidopsis*. Plant Physiol 132:768–778

Petkova M, Dimova M, Dimova D, Bistrichanov S (2014) Effect of gamma-irradiation on the fatty acid composition and susceptibility to powdery mildew (*Erysiphe cruciferarum*) of oilseed rape plants. Agric Sci Technol 6(4):413–416

Pieterse CM, Van der Does D, Zamioudis C, Leon-Reyes A, Van Wees SC (2012) Hormonal modulation of plant immunity. Annu Rev Cell Dev Biol 28:489–521. https://doi.org/10.1146/annurev-cellbio-092910-154055

Pinosa F, Buhot N, Kwaaitaal M, Fahlberg P, Thordal-Christensen H, Ellerstrom M, Andersson MX (2013) *Arabidopsis* phospholipdase is involved in basal defense and nonhost resistance to powdery mildew fungi. Plant Physiol 163:896–906

Pinot F, Benveniste IJ, Salau P, Loreau O, Noe J, Schreiber L, Durst F (1999) Production in vitro by the cytochrome P450 CYP94A1 of major C18 cutin monomers and potential messengers in plant–pathogen interactions: enantioselectivity studies. Biochem J 342:27–32

Pleines S, Friendt W (1989) Genetic control of linolenic acid concentrations in seed oil of rapeseed (*Brassica napus* L.). Theor Appl Genet 78:793–797

Plotnikova JM, Reuber TL, Ausubel FM, Pfister DH (1998) Powdery mildew pathogenesis of *Arabidopsis thaliana*. Mycologia 90:1009–1016

Poraty-Gavra L, Zimmermann P, Haigis S, Bednarek P, Hazak O, Stelmakh OR, Sadot E, Schulze-Lefert P, Gruissem W, Yalovsky S (2013) The *Arabidopsis* Rho of plants GTPase AtROP6functions in developmental and pathogen response pathways. Plant Physiol 161:1172–1188

Prerostova S, Dobrev P, Konradyova V, Knirsch V, Gaudinova A, Kramna B, Kazda J, Ludwig-Muller J, Vankova R (2018) Hormonal responses to *Plasmodiophora brassicae* infection in *Brassica napus* cultivars differing in their pathogen resistance. Int J Mol Sci 19:4024. https://doi.org/10.3390/ijms19124024

Qiu JL, Fiil BK, Petersen K, Nielsen HB, Botanga CJ, Thorgrimsen S, Palma K, Suarez-Rodriguez MC, Sandbech-Clausen S, Lichota J, Brodersen P, Grasser KD, Mattsson O, Glazebrook J, Mundy J, Petersen M (2008) *Arabidopsis* MAP kinase 4 regulates gene expression through transcription factor release in the nucleus. EMBO J 27:2214–2221

Quentin M, Baures I, Hoefle C, Caillaud MC, Allasia V, Panabieres F, Abad P, Huckelhoven R, Keller H, Favery B (2016) The *Arabidopsis* microtubule-associated protein MAP65-3 supports infection by filamentous biotrophic pathogens by down-regulating salicylic acid-dependent defences. J Exp Bot 67(6):1731–1743

Quilis J, Penas G, Messeguer J, Brugidou C, Segundo BS (2008) The *Arabidopsis* AtNPR1 inversely modulates defense responses against fungal, bacterial, or viral pathogens while conferring hypersensitivity to abiotic stresses in transgenic rice. Mol Plant Microbe Interact 21:1215–1231. https://doi.org/10.1094/MPMI-21-9-1215

Rahimi M, Bahrani A (2011) Effect of gamma irradiation on qualitative and quantitative characteristics of canola. Middle East J Sci Res 8:519–525

Rajniak J, Barco B, Clay NK, Sattely ES (2015) A new cyanogenic metabolite in *Arabidopsis* required for inducible pathogen defence. Nature 525:376–379. https://doi.org/10.1038/nature14907

Ramonell KM, Zhang B, Ewing RM, Chen Y, Xu D, Stacey G, Somerville S (2002) Microarray analysis of chitin elicitation in *Arabidopsis thaliana*. Mol Plant Pathol 3:301–311

Ramonell K, Berrocal-Lobo M, Koh S, Wan J, Edwards H, Stacey G, Somerville S (2005) Loss-of-function mutations in chitin responsive genes show increased susceptibility to the powdery mildew pathogen *Erysiphe cichoracearum*. Plant Physiol 138:1027–1036

Rauhut T, Glawischnig E (2009) Evolution of camalexin and structurally related indolic compounds. Phytochemistry 70:1638–1644

Rawlinson CJ, Doughty KJ, Bock CH, Church VJ, Milford GFJ, Fieldsend JK (1989) Diseases and responses to disease and pest control on single- and double-low cultivars of winter oilseed rape. Aspects Appl Biol 23:393–400

Reiner T, Hoefle C, Huesmann C, Menesi D, Feher A, Huckelhoven R (2014) The *Arabidopsis* ROP-activated receptor-like cytoplasmic kinase RLCK VI_A3 is involved in control of basal resistance to powdery mildew and trichome branching. Plant Cell Rep 34(3):457–468. https://doi.org/10.1007/s00299-014-1725-1

Rekhter D, Mohnike L, Feussner K, Zienkiewicz K, Zhang Y, Feussner I (2019) Enhanced disease susceptibility 5 (EDS5) is required for N-hydroxy pipecolic acid formation. bioRxiv: 630723.

Reuber TL, Plotnikova JM, Dewdney J, Rogers EE, Wood W, Ausubel FM (1998) Correlation of defense gene induction defects with powdery mildew susceptibility in *Arabidopsis* enhanced disease susceptibility mutants. Plant J 16:473–485

Ride JP (1983) Cell walls and other structural barriers in defense. In: Callow JA (ed) Biochemical plant pathology. Wiley, Chichester, pp 215–236

Rietz S, Stamm A, Malonek S, Wagner S, Becker D, Medina-Escobar N, Vlot AC, Feys BJ, Niefind K, Parker JE (2011) Different roles of enhanced disease susceptibility1 (EDS1) bound to and dissociated from phytoalexin Deficient4 (PAD4) in *Arabidopsis* immunity. New Phytol 191:107–119. https://doi.org/10.1111/j.1469-8137.2011.03675.x

Rolland F, Baena-Gonzalez E, Sheen J (2006) Sugar sensing and signaling in plants: conserved and novel mechanisms. Annu Rev Plant Biol 57:675–709

Roux F, Voisin D, Badet T, Balague C, Barlet X, Huard-Chauveau C, Roby D, Raffaele S (2014) Resistance to phytopathogens e tutti quanti: placing plant quantitative disease resistance on the map. Mol Plant Pathol 15(5):427–432. https://doi.org/10.1111/mpp.12138

Rouxel T, Kollmann A, Boulidard L, Mithen R (1991) Abiotic elicitation of indole phytoalexins and resistance to *Leptosphaeria maculans* within Brassiceae. Planta 184(2):271–278. https://www.jstor.org/stable/23380984

Rushton PJ, Somssich IE (1998) Transcriptional control of plant genes responsive to pathogens. Curr Opin Plant Biol 1:311–315

Russo VM, Bushnell WR (1989) Responses of barley cells to puncture by micro needles and to attempted penetration by *Erysiphe graminis* f. sp. *hordei*. Can J Bot 67:2912–2921

Rusterucci C, Aviv DH, Holt BF, Dangl JL, Parker JE (2001) The disease resistance signaling components EDS1 and PAD4 are essential regulators of the cell death pathway controlled by LSD1 in *Arabidopsis*. Plant Cell 13:2211–2224

Ryder TB, Hedrick SA, Bell JN, Liang X, Clouse SD, Lamb CJ (1987) Organization and differential activation of a gene family encoding the plant defense enzyme chalcone synthase in *Phaseolus vulgaris*. Mol Genet Genomics 210:219–233

Saenz GS, Taylor JW (1999) Phylogeny of the Erysiphales (powdery mildews) inferred from internal transcribed spacer (ITS) ribosomal DNA sequences. Can J Bot 77:150–169

Sagi M, Fluhr R (2001) Superoxide production by plant homologues of the gp91phox NADPH oxidase. Modulation of activity by calcium and by tobacco mosaic virus infection. Plant Physiol 126:1281–1290. https://doi.org/10.1104/pp.126.3.1281

Saharan GS, Naresh M, Sangwan MS (2005) Development of disease resistance in rapeseed-mustard. In: Saharan GS, Naresh M, Sangwan MS (eds) Diseases of oilseed crops. Indus Publishing Co., New Delhi, pp 561–617

Saharan GS, Verma PR, Meena PD, Kumar A (2014) White rust of crucifers: Biology, ecology and disease management. Springer Verlag, Frankfurt, p 244, ISBN 978-81-322-1791-6

Saharan GS, Naresh M, Meena PD (2016) Alternaria blight of crucifers: Biology, ecology and disease management, vol 326. Springer Verlag, Singapore, ISBN 978-981-10-0019-5

Saharan GS, Naresh M, Meena PD (2017) Downy mildew disease of crucifers: biology, ecology and disease management. Springer Verlag, Singapore, LVI, 357, ISBN 978-981-10-7499-8

Saharan GS, Naresh M, Meena PD (2019) Powdery mildew disease of crucifers: Biology, ecology and disease management. Springer Verlag, Singapore, p 362, ISBN 978-981-13-9852-0

von Schaewen A, Stitt M, Schmidt R, Sonnewald U, Willmitzer L (1990) Expression of a yeast-derived invertase in the cell wall of tobacco and *Arabidopsis* plants leads to accumulation of carbohydrate and inhibition of photosynthesis and strongly influences growth and phenotype of transgenic tobacco plants. EMBO J 9:3033–3044

Scheideler M, Schlaich NL, Fellenberg K, Beissbarth T, Hauser NC, Vingron M, Slusarenko AJ, Hoheisel JD (2002) Monitoring the switch from housekeeping to pathogen defense metabolism in *Arabidopsis thaliana* using cDNA arrays. J Biol Chem 277:10555–10561

Scheler B, Schnepf V, Galgenmuller C, Ranf S, Huckelhoven R (2016) Barley disease susceptibility factor RACB acts in epidermal cell polarity and positioning of the nucleus. J Exp Bot 67 (11):3263–3275. https://doi.org/10.1093/jxb/erw141

Schenk PM, Kazan K, Wilson I, Anderson JP, Richmond T, Somerville SC, Manners JM (2000) Coordinated plant defense responses in *Arabidopsis* revealed by microarray analysis. Proc Natl Acad Sci U S A 97(11):655–660. https://doi.org/10.1073/pnas.97.21.11655

Schlaeppi K, Abou-Mansour E, Buchala A, Mauch F (2010) Disease resistance of *Arabidopsis* to *Phytophthora brassicae* is established by the sequential action of indole glucosinolates and camalexin. Plant J 62:840–851. https://doi.org/10.1111/j.1365-313X.2010.04197.x

Schmelzer E (2002) Cell polarization, a crucial process in fungal defence. Trends Plant Sci 7:411–415

Schmidt SM, Panstruga R (2008) Cytoskeleton functions in plant-microbe interactions. Physiol Mol Plant Pathol 716(1):135–148. https://doi.org/10.1016/j.pmpp.2008.01.001

Schon M, Toller A, Diezel C, Roth C, Westphal L, Wiermer M, Somssich IE (2013) Analyses of *wrky18 wrky40* plants reveal critical roles of SA/EDS1 signaling and indole-glucosinolate biosynthesis for *Golovinomyces orontii* resistance and a loss-of resistance towards *Pseudomonas syringae* pv. *tomato* AvrRPS4. Mol Plant Microbe Interact 26:758–767

Schuhegger R, Nafisi M, Mansourova M, Petersen BL, Olsen CE, Svatos A, Halkier BA, Glawischnig E (2006) CYP71B15 (PAD3) catalyzes the final step in camalexin biosynthesis. Plant Physiol 141:1248–1254

Schuhegger R, Nafisi M, Mansourova M, Petersen BL, Olsen CE, Svatos A, Halkier BA, Glawischnig E (2007a) CYP71 B15 (PAD3) catalyzes the final step in camalexin biosynthesis-correction. Plant Physiol 145:1086–1086

Schuhegger R, Rauhut T, Glawischnig E (2007b) Regulatory variability of camalexin biosynthesis. J Plant Physiol 164:636–644. https://doi.org/10.1016/j.jplph.2006.04.012

Schuller A, Kehr J, Ludwig-Muller J (2013) Laser microdissection coupled to transcriptional profiling of *Arabidopsis* roots inoculated by *Plasmodiophora brassicae* indicates a role for brassinosteroids in clubroot formation. Plant Cell Physiol 55:392–411. https://doi.org/10.1093/pcp/pct174

Schuller A, Kehr J, Ludwig-Muller J (2014) Laser microdissection coupled to transcriptional profiling of Arabidopsis roots inoculated by *Plasmodiophora brassicae* indicates a role for brassinosteroids in clubroot formation. Plant Cell Physiol 55:392–411. https://doi.org/10.1093/pcp/pct174

Schultheiss H, Dechert C, Kogel KH, Huckelhoven R (2002) A small GTP-binding host protein is required for entry of powdery mildew fungus into epidermal cells of barley. Plant Physiol 128:1447–1454

Schultheiss H, Dechert C, Kogel KH, Huckelhoven R (2003) Functional analysis of barley RAC/ROP G-protein family members in susceptibility to the powdery mildew fungus. Plant J 36(5):589–601

Schuster A, Schmoll M (2010) Biology and biotechnology of *Trichoderma*. Appl Microbiol Biotechnol 87:787–799

Schweizer P (2007) Non-host resistance of plants to powdery mildew-new opportunities to unravel the mystery. Physiol Mol Plant Pathol 70:3–7

Schweizer P, Kmecl A, Carpita N, Dudler R (2000) A soluble carbohydrate elicitor from *Blumeria graminis* f. sp *tritici* is recognized by a broad range of cereals. Physiol Mol Plant Pathol 56:157–167

Seskar M, Shulaev V, Raskin I (1998) Endogenous methyl salicylate in pathogen-inoculated tobacco plants. Plant Physiol 116:387–392

Seyfferth C, Tsuda K (2014) Salicylic acid signal transduction: the initiation of biosynthesis, perception and transcriptional reprogramming. Front Plant Sci 5:697. https://doi.org/10.3389/fpls.2014.00697

Shah N, Sun J, Yu S, Yang Z, Wang Z, Huang F, Dun B, Gong J, Liu Y, Li Y, Li Q, Yuan L, Baloch A, Li G, Li S, Zhang C (2019) Genetic variation analysis of field isolates of clubroot and their responses to *Brassica napus* lines containing resistant gene CRb, PbBa8.1 and their combination in homozygous and heterozygous state. Mol Breed 39:153. https://doi.org/10.1007/s11032-019-1075-3

Shah N, Li Q, Xu Q, Liu J, Huang F, Zhan Z, Qin P, Zhou X, Yu W, Zhu L, Zhang C (2020)*CRb* and *PbBa8.1* synergically increases resistant genes expression upon infection of *Plasmodiophora brassicae* in *Brassica napus*. Genes (Basel) 11(2):202. https://doi.org/10.3390/genes11020202

Shapiro AD, Zhang C (2001) The role of *NDR1* in avirulence gene-directed signaling and control of programmed cell death in *Arabidopsis*. Plant Physiol 127:1089–1101

Sharma P, Sain SK (2004) Induction of systemic resistance in tomato and cauliflower by *Trichoderma* spp. against stalk rot pathogen, *Sclerotinia sclerotiorum* Lib de Bary. J Biol Control 18(1):21–27

Shen QH, Saijo Y, Mauch S, Biskup C, Bieri S, Keller B, Seki H, Ulker B, Somssich IE, Schulze-Lefert P (2007) Nuclear activity of MLA immune receptors links isolate-specific and basal disease-resistance responses. Science 315:1098–1103

Shirasu K, Nakajima H, Rajasekhar VK, Dixon RA, Lamb C (1997) Salicylic acid potentiates an agonist-dependent gain control that amplifies pathogen signals in the activation of defense mechanisms. Plant Cell 9:261–270

Shirasu K, Nakajima H, Rajasekhar VK, Dixon RA, Lamb C (1999) Salicylic acid potentiates an agonist-dependent gain control that amplifies pathogen signals in the activation of defense mechanisms. Plant Cell 9:261–270

Shiu SH, Karlowski WM, Pan R, Tzeng YH, Mayer KF, Li WH (2004) Comparative analysis of the receptor-like kinase family in *Arabidopsis* and rice. Plant Cell 16:1220–1234

Shoresh M, Harman GE, Mastouri F (2010) Induced systemic resistance and plant responses to fungal biocontrol agents. Annu Rev Phytopathol 48:21–43

Siemens J, Nagel W, Ludwig-Muller J, Sarcristán MD (2002) The interaction of *Plasmodiophora brassicae* and *Arabidopsis thaliana*: parameters for disease quantification and screening mutant lines. J Phytopathol 150:592–605

Siemens J, Keller I, Sarx J, Kunz S, Schuller A, Nagel W, Schmulling T, Parniske M, Ludwig-Muller J (2006) Transcriptome analysis of *Arabidopsis* clubroots indicate a key role for cytokinins in disease development. Mol Plant Microbe Interact 19:480–494. https://doi.org/10.1094/MPMI-19-0480

Siemens J, Bulman S, Rehn F, Sundelin T (2009a) Molecular biology of *Plasmodiophora brassicae*. J Plant Growth Regul 28:245–251

Siemens J, Graf H, Bulman S, In O, Ludwig-Muller J (2009b) Monitoring expression of selected *Plasmodiophora brassicae* genes during clubroot development in *Arabidopsis thaliana*. Plant Pathol 58:130–136

Singh A (2017) Glucosinolates and plant defense. In: Mérillon JM, Ramawat KG (eds) Glucosinolates. Springer International Publishing, Cham, pp 237–246

Singh SB, Singh DV, Bains BS (1980) *In vivo* cellulase and pectinase production by *Albugo candida* and *Peronospora parasitica*. Indian Phytopathol 33:370–371

Skoric D, Jocic S, Sakac Z, Lecic N (2008) Genetic possibilities for altering sunflower oil quality to obtain novel oils. Can J Physiol Pharmacol 86:215–221

Slusarenko AJ (1996) The role of lipoxygenase in plant resistance to infection. In: Piazza GJ (ed) Lipoxygenase and lipoxygenase pathway enzymes. American Oil Chemists Society Press, Champaign, pp 176–197

Slusarenko AJ, Mauch-Mani B (1991) Downy mildew of *Arabidopsis thaliana* caused by *Peronospora parasitica*: a model system for the investigation of the molecular biology of host-pathogen interactions. In: Hennecke H (ed) Advances in the molecular genetics of plant–microbe interactions, vol 1. Kluwer Academic, Dordrecht, pp 280–283

Slusarenko AJ, Schlaich NL (2003) Downy mildew of *Arabidopsis thaliana* caused by *Hyaloperonospora parasitica* (formerly *Peronospora parasitica*). Mol Plant Pathol 4:159–170

Sonderby IE, Geu-Flores F, Halkier BA (2010) Biosynthesis of glucosinolates—gene discovery and beyond. Trends Plant Sci 15:283–290. https://doi.org/10.1016/j.tplants.2010.02.005

Song T, Chu M, Lahlali R, Yu F, Peng G (2016) Shotgun label-free proteomic analysis of clubroot (*Plasmodiophora brassicae*) resistance conferred by the gene Rcr1 in *Brassica rapa*. Front Plant Sci 7:1013. https://doi.org/10.3389/fpls.2016.01013

Spencer-Phillips PTN, Gay JL (1981) Domains of ATPase in plasma membrane and transport through infected plant cells. New Phytol 896(1):393–400

Speth EB, Lee YN, He SY (2007) Pathogen virulence factors as molecular probes of basic plant cellular functions. Curr Opin Plant Biol 10:580–586

Spoel SH, Koornneef A, Claessens SM, Korzelius JP, Van-Pelt JA, Mueller MJ, Buchala AJ, Métraux JP, Brown R, Kazan K, Van Loon LC, Dong X, Pieterse CMJ (2003) NPR1 modulates cross-talk between salicylate- and jasmonate-dependent defense pathways through a novel function in the cytosol. Plant Cell 15:760–770. https://doi.org/10.1105/tpc.009159

Sreeramulu S, Mostizky Y, Sunitha S, Shani E, Nahum H, Salomon D, Hayun LB, Gruetter C, Rauh D, Ori N, Sessa G (2013) BSKs are partially redundant positive regulators of brassinosteroid signaling in *Arabidopsis*. Plant J 74:905–919

Stein M, Dittgen J, Sanchez-Rodriguez C, Hou BH, Molina A, Schulze-Lefert P (2006) *Arabidopsis* PEN3/PDR8, an ATP binding cassette transporter, contributes to nonhost resistance to inappropriate pathogens that enter by direct penetration. Plant Cell 18:731–746. https://doi.org/10.1105/tpc.105.038372

Stein E, Molitor A, Kogel KH, Waller F (2008) Systemic resistance in *Arabidopsis* conferred by the mycorrhizal fungus *Piriformospora indica* requires jasmonic acid signaling and the cytoplasmic function of NPR1. Plant Cell Physiol 49:1747–1751. https://doi.org/10.1093/pcp/pcn147

Stitt M, von Schaewen A, Willmitzer L (1990) "Sink" regulation of photosynthetic metabolism in transgenic tobacco plants expressing yeast invertase in their cell wall involves a decrease of the Calvin-cycle enzymes and an increase in glycolytic enzymes. Planta 183:40–50

Stotz HU, Sawada Y, Shimada Y, Hirai MY, Sasaki E, Krischke M, Brown PD, Saito K, Kamiya Y (2011) Role of camalexin, indole glucosinolates, and side chain modification of glucosinolate-derived isothiocyanates in defense of *Arabidopsis*against *Sclerotinia sclerotiorum*. Plant J 67:81–93

Strack D, Vogt T, Schliemann W (2003) Recent advances in betalain research. Phytochemistry 62:247–269

Strawn MA, Marr SK, Inoue K, Inada N, Zubieta C, Wildermuth MC (2007) *Arabidopsis* isochorismate synthase functional in pathogen-induced salicylate biosynthesis exhibits properties consistent with a role in diverse stress responses. J Biol Chem 282:5919–5933

Strelkov SE, Hwang SF, Manolii VP, Cao T, Feindel D (2016) Emergence of new virulence phenotypes of *Plasmodiophora brassicae* on canola (*Brassica napus*) in Alberta, Canada. Eur J Plant Pathol 145:517–529. https://doi.org/10.1007/s10658-016-0888-8

Strelkov SE, Hwang SF, Manolii VP, Cao T, Fredua-Agyeman R, Harding MW, Peng G, Gossen BD, McDonald MR, Feindel D (2018) Virulence and pathotype classification of *Plasmodiophora brassicae* populations collected from clubroot resistant canola (*Brassica napus*) in Canada. Can J Plant Pathol 40:284–298

Sun J, Zhang J, Fang H, Peng L, Wei S, Li C, Zheng S, Lu J (2019) Comparative transcriptome analysis reveals resistance-related genes and pathways in *Musa acuminata* banana 'Guijiao 9' in response to Fusarium wilt. Plant Physiol Biochem 141:83–94. https://doi.org/10.1016/j.plaphy.2019.05.022

Sundaresha S, Rohini S, Appanna VK, Arthikala MK, Shanmugam NB, Shashibhushan NB, Hari Kishore CM, Pannerselvam R, Kirti PB, Udayakumar M (2016) Co-overexpression of *Brassica juncea* NPR1 (BjNPR1) and *Trigonella foenum*-graecum defensin (Tfgd) in transgenic peanut provides comprehensive but varied protection against *Aspergillus flavus* and *Cercospora arachidicola*. Plant Cell Rep 35:1189–1203. https://doi.org/10.1007/s00299-016-1945-7

Suwabe K, Tsukazaki H, Iketani H, Hatakeyama K, Fujimura M, Nunome T, Fukuoka H, Matsumoto S, Hirai M (2003) Identification of two loci for resistance to clubroot (*Plasmodiophora brassicae* Woronin) in *Brassica rapa* L. Theor Appl Genet 107:997–1002

Swarbrick PJ, Schulze-Lefert P, Scholes JD (2006) Metabolic consequences of susceptibility and resistance (race-specific and broad-spectrum) in barley leaves challenged with powdery mildew. Plant Cell Environ 29:1061–1076

Swiderski MR, Innes RW (2001) The *Arabidopsis* PBS1 resistance gene encodes a member of a novel protein kinase subfamily. Plant J 26:101–112

Szadkowski E, Eber F, Huteau V, Lode M, Coriton O, Jenczewski E, Chevre AM (2011) Polyploid formation pathways have an impact on genetic rearrangements in resynthesized *Brassicanapus*. New Phytol 191:884–894. https://doi.org/10.1111/j.1469-8137.2011.03729.x

Tada Y, Spoel SH, Pajerowska-Mukhtar K, Mou Z, Song J, Wang C, Zou J, Dong X (2008) Plant immunity requires conformational charges of NPR1 via S-nitrosylation and thioredoxins. Science 321:952–956. https://doi.org/10.1126/science.1156970

Takemoto D, Jones DA, Hardham AR (2003) GFP-tagging of cell components reveals the dynamics of sub-cellular re-organization in response to infection of *Arabidopsis* by oomycete pathogens. Plant J 33:775–792

Takemoto D, Jones D, Hardham A (2006) Re-organisation of the cytoskeleton and endoplasmic reticulum in the *Arabidopsis pen1-1* mutant inoculated with the non-adapted powdery mildew *Blumeria graminis* f. sp. *hordei*. Mol Plant Pathol 76(1):553–563

Tan ZY, Wada Y, Chen JS, Ohshima K (2004) Inter- and intralineage recombinants are common in natural populations of Turnip mosaic virus. J Gen Virol 85:2683–2696

Tang XY, Xie MT, Kim YJ, Zhou JM, Klessig DF, Martin GB (1999) Overexpression of Pto activates defense responses and confers broad resistance. Plant Cell 11:15–29

Tang D, Ade J, Frye CA, Innes RW (2005a) Regulation of plant defense responses in *Arabidopsis* by EDR2, a PH and START domain-containing protein. Plant J 446(1):245–257

Tang D, Christiansen KM, Innes RW (2005b) Regulation of plant disease resistance, stress responses, cell death and ethylene signaling in *Arabidopsis* by the EDR1 protein kinase. Plant Physiol 1386(1):1018–1026

Tang H, Wang X, Bowers JE, Ming R, Alam M, Paterson AH (2008) Unraveling ancient hexaploidy through multiply-aligned angiosperm gene maps. Genome Res 18(12):1944–1954. https://doi.org/10.1101/gr.080978.108

Thaler JS, Humphrey PT, Whiteman NK (2012) Evolution of jasmonate and salicylate signal crosstalk. Trends Plant Sci 17:260–270. https://doi.org/10.1016/j.tplants.2012.02.010

Thomma BP, Eggermont K, Penninckx IA, Mauch-Mani B, Vogelsang R, Cammue BP, Broekaert WF (1998) Separate jasmonate-dependent and salicylate-dependent defense-response pathways in *Arabidopsis* are essential for resistance to distinct microbial pathogens. Proc Natl Acad Sci U S A 95(25):15107–15111. https://doi.org/10.1073/pnas.95.25.15107

Thomma BP, Eggermont K, Tierens KF, Broekaert WF (1999) Requirement of functional ethylene-insensitive 2 gene for efficient resistance of *Arabidopsis* to infection by *Botrytis cinerea*. Plant Physiol 121:1093–1102

Thomma BP, Penninckx IA, Broekaert WF, Cammue BP (2001) The complexity of disease signaling in *Arabidopsis*. Curr Opin Immunol 13:63–68. https://doi.org/10.1016/S0952-7915(00)00183-7

Thordal-Christensen H (2003) Fresh insights into processes of non-host resistance. Curr Opin Plant Biol 6:351–357

Thornton JD, Cooke RC (1974) Changes in respiration, chlorophyll content and soluble carbohydrates of detached cabbage cotyledons following infection with *Peronospora parasitica* (Pers. ex. Fr.) Fr. Physiol Plant Pathol 4:117–125

Timonen KL, Vanninen E, de Hartog J, Ibald-Mulli A, Brunekreef B, Gold DR, Heinrich J, Hoek G, Lanki T, Peters A, Tarkiainen P, Kreyling W, Pekkanen J (2006) Effects of ultrafine and fine particulate and gaseous air pollution on cardiac autonomic control in subjects with coronary artery disease-the ULTRA study. J Expo Sci Environ Epidemiol 16(4):332–341

Torrens-Spence MP, Bobokalonova A, Carballo V, Glinkerman CM, Pluskal T, Shen A, Weng JK (2019) PBS3 and EPS1 complete salicylic acid biosynthesis from isochorismate in *Arabidopsis*. bioRxiv: 601948

Torres MA, Jones JDG, Dangl JL (2006) Reactive oxygen species signaling in response to pathogens. Plant Physiol 141:373–378. https://doi.org/10.1104/pp.106.079467

Trujillo M, Troeger M, Niks RE, Kogel KH, Huckelhoven R (2004) Mechanistic and genetic overlap of barley host and non-host resistance to *Blumeria graminis*. Mol Plant Pathol 5:389–396. https://doi.org/10.1111/j.1364-3703.2004.00238.x

Tsuda K, Somssich IE (2015) Transcriptional networks in plant immunity. New Phytol 206:932–947

Tsuji J, Jackson EP, Gage DA, Hammerschmidt R, Sommerville SC (1992) Phytoalexin accumulation in *Arabidopsis thaliana* during the hypersensitive response to *Pseudomonas syringae* pv *syringae*. Plant Physiol 98:1304–1309

Ueno H, Matsumoto E, Aruga D, Kitagawa S, Matsumura H, Hayashida N (2012) Molecular characterization of the CRa gene conferring clubroot resistance in *Brassica rapa*. Plant Mol Biol 80:621–629. https://doi.org/10.1007/s11103-012-9971-5

Ulker B, Shahid Mukhtar M, Somssich IE (2007) The WRKY70 transcription factor of *Arabidopsis* influences both the plant senescence and defense signaling pathways. Planta 226:125–137

Underwood W, Somerville SC (2008) Focal accumulation of defences at sites of fungal pathogen attack. J Exp Bot 59:3501–3508

Underwood W, Somerville SC (2013) Perception of conserved pathogen elicitors at the plasma membrane leads to relocalization of the *Arabidopsis* PEN3 transporter. Proc Natl Acad Sci U S A 110:12492–12497

Van der Does D, Leon-Reyes A, Koornneef A, Van Verk MC, Rodenburg N, Pauwels L, Goossens A, Korbes AP, Memelink J, Ritsema T, Van Wees SCM, Pieterse CMJ (2013) Salicylic acid suppresses jasmonic acid signaling downstream of SCFCOI1-JAZ by targeting GCC promoter motifs via transcription factor ORA59. Plant Cell 25:744–761. https://doi.org/10.1105/tpc.112.108548

Van der Luit AH, Piatti T, van Doom A, Musgrave A, Felix G, Boiler T, Munnik T (2000) Elicitation of suspension-cultured tomato cells triggers the formation of phosphatidic acid and diacylglycerol pyrophosphate. Plant Physiol 123:1507–1516

Van Schie CC, Takken FL (2014) Susceptibility genes: how to be a good host. Annu Rev Phytopathol 52:551–581

Vance CP, Kirk TK, Sherwood RT (1980) Lignification as a mechanism of disease resistance. Annu Rev Phytopathol 18:259–288. https://doi.org/10.1146/annurev.py.18.090180.001355

VanEtten HD, Mansfield JW, Bailey JA, Farmer EE (1994) Two classes of plant antibiotics: phytoalexins versus "phytoanticipins". Plant Cell 6:1191–1192

Vassileva VN, Kouchi H, Ridge RW (2005) Microtubule dynamics in living root hairs: transient slowing by lipochitin oligosaccharide nodulation signals. Plant Cell 17:1777–1787

Verhagen BWM, Glazebrook J, Zhu T, Chang HS, van Loon LC, Pieterse CMJ (2004) The transcriptome of rhizobacteria-induced systemic resistance in *Arabidopsis*. Mol Plant Microbe Interact 17:895–908

Veronese P, Chen X, Bluhm B, Salmeron J, Dietrich RA, Mengiste T (2004) The BOS loci of *Arabidopsis* are required for resistance to *Botrytis cinerea* infection. Plant J 40:558–574

Veronese P, Nakagami H, Bluhm B, Abuqamar S, Chen X, Salmeron J, Dietrich RA, Hirt H, Mengiste T (2006) The membrane-anchored BOTRYTIS-INDUCED KINASE1 plays distinct roles in *Arabidopsis* resistance to necrotrophic and biotrophic pathogens. Plant Cell 18 (1):257–273. https://doi.org/10.1105/tpc.105.035576

Vivancos J, Labbe C, Menzies JG, Belanger RR (2015) Silicon-mediated resistance of *Arabidopsis* against powdery mildew involves mechanisms other than the salicylic acid (SA)-dependent defence pathway. Mol Plant Pathol 16:572–582

Vlieghe K, Boudolf V, Beemster GTS, Maes S, Magyar Z, Atanassova A, de Almeida EJ, De Groodt R, Inze D, De Veylder L (2005) The DP-E2F-like gene *DEL1* controls the Endocycle in *Arabidopsis thaliana*. Curr Biol 15:59–63

Voegele RT, Stuck C, Hahn M, Mendgen K (2001) The role of haustoria in sugar supply during infection of broad bean by the rust fungus *Uromyces fabae*. Proc Natl Acad Sci U S A 98:8133–8138

Vogel JP, Raab TK, Somerville CR, Somerville SC (2004) Mutations in PMR5 result in powdery mildew resistance and altered cell wall composition. Plant J 40(6):968–978

Voit OE (2003) Biochemical and genomic regulation of the trehalose cycle in yeast: review of observations and canonical model analysis. J Theor Biol 223:55–78

Vorwerk S, Somerville S, Somerville C (2004) The role of plant cell wall polysaccharide composition in disease resistance. Trends Plant Sci 9:203–209

Vorwerk S, Schiff C, Santamaria M, Koh S, Nishimura M, Vogel J, Somerville C, Somerville S (2007) *EDR2* negatively regulates salicylic acid-based defences and cell death during powdery mildew infections of *Arabidopsis thaliana*. BMC Plant Biol 7:35. https://doi.org/10.1186/1471-2229-7-35

Vos IA, Verhage A, Schuurink RC, Watt LG, Pieterse CMJ, Van Wees SCM (2013) Onset of herbivore-induced resistance in systemic tissue primed for jasmonate-dependent defenses is activated by abscisic acid. Front Plant Sci 4:539. https://doi.org/10.3389/fpls.2013.00539

Wagner S, Stuttmann J, Rietz S, Guerois R, Brunstein E, Bautor J, Niefind K, Parker JE (2013) Structural basis for signaling by exclusive EDS1 heteromeric complexes with SAG101 or PAD4 in plant innate immunity. Cell Host Microbe 14:619–630. https://doi.org/10.1016/j.chom.2013.11.006

Wagner G, Laperche A, Lariagon C, Marnet N, Renault D, Guitton Y, Bouchereau A, Delourme R, Manzanares-Dauleux M, Gravot A (2019) Resolution of quantitative resistance to clubroot into QTL-specific metabolic modules. J Exp Bot 70:5375–5390

Wally O, Jayaraj J, Punja ZK (2009) Comparative resistance to foliar fungal pathogens in transgenic carrot plants expressing genes encoding for chitinase, beta-1, 3-glucanase and peroxidase. Eur J Plant Pathol 123:331–342. https://doi.org/10.1007/s10658-008-9370-6

Walters DR, Ratsep J, Havis Neil D (2013) Controlling crop diseases using induced resistance: challenges for the future. J Exp Bot 64:1263–1280

Wan JR, Zhang XC, Neece D, Ramonell KM, Clough S, Kim SY, Stacey MG, Stacey G (2008) A LysM receptor-like kinase plays a critical role in chitin signaling and fungal resistance in *Arabidopsis*. Plant Cell 206(1):471–481

Wang X (2004) Lipid signaling. Curr Opin Plant Biol 7:329–336

Wang D, Amornsiripanitch N, Dong X (2006) A genomic approach to identify regulatory nodes in the transcriptional network of systemic acquired resistance in plants. PLOS Pathog 2:e123. https://doi.org/10.1371/journal.ppat.0020123

Wang L, Tsuda K, Truman W, Sato M, Nguyen LV, Katagiri F, Glazebrook J (2011a) CBP60g and SARD1 play partially redundant critical roles in salicylic acid signaling. Plant J 3000:1029–1041. https://doi.org/10.1111/j.1365-313X.2011.04655.x

Wang YH, Gehring C, Irving HR (2011b) Plant natriuretic peptides are apoplastic and paracrine stress response molecules. Plant Cell Physiol 52:837–850. https://doi.org/10.1093/pcp/pcr036

Wang Y, Cheng X, Shan Q, Zhang Y, Liu J, Gao C, Qiu JL (2014) Simultaneous editing of three homoalleles in hexaploid bread wheat confers heritable resistance to powdery mildew. Nat Biotechnol 32:947–951

Wang Z, Bao L-L, Zhao F-Y, Tang M-Q, Chen T, Li Y, Wang B-X, Fu B, Fang H, Li G-Y, Cao J, Ding L-N, Zhu K-M, Liu S-Y, Tan X-L (2019) BnaMPK3 Is a key regulator of defense responses to the devastating plant pathogen *Sclerotinia sclerotiorum* in Oilseed rape. Front Plant Sci 10:91. https://doi.org/10.3389/fpls.2019.00091

Wasson AP, Ramsay K, Jones MGK, Mathesius U (2009) Differing requirements for flavonoids during the formation of lateral roots, nodules and root knot nematode galls in *Medicago truncatula*. New Phytol 183:167–179

Wasteneys GO (2004) Progress in understanding the role of microtubules in plant cells. Curr Opin Plant Biol 7:651–660

Wasteneys GO, Galway ME (2003) Remodeling the cytoskeleton for growth and form: an overview with some new views. Annu Rev Plant Biol 54:691–722

Wasternack C, Hause B (2013) Jasmonates: biosynthesis, perception, signal transduction and action in plant stress response, growth and development. An update to the 2007 review in Annals of Botany. Ann Bot 111:1021–1058. https://doi.org/10.1093/aob/mct067

Wawrzynska A, Rodibaugh NL, Innes RW (2010) Synergistic activation of defense responses in *Arabidopsis* by simultaneous loss of the gsl5 callose synthase and the edr1 protein kinase. Mol Plant Microbe Interact 23:578–584

Weis C, Huckelhoven R, Eichmann R (2013) LIFEGUARD proteins support plant colonization by biotrophic powdery mildew fungi. J Exp Bot 64:3855–3867

Weis C, Hildebrandt U, Hoffmann T, Hemetsberger C, Pfeilmeier S, Konig C, Schwab W, Eichmann R, Huckelhoven R (2014) CYP83A1 is required for metabolic compatibility of *Arabidopsis* with the adapted powdery mildew fungus *Erysiphe cruciferarum*. New Phytol 202:1310–1319

Weymann K, Hunt M, Uknes S, Neuenschwander U, Lawton K, Steiner HY, Ryals J (1995) Suppression and restoration of lesion formation in *Arabidopsis* lsd mutants. Plant Cell 7:2013–2022

Whetten R, Sederoff R (1995) Lignin biosynthesis. Plant Cell 7:1001–1013

Widarto HT, Van der Meijden E, Lefeber AWM, Erkelens C, Kim HK, Choi YH, Verpoorte R (2006) Metabolomic differentiation of *Brassica rapa* following herbivory by different insect instars using two-dimensional nuclear magnetic resonance spectroscopy. J Chem Ecol 32:2417–2428

Wiermer M, Feys BJ, Parker JE (2005) Plant immunity: the EDS1 regulatory node. Curr Opin Plant Biol 8:383–389. https://doi.org/10.1016/j.pbi.2005.05.010

Wildermuth MC, Dewdney J, Wu G, Ausubel FM (2001) Isochorismate synthase is required to synthesize salicylic acid for plant defense. Nature 414:562–565

Winge P, Brembu T, Kristensen R, Bones AM (2000) Genetic structure and evolution of RAC-GTPases in *Arabidopsis thaliana*. Genetics 156:1959–1971

Winter D, Vinegar B, Nahal H, Ammar R, Wilson GV, Provart NJ (2007) An "electronic fluorescent pictograph" browser for exploring and analyzing large-scale biological data sets. PLoS One 2:e718. https://doi.org/10.1371/journal.pone.0000718

Wittstock U, Halkier BA (2002) Glucosinolate research in the *Arabidopsis* era. Trends Plant Sci 7:263–270

Wright DP, Baldwin BC, Shephard MC, Scholes JD (1995a) Source-sink relationships in wheat leaves infected with powdery mildew. II. Changes in the regulation of the Calvin cycle. Physiol Mol Plant Pathol 47:255–267

Wright DP, Baldwin BC, Shephard MC, Scholes JD (1995b) Source-sink relationships in wheat leaves infected with powdery mildew. I. Alterations in carbohydrate metabolism. Physiol Mol Plant Pathol 47:237–253

Wu Y, Zhang D, Chu JY, Boyle P, Wang Y, Brindle ID, Luca VD, Despres C (2012) The *Arabidopsis* NPR1 protein is a receptor for the plant defense hormone salicylic acid. Cell Rep 1:639–647. https://doi.org/10.1016/j.celrep.2012.05.008

Wu G, Liu S, Zhao Y, Wang W, Kong Z, Tang D (2015) ENHANCED DISEASE RESISTANCE4 associates with CLATHRIN HEAVY CHAIN2 and modulates plant immunity by regulating relocation of edr1 in *Arabidopsis*. Plant Cell 27:857–873

Xiao SY, Ellwood S, Calis O, Patrick E, Li TX, Coleman M, Turner JG (2001) Broad-spectrum mildew resistance in *Arabidopsis thaliana* mediated by *RPW8*. Science 2916(1):118–120

Xiao S, Brown S, Patrick E, Brearley C, Turner JG (2003) Enhanced transcription of the *Arabidopsis* disease resistance genes *RPW8.1* and *RPW8.2* via a salicylic acid-dependent amplification circuit is required for hypersensitive cell death. Plant Cell 15(1):33–45. https://doi.org/10.1105/tpc.006940

Xiong X, Sun S, Li Y, Zhang X, Sun J, Xue F (2019) The cotton WRKY transcription factor GhWRKY70 negatively regulates the defense response against *Verticillium dahliae*. Crop J 7:393–402. https://doi.org/10.1016/j.cj.2018.10.005

Xu X, Kanbara K, Azakami H, Kato A (2004) Expression and characterization of *Saccharomyces cerevisiae* Cne1p, a calnexin homologue. J Biochem 135(5):615–618

Xu T, Wen M, Nagawa S, Fu Y, Chen JG, Wu MJ, Perrot-Rechenmann C, Friml J, Jones AM, Yang Z (2010) Cell surface-and rho GTPase-based auxin signaling controls cellular inter-digitation in *Arabidopsis*. Cell 143:99–110

Xu G, Guo C, Shan H, Kong H (2012) Divergence of duplicate genes in exon–intron structure. Proc Natl Acad Sci 109(4):1187–1192

Yan S, Dong X (2014) Perception of the plant immune signal salicylic acid. Curr Opin Plant Biol 20:64–68

Yang Y, Shah J, Klessig DF (1997) Signal perception and transduction in plant defense responses. Genes Dev 11:1621–1639

Yang L, Qin L, Liu G, Peremyslov VV, Dolja VV, Wei Y (2014) Myosins XI modulates host cellular responses and penetration resistance to fungal pathogens. Proc Natl Acad Sci U S A 111:13996–14001

Yedidia I, Benhamou N, Chet I (1999) Induction of defense responses in cucumber plants (*Cucumis sativus* L.) by the biocontrol agent *Trichoderma harzianum*. Appl Environ Microbiol 65:1061–1070

Yoshimoto K, Jikumaru Y, Kamiya Y, Kusano M, Consonni C, Panstruga R, Ohsumi Y, Shirasu K (2009) Autophagy negatively regulates cell death by controlling npr1-dependent salicylic acid signaling during senescence and the innate immune response in *Arabidopsis*. Plant Cell 21:2914–2927

You J, Chan Z (2015) ROS regulation during abiotic stress responses in crop plants. Front Plant Sci 6:1092. https://doi.org/10.3389/fpls.2015.01092

Yu D, Chen C, Chen Z (2001) Evidence for an important role of WRKY DNA binding proteins in the regulation of NPR1 gene expression. Plant Cell 13:1527–1540

Yu F, Zhang X, Huang Z, Chu M, Song T, Falk KC, Deora A, Chen Q, Zhang Y, McGregor L, Gossen BD, McDonald MR, Peng G (2016) Identification of genome-wide variants and discovery of variants associated with *Brassica rapa* clubroot resistance gene Rcr1 through bulked segregant RNA sequencing. PLoS One 11:e0153218. https://doi.org/10.1371/journal.pone.0153218

Yun BW, Atkinson HA, Gaborit C, Greenland A, Read ND, Pallas JA, Loake GJ (2003) Loss of actin cytoskeletal function and EDS1 activity, in combination, severely compromises non-host resistance in Arabidopsis against wheat powdery mildew. Plant J 346(1):768–777

Zablackis E, Huang J, Muller B, Darvill AG, Albersheim P (1995) Characterization of the cell wall polysaccharides of *Arabidopsis thaliana* leaves. Plant Physiol 107:1129–1138

Zander M, La Camera S, Lamotte O, Metraux JP, Gatz C (2010) *Arabidopsis thaliana* class-II TGA transcription factors are essential activators of jasmonic acid/ethylene-induced defense responses. Plant J 61:200–210. https://doi.org/10.1111/j.1365-313X.2009.04044.x

Zander M, Chen S, Imkampe J, Thurow C, Gatz C (2012) Repression of the *Arabidopsis thaliana* jasmonic acid/ethylene-induced defense pathway by TGA-interacting glutaredoxins depends on their C-terminal ALWL motif. Mol Plant 5:831–840. https://doi.org/10.1093/mp/ssr113

Zander M, Thurow C, Gatz C (2014) TGA transcription factors activate the salicylic acid-suppressible branch of the ethylene-induced defense program by regulating ORA59 expression. Plant Physiol 165:1671–1683. https://doi.org/10.1104/pp.114.243360

Zeyen RJ, Kruger WM, Lyngkjær MF, Carver TLW (2002) Differential effects of D-mannose and 2-deoxym-D-glucose on attempted powdery mildew fungal infection of inappropriate and appropriate Gramineae. Physiol Mol Plant Pathol 61:315–323

Zhang C, Shapiro AD (2002) Two pathways act in an additive rather than obligatorily synergistic fashion to induce systemic acquired resistance and *PR* gene expression. BMC Plant Biol 2:9. https://doi.org/10.1186/1471-2229-2-9

Zhang Y, Tessaro MJ, Lassner M, Li X (2003) Knockout analysis of *Arabidopsis* transcription factors TGA2, TGA5, and TGA6 reveals their redundant and essential roles in systemic acquired resistance. Plant Cell 15:2647–2653. https://doi.org/10.1105/tpc.014894

Zhang J, Shao F, Li Y, Cui H, Chen L, Li H, Zou Y, Long C, Lan L, Chai J, Chen S, Tang X, Zhou JM (2007) A *Pseudomonas syringae* effector inactivates MAPKs to suppress PAMP-induced immunity in plants. Cell Host Microbe 1:175–185

Zhang X, Francis MI, Dawson WO, Graham JH, Orbovic V (2010) Over-expression of the *Arabidopsis* NPR1 gene in citrus increases resistance to citrus canker. Eur J Plant Pathol 128:91–100. https://doi.org/10.1007/s10658-010-9633-x

Zhang H, Feng J, Zhang S, Zhang S, Li F, Strelkov SE, Sun R, Hwang SF (2015) Resistance to *Plasmodiophora brassicae* in *Brassica rapa* and *Brassica juncea* genotypes from China. Plant Dis 99:776–779. https://doi.org/10.1094/PDIS-08-14-0863-RE

Zhang H, Feng J, Hwang SF, Strelkov SE, Falak I, Huang X, Sun R (2016) Mapping of clubroot (*Plasmodiophora brassicae*) resistance in canola (*Brassica napus*). Plant Pathol 65:435–440. https://doi.org/10.1111/ppa.12422

Zhao J, Last RL (1996) Coordinate regulation of the tryptophan biosynthetic pathway and indolic phytoalexin accumulation in *Arabidopsis*. Plant Cell 8:2235–2244

Zhao Y, Hull AK, Gupta NR, Goss KA, Alonso J, Ecker JR (2002) Trp-dependent auxin biosynthesis in *Arabidopsis*: involvement of cytochrome P450s CYP79B2 and CYP79B3. Genes Dev 16:3100–3112. https://doi.org/10.1101/gad.1035402

Zhao N, Ferrer JL, Ross J, Guan J, Yang Y, Pichersky E, Noel JP, Chen F (2008) Structural, biochemical, and phylogenetic analyses suggest that indole-3-acetic acid methyltransferase is an evolutionarily ancient member of the SABATH family. Plant Physiol 146:455–467

Zhao C, Nie H, Shen Q, Zhang S, Lukowitz W, Tang D (2014) EDR1 physically interacts with MKK4/MKK5 and negatively regulates a MAP kinase cascade to modulate plant innate immunity. PLoS Genet 10:e1004389. https://doi.org/10.1371/journal.pgen.1004389

Zhao Y, Bi K, Gao Z, Chen T, Liu H, Xie J, Cheng J, Fu Y, Jiang D (2017) Transcriptome analysis of *Arabidopsis thaliana* in response to *Plasmodiophora brassicae* during early infection. Front Microbiol 8:673

Zheng XY, Spivey NW, Zeng WQ, Liu PP, Fu ZQ, Klessig DF, He SY, Dong XN (2012) Coronatine promotes Pseudomonas syringae virulence in plants by activating a signaling cascade that inhibits salicylic acid accumulation. Cell Host Microbe 11:587–596

Zheng X, McLellan H, Fraiture M, Liu X, Boevink PC, Gilroy EM, Chen Y, Kandel K, Sessa G, Birch PR, Brunner F (2014) Functionally redundant RXLR effectors from *Phytophthora infestans* act at different steps to suppress early flg22-triggered immunity. PLoS Pathog 10: e1004057

Zhong X, Xi L, Lian Q, Luo X, Wu Z, Seng S, Yuan X, Yi M (2015) The NPR1 homolog GhNPR1 plays an important role in the defense response of *Gladiolus hybridus*. Plant Cell Rep 34:1063–1074. https://doi.org/10.1007/s00299-015-1765-1

Zhou N, Tootle TL, Tsui F, Klessig DF, Glazebrook J (1998) PAD4 functions upstream from salicylic acid to control defense responses in *Arabidopsis*. Plant Cell 10:1021–1030

Zierold U, Scholz U, Schweizer P (2005) Transcriptome analysis of mlo-mediated resistance in the epidermis of barley. Mol Plant Pathol 6:139–151. https://doi.org/10.1111/j.1364-3703.2005.00271.x

Zimmerli L, Stein M, Lipka V, Schulze-Lefert P, Somerville S (2004) Host and non-host pathogens elicit different jasmonate/ethylene responses in *Arabidopsis*. Plant J 40:633–646. https://doi.org/10.1111/j.1365-313X.2004.02236.x

Zipfel C, Robatzek S, Navarro L, Oakeley EJ, Jones JDG, Felix G, Boller T (2004) Bacterial disease resistance in *Arabidopsis* through flagellin perception. Nat Int J Sci 428:764–767

Biometabolomics of Host Resistance to Hemi-biotrophs and Necrotrophs

5

Abstract

The mechanisms of nonhost resistance (NHR) in plants to *Alternaria* operate at preinvasion stage mainly by both preformed and inducible defense responses. Preformed defenses involved structural features like trichomes, chemical compounds (antimicrobial enzymes, secondary metabolites), cuticular waxes, formation of phytoalexins, and detoxification of toxins. Defense responses mainly involve ROS accumulation and defense-induced cell death where fungal proliferation is restricted by rapid hypersensitive reaction (HR). Resistance in *Brassica* to *Alternaria* depends on phytoalexins, detoxification, and simultaneous phytoalexins (camalexin) elicitation. MAPK 4 gene activates defense in *Brassica* to *Alternaria*. The biomolecules like phenols, polyphenol oxidase, peroxidase, PAL, and catalase are higher in Alternaria-tolerant genotypes of *Brassica*. *GLIP1* gene in association with ethylene signaling is a critical factor in host resistance to *Alternaria*. β-Aminobutyric acid (BABA) and zeatin upregulated host immunity via elevation of MAPK 5 which antagonizes the effect of *A. brassicae*. Resistant *Brassica* genotypes accumulate phytoalexins like camalexin, brassinin, allyl (AI ITC), and benzyl (Bz ITC) isothiocynates to inhibit *Alternaria* infection. Foliar sprays of calcium compounds sequester the organic acids at the site of *Alternaria* infection to boost calcium content of plants to provide resistance. The MAPK signaling cascade leading to biosynthesis of camalexin triggers defense responses in *B. rapa* to *Alternaria*. The upregulation of genes such as WARKY, peroxidase, p450 oxidase, and chitinase in *Brassica* mediated with JA and SA pathways enhances defense response to *Alternaria*. Differential defense signaling pathways of JA and SA operates in different species of *Brassica* to *Alternaria*.

Keywords

Biometabolomic resistance to Hemibiotrophs · Necrotrophs · NHR mechanisms · Detoxification for host defense · MAPK interaction with toxin · Bimolecular basis

G. S. Saharan et al., *Molecular Mechanism of Crucifer's Host-Resistance*,
https://doi.org/10.1007/978-981-16-1974-8_5

of resistance · Proteomics of disease resistance · Induction of resistance · Elicitation of phytoalexins · Calcium sequestration · MAPK signaling pathways · Defense genes identification · Differential defense signaling pathways · Molecular events in *Colletotrichum* interaction · Induction of phytohormones · Cellular resistance to *Leptosphaeria* · Heightened defense response in resistant hosts · *Rlm2*-mediated resistant · NADPH oxidases to *Rhizoctonia* · Glucosinolates in the defense · Biomolecular mechanism to *Sclerotinia* · *GDSL1* overexpression to *Sclerotinia* · *BnaMPK3* gene · Crosstalk of signaling molecules · Nitric oxide in defense-related signaling · Metabolites of *Arabidopsis* to *Verticillium* · Tryptophan-derived secondary metabolites · Biomolecular resistance to *Verticillium* · Proteomics of defense to *Xanthomonas* · Biometabolomics of resistance to *Xanthomonas* · Endochitinase gene · Overexpression of CH1-B4 like protein

Mutation of lysine histidine transporter 1 (lhr1) in *Arabidopsis* reduces contents of glutamine, alanine, and proline to enhance resistance against Colletotrichum mediated with SA signaling and increased callose deposition and PR1. Defense responses in *Arabidopsis* to Colletotrichum are activated through SA signaling at early stage of interaction and by JA signaling subsequently.

TF homologs associated with JA (MYC3, JAM1), ABA (AKS2, ABF3), and calcium (CAMTA5, CAMTA 3/SR1) signaling regulate defense response processes in resistant host of *Brassica* to *Leptosphaeria.* At the site of host–pathogen interface, DEGs are downregulated in resistant plants in association with response to cytokinin, photosynthesis, amino acids biosynthesis, metabolisms, and unsaturated fatty acids biosynthesis. Defense response in resistant cvs. is faster activated by complex network of WRKY, BZIP, bHLH, CAMTA, and host shock TFs. Calcium signaling is regulated for resistance and basal defense against *L. maculans*. There is differential response of resistant and susceptible cvs. of *Brassica* for expression, biosynthesis, and signaling pathways of different hormones in response to *L. maculans*. Genes associated with surface receptors and MAPK activities are upregulated in *Leptosphaeria*-resistant hosts but not in susceptible ones.

NADPH oxidases are critical components of resistance to *R. solani* AG8 in *Arabidopsis.* There is a role of glucosinolates in the defense of *Brassica* against pathogens. Defense response against *Sclerotinia* are regulated by a complex network of hormonal and signaling pathways via JA, ET, and abscisic acid in promoting the production of glucosinolates (GSLs) and other metabolites working together to enhance host resistance. Several pathways and biological processes are associated with the defense against *Sclerotinia*, such as oxidative burst, cell wall enforcement or modification, secondary metabolisms and calcium binding, signaling, antioxidation, glutathione metabolism, glucosinolate metabolisms, and biosynthesis of lignin. In *Brassica*, resistance is expressed by quick accumulation of ROS via activating Ca^{2+} signaling and suppression of oxalic acid generation in *Sclerotinia*. Overexpression of AtGDSL1 in *B. napus* enhances resistance to *Sclerotinia* with increased ROS and SA levels and reduced JA level. In *B. napus*, mitogen SA-activated protein kinase gene *BnaMPK3* plays an important role in the defense against *Sclerotinia*. Host resistance to *Sclerotinia* involves various biological process like redox homeostasis; signaling by lipids, calcium, SA, JA, ET, and ABA; histone, DNA methylation-mediated transcription regulation, defensin-like proteins, cyanatelyase, miRNA, PTGs, and glycolate oxidase genes encoding crucial enzymes in photorespiration. The crosstalk of signaling molecules, namely, JA, SA, ET, ABA, Auxin, NO, ROSs, and OA, is very complex to induce defense against *Sclerotinia* on *Brassica*, and it varies with interaction of pathogen with different genotypes. Differential alternative slicing genes and isoform regulate resistance in *Brassica* to *Sclerotinia*.

The phenylpropanoids are involved in defense responses of *Arabidopsis* against *Verticillium*. Tryptophan-derived secondary metabolites like camalexin provide defense against *Verticillium* to *Brassica*. The application of BABA induces biomolecules to confer resistance in *Brassica* to *Verticillium*. The increased level of proteins like annexin, AIG2, ferredoxin, ferredoxin–NADP, and epithiospecifier like in resistant cvs. of *Brassica* provides resistance to *Xanthomonas*. The role of several proteins, in resistant and susceptible cvs. of *Brassica*, has been characterized with several signal molecules against *Xanthomonas* and responses indicated up- and downregulated levels. *CHI-B4*-like gene encoding an endochitinase has been identified in *Arabidopsis* to confer resistance to *Xanthomonas*. The overexpression of *BoCHB4* in *Brassica* confers resistance to *Xanthomonas*.

5.1 Introduction

Biometabolomics and biomolecular mechanisms of host resistance is expressed in the form of different metabolites and biomolecules involved in the induction of biochemical changes and pathways in the host to confer resistance against biotic and abiotic stresses. Biometabolomics and systems biology determines a variety of molecular targets for breeding disease resistance cvs. in *Brassica* crops. Transcriptomics and proteomics help to discover novel biochemical pathways involved in conferring resistance to virulent pathogens and reveal mechanisms of biometabolomics host resistance. With the use of multi-omics supplemented with functional studies, key resistance pathways involved in *Brassica*-hemibiotrophs pathosystem has been discovered, where the induction of JA signaling elevated ROS with reprogramming of the phenyl propanoid pathways to impart resistance. The hormone SA plays a major role in plant resistance to hemibiotrophs. In *A. thaliana*, SA synthesis occurs in plastids via isochorismate synthesis (ICS1 or SID2) and is triggered by pathogens. SA accumulated in the cytoplasm can be converted to SA glycoside (SAG) which is stored in the vacuole and hydrolyzed back to SA when required. Elevated levels of total SA (free SA plus SAG) are correlated with induction of defense gene expression and enhanced host resistance. Different from SA, JA and ET accumulate mainly in response to necrotrophs. Loss of susceptibility is frequently caused by deregulation of SA- or JA-dependent host defense signaling by the impairment of cellular rearrangement or by the limitation of nutrient supply for the pathogens. There is role of MAPK 4 gene in host defense against *Alternaria*. MAP kinase signaling cascade is known to be involved in the synthesis of camalexin which contributes to disease resistance against necrotrophs. The lifestyle transition of the pathogens from biotrophics to necrotrophics is regulated by signaling molecules from SA to JA and ET.

5.2 *Brassica–Alternaria*: Biometabolomic Resistance

5.2.1 Nonhost Resistance (NHR) Mechanisms in Nonhost Plants to *Alternaria brassicae*

NHR to fungi is highly conserved among different plant species and seems to involve multiple layers of resistance. NHR mainly comprises preinvasion and post-invasion defenses (Lee et al. 2017; Fonseca and Mysore 2018). Based on the current understanding of the NHR mechanisms in other related nonhost plant pathosystems such as *Arabidopsis* and powdery mildew fungus, barley and powdery mildew fungus, and Medicago and the rust fungal pathogen, Fatima et al. (2019) have proposed the possible NHR mechanisms in a nonhost plant against *A. brassicae* (Fig. 5.1). The NHR mechanisms operating at the preinvasion level mainly involve both preformed and inducible defense responses. The germination of *A. brassicae* spores requires host stimuli, high humidity, and surface water to increase wetness on the leaf surface (Verma and Saharan 1994; Saharan et al. 2016). Preformed defenses may involve structural features such as the abundance of leaf trichomes, which reduces leaf wettability, as well as the lack of stimuli from the nonhost plant, which eventually causes irregular spore germination or a complete lack thereof. Preformed barriers also include chemical compounds present on the leaf surface, for example, antimicrobial enzymes and secondary metabolites, which might inhibit spore germination. The ability of *A. brassicae* spores to germinate on nonhost tomato and wheat plants was tested by McRoberts and Lennard (1996). They found that the spores were able to germinate with the same efficiency on the nonhost and host plants. These spores may have overcome the preformed chemical barriers, thereby being able to germinate on the leaf surface. Furthermore, the entry of germinated *A. brassicae* spores inside the plants is achieved either by targeting the stomata, which is a more predominant mode for entry, or directly through the cuticle.

Stomatal recognition by a pathogen requires specific plant topographical cues, which allow germ tube development and eventually facilitate the penetration inside the plant (Niks and Rubiales 2002). A study on tomato and wheat plants showed that germ tube development of *A. brassicae* was not affected, but the pathogens failed to enter the stomata of the nonhost plants, unlike in the host plants. It is possible that in the nonhost plants, stomata are not accurately recognized by the pathogen because the surface topography may significantly differ from that of the host leaf. The lack of stomatal recognition by the pathogen may lead to the development of the germ tube, albeit away from the location of the stomatal opening. This sort of disruption in the orientation of the germ tube may not allow the pathogen to enter inside the leaf tissue. Another preformed structural barrier includes cuticular waxes, which retard

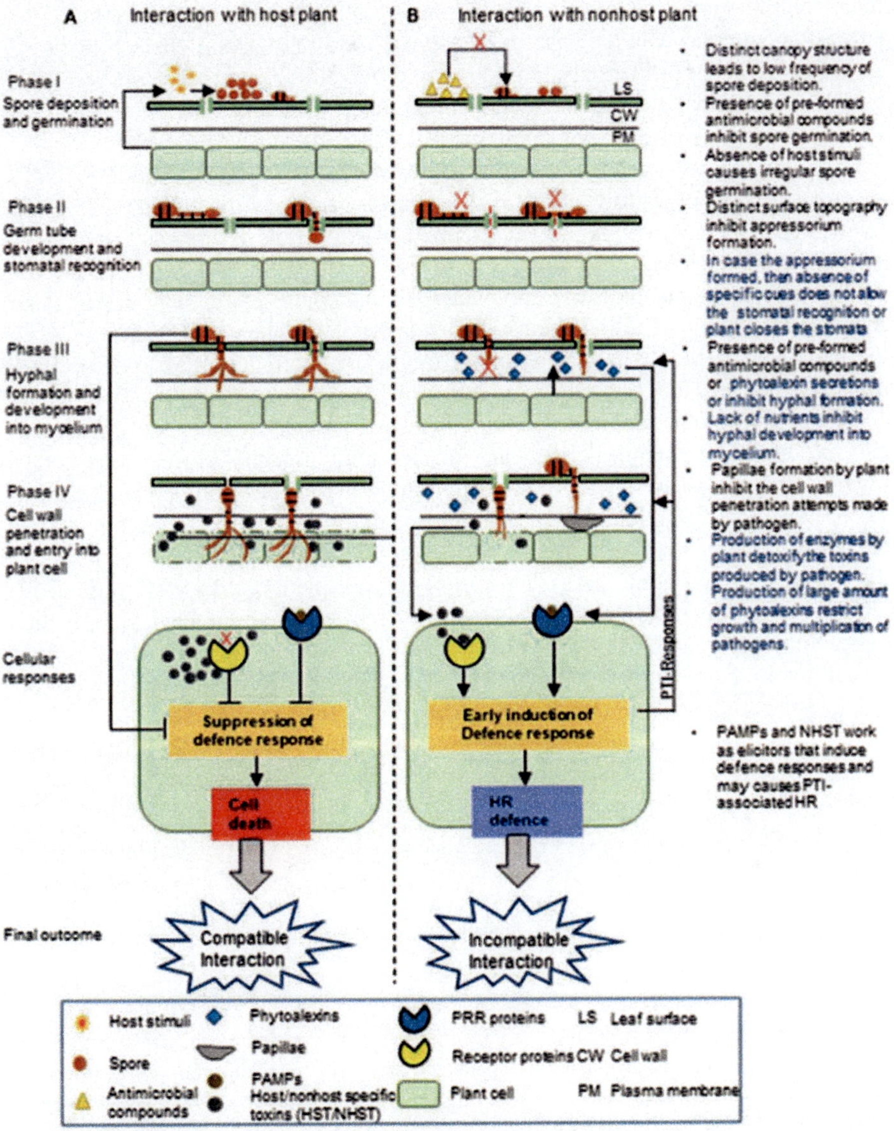

Fig. 5.1 Proposed stages of nonhost resistance against *Alternaria brassicae*. This schematic representation depicts the possible events during the interaction of *Alternaria brassicae* with the host (**a**) and nonhost plants (**b**), which is divided into four different phases. The NHR toward *A. brassicae* may involve the preformed and inducible defense barrier at different phases. I. The first phase involves the deposition and germination of *A. brassicae* spores on the leaf surface. Nonhost plants have a low frequency of spore deposition on the leaf surface due to the differences in their canopy structure compared to the host plant. Moreover, spore germination might be inhibited due to the presence of preformed antimicrobial compounds on the leaf surface and irregular germination of spores due to the absence of host stimuli. II. The second phase involves germ tube development from the spores and the recognition of the stomata. The difference in the leaf surface topography of the nonhost plant compared to the host plant may not allow the proper development of the germ tube into the appressorium. At the next level, even if the appressorium is formed, either stomata are

pathogen entry into the leaf tissues (Serrano et al. 2015). For example, barley has excessive cuticular wax, consisting of approximately 54% long-chain β-diketones, 20% hydroxylated β-diketones, and other hydrocarbons. Together, these form thick cuticular layers that occlude the stomatal features involved in germ tube penetration inside the leaf tissue (Uppalapati et al. 2012). Besides, *Alternaria*, being a necrotrophic pathogen, sometimes directly enters the host plant through the cuticle by enzymatically digesting cutin waxes by means of cutinolytic enzymes. For a few tolerant host species, a high content of cuticular waxes acts as a physical barrier that provides resistance to *A. brassicae* (Conn and Tewari 1989; Saharan et al. 2016).

The nonhost plants may have appreciably high amounts of cuticular waxes, which may restrict pathogens from penetrating inside the plant. The penetration attempts by the pathogen on the nonhost plant may trigger inducible defense responses in the plant.

The nonhost plant can induce stomatal closure, prevent the entry of *A. brassicae*, and mount an inducible chemical barrier involving the rapid formation of phytoalexins antimicrobial compounds inhibiting hyphal development and differentiation. The differentiation of hyphae is important for penetration through intercellular spaces, and certain nutrients are essential for hyphal development and differentiation (Giri et al. 2013). Thus, in a nonhost plant, the lack of nutrients and presence of antimicrobial compounds in the apoplast may inhibit the development of hyphae into mycelium. Moreover, the defense responses also involve the formation of a structural barrier to arrest fungal growth and penetration in the cell wall (Voigt 2015). The pathogen *A. brassicae* fails to penetrate the cell wall in tomato and wheat plants, and callose-containing papilla formation is observed in the cell wall regions where the pathogen attempts to penetrate such plants. The pathogen also produces nonhost-specific or general toxins that can damage the plant cells, which eventually leads to necrosis (Parada et al. 2007). To avoid this, a nonhost plant may recognize these toxins and employ defense mechanisms to detoxify these toxins. Some NHR mechanisms operate at the post-invasion level, which might restrict the pathogen from invading the epidermal cells in the nonhost plant. These defense responses

---

Fig. 5.1 (continued) closed by plant or the pathogen may fail to recognize the stomata due to the absence of specific cues, generally derived from the distinct composition and amount of cuticular waxes in the nonhost leaf, which might be entirely different in the case of the host leaf. III. The third phase involves fungal hyphal formation and its development into mycelium. The nonhost plant may inhibit hyphal formation and its development into mycelium by secreting inducible chemicals by preformed antimicrobial compounds or by limiting nutrient availability in the intercellular spaces. IV. The fourth phase involves the penetration of the fungal pathogen into the plant cell wall and its subsequent entry inside the plant cell. The nonhost plant may inhibit the penetration of the fungal hyphae into the plant cell wall by inducing the formation of papillae. The defense responses are induced by nonhost plants after recognizing PAMPs or elicitor molecules. This may lead to the production of enzymes and a large amount of phytoalexins by the nonhost plants, which detoxify the toxins, produced by the fungal pathogen and restrict fungal growth and multiplication. Note: The text on the right in the figure indicates the NHR mechanisms, and the highlighted blue text indicates the most exploitable NHR mechanisms to develop durable disease-resistant *Brassica* crops (Fatima et al. 2019) mainly involve ROS accumulation and defense-induced cell death, which do not allow the mycelium to ramify through and between the cells. Eventually, fungal proliferation might be restricted by rapid HR (Fatima et al. 2019).

5.2.2 Detoxification for Host Defense Against *Alternaria* Species

Destruxin B is a host-specific *Alternaria* phytotoxin, whose selective phytotoxicity is due to the detoxification in tissues of resistant species (Pedras et al. 2001). In addition, the hydroxylation product that is generated during detoxification was found to induce phytoalexin biosynthesis in resistant, but not in susceptible, plant species. This suggests that resistance depends on phytotoxin detoxification and simultaneous phytoalexin elicitation (Pedras et al. 2001). Phytoalexins have also been implicated in pathogen containment for other *Alternaria* host interactions. In *Camelina sativa* (false flax), resistance against *A. brassicae* has been correlated with the production of the phytoalexin camalexin (Jejelowo et al. 1991). Moreover, *Arabidopsis* synthesizes camalexin upon pathogen challenge. Since *Arabidopsis* is amendable to phytopathological studies as well as molecular genetics, it has been established as a model for dissecting plant defense mechanisms. The *Arabidopsis pad3* mutant (phytoalexins deficient) was found to be deficient in camalexin biosynthesis and to display severely enhanced susceptibility to *A. brassicicola*, demonstrating that camalexin contributes to *Alternaria* resistance (Thomma et al. 1999b). Although camalexin was found to exhibit direct antimicrobial activity to this fungus in vitro (Thomma et al. 1999b), it cannot be excluded that camalexin also contributes to *Alternaria* resistance in an indirect way, as camalexin was found to inhibit production of the host-specific toxin destruxin B in *A. brassicae* (Pedras et al. 1998).

In a study of 24 different *Arabidopsis* ecotypes, it was not possible to discover a direct relationship between camalexin production upon *A. brassicicola* inoculation on the one hand and resistance against this pathogen on the other (Kagan and Hammerschmidt 2002). This is perhaps not surprising, since defense against *A. brassicicola* in *Arabidopsis* appears to depend on multiple defense components, the relative importance of which can vary between ecotypes. Previously, jasmonate has also been implicated in defense against *Alternaria*, as the jasmonate insensitive *Arabidopsis* mutant *coi1* was found to display enhanced susceptibility to this pathogen (Thomma et al. 1998, 1999a, 2000). Jasmonate is a plant hormone that, along with salicylic acid and ethylene, plays a role in the activation of defense responses in plants (Thomma et al. 2001). However, mutants affected in either salicylic acid or ethylene signaling are as resistant as wild-type plants, leading to the conclusion that neither salicylic acid nor ethylene is directly involved in resistance against *Alternaria* in *Arabidopsis* (Thomma et al. 1998, 1999a). Jasmonate was found not to directly induce camalexin biosynthesis and, in addition, jasmonate-insensitivity does not lead to camalexin deficiency. It was therefore concluded that both camalexin and jasmonate mediate host resistance against *Alternaria* through separate but parallel pathways in *Arabidopsis* (Thomma et al. 1998, 1999b, 2001). Further, support for this observation came from the finding that the *Arabidopsis* mutation *esa1* affects resistance against *A. brassicicola* through a severe reduction in both camalexin production as well as jasmonate-dependent gene induction (Tierens et al. 2002). It is currently not clear, however, which jasmonate-inducible effect or molecules are responsible for the containment of *Alternaria* in *Arabidopsis*.

In a cDNA microarray analysis on a set of 2375 *Arabidopsis* genes, 168 genes were found to be upregulated and 39 were found to be repressed 72 h after inoculation with *A. brassicicola*. About 10% of these 207 genes can be implicated in plant defense, and about one-third in cell maintenance and development (Schenk et al. 2000). However, a precise role for any of the responsive genes in the defense of *Arabidopsis* against *A. brassicicola* has not yet been established. From studies using ethylene- and jasmonate-insensitive *Arabidopsis* mutants it can be concluded that a set of PR genes encoding plant defensin, and hevein-like protein, and basic chitinase are not markedly involved in *Alternaria* resistance (Thomma et al. 1998, 1999a). In tomato, PR-proteins have been implicated in *Alternaria* resistance through the observation that resistance in tomato against *A. solani* correlates with high and rapid accumulation of PR-proteins. However, it is speculated that those PR-proteins, many of which possess hydrolytic activity, are not necessarily directly involved in arresting the pathogen, but rather release elicitors from the pathogen cell wall, thus triggering a HR (Lawrence et al. 2000).

The production of reactive oxygen species (ROS) is one of the earliest defense reactions of plants against pathogen attack. In the presence of tobacco plant extracts, *A. alternata* was found to synthesize mannitol, which can act as a potent quencher of reactive oxygen species. This suggests that *Alternaria* uses the active oxygen quencher mannitol to overcome host defense. On the other hand, pathogen attack was found to induce mannitol dehydrogenase-expression in tobacco, suggesting that plants counter this fungal defense system by catabolizing fungal mannitol (Jennings et al. 1998). Support for this hypothesis also came from the fact that the constitutive expression of a mannitol dehydrogenase gene in tobacco confers resistance against *A. alternata*, but not against non-mannitol secreting fungi (Jennings et al. 2002).

Using a transgenic approach, enhanced levels of resistance against *Alternaria* species were also obtained in a number of cases. The expression of antiapoptotic genes provides disease resistance, against *Alternaria* species among others, through blocking a programmed cell death response (Dickman et al. 2001; Lincoln et al. 2002). Constitutive expression of a radish defensin gene, a rubber tree chitin-binding lectin, and a human lysozyme, respectively, conferred enhanced resistance against *Alternaria* spp. in tobacco, Indian mustard, and carrot (Kanrar et al. 2002; Takaichi and Oeda 2000; Terras et al. 1995). In addition, constitutive expression of an endochitinase gene isolated from the biocontrol fungus *Trichoderma harzianum* has resulted in enhanced *Alternaria* resistance in transgenic tobacco, potato, and broccoli (Lorito et al. 1998; Mora and Earle 2001). In addition, the treatment of plants with biocontrol organisms has in some cases resulted in enhanced resistance against *Alternaria* pathogens (Morita et al. 2003; Ton et al. 2002). Many aspects of the interaction between *Alternaria* species and their respective hosts are currently under investigation. The function and biosynthesis of host-specific toxins, and melanins in particular, has gained much attention, and considerable advances have

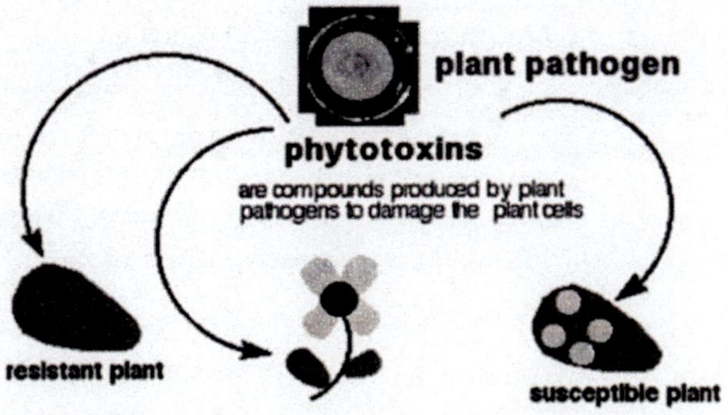

Fig. 5.2 Phytotoxin are produced by microbial plant pathogens (Verma and Saharan 1994; Saharan et al. 2016)

been made. Finally, although the general strategies employed by the pathogenic strains to attack their hosts are similar, the responses of the various hosts against these attacks are very diverse (Thomma 2003).

White mustard cells can transform a very toxic compound called destruxin B (^{14}C-labeled), produced by the microorganism that causes the plant disease, into a nontoxic compound called hydroxydestruxin B. Hydroxydestruxin B (^{14}C-labeled) was further biotransformed to the β-D-glucosyl derivative at slower rates. The structures of hydroxydestruxin B and β-D-glucosyl hydroxydestruxin B were deduced from their spectroscopic data (NMR, HRMS, FTIR) and confirmed by total chemical synthesis (Fig. 5.2). Although these hydroxylation and glucosylation reactions occurred in both resistant (*S. alba*) and susceptible (*B. napus*, *B. juncea*, and *B. rapa*) species, hydroxylation was the rate-limiting step in the susceptible species, whereas glucosylation was the rate-limiting step in the resistant species. Remarkably, it was observed that the hydroxyl destruxin B induced the biosynthesis of phytoalexins in black spot resistant, but not in susceptible species. This appears to be a unique example of phytotoxin detoxification and simultaneous phytoalexin elicitation by the detoxification product. *Sinapis alba* can overcome the fungal invader through detoxification of destruxin B coupled with production of phytoalexins (Figs. 5.3 and 5.4). Since the hydroxylation reaction happens at a very fast rate and it leads to a less toxic product, it may prevent the plant cells from being killed by the pathogen. Canola plants can become resistant to the black spot disease upon incorporation of this detoxification trait (Pedras et al. 1999, 2001). In unique interactions, three different rapeseed pathogens convert the phytoalexin cyclobrassinin into three different phytoalexins, brassilexin, dioxibrassinin, and brassicanal A, and ultimately each phytoalexin into nontoxic products. Such pathogen may have acquired a more effective mechanism for overcoming this chemical by adopting phytoalexin biosynthetic pathways operating in planta.

This strategy appears quite plausible, especially considering that most fungal pathogens have been coevolving with plants for numerous generations (Pedras and Okanga 1999; Fig. 5.5). Each detoxification pathway is (1) specific to a group of pathogens and that (2) the enzymes involved in these transformations are quite selective (Pedras 1998; Pedras et al. 2000). Pedras et al. (2009) have established that *A. brassicicola* produces brassicicolin A (1) (a mixture of epimers, Gloer et al.

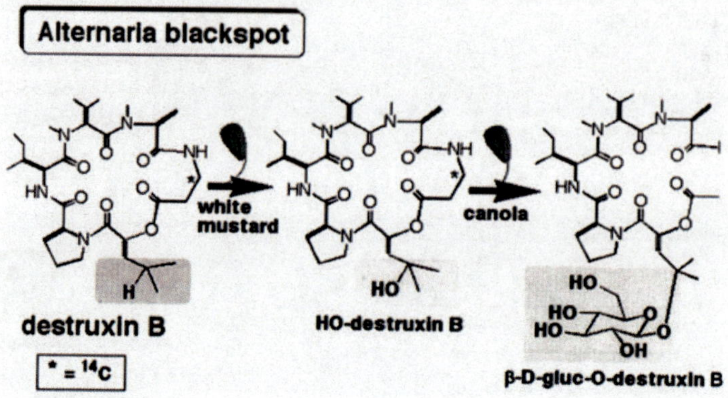

Fig. 5.3 Enzymatic reactions of plants to the pathogen causing Alternaria black spot (Verma and Saharan 1994; Saharan et al. 2016)

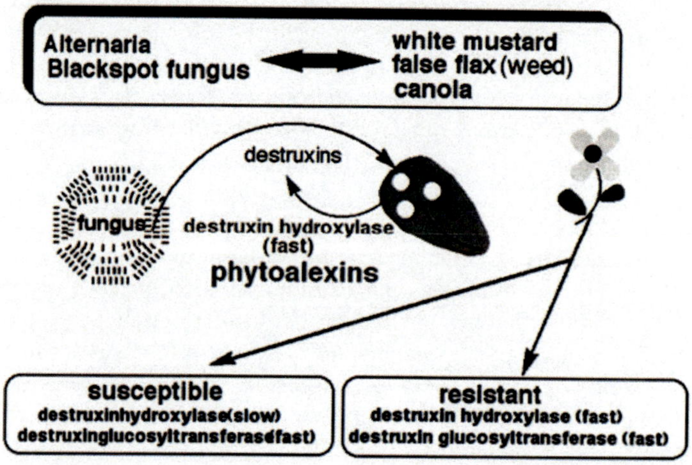

Fig. 5.4 Defense responses of plants resistant and susceptible to Alternaria black spot (Verma and Saharan 1994; Saharan et al. 2016)

1988) as the major host-selective phytotoxin. Brassicicolin A (1) appeared to be more phytotoxic to the susceptible cruciferous species *B. juncea* than to the tolerant *B. napus.*

Previous claims that *A. brassicicola* produced destruxins (Cooke et al. 1997; Evans et al. 1996) could not be confirmed; in spite of the availability of several destruxins in metabolite library (Pedras et al. 2000), none of these phytotoxins could be detected in either extracts of liquid cultures or in infected leaf tissues. In addition, the four polyketides phomapyrones A (3), F (4), and G (5), and infectopyrone (6), previously isolated from *Leptosphaeria maculans* and *L. bigloboso* (Pedras et al. 1994; Pedras and Biesenthal 2001; Pedras and Chumala 2005), were also isolated from *A. brassicicola.* Clones of *A. brassicicola* (altr203 × n03 and altrO14 × eO1)

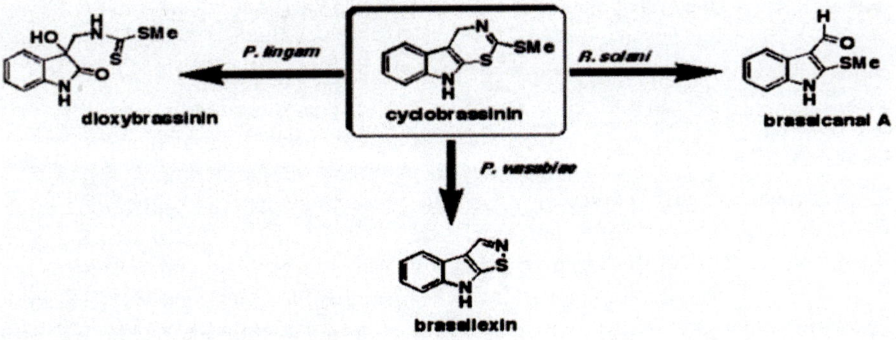

Fig. 5.5 Transformation of the phytoalexin cyclobrassinin by different pathogens of Brassicas (Verma and Saharan 1994; Saharan et al. 2016)

were recently reported (Cramer et al. 2006) to encode enzymes involved in the biosynthesis of undetermined polyketides, which could be the phomapyrones 3 and 5. In this context, it is pertinent to point out that genome sequences currently available in public databases have indicated the existence of numerous "orphan pathways," that is, gene clusters for which the encoded natural product is unknown (Gross 2007). Therefore, it is expected that further new metabolites and/or phytotoxins could potentially be isolated from fungal species previously investigated, such as *A. brassicicola* (Pedras et al. 2009).

Phytoalexin production in leaves of *B. juncea* infected with *A. brassicicola* led to the detection and isolation of two known phytoalexins and a phytoanticipin: cyclobrassinin (11) and spirobrassinin (12), and indolyl-3-acetonitrile (15), respectively. Since indolyl-3-acetonitrile (15) was occasionally detected in control leaves, it is most likely a phytoanticipin. In addition, other minor phytoalexins (13 and 14) and phytoanticipins (16 and 17) were detected in fractions of leaf extracts. Since the product of detoxification of brassinin (10), and *N*′-acetyl-3-indolylmethanamine (19), was also isolated from infected, but never detected in control leaves, it is suspected that brassinin (10) do not accumulate in enough quality in infected tissues because it is quickly detoxified to 19, similar to the detoxification that occurs in *A. brassicicola* cultures. To support this hypothesis, Pedras et al. (2009) determined that *A. brassicicola* detoxifies brassinin (10) to indolyl-3-methanamine (18) and *N*′-acetyl-3-indolylmethanamine (19) in fungal cultures. Although brassinin (10) is not a strong antifungal against *A. brassicicola*, this phytoalexin is a precursor of brassilexin (14) and other phytoalexins of *B. juncea* (Figures in the parenthesis are phytoalexin structures as per list of Pedras et al. 2007). It is very probably that its detoxification makes the plant more vulnerable to fungal invasion, and brassinin-detoxifying enzymes produced by *A. brassicicola* might contribute to its virulence. Comparison of the in vitro antifungal activity of the phytoalexins present in infected tissues of *B. juncea* and that of camalexin (20, not produced in *Brassica* species) suggest that camalexin might be a more effective plant defense against *A. brassicicola*.

5.2.3 Role of MAPK Cascade for Interaction with Toxin

Since cell perceives the signal through the cell surface receptors including receptors like kinases (RLK), like proteins (RLP), and like extracellular binding proteins, they are known to interact with MAMPs (Microbes-Associated Molecular Pattern) or PAMPS (pathogen-associated molecular pattern), and effectors. Receptor-like kinases interact with pathogen-associated molecular pattern, which are small peptides like chitin-oligosaccharide (Hamel and Beaudoin 2010) and flagellin (Pitzschke et al. 2009). Destruxin B, which is a small peptide, produced by *A. brassicae*, is found to interact with Lys M receptor-like kinase and WAK receptor through in silico interaction. The results of docking interaction of Destruxin B with different receptor-like kinases from different species indicate that Lys M receptor-like kinase in *Arabidopsis* showed better interaction with it. A mutation in Lys M receptor-like protein (Lys M RLK I) gene blocked the induction of almost all chito-oligosaccharide-responsive genes and led to increased susceptibility to fungal pathogens, indicating that Lys M RLK1 is essential for chitin signaling in plants (Wan et al. 2008). *Arabidopsis* genome has more than 80 MAPKKKs, but very few members of MAPKKKs have been studied. To continue the downstream cascade, and to take part in the defense, docking was performed with MAPKKKs (MAPKKK I, MAPKKK 2, MAPKKK I2, and MAPKKK 19) and receptor kinases which showed better interaction with destruxin B. It was found that MAPKKK 19 showed better interaction with Lys M receptor-like kinase (*Ricinus communis*); WAK receptor-like kinase *(A. thaliana)* and MAPKKK 2 were also found to interact better with CHRK 1 receptor-like kinase (*Senecio squalidus*).

Activity of MAPK 4 has been identified in plant defense against biotic stress (Desikan et al. 2001). Out of all ten MAPKKs known, activity of some MAPKKs in plant defense has already been identified against biotic stress, namely, MAPKK 3 and MAPKK 5 (Asai et al. 2002). Docking results of all ten MAPKKs with MAPK 4 showed that MAPKK 5, MAPKK 8, MAPKK 3, and MAPKK 9 have better interaction with MAPK 4. There are several evidences in the literature that showed MAPKK 3, MAPKK 5, MAPKK 1, and MAPKK 2 are the upstream kinase of MAPK 4 (Popescu et al. 2009a, b; Andreasson and Ellis 2010), and MAPKK 9 is the upstream kinase of MAPK 3/MAPK 6 (Tena et al. 2011). According to the superimposition results, MAPKK 9 and MAPKK 5 also belong to the same group C. So it could be hypothesized that MAPKK 9 might be the upstream kinase not only for MAPK 3/MAPK 6 but also for MAPK 4. Recently, one group also tried to determine the interaction of all MAPKs with all upstream MAPKKs through functional protein microarrays and found that MAPK 4 interacts with MAPKK 1, MAPKK 2, MAPKK 3, MAPKK 5, and MAPKK 7 (Popescu et al. 2008). Tena et al. (2011) showed that activation of MAPK 4 cascade is required for sustainable activation of defense pathway. Since our docking results do not show the better interaction of MAPK 4 with MAPKK 1/MAPKK 2, it implies that this interaction occurs through scaffold protein, or after post-translational modification of the protein. At the early stage of infection of *Alternaria* blight, all three MAPKs (MAPK 3, MAPK 6, and MAPK 4) were expressed. It should be further investigated by the wet lab experiment if all these three kinases (MAPK 3, MAPK 4, and MAPK 6) are activated by specific MAPKKs or by a common MAPKK. To determine the upstream members of MAPKKs in the signaling module, the docking was again performed with same members of MAPKKKs family which have already been docked with receptor-like kinases, and results suggested that MAPKK 9 interacts with MAPKKK 2/MAPKKK I9. To evaluate the downstream targets of MAPK 4, the docking performed with some WRKY transcription factors in plant defense, namely, WRKY 21, 25, 29, 33, 40, 53, 69 (Popescu et al. 2008; Tena et al. 2011; Asai et al. 2002; Fill et al. 2009; Cristna et al. 2010; Anderson et al. 2005), indicated that WRKY 25 and 40 could be the better interacting partner of MAPK 4. Several molecular and biochemical studies have shown that MAPK 4 interacts with all above listed WRKY proteins, except WRKY 29, a downstream partner of MAPK 3/MAPK 6; it can be assumed that this might be due to the specificity provided by the scaffold proteins. The best-studied example of scaffold protein is interaction of MAPK 4 with WRKY 33 via MKS I protein (Fill et al. 2009). By protein microarray and Y2H studies, WRKY 40 and WRKY 25 were reported to be the downstream interacting protein of MAPK 4 (Popescu et al. 2008; Anderson et al. 2005). Superimposition results also group WRKY 40 in a same group which is also supported by the docking results. Marmath et al. (2013) described the role of MAPK 4 in plant defense against Alternaria blight and predicted some potential candidate for MAPK 4 cascade. Observations lead to propose the model of MAPK cascade involving MAPK 4 in pathogenesis of Alternaria blight disease.

5.2.4 Bimolecular Basis of Resistance

The *Alternaria*-resistant *B. juncea* genotype RC-781 is characterized by relatively high concentrations of phenols and low concentrations of sugars and nitrogen compared to a susceptible genotype Prakash (Gupta et al. 1984). The leaf surface constituents including wax, total phenols, soluble nitrogen, and total soluble and reducing sugars of susceptible genotypes BSH-1, YSPB-24 (*B. rapa*), and RH-30 (*B. juncea*) and resistant genotypes Tower, HNS-3 (*B. napus*), AC-2 (*B. carinata*), and local (*B. alba*) have been determined by Gupta et al. (1987a, b, c), at 30, 50, 70, and 90 days of plant growth. The concentration of the phenolic compounds is significantly higher in resistant compared to susceptible species at all plant growth stages. Total soluble sugars, reducing sugars, and soluble nitrogen levels are, however, lower in resistant genotypes.

The analysis of total phenols and specific activities of polyphenol oxidase (PPO), peroxidase (PO), and catalase in *A. brassicae* and *A. brassicicola* infected mustard leaves of tolerant (RC-781, RH-8113) and susceptible (RH 30, Prakash) cultivars of mustard were determined at 20 days interval after 40 days of sowing showed that the total phenols increased initially and decreased with the age of the plant; these were markedly higher in tolerant as compared to the susceptible ones at all growth stages. After infection, their amount decreased in all genotypes. The specific activities of polyphenol oxidase remained higher while that of peroxidase lower in tolerant than

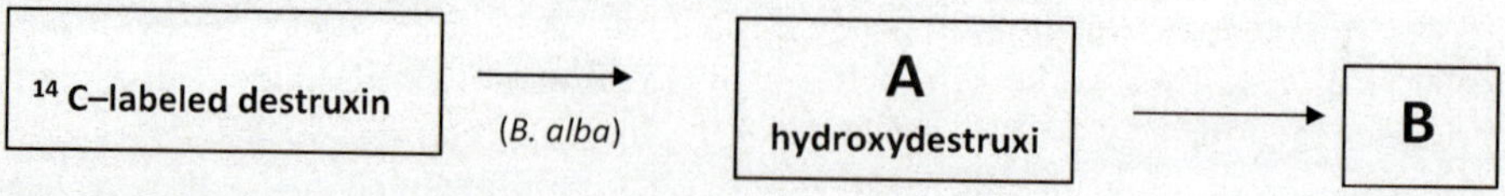

Fig. 5.6 Metabolism of destruxin B by white mustard and rapeseed (Verma and Saharan 1994; Saharan et al. 2016)

in the susceptible cultivars. In response to infection, although the activities of both the enzymes increased sharply in the cultivars, this increase was considerably higher in susceptible than in tolerant cultivars. Catalase activity was appreciably higher at initial stage and dropped markedly at later stages (Gupta et al. 1990).

Incorporation of *A. brassicae* culture filtrate into subcultures of hypocotyl explants callus on MS medium supplemented with 1 mg L^{-1} NAA and 1 mg L^{-1} BAP (MS N_1B_1) caused greater accumulation of total soluble and reducing sugars in susceptible *B. juncea* cv. Kranti than in resistant *B. napus* cv. GSH-1 and *B. juncea* cv. RH-781. Phenolic content and flavonols were higher in GSH-1 and RH-781 calli than that of cv. Kranti. Protein content though increased in all three cultivars, increase was more in resistant than the susceptible cultivars (Kiran et al. 2003). Destruxin B detoxification of white mustard (*Sinapis alba*) through biotransformation can help discover the specific disease resistance enzyme(s)/gene(s). If the black spot disease resistance of white mustard is related with the presence of specific detoxifying enzyme(s), this trait can be transferred to susceptible Brassicas and may improve their resistance to this destructive fungal disease (Fig. 5.6; Pedras et al. 2003).

5.2.5 Proteomics of Disease Resistance

Two *Brassica* lines derived from an inter-specific cross between *B. napus* and *B. carinata* were evaluated for tolerance to *A. brassicae*. Pathogen-induced chlorosis and necrosis spread significantly in one line whereas it remains localized in the other. Proteome level changes in response to the fungal pathogen invasion were investigated using two-dimensional electrophoresis. Levels of 48 proteins were significantly affected at various time points in the tolerant line (41 upregulated and seven downregulated). In contrast, in the susceptible line, the levels of 23 proteins were significantly affected with four increasing and 19 decreasing. The identities of 38 proteins and those identified from the tolerant line included enzymes involved in the generation of reactive oxygen species (ROS), ROS-mediated signaling, auxin signal transduction, and metabolic pathways. Proteome level changes results suggesting a role for ROS-mediated auxin signaling in this pathosystem was further investigated and confirmed using quantitative real-time PCR (Sharma et al. 2007). Chitinase modifying proteins (*Cmps*) are secreted by fungal pathogens of crucifers, which interfere with fungalysin *Cmp* activity to improve plant resistance to multiple fungal diseases (Naumann and Wicklow 2013).

The *Arabidopsis thaliana* secretome analyzed by the proteomic approach led to the identification of secreted proteins implicated in many aspects of cell biology. Oh et al. (2005) investigated the change in the *Arabidopsis* secretome in response to salicylic acid and identified several proteins involved in pathogen response. One of these, a secreted lipase with a GDSL-like motif designated as GDSL LIPASE 1 (*GLIP1*), was further characterized for its function in disease resistance; *glip1* plants are markedly more susceptible to infection by *A. brassicicola* compared with the parental wild plants. The recombinant *GLIP1* protein possessed lipase and antimicrobial activities that directly disrupt fungal spore integrity. *GLIP1* appears to trigger systemic resistance signaling in plants when challenged with *A. brassicicola*, because pretreatment of *glip1* mutant with recombinant *GLIP1* protein inhibited *A. brassicicola*-induced cell death in both peripheral and distal leaves. Moreover, *glips1* shows altered expression of defense- and ethylene-related genes. *GLIP1* transcription was increased by ethephon, the ethylene releaser, but not by salicylic acid or jasmonic acid. *GLIP1* in association with ethylene signaling may be a critical component in plant resistance to *A. brassicicola*.

Chitinase, capable of degrading the cell walls of invading phytopathogenic fungi, plays an important role in plant defense response, particularly when this enzyme is overexpressed through genetic engineering. *Brassica* plant (*B. juncea* L.) was transformed with chitinase gene tagged with an overexpressing promoter 35 S *CaMV*. In in vitro fungal growth inhibition assays, chitinase inhibits the fungal colony size by 12–56% over the non-transgenic control. The bioassay under artificial epiphytotic conditions reveals the delay in the onset of disease as well as reduces lesion number, and size in 35 S chitinase *Brassica* as compared to the untransformed control plants (Mondal et al. 2003).

Genotypes EC-399299, CE-399296, EC-399313, and PHR-2 are comparatively more resistant to infection by *A. brassicae*. Increased level of PAL, PPO, and peroxidase may play an important role in undertaking the defense mechanisms of *B. juncea* genotypes against Alternaria blight pathogenesis (Parihar et al. 2012). Spary of chitosan with 1 ppm concentration on the leaves of *Brassica juncea*, superoxide dismutase and with 5 ppm chitosan peroxide and phenylalanine ammonia lyase significantly increased resistance to Alternaria blight with respect to control. Phenolic content in the leaves of treated plants also increased with both the treatments of chitosan. Defense-related enzymes and phenolic content showed positive correlations, that is, increase with the elicitor application. Since these enzymes and phenolics increased resistance to pathogens, they may provide protection against Alternaria blight (Sood and Sohal 2012). After inoculation of *B. juncea* leaves with *A. brassicae*, activities of the cell wall-degrading enzymes, polygalacturonase (EC-3.2.1.15) and cellulase (EC-3.2.1.4), decreased in leaf blight-resistant cultivar RC-781 and increased in the susceptible cultivar Varuna up to 3 days. In both cultivars, 11 polypeptides are observed in leaves in the absence of *A. brassicae* inoculation. After inoculation, although, there was no change in the polypeptide pattern in the resistant cv. RC-781, but in the susceptible cv. Varuna, four polypeptides (43.7–58.8 kDa) disappeared on the third day after inoculation (Garg et al. 1999). Occurrence of a lipase in *A. brassicicola* spores plays a crucial

Table 5.1 Specific activity of peroxidase enzyme in hypocotylar calli of *Brassica* species raised on MS medium supplemented with or without fungal culture filtrate (FCF) of *Alternaria brassicae* (Kiran et al. 2004)

Species	Activity of peroxidase enzyme* at different concentration of FCF (%)		
	Control	5	10
Brassica juncea cv. Kranti	11.21	13.77	16.88
Brassica juncea cv. RH-781	3.57	13.41	21.73
Brassica juncea cv. GSH-1	0.18	0.37	1.89

*Enzymes unit = Change in absorbance/3 min/mg/protein

Table 5.2 Specific activity of enzyme catalase in hypocotylar calli of *Brassica* species raised on MS medium supplemented with or without fungal culture filtrate (FCF) of *Alternaria brassicae* (Kiran et al. 2004)

Species	Activity of enzyme* catalase at different concentration of FCF (%)		
	Control	5	10
Brassica juncea cv. Kranti	0.79	0.83	0.87
Brassica juncea cv. RH-781	4.21	3.57	1.72
Brassica juncea cv. GSH-1	1.43	1.21	0.45

*Enzymes unit = Change in absorbance/5 min/mg/protein

role in the infection of cauliflower leaves. Addition of anti-lipase antibodies into conidial suspension prior to inoculation resulted in reducing black spot lesions by 90% on intact cauliflower leaves, but not on leaves from which surface wax had been removed. Spore surface bound lipase is thought to interact closely with epicuticular leaf waxes for adhesion, and/or penetration of fungal propagules during the early stages of host parasite interactions (Berto et al. 1999).

Dhingra et al. (2004) reported higher peroxidase activities in two *B. juncea* cultivars than in *B. napus* cv. GSH-1. Between two *B. juncea* cultivars, enzyme activity in Kranti was more than three times than in RH-781. Incorporation of fungal culture filtrate (FCF) to the culture medium although increased peroxidase activity in all cases, further increase in the FCF concentration, the peroxidase activity increase was much lesser in Kranti than in RH-781 and GSH-1 (Table 5.1). The catalase activity is low in the calli of *B. juncea* cv. Kranti than in *B. juncea* cv. RH-781 and *B. napus* cv. GSH-1. Between the latter two cultivars, the former showed three times more activity than the latter. Catalase activity was not affected by FCF in the calli of Kranti, while activity decreased in other two cultivars; the decrease was more in GSH-1 than in RH-781 (Table 5.2). Poly-phenol oxidase activity was low in the calli of *B. juncea* cv. Kranti than in *B. juncea* cv. RH-781 and *B. napus* cv. GSH-1 (Table 5.3). Addition of FCF to the culture medium, in general, increased PPO activity; the increase was much lesser in cv. Kranti than in the other two cultivars. Calli of GSH-1 showed more than twofold increase in PPO activity over that of RH-781 (Table 5.3; Kiran et al. 2003). Inoculation of leaves with *A. brassicae*

Table 5.3 Specific activity of enzyme polyphenol oxidase in hypocotylar calli of *Brassica* species raised on MS medium supplemented with or without fungal culture filtrate (FCF) of *Alternaria brassicae* (Kiran et al. 2004)

	Activity of enzyme* polyphenol oxidase at different concentration of FCF (%)		
Species	Control	5	10
Brassica juncea cv. Kranti	4.12	5.37	6.54
Brassica juncea cv. RH-781	5.70	8.49	10.56
Brassica juncea cv. GSH-1	6.19	10.21	14.63

*Enzymes unit = Change in absorbance/5 min/mg/protein

increased PPO activity in both resistant and susceptible cvs. of *B. juncea*, but the increase was much higher in RC-781 than in Varuna (Gupta 1995; Saharan et al. 2016).

5.2.6 Induction of Resistance

Plants possess a very vast array of defense strategies to combat pathogen's growth and colonization. One of the most efficient and immediate defense reactions of plants against the pathogen is the hypersensitive response, which is defined as "the rapid death of plant cells in association with restriction of pathogen growth" (Goodman and Novacky 1994). It is characterized by the presence of brown, dead cells at the site of infection which limit the growth of biotrophic pathogens. Hypersensitive reaction is a highly concerted complex defense response which is followed by the production of reactive oxygen species (Lamb and Dixon 1997), modification of ion fluxes (Levine et al. 1996), and activation of defense through the synthesis of signaling molecules such as jasmonic acid, salicylic acid, and protein kinase (Dangl et al. 1996; Dixon et al. 1994). Salicylic acid accumulation leads to the onset of systemic acquired resistance in the distal plant tissues (Ryals et al. 1996). These events are accompanied by the activation of several plant defense genes, local accumulation of pathogenesis-related (PR) proteins, activation of transcription factors, degradation of proteins by the polyubiquitin system, and programmed cell death (Mishra et al. 2010).

A pre-inoculative foliar application of 5 mM β-amino-butyric acid (BABA) significantly inhibits the colonization of *A. brassicae* on leaves of *B. carinata* susceptible cultivar Car 6. BABA treatment leads to transient but significant increase in H_2O_2 level during early stages of pathogen colonization. A significant increase in superoxide dismutase (SOD), ascorbate peroxidase (APX), and guaracol-dependent peroxidase (GDP) is known to inhibit the oxidative stress in BABA-treated plants in response to pathogen infection. BABA treatment also leads to proper balance of oxidants and antioxidants suitable for expression of resistance, resulting in curtail of pathogen ingress during early stages of colonization (Chavan et al. 2013). The studies have revealed an important role of hormones in plant immunity. Cytokinins are phytohormones that are involved in various regulatory processes including plant defense. Zeatin, a cytokinin, antagonizes the effect of *A. brassicae* pathotoxin in cell culture of *B. juncea*. Phytohormones are also the inducers of MAP kinase signaling pathways, which are the important signaling modules in eukaryotic cells. Marmath et al. (2013) observed the role of exogenous application of zeatin on disease score, and infection behavior of *A. brassicae*, expression pattern of MAPK 4 in the nonhost *Sinapsis alba*, susceptible *B. juncea* cv. Varuna, moderately tolerant *B. juncea* cv. Divya, and transgenic *Brassica* to confer its role in plant immunity. High concentration of zeatin leads to increased defense responses by delaying the infection process as well as significantly reducing the disease score. Semi-quantitativc RT-PCR reveals that zeatin also increases the expression of MAPK 4 at early hours of infection. Zeatin upregulates plant immunity via an elevation of MAPK 4, and clearly reflects that it antagonizes the effect of *A. brassicae*. The cross talk between zeatin and MAPK 4 signaling pathways may help plants fine-tune defense responses against *A. brassicae* in Alternaria blight (Saharan et al. 2016).

5.2.7 Elicitation of Phytoalexins for Host Resistance

The accumulation of phytoalexins in Brassicas after exposure to microorganisms, and their role in disease resistance have been demonstrated by many workers. These are low molecular weight, antimicrobial compounds that are both synthesized by and accumulated in plants after exposure to microorganisms (Paxton 1981). So far, approximately a dozen different phytoalexins have been reported from crucifers (Bains and Tewari 1985; Browne et al. 1991; Devys et al. 1988; Takasugi et al. 1988). All phytoalexins identified from Brassicas contain sulfur in their molecules. Conn et al. (1987, 1991) and Tewari et al. (1987, 1988) described elicitation of phytoalexins in var. *Oleifera* (canola), *B. rapa* ssp. *rapifera* (an accession of turnip), *B. napus*, *Camelina sativa*, *Eruca sativa*, and *Capsella bursa-pastoris* similar to that of cyclobrassinin. However, the quantity produced in turnip is much greater. Jejelowo et al. (1991) studied the kinetics of phytoalexin elicitation in *C. sativa.* An accession of *E. sativa* was reported to show hypersensitive response to *A. brassicae* (Conn and Tewari 1986). Browne et al. (1991) described camalexin and methoxy-camalexin phytoalexins from *C. sativa*, which are thiazoyl-substituted indole phytoalexins and showed strong structural resemblance to the fungicide thiabendazole.

The antifungal effects of camalexin, brassinin, allyl-(AI ITC), and benzyl–(Bz ITC) isothiocyanates, at different development stages of *A. brassicae* and *A. brassicicola* have been observed by Sellam et al. (2007). Irrespective of the isolate, the phytoalexin camalexin exhibits the greater inhibitory effect with mean EC_{50} values ranging from 34 μm (germ-tube elongation) to 183 μm (mycelial growth). Germ-tube elongation is more sensitive compared to conidial germination, and mycelial growth, with mean EC_{50} values of the former of 81 μm, 520 μm, and 870 μm for brassinin, Bz ITC, and AI ITC, respectively. However, the role of camalexin in disease resistance varies among different *Arabidopsis* population in

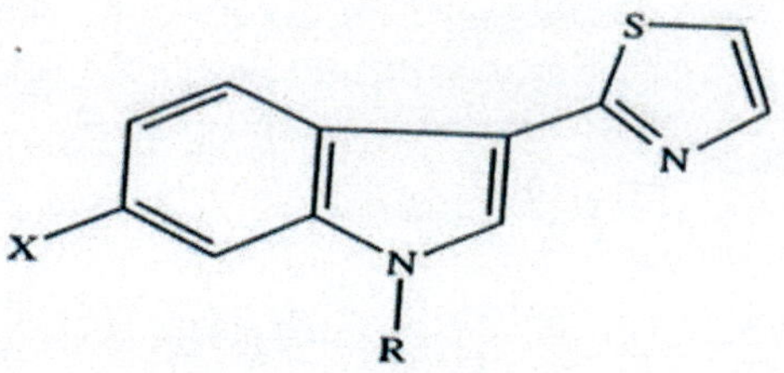

Fig. 5.7 Structures of camalexin (1), 6-methoxycamalexin (2), and *N*-methylcamalexin (3) (Verma and Saharan 1994; Saharan et al. 2016)

nature and provides some clue to other possible determinants of resistance to *A. brassicicola* (Kagan and Hammerschmidt 2002).

The relationship between conidial concentration, germling growth, and phytoalexin production by *C. sativa* leaves inoculated with *A. brassicae* has been studied by Jejelowo et al. (1991). The rapid rate of phytoalexin accumulation shortly after inoculation results into inhibition of fungal growth on the leaf surface. The phytoalexin slows germination and inhibits germ-tube growth of *A. brassicae* conidia in vitro. Three phytoalexins, camalexin ($C_{11}H_8N_2S$), 6-methoxycamalexin ($C_{12}H_{10}N_2SO$), and *N*-methylcamalexin ($C_{12}H_{10}N_2S$), have been isolated from shepherd's purse (*Capsella bursa-pastoris*) challenged by *A. brassicae* (Fig. 5.7). Phytoalexin elicitation in shepherd's purse is associated with its resistance to *A. brassicae* (Jimenez et al. 1997). *Capsella bursa-pastoris* shares one phytoalexin with *C. sativa* and the other one with rapeseed and is also highly resistant to *A. brassicae* (Conn et al. 1988). The highly resistant *Camelina sativa* elicits large quantities of two novel phytoalexins; it appears that in crucifers, both the quantity and quality of phytoalexins are important in regulating resistance to *A. brassicae.*

These phytoalexins are highly antimicrobial, as shown by the *Cladosporium* sp. thin-layer chromatography bioassay (Tewari 1991a, b). Antifungal compounds have been found to accumulate when root slices of radish are inoculated with the spore suspension of *A. alternata.* Antifungal activity of the diffusate shows inhibitory action to spore germination, germ tube elongation, and radial mycelial growth. The phytoalexins produced are fungistatic in nature (Kulshreshtha and Chauhan 1985). Chinese cabbage lines/hybrids on *A. brassicae* inoculation produce phytoalexins (Saharan et al. 2003, 2016).

5.2.8 Calcium Sequestration for Host Resistance

The plant cell walls are rich in calcium, which is tightly bound to pectins. Calcium is known to be a factor in disease resistance. The insoluble calcium polypectates are resistant to hydrolysis by pectolytic enzymes produced by the pathogens (Vidhyasekaran 1988). Oxalic acid and possibly other organic acids produced by various pathogens sequester calcium, which may overcome this form of resistance (Punja and Jankins 1984; Rao and Tewari 1987). Examination of black spot lesions on rapeseed leaves by scanning electron microscopy in conjunction with energy-dispersive X-ray microanalysis has revealed sequestration of calcium by *A. brassicae*. Therefore, there are possibilities of enhancing resistance to *A. brassicae* in rapeseed by soil or foliar application of calcium compounds (Tewari 1991a, b). Foliar sprays of calcium compounds sequester the organic acids at the site of infection, and soil application has the potential of boosting calcium content of the plant (Saharan et al. 2016).

5.2.9 MAPK Signaling Pathways for Camalexin Biosynthesis in *Brassica* Triggers *Alternaria* Resistance

The host–pathogen interaction is precisely driven by selection to acquire improved defense against pathogens; therefore, during the course of evolution, plants have adapted to the processes that trigger the synthesis and accumulation of a vast array of structurally diverse antimicrobial secondary metabolites. Diversity among the secondary metabolites is considered to be partly originating from this host–pathogen interaction as a byproduct of the evolutionary process. This notion reflects that each particular class of secondary metabolites is restricted to a narrow phylogenetic lineage, which is responsible for developing disease resistance against specific pathogens. Among these antimicrobial secondary metabolites, phytoalexins represent an important class which can be defined as small molecular weight antimicrobial compounds that are synthesized de novo by plants in response to an attack and progression of the fungal pathogens and are considered to be the major determinants of induced plant resistance against these pathogens. However, the regulatory mechanisms for phytoalexin biosynthesis are still largely unknown, but previous investigations have suggested that the mitogen-activated protein kinase (MAPK) signaling pathways are the core of the defense mechanisms that regulate phytoalexin biosynthesis, often orchestrated in microbe-associated molecular patterns (MAMP)-triggered immunity. It has been observed that the MAPK cascade activates transcription factors to regulate the phytoalexin biosynthesis in MAMP-triggered immunity. MAMP-triggered immunity exerts a number of defense responses upon recognition of reactive oxygen species (ROS) accumulation, including the synthesis of antimicrobial secondary metabolites or phytoalexins along with transcriptional activation of pathogenesis-related genes. The MAMP signal is received by plant pattern recognition receptors (PRRs) that are shortly followed by the activation of plant MAPK signaling pathways consisting of three kinase-signaling modules (MAPK, MAPK kinase (MAPKK), and MAPKK kinase (MAPKKK)).

There is a typical indole-derived phytoalexin known as "Camalexin" that is synthesized and accumulated in *Arabidopsis thaliana* to trigger defense against a fungal pathogen *Botrytis cinerea*. Infection to *Arabidopsis thaliana* with *Botrytis cinerea* is recognized by the pattern recognition receptors (PRRs) of the host which then express the genes involved in the downstream processing of the MAPK signaling cascade to synthesize and accumulate camalexin in order to trigger defense against the pathogen. The biosynthesis of camalexin is regulated by MAPK signaling pathway via expression of AtMPK3, AtMPK4, and AtMPK6 promoters in *Arabidopsis thaliana*. AtWRKY33, which is a DNA-binding protein, and OsTGAP1, which is a b ZIP transcription factor, have both been identified to be the inducers of diterpenoid phytoalexin biosynthesis in *Arabidopsis*.

Phytoalexins are small antimicrobial molecules synthesized and accumulated by plants upon exposure to pathogens. Camalexin is an indole-derived phytoalexin, which is accumulated in plants including *Arabidopsis thaliana*, and other Brassicaceae, which plays a major role in disease resistance against fungal pathogens. In *Arabidopsis thaliana*, MAP kinase signaling cascade is known to be involved in the synthesis of camalexin, which contributes to disease resistance against a necrotrophic fungal pathogen, *Botrytis cinerea*. The MAPK signaling cascade leading to biosynthesis of camalexin that triggers defense responses in *B. rapa* upon exposure to the most devastating necrotrophic fungus, Alternaria brassicae, has been elucidated with the help of MAPK cascade in Arabidopsis thaliana. Molecular modeling, docking, and protein-protein interaction analysis of MAP kinases retrieved from *Brassica rapa* genome have been carried out to reveal the above cascade. The tertiary structure prediction of MAPKs obtained through molecular modeling revealed that all protein models fulfill the criteria of being the stable structures. The molecular docking of predicted models for elucidating potential partners of MAPKs revealed strong interactions between MKK1, MKK4, MKK5, MAPK3, and MAPK6 with MKK9. The MAPK signaling cascade also shows different genes that express and play major role in camalexin biosynthesis in *B. rapa* during defense response to *A. brassicae*. The understanding of MAPK defense signaling pathway in *B. rapa* against devastating fungal pathogen Alternaria brassicae would help in devising strategies to develop disease resistance in *Brassica* crops (Fig. 5.8; Gaur et al. 2018).

5.2.10 Identification and Characterization of Defense Genes in *Brassica* to *Alternaria*

Efforts have been made to identify and characterize resistant and defense-related genes triggered during resistance toward Alternaria blight, a recalcitrant disease caused by *Alternaria brassicicola* and Alternaria brassicae in *Arabidopsis* and *Brassica*. Besides, the key components of JA-SA-ET involved in defense response were also investigated. The identified top ten resistant and defense-related genes are listed. These genes could be utilized for development of molecular markers linked with disease resistance which can further be utilized in molecular breeding program. Moreover, the results can also be utilized for transgenesis, directed mutagenesis, cisgenesis, and gene editing for development of resistant *Brassica* plants against Alternaria blight.

The identification of resistance (R) and defense-related genes unlocked interesting possibilities for prevention and management of diseases caused by several

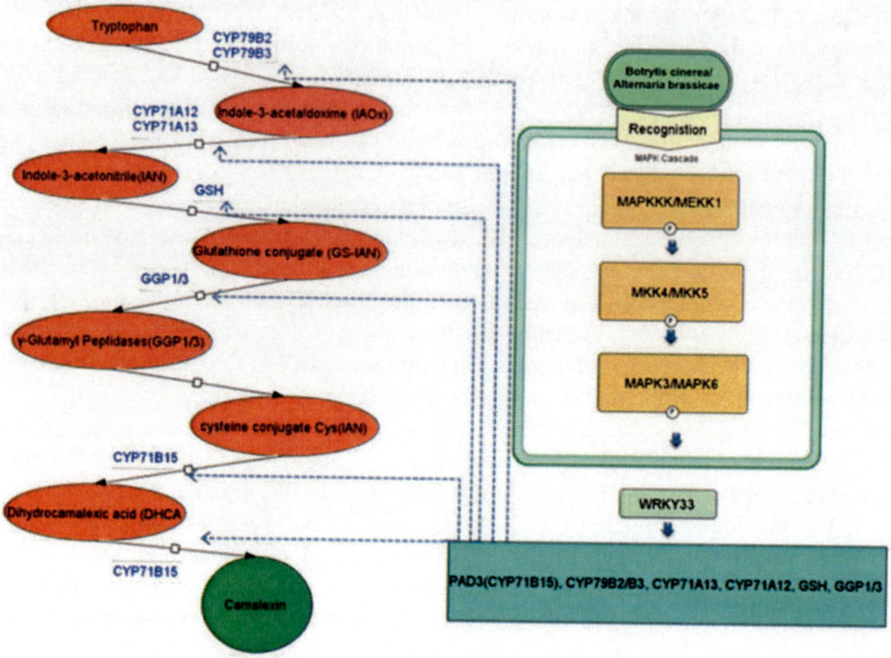

Fig. 5.8 Camalexin biosynthesis pathway and regulatory gene expressed by MAPK cascade designed using cell designer tool (Gaur et al. 2018)

pathogens. However, such genes are available in limited numbers which can be deployed in plants to engineer defense against limited number of pathogens. On the other hand, efficient application of microarray technology and functional genomics tools allow us to discover important candidate genes through stimulating better understanding of disease resistance and plant defense signaling. It could disclose novel insights on the interactions among signaling pathways and other processes of plant systems involved in plant–pathogen interactions. Various studies conducted in the recent past on the signaling machinery toward necrotrophic fungal pathogens have helped to dissect various components.

The knowledge on molecular mechanism of host–pathogen interaction is considered to be prerequisite for engineering disease-resistant varieties of Brassica against Alternaria blight disease. Throughout plant–pathogen interactions, our knowledge of responses has taken a big leap forward. Nonetheless, over the course of this decade, we still have several aspects and challenges to address different questions associated to these interactions. It is believed that the huge data on expression of resistance and defense-related genes with respect to plant–pathogen interaction can be analyzed to identify key candidate gene(s) which can be modified by genetic engineering or molecular breeding approaches to engineer disease resistance in *Brassica* (Pandolfi et al. 2017; Imam et al. 2016).

The efforts have also been made in dissecting the different components of defense signal transduction pathways activated toward different pathogens. Jasmonic acid/ ethylene and salicylic acid-mediated signaling pathways are activated against necrotrophic and biotrophic fungal pathogens. *A. thaliana* has already been demonstrated as host for Alternaria blight disease of *Brassica*. Therefore, it is being felt that *Alternaria brassicicola-Arabidopsis thaliana* could be used as one of the excellent model system for deciphering the intricacy of Alternaria blight in *Brassica*. Preprocessing of microarray data is the phenomenon of extracting and transforming the intensities of raw fluorescence into a signal normalized for biological variations and experimental errors. GCRMA (Guanine Cytosine Robust Multi-array Analysis) method was used for background correction of downloaded microarray data from NCBI GEO (Wu and Irizarry 2010). It converts background adjusted probe intensities into expression measures as same has been used by RMA (Robust Multi-array Average) for normalization and summarization of data. It performs much better than the other commonly used methods for normalization to identify upregulated and downregulated DEGs. It has demonstrated different sets of differentially expressed JA, SA, and ET responsive genes (DEGs) at 9 and 24 h after infection of *A. brassicicola*. All the upregulated and downregulated DEGs given in parenthesis for each contrast, that is, wild-type pathogen-treated plant with control plant (WT9C9: 1327up, 1527down; WT24C24: 1510up, 1455down); jasmonic acid mutant pathogen-treated plant with control (JAM9C9: 809up, 350down; JAM24C24: 2201up, 2397down); salicylic acid mutant pathogen-treated plant with control (SAM9C9: 1355up, 1161down; SAM24C24: 1819up, 1984down); ethylene mutant-treated plant with control (ETM9C9: 917up, 650down; ETM24C24: 1895up, 2269down) at 9 and 24 h after *Alternaria brassicicola* infection. During data analysis and annotation, among many defenserelated genes, NHL10 and HCHIB were recognized as resistance genes. The NHL10 is non-race-specific disease resistance gene (NDR1), and HCHIB is involved in ethylene-/ jasmonate-mediated systemic acquired resistance during pathogenesis. They are involved in defense responses during pathogenesis of Alternaria blight in *Arabidopsis thaliana*. Besides, enrichment analyses of the DEGs led to formation of gene ontology to define the significant genes involved in biological processes, molecular function, cellular components, and pathways. The genes involved in defense response toward fungi in *A. thaliana* were mapped on *Brassica rapa* genome sequences to identify and characterize the similar *Brassica* sequences. Characterization and comparative analysis of identified genes were carried through molecular phylogeny analysis and domain prediction to identify resistance and defense-related genes in *Brassica*. Infection of *A. brassicicola* led to upregulation of various genes such as WRKY, peroxidase, p450 oxidases, and chitinase which mediate defense response in *Arabidopsis* and *Brassica* upon infection with pathogen. It has been observed that the expression of these genes increase more in the presence of JA and SA than the wild-type plants at 24 h post infection. This indicates that expression of defense-related genes increase post infection of *A. brassicicola* to combat pathogen's spread, and this effect is enhanced by JA and SA which are well-known defense inducers. However, there were observed few genes which were downregulated in the presence of JA and SA. These genes appear to be the ones which are involved in pathogenesis process and that they are downregulated by JA or SA to trigger defense response against the pathogen. Protein-protein interaction networks of genes involved in defense response toward fungi were constructed from contrast JAMC, SAMC, and ETMC to determine the key components of JA-SA-ET-mediated pathway involved in disease resistance through network analysis. The genes, CZF1; WRKY; Movement Protein (MP); INDOLE-3-ACETIC ACID INDUCIBLE 19 (IAA19); Auxin-responsive gene (AXR3); INDOLE-3-ACETIC ACID INDUCIBLE 1 (IAA1); auxin response factor 6 (ARF6); and extra-large G-protein 2 (XLG2) involved in DNA-binding transcription factor activity, GTP binding, GTPase activity, and protein binding as well as defense response toward fungi, were investigated. However, XLG2 which is found in all contrast in the category of hubs is a well-characterized gene playing significant role in disease resistance (Shinde et al. 2018; Kaman-Tooth et al. 2019; Berardini et al. 2015; Zhu et al. 2009; Pathak et al. 2020).

5.2.11 Differential Defense Signaling Pathways in *Brassica* to *Alternaria*

A comparative study on defense gene induction was undertaken to explore the signaling mechanisms that govern the defense responses of susceptible Brassica juncea and moderately resistant *Sinapsis alba*, belonging to the family Brassicaceae, to Alternaria leaf spot disease. The expression of five selected defense-related genes, namely, pathogenesis related (PR)-1, PR-2, PR-3, non-expresser of PR-1 (NPR-1), and plant defensin (PDF1.2), was examined after inoculation of the seedlings of *B. juncea* and *S. alba*. Transcripts of all five defense-related genes accumulated at a greater level and earlier in *S. alba* than in *B. juncea* upon challenge inoculation with *A. brassicae* locally as well as systemically. Although PDF1.2 is known to be induced only by jasmonic acid (JA) and PR-1 by salicylic acid (SA) in the model plant *Arabidopsis thaliana* belonging to the family Brassicaceae, PDF1.2 was induced also by SA and PR-1 by JA in both *B. juncea* and *S. alba*. These results suggest that defense signaling pathways in *B. juncea* and *S. alba* are different from *A. thaliana*. Although JA responsive genes are predominantly induced in most crop plants and model plants against necrotrophic fungal pathogens, both JA and SA responsive genes seem to play an important role in *S. alba* conferring some resistance against *A. brassicae* (Nayanakantha et al. 2016).

5.3 *Arabidopsis–Colletotrichum*: Biometabolomics Resistance

5.3.1 Molecular Events During Host–Pathogen Interaction

Plants usually defend against pathogens by activating both localized and systemic resistance responses. These responses include hypersensitive response, cell-wall fortification, synthesis of phytoalexins, and production of other antimicrobial secondary metabolites or pathogenesis-related proteins (PR-proteins). Signaling molecules implicated in these inducible defense systems include salicylic acid (SA), jasmonic acid (JA), ethylene (ET), abscisic acid (ABA), auxin, gibberellins (GAs), cytokinins (CKs), brassinosteroids (BRs), and reactive oxygen species (ROS). These phytohormones can induce defense responses individually and also interact synergistically or antagonistically to further orchestrate downstream signaling. A large number of fungal and pathogens have been reported to infect the model plant *A. thaliana*, either naturally or in the laboratory. As a typical hemibiotrophic fungus, *C. higginsianum* develops a series of specialized infection structures. In particular, the intracellular BH of *C. higginsianum* are equivalent to haustoria, and hence this pathosystem can provide insights into the molecular basis of biotrophy in obligatory parasitic organisms, such as rusts, powdery mildews, and downy mildews, all of which are not readily culturable or genetically manipulated (Shigenaga and Argueso 2016; Yan et al. 2018).

5.3.1.1 Primary Metabolic Pathways

For successful establishment in host plants, biotrophic and hemibiotrophic fungi need to obtain nutrients from living host cells and effectively evade the host defense system. Colonization by a fungal pathogen is associated with multiple metabolic changes in the plant host, notably increases in the expression of several genes involved in primary metabolic pathways, synthesis or degradation of carbohydrates, amino acids, lipids, and mineral transport. It has been suggested that the role of primary metabolism during plant–pathogen interactions is to support cellular energy requirements for plant defense responses. The mutation of lht1 (lysine histidine transporter 1) in *Arabidopsis* can significantly reduce contents of glutamine, alanine, and proline, resulting in enhanced resistance not only to *C. higginsianum* but also to diverse bacterial and fungal pathogens. After inoculation with these pathogens, the lht1 mutant also exhibited increased callose deposition, higher accumulation of SA, and constitutive expression of PR-1 (Liu et al. 2010; Rojas et al. 2014).

Evidence has suggested that components of primary metabolism also can act as signals regulating various aspects of plant defense. The fatty acids and lipids play important roles in plant defense and biosynthesis of the major defense hormone JA. Little is known about the role of plant primary metabolism in defense against attack by this hemibiotroph. To date, several metabolic functions have been identified that influence compatibility of *C. higginsianum* with the plant host. Glycerol-3-phosphate (G3P) is an important component in carbohydrate and lipid metabolic processes. Infection of *Arabidopsis* by *C. higginsianum* leads to an increase in G3P levels and a simultaneous decrease in glycerol levels in the plant.

Cells impaired in the utilization of G3P accumulated higher levels of pathogen-induced G3P and exhibited enhanced resistance. The NADP-malic enzyme catalyzes the oxidative decarboxylation of L-malate using NADP+ as coenzyme, producing pyruvate, CO_2, and NADPH present as a multigene family. In *A. thaliana*, loss of cytosolic NADP-ME2 leads to increased susceptibility to infection by pathogens such as *C. higginsianum*. The data suggest that NADP-ME2 has a function during the basal defense response, where it may be required for ROS production after pathogen recognition. Since primary metabolism is essential for survival, associated genes are unlikely to be eliminated during natural selection, in contrast to most R-genes which are only periodically important for survival (i.e., when pathogens attack). Therefore, engineering resistance against pathogens by selection of resistance-related genes that also have a primary metabolic function is expected to provide more durable resistance (Lim et al. 2017; Voll et al. 2012).

5.3.1.2 Induction of Phytohormones

Genetic studies with *Arabidopsis*-signaling mutants have shown that SA-dependent responses are deployed against biotrophic pathogens, whereas ethylene- or JA-dependent responses are more important for induced resistance to necrotrophic pathogens, suggesting that resistance to hemibiotrophic pathogens such as Colletotrichum may require a combination of these pathways. Genome-wide studies using cDNA arrays in *Arabidopsis* infected with the *C. higginsianum* revealed that defense reactions activate the SA-dependent signaling pathway at the early stage of the interaction between *Arabidopsis* and *C. higginsianum*, and the subsequent defense reaction may depend on the JA-dependent signaling pathway because the correlation with SA signaling decreased rapidly and that of JA-signaling increased relatively. Differential defense signaling crosstalk and PR gene expression are involved in cultivar-specific resistance of kimchi cabbage plants to anthracnose, black spot, and black rot diseases, and the resistance is strongly associated with the hormone-dependent transcriptional induction of defense genes (Lee and Hong 2015; Thomma et al. 2001).

5.4 *Brassica–Leptosphaeria*: Biometabolomics Resistance

5.4.1 Jasmonic Acid (JA) and Calcium-Associated Transcription Factors Influence the Cellular Resistance Response in Host to *Leptosphaeria maculans*

Gene regulatory networks from co-expressed gene sets were studied using SeqEnrich (Becker et al. 2019). A putative gene regulatory network was identified in resistant plants early in the defense response (Fig. 5.9c). Data predict TF homologs canonically associated with JA (MYC3, JAM1), ABA (AKS2, ABF3), and calcium (CAMTA5, CAMTA3/SR1) signaling that regulate essential defense processes at the inoculation site in resistant plants. This includes overarching regulation of calcium ion binding, response to ABA, JA-mediated signaling,

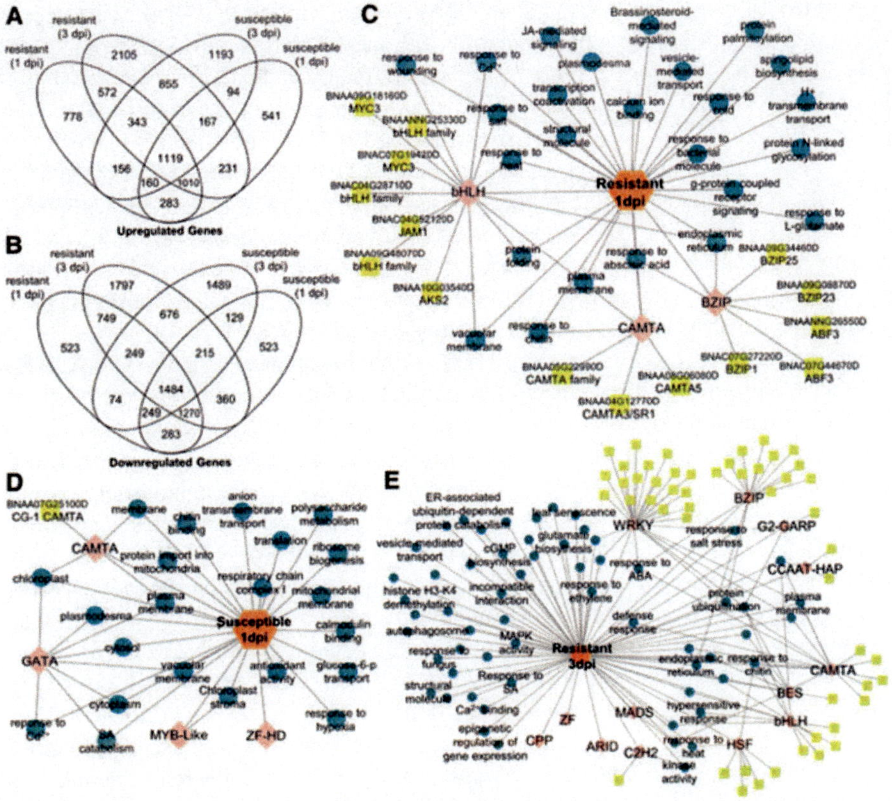

Fig. 5.9 Differential gene expression analyses and predictive transcription factor networks. (**a**) Significantly ($P < 0.05$) upregulated genes in each treatment as compared to their mock inoculated controls. Four-way VENN diagram shows number of unique or shared genes between treatments. (**b**) Significantly ($P < 0.05$) downregulated genes in each treatment as compared to their mock inoculated controls. Four-way VENN diagram shows number of unique or shared genes between treatments. (**c**) Predicted network showing transcription factors (yellow squares), DNA binding motifs (pink diamonds), and gene ontology terms (teal circles) enriched in resistant-specific differentially expressed genes at 1 day post-inoculation (orange hexagon). Gray lines show predicted connections between transcription factors, motifs, gene ontology terms, or the resistant-specific gene set. (**d**) Predicted network showing genes specifically upregulated in susceptible hosts at 1 day post-inoculation. (**e**) Predicted network showing genes specifically upregulated in resistant hosts at 3 days post-inoculation. 190 × 180 mm (300 × 300 dpi) (Becker et al. 2019)

brassinosteroid-mediated signaling, response to chitin, and G-protein coupled receptor activity (Fig. 5.9c). To validate gene activity, Becker et al. (2019) measured mRNA levels of *JAM1* and *MYC3* in all treatments using LMD-qPCR and found the expression of *JAM1* and *MYC3* 8.8-fold and 1.8-fold higher in resistant plants compared to their susceptible counterparts respectively. Both genes were upregulated in susceptible and resistant lines when compared to mock.

In *Arabidopsis*, CAMTA3 and CAMTA5 transcriptionally activate genes associated with ABA signaling including the master regulator *DREB1* (Kidokoro et al. 2017). In *Arabidopsis, BZIP1* is transcriptionally activated by calcium signaling cascades and initiates ABA-mediated changes to glucose and amino acid metabolism (Hartmann et al. 2015). In *Arabidopsis*, MYC3 is a positive regulator of JA (Cheng et al. 2011) and JAM1 a modulator of JA signaling (Sasaki-Sekimoto et al. 2013) that coordinates JA defense responses. The important role of bHLH TFs in RLP-SOBIR1 immunity is also supported by the discovery that bHLH TFs can be directly activated by SOBIR1 in cotton (Zhou et al. 2018). Together, this provides a putative regulatory framework activated during *Rlm2*-mediated immunity that reprograms cells to respond to the hemibiotrophic pathogen.

Marked distinctions in the expression of JA biosynthesis genes and markers suggest JA is repressed in susceptible cotyledons and is consistent with reports showing repression of JA markers during compatible *L. maculans–B. napus* interactions (Becker et al. 2017; Lowe et al. 2014), despite endogenous JA levels never being assayed. JA signaling plays a central role in defense against necrotrophs by limiting cell death (Rossi et al. 2011) and favoring accumulation of antifungal compounds (Antico et al. 2012). As *L. maculans* has been described as a primarily necrotrophic fungus, an effective response to the pathogen would likely involve extensive JA signaling. The ability of JA to antagonize cell death provides an explanation for the delayed onset of cell death commonly observed in *B. napus* cotyledons during the incompatible response to *L. maculans* (Stotz et al. 2014).

Rlm2-mediated immunity may fortify defense and protect signaling pathways from pathogen interference through the production of calcium signals, such as those initiated through the *CPKs/CDPKs* upregulated. CPK5 stimulates hormone production in response to fungi (Gravino et al. 2015) and phosphorylates RBOHD to produce ROS (Dubiella et al. 2013). Additionally, core calcium sensors upregulated in resistant plants activate strong defensive cues; CAM1 increases nitric oxide (Choi et al. 2009; Zhou et al. 2016) and ROS (Choi et al. 2009) production, which is amplified in the presence of CBL1 (Drerup et al. 2013). Thus, calcium signaling in concert with Rlm2/SOBIR1-mediated signaling is likely to direct and amplify the immune response.

When the calcium chelator EGTA was applied to the inoculation site, *Rlm2*-mediated immunity was abolished and infection progressed rapidly in both lines, suggesting calcium signaling is required to initiate resistance. The ability of EGTA to affect disease progression has been demonstrated previously. Pretreatment of wheat leaves with EGTA prevented NO production and delayed the peak hypersensitive response to *Puccinia triticina* by 24 h (Qiao et al. 2015), and also increased *Tobacco Mosaic Virus* load in *Arabidopsis* leaves by obstructing systemic acquired resistance and preventing *PR1* accumulation (Zhao et al. 2018). The accumulation of *PR1* suggests SA signaling was not compromised by EGTA treatment and occurred upstream or independent of calcium influx. Additionally, as calcium deficiency in cotyledons initiates abiotic stress responses, SA production was likely stimulated by calcium chelation in combination with *L. maculans* inoculation. SA production may have induced cell death and contributed to observed susceptibility. Although EGTA was applied locally at the inoculation site, and its effect is transient, *L. maculans* colonized the entire cotyledon despite the presence of *Rlm2*. This suggests early signaling events that prevent fungal establishment are likely critical for determining host fate. The activity of *Rlm2* and other *R*-genes likely protects the integrity of MAPK signaling pathway, allowing the host to maintain JA signaling and respond appropriately (Becker et al. 2019).

5.4.2 Contrasting Expression Profiles of Genes at the Inoculation Site of Susceptible and Resistant Plants

Genes at the host–pathogen interface contributes to resistance and susceptibility. A putative regulatory network specific to susceptible plants at 1 dpi associated with development, energy production, and plastid maintenance (Fig. 5.9d) has been described. Specifically, four DNA sequence motifs (MYB-like, ZF, CAMTA, GATA) and one CG-1 CAMTA TF were predicted to coordinate bioprocesses associated with the chloroplast, mitochondrial membrane, respiratory chain complex I, polysaccharide metabolism, ribosome biogenesis, and glucose 6-p transport (Fig. 5.9d). In contrast, genes associated with photosynthesis and primary metabolism were identified in a putative regulatory network downregulated in resistant plants. Downregulated DEGs were associated with response to cytokinin, photosynthesis, amino acid biosynthesis, metabolism, and unsaturated fatty acid biosynthesis, indicating these processes may be inhibited in resistant plants (Becker et al. 2019).

5.4.3 A Heightened Defense Response in Resistant Hosts Coincides with Attenuated Defense in Susceptible Cotyledons

A complex predictive network active specifically in resistant plants at 3 dpi composed of WRKY, BZIP, bHLH, CAMTA, and heat-shock TFs have been observed (Becker et al. 2019) (Fig. 5.9e). These TFs were predicted to regulate plant immune responses including protein ubiquitination, hypersensitivity, auto-phagosome, epigenetic regulation, MAPK activity, response to ethylene, SA, ABA, and calcium binding. They have also identified a putative regulatory network active in all inoculated plants except susceptible plants at 3 dpi suggesting their expression may become attenuated in susceptible cotyledons. This network contains WRKY, heat-shock, ABA INSENSITVE GROWTH 1, and CAMTA TFs that putatively regulate genes associated with kinase activity, JA-mediated signaling, PAMP-receptor signaling, and a diverse set of other plant defense responses.

5.4.4 Calcium Signaling is Required for Resistance and Basal Defense Against *Leptosphaeria maculans*

In the network analyses-associated resistance with CAMTAs and calcium-related biological functions, Becker et al. (2019) investigated for mRNA levels of calcium transporters, signal transducers, and binding proteins (Fig. 5.9a). Homologs of calcium sensors *CAM1*, *CAM5*, *CAM7*, and *CAM-LIKE 40* were 2.15-fold more expressed in resistant plants at 1 dpi. Similarly, calcium transporters (*CYCLIC NUCLEOTIDE GATED CALCIUM CHANNEL 3, 12, 19; CALCIUM ATPASE 2, 11*) and calcium-dependent signal transducers (*CALMODULIN-DOMAIN PROTEIN KINASE 5, 9; CALCIUM-DEPENDENT PROTEIN KINASE 6, 28*; and *CALCINEURIN B-LIKE GENE 1*) were an average of 1.41-fold higher in resistant plants at 1 dpi, increasing to 4.15-fold by 3 dpi. Previous studies have used the calcium chelator EGTA to explore the role of calcium in bioprocesses such as defense (Ma et al. 2017; Qiao et al. 2015; Hu et al. 2007; Lee and Back 2016). Becker et al. (2019) applied EGTA before inoculation with *L. maculans* to explore how the interaction would occur at a calcium-depleted interface and to provide physiological support that calcium is required for *Rlm2*-mediated immunity. EGTA alone caused minor chlorosis around the inoculation site at 3 dpi (Fig. 5.10a–c). EGTA pretreatment abolished resistance and caused rapid pathogen proliferation in both lines (Fig. 5.10c). By 3 dpi, fungal load was 19.5-fold higher in EGTA-pretreated resistant cotyledons versus *L. maculans* alone, and similar to pretreated susceptible cotyledons. Lesions continued to spread aggressively in EGTA-pretreated plants, with complete cotyledon coverage and visible pycnidia by 11 dpi. *PR1* mRNA level was used as an indicator of SA-mediated defense activation in all treatments at 1 and 3 dpi (Fig. 5.10d). Inoculation without EGTA application induced *PR1* gene activity at both time points, with higher expression in resistant lines (1.48-fold higher at 1 dpi, 8.45-fold at 3 dpi). EGTA application without *L. maculans* inoculation transiently activated host defense, with a 149-fold increase in *PR1* mRNA level in both treatments at 1 dpi that diminished by 3 dpi that may be related to observed chlorosis. EGTA pretreatment preceding inoculation with *L. maculans* strongly increased *PR1* accumulation at both time points with no difference between genotypes, indicating SA signaling was not compromised by apoplastic calcium depletion (Fig. 5.10d).

5.4.5 Hormonal Signaling During *Rlm2*-Mediated Resistance

As hormone signaling is essential for host defense programming, The expression of biosynthetic genes and markers directly at the inoculation site for ET, JA, SA, pipecolic acid (PiP), auxin, and ABA have been investigated (Fig. 5.11). *ACC OXIDASE 2* (*ACO2*) and *ACC SYNTHASE 6* (*ACS6*), essential for ET production, were an average of 1.97-fold and 2.01-fold higher in resistant cotyledons at 1 dpi and 3 dpi, respectively. Primary JA biosynthesis genes *ALLENE OXIDE SYNTHASE* (*AOS*) and *ALLENE OXIDE CYCLASE 3* (*AOC3*) had similar expression levels

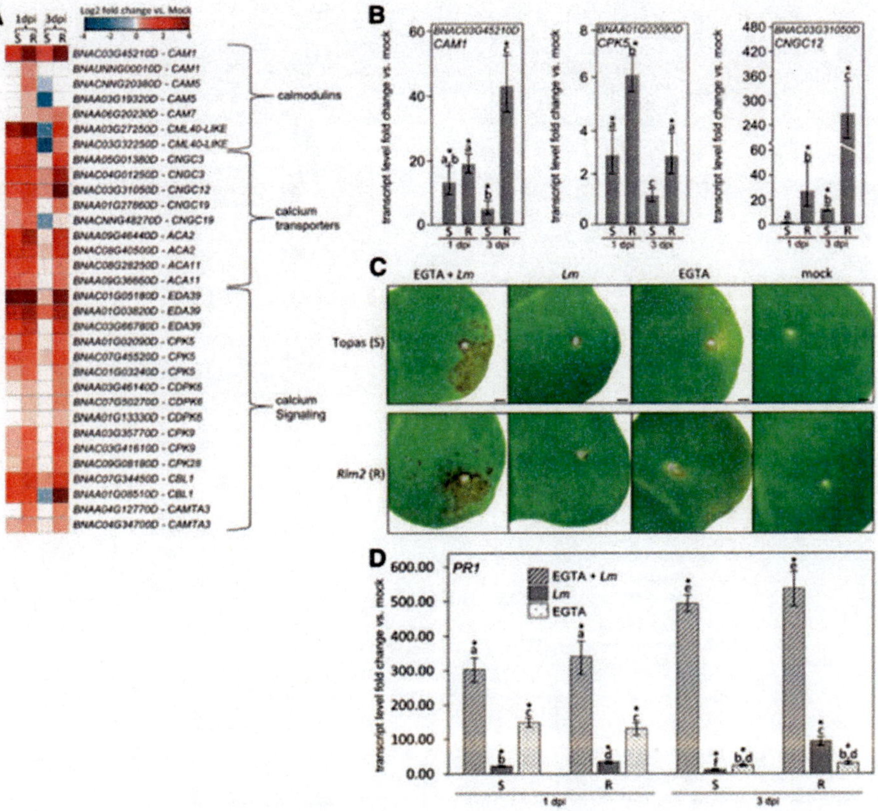

Fig. 5.10 Role of calcium signaling in resistance and inhibition of pathogen establishment. (**a**) Log2 FPKM fold change of calcium transporters and signal transducers. A more intense red color denotes upregulation and blue downregulation vs. mock (**b**) Relative mRNA abundance of select genes from A as measured by qPCR. (**c**) Treatment of cotyledons with calcium chelator EGTA prior to infection at 3 dpi. EGTA + Lm = co-inoculation with calcium chelator and pathogen *Leptosphaeria maculans*; *Lm* = treatment with pathogen alone; EGTA = treatment with chelator alone; mock = inoculation with water. Scale = 1 mm. (**d**) relative mRNA abundance of PR1 in following EGTA application. Significance of qPCR results was determined with a one-way ANOVA ($P < 0.05$) between the different treatments, where significance is indicated by different letters. Significant deviation from mock controls is denoted by an (*), as determined by a Student's *t*-test ($P < 0.05$). 190 × 170 mm (300 × 300 dpi) (Becker et al. 2019)

between lines at 1 dpi; however, at 3 dpi they were downregulated specifically in susceptible cotyledons (Fig. 5.11a). Additionally, a marker of JA/ET signaling, *PLANT DEFENSIN 1.2* (*PDF1.2*), was 3.2-fold downregulated in susceptible plants, in contrast to a 9.6-fold increase in resistant plants. *COI1* was the only JA marker tested with higher overall expression in susceptible plants, with 59% higher expression. Other markers of JA activity (*ETR1*, *JAZ1*, *JAR1*, *MPK5*) were an average of 83% higher in resistant hosts. Together, data indicated a robust JA response at the host–pathogen interface at 3 dpi specifically in resistant plants. Because of the higher

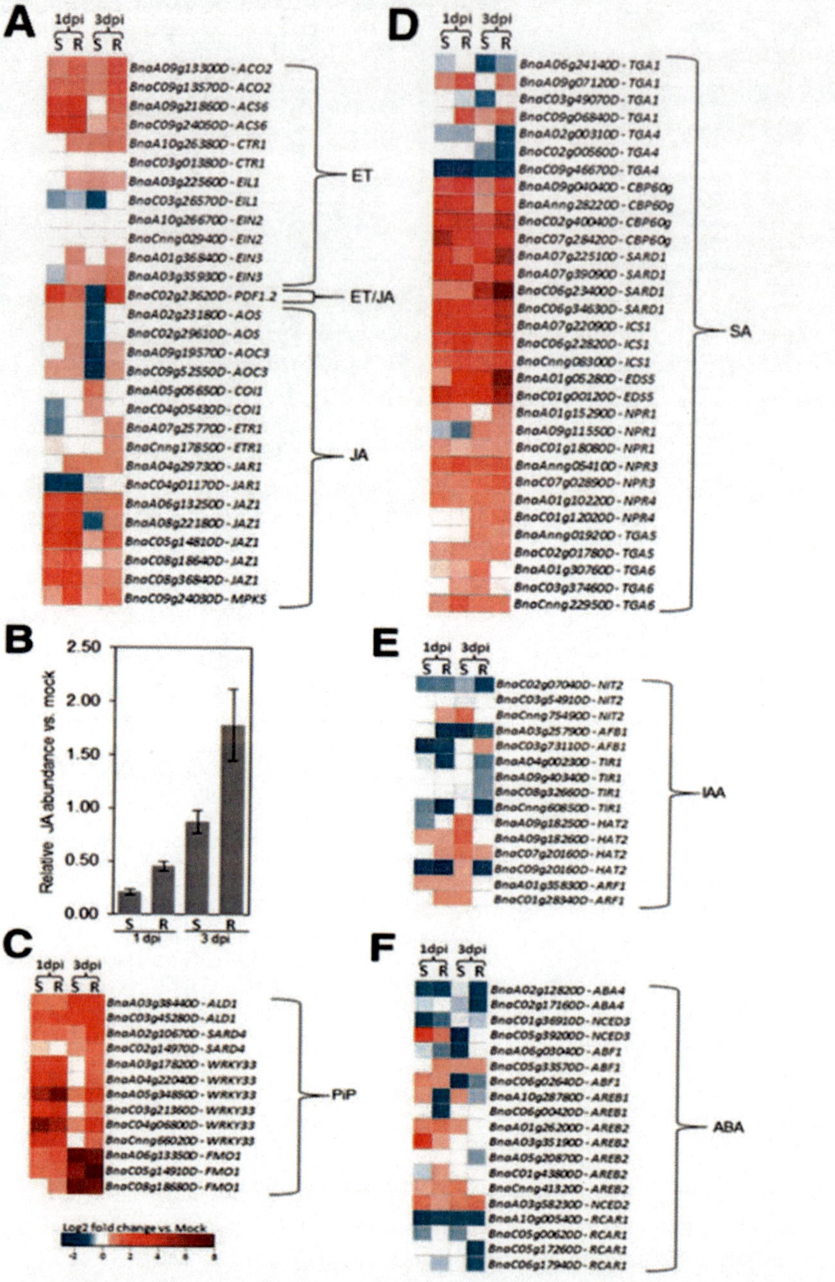

Fig. 5.11 Expression of hormone biosynthesis genes and signaling markers, and JA concentration in cotyledons, in response to *Leptosphaeria maculans*. (**a**) Relative Log2 FPKM fold change of biosynthetic genes and signaling markers for ethylene and jasmonic acid (ET/JA) in resistant and susceptible hosts. (**b**) Concentration of JA in hand-dissected infection sites, as compared to mock controls. Error bars show standard error. (**c**) Relative Log2 FPKM fold change of biosynthetic genes and signaling markers for pipecolic acid, (**d**) salicylic acid, (**e**) auxin, and (**f**) abscisic acid. 61 × 110 mm (600 × 600 dpi) (Becker et al. 2019)

expression of JA-related biosynthesis genes and markers in resistant plants, Becker et al. (2019) measured the physiological JA concentration in all treatments (Fig. 5.11b). At both time points JA levels were higher in resistant cotyledons. Interestingly, JA decreased in both susceptible and resistant cotyledons at 1 dpi as compared to mock controls (Fig. 5.11b). At 3 dpi, JA concentrations returned to near control levels in susceptible cotyledons and increased 1.8 times in resistant cotyledons.

Homologs of *ISOCHORISMATE SYNTHASE 1* (*ICS1*), critical to SA production, were an average of 1.6-fold and 3.1-fold higher in resistant cotyledons at 1 dpi and 3 dpi, respectively, as compared to their susceptible counterpart (Fig. 5.11d). In addition, regulators of ISC1 gene activity (*TGA1*, *4*, *SARD1*, *CBP60g*) were an average of 2.64-fold higher in resistant plants at 3 dpi; however, the receptors *NPR1*, *3*, *4* and transcription factors *TGA5*, *6* showed no overall difference in expression between the two lines. EDS5, another downstream regulator of SA activity, was fourfold higher in resistant plants at 3 dpi.

PiP biosynthesis branches from the SA pathway following SARD1 and CBP60g signaling and has emerged as an important signaling hormone. Markers for PiP biosynthesis and activity were an average of 2.45-fold higher in resistant plants at 3 dpi, suggesting this pathway may play a role in resistance (Fig. 5.11c). Auxin markers showed similar expression in both hosts at 1 dpi. At 3 dpi, expression of biosynthetic gene *NITILASE 2* (*NIT2*) along with markers (*AFB1*, *TIR1*, *HAT2*, *ARF1*) was an average of 53% higher in susceptible hosts (Fig. 5.11e). ABA markers showed no clear pattern between the two hosts at 1 and 3 dpi, with mixed expression (Fig. 5.11f). For example, putative homologs of markers *ABSCISIC ACID RESPONSIVE ELEMENTS-BINDING FACTOR 1* (*AREB1*) and *AREB2* were upregulated only in susceptible plants, whereas a homolog of key biosynthetic enzyme *NCED3* was specifically downregulated (Becker et al. 2019).

5.4.6 Pathogen Detection and Signaling is Compromised in Susceptible Plants and Maintained in *Rlm2*-Mediated Resistant

Cohorts of genes expressed specifically in resistant *Rlm2* plants at 3 dpi (Fig. 5.9e) or attenuated in susceptible plants at the same time point directly at the host–pathogen interface has been identified. These genes coded for core signaling machinery involved in fungal detection and defense elicitation through the receptors *SOBIR1* and *CERK1*. The mRNA levels of these receptors, along with their co-receptors and downstream signal transduction machinery has been plotted (Fig. 5.12a, b). With the exception of *MAPK KINASE 9* (*MKK9*) and *BOTRYTIS-INDUCED*

KINASE 1 (*BIK1*), the entire pathway was more highly expressed in resistant plants at 3 dpi (Fig. 5.12b). This includes homologs of *SOBIR1* and *CERK1*, along

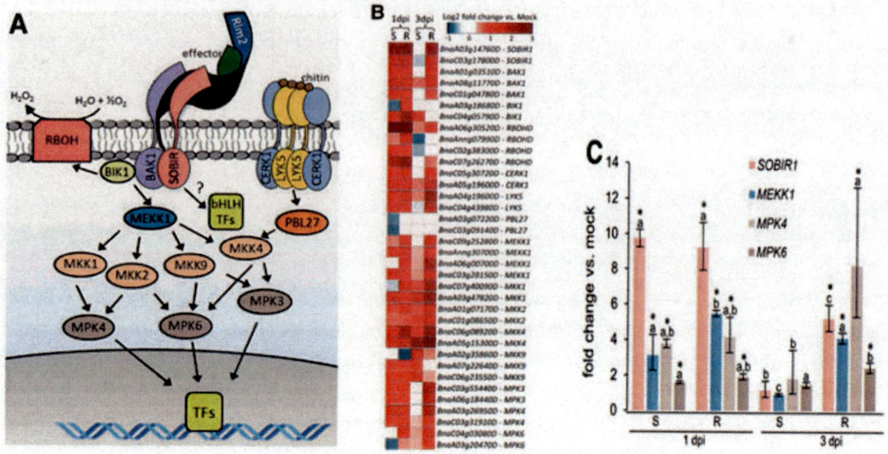

Fig. 5.12 Expression levels of cell-surface receptors and signal transduction machinery responsible for initiating host defense. (**a**) Partial model of signaling events following activation of cell surface receptors SOBIR and CERK1. Activation of signal transduction cascades results in regulation of gene transcription. (**b**) Gene expression changes (Log2 fold change vs. mock) of signaling pathway components as measured by RNA-seq. Genes associated with surface receptors and MAPK activity are upregulated in resistant hosts, but not susceptible, at 3 days post inoculation. (**c**) Gene expression of selected genes measured by qPCR. 66 × 240 mm (600 × 600 dpi) (Becker et al. 2019)

with functional partners *BAK1* and *LYK5*, and signal transducers *MAPK/ERK KINASE KINASE 1* (*MEKK1*); *MKK1*, *2*, *4*; and *MPK3*, *4*, *6* that were an average of fourfold higher at the host–pathogen interface during *Rlm2*-mediated immunity. The targeted transcript profiling was used to measure mRNA abundance of *SOBIR1*, *MAPK/ERK KINASE KINASE 1* (*MEKK1*), *MPK4*, and *MPK6* directly at the inoculation site (Fig. 5.12c). At 1 dpi, all targets were upregulated during infection and showed similar expression profiles between genotypes (Fig. 5.12c). At 3 dpi, data showed no difference between susceptible interactions and mock controls; however, all genes were significantly upregulated in resistant plants with mRNA levels 2.4- to 8.2-fold those observed in mock treatments (Fig. 5.12c). As intracellular NLRs also play a role in amplifying the host immune response, plotted expression of all differentially expressed NLRs along with known NLR lipase-like proteins involved in signal transduction or yet to be identified functions were plotted. The observed patterns of gene activity with similar expression of NLRs and markers for NLR-related signaling between hosts at 1 dpi, and a 93% overall expression increased in resistant hosts at 3 dpi (Becker et al. 2019).

5.5 *Brassica–Rhizoctonia*: Biometabolomics Resistance

5.5.1 Resistance to *Rhizoctonia* in *Arabidopsis* is Mediated Through NADPH Oxidases

R. solani has been characterized and grouped into 13 anastomosis groups (AG) that vary in pathogenicity, physical characteristics, and sequence variations. *R. solani* is well known for causing rice sheath blight which is one of the most prevalent rice diseases. Isolates of AG8 cause bare patch disease of cereals and legumes and can cause severe root rot in canola. It is pathogenic on canola and causes severe hypocotyl rot on mustard. There is little variation in the response to either AG8 or AG2-1 among *Arabidopsis* ecotypes, and resistance to AG8 is independent of common defense signaling pathways. Whole genome gene expression analysis at 7 days post-infection identified genes that were differentially expressed in either the resistant or susceptible response including cell-wall-associated genes, heat shock protein genes, and other genes associated with oxidative stress responses. Key genes involved in reactive oxygen species (ROS) production demonstrated that specific NADPH oxidases are a critical component of resistance to *R. solani* AG8 in *Arabidopsis*, opening up new avenues to potentially control this destructive pathogen in various crop plants.

Arabidopsis is resistant to *R. solani* AG8 but susceptible to *R. solani* AG2-1. A screen of 36 *Arabidopsis* ecotypes and mutants affected in the auxin, camalexin, salicylic acid, abscisic acid, and ethylene/jasmonic acid pathways did not reveal any variation in response to *R. solani* and demonstrated that resistance to AG8 was independent of these defense pathways. The *Arabidopsis* Affymetrix ATH1 Genome array was used to assess global gene expression changes in plants infected with AG8 and AG2-1 at 7 days post-infection. While there was considerable overlap in the response, some gene families were differentially affected by AG8 or AG2-1 and included those involved in oxidative stress, cell-wall-associated proteins, transcription factors, and heat shock protein genes. Since a substantial proportion of the gene expression changes were associated with oxidative stress responses, Foley et al. (2013) analyzed the role of NADPH oxidases in resistance. While single NADPH oxidase mutants had no effect, a NADPH oxidase double mutant *atrbohf atrbohd* resulted in an almost complete loss of resistance to AG8, suggesting that reactive oxidative species play an important role in *Arabidopsis's* resistance to *R. solani* (Foley et al. 2013).

5.6 *Brassica–Sclerotinia*: Biometabolomics Resistance

5.6.1 The Role of Glucosinolates in the Defense of *Brassica* Against Pathogens

Plants defend themselves against pathogens by synthesizing and accumulating chemically diverse metabolites with antibiotic properties (Pastorczyk and Bednarek 2016). Constitutive synthesis and accumulation of defensive metabolites (phytoanticipins) is resource-demanding; therefore, de novo synthesis can be induced after plant tissue is exposed to pathogen infection (phytoalexins) (Pastorczyk and Bednarek 2016). Brassicaceae species accumulate up to 1% of dry weight as the sulfur-containing metabolites, glucosinolates (GSLs) (Calmes et al. 2015). GSLs can be classified into three chemical classes, depending on the side chain of their precursor amino acid. Methionine, alanine, valine, leucine, and isoleucine are precursors of aliphatic GSLs. Phenylalanine and tyrosine render aromatic GSLs, and tryptophan is the precursor of indolic GSLs (Zukalova and Vasak 2002). They are constitutively synthesized and stored in plant cells, functioning mainly as phytoanticipins, although they can be induced in response to a variety of stimuli, as well as following exposure to insect damage and pathogens infection (Brader et al. 2006; Buxdorf et al. 2013). De novo synthesis of indolic GSLs in *Arabidopsis* can be triggered by plant pathogens, at the same time inducing synthesis of the phytoalexin camalexin through the upregulation of the genes CYP79B2 and CYP79B3, which are shared by several indolic pathways (Frerigmann et al. 2016; Stahl et al. 2016).

Myrosinase enzymes hydrolyze GSLs into several chemically diverse active products. Myrosinases and GSLs are stored in different plant compartments, and after tissue damage caused by injuries, pests, or pathogens, these constituents mix, and hydrolysis of GSLs occurs. Hydrolytic products exhibit antimicrobial activities against a wide range of bacterial and fungal plant pathogens in vitro (Calmes et al. 2015; Dufour et al. 2015; Li et al. 2013; Sotelo et al. 2015). Antimicrobial effects depend on the dose, pathogen, and the specific race or isolate (Sotelo et al. 2015). Besides, the chemical structure of the breakdown products determine their activity in vitro, with isothiocyanates (ITCs) being the most toxic, even at low concentrations. Toxicity of ITCs is also related to the length of their side chain (Li et al. 2013; Pastorczyk and Bednarek 2016).

In planta activation of GSLs is dependent on tissue breakdown; however, this does not happen in the interaction with biotrophic pathogens. Production and liberation of active indole compounds to the apoplast can also be mediated by the action of the genes pen2 and pen3 as described with the biotroph fungus *Blumeria graminis* fsp. *hordei* (Bgh) (Johansson et al. 2014). To demonstrate the role of GSLs in the immunity of Brassicaceae plants to various pathogens, knockout mutants of genes implicated in the synthesis and hydrolysis of GSLs in *Arabidopsis* have been tested against various bacteria and fungi. The double mutant CYP79B2/B3, which does not accumulate indolic GSLs or camalexin, and the double mutant MYB28/29, which does not accumulate aliphatic GSLs, modified the sensitivity of plants against isolates of the fungi *Alternaria brassicicola*, *Botrytis cinerea* (Buxdorf et al. 2013), and *Sclerotinia sclerotiorum* (Stotz et al. 2011). However, transgenic plants of *Arabidopsis* accumulating specific GSLs can increase their resistance to pathogens. Expression of the CYP79D2 gene from cassava in *Arabidopsis* causes accumulation of aliphatic isopropyl and methylpropyl GSLs that enhance resistance to the bacteria *Erwinia carotovora*. *Arabidopsis* expressing the sorghum CYP79A1 or overexpressing the endogenous CYP79A2 accumulates *p*-hydroxybenzyl or benzyl GSLs, respectively, increasing resistance to *Pseudomonas syringae* (Brader et al. 2006). The contribution of GSLs to the defense of *Arabidopsis* is dependent on the pathogen species and even on the isolate under consideration. Specialist pathogens can adapt themselves by detoxifying ITCs and/or suppressing the induction of GSLs biosynthesis. Variability in the response among pathogens may also be due to different life styles (i.e., biotrophic versus necrotrophic) or differential host ranges (Buxdorf et al. 2013; Pastorczyk and Bednarek 2016).

The role of specific metabolites in defense of *Brassica* crops against pathogens is not well characterized. The content of GSLs in several varieties of *Brassica oleracea* and *B. napus* has been studied in relation to resistance to *Xanthomonas campestris* pv. *campestris* (Pammel) Dowson, *S. sclerotiorum* and *B. cinerea*. But it should be noted that the host genotypes had different GSL profiles and concentrations, as well as different genetic backgrounds. The lines of *B. napus*, which overexpressed UGT74B, an indolic GSL pathway gene, enhanced not only the production of total indolic GSLs but also the resistance to *S. sclerotiorum* and *B. cinerea* (Zhang et al. 2015). Modulating the content of GSLs may affect *Brassica* plant resistance to pathogens. Sotelo et al. (2016) used mass selection to develop a set of *B. oleracea* var. *acephala* L. (kale) genotypes differing in the content of three GSLs: the aliphatics sinigrin (2-propenyl, SIN), glucoiberin (3-methylsulfinylpropyl, GIB), and the indolic glucobrassicin (3-indolylmethyl, GBS). These genotypes were evaluated and concluded that different alleles at the GSL-ALK locus were responsible for variation in the content of SIN and GIB, whereas the CYP81F2 gene may be responsible for the variation in concentration of GBS. This material can be used to study the effects of major GSLs in kales, with the advantage that the genotypes all have the same genetic background. The role of SIN, GIB, and GBS in the defense of kale plants against the necrotrophic pathogens, *S. sclerotiorum* (*Ss*) and *X. campestris* pv. *campestris* (*Xcc*), was assessed (Madloo et al. 2019).

The genotypes selected for higher GSL content invariably had a greater concentration of the target GSL compared with the corresponding genotypes selected for low GSL content. It is established that changes in a single GSL gene may lead to pleiotropic effects on the concentrations of other GSLs. When divergent selection modifies the content of SIN, effects on the content of GIB, GBS, or other GSLs may also occur (Sotelo et al. 2016). The minor GSLs present in kale leaves for each selection in studies of the relationship with disease severity. The differential performance of the divergent selections after inoculation with *Ss* or *Xcc* was recorded. Genotypes selected for high SIN content were more resistant to *Ss* and genotypes selected for high GBS content were more resistant to *Xcc*. However, the effect of GIB content was not clear, since the relationship with lesion area was time dependent in the *Ss* experiment, and there was no relationship with lesion area when genotypes were inoculated with *Xcc*. The results showed that the aliphatic GSL, SIN, inhibited the growth of *S. sclerotiorum*, that is, HSIN genotypes had smaller lesions (less severe disease). These results are congruent with those of a previous report showing a role of SIN in reducing severity of disease (Stotz et al. 2011). The content of aliphatic GSLs in *Arabidopsis* was correlated with changes in susceptibility to *Ss*. In addition, when genotypes of the SIN selection were inoculated with *S. sclerotiorum*, the concentrations of two indolic GSLs (NEOGBS and MEGBS) were significantly higher in HSIN than in LSIN. The resistance observed with the SIN selection against *S. sclerotiorum* may be attributed to a combined effect of the aliphatic SIN and the indolic GSLs that together enhanced the plant defense response.

The effect of GIB selection was not clear, since the response of the HGIB and LGIB genotypes was dependent upon time when plants were inoculated with *S. sclerotiorum*. Earlier reports indicate that different alleles at the GSL-ALK locus are selected for in the divergent selections for SIN and GIB (Sotelo et al. 2016). GSL-ALK alkenized GIB to render SIN. Therefore, when the content of one increases, the content of the other decreases. SIN inhibited the growth of *S. sclerotiorum* as demonstrated in SIN selection. HGIB had lower concentration of SIN than LGIB. This could explain why the selection to modify GIB content did not have a clear effect on lesion area as a result of infection by *S. sclerotiorum*. Several reports indicate that indolic GSLs are inhibitory against *S. sclerotiorum* in vivo (Zhang et al. 2015). However, Madloo et al. (2019) did not find a clear effect of the HGBS genotypes reducing lesion area caused by *Ss*. Lack of an effect may be due to variations in the GBS content in the divergent selection used. Such variations are reported to occur in expression of the CYP81F2 gene (Sotelo et al. 2016). The CYP81F2 gene converts GBS into OHGBS, and should not affect other metabolic pathways. However, research that relates indolic GSLs with resistance to *S. sclerotiorum* in *Arabidopsis* through mutations in the UGT74B1 or CYP79B genes indicate they affect not only the content of indolic GSLs but also other plant immune system metabolites (Zhang et al. 2015). Therefore, it is not clear if the effect can be attributed to an increase in total indolic GSLs or to the induction of other plant immune system metabolites (i.e., camalexin or auxins). Thus, defense responses against *S. sclerotiorum* are probably regulated by a complex network of hormonal and signaling pathways (i.e., jasmonic acid, ethylene, abscisic acid) promoting the production of GSLs and other metabolites, working together to enhance plant resistance. Little is known about the mechanisms of resistance to *Xcc* in *Brassica* crops. Aires et al. (2011) concluded that both aliphatic and indolic GSLs may play a complex role in the defense, depending on the species of *Brassica*. Velasco et al. (2013) reported that the aliphatic gluconapin (3-butenyl) can protect *B. rapa* against *Xcc*. Madloo et al. (2019) found that HGBS plants were more resistant to *Xcc*, whereas the aliphatics SIN and GIB did not have an effect in reducing lesion area. Overall, these findings suggest that an increment of individual GSLs does not always lead to enhanced disease resistance.

GSLs are present in all tissues of the plants. They are produced constitutively, but are also inducible in response to biotic stress. To explore the relationship between the endogenous concentration of GSLs and the induced component in GSLs as a result of infection by *S. sclerotiorum*, the content of GSLs in the high and low selections of each genotypes was compared to that of the controls. Aires et al. (2011) related the suppression of de novo synthesis of defense compounds and the inhibition of myrosinase activity with the mode of infection by a necrotrophic pathogen. Whereas hypersensitive cell death can be an effective defense strategy against biotrophs, it may not be efficient against necrotrophic pathogens. However, a reduction in lesion size caused by *S. sclerotiorum* in oilseed rape (*B. napus*) was related to induction of indolic and aromatic GSLs (Li et al. 1999). Induced synthesis of GSLs may be related to the plant genotype and to its interaction with specific pathogens. Similar to the observations of Li et al. (1999), it was found that indolic GSLs were induced after inoculation with *Ss* in all three selections, but it was most pronounced with the GIB genotype. However, aliphatic GSLs were repressed in the GIB genotype. These apparently opposing responses among GSLs could contribute to the poor performance of GIB against *S. sclerotiorum*.

The process of GSL hydrolysis may be related to the differential performance of the divergent selections after infection with *S. sclerotiorum* and *Xcc*. Myrosinases and GSLs are localized in two different specialized cells. When there is tissue destruction, for example, if a necrotroph infects, the GSLs are exposed to the myrosinases which hydrolyze the β-D-thioglucosidic bond of the GSL to release β-D-glucose and an unstable aglucone (Calmes et al. 2015; Shirakawa and Hara-Nishimura 2018). The aglucone spontaneously forms an ITC by rearrangement or decomposes to a nitrile. Formation of ITCs is favored at pH values of 5–6, while nitriles are preferentially formed at a pH <4 or at increased Fe^{2+} concentrations (Hanschen et al. 2018). In addition, epithio-specific proteins may promote the formation of nitriles and epithionitriles by modulating myrosinase activity. As a consequence, tissue disruption typically leads to the release of complex mixtures of GSL breakdown products (Hanschen et al. 2018). Madloo et al. (2019) measured the quantity of intact GSLs and not their degradation products after inoculation. A reduction in the concentration of GSLs was expected when comparing inoculated plants with controls as a part of the process of degradation. However, this does not always happen, as several GSLs were induced upon *Ss* infection.

Therefore, the concentrations of GSLs in inoculated plants are the product of their degradation and of their induced synthesis upon infection. When comparing the three divergent selections, the HSIN and HGBS plants had greater resistance to *S. sclerotiorum* and *Xcc*, respectively. However, the resistance is not only related to the content of GSLs: in the SIN divergent selection group, the content of SIN was related to resistance. But the relationship with resistance was not observed in the GBS selection, where the HGBS plants had a higher content of SIN compared to the LGBS plants. Other processes in the plant could modify the plant response as the three divergent selections were performed independently, resulting in different genetic backgrounds for each GSL selection. Modifying the content of GSLs could affect other processes in the plant (Madloo et al. 2019).

5.6.2 Biomolecular Mechanism of *Brassica* Resistance to *Sclerotinia*

S. sclerotiorum secretes cell wall-degrading enzymes (CWDEs) (pectinases, cellulases, hemicellulases) to facilitate plant cell wall degradation, hyphae colonization, and lesion expansion. These processes are accompanied by the production of the nonhost-selective toxin oxalic acid (OA) to manipulate the host redox environment, which is a benefit to colonization of the host plant, suppression of host autophagy, and inhibition of plant defense responses. A large number of secretory proteins serving as potential effectors were predicted in *Sclerotinia* with use of bioinformatics tools, and several genes encoding effector have been identified (Zhu et al. 2013; Lyu et al. 2016; Yu et al. 2017; Yang et al. 2018).

In rapeseed, a number of QTLs and expressed sequence tags (ESTs) in association with resistance were identified in the partial resistant accessions such as 'Zhongshuang 9' and 'Zhongyou 821' (Wu et al. 2016). Several pathways and biological processes were revealed to associate with the defense against *Sclerotinia*, such as oxidative burst, cell wall enforcement or modification, secondary metabolism and calcium binding, signaling, antioxidation, glutathione metabolism, glucosinolate metabolism, and biosynthesis of lignin (Joshi et al. 2016; Wu et al. 2016; Wei et al. 2016). However, these studies only emphasized the response from the pathogen or the host during infection, producing an incomplete view of the interaction. For a comprehensive understanding of plant–pathogen interactions, it is valuable to analyze gene expression alterations in both pathogen and host (Westermann et al. 2012; Kawahara et al. 2012; Yazawa et al. 2013; Zhuang et al. 2012). The lack of an available host genotype with high resistance has limited the depth of understanding the mechanistic interaction between *S. sclerotiorum* and its host. A genotype cultivar 'C01' of *B. incana* (a wild *B. oleracea* species) related to rapeseed, with nearly 20-fold higher stem resistance than a partial resistant *B. napus* cultivar 'Zhongyou 821' was identified, and resistance QTLs were mapped in a resistance segregation population derived from a cross between 'C01' and a susceptible *B. oleracea* (Mei et al. 2011, 2013). The stems of extreme disease-resistant (R) and susceptible (S) lines in this segregation population were inoculated with *Sclerotinia*, and the transcriptome changes in both host and pathogen in lesions were analyzed. The data suggested the resistant *B. oleracea* quickly perceive *S. sclerotiorum* secretory proteins and/or pathogen-associated molecular pattern (PAMPs), limit the expression of virulence genes, and quickly accumulate ROS to inhibit the *S. sclerotiorum* establishment by activating Ca^{2+} signaling and suppressing pathogen OA generation. Meanwhile, the gene expression of pectin methylesterase inhibitors (*PMEIs*) and pectinesterase inhibitors (*PEIs*) in resistant *B. oleracea* was promoted to suppress the cell wall degradation and to prevent hyphae expansion (Ding et al. 2019).

The transcriptome changes from lesions produced by *S. sclerotiorum* on disease-resistant (R) and susceptible (S) *Brassica oleracea* pools bulked from a resistance-segregating F2 population have been assessed (Ding et al. 2019). Virulence genes of *S. sclerotiorum*, including polygalacturonans, chitin synthase, secretory proteins, and oxalic acid biosynthesis, were significantly repressed in lesions of R *B. oleracea* at 12 hpi but exhibited similar expression patterns in R and S *B. oleracea* at 24 hpi. Resistant *B. oleracea* induced expression of receptors potentially to perceive *Sclerotinia* signals during 0–12 hpi and deployed complex strategies to suppress the pathogen establishment including the quick accumulation of reactive oxygen species (ROS) via activating Ca^{2+} signaling and suppressing pathogen oxalic acid (OA) generation in *S. sclerotiorum*. In addition, cell wall degradation was inhibited in the resistant *B. oleracea* potentially to prevent the expansion of *Sclerotinia* hyphae. The transcriptome changes in *S. sclerotiorum* and host revealed that resistant *B. oleracea* produce strong responses against *S. sclerotiorum* during early infection (Ding et al. 2019).

5.6.3 *Arabidopsis GDSL1* Overexpression Enhances *Sclerotinia* Resistance in *Brassica napus*

GDSL-type lipases/esterases represent a subfamily of lipolytic enzymes in which the amino acid sequence contains a GDSL motif, GxSxxxxG, with the active site serine (S) located near the N-terminus (Lee et al. 2009; Updegraff et al. 2009). Compared with the current advanced understanding of the roles of these enzymes in animals and microorganisms, there is still a limited understanding of plant GDSL lipases, which were discovered relatively late. The plant GDSL lipases have received more attention because of their multifunctional properties, especially in relation to plant disease resistance and stress responses (Dong et al. 2016). In *Arabidopsis*, two GDSL LIPASEs (GLIP), GLIP1 and GLIP2, play key roles in resistance to *Alternaria brassicicola* and *Erwinia carotovora*, respectively (Lee et al. 2009; Oh et al. 2005).

AtGLIP1 can modulate ET-associated systemic immunity through the regulation of ET signaling (Kim et al. 2013, 2014). The overexpression of *A. thaliana* Li-tolerant lipase 1 increases salt tolerance in yeast and *Arabidopsis* (Naranjo et al. 2006). Two closely related GDSL motif-containing lipases, *BnGDSL1* and *AtGDSL1*, were involved in regulating seed germination and seed oil content in *B. napus* (Ding et al. 2019). Based on their different functions in SSR resistance, Ding et al. (2020) showed that *AtGDSL1*, but not *BnGDSL1*, appears to be a critical component of the incompatible interaction with *S. sclerotiorum*. Moreover, the increased resistance to disease of *AtGDSL1* transgenic rapeseed plants was correlated with enhanced phosphatidic acid (PA) production, which might, in turn, activate downstream ROS/SA defense signaling events that were critical for resistance. Further studies demonstrated that the counterpart of *AtGDSL1* with respect to *S. sclerotiorum* resistance in *B. napus* was the *BnGLIP1* gene harbored in selective regions on chromosome C07 during breeding. There was significant linkage among single-nucleotide polymorphisms (SNPs) within the *BnGLIP1* gene, and the alleles harboring SNPs exist in fewer varieties (36%) in the natural population, suggesting that *BnGLIP1* could be used as a potential target to improve *S. sclerotiorum* resistance in rapeseed breeding (Ding et al. 2020).

Arabidopsis thaliana GDSL1 was characterized as an extracellular *GDSL* lipase gene functioning in *Sclerotinia* resistance. Loss of *AtGDSL1* function resulted in enhanced susceptibility to *S. sclerotiorum*. Conversely, overexpression of *AtGDSL1* in *B. napus* enhanced resistance, which was associated with increased reactive oxygen species (ROS) and salicylic acid (SA) levels and reduced jasmonic acid levels. In addition, *AtGDSL1* can cause an increase in lipid precursor phosphatidic acid levels, which may lead to the activation of downstream ROS/SA defense-related pathways. However, the rapeseed *BnGDSL1* with highest sequence similarity to *AtGDSL1* had no effect on SSR resistance. A candidate gene association study revealed that only one *AtGDSL1* homolog from rapeseed, *BnaC07g35650D* (*BnGLIP1*), significantly contributed to resistance traits in a natural *B. napus* population, and the resistance function was also confirmed by a transient expression assay in tobacco leaves. Moreover, genomic analyses revealed that *BnGLIP1* locus was embedded in a selected region associated with SSR resistance during the breeding process, and its elite allele type belonged to a minor allele in the population. Thus, *BnGLIP1* is the functional equivalent of *AtGDSL1* and has a broad application in rapeseed *S. sclerotiorum* resistance breeding (Ding et al. 2020).

5.6.4 *BnaMPK3* Gene Regulates Defense Responses to *Sclerotinia* in *Brassica napus*

It has been shown that mitogen-activated protein kinases (MAPKs or MPKs) play an important role in signal transduction in response to hormones and environmental stresses. A MAPK activity is controlled by sequential activation of two protein kinases, by which an MAPK kinase kinase activates an MAPK kinase that in turn activates an MAPK. Evidence is now accumulating that some MAPK family members have been implicated in plant defense as a component of defense signaling pathways. For example, in the model plant *Arabidopsis thaliana*, microbe-associated molecular patterns (MAMPs), such as the bacterial flagellum-derived flg22 peptide, trigger the activation of AtMAPKs, including AtMPK1, AtMPK3, AtMPK4, AtMPK6, AtMPK11, and AtMPK13 (Meng and Zhang 2013; Nitta et al. 2014). Of these MAPKs, AtMPK4 was first described as a negative regulator of plant immunity, whereas some others, such as AtMPK3, were believed to play positive roles in plant immunity based on their rapid activation by flg22 and pathogen inoculation (Zhang and Klessig 2001). Further, the *Arabidopsis thaliana* MPK3 loss-of-function mutants showed significantly lower bacterial titers of *Pseudomonas syringae*, and in expression of genes involved in SA biosynthesis and signaling, either control or flg22-treated AtMPK3 loss-of-function mutants have no difference from AtMPK4 mutants that display enhanced resistance to biotrophic pathogens *Hyaloperonospora arabidopsidis* and *P. syringae* (Petersen et al. 2000; Freidit Frey et al. 2014). However, another trial revealed no clear susceptible phenotype in the AtMPK3 mutants when sprayed or infiltration-inoculated with *P. syringae* (Su et al. 2017). MPK3 genes have been identified in various crop species and implicated in defense responses and other physiological processes. Diseases caused by

S. sclerotiorum have traditionally been difficult to control. Breeding *S. sclerotiorum*-resistant oilseed rape cultivars using traditional methods is difficult since no immune or highly resistant germplasm in *B. napus* is available. Molecular breeding is pursued as an important strategy for controlling Sclerotinia diseases. Wang et al. (2019) used both gain- and loss-of-function approaches to investigate the specific role of *B. napus.MPK3* (*BnaMPK3*) in defense responses against *S. sclerotiorum*, and a candidate gene association analysis was used to validate the contribution of the genomic loci of the gene to resistance to the pathogen in a natural *B. napus* population.

Identification of important genes in the defense responses is critical for molecular breeding, an important strategy for controlling the disease. A *B. napus* mitogen-activated protein kinase gene, *BnaMPK3*, plays an important role in the defense against *S. sclerotiorum* in oilseed rape. *BnaMPK3* is highly expressed in the stems, flowers, and leaves, and its product is localized in the nucleus. Furthermore, *BnaMPK3* is highly responsive to infection by *S. sclerotiorum* and treatment with jasmonic acid (JA) or the biosynthesis precursor of ethylene (ET), but not to treatment with salicylic acid (SA) or abscisic acid. Moreover, overexpression (OE) of *BnaMPK3* in *B. napus* and *Nicotiana benthamiana* results in significantly enhanced resistance to *S. sclerotiorum*, whereas resistance is diminished in RNAi transgenic plants. After *S. sclerotiorum* infection, defense responses associated with ET, JA, and SA signaling are intensified in the *BnaMPK3*-OE plants but weakened in the *BnaMPK3*-RNAi plants when compared to those in the wild-type plants; by contrast the level of both H_2O_2 accumulation and cell death exhibits a reverse pattern. The candidate gene association analyses show that the BnaMPK3-encoding *BnaA06g18440D* locus is a cause of variation in the resistance to *S. sclerotiorum* in natural *B. napus* population. These results suggest that *BnaMPK3* is a key regulator of multiple defense responses to *S. sclerotiorum*, which may guide the resistance improvement of oilseed rape and related economic crops (Wang et al. 2019).

5.6.5 Multifaceted Molecular Mechanisms of Defense Against *Sclerotinia sclerotiorum*

Facing the attack *S. sclerotiorum*, plants need to accurately detect and send timely signal and then appropriately respond to each of the different pathogenic strategies by deploying a range of multifaceted defense response mechanisms (Fig. 5.13). Defense responses often begin with either recognition of pathogen (or microbe)-associated molecular patterns (PAMPs or MAMPs) by plant-cell-surface pattern recognition receptors (PRRs) or recognition of the pathogen's virulence molecules, termed effectors, by a set of plant intracellular resistance (R)-gene products, which results in MAMP (or PAMP)-triggered immunity (MTI or PTI) or effector-triggered immunity (ETI), respectively (Jones and Dangl 2006). ETI is also named as gene-for-gene interaction. In the plant–*Sclerotinia* pathosystem, ETI (or R-gene-mediated resistance) has not been observed (Yang et al. 2018). In contrast, the presence of SCFE1 (*Sclerotinia* culture filtrate elicitor1), an elicitor in *S. sclerotiorum* evoking

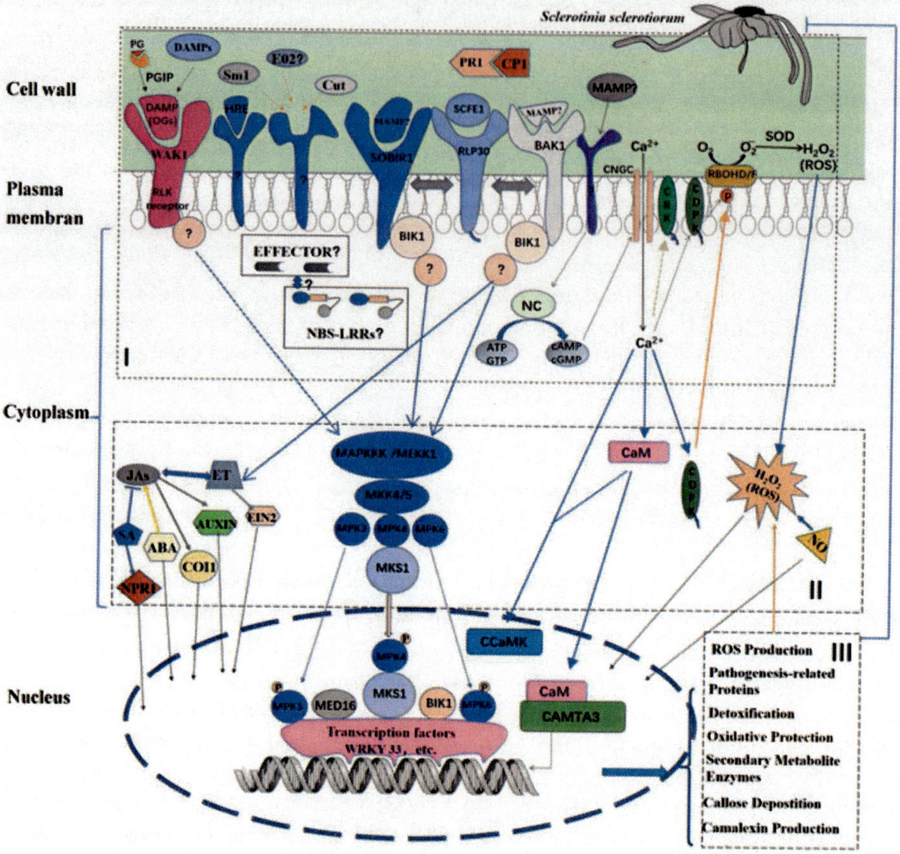

Fig. 5.13 Schematic model of plant innate immune defense process against *Sclerotinia sclerotiorum* infection. In the model, plant defense process can be split into (I) recognition, (II) signal-transduction, and (III) defense response. (I) Plant recognition of MAMPs (or PAMPs) and DAMPs are detected by PRRs in apoplast. The SCFE1–RLP30/SOBIR1/BAK1 recognition, a classical PAMP–PRR recognition mode, is shown. PRRs of other MAMPs and the recognition of OGs, typical DAMPs, by WAK1 remain to be identified. Additionally, potential effectors secreted by *S. sclerotiorum* may be recognized by potential NBS-LRRs R-protein in cytoplasm. (II) The signal-transduction is performed by the MAP kinase cascade or MAPK-independent pathways. In response to *S. sclerotiorum* infection, the MAPK-independent pathways, such as calcium (Ca) signaling pathway, are also involved. In the Ca signaling pathway, firstly the NC may be activated by unknown PAMP(s) to produce cAMP and cGMP. Subsequently, the two cyclic nucleotides activate Ca^{2+} channels such as CNGCs, resulting in a Ca^{2+} elevation in the cytoplasm. Then the signal of the cytosolic Ca^{2+} elevation is transduced by Ca^{2+} sensor proteins, such as CaMs, CDPKs, as well as CRKs. Further, CDPKs phosphorylate and activate RBOHD/F, resulting in ROS accumulation. However, substrates of CRKs are not yet identified under the pathogen infection. Importantly, many signal molecules, including SA, JA, ET, auxin, ABA, NO, as well as ROS, play key roles in the signal transduction. (III) These signals direct various nuclear proteins including transcription factors, transcription activators, as well as protein kinases, such as WRKY33, CAMTA3, and CCaMK, to activate specific defense responses, including pathogenesis-related proteins, ROS production, detoxification, oxidative protection, callose deposition camalexin production, and other specialized secondary metabolites, and the model is expounded in the following text. *PR1* pathogenesis-related protein1, *MAMPs* microbial-associated molecular patterns, *DAMPs* damage-associated molecular patterns, *PG* polygalacturonases, *RLPs* receptor-like proteins, *BAK1*

MAMP-triggered immune responses and sensed by RLP30 (Receptor-like protein30), demonstrates the relevance of MTI in resistance to *S. sclerotiorum* (Zhang et al. 2013). RLP30, belonging to a specific class of plasma membrane (PM)-localized receptors, presents all the hallmarks of a cell surface-located RLP with an N-terminal signal peptide, an extracellular domain containing 21 LRRs (which possibly act as the SCFE1 binding site), a single transmembrane domain, and a short cytoplasmic tail of 25 amino acid residues (Zhang et al. 2013). Other elicitors, such as HRE (A heat-released elicitor) (Bertinetti and Ugalde 1996), *S. sclerotiorum* Cut (cutinase) (Zhang et al. 2014), and SsSm1 (a Cerato-platanin family protein) (Pan et al. 2018), have also been identified from *S. sclerotiorum*.

Upon infection with *S. sclerotiorum*, these elicitors can inspire the host plant to produce an immune response against this invading pathogen. Interestingly, a fungus-secreted protein, the cerato-platanin protein (CP), acts as a MAMP to elicit plant defense responses (Yang et al. 2009; Frias et al. 2011), but *S. sclerotiorum* CP1 (SsCP1) interacts with PR1, overexpression of which shows more resistance to *S. sclerotiorum*, in the apoplast to facilitate infection by *S. sclerotiorum* (Yang et al. 2018). CP also induces synthesis of ROS and triggers local cell death, suggesting that the MTI triggered by CP may be involved in the production of HR, which may also facilitate the infection by *S. sclerotiorum* with necrotrophic lifestyle. In addition, a study showed that treatment of *N. benthamiana* with VmE02 (a small cysteine-rich protein), a novel PAMP from the necrotrophic fungus *Valsa mali*, enhances plant resistance to *S. sclerotiorum* and *Phytophthora capsici* (Nie et al. 2019).

Host damage-associated molecular patterns (DAMPs) also activate protective immune signaling in plants. Oligogalaturonides (OGs) are representative DAMPs. They are released from plant cell wall after infection of pathogens like *Botrytis cinerea* and *S. sclerotiorum*, function as DAMPs, and induce plant defense response (Ferrari et al. 2013). OGs are recognized by WAK1 (WALL-ASSOCIATED KINASE 1) which is capable of activating the kinase domain of the elongation factor Tu (EF-TU) receptor (EFR). So OGs seem to be involved in MTI responses to necrotrophic pathogens (Mengiste 2012). Interestingly, polygalacturonase inhibitor proteins (PGIPs) seem to shift a breakdown process toward generating OGs. In

Fig. 5.13 (continued) BR insensitive1-associated receptor kinase1, *SOBIR1* suppressor of BIR1 BIR1-1, *WAK1* wall-associated kinase 1, *RLK* receptor-like kinase, *BIK1* botrytis-induced kinase1, *NBS-LRR* nucleotide-binding site-leucine-rich repeat, *MAPK* mitogen-activated protein kinase, *MAPKKKs* MAPK kinase kinases, *MAPKKs* MAPK kinases, *MKS1* MPK4 substrate 1, *SA* salicylate, *JAs* jasmonate, *ET* ethylene, *EIN2* ethylene insensitive 2, *ABA* abscisic acid, *NO* nitric oxide, *NPR1* non-expresser of *PR1*, *COI1* the coronatine-insensitive protein 1, *MED16* mediator complex subunit, *NC* nucleotidyl cyclase, *CNGC* cyclic nucleotide gated channel, *ATP* adenosine triphosphate, *GTP* guanosine triphosphate, *cAMP* cyclic adenosine monophosphate, *cGMP* cyclic guanosine monophosphate, *NC* nucleotidyl cyclase, *CaM* calmodulin, *CDPK* calcium-dependent protein kinase, *CRK* CDPK-related kinase, *RBOH* respiratory burst oxidase homologue, *SOD* superoxide dismutase, *CCaMK* Ca and CaM-dependent protein kinase, *CAMTA3* CaM-binding transcription activator3 (Wang et al. 2019)

A. thaliana, the overexpression of *B. napus PGIP2* inhibited necrotic lesions, but had no long-term effects on *S. sclerotiorum* disease progression (Bashi et al. 2013). The ectopic expression of *OsPGIP2* in rapeseed conferred the resistance to *S. sclerotiorum* at both the seedling and adult stages (Wang et al. 2018).

RLP30 belongs to a specific class of plasma membrane (PM)-localized receptors that carry an extracellular ligand-binding domain, but lack any obvious cytoplasmic signaling-competent moiety (Zhang et al. 2013; Liebrand et al. 2014), so following recognition of MAMPs by RLP30, the transduction of the extracellular signals to intracellular targets is dependent on additional signaling partners. Two transmembrane RLKs (receptor-like kinases), BAK1 (Brassinosteroid insensitive1-associated receptor kinase1) and SOBIR1/EVR (SUPPRESSOR OF BIR1-1/EVERSHED), are identified as RLP30 interacting and signaling partners and are responsible for subsequent signal transduction to cytoplasmic targets (Zhang et al. 2013). The activation of mitogen-activated protein kinases (MAPKs), including MAPK 3, 4, and 6, by the SCFE1-containing fraction was impaired in *bak* and *sobir1* mutant plants, indicating that *BAK1* and *SOBIR1* are required for the signal transduction from RLP30 to the intracellular MAPK kinase kinase(s) (MAPKKK).

MAPKKK activates the MAP kinase kinase(s) (MAPKK), by which the MAP kinase(s) (MAPK or MPK) is sequential activated by MAPKK. Additionally, SCFE1-dependent production of ethylene (ET), an important signaling molecule, was abolished in *bak1-5* mutant plants, and mutants of *RLP30*, *BAK1*, and *SOBIR1* are more susceptible to *S. sclerotiorum* and the related fungus *Botrytis cinerea* (Zhang et al. 2013). Further, the signaling from RLP30/SOBIR1/BAK1 to MAPKKK might require the RLCK (receptor-like cytoplasmic kinase), because Botrytis-induced kinase1 (BIK1), a RLCK, can cooperate with BAK1 to regulate constitutive immunity and cell death in *A. thaliana* (Eckardt 2011), which need to be further identified in the pathosystem between *S. sclerotiorum* and its hosts. BIK1 was initially identified as a plasma membrane localized protein (Veronese et al. 2006). Study have shown that BIK1 can also localize to the nucleus and interact directly with WRKY transcription factors, such as WRKY33, to be involved in the jasmonic acid (JA) and salicylic acid (SA) regulation (Lal et al. 2018).

MAPKs play an important role in signal transduction from cytoplasmic to nucleus, leading to the rapid changes of transcription depending on a series of phosphorylation. It has been reported that the MAP kinase 4 (MPK4) and the WRKY33 transcription factor play an important role in resistance to *S. sclerotiorum* as well as *Botrytis cinerea* (Wang et al. 2009, 2014, 2015a). MPK4 is localized in cytoplasm and nuclei, and the MPK4 substrate MKS1 is primarily localized in nuclei (Anderson et al. 2005). In the absence of pathogens, depending on MKS1, MPK4 can exist in nuclear composited with WRKY33 (Qiu et al. 2008; Wang et al. 2014). Challenge with pathogens or MAMPs leads to the activation of MPK4 and phosphorylation of MKS1. Subsequently, the MKS1–WRKY33 complex is released from MPK4, and WRKY33 activates the expression of *PHYTOALEXIN DEFICIENT 3* (*PAD3*) and *Cytochrome P450 71A13* (*CYP71A13*) (Qiu et al. 2008), both of which are involved in antimicrobial camalexin synthesis (Birkenbihl et al. 2012; Liu et al. 2015). WRKY33 also positively regulates the expression of JA/ET defense pathway marker gene *PLANT DEFENSIN1.2* (*PDF1.2*) and *OCTADECANOID-RESPONSIVE ARABIDOPSIS AP2/ERF59* (*ORA59*) (Birkenbihl et al. 2012; Wang et al. 2014, 2015a). However, how WRKY33 is released from the MKS1–WRKY33 complex remains to be determined. The MKS1–WRKY33 complex can exist both before and after phosphatase treatments of MKS1 in *A. thaliana*, and phospho-mimics, non-phosphorylatable, and wild-type forms of MKS1 bind WRKY33 equally well in yeast, suggesting that the complex exists independently of MKS1 phosphorylation. It has been shown that WRKY33 can be phosphorylated by MPK3/MPK6 in vivo in response to the pathogen infection in *A. thaliana*, which may be a hint for the release of WRKY33. *B. napus* MPK3 is a key regulator in defense responses to *S. sclerotiorum* (Wang et al. 2019).

The *A. thaliana* mediator complex subunit MED16 can also physically interact with WRKY33 in yeast and in planta (Wang et al. 2015a). The mediator complex, a multiprotein co-activator scaffold acting as a bridge between RNA polymerase II (RNAPII) and transcription factors, is involved in host transcriptional reprogramming after pathogen challenge, being paramount in the establishment of plant defense (Samanta and Thakur 2015). Mutations in MED16 subunit abolished induction of JA/ET cross-talk genes and reduced resistance to *S. sclerotiorum* and *B. cinerea* (Zhang et al. 2012; Wang et al. 2015a). Thus, these data suggest that WRKY33 seems to be a recruitment center in defense to *S. sclerotiorum*, as well as *B. cinerea*.

In addition to the MAPK pathway, calcium (Ca) signaling pathway has been reported to play important roles in plant defense against *S. sclerotiorum*. A Ca^{2+} elevation in the cytoplasm is known to be an important early event in plant cell perception of pathogen invasion (Dangl et al. 1996; Ma and Berkowitz 2011). Plant cyclic nucleotide-gated ion channels (CNGCs) provide a pathway for Ca^{2+} conductance across the plasma membrane (PM) and facilitate cytosolic Ca^{2+} elevation in response to pathogen signals. In the case of the interaction of plant host with *S. sclerotiorum*, pharmacological assays showed that the putative CNGC activators cGMP (cyclic adenosine monophosphate) and cAMP (cyclic guanosine monophosphate) enhanced resistance of tomato (*Solanum lycopersicum*) against this pathogen. Further, two tomato CNGC genes, *SlCNGC1* and *SlCNGC6*, were reported to play a positive role in tomato resistance to *S. sclerotiorum* (Saand et al. 2015a, b). Interestingly, other CNGC genes such as *SlCNGC17* and *SlCNGC18* exhibited negative roles in this resistance (Saand et al. 2015a). Transient elevations of the Ca^{2+} concentration in the cytoplasm can be sensed by various Ca^{2+} sensor proteins including calmodulins (CaMs) and calcium-dependent protein kinases (CDPKs) (Ranty et al. 2006; Hamel et al. 2014). It has been reported that a CaM-binding transcription activator CAMTA3 negatively regulated PTI probably by directly targeting *BAK1* under the *S. sclerotiorum* infection (Rahman et al. 2016). In contrast, Ca and CaM-dependent protein kinase (CCaMK), a nuclear-localized protein, positively regulates resistance to *via* promoting H_2O_2 accumulation (Wang et al. 2015b). The plasma membrane- and the cytosol-localized CDPKs can also positively regulate H_2O_2 accumulation *via* phosphorylating and activating RBOHD/

F (Liu and He 2016). However, knock-down of a set of *CDPKs* did not affect resistance to *S. sclerotiorum* in tomato (Wang et al. 2016). In contrast, *SlCRK6*, a CDPK-related kinase (CRK) that probably localizes at the plasma membrane, plays a positive role in this resistance (Wang et al. 2016). The CRK is another type of protein kinase closely related to CDPKs in structure. Unlike CDPKs, however, substrates of CRKs are not yet identified under the pathogen infection. Together, the increasing evidence indicated roles for Ca^{2+} signaling in resistance to *S. sclerotiorum*. Activation of Ca^{2+} channels, such as CNGC, requires the involvement of nucleotidyl cyclase (NC) which generates Ca^{2+} channel activators: cAMP and cGMP, together referred to as the cyclic nucleotide signaling system (Swiezawska et al. 2018).

In fact, a large body of data has implied that plant defense to *S. sclerotiorum* is complex and involves multiple signaling pathways. This is supported by a study on the protein profile of *B. napus* in response to *S. sclerotiorum*. The study indicated that plant defense to this pathogen involves various biological processes including redox homeostasis, lipid signaling, calcium signaling, histone, and DNA methylation-mediated transcription regulation and defense-related proteins such as defensin and defensin-like proteins as well as cyanate lyase. Moreover, it was reported plant defense to *S. sclerotiorum* is also regulated at miRNA level and probably involves PTGS (Post-Transcriptional Gene Silencing) (Cao et al. 2016a, b). Glycolate oxidase genes, encoding crucial enzymes in photorespiration, were reported to be involved in resistance to *S. sclerotiorum* (Xu et al. 2018b), suggesting a potential role for photorespiration in this resistance. In addition, defense against *S. sclerotiorum* was found to be stage-/phase-associated, and the phytohormones SA, ET, JA, and abscisic acid (ABA) likely play an essential, but pathosystem-dependent, role in leaf stage-associated resistance (Xu et al. 2018a; Wang et al. 2019).

5.6.6 Crosstalk of Signaling Molecules in Plants for Defense Against *Sclerotinia sclerotiorum*

Initiation of MTI, as well as ETI, is correlated with a complex network of defense signaling pathways, resulting in defensive cellular responses and changes in expression of thousands of host genes (Zipfel et al. 2006; Zipfel 2014; Lewis et al. 2015). The hormones, such as JA, SA, ET, ABA, and auxin, function as signaling molecules to transduce the perception signal through a huge and complex signaling network, and then complex defense responses are activated. Early studies on global profiles of host gene expression in response to infection with *S. sclerotiorum* indicated that genes associated with JA and ET signal are induced, but few of SA responsive genes are identified (Zhao et al. 2007, 2009). In plant immune responses, SA can antagonize the JA signaling pathway, which has already been identified in plants during *S. sclerotiorum* infection (Wang et al. 2012), and JA and ET are well known to synergistically induce defense responses and cooperate in resistance against pathogens. However, a study on *Arabidopsis* mutants, including *npr1* (*non-expresser of PR genes1*) being defective in SA perception, *coi1-2* impaired in JA perception, and *ein2* (ethylene *insensitive*), revealed that defense against *S. sclerotiorum* in *A. thaliana* is dependent on all of SA-, JA-, and ET-mediated signaling (Guo and Stotz 2007). The conclusion is supported by results from *B. napus* (Wang et al. 2012; Novakova et al. 2014), and it is suggested that defense against *S. sclerotiorum* in *B. napus* is associated with a sequential activation of SA and JA signaling (Wang et al. 2012). This can be explained by the new model depicting the lifestyle transition of the pathogen from biotrophic to necrotrophic growth (Kabbage et al. 2015), because it has been suggested that the SA signaling protects against biotrophs, while JA/ET signaling activates defense responses against necrotrophs, as well as herbivorous insects. In favor of a role of ET signaling in this defense, is a discovery that *B. napus* MPK3 is a positive regulator of ET signaling which positively regulates resistance to *S. sclerotiorum* (Wang et al. 2019).

Roles of SA and JA signaling in defense against *S. sclerotiorum* are challenging. By using three different mutants or transgenic lines impaired in SA production (*sid1/eds5*, *sid2/eds16*, and *nahG*) and one impaired in SA signaling (*npr1-1*), Perchepied et al. (2010) suggested that resistance to *S. sclerotiorum* is not dependent on SA. It was thought that the availability of bioactive jasmonates is not essential for resistance to *S. sclerotiorum*, because *jar1-1*, impaired in biosynthesis of jasmonoyl-L-amino acid, showed a completely wild-type phenotype in response to *S. sclerotiorum*, although *coi1-1* was found to be highly susceptible to *S. sclerotiorum* (Guo and Stotz 2007). However, studies have shown that SA, BTH (benzothiadiazole, a SA functional analogue), and MeJA application to *B. napus* leaves significantly results in increased resistance to *S. sclerotiorum* (Wang et al. 2012; Novakova et al. 2014). These results suggest that the role of SA or JA signaling in defense against *S. sclerotiorum* may vary depending on the plant species or that possible crosstalk between these signaling pathways and other regulatory pathways is likely to play a role in this defense (Perchepied et al. 2010). For example, an analysis on both jasmonate-dependent and COI1-independent defense responses against *S. sclerotiorum* showed that auxin signaling regulates the COI1-independent defense response pathway and plays an important role in resistance to this pathogen in *A. thaliana* (Stotz et al. 2011).

It has been suggested that ABA signaling would interlink with SA, JA, and ET signaling to affect pathogen resistance *via* a complex interplay of mechanisms (Mauch-Mani and Mauch 2005; Adie et al. 2007; Kaliff et al. 2007). ABA is required for biosynthesis of JA and expression of JA signaling defense gene in response to infection with *Pythium* spp., a necrotrophic pathogen. In the case of resistance to *S. sclerotiorum*, *A. thaliana* mutants (*aba3-2*, *abi1-1*, and *abi2-1*), being defective in ABA biosynthesis *or* perception, showed a complete loss of resistance to *S. sclerotiorum* (Guimaraes and Stotz 2004; Perchepied et al. 2010). ABA-mediated regulation of guard cells is a major mechanism of this defense in these interactions, which is compatible with two views. One is that OA, the essential pathogenicity factor for *S. sclerotiorum*, was concluded to favor infection by both inducing stomatal opening and inhibiting ABA-mediated stomatal closure (Guimaraes and Stotz 2004; Billon-Grand et al. 2012). Another is that the decreased ambient pH caused by *S. sclerotiorum* infection results in increased synthesis of photo-protective compounds of the xanthophyll cycle that serve as precursors for ABA synthesis, and thus the decrease in de novo ABA biosynthesis through decreasing xanthophyll precursors was suggested to account for enhanced plant susceptibility to *S. sclerotiorum* (Zhou et al. 2015). These observations also suggest that ABA signaling may be bidirectionally regulated by both *S. sclerotiorum* and its host plants, which needs to be fully understood by further studies. However, in other two *A. thaliana* mutants (*aba2-3* or *abi5*), they didn't show any significant changes in their response to *S. sclerotiorum* compared to control plants (Perchepied et al. 2010), suggesting that it remains unclear whether ABA signaling is required for this resistance.

In addition to these hormones, NO (nitric oxide) and ROSs (reactive oxygen species) are important signaling molecules (Kunkel and Brooks 2002). Nitric oxide (NO) is rapidly generated after recognition of pathogens (Wendehenne et al. 2004). The NO-impaired mutants showed an extremely susceptible phenotype to *S. sclerotiorum*, revealing the role of NO in resistance to this pathogen (Perchepied et al. 2010). It is suggested that the major role of NO in this resistance is multifaceted, such as regulating defense gene expression, interfering with the ROS signaling pathway, or modulating cell death (Perchepied et al. 2010). It has been shown that inhibition of host cell death by expressing negative regulators of mammalian apoptosis in transgenic tobacco plants leads to markedly enhanced resistance to *S. sclerotiorum* (Dickman et al. 2001), when ROS induction is inhibited, apoptotic-like cell death induced by oxalic acid does not occur and the PCD (programmed cell death) response is required for disease development (Kim et al. 2008). Further, the control of cell death governs the outcome of the *S. sclerotiorum*–plant interaction (Kabbage et al. 2013), and once infection is established, the necrotrophic *S. sclerotiorum* induces the generation of plant ROS, leading to PCD of host tissue, the result of which is of direct benefit to the pathogen (Williams et al. 2011; Wang et al. 2019).

5.6.7 Role of Nitric Oxide in Defense-Related Signaling Pathways to *Sclerotinia*

Studies of the interaction between *Arabidopsis thaliana* and the necrotrophic fungal pathogen Sclerotinia sclerotiorum have been hampered by the extreme susceptibility of this model plant to the fungus. In addition, analyses of the plant defense response suggested the implication of a complex interplay of hormonal and signaling pathways. To get a deeper insight into this host–pathogen interaction, the natural variation in *Arabidopsis* for resistance to *S. sclerotiorum* was analyzed. The results revealed a large variation of resistance and susceptibility in *Arabidopsis*, with some ecotypes, such as Ws-4, Col-0, and Rbz-1, being strongly resistant, and others, such as Shahdara, Ita-0, and Cvi-0, exhibiting an extreme susceptibility. The role of different signaling pathways in resistance was then determined by assessing the symptoms of mutants affected in the perception, production, or transduction of hormonal signals after inoculation with *S. sclerotiorum*. This analysis led to the conclusions that (1) signaling of inducible defenses is predominantly mediated by jasmonic acid and abscisic acid, influenced by ethylene, and independent of salicylic acid, and (2) nitric oxide (NO) and reactive oxygen species are important signals required for plant resistance to *S. sclerotiorum*. Defense gene expression analysis supported the specific role of NO in defense activation (Perchepied et al. 2010).

5.6.8 Differential Alternative Slicing Genes and Isoform Regulate Resistance to *Sclerotinia*

Identification of genetic factors and specific response mechanisms that influence resistance to SSR by analyzing the RNA-seq data of infected and mock-treated resistant and susceptible cultivars was analyzed by Ma et al. (2020). Two types of analysis were performed: post-transcriptional differences were explored using a DAS analysis, and isoform-level transcriptional regulation was elucidated using a co-expression network analysis. A total of 79 DAS genes were identified, which was fewer than expected. However, a previous study showed that the use of ten differential splicing analysis tools resulted in the identification of different numbers of DAS genes. The number of identified DAS genes ranged from 0 (cuttdiff2) to 4506 (edgeR) in a human prostate cancer dataset, with 2962 and 0 DAS genes identified using rMATS in human and mouse datasets, respectively. In addition, 252, 171, and 42 DAS genes were identified in comparisons between control and intolerant, control and tolerant, and intolerant and tolerant catfish following heat stress. Additionally, rMATS was used to identify 24, 40, and 74 DAS genes in the root, juvenile leaves and old leaves of *B. napus*, respectively, in response to boron deficiency (Gu et al. 2019).

The 79 DAS genes identified here were mainly involved in the nuclear retention of unspliced prem RNA and the nuclear mRNA surveillance of the spliceosome. Using the same double-end RNA-seq data here, it has been reported that 13,276 differentially expressed genes (DEGs) participated in metabolic processes and responses to stimuli such as response to chitin, RNA methylation, secondary cell wall biogenesis, and gravitropism (Wu et al. 2016). However, these genes had little overlap with the 79 DAS genes identified here. A small overlap was identified between the DEGs and DAS genes in rice plants exposed to mineral (phosphorus, zinc, manganese, copper, and iron) deficiencies. These comparisons suggested that AS is an additional layer of regulation, complementary to DEGs, that enables plants to respond to their environment (Dong et al. 2018).

S. sclerotiorum infects *B. napus* using mycelia or through the production of apothecia which are laden with fungal spores. The fungus produces multiple pathogenicity factors that attack *B. napus*, such as oxalic acid and an array of lytic enzymes. When attacked by fungi, plants undergo a three-step process: pathogen perception, signal transduction, and the defense response. In the plant–*Sclerotinia* system, pathogen detection is based on the recognition of pathogen-associated molecular patterns (PAMPs) by pattern recognition receptors, such as RLP30 (receptor-like protein 30) and other receptor-like kinases, or via cytoplasmic leucine-rich proteins (LRRs). The subsequent signal transduction involves a MAPK (mitogen-activated protein kinase) cascade or MAPK-independent pathways involving many signaling molecules, including reactive oxygen species (ROS), nitric oxide, salicylic acid, jasmonic acid, and ethylene. These signals drive plant PAMP-triggered immunity through the production of nuclear proteins including transcription factors and protein kinases that activate ROS production, detoxification, oxidative protection, callose deposition, camalexin production, and the production of other specialized metabolites. The DAS genes identified such as the immune-associated genes and those encoding important enzymes might play a major role in the response to *S. sclerotiorum* (Ma et al. 2020).

Among the 79 DAS genes, we identified several genes, including SR, that are involved in RNA splicing. As constitutive and AS regulators, SR proteins consist of RNA recognition motif and a Ser/Arg-rich domain, which, respectively, recognize RNA and the spliceosome. It has been reported that genes encoding SR proteins in plants appear to function mainly in stress responses; for example, in *Arabidopsis*, the *sr34b* T-DNA insertion mutant had a lower cadmium tolerance with a splicing defect in its iron-regulated transporter 1, whereas the *sr45* mutant was hypersensitive to glucose and abscisic acid. In rice, the SR40, SCL57, and SCL25 proteins were important for phosphorus uptake and remobilization. Studies have reported the stress-triggering AS pattern of certain *SR* transcripts, which may contain premature termination codons, resulting in their degradation by the nonsense-mediated decay pathway. For example, the abundance of an unproductive premature termination codon-containing AS isoform of *Arabidopsis* SR30 decreases at elevated temperatures. The AS of SR genes can increase plant tolerance to hormones and stress treatments in rice and *B. napus*. It was interesting to discover that the gene encoding the splicing factor PRP18 protein (*BnaA09g50970D*) is an overlapping DAS gene, as the homologous mutant in *A. thaliana* produces short roots and small siliques following its induction of large genome-wide IR events. Additionally, *BnaCnn1978*, encoding an RNA-binding protein, was identified as an infection-induced DAS gene. The splicing-related gene *SR34a* was also involved in two defense-related modules, and the increased number of ES events at 48 hpi with *S. sclerotiorum* suggests that this crucial splicing factor might be activated in response to pathogen stress to regulate downstream networks (Gu et al. 2019; Peng et al. 2013).

In addition to SR34a, other identified five DAS genes (*APX1*, *NIT2*, *AAT1*, *V-ATPase* F, and a gene encoding an unknown protein) as crucial candidate genes in the response to pathogen defense. APX isoenzymes function in controlling ROS metabolism and eliminating H_2O_2. APX can respond to environmental stresses such as oxidative stresses, light stress, and pathogen infection (*Trichoderma harzianum*). The identified APX1 as a cultivar-related DAS with isoforms is involved in different co-expression networks in the response to pathogen stress. NIT2 in *A. thaliana* is able to hydrolyze a broad substrate range, whereas a study revealed that AtNIT2 expression was de novo-induced by a plant pathogen (*Pseudomonas syringae* pv. *tomato*) and responsible for R-gene-mediated resistance responses. The

BnNIT2 was differently spliced in the susceptible cultivar at 24 hpi compared with the mock-treated plants, and seemed to reduce the ES events in the susceptible cultivar in response to *S. sclerotiorum*. V-ATPase F proton pumps are responsible for the ATP hydrolysis that drives ion transport. It has been reported that V-ATPase functions in response to phosphorus deficiency, copper stress, and fungal infection (the dermatophyte *Trichophyton rubrum*), among other stresses. The identified V-ATPase F is a pathogen-induced and cultivar-related DAS gene, and its AS frequency increased at 48 hpi. AATs are responsible for the long-distance transport of amino acids from the root to the shoot. The overexpression of AtAAT1 resulted in improved resistance against the pathogen *Pseudomonas syringae*. The identified AAT1 is both a pathogen-induced and cultivar-related DAS gene, which might lead to the generation of more AS events at 24 hpi with *S. sclerotiorum* to increase plant resistance to this pathogen. A total of 79 DAS genes were identified in *B. napus* after inoculation with *S. sclerotiorum*, and a combination of a DAS analysis and an isoform WGCNA revealed six candidate genes with important putative functions in pathogen resistance. These observations enhance our understanding of the post-transcriptional response of *B. napus* to *S. sclerotiorum* infection and pave the way to developing rapeseed lines with improved resistance to this devastating pathogen. Continuous advances in molecular biological experiments in exploring the function of different isoforms of one single gene should facilitate our findings of the candidate isoforms in pathogen defense (Ma et al. 2020).

5.7 *Brassica–Verticillium*: Biometabolomic Resistance

5.7.1 Metabolites Involved in Defense of *Arabidopsis* to *Verticillium*

It has been established that *Arabidopsis* is a suitable host to study plant reactions against *V. longisporum* (Reusche et al. 2012; Tran et al. 2013). In *Arabidopsis*, infection leads to stunted growth, hyper-plastic xylem formation, and premature senescence. Plant–pathogen interactions are generally characterized by a reprogramming in plant metabolism. In general, growth is arrested and defense mechanisms are activated. In *Arabidopsis*, several secondary metabolites with proven or potential antimicrobial effects are synthesized upon infection: indolic compounds including the phytoalexin camalexin, glucosinolates, phenylpropanoids, terpenes, benzenoids, and fatty acid derivatives (D'Auria and Gershenzon 2005). The secondary metabolites derived from the indole pathway are induced in *Arabidopsis* roots upon attempted entry of *V. longisporum* (Iven et al. 2012). Konig et al. (2014) identified many metabolic patterns that are altered in *Arabidopsis* when *V. longisporum* has already colonized the xylem vessels of the aerial parts. Metabolite fingerprinting and subsequent targeted analysis identified with 20 compounds that were derived from the phenylpropanoid pathway, as well as several amino acids. To test the contribution of the phenylpropanoid pathway on the interaction of *Arabidopsis* with *V. longisporum*, different mutants in the pathway were analyzed. Comparison of cell wall amount suggested that lignin is not a major factor in the defense response, but that soluble phenylpropanoids such as sinapoyl glucose, coniferyl alcohol, and coniferin do directly inhibit fungal growth.

Using the model plant *Arabidopsis*, Konig et al. (2014) analyzed metabolic changes upon *V. longisporum* infection in order to identify possible defense strategies of Brassicaceae against this fungus. Metabolite fingerprinting identified infection-induced metabolites derived from the phenylpropanoid pathway. Targeted analysis confirmed the accumulation of sinapoyl glucosides, coniferin, syringing, and lignans in leaves from early stages of infection. At later stages, the amounts of amino acids increased. To test the contribution of the phenylpropanoid pathway, mutants in the pathway were analyzed. The sinapate-deficient mutant fah1-2 showed stronger infection symptoms than wild-type plants, which is most likely due to the lack of sinapoyl esters. Moreover, the coniferin accumulating transgenic plant UGT72E2-OE was less susceptible. Consistently, sinapoyl glucose, coniferyl alcohol, and coniferin inhibited pathogen growth and melanization in vitro, whereas sinapoyl alcohol and syringin did not. The amount of lignin was not significantly altered supporting the notion that soluble derivatives of the phenylpropanoid pathway contribute to defense (Fig. 5.14). These data show that soluble phenylpropanoids are important for the defense response of *Arabidopsis* against *V. longisporum* and that metabolite fingerprinting is a valuable tool to identify infection-relevant metabolic markers (Konig et al. 2014).

5.7.2 Tryptophan-Derived Secondary Metabolites in *Arabidopsis* Roots Contribute to the Defense Against *Verticillium*

The soil-borne pathogen *Verticillium longisporum* causes vascular disease on Brassicaceae host plants such as oilseed rape. The fungus colonizes the root xylem and moves upward to the foliage where disease symptoms become visible. Using *Arabidopsis* as a model for early gene induction, Iven et al. (2012) performed root transcriptome analyses in response to hyphal growth immediately after spore germination and during penetration of the root cortex, respectively.

Infected roots showed a rapid reprogramming of gene expression such as activation of transcription factors and stress- and defense-related genes. The highly coordinated gene induction was observed resulting in the production of tryptophan-derived secondary metabolites. Earlier leaves showed that enzymes encoded by *CYP81F2* and *PEN2* (*PENETRATION2*) execute the formation of antifungal indole glucosinolate (IGS) metabolites. In *Verticillium*-infected roots, Iven et al. (2012) found transcriptional activation of *CYP81F2* and the *PEN2* homolog *PEL1* (*PEN2-LIKE1*), but no increase in antifungal IGS breakdown products. In contrast, indole-3-carboxylic acid (I3CA) and the phytoalexin camalexin accumulated in infected roots but only camalexin inhibited *Verticillium* growth in vitro. Whereas genetic disruption of the individual metabolic pathways leading to either camalexin or CYP81F2-dependent IGS metabolites did not alter *Verticillium*-induced disease symptoms, a *cyp79b2 cyp79b3* mutant impaired in both branches resulted in significantly enhanced susceptibility. The data provide an

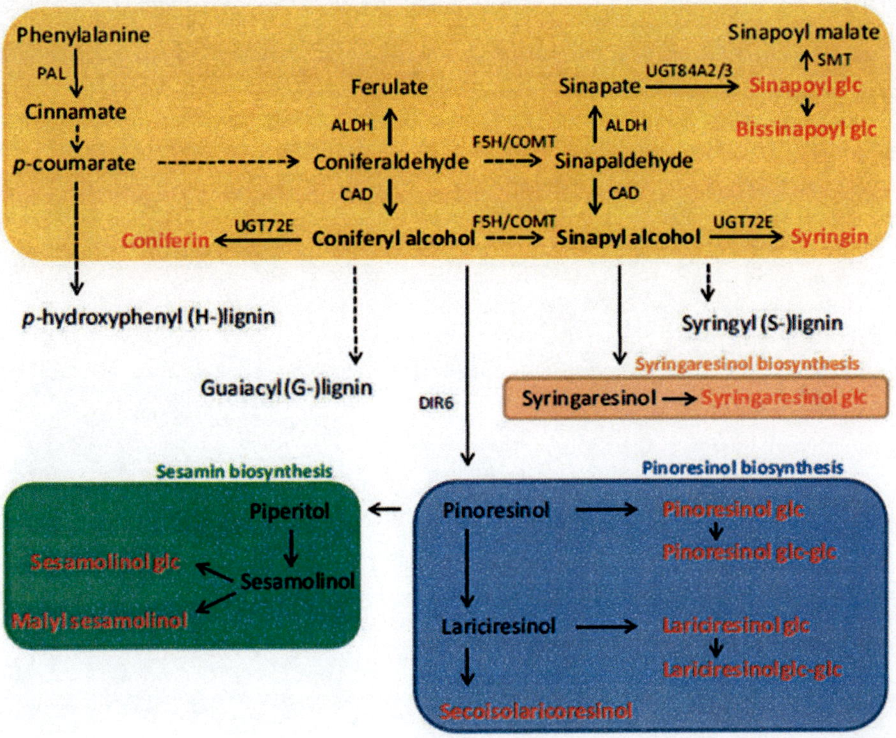

Fig. 5.14 Schematic overview of the phenylpropanoid biosynthesis pathway. Infection markers detected by metabolite fingerprinting are marked in red. *PAL* phenylalanine ammonia lyase, *F5H* ferulate-5-hydroxylase, *COMT* caffeic acid/5-hydroxyferulic acid *O*-methyl transferase, *CAD* cinnamyl alcohol dehydrogenase, *DIR* dirigent protein, *ALDH* aldehyde dehydrogenase, *SMT* sinapoyl glucose: malate sinapoyl transferase, *UGT* UDP-glycosyl transferase (Konig et al. 2014)

insight into root-specific early defenses and suggest tryptophan-derived metabolites as active antifungal compounds against a vascular pathogen (Fig. 5.15; Iven et al. 2012).

5.7.3 Induction of Biomolecular Resistance to *Verticillium* in *Brassica* by BABA

Plants have developed a wealth of defense strategies to protect themselves against the wide range of pathogens and pests that have co-evolved. Defense responses are expressed through complex signaling networks involving signaling compounds such as reactive oxygen species (ROS), salicylic acid (SA), jasmonate (JA), ethylene (ET), and abscisic acid (ABA). These signaling intermediates regulate the expression of pathogenesis-related (PR) genes that encode for proteins playing either a role in the structural and physiological manifestation of defense through cell wall

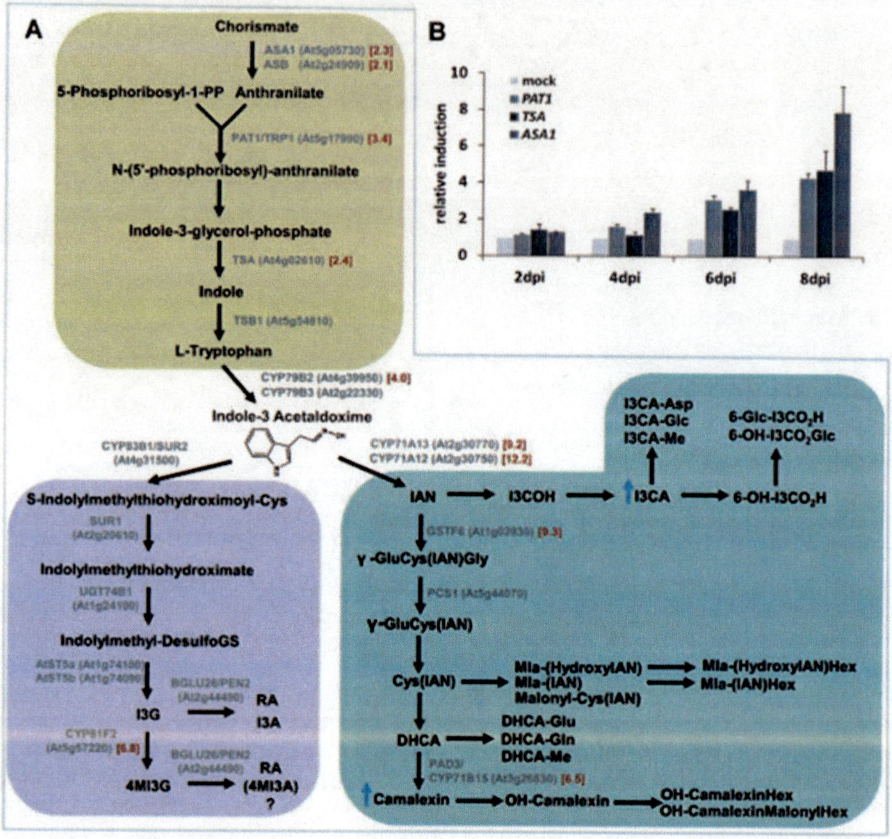

Fig. 5.15 Coordinated transcriptional induction of the tryptophan-derived secondary metabolism. (**a**) Schematic representation of the biosynthetic pathway leading to tryptophan (light green), IGS (blue), and camalexin (dark green). Depicted are the major intermediates and enzymes. The *Verticillium longisporum*-induced genes and the corresponding *Arabidopsis* Gene Identifier (AGI) are given in brackets. The (Log2) induction values from the microarray experiment are shown in red square brackets for the 3-dpi time point. Metabolites found to be increased after infection is marked by a blue arrow. *IAN* indole-3-acetonitrile, *I3CHO* indole-3-carbaldehyde, *I3CA* indole-3-carboxylic acid, *DHCA* dihydrocamalexic acid, *I3G* indole-3-ylmethyl glucosinolate, *I3A* indole-3-ylmethylamine, *4MI3G* 4-methoxyindole-3-ylmethylglucosinolate, *RA* raphanusamic acid. (**b**) qPCR analysis of relative transcript abundance in *Verticillium longisporum*-infected *Arabidopsis thaliana* Col-0 Wt roots. Expression of *PAT1* (At5g17990), *TSA* (At4g02610), and *ASA1* (At5g05730) was determined at 2, 4, 6, and 8 dpi. The relative transcript abundance was normalized to mock inoculated control (set to one). The mean values (+SE) are obtained from two biological and four technical replicates. For each replicate and condition RNA from pooled roots of three Petri dishes was prepared (>100 seedling roots) (Iven et al. 2012)

strengthening, lignification, phytoalexin synthesis, and expression of localized cell death or have direct antifungal properties. In addition to basal resistance, plants are capable of developing an induced resistance (IR) that is a physiological state of enhanced defensive capacity elicited by specific environmental stimuli, by which plants' innate defenses are potentiated against subsequent biotic challenge. The best-known systems of IR are systemic acquired resistance (SAR) and induced systemic resistance (ISR). In addition to these two types, treatment with a non-protein amino acid, BABA (β-aminobutyric acid), may induce resistance in a number of plant species against a wide range of pathogens. However, the mode of action of BABA in inducing resistance remains unresolved. In some plant–pathogen systems, BABA was reported to induce resistance via the SA signaling pathway (Zimmerli et al. 2001). In some other plant–pathogen systems, BABA IR was shown to be expressed through an SA-independent pathway (Zimmerli et al. 2000). There is also some evidence of the involvement of JA in BABA IR because the inhibition of lipoxygenase (LOX) activity, which causes JA deficiency, was seen to reduce BABA IR (Hamiduzzaman et al. 2005). Finally, BABA has also been reported to induce resistance through ABA-primed callose deposition in the *Arabidopsis–Alternaria* interaction, a resistance reaction which was expressed even in mutants defective in SA, JA, or ET signaling (Ton and Mauch-Mani 2004).

Some characteristics of the genotype-related defense responses in *B. napus* against *V. longisporum* have been partly elucidated (Eynck et al. 2009), while resistance responses during induced resistance are completely unexplored. Kamble et al. (2013) explored the potential of induction of resistance in oilseed rape with a known chemical inducer, BABA, and to compare induced defense responses with genotype-related resistance reported previously (Eynck et al. 2009). First, the effective concentration of BABA was estimated based on the alleviating effect on disease severity in terms of stunting caused by the pathogen, followed by attempts to identify some of the components of induced resistance. Ramification of the pathogen was followed by using quantitative real-time PCR. Further parameters included the role of physical barriers in vascular tissues, endogenous H_2O_2 and SA levels, PAL enzyme activity, and expression of defense-related genes, *PR*-1 and *PDF1.2*, as markers for the involvement of SA- and JA-mediated signaling pathways, respectively.

A pre-inoculated soil drench application of 0–5 mM β-aminobutyric acid (BABA) significantly inhibited colonization of oilseed rape (*Brassica napus*, susceptible cultivar Falcon) by *Verticillium longisporum* and also prevented plant stunting caused by the pathogen. To better understand the defense responses induced by BABA, the presence of occlusions in the plant hypocotyl, levels of salicylic acid (SA) and hydrogen peroxide (H_2O_2), phenylalanine ammonia lyase (PAL) activity, and expression of *PR*-1 and *PDF1.2* genes were examined. Transverse sections through the hypocotyl region of BABA-treated plants showed clear vessels surrounded by phenol-storing cells, in contrast to numerous obstructed vessels in water-treated plants, in response to the pathogen. A significant increase in SA levels was observed in the hypocotyls of both water- and BABA-treated plants in response to the pathogen; however, SA levels were unrelated to disease resistance. The level of H_2O_2 decreased in both treatments in response to the pathogen. A significant increase in PAL activity was observed in hypocotyl tissues of BABA-treated plants. The expression patterns of *PR*-1 and *PDF1.2* were similar in the two treatments in response to the pathogen, indicating no involvement of these genes in resistance. The results indicate a similar organ specificity of the plant hypocotyl for chemically induced internal resistance as for genotype-related resistance, two phenomena which, however, are based on contrasting cytological responses in the vascular tissues. Nonetheless, evidence is provided that the activity of the phenylpropanoid pathway plays a crucial role in both types of resistance (Kamble et al. 2013).

5.7.4 Role of Biophenols in Resistance of *Brassica* to *Verticillium*

Verticillium longisporum is a vascular fungal pathogen presently threatening oilseed rape production in Europe. Systemic spread and vascular responses were studied in a susceptible ('Falcon') and a resistant genotype (SEM 05-500256) of *Brassica napus*. Colonization of both genotypes after dip-inoculation of the roots followed by quantitative polymerase chain reaction revealed similarities only in the initial stages of root penetration and colonization of the hypocotyl, while a substantial invasion of the shoot was only recorded in 'Falcon'. It is concluded that the type of resistance represented in SEM 05-500256 does not prevent the plant base from being invaded as it is internally expressed well after root penetration and colonization of the plant base. The morphological and biochemical nature of barriers induced in the hypocotyl tissue upon infection was studied with histochemical methods accompanied by biochemical analyses. Histochemical studies revealed the build-up of vascular occlusions and the reinforcement of tracheary elements through the deposition of cell wall-bound phenolics and lignin. Furthermore, the accumulation of soluble phenolics was observed. Although these responses were found in vascular tissues of both genotypes, they occurred with a significantly higher intensity in the resistant genotype and corresponded with the disease phenotype. In the resistant genotype, phenols were differentially expressed in a time-dependent manner with preformed soluble and cell wall-bound phenolics at earlier time points and de novo formation of lignin and lignin-like polymers at later stages of infection. A crucial role of phenol metabolism in internal defense of *B. napus* against *V. longisporum* and locating the crucial defense responses in the plant hypocotyl has been observed (Eynck et al. 2009).

5.8 *Brassica–Xanthomonas*: Biometabolomics Resistance

5.8.1 Proteomics of Defense Response in *Brassica* to *Xanthomonas*

The disease/defense response proteins identified were increased in both cultivars and were involved with oxidative stress challenged with *Xanthomonas* (Santos et al. 2019). However, a higher number of pathogen-related proteins associated with plant responses showed increased abundance in the resistant cultivar (16%), whereas only a few (5%) were increased in the susceptible cultivar. The increased defense proteins identified in both cultivars were annexin (I3Y171), AIG2 (A0A0D3BZV5), ferredoxin (A0A0D3BV84), ferredoxin-NADP leaf isozyme 1 and 2 chloroplastic (A0A0D3E2R1; A0A0D3DQI2), and mitochondrial outer membrane protein 4 (A0A0D3B2Z9).

Several increased proteins in susceptible plants were involved with responses to abscisic acid (ABA), while in resistant plants these proteins showed decreased abundance. ABA is a phytohormone, known as a signaling molecule, responsible for the regulation of abiotic stress response (Taiz et al. 2017). Studies have shown that ABA can suppress the plant immune response (Kim et al. 2011; Desclos-Theveniau et al. 2012), and in many pathosystems, this phytohormone can act antagonistically to the salicylic acid (SA) pathway. SA is another important phytohormone that can confer plant resistance against pathogens (Audenaert et al. 2002; Jiang et al. 2010). Furthermore, ABA can suppress the MAPK pathway, causing immune-suppression in *A. thaliana* and possibly in other cruciferous plants (Mine et al. 2017). ABA's effect during plant–pathogen interaction is considered complex; however, it is possible that the increased abundance of proteins involved with ABA response can be a mechanism, which can favor susceptibility (Kim et al. 2011; Desclos-Theveniau et al. 2012).

The increase of ABA can also lead to the accumulation of other proteins such as aquaporins. The aquaporins PIP3 (Q9FUC0) and BoPIP1b1 (Q9FUL1) were differentially abundant; the second was evaluated by qRT-PCR and showed a downregulation trend in both cultivars. These aquaporins were identified in chloroplast samples with increased abundance only in the susceptible plants. PIP aquaporins are intramembrane channels important for the transport of water and CO_2 in the plant tissues (Luu and Maurel 2005; Verkman 2013). The detection of these proteins in chloroplast samples was not expected; however, since sampled intact and broken chloroplasts, it is possible that some non-chloroplastic proteins were also isolated. The results suggest that the accumulation of these proteins may be related to ABA, as observed by Aroca et al. (2006), in leaves of *Phaseolus vulgaris* after ABA treatment. Aquaporins are multifunctional, and some isoforms are able to detect pathogen molecular patterns (PAMPs) such as harpins (Zhu et al. 2000; Flexas et al. 2007). The transport of hydrogen peroxide has also been associated to aquaporins (Taiz et al. 2017). It was demonstrated that the loss of function of the gene locus AtPIP1; four in *Arabidopsis* cancels the import of apoplastic H_2O_2 induced by the pathogen and consequently blocks the plant immune response (Tian et al. 2016).

Curiously, another protein named epithiospecifier-like (ESP-like; A0A0D3CQU9) involved in defense response was decreased in susceptible plants and unchanged in resistant plants. Conversely, the expression of ESP gene was upregulated according to qRT-PCR results in both cultivars (23-fold change in the resistant plant and threefold change in the susceptible). In the previous study performed by Ribeiro et al. (2018), another resistant cabbage cultivar (Uniao) was analyzed and ESP protein was exclusively identified in the resistant cultivar infected with *Xcc* by 2-DE analysis, demonstrating that the regulation of ESP protein can be important for the plant defense against *Xcc*. The ESP protein is related to the glucosinolate pathway involved in plant protection against herbivory pests.

Glucosinolates are secondary metabolites, known as phytoanticipins (preformed antimicrobial compounds), representing one of the first chemical barriers against pathogen attack (Osbourn 1996). These metabolites can be found extensively in Brassicaceae plants (i.e., broccoli, cabbage, mustard) and are biologically active compounds reported in some processes of plant defense including stress response and antioxidant activities (Bennett and Wallsgrove 1994; Halkier and Gershenzon 2006; Santos et al. 2019).

5.8.2 Biometabolomics of *Brassica oleracea–Xanthomonas* Interaction as a Model of Resistance

The protein profiles of the resistant and susceptible plants were similar, especially regarding the predominant GO terms (Santos et al. 2019). However, a higher number of pathogen-related proteins were identified in the resistant plants as in a proposed model of this interaction based on protein localization and their role in the cell (Fig. 5.16; Table 5.4). This model can help better understand the plant response to Xcc infection and provide candidate genes for the development of more efficient pathogen control strategies. The classical mechanism of bacterial recognition occurs at the beginning of infection, in an attempt to neutralize the effectors released by the bacterium and repair the damage caused in the cell. In the resistant cultivar Astrus, several proteins with increased abundance were recorded probably involved in this initial response (see step 1 in the model), such as lectin-like protein (A0A0D3CJY3), leucine-rich repeat receptor kinase (A0A0D3D099), and leucine-rich repeat receptor-like serine threonine kinase (A0A0D3CSX5). These transmembrane signaling proteins, together with other proteins, such as NB-LRR, can be essential to sense the pathogen and promote systemic immunity (O'Neill and Bowie 2007; Couto and Zipfel 2016). Other signaling proteins identified showing increased abundance were the universal stress protein A (A0A0D3CN30) and universal stress PHOS34-like (A0A0D3CTQ3). Although the exact role of PHOS34 in plant defense is not known, studies have reported that this protein can be phosphorylated by MPK3 and MPK6, and after treatment with the flagellin flg22 peptide (Merkouropoulos et al. 2008), suggesting that this protein may be related with cellular signaling in the presence of the bacterium.

Several antioxidant/detoxification proteins were also increased, including superoxide dismutase Fe (F8U7Z7), peroxidase (A0A0D3C7R8), peroxiredoxin (A0A0D3DSN3; A0A0D2ZRQ6; A0A0D3BYD5), glutathione *S*-transferase U5, and U19 (A0A0D3B771; A0A0D3CVZ5). The accumulation of ROS can be toxic to the pathogen by inhibiting and/or reducing its survival (Jones and Dangl 2006; Zhang and Zhou 2010). However, ROS accumulation can lead to the oxidation of important cell components like lipids and genetic material (Sharma et al. 2016). An intriguing result obtained was the decrease of other extra and intracellular antioxidant proteins including glutathione peroxidase mitochondrial (A0A0D3AT05; A0A0D3DQE3); peroxidase 3-like (A0A0D3C7R9); peroxiredoxin-mitochondrial (A0A0D3CMF0); and peroxidase 32 (A0A0D3E2V6) in the resistant interaction.

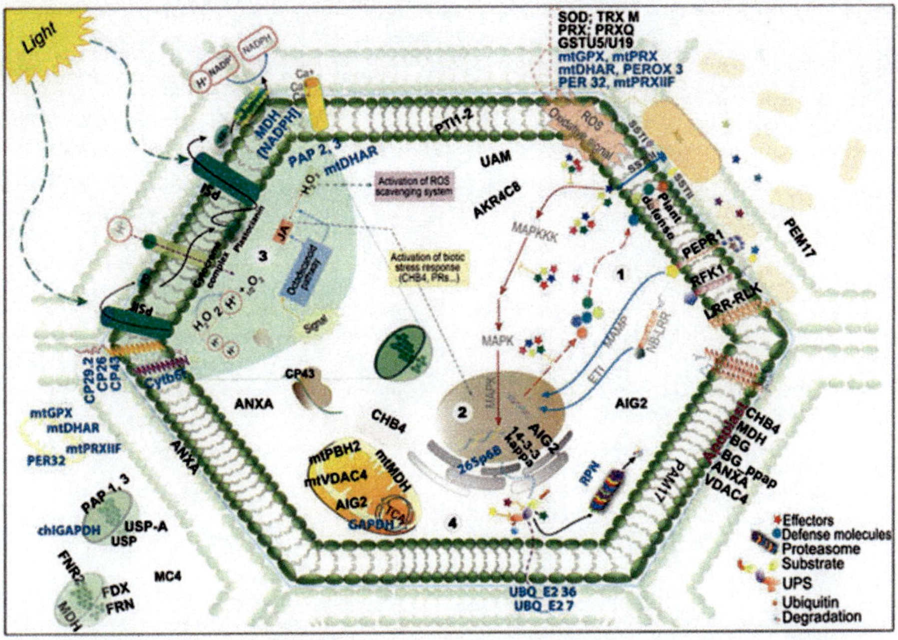

Fig. 5.16 Schematic view of a model proposed with the proteins identified in the resistant *Brassica oleracea–Xanthomonas campestris* pv. *campestris* (*Xcc*) infected leaf and chloroplast-enriched proteomes. The figure shows the localization, as proposed by UniProt database; detailed information on the proteins is presented in Table 5.4. The names in black and blue indicate proteins with increased and decreased abundance, respectively. The steps begin at the recognition of the pathogen, involving important signaling proteins, activation of molecular defense response pathways and oxidative stress response (steps 1–3), followed by UPS (ubiquitin pathway system) modulation and repair proteins (step 4), as well as alteration of metabolic and photosynthetic pathways (steps 3 and 4) (Santos et al. 2019)

This result may indicate that a balance in the abundance of proteins related to oxidative stress, maintaining some proteins with increased abundance and others with decreased abundance may be important for an efficient control of the pathogen without extensive damage to the plant tissue.

Once the pathogen overcomes the first line of defense, other events occur in response to the effector delivery into the cell by the type III secretion system. At this stage, proteins such as NB-LRR proteins interact with the pathogen effectors (Spallek et al. 2009; Marino et al. 2012). Santos et al. (2019) identified a leucine-rich repeat receptor kinase PEPR1 (A0A0D3D099), which has been reported as the receptor for AtPep1, a peptide elicitor from *Arabidopsis* that signals activation of innate immune response against pathogens (Yamaguchi et al. 2006) as well as a probable LRR receptor-like serine/threonine-protein kinase At1g29720 RFK1-like (A0A0D3CSX5). Both proteins are integral components of the membrane (model-step 1) and may be involved in triggering a defense response. Ubiquitination pathway also seems to play an important role in the resistant interaction. The protein

Table 5.4 Differential proteins and encoding genes analyzed by qRT-PCR analysis (RI: RC and SI: SC interactions) and discussed in the proposed interaction model (Santos et al. 2019)

Gene[a]	Gene product (full name)	Protein (SN)[b]	UniProt accession#	Protein fold change		Gene fold change (Log2)		Gene ontology (biological process)[c]
Differential genes **analyzed by qRT-PCR and included in the model**				**R**	**S**	**R**	**S**	
BoAKR2A	Ankyrin repeat domain-containing protein 2-like	AKR2	A0A0D3BK51	ni	2	−0.37	−0.003	Protein targeting chloroplast
BoAMC4	Metacaspase-4	MC4	A0A0D3D1T5[d]	2	2	2.88	1.172	Positive regulation of programmed cell death; protein autoprocessing
BoANN2	Annexin	ANN2	I3Y171	2	ni	3.80	2.357	Calcium ion transmembrane transport; response to oxidative stress
BoAPY5	Apyrase 5	APY5	A0A0D3CA22	nd	10	0.73	0.179	None predicted
At5g02240	Uncharacterized protein At5g02240	At5g02240	A0A0D3EID2[d]	nd	3	−0.31	0.897	Response to abscisic acid
BoATPC1	ATP synthase gamma chain 1, chloroplastic	ATPG1	A0A0D3E873	nd	11	−0.23	−0.188	ATP synthesis-coupled proton transport
BoBAG7	BAG family molecular chaperone regulator 7-like	BAG7	A0A0D3A4W0	ni	81	1.21	0.452	Cellular response to unfolded protein; protein folding; cellular response to heat
BoCHB4	Basic endochitinase CHB4-like	CHB4	A0A0D3BPL2	6	ni	9.17	4.495	Cell wall macromolecule catabolic process; chitin catabolic process; response to virus; systemic acquired resistance
BoCLPP	ATP-dependent Clp protease proteolytic subunit	CLPP	A0A0D3A7F3	ni	2	1.18	−0.385	Protein quality control for misfolded or incompletely synthesized proteins
BoDEGP8	Protease do-like 8, chloroplastic	DEGP8	A0A0D3BZW4[d]	3	ni	0.35	6.457	Photosystem II repair

BoDTX	Protein detoxification	DTX	A0A0D3BJ77	ni	2	0.54	0.268	Abscisic acid transport; drug transmembrane transport; regulation of response to water deprivation
BoENH1	Rubredoxin_like, 1	ENNH1	A0A0D3AJG1	nd	2	−1.58	−0.639	None predicted
BoERF1-3	Eukaryotic peptide chain release factor subunit 1–3	eRF1-3	A0A0D3E0E3[d]	3	nd	1.05	0.317	Cytoplasmic translational termination; regulation of growth; translational termination
BoESP	Epithiospecifier-like	ESP	A0A0D3CQU9	nd	2	4.56	1.607	Defense response to bacterium; catabolic process; nitrile biosynthetic process; response to jasmonic acid
BoFSD1	Superoxide dismutase	SOD	F8U7Z7	3	nd	0.71	−0.017	Cellular response to oxidative stress; defense response to bacterium; cellular response to salt stress; cellular response to UV-B
BoGASA1	Gibberellin-regulated protein 1	GAST1	A0A0D3D0Y9[d]	2	nd	0.01	−1.905	Response to abscisic acid; response to brassinosteroid; response to gibberellin
BoIF-2	Translation initiation factor IF-2	IF2	A0A0D3CAZ7	2	nd	−1.30	0.110	Translational initiation; translation; nucleotide binding
BoLLP	Lectin-like protein At3g16530	LLP	A0A0D3CJY3	3	ni	−6.51	3.470	Response to chitin; response to oomycetes
BoMLP31	MLP-like protein 31	MLP31	A0A0D3APR4	4	ni	−0.83	−0.860	Defense response
BoPER32	Peroxidase 32	PEROX32	A0A0D3E2V6	2	nd	1.92	0.239	Response to oxidative stress; hydrogen peroxide catabolic process; response to cytokinin
BoPIP1-1	Aquaporin PIP1b1	PIP1b1	Q9FUL1[d]	ni	2	−0.46	−0.003	Water transport; response to water deprivation
BoPPD4	psbP domain-containing protein 4, chloroplastic	PSBP4	A0A0D3D1B1[d]	nd	9	−0.64	−0.349	Photosynthesis
BoPRX2F		PRXIIF	A0A0D3CMF0[d]	2	2	−0.33	0.920	

(continued)

Table 5.4 (continued)

Gene[a]	Gene product (full name)	Protein (SN)[b]	UniProt accession#	Protein fold change		Gene fold change (Log2)		Gene ontology (biological process)[c]
	Peroxiredoxin IIF, mitochondrial							Cell redox homeostasis; response to cadmium ion; response to oxidative stress
BoPSBE	Cytochrome b559 subunit alpha	PSBE	A0A0H3Y313[d]	nd	2	−0.44	−0.383	Photosynthetic electron transport chain
BoPSBD	Photosystem II D2 protein	PSII D2	A0A191SEU8	nd	2	−0.13	−0.223	Photosynthetic electron transport in photosystem II; protein-chromophore linkage
BoRFK1	LRR receptor-like serine/threonine-protein kinase At1g29720	LRR-RLK	A0A0D3CSX5	2	nd	−0.71	1.010	Protein autophosphorylation; regulation of innate immune response; jasmonic acid and ethylene-dependent systemic resistance
BoRGP1	UDP-arabinopyranose mutase 1-like	UAM	A0A0D3B9D8	9	ni	2.20	1.968	Plant-type cell wall organization or biogenesis
BoTL17	Thylakoid lumenal 17.4 kDa protein, chloroplastic	P17.4	A0A0D3EAP2[d]	3	6	−0.69	1.944	Protein binding
BoYCF54	Uncharacterized protein Ycf54	YCF54	A0A0D3ECB6[d]	3	3	−0.38	−0.436	None predicted
BoUSPA	Universal stress protein YxiE-like	YXIE	A0A0D3CTQ3	2	2	0.73	0.642	None predicted
Additional proteins **included in the model**				**R**	**S**			
BoTRXM	Thioredoxin M chloroplastic	TRXM	A0A0D3DZ53[d]	2	2	Cell redox homeostasis; glycerol ether metabolic process; regulation of carbohydrate metabolic process		
BoPRXQ	Peroxiredoxin Q, chloroplastic isoform X2	PRXQ	A0A0D2ZRQ6[d]	2	nd	Cell redox homeostasis		

BoPRX	Peroxiredoxin-chloroplastic	PRX	A0A0D3BYD5[d]	2	nd	Cell redox homeostasis
			A0A0D3DSN3[d]	2	nd	
106329510	Pectinesterase	PEM17	A0A0D3B6U2[d]	4	2	Cell wall modification; pectin catabolic process
BoGSTU5	Glutathione *S*-transferase U5	GSTU5	A0A0D3B771	2	nd	Response to oxidative stress; response to toxic substance; toxin catabolic process
BoGSTU19	Glutathione *S*-transferase U19	GSTU19	A0A0D3CVZ5	2	−2	Glutathione metabolic process; response to oxidative stress; cellular response to water deprivation
BoBoAIG2	Aig2 protein	AIG2	A0A0D3BZV5[d]	2	2	Response to bacterium
106337169	Ferredoxin	FDX	A0A0D3BV84[d]	2		Transport; electron transport
BoLFNR	Ferredoxin–NADP leaf isozyme 1 chloroplastic	FNR	A0A0D3E2R[d]	1.5	nd	Defense response to; defense response to fungus, incompatible interaction; photosynthesis; response to cytokinin
BoLFNR2	Ferredoxin–NADP leaf isozyme 2 chloroplastic	FNR2	A0A0D3DQI2[d]	2	nd	Defense response to; defense response to fungus, incompatible interaction; photosynthesis; response to cytokinin
BoGF14 BoKAPPA	14-3-3 GF14 kappa isoform X1	GF14 kappa	A0A0D3BET0	2	nd	Regulation of metabolic; response to freezing
BoAKR4C8	Aldo-keto reductase family 4 member C8	AKR4C8	A0A0D3BR44	2	−2	Oxidation-reduction process; response to cadmium ion; response to toxic substance; response to cold; response to salt stress
BoPBH2	Prohibitin 2, mitochondrial-like	mtPBH2	A0A0D3C7E7[d]	2	2	Mitochondrion organization; cell division; defense response to bacterium; negative regulation of cell division; response to auxin
BoVDAC4	Mitochondrial outer membrane porin 4	mtVDAC4	A0A0D3B2Z9[d]	2	ni	Regulation of growth; response to bacterium
BoMDH	Malate dehydrogenase mitochondrial	mtMDH	A0A0D3CQE1[d]	3	2	Carbohydrate metabolic process; malate metabolic process; tricarboxylic acid cycle
			A0A0D3CQN2[d]	2	2	

(continued)

Table 5.4 (continued)

Gene[a]	Gene product (full name)	Protein (SN)[b]	UniProt accession#	Protein fold change	Gene fold change (Log2)	Gene ontology (biological process)[c]
			A0A0D3BMU9	4	−2	
BoMDH	Malate dehydrogenase, chloroplastic	chlMDH	A0A0D3CGY3	2	1.5	Carbohydrate metabolic process; malate metabolic process; tricarboxylic acid cycle
106341843	Glucan endo-1,3-beta-glucosidase-like	BG	A0A0D3BXB6	7	3	Carbohydrate metabolic process
106300472	Glucan endo-1,3-beta-glucosidase-like (beta-1,3-glucanase)	BG_ppap	A0A0D3CTF1[d]	2	ni	Carbohydrate metabolic process; cell communication
BoGAPDH	Glyceraldehyde-3-phosphate dehydrogenase, chloroplastic	chlGAPDH	A0A0D3DXN8	2	nd	Glucose metabolic process
BoPAP 1	Plastid-lipid-associated 1, chloroplastic	chlPAP 1	A0A0D3E8B6[d]	2	2	Photoinhibition; response to abscisic acid; response to cold
			A0A0D3E8B7[d]	2	2	
			A0A0D3B8J8	2	3	
BoPAP 2	Plastid-lipid-associated 2	PAP 2	A0A0D3A546	2	nd	None predicted
BoPAP 3	Plastid-lipid-associated 3	PAP 3	A0A0D3BRT9[d]	2	2	None predicted
BoUSPA	Universal stress A	USP-A	A0A0D3CN30[d]	2	ni	None predicted
BoPEPR1	Leucine-rich repeat receptor kinase Pepr1	PEP1	A0A0D3D099[d]	3	2	Immune response; innate immune response; response to jasmonic acid; response to wounding
NA	Leucine-rich repeat receptor-like serine threonine kinase At3g14840	LRR-RLK	A0A0D3BBD1[d]	2	ni	Protein autophosphorylation; regulation of innate immune response; jasmonic acid and ethylene-dependent systemic resistance
BoPTI12	PTI1-like tyrosine-protein kinase 2	PTI1-2	A0A0D3B4A1	1.5	nd	Defense response; protein phosphorylation

BoGAPC	Glyceraldehyde-3-phosphate dehydrogenase	GAPDH	A0A0D2ZPE9	2	nd	Glucose metabolic process
			A0A0D3C9I2[d]	2	nd	
106335373	Malate dehydrogenase [NADP] chloroplastic-like	chlMDH	A0A0D3B2U9	2	nd	Carbohydrate metabolic process; malate metabolic process
BoGPX1	Glutathione peroxidase mitochondrial	mtGPX	A0A0D3AT05	2	nd	Response to oxidative stress
BoDHAR1	Glutathione *S*-transferase DHAR1, mitochondrial-like	mtDHAR	A0A0D3DQE3	2	nd	Cellular response to hydrogen peroxide; defense response; positive regulation of salicylic acid-mediated signaling pathway; response to jasmonic acid
BoPER3	Peroxidase 3-like	PEROX3	A0A0D3C7R9[d]	2	nd	Hydrogen peroxide catabolic process; response to oxidative stress
BoPER32	Peroxidase 32	PER32	A0A0D3E2V6	2	nd	Hydrogen peroxide catabolic process; response to cytokinin; response to oxidative stress
BoLHCB5	Chlorophyll a-b binding protein CP26, chloroplastic	CP26	A0A0D3B7Z5	4	ni	Non-photochemical quenching; photosynthesis, light harvesting; photosystem II assembly; protein-chromophore linkage
BoLHCB4.2	Chlorophyll a-b binding protein CP29.2, chloroplastic	CP29.2	A0A0D3CLT0	2	2	Photosynthesis, light harvesting; protein-chromophore linkage; response to blue light; response to cytokinin
NA	Chlorophyll a-b binding protein CP43, chloroplastic	CP43	A0A0D3CFB6	2	ni	Photosynthetic electron transport in photosystem II; protein-chromophore linkage
BoRPN	26S proteasome non-ATPase regulatory subunit 5	RPN5	A0A0D3CJZ8	2	nd	Proteasome assembly; translation
BoRPT3	26S protease regulatory subunit 6B homolog	26Sp6B	A0A0D3AL95	2	ni	Protein catabolic process

(continued)

Table 5.4 (continued)

Gene[a]	Gene product (full name)	Protein (SN)[b]	UniProt accession#	Protein fold change		Gene fold change (Log2)	Gene ontology (biological process)[c]
BoUBC36	Ubiquitin-conjugating enzyme E2 36	UBC36	A0A0D3ARJ5	2	nd		Post replication repair; protein K63-linked ubiquitination
BoUBC7	Ubiquitin-conjugating enzyme E2 7	UBC7	A0A0D3ECQ3	2	nd		Protein ubiquitination

"N/A" refers to non-annotated. The positive numbers indicate gene fold change upregulated or increased abundance of gene product and the negative downregulated or decreased abundance. "*Differential gene expression statistically validated*" ($p \leq 0.05$)

nd, non-differential protein (fold change <1.5); ni, non-identified proteins in one of the cultivars; Log2, results presentation of fold change values for gene relative expression evaluated by qRT-PCR; R, resistant; S, susceptible; mt, mitochondrial; chl, chloroplastic—prefix used here for differentiation of gene product localization

[a] *Brassica oleracea* (Bo) gene name based on the homolog in *Arabidopsis thaliana*

[b] Short name of gene product

[c] Gene ontology, summerizing the principal biological process terms showed in database

[d] Gene product identified in enrichment chloroplast simple

14-3-3-GF14 kappa (A0A0D3BET0), known as a metabolism regulator associated with abiotic stress, was identified and can modulate other proteins by facilitating their degradation by ubiquitins (Fuller et al. 2006; Chang et al. 2009; Liu et al. 2017). The ubiquitin pathway is necessary to tag proteins that should be degraded; however, bacterial effectors may also interact with ubiquitin proteasome system (UPS) as a false system protein (Fig. 5.16). The bacterial effectors can be ubiquitinated and degraded by proteasomes; they can also interfere in the system, act as a ubiquitin ligase or inhibit the specific UPS steps (Collins and Brown 2010). Proteins related to ubiquitination showed reduced abundance in the present work (A0A0D3CJZ8; A0A0D3ARJ5, A0A0D3ECQ3, A0A0D3BLH4). In a highly resistant plant, ubiquitination proteins also showed reduced abundance at 24 hai (Ribeiro et al. 2018), which may indicate a negative regulation of this pathway, leading to cell death and consequently resulting in limitation of bacterial growth (Spallek et al. 2009; Marino et al. 2012).

Proteins involved with defense against pathogens were also increased including Ferredoxin–NADP leaf isozyme 1 and 2, chloroplastic (A0A0D3E2R1; A0A0D3DQI2), and annexin (I3Y171). Annexins are members of a well-known family of proteins involved in tolerance against environmental stresses and have been studied in tobacco, cotton, *Brassica*, and *Arabidopsis* plants (Jami et al. 2008; Konopka-Postupolska et al. 2009; Clark et al. 2012; Szalonek et al. 2015). AIG2 (A0A0D3BZV5) is another defense protein, which has not been functionally characterized yet; however, it is known that the corresponding gene is induced in *Arabidopsis* by the avirulent gene avrRpt2 of *Pseudomonas syringae* (Reuber and Ausubel 1996). Defense response is also highly correlated with the levels of phytohormones, such as JA, ethylene (ET), ABA, and cytokinin. JA is important in the plant defense against various stresses. As seen in the model, the indirect activation of JA by the octadecanoid pathway and H_2O_2 accumulation can result in the activation of biotic stress response. The JA pathway can activate other pathways such as the signal transduction pathway, inducing the formation of chemical and physical barriers against pathogen or herbivore attacks (Kazan and Manners 2008). In addition to the lectin proteins identified, plastid lipid-associated protein 2 and 3, chloroplastic (A0A0D3A546; A0A0D3BRT9) were also identified.

Another phytohormone involved in defense signaling is cytokinin which is involved in plant development, cellular differentiation, and senescence (Hwang et al. 2012). It has been reported that high levels of this hormone increased plant immunity (Swartzberg et al. 2008; Choi et al. 2010; Argueso et al. 2012). Some proteins responsive to cytokinins were increased including 50S ribosomal chloroplastic protein (A0A0D3C0C3); binding partner of ACD11 1 isoform X2 (A0A0D3AJE9); and succinate dehydrogenase subunit 5 mitochondrial-like isoform X2 (A0A0D3DK02). Based on these results, it seems that the regulation of proteins responsive to these phytohormones may play an important role in resistance against *Xcc*.

Secondary metabolites also play an important role in plant defense. Chloroplastic plastid-lipid-associated proteins (1, 2, and 3) were identified with increased abundance and are associated with the storage of carotenoids in plants and sequestration of hydrophobic compounds (Ting et al. 1998; Laizet et al. 2004; Leitner-Dagan et al. 2006). Finally, when the initial defense mechanisms are not enough to contain the pathogen, cell death can also be activated. Metacaspase 4 (A0A0D3D1T5), an important protein reported as a participant in the cell death mechanism (Kwon and Hwang 2013), showed increased abundance in both cultivars. qRT-PCR analysis showed that the corresponding gene was upregulated sevenfold in the resistant cultivar and twofold in the susceptible one. Taken together, the model presented can represent a step-by-step process of the defense mechanism in resistant *Brassica* plants, beginning at the recognition of the pathogen, with the activation of important signaling proteins, molecular defense response pathways, and oxidative stress response (steps 1–3 in the model), followed by UPS modulation and repair proteins (step 4), and alteration of metabolic and photosynthetic pathways (steps 3 and 4). The model may also contribute to better understand the molecular responses during the plant–pathogen interaction reflected by the differential abundance of proteins under *Xcc* infection (Santos et al. 2019).

5.8.3 Functional Characterization of Endochitinase Gene in *Brassicas* for Resistance to *Xanthomonas*

Functional genomics and proteomics techniques have been important tools for exploring and understanding plant–pathogen interaction mechanisms. Proteomic studies can provide the link between gene expression and protein abundance and help identify key proteins involved in plant defense and resistance (Kamal et al. 2010a, b; Komatsu and Hossain 2017). Although mass spectrometer sensitivity and software development have improved protein identification, there are still some limitations in the detection of low abundant proteins. One alternative to overcome this problem is the analysis of subcellular proteomes. This strategy can reduce sample complexity and provide the identification of a high amount of additional proteins contributing to a better understanding of the metabolic pathways involved (Stekhoven et al. 2014; Wang and Komatsu 2016). Indeed, analyses of subcellular proteomes have been widely performed and presented a better picture of differential protein abundance under different stress conditions (Peltier et al. 2000; Uberegui et al. 2015). Chloroplasts have an important role in stress response and therefore the study of the chloroplast proteome can bring important contributions for the elucidation of plant defense, especially since this organelle participates actively in plant immune response (Audran et al. 2016). Ribeiro et al. (2018) analyzed *B. oleracea* leaves inoculated with Xcc by 2-DE. Although differential protein spots were detected, the 2-DE technique is highly limited, particularly in the detection of low abundant proteins (Corthals et al. 2000; Wang and Hanash 2003). Therefore, Santos et al. (2019) performed bottom-up proteomics of inoculated leaves at the same time point (24 h after infiltration), in order to understand protein abundance at an early stage of infection. The total leaf proteome was further complemented with the analysis of chloroplast-enriched samples and the expression levels for 30 genes encoding the identified differential proteins were determined by qRT-PCR.

Additionally, one protein was selected for overexpression in *Arabidopsis thaliana* to verify its involvement in resistance to Xcc.

Although the use of resistant cultivars is the main strategy to control this disease, there are limited sources of resistance. Santos et al. (2019) used the LC-MS/MS technique to analyze young cabbage leaves and chloroplast-enriched samples at 24 h after infection by Xcc, using both susceptible (Veloce) and resistant (Astrus) cultivars. A comparison between susceptible Xcc-inoculated plants and the control condition, as well as between resistant Xcc-inoculated plants with the control, was performed, and more than 300 differentially abundant proteins were identified in each comparison. The chloroplast enriched samples contributed with the identification of 600 additional protein species in the resistant interaction and 900 in the susceptible one, which were not detected in total leaf sample. The expression levels for 30 genes encoding the identified differential proteins were observed by qRT-PCR. CHI-B4 like gene, encoding an endochitinase showing a high increased abundance in resistant Xcc-inoculated leaves, was selected for functional validation by overexpression in *Arabidopsis thaliana*. Compared to the wild type (Col-0), transgenic plants were highly resistant to Xcc indicating that CHI-B4 like gene could be an interesting candidate to be used in genetic breeding programs aiming at black rot resistance (Santos et al. 2019).

5.8.4 Functional Characterization of Overexpression of CH1-B4 Like Protein in *Arabidopsis*

Several candidate proteins were identified, potentially involved in the resistance response to *Xcc* (Santos et al. 2019). One of these proteins, the CHI-B4 like protein, showed increased abundance in the proteomic analysis and high gene expression levels. Endochitinase-like proteins are members of the chitinase family that participate in the catabolic process in the cell (Stintzi et al. 1993). The chitinases are also classified as pathogenesis-related proteins (PRs), and endochitinases belong to group 3 of PRs that cleave chitin molecules. In general, studies have related chitinases to plant–fungus interaction (Jalil et al. 2015). *A. thaliana* plants overexpressing cabbage BoCHB4 gene under the control of CaMV 35S promoter were obtained through transformation. The presence of the transgene was confirmed in five homozygote plants, and single copy insertions were observed by Southern blot analysis for each positive event. The transcript level of the transgene was also assessed, and the transgenic lines showed a relative expression level 588-fold higher than the wild type (Col-0). Since *CHIB4* like can also be found in *A. thaliana*, an alignment of the nucleotide sequences of both genes (endogenous and transgenic) was performed and showed 84% identity with the *Arabidopsis* gene, and therefore the expression levels detected were probably related to the transgene and not to the endogenous gene.

A phenotypic evaluation of *A. thaliana* plants overexpressing *BoCHB4* was performed, and at 48 hai the WT plants began to show the first symptoms. The phenotypic analysis of the WT and transgenic lines after inoculation of *Xcc* showed that at 5 days after inoculation (dai), among the surviving plants, the WT replicates were almost totally necrotic and most leaves were dead, while the transgenic plants showed no symptoms even at 15 days after inoculation. According to the disease scoring index based on Meyers et al. (2005), *A. thaliana* WT Col-0 was highly susceptible, showing severe necrosis and leaf death, while the transformed plants were highly resistant, since no symptoms were observed. Based on these results, it was concluded that the overexpression of *BoCHB4* can confer resistance against the bacterial pathogen Xcc. This work provides an important contribution regarding the comprehension of resistance mechanisms and offers candidate genes to be used in genetic breeding programs aiming at the development of more efficient strategies for black rot disease control.

References

Adie BA, Pérez-Pérez J, Pérez-Pérez MM, Godoy M, Sánchez-Serrano JJ, Schmelz EA, Solano R (2007) ABA is an essential signal for plant resistance to pathogens affecting JA biosynthesis and the activation of defenses in *Arabidopsis*. Plant Cell 19(5):1665–1681. https://doi.org/10.1105/tpc.106.048041

Aires A, Dias CSP, Carvalho R, Oliveira MH, Monteiro AA, Simoes MV, Rosa EAS, Bennett RN, Saavedra MJ (2011) Correlations between disease severity, glucosinolate profiles and total phenolics and *Xanthomonas campestris* pv. *campestris* inoculation of different Brassicaceae. Sci Hortic 129:503–510

Anderson JP, Thatcher LF, Singh KB (2005) Plant defense responses: conservation between models and crops. Funct Plant Biol 32:21–34. https://doi.org/10.1071/FP04136

Andreasson E, Ellis B (2010) Convergence and specificity in the *Arabidopsis* MAPK nexus. Trends Plant Sci 15:106–113

Antico CJ, Colon C, Banks T, Ramonell KM (2012) Insights into the role of jasmonic acid-mediated defenses against necrotrophic and biotrophic fungal pathogens. Front Biol 7:48–56. https://doi.org/10.1007/s11515-011-1171-1

Argueso CT, Ferreira FJ, Epple P, To JP, Hutchison CE, Schaller GE, Dangl JL, Kieber JJ (2012) Two-component elements mediate interactions between cytokinin and salicylic acid in plant immunity. PLoS Genet 8:e1002448. https://doi.org/10.1371/journal.pgen.1002448

Aroca R, Ferrante A, Vernieri P, Chrispeels MJ (2006) Drought, abscisic acid and transpiration rate effects on the regulation of PIP aquaporin gene expression and abundance in *Phaseolus vulgaris* plants. Ann Bot 98:1301–1310. https://doi.org/10.1093/aob/mcl219

Asai T, Tena G, Plotnikova J, Willmann MR, Chiu WL, Gomez-Gomez L, Boller T, Ausubel FM, Sheen J (2002) MAP kinase signalling cascade in Arabidopsis innate immunity. Nature 415:977–983

Audenaert K, De Meyer GB, Höfte MM (2002) Abscisic acid determines basal susceptibility of tomato to *Botrytis cinerea* and suppresses salicylic acid-dependent signaling mechanisms. Plant Physiol 128:491–501. https://doi.org/10.1104/pp.010605

Audran C, Serrano I, Rivas S (2016) Chloroplasts at work during plant innate immunity. J Exp Bot 67:3845–3854. https://doi.org/10.1093/jxb/erw088

Bains PS, Tewari JP (1985) Purification and properties of the phytoalexin(s) produced by *Alternaria brassicae*. Phytopathology 75:1298. (Abstr)

Bashi ZD, Rimmer SR, Khachatourians GG, Hegedus DD (2013) *Brassica napus* polygalacturonase inhibitor proteins inhibit *Sclerotinia sclerotiorum* polygalacturonase enzymatic and necrotizing activities and delay symptoms in transgenic plants. Can J Microbiol 59:79–86

Becker MG, Zhang X, Walker PL, Wan JC, Millar JL, Khan D, Granger MJ, Cavers JD, Chan AC, Fernando DW, Belmonte MF (2017) Transcriptome analysis of the *Brassica napus*-

Leptosphaeria maculans pathosystem identifies receptor, signalling and structural genes underlying plant resistance. Plant J 90(3):573–586. https://doi.org/10.1111/tpj.13514

Becker MG, Haddadi P, Wan J, Adam L, Walker P, Larkan NJ, Daayf F, Borhan MH, Belmonte MF (2019) Transcriptome analysis of *Rlm2*-mediated host immunity in the *Brassica napus–Leptosphaeria maculans* pathosystem. Mol Plant-Microbe Interact 32(8):1001–1012. https://doi.org/10.1094/MPMI-01-19-0028-R

Bennett RN, Wallsgrove RM (1994) Tansley review no. 72. Secondary metabolites in plant defence mechanisms. New Phytol 127:617–633

Berardini TZ, Reiser L, Li D, Mezheritsky Y, Muller R, Strait E, Huala E (2015) The *Arabidopsis* information resource: making and mining the "gold standard" annotated reference plant genome. Genesis 53:474

Bertinetti C, Ugalde RA (1996) Studies on the response of carrot cells to a *Sclerotinia sclerotiorum* elicitor: induction of the expression of an extracellular glycoprotein mRNA. Mol Plant-Microbe Interact 9(7):658–663. https://doi.org/10.1094/MPMI-9-0658

Berto P, Commenil P, Belingheri L, Dehorter B (1999) Occurrence of a lipase in spores of Alternaria brassicicola with a crucial role in the infection of cauliflower leaves. FEMS Microbiol Lett 180:183–189

Billon-Grand G, Rascle C, Droux M, Rollins JA, Poussereau N (2012) pH modulation differs during sunflower cotyledon colonization by the two closely related necrotrophic fungi *Botrytiscinerea* and *Sclerotinia sclerotiorum*. Mol Plant Pathol 13(6):568–578. https://doi.org/10.1111/j.1364-3703.2011.00772.x

Birkenbihl RP, Diezel C, Somssich IE (2012) *Arabidopsis* WRKY33 is a key transcriptional regulator of hormonal and metabolic responses toward Botrytis cinerea infection. Plant Physiol 159(1): 266–285. https://doi.org/10.1104/pp.111.192641

Brader G, Mikkelsen MD, Halkier BA, Palva ET (2006) Altering glucosinolate profiles modulates disease resistance in plants. Plant J 46(5):758–767. https://doi.org/10.1111/j.1365-313X.2006.02743.x

Browne LM, Conn KL, Ayer WA, Tewari JP (1991) The camalexines: new phytoalexins produced in the leaves of *Camelina sativa*. Tetrahdron 47:3909–3914

Buxdorf K, Yaffe H, Barda O, Levy M (2013) The effects of glucosinolates and their breakdown products on necrotrophic fungi. PLoS One 8:e70771

Calmes B, N'Guyen G, Dumur J, Brisach CA, Campion C, Iacomi B, Pigne S, Dias E, Macherel D, Guillemette T, Simoneau P (2015) Glucosinolate derived isothiocyanates impact mitochondrial function in fungal cells and elicit an oxidative stress response necessary for growth recovery. Front Plant Sci 6:414. https://doi.org/10.3389/fpls.2015.00414

Cao JY, Xu YP, Cai XZ (2016a) TMT-based quantitative proteomics analyses reveal novel defense mechanisms of Brassica napus against the devastating necrotrophic pathogen Sclerotinia sclerotiorum. J Proteome 143:265–277. https://doi.org/10.1016/j.jprot.2016.03.006

Cao JY, Xu YP, Zhao L, Li SS, Cai XZ (2016b) Tight regulation of the interaction between Brassica napus and Sclerotinia sclerotiorum at the microRNA level. Plant Mol Biol 92(1–2):39–55. https://doi.org/10.1007/s11103-016-0494-3

Chang IF, Curran A, Woolsey R, Quilici D, Cushman JC, Mittler R et al (2009) Proteomic profiling of tandem affinity purified 14-3-3 protein complexes in *Arabidopsis thaliana*. Proteomics 9:2967–2985. https://doi.org/10.1002/pmic.200800445

Chavan V, Bhargava S, Kamble A (2013) Temporal modulation of oxidant and antioxidative responses in *Brassica carinata* during β-aminobutyric acid-induced resistance against *Alternaria brassicae*. Physiol Mol Plant Pathol 83:35–39. https://doi.org/10.1016/j.pmpp.2013.03.002

Cheng F, Liu S, Wu J, Fang L, Sun S, Liu B, Li P, Hua W, Wang X (2011) BRAD, the genetics and genomics database for *Brassica* plants. BMC Plant Biol 11:136. https://doi.org/10.1186/1471-2229-11-136

Choi HW, Lee DH, Hwang BK (2009) The pepper calmodulin gene *CaCaM1* is involved in reactive oxygen species and nitric oxide generation required for cell death and the defense response. Mol Plant-Microbe Interact 22:1389–1400

Choi J, Huh SU, Kojima M, Sakakibara H, Paek KH, Hwang I (2010) The cytokinin-activated transcription factor ARR2 promotes plant immunity via TGA3/NPR1-dependent salicylic acid signaling in *Arabidopsis*. Dev Cell 19:284–295. https://doi.org/10.1016/j.devcel.2010.07.011

Clark GB, Morgan RO, Fernandez MP, Roux SJ (2012) Evolutionary adaptation of plant annexins has diversified their molecular structures, interactions and functional roles. New Phytol 196:695–712. https://doi.org/10.1111/j.1469-8137.2012.04308.x

Collins CA, Brown EJ (2010) Cytosol as battleground: ubiquitin as a weapon for both host and pathogen. Trends Cell Biol 20:205–213. https://doi.org/10.1016/j.tcb.2010.01.002

Conn KL, Tewari JP (1986) Hypersensitive reaction induced by *Alternaria brassicae* in *Eruca sativa*, an oil yielding crucifer. Can J Plant Pathol 8:348

Conn KL, Tewari JP (1989) Interactions of *Alternaria brassicae* conidia with leaf epicuticular wax of canola. Mycol Res 93:240–242. https://doi.org/10.1016/S0953-7562(89)80126-1

Conn KL, Tewari JP, Dahiya JS (1987) Resistance to *Alternaria brassicae* and phytoalexin-elicitation in rapeseed and some other crucifers. Phytopathology 77:1703. (Abstr)

Conn KL, Tewari JP, Dahiya JS (1988) Resistance to *Alternaria brassicae* and phytoalexins-elicitation in rapeseed and other crucifers. Plant Sci 56:21–25. https://doi.org/10.1016/0168-9452(88)90180-X

Conn KL, Jejelowa OA, Tewari JP, Bains PS (1991) *Camelina sativa* phytoalexins (Calmexin and methoxy camalexin) provide resistance against *Alternaria brassicae*. Can J Plant Pathol 13: Abstr 274

Cooke DEL, Jenkins PD, Lewis DM (1997) Production of phytotoxic spore germination liquids by *Alternaria brassicae* and *A. brassicicola* and their effect on species of the family Brassicaceae. Ann Appl Biol 131:413–426

Corthals GL, Wasinger VC, Hochstrasser DF, Sanchez JC (2000) The dynamic range of protein expression: a challenge for proteomic research. Electrophoresis 21:1104–1115. https://doi.org/10.1002/(SICI)1522-2683(20000401)21:6<1104::AID-ELPS1104>3.0.CO;2-C

Couto D, Zipfel C (2016) Regulation of pattern recognition receptor signalling in plants. Nat Rev Immunol 16:537–552. https://doi.org/10.1038/nri.2016.77

Cramer RA, La Rota CM, Cho Y, Thon M, Craven KD, Knudson DL, Mitchell TK, Lawrence CB (2006) Bioinformatic analysis of expressed sequence tags derived from a compatible *Alternaria brassicicola–Brassica oleracea* interaction. Mol Plant Pathol 7(2):113–124. https://doi.org/10.1111/J.1364-3703.2006.00324.X

Cristna M, Rodriguez S, Pitersen M, Mundy J (2010) Mitogen activated protein kinase signaling in plant. Annu Rev Plant Biol 61:621–649

D'Auria JC, Gershenzon J (2005) The secondary metabolism of *Arabidopsis thaliana*: growing like a weed. Curr Opin Plant Biol 8:308–316

Dangl JL, Dietrich RA, Richberg MH (1996) Death don't have no mercy: cell death programs in plant-microbe interactions. Plant Cell 8(10):1793–1807. https://doi.org/10.1105/tpc.8.10.1793

Desclos-Theveniau M, Arnaud D, Huang TY, Lin GJ, Chen WY, Lin YC, Zimmerli L (2012) The *Arabidopsis* lectin receptor kinase LecRK-V.5 represses stomatal immunity induced by *Pseudomonas syringae* pv. *tomato* DC3000. PLoS Pathog 8:e1002513. https://doi.org/10.1371/journal.ppat.1002513

Desikan R, Hancock JT, Ichimura K, Shinozaki K, Neill SJ (2001) Harpin induces activation of the *Arabidopsis* mitogen-activated protein kinases AtMPK4 and AtMPK6. Plant Physiol 126:1579–1587

Devys M, Barbier M, Loiselet I, Rouxel T, Sarniguet A, Kollmann A, Bousquet JF (1988) Brassilexin a noval Sulphur containing phytoalexin from *Brassica juncea* L. Crucif Tetrahed Lett 29(49):6447–6448

Dhingra HR, Kiran MN, Sangwan MS (2004) Effect of culture filtrate of *Alternaria brassicae* on the activities of oxidative enzymes in calli of *Brassica* species. J Mycol Plant Pathol 34:110–112

Dickman MB, Park YK, Oltersdorf T, Li W, Clemente T, French R (2001) Abrogation of disease development in plants expressing animal antiapoptotic genes. Proc Natl Acad Sci U S A 98 (12):6957–6962. https://doi.org/10.1073/pnas.091108998

Ding Y, Mei J, Wu Q, Xiong Z, Li Y, Shao C, Wang L, Qian W (2019) Synchronous improvement of subgenomes in allopolyploid: a case of Sclerotinia resistance improvement in Brassica napus. Mol Breed 39:10

Ding LN, Li M, Guo XJ, Tang MQ, Cao J, Wang Z, Liu R, Zhu KM, Guo L, Liu SY, Tan XL (2020) *Arabidopsis* GDSL1 over expression enhances rapeseed *Sclerotinia sclerotiorum* resistance and the functional identification of its homolog in *Brassica napus*. Plant Biotechnol J 18:1255–1270. https://doi.org/10.1111/pbi.13289

Dixon RA, Harrison MJ, Lamb CJ (1994) Early events in the activation of plant defense responses. Annu Rev Phytopathol 32:479–501

Dong X, Yi H, Han CT, Nou IS, Hur Y (2016) GDSL esterase/lipase genes in *Brassica rapa* L.: genome-wide identification and expression analysis. Mol Gen Genomics 291:531–542

Dong C, He F, Berkowitz O, Liu J, Cao P, Tang M, Shi H, Wang W, Li Q, Shen Z (2018) Alternative splicing plays a critical role in maintaining mineral nutrient homeostasis in Rice (*Oryza sativa*). Plant Cell 30:2267–2285

Drerup MM, Schlücking K, Hashimoto K, Manishankar P, Steinhorst L, Kuchitsu K, Kudla J (2013) The Calcineurin B-like calcium sensors CBL1 and CBL9 together with their interacting protein kinase CIPK26 regulate the *Arabidopsis* NADPH oxidase RBOHF. Mol Plant 6:559–569

Dubiella U, Seybold H, Durian G, Komander E, Lassig R, Witte CP, Schulze WX, Romeis T (2013) Calcium-dependent protein kinase/NADPH oxidase activation circuit is required for rapid defense signal propagation. Proc Natl Acad Sci U S A 110:8744–8749

Dufour V, Stahl M, Baysse C (2015) The antibacterial properties of isothiocyanates. Microbiol SGM 161:229–243

Eckardt NA (2011) BIK1 function in plant growth and defense signaling. Plant Cell 23(8):2806. https://doi.org/10.1105/tpc.111.230811

Evans N, Mc Roberts N, Hill RA, Marshall G (1996) Phytotoxin production by *Alternaria linicola* and phytoalexin production by linseed host. Ann Appl Biol 129:415–431

Eynck C, Koopmann B, Karlovsky P, von Tiedemann A (2009) Internal resistance in winter oilseed rape inhibits systemic spread of the vascular pathogen *Verticillium longisporum*. Phytopathology 99:803–811

Fatima U, Bhorali P, Borah S, Senthil-Kumar M (2019) Perspectives on the utilization of resistance mechanisms from host and nonhost plants for durable protection of *Brassica* crops against Alternaria blight. Peer J 7:e7486. https://doi.org/10.7717/peerj.7486

Ferrari S, Savatin DV, Sicilia F, Gramegna G, Cervone F, Lorenzo GD (2013) Oligogalacturonides: plant damage-associated molecular patterns and regulators of growth and development. Front Plant Sci 4:49. https://doi.org/10.3389/fpls.2013.00049

Fill BK, Petersen K, Petersen M, Mundy J (2009) Gene regulation by MAP kinase cascades. Curr Opin Plant Biol 12:615–621

Flexas J, Ortuno MF, Ribas-Carbo M, Diaz-Espejo A, Florez-Sarasa ID, Medrano H (2007) Mesophyll conductance to CO_2 in *Arabidopsis thaliana*. New Phytol 175:501–511. https://doi.org/10.1111/j.1469-8137.2007.02111.x

Foley RC, Gleason CA, Anderson JP, Hamann T, Singh KB (2013) Genetic and genomic analysis of *Rhizoctonia solani* interactions with *Arabidopsis*; evidence of resistance mediated through NADPH oxidases. PLoS One 8(2):e56814. https://doi.org/10.1371/journal.pone.0056814

Fonseca JP, Mysore KS (2018) Genes involved in nonhost disease resistance as a key to engineer durable resistance in crops. Plant Sci 279:108–116

Freidit Frey N, Garcia AV, Bigeard J, Zaag R, Bueso E, Garmier M, Pateyron S, de Tauzia-Moreau ML, Brunaud V, Balzergue S, Colcombet J, Aubourg S, Martin-Magniette ML, Hirt H (2014) Functional analysis of *Arabidopsis* immune-related MAPKs uncovers a role for MPK3 as negative regulator of inducible defenses. Genome Biol 15:R87. https://doi.org/10.1186/gb-2014-15-6-r87

Frerigmann H, Pislewska-Bednarek M, Sanchez-Vallet A, Molina A, Glawischnig E, Gigolashvili T, Bednarek P (2016) Regulation of pathogen-triggered tryptophan metabolism in *Arabidopsis thaliana* by MYB transcription factors and indole glucosinolate conversion products. Mol Plant 9:682–695

Frias M, Gonzalez C, Brito N (2011) BcSpl1, a cerato-platanin family protein, contributes to *Botrytiscinerea* virulence and elicits the hypersensitive response in the host. New Phytol 192 (2):483–495. https://doi.org/10.1111/j.1469-8137.2011.03802.x

Fuller B, Stevens SM Jr, Sehnke PC, Ferl RJ (2006) Proteomic analysis of the 14-3-3 family in *Arabidopsis*. Proteomics 6:3050–3059. https://doi.org/10.1002/pmic.200500729

Garg S, Dhawan K, Chawla HKL, Nainawatee HS (1999) *Alternaria brassicae* induced changes in title activity of cell wall degrading enzymes in leaves of *Brassica juncea*. Biol Plant 30:387–392

Gaur M, Tiwari A, Chauhan RP, Pandey D, Kumar A (2018) Molecular modeling, docking and protein-protein interaction analysis of MAPK signaling cascade involved in Camalexin biosynthesis in *Brassica rapa*. Bioinformation 14(4):145–152

Giri P, Taj G, Kumar A (2013) Comparison of artificial inoculation methods for studying pathogenesis of *Alternaria brassicae* (Berk.) Sacc. on *Brassica juncea* (L.) Czern.(Indian mustard). Afr J Biotechnol 12:2422–2426

Gloer JB, Poch GK, Short DM, Mc Closkey DV (1988) Structure of Brassicicolin A: isocyanide antibiotic from the phyloplane fungus *Alternaria brassicicola*. J Org Chem 53:3758–3761

Goodman RN, Novacky AJ (1994) The hypersensitive response reaction in plants to pathogens, a resistance phenomenon. Amer Phytopathol Sco Press, St Paul, MN

Gravino M, Savatin DV, Macone A, Lorenzo GD (2015) Ethylene production in *Botrytis cinerea*- and oligogalacturonide-induced immunity requires calciumdependent protein kinases. Plant J 84:1073–1086. https://doi.org/10.1111/tpj.13057

Gross H (2007) Strategies to unravel the function of orphan biosynthesis pathways: recent examples and future prospects. Appl Microbiol Biotechnol 75:267–277

Gu J, Li W, Wang S, Zhang X, Coules A, Ding G, Xu F, Ren J, Lu C, Shi L (2019) Differential alternative splicing genes in response to boron deficiency in *Brassica napus*. Genes 10:224

Guimaraes RL, Stotz HU (2004) Oxalate production by *Sclerotinia sclerotiorum* deregulates guard cells during infection. Plant Physiol 136(3):3703–3711. https://doi.org/10.1104/pp.104.049650

Guo X, Stotz HU (2007) Defense against *Sclerotinia sclerotiorum* in *Arabidopsis* is dependent on jasmonic acid, salicylic acid, and ethylene signaling. Mol Plant-Microbe Interact 20 (11):1384–1395. https://doi.org/10.1094/MPMI-20-11-1384

Gupta V (1995) Changes in activities of oxidative enzymes and polypeptide patterns of *Brassica juncea* in relation to Alternaria blight. MSc thesis, HAU, Hisar

Gupta SK, Kumar P, Yadav TP, Saharan GS (1984) Changes in phenolic compounds, sugars and total nitrogen in relation to Alternaria leaf blight in Indian mustard. Haryana Agric Univ J Res 14:535–537

Gupta SK, Gupta PP, Yadava TP (1987a) Leaf surface constituents of *Brassica* species in relation to Alternaria leaf blMht (*Alternaria brassicae* (Berk.) Sacco and *A. brassicicola* Schw. Wilts.). In: Proc.7 Int. Rapeseed Congr, Poznan, Poland, pp 1241–1247

Gupta SK, Gupta PP, Yadava TP (1987b) Leaf surface constituents of *Brassica* species in relation to Alternaria leaf blight (*Alternaria brassicae* (Berk.) Sacc and *A. brassicicola* Schw. Wilts.). In: 7th Intl Rapeseed Congr, Poznan, Poland, p 252

Gupta SK, Gupta PP, Yadava TP (1987c) Leaf surface constituents of mustard genotypes in relation to Alternaria leaf blight disease. Indian J Mycol Plant Pathol 17:332–334

Gupta SK, Gupta PP, Yadav TP, Kaushik CD (1990) Metabolic changes in mustard due to *Alternaria* leaf blight. Indian Phytopathol 43:64–69

Halkier BA, Gershenzon J (2006) Biology and biochemistry of glucosinolates. Annu Rev Plant Biol 57:303–333. https://doi.org/10.1146/annurev.arplant.57.032905.105228

Hamel LP, Beaudoin N (2010) Chit oligosaccharide sensing and downstream signaling: contrasted outcomes in pathogenic and beneficial plant-microbe interactions. Planta 232:787–806

Hamel LP, Sheen J, Seguin A (2014) Ancient signals: comparative genomics of green plant CDPKs. Trends Plant Sci 19(2):79–89. https://doi.org/10.1016/j.tplants.2013.10.009

Hamiduzzaman M, Jakab G, Barnavon L, Neuhaus J, Mauch-Mani B (2005) β-aminobutyric acid-induced resistance against downy mildew in grapevine acts through the potentiation of callose formation and jasmonic acid signaling. Mol Plant-Microbe Interact 18:819–829

Hanschen FS, Pfitzmann M, Witzel K, Stuetzel H, Schreiner M, Zrenner R (2018) Differences in the enzymatic hydrolysis of glucosinolates increase the defense metabolite diversity in 19 *Arabidopsis thaliana* accessions. Plant Physiol Biochem 124:126–135

Hartmann L, Pedrotti L, Weiste C, Fekete A, Schierstaedt J, Gottler J, Kempa S, Krischke M, Dietrich K, Mueller MJ, Vicente-Carbajosa J, Hanson J, Droge-Laser W (2015) Crosstalk between two bZIP signaling pathways orchestrates salt-induced metabolic reprogramming in *Arabidopsis* roots. Plant Cell 27:2244–2260

Hu X, Jiang M, Zhang J, Lin F, Tan M (2007) Calcium-calmodulin is required for abscisic acid-induced antioxidant defense and functions both upstream and downstream of H_2O_2 production in leaves of maize (*Zea mays*) plants. New Phytol 173:27–38

Hwang SF, Strelkov SE, Feng J, Gossen BD, Howard RJ (2012) *Plasmodiophora brassicae*: a review of an emerging pathogen of the Canadian canola (*Brassica napus*) crop. Mol Plant Pathol 13:105–113. https://doi.org/10.1111/j.1364-3703.2011.00729.x

Imam J, Singh PK, Shukla P (2016) Plant microbe interactions in post genomic era: perspectives and applications. Front Microbiol 7:1488

Iven T, Konig S, Singh S, Braus-Stromeyer SA, Bischoff M, Tietze LF, Braus GH, Lipka V, Feussner I, Droge-Laser W (2012) Transcriptional activation and production of tryptophan-derived secondary metabolites in *Arabidopsis* roots contributes to the defense against the fungal vascular pathogen *Verticillium longisporum*. Mol Plant 5:1389–1402. https://doi.org/10.1093/mp/sss044

Jalil SU, Mishra M, Ansari MI (2015) Current view on chitinase for plant defense. Trends Biosci 8:6733–6743

Jami SK, Clark GB, Turlapati SA, Handley C, Roux SJ, Kirti PB (2008) Ectopic expression of an annexin from *Brassica juncea* confers tolerance to abiotic and biotic stress treatments in transgenic tobacco. Plant Physiol Biochem 46:1019–1030

Jejelowo OA, Conn KL, Tewari JP (1991) Relationship between conidial concentration, germling growth, and phytoalexin production by *Camelina sativa* leaves inoculated with *Alternaria brassicae*. Mycol Res 95:928–934. https://doi.org/10.1016/S0953-7562(09)80089-0

Jennings DB, Ehrenshaft M, Pharr DM, Williamson JD (1998) Roles for mannitol and mannitol dehydrogenase in active oxygen-mediated plant defense. Proc Natl Acad Sci U S A 95:15129–15133

Jennings DB, Daub ME, Pharr DM, Williamson JD (2002) Constitutive expression of a celery mannitol dehydrogenase in tobacco enhances resistance to the mannitol-secreting fungal pathogen *Alternaria alternata*. Plant J 32:41–49

Jiang CJ, Shimono M, Sugano S, Kojima M, Yazawa K, Yoshida R, Inoue H, Hayashi N, Sakakibara H, Takatsuji H (2010) Abscisic acid interacts antagonistically with salicylic acid signaling pathway in rice-Magnaporthe grisea interaction. Mol Plant-Microbe Interact 23:791–798. https://doi.org/10.1094/MPMI-23-6-0791

Jimenez LD, Ayer WA, Tewari JP (1997) Phytoalexins produced in the leaves of *Capsellabursa-pastoris* (shepherd's purse). Phytoprotection 78:99–103

Johansson ON, Fantozzi E, Fahlberg P, Nilsson AK, Buhot N, Tor M, Andersson MX (2014) Role of the penetration-resistance genes PEN1, PEN2 and PEN3 in the hypersensitive response and race-specific resistance in *Arabidopsis thaliana*. Plant J 79:466–476

Jones JD, Dangl JL (2006) The plant immune system. Nature 444:323–329

Joshi RK, Megha S, Rahman MH, Basu U, Kav NNA (2016) Global study of transcriptome dynamics in canola (Brassica napus L.) responsive to *Sclerotinia sclerotiorum* infection using RNA-Seq. Gene 590:57–67

Kabbage M, Williams B, Dickman MB (2013) Cell death control: the interplay of apoptosis and autophagy in the pathogenicity of *Sclerotinia sclerotiorum*. PLoS Pathog 9(4):e1003287. https://doi.org/10.1371/journal.ppat.1003287

Kabbage M, Yarden O, Dickman MB (2015) Pathogenic attributes of *Sclerotinia sclerotiorum*: switching from a biotrophic to necrotrophic lifestyle. Plant Sci 233:53–60

Kagan IA, Hammerschmidt R (2002) *Arabidopsis* ecotype variability in camalexin production and reaction to infection by *Alternaria brassicicola*. J Chem Ecol 28:2121–2140

Kaliff M, Staal J, Myrenas M, Dixelius C (2007) ABA is required for *Leptosphaeria maculans* resistance *via* ABI1- and ABI4-dependent signaling. Mol Plant-Microbe Interact 20 (4):335–345. https://doi.org/10.1094/MPMI-20-4-0335

Kamal AHM, Kim KH, Shin KH, Choi JS, Baik BK, Tsujimoto H, Heo HY, Park CS, Woo SH (2010a) Abiotic stress responsive proteins of wheat grain determined using proteomics technique. Aust J Crop Sci 4:196–208

Kamal AHM, Kim KH, Shin KH, Kim DE, Oh MW, Choi JS, Hirano H, Heo HY, Woo SH (2010b) Proteomics-based dissection of biotic stress responsive proteins in bread wheat (*Triticum aestivum* L.). Afr J Biotechnol 9:7239–7255

Kaman-Tooth E, Danko T, Gullner G, Bozso Z, Palkovics L, Pogany M (2019) Contribution of cell wall peroxidase-and NADPH oxidase-derived reactive oxygen species to *Alternaria brassicicola*-induced oxidative burst in *Arabidopsis*. Mol Plant Pathol 20(4):485–499

Kamble A, Koopmanna B, von Tiedemann A (2013) Induced resistance to *Verticillium longisporum* in *Brassica napus* by b-aminobutyric acid. Plant Pathol 62:552–561. https://doi.org/10.1111/j.1365-3059.2012.02669.x

Kanrar S, Venkateswari JC, Kirti PB, Chopra VL (2002) Transgenic expression of hevein, the rubber tree lectin, in Indian mustard confers protection against *Alternaria brassicae*. Plant Sci 162:441–448

Kawahara Y, Oono Y, Kanamori H, Marsumoto T, Itoh T, Minami E (2012) Simultaneous RNA-Seq analysis of a mixed transcriptome of rice and blast fungus interaction. PLoS One 7: e49423

Kazan K, Manners JM (2008) Jasmonate signaling: toward an integrated view. Plant Physiol 146:1459–1468. https://doi.org/10.1104/pp.107.115717

Kidokoro S, Yoneda K, Takasaki H, Takahashi F, Shinozaki K, Yamaguchi-Shinozakia K (2017) Different cold-signaling pathways function in the responses to rapid and gradual decreases in temperature. Plant Cell 29(4):760–774. https://doi.org/10.1105/tpc.16.00669

Kim KS, Min JY, Dickman MB (2008) Oxalic acid is an elicitor of plant programmed cell death during Sclerotinia sclerotiorum disease development. Mol Plant-Microbe Interact 21 (5):605–612. https://doi.org/10.1094/MPMI-21-5-0605

Kim S, Song YH, Lee JY, Choi SR, Dhandapani V, Jang CS, Lim YP, Han T (2011) Identification of the BrRHP1 locus that confers resistance to downy mildew in Chinese cabbage (*Brassica rapa* ssp. pekinensis) and development of linked molecular markers. Theor Appl Genet 123:1183–1192. https://doi.org/10.1007/s00122-011-1658-9

Kim J, Kang WH, Yang HB, Park S, Jang C, Yu HJ, Kang BC (2013) Identification of a broad spectrum recessive gene in *Brassica rapa* and molecular analysis of the eIF4E gene family to develop molecular markers. Mol Breed 32:385–398

Kim H, O'Connell R, Maekawa-Yoshikawa M, Uemura T, Neumann U, Schulze-Lefert P (2014) The powdery mildew resistance protein RPW8.2 is carried on VAMP721/722 vesicles to the extrahaustorial membrane of haustorial complexes. Plant J 79:835–847

Kiran DHR, Mehta N, Sangwan MS (2003) Effect of culture filtrate of *Alternaria brassicae* on biochemical constituents of calli of Brassicas. J Mycol Plant Pathol 33:51–55

Kiran DHR, Mehta N, Sangwan MS (2004) Effect of culture filtrate of *Alternaria brassicae* on the activities of oxidative enzymes in calli of *Brassica* species. J Mycol Plant Pathol 34:110–112

Komatsu S, Hossain Z (2017) *Preface—plant proteomic research*. Multidisciplinary Digital Publishing Institute, Basel

Konig S, Feussner K, Kaever A, Landesfeind M, Thurow C, Karlovsky P, Gatz C, Polle A, Feussner I (2014) Soluble phenylpropanoids are involved in the defense response of *Arabidopsis* against *Verticillium longisporum*. New Phytol 202:823–837. https://doi.org/10.1111/nph.12709

Konopka-Postupolska D, Clark G, Goch G, Debski J, Floras K, Cantero A, Fijolek B, Roux S, Hennig J (2009) The role of annexin 1 in drought stress in *Arabidopsis*. Plant Physiol 150:1394–1410. https://doi.org/10.1104/pp.109.135228

Kulshreshtha BM, Chauhan RKS (1985) Production of phytoalexins in the roots of radish (*Raphanus sativus*) after inoculation with *Alternaria alternata*. Indian Phytopathol 38:286–291

Kunkel BN, Brooks DM (2002) Cross talk between signaling pathways in pathogen defense. Curr Opin Plant Biol 5(4):325–331. https://doi.org/10.1016/S1369-5266(02)00275-3

Kwon SI, Hwang DJ (2013) Expression analysis of the metacaspase gene family in Arabidopsis. J Plant Biol 56:391–398. https://doi.org/10.1007/s12374-013-0290-4

Laizet Y, Pontier D, Mache R, Kuntz M (2004) Subfamily organization and phylogenetic origin of genes encoding plastid lipid-associated proteins of the fibrillin type. J Genome Sci Technol 3:19–28. https://doi.org/10.1166/gl.2004.038

Lal NK, Nagalakshmi U, Hurlburt NK, Flores R, Bak A, Sone P, Ma X, Song G, Walley J, Shan L, He P, Casteel C, Fisher AJ, Dinesh-Kumar SP (2018) The receptor-like cytoplasmic kinase bik1 localizes to the nucleus and regulates defense hormone expression during plant innate immunity. Cell Host Microbe 23(4):485. https://doi.org/10.1016/j.chom.2018.03.010

Lamb C, Dixon RA (1997) The oxidative burst in plant disease resistance. Annu Rev Plant Physiol Plant Mol Biol 48:251–275

Lawrence CB, Singh NP, Qiu J, Gardner RG, Tuzun S (2000) Constitutive hydrolytic enzymes are associated with polygenic resistance of tomato to *Alternaria solani* and may function as an elicitor release mechanism. Physiol Mol Plant Pathol 57:211–220

Lee HY, Back K (2016) Mitogen-activated protein kinase pathways are required for melatonin-mediated defense responses in plants. J Pineal Res 60:327–335

Lee YH, Hong JK (2015) Differential defence responses of susceptible and resistant kimchi cabbage cultivars to anthracnose, black spot and black rot diseases. Plant Pathol 64:406–415. https://doi.org/10.1111/ppa.12262

Lee DS, Kim BK, Kwon SJ, Jin HC, Park OK (2009) *Arabidopsis* GDSL lipase 2 plays a role in pathogen defense via negative regulation of auxin signaling. Biochem Biophys Res Commun 379:1038–1042

Lee HA, Lee HY, Seo E, Lee J, Kim SB, Oh S, Choi E, Choi E, Lee SE, Choi D (2017) Current understandings of plant nonhost resistance. Mol Plant-Microbe Interact 30:5–15. https://doi.org/10.1094/MPMI-10-16-0213-CR

Leitner-Dagan Y, Ovadis M, Shklarman E, Elad Y, David DR, Vainstein A (2006) Expression and functional analyses of the plastid lipid-associated protein CHRC suggest its role in chromoplastogenesis and stress. Plant Physiol 142:233–244. https://doi.org/10.1104/pp.106.082404

Levine A, Pennell RI, Alvarez ME, Palmer R, Lamb C (1996) Calcium-mediated apoptosis in a plant hypersensitive disease resistance response. Curr Biol 6:427–437

Lewis LA, Polanski K, de Torres-Zabala M, Jayaraman S, Bowden L, Moore J, Penfold CA, Jenkins DJ, Hill C, Baxter L, Kulasekaran S, Truman W, Littlejohn G, Prusinska J, Mead A, Steinbrenner J, Hickman R, Rand D, Wild DL, Ott S, Buchanan-Wollaston V, Smirnoff N, Beynon J, Denby K, Grant M (2015) Transcriptional dynamics driving mamp-triggered immunity and pathogen effector-mediated immunosuppression in *Arabidopsis* Leaves following infection with pseudomonas syringae pv tomato DC3000. Plant Cell 27(11):3038–3064. https://doi.org/10.1105/tpc.15.00471

Li Y, Chen J, Bennett R, Kiddle G, Wallsgrove R, Huang Y, He Y (1999) Breeding, inheritance, and biochemical studies on *Brassica napus* cv. Zhongyou 821: tolerance to *Sclerotinia sclerotiorum* (stem rot). In: Wratten N, Salisbury PA (eds) Proc 10th. Intl Rapeseed Cong, Canberra, Australia, September 26-29, p 61

Li DG, Shu YA, Li PL, Zhang WB, Ni HW, Cao YS (2013) Synthesis and structure-activity relationships of aliphatic isothiocyanate analogs as antibiotic agents. Med Chem Res 22:3119–3125

Liebrand TW, van den Burg HA, Joosten MH (2014) Two for all: receptor-associated kinases SOBIR1 and BAK1. Trends Plant Sci 19:123–132. https://doi.org/10.1016/j.tplants.2013.10.003

Lim GH, Singhal R, Kachroo A, Kachroo P (2017) Fatty acid–and lipid-mediated signaling in plant defense. Annu Rev Phytopathol 55:505–536

Lincoln JE, Richael C, Overduin B, Smith K, Bostock R, Gilchrist DG (2002) Expression of the antiapoptotic baculovirus p35 gene in tomato blocks programmed cell death and provides broad-spectrum resistance to disease. Proc Natl Acad Sci U S A 99:15217–15221

Liu Y, He C (2016) Regulation of plant reactive oxygen species (ROS) in stress responses: learning from AtRBOHD. Plant Cell Rep 35(5):995–1007. https://doi.org/10.1007/s00299-016-1950-x

Liu LJ, Zhang YY, Tang SY, Zhao QZ, Zhang ZH, Zhang HW, Dong L, Guo H, Xie Q (2010) An efficient system to detect protein ubiquitination by agroinfiltration in *Nicotiana benthamiana*. Plant J 61:893–903. https://doi.org/10.1111/j.1365-313X.2009.04109.x

Liu S, Kracher B, Ziegler J, Birkenbihl RP, Somssich IE (2015) Negative regulation of ABA signaling by WRKY33 is critical for *Arabidopsis* immunity towards Botrytis cinerea 2100. elife 4:e07295. https://doi.org/10.7554/eLife.07295

Liu Z, Jia Y, Ding Y, Shi Y, Li Z, Guo Y, Gong Z, Yang S (2017) Plasma membrane CRPK1-mediated phosphorylation of 14-3-3 proteins induces their nuclear import to fine-tune CBF signaling during cold response. Mol Cell 66:117–128. https://doi.org/10.1016/j.molcel.2017.02.016

Lorito M, Woo SL, Garcia I, Colucci G, Harman GE, Pintor-Toro JA, Filippone E, Muccifora S, Lawrence CB, Zoina A, Tuzun S, Scala F, Fernandez IG (1998) Genes from mycoparasitic fungi as a source for improving plant resistance to fungal pathogens. Proc Natl Acad Sci U S A 95:7860–7865

Lowe RGT, Cassin A, Grandaubert J, Clark BL, Van de Wouw AP, Rouxel T, Howlett BJ (2014) Genomes and transcriptomes of partners in plant-fungal-interactions between canola (*Brassica napus*) and two *Leptosphaeria* species. PLoS One 9(7):e103098. https://doi.org/10.1371/journal.pone.0103098

Luu DT, Maurel C (2005) Aquaporins in a challenging environment: molecular gears for adjusting plant water status. Plant Cell Environ 28:85–96. https://doi.org/10.1111/j.1365-3040.2004.01295.x

Lyu X, Shen C, Fu Y, Xie J, Jiang D, Li G, Cheng J (2016) A small secreted virulence-related protein is essential for the necrotrophic interactions of *Sclerotinia sclerotiorum* with its host plants. PLoS Pathol 12:e1005435

Ma W, Berkowitz GA (2011) Ca2+ conduction by plant cyclic nucleotide gated channels and associated signaling components in pathogen defense signal transduction cascades. New Phytol 190(3):566–572. https://doi.org/10.1111/j.1469-8137.2010.03577.x

Ma Q, Grones P, Robert S (2017) Auxin signaling: a big question to be addressed by small molecules. J Exp Bot 69:313–328. https://doi.org/10.1093/jxb/erx375

Ma JQ, Xu W, Xu F, Lin A, Sun W, Jiang HH, Lu K, Li JN, Wei LJ (2020) Differential alternative splicing genes and isoform regulation networks of rapeseed (*Brassica napus* L.) infected with *Sclerotinia sclerotiorum*. Genes 11:784. https://doi.org/10.3390/genes11070784

Madloo P, Lema M, Francisco M, Soengas P (2019) Role of major glucosinolates in the defense of kale against *Sclerotinia sclerotiorum* and *Xanthomonas campestris* pv. *campestris*. Phytopathology 109(7):1246–1256. https://doi.org/10.1094/PHYTO-09-18-0340-R

Marino D, Peeters N, Rivas S (2012) Ubiquitination during plant immune signaling. Plant Physiol 160:15–27. https://doi.org/10.1104/pp.112.199281

Marmath KK, Giri P, Taj G, Pandey D, Kumar A (2013) Effect of zeatin on the infection process and expression of MAPK-4 during pathogenesis of *Alternaria brassicae* in non-host and host *Brassica* plants. Afr J Biotechnol 12(17):2164–2174. https://doi.org/10.5897/AJB12.2725

Mauch-Mani B, Mauch F (2005) The role of abscissic acid in plant–pathogen interactions. Curr Opin Plant Biol 8:409–414

McRoberts N, Lennard JH (1996) Pathogen behaviour and plant cell reactions in interactions between *Alternaria* species and leaves of host and nonhost plants. Plant Pathol 45(4):742–752

Mei J, Qian L, Disi JO, Yang X, Li Q, Li J, Frauen M, Cai D, Qian W (2011) Identification of resistant sources against *Sclerotinia sclerotiorum* in *Brassica* species with emphasis on *B. oleracea*. Euphytica 177:393–399

Mei J, Ding Y, Lu K, Wei D, Liu Y, Disi JO, Li J, Liu L, Liu S, McKay J, Qian W (2013) Identification of genomic regions involved in resistance against *Sclerotinia sclerotiorum* from wild *Brassica oleracea*. Theor Appl Genet 126:549–556. https://doi.org/10.1007/s00122-012-2000-x

Meng X, Zhang S (2013) MAPK cascades in plant disease resistance signalling. Annu Rev Phytopathol 51:245–266

Mengiste T (2012) Plant immunity to necrotrophs. Annu Rev Phytopathol 50:267–294

Merkouropoulos G, Andreasson E, Hess D, Boller T, Peck SC (2008) An *Arabidopsis* protein phosphorylated in response to microbial elicitation, AtPHOS32, is a substrate of MAP kinases 3 and 6. J Biol Chem 283:10493–10499. https://doi.org/10.1074/jbc.M800735200

Meyers BC, Kaushik S, Nandety RS (2005) Evolving disease resistance genes. Curr Opin Plant Biol 8:129–134. PMID: 15752991

Mine A, Berens ML, Nobori T, Anver S, Fukumoto K, Winkelmuller TM, Takeda A, Becker D, Tsuda K (2017) Pathogen exploitation of an abscisic acid-and jasmonate-inducible MAPK phosphatase and its interception by *Arabidopsis* immunity. Proc Natl Acad Sci 114:7456–7461. https://doi.org/10.1073/pnas.1702613114

Mishra A, Pandey D, Goel A, Kumar A (2010) Molecular cloning and in-silico analysis of functional homologous of hypersensitive response genes induced during pathogenesis of Alternaria Blight in two genotypes of *Brassica*. J Proteom Bioinformatics 3:244–248

Mondal KK, Chatterjee SC, Viswakarma N, Bhattacharya RC, Grover A (2003) Chitinase mediated inhibitory activity of *Brassicas* transgenic on growth of *Alternaria brassicae*. Curr Microbiol 47:171–173. https://doi.org/10.1007/s00284-002-3980-6

Mora MM, Earle ED (2001) Resistance to *Alternaria brassicicola* in transgenic broccoli expressing a *Trichoderma harzianum* endochitinase gene. Mol Breed 8:1–9

Morita S, Azuma M, Aoba T, Satou H, Narisawa K, Hashiba T (2003) Induced systemic resistance of Chinese cabbage to bacterial leaf spot and Alternaria leaf spot by the root endophytic fungus, *Heteroconium chaetospira*. J Gen Plant Pathol 69:71–75

Naranjo MA, Forment J, Roldan M, Serrano R, Vicente O (2006) Overexpression of *Arabidopsis thaliana* LTL1, a salt-induced gene encoding a GDSL-motif lipase, increases salt tolerance in yeast and transgenic plants. Plant Cell Environ 29(10):1890–1900

Naumann TA, Wicklow DT (2013) Chitinase modifying proteins from phylogenetically distinct lineages of *Brassica* pathogens. Physiol Mol Plant Pathol 82:1–9

Nayanakantha NMC, Rawat S, Ali S, Grover A (2016) Differential expression of defense related genes in *Sinapsis alba* and *Brassica juncea* upon the infection of *Alternaria brassicae*. Trop Agric Res 27(3):123–136

Nie J, Yin Z, Li Z, Wu Y, Huang L (2019) A small cysteine-rich protein from two kingdoms of microbes is recognized as a novel pathogen-associated molecular pattern. New Phytol 222 (2):995–1011. https://doi.org/10.1111/nph.15631

Niks RE, Rubiales D (2002) Potentially durable resistance mechanisms in plants to specialized fungal pathogens. Euphytica 124:201–216. https://doi.org/10.1023/A:1015634617334

Nitta Y, Ding P, Zhang Y (2014) Identification of additional MAP kinases activated upon PAMP treatment. Plant Signal Behav 9:e976155. https://doi.org/10.4161/15592324.2014.976155

Novakova M, Sasek V, Dobrev PI, Valentova O, Burketova L (2014) Plant hormones in defense response of Brassica napus to *Sclerotinia sclerotiorum* - reassessing the role of salicylic acid in the interaction with a necrotroph. Plant Physiol Biochem 80:308–317. https://doi.org/10.1016/j.plaphy.2014.04.019

O'Neill LAJ, Bowie AG (2007) The family of five: TIR-domain-containing adaptors in Toll-like receptor signalling. Nat Rev Immunol 7:353–364. https://doi.org/10.1038/nri2079

Oh IS, Park AR, Bae MS, Kwon SJ, Kim YS, Lee JE, Kang NY, Lee S, Cheong H, Park OK (2005) Secretome analysis reveals an *Arabidopsis* lipase involved in defense against *Alternaria brassicicola*. Plant Cell 17:2832–2847

Osbourn AE (1996) Preformed antimicrobial compounds and plant defense against fungal attack. Plant Cell 8:1821–1831. https://doi.org/10.1105/tpc.8.10.1821

Pan Y, Wei J, Yao C, Reng H, Gao Z (2018) SsSm1, a Cerato-platanin family protein, is involved in the hyphal development and pathogenic process of *Sclerotinia sclerotiorum*. Plant Sci 270:37–46. https://doi.org/10.1016/j.plantsci.2018.02.001

Pandolfi V, Neto JRCF, da Silva MD, Amorim LLB, Wanderley-Nogueira AC, de Oliveira Silva RL, Kido EA, Crovella S, Iseppon AMB (2017) Resistance (R) genes: applications and prospects for plant biotechnology and breeding. Curr Protein Pept Sci 18(4):323–334

Parada RY, Oka K, Yamagishi D, Kodama M, Otani H (2007) Destruxin B produced by *Alternaria brassicae* does not induce accessibility of host plants to fungal invasion. Physiol Mol Plant Pathol 71:48

Parihar PS, Prakash O, Punetha H (2012) Investigation on defensive enzymes activity of Brassica juncea genotypes during pathogenesis of Alternaria blight. Nat Sci 10(2):63–68

Pastorczyk M, Bednarek P (2016) The function of glucosinolates and related metabolites in plant innate immunity. In: Kopriva S (ed) Advances in botanical research, glucosinolates, vol 80. Elsevier, London, pp 171–198

Pathak RK, Baunthiyal M, Pandey D, Kumar A (2020) Computational analysis of microarray data of *Arabidopsis* thaliana challenged with *Alternaria brassicicola* for identification of key genes in *Brassica*. J Genet Eng Biotechnol 18:17. https://doi.org/10.1186/s43141-020-00032-y

Paxton JD (1981) Phytoalexins - a working redefinition. J Phytopathol 101:106–109

Pedras MSC (1998) Phytoalexins from crucifers: mimicking or overcoming plant chemical defenses? Recent Res Dev Phytochem 2:259–267

Pedras MSC, Biesenthal CJ (2001) Isolation, structure determination, and phytotoxicity of unusual dioxopiperazines from the phytopathogenic fungus *Phoma lingam*. Phytochemistry 58:905–909

Pedras MSC, Chumala PB (2005) Phomapyrones from blackleg causing phytopathogenic fungi: isolation, structure determination, biosyntheses and biological activity. Phytochemistry 66:81–87

Pedras MSC, Okanga FI (1999) Strategies of cruciferous pathogenic fungi: detoxification of the phytoalexin cyclobrassinin by mimicry. J Agric Food Chem 47:1196–1202

Pedras MSC, Morales VM, Taylor JL (1994) Phomapyrones: three metabolites from the blackleg fungus. Phytochemistry 36:1315–1318

Pedras MSC, Kahn AQ, Taylor JL (1998) The phytoalexin camalexin is not metabolized by *Phoma lingam*, *Alternaria brassicae*, or phytopathogenic bacteria. Plant Sci 139:1–8

Pedras MSC, Zaharia JL, Smith KC, Gai Y, Ward DE (1999) Alternaria blackspot phytotoxins: new strategies for determining specific disease resistance traits. In: Proc of the 10th Intl Rapeseed Cong, Caberra, Australia, 26–29 Sep 1999

Pedras MSC, Biesenthal CJ, Zaharia IL (2000) Comparison of the phytotoxicity of the phytotoxin destruxin B and four natural analogs. Plant Sci 156:185–192

Pedras MSC, Zaharia IL, Gai Y, Zhou Y, Ward DE (2001) In planta sequential hydroxylation and glycosylation of a fungal phytotoxin: avoiding cell death and overcoming the fungal invader. Proc Natl Acad Sci U S A 98:747–752. https://doi.org/10.1073/pnas.98.2.747

Pedras MSC, Chumala PB, Suchy M (2003) Phytoalexins from Thlaspiarvense, a wild crucifer resistant to virulent *Leptosphaeria maculans*: structures, syntheses and antifungal activity. Phytochemistry 64:949–956

Pedras MSC, Zheng Q, Sarma-Mamillapalle Vijay K (2007) The phytoalexins from brassicaceae: structure, biological activity, synthesis and biosynthesis. Nat Prod Commun 2:319–330

Pedras MSC, Chumala PB, Jin W, Islam MS, Hauck DW (2009) The phytopathogenic fungus *Alternaria brassicicola*: phytotoxin production and phytoalexin elicitation. Phytochemistry 70:394–402

Peltier JB, Friso G, Kalume DE, Roepstorff P, Nilsson F, Adamska I et al (2000) Proteomics of the chloroplast: systematic identification and targeting analysis of lumenal and peripheral thylakoid proteins. Plant Cell 12:319–341. https://doi.org/10.1105/tpc.12.3.319

Peng G, Pageau D, Strelkov SE, Lahlali R, Hwang SF, Hynes RK (2013) Assessment of crop rotation, cultivar resistance and *Bacillus subtilis* biofungicide for control of clubroot on canola. Acta Hortic 1005:591–598

Perchepied L, Balague C, Riou C, Claudel-Renard C, Riviere N, Grezes-Besset B, Roby D (2010) Nitric oxide participates in the complex interplay of defense-related signaling pathways controlling disease resistance to *Sclerotinia sclerotiorum* in *Arabidopsis thaliana*. Mol Plant-Microbe Interact 23:846–860. https://doi.org/10.1094/MPMI-23-7-0846

Petersen M, Brodersen P, Naested H, Andreasson E, Lindhart U, Johansen B, Nielsen HB, Lacy M, Austin MJ, Parker JE, Sharma SB, Klessig DF, Martienssen R, Mattsson O, Jensen AB, Mundy J (2000) *Arabidopsis* map kinase 4 negatively regulates systemic acquired resistance. Cell 103 (7):1111–1120. https://doi.org/10.1016/S0092-8674(00)00213-0

Pitzschke A, Schikora A, Hirt H (2009) MAPK cascade signalling networks in plant defense. Curr Opin Plant Biol 12:421–426

Popescu SC, Popescu GV, Bachan S, Zhang Z, Gerstein M, Snyder M, Dinesh-Kumar SP (2008) MAPK target networks in *Arabidopsisthaliana* revealed using functional protein microarrays. Genes Dev 23:80–92. https://doi.org/10.1101/gad.1740009

Popescu SC, Popescu GV, Bachan S, Zhang Z, Gerstein M, Snyder M, Dinesh-Kumar SP (2009a) MAPK target networks in *Arabidopsis thaliana* revealed using functional protein microarrays. Genes Dev 23:80–92

Popescu SC, Popescu GV, Snyder M, Dinesh SP (2009b) Integrated analysis of co-expressed MAP kinase substrates in *Arabidopsis thaliana*. Plant Signal Behav 4:524–527

Punja ZK, Jankins SF (1984) Light and scanning electron microscopic observations of calcium oxalate crystals produced during growth of Sclerotium rolfsii in culture and in infected tissue. Can J Bot 62:2028–2032

Qiao M, Sun J, Liu N, Sun T, Liu G, Han S, Hou C, Wang D (2015) Changes of nitric oxide and its relationship with H2O2 and Ca2+ in defense interactions between wheat and *Puccinia triticina*. PLoS One 10(7):e0132265. https://doi.org/10.1371/journal.pone.0132265

Qiu JL, Fiil BK, Petersen K, Nielsen HB, Botanga CJ, Thorgrimsen S, Palma K, Suarez-Rodriguez MC, Sandbech-Clausen S, Lichota J, Brodersen P, Grasser KD, Mattsson O, Glazebrook J, Mundy J, Petersen M (2008) Arabidopsis MAP kinase 4 regulates gene expression through transcription factor release in the nucleus. EMBO J 27:2214–2221

Rahman M, Mamidi S, Del Rio L, Ross A, Kadir MM, Rahaman MM, Arifuzzaman M (2016) Association mapping in *Brassica napus* (L.) accessions identifies a major QTL for blackleg disease resistance on chromosome A01. Mol Breed 36:90. https://doi.org/10.1007/s11032-016-0513-8

Ranty B, Aldon D, Galaud JP (2006) Plant calmodulins and calmodulin-related proteins: multifaceted relays to decode calcium signals. Plant Signal Behav 1(3):96–104. https://doi.org/10.4161/psb.1.3.2998

Rao DV, Tewari JP (1987) Production of oxalic acid by *Mycena citricolor*, causal agent of the American leaf spot of coffee. Phytopathology 77:780–785

Reuber TL, Ausubel FM (1996) Isolation of *Arabidopsis* genes that differentiate between resistance responses mediated by the RPS2 and RPM1 disease resistance genes. Plant Cell 8:241–249. https://doi.org/10.1105/tpc.8.2.241

Reusche M, Thole K, Janz D, Truskina J, Rindfleisch S, Drübert C, Polle A, Lipka V, Teichmanna T (2012) *Verticillium* infection triggers VASCULAR-RELATED NAC DOMAIN7–dependent de novo xylem formation and enhances drought tolerance in *Arabidopsis*. Plant Cell 24:3823–3837

Ribeiro DG, Cunha GCR, Santos C, Silva LP, Oliveira Neto OB, Labuto LBD, Mehta A (2018) *Brassica oleracea* resistance-related proteins identified at an early stage of black rot disease. Physiol Mol Plant Pathol 104:9–14. https://doi.org/10.1016/j.pmpp.2018.06.002

Rojas CM, Senthil-Kumar M, Tzin V, Mysore KS (2014) Regulation of primary plant metabolism during plant-pathogen interactions and its contribution to plant defense. Front Plant Sci 5:17. https://doi.org/10.3389/fpls.2014.00017

Rossi FR, Gárriz A, Marina M, Romero FM, González ME, Collado IG, Pieckenstain FL (2011) The sesquiterpene botrydial produced by Botrytis cinerea induces the hypersensitive response on plant tissues and its action is modulated by salicylic acid and jasmonic acid signaling. Mol Plant-Microbe Interact 24:888–896

Ryals JA, Neuenschwander UH, Willits MG, Molina A, Steiner HY, Hunt MD (1996) Systemic acquired resistance. Plant Cell 8:1809–1819

Saand MA, Xu YP, Li W, Wang JP, Cai XZ (2015a) Cyclic nucleotide gated channel gene family in tomato: genome-wide identification and functional analyses in disease resistance. Front Plant Sci 6:303. https://doi.org/10.3389/fpls.2015.00303

Saand MA, Xu YP, Munyampundu JP, Li W, Zhang XR, Cai XZ (2015b) Phylogeny and evolution of plant cyclic nucleotide-gated ion channel (CNGC) gene family and functional analyses of tomato CNGCs. DNA Res 22(6):471–483. https://doi.org/10.1093/dnares/dsv029

Saharan GS, Mehta N, Sangwan MS (2003) Nature and mechanism of resistance to Alternaria blight in rapeseed-mustard system. Annu Rev Plant Pathol 2:85–128

Saharan GS, Mehta N, Meena PD (2016) Alternaria blight of crucifers: biology, ecology and disease management, vol 326. Springer, Singapore. ISBN 978-981-10-0019-5

Samanta S, Thakur JK (2015) Importance of Mediator complex in the regulation and integration of diverse signaling pathways in plants. Front Plant Sci 6:757. https://doi.org/10.3389/fpls.2015.00757

Santos C, Nogueira FCS, Domont GB, Fontes W, Prado GS, Habibi P, Santos VO, Oliveira-Neto OB, Grossi-de-Sa MF, Jorrin-Novo JV, Franco OL, Mehta A (2019) Proteomic analysis and functional validation of a *Brassica oleracea* endochitinase involved in resistance to Xanthomonas campestris. Front Plant Sci 10:414. https://doi.org/10.3389/fpls.2019.00414

Sasaki-Sekimoto Y, Jikumaru Y, Obayashi T, Saito H, Masuda S, Kamiya Y, Ohta H, Shirasu K (2013) Basic helix-loop-helix transcription factors JASMONATE-ASSOCIATED MYC2-LIKE1 (JAM1), JAM2, and JAM3 are negative regulators of jasmonate responses in *Arabidopsis*. Plant Physiol 163:291–304

Schenk PM, Kazan K, Wilson I, Anderson JP, Richmond T, Somerville SC, Manners JM (2000) Coordinated plant defense responses in *Arabidopsis* revealed by microarray analysis. Proc Natl Acad Sci U S A 97:11655–11660

Sellam A, Dongo A, Guillemette T, Hudhomme P, Simoneau P (2007) Transcriptional responses to exposure to the brassicaceous defence metabolites camalexin and allyl-isothiocyanate in the necrotrophic fungus *Alternaria brassicicola*. Mol Plant Pathol 8:195–208. https://doi.org/10.1111/j.1364-3703.2007.00387.x

Serrano M, Coluccia F, MarthaTorres FLH, Metraux JP (2015) The cuticle and plant defense to pathogens. Front Plant Sci 5:274

Sharma N, Rahman MH, Strelkov S, Thiagarajah M, Vipan K, Bansal VK, Kav NNV (2007) Proteome-level changes in two *Brassica napus* lines exhibiting differential responses to the fungal pathogen *Alternaria brassicae*. Plant Sci 172:95–110

Sharma BB, Kalia P, Yadava DK, Dingh D, Sharma TR (2016) Genetics and molecular mapping of black rot resistance locus *Xca1bc* on chromosome B-7 in Ethiopian mustard (*Brassica carinata* A. Braun). PLoS One 11:e0152290. https://doi.org/10.1371/journal.pone.0152290

Shigenaga AM, Argueso CT (2016) No hormone to rule them all: interactions of plant hormones during the responses of plants to pathogens. Semin Cell Dev Biol 56:174–178. https://doi.org/10.1016/j.semcdb.2016.06.005

Shinde BA, Dholakia BB, Hussain K, Aharoni A, Giri AP, Kamble AC (2018) WRKY1 acts as a key component improving resistance against *Alternaria solani* in wild tomato, *Solanum arcanum* Peralta. Plant Biotechnol J 16(8):1502–1513

Shirakawa M, Hara-Nishimura I (2018) Specialized vacuoles of myrosin cells: chemical defense strategy in Brassicales plants. Plant Cell Physiol 59:1309–1316

Sood N, Sohal BS (2012) Studies in relation to chitosan and its potential to enhance host resistance in *Brassica juncea* L. against Alternaria blight. Plant Dis Res 27(2):144–148

Sotelo T, Lema M, Soengas P, Cartea ME, Velasco P (2015) In vitro activity of glucosinolates and their degradation products against *Brassica*-pathogenic bacteria and fungi. Appl Environ Microbiol 81:432–440

Sotelo T, Velasco P, Soengas P, Rodriguez VM, Cartea ME (2016) Modification of leaf glucosinolate contents in *Brassica oleracea* by divergent selection and effect on expression of genes controlling glucosinolate pathway. Front Plant Sci 7:1012. https://doi.org/10.3389/fpls.2016.01012

Spallek T, Robatzek S, Gohre V (2009) How microbes utilize host ubiquitination. Cell Microbiol 11:1425–1434. https://doi.org/10.1111/j.1462-5822.2009.01346.x

Stahl E, Bellwon P, Huber S, Schlaeppi K, Bernsdorff F, Vallat-Michel A, Mauch F, Zeier J (2016) Regulatory and functional aspects of indolic metabolism in plant systemic acquired resistance. Mol Plant 9:662–681

Stekhoven DJ, Omasits U, Quebatte M, Dehio C, Ahrens CH (2014) Proteome-wide identification of predominant subcellular protein localizations in a bacterial model organism. J Proteome 99:123–137. https://doi.org/10.1016/j.jprot.2014.01.015

Stintzi A, Heitz T, Prasad V, Wiedemann-Merdinoglu S, Kauffmann S, Geoffroy P, Legrand M, Fritig B (1993) Plant 'pathogenesis-related' proteins and their role in defense against pathogens. Biochimie 75:687–706. https://doi.org/10.1016/0300-9084(93)90100-7

Stotz HU, Sawada Y, Shimada Y, Hirai MY, Sasaki E, Krischke M, Brown PD, Saito K, Kamiya Y (2011) Role of camalexin, indole glucosinolates, and side chain modification of glucosinolate-derived isothiocyanates in defense of Arabidopsis against Sclerotinia sclerotiorum. Plant J 67:81–93

Stotz HU, Mitrousia GK, de Wit PJGM, Fitt BDL (2014) Effector-triggered defense against apoplastic fungal pathogens. Trends Plant Sci 19:491–500

Su J, Zhang M, Zhang L, Sun T, Liu Y, Lukowitz W, Xu J, Zhang S (2017) Regulation of stomatal immunity by interdependent functions of a pathogen-responsive MPK3/MPK6 cascade and abscisic acid. Plant Cell 29:526. https://doi.org/10.1105/tpc.16.00577

Swartzberg D, Kirshner B, Rav-David D, Elad Y, Granot D (2008) *Botrytis cinerea* induces senescence and is inhibited by auto regulated expression of the IPT gene. Eur J Plant Pathol 120:289–297. https://doi.org/10.1007/s10658-007-9217-6

Swiezawska B, Duszyn M, Jaworski K, Szmidt-Jaworska A (2018) Downstream targets of cyclic nucleotides in plants. Front Plant Sci 9:1428. https://doi.org/10.3389/fpls.2018.01428

Szalonek M, Sierpien B, Rymaszewski W, Gieczewska K, Garstka M, Lichocka M, Sass L, Paul K, Vass I, Vankova R, Dobrev P, Szczesny P, Marczewski W, Krusiewicz D, Strzelczyk-Zyta D, Hennig J, Konopka-Postupolska D (2015) Potato annexin STANN1 promotes drought tolerance and mitigates light stress in transgenic *Solanum tuberosum* L. plants. PLoS One 10:e0132683. https://doi.org/10.1371/journal.pone.0132683

Taiz L, Zeiger E, Moller IM, Murphy A (2017) Fisiologia e desenvolvimento vegetal. Artmed Editora, Porto Alegre

Takaichi M, Oeda K (2000) Transgenic carrots with enhanced resistance against two major pathogens, *Erysiphe heraclei* and *Alternaria dauci*. Plant Sci 153:135–144

Takasugi M, Monde K, Katsui N, Shirata A (1988) Novel sulphur containing phytoalexins from the chinese *Brassica campestris* L. spp. *pekinensis* (cruciferae). Bull Chem Soc Japan 61:285–289

Tena G, Boudsocq M, Sheen J (2011) Protein kinase signaling networks in plant innate immunity. Curr Opin Plant Biol 14:519–529

Terras FRG, Eggermont K, Kovaleva V, Raikhel NV, Torrekens S, Van Leuven F, Osborn RW, Kester A, Rees SB, Vanderleyden J, Cammue BPA, Broekaert WF (1995) Small cysteine-rich antifungal proteins from radish. Their role in host defense. Plant Cell 7:573–588

Tewari JP (1991a) Structural and biochemical bases of black spot disease of crucifers. Adv Struct Biol 1:325–349

Tewari JP (1991b) Resistance to *Alternaria brassicae* in crucifers. Bull SROP 14(6):154–161

Tewari JP, Conn KL, Dahiya JS (1987) Resistance to *Alternaria brassicae* in crucifer. In: Proc. 7th IRC, Poznan, vol 48, pp 1085–1089

Tewari JP, Conn KL, Dahiya JS (1988) Dynamic response of rapeseed and other crucifers to *Alternaria brassicae* and *Rhizoctonia solani*. In: Proc. 5th Intl Cong Plant Pathol, Koyoto, p 242

Thomma BPHJ (2003) *Alternaria* spp.: from general saprophyte to specific parasite. Mol Plant Pathol 4(4):225–236. https://doi.org/10.1046/j.1364-3703.2003.00173.x

Thomma BP, Eggermont K, Penninckx IA, Mauch-Mani B, Vogelsang R, Cammue BP, Broekaert WF (1998) Separate jasmonate-dependent and salicylate-dependent defense-response pathways in *Arabidopsis* are essential for resistance to distinct microbial pathogens. Proc Natl Acad Sci U S A 95(25):15107–15111. https://doi.org/10.1073/pnas.95.25.15107

Thomma BP, Eggermont K, Tierens KF, Broekaert WF (1999a) Requirement of functional ethylene-insensitive 2 gene for efficient resistance of Arabidopsis to infection by Botrytis cinerea. Plant Physiol 121:1093–1102

Thomma BPHJ, Nelissen I, Eggermont K, Broekaert WF (1999b) Deficiency in phytoalexin production causes enhanced susceptibility of *Arabidopsis thaliana* to the fungus *Alternaria brassicicola*. Plant J 19:163–171

Thomma BPHJ, Eggermont K, Broekaert WF, Cammue BPA (2000) Disease development of several fungi on *Arabidopsis* can be reduced by treatment with methyl jasmonate. Plant Physiol Biochem 38:421–427

Thomma BP, Penninckx IA, Broekaert WF, Cammue BP (2001) The complexity of disease signaling in Arabidopsis. Curr Opin Immunol 13:63–68. https://doi.org/10.1016/S0952-7915(00)00183-7

Tian S, Wang X, Li P, Wang H, Ji H, Xie J, Qiu Q, Shen D, Dong H (2016) Plant aquaporin AtPIP1;4 links apoplastic H2O2 induction to disease immunity pathways. Plant Physiol 171:1635–1650. https://doi.org/10.1104/pp.15.01237

Tierens FMJ, Thomma BPHJ, Bari RP, Garnier M, Eggermont K, Brouwer M, Penninckx IAMA, Broekaert WF, Cammue BPA (2002) Esa1, an *Arabidopsis* mutant with enhanced susceptibility to a range of necroptrophic fungal pathogens, shows a distorted induction of defense responses by reactive oxygen generated compounds. Plant J 29:131–140

Ting JTL, Wu SSH, Ratnayake C, Huang AHC (1998) Constituents of the tapetosomes and elaioplasts in *Brassica campestris* tapetum and their degradation and retention during microsporogenesis. Plant J 16:541–551. https://doi.org/10.1046/j.1365-313x.1998.00325.x

Ton J, Mauch-Mani B (2004) b-aminobutyric acid-induced resistance against necrotrophic pathogen is based on ABA-dependent priming for callose. Plant J 38:119–130

Ton J, Van Pelt JA, Van Loon LC, Pieterse CM (2002) Differential effectiveness of salicylate-dependent and jasmonate/ethylene-dependent induced resistance in *Arabidopsis*. Mol Plant-Microbe Interact 15:27–34

Tran VT, Braus-Stromeyer SA, Timpner C, Braus GH (2013) Molecular diagnosis to discriminate pathogen and a pathogen species of the hybrid *Verticillium longisporum* on the oilseed crop Brassica napus. Appl Microbiol Biotechnol 97:4467–4483. https://doi.org/10.1007/s00253-012-4530-1

Uberegui E, Hall M, Lorenzo O, Schroder WP, Balsera M (2015) An *Arabidopsis* soluble chloroplast proteomic analysis reveals the participation of the Executer pathway in response to increased light conditions. J Exp Bot 66:2067–2077. https://doi.org/10.1093/jxb/erv018

Updegraff EP, Fang Z, Preuss D (2009) The extracellular lipase EXL4 is required for efficient hydration of *Arabidopsis* pollen. Sex Plant Reprod 22:197–204

Uppalapati SR, Ishiga Y, Doraiswamy V, Bedair M, Mittal S, Chen JH, Nakashima J, Tang YH, Tadege M, Ratet P, Chen RJ, Schultheiss H, Mysore KS (2012) Loss of abaxial leaf epicuticular wax in *Medicago truncatula* irg1/palm1 mutants results in reduced spore differentiation of anthracnose and nonhost rust pathogens. Plant Cell 24:353–370. https://doi.org/10.1105/tpc.111.093104

Velasco P, Lema M, Francisco M, Soengas P, Elena Cartea M (2013) In vivo and in vitro effects of secondary metabolites against *Xanthomonas campestris* pv. *campestris*. Molecules 18:11131–11143

Verkman AS (2013) Aquaporins. Curr Biol 23:R52–R55. https://doi.org/10.1016/j.cub.2012.11.025

Verma PR, Saharan GS (1994) Monograph on *Alternaria* diseases of crucifers. In: Saskatoon Research Centre Technical Bulletin 1994–6E. Agriculture and Agri-Food Canada, Saskatoon, SK

Veronese P, Nakagami H, Bluhm B, Abuqamar S, Chen X, Salmeron J, Dietrich RA, Hirt H, Mengiste T (2006) The membrane-anchored BOTRYTIS-INDUCED KINASE1 plays distinct roles in *Arabidopsis* resistance to necrotrophic and biotrophic pathogens. Plant Cell 18 (1):257–273. https://doi.org/10.1105/tpc.105.035576

Vidhyasekaran P (1988) Physiology of disease resistance in plants, vol II. CRC Press, Boca Raton, FL. 128 p

Voigt CA (2015) Callose-mediated resistance to pathogenic intruders in plant defense-related papillae. Front Plant Sci 5:168

Voll LM, Zell MB, Engelsdorf T, Saur A, Wheeler MG, Drincovich MF, Weber AP, Maurino VG (2012) Loss of cytosolic NADP-malic enzyme 2 in *Arabidopsis thaliana* is associated with enhanced susceptibility to Colletotrichum higginsianum. New Phytol 195:189–202

Wan JR, Zhang XC, Neece D, Ramonell KM, Clough S, Kim SY, Stacey MG, Stacey G (2008) A LysM receptor-like kinase plays a critical role in chitin signaling and fungal resistance in Arabidopsis. Plant Cell 206(1):471–481

Wang H, Hanash S (2003) Multi-dimensional liquid phase based separations in proteomics. J Chromatogr B 787:11–18. https://doi.org/10.1016/S1570-0232(02)00335-5

Wang X, Komatsu S (2016) Plant subcellular proteomics: application for exploring optimal cell function in soybean. J Proteome 143:45–56. https://doi.org/10.1016/j.jprot.2016.01.011

Wang Z, Mao H, Dong C, Ji R, Cai L, Fu H, Liu S (2009) Over expression of *Brassica napus MPK4* enhances resistance to *Sclerotinia sclerotiorum* in oilseed rape. Mol Plant-Microbe Interact 22:235–244

Wang Z, Tan X, Zhang Z, Gu S, Li G, Shi H (2012) Defense to *Sclerotinia sclerotiorum* in oilseed rape is associated with the sequential activations of salicylic acid signaling and jasmonic acid signaling. Plant Sci 184:75–82. https://doi.org/10.1016/j.plantsci.2011.12.013

Wang Y, Cheng X, Shan Q, Zhang Y, Liu J, Gao C, Qiu JL (2014) Simultaneous editing of three homoalleles in hexaploid bread wheat confersheritable resistance to powdery mildew. Nat Biotechnol 32:947–951

Wang C, Yao J, Du X, Zhang Y, Sun Y, Rollins JA, Mou Z (2015a) The *Arabidopsis* mediator complex subunit16 is a key component of basal resistance against the necrotrophic fungal pathogen *Sclerotinia sclerotiorum*. Plant Physiol 169(1):856–872. https://doi.org/10.1104/pp.15.00351

Wang JP, Munyampundu JP, Xu YP, Cai XZ (2015b) Phylogeny of plant calcium and calmodulin-dependent protein kinases (ccamks) and functional analyses of tomato ccamk in disease resistance. Front Plant Sci 6:1075. https://doi.org/10.3389/fpls.2015.01075

Wang GX, Lv J, Zhang J, Han S, Zong M, Guo N, Zeng XY, Zhang YY, Wang YP, Liu F (2016) Genetic and epigenetic alterations of *Brassica nigra* introgression lines from somatic hybridization: a resource for cauliflower improvement. Front Plant Sci 7:1258. https://doi.org/10.3389/fpls.2016.01258

Wang Z, Yang C, Chen H, Wang P, Wang P, Song C, Zhang X, Wang D (2018) Multi-gene co-expression can improve comprehensive resistance to multiple abiotic stresses in *Brassica napus* L. Plant Sci 274:410–419. https://doi.org/10.1016/j.plantsci.2018.06.014

Wang Z, Bao LL, Zhao FY, Tang MQ, Chen T, Li Y, Wang BX, Fu B, Fang H, Li GY, Cao J, Ding LN, Zhu KM, Liu SY, Tan XL (2019) BnaMPK3 Is a key regulator of defense responses to the devastating plant pathogen Sclerotinia sclerotiorum in Oilseed rape. Front Plant Sci 10:91. https://doi.org/10.3389/fpls.2019.00091

Wei L, Jian H, Lu K, Filardo F, Yin N, Liu L, Qu C, Li W, Du H, Li J (2016) Genome-wide association analysis and differential expression analysis of resistance to Sclerotinia stem rot in *Brassica napus*. Plant Biotechnol J 14:1368–1380. https://doi.org/10.1111/pbi.12501

Wendehenne D, Durner J, Klessig DF (2004) Nitric oxide: a new player in plant signalling and defence responses. Curr Opin Plant Biol 7(4):449–455. https://doi.org/10.1016/j.pbi.2004.04.002

Westermann AJ, Gorski S, Vogel J (2012) Dual RNA-seq of pathogen and host. Nat Rev Microbiol 10:618–630

Williams B, Kabbage M, Kim HJ, Britt R, Dickman MB (2011) Tipping the balance: *Sclerotinia Sclerotiorum* secreted oxalic acid suppresses host defences by manipulating the host redox environment. PLoS Pathog 7(6):e1002107. https://doi.org/10.1371/journal.ppat.1002107

Wu, ZJ, Irizarry R (2010) Description of gcrma package. https://www.bioconductor.org/packages/3.7/bioc/vignettes/gcrma/inst/doc/gcrma2.0.pdf

Wu J, Zhao Q, Liu S, Shahid M, Lan L, Cai G, Zhang C, Fan C, Wang Y, Zhou Y (2016) Genome-wide association study identifies new loci for resistance to *Sclerotinia* stem rot in *Brassica napus*. Front Plant Sci 7:1418. https://doi.org/10.3389/fpls.2016.01418

Xu YP, Lv LH, Xu YJ, Yang J, Cao JY, Cai XZ (2018a) Leaf stage-associated resistance is correlated with phytohormones in a pathosystem-dependent manner. J Integr Plant Biol 60 (8):703–722. https://doi.org/10.1111/jipb.12661

Xu YP, Yang J, Cai XZ (2018b) Glycolate oxidase gene family in *Nicotiana benthamiana*: genome-wide identification and functional analyses in disease resistance. Sci Rep 8(1):8615. https://doi.org/10.1038/s41598-018-27000-4

Yamaguchi Y, Pearce G, Ryan CA (2006) The cell surface leucine-rich repeat receptor for AtPep1, an endogenous peptide elicitor in *Arabidopsis*, is functional in transgenic tobacco cells. Proc Natl Acad Sci U S A 103:10104–10109. https://doi.org/10.1073/pnas.0603729103

Yan Y, Yuan Q, Tang J, Huang J, Hsiang T, Wei Y, Zheng L (2018) *Colletotrichum higginsianum* as a model for understanding host-pathogen interactions: a review. Int J Mol Sci 19(7):2142. https://doi.org/10.3390/ijms19072142

Yang Y, Zhang H, Li G, Li W, Wang X, Song F (2009) Ectopic expression of MgSM1, a Ceratoplatan in family protein from *Magnaporthe grisea*, confers broad-spectrum disease resistance in *Arabidopsis*. Plant Biotechnol J 7(8):763–777. https://doi.org/10.1111/j.1467-7652.2009.00442.x

Yang K, Nath UK, Biswas MK, Kayum MA, Yi G, Lee J, Yang TJ, Nou IS (2018) Whole-genome sequencing of Brassica oleracea var. capitata reveals new diversity of the mitogenome. PLoS One 13:e01194356

Yazawa T, Kawahigashi H, Matsumoto T, Mizuno H (2013) Simultaneous transcriptome analysis of sorghum and *Bipolaris sorghicola* by using RNA-seq in combination with de novo transcriptome assembly. PLoS One 8:e62460

Yu F, Zhang X, Peng G, Falk KC, Strelkov SE, Gossen BD (2017) Genotyping-by-sequencing reveals three QTL for clubroot resistance to six pathotypes of *Plasmodiophora brassicae* in *Brassica rapa*. Sci Rep 7:4516. https://doi.org/10.1038/s41598-017-04903-2

Zhang SQ, Klessig DF (2001) MAPK cascades in plant defense signaling. Trends Plant Sci 6:520–527

Zhang J, Zhou JM (2010) Plant immunity triggered by microbial molecular signatures. Mol Plant 3:783–793. https://doi.org/10.1093/mp/ssq035

Zhang L, Li H, Wang J (2012) The statistical power of inclusive composite interval mapping in detecting digenic epistasis showing common F2 segregation ratios. J Integr Plant Biol 54 (4):270–279

Zhang W, Fraiture M, Kolb D, Loffelhardt B, Desaki Y, Boutrot FF, Tor M, Zipfel C, Gust AA, Brunner F (2013) *Arabidopsis* receptor-like protein30 and receptor-like kinase suppressor of BIR1-1/EVERSHED mediate innate immunity to necrotrophic fungi. Plant Cell 25:4227–4241

Zhang X, White RP, Demir E, Jedryczka M, Lange RM, Islam M, Li ZQ, Huang YJ, Hall AM, Zhou G, Wang Z, Cai X, Skelsey P, Fitt BDL (2014) *Leptoshaeria* spp. Phoma stem canker and potential spread of *L. maculans* on oilseed rape crops in China. Plant Pathol 63:598–612

Zhang Y, Huai D, Yang Q, Cheng Y, Ma M, Kliebenstein DJ, Zhou Y (2015) Overexpression of three glucosinolate biosynthesis genes in *Brassica napus* identifies enhanced resistance to *Sclerotinia sclerotiorum* and *Botrytis cinerea*. PLoS One 10:e0140491

Zhao J, Wang J, An L, Doerge RW, Chen ZJ, Grau CR, Meng J, Osborn TC (2007) Analysis of gene expression profiles in response to *Sclerotinia sclerotiorum* in *Brassica napus*. Planta 227:13–24. https://doi.org/10.1007/s00425-007-0586-z

Zhao J, Buchwaldt L, Rimmer R, Sharpe A, McGregor L, Bekkaoui D, Hegedus D (2009) Patterns of differential gene expression in *Brassica napus* cultivars infected with *Sclerotinia sclerotiorum*. Mol Plant Pathol 10:635–649. https://doi.org/10.1111/j.1364-3703.2009.00558.x

Zhao L, Chen Y, Yang W, Zhang Y, Chen W, Feng C, Wang Q, Wu Y (2018) Polysaccharide peptide–induced virus resistance depends on ca2+ in flux by increasing the salicylic acid content and upregulating the leucine-rich repeat gene in *Arabidopsis thaliana*. MPMI 31(5):516–524. https://doi.org/10.1094/MPMI-10-17-0242-R

Zhou J, Zeng L, Liu J, Xing D (2015) Manipulation of the xanthophyll cycle increases plant susceptibility to *Sclerotinia sclerotiorum*. PLoS Pathog 11(5):e1004878. https://doi.org/10.1371/journal.ppat.1004878

Zhou LZ, Howing T, Muller B, Hammes UZ, Getl C, Dresselhaus T (2016) Expression analysis of KDEL CysEPs programmed cell death markers during reproduction in *Arabidopsis*. Plant Rep 29:265–272

Zhou Y, Sun L, Wassan GM, He X, Shaban M, Zhang L, Zhu L, Zhang X (2018) GbSOBIR 1 confers Verticillium wilt resistance by phosphorylating the transcriptional factor GbbHLH171 in *Gossypium barbadense*. Plant Biotechnol J 17:152–163

Zhu W, Magbanua MM, White FF (2000) Identification of two novel hrp-associated genes in the hrp gene cluster of *Xanthomonas oryzae* pv. *oryzae*. J Bacteriol 182:1844–1853. https://doi.org/10.1128/JB.182.7.1844-1853.2000

Zhu H, Li GJ, Ding L, Cui X, Berg H, Assmann SM, Xia Y (2009) *Arabidopsis* extra large G-protein 2 (XLG2) interacts with the Gβ subunit of heterotrimeric G protein and functions in disease resistance. Mol Plant 2(3):513–525

Zhu W, Wei W, Fu Y, Cheng J, Xie J, Li G, Yi X, Kang Z, Dickman MB, Jiang D (2013) A secretory protein of necrotrophic fungus *Sclerotinia sclerotiorum* that suppresses host resistance. PLoS One 8(1):e53901. https://doi.org/10.1371/journal.pone.0053901

Zhuang X, McPhee K, Coram T, Peever T, Chilvers M (2012) Rapid transcriptome characterization and parsing of sequences in a non-model host-pathogen interaction: Pea-*Sclerotinia sclerotiorum*. BMC Genomics 13:668

Zimmerli L, Jakab G, Metraux JP, Mauch-Mani B (2000) Potentiation of pathogen-specific defense mechanisms in *Arabidopsis* by b-aminobutyric acid. Proc Natl Acad Sci U S A 97:12920–12925

Zimmerli L, Metraux J, Mauch-Mani B (2001) b-aminobutyric acid-induced protection of *Arabidopsis* against the necrotrophic fungus *Botrytis cinerea*. Plant Physiol 126:517–523

Zipfel C (2014) Plant pattern-recognition receptors. Trends Immunol 35(7):345–351. https://doi.org/10.1016/j.it.2014.05.004

Zipfel C, Kunze G, Chinchilla D, Caniard A, Jones JDG, Boller T, Felix G (2006) Perception of the bacterial PAMP EF-Tu by the receptor EFR restricts Agrobacterium-mediated transformation. Cell 125:749–760

Zukalova H, Vasak J (2002) The role and effects of glucosinolates of Brassica species – a review. Rostlinia Viroba 48:175–180

Glimpses of Host Resistance Genomics

6

Abstract

Cruciferous crops are challenged by numerous biotic and abiotic stresses under field conditions. Some of the biotic stresses are of major consequences causing heavy losses to the economically very important *Brassica* crops at global level. The management practices employed to control these stresses are delivered through agronomic practices, chemical application, use of biological agents, and integration of all with growing resistant cultivars. The use of resistant cultivars is most economical, ecofriendly practices to get higher yield with better quality of *Brassica* produce under the threat of various biotic stresses of fungal, bacterial, viral, and nematodes origin. The breeding of resistant cultivars of *Brassica* crops has received significant attention of scientists all over the world after the advances made in genomics of host-resistance using multiomics technology and analysis of transcriptomics profiles through next-generation sequencing approaches to reveal molecular mechanisms of host-parasitic interaction and expression of host resistance. The resistance conferred by R-genes products is of five major kinds, that is, NTL (TIR-NBS-LRR), CNL (CC-NBS-LRR), RLK (receptor-like kinase), RLP (receptor-like proteins), and Pto (a Ser/Thr kinase proteins). Most of the R-genes in plants are members of NBS-LRR proteins. Identification of R-genes sources is the most essential and crucial step to breed *Brassica* crops with strong, durable, and effective resistance against all pathogens and their pathotypes. Several R-genes effective against major pathogens of *Brassica* with quantitative and qualitative resistance have been identified, cloned, and molecularly mapped on different chromosomes of *Brassica* species. These R-genes have been functionally characterized to use them for breeding resistant cultivars of *Brassica* through different breeding approaches including molecular technology. R-genes governing resistance to multiple pathogens and seeding as well as adult plant stages have been characterized. Pathotype/isolate/strain specific R-genes effective against fungal, bacterial, and viral pathogens of *Brassica*

G. S. Saharan et al., *Molecular Mechanism of Crucifer's Host-Resistance*,
https://doi.org/10.1007/978-981-16-1974-8_6

have been identified and molecularly characterized to use through appropriate introgression approaches. The genetics of resistance in the form of inheritance of R-genes in *Brassica* against different pathogens have been determined. Inheritance of resistance in crosses of *B. juncea* × *B. carinata* to *Alternaria brassicae* is governed by adaptive genes, dominant genes, additive × additive type epistatic genes, and additive × dominance and dominance × dominance type of nonallelic interaction genes. Partial resistance has been identified against *Alternaria* and *Sclerotinia* pathogens. The resistance to *Albugo* in *Brassica* crops is governed by one or two dominant genes with epistatic effects, additive dominant with epistatic effects and partial-resistant single recessive genes *wpr*. In interspecific crosses of *B. juncea* × *B. carinata*, single resistance gene with complete dominance confers resistance to *Erysiphe*. In *Brassica*, resistance to *Fusarium* is conferred by a dominant gene. Inheritance to *Hyaloperonospora* is governed by one to four dominant complementary genes, additive genes, one to two partial dominant genes, partial dominant epistatic, and gene dosage effects along with quantitative genes depending on host species and pathotype interaction. In *Brassica*, most resistant cultivars contain major genes/QTLs' conferring resistance to *Leptosphaeria* pathotypes. Resistant to *Plasmodiophora* in *Brassica* is governed by two to four dominant genes, additive and recessive genes and QTLs. *Plasmodiophora brassicae* pathotype 5× identified two independent QTLs, *Rer* 8 and *Lc* 9, in the A-genome on chromosomes AO2 and AO8 whereas a single dominant gene, *Rcr* 7, has been identified to provide resistance to pathotypes 3 h and 5× in the C-genome on chromosome Co7. In *Brassica*, partial dominant, polygenic, additive epistatic interactions confer resistance to *Sclerotinia*. In leaf and stem, resistance is conferred by QTLs' dominant and additive x additive epistatic interactions of genetic effects. In *B. rapa*, resistance to *TuMV* is controlled by dominant and recessive genes. In *B. oleracea*, resistance to *TuMV* is controlled by both dominant genes and multiple loci. In *B. oleracea*, *B. carinata*, and *B. napus*, resistance to *Xanthomonas* is controlled by single dominant genes. Substantial progress has been achieved in the transfer of R-genes to develop disease-resistant cultivars of *Brassica* crops against major pathogens using conventional and molecular techniques. To develop durable *Alternaria*-resistant *Brassica* cultivars, inter- or intraspecific introgression of R-genes from nonhost, microarray-based gene identification, and transfer by gene silencing, and transfer of an endo-chitinase gene from *Trichoderma* approaches have been used. R-genes sources of multiple disease resistance have been used to breed resistant cultivars of *Brassica* to *Albugo*, *Alternaria*, *Erysiphe*, *Hyaloperonospora*, *Leptosphaeria*, and *Xanthomonas*. Clubroot-resistant cultivars of *B. rapa*, *B. oleracea*, *B. napus*, *Raphanus sativus*, and *A. thaliana* have been developed. *Sclerotinia*-resistant *Brassica* cultivar has been obtained by introducing QTLs from wild relatives using marker-assisted introgression approach. Resistance to *Xanthomonas* has been introgressed from *B. carinata* to *B. oleracea* through embryo rescue. Crucifers' interactions with multiple stresses trigger a wide range of defense reactions at molecular and cellular levels to ensure that host cells are insulated from these challenges they encounter. The production of ROS, MAP kinase cascades, and hormonal signaling are triggered by multiple stresses and play crucial roles as common components of the immune system. GbRLK, a receptor-like kinase (RLK) gene, TF's, MYBTF's, ATAF1, HSPs gene, miRNA, SnRK, and calcium are involved in defense signaling pathways to multiple stresses.

Keywords

Functional characterization of R-genes · Identification of R-genes · Inheritance of disease resistance · Transfer of disease resistance · Molecular mechanisms · Management of R-genes · Use of R-genes · Exploitation of R-genes · Sources of resistance · Genetics of disease resistance · Deployment of R-genes · Introgression and pyramiding of major R-QTL's · Distant hybridization for *Xcc* resistance · Regulation of genomics · Biometabolomics for host resistance

Management of crucifers' disease resistance for its effective utilization under field conditions is very important and challenging. Effective utilization of R-loci and their allelic variations enhances the durability of resistance cultivars against different pathogens and their pathotypes. The deployment of R-genes cultivars based on quantitative and qualitative pyramiding along with integrated disease management is the key lever to increase the durability of R-genes against rapidly evolving pathogen populations with new virulences. In addition to major gene-resistant cultivars, use of horizontal resistance, nonhost resistance, multilayered and multi-component resistance; use of bacterial and fungal biocontrol agents; use of Proca, nitrogen, and arginase gene to modulate resistance positively; and use of genes like WRKY and CAMA should be exploited to fortify the resistance of cvs. Under field conditions. Standardized techniques on germplasm screening, identification of R-genes, introgression of R-genes, and methodology for breeding disease-resistant cultivars against different pathogen included here will be helpful for students, teachers, and researchers to plan their research projects on crucifers' host–pathogen interactions.

6.1 Introduction

Crucifers are significant component of cultivated crop biodiversity of the world. Out of total cultivated area in the world, crucifers occupy 2.69% area with 1384.8 mha total cultivated area producing 147.69 mmt vegetables, oil yielding food, and fodder crop for feeding human and animal population at global level. Large number of biotic and abiotic stresses is known to deteriorate quantity and quality of crops in the largest group of crucifers encompassing Brassicaceae family. The biostresses have been managed with several approaches including disease-resistant cvs. as a major component to save resources and environment. The genomics studies accompanied with molecular omics approaches have strengthened our understanding on molecular mechanisms of host–parasite interactions and functional characterization of valuable genes to breed better cvs. of *Brassica* crops with high yield, quality, and biotic and abiotic stress management attributes. During this century, multiomics approaches like genomics, proteomics, transcriptomics, biometabolomics, pangenomics, and bioinformatics have been extensively used to comprehend the genomics of host resistance through identification, isolation, cloning, and molecular mapping of R-genes with functional characterization to confer high level of resistance to *Brassica* pathogens. It has facilitated breeding of disease resistance *Brassica* crop with strong and durable resistance against major pathogens with high level of precision and efficacy.

6.2 Principles of Host Resistance

Cruciferous crop plants are always challenged by various biotic and abiotic stresses. In agriculture cropping system, plant protection is being delivered using different approaches such as chemical control, various agronomic practices, biological control, integrated pest management (IPM), and cultivation of resistance cultivars. Among these approaches, resistance cultivars are the most economical, environmentally sustainable solution to control different diseases in *Brassica* species. Previous studies suggested that the inheritance of disease resistance is either qualitative or quantitative in *Brassica* species. Recently mapped disease resistance genetic loci and closely linked molecular marker to these loci can be used for marker-assisted selection in cruciferous resistance breeding of *Brassica* species. Extensive use of recently available resistance sources can be combined with molecular tools and new technologies such as gene/QTL mapping, fine mapping, gene cloning, comparative genomics, and analysis of transcriptomic profiles through next-generation sequencing approaches, to enhance our understanding of resistance mechanism. Novel information can help control diseases in an effective way, so yield losses would be reduced and the quality of Brassica crop product would be improved. The principles involved in the effective utilization of genomics technology largely depend on the following approaches including molecular genetics.

In natural environment, cruciferous plants are challenged by 167 biotic (44 pathogen, 87 pests, and 36 weeds) and 19 abiotic stresses (environmental, soil salinity and alkalinity, nutrients' deficiency, and hormonal changes), and some of them have successfully invaded crop plants to cause diseases which resulted in deterioration of crop quality and yield. The major pathogens of crucifers can be ranked at global level on the basis of host range, geographical distribution, yield losses, and emphasis given by the *Brassica* scientists on genomic investigations for improving the yield and quality of *Brassica* crops. On the basis of average rank at global level eight pathogens (1—*Plasmodiophora brassicae*; 2—*Alternaria brassicae*; 3—*Sclerotinia sclerotiorum*; 4—*Leptosphaeria maculans*; 5—*Erysiphe cruciferarum*; 6—*Albugo candida*; 7—*Turnip mosaic virus*; 8—*Xanthomonas campestris* pv. *campestris*) can be considered as of major consequences with 18.25 to 68.25 rank (Table 1.1, Chap. 1). In order to cope with disease attacks, the plants have developed multiple layers of defense mechanisms. Plant disease resistance (R)-genes which specifically interact/recognize with corresponding pathogen avirulence (*avr*) genes are considered as plant genetic factors of a major layer. The interactions of this gene-for-gene (or genes-for-genes) manner activate the signal transduction cascades that turn on complex defense responses against pathogen attack and this is called incompatible interaction. The interaction between a host species and a pathogenic species is dynamic where a host variety often lost the R-;gene-dependent resistance due to its pathogen race evolution for a virulent gene, and thus a new R-gene was selected against this new race. R-genes provide innate immunity whereas outcomes of defense responses lacking R-genes are partial resistance. Therefore, identification of R genes is crucial for resistant variety development and relevant mechanism investigation.

More than one hundred R-genes, which are reported in PRGdb (http://prgdb.crg.eu/wiki), were functionally identified and comprise a super family in plants. Sequence composition analysis of R-genes indicate that they share high similarity and contain seven different conserved domains like NBS (nucleotide-binding site), LRR (leucine-rich repeat), TIR (Toll/Interleukin-1 receptor), CC (coiled coil), LZ (leucine zipper), TM (transmembrane), and STK (serine-threonine kinase). Based on domain organization, R gene products can be categorized into five major types: TNL (TIR-NBS-LRR), CNL (CC-NBS-LRR), RLK (receptor-like kinases), RLP (receptor-like proteins), and Pto (a Ser/Thr kinase protein). Most of the R-genes in plant kingdom are members of NBS-LRR (nucleotide-binding site–leucine-rich repeat) proteins. "NBS" and "LRR" domains play different roles in plant–microbe interaction, where the former have the ability to bind and hydrolyze ATP or GTP and the latter is involved in protein–protein interactions. NBS-LRR proteins in plants share sequence similarity with the mammalian NOD-LRR containing proteins which play a role in inflammatory and immune responses. On the basis of presence or absence of N-terminal domains (TOLL/ interleukin-1 receptor (TIR) and the coiled-coil (CC) motif), NBS-LRR class can be further divided into two major types, TNL (TIR-NBS-LRR) and CNL (CC-NBS-LRR). TNL type shares homology with the Drosophila toll and human interleukin-1 receptor (TIR). The two types show divergence in their sequence and signaling pathways. Several partial NBS-LRR variants like TIR, TIR-NBS (TN), CC, CC-NBS (CN), and NBS (N) have also been identified in plant species (Wan et al. 2012; Inohara et al. 2005; Meyers et al. 2002, 2003).

The whole genome sequence data enabled the genome-wide identification, mapping, and characterization of candidate NBS-containing R-genes in economically important plants. For example, the approximate arrays of 159 NBS-encoding R-genes in *Arabidopsis thaliana* have been identified. Earlier genome-wide studies have demonstrated that TNL subfamily is abundant in dicots while absent in cereals (monocots). The presence of the full length of TNL and CNL types in the common ancestor (mosses) of both angiosperms and gymnosperms and exceptional presence of truncated domains of TN- or TX-type proteins in cereals indicate that the TNL class might have been lost in monocot plants. On the chromosomes, the NBS-LRR R-genes are arranged in clusters. The genes in the clusters could be homogenous (often tandem duplicated from single ancestor gene) or heterogeneous (with different protein domains). However, the variation of the number and sequences of the R-genes presented in the *Brassica* lineage since split from the *Arabidopsis* lineage and their distributions in chromosomes are unknown.

The genera *Brassica* and *Arabidopsis* both belong to the mustard family Brassicaceae (Cruciferae) and are a model plant and a model crop, respectively. The two genera shared a latest and obviously detectable alpha genome duplication event before their divergence ~20 million years ago (MYA), and subsequently *Brassica* ancestor underwent a whole genome triplication event (common to the tribe Brassicaceae) ~16 MYA. In *Brassica*, interspecific cytogenetic relationship between important crops (oilseed and vegetables) is well-described by a "U" triangle where each two diploid species [*Brassica rapa* (AA, $2n = 20$), *Brassica oleracea* (CC, $2n = 18$), and *B. nigra* (BB, $2n = 16$)] formed a tetraploidy species [*B. napus* (AACC, $2n = 38$), *B. juncea* (AABB, $2n = 36$) or *B. carinata* (BBCC, $2n = 34$)]. This well-established phylogenetic relationship provides a chance to trace evolution of the R-genes between wild plants and their relative crops. Yu et al. (2014) identified R-genes on genome-wide scale in *B. oleracea* and *B. rapa* and provided insights into their evolutionary history and disease resistance in crucifers.

Plant disease resistance (R) genes with the nucleotide-binding site (NBS) play an important role in offering resistance to pathogens. The availability of complete genome sequences of *B. oleracea* and *Brassica rapa* provides an important opportunity for researchers to identify and characterize NBS-encoding R-genes in *Brassica* species and to compare with analogues in *A. thaliana* based on a comparative genomics approach. However, little is known about the evolutionary fate of NBS-encoding genes in the *Brassica* lineage after split from *A. thaliana*. Yu et al. (2014) presented genome-wide analysis of NBS-encoding genes in *B. oleracea*, *B. rapa*, and *A. thaliana*. Through the employment of HMM search and manual curation, they have identified 157, 206, and 167 NBS-encoding genes in *B. oleracea*, *B. rapa*, and *A. thaliana* genomes, respectively. Phylogenetic analysis among three species classified NBS-encoding genes into six subgroups. Tandem duplication and whole genome triplication (WGT) analyses revealed that after WGT of the *Brassica* ancestor, NBS-encoding homologous gene pairs on triplicated regions in *Brassica* ancestor were deleted or lost quickly, but NBS-encoding genes in *Brassica* species experienced species-specific gene amplification by tandem duplication after divergence of *B. rapa* and *B. oleracea*. Expression profiling of NBS-encoding orthologous gene pairs indicated the differential expression pattern of retained orthologous gene copies in *B. oleracea* and *B. rapa*. Furthermore, evolutionary analysis of CNL type NBS-encoding orthologous gene pairs among three species suggested that orthologous genes in *B. rapa* species have undergone stronger negative selection than those in *B. oleracea* species. But for TNL type, there are no significant differences in the orthologous gene pairs between the two species. The identification and characterization of NBS-encoding genes in *B. rapa*, and *B. oleracea* based on whole genome sequences will be landmark in genomics research. Through tandem duplication and whole genome triplication analysis in *B. oleracea*, *B. rapa*, and *A. thaliana* genomes, the study provides insight into the evolutionary history of NBS-encoding genes after divergence of *A. thaliana* and the *Brassica* lineage. These results together with expression pattern analysis of NBS-encoding orthologous genes provide useful resource for functional characterization of these genes and genetic improvement of relevant crops (Yu et al. 2014).

Molecular genetics investigations of disease resistance have been at the vanguard in *Arabidopsis* research of utilizing natural genetic variation. Numerous disease resistance genes (so-called R-genes) have been revealed that confer genotype or race-specific resistance to a wide assortment of pathogens. A majority of these genes are of the same structural class, now widely known by the abbreviated name of NB-LRR (**n**ucleotide-**b**inding site and **l**eucine-**r**ich **r**epeat domains). Genes of this class have also been described in a wide diversity of crops such as maize, lettuce, sugar beet, tomato, and tobacco. An obvious spin-off, therefore, would be to use DNA sequence from NB-LRR genes and other known R-gene classes to develop more effective genetic markers for traditional breeding of resistance in crops. The prospect of using R-genes themselves to confer resistance in crops has also been demonstrated. Most *Arabidopsis* research has been advanced using genetic variation that was derived artificially from mutagenesis induced chemically, by irradiation, or by transposon or T-DNA insertion. For investigation of disease resistance, numerous mutants classified broadly as ones that enhance susceptibility to avirulent pathogens or else enhance resistance to virulent pathogens have been molecularly characterized. A complex impression of defense signaling has emerged from these mutational analyses and consequently redefines the frontline for *Arabidopsis* pathology as an effort to identify the intermediate components of defense regulation and dynamic relationships among distinct responses.

The underlying defense responses appear to be widely conserved among plant species. In some cases, genes have been conserved across Kingdoms, such as *SGT1* which has recently been identified as a major regulator of powdery mildew resistance in barley and downy mildew resistance in *Arabidopsis*, having first been described as a critical gene for cell cycling and proteolysis in yeast. Disease resistance can potentially be improved in crops by altering the expression of defense regulation, as suggested by broad spectrum resistance that was achieved by overexpression in *Arabidopsis* of *NPR1*, a regulator of acquired disease resistance.

Wild plant species represent a gene pool that can potentially provide a rich source of durable, broad spectrum disease resistance for use in crops. Species that are interfertile with a particular crop have often been used for transferring monogenic resistance that is often genotype or race specific. Unfortunately, such resistance derived from either crop or wild germplasm often lacks durability when deployed in field production because of selection it imposes that force the pathogen to adapt and overcome the resistance. Nondurable resistance is potentially still useful; however, plant breeders are usually challenged to continue the search for new sources of resistance. Broad spectrum resistance from wild germplasm, which may be polygenic and inherently more difficult to transfer and reconstruct in a crop, will be increasingly more accessible as a result of recent advances in our knowledge of candidate resistance genes. The underlying basis for this resistance is that the pathogen as a taxonomic group (species or subspecies) is highly specialized to cause disease in the crop, but not in the wild crop relative. Broad spectrum resistance that occurs universally in the wild relative is referred to below as *species level ("nonhost") resistance* to the pathogen.

Some of the advances demonstrate clearly that signaling mechanisms governing genotype-specific resistance can also govern species level resistance. For example, mutations in one of the defense regulators discovered in *Arabidopsis*, designated *EDS1* (**e**nhanced disease **s**usceptibility), exhibit broad spectrum susceptibility to several pathogens including isolates of *Albugo candida* (white rust) and *Hyaloperonospora parasitica* (downy mildew) from *B. oleracea*. Isolates of both pathogens collected from either *Brassica* or *Arabidopsis* represent distinct subspecies, based on their restrictive host specialization and molecular divergence. Therefore, *EDS1* is an essential regulator of species level resistance to at least two *Brassica* pathogens. The highly conserved protein, *SGT1* has also recently been reported as a key regulator for species level resistance.

Two bacterial R-genes, *RPS4* and *RPS5*, have been described in *Arabidopsis* which demonstrate that NBLRR genes can provide necessary components of species level resistance. Both of these genes were characterized using corresponding avirulence determinants that were derived from legume isolates of *Pseudomonas syringae*. Arabidopsis exhibits species level resistance, at least in symptomology, to these pathogens. The avirulence determinants, however, were introduced into an *Arabidopsis*-virulent isolate of *Ps. syringae* to facilitate molecular characterization of the resistance genes. Transient suppression of species level resistance has also been observed in *Arabidopsis*. For example, *Bremia lactuca* (lettuce downy mildew) can reproduce asexually in cotyledons or true leaves following preinfection with *Albugo candida* (Holub et al. 2002).

Crucifers include genus *Brassica* in the Brassicaceae family with 325 genera with 3740 species with 75 accepted species. Most of the species in the genus are cultivated for their edible roots, leaves, stems, buds, flower, and seeds for human consumption as raw, cooked, edible oil, and as fodder for animals. Genetic diversity within *Brassica* species has been broadly studied with a special focus on six crop species widely cultivated for vegetables and oilseed production. Out of 44 pathogens known to infect crucifers, 16 pathogens have received attention for genomics investigations. The principles involved in the effective utilization of genomics technology largely depend on several omics approaches for identification, inheritance, transfer, and management of R-genes for developing durable-resistant cultivars of crucifers. Wild plant species represent a gene pool that can potentially provide a rich source of durable, broad spectrum disease resistance for use in crops. Hybridization between *Brassica* species and wild relatives' coenospecies and cytodemes can be possible under controlled conditions with molecular technology. Interspecific hybridization for developing disease resistance in *Brassica* crops has potential since they share genomics in common. Genes containing a nucleotide-binding site (NBS) constitute one of the largest R-genes families which are co-located with resistance QTLs and facilitate multiple disease resistance breeding in *Brassica* crops. Elevated resistance against biotrophs is often related with increased susceptibility to necrotrophs and vice versa. Plants have evolved intracellular receptors with nucleotide-binding leucine-rich repeat domain (NLRs) to detect pathogen effectors or their activity to trigger immunity. Understanding the interspecific and intraspecific relationship among cultivated *Brassica* species will be useful for better parental selection to widen the genetic base for multiple disease resistance breeding programs of *Brassica* crops. Despite the genome complexity of *Brassica* species, homoeologous rearrangements/recombination occurs more frequently in resynthesized species contributing toward genetic diversity to provide good sources of R-genes for breeding resistant cultivars. The complexity of R-genes resistance levels varies in *Brassica* species for major pathogens due to evolution of pathogenic variability. The genomic era is characterized by high-efficiency integrated breeding (HIB) in which multiple methods are combined including traditional ways such as microspore culture, backcrossing, and distant introgression with modern ways such as MAS, gene editing, and genome design. During HIB genome background analysis is helpful in eliminating undesirable linkage drags and rapidly identifying desirable individuals. Several numbers of genes have well documented that they are differentially expressed against different pathogens. Several omics technologies and approaches have provided tool to look at the differences in DNA, RNA, proteins, and other cellular molecules between species and among individuals of species to get better insight into the genomics of crucifers' disease resistance breeding programs. Using the various omics and multiomics tools, supplemented with further fine tuning of the bioinformatics methods will speed up the screening of favorable alleles in *Brassica* germplasm promoting resistance against major pathogens along with identification and cloning of favorable genes with increased precision.

6.3 Identification of R-Genes Sources

Identification of R-genes sources is the first most essential and crucial step toward breeding of disease resistance cultivars through conventional and molecular approaches. Crucifer's pathogens are of diverse nature from biotrophs and hemibiotrophs to necrotrophs with a lot of pathogenic variability in the form of races, pathotypes, and strains. For introgression of durable resistance both qualitative and quantitative, R-genes sources are required to plan a breeding of cultivars with desired traits of yield and quality in *Brassica* crops. Considerable efforts have been made to take the R-genes from cultivated diploid and amphidiploid *Brassica* species in addition to *A. thaliana*, *Brassica* wild relatives, and land races. With the use of omics approaches, several R-genes effective against major pathogens of crucifers have been identified, cloned, and mapped on chromosomes of *Brassica* with functional characterization using the various omics or multi-omics tools, supplemented with further fine tuning of the bioinformatics method, will not only speed up the screening of favorable alleles in *Brassica* germplasm promoting resistance against the major pathogens but also expedite the identification and cloning of favorable genes with increased precision. Efforts have been made to incorporate and introgress desired R-genes in *Brassica* crops through different breeding approaches including molecular technology. Now, after genome sequencing of both host and pathogen, it has become easy to use the R-genes pyramiding in a cultivar with desirable gene editing and gene silencing approach.

Identification of effective and durable sources of R-genes with amicable introgression in the crucifers' genotypes using conventional as well as molecular/biotechnological approaches is essential to breed disease-resistant cultivars of *Brassica* to ameliorate production and productivity at global level. R-genes identification is the first and foremost step toward the disease-resistant hybrid/varieties development through appropriate breeding programs. R-genes conferring qualitative and quantitative (QTLs) loci have been identified from *Brassica* species, cruciferous wild relatives, land races, and *Brassica* coenospecies against major pathogens of crucifers. Several sources of *Brassica* species with R-genes and QTLs for inter- and intraspecific hybridization disease resistance cultivars development with multiple pathogens protection like *Albugo*, *Alternaria*, *Erysiphe*, and *Hyaloperonospora* have been identified. R-genes governing seedling and adult plant resistance have been determined. Some R-genes show differential interaction on *Brassica* genotypes at different locations because of environmental and pathogenic variations. Multi R-genes genotypes have been identified from *Brassica* species against *Albugo candida*. R-genes in *Brassica* species to Colletotrichum *higginsianum* have been characterized using molecular and biochemical approaches. Several genotypes of *Brassica* species has slow mildewing attributes to provide protection against *Erysiphe cruciferarum*. In *B. oleracea*, a single R-genes and polygenic resistance to Fusarium wilt has been characterized. Several R-genes sources in *A. thaliana* have been characterized as RPP genes against *H. arabidopsidis*.

In *Brassica* species, several Leptosphaeria qualitative race-specific R-genes and quantitative QTLs have been identified. However, it is uncertain if some of these genes are same genes with different nomenclature or allelic variants of the same gene, since researchers have used different crosses, different isolates, and different markers system. R-genes effective against clubroot (Plasmodiophora *brassicae*) have been identified from cruciferous species as pathotypes specific as well as QTLs. GBS identified QTLs provide resistance to multiple pathotypes. Multiple wild *Brassica* species have shown high levels of resistance against stem rot pathogen (Sclerotinia). Isolate-specific-resistant sources to TuMV have been identified from *Brassica* species. Race-specific sources of resistance to Xanthomonas have been identified from cultivated *Brassica* species and *A. thaliana*. Black rot (*Xanthomonas*) R-genes have been identified in cauliflower through molecular approaches. In general, the *Brassica* A-genome is a rich source for resistance to Turnip mosaic virus, Black rot, downy mildew, and clubroot, whereas B-genome possesses resistance against Black rot and Blackleg, and C-genome harbors stem rot, Fusarium wilt, and Downy mildew resistance. Well-characterized R-genes/QTLs in *Brassica* crops have been identified and genetically mapped against clubroot, Sclerotinia stem rot, downy mildew, and *Leptosphaeria*. However, only limited R-genes/alleles have been cloned from *Brassica* species. Eight different bonafide R-genes like (RGL) PCR fragments have been obtained from *A. thaliana* accessions *Col* and L*er*. The application of omics technology in R-genes identification has widened the scope of *Brassica* species genome sequencing, use of pangenomics, use of NGS-based SNP methods, in silico method, NGS-based bulked segregant analysis (BSA), resistance gene enrichment and sequencing (Ren Seq), and effectromics approaches to detect R-genes.

Genetic resistance is one of the most important attributes of the host defense against pathogens. Host resistance provides an economical, environmentally benign, and widely accepted method of managing disease especially downy mildew in crucifers. Present day's emphasis on organic agriculture relies mainly on large-scale deployment of disease-resistant crop cultivars in addition to the use of techniques such as cultural practices (sowing time, soil solarization, crop rotation, use of organic manures, plant population, clean cultivation, soil and plant moisture regulation) and biological disease control. Hyaloperonospora–*Brassica* interaction can be classified into two types, compatible and incompatible interaction, leading to the phenotypes of susceptibility, and resistance to certain pathogens. Obviously, the incompatible interaction has been extensively exploited by the *Brassica* breeders to develop resistant cultivars for increasing crop production. Large-scale deployment of elite cultivars carrying an R-gene (monoculture) imposes higher selection pressure on the pathogen carrying the cognate *Avr* gene to survive, resulting in the modification or deletion of the recognized *Avr* gene or generation of novel effector genes that can escape the recognition of the old 'R'-gene. This *R-Avr* interactive co-evolution explains why many resistant cultivars in the field lose their resistance in a relatively short period of time. Hence, knowledge on the molecular mechanisms of *Hyaloperonospora-Brassica*, that is, host–pathogen interaction, and use of horizontal, durable, and broad-spectrum resistance, has become very important in disease resistance breeding programs. The interplay of host resistance and pathogen virulence in latter system is likely to be quite different to major gene–based system. Major gene resistance in crucifers to downy mildew has been exploited for breeding resistant cultivars.

To assess the nature of powdery mildew resistance in crops, particularly in context to slow mildewing components. The studies revealed that the incubation period ranged from 3 to 4 days. The maximum incubation period of 4 days was recorded on the varieties RH-9304 and RH-9801, whereas in other cases, it was 3 days. The latent period of test cvs. Ranged between 1 and 3 days. The latent period was 1–2 days in varieties belonging to the *Brassica juncea*. However, it was 3 days in case of variety GSL-1. The results also revealed that there was no significant difference in the latent period in the varieties belonging to *Brassica juncea*. However, it differed in case of *Brassica napus* where it was slightly higher. The number of powdery mildew specks/leaf was also recorded and revealed that the variety GSL-1 showed minimum number of specks/leaf (5.52) whereas the variety RH-9801 contracted maximum number of the specks/leaf (39.40). The variety GSL-1 contracted less number of specks per leaf which significantly differed from the other varieties (less than ten specks /leaf) which may be considered as resistant, whereas other varieties such as RH-9801, RH-9304, and RH-30 had higher number of specks/leaf (more than 30), and may be termed as susceptible. The other varieties such as RH-9901, RH-8812, RC-781, and Purple Mutant contracted the powdery mildew specks ranging between 10 and 30 specks/leaf and considered as moderately susceptible.

The minimum number of conidia/speck was recorded on the variety GSL-1 (10 conidia/speck). It was followed by the variety RC-781 where it was 41.58 conidia/speck. In other varieties such as RH-30, RH-9801, RH-9304, RH-8812, RH-9901, and Purple Mutant, the conidial production ranged between 50.08 and 83.75 conidia per speck being maximum on the variety RH-30 (83.8) and minimum on RH-8812 (50.1). The considerable amount of conidia per speck were produced in all the susceptible varieties belonging to except GSL-1, and HC-9603. The variety GSL-1 and HC-9603 appeared as resistant to powdery mildew with the expression of slow mildewing components, whereas other varieties belonging to group appeared as susceptible to the disease showing faster powdery mildew development under field conditions (Singh 1994; Meena et al. 2018).

6.4 Inheritance of Disease Resistance

In the absence of host resistance, intensive cultivation has resulted in perpetuation, build-up, and dissemination of *Alternaria* species on crucifers in areas where these crops are grown. The main factors responsible for the increase of inoculum of pathogenic Alternaria to the present level include the lack of genetic sources of resistance, cultivation of susceptible crops with high plant density, use of irrigation with high rates of nitrogen fertilizers, contiguous areas under monoculture, staggered sowing dates, poor weed management, and poor plant protection strategies. The information gathered on *Alternaria* diseases of crucifers indicates that some gaps and bottle necks still exist in our comprehensive understanding of various dimensions of Alternaria–host pathosystem. Although some information is available on genetic sources of resistance, the information on the nature, mechanism and inheritance of resistance are still not satisfactory and/or lacking. The results of this area of investigation are an important prerequisite to incorporate stable and effective sources of resistance in breeding programs. In diseases caused by *Alternaria* species where major gene resistance is not known, the potential of minor genes/horizontal resistance, tolerance, and slow blighting should be exploited. Feasibility of induced resistance needs attention. Biotechnology or molecular biology and genetic engineering techniques can be used to transfer resistant genes from distantly related sources. Identification and sequencing of resistant genes along with identification of suitable gene combinations may be taken up. Hybrids and GM cultivars with multiple disease resistance are need of the hour.

Characterization of secondary metabolite biosynthetic genes and signal transduction and their role in pathogenicity and fungal development are important areas of investigation. Disruption of Aso-1, a gene required for hyphal fusion (anastomosis), is important for *Alternaria* spores to become pathogenic. Studies on the nature and mechanism of inheritance of virulence genes in different species of *Alternaria* are required. Determination of virulence spectrum of pathotypes is another aspect to be looked into. Exploitation of morphological, structural, and biochemical basis of resistance needs in-depth study in crucifer since lot of genetic variability is available in related and distantly related crucifers all over the world. Comparative studies on all aspects of host–parasite interaction in crucifers with all the four species of *Alternaria* may reveal some fruitful results for comprehensive and management of disease.

Host resistance in crucifers against *Alternaria* species has various components. Inheritance of resistance in inter- and intraspecific crosses of *B. juncea* and *B. carinata* to *Alternaria brassicae* is governed by adaptive genes, dominant genes, additive × additive type epistatic genes, additive × dominance, and dominance × dominance type of nonallelic interaction genes (Singh and Singh 1989; Krishnia et al. 2000). Inter-mating between resistance plants helps in increasing the level of resistance against *A. brassicae* through pyramiding of resistant genes. Elevated horizontal resistance in oilseed *Brassicas* genotypes has been recorded (Saharan and Kadian 1983; Saharan and Krishnia 2001). *Brassica* genotypes, Rajat, Kranti, RH-781, and RL-1359, have been identified with disease tolerance attributes (Gupta et al. 2002). Genotypes PR-8988 and PR-9024 show higher degree of partial resistance or slow blighting (Kumar and Kolte 2001). Epicuticular wax (Candle, Tobin, Altex, Midas, Tower), low number, and narrow stomatal aperture (Tower, RC-781) provide resistance to *Alternaria* infection in *Brassica* species (Saharan and Kadian 1983; Conn et al. 1984; Tewari 1991). The concentration of phenolic compounds and activation of polyphenol oxidase and catalase are higher in tolerance genotype of mustard (Gupta et al. 1990).

Chitinase-modifying proteins, *Cmps*, are secreted by fungal pathogens of crucifers, which hamper with fungalysin *cmp* activity to develop plant resistance to multiple fungal diseases (Naumann and Wicklow 2013). *GLIP1* in association with ethylene signaling may be a critical component in plant resistance (*A. thaliana*) to *A. brassicicola* (Oh et al. 2005). *B. juncea* plants transformed with chitinase gene tagged with an overexpressing promoter, 35S CaMV, gave defense response by degrading the cell walls of invading fungi (Mondal et al. 2003). Increased levels of PAL, PPO, and peroxidase play an important role in the defense mechanisms of *B. juncea* genotypes against *Alternaria* pathogenesis (Parihar et al. 2012). β–Aminobutyric acid treatment leads to proper balance of oxidant and antioxidants suitable for expression of resistance in *B. carinata* against *A. brassicae* by curtailing pathogen entrance during early stages of colonization (Chavan et al. 2013). Zeatin, a cytokinin, upregulates plant immunity via an elevation of *MAPK-4* and antagonizes the effects of *A. brassicae* (Marmath et al. 2013). Transgenic expression of hevein, the rubber tree lectin, in *B. juncea* cv. RLM-198 confers defense against *A. brassicae* (Kanrar et al. 2002), β–Aminobutyric acid pretreatment of *B. juncea* plants induces *A. brassicae* resistance mediated through an enhanced expression of pathogenesis-related protein genes, independent of SA and JA accumulation (Kamble and Bhargava 2007). Hypersensitive response gene (hsr 203J) like homologues of *Brassica* play important role in differential defense response against *A. brassicae* (Mishra et al. 2010). A highly contiguous genome assembly of *A. brassicae* has been generated using Nanopore MinION sequencing with an N50 of 2.98 Mb, yielding nine full chromosome-level sequences. This study adds to the current genomic resources available for the genus *Alternaria* and will provide opportunities for further analyses to unravel the mechanisms underlying pathogenicity of this important necrotrophic pathogen (Rajarammohan et al. 2019).

The cDNA encoding Pm AMP 1 has been successfully incorporated into the genome of *B. napus*, and it is in planta expression that confers greater protection against *A. brassicae*. Pm AMP1 is a cysteine-rich antimicrobial peptide isolated from *Pinus monticola* (Verma et al. 2012). Combined expression of a barley class II chitinase and Type I ribosome inactivating protein in transgenic *B. juncea* provides resistance against *A. brassicae* (Chhikara et al. 2012). Transcriptional responses to exposure to the brassicaceous defense metabolites camalexin and allyl-isothiocyanate in *A. brassicicola* have been recorded (Sellam et al. 2007).

Elicitation and accumulation of phytoalexins in crucifers after exposure to Alternaria and their role in disease resistance have been demonstrated (Verma and Saharan 1994). Calcium sequestration property of *A. brassicae* can be used in enhancing resistance to this pathogen in rapeseed by soil or foliar application of calcium compounds (Tewari 1991). Many sources of resistance to *Alternaria* species have been identified from different *Brassica* species, but very few have been utilized to develop resistant cultivars (Verma and Saharan 1994; Sharma et al. 2002). Strong and positive correlation in the increased level of resistance to different pathogens reported in *Brassica* species will facilitate improvement in accumulation of resistance to multiple disease resistance (Kumar and Saharan 2002). In spite of several bottlenecks in the development of resistance cultivars, various methods/techniques together with conventional as well as biotechnological approaches have been utilized to incorporate desired traits in cruciferous crops against black spot disease (Nowicki et al. 2012; Aneja and Agnihotri 2013). The resistance sources available in host germplasm, the exploitation of NHR mechanisms, and their applications in protecting *Brassica* crops from Alternaria blight need attention promptly. The insights might also be applicable to a wider repertoire of plant pathogens (Fatima et al. 2019).

Sclerotinia sclerotiorum (Lib.) de Bary, the causal organism of stem rot of *Brassica* along with over 500 host plants are distributed worldwide. *Sclerotinia* rot is menace to cultivation of oilseed *Brassica* crops all over the world. Infection occurs on leaves, stems, and pods at different developmental stages, causing seed yield losses of up to 80%, as well as significant reductions in oil content and quality. The initial mycelial infection at the base of the stem is an appearance of elongated water-soaked lesions that expand rapidly. Ascosporic (carpogenic) infection is quite general and occurs on the leaves or leaf axil. Effective pathogenesis by *S. sclerotiorum* requires the secretion of pathogenicity factors including oxalic acid and extracellular lytic enzymes. Germination of overwintered sclerotia and release, survival, and germination of ascospores are important factors for the development of disease and in the life cycle of this pathogen. Isolates of *S. sclerotiorum* show a high level of morphological variability and molecular diversity. Management of *S. sclerotiorum* is a major challenge, and the best being the integration of various IPM measures. Partial resistance has been identified in some *Brassica napus* and,

B. juncea, genotypes, though; wild *Brassicas* show better resistant reactions (Sharma et al. 2015).

Identification of effective R-genes sources along with the nature of inheritance in *Brassica* genotypes is the basic impact in breeding of disease resistance cultivars of crucifers against different range of pathogens. There are two different genetic mechanisms for disease resistance traits being inherited in generations. One is monogenic resistance which is based on single gene and second is quantitative which is governed by two or more genes. In most cases, single R-genes confer complete resistance but are active against few races of the pathogen. R-genes are clustered at some loci in the genome or exist as different alleles conferring resistance toward specific pathogen races. Quantitative resistance shows efficacy against more number of pathogenic races. Resistance breeding in crucifers depends on both types of resistance. Monogenic resistance is easy to incorporate but it is not durable. Mostly quantitative resistance is preferred. The use of molecular markers has allowed the isolation, cloning, and gene pyramiding to develop more durable resistance cvs. of *Brassica* crops. Qualitative resistance mostly provides defense against biotrophs, whereas quantitative resistance is involved in defense response to all plant pathogens from biotrophs, hemibiotrophs to necrotrophs pathogenic to crucifers.

The inheritance of disease resistance in *Brassica* crops is by both qualitative one gene and polygenes and quantitative conferred by QTLs. The resistance to white rust (*Albugo*) in different *Brassica* species is governed by single dominant gene, one or two dominant genes with epistatic effects, additive dominant genes with epistatic effects, and partial-resistant single recessive genes *wpr*. In *A. thaliana*, *WRR* 4 genes confer broad-spectrum resistance to races AC-2, 4, 7, and 9. In *Brassica* genotypes, significant additive gene effects and dominant gene effects govern resistance to Alternaria blight disease. In some crosses, additive x dominance epistatic effects are significant. In *Brassica juncea* genotypes, *Alternaria* tolerance with components of horizontal resistance attributes has been identified. In the interspecific crosses of *B. juncea* × *B. carinata*, single resistance gene with complete dominance confers resistance to *Erysiphe*. In *A. thaliana*, accessions resistance to powdery mildew is either polygenic based on atypical R-gene RPW8 or combination thereof with RPW8 representing major QTL. The inheritance of enhanced R-genes in *edr* 1 mutant is conferred by recessive genes to powdery mildew. In *Brassica*, resistance to *Fusarium* is conferred by a dominant gene. Inheritance of resistance in crucifers to downy mildew has been observed to be governed by one to four dominant complementary genes, additive genes, one to two partial dominant genes, partial dominant epistatic and gene dosage effects along with quantitative genes depending on host species, and pathotype interactions. In *Brassica*, most resistant cultivars contain major genes (QTLs) conferring resistance to specific blackleg pathotypes (*Leptosphaeria maculans*). Resistant loci to *L. maculans* are distributed in A- and B-genome of *B. napus*. Clubroot (CR) resistance in *B. oleracea* is governed by 2 to 4 dominant genes, additive and recessive genes, and more than 22 QTLs. In *B. napus*, clubroot resistance is governed by one to two dominant genes and more than 19 QTLs, whereas in *B. rapa*, two dominant genes interact in a nonadditive manner to confer resistance to different *Plasmodiophora brassicae* pathotypes. Across all *Brassica* spp. at least 15 CR loci from *B. rapa*, 22 QTLs from *B. oleracea*, and 16 QTLs from *B. napus* have been identified. In *B. rapa*, chromosomes AO3 carry the highest number of loci. *P. brassicae* pathotype 5× identified two independent QTLs, *Rcr* 8 and *Lc9*, in the A-genome on chromosomes AO2 and AO8. A single dominant gene, *Rcr* 7, has been identified to provide resistance to pathotypes 3H and 5X in the C-genome on chromosome CO7. A modifier gene has been identified to increase efficiency of a dominant CR gene against virulent pathotypes. The inheritance of resistance to *Sclerotinia* is complex since no single consistent mode exists to provide resistance to *Brassica* crops. In *Brassica*, partial dominant, polygenic, additive epistatic interactions confer resistance to *Sclerotinia*. Quantitative resistance in both leaf and stem by dominant and additive × additive epistatic interactions controls the genetic effects. In *B. rapa*, resistance to Turnip mosaic virus (TuMV) is controlled by dominant and recessive genes. In *B. oleracea*, resistance to TuMV has been observed to be controlled by both dominant genes and multiple loci. The inheritance of resistance in *Brassica* species to black rot (*Xanthomonas*) is very complex and depends on genotype-race interactions. In *B. oleracea*, resistance is controlled by single dominant gene to race 1 and 4. It is polygenic recessive to races 1 and 3 in *B. oleracea* line Badger Inbred-16 and high level of resistance to races 1 and 4 in *B. carinata* controlled by a single dominant locus (Xca1). Resistance to race 4 in three *B. napus* lines is controlled by a single dominant locus (Xca4).

Latent period and sporulation capacity are two major components of partial resistance and governed by the same mechanism in the interaction between broccoli (*B. oleracea*) and *H. parasitica*. Twenty doubled haploid broccoli lines from breeding material were evaluated for resistance to *H. parasitica* at the seedling stage. All lines supported sporulation of the pathogen, but to varying extents and intensities. Partial resistance of the more resistant lines 'br8' and 'br9' reduced conidia production on cotyledons by 50–70% compared with the most susceptible lines. Inoculation of the two most resistant lines with 13 isolates of different geographical origins revealed that 'br9' showed a rather uniform level of resistance to all isolates while 'br8' showed some isolate specificity. Partial resistance was evaluated in six of the broccoli lines in a half diallel set of crosses. Disease assessment of seedlings showed that additive genetic effects explained 45.8% and 31.8% of the total variation of sporulation score and conidia production, respectively. This suggests that recurrent selection for partial resistance to *H. parasitica* in early generation inbreds or in populations of broccoli will be efficient to obtain cotyledon resistance (Jensen et al. 1999).

Seedling resistance conditioned by a single major gene is isolate specific and may lose the effect through selection of virulent types of the pathogen, if resistant cultivars are produced on large areas (Lucas et al. 1988). Partial quantitative disease resistance by definition provides reduced host invasion without complete protection and is generally anticipated to be based on several genes. It may be a source of durable resistance with less isolate specificity than resistance based on single genes (Parlevliet 1992). In *B. oleracea*, partial resistance to *H. parasitica* at the seedling stage has been reported in Portuguese coles (Monteiro and Williams 1989; Dias et al. 1993) and in gene bank accessions of cabbage and broccoli (Thomas and Jourdain 1990, 1992). Jensen et al. (1999) characterize partial resistance to *H. parasitica* at the cotyledon stage within 20 doubled haploid broccoli lines originating from a wide range of breeding material. Inheritance of the resistance was analyzed in progeny from a half diallel set of crosses.

6.5 Transfer of Disease Resistance in *Brassica* Crops

Cruciferous crop plants (canola/oilseed rape/rapeseed/mustard/cabbage/cauliflower) are economically important oilseed crop, widely used as edible oil. These crops are mostly cultivated in temperate and subtropical regions. To advance further comprehensions into the *Brassica*–pathogen interactions, it is critical to have current knowledge about the advanced skills which can offer encouraging consequences and help to impetus limitations in understanding of the resistance mechanisms. Resistance development is the utmost economical, environmentally sustainable resolution to cope diseases in *Brassica* species. Transfer of disease resistance remains attained once sources of R-genes recognized with sound characterized mode of inheritance of genes in the progenies with all desired traits of yield and quality. To develop stress tolerance and/or resistant *Brassica* cultivars, there is need to apprehend both of the plant and pathogenic disease development mechanisms.

The transfer of disease resistance is achieved after identification of R-genes sources with well-characterized mode of inheritance of genes in the progenies to get a better output as disease resistance cultivars with all desired traits of yield and quality. Various methods including conventional and molecular approaches have been used either alone or in combination to incorporate desired traits to develop improved cultivars of crucifers. Hybridization combined with backcrossing either singly or in combination with pedigree selection is the most widely used and accepted conventional method to transfer disease resistance in crucifers. The level of resistance in developed cultivars has been achieved through multiple crosses, recurrent selection, diallele crossing, and selective mating system. The resistance has been introgressed with intraspecific pedigree, backcross, modified recurrent mass selection, interspecific genome substitution, chromosome substitution, and gene introgression. The molecular and genetic engineering approaches like genome manipulation, cytoplasmic genome manipulation, transformation, foreign gene expression, and embryo rescue for wide hybridization or distant hybridization have been exploited. The introgression and pyramiding of major R-QTLs have been attempted including transgenic development. In crucifers, substantial problems have been faced in the introgression and pyramiding of major QTLs including transgenic development because of diverse nature of pathogens infecting these crops with different modes of infection, pathogenesis, and survival as biotrophs, hemi-biotrophs, and necrotrophs.

Substantial progress has been made in transfer of R-genes from different sources and developing of disease-resistant cultivars of major cruciferous crops against major pathogens using conventional and molecular biotechnological approaches. However, problems in transfer of R-genes arise due to diverse nature of pathogens which are physiologically and taxonomically highly variable; therefore no single gene product and approach can be expected to have a direct effect on all types of pathogens. The advent of molecular genetics has provided new opportunities for mapping and tracking of genes of disease resistance leading to more efficient marker-assisted selection. The improvement in plant transformation protocols and high-throughput gene attenuation methods like RNA interference and virus induced gene silencing (VIGS) will accelerate the identification and transfer of novel genes for durable resistance in crucifers. The introgression of R-genes to major pathogens have been widely used to breed disease-resistant cultivars in *Brassica* crops through inter- or intraspecific crosses and embryo rescue. The conventional and modern approaches have been used to exploit R-genes sources from wild relatives to breed disease-resistant cultivars of *Brassica* crops. For developing durable Alternaria blight–resistant *Brassica* crops, strategies like inter- or intraspecific introgression of R-genes from nonhost plant to *Brassica* crops, microarray-based gene identification and transfer by gene silencing, and transfer of an endo-chitinase gene from *Trichoderma* approaches have been used. R-genes sources of multiple disease resistance have been utilized to breed resistant cultivars of *Brassica* to *Albugo*, *Alternaria*, *Erysiphe*, *Hyaloperonospora*, *Leptosphaeria*, and *Xanthomonas*. To combine the traits from distantly related *Brassica* species for development of superior, durable, and tolerant cultivars of *Brassica* to major diseases, approaches like in vitro embryo rescue, somatic hybridization, interspecific hybridization, distant hybridization, somaclonal variations, genetic transformations, molecular markers, mutagenic resistance, induction of systemic resistance, and genetic engineering have been exploited. Substantial progress has been made in breeding for clubroot resistance genotypes/cultivars of *B. rapa*, *B. oleracea*, *B. napus*, *Raphanus sativus*, and *A. thaliana*. Several clubroot loci or QTLs have been mapped on the chromosomes of these species. There is substantial genetic and pathotype variation in *Plasmodiophora brassicae* populations all over the world, and the deployment of a cultivar with single gene resistance imposes a strong selection pressure to evolve new virulences to overcome this resistance. The successful development of *Sclerotinia*-resistant *Brassica* cultivars has been obtained by introducing multiple-resistant loci (QTLs) from wild relatives using marker-assisted introgression approach. Black rot (*Xanthomonas*) resistance has been introgressed from *B. carinata* to *B. oleracea* through embryo rescue. Somatic hybrids by protoplast fusion between rapid cycling *B. oleracea* and *Xanthomonas*-resistant lines of *B. napus* have been developed. Understanding the evolutionary potential between the pathogen and the host is essential in developing and transfer of R-genes for durable *Brassica*-resistant cultivars. In the present era, the breeding for disease-resistant *Brassica* cultivars, application of molecular approaches like marker-assisted selection, genome wide marker-assisted selection, genetic diversity analysis, and comparative genomics have opened new avenues for superior, durable, and multiple disease-resistant cultivars in different *Brassica* crops (Saharan et al. 2021).

6.6 Host Resistance Signaling Network System to Multiple Stresses

Cruciferous crops can suffer from stress too due to various biotic and abiotic. Biotic stress is caused by living organisms like fungi, bacteria, oomycetes, viruses, insects, weeds, and parasitic animals and plants. These organisms can infect plants to draw plant-produced nutrients thus causing stress in plants. These stresses can significantly reduce crop yield and sometimes they kill the plants. Abiotic stresses are caused by nonliving factors such as high temperature, low temperature (frost), excess water, drought, salinity, mineral toxicity, and lack of nutrients and minerals. These stresses can affect plant growth and development, causing significant reductions in crop yield. Some of these stresses can coexist increasing the overall magnitude of stress on plants. In fact, stress can seriously affect the plant growth, which can have a severe implication which further reduces the growth and ultimately the loss in food production. With the world's growing population, there is need to increase food production. One significant way to do this is to find ways to reduce stress in plants. There is need to develop or engineer the plant for multiple stress tolerance. The molecular plant breeding can help in the development of stress-tolerant plants with the help of microbe interactions, particularly with the help of nonhost resistance and for developing genetic resources for plant functional genomics.

Multiple stresses in the form of biotic and abiotic stresses, as a part of the natural ecosystem, seriously impact crucifer's crop productivity and threaten global oilseed and vegetable food security. Plants in their natural habitats are persistently and simultaneously confronted with a range of biotic and other internal adverse changes imposed by multiple interactions of soil environment. Environmental pollution, global warming, and climate change exacerbate the impact of biotic and abiotic stresses on plant growth and yield. Plants have evolved sophisticated defense network, also called innate immune system, in response to ever-changing environmental conditions. Recent evidence highlights the complex nature of interactions between biotic and abiotic stress responses, significant aberrant signaling crosstalk in response to combined stresses, and a degree of overlap, but unique response to each environmental stimulus. Further, the results of simultaneous combined biotic and abiotic stress studies indicate that abiotic stresses particularly heat and drought enhance plant susceptibility to plant pathogens. Plant biotic and/or abiotic stress interactions trigger a wide range of defense reactions at molecular and cellular levels to ensure that plant cells are insulated from the challenges they encounter (Kim et al. 2014).

A sophisticated defense network is activated to orchestrate transcriptional reprogramming using a two-tiered defense approach (Dodds and Rathjen 2010; Suda and Katagiri 2010). Pathogen−/microbe-associated molecular patterns (PAMP/MAPM)-triggered immunity (PTI) is the first line of defense. It is a general and nonspecific defense response and provides basal resistance not only against entire classes of microbial pathogens but also abiotic stresses. PTI constitutes the inducible defense through an array of early cellular responses comprising of ion flux across the membrane, production of reactive oxygen species (ROS) and mitogen-activated protein kinase (MAPK) cascades phosphorylation, and long-term later responses that induce callose deposition (Nakagami et al. 2005; Dodds and Rathjen 2010; Suda and Katagiri 2010; Schwessinger et al. 2015).

Defense-related phytohormones comprise salicylic acid (SA), jasmonic acid (JA), ethylene (ET), and abscisic acid (ABA) as the core immune signaling components. They are produced in response to both, biotic and abiotic stresses (Dodds and Rathjen 2010; Suda and Katagiri 2010). Among those, ROS, MAPkinase cascades, and hormone signaling pathways are of paramount importance as they lead the defense responses to biotic and abiotic stresses (Nakagami et al. 2005; Fujita et al. 2006; Atkinson and Urwin 2012; Kissoudis et al. 2014; Perez and Brown 2014; Schwessinger et al. 2015). Nejat and Mantri (2017) proposed that PTI which is activated either by PAMP or MAPM is the "stress-triggered immunity" as it has evolutionary conserved molecular signatures across similar types of microbes or abiotic stresses. The effector-triggered immunity (ETI) is the second, prolonged and robust layer of plant immune response that is specific to the pathogen infection. It is activated by resistance R-genes when pathogen virulence factors, called effectors, are released into plant cells (Cui et al. 2015). If the first line of inducible nonspecific defense, so-called PTI, is successfully overcome by pathogens through suppression of PTI response, then the second line of plant immunity is activated to produce defense proteins that cause programmed cell death. This is also referred to as the hypersensitive response (HR) (Thomma et al. 2011). Hence, after stress recognition, the immune system triggers a wide range of defense mechanisms to orchestrate transcriptional reprogramming through receptor proteins, signal transduction cascades, kinase cascades, reactive oxygen species, hormone signaling pathways, heat shock proteins, and transcription factors to protect plants against multiple stresses. However, recent advances suggest significant aberrant signaling crosstalk in plant resistance response to multiple stresses compared with an individual stress (Atkinson et al. 2013; Prasch et al. 2013; Suzuki et al. 2014).

Cruciferous plants have developed numerous molecular mechanisms to respond to various environmental and multiple biotic and abiotic stresses. Whenever plants encounter any kind of stress, first of all there is perception of the signals by the potential sensors. This further generates secondary messengers (e.g., ABA, ROS, Ca^{2+}, inositol phosphates). The secondary messengers can alter the levels of intracellular Ca^{2+}. Any change in cytoplasmic Ca^{2+} is sensed by Ca^{2+} sensors, that is, calcium-binding proteins, and after interacting with corresponding interacting partners, they initiate a protein phosphorylation cascade. This signaling cascade ultimately regulates the transcription factors that manage specific suite of stress-regulated genes and the functional proteins involved in protecting the cell against stress. The expression of stress-regulated genes leads to the production of various kinds of regulatory molecules like the phytohormones ABA, ET, and SA and transcription factors. A second round of signaling can get started by these regulatory molecules which may further enhance the initial signaling pathway. These signaling networks are very complex and interconnected at many levels for transferring information to generate robust host resistance. Moreover, during signaling other components are also involved. A successful signal transduction pathway demands proper coordination of all signaling molecules. This is brought about by specific molecules whose functions are to transport, bring together, or modify the signaling molecules. They themselves are not involved in relaying the signal. These proteins consist of scaffolding proteins, protein modifiers (e.g., enzymes needed for protein glycosylation, methylation, ubiquitination, and lipidation), and adaptors. Cruciferous interactions with multiple stresses trigger a wide range of defense reactions at molecular and cellular levels to ensure that host cells are insulated from these challenges they encounter. The production of ROS, MAP kinase cascades, and hormone signaling are triggered by multiple stresses and play crucial roles as common components of the host immune system. GbRLK, a receptor-like kinase (RLK) gene, is involved in response to multiple stresses as positive regulator transcription factors (TFs) are the central core of the genes as defense signaling pathways to multiple stress. MYB TFs play essential roles in multiple stress responses. ATAF1 is a multifaceted TF with versatile capabilities and plays an important role in mediating crosstalk between multiple stress responses. HSPs gene regulate the function of NBS-LRR genes in response to multiple stresses. miRNAs may target only one locus or several loci in response to multiple stresses. Cell signaling in response to multiple stresses largely depends on the SnRK family of protein kinases in plants. Calcium is a key signaling pathway in host responses to multiple stresses through calcium sensor proteins like calcineurin-B-like protein (CBls), the calmodulin (CaM), and calmodulin-like protein (CMLs).

Transcriptome analyses based on microarrays and proteomic studies have given rise to rapid progress in the field of plant signal transduction and gene regulation. GeneChip and cDNA microarrays together with massive whole genome sequencing proved to be useful in identifying novel signaling determinants on a whole-genome scale in response to various stresses. This has increased our understanding toward the complex mechanisms involved in various aspects of plant responses to environmental cues. Proteomics approach is useful in studying post-translational modifications of the proteins. It can also be used to clone unique genes by means of differential analysis which will expedite our understanding of stress signaling mechanisms in plants. Therefore, stress-tolerant plants can be genetically engineered by the combined efforts of plant molecular biologists, physiologists, and breeders.

6.7 Molecular Mechanisms of Host Resistance

Cruciferous crops are attacked by several pathogens which deteriorate yield and quality of major *Brassica* crops affecting global food security. The pathogens infecting these crops have diverse lifestyle, genomic plasticity with rapid evolutionary mechanisms, prolific reproduction, and longer survival nature which is impedimental to manage efficiently under field conditions. Moreover, pathogens secrete small protein known as effector genes, which counter plant defense mechanisms and modulate host physiology to promote pathogen growth and reproduction. However, plants evolve disease resistance (R)-genes encoding receptor proteins that detect and respond to pathogen effector gene, leading to the activation of effector-triggered immunity. This discovery with the use of multiomics of approaches has revealed molecular mechanisms of crucifers' host resistance to manage major pathogens of *Brassica* crops. Host resistance in crucifers to biotic and abiotic stresses is multilayered and multicomponent. Resistance is activated either through SA signaling and simultaneous perception of ethylene and jasmonic acid depending on biotrophic, hemibiotrophic, and necrotrophic *Brassica* host–pathogen interactions. Molecular mechanisms of host resistance against biotrophs have been revealed through identification, functionally characterization, and fine mapping of R-genes and QTLs on different chromosomes of *Brassica* species. The molecular mechanisms of R-genes resistance imparted by NBS-LRR are effective against biotrophs or hemibiotrophs but are not effective against necrotrophs. JA- and ET-mediated signaling is predominantly involved in general defense to necrotrophs. Similarly, R-genes and QTLs have been identified from *B. rapa*, *B. juncea*, *B. nigra*, and *B. napus* to provide resistance to hemibiotrophs and necrotrophs. Cysteine-rich protein kinase genes and multi-environment stable six QTLs suitable for quantitative blackleg resistance in *B. napus* have been identified. Overexpression of *BnNPR1*, *OsPGIP2*, *BnWRKY33*, and *BnMPK4* genes enhances resistance in *B. napus* to Sclerotinia. Seven NBS-LRR encoding genes linked to black rot resistance in *B. oleracea* have been identified on four QTL regions and chromosomes Co1, Co3, CO6, CO7, and CO8, respectively. Biometabolomics mechanisms of *Brassica* host resistance is initiated soon after perception of pathogens by signaling molecules and results in execution of broad range of defense responses to stop the infection and pathogenesis of invading pathogens. Biometabolomics and biomolecular mechanisms of host resistance is expressed in the form of different metabolites (primary and secondary) and biomolecules involved in the induction of biochemical changes and pathways in the host to confer resistance at pre- and post-invasion stages to biotrophs, hemibiotrophs, and necrotrophs. Transcriptomics and proteomics have discovered novel biochemical pathways and proteins/genes involved in conferring resistance to virulent pathotypes. With the use of multiomics supplemented with functional genomics studies, key resistance pathways involved in *Brassica* host-pathosystem and induction of differential signaling molecules like SA, JA, and ET in response to diverse pathogens have been revealed.

Novel molecular approaches have been exploited to breed disease-resistant cultivars of *Brassica*. Through introgression breeding high level of resistance has been obtained from *B. nigra* and *B. carinata* to *B. napus* against black leg (*Leptosphaeria*). Similarly transfer of resistance to black rot from *B. carinata* to *B. oleracea*, from *B. hirta* to *B. juncea* against *Alternaria*, and from *B. carinata* to *B. oleracea* against powdery mildew has been obtained. The approaches like in vitro embryo rescue, somatic hybridization, somaclonal variations, genetic transformation, molecular markers, use of systemic resistance induction, and genetic engineering have great promise but have been used against few pathogens. Clubroot resistance in *B. napus* and *B. rapa* has been transferred through interspecific and distant hybridization using molecular markers. R-genes from European Clubroot-Differentials (ECD) set have been introduced into *B. napus*, and other species, to provide resistance to clubroot. Powdery mildew resistance in *B. oleracea* has been obtained from *B. carinata* through embryo rescue. The mutagenic agent gamma radiation increases resistance to *E. cruciferarum* in *Brassica*. Durable resistance in *Brassica* species to *Alternaria* can be obtained through introgression of genes from nonhost plants by inter- or intraspecific crosses, identification and transfer of NHR based on microarray, identification and transfer of NHR genetic resources from model host plants, identification of functional genes based on bioinformatics, and use of endo-chitinase gene from *Trichoderma virens*. Marker-assisted introgression of resistance from wild relatives has been achieved in *B. juncea* to *Sclerotinia*. Resistant *B. napus* lines pyramided with multiple R-QTL derived from wild *B. oleracea* have been developed via hexaploidy strategy. Introgression of black rot resistance from *B. carinata* to *B. oleracea* through embryo rescue has been attempted successfully. The *B. carinata* genotypes, NPC-9 and HCA-6, having very high resistance to *XCC* race 1 and 4 have been used to introgress resistance in Indian cauliflower through embryo rescue approaches.

The advances made in the molecular genetics through whole genome sequencing of *Brassica* crops have led to quantum jump not only in understanding plant genetic diversity but also for accessing these enormous resources. Novel sources of R-genes for blackleg has been tapped for other A- and C-genomes of *Brassica* species from the B-genomes in *B. nigra*, *B. juncea*, *B. carinata*, and a wild species *Sinapis arvensis*. Several sources of resistance to 15 major pathogens of crucifers except nematodes have been identified from different *Brassica* species and wild relatives. Some of these sources have been utilized to breed disease-resistant cultivars of *Brassica* through conventional and molecular approaches. From *Brassica* species, multiple pathogen resistance sources have been identified which provide resistance to Albugo, Alternaria, Erysiphe, and Hyaloperonospora. There is strong evidence for positive genetic correlation among *Albugo*, *Alternaria*, *Erysiphe*, *Hyaloperonospora*, Leptosphaeria, and Xanthomonas to develop multiple disease-resistant cultivars of *Brassica* crops. Clubroot-resistant cultivars with R-genes and QTLs have been developed in *B. rapa* and *B. napus* with pathotype specificity at different counties, but most of them have scummed through evaluation of new virulences of Plasmodiophora. In *Brassica* crops, transfer of resistance through different approaches is facing problems because of diverse nature of infecting pathogens for which there is no single gene product to have a direct toxic effect on all kinds of pathogens. Consequently host has to develop quite different strategies to deal with these pathogens.

6.8 Management of Disease Resistance

Management of crucifers' disease resistance for its effective utilization under field conditions is very important and challenging. The *Brassica* genomes (A-, B-, and C-genome) are crucially important to provide novel genetic inheritance for economically important traits that can be used for the overall improvement of crop production and quality. For example, single genome of diploid *B. rapa* (Agenome) holds more than 230 R-gene sequences in 16 gene families (Cheng et al. 2011), among which, over 8 genetic loci have been identified in different research studies, which have functional properties for crucifers' disease resistance. There are possibilities of the existence of more R-genes specific to the disease resistance and their allelic variations in different genetic pools or wild relative species. Effective utilization of resistance loci and their allelic variations may enhance the durability of resistance against different diseases in different *Brassica* species. As a long history of host and pathogen association, it has been revealed relatively high evolutionary patterns of the pathogens. In various cultivating geographical regions of *Brassica* crops, persistence of different pathotypes with high levels of pathogenicity poses challenges to breed durable disease-resistant cultivars. The breakdown of disease-resistant cultivars has become a serious problem in various countries of the world. Effective management of resistance genetic resources in breeding novel cultivars could enhance the performance of resistance loci in different *Brassica* species for sustainable, more durable, and cost-effective control of the crucifers' diseases. Understanding the evolutionary potential between the pathogen and the host is essential in developing durable-resistant *Brassica* crops.

In the integrated disease management strategies of cruciferous crops, use of R-genes through development of disease resistance cultivars is one of the most important ecofriendly and cost-effective components. Major R-genes impose strong selection pressure on pathogen populations and in combination with a number of factors can act to cause "boom-and-bust" cycle of pathotype diversification within a population leading to breakdown or erosion of R-genes. Strong selection imposed by widespread deployment of one R-gene also may lead to breakdown of resistance conferred by another gene, if the two complementary avirulence genes are closely linked. Therefore, the effective and sustainable deployment of resistant cultivars to manage crucifers' disease requires the integrated approaches to manage disease resistance with other control measures. The deployment of R-genes cultivars based on qualitative and quantitative pyramiding is considered as a key lever to increase the durability of R-genes against rapidly evolving pathogen populations. The disease resistance can be effectively managed in crucifers' crops against different pathogens under field conditions by fortification with different approaches. To manage white rust under field conditions, in addition to major gene, resistant cvs. With slow white rusting and systemic resistance induction approaches may be exploited. Nonhost resistance (NHR) is the most durable form of plant immunity effective against all genetic variants of all pathogens and it is multilayered. NHR-related genes can be transferred to host plants to confer durable resistance to *Alternaria brassicae* and other pathogens of crucifers. Multilayered and multicomponent mechanisms of resistance to crucifers' powdery mildew pathogen may be more durable along with major gene resistance under field conditions. The competitive ability of the disease resistance genotype increases significantly in the presence of *Hyaloperonospora parasitica* isolate when competing with susceptible genotype. Small RNA inhibits infection of *Hyaloperonospora*. The deployment of different R-genes cvs. of *Brassica* against *Leptosphaeria* through space and time derives the genetic composition of the pathogen populations. The deployment of R-genes cvs. of *Brassica* to manage clubroot has been the dominant component to manage *Plasmodiophora* in several countries. The evolution of new virulence in the pathogen has eroded resistance in these countries. R-genes pyramiding in *Brassica* cvs. May provide protection against different *P. brassicae* pathotypes. The erosion of R-genes in *Brassica* may be managed through integrated approaches such as use of nonhost and basal resistance, use of bacterial and fungal biocontrol agents, use of ProCa, nitrogen, and arginase gene to modulate resistance positively. Introgression of R-genes from *B. carinata* against *Psuedocercosporaella* (white leaf spot) in *Brassica* crops has enormous potential. Hypovirulent isolates of *Sclerotinia* with the potential in reducing virulence in pathogen populations can help to manage stem rot disease. The breeding of TuMV *Brassica* cvs. can be achieved through transformation of R-genes into crops, molecular markers-assisted selection (MAs), pyramiding of R-genes, and host-induced gene silencing (HIGS). Interspecific hybridization with related or distant species followed by embryo rescue or somatic hybridization for the development of black rot (*Xanthomonas*)-resistant hybrids has been accomplished. Beet cyst nematode has been managed by pyramiding of R-genes in *Brassica*. Multiple pathogen disease resistance of *Brassica* has been managed by development of transgenics, use of fungal and bacterial biocontrol agents, and use of other defense-related genes like WRKY and CAMTA. The breakdown of resistance in *Brassica* can be assessed with molecular approaches to prevent another breakdown of disease resistance. The phenomena of net cost of R-genes resistance in the diseased crucifers operates under natural cultivation system.

6.9 Development of Resistance Cultivars' Techniques

Techniques, methodology, procedures, and protocols developed are essential and prerequisites for research and teaching in science and technology as well as for the development of resistant varieties of crucifers. To initiate any research project, the first and foremost requirement is a standardized, validated, and reproducible technique. A scientific technique is a systematic way of obtaining information about a scientific nature in the areas of explored and unexplored material or product or biotic and abiotic interactions. Scientific techniques can be divided into preparative (cultures purification, growth, and analysis) and measurement techniques like molecular characterization, isolation, cloning, and other genomics protocols. Some selected techniques have been included in this book to facilitate research workers to use, refine, and obtain desired results in the areas of genomics of crucifers and host resistance. For the benefit of graduate students and research scientists of *Brassica*, these selected techniques with methodology, procedures, and protocols will be helpful in planning disease resistance breeding programs. These techniques are standardized and have been developed by experts working on genomics of crucifers' host-pathosystem. They can be used for identification of sources of resistance with R-genes by screening germplasm from different sources. The procedures and methodology for transfer of resistance through various approaches have also been included for developing and breeding disease resistance cultivars of *Brassica* against range of pathogens. To get reliable and widely acceptable results in research pursuits is the primary object of any research programs which can be met only through standardized techniques included in the chapter.

6.10 Exploitation of Novel Protocols for Breeding Disease-Resistant Cultivars of Crucifers

Cruciferous crop plants (canola/oilseed rape/rapeseed/mustard/cabbage/cauliflower) are economically important crops, widely used as edible oil vegetables. These crops are mostly cultivated in temperate and subtropical regions. *Brassica* crops are always challenged by major diseases such as Sclerotinia stem rot, white rust, downy mildew, Alternaria blight, blackleg, and clubroot. These diseases cause significant yield and economic losses in rapeseed-mustard and vegetable crops worldwide, which has been aggravated by global climate change. If it is not resolved meritoriously, it will impend global food security. To advance further comprehensions into the *Brassica*–pathogen interactions, it is advisable to analyze critically the current knowledge in this area and how advanced skills and technologies can offer encouraging consequences to understand the resistance mechanisms. Resistant varieties development is the utmost important, economical, environmentally sustainable solution and effective to cope with diseases management in *Brassica* species. Transfer of disease resistance remains attained once sources of R-genes have recognized with sound characters, mode of inheritance of genes in the progenies along with all desired traits of yield and quality. Conventional and molecular approaches have been used either alone or in combination to incorporate desired and popular traits to develop improved cultivars of crucifers. Hybridization combined with backcrossing either singly or in combination with pedigree selection is the most widely used and accepted conventional method to transfer disease resistance in crucifers. The level of resistance in developed cultivars has been accomplished through multiple crosses, recurrent selection, diallele crossing, and selective mating system. The resistance has been introgressed with intraspecific pedigree, backcross, modified recurrent mass selection, interspecific genome substitution, chromosome substitution, and gene introgression against different pathogens. The molecular and genetic engineering techniques like genome manipulation, cytoplasmic genome manipulation, transformation, foreign gene expression, and embryo rescue for wide hybridization have been exploited in *Brassica*. The introgression and pyramiding major RQTLs have been attempted including transgenics development. To develop stress tolerance and/or resistant *Brassica* cultivars, there is need to apprehend both of the plant and pathogenic disease development mechanisms.

6.11 Transfer of Disease Resistance from Germplasm Sources

Several attempts have been made to exploit resistance sources from the wild relatives of *Brassica* and transfer them into cultivated Brassica (Aneja and Agnihotri 2013). Efforts have been made in screening blight-resistant cultivars; however, high levels of resistance to *A. brassicae* have not been observed among cultivated *Brassica* species. The species show varying degrees of resistance responses against Alternaria blight, and accordingly, *B. rapa* and *B. juncea* are classified as being more susceptible to the disease than *B. napus* and *B. carinata* (Skoropad and Tewari 1977). These related wild species have been found to harbor genes conferring resistance to Alternaria blight that can be used for resistance breeding (Ghose et al. 2008; Chatterjee et al. 2013).

In the past few decades, various conventional and modern approaches have been used to develop blight-resistant cultivars by exploiting resistance sources from host germplasm. Conventional approaches involve the screening of germplasm for blight-resistant plant materials, followed by the transfer of resistance traits from tolerant species into susceptible cultivars by hybridization combined with multiple crosses. Inter- and intraspecific crosses of *B. juncea* and *B. carinata* were established to understand the inheritance pattern of resistance to Alternaria blight (Krishnia et al. 2000), indicating that dominant and additive genes govern the inheritance of resistance to Alternaria blight. Thus, a successful breeding method should utilize both the gene effects. However, there is a limitation in such breeding approaches owing to the low availability of suitable resistance sources/genes within the available germplasm of cultivated *Brassica* and insurmountable self-incompatibility barriers in combining the resistance traits from distantly related wild *Brassica* species. Modern biotechnological interventions comprising tissue culture and genetic transformation approaches have also been recognized for the development of improved varieties. In vitro ovary culture and ovule culture have been attempted for transferring resistance traits from *S. alba* cv. Carine to *B. napus* cv. Brutor (Chevre et al. 1994) and *Erucastrum cardaminoides* to *B. oleracea* var. *alboglabra* (Mohanty et al. 2009). Intergeneric hybrids of *B. campestris* and *B. spinescens* have also been developed (Agnihotri et al. 1991). Moreover, the somatic hybridization technique has been applied for transferring the resistance traits from blight-tolerant species to susceptible ones, for example, *S. alba* to *B. napus* (Primard et al. 1988), *B. carinata* to *B. juncea* (Sharma and Singh 1992), and *B. nigra* and *B. oleracea* (Jourdan and Salazar 1993). Attempts have also been made for the introduction of somaclonal variations for incorporation of disease resistance/tolerance to Alternaria blight through mutagenesis using gamma rays, ethyl methanesulfonate, and ethyl nitrosourea (Verma and Rai 1980; Agnihotri et al. 2009, Meena et al. 2020). In addition, the defense response might be activated by treatment with biotic and abiotic agents for combating *A. brassicae* infection in oilseed Brassicas. Vishwanath et al. (1999) showed that resistance in a susceptible cultivar of *B. juncea* against the extremely virulent *A. brassicae* isolate A and reasonably virulent isolate C, both from *B. carinata*, was induced after inoculating the plants with the avirulent *A. brassicae* isolate D. Similarly, the application of β-aminobutyric acid on the leaves of *B. carinata* was found to induce resistance to *A. brassicae* (Chavan et al. 2013). In future, the activation of the defense response without gene manipulation may act as a better substitute for the conventional use of fungicides. Unfortunately, till date, none of the conventional or modern biotechnological methods have been found to be feasible or stable in developing blight-resistant *Brassica* species. Under such circumstances, it is pragmatic to use the novel sources of resistance from plant species that are not hosts to the pathogen. The transfer of genes from nonhost plants conferring resistance to the pathogen is a promising strategy in parallel to the current efforts toward the development of oilseed *Brassica* varieties resistant to Alternaria blight (Fatima et al. 2019).

6.12 Sources of Resistance

Digenomic *Brassica* species such as *B. napus* and *B. juncea* have better sources of resistance than the monogenomic species like *B. rapa.* Some lines of *B. alba*, *B. carinata*, *B. spinescenes*, *B. maurorum*, *Eruca sativa*, *Camelina sativa*, *Capsella bursa-pastoris*, *Diplotaxis* sp., and *N. paniculata* have also shown resistance to *A. brassicae* in different areas of the world (Brun et al. 1987a, b; Conn et al. 1991; Grontoft 1986; Jejelowa et al. 1991; Saharan 1992, 1997; Tewari and Conn 1993; Hansen and Earle 1997; Chrungu et al. 1999; Sharma et al. 2002). Sources of resistance to *A. brassicae* and *A. brassicicola* in different host species from different areas of the world have been reported. In India, *Brassica* genotypes, namely, CSR 43, CSR I42, CSR-I42–2, CSR 343, CSR 448, CSR 622, CSR 741, Gulivar, KRV-Tall, Midas, PHR 1, RC-781, TMV2, Tower, and YRT3, were consistently found to have field resistance proven after testing for several years at different locations under uniform disease nursery trials (Kolte 1985b, 1987; Saharan 1984, 1992, 1997; Saharan and Chand 1988; Saharan et al. 2003, 2005, 2014, 2016, 2018, 2019, 2021; Kumar et al. 2015; Verma and Saharan 1994; Meena et al. 2019c). *Brassica* coenospecies are rich reservoir for genetic resistance to *A. brassicae* (Sharma et al. 2002). *Alternaria brassicae* pathotype-specific resistance has been recorded in genotype, GS-05-1 of *B. napus* to pathotypes, Abr 1 and Abr 5, whereas genotypes, RH-8544, Pusa Swarnim, and HC-9605 of *B. juncea* showed moderately resistant reaction to three pathotypes (Kumar et al. 2014). Pravel et al. (2015) reported tolerant to Alternaria blight genotypes of *B. juncea* DRMR 2805 and DRMR 2806. Tolerance to Sclerotinia rot was reported in *B. juncea* genotypes, EC 597317, EC 597328, RH 1222–28, DRMR 261, DRMR 360, DRMR 1493, DRMR 1034, and DRMR 2035 (Sharma et al. 2018). Resistant to white rust was reported by Meena et al. (2019a, b) in DRMR 2019 and DRMR 2035 genotypes of *B. juncea* while DRMR 2035 also showed tolerance to Sclerotinia rot.

6.12.1 Sources of Disease Resistance from Cruciferous Relatives

Sources with high level of resistance against *A. brassicicola* and *A. brassicae* have not been identified among the cultivated *Brassica* species, although individual cabbage varieties differ considerably in their level of susceptibility to black spot (Otani et al. 2001). The highest level of *Alternaria* resistance among the oilseed *Brassica* crops is displayed by the Ethiopian mustard (*B. carinata*). Among the wild cruciferous plants closely related to the *Brassica* genus, the highest resistance against *A. brassicae* is found in white mustard (*Sinapis alba*) (Kolte 1985a; Brun et al. 1987a, b; Ripley et al. 1992; Sharma and Singh 1992; Verma and Saharan 1994; Hansen and Earle 1995, 1997). The highest overall *Alternaria* spp. resistance, however, has been identified in the crucifers more distant from the *Brassica*, such as *Camelina* (*Camelina sativa*; false flax), shepherd's purse (*Capsella bursa-pestoris*), taramira (*Eruca sativa*), and ball mustard (*Neslia paniculata*) (Conn and Tewari 1986; Conn et al. 1988; Tewari 1991). Resistance against *A. brassicae* and *A. brassicicola* has been reported among other wild members of the Brassicaceae family (Sharma et al. 2002; Tewari and Conn 1993; Warwick 2011), namely, *Alliaria petiolata*, *Barbarea vulgaris*, *B. elongate*, *B. desnottessi*, *B. fruticulosa*, *B. maurorum*, *B. nigra*, *B. souliei*, *B. spinescens*, *C. sativa*, *C. bursa-pestoris*; *Coincya* spp., *Diplotaxis catholica*, *D. berthautii*, *D. creacea*, *D. erucoides*, *D. tenuifolia*; *Erucastrum gallicum*, *E. vesicaria* subsp. *sativa*; *Hemicrambe fruticulosa*, *H. matronalis*, *N. paniculata*, *R. sativus*, *S. alba*, and *S. arvensis*. The completely immune plants remain symptom free both under natural field infection and under controlled artificial inoculation (Sharma et al. 2002). Comparatively, broccoli and cauliflower varieties exhibit moderate *Alternaria* resistance, while the cabbage as susceptible.

6.12.2 Sources of Multiple Disease Resistance

Brassica alba and genotypes, namely, HC-1, PCC-2 (*B. carinata*), GSL-1501, GS-7027, Midas, Tower (*B. napus*), and DIR-1507, DIR-1522 (*B. juncea*), have been identified as sources of multiple disease resistance against white rust, Alternaria blight, downy mildew, and powdery mildew diseases (Dang et al. 2000; Kumar and Saharan 2002). Earlier *B. napus* genotype, GSL-1501, was reported resistant to white rust and powdery mildew (Gupta and Singh 1994). The variations in the incidence of powdery mildew scores among different *Brassica* genotypes differ considerably due to variations in date of sowing, stage of the crop at the time of observations, and prevailing environmental conditions. Hence, early maturing genotypes or the genotypes planted early in the season would help in escaping the powdery mildew infection.

The genotypes EC-322090, EC-322092, EC-322093, EC-399301, EC-399296, EC 399299, PHR-2, RC-781 and of *B. juncea*, HC-1 and PCC-2 of *B. carinata*, and GS-7027, GSL-1, GSL-1501, Gulivar 1, HNS -4 and Tower sel. 1 of *B. napus* are resistant to *A. candida* and *A. brassicae* (Awasthi et al. 2012; Meena et al. 2014;

Sandhu et al. 2015). Whereas genotypes EC-129126-1, EC-129126-2, PR-8805 of *B. juncea*, Candle, Tobin, Torch of *B. rapa*, DHC-1, HDC-4, DHC-1960, HC-1, PCC-2 of *B. carinata*, and GSL-1501, N 20–12-2 and Wester of *B. napus* are resistant to *A. candida* and *Erysiphe cruciferarum*. However, HC-1, PCC-2 of *B. carinata* and GSL-1501 (*B. napus*) are resistant to both *A. brassicae* and *E. cruciferarum*. GSL-1501 and Gulivar-1 of *B. napus* were found resistant to *A. candida* and *E. cruciferarum* by Gupta and Singh (1994). The genotypes Wester (*B. napus* Tobin and Candle (*B. rapa*) are reported completely free from *A. candida* and *E. cruciferarum* (Shivpuri et al. 1997). The variations in the observations may be attributed to the occurrence of new races of the pathogen and congenial weather factors (Kumar and Saharan 2002).

6.13 Relationship Between Major Foliar Diseases

The increase in level of resistance to different pathogens is positively correlated, the improvement in the accumulation of resistance to multiple diseases may be facilitated easily in agricultural or natural populations; negative association among the level of resistance will hinder plant defense against pathogens attack and leads to difficulties in development of multiple disease resistance. The nonavailability of single resistance source line with different mode of inheritance further complicates the work of genetic amelioration for multiple disease resistance. The significant and positive association between disease score of white rust, Alternaria blight, and powdery mildew has been recorded except, white rust stag head and powdery mildew infection in genotypes of different *Brassica* species. A high degree of correlation is observed between leaf phases of white rust and Alternaria blight. Similarly, the positive genetic correlation is observed between *Hyaloperonospora parasitica* and *Leptosphaeria maculans* and *Hyaloperonospora parasitica* and *Albugo candida* in *B. rapa* (Mitchell-Olds et al. 1995; Mehta et al. 2018). Sharma et al. (1991) reported significant positive correlation between Alternaria blight and black rot in cauliflower. The strong evidence for a positive genetic correlation in the level of resistance against *A. candida*, *A. brassicae*, and *Erysiphe cruciferarum* may indicate the presence of plant resistance genes that provide defense against some fundamental characteristics common to different taxonomic orders of fungal pathogens as observed by Bruns et al. (1991). Genetic correlations may arise from several factors like (1) linkage disequilibrium, a nonrandom association among genotypes at several loci, which is rarely found in cross pollinated species (Crow and Kimura, 1970). However, the linkage disequilibrium may arise from previous generations of selection or small population size, (2) Pleiotropy, and (3) tight linkage between resistance genes (Martin et al. 1993).

6.14 Development of Resistant Cultivars

Resistant cultivars are the only solution for protection against *P. brassicae* in combination with suitable crop rotation and management strategies (Peng et al. 2014).

6.14.1 Genetics and Breeding

The clubroot host-plant resistance can be found in main cruciferous crops (except *Brassica juncea* and *B. carinata*) (Diederichsen et al. 2009). The resistance against *P. brassicae* is polygenic (quantitative) and also monogenic (qualitative) (Crute et al. 1980; Voorrips 1996). Many of the resistance types are pathotype-specific and do not cover all pathotypes of *P. brassicae* (Diederichsen et al. 2009). The studies proved the existence of eight resistance loci, some of them are pathotype-specific major genes, and some are quantitative loci (QTL) with specific or nonspecific resistance responses (Voorrips 1996; Hwang et al. 2012). Some studies also proved that resistance is qualitative and is controlled by dominant and recessive genes (Chiang and Crete 1983; Yoshikawa 1993). The oilseed rape–resistant cultivar Mendel was bred from ECD hosts-resistant turnip (*B. rapa*, ECD-04) and kale (*B. oleracea* ECD-15) through backcross with oilseed rape (Diederichsen et al. 2009). The F_1 backcrosses were inoculated with *P. brassicae*, and the most resistant ones were selected for further breeding and genetic analysis (Diederichsen et al. 2009). Later the presence of resistant genes was confirmed. There is one major locus and two recessive genes in the cultivar Mendel, whereas other genes were lost in the breeding process and negative selection (Diederichsen et al. 2009). This is why cv. Mendel is recommended to use wisely together with other management strategies (Diederichsen et al. 2009; Dixon 2009). The same situation exists in other resistant cultivars.

6.14.2 Mechanism of Resistance

The resistance mechanisms are based on suppressing the pathogen development in a host rather than on infection avoidance. The term "resistance" is used because the infection inhibition is based on resistance genes, not on tolerance genes. The zoospore penetration occurs in the root hairs of a resistant host (the primary phase of the pathogen life) (Hwang et al. 2012). The secondary phase of infection in the root cortex also follows, but formed plasmodia do not mature and resting spores are not created (Hwang et al. 2011; Deora et al. 2013). Slight suppression of primary infection and significant suppression of secondary infection occur during the resistance response, the disruption of the cell walls is lesser, and the secondary thickening of the stem is not stopped (Donald et al. 2008). The immune response includes the activation of metabolic pathways for synthesis of phytoalexins, phenolic compounds, lignin, and other substances that block the pathogen invasion (Siemens et al. 2006).

6.14.3 R-Genes Cultivars in the Field

The first resistant cultivars were Mendel and Tosca, marketed at the beginning of the new millennium. These cultivars are heterozygous hybrids with one gene of pathotype-specific resistance. Unfortunately, the Mendel host plant resistance was broken in some parts of Germany and France, where it was grown since 2000 (Diederichsen et al. 2014). In Great Britain, resistance was broken quite rapidly in the field, where cultivars were grown under strong selection pressure (Diederichsen et al. 2003). These cultivars were also tested on local pathotypes in Canada. Cultivar Mendel showed resistance against pathotypes 3, 5, 6, and 8 (Williams set) (Rahman et al. 2014). Cultivar Tosca was susceptible to pathotype 3 and resistant to pathotype 2 (Williams set) (Peng et al. 2013). These results showed that the host-plant resistance is quite variable and some genes may be ineffective in some regions with specific pathotypes (Peng et al. 2014). New resistant cultivars of spring rape with genes from European cultivars were introduced into Canada in 2009 (Peng et al. 2014). In the European market, the available resistant cultivars are derived from the first resistant cultivar Mendel (Diederichsen et al. 2014). It is expected that a wider range of resistant varieties will come to the market in the next years (Diederichsen et al. 2014). However, these varieties cannot be grown without suitable agronomic measures and pathogen-resistance management strategies (Diederichsen et al. 2014). Field experiments with resistant cultivars in infested fields were also conducted in the Czech Republic. Eight resistant cultivars were tested, some of them are normally available on the market and some were test cultivars in the breeding process (new breeds). Normally traded cultivars had stable performance in resistance results and good yield. Cultivars from new breeding showed some variations in resistance performance and yield (Ricarova et al. 2016).

6.14.4 Deployment of R-Genes Cultivars

Clubroot-resistant canola cultivars have been the dominant component of the clubroot management strategy in western Canada in recent years. The first resistant cultivar for western Canada, Pioneer '45H29', was registered in 2009, and its release was quickly followed by several other resistant cultivars (Strelkov et al. 2011; Gossen et al. 2013). However, the durability of this resistance is not known. The genetic basis for the resistance in these cultivars also is proprietary knowledge that is not available to the public. This has complicated efforts aimed at resistance stewardship, since it is not possible to develop rational strategies for the rotation of resistance sources in infested fields. The number of resistance genes also that are currently available to breeders is limited (Hirai 2006). Recent studies indicated that there were no detectable differences in pathotype reactions among clubroot-resistant canola cultivars available in western Canada (Deora et al. 2012, 2013). This represents one line of evidence that resistance genes in all these lines originated from a similar source. In contrast, another study by LeBoldus et al. (2012) revealed that pathogen populations that had been repeatedly cycled on and become adapted to one *Brassica* host did not exhibit a similar adaptation to other hosts, when the cycled populations were inoculated on the latter, indicating that distinct sources of resistance may occur in at least some genotypes. Regardless of the relationships between resistance sources, there is substantial genetic and pathotype variation present in the *P. brassicae* populations in western Canada (Strelkov et al. 2006; Xue et al. 2008; Cao et al. 2009), and the deployment of a cultivar with single-gene resistance against a genetically diverse pathogen on a large scale imposes a strong selection pressure for pathogen genotypes that are able to overcome this resistance (Hwang et al. 2012). Similar mechanisms may operate in other countries also.

6.15 Introgression and Pyramiding of Major R-QTLs into *Brassica napus* Against *Sclerotinia*

To maintain stable production in the face of damage from diseases, the development of resistant lines or cultivars is a cost-effective strategy. However, due to the lack of highly resistant resource in *B. napus* gene pool, the breeding of *Sclerotinia*-resistant rapeseed has been difficult and rather unsuccessful. Alternatively, diploid progenitors which always provide precious gene resources could be considered in rapeseed improvement. *Sclerotinia*-resistant sources were reported in *B. oleracea* germplasm (Mei et al. 2011). In particular, a wild *B. oleracea* genotype (*B. incana*) identified as strong resistance to *S. sclerotiorum* (Mei et al. 2011), provided a promising basis to improve *Sclerotinia* resistance in present rapeseed cultivars. However, this raised problems in utilization of the *B. oleracea* resource by resynthesizing rapeseed technique, that is, resynthesized allotetraploid lines are genetically unstable (Hurgobin et al. 2018; Wu et al. 2018; Zou et al. 2018), and the synthetic rapeseed lines contain A-genome from *B. rapa* which is highly susceptible to *S. sclerotiorum* (Ding et al. 2013; Mei et al. 2011). Therefore, novel breeding strategies are needed to avoid these problems (Mei et al. 2020). There is another bottleneck in *Sclerotinia*-resistant breeding of rapeseed, as it is a typical quantitative trait controlled by multiple genetic loci (Mei et al. 2013; Wei et al. 2016; Wu et al. 2013, 2016). Till date, no resistant gene has been cloned from any of those reported QTLs, and most importantly, a single locus always provides very limited contribution to resistance. These factors have largely limited the molecular marker-assisted breeding for *Sclerotinia*-resistant rapeseed.

Resistance QTLs were identified from *Sclerotinia*-resistant wild *B. oleracea* (Mei et al. 2013), and several molecular markers were proved effective in selecting resistant individuals (Ding et al. 2015, 2019; Mei et al. 2015). A hexaploidy strategy was preliminarily proposed to introduce *Sclerotinia* resistance from the wild *B. oleracea* into rapeseed (Mei et al. 2015). The genotypic (marker-assisted selection) and phenotypic selection in high generations generated from the hexaploidy strategy successfully yielded resistant rapeseed lines with multiple resistant QTLs, improved *Sclerotinia* resistance, and high seed yield and quality. Resistant rapeseed lines pyramided with multiple resistant QTLs derived from *B. oleracea* were developed via a hexaploidy strategy. Interspecific hexaploids were generated between rapeseed variety 'Zhongshuang 9' and a wild *B. oleracea* which was highly resistant to *S. sclerotiorum*, followed by backcrossing with Zhongshuang 9 and successive selfing. By molecular marker-assisted selection, three major resistant QTLs were transferred and pyramided from *B. oleracea* into two BC1F8 lines which exhibited ~35% higher resistance level than Zhongshuang 9 and produced good seed yield and seed quality. The successful development of *Sclerotinia*-resistant rapeseed lines by introducing multiple resistant loci from wild *B. oleracea* has been obtained. This revealed the effectiveness of pyramiding multiple QTLs in improving *Sclerotinia* resistance in rapeseed and provided a novel breeding strategy on utilization of *B. oleracea* in rapeseed improvement (Wang et al. 2020).

6.16 Introgression of Resistance from Wild *Brassica* Species into *Brassica juncea* to *Sclerotinia*

A wild *Brassicaceae* species with excellent resistance against stem rot to develop a set of *B. juncea–B. fruticulosa* introgression lines (ILs) was used by Atri et al. (2019). These were assessed for resistance using a highly reproducible stem inoculation technique against a virulent pathogen isolate. Over 40% of ILs showed higher levels of resistance. IL-43, IL-175, IL-215, IL-223, and IL-277 were most resistant ILs over three crop seasons. Sequence reads (21×) from the three most diverse ILs were then used to create *B. juncea* pseudo molecules, by replacing SNPs of reference *B. juncea* with those of re-sequenced ILs. Genotyping by sequencing (GBS) was also carried out for 88 ILs. Resultant sequence tags were then mapped on to the *B. juncea* pseudo molecules, and SNP genotypes prepared for each IL. Genome-wide association studies helped to map resistance responses to stem rot. A total of 13 significant loci were identified on seven *B. juncea* chromosomes (A01, A03, A04, A05, A08, A09, and B05). Annotation of the genomic region around identified SNPs allowed identification of 20 candidate genes belonging to major disease resistance protein families, including *TIR-NBS-LRR* class, chitinase, malectin/receptor-like protein kinase, defensin-like (*DEFL*), desulfoglucosinolate sulfotransferase protein, and lipoxygenase. A majority of the significant SNPs could be validated using whole genome sequences (21×) from five advanced generation lines being bred for *Sclerotinia* resistance as compared to three susceptible *B. juncea* germplasm lines. The understanding of the defensive pathway of *B. fruticulosa* resistance will also enable development of marker candidates for assisted transfer of introgressed resistant loci into agronomically superior cultivars of crop *Brassica* (Atri et al. 2019).

6.17 Introgression of Black Rot Resistance from *Brassica carinata* to *B. oleracea* Through Embryo Rescue

The black rot resistance sources have been reported in A- and B-genomes of *Brassica* species (Westman et al. 1999; Taylor et al. 2002), and genes present in these genomes can be pyramided in Cole crops to provide a durable solution to this malady. The monogenomic and digenomic species in Uthant triangle carry resistance gene(s) for Xcc, which can be utilized for guarding *B. oleracea* genotypes against this disease using embryo rescue and somatic hybridization techniques. Genetic transformation option can be explored when resistance is not available within and related gene pools. Besides, a number of disease resistance genes in *Brassicas* and *A. thaliana* have been postulated and mapped, but none has been cloned as yet (Vicente and Holub 2013). Therefore, there is no information on genetic transformation for resistance to black rot disease in *B. oleracea* group.

Brassica carinata (Ethiopian mustard, $2n = 4x = 34$, BBCC-genome) is an important oilseed crop thought to have originated in the Ethiopian plateau (Warwick 2011). Its leaves are also consumed as vegetable in Northeast Africa. The *Brassica* B-genome is known to carry several important traits, yet there has been limited analyses of its underlying genome structure (Navabi et al. 2013). *B. carinata* harbors genes controlling valuable resistance/tolerance traits for biotic and abiotic stresses, namely, resistance to black leg and black rot and tolerance to aluminum, salinity, heat, and drought (Pan et al. 2012; Enjalbert et al. 2013). Moreover, strong resistance to Xcc race 1 and 4 has been reported in high proportion of *B. nigra* (B-genome) and *B. carinata* (BC-genome) accessions (Taylor et al. 2002; Vicente et al. 2002). A single dominant gene conferring high resistance to *Xcc* race 1 was reported by Tonguc et al. (2003). A single dominant gene (Xca1bc) controlling black rot resistance was mapped with intron length polymorphic markers at B-7 genome of *B. carinata* (Sharma et al. 2016). Therefore, it is important and feasible to transfer this black rot resistance gene into cauliflower cultivars. Alien gene transfer in crop plants has emerged as a potent approach to broaden gene pools and increase genetic variability in existing germplasm. Hybridization between a diploid and a tetraploid species is difficult and failure occurs at many stages starting from pollination incompatibility to pre–/post-fertilization barriers.

Search for new gene(s) for disease resistance in closer genome-related species, namely, *B. napus* (AACC) and *B. carinata* (BBCC), for incorporation into *B. oleracea* group via interspecific (Tonguc et al. 2003) or somatic hybridization using embryo rescue (Hansen and Earle 1995) has to be a continuous process. Of the three *Brassica* genomes, A and C are very closely related while the B-genome is phylogenetically distant. Zhou et al. (1997) transferred resistance to Xcc from *B. napus* to broccoli (*B. oleracea*) through protoplast fusion. Tonguc and Griffiths (2004b), however, explored the half B-genome *B. juncea* (AABB) for the introgression of black rot resistance into *B. oleracea* using in vitro embryo rescue. Somatic hybridization was attempted to transfer black rot resistance from B. nigra into *B. oleracea* (Wang et al. 2011, 2016).

Utilization of embryo rescue or somatic hybridization techniques, or combination of both, could help to overcome natural reproductive barriers in the development of interspecific hybrids in *Brassica* (Ayotte et al. 1987; Hansen and Earle 1995; Momotaz et al. 1998; Weerakoon et al. 2009; Niemann et al. 2013). Wide crosses between crop plants and their wild relatives have now become routinely possible through the use of the embryo rescue technique. Different techniques of plant cell and tissue culture, such as ovary, ovule, and embryo culture as well as protoplast fusion, have proved useful for production of interspecific hybrids. Rescue of hybrid embryos and their culture in vitro helps to overcome post-fertilization barriers in interspecific crosses. Since its first use in *Brassica* by Nishi et al. (1959), extensive investigations have been carried out to improve the techniques for obtaining higher seed set (Inomata 1993, 2002; Zhang et al. 2003, 2004). The successful application of this technique depends on the stage of the embryo being rescued and cultured in vitro. Several attempts were made to transfer desirable gene(s) from alien *Brassica* spp. to *B. oleracea* such as powdery mildew resistance (Tonguc and Griffiths 2004a), downy mildew (Chiang et al. 1977), male sterility (Chiang and Crete 1987), and atrazine resistance (Jourdan et al. 1989). The progress for marker-assisted Xcc resistance gene transfer from *B. carinata* to cauliflower has been very slow (Tonguc et al. 2003; Tonguc and Griffiths 2004b). The transfer of monogenic resistance to Xcc race 1 from an accession "NPC-9" of *B. carinata* into a commercial susceptible cauliflower variety "Pusa Sharad" using embryo rescue technique has been attempted by Sharma et al. (2017).

Black rot caused by *X. campestris* pv. *campestris* (*Xcc*) is a very important disease of cauliflower (*B. oleracea botrytis* group) resulting in 10–50% yield losses every year. Since there is a dearth of availability of resistance to black rot disease in *B. oleracea* (C-genome), exploration of A- and B-genomes was inevitable as they have been reported to be potential reservoirs of gene(s) for resistance to black rot. To utilize these sources, interspecific hybrid and backcross progeny (B1) were generated between cauliflower "Pusa Sharad" and Ethiopian mustard "NPC-9" employing in vitro embryo rescue technique. Direct ovule culture method was better than siliqua culture under different temperature regime periods. Hybridity testing of F1 interspecific plants was carried out using co-dominant SSR marker and *Brassica* B- and C-genome-specific (DB and DC) primers. Meiosis in the di-genomic (BCC) interspecific hybrid of *B. oleracea botrytis* group ($2n = 18$, CC) × *B. carinata* ($2n = 4x = 34$, BBCC) was highly disorganized and cytological analysis of pollen mother cells revealed chromosomes $2n = 26$ at metaphase-I. Fertile giant pollen grain formation was observed frequently in interspecific F1 hybrid and BC1 plants. The F1 interspecific plants were found to be resistant to *Xcc* race 1. Segregation distortion was observed in BC1 generation for black rot resistance and for different morphological traits. The At1g70610 marker analysis confirmed successful introgression of black rot resistance in interspecific BC1 population. This effort will go a long way in pyramiding gene(s) for resistance against black rot in Cole crops, especially cauliflower and cabbage for developing durable resistance, thus minimizing dependency on bactericides (Sharma et al. 2017).

6.18 Distant Hybridization for *Xcc* Resistance Through Somatic Hybridization and Embryo Rescue

Wide hybridization or distant hybridization is a powerful approach to combine the diverged genomes into a single nucleus. Wide hybridization is an important driving force in broadening the gene pools, the evolution of novel plant forms and speciation, transferring the genomes of one species to another, enhancing the genetic diversity, and improving the quality of crops (Soltis and Soltis 2000; Barton 2001; Jiao et al. 2011). The *Brassica* vegetables and oilseed crops in the Cruciferae family have undergone multiple genomic duplications (Schranz et al. 2006). The classic genome polyploidization event is the interspecific hybridization of *B. rapa* (A-genome), *B. nigra* (B-genome), and *B. oleracea* (C-genome) to form the allotetraploids *B. juncea* (AB-genome), *B. napus* (AC-genome), and *B. carinata* (BC-genome) (Nagaharu 1935). Since the first intergeneric hybrid between *R. sativus* and *B. oleracea*, '*Raphanobrassica*', reported by Karpechenko (1928), numerous intergeneric and interspecific hybrids have been produced between *Brassica* species of the U triangle (Nagaharu 1935) and *R. sativus*. The related or wild species are a rich reservoir of several useful genes having resistance to various biotic or abiotic stresses. In the family Brassicaceae, interspecific hybridization is a potential approach for transfer of desirable genes from one species to another, enhancing genetic diversity and creation of hexaploids for the improvement of *Brassica* crops (Prakash 1973; Prakash and Chopra 1988; Choudhary et al. 1999; Siemens et al. 2002; Niemann et al. 2012; Kaneko and Bang 2014). The sexual incompatibility or reproductive barriers can be overcome by utilization of embryo rescue or somatic hybridization techniques, or a combination of both, in the development of interspecific hybrids in crucifer crops (Hansen and Earle 1995; Weerakoon et al. 2009; Niemann et al. 2012; Dey et al. 2015). The development of black rot resistant hybrids is one of the major objectives of *Brassica* breeders. Since the resistance sources in *B. oleracea* are insufficient, interspecific hybridization with related or distant species followed by embryo rescue or somatic hybridization for the development of black rot resistant hybrids has been practiced and accomplished by many workers.

The lines of *B. napus* have very high resistance to black rot (*X. campestris* pv. *campestris*) and resistance is governed by a dominant gene, but the successful crossing between two species (*B. oleracea* and *B. napus*) is very difficult (Hansen and Earle 1995). Therefore, development of somatic hybrids by protoplast fusion between rapid cycling *B. oleracea* and *X. campestris* pv. *campestris* resistant line of *B. napus* has been attempted. Hybrid identity was confirmed by morphological studies, flow cytometric estimation of nuclear DNA content, and analysis of RAPDs. Wang et al. (2011) generated new alien addition lines having high resistance to black rot (*X. campestris* pv. *campestris*) by employing a somatic hybridization technique between cauliflower and black mustard. It has been evident that success of interspecific hybridization is also influenced by the genotype of the parents (Meng and Lu 1993; Tonguc and Griffiths 2004a, b; Mason et al. 2011). Tonguc and Griffiths (2004a, b) utilized two *B. juncea* accessions (A 19182 and A

19183) having resistance to Xcc race 1 and 4, as maternal and paternal parents to develop interspecific hybrids with *B. oleracea* cultivars. They employed embryo rescue to recover interspecific hybrids and further confirmed by inheritance of paternal molecular markers. They obtained 26 interspecific hybrid plants between A 19182 and *B. oleracea* cultivars, but no interspecific hybrids were obtained using A 19183, suggesting influence of genotype in the success of interspecific hybridization. The interspecific hybrid plants were male-sterile, and they were used successfully as maternal parents to develop backcross plants employing embryo rescue. All hybrid and BC1 plants were found resistant to races 1 and 4 of Xcc. Dey et al. (2015) reported introgression of black rot resistance genes into Indian cauliflower (*B. oleracea* var. *botrytis* L.) through interspecific hybridization (*B. carinata* 9 *B. oleracea*) followed by embryo rescue. They utilized two *B. carinata* genotypes, NPC-9 and HCA-6, having very high resistance to Xcc race 1 and 4, for interspecific hybridization to introgress resistance genes from *B. carinata* into Indian cauliflower employing three different embryo rescue approaches. They reported that all the F1 hybrids and back cross were resistant to *Xcc* races 1 and 4 (Singh et al. 2018).

6.19 Novel Sources and Transfer of R-Genes

Novel sources of R-genes for blackleg can be sourced from other A- and C-genomes of *Brassica* species (Chen et al. 2010) or from the B-genomes in *B. nigra* (Chevre et al. 1996), *B. juncea* (Christianson et al. 2006), and *B. carinata* (Plieske et al. 1998). Wild Brassicaceae species, such as *Sinapis arvensis*, also serve as a good source of R-genes for blackleg (Snowdon et al. 2000). However, R-genes from other wild species remain to be tapped. Besides identifying candidate R-genes through comparative genomic analysis, more screening work on wild and land races, including species from the Brassicaceae family, should be performed against the main pathogens to identify resistant plants. This would broaden the gene pool for introgression of novel R genes or alleles into *Brassica* cultivars. Combining this information with genome sequencing will enable characterization of these novel resistance genes. Although effector genes are usually unique and do not often share structural similarity but encode very diverse proteins, it is still possible to identify R-gene that could recognize common or near identical effector gene(s) in different pathogen pathotypes/races/isolates, which already exists naturally in some plants/crops (Dangl et al. 1992). This type of R-gene is useful and can be transferred between hosts that are closely related or within the same genus or family. Some R-genes can also be transferred from host to nonhost using GM technologies, providing broad spectrum immunity against one or more pathogens. One such successful example is the pepper *Bs2* gene against bacterial spot disease transferred from its host pepper into nonhost tomato (both members of Solanaceae family; Tai et al. 1999). The Bs2 gene encodes the R-gene NBS-LRR class (Tai et al. 1999), while the gene avrBs2 in *Xanthomonas campestris* pv. *vesicatoria*, causal pathogen of bacterial spot disease, is essential for virulence activity and is highly conserved between and within the pathovars of *X. campestris* (Kearney and Staskawicz 1990). This has opened up new possibilities to protect economically valuable crops, which lack R-gene resources, from disastrous pathogens (Neik et al. 2017).

6.20 Factors Affecting Transfer of Plant Disease Resistance

The different cultivars in biology of the several types of plant pathogens (fungi, bacteria, viruses, and nematodes) infecting crucifers present substantial problems in transfer of R-genes and developing GM-resistant plants. Firstly, the kind of organisms causing disease is taxonomically highly diverse; the major groups include cellular pathogens (bacteria, fungi, and the algal Oomycetes) and molecular pathogens (viruses). These are physiologically very different from each other, and therefore, no single gene product can be expected to have a direct toxic effect on all types of pathogens. Secondly, pathogens use three major life strategies, namely biotrophic, hemibiotrophic, and necrotrophic. Biotrophic pathogens essentially act as a sink for the host's anabolic assimilates, and therefore keep it alive. Meanwhile, necrotrophic pathogens consume the host tissues on invasion. Hemi-biotrophs combine both strategies in their life cycle. Consequently, plants have developed quite different ways for dealing with these strategies.

References

Agnihotri A, Lakshmikumaran MS, Jagannathan V, Shivanna KR (1991) Wide hybridization for improvement in cultivated brassicas. Acta Hortic (ISHS) 289:213–214

Agnihotri A, Gupta K, Prem D, Sarkar D, Mehra VS, Zargar SM (2009) Genetic enhancement in rapeseed-mustard for quality and disease resistance through in vitro techniques. In: Proceedings of 16th Australian Research Assembly on Brassicas, Ballarat, Australia, Sept 14–16, p 28

Aneja JK, Agnihotri A (2013) Alternaria blight of oilseed *brassicas*: epidemiology and disease control strategies with special reference to use of biotechnological approaches for attaining host resistance. J Oilseed Brassica 4:1–10

Atkinson NJ, Urwin PE (2012) The interaction of plant biotic and abiotic stresses: from genes to the field. J Exp Bot 63:3523–3544

Atkinson NJ, Lilley CJ, Urwin PE (2013) Identification of genes involved in the response of Arabidopsis to simultaneous biotic and abiotic stresses. Plant Physiol 162:2028–2041

Atri C, Akhatar J, Gupta M, Gupta N, Goyal A, Rana K, Kaur R, Mittal M, Sharma A, Singh MP, Sandhu PS, Barbetti MJ, Banga SS (2019) Molecular and genetic analysis of defensive responses of *Brassica juncea*—*B. fruticulosa* introgression lines to *Sclerotinia* infection. Sci Rep 9:17089. https://doi.org/10.1038/s41598-019-53444-3

Awasthi RP, Nashaat NI, Kolte SJ, Tewari AK, Meena PD, Renu B (2012) Screening of putative resistant sources against Indian and exotic isolates of *Albugo candida* inciting white rust in rapeseed-mustard. J Oilseed Brassica 3:27–37

Ayotte R, Harney PM, Souza MV (1987) The transfer of atriazine resistance from *Brassica napus* L. to *B. oleracea* L. production of F1 hybrids through embryo rescues. Euphytica 36:615–624. https://doi.org/10.1007/BF00041511

Barton NH (2001) The role of hybridization in evolution. Mol Ecol 10:551–568

Brun H, Plessis J, Renard M (1987a) Resistance of some crucifers to *Alternaria brassicae* (Berk.) Sacc. In: Proceedings of the 7th International rapeseed Congress Poznan, Poland, pp 1222–1227

Brun H, Renard M, Plessis J (1987b) Resistance of some crucifers to *Alternaria brassicae* (Berk.) Sacc. In: Proceedings of the 7th International rapeseed Congress Poznan, Poland, p 247

Bruns TD, White TJ, Taylor JW (1991) Fungal molecular systematic. Annu Rev Ecol Syst 22:525–564

Cao T, Manolii VP, Hwang SF, Howard RJ, Strelkov SE (2009) Virulence and spread of *Plasmodiophora brassicae* [clubroot] in Alberta, Canada. Can J Plant Pathol 31:321–329

Chatterjee M, Mazumder M, Basu D (2013) Functional analysis of the promoter of a glycosyl hydrolase gene induced in resistant *Sinapis alba* by *Alternaria brassicicola*. Phytopathology 103:841–850. https://doi.org/10.1094/PHYTO-11-12-0303-R

Chavan V, Bhargava S, Kamble A (2013) Temporal modulation of oxidant and anti-oxidative responses in *Brassica carinata* during β-aminobutyric acid-induced resistance against *Alternaria brassicae*. Physiol Mol Plant Pathol 83:35–39. https://doi.org/10.1016/j.pmpp.2013.03.002

Chen S, Zou J, Cowling WA, Meng J (2010) Allelic diversity in a novel gene pool of canola-quality *Brassica napus* enriched with alleles from *B. rapa* and *B. carinata*. Crop Pasture Sci 61:483–492. https://doi.org/10.1071/CP09327

Cheng F, Liu SY, Wu J, Fang L, Sun S, Liu B, Li P, Hua W, Wang X (2011) BRAD, the genetics and genomics database for Brassica plants. BMC Plant Biol 11:136. https://doi.org/10.1186/1471-2229-11-136

Chevre AM, Eber F, Margale E, Kerlan MC (1994) Comparison of somatic and sexual *Brassica napus-Sinapis alba* hybrids and their progeny by cytogenetic studies and molecular characterization. Genome 37:367–371

Chevre AM, Eber F, This P, Barret P, Tanguy X, Brun H, Dlseny M, Renard M (1996) Characterization of *Brassica nigra* chromosomes and of blackleg resistance in *B. napus–B. nigra* addition lines. Plant Breed 115:113–118. https://doi.org/10.1111/j.1439-05231996.tb00884.x

Chhikara S, Chaudhury D, Dhankher OP, Jaiwal PK (2012) Combined expression of a barley class II chitinase and type I ribosome inactivating protein in transgenic *Brassica juncea* provides protection against *Alternaria brassicae*. Plant Cell Tissue Org Cult 108:83–89

Chiang MS, Crete R (1983) Transfer of resistance to race 2 of *Plasmodiophora brassicae* from *Brassica napus* to cabbage (*B. oleracea* spp. *capitata*). V. the inheritance of resistance. Euphytica 32:479–483

Chiang MS, Crete R (1987) Cytoplasmic male sterility in *Brassica oleracea* induced by *B*. napus cytoplasm-female fertility and restoration of male fertility. Can J Plant Sci 672:891–897. https://doi.org/10.4141/cjps87-126

Chiang MS, Chiang BY, Grant WF (1977) Transfer of resistance to race 2 of *Plasmodiophora brassicae* from *Brassica napus* to cabbage (*B. oleracea* var. *capitata*). I. Interspecific hybridization between *B. napus* and *B. oleracea* var. *capitata*. Euphytica 26:319–336

Choudhary BR, Joshi P, Ramarao S (1999) Interspecific hybridization between *Brassica carinata* and *Brassica rapa*. Plant Breed 119:417–420

Christianson JA, Rimmer SR, Good AG, Lydiate DJ (2006) Mapping genes for resistance to *Leptosphaeria maculans* in *Brassica juncea*. Genome 49:30–41. https://doi.org/10.1139/g05-085

Chrungu B, Verma N, Mohanty A, Pradhan A, Shivanna KR (1999) Production and characterization of interspecific hybrids between *Brassica maurorum* and crop brassicas. Theor Appl Genet 98:608–613

Conn KL, Tewari JP (1986) Hypersensitive reaction induced by *Alternaria brassicae* in *Eruca sativa*, an oil-yielding crucifer. Can J Plant Pathol 8:348. (Abstr)

Conn KL, Tewari JP, Hadziyev D (1984) The role of epicuticular wax in canola in resistance to *Alternaria brassicae*. Phytopathology 74:851

Conn KL, Tewari JP, Dahiya JS (1988) Resistance to *Alternaria brassicae* and phytoalexin-elicitation in rapeseed and other crucifers. Plant Sci 55:21–25. https://doi.org/10.1016/0168-9452(88)90037-4

Conn KL, Jejelowa OA, Tewari JP, Bains PS (1991) *Camelina sativa* phytoalexins (Calmexin and methoxy camalexin) provide resistance against *Alternaria brassicae*. Abstract 274. Can J Plant Pathol:13

Crow JF, Kimura M (1970) An introduction to population genetics theory. Harper and Row, New York

Crute IR, Gray AR, Crisp P, Buczacki ST (1980) Variation in *Plasmodiophora brassicae* and resistance to clubroot disease in brassicas and allied crops-a critical review. Plant Breed Abstr 50:91–104

Cui H, Tsuda K, Parker JE (2015) Effector-triggered immunity: from pathogen perception to a robust defense. Annu Rev Plant Biol 66:487–511

Dang JK, Sangwan MS, Mehta N, Kaushik CD (2000) Multiple disease resistance against four fungal foliar diseases of rapeseed-mustard. Indian Phytopathol 53:455–458

Dangl JL, Ritter C, Gibbon MJ, Mur LA, Wood JR, Goss S, Mansfield J, Taylor JD, Vivian A (1992) Functional homologs of the Arabidopsis RPM1 disease resistance gene in bean and pea. Plant Cell 4:1359–1369. https://doi.org/10.1105/tpc.4.11.1359

Deora A, Gossen BD, Mcdonald MR (2012) Infection and development of *Plasmodiophora brassicae* in resistant and susceptible canola cultivars. Can J Plant Pathol 34(2):239–247. https://doi.org/10.1080/07060661.2012.681071

Deora A, Gossen BD, McDonald MR (2013) Expression of resistance to *Plasmodiophora brassicae* in four resistant canola cultivars. Can J Plant Pathol 35

Dey SS, Sharma K, Dey RB, Kumar GMS, Singh D, Kumar R, Parkash C (2015) Inter specific hybridization (*Brassica carinata* × *Brassica oleracea*) for introgression of black rot resistance genes into Indian cauliflower (*B. oleracea* var. botrytis L.). Euphytica 204:149–162. https://doi.org/10.1007/s10681-015-1352-0

Dias JS, Ferreira ME, Williams PH (1993) Screening of Portuguese cole landraces (*Brassica oleracea* L.) with *Peronospora parasitica* and *Plasmodiophora brassicae*. Euphytica 67:135–141

Diederichsen E, Deppe U, Sacristan MD (2003) Characterization of clubroot resistance in recent winter oilseed rape material. In: Proceedings of the 11th international rapeseed congress, Copenhagen, Denmark, vol 1, pp 68–70

Diederichsen E, Frauen M, Linders EGA, Hatakeyama K, Hirai M (2009) Status and perspectives of clubroot resistance breeding in crucifer crops. J Plant Growth Regul 28:265–281. https://doi.org/10.1007/s00344-009-9100-0

Diederichsen E, Frauen M, Ludwig-Muller J (2014) Clubroot disease management challenges from a German perspective. Can J Plant Pathol 36:85–98. https://doi.org/10.1080/07060661.2013.861871

Ding Y, Mei J, Li Q, Liu Y, Wan H, Wang L, Becker HC, Qian W (2013) Improvement of *Sclerotinia sclerotiorum* resistance in *Brassica napus* by using *B. oleracea*. Genet Resour Crop Evol 60:1615–1619. https://doi.org/10.1007/s10722-013-9978-z

Ding Y, Mei J, Liu Y, Wang L, Li Y, Wan H, Li J, Qian W (2015) Transfer of Sclerotinia stem rot resistance from wild *Brassica oleracea* into *B. rapa*. Mol Breed 35:225

Ding Y, Mei J, Wu Q, Xiong Z, Li Y, Shao C, Wang L, Qian W (2019) Synchronous improvement of sub genomes in allopolyploid: a case of *Sclerotinia* resistance improvement in *Brassica napus*. Mol Breed 39:10

Dixon GR (2009) Husbandry—the sustainable means of controlling soil borne pathogens: a synoptic review. In: Proceedings of the symposium of horticulture in Europe, Dixon GR (ed). Vienna, Austria, 17–21. February 2008. Acta Hort 817:233–242

Dodds PN, Rathjen JP (2010) Plant immunity: towards an integrated view of plant-pathogen interactions. Nat Rev Genet 11:539–548

Donald EC, Jaudzems G, Porter IJ (2008) Pathology of cortical invasion by *Plasmodiophora brassicae* in clubroot resistant and susceptible *Brassica oleracea* hosts. Plant Pathol 57:201–209. https://doi.org/10.1111/j.1365-3059.2007.01765.x

Enjalbert JN, Zheng S, Johnson JJ, Mullen JL, Byrne PF, McKay JK (2013) Brassicaceae germplasm diversity for agronomic and seed quality traits under drought stress. Ind Crop Prod 47:176–185

Fatima U, Bhorali P, Borah S, Senthil-Kumar M (2019) Perspectives on the utilization of resistance mechanisms from host and nonhost plants for durable protection of *Brassica* crops against Alternaria blight. Peer J 7:e7486. https://doi.org/10.7717/peerj.7486

Fujita M, Fujita Y, Noutoshi Y, Takahashi F, Narusaka Y, Yamaguchi-Shinozaki K, Shinozaki K (2006) Crosstalk between abiotic and biotic stress responses: a current view from the points of convergence in the stress signaling networks. Curr Opin Plant Biol 9:436–442

Ghose K, Dey S, Barton H, Loake GJ, Basu D (2008) Differential profiling of selected defense-related genes induced on challenge with *Alternaria brassicicola* in resistant white mustard and their comparative expression pattern in susceptible India mustard. Mol Plant Pathol 9 (6):763–775. https://doi.org/10.1111/j.1364-3703.2008.00497.x

Gossen BD, McDonald MR, Hwang SF, Strelkov SE, Peng G (2013) A comparison of clubroot development and management on canola and *Brassica* vegetables. Can J Plant Pathol 35 (2):175–191. https://doi.org/10.1080/07060661.2013.763293

Grontoft M (1986) Resistance to *Alternaria* spp. in oil crops. Sveriges Utsadesforenings Tidskrift 96(3):263

Gupta RBL, Singh M (1994) Source of resistance to white rust and powdery mildew of mustard. Intl J Trop Plant Dis 12:225–227

Gupta SK, Gupta PP, Yadav TP, Kaushik CD (1990) Metabolic changes in mustard due to Alternaria leaf blight. Indian Phytopath 43:64–69

Gupta K, Saharan GS, Singh D (2002) Effective selection criteria to assess disease tolerance in Indian mustard [*Brassica juncea* (L.) Czern and Coss.]. J Mycol Plant Pathol 32:72–80

Hansen LN, Earle ED (1995) Transfer of resistance to *Xanthomonas campestris* pv. *Campestris* into *Brassica oleracea* L. by protoplast fusion. Theor Appl Genet 91:1293–1300. https://doi.org/10.1007/BF00220944

Hansen LH, Earle ED (1997) Somatic hybrids between *Brassica oleracea* (L.) and *Sinapis alba* (L.) with resistance to *Alternaria brassicae* (Berk) Sacc. Theor Appl Genet 94:1078–1085

Hirai M (2006) Genetic analysis of clubroot resistance in *Brassica* crops. Breed Sci 56:223–229

Holub EB, Tör M, Cooper A, Gordon P, Gunn N (2002) Genetics of disease resistance in Arabidopsis to crop pathogens. In: Proc 6th Conf. EFPP 2002, Prague. Plant Protect Sci 38 (Spl 1), pp 151–154

Hurgobin B, Golicz AA, Bayer PE, Chan CKK, Tirnaz S, Dolatabadian A, Schiessl SV, Samans B, Montenegro JD, Parkin IAP, Pires JC, Chalhoub B, King GJ, Snowdon R, Batley J, Edwards D (2018) Homoeologous exchange is a major cause of gene presence/absence variation in the amphidiploid *Brassica napus*. Plant Biotechnol J 16:1265–1274

Hwang SF, Ahmed HU, Zhou Q, Strelkov SE, Gossen BD, Peng G, Turnbull D (2011) Influence of cultivar resistance and inoculum density on root hair infection of canola (*Brassica napus*) by *Plasmodiophora brassicae*. Plant Pathol 60:820–829

Hwang SF, Strelkov SE, Feng J, Gossen BD, Howard RJ (2012) *Plasmodiophora brassicae*: a review of an emerging pathogen of the Canadian canola (*Brassica napus*) crop. Mol Plant Pathol 13:105–113. https://doi.org/10.1111/j.1364-3703.2011.00729.x

Inomata N (1993) Cross ability and cytology of hybrid progenies in the cross between *Brassica campestris* and three wild relatives of *B. oleracea, B. bourgeaui, B. cretica* and *B. montana*. Euphytica 69:7–17. https://doi.org/10.1007/BF00021721

Inohara N, Chamaillard M, McDonald C, Nunez G (2005) NOD-LRR proteins: role in host-microbial interactions and inflammatory disease. Annu Rev Biochem 74:355–383. https://doi.org/10.1146/annurev.biochem.74.082803.133347

Inomata N (2002) A cytogenetic study of the progenies of hybrids between *Brassica napus* and *B. oleracea, B. bourgeaui, B. cretica* or *B. montana*. Plant Breed 121:174–176. https://doi.org/10.1046/j.1439-0523.2002.00695.x

Jejelowa OA, Conn KL, Tewari JP (1991) Relationship between conidial concentration, germling growth, and phytoalexin production by *Camelina sativa* leaves inoculated with *Alternaria brassicae*. Mycol Res 95:928–934. https://doi.org/10.1016/S0953-7562(09)80089-0

Jensen BD, Værbak S, Munk L, Andersen SB (1999) Characterization and inheritance of partial resistance to downy mildew, *Peronospora parasitica*, in breeding material of broccoli, *Brassica oleracea* convar. *botrytis* var. *italica*. Plant Breed 118:549–554

Jiao Y, Wickett NJ, Ayyampalayam S, Chanderbali AS, Landherr L, Ralph PE, Tomsho LP, Hu Y, Liang H, Soltis PS, Soltis DE, Clifton SW, Schlarbaum SE, Schuster SC, Ma H, Leebens-Mack J, de Pamphili CW (2011) Ancestral polyploidy in seed plants and angiosperms. Nature 473:97–100

Jourdan PS, Salazar E (1993) *B. carinata* resynthesized by protoplast fusion. Theor Appl Genet 86:567–572

Jourdan PS, Earle ED, Mutscher MA (1989) Atrazine-resistant cauliflower obtained by somatic hybridization between *Brassica oleracea* and ATR- *B. napus*. Theor Appl Genet 78:271–279. https://doi.org/10.1007/BF00288810

Kamble A, Bhargava S (2007) β–aminobutyric acid-induced resistance in *Brassica juncea* against the necrotrophic pathogen *Alternaria brassicae*. J Phytopathol 155:152–158

Kaneko Y, Bang SW (2014) Interspecific and intergeneric hybridization and chromosomal engineering of brassicaceae crops. Breed Sci 64:14–22

Kanrar S, Venkateswari JC, Kirti PB, Chopra VL (2002) Transgenic expression of hevein, the rubber tree lectin, in Indian mustard confers protection against *Alternaria brassicae*. Plant Sci 162:441–448

Karpechenko GD (1928) Polyploid hybrids of *Raphanus sativus* L. × *Brassica oleracea* L. Z Indukt Abstamm Vererbungsl 48:1–85

Kearney B, Staskawicz BJ (1990) Widespread distribution and fitness contribution of *Xanthomonas campestris* avirulence gene avrBs2. Nature 346:385–386. https://doi.org/10.1038/346385a0

Kim Y, Tsuda K, Igarashi D, Hillmer RA, Sakakibara H, Myers CL, Katagiri F (2014) Signaling mechanisms underlying the robustness and tunability of the plant immune network. Cell Host Microbe 15:84–94

Kissoudis C, van de Wiel C, Visser RGF, van der Linden G (2014) Enhancing crop resilience to combined abiotic and biotic stress through the dissection of physiological and molecular crosstalk. Front Plant Sci 5:207

Kolte SJ (1985a) Disease management strategies for rapeseed-mustard crops in India—a review. Agric Rev 6:81–88

Kolte SJ (1985b) Diseases of annual edible oilseed crops. In: Rapeseed-mustard and sesame diseases, vol 2. CRC, Boca Raton, p 135

Kolte SJ (1987) Important diseases of rapeseed and mustard in oilseed crops: Niger and rapeseed mustard. In: Proceedings of the Third Oil Crops Network, held in Addis Ababa, Ethiopia, 6–10 October, 1986, pp 99–106

Krishnia SK, Saharan GS, Singh D (2000) Genetics of Alternaria blight resistance in inter and intraspecific crosses of *Brassica juncea* and *B. carinata*. Ann Biol 16(2):212–216

Kumar B, Kolte SJ (2001) Progression of Alternaria blight of mustard in relation to components of resistance. Indian Phytopath 54:329–331

Kumar S, Saharan GS (2002) Sources of multiple disease resistance in *Brassica* spp. J Mycol Plant Pathol 32:184–188

Kumar D, Maurya N, Bharati YK, Kumar A, Kumar K, Srivastava K, Chand G, Singh SK, Mishra RK, Kumar A (2014) Alternaria blight of oilseed *brassicas*: a comprehensive review. Afr J Microbiol Res 8:2816–2829. https://doi.org/10.5897/AJMR2013.6434

Kumar A, Banga SS, Meena PD, Kumar PR (2015) Brassica oilseeds breeding and management. CABI International, London, p 280. ISBN-13: 9781780644837

LeBoldus JM, Manolii VP, Turkington TK, Strelkov SE (2012) Adaptation to Brassica host genotypes by a single-spore isolate and population of *Plasmodiophora brassicae* (clubroot). Plant Dis 96:833–838

Lucas JA, Crute IR, Sherriff C, Gordon PL (1988) The identification of a gene for race-specific resistance to *Peronospora parasitica* (downy mildew) in *Brassica napus* var. *oleifera* (oilseed rape). Plant Pathol 37:538–545. https://doi.org/10.1111/j.1365-3059.1988.tb02112.x

Marmath KK, Giri P, Sharma S, Taj G, Kumar A (2013) In-silico interaction studies of *Alternaria brassicae* toxin destruxin B and potential partners of MAPK4 cascade. Int J Agric Env Biotechnol 6:203–210

Martin GB, Brommonschenkel SH, Chunwongse J, Frary A, Ganal MW, Spivy R, Wu T, Earle ED, Tanksley SD (1993) Map-based cloning of a protein kinase gene conferring disease resistance in tomato. Science 262:1432–1436

Mason AS, Nelson MN, Yan G, Cowling WA (2011) Production of viable male unreduced gametes in Brassica interspecific hybrids is genotype specific and stimulated by cold temperatures. BMC Plant Biol 11:103. https://doi.org/10.1186/1471-2229-11-103

Meena PD, Manju B, Sharma P, Singh D (2014) Interaction and tolerance to *Alternaria brassicae* in Indian mustard (*Brassica juncea*) genotypes. J Oilseeds Res 31:136–139

Meena SS, Meena PD, Singh VV, Meena HS, Singh D, Yadav R, Singh KH, Sharma P, Nanjundan J, Singh BK, Singh YP, Ram B (2019a) DRMR-2019 (IC0598622; INGR17077), an Indian mustard (*Brassica juncea*) germplasm with White rust resistance. Indian J Plant Genet Resour 32(2):281

Meena SS, Meena PD, Singh VV, Meena HS, Singh D, Yadav R, Singh KH, Sharma P, Nanjundan J, Singh BK, Singh YP, Bhagirath R (2019b) DRMR-2035 (IC0598623; INGR17078), an Indian mustard (*Brassica juncea*) germplasm with white rust resistance. Indian J Plant Genet Resour 32(2):281

Meena PD, Mehta N, Saharan GS (2019c) *Alternaria-Brassica* associations: research progress and prospects. Agric Res J 56(4):600–612

Meena PD, Sujith Kumar MS, Meena HS, Jambhulkar S, Gohartaj, Pathak D, Srivastava S, Gupta R, Singh D, Gurung B, Rai PK (2020) Confirmation of induced tolerance to Alternaria blight disease in Indian mustard (*Brassica juncea* L.). Appl Biochem Biotechnol 192 (3):965–978. https://doi.org/10.1007/s12010-020-03362-2

Mehta N, Saharan GS, Meena PD (2018) Expression of disease resistance in *Brassica-Hyaloperonospora* host-patho system—a review. Plant Dis Res 33(2):112–141

Mei J, Qian L, Disi JO, Yang X, Li Q, Li J, Frauen M, Cai D, Qian W (2011) Identification of resistant sources against *Sclerotinia sclerotiorum* in *Brassica* species with emphasis on *B. oleracea*. Euphytica 177:393–399

Mei J, Ding Y, Lu K, Wei D, Liu Y, Disi JO, Li J, Liu L, Liu S, McKay J, Qian W (2013) Identification of genomic regions involved in resistance against *Sclerotinia sclerotiorum* from wild *Brassica oleracea*. Theor Appl Genet 126:549–556. https://doi.org/10.1007/s00122-012-2000-x

Mei J, Liu Y, Wei D, Wittkop B, Ding Y, Li Q, Li J, Wan H, Li Z, Ge X, Frauen M, Snowdon RJ, Qian W, Friedt W (2015) Transfer of Sclerotinia resistance from wild relative of *Brassica oleracea* into *Brassica napus* using a hexaploidy step. Theo Appl Genet 128:639–644

Mei J, Shao C, Yang R, Feng Y, Gao Y, Ding Y, Li J, Qian W (2020) Introgression and pyramiding of genetic loci from wild *Brassica oleracea* into *B. napus* for improving *Sclerotinia* resistance of rapeseed. Theor Appl Genet 133:1313–1319. https://doi.org/10.1007/s00122-020-03552-w

Meng J, Lu M (1993) Genotype effects of *Brassica napus* on its reproductive behaviour after pollination with *B. juncea*. Theor Appl Genet 87:238–242. https://doi.org/10.1007/BF00223771

Meyers BC, Morgante M, Michelmore RW (2002) TIR-X and TIR-NBS proteins: two new families related to disease resistance TIR-NBS-LRR proteins encoded in *Arabidopsis* and other plant genomes. Plant J 32:77–92. https://doi.org/10.1046/j.1365-313X.2002.01404.x

Meyers BC, Kozik A, Griego A, Kuang H, Michelmore RW (2003) Genome-wide analysis of NBS-LRR-encoding genes in *Arabidopsis*. Plant Cell 15:809–834. https://doi.org/10.1105/tpc.009308

Mishra A, Pandey D, Goel A, Kumar A (2010) Molecular cloning and in-silico analysis of functional homologues of hypersensitive response gene(s) induced during pathogenesis of Alternaria blight in two genotypes of Brassica. J Proteomics Bioinform 3:244–248

Mitchell-Olds T, James RV, Palmer MJ, Williams PH (1995) Genetics of *Brassica rapa* (syn. *Campestris*). 2. Multiple disease resistance to three fungal pathogens: *Peronospora parasitica*, *Albugo candida* and *Leptosphaeria maculans*. Heredity 75:362–369

Mohanty A, Chrungu B, Verma N, Shivanna KR (2009) Broadening the genetic base of crop brassicas by production of new intergeneric hybrid. Czech J Genet Plant Breed 45(3):117–122

Momotaz A, Kato M, Kakihara F (1998) Production of inter-generic hybrids between Brassica and *Sinapsis* species by means of embryo rescue techniques. Euphytica 103:123–130. https://doi.org/10.1023/a:1018331528368

Mondal KK, Chatterjee SC, Viswakarma N, Bhattacharya RC, Grover A (2003) Chitinase-mediated inhibitory activity of Brassica transgenic on growth of *Alternaria brassicae*. Curr Microbiol 47:171–173

Monteiro AA, Williams PH (1989) The exploration of genetic resources of Portuguese cabbage and kale for resistance to several Brassica diseases. Euphytica 41:215–225

Nagaharu U (1935) Genome analysis in *Brassica* with special reference to the experimental formation of *B. napus* and peculiar mode of fertilization. Japan J Bot 7:389–452

Nakagami H, Pitzchke A, Hirt H (2005) Emerging MAP kinase pathways in plant stress signaling. Trends Plant Sci 10:339–346

Naumann TA, Wicklow DT (2013) Chitinase modifying proteins from phylogenetically distinct lineages of Brassica pathogens. Physiol Mol Plant Pathol 82:1–9

Navabi ZK, Huebert T, Sharpe AG, Carmel M, Bancroft I, Isobel Parkin AP (2013) Conserved microstructure of the Brassica B genome of *Brassica nigra* in relation to homologous regions of *Arabidopsis thaliana, B. rapa* and *B. oleracea*. BMC Genomics 14:250. https://doi.org/10.1186/1471-2164-14-250

Neik TX, Barbetti MJ, Batley J (2017) Current status and challenges in identifying disease resistance genes in *Brassica napus*. Front Plant Sci 8:1788. https://doi.org/10.3389/fpls.2017.01788

Nejat N, Mantri N (2017) Plant immune system: crosstalk between responses to biotic and abiotic stresses the missing link in understanding plant defence. Curr Issues Mol Biol 23:1–16. https://doi.org/10.21775/cimb.023.001

Niemann J, Wojciechowski A, Janowicz J (2012) Broadening the variability of quality traits in rapeseed through interspecific hybridization with an application of immature embryo culture. J Biotechnol Comput Biol Bionanotechnol 93:109–115

Niemann J, Wojciechowski A, Jedryczka M, Kaczmarek J (2013) Interspecific hybridization as a tool for broadening the variability of useful traits in rapeseed (*Brassica napus* L.). In: Branca F, Tribulato A (eds) Proceedings of the VIth IS on brassicas and XVIIIth crucifer genetics workshop. Acta Hort, Catania, p 1005

Nishi S, Wata JK, Toda M (1959) On the breeding of interspecific hybrids between two genomes. 'C' and 'B' of *Brassica* through the application of embryo culture techniques. Japan J Breed 8:215–222

Nowicki M, Nowakowska M, Niezgoda A, Kozik EU (2012) Alternaria black spot of crucifers: symptoms, importance of disease, and perspectives of resistance breeding. Veg Crops Res Bull 76:5–19

Oh IS, Park AR, Bae MS, Kwon SJ, Kim YS, Lee JE, Kang NY, Lee S, Cheong H, Park OK (2005) Secretome analysis reveals an Arabidopsis lipase involved in defense against *Alternaria brassicicola*. Plant Cell 17:2832–2847

Otani H, Kohnobe A, Narita M, Shiomi H, Kodama M, Kohmoto K (2001) A new type of host-selective toxin, a protein from *Alternaria brassicicola*. In: Keen NT, Mayama S, Leach JE, Tsuyumu S (eds) Delivery and perception of pathogen signals in plants. APS Press, St. Paul, MN, pp 68–76

Pan X, Caldwell CD, Falk KC, Lada R (2012) The effect of cultivar, seeding rate and applied nitrogen on *Brassica carinata* seed yield and quality in contrasting environments. Can J Plant Sci 92:961–971. https://doi.org/10.4141/cjps2011-169

Parihar PS, Prakash O, Punetha H (2012) Changes in metabolites of *Brassica juncea* (Indian mustard) during progressive infection of *Alternaria brassicae*. Nat Sci 10:39–42

Parlevliet JE (1992) Selecting components of partial resistance. In: Stalker HT, Murphy JP (eds) Plant breeding in the 1990s, pp 281–302

Peng G, Pageau D, Strelkov SE, Lahlali R, Hwang SF, Hynes RK (2013) Assessment of crop rotation, cultivar resistance and *Bacillus subtilis* biofungicide for control of clubroot on canola. Acta Hortic 1005:591–598

Peng G, Falk KC, Gugel RK, Franke C, Yu F, James B, Strelkov SE, Hwang SF, McGregor L (2014) Sources of resistance to *Plasmodiophora brassicae* (clubroot) pathotypes virulent on canola. Can J Plant Pathol 36:89–99. https://doi.org/10.1080/07060661.2013.863805

Perez IB, Brown P (2014) The role of ROS signaling in cross-tolerance: from model to crop. Front Plant Sci 5:754

Plieske J, Struss D, Robbelen G (1998) Inheritance of resistance derived from the B-genome of Brassica against *Phoma lingam* in rapeseed and the development of molecular markers. Theor Appl Genet 97:929–936. https://doi.org/10.1007/s001220050973

Prakash S (1973) Non-homologous meiotic pairing in the A and B genomes of Brassica: its breeding significance in the production of variable amphidiploids. Genet Res 21:133–137

Prakash S, Chopra VL (1988) Introgression of resistance to shattering in *Brassica napus* from *Brassica juncea* through non-homologous recombination. Plant Breed 101:167–168

Prasch CM, Sonnewald U (2013) Simultaneous application of heat, drought and virus to Arabidopsis plants reveals significant shifts in signaling networks. Plant Physiol 162:1849–1866

Pravel P, Thakur AK, Meena PD, Meena HS, Sharma P, Singh D, Majumdar R (2015) Genetic diversity assessment in Indian mustard (*Brassica juncea* L.) for Alternaria blight tolerance using SSR markers. J Oilseed Brassica 6(1):175–182

Primard C, Vedel F, Mathieu C, Pelletier G, Chevre AM (1988) Interspecific somatic hybridization between *Brassica napus* and *Brassica hirta* (*Sinapis alba* L.). Theor Appl Genet 75:546–552

Rahman H, Peng G, Yu F, Falk KC, Kulkarni M, Selvaraj G (2014) Genetics and breeding for clubroot resistance in Canadian spring canola (*Brassica napus* L.). Can J Plant Pathol 36:122–134. https://doi.org/10.1080/07060661.2013.862571

Rajarammohan S, Pental D, Kaur J (2019) Near-complete genome assembly of *Alternaria brassicae*—a necrotrophic pathogen of Brassica crops. MPMI 32:928–930

Ricarova V, Kazda J, Singh K, Ryšánek P (2016) Clubroot caused by *Plasmodiophora brassicae* worn.: a review of emerging serious disease of oilseed rape in the Czech Republic. Plant Protect Sci 52:71–86. https://doi.org/10.17221/87/2015-PPS

Ripley V, Thorpe M, Iler S, Mizier K, Beversdorf WD (1992) Isozyme analysis as a tool for introgression of *Sinapis alba* germplasm into Brassica napus. Theor Appl Genet 84:403–410. https://doi.org/10.1007/BF00229500

Saharan GS (1984) A review of research on rapeseed-mustard pathology in India. In: Annual Workshop AICORPO, ICAR, Jaipur, 6–10 August 1984

Saharan GS (1992) Disease resistance. In: Labana KS, Banga SS, Banga SK (eds) Breeding oilseed brassicas. Narosa Publishing House, New Delhi, pp 181–200

Saharan GS (1997) Disease resistance. In: Kalia HR, Gupta SC (eds) Recent advances in oilseed brassicas, vol 12. Kalyani Pub, Ludhiana, pp 233–259

Saharan GS, Chand JN (1988) Diseases of rapeseed and mustard. Directorate of Oilseed Crops (in Hindi). Directorate of Publication, Haryana Agric Univ Press, Hisar. Chapter 3, pp 84–91

Saharan GS, Kadian AK (1983) Analysis of components of horizontal resistance in rapeseed and mustard cultivars against *Alternaria brassicae*. Indian Phytopath 36:503–507

Saharan GS, Krishnia SK (2001) Multiple disease resistance in rapeseed and mustard. In: Nagarajan S, Singh DP (eds) Role of resistance in intensive agriculture. Kalyani Publications, New Delhi, pp 98–108

Saharan GS, Mehta N, Sangwan MS (2003) Nature and mechanism of resistance to Alternaria blight in rapeseed-mustard system. Ann Rev Plant Pathol 2:85–128

Saharan GS, Mehta N, Sangwan MS (2005) Development of disease resistance in rapeseed-mustard. In: Saharan GS, Mehta N, Sangwan MS (eds) Diseases of oilseed crops. Indus Publishing, New Delhi, pp 561–617

Saharan GS, Verma PR, Meena PD, Kumar A (2014) White rust of crucifers: biology, ecology and disease management. Springer Verlag, Frankfurt, p 244. ISBN 978-81-322-1791-6

Saharan GS, Mehta N, Meena PD (2016) Alternaria blight of crucifers: biology, ecology and disease management. Springer Verlag, Singapore, p 326. ISBN: 978-981-10-0019-5

Saharan GS, Mehta N, Meena PD (2018) Downy mildew disease of crucifers: biology, ecology and disease management, vol 56. Springer Verlag, Singapore, p 357. ISBN 978-981-10-7499-8

Saharan GS, Mehta N, Meena PD (2019) Powdery mildew disease of crucifers: biology, ecology and disease management. Springer Verlag, Singapore, p 362. ISBN 978-981-13-9852-0

Saharan GS, Mehta N, Meena PD (2021) Clubroot disease of crucifers: biology, ecology and disease management. Springer Verlag, Singapore. (In Press)

Sandhu PS, Brar KS, Chauhan JS, Meena PD, Awasthi RP, Rathi AS, Kumar A, Gupta JC, Kolte SJ, Manhas SS (2015) Host–pathogen interactions of *Brassica* genotypes for white rust (*Albugo candida*) disease severity under aided epiphytotic conditions in India. Phytoparasitica 43 (2):197–207. https://doi.org/10.1007/s12600-014-0432-3

Schranz ME, Lysak MA, Mitchell-Olds T (2006) The ABC's of comparative genomics in the Brassicaceae: building blocks of crucifer genomes. Trends Plant Sci 11:535–542. https://doi.org/10.1016/j.tplants.2006.09.002

Schwessinger B, Bart R, Krasileva K, Coaker G (2015) Focus issue on plant immunity: from model systems to crop species. Front Plant Sci 6:195

Sellam A, Dongo A, Guillemette T, Hudhomme P, Simoneau P (2007) Transcriptional responses to exposure to the brassicaceous defence metabolites camalexin and allyl-isothiocyanate in the necrotrophic fungus *Alternaria brassicicola*. Mol Plant Pathol 8:195–208

Sharma TR, Singh BM (1992) Transfer of resistance to *Alternaria brassicae* in *Brassica juncea* through interspecific hybridization among brassicas. J Genet Breed 46(4):373–378

Sharma BR, Dhiman JS, Thakur JC, Singh A, Bajaj KL (1991) Multiple disease resistance in cauliflower. Adv Hort Sci 5:30–34

Sharma G, Kumar DV, Haque A, Bhat SR, Shyam P, Chopra VL (2002) *Brassica* coenospecies: a rich reservoir for genetic resistance to leaf spot caused by *Alternaria brassicae*. Euphytica 125:411–419. https://doi.org/10.1023/A:1016050631673

Sharma P, Meena PD, Verma PR, Saharan GS, Mehta N, Singh D, Kumar A (2015) *Sclerotinia sclerotiorum* (Lib) de Bary causing Sclerotinia rot in oilseed Brassicas: a review. J Oilseed Brassica 6:1–44

Sharma BB, Kalia P, Yadava DK, Dingh D, Sharma TR (2016) Genetics and molecular mapping of black rot resistance locus Xca1bc on chromosome B-7 in Ethiopian mustard (*Brassica carinata* a. Braun). PLoS One 11:e0152290. https://doi.org/10.1371/journal.Pone.0152290

Sharma BB, Kalia P, Singh D, Sharma TR (2017) Introgression of black rot resistance from *Brassica carinata* to cauliflower (*Brassica oleracea botrytis* group) through embryo rescue. Front Plant Sci 8:1255. https://doi.org/10.3389/fpls.2017.01255

Sharma P, Singh S, Singh VV, Gupta NC, Prasad L, Meena PD, Ambawat S, Rai PK (2018) Identification and characterization of new resistance sources against Sclerotinia stem rot (*Sclerotinia sclerotiorum*) in oilseed Brassica. J Mycol Pl Pathol 48(1):35–44

Shivpuri A, Chippa HP, Gupta RLB, Sharma KN (1997) Field evaluation of mustard genotypes against white rust, powdery mildew, and stem rot. Ann Arid Zone Res 36:387–390

Siemens J, Nagel W, Ludwig-Muller J, Sarcristán MD (2002) The interaction of *Plasmodiophora brassicae* and *Arabidopsis thaliana*: parameters for disease quantification and screening mutant lines. J Phytopathol 150:592–605

Siemens J, Keller I, Sarx J, Kunz S, Schuller A, Nagel W, Schmulling T, Parniske M, Ludwig-Muller J (2006) Transcriptome analysis of *Arabidopsis* clubroots indicate a key role for cytokinins in disease development. Mol Plant Microbe Interact 19:480–494. https://doi.org/10.1094/MPMI-19-0480

Singh K (1994) Studies on the ecofriendly management of powdery mildew (*Erysiphe cruciferarum* Opiz ex. Junell) of mustard [*Brassica juncea* (L) Czern & Coss.]. MSc Thesis, Department of Plant Pathology, CCS HAU, Hisar. Pages 106+ xvii

Singh H, Singh D (1989) Studies on genetic control of resistance to *Alternaria brassicae* in Indian mustard. In: Proc XIIth Eucarpia Congress, 11–13, Abstr

Singh S, Dey SS, Bhatia R, Batley J, Kumar R (2018) Molecular breeding for resistance to black rot [*Xanthomonas campestris* pv. *Campestris* (Pammel) Dowson] in *brassicas*: recent advances. Euphytica 214:196

Skoropad WP, Tewari JP (1977) Field evaluation of the epicuticular wax in rapeseed and mustard in resistance to *Alternaria brassicae*. Can J Plant Sci 57:1001–1003. https://doi.org/10.4141/cjps77-146

Snowdon RJ, Winter H, Diestel A, Sacristan MD (2000) Development and characterization of Brassica napus-Sinapis arvensis addition lines exhibiting resistance to *Leptosphaeria maculans*. Theor Appl Genet 101:1008–1014. https://doi.org/10.1007/s001220051574

Soltis PS, Soltis DE (2000) The role of genetic and genomic attributes in the success of polyploids. Proc Natl Acad Sci U S A 97:7051–7057

Strelkov SE, Tewari JP, Smith-Degenhardt E (2006) Characterization of *Plasmodiophora brassicae* populations from Alberta, Canada. Can J Plant Pathol 28:467–474. https://doi.org/10.1080/07060660609507321

Strelkov SE, Hwang SF, Howard RJ, Hartman M, Turkington TK (2011) Progress towards the sustainable management of clubroot [*Plasmodiophora brassicae*] of canola on the Canadian prairies. Prairie Soils Crops J 4:114–121

Suda K, Katagiri F (2010) Comparing signaling mechanisms engaged in pattern-triggered and effector-triggered immunity. Curr Opin Plant Biol 13:459–465

Suzuki N, Rivero RM, Shulaev V, Blumwald E, Mittler R (2014) Abiotic and biotic stress combinations. New Phytol 203:32–43

Tai TH, Dahlbeck D, Clark ET, Gajiwala P, Pasion R, Whalen MC, Stall RE, Staskawicz BJ (1999) Expression of the Bs2 pepper gene confers resistance to bacterial spot disease in tomato. Proc Natl Acad Sci U S A 96:14153–14158. https://doi.org/10.1073/pnas.96.24.14153

Taylor JD, Conway J, Roberts SJ, Astley D, Vicente JG (2002) Source and origin of resistance to *Xanthomonas campestris* pv. Campestris in Brassica genomes. Phytopathology 92:105–111

Tewari JP (1991) Structural and biochemical bases of black spot disease of crucifers. Adv Struct Biol 1:325–349

Tewari JP, Conn KL (1993) Reaction of some wild crucifers to *Alternaria brassicae*. Bull OILB SROP 16:53–58

Thomas CE, Jourdain EL (1990) Evaluation of broccoli and cauliflower germplasm for resistance to race 2 of *Peronospora parasitica*. Hortic Sci 25:1429–1431

Thomas CE, Jourdain EL (1992) Resistance to race 2 of *Peronospora parasitica* in U.S. plant introductions of *Brassica oleracea* var. *capitata*. Hortic Sci 27:1120–1122

Thomma BP, Nürnberger T, Joosten MH (2011) Of PAMPs and effectors: the blurred PTI-ETI dichotomy. Plant Cell 23:4–15

Tonguc M, Griffiths PD (2004a) Transfer of powdery mildew resistance from *Brassica carinata* to *Brassica oleracea* through embryo rescue. Plant Breed 123:587–589. https://doi.org/10.1111/j.1439-0523.2004.00987.x

Tonguc M, Griffiths PD (2004b) Development of black rot resistant interspecific hybrids between *Brassica oleracea* L. cultivars and *Brassica* accession a 19182, using embryo rescue. Euphytica 136:313–318. https://doi.org/10.1023/B:EUPH.0000032733.47031.5f

Tonguc M, Earle ED, Griffiths PD (2003) Segregation distortion of *Brassica carinata* derived black rot resistance in *Brassica oleracea*. Euphytica 134:269–276

Verma VD, Rai B (1980) Note on induced mutagenesis for spotting out the sources of resistance to Alternaria leaf spot in Indian mustard. Indian J Agric Sci 50(3):278

Verma PR, Saharan GS (1994) Monograph on Alternaria diseases of crucifers. In: Agriculture and agri-food Canada, pp 1–146

Verma SS, Yajima WR, Rahman MH, Shah S, Liu JJ, Ekramoddoullah AK, Kav NN (2012) A cysteine-rich antimicrobial peptide from *Pinus monticola* (PmAMP1) confers resistance to multiple fungal pathogens in canola (*Brassica napus*). Plant Mol Biol 79:61–74

Vicente JG, Holub EB (2013) *Xanthomonas campestris* pv. *Campestris* (cause of black rot of crucifers) in the genomic era is still a worldwide threat to *Brassica* crops. Mol Plant Pathol 14:2–18. https://doi.org/10.1111/j.1364-3703.2012.00833.x

Vicente JG, Taylor JD, Sharpe AG, Parkin IA, Lydiate PDJ, King GJ (2002) Inheritance of race-specific resistance to *Xanthomonas campestris pv. Campestris* in *Brassica* genomes. Phytopathology 92:1134–1141. https://doi.org/10.1094/PHYTO.2002.92.10.1134

Vishwanath, Kolte SJ, Singh MP, Awasthi RP (1999) Induction of resistance in mustard (*Brassica juncea*) against Alternaria black spot with an avirulent *Alternaria brassicae* isolate-D. Eur J Plant Pathol 105:217–220. https://doi.org/10.1023/A:1008717323002

Voorrips RE (1996) Production, characterization and interaction of single-spore isolates of *Plasmodiophora brassicae*. Eur J Plant Pathol 102:377–383

Wan H, Yuan W, Ye Q, Wang R, Ruan M, Li Z, Zhou G, Yao Z, Zhao J, Liu S, Yang Y (2012) Analysis of TIR- and non-TIR-NBS-LRR disease resistance gene analogous in pepper: characterization, genetic variation, functional divergence and expression patterns. BMC Genomics 13:502. https://doi.org/10.1186/1471-2164-13-502

Wang X, Wang H, Wang J, Sun R, Wu J, Liu S, Bai Y, Mun JH, Bancroft I, Cheng F, Huang S, Li X, Hua W, Wang J, Wang X, Freeling M, Pires JC, Paterson AH, Chalhoub B, Wang B, Hayward A, Sharpe AG, Park BS, Weisshaar B, Liu B, Li B, Liu B, Tong C, Song C, Duran C, Peng C, Geng C, Koh C, Lin C, Edwards D, Mu D, Shen D, Soumpourou E, Li F, Fraser F, Conant G, Lassalle G, King GJ, Bonnema G, Tang H, Wang H, Belcram H, Zhou H, Hirakawa H, Abe H, Guo H, Wang H, Jin H, Parkin IA, Batley J, Kim JS, Just J, Li J, Xu J, Deng J, Kim JA, Li J, Yu J, Meng J, Wang J, Min J, Poulain J, Wang J, Hatakeyama K, Wu K, Wang L, Fang L, Trick M, Links MG, Zhao M, Jin M, Ramchiary N, Drou N, Berkman PJ, Cai Q, Huang Q, Li R, Tabata S, Cheng S, Zhang S, Zhang S, Huang S, Sato S, Sun S, Kwon SJ, Choi SR, Lee TH, Fan W, Zhao X, Tan X, Xu X, Wang Y, Qiu Y, Yin Y, Li Y, Du Y, Liao Y, Lim Y, Narusaka Y, Wang Y, Wang Z, Li Z, Wang Z, Xiong Z, Zhang Z (2011) *Brassica rapa* genome sequencing project consortium. The genome of the mesopolyploid crop species *Brassica rapa*. Nat Genet 43:1035–1039. https://doi.org/10.1038/ng.919

Wang GX, Lv J, Zhang J, Han S, Zong M, Guo N, Zeng XY, Zhang YY, Wang YP, Liu F (2016) Genetic and epigenetic alterations of *Brassica nigra* introgression lines from somatic hybridization: a resource for cauliflower improvement. Front Plant Sci 7:1258. https://doi.org/10.3389/fpls.2016.01258

Wang Z, Zhao FY, Tang MQ, Chen T, Bao LL, Cao J, Li YL, Yang YH, Zhu KM, Liu S, Tan XL (2020) *BnaMPK6* is a determinant of quantitative disease resistance against *Sclerotinia sclerotiorum* in oilseed rape. Plant Sci 291:110362

Warwick SI (2011) Brassicaceae in agriculture. In: Bancroft I, Schmidt R (eds) Genetics and genomics of the Brassicaceae. Springer, New York, pp 33–67. https://doi.org/10.1007/978-1-4419-7118-0_2

Weerakoon SR, Si P, Zili W, Meng J, Yan G (2009) Production and confirmation of hybrids through interspecific crossing between tetraploid *B. juncea* and diploid *B. oleracea* towards a hexaploid Brassica population. In: 16th Australian Research Assembly on Brassicas (Ballarat, VIC)

Wei L, Jian H, Lu K, Filardo F, Yin N, Liu L, Qu C, Li W, Du H, Li J (2016) Genome-wide association analysis and differential expression analysis of resistance to Sclerotinia stem rot in *Brassica napus*. Plant Biotechnol J 14:1368–1380. https://doi.org/10.1111/pbi.12501

Westman AL, Kresovich S, Dickson MH (1999) Regional variation in *Brassica nigra* and other weedy crucifers for disease reaction to *Alternaria brassicicola* and *Xanthomonas campestris* pv. *Campestris*. Euphytica 106:253–259. https://doi.org/10.1023/A:1003544025146

Wu J, Cai G, Tu J, Li L, Liu S, Luo X, Zhou L, Fan C, Zhou Y (2013) Identification of QTLs for resistance to Sclerotinia stem rot and BnaC.IGMT5.A as a candidate gene of the major resistant QTL SRC6 in *Brassica napus*. PLoS One 8:e67740. https://doi.org/10.1371/journal.pone.0067740

Wu J, Zhao Q, Yang Q, Liu H, Li Q, Yi X, Cheng Y, Guo L, Fan C, Zhou Y (2016) Comparative transcriptomic analysis uncovers the complex genetic network for resistance to *Sclerotinia sclerotiorum* in *Brassica napus*. Sci Rep 6:19007. https://doi.org/10.1038/srep19007

Wu J, Lin L, Xu M, Chen P, Liu D, Sun Q, Ran L, Wang Y (2018) Homoeolog expression bias and expression level dominance in resynthesized allopolyploid *Brassica napus*. BMC Genomics 19 (1):586

Xue S, Cao T, Howard RJ, Hwang SF, Strelkov SE (2008) Isolation and variation in virulence of single-spore isolates of *Plasmodiophora brassicae* from Canada. Plant Dis 92:456–462

Yoshikawa H (1993) Studies on breeding of clubroot resistance in cole crops. Bull Natl Res Inst Veg Ornam Plants Tea Jpn Ser A 7:1–165

Yu J, Tehrim S, Zhang F, Tong C, Huang J, Cheng X, Dong C, Zhou Y, Qin R, Hua W, Liu S (2014) Genome-wide comparative analysis of NBS-encoding genes between *Brassica* species and *Arabidopsis thaliana*. BMC Genomics 15:3. https://doi.org/10.1186/1471-2164-15-3

Zhang Y, Tessaro MJ, Lassner M, Li X (2003) Knockout analysis of *Arabidopsis* transcription factors TGA2, TGA5, and TGA6 reveals their redundant and essential roles in systemic acquired resistance. Plant Cell 15:2647–2653. https://doi.org/10.1105/tpc.014894

Zhang GQ, Tang GX, Song WJ, Zhou WJ (2004) Resynthesizing *Brassica napus* from interspecific hybridization between *Brassica rapa* and *B. oleracea* through ovary culture. Euphytica 140:181–187. https://doi.org/10.1007/s10681-004-3034-1

Zhou JM, Tang XY, Martin GB (1997) The Pto kinase conferring resistance to tomato bacterial speck disease interacts with proteins that bind a cis-element of pathogenesis-related genes. EMBO J 16:3207–3218. https://doi.org/10.1093/emboj/16.11.3207

Zou P, Lu X, Jing C, Yuan Y, Lu Y, Zhang C, Meng L, Zhao H, Li Y (2018) Low-molecular-Weightt polysaccharides from *Pyropia yezoensis* enhance tolerance of wheat seedlings (*Triticum aestivum L.)* to salt stress. Front Plant Sci 9:427. https://doi.org/10.3389/fpls.2018.00427

Molecular Mechanisms of Host Resistance at a Glance

7

Abstract

Cruciferous crops are attacked by several pathogens which deteriorate yield and quality of major *Brassica* crops affecting global food security. The pathogens infecting these crops have diverse lifestyle, genomic plasticity with rapid evolutionary mechanisms, prolific reproduction, and longer survival nature which is impedimental to manage efficiently under field conditions. Moreover pathogens secrete small protein known as effector genes, which counter plant defense mechanisms and modulate host physiology to promote pathogen growth and reproduction. However, plants evolve disease resistance (R)-genes encoding receptor proteins that detect and respond to pathogen effector gene, leading to the activation of effector-triggered immunity. This discovery with the use of multiomics approaches has revealed molecular mechanisms of crucifers' host resistance to manage major pathogens of *Brassica* crops. Host resistance in crucifers to biotic and abiotic stresses is multilayered and multicomponent. Resistance is activated through SA signaling and simultaneous perception of ethylene and jasmonic acid depending on biotrophic, hemibiotrophic, and necrotrophic *Brassica* host–pathogen interactions. Molecular mechanisms of host resistance against biotrophs have been revealed through identification, functionally characterization, and fine mapping of R-genes and QTLs on different chromosomes of *Brassica* species. The molecular mechanisms of R-genes resistance imparted by NBS-LRR are effective against biotrophs or hemibiotrophs but are not effective against necrotrophs. JA- and ET-mediated signaling is predominantly involved in general defense to necrotrophs. Similarly, R-genes and QTLs have been identified from *B. rapa*, *B. juncea*, *B. nigra*, and *B. napus* to provide resistance to hemibiotrophs and necrotrophs. Cysteine-rich protein kinase genes and multienvironment stable six QTLs suitable for quantitative blackleg resistance in *B. napus* have been identified. Overexpression of *BnNPR1*, *OsPGIP2*, *BnWRKY33*, and *BnMPK4* genes enhances resistance in *B. napus* to *Sclerotinia*.

G. S. Saharan et al., *Molecular Mechanism of Crucifer's Host-Resistance*, https://doi.org/10.1007/978-981-16-1974-8_7

Seven NBS-LRR-encoding genes linked to black rot resistance in *B. oleracea* have been identified on four QTL regions and chromosomes CO1, CO3, CO6, CO7, and CO8, respectively. Biometabolomics mechanisms of *Brassica* host resistance is initiated soon after perception of pathogens by signaling molecules and results in execution of broad range of defense responses to stop the infection and pathogenesis of invading pathogens. Biometabolomics and biomolecular mechanisms of host resistance are expressed in the form of different metabolites (primary and secondary) and biomolecules involved in the induction of biochemical changes and pathways in the host to confer resistance at pre- and post-invasion stages to biotrophs, hemibiotrophs, and necrotrophs. Transcriptomics and proteomics have discovered novel biochemical pathways and proteins/genes involved in conferring resistance to virulent pathotypes. With the use of multiomics supplemented with functional genomics studies, key resistance pathways involved in *Brassica* host-pathosystem and induction of differential signaling molecules like SA, JA, and ET in response to diverse pathogens have been revealed.

Keywords

Molecular mechanism of host resistance · Biotrophs · Necrotrophs · Hemibiotrophs · Crucifers' pathogens · Biomolecular mechanisms of resistance · Molecular approaches for breeding disease resistance · Durable resistance · Novel molecular approaches for resistance · Development of transgenics · Induction of systemic resistance · NHR by gene silencing · Endo-chitinase gene from *Trichoderma virens*

Novel molecular approaches have been exploited to breed disease-resistant cultivars of *Brassica*. Through introgression breeding high level of resistance has been obtained from *B. nigra* and *B. carinata* to *B. napus* against black leg (*Leptosphaeria*). Similarly transfer of resistance to black rot from *B. carinata* to *B. oleracea*, from *B. hirta* to *B. juncea* against *Alternaria*, and from *B. carinata* to *B. oleracea* against powdery mildew has been obtained. The approaches like in vitro embryo rescue, somatic hybridization, somaclonal variations, genetic transformation, molecular markers, use of systemic resistance induction, and genetic engineering have great promise but have been used against few pathogens. Clubroot resistance in *B. napus* and *B. rapa* has been transferred through interspecific and distant hybridization using molecular markers. R-genes from European Clubroot-Differentials (ECD) set have been introduced into *B. napus*, and other species, to provide resistance to clubroot. Powdery mildew resistance in *B. oleracea* has been obtained from *B. carinata* through embryo rescue. The mutagenic agent gamma radiation increases resistance to *E. cruciferarum* in *Brassica*. Durable resistance in *Brassica* species to *Alternaria* can be obtained through introgression of genes from nonhost plants by inter- or intraspecific crosses, identification and transfer of NHR based on microarray, identification and transfer of NHR genetic resources from model host plants, identification of functional genes based on bioinformatics, and use of endo-chitinase gene from *Trichoderma virens*. Marker-assisted introgression of resistance from wild relatives has been achieved in *B. juncea* to *Sclerotinia*. Resistant *B. napus* lines pyramided with multiple R-QTL derived from wild *B. oleracea* have been developed via hexaploidy strategy. Introgression of black rot resistance from *B. carinata* to *B. oleracea* through embryo rescue has been attempted successfully. The *B. carinata* genotypes, NPC-9 and HCA-6, having very high resistance to *XCC* race 1 and 4 have been used to introgress resistance in Indian cauliflower through embryo rescue approaches.

The advances made in the molecular genetics through whole genome sequencing of *Brassica* crops have led to quantum jump not only in understanding plant genetic diversity but also for accessing these enormous resources. Novel sources of R-genes for blackleg has been tapped for other A- and C-genomes of *Brassica* species from the B-genomes in *B. nigra*, *B. juncea*, *B. carinata*, and a wild species *Sinapis arvensis*. Several sources of resistance to 15 major pathogens of crucifers except nematodes have been identified from different *Brassica* species and wild relatives. Some of these sources have been utilized to breed disease-resistant cultivars of *Brassica* through conventional and molecular approaches. From *Brassica* species, multiple pathogen resistance sources have been identified which provide resistance to Albugo, Alternaria, Erysiphe, and Hyaloperonospora. There is strong evidence for positive genetic correlation among *Albugo*, *Alternaria*, *Erysiphe*, *Hyaloperonospora*, Leptosphaeria, and Xanthomonas to develop multiple disease-resistant cultivars of *Brassica* crops. Clubroot-resistant cultivars with R-genes and QTLs have been developed in *B. rapa* and *B. napus* with pathotype specificity at different counties, but most of them have scummed through evaluation of new virulences of Plasmodiophora. In *Brassica* crops, transfer of resistance through different approaches is facing problems because of diverse nature of infecting pathogens for which there is no single gene product to have a direct toxic effect on all kinds of pathogens. Consequently host has to develop quite different strategies to deal with these pathogens.

7.1 Introduction

Cruciferous crop plants (canola/oilseed rape/rapeseed/mustard/cabbage/cauliflower) are economically important crops, widely used as edible oil vegetables. These crops are mostly cultivated in temperate and subtropical regions. Brassica crops are always challenged by major diseases such as Sclerotinia stem rot, white rust, downy mildew, Alternaria blight, blackleg, and clubroot. These diseases cause significant yield and economic losses in rapeseed-mustard and vegetable crops worldwide, which has been aggravated by global climate change. If it is not resolved meritoriously, it will impend global food security. To advance further comprehensions into the *Brassica*–pathogen interactions, it is advisable to analyze critically the current knowledge in this area and how advanced skills and technologies can offer encouraging consequences to understand the resistance mechanisms. Resistant varieties development is the utmost important, economical, environmentally sustainable solution and effective to cope with diseases management in *Brassica* species. Transfer of disease resistance remains attained once sources of R-genes have recognized with sound characters, mode of inheritance of genes in the progenies along with all desired traits of yield and quality. Conventional and molecular approaches have been used either alone or in combination to incorporate desired and popular traits to develop improved cultivars of crucifers. Hybridization combined with backcrossing either singly or in combination with pedigree selection is the most widely used and accepted conventional method to transfer disease resistance in crucifers. The level of resistance in developed cultivars has been accomplished through multiple crosses, recurrent selection, diallele crossing, and selective mating system. The resistance has been introgressed with intraspecific pedigree, backcross, modified recurrent mass selection, interspecific genome substitution, chromosome substitution, and gene introgression against different pathogens. The molecular and genetic engineering techniques like genome manipulation, cytoplasmic genome manipulation, transformation, foreign gene expression, and embryo rescue for wide hybridization have been exploited in *Brassica*. The introgression and pyramiding major RQTLs have been attempted including transgenics development. To develop stress tolerance and/or resistant *Brassica* cultivars, there is need to apprehend both of the plant and pathogenic disease development mechanisms.

The genomics of crucifers host resistance has been strengthened significantly after the sequencing of all the six *Brassica* species genome and major crucifers' pathogen genome. The development and application of omics technology and approaches have facilitated to discover molecular mechanisms of host–pathogen interaction and host resistance in major crucifers' host pathosystem. During the course of literature search, it was revealed that voluminous research data have been generated on genomics of crucifers' host-pathosystem which needs separate treatment on the genomics of host resistance and molecular mechanisms of host resistance. In this chapter to make the story of host resistance complete, briefly molecular mechanisms of host resistance has been covered. The readers are advised to consult the other book entitled *Genomics of Crucifers Host Resistance* for detailed information.

7.2 Pathogen Effector Gene Regulates Molecular Mechanisms of Host Defense

Biotic and abiotic stresses of crucifer crop plants cause significant yield losses and affect global food security, particularly in the face of increased demand due to a growing world population (Fisher et al. 2012). Diverse lifestyle, genome plasticity and evolution, prolific reproduction, and longevity of spores under harsh environment impede the efficient control of pathogen causing crucifers crop diseases. Bacterial and fungal pathogens secrete small proteins, known as effectors, which counter plant defense and modulate plant physiology to promote pathogen growth and reproduction (Presti et al. 2015; Stergiopoulos and de Wit 2009). Accumulating evidences show that the secreted effectors are key players in suppressing pathogen-associated molecular patterns (PAMPs)-triggered immunity (PTI) induced upon fungal recognition by plants (Presti et al. 2015). The lysine motif (LysM) effector ECP6 secreted by the tomato pathogen *Cladosporium fulvum* suppresses chitin-triggered immunity. ECP6 binds chitin via the intramolecular chitin-binding groove formed by the LysM domains, resulting in the sequestering of chitin released from the cell walls of invading hyphae (de Jonge et al. 2010; Sanchez-Vallet et al. 2013). Plants, on the other hand, have evolved disease resistance (R)-genes encoding receptor proteins that detect and respond to pathogen effector proteins, leading to the activation of effector-triggered immunity (ETI) (Jones and Dangl 2006). Effectors that trigger ETI are referred to as avirulence (*Avr*) genes. Discovery of the host targets of effectors is essential for our understanding of effector activity and molecular mechanisms of plant defense (Win et al. 2012). There are many examples of plant proteins that are targeted by bacterial effectors secreted by the type III secretion system (Block and Alfano 2011; Deslandes and Rivas 2012). However, despite the economic importance of diseases caused by the fungal plant pathogens, for most fungal effectors, the host targets and underlying molecular mechanisms remain unknown. Mitogen-activated protein (MAP) kinase cascades are highly conserved signaling modules in eukaryotes and have a central role in plant immunity against pathogen attack (Meng and Zhang 2013; Pitzschke et al. 2009). A MAP kinase cascade is commonly composed of three types of kinases: a MAP kinase kinase kinase (MAPKKK), a MAP kinase kinase (MKK), and a MAP kinase (MPK). In *Arabidopsis*, pathogen-responsive MPK signaling has been reported to be involved in both PTI and ETI, in which the best characterized MPKs are MPK3, MPK4, and MPK6 (Asai et al. 2002; Rasmussen et al. 2012). PAMPs, such as flg22 (a conserved 22-amino acid flagellin peptide) and elf18 (elongation factor-Tu peptide), activate the kinase signaling cascades involving MPK3, MPK4, and MPK6 (Felix et al. 1999; Gao et al. 2008; Zipfel et al. 2006). In contrast, plant pathogens have evolved mechanisms to target the MPK signaling pathways to enhance plant susceptibility. *Pseudomonas syringae* HopAI1 targets MPK3 and MPK6 and inactivates their kinase function to suppress plant defense responses (Zhang et al. 2007). *P. syringae* HopF2 targets MKK5 and can inactivate MKK5 via ADP-ribosylation of the C terminus of MKK5 in vitro (Wang et al. 2010). AvrB is a *P. syringae* effector that interacts with MPK4 to perturb hormone signaling and promote infection (Cui et al. 2010). *Phytophthora infestans* RXLR effector PexRD2 as a virulence factor interacts with the kinase domain of the host MAPKKKε to suppress MAPKKKε-dependent phosphorylation of MPKs to modulate plant immunity (King et al. 2014). These findings highlight the importance of MPK pathways in plant immunity and as targets of bacterial and biotrophs pathogen effectors. However, to date no MAPKs have been identified as targets of effectors from fungal plant pathogens (Ma et al. 2018).

Plant MPKs can be classified into four groups (A, B, C, D) based on the conserved amino acid sequences of the T × Y motif present in the activation loop of MPKs (Ichimura et al. 2002). Among the best studied MPKs, MPK3 and MPK6 are in group A and MPK4 belongs to group B. However, MPK9 belongs to the D group with the plant-specific TDY phosphorylation motif and is distinct from the A, B, and C groups. In *Arabidopsis*, studies have reported that AtMPK9, together with AtMPK12, is involved in stomatal closure (Lee et al. 2016). Both abscisic acid (ABA)- and methyl jasmonate-induced stomatal closures are impaired in *Arabidopsis* double mpk9/mpk12 mutants, but not in mpk9 or mpk12 single mutants, indicating a functional redundancy (Jammes et al. 2009; Khokon et al. 2015). The mpk9/mpk12 double mutant is highly susceptible to *P. syringae* pv. tomato (Pst) DC3000 infection (Jammes et al. 2011). However, Montillet et al. (2013) reported that ABA signaling, including open stomata 1 (OST1) protein kinase, AtMPK9, or AtMPK12, plays a limited role in response to Pst early infection. This is based on the observations that there is no difference between double mpk9/mpk12 mutants and wild-type *Arabidopsis* Col in response to flg22-induced stomatal closure and expression of ABA-specific marker genes is not affected by Pst treatment (Montillet et al. 2013). Thus it is unclear whether MPK9 acts in ABA-mediated guard cell immune signaling in response to biotic stresses. A study by Nagy et al. (2015) revealed that AtMPK9 is activated through intramolecular autophosphorylation independent of any upstream MAPKKs, which is similar to MAPKK-independent activation mechanisms reported for the mammalian atypical MAPKs, such as extracellular signal-regulated kinase (ERK) 7/8 (Klevernic et al. 2006; Nagy et al. 2015).

Host resistance in crucifers to biotic and abiotic stresses is multilayered and multicomponent. The perception of defense response and molecular mechanisms in a host against the pathogens are of great significance in the breeding of resistant cultivars. The sequencing of host and pathogen genome during the last decade has facilitated the researchers using modern omics approaches to reveal molecular mechanisms of host–pathogen interaction and induction of signal molecules for expression of R-genes in the host at different levels of host growth. The intricate immune responses are evolved through accumulation of ROS, H_2O_2, deposition of callose, pectin, cellulose, waxes, silicon, ion-influxes, formation of papilla, cell wall apposition, phenolic compounds, overexpression of R-genes, PR proteins, protein phosphorylation, biosynthesis of phytoalexins, fungal enzymes inhibitors, triggering of HR, induction of SAR, nonhost resistance mechanisms, the production of secondary metabolites, calcium and hormones signaling and homeostasis, and the induction of transcriptional factors and mitogen-activated protein kinases. Resistance is activated through SA signaling and simultaneous perception of ethylene and jasmonic acid depending on biotrophic and necrotrophic host–pathogen interactions. The majority of plant R-genes encode nucleotide-binding site leucine-rich repeat (NB-LRR)–type protein, which can be further, grouped into two subclasses based on their N-terminal sequence. First, those containing coiled coil (CC) domain (CC-NB-LLR) and second those containing a domain with interleukin-1 receptor (TIR) (TIR-NB-LRR). The NBS-LRR proteins are thought to recognize effectors or effector-specific signals from the pathogen and activate the host immune system to provide hypersensitive responses.

Cruciferous plants evolve R-genes in response to pathogen attack which encodes receptor proteins that detect and respond to pathogen effector protein leading to the activation of effector-triggered host immunity. Effectors that trigger ETI are referred to as avirulence gene (*Avr*). Mitogen-activated protein (MAP) kinase cascades are highly conserved signaling molecules in eukaryotes and have a central role in plant immunity against pathogen attack. Plants are very often exposed to a variety of biotic stresses and thus have evolved multidimensional defense approaches to survive or retain their fitness. Plants use both preformed and inducible defense mechanisms to overcome pathogen challenges. Much stronger and long-lasting is inducible defense response such as systemic acquired resistance (SAR), which confers enhanced disease resistance to broad range of pathogens. NPR1 gene in *Brassica* plants mediates the SA-induced expression of pathogenesis-related (PR) genes and SAR. Overexpression of NPR1 results in the increased transcript levels of antifungal genes like PR1, PR2 (glucanase), and PR5 (thaumatin). Many of the underlying resistance genes have been cloned, and their downstream signaling pathways have been characterized using *Arabidopsis* host-pathosystem in molecular mechanisms of host resistance. Plants have evolved intracellular receptors with nucleotide-binding, leucine-rich repeat domains (NLRs) to detect fungal effectors, or their activity to trigger immunity. The high polymorphic information content (PIC) and number of phylogenetically informative bands established RAPD as a useful tool for phylogenetic reconstruction, quantification of genetic diversity for conservation, cultivar classification, and molecular disease resistance cultivars breeding in *Brassica*. The clustering and duplication of R-genes can cause problems in R-genes identification as these genes tend to collapse in genome sequence assemblies. The outcome of host–pathogen interactions is determined by R-genes that enable host to recognize invading pathogens and activate inducible defense mechanisms. A typical R-genes allele encodes “race-specific” resistance to only one or a few strains of a single pathogen species. R-genes loci are functionally polymorphic within a plant species and encode alternate alleles that either recognize different strains of the same pathogen or do not recognize any tested pathogen. The molecular studies have revealed that R-genes often reside in complex loci consisting of R-genes and lightly linked homologs. The cloning of R-genes against diverse pathogen provides the tools for comparative analysis of R-genes alleles within and between species, which will provide insight into the evolutionary history of R-genes. The functional polymorphism at R-genes loci can arise from gene deletions. The majority of plant PRRs identified belongs to the family of leucine-rich repeat receptor-like kinase (LRR-RLKs) or receptors-like protein (LRR-RLPs) that lacks a C-terminal kinase domain and interacts with RLKs for transducing signals. Several numbers of genes have well documented that they are differentially expressed against different biotic stresses. Several techniques and approaches have been used to reveals molecular mechanisms of crucifers’ host resistance. These techniques include genomics, bioinformatics, epigenetics, transcriptomics, proteomics, metabolomics, metagenomics, pathogenomics, phylogenomics, and pangenomics.

7.3 Molecular Host Defense Responses to Biotrophs

The genome sequencing of a small cruciferous weed, *Arabidopsis thaliana*, and its establishment as a model for host pathosystem to reveal molecular mechanisms of host–pathogen interactions have transformed traditional sciences like Genetics, Plant Breeding, and Plant Pathology into Molecular Genetics, Molecular Plant Breeding, and Molecular Plant Pathology under the umbrella of Molecular Biology. During the last three decades, significant investigations have been carried out on molecular mechanisms of crucifers host resistance. Out of 44 pathogens known to infect crucifers, 16 pathogens are of significant importance at global level on the basis of geographical distribution, host range, yield losses, and investigations on the molecular studies. On the basis of mode of acquisition of nutrients from the host (life style) and mode of infection, differences in molecular mechanisms of host–pathogen interactions can be divided as biotrophs, hemibiotrophs, and necrotrophs. Biotrophs essentially act as a sink for the host, enable to derive assimilates, and keep it alive. Necrotrophs consume the host tissues on invasion. Hemibiotrophs combine both strategies in their life cycle. Therefore, host plants have quite different molecular mechanisms and strategies to deal with these groups of pathogens. Pathogen infection produces a variety of phytohormones depending mainly on the pathogen lifestyle and their mode of infection. Salicylic acid pathway generally provides resistance to biotrophic pathogens, whereas JA/ET pathways are commonly associated with resistance to necrotrophic pathogens, and to herbivorous pests. Generally, SA and JA signaling pathways operate antagonistically, and thus, elevated resistance against biotrophs is often related with increased susceptibility to necrotrophs and vice versa.

Molecular mechanisms of crucifers' host resistance against biotrophs have been revealed in different *Brassica* species during host–pathogen interactions at different stages of host growth and disease development. The R-genes and QTLs have been identified, functionally characterized, and molecularly mapped on the chromosomes of *Brassica* species expressed during pathogens interactions. Timing of the expression of defense-related genes plays a crucial role during pathogenesis and host resistance for mounting a successful defense response. A single gene (*Acr*) responsible for conferring resistance to *Albugo* was mapped on a densely populated *B. juncea* RFLP map. A single dominant *RAC1* gene in accession Ksk 1 of *Arabidopsis thaliana* confers resistance to *A. candida* isolate ACeM1. Two accessions of *A. thaliana* (Ksk1 and Ksk2) were used to identify and map three loci (RAC1, RAC2, and RAC3) of genes that confer resistance to *A. candida*. In *B. rapa*, the ACA1 locus was mapped to linkage group 4 and was flanked by RFLP marker loci. *WRR* 4 genes encode a cytoplasmic TIR-NBL-LRR receptor like protein in Columbia *A. thaliana* and confer broad-spectrum white rust resistance to four races of *A. candida*. Three WRR (*WRR4BCol*-0, *WRR8Sf*-2, and *WRR9Hi*-0) genes against *Ac2V*, and a gene *WRR12* (*SOC3*), conferring NHR to AiBoT have been identified. Analysis of allelic variants of BjuA046215 in Indian, east Europe, and Chinese gene pool suggested the presence of three types of alleles. Allele 1 is present in the Chinese Tumida, Allele 2 in Indian, and Allele 3 in east European gene pools.

Host resistance in *Brassica* species to powdery mildew is multilayered and multicomponent at pre- and post-penetration both stages. Host resistance is activated either through SA signaling and simultaneous perception of ethylene and jasmonic acid. MLO genes encoding seven transmembrane calmodulin-binding proteins confer broad spectrum resistance to *Arabidopsis* powdery mildew. The Pmr mutants confer resistance to powdery mildew through altered cell wall composition of host. Salicylic acid enhances the expression of RPW 8.1, and RPW 8.2 leading to HR or SHL and resistance. *BjNPR1* gene activates SAR to confer broad spectrum resistance to powdery mildew of *B. juncea*. There is role of WRKY transcription factors and overexpression of R-genes like PMR, MLO, PEN, EDR, MAPK, MAPK 65-3, NPR1, PAD3, PAD4, ED5, SNARE, RLKs, and KDL (AtCEP1) to confer resistance to powdery mildew of crucifers. Powdery mildew resistance genes of *Arabidopsis* have been mapped on chromosomes II (RPW 1), III (RPW 2, 3, 7, 8), IV (RPW 4), and V (RPW 5, 6). The papain-type KDEL Cys EP CEP1 fulfills function in plant defense during late development of *Erysiphe cruciferarum* in close spatial association with the fungal haustorium, and haustorial callose encasement.

Resistance to Hyaloperonospora in *Brassica* is controlled by dominant gene at seedling and adult plant stages. White rust R loci on three *Arabidopsis* chromosomes are closely linked to downy mildew R loci. Downy mildew R-genes are differentially expressed. The identified 129 genes of which 121 TDFs' are upregulated and eight are down-regulated. R-genes of *Arabidopsis* are mapped on five chromosomes in MRC-A, MRC-B, MRC-F, MRC-H, and MRC-J as RPP1 to RPP 27. Single transduction pathways used by the different RPP genes are complex and diverse. Stable and transient gene silencing approaches have been applied in downy mildew of crucifers. Strong dependence on genes EDS1 and NDR1 is governed by R-genes structural type rather than pathogens of *Arabidopsis*. Several clubroot resistance (CR) genes have been identified and mapped on *B. rapa*, *B. oleracea*, *B. nigra*, *B. juncea*, *B. napus*, and *A. thaliana* chromosomes. *Brassica* species genome sequencing has allowed rapid identification of genome wide QTLs and GWAS regions for R-genes of clubroot. With the use of transcriptomic and proteomic approaches, more than 1000 *Arabidopsis* genes have been identified which are differentially expressed in Plasmodiophora *brassicae* infected roots. More than ten CR-genes in *B. rapa* have been identified with pathotypes specificity. The genetic analysis, genetic mapping, and fine mapping on chromosomes of *Brassica* species with molecular markers and pathotypes specificity have been studied in *B. rapa*, *B. oleracea*, *B. napus*, *B. napus* var. *napobrassicae*, and *A. thaliana*. The genes encoding TIR-NBS-LRR class proteins impart resistance against *P. brassicae*. The R-genes are present in clusters on the chromosomes of *Brassica* species. R-genes QTLs using dd RAD-seq has been mapped in *B. rapa* against clubroot pathotypes. The CR QTLs have been identified through GWAS analysis in the natural populations of *B. napus*. Bioinformatic analysis has been used to predict CR QTLs of *Brassica*. For the determination of loss of R-genes, linkage markers of CR in *Brassica* have been identified. Marker assisted selection of CR genes has reinforced the accumulation of varied R-genes in *Brassica*. The genetic origin of CR genes has been revealed in *Brassica*. Quantitative resistance to clubroot is mediated by transgenerational epigenetic variation in *Arabidopsis*. The use of SNP microarray, proteomic, and transcriptome approaches in *Brassica* have revealed multicomponent molecular mechanisms conferred by CR QTLs to *P. brassicae*. In *Brassica*, R-genes and QTLs have been identified with molecular markers and mapped on chromosomes to TuMV isolates. The retr01/retr02/retr03 genes have broad spectrum resistance to TuMV. Alternative oxidase gene (*BjAOX1a*) in *Brassica* enhances resistance to Turnip mosaic virus. The techniques such as yeast two hybrid (Y2H), biomolecular fluorescence complement (BiFC), and co-immune precipitation (COIP) methods to analyze the resistance interaction between host and TuMV isolates have been employed.

7.4 Molecular Host Defense Responses to Hemi-Biotrophs and Necrotrophs

The development in the science and technology through modern approaches like omics and genomic have facilitated to investigate molecular mechanisms of crucifer's host resistance against diverse nature of biotic and abiotic stresses. The molecular mechanisms of host resistance have been investigated in *Brassica* crops using three different categories of host-pathosystem, namely, biotrophs, hemibiotrophs, and necrotrophs. The molecular mechanisms of R-genes resistance imparted by NBS-LRR are effective against biotrophs or hemibiotrophs but are not effective against necrotrophs. Salicylic acid (SA), jasmonic acid (JA), and abscisic acid (ABA) are the key modulators involved in signaling pathways during host defense. Jasmonic acid- and ethylene-mediated signaling is predominantly involved in general defense against necrotrophs. NBS-LRR genes are intracellular receptors that can recognize the pathogen invasion by binding to pathogen effector proteins in host plant and trigger various defense signal transduction to yield hypersensitive response and inhibit pathogen development. The NBS domain binds and hydrolyzes ATP and GTP, while LRR domain is liable for protein–protein interaction. NBS domains are of various motifs like P-loop and kinase 2 motif. NBS-LRR genes are grouped into TIR-BSS-LRR (TNL) and CC-NBS-LRR (ONL). Many NBS-LRR genes have been identified in *Brassica* species for the defense against different pathogens infecting *Brassica* crops. The first line of plant defense (pre-penetration) against fungal pathogen relies on cell surface-mediated defense signaling initiated by recognition of PAMPs, and secondary (post-penetration) defense responses are often induced by the salicylic acid (SA) or jasmonic acid/ethylene (JA/ET) phytohormone signaling pathways.

The molecular mechanisms of nonhost resistance to Alternaria in *Brassica* indicated activation of defense-related genes earlier in nonhost plants than the host plants. Six QTLs governing resistance to *Alternaria brassicae* have been identified, five of them are population specific, while one QTL is common between all the three mapping populations. In *Brassica*, defense-related genes (PR proteins) including defensins (β-glucanases and chitinases) have been identified against *Alternaria*. Different defensin genes in *B. juncea* have been characterized for their structures, evolution, cellular location, and regulation in response to SA, JA, and *Alternaria* infection. Chitinase genes in *Brassica* have been identified to breed *Alternaria*-resistant cultivars. A single R-locus RCH-1 has been identified at the tip of chromosomes 4 in *Arabidopsis thaliana* ecotype Eil-O against Colletotrichum *higginsianum.* The locus RCHz maps to an extensive cluster of R-loci known as MRC-J in the *Arabidopsis* ecotype Ws-O. In *B. oleracea*, two types of R-genes have been identified. A type with single dominant genes is effective against Fusarium race 1 and stable under high or low temperatures. B type is polygenic resistance, and it is unstable under high temperature above 24 °C. Type A resistance to *FOC* 1 race 1 conferred by a dominant single gene, *FOC1*, has been mapped and molecular markers have been developed. R-genes and QTLs have been identified from *B. rapa*, *B. juncea*, *B. nigra*, and *B. napus* to provide resistance to Leptosphaeria. The LepR1 and lepR2 genes provide cotyledon and adult plant resistance in *B. napus* to *Leptosphaeria*. The background is useful markers to select the most suitable genotype as a recipient parent while developing *Brassica* cvs. with *Lm* resistance. The combined effect of R-genes and host genetic background generates stronger defense responses through induction of genes related to callose deposition, upregulation of ROS, and SA and JA pathways.

Expression of *Cf9* and *Avr* 9 genes in *Brassica* induces resistance to *Leptosphaeria maculans*. (*Lm*) R-genes and QTLs from *B. rapa*, *B. juncea*, and *B. napus* have been molecularly mapped with markers to provide resistance to *Lm*. Two R-genes, *LepR3* and *Rlm2*, have been cloned along with seven corresponding avirulence (*Avr*) genes (*Avr lm1*, *AvrLm6*, *AvrLm3*, *AvrLm5/AvrLmJ1*, *AvrLm4-7*, *Avr Lm6*, and *AvrLm11*). Molecular basis of assessment of R-genes breakdown in *Brassica* cvs. can be predicted by frequencies of avirulence/virulence alleles in populations of *Lm*. In resistant and susceptible lines of *B. napus*, gene expression profile differs to confer resistance to *Lm*. Cysteine-rich protein kinase genes and multienvironment stable six QTLs suitable for quantitative blackleg resistance in *B. napus* have been identified. Marker-associated mapping of QTLs in *Brassica* will help identification of desirable alleles and facilitate QTL introgression. For stable resistance to *Lm*, eight QTL regions for resistance on AO2, AO9, AO10, CO1, and CO9 have been identified in *B. napus* under diverse growing environments. Nine of the 18 blackleg R-genes from *Brassica* species correspond to genetically defined *Avr* genes from *Lm*. Mechanisms of resistance to light leaf spot (*Pyrenopeziza*) in *Brassica* is controlled by two R-genes. QTLs controlling partial resistance in *Brassica* at leaf and stem infection stages by *Sclerotinia* have been identified. Resistance to *Sclerotinia* operates at host tissue breakdown, preformed antifungal compounds, and production of phytoalexins. Resistance in *Brassica* to *Sclerotinia* is governed by a complex interplay of minor effect genes designated as quantitative disease resistance (QDR). The candidates' defense-related genes have been mapped by genome-wide association combined with other omics approaches. Overexpression of genes like *BnNPR1*, *OsPGIP2*, *BnWRKY33*, and *BnMPK4* enhances resistance in *B. napus* to *Sclerotinia*, whereas overexpression of *AtWRKY28* and AtWRKY75 genes in *Arabidopsis* enhanced resistance to oxalic acid and *Sclerotinia*. In *B. napus* the microsatellite marker for genome-wide associations has identified three loci DSRC4, DSRC6, and DSRC8 for Sclerotinia stem rot resistance. SSR-based association mapping identified six markers loci associated to *Sclerotinia* resistance in A- and B-genomes of *Brassica*. Genome-wide association analysis of 84ILS recognized a large number of SNPs associated to *Sclerotinia* resistance in chromosomes AO3, AO6, and BO3.

In *B. napus*, one major RQTL contributed by R53 on chromosome C5 and minor QTL contributed by Express 617 on chromosome C1 confer resistance to *Verticillium longisporum*. Black rot (*XCC*) resistance genes and QTLs in different *Brassica* species have been identified on different chromosomes to *Xanthomonas* races 1 and 4. Seven NBS-LRR-encoding genes linked to black rot resistance in *B. oleracea* have been identified on four QTL regions and chromosomes CO1, CO3, CO6, CO7, and CO8, respectively. Micro RNAs (miR156, miR 167, miR169, and miR 90) play a role in *B. oleracea* resistance enhancement to *XCC*. R-gene1 conferring resistance to race 1 has been identified in *B. carinata*, *B. juncea*, and *B. nigra*.

7.5 Biomolecular Mechanisms of Host Defense to Biotrophs

Biometabolomics of disease resistance plays a significant role in induction of immunity signaling pathways and defense responses to biotrophs of crucifers. With the perception of pathogens, signaling is initiated that result in the execution of a broad range of defense responses to stop the infection and pathogenesis of invading pathogens. Polar metabolites of *Brassica* plants including glucosinolates, polar indole metabolites, soluble compounds, and wall-bound phenolics are induced to confer resistance to *Albugo*. The phytoalexins cyclobrassinin, rapalexin, and brassilexin are inhibitory to *Albugo* zoospores release.

Strong and accelerated activation of defense-related transcriptional programs in *mlo2 mlo6 mlo12* triple mutants of *Arabidopsis* has been observed against powdery mildew with jasmonic acid/ethylene (JA/ET) signaling and biosynthesis of indole-derived antimicrobial compounds. Defense compounds like camalexin, indole glucosinolates, and indole cyanogenic metabolites provide full immunity in *mlo2 mlo6 mlo12* mutant to powdery mildew. Salicylic acid (SA) plays an important role in resistance to powdery mildew of *Arabidopsis* required for signaling of mutant's edr1, edr2, and edr4. Genes encoding camalexin biosynthesis for powdery mildew resistance have been determined. Pre-penetration resistance to powdery mildew of *Arabidopsis* operates through cuticle and epicuticular wax when pathogen exerts pressure to penetrate cell wall, host activates, mechano-sensors to activate immune signaling gene *CERK1* to contribute basal resistance. Resistance is expressed through papilla formation which contains callose, silicon, ROS, and phenolic compounds. Induction of salicylate, NPR1, PAD4, and EDS5 in *Arabidopsis* provides resistance to powdery mildew. Chito-oligomers in *Brassica* induce defense responses to powdery mildew. Salicylic acid-dependent transcription of *RPW* 8.1 and *RPW* 8.2 genes leads to accumulation of SA and formation of SHL or HR in *Brassica* to confer powdery mildew resistance. *BjNPR1* transgenic *B. juncea* plants exhibit partial resistance to powdery mildew through activation of SAR. The expression of gene encoding AtMAP 65-3 in *A. thaliana* downregulates SA signaling for powdery mildew infection. Loss of AtMAP 65-3 function in the map65-3 mutant reduces infection at the stage of leaf penetration. There is role of AtROP-regulated AtRLCK VI A3 in basal resistance to *Erysiphe cruciferarum*. The PAD 3 overexpressing plants accumulate higher camalexin and MiMick-resistant eyp 8301-3 mutant to powdery mildew. Powdery mildew resistance of atg 2 plants depends on SA signaling. Autophagy contributes to suppression of cell death and defense response to powdery mildew. JA/Et-induced defense responses are effective against virulent powdery mildew fungi if stimulated constitutively, artificially, or systemically despite biotrophic nature of the interaction. Expression of WRKY TFs in *Brassica* enhances resistance to powdery mildew in edr1 plants. SR 1 plays a critical role in powdery mildew resistance by regulating EIN 3 and NDR1 expression. Identification of defense-related TFs/genes in the regulation indicates recognition of pathogen, response initiation and defense execute an efficient immune output. In *Arabidopsis*, WRKY 70 is transcriptionally induced by SA signaling and mediates basal resistance to powdery mildew. Transcriptional regulation and expression of several genes in response to powdery mildew interaction with host and nonhosts is regulated by signaling molecules like SA, JA, and ET pathways differentially. The application of *Trichoderma harzianum* and cultural filtrate provides systemic resistance to powdery mildew in *Brassica*. Phytohormone abscisic acid (ABA) is involved in NHR against powdery mildew of *Brassica*. Two distinct pre-haustorial nonhost defense mechanisms mediated by genes *PEN1*, *SNAP 33*, *VAMP 721*, *VAMP 722*, *PEN 2*, and *PEN* 3 are subject to pathogen-induced cell polarization and mediate the induction of antimicrobial compounds, which is common in both. *Arabidopsis* NHR to powdery mildew is based on multicomponents and independently effective defense layers. *PEN* gene mediated pre-invasion resistance and post-invasion immunity is controlled by *EDS1*, *PAD4*, and *SAG* 101 genes. Mutagenic resistance to powdery mildew is expressed by biochemical changes in the form of fatty acids, enzymes, and biosynthetic pathways. Induction of several biomolecules in *Brassica* confers resistance to downy mildew. The synthesis of phytoalexins in *A. thaliana* is mediated by genes *PAD1* to *PAD4* to provide resistance against downy mildew. In resistant genotypes of *Brassica*, cell wall lignification provides resistance to downy mildew. Salicylic acid is important for defense against most *Hyaloperonospora parasitica* races as signal amplifier to induce antipathogenic molecules to provide downy mildew resistance. Downy mildew resistance signal pathways for the expression of R-gens are very complex.

R-genes pyramiding in *Brassica* for clubroot resistance strongly trigger multiple resistance pathways by hormonal signal transduction, secondary metabolites, transcription factors, and other process via multigene network controlled by SA, ROS, PCD, and R line *ZHE-226* gene. The pyramided line 618 R showed higher number of DEGs than the parental lines. PbBa 8.1 shared higher number of DEGs than CRb. SA and ROS signaling pathways are significant in pyramided and parental lines. There is significant indication of PR proteins, thaumatin, and lipid transfer proteins in the pyramided line as compared to parental lines. Lignin biosynthesis enhances resistance to *Plasmodiophora brassicae* at early stage of infection. Flavonoids act as defense compounds or antioxidants for clubroot. MET serves as the major energy source required for defense responses mediated by Rcr1 against clubroot. To activate defense response in *Brassica* to clubroot, there is induction of signal molecules like SA, BA, SABATH enzymes, JA, IAA, GA, and MeSA. At the later stages of *P. brassicae* infection, the gene expression is different in *B. napus* cvs. differing in resistance for induction of hormones like CK, IAA, SA, JA, ABA, and ET. In general gene expression was low in resistant cultivar and higher in susceptible ones. There is differential expression of hormones in leaves verses roots and formation of galls in resistant and susceptible cultivars indicating organ and host resistance specificity in production of hormones. However, SA signal transduction is up-regulated mainly in the R-genotypes in comparison to susceptible ones. The biometabolomic resistance to clubroot in *Brassica* is expressed by NBS-LRR proteins of PbBa 1 and CRb which recognize the elicitors released by *P. brassicae* and trigger defense signal transduction. The key pathways are SA, JA, MAPK, ET, Ca^{2+}, and ROS. RBOH and phenylpropanoid biosynthesis activate ROS, SA signaling, PCD, HR, and antioxidant enzymes to enhance defense responses in *Brassica* to clubroot. Numerous genes for immune-related functional categories are associated with SA-mediated responses. JA-mediated responses are mostly inhibited especially in clubroot-resistant genotypes.

The biometabolomics of disease resistance plays a significant role in induction of immunity signaling pathways and defense responses. Following the perception of pathogens, signaling is initiated that ultimately results in the execution of a broad range of defense responses that stop the invading pathogens. In *Arabidopsis*, many genetic screens have been performed to identify genes that are important for defense signaling. Some of these screens were focused on signaling downstream of specific R-proteins, while others were aimed to identify more general signal transduction components. The important plant proteins, EDS1 and NPR1, play crucial roles in the interaction with pathogens and are well studied in *Arabidopsis*. EDS1 and NPR1 are closely linked to the immunity-related hormone salicylic acid (SA) that is also key for resistance to several pathogens. Multi-omics approaches combining genomics, transcriptomics, proteomics, and biometabolomics using computational strategies will allow us to identify suitable mimicking molecules in the pathogen and/or host species that trigger stronger plant defense system during plant–pathogen interaction.

EDS1 (ENHANCED DISEASE SUSCEPTIBILITY 1) is a lipase-like protein that is required for the function of a subgroup of NLRs, those with a Toll-interleukin 1 receptor domain called *TIR-NLR*. In addition, EDS1 has a function in oxidative stress responses and as a positive regulator in basal resistance to virulent pathogens making the *eds1* mutant-enhanced disease-susceptible (Wiermer et al. 2005). Two sequence-related signaling partners, PAD4 (PHYTOALEXIN-DEFICIENT 4) and SAG101 (SENESCENCE-ASSOCIATED GENE 101), form complexes with EDS1 and have signaling functions in the cytoplasm and nucleus. The NLR helper protein NRG1 also resides in this complex and plays a role in cell death signaling. Interestingly, EDS1 was found to associate with AvrRps4, a *Pseudomonas* effector, and to interact with TNLs, like RPS4 that mediate recognition of AvrRps4. It was hypothesized that EDS1, being important for basal resistance, could be a target for pathogen effectors and thus serve as a molecular bridge to mediate recognition by R-proteins that guard the EDS1 protein in addition to its signaling functions. This hypothesis is currently debated and under further investigation. The *eds1* mutant was originally identified in a screen for loss of resistance to downy mildew (Parker et al. 1996). EDS1 was shown to be required for the function of *TIR-NLR* genes such as *RPP1*, *RPP4*, *RPP5*, and *WRR4*. Virulent strains of *H. parasitica* and *Albugo* are also able to cause more severe infections on *eds1* mutant plants because of the reduced level of basal resistance. *Arabidopsiseds1* mutants also enhance susceptibility to a number of pathogens including *H. parasitica*. The lowered basal immunity is associated with reduced production of the defense hormone SA that is essential for defense against biotrophic pathogens. EDS1 positively regulates SA accumulation, acting upstream of SA, but the *EDS1* gene is also activated at the transcriptional level by SA. EDS1 thus seems to be a part of an SA-associated positive feedback loop of plant defense (Lapin et al. 2019).

SA is not only crucial for resistance to downy mildew but also affects basal and R-gene-mediated resistance to many other pathogens of crucifers. Exogenous application of the hormone SA is sufficient to trigger efficient defense responses to a broad range of pathogens, and in particular to downy mildew and hemi-biotrophic species. SA (2-hydroxybenzoic acid) is synthesized via two main routes, the isochorismate (IC) and phenylalanine ammonia lyase (PAL) routes. The *Arabidopsis* IC route goes via two IC synthases (ICS) and downstream steps that differ from the canonical pathway defined in bacteria (Torrens-Spence et al. 2019). The *ics1* (*sid2*) mutant has received the majority of experimental attention to date. *Ics1* mutants exhibit 90% lower levels of SA, and levels are ~95% lower in the *ics1 ics2* double mutant following powdery mildew infection. The *ICS1* gene is induced by MAMPs and by NLR-mediated pathogen recognition leading to high local levels of SA that activate defense. Besides local defenses, SA plays an important role, together with pipecolic acid, in systemic acquired resistance (SAR) that enhances immunity of distant plant tissues to protect against future infections. The export protein EDS5 has been implicated in synthesis of pipecolic acid, in addition to its well-known role in SA biosynthesis, revealing a surprising instance of immune convergence (Rekhter et al. 2019).

Downstream of SA, the central NPR1 (NONEXPRESSOR of PR GENES 1) protein plays a crucial role in the transcriptional activation of SA-induced defense in *Arabidopsis*. Mutation of *NPR1* leads to loss of basal resistance to (hemi-)biotrophic pathogens and of SAR. NPR1 proteins form a multimeric complex in the plant cell cytoplasm. Upon triggering of the plant immune system, SA accumulates and subsequently elicits the thioredoxin-mediated reduction of a cysteine residue in NPR1. This results in the release of monomeric NPR1 that can travel into the nucleus where it interacts with TGA transcription factors to activate transcription of defense genes, like pathogenesis-related (*PR*) genes. NPR1 and the two paralogous *Arabidopsis* proteins NPR3 and NPR4 have been shown to bind SA and could thus constitute the SA immune receptors (Ding et al. 2018; Innes 2018). The RPP proteins vary significantly in their requirements for functionality of these signaling components. For example, resistance mediated by the TIR-NLR protein RPP5 is strongly compromised by mutations that nullify EDS1, NPR1, or SA biosynthesis. On the other hand, the CC-NLR proteins RPP7 and RPP13 retain almost full functionality in these backgrounds. These differences underscore the complexity of ETI regulation. It is well known that many proteins are specifically produced during plant defense. In particular, many secreted and vacuolar proteins have been classified as pathogenesis-related proteins. Also in *Arabidopsis* many PR proteins have been identified, for example, PR1, PR2, and PR5 that were found in apoplastic fluid of *Arabidopsis* treated with 2,6-dichloroisonicotinic acid (INA), which is a mimic of SA. Several of these PR proteins have anti-pathogenic activities, while others contribute to resistance. One of the most used marker genes of SA-induced plant defense is *PR1*. Plants have multiple *PR1* genes encoding different protein variants with different activities. Overexpression of several *PR1* genes in different plant species also increased their resistance to pathogens. PR1 has been shown to bind sterols, indicative of defensive mode of action that is based on restriction of this important nutrient for pathogens (Gamir et al. 2017). The PR2 proteins are β-1, 3-beta-glucanases that are thought to attack the cell wall of invading fungal pathogens. Several cases of enhanced resistance of plants overexpressing *PR2* to pathogens are well known.

Other nonproteinaceous compounds that are important for resistance to pathogens in *Arabidopsis* are small secondary metabolites. Important anti-pathogenic compounds produced by *Arabidopsis* are, for instance, indole-glucosinolates (iGS) and the phytoalexin camalexin. Mutant plants that no longer produce either iGS or camalexin are only slightly more susceptible to *P. brassicae* and other pathogens. A very strong gain in susceptibility was observed in the double mutant *cyp79b2 cyp79b3* that is blocked in the production of indole-3-aldoxime, which is a common precursor for iGS and camalexin. Camalexin appears to contribute to resistance to *Arabidopsis* downy mildew, as several *phytoalexin-deficient* (*pad*) mutants are slightly more susceptible to *H. parasitica*.

A combination of metabolomics with quantitative genetics was used to discover the potential role of gluconasturtiin in the *B. napus* resistance response against Clubroot and the underlying QTL controlling the trait on chromosome C03 and C09. Gluconasturtiin is a type of glucosinolate compound associated with the biotic resistance responses of *Brassica* species. With the use of multi-omics supplemented with functional studies, key resistance pathways involved in the *Brassica* host pathosystem have been revealed (Fig. 4.1, Chap. 4) (Singh 2017; Wagner et al. 2019).

7.6 Biomolecular Mechanisms of Host Defense to Hemi-Biotrophs and Necrotrophs

The mechanisms of nonhost resistance (NHR) in plants to *Alternaria* operate at pre-invasion stage mainly by both preformed and inducible defense responses. Preformed defenses involved structural features like trichomes, chemical compounds (antimicrobial enzymes, secondary metabolites), cuticular waxes, formation of phytoalexins, and detoxification of toxins. Defense responses mainly involve ROS accumulation and defense-induced cell death where fungal proliferation is restricted by rapid hypersensitive reaction (HR). Resistance in *Brassica* to *Alternaria* depends on phytoalexins, detoxification, and simultaneous phytoalexins (camalexin) elicitation. MAPK4 gene activates defense in *Brassica* to *Alternaria*. The biomolecules like phenols, polyphenol oxidase, peroxidase, PAL, and catalase are higher in Alternaria-tolerant genotypes of *Brassica*. *GLIP1* gene in association with ethylene signaling is a critical factor in host resistance to *Alternaria*. β-Aminobutyric acid (BABA) and zeatin upregulated host immunity via elevation of MAPK 5 which antagonizes the effect of *A. brassicae*. Resistant *Brassica* genotypes accumulate phytoalexins like camalexin, brassinin, allyl (Al TIC), and benzyl (Bz ITC) isothiocynates to inhibit *Alternaria* infection. Foliar sprays of calcium compounds sequester the organic acids at the site of *Alternaria* infection to boost calcium content of plants to provide resistance. The MAPK signaling cascade leading to biosynthesis of camalexin triggers defense responses in *B. rapa* to *Alternaria*. The upregulation of genes such as WARKY, peroxidase, p450 oxidase, and chitinase in *Brassica* mediated with JA and SA pathways enhances defense response to *Alternaria*. Differential defense signaling pathways of JA and SA operates in different species of *Brassica* to *Alternaria*.

Mutation of lysine histidine transporter 1 (lhr1) in *Arabidopsis* reduces contents of glutamine, alanine, and proline to enhance resistance against Colletotrichum mediated with SA signaling and increased callose deposition and PR1. Defense responses in *Arabidopsis* to Colletotrichum are activated through SA signaling at early stage of interaction and by JA signaling subsequently.

The TF homologs associated with JA (MYC3, JAM1), ABA (AKS2, ABF3), and calcium (CAMTA5, CAMTA 3/SR1) signaling regulate defense response processes in resistant host of *Brassica* to *Leptosphaeria*. At the site of host–pathogen interface, DEGs are downregulated in resistant plants in association with response to cytokinin, photosynthesis, amino acids biosynthesis, metabolisms, and unsaturated fatty acids biosynthesis. Defense response in resistant cvs. is faster activated by complex network of WRKY, BZIP, bHLH, CAMTA, and host shock TFs. Calcium signaling is regulated for resistance and basal defense against *L. maculans*. There is differential response of resistant and susceptible cvs. of *Brassica* for expression, biosynthesis, and signaling pathways of different hormones in response to *L. maculans*. Genes associated with surface receptors and MAPK activities are upregulated in *Leptosphaeria*-resistant hosts but not in susceptible ones.

NADPH oxidases are critical components of resistance to *R. solani* AG8 in *Arabidopsis*. There is a role of glucosinolates in the defense of *Brassica* against pathogens. Defense response against *Sclerotinia* are regulated by a complex network of hormonal and signaling pathways via JA, ET, and abscisic acid in promoting the production of GSLs and other metabolites working together to enhance host resistance. Several pathways and biological processes are associated with the defense against *Sclerotinia*, such as oxidative burst, cell wall enforcement or modification, secondary metabolisms and calcium binding, signaling, antioxidation, glutathione metabolism, glucosinolate metabolisms, and biosynthesis of lignin. In *Brassica*, resistance is expressed by quick accumulation of ROS via activating Ca^{2+} signaling and suppression of oxalic acid generation in *Sclerotinia*. Overexpression of AtGDSL1 in *B. napus* enhances resistance to *Sclerotinia* with increased ROS and SA levels and reduced JA level. In *B. napus*, mitogen SA-activated protein kinase gene *BnaMPK3* plays an important role in the defense against *Sclerotinia*. Host resistance to *Sclerotinia* involves various biological process like redox homeostasis; signaling by lipids, calcium, SA, JA, ET, and ABA; histone, DNA methylation-mediated transcription regulation, defensin-like proteins, cyanate lyase, miRNA, PTGs, and glycolate oxidase genes encoding crucial enzymes in photorespiration. The crosstalk of signaling molecules, namely, JA, SA, ET, ABA, Auxin, NO, ROSs, and OA, is very complex to induce defense against *Sclerotinia* on *Brassica*, and it varies with interaction of pathogen with different genotypes. Differential alternative slicing genes and isoform regulate resistance in *Brassica* to *Sclerotinia*. The phenylpropanoids are involved in defense responses of *Arabidopsis* against *Verticillium*. Tryptophan-derived secondary metabolites like camalexin provide defense against *Verticillium* to *Brassica*. The application of BABA induces biomolecules to confer resistance in *Brassica* to *Verticillium*. The increased level of proteins like annexin, AIG2, ferredoxin, ferredoxin–NADP, and epithiospecifier like in resistant cvs. of *Brassica* provides resistance to *Xanthomonas*. The role of several proteins, in resistant and susceptible cvs. of *Brassica* has been characterized with several signal molecules against *Xanthomonas* and responses indicated up- and downregulated levels. *CHI-B4*-like gene encoding an endochitinase has been identified in *Arabidopsis* to confer resistance to *Xanthomonas*. The overexpression of *BoCHB4* in *Brassica* confers resistance to *Xanthomonas*.

Biometabolomics and biomolecular mechanisms of host resistance is expressed in the form of different metabolites and biomolecules involved in the induction of biochemical changes and pathways in the host to confer resistance against biotic and abiotic stresses. Biometabolomics and systems biology determines a variety of molecular targets for breeding disease resistance cvs. in *Brassica* crops. Transcriptomics and proteomics help to discover novel biochemical pathways involved in conferring resistance to virulent pathogens and reveal mechanisms of biometabolomics host resistance. With the use of multi-omics supplemented with functional studies, key resistance pathways involved in *Brassica*-hemibiotrophs pathosystem has been discovered, where the induction of JA signaling elevated ROS with reprogramming of the phenyl propanoid pathways to impart resistance. The hormones SA plays a major role in plant resistance to hemibiotrophs. In *A. thaliana*, SA synthesis occurs in plastids via isochorismate synthesis (ICS1 or SID2) and is triggered by pathogens. SA accumulated in the cytoplasm can be converted to SA glycoside (SAG) which is stored in the vacuole and hydrolyzed back to SA when required. Elevated levels of total SA (free SA plus SAG) are correlated with induction of defense gene expression and enhanced host resistance. Different from SA, JA, and ET accumulate mainly in response to necrotrophs. Loss of susceptibility is frequently caused by deregulation of SA- or JA-dependent host defense signaling by the impairment of cellular rearrangement or by the limitation of nutrient supply for the pathogens. There is role of MAPK 4 gene in host defense against *Alternaria*. MAP kinase signaling cascade is known to be involved in the synthesis of camalexin which contributes to disease resistance against necrotrophs. The lifestyle transition of the pathogens from biotrophics to necrotrophics is regulated by signaling molecules from SA to JA and ET.

7.7 Techniques to Study Molecular Mechanism of Host Resistance

To understand the molecular mechanisms of crucifers host resistance, useful protocols have been developed to analyze the host–pathogen interactions through omics approaches like genomics, proteomics, transcriptomics, and bioinformatics. Novel R-genes have been identified, molecularly characterized, and have been fine mapped on the chromosomes of different *Brassica* species. Techniques have been developed to introgress high yielding and disease resistance genes through traditional as well as transgenic approaches. Protocols for molecular characterization of R-genes in *Brassica* species, DNA extraction, purification, and quantification, construction of a linkage map and mapping of R-loci and proteome analysis of *Brassica-Albugo* pathosystem have been developed. The procedures for analysis of molecular and biochemical mechanisms of resistance in *Brassica* to Alternaria and molecular characterization of *Alternaria* genes showing fungicidal resistance have been described. Genome-wide identification of defensin genes in *B. juncea* and *Camelina sativa*; analysis of expression of defensin genes; and identification of distribution of chitinase genes against *Alternaria* are the techniques available. *Brassica-Erysiphe* molecular techniques include identification of molecular markers linked powdery mildew R-genes and DNA sequence analysis of *Erysiphe* isolates. In *Brassica–Hyaloperonospora* interactions, assessment of small RNA role; R-genes overexpression; and c-DNA-AFLP analysis to reveal gene expression have been developed. *Brassica-Leptosphaeria* pathosystem has revealed phylogenetic relationship of R-loci; transcriptome analysis; cloning and transformation of *Leptosphaeria* avirulence genes; identification of QTLs; molecular mapping of R-genes; and identification of NBS-encoding genes in *Brassica* species. Molecular techniques developed with the *Brassica-Plasmodiophora* system are mapping of R-genes and QTLs; SNP array, mapping, population structure, and linkage disequilibrium analysis; genome-wide association study; and the methodology for transcriptome analysis of *Brassica*-Sclerotinia pathosystem has been developed. The protocols for TuMV detection; preservation; inoculation; and identification by ELISA have been developed. In *Brassica* species RNA sequencing and NBS domain and NBS-associated conserved domains protocols have been developed.

In science and technology to get reliable and widely acceptable results, one of the basic impacts is use of standard, repeatable, and high throughput methodology, procedure, protocols, and techniques. The techniques included in this chapter have been developed by the experts of molecular biology who have long experience of working with crucifers' host-pathosystem. The development of molecular techniques was initiated during the 1990s after the sequencing of *Arabidopsis thaliana* genome using *Arabidopsis* host-pathosystem. With the use of this system, rapid progress has been made on molecular mechanisms of host resistance through analysis of crucifers' host–pathogen genomics, proteomics, and transcriptomics. These techniques can be used for demonstrating biomolecular events by the teachers for the graduate students and by research scientists to pursue their research programs to get widely acceptable results on molecular mechanisms of host resistance. Only selected techniques required to study molecular mechanisms of host resistance have been described. For more details on procedures, readers are advised to consult cross-references given for each technique.

7.8 Novel Molecular Approaches for Breeding Disease Resistance Cultivars of *Brassica*

7.8.1 R-Genes Introgression Approach

The introgression of genes for disease resistance between species has been widely studied in *Brassica*. The utilization of the B-genome as a source of resistance against *Leptosphaeria maculans* (blackleg) from diploid and tetraploid species has been studied. For instance, chromosome B4 from *B. nigra* was introgressed into rapeseed variety "Darmor", which exhibited high resistance with the addition of this chromosome. Similarly, the high resistance from *B. juncea* was obtained in selected recombinant lines of *B. napus* carrying a resistance gene located on chromosome B8. Alike, successfully introgressed a B-genome chromosome from *B. carinata* to *B. napus*, with plants carrying this chromosome showing variation in traits such as blackleg resistance, days to flowering, days of maturity, and fatty acid composition. Transfer of black rot resistance from *B. carinata* to *B. oleracea*, resistance to Alternaria leaf blight caused by *Alternaria brassicae* from *B. hirta* to *B. juncea*, and transfer of powdery mildew resistance from *B. carinata* to *B. oleracea* through embryo rescue followed by backcrossing to *B. oleracea* have been successfully demonstrated (Saharan et al. 2019).

Besides utilizing conventional methods of selection and sexual reproduction, biotechnological techniques involving tissue culture and genetic transformation-based methods have revolutionized the search and development of economically promising disease-resistant crops. These tools have enabled the researchers to combine the traits from distantly related *Brassica* species for development of superior Alternaria blight-tolerant varieties of *Brassica* (Aneja et al. 2012; Aneja and Agnihotri 2013; Saharan et al. 2003, 2016).

7.8.2 Using In Vitro Embryo Rescue Technique

Embryo rescue, documented for the first time in the eighteenth century, is one of the most successful techniques for the production of interspecific and intergeneric hybrids from naturally incompatible crosses that do not result in the production of viable offspring. The technique involves excision of the ovary/ovule/embryo resulting from an interspecific cross, its culturing and maturing on suitable media for the production of a hybrid plant with desired traits. The technique has proven very efficient in transferring Alternaria blight tolerance in oilseed *Brassicas.* In vitro ovary culture was used to transfer the resistance against Alternaria blight from *B. tourniforti* to *B. juncea* cv. RH 30 (Yadav et al. 1991). Agnihotri et al. (1991) developed intergeneric hybrids of *B. campestris* and *B. spinenscens* through sequential ovary, ovule, and embryo culture. The hybrid plants were then multiplied using somatic embryogenesis.

The resistance trait from *Sinapis alba* cv. Carine was transferred to *B. napus* cv. Brutor using in vitro fertilized ovary culture (Chevre et al. 1994). Intergeneric hybrids with Alternaria blight resistance were developed between *Erucastrum cardaminoides* and *B. oleracea* var. *Alboglabra* using sequential ovary and ovule culture (Mohanty et al. 2009). Gupta et al. (2010) also used interspecific hybridization in combination with in vitro ovule culture to incorporate high tolerance to Alternaria blight and white rust from *B. carinata* cv. Kiran to low erucic acid TERl (OE) M21 lines; hybrids with high tolerance to Alternaria blight and white rust contained negligible amount of erucic acid (1.3%) and high oleic acid (42.5%).

7.8.3 Use of Somatic Hybridization

Somatic hybridization refers to the enzymatic removal of the cell wall and isolation of protoplasts which are then fused together to produce hybrids. Due to the absence of the cell wall, there occurs a nonspecific fusion between the two protoplasts without any barrier to intergeneric/interspecific crosses. It also bypasses both the pre- and post-fertilization barriers. This technique has been utilized effectively in *Brassicas*, but the studies are limited to *B. napus* and *B. oleracea.* Efforts have been made to transfer *Alternaria*-resistant trait from *Moricandida arvensis* to *B. oleracea* (Toriyama et al. 1987), and from *S. alba* to *B. napus* (Primard et al. 1988). Sharma and Singh (1992) attempted the transfer of *Alternaria* resistance trait from *B. carinata* to *B. juncea.* Leaf mesophyll protoplasts from *M. arvensis* and *B. napus* were hybridized to produce disease-resistant hybrid plants (O'Neill et al. 1996). Jourdan and Salazar (1993) resynthesized *B. carinata* by protoplast fusion between *B. nigra* and *B. oleracea*, and the hybrids thus obtained were fertile and grew into robust plants. Hybridization between *S. alba* and *B. oleracea*, and between *Camelina sativa* and *B. oleracea* has also shown to produce resistant hybrids (Hansen and Earle 1997; Hansen 1998). Sigareva and Earle (1999) developed somatic hybrids between *S. alba* × *B. oleracea* by protoplast fusion followed by embryo rescue and recovered four highly resistant hybrid progenies after repeated back crosses.

7.8.4 Use of Somaclonal Variations

These are regarded as the impulsive, inheritable changes in gene expression which often appear under in vitro conditions especially in tissue culture experiments. Sharma and Singh (1995) used somaclonal variations for incorporation of disease resistance/tolerance against Alternaria blight. Resistance in Alternaria blight susceptible *B. juncea* varieties was achieved by inducing variations through mutagenesis (exposing seeds to gamma radiations) (Verma and Rai 1980) and by treating the microspores derived embryos with chemical mutagens (Agnihotri et al. 2009) including ethyl methane-sulfonate (EMS) and ethyl nitrosourea (ENU).

7.8.5 Development of Transgenics

Production of transgenic plants through genetic transformation forms another biotechnological approach to integrate the disease resistance genes from resistant/tolerant genotypes to the economically important susceptible ones. The genetic transformations in *B. juncea* have been reported to achieve delayed onset of the Alternaria blight. Kanrar et al. (2002) isolated cDNA sequence for "hevein-rubber tree lectin" from *Hevea brasiliensis* and transferred it into *B. juncea* cv. RLM-198 through binary vector *pBinAR*. The transgenic plants thus produced supported delayed onset of Alternaria blight with considerably low disease intensity. Mondal et al. (2003) identified the role of "chitinase" in plant defense responses and transformed *B. juncea* (cv. RLMI98) with chitinase gene tagged with overexpressing promoter 35S *CaMV*. The transformed plants resulted in delayed disease onset and about 12–56% reduction in fungal colony size. Osmotin protein, known for its role in signal transduction, when transferred to *B. juncea* also showed tolerance to *A. brassicae* in transgenic plants (Taj et al. 2004). In another study, transgenic Indian mustard expressing "*class I glucanase*" gene under the control of *CaMV* 35S promoter was developed (Mondal et al. 2007). For undertaking the genetic transformation, cDNA encoding "*classI glucanase*" gene was isolated from tomato and cloned into *Bam*HI-*Sal I* restriction site of binary vector, *pBinAR* flanking *CaMV* 35S constitutive promoter, and an *ocs* terminator region. The resulting recombinant $pBinG_B$ clone was transferred into *Agrobacterium tumefaciens* strain GV 2260, which was further used to transform *B. juncea* (cv. RLM 198). A study by Verma et al. (2012) reported the introgression of cysteine rich-antimicrobial peptide, *PmAMP1* from *Pinus monticola* into *B. napus*, thus conferring enhanced protection against *A. brassicae*, *Leptosphaeria maculans*, and *Sclerotinia sclerotiorum*. Combined expression of type I ribosome inactivating protein (RIP; *AAA32951*) and Barley "*class II chitinase*" gene in *B. juncea* not only reduced the growth of fungal hyphae by 44% but also affected the onset of disease (Chhikara et al. 2012). Efforts have also been made to study the signal transduction pathways in *Brassica* species. Taj et al. (2011) reported interaction of MAPK3 and Lox genes during pathogenesis of *A. brassicae*, which is suggested to play a role in biosynthesis of jasmonic acid and jasmonic acid-induced expression of defense genes in *B. juncea*.

7.8.6 Molecular Marker-Assisted Breeding

As the efficient management of plant diseases relies on the accurate identification of the plant pathogen, knowledge of the variability in the pathogen population, both at the pathological and genetic levels, is very important. Characterization of variability in the pathogens including *Alternaria* spp. has been done earlier using traditional methods of morphological markers (Meena et al. 2012), biochemical tests (Vishwanath and Kolte 1997), and cultural and pathogenic assays (Meena et al. 2012; Singh et al. 2011). These markers differentiate the pathogens on the basis of their growth behavior, physiological characters, and microscopic appearance. However, these criteria are greatly influenced by different factors including age and quality of the inoculum, age and type of host, culture media used, and incubation conditions. Also, these techniques are more laborious, time-consuming, and often give nonreproducible results. To overcome these limitations of traditional markers, detection of genetic variability through DNA fingerprinting technology by the use of molecular markers has come to the aid of plant breeders and pathologists. The main advantage of using these fingerprinting techniques is that, they do not require any prior knowledge or availability of variable sequences in the database, as they deal with the genome-wide biodiversity.

Various molecular markers being used for assessing variability include internal transcribed spacer regions (ITS), restriction fragment length polymorphism (RFLP), randomly amplified polymorphic DNA (RAPD), amplified fragment length polymorphism (AFLP), inter-simple sequence repeats (ISSRs), microsatellite, sequence tagged sites (STS), and single nucleotide polymorphism (SNPs). Due to the presumed dependency on traditional characterization, only a few studies are reported on molecular characterization of *Alternaria* species pathogenic to oilseed Brassicas. The ITS regions (also called the conserved regions) in the fungal genome are considered the most popular locus for DNA-based mycological studies at species level. Chou and Wu (2002) and Berbee et al. (2003) studied the ITS region of rDNA of *A. brassicae* and *A. brassicicola* to evaluate the phylogeny of the pathogen. Molecular techniques like RAPD that do not require any prior knowledge of the DNA sequences have been efficiently used to analyze the genetic variations in *Alternaria* species (Cooke et al. 1998; Sharma and Tewari 1995, 1998; Gherbawy 2005). However, due to the limitation of reproducibility in RAPD-generated data, the assaying of genetic variability in *Alternaria* species has shifted to more sensitive techniques like AFLP (Bock et al. 2002) and microsatellite markers. Twelve polymorphic microsatellite markers were isolated from *A. brassicicola*. The number of alleles detected in 12 loci range from 2 to 10 (mean 3.5). Cross-species amplification shows that the designated primers are specific to the pathogen (Avenot et al. 2005). Though, these molecular tools have revealed huge amount of genetic variations in *Alternaria* at the sub-generic as well as sub-species level, this field still remains largely unexplored and demands an extensive study into the variability at different geographic levels among different hosts.

7.8.7 Induction of Systemic Resistance

Induced host resistance refers to activation of resistance in otherwise susceptible plants, through biotic as well as abiotic agents (Stitcher et al. 1997; Kessman et al. 1994), without changing their basic genetic make-up. Foliar application of β-aminobutyric acid induces resistance against *A. brassicae* in *B. carinata* (Chavan et al. 2013). *GLIP1* protein in association with ethylene signaling is a critical component in plant resistance to *A. brassicicola* (Oh et al. 2005). The resistance/defense-related genes in the vulnerable plants can be activated by inoculating the plant either by an avirulent form of the pathogen or by limited inoculation with the pathogen (Deverall 1995). The infecting avirulent pathogen triggers natural defense responses in the plant through the release of the elicitors, which then result in the expression of novel anti-pathogenic proteins. In order to harp the benefits of induced host resistance and build up a stable, long-term resistance mechanism in the host plant against the pathogen, there is a need to identify the pathogen and understand its behavior under diversified conditions. In case of Alternaria blight, considerable morphological and pathological variations have been observed in *A. brassicae* under different environmental conditions (Kaur et al. 2007; Singh et al. 2007; Patni et al. 2005; Meena et al. 2005, 2012; Ansari et al. 1989; Awasthi and Kolte 1989; Saharan et al. 2015, 2016). Vishwanath et al. (1999) reported the induction of resistance in susceptible *B. juncea* cv. PR-15 against extremely virulent *A. brassicae* isolate A (AbA) and moderately virulent isolate C (AbC) from *B. carinata* cv. PPCS-1 by inoculating with an avirulent isolate D (AbD) isolated from same *B. carinata* cultivar. About 60% reduction in the disease severity was observed against AbA and AbC by the induction of resistance. However, due to the limited studies in molecular characterization of the pathogen, the clear outcome is still awaited.

Variations in *A. brassicae* isolates have been recorded in traits including colony morphology, colony diameter ranging from 32 to 68 mm, surface texture from velvety to wooly, color from cream to dark olive green, and sporulation from very less to intense. The host–pathogen interaction studies using four varieties of *B. juncea* and two of *B. rapa* have shown pathological variation among the isolates that differ in their aggressiveness from avirulent (few) to highly virulent (Aneja et al. 2012; Aneja and Agnihotri 2013). A thorough knowledge of diversity in the pathogen will not only help in characterizing and analyzing it phylogenetically but also in inducing durable resistance against Alternaria blight, which can act as an environment friendly substitute to the traditional use of fungicides.

7.8.8 Use of Genetic Modification Technique

The development of *Alternaria*-resistant genetically modified plants (GM) is an excellent approach to manage Alternaria diseases of crucifers. Resistant genes from other plants or bacteria-encoding enzymes like chitinase or glucanase can be introduced in crucifers. Both chitinase and glucanase enzymes break down chitin and glucan, which are essential components of fungal cell wall. Introduction of plant genes to enhance innate plant defense mechanisms like activating phytoalexins, proteinase inhibitors, and detoxification of phytotoxins is another useful way of developing resistant varieties. Invoking hypersensitive reaction when a fungal pathogen is attacking can also be an innovative approach.

7.9 R-Genes Transfer by Molecular Marker

To develop resistant *Brassica* cultivars, several resistant sources such as European turnips, Chinese cabbage, and rutabaga cultivars are available, and the resistance in these sources is dominant, which makes it easier to transfer clubroot resistance through interspecific and intraspecific hybridization. Rutabaga cultivars have been identified as clubroot-resistant sources. However, the genetics of clubroot resistance in rutabaga is complicated, so it is difficult to develop molecular markers that are closely linked to the dominant clubroot resistance genes (Ayers and Lelacheur 1972).

In Chinese cabbage, the dominant clubroot resistance from European turnips has been successfully used to develop clubroot-resistant Chinese cabbage. Since the gene mapping has been performed extensively in Chinese cabbage, molecular markers closely linked to clubroot resistance loci that are used in gene mapping can be easily selected to transfer clubroot resistance genes in the development of Chinese cabbage cultivars through molecular marker-assisted selection. Since the amphidiploid *B. napus* has a very complex genome, most of the molecular markers developed in *B. rapa* may not be polymorphic and cannot be directly used in *B. napus*. Additional efforts are required to develop molecular markers in *B. napus* when the mapped clubroot resistance loci in Chinese cabbage are transferred into *B. napus*. Currently, most clubroot resistance genes in European turnips have not been intensively investigated, and mapping and cloning of these clubroot resistance genes in European turnips will allow using these genes effectively and efficiently in *Brassica* breeding (Hirani and Li 2015).

7.10 R-Genes Transfer by Interspecific Hybridization

A dominant clubroot resistance gene from *B. rapa* (Chinese cabbage) was fine mapped, and molecular markers were developed in Chinese cabbage that could be used for marker-assisted selection (MAS) in other *Brassica* crops. To transfer this clubroot resistance gene to *B. napus* (canola), an interspecific hybridization was made between *B. napus* (canola) and *B. rapa* (Chinese cabbage). Subsequently, the F_1 was backcrossed to the canola recurrent parent for three generations to produce BC_1, BC_2, and BC_3 progenies. Using these populations, simple sequence repeat (SSR) markers flanking the clubroot resistance gene were used to perform MAS in canola. These molecular markers were then evaluated in 13 different canola and rapeseed quality genotypes in *B. napus* and *B. rapa*. These markers exhibited high reliability in identifying clubroot resistance in this diverse set of *Brassica* genotypes.

Clubroot resistance also co-segregated with the SSR markers flanking the clubroot resistance gene in the BC_3, and BC_3S1. The segregation ratio of resistant and susceptible individuals in the BC_3 supported the expected 1:1 ratio for the segregation of a single Mendelian gene. BC_3S1 families with homozygous clubroot resistance were developed during this process and will be valuable sources of clubroot resistance in *B. napus* breeding activities (Hirani et al. 2016).

7.11 R-Genes Transfer by Distant Hybridization

The three clubroot-resistant Chinese cabbages (1003, 1007, and 1008) were used to transfer the clubroot resistance genes to *B. napus* by distant hybridization combined with embryo rescue. Three methods including morphological identification, cytology identification, and molecular marker assisted selection were used to determine hybrid authenticity, and 0, 2, and 4 false hybrids were identified by these three methods. In total, 297 true hybrids were identified. Clubroot resistance markers and artificial inoculation were utilized to determine the source of clubroot resistance in the true hybrids. Two simple sequence repeat (SSR) and two intron polymorphic (IP) markers linked to clubroot resistance genes were identified. The clubroot resistance genes 1007 and 1008 were mapped to A03. At last, 159 clubroot-resistant hybrids were obtained by clubroot resistance markers and artificial inoculation. Four molecular markers associated with clubroot resistance were identified, which were located in a region from 23.77 to 25.56 Mb on A03 of *B. rapa*. In the adjacent area of this region, two clubroot resistant genes *CRa* (Ueno et al. 2012) and *CRb* (Hatakeyama et al. 2017) have been cloned, in which both genes encode the TIR-NBS-LRR (TNL) protein. Through analyzing the genes structure of the region surrounding the *CRa* and *CRb*, it was found that six genes *Bra012540*, *Bra012541*, *Bra019409*, *Bra019410*, *Bra019412*, and *Bra019413* have the structure of TNL (Liu et al. 2018).

7.12 R-Genes Transfer from European Clubroot Differential (ECD) Set

European turnips are the most resistant sources and their resistance genes have been introduced into other crops such as oilseed rape (*Brassica napus* L.), Chinese cabbage, and other *Brassica* vegetables. The European clubroot differential (ECD) set contains four turnip accessions (ECD1–4). These ECD turnips exhibit high levels of resistance to clubroot when they are tested under controlled environmental conditions with Canadian field isolates. Gene mapping of the clubroot resistance genes in ECD1–4 were performed and three independent dominant resistance loci were identified. Two resistance loci were mapped on chromosome A03, and the third on chromosome A08. Each ECD turnip accession contained two of these three resistance loci. Some resistance loci were homozygous in ECD accessions, while others showed heterozygosity based on the segregation of clubroot resistance in

20 BC_1 families derived from ECD1 to 4. Molecular markers were developed linked to each clubroot resistance loci for the resistance gene introgression in different germplasm (Hirani et al. 2018).

7.13 R-Genes Transfer Through Embryo Rescue for Powdery Mildew Resistance

Interspecific hybrid plants and backcross 1 (BC_1) progeny were produced through sexual crosses, and embryo rescue between *Brassica carinata* accession PI 360883, and *B. oleracea* cvs. Titleist and Cecile to transfer resistance against powdery mildew of *B. oleracea*. Four interspecific hybrids were obtained through application of embryo rescue from crosses with *B. carinata* as the maternal parent, and their interspecific nature confirmed through plant morphology and random amplified polymorphic DNA (RAPD) analysis. Twenty-one BC_1 plants were obtained through sexual crosses and embryo rescue, although embryo rescue was not necessary to produce first backcross generation plants between interspecific hybrids and *B. oleracea*. All interspecific hybrids, and eight of the BC_1 plants were resistant to powdery mildew (Tonguc and Griffiths 2004).

7.14 Use of Mutagenic Approach for Resistance to Powdery Mildew

The method of induced mutagenesis is a powerful tool for creating new genotypes with an increased yield, quality, and resistance to diseases and pests (Ahloowalia et al. 2004). Upon irradiation with gamma rays, the variation of the resulting genetic alterations has high frequency, and wider range of morphological and biochemical changes (Skoric et al. 2008). There is evidence that the mutations may also change the expression of certain enzymes in the biosynthetic pathway of fatty acids (Auld et al. 1992; Pleines and Friendt 1989). The irradiation of rapeseed with low doses of gamma rays of 100 and 150 Gy has a stimulating effect on the synthesis of linoleic and linolenic acids. Conversely, at higher doses of the mutagen factor, the concentration of these fatty acids decreases while the amount of oleic acid increases (Rahimi and Bahrani 2011). The fatty acids synthesized in plants play an important role by participating in the composition of tissue hormones, in cell, and nuclear membrane structures (Kolattukudy 1974). It is assumed that unsaturated fatty acids in plant cells are involved in cutin and suberin synthesis (Pinot et al. 1999). Cutin, which acts as first barrier and protects plant organs (leaves and fruits) from dehydration providing protection against pathogens and pests, is composed of C16 and C18 fatty acids. Suberin, which is predominantly composed of hydroxyl and epoxy unsaturated fatty acids, is a structural component of the plant underground part roots and tubers (Kolattukudy 1980). C18 fatty acids in plants play a role in building the defense mechanism against diseases, and pests (Kachroo and Kachroo 2009). It has been found that hydroxyl fatty acids also inhibit the development of *Erysiphe graminis*, *Puccinia recondita*, and *Phytophthora infestans* (Hou and Forman 2000). The powdery mildew development is favored by meteorological factors, such as temperature and high humidity (Penaud 1999; Enright and Cipollini 2007). Fungicide application is often not efficient enough, due to the pathogen's rapid spread in favorable conditions. The effects of gamma radiation on the fatty acid composition of rapeseed oil and the increased resistance to *E. cruciferarum* have been determined by Petkova et al. (2014). Biochemical changes in the oil after gamma ray irradiation are also associated with changes in the susceptibility to powdery mildew.

Petkova et al. (2014) assessed through the gas chromatography, the difference in the content of saturated and unsaturated fatty acids in the seeds of the nonirradiated variants, and the ones irradiated with two doses of gamma radiation. The data for 3 years of the fatty acid composition of the Abacus hybrid oil, and its M0, M1, and M2 was decomposing the populations. The chemical analysis showed generally that the irradiated plants showed a lower content of oleic acid and a higher content of linoleic acid and linolenic acid compared with the nonirradiated variants. There was tendency toward increase in the total content of raw fats in mutant populations of M1 and M2 of Abacus as compared with nonirradiated plants. This, in turn, is associated with a reduction in the total content of saturated fatty acids and an increase in the content of unsaturated fatty acids. There is an inverse relationship between the values of the total fat content and the content of saturated and unsaturated fatty acids. The rapeseed oil of the Abacus hybrid showed higher content of saturated fatty acids as compared with the M2 and M3 generations of the irradiated variants. In contrast, the percentage content of unsaturated fatty acids is influenced positively by irradiation with gamma rays and is comparatively high in M2 A1-100, and M2 A-150 92.7%, and 93.7%, compared with the baseline 89.7%. The extent of damage by the powdery mildew on plants of the Abacus variety and its gamma ray irradiation-treated variants shows the arithmetic mean values, and their errors, accuracy indicators, and variation coefficients. The highest extent of damage on the stems and branches in the range of 98.9–99% ($p = 0.1\%$) was observed with the M0 Abacus generation, irradiated with absorbed dose 100 Gy, and 150 Gy gamma rays, compared with the nonirradiated variant. The irradiation did not have a positive impact on the sustainability of pods to *E. cruciferarum*. The differences found are insignificant compared with the corresponding baseline group. In contrast, the M1 and M2 generations were shown to have a proven lower extent of damage by the phytopathogen as compared with the baseline. It is assumed that the increased resistance demonstrated is due to the higher content of unsaturated fatty acids in the seeds of the irradiated plants. Unsaturated fatty acids can be used as a means of biological control of plant diseases on global scales. Increasing the dose of the mutagen leads to an increase in the concentration of unsaturated fatty acids with 18 carbon atoms and to a decrease in the extent of damage by *Erysiphe* cruciferarum (Petkova et al. 2014).

7.15 Introgression of Durable Resistance in *Brassica* to *Alternaria*

Based on a literature survey on intercropping systems, Fatima et al. (2019) have shortlisted the nonhost plants for *A. brassicae* that can be chosen for dissecting the NHR mechanisms. The NHR to *A. brassicae* is based on fungal invasion and penetration events. Prior to any specific study, it is important to classify the nonhost plants based on these four phases by using pathological assays. This will provide a clear idea about whether the NHR operates at the pre-invasion or post-invasion level. The following strategies and approaches can be used for developing durable Alternaria blight-resistant *Brassica* crops.

7.15.1 Introgression of Genes from Nonhost Plants to *Brassica* Crops by Inter- or Intraspecific Crosses

The identification of genes conferring NHR by utilizing conventional or molecular breeding approaches, followed by the introgression of candidate genes from nonhost plants to *Brassica* crops is essential. This can be achieved by interspecific crosses of the host with a nonhost by transfer of resistance traits from the nonhost plant to a host plant. The unwanted background in the progeny can be removed by crossing it with wild-type host plants and selecting for the targeted traits only. Then, the progeny can be screened for response to *A. brassicae*. Plants showing resistance to the pathogen can be subsequently subjected to QTL mapping to identify the genes involved in NHR. Once the genes are identified, they can be successfully introgressed into the desired varieties to provide resistance to *A. brassicae*. The major limitation with this approach, however, is that the progeny generally suffers from sterility and abnormal segregation, which can hinder the identification of genes involved in resistance. The only available example of an interspecific cross made between a host and nonhost plant is that of *Lactuca* and *Bremia* (Jeuken et al. 2008).

Secondly by crossing two different varieties of nonhost plants (intraspecific cross: completely resistant genotype × moderately resistant genotype) that differ in the degree of resistance toward *A. brassicae*. This will allow the inheritance of NHR-specific QTL of the resistant donor into the moderately susceptible genotypes of the same species of the nonhost plant (Jafary et al. 2008; Niks and Marcel 2009). The progeny obtained can be similarly studied for identifying the genes involved in NHR and transferring them into target crops. However, if the nonhost plant does not have significant variation in the degree of resistance toward the pathogen, the inheritance of NHR-specific genes might be difficult (Table 7.1; Fatima et al. 2019).

Table 7.1 List of the potential nonhost plants toward *Alternaria brassicae* (Fatima et al. 2019)

Name of the nonhost plant	Name of the host plant intercropped with nonhost plants	Criteria for categorization of nonhost plant	Reference
Chickpea (*Cicer arietinum* and *C. kabulium*)	*B. juncea, B. napus, B. rapa*	Intercropping system	Ahlawat and Sharma (2002), Ahlawat et al. (2005), Singh et al. (2010), Shekhawat et al. (2012), Jeromela et al. (2017)
Lentils (*Lens culinaris*)	*B. juncea, B. rapa*	Intercropping system	Shyam et al. (2007), Shekhawat et al. (2012), Ao and Saud (2016), Jeromela et al. (2017)
Barley (*Hordeum vulgare*)	*B. juncea, B. rapa*	Intercropping system	Singh et al. (2010), Shekhawat et al. (2012)
Wheat (*Triticum aestivum*)	*B. juncea, B. napus, B. rapa*	Intercropping system	Singh et al. (2010), Shekhawat et al. (2012), Ebrahimi et al. (2016)
		Literature	
Tomato (*Solanum lycopersicum*)		Literature	McRoberts and Lennard (1996)
Potato (*Solanum tuberosum*)	*B. juncea B. napus*	Intercropping system	Singh et al. (2010), Shekhawat et al. (2012)
Sugarcane (*Saccharum officinarum*)	*B. juncea*	Intercropping system	Singh et al. (1986), Toor et al. (2000), Suman et al. (2006);
	B. rapa, B. napus		Ahlawat and Sharma (2002);
	B. campestris		Singh et al. (2010)

7.15.2 Identification and Transfer of NHR Based on Microarray Analysis

The strategy based on microarray analysis after infecting the nonhost plant (e.g., chickpea or barley) with *A. brassicae* then identifies the candidate genes from the list of upregulated or downregulated genes. If a homologous gene is identified, mutant or overexpression lines can be generated for the identified genes in the susceptible model plant. Alternatively, the candidate genes can be cloned from the nonhost plant and expressed in the susceptible model plant. These transgenic plants can then be tested for their response to *A. brassicae* infection. Finally, the genes imparting NHR to *A. brassicae* can be successfully transferred to *Brassica* crops.

7.15.3 Identification of NHR by Gene Silencing

The approach to identify genes underlying NHR involves gene silencing in the nonhost plant. Virus-induced gene silencing (VIGS)-based forward genetic screening can be performed in nonhost plants that have established VIGS protocols and cDNA libraries to identify the genes involved in NHR to *A. brassicae*. For example, for wheat and barley, the Barley stripe mosaic virus (BSMV)-based vector and a well-developed VIGS protocol for gene silencing are available (Ma et al. 2012). Such a strategy will be useful in identifying important genes involved in NHR by screening a VIGS population for its response to *A. brassicae*. Furthermore, high-throughput transient expression and gene silencing methods, for example, transient-induced gene silencing (TIGS), have been used in a nonhost plant such as barley. This technique is based on the transformation of leaf epidermal cells by the biolistic method (Douchkov et al. 2014a, b). TIGS constructs may be made for the genes that are upregulated in the nonhost plant after *A. brassicae* infection. These TIGS constructs can be bombarded into leaf epidermal cells of the nonhost plant to induce gene silencing. After bombardment, leaf discs can be infected with *A. brassicae* and screened for the response. TIGS will allow the identification of potentially important candidate genes involved in NHR to *A. brassicae*. The genes identified using any of these functional genomic methods can be successfully transferred to Brassica crops to ensure resistance to *A. brassicae*.

7.15.4 Identification and Transfer of NHR Genetic Resources from Model Host Plants

The model plants that have well-developed molecular and genetic resources and a large population of mutants as nonhosts for *A. brassicae* are good sources to transfer nonhost resistance. For example, *Brachypodium distachyon* can be directly utilized for identifying genes and dissecting the pathways involved in NHR (An et al. 2016). If the mutants are not readily available, then with the help of well-developed genetic tools, transposon or T-DNA insertional mutagenesis can be attempted in the model nonhost plant. The mutant population can be screened for the response to *A. brassicae* to identify the candidate genes involved in NHR. These candidate genes can then be successfully transferred to Brassica crops for durable resistance to *A. brassicae*. *Arabidopsis* is a well-established, easy-to-manipulate model plant with ample genetic resources. Another approach can be screening the response of all available *Arabidopsis* accessions to *A. brassicae*. This may help identify *Arabidopsis* accessions imparting resistance to *A. brassicae* (Rajarammohan et al. 2017; Mandal et al. 2018).

7.15.5 Identification of Targeted Functional Genomics Based on Bioinformatics

The literature mining and targeted functional genomics to dissect the different layers of NHR involved in pre- and postinvasive defense responses of *Brassica* to pathogens have wide scope. The available literature provides hints about possible candidates specifically involved in pre- and postinvasive defenses. This will provide the opportunity to directly start studying these candidates for their role in NHR.

Another possibility is that the postinvasive defenses may play a more important role in restricting the fungal pathogen rather than preinvasive defenses against *A. brassicae*, which can be determined by performing pathological assays. In such cases, we can use mutants with mutations in the genes important for preinvasion defense to identify the genes involved in postinvasion defenses. For example, the mutant of the barley gene Ror1, a homolog of the *Arabidopsis* gene Pen1, is important for preinvasive defense against many nonhost pathogens, including rust pathogens (Collins et al. 2003; Trujillo et al. 2004). This mutant line can be used for identifying genes involved in the postinvasion defense response. The response of a ror1 mutant line should be tested toward *A. brassicae* to ascertain whether this gene is involved in preinvasive defense against this pathogen. Subsequently, the NHR-specific candidate genes can be identified by transcriptomic analysis in wild-type barley after *A. brassicae* infection. Gene silencing approaches can be applied to silence the candidate genes in the ror1 mutant line for identifying their role in postinvasion immunity by screening them against *A. brassicae*. The validation experiments can be carried out by constructing double mutants of candidate genes and Ror1 in barley. The double mutants compromising NHR and showing a higher degree of susceptibility to *A. brassicae* compared to the single mutant ror1 can be likely candidates that contribute to NHR at the postinvasion level. The important genes identified can be further transferred, or their homologs can be overexpressed in the susceptible *Brassica* crops to test their resistance to *A. brassicae* (Fatima et al. 2019).

7.15.6 Use of an Endo-Chitinase Gene from *Trichoderma virens* for Tolerance to *Alternaria* in Transgenic *Brassica*

Genetic engineering is an alternative for developing disease-resistant plants where the resistance source is not available or there are sexual compatibilities between the source and the cultivar (Grover and Gawthaman 2003; Chhikara et al. 2012). Genes from the mycoparasitic *Trichoderma* spp. have been used to impart tolerance to biotic and abiotic stresses in many crops (Nicolas et al. 2014). Among the *Trichoderma* genes transferred to plants, an endochitinase of 42 kDa has been very popular due to its broad spectrum of activity against many fungal pathogens. Endochitinases are known to degrade fungal chitin by hydrolyzing the glycosidic bond between chitin monomers. In addition, the endochitinases also release chitin oligomers which act as elicitor molecules triggering broad spectrum plant defense. There are very few reports on development of transgenic *Brassica* species for resistance to fungal pathogens. For example, *B. napus* transformed with two genes *sporamin* and *chitinasePjChi*-1 (derived from sweet potato and *Paecliomyces javanicus*, respectively) showed enhanced resistance against both *Plutella xylostella* and *Sclerotinia sclerotiorum* (Liu et al. 2011). Wu et al. (2009) introduced a plant defensin gene Ovd, cloned from *Orychophragmus violaceus* (L.) into *B. napus*, and the transgenic plants showed resistance to *S. sclerotiorum*. Transgenic *B. napus* expressing *B. napus* mitogen-activated protein kinase (*BnMPK4*) showed resistance to *S. sclerotiorum* (Wang et al. 2009). Chitinase and glucanase gene from tomato have been introduced into *B. juncea* and the transgenic plants showed resistance to *A. brassicae* (Mondal et al. 2003, 2007). Kamble et al. (2016) observed expression of an endochitinase '*ech42*' gene from *Trichoderma virens* in transgenic *B. juncea* lines and also assessed tolerance of transgenic lines against two fungal pathogens, *A. brassicicola* and *A. brassicae*, causing Alternaria blight.

An endochitinase gene '*ech42*' from the biocontrol fungus *Trichoderma virens* was introduced to *Brassica juncea* via *Agrobacterium tumefaciens*-mediated genetic transformation method. Integration and expression of the '*ech42*' gene in transgenic lines were confirmed by PCR, RT-PCR, and southern hybridization. Transgenic lines (T1) showed expected 3:1 Mendelian segregation ratio when segregation analysis for inheritance of transgene 'hpt' was carried out. Fluorimetric analysis of transgenic lines (T0 and T1) showed sevenfold higher endochitinase activities than the non-transformed plant. *Fluorimetric zymogram* showed presence of endochitinase (42 kDa) in crude protein extract of transgenic lines. In detached leaf bioassay with fungi *Alternaria brassicae* and *Alternaria brassicicola*, transgenic lines (T0 and T1) showed delayed onset of lesions as well as 30–73% reduction in infected leaf area compared to non-transformed plant (Kamble et al. 2016).

7.16 *Brassica* Genetic Diversity Accessing and Exploiting

The advent of molecular genetics provided new opportunities for mapping and tracking genes of agronomic interest, leading to more efficient marker-assisted selection. Whole genome sequencing, starting with *Arabidopsis* and rice as models for dicotyledons and monocotyledons, respectively, and followed by a rapidly increasing number of crop plant genomes, has led to a quantum leap not only in understanding plant genetic diversity but also for accessing this enormous resource. For many crop plant species, for example, tomato, barley, maize, wheat, and various *Brassica* species, either the entire genome or the generic parts of the genome are now emerging. As bioinformatics tools for analyzing the exponential increase in genome data improve, the practical utility of such data will also be enhanced. This will extend the options for breeding disease resistance.

The presence of conserved motifs in plant resistance (R-genes), such as the nucleotide-binding site leucine-rich repeat (NBS-LRR) domains, has facilitated the identification of gene families and resistance gene analogues in other plants. The *Arabidopsis* genome has around 150 NBS-LRR-encoding genes and rice c. 400 (McHale et al. 2006). Studies of genome structure have shown that many putative R-genes are clustered and have undergone duplication and evolution due to diversifying selection. Functional analysis of these entire candidate genes is a demanding task, but improvement in plant transformation protocols and high-throughput gene attenuation methods, such as RNA interference (RNAi) and virus-induced gene silencing (VIGS), should accelerate the identification of novel genes of practical utility.

7.17 Introgression of *Sclerotinia* Resistance into *Brassica juncea* Through Marker-Assisted Breeding

A set of 96 *Brassica juncea–Erucastrum cardaminoides* introgression lines (ILs) were developed with genomic regions associated with Sclerotinia stem rot (*Sclerotinia sclerotiorum*) resistance from a wild Brassicaceous species *E. cardaminoides*. ILs were assessed for their resistance responses to stem inoculation with *S. sclerotiorum*, over three crop seasons. Initially, ILs were genotyped with transferable SSR markers and subsequently through genotyping by sequencing. SSR-based association mapping identified six marker loci associated to resistance in both A- and B-genomes. Subsequent genome-wide association analysis (GWAS) of 84 ILs recognized a large number of SNPs associated to resistance, in chromosomes A03, A06, and B03. Chromosomes A03 and A06 harbored the maximum number of resistance related SNPs. Annotation of linked genomic regions highlighted an array of resistance mechanisms in terms of signal transduction pathways, hypersensitive responses, and production of antifungal proteins and metabolites. Of major importance was the clustering of SNPs, encoding multiple resistance genes on small regions spanning approximately 885 kb region on chromosome A03 and 74 kb on B03. Five SNPs on chromosome A03 (6,390,210-381) were associated with LRR-RLK (receptor like kinases) genes that encode LRR-protein kinase family proteins. Genetic factors associated with pathogen-associated molecular patterns (PAMPs) and effector-triggered immunity (ETI) were predicted on chromosome A03, exhibiting 11 SNPs (6,274,763-994). These belonged to three R-Genes encoding TIR-NBS-LRR proteins. Marker trait associations (MTAs) identified will facilitate marker-assisted introgression of these critical resistances into new cultivars of *B. juncea* initially and, subsequently, into other crop *Brassica* species (Rana et al. 2019). Association mapping allowed detection of ten marker trait associations (MTAs) after Benferroni correction. These were CNU-M-157-2, RA2-G05, CNU-M-353-3, CNU-M-442-5, ACMP00454-2, ACMP00454-3, EIN2-3-1, M 641-1, Na10D09-1, and Na 10D11-1 (Rana et al. 2017).

References

Agnihotri A, Lakshmikumaran MS, Jagannathan V, Shivanna KR (1991) Wide hybridization for improvement in cultivated Brassicas. Acta Hortic 289:213–214

Agnihotri A, Gupta K, Prem D, Sarkar G, Mehra VS, Zargar SM (2009) Genetic enhancement in rapeseed-mustard for quality and disease resistance through *in vitro* techniques. In: Proceedings of 16th Australian Research Assembly on Brassicas, Ballarat, Australia, Sept 14–16, p 28

Ahlawat IPS, Sharma RP (2002) Agronomic terminology, vol 50. Indian Society of Agronomy, Division of Agronomy, Indian Agriculture Research Institute, New Delhi, p 132

Ahlawat IPS, Gangaiah B, Ompal S (2005) Production potential of chickpea (*Cicer arietinum*) based intercropping systems under irrigated conditions. Indian J Agron 50:27–30

Ahloowalia BS, Maluszynski M, Nichterlein K (2004) Global impact of mutation-derived varieties. Euphytica 135:187–201

An T, Cai Y, Zhao S, Zhou J, Song B, Bux H, Qi X (2016) *Brachypodium distachyon* TDNA insertion lines: a model pathosystem to study nonhost resistance to wheat stripe rust. Sci Rep 6:25510. https://doi.org/10.1038/srep25510

Aneja JK, Agnihotri A (2013) Alternaria blight of oilseed *Brassicas*: epidemiology and disease control strategies with special reference to use of biotechnological approaches for attaining host resistance. J Oilseed Brassica 4:1–10

Aneja JK, Goel P, Agnihotri A (2012) Diversity assessment in *Alternaria brassicae*. Abstracts. In: Proc. of the 1st National *Brassica* Conference on production barriers and technological options in Oilseed Brassica, organized by Society for Rapeseed-Mustard Research, Bharatpur and Indian Council of Agricultural Research, New Delhi, at CCS Haryana Agriculture University Hisar, Mar 2–3, 2012, pp 149

Ansari NA, Khan MW, Muheet A (1989) Effect of some factors on growth and sporulation of *Alternaria brassicae* causing Alternaria blight of rapeseed and mustard. Acta Bot Ind 17:49–53

Ao E, Saud KTR (2016) Production potential of rapeseed (*Brassica rapa* var. *dichotoma*)— based intercropping systems under rain fed conditions. J Oilseed Brassica 1:91–97

Asai T, Tena G, Plotnikova J, Willmann MR, Chiu WL, Gomez-Gomez L, Boller T, Ausubel FM, Sheen J (2002) MAP kinase signalling cascade in Arabidopsis innate immunity. Nature 415:977–983

Auld D, Heikkinen M, Erickson D, Sernyk J, Romero J (1992) Rapeseed mutants with reduced levels of polyunsaturated fatty acids and increased levels of oleic acid. Crop Sci 32:657–662

Avenot H, Dongo A, Simoneau NB, Iacomivasilescu B, Hamon B, Peltiers D, Simoneau P (2005) Isolation of 12 polymorphic microsatellite loci in the phytopathogenic fungus *Alternaria brassicicola*. Mol Ecol Notes 5:948–950

Awasthi RP, Kolte SJ (1989) Variability in *Alternaria brassicae* affecting rapeseed and mustard. Indian Phytopathol 42:275

Ayers GW, Lelacheur KE (1972) Genetics of resistance in rutabaga to races of *Plasmodiophora brassicae*. Can J Plant Sci 52:897–900

Berbee ML, Payne BP, Zhang G, Roberts RG, Turgeon BG (2003) Shared ITS DNA substitutions in isolates of opposite mating type reveal a recombining history for three presumed species in the filamentous ascomycete genus Alternaria. Mycol Res 107(2):169–182

Block A, Alfano JR (2011) Plant targets for Pseudomonas syringae type III effectors: virulence targets or guarded decoys? Curr Opin Microbiol 14:39–46

Bock CH, Thrall PH, Brubaker CL, Burdon JJ (2002) Detection of genetic variation in *Alternaria brassicicola* using AFLP fingerprinting. Mycol Res 106(4):428–434

Chavan V, Bhargava S, Kamble A (2013) Temporal modulation of oxidant and antioxidative responses in *Brassica carinata* during β-aminobutyric acid-induced resistance against *Alternaria brassicae*. Physiol Mol Plant Pathol 83:35–39. https://doi.org/10.1016/j.pmpp.2013.03.002

Chevre AM, Eber F, Margale E, Kerlan MC (1994) Comparison of somatic and sexual *Brassica napus-Sinapis alba* hybrids and their progeny by cytogenetic studies and molecular characterization. Genome 37:367–371

Chhikara S, Chaudhury D, Dhankher OP, Jaiwal PK (2012) Combined expression of a barley class II chitinase and type I ribosome inactivating protein in transgenic *Brassica juncea* provides protection against *Alternaria brassicae*. Plant Cell Tissue Organ Cult 108:83–89

Chou HH, Wu WS (2002) Phylogenetic analysis of internal transcribed spacer regions of the genus *Alternaria*, and the significance of filament-beaked conidia. Mycol Res 106:164–169

Collins NC, Thordal-Christensen H, Lipka V, Bau S, Kombrink E, Qiu JL (2003) SNARE-protein-mediated disease resistance at the plant cell wall. Nature 425:973–977. https://doi.org/10.1038/nature02076

Cooke DEL, Forster JW, Jenkins PD, Gareth DJ, Lewis DM (1998) Analysis of intraspecific and interspecific variation in the genus *Alternaria* by the use of RAPD-PCR. Ann Appl Biol 132:197–209

Cui H, Wang Y, Xue L, Chu J, Yan C, Fu J, Chen M, Innes RW, Zhou JM (2010) *Pseudomonas syringae* effector protein AvrB perturbs *Arabidopsis* hormone signaling by activating MAP kinase 4. Cell Host Microbe 7:164–175

Deslandes L, Rivas S (2012) Catch me if you can: bacterial effectors and plant targets. Trends Plant Sci 17:644–655

Deverall BJ (1995) Plant protection using natural defense systems of plants. Adv Plant Pathol 11:211–228

Ding Y, Sun T, Ao K, Peng Y, Zhang Y, Li X, Zhang Y (2018) Opposite roles of salicylic acid receptors NPR1 and NPR3/NPR4 in transcriptional regulation of plant immunity. Cell 173:1454–1467.e1415

Douchkov D, Luck S, Johrde A, Nowara D, Himmelbach A, Rajaraman J, Schweizer P (2014a) Discovery of genes affecting resistance of barley to adapted and non-adapted powdery mildew fungi. Genome Biol 15(12):518. https://doi.org/10.1186/gb-2014-15-1-r1

Douchkov D, Baum T, Ihlow A, Schweizer P, Seiffert U (2014b) Microphenomics for interactions of barley with fungal pathogens. In: Tuberosa R, Graner A, Frison E (eds) Genomics of plant genetic resources, 2nd edn. Springer, Heidelberg, pp 123–148

Ebrahimi E, Kaul HP, Neugschwandtner RW, Dabbagh Mohammadi Nassab A (2016) Productivity of wheat (*Triticum aestivum* L.) intercropped with rapeseed (*Brassica napus* L). Can J Plant Sci 97:557–568

Enright S, Cipollini D (2007) Infection by powdery mildew *Erysiphe cruciferarum* (Erysiphaceae) strongly affects growth and fitness of *Alliaria petiolata* (Brassicaceae). Am J Bot 94:1813–1820

Fatima U, Bhorali P, Borah S, Senthil-Kumar M (2019) Perspectives on the utilization of resistance mechanisms from host and nonhost plants for durable protection of *Brassica* crops against Alternaria blight. Peer J 7:e7486. https://doi.org/10.7717/peerj.7486

Felix G, Duran JD, Volko S, Boller T (1999) Plants have a sensitive perception system for the most conserved domain of bacterial flagellin. Plant J 18:265–276

Fisher MC, Henk DA, Briggs CJ, Brownstein JS, Madoff LC, McCraw SL, Gurr SJ (2012) Emerging fungal threats to animal, plant and ecosystem health. Nature 484:186–194

Gamir J, Darwiche R, Van't Hof P, Choudhary V, Stumpe M, Schneiter R, Mauch F (2017) The sterol-binding activity of PATHOGENESIS-RELATED PROTEIN 1 reveals the mode of action of an antimicrobial protein. Plant J 89:502–509

Gao M, Liu J, Bi D, Zhang Z, Cheng F, Chen S, Zhang Y (2008) MEKK1, MKK1/MKK2 and MPK4 function together in a mitogen-activated protein kinase cascade to regulate innate immunity in plants. Cell Res 18:1190–1198

Gherbawy YAMH (2005) Genetic variation among isolates of *Alternaria* spp. from selected Egyptian crops. Arch Phytopathol Plant Protect 38(2):77–89

Grover A, Gawthaman R (2003) Strategies for development of fungus-resistant transgenic plants. Curr Sci 84:330–340

Gupta K, Prem D, Agnihotri A (2010) Pyramiding white rust and Alternaria resistance in low erucic *Brassica juncea* using *B. carinata*. J Oilseed Brassica 1(2):55–65

Hansen LN (1998) Intertribal somatic hybridization between rapid cycling *Brassica oleracea* and *Camelina sativa*. Euphytica 104:173–179

Hansen LH, Earle ED (1997) Somatic hybrids between *Brassica oleracea* (L.) and *Sinapis alba* (L.) with resistance to *Alternaria brassicae* (Berk) Sacc. Theor Appl Genet 94:1078–1085

Hatakeyama K, Niwa T, Kato T, Ohara T, Kakizaki T, Matsumoto S (2017) The tandem repeated organization of *NB-LRR* genes in the clubroot-resistant *CRb* locus in *Brassica rapa* L. Mol Genet Genomics 292:397–405. https://doi.org/10.1007/s00438-016-1281-1

Hirani AH, Li G (2015) Understanding the genetics of clubroot resistance for effectively controlling this disease in *Brassica* species. In: Plants for the future. Intech Open, London. https://doi.org/10.5772/60936

Hirani AH, Gao F, Liu J, Fu G, Li G (2016) Transferring clubroot resistance from Chinese cabbage (*Brassica rapa*) to canola (*B. napus*). Can J Plant Pathol 38(1):82–90. https://doi.org/10.1080/07060661.2016.1141799

Hirani AH, Gao F, Liu J, Fu G, Wu C, McVetty PBE, Duncan RW, Li G (2018) Combinations of independent dominant loci conferring clubroot resistance in all four turnip accessions (*Brassica rapa*) from the European clubroot differential set. Front Plant Sci 9:1628. https://doi.org/10.3389/fpls.2018.01628

Hou CT, Forman RJ (2000) Growth inhibition of plant pathogenic fungi by hydroxy fatty acids. J Indust Microbiol Biochem 24:275–276

Ichimura K, Shinozaki K, Tena G, Sheen J, Henry Y, Champion A, Kreis M, Zhang S, Hirt H (2002) Mitogen-activated protein kinase cascades in plants: a new nomenclature. Trends Plant Sci 7:301–308

Innes R (2018) The positives and negatives of NPR: a unifying model for salicylic acid signaling in plants. Cell 173:1314–1315

Jafary H, Albertazzi G, Marcel TC, Niks RE (2008) High diversity of genes for nonhost resistance of barley to heterologous rust fungi. Genetics 178:2327–2339. https://doi.org/10.1534/genetics.107.077552

Jammes F, Song C, Shin D, Munemasa S, Takeda K, Gu D, Cho D, Lee S, Giordo R, Sritubtim S (2009) MAP kinases MPK9 and MPK12 are preferentially expressed in guard cells and positively regulate ROS-mediated ABA signaling. Proc Natl Acad Sci U S A 106:20520–20525

Jammes F, Yang X, Xiao S, Kwak JM (2011) Two Arabidopsis guard cell-preferential MAPK genes, MPK9 and MPK12, function in biotic stress response. Plant Signal Behav 6:1875–1877

Jeromela AM, Mikic AM, Vujic S, Cupina B, Krstic D, Dimitrijevic A, Vasiljevic S, Mihailovic V, Cvejic S, Miladinovic D (2017) Potential of legume–*Brassica* intercrops for forage production and green Manure: encouragements from a temperate southeast European environment. Front Plant Sci 8:312

Jeuken MJW, Pelgrom K, Stam P, Lindhout P (2008) Efficient QTL detection for nonhost resistance in wild lettuce: backcross inbred lines versus F2 population. Theor Appl Genet 116:845–857. https://doi.org/10.1007/s00122-008-0718-2

Jones JDG, Dangl JL (2006) The plant immune system. Nature 444:323–329

de Jonge R, Peter van Esse H, Kombrink A, Shinya T, Desaki Y, Bours R, van der Krol S, Shibuya N, Joosten MHAJ, Thomma BPHJ (2010) Conserved fungal LysM effector Ecp6 prevents chitin-triggered immunity in plants. Science 329:953–955

Jourdan PS, Salazar E (1993) *B. carinata* resynthesized by protoplast fusion. Theor Appl Genet 86:567–572

Kachroo A, Kachroo P (2009) Fatty acid-derived signals in plant defense. Annu Rev Phytopathol 47:153–176

Kamble S, Mukherjee PK, Eapen S (2016) Expression of an endochitinase gene from *Trichoderma virens* confers enhanced tolerance to Alternaria blight in transgenic *Brassica juncea* (L.) Czern & Coss lines. Physiol Mol Biol Plants 22(1):69–76. https://doi.org/10.1007/s12298-016-0340-8

Kanrar S, Venkateswari JC, Kirti PB, Chopra VL (2002) Transgenic expression of hevein, the rubber tree lectin, in Indian mustard confers protection against *Alternaria brassicae*. Plant Sci 162:441–448

Kaur S, Singh G, Banga SS (2007) Documenting variation in *Alternaria brassicae* isolates based on conidial morphology, fungicidal sensitivity and molecular profile. In: Proceeding of the 12th International Rapeseed Congress, 26–30 March, Wuhan, China, vol 4, pp 87–89

Kessman H, Staub T, Ligon J, Oostendorp M, Ryals J (1994) Activation of systemic acquired disease resistance in plants. Eur J Plant Pathol 100:359–369

Khokon MAR, Salam MA, Jammes F, Ye W, Hossain MA, Uraji M, Nakamura Y, Mori IC, Kwak JM, Murata Y (2015) Two guard cell mitogen-activated protein kinases, MPK9 and MPK12, function in methyl jasmonate-induced stomatal closure in Arabidopsis thaliana. Plant Biol (Stuttg) 17:946–952

King SRF, McLellan H, Boevink PC, Armstrong MR, Bukharova T, Sukarta O, Win J, Kamoun S, Birch PRJ, Banfield MJ (2014) *Phytophthora infestans* RXLR effector PexRD2 interacts with host MAPKKKε to suppress plant immune signalling. Plant Cell 26:1345–1359

Klevernic IV, Stafford MJ, Morrice N, Peggie M, Morton S, Cohen P (2006) Characterization of the reversible phosphorylation and activation of ERK8. Biochem J 394:365–373

Kolattukudy PE (1974) Biosynthesis of a hydroxy fatty acid polymer, cutin. Identification and biosynthesis of 16-oxo-9- or 10- hydroxypalmitic acid, a novel compound in *Vicia faba*. Biochemistry 13:1354–1363

Kolattukudy PE (1980) Biopolyester membranes of plants: cutin and suberin. Science 208:990–1000

Lapin D, Kovacova V, Sun X, Dongus J, Bhandari DD, Born PV, Bautor J, Guarneri N, Stuttmann J, Beyer A, Parker JE (2019) A coevolved EDS1-SAG101-NRG1 module mediates cell death signaling by TIR-domain immune receptors. BioRxiv:572826

Lee J, Izzah NK, Choi BS, Joh HJ, Lee SC, Perumal S, Seo J, Ahn K, Jo EJ, Choi GJ, Nou IS, Yu Y, Yang TJ (2016) Genotyping-by-sequencing map permits identification of clubroot resistance QTLs and revision of the reference genome assembly in cabbage (Brassica oleracea L.). DNA Res 23:29–41. https://doi.org/10.1093/dnares/dsv034

Liu H, Guo X, Naeem MS, Liu D, Xu L, Zhang W, Tang G, Zhou W (2011) Transgenic *Brassica napus* L. lines carrying a two gene construct demonstrate enhanced resistance against *Plutella xylostella* and *Sclerotinia sclerotiorum*. Plant Cell Tissue Organ Cult 106(1):143–151. https://doi.org/10.1007/s11240-010-9902-6

Liu C, Yang Z, He P, Munir S, Wu Y, Ho H, He Y (2018) Deciphering the bacterial and fungal communities in clubroot-affected cabbage rhizosphere treated with *Bacillus subtilis* XF-1. Agric Ecosyst Environ 256:12–22

Ma M, Yan Y, Huang L, Chen M, Zhao H (2012) Virus-induced gene-silencing in wheat spikes and grains and its application in functional analysis of HMW-GS-encoding genes. BMC Plant Biol 12:1. https://doi.org/10.1186/1471-2229-12-1

Ma L, Djavaheri M, Wang H, Larkan NJ, Haddadi P, Beynon E, Gropp G, Borhan MH (2018) Leptosphaeria maculans effector protein AvrLm1 modulates plant immunity by enhancing MAP Kinase 9 phosphorylation. iScience 3:177–191

Mandal S, Rajarammohan S, Kaur J (2018) *Alternaria brassicae* interactions with the model Brassicaceae member *Arabidopsis thaliana* closely resembles those with Mustard (*Brassica juncea*). Physiol Mol Biol Plants 24:51–59. https://doi.org/10.1007/s12298-017-0486-z

McHale L, Tan X, Koehl P, Michelmore RW (2006) Plant NBS-LRR proteins: adaptable guards. Genome Biol 7:212. https://doi.org/10.1186/gb-2006-7-4-212

McRoberts N, Lennard JH (1996) Pathogen behaviour and plant cell reactions in interactions between *Alternaria* species and leaves of host and nonhost plants. Plant Pathol 45(4):742–752

Meena PD, Chattopadhyay C, Kumar VR, Meena RL, Rana US (2005) Spore behaviour in atmosphere and trends in variability of *Alternaria brassicae* population in India. J Mycol Plant Pathol 35:511

Meena PD, Rani A, Meena R, Sharma P, Gupta R, Chowdappa P (2012) Aggressiveness, diversity and distribution of *Alternaria brassicae* isolates infecting oilseed *Brassica* in India. Afr J Microbiol Res 6:5249–5258

Meng X, Zhang S (2013) MAPK cascades in plant disease resistance signaling. Annu Rev Phytopathol 51:245–266

Mohanty AL, Chrungu B, Verma N, Shivanna KR (2009) Broadening the genetic base of crop Brassicas by production of new intergeneric hybrid. Czech J Genet Plant Breed 45(3):117–122

Mondal KK, Chatterjee SC, Viswakarma N, Bhatacharya RC, Grover A (2003) Chitinase mediated inhibitory activity of *Brassicas* transgenic on growth of *Alternaria brassicae*. Curr Microbiol 47:171–173

Mondal KK, Bhattacharya RC, Koundal KR, Chatterjee SC (2007) Transgenic Indian mustard (*Brassica juncea*) expressing tomato glucanase leads to arrested growth of *Alternaria brassicae*. Plant Cell Rep 26:247–252. https://doi.org/10.1007/s00299-006-0241-3

Montillet JL, Leonhardt N, Mondy S, Tranchimand S, Rumeau D, Boudsocq M, Garcia AV, Douki T, Bigeard J, Laurière C (2013) An abscisic acid-independent oxylipin pathway controls stomatal closure and immune defense in Arabidopsis. PLoS Biol 11:e1001513

Nagy SK, Darula Z, Kállai BM, Bögre L, Bánhegyi G, Medzihradszky KF, Horváth GV, Mészáros T (2015) Activation of AtMPK9 through autophosphorylation that makes it independent of the canonical MAPK cascades. Biochem J 467:167–175

Nicolas C, Hermosa R, Rubio B, Mukherjee P, Monte E (2014) *Trichoderma* genes in plants for stress tolerance – status and prospects. Plant Sci 228:71–78

Niks RE, Marcel TC (2009) Nonhost and basal resistance: how to explain specificity? New Phytol 182:817–828. https://doi.org/10.1111/j.1469-8137.2009.02849.x

O'Neill CM, Murata T, Morgan CL, Mathia RJ (1996) Expression of the C3-C4 intermediate character in somatic hybrids between B. napus and the C3-C4 species *Moricandida arvensis*. Theor Appl Genet 93:1234–1241

Oh IS, Park AR, Bae MS, Kwon SJ, Kim YS, Lee JE, Kang NY, Lee S, Cheong H, Park OK (2005) Secretome analysis reveals an Arabidopsis lipase involved in defense against *Alternaria brassicicola*. Plant Cell 17:2832–2847

Parker JE, Holub EB, Frost LN, Falk A, Gunn ND, Daniels MJ (1996) Characterization of eds1, a mutation in Arabidopsis suppressing resistance to Peronospora parasitica specified by several different RPP genes. Plant Cell 8:2033–2046

Patni CS, Kolte SJ, Awasthi RP (2005) Cultural variability of *Alternaria brassicae*, causing Alternaria blight of mustard. Ann Plant Physiol 19:231–242

Penaud A (1999) Chemical control and yield losses caused by Erysiphe cruciferarum on oilseed rape in France. In: Proceeding of the 10 International Rapeseed Congress, Australia

Petkova M, Dimova M, Dimova D, Bistrichanov S (2014) Effect of gamma-irradiation on the fatty acid composition and susceptibility to powdery mildew (*Erysiphe cruciferarum*) of oilseed rape plants. Agric Sci Technol 6(4):413–416

Pinot F, Benveniste IJ, Salau P, Loreau O, Noe J, Schreiber L, Durst F (1999) Production in vitro by the cytochrome P450 CYP94A1 of major C18 cutin monomers and potential messengers in plant–pathogen interactions: enantioselectivity studies. Biochem J 342:27–32

Pitzschke A, Schikora A, Hirt H (2009) MAPK cascade signalling networks in plant defence. Curr Opin Plant Biol 12:421–426

Pleines S, Friendt W (1989) Genetic control of linolenic acid concentrations in seed oil of rapeseed (*Brassica napus* L.). Theor Appl Genet 78:793–797

Presti LL, Lanver D, Schweizer G, Tanaka S, Liang L, Tollot M, Zuccaro A, Reissmann S, Kahmann R (2015) Fungal effectors and plant susceptibility. Annu Rev Plant Biol 66:513–545

Primard C, Vedel F, Mathieu C, Pelletier G, Chevre AM (1988) Interspecific somatic hybridization between *Brassica napus* and *Brassica hirta* (*Sinapis alba* L.). Theor Appl Genet 75:546–552

Rahimi M, Bahrani A (2011) Effect of gamma irradiation on qualitative and quantitative characteristics of canola. Middle-East J Sci Res 8:519–525

Rajarammohan S, Kumar A, Gupta V, Pental D, Pradhan AK, Kaur J (2017) Genetic architecture of resistance to *Alternaria brassicae* in *Arabidopsis thaliana*: QTL mapping reveals two major resistance-conferring loci. Front Plant Sci 8:260

Rana K, Atri C, Gupta M, Akhatar J, Sandhu PS, Kumar N, Jaswal R, Barbetti MJ, Banga SS (2017) Mapping resistance responses to *Sclerotinia* infestation in introgression lines of *Brassica juncea* carrying genomic segments from wild Brassicaceae *B. fruticulosa*. Sci Rep 7:5904. https://doi.org/10.1038/s41598-017-05992-9

Rana K, Atri C, Akhatar J, Kaur R, Goyal A, Singh MP, Kumar N, Sharma A, Sandhu PS, Kaur G, Barbetti MJ, Banga SS (2019) Detection of first marker trait associations for resistance against *Sclerotinia sclerotiorum* in *Brassica juncea–Erucastrum cardaminoides* introgression lines. Front Plant Sci 10:1015. https://doi.org/10.3389/fpls.2019.01015

Rasmussen MW, Roux M, Petersen M, Mundy J (2012) MAP kinase cascades in *Arabidopsis* innate immunity. Front Plant Sci 3:169

Rekhter D, Mohnike L, Feussner K, Zienkiewicz K, Zhang Y, Feussner I (2019) Enhanced disease susceptibility 5 (EDS5) is required for N-hydroxy pipecolic acid formation. BioRxiv:630723

Saharan GS, Mehta N, Sangwan MS (2003) Nature and mechanism of resistance to Alternaria blight in rapeseed-mustard system. Annu Rev Plant Pathol 2:85–128

Saharan GS, Mehta N, Verma PR, Meena PD (2015) Analysis of variability determinants in *Alternaria-Brassica*-Interactions. J Oilseed Brassica 6:126–141

Saharan GS, Mehta N, Meena PD (2016) Alternaria blight of crucifers: biology, ecology and disease management, vol 326. Springer, Singapore. ISBN: 978-981-10-0019-5

Saharan GS, Mehta N, Meena PD (2019) Powdery mildew disease of crucifers: biology, ecology and disease management. Springer, Singapore, p 362. ISBN 978-981-13-9852-0

Sanchez-Vallet A, Saleem-Batcha R, Kombrink A, Hansen G, Valkenburg DJ, Thomma BPHJ, Mesters JR (2013) Fungal effector Ecp6 outcompetes host immune receptor for chitin binding through intrachain LysM dimerization. elife 2:e00790

Sharma TR, Singh BM (1992) Transfer of resistance to *Alternaria brassicae* in *Brassica juncea* through interspecific hybridization among brassicas. J Genet Breed 46(4):373–378

Sharma TR, Singh BM (1995) Generations and evaluations of somaclones of *Brassica juncea* for resistance to *Albuga candida* and *Alternaria brassicae*. Proc Ind Natl Sci Acad B Biol Sci 61 (2):155–161

Sharma TR, Tewari JP (1995) Detection of genetic variation in *Alternaria brassicae* by RAPD fingerprints. J Plant Biochem Biotechnol 4:105–107

Sharma TR, Tewari JP (1998) RAPD analysis of three *Alternaria* species pathogenic to crucifers. Mycol Res 102:807–814

Shekhawat K, Rathore SS, Premi OP, Kandpal BK, Chauhan JS (2012) Advances in agronomic management of Indian mustard (Brassica juncea (L.) Czernj. Cosson): an overview. Int J Agron 2012:1–14

Shyam SY, David L, Philip C (2007) Lentil an ancient crop for modern times, vol 472. Springer, Dordrecht, p 259

Sigareva MA, Earle ED (1999) Production and characterization of somatic hybrids between *Camilina sativa* and rapid-cycling *B. oleracea*. Theor Appl Genet 98:164–170

Singh A (2017) Glucosinolates and plant defense. In: Mérillon JM, Ramawat KG (eds) Glucosinolates. Springer International Publishing, Cham, pp 237–246

Singh VS, Kothari K, Tripathi HN (1986) Studies on intercropping in sugarcane in central Uttar Pardesh. Indian Sugarc J 35:559–562

Singh D, Singh R, Singh H, Yadav RC, Yadav N, Barbetti MJ, Salisbury P, Nimbal S, Chattopadhyay C, Kumar A (2007) Cultural and morphological variability in *Alternaria brassicae* isolates of Indian mustard (*Brassica juncea* L. Czern & Coss.). In: Proc 12th Inter Rapeseed Congress, 26–30 March, Wuhan, China, vol 4, pp 158–160

Singh RK, Kumar H, Singh AK (2010) *Brassica* based intercropping systems—a review. Agric Rev 3:253–266

Singh R, Singh D, Singh H, Caxia L, Sivasithampram K, Salisbury P, Barbetti MJ (2011) Pathogenic variability in *Alternaria brassicae* incitant of Alternaria blight of *Brassica* species. In: Proc 13th International Rapeseed Congress, 5–9 June, Prague, Czech Republic, pp 1265–1270

Skoric D, Jocic S, Sakac Z, Lecic N (2008) Genetic possibilities for altering sunflower oil quality to obtain novel oils. Can J Physiol Pharmacol 86:215–221

Stergiopoulos I, de Wit PJ (2009) Fungal effector proteins. Annu Rev Phytopathol 47:233–263

Stitcher L, Mauch-Mani B, Metraux JP (1997) Systemic acquired resistance. Annu Rev Phytopathol 35:235–270

Suman A, Lal M, Singh AK, Gaur A (2006) Microbial biomass turnover in Indian subtropical soils under different sugarcane intercropping systems. Agron J 98:698–704. https://doi.org/10.2134/agronj2005.0173

Taj G, Kumar A, Bansal KC, Garg GK (2004) Introgression of osmotin gene for creation of resistance against Alternaria blight by perturbation of cell cycle machinery. Indian J Biotechnol 3:291–198

Taj G, Agarwal P, Grant M, Kumar A (2011) Co-expression and in-silico interaction studies for inter-linking the activation of MAPK3 and LOX genes during pathogenesis of *Alternaria brassicae* in *Brassica juncea*. J Oilseed Brassica 2:13–20

Tonguc M, Griffiths PD (2004) Transfer of powdery mildew resistance from *Brassica carinata* to *Brassica oleracea* through embryo rescue. Plant Breed 123:587–589. https://doi.org/10.1111/j.1439-0523.2004.00987.x

Toor SS, Singh S, Garch AIS, Kapur ML, Singh A (2000) Gobhi sarson as an intercrop in Autumn Sugarcane for higher returns. Intensiv Agric 38:29–30

Toriyama K, Hinata K, Kameya T (1987) Production of somatic hybrid plants "*Brassica moricandia*", through protoplast fusion between *Moricandida arvensis* and *Brassica oleracea*. Plant Sci 48(2):123–128

Torrens-Spence MP, Bobokalonova A, Carballo V, Glinkerman CM, Pluskal T, Shen A, Weng JK (2019) PBS3 and EPS1 complete salicylic acid biosynthesis from isochorismate in *Arabidopsis*. Mol Plant 12(12):1577–1586. https://doi.org/10.1016/j.molp.2019.11.005

Trujillo M, Troeger M, Niks RE, Kogel KH, Huckelhoven R (2004) Mechanistic and genetic overlap of barley host and non-host resistance to *Blumeria graminis*. Mol Plant Pathol 5:389–396. https://doi.org/10.1111/j.1364-3703.2004.00238.x

Ueno H, Matsumoto E, Aruga D, Kitagawa S, Matsumura H, Hayashida N (2012) Molecular characterization of the CRa gene conferring clubroot resistance in *Brassica rapa*. Plant Mol Biol 80:621–629. https://doi.org/10.1007/s11103-012-9971-5

Verma VD, Rai B (1980) Note on induced mutagenesis for spotting out the sources of resistance to Alternaria leaf spot in Indian mustard. Indian J Agric Sci 50(3):278

Verma S, Yajima WR, Rahman MH, Shah S, Liu JJ, Ekramoddoullah AKM, Kav NNV (2012) A cysteine-rich antimicrobial peptide from *Pinus monticola* (PmAMP1) confers resistance to multiple fungal pathogens in canola (*Brassica napus*). Plant Mol Biol 79(1–2):61–74

Vishwanath, Kolte SJ (1997) Biochemical variation among three isolates of *Alternaria brassicae*. Indian Phytopathol 50(3):437–438

Vishwanath, Kolte SJ, Singh MP, Awasthi RP (1999) Induction of resistance in mustard (*Brassica juncea*) against Alternaria black spot with an avirulent *Alternaria brassicae* isolate-D. Eur J Plant Pathol 105:217–220. https://doi.org/10.1023/A:1008717323002

Wagner G, Laperche A, Lariagon C, Marnet N, Renault D, Guitton Y, Bouchereau A, Delourme R, Manzanares-Dauleux MJ, Gravot A (2019) Resolution of quantitative resistance to clubroot into QTL-specific metabolic modules. J Exp Bot 70:5375–5390

Wang Z, Mao H, Dong C, Ji R, Cai L, Fu H, Liu S (2009) Over expression of *Brassica napus MPK4* enhances resistance to *Sclerotinia sclerotiorum* in oilseed rape. Mol Plant-Microbe Interact 22:235–244. https://doi.org/10.1094/MPMI-22-3-0235

Wang Y, Li J, Hou S, Wang X, Li Y, Ren D, Chen S, Tang X, Zhou JM (2010) A *Pseudomonas syringae* ADP-ribosyltransferase inhibits *Arabidopsis* mitogen-activated protein kinase kinases. Plant Cell 22:2033–2044

Wiermer M, Feys BJ, Parker JE (2005) Plant immunity: the EDS1 regulatory node. Curr Opin Plant Biol 8:383–389. https://doi.org/10.1016/j.pbi.2005.05.010

Win J, Chaparro-Garcia A, Belhaj K, Saunders DGO, Yoshida K, Dong S, Schornack S, Zipfel C, Robatzek S, Hogenhout SA, Kamoun S (2012) Effector biology of plant-associated organisms: concepts and perspectives. Cold Spring Harb Symp Quant Biol 77:235–247

Wu XM, Chen BY, Lu G, Wang HZ, Xu K, Guizhan G, Song Y (2009) Genetic diversity in oil and vegetable mustard (*Brassica juncea*) landraces revealed by SRAP markers. Genet Resour Crop Evol 56:1011–1022. https://doi.org/10.1007/s10722-009-9420-8

Yadav RC, Sareen PK, Chowdhury JB (1991) Interspecific hybridization in *Brassica juncea* x *Brassica tournifortii* using ovary culture. Crucif Newsl 14/15:84

Zhang J, Shao F, Li Y, Cui H, Chen L, Li H, Zou Y, Long C, Lan L, Chai J (2007) A *Pseudomonas syringae* effector inactivates MAPKs to suppress PAMP-induced immunity in plants. Cell Host Microbe 1:175–185

Zipfel C, Kunze G, Chinchilla D, Caniard A, Jones JDG, Boller T, Felix G (2006) Perception of the bacterial PAMP EF-Tu by the receptor EFR restricts Agrobacterium-mediated transformation. Cell 125:749–760

Techniques for Molecular Mechanism of Host Resistance

8

Abstract

To understand the molecular mechanisms of crucifers host resistance, useful protocols have been developed to analyze the host–pathogen interactions through omics approaches like genomics, proteomics, transcriptomics, and bioinformatics. Novel R-genes have been identified, molecularly characterized, and have been fine mapped on the chromosomes of different *Brassica* species. Techniques have been developed to introgress high yielding and disease resistance genes through traditional as well as transgenic approaches. Protocols for molecular characterization of R-genes in *Brassica* species, DNA extraction, purification, and quantification, construction of a linkage map and mapping of R-loci and proteome analysis of *Brassica-Albugo* pathosystem have been developed. The procedures for analysis of molecular and biochemical mechanisms of resistance in *Brassica* to *Alternaria* and molecular characterization of *Alternaria* genes showing fungicidal resistance have been described. Genome wide identification of defensin genes in *B. juncea* and *Camelina sativa*; analysis of expression of defensin genes; and identification of distribution of chitinase genes against *Alternaria* are the techniques described. *Brassica-Erysiphe* molecular techniques include identification of molecular markers linked powdery mildew R-genes and DNA sequence analysis of *Erysiphe* isolates. In *Brassica–Hyaloperonospora* interactions, assessment of small RNA role; R-genes overexpression; and c-DNA-AFLP analysis to reveal gene expression have been developed. *Brassica-Leptosphaeria* pathosystem has revealed phylogenetic relationship of R-loci; transcriptome analysis; cloning and transformation of *Leptosphaeria* avirulence genes; identification of QTLs; molecular mapping of R-genes; and identification of NBS-encoding genes in *Brassica* species. Molecular techniques developed with the *Brassica-Plasmodiophora* system are mapping of R-genes and QTLs; SNP array, mapping, population structure, and linkage disequilibrium analysis; genome-wide association study; and the methodology for transcriptome analysis

G. S. Saharan et al., *Molecular Mechanism of Crucifer's Host-Resistance*, https://doi.org/10.1007/978-981-16-1974-8_8

of *Brassica-Sclerotinia* pathosystem has been developed. The protocols for TuMV detection; preservation; inoculation; and identification by ELISA have been developed. In *Brassica* species RNA sequencing and NBS domain and NBS-associated conserved domains protocols have been developed.

Keywords

Molecular technology for resistance development · DNA extraction · Purification · Quantification · Development of isogenic lines · Gene expression and cDNA analysis · Proteome analysis · Protein extraction · Quantitative real-time PCR · Physical mapping and gene characterization · Biochemical assays · Genome wide identification of defensin genes · Gene structure and prediction · Phylogenetic analysis · Enzyme assay's · DNA sequence analysis · Identification and overexpression of R-genes · Gene cloning and DNA sequence analysis · Linkage map construction and QTL analysis · CDNA-AFLP analysis · Phylogenetic relationship · Cloning and transformation · Identification of NBS-encoding genes · Identification and pyramiding of R-QTL's

8.1 Introduction

In science and technology to get reliable and widely acceptable results, one of the basic impacts is use of standard, repeatable, and high throughput methodology, procedure, protocols, and techniques. The techniques included in this chapter have been developed by the experts of molecular biology who have long experience of working with crucifers' host-pathosystem. The development of molecular techniques was initiated during 1990s after the sequencing of Arabidopsis thaliana genome using *Arabidopsis* host-pathosystem. With the use of this system, rapid progress has been made on molecular mechanisms of host resistance through analysis of crucifers' host–pathogen genomics, proteomics, and transcriptomics. These techniques can be used for demonstrating biomolecular events by the teachers for the graduate students and by research scientists to pursue their research programs to get widely acceptable results on molecular mechanisms of host resistance. Only selected techniques required to study molecular mechanisms of host resistance have been described. For more details on procedures, readers are advised to consult cross-references given for each technique.

8.2 *Brassica-Albugo*: Molecular Techniques

8.2.1 DNA Extraction by CTAB Method

For DNA extraction, the methodology described by Lee and Taylor (1990) was followed with the following steps: Mycelium is harvested from liquid cultures by filtration through Whatman no.3 filter paper and damp dried. Grind freeze-dried mycelium into a fine powder in liquid nitrogen and place 50 mg in eppendorf centrifuge tubes. The homogenized material is transferred to pre-warmed 2× CTAB DNA extraction buffer (100 mM Tris (pH 8.0), 20 mM EDTA (pH 8.0), 1.4 M NaCl, 2% CTAB, 2 μl/ml β-merceptoethenol). Incubate for 1 h in a 60 °C water bath, occasionally mixing gently. After removing from water bath, equal volume of chloroform-isoamylalchohol (24:1) is added, and mixed by inversion for 15 min. Spin at 15,000 rpm (10,000 × *g*) for 15 min at 4 °C. The aqueous phase is taken and transferred to another tube. Pre-chilled equal volume of absolute alcohol or 6 vol. of isopropanol is added to precipitate the DNA. DNA-CTAB complex is precipitated as a fibrous network, lifted by a pasture pipette, and washed in solution. Centrifuge at 15,000 rpm for 15 min at 4 °C. The pellet is washed with 70% alcohol, micro centrifuge (10,000 rpm) three times, each for 10 min at 4 °C, and dried overnight. The dried DNA pellet is re-dissolved in TE buffer (pH 8.0) by keeping overnight at room temperature.

1. **Purification of DNA:** RNA is removed by treating the sample with DNase free RNase. Protein including RNase is removed by treating with Chloroform: Isoamylalchohol (24:1). The purification is carried out by following the steps of method given by White et al. (1990): 2.5 μl of RNase is added to 0.5 ml of crude DNA (2.5 μl of RNase = 25 μg of RNase, so treatment is 50 μg/ml of DNA preparation). Mix gently and incubate at 37 °C for 1 h. After 1 h, a mixture of 0.3–0.4 ml of chloroform:isoamylalchohol (24:1) is added, and mixed thoroughly for 15 min till an emulsion is formed. Centrifuge for 15 min at 15,000 rpm. The supernatant is taken avoiding whitish layer at interface. The DNA is re-precipitated by adding double the quantity of absolute alcohol. To pellet DNA, the tube is centrifuged for 5 min at 5000–10,000 rpm. The pellet is washed with 70% alcohol and dried overnight. The DNA is re-dissolved in TE buffer.
2. **Gel Analysis by Agarose Gel Electrophoresis:** The integrity of DNA is judged through gel analysis by the method of Hudspeth et al. (2000) in following steps: Cast 150 ml of 0.8% agarose gel in 1× TBE buffer containing ethidium bromide (0.5 μg/ml). Load 2 μl of DNA with 6× loading dye (Bromophenol blue). Load a known amount of uncut Lambda phage DNA as control in the adjacent well. Run the gel at 50 V for 1 h. Visualize gel under UV light. Presence of single compact band at the corresponding position to phage DNA indicates high molecular weight of isolated DNA.
3. **Quantification of DNA:** The quantification of DNA is done by observing it at 260 and 280 nm wavelength by using a spectrophotometer (UV visible) method of Hudspeth et al. (2000): Take 3 ml TE buffer in a cuvette and calibrate the spectrophotometer at 260 nm as well as at 280 nm. Add 3 μl of DNA, mix properly, and take reading of optical density (OD) at both 260 and 280 nm. Estimate the DNA concentration employing the following formula: Amount of DNA (μg/ml) = (OD) 260× 50× dilution factor/l000. Check the quality of DNA from the ratio of OD value recorded at 260 and 280 nm.

8.2.2 Molecular Characterization of R-Genes in *Brassica* to *Albugo*

1. **Development of AcB1-A5.1 Near Isogenic Lines (NILs):** NILs for white rust resistance-conferring locus AcB1-A5.1 were developed in susceptible Indian lines Varuna, Pusa Bold, Pusa Jai Kisan, and Rohini by marker-assisted back-cross breeding (MAB). Locus AcB1-A5.1 was introgressed from the donor parent, Donskaja-IV into susceptible lines through positive selection by a PCR-based Intron polymorphism (IP) marker At2g36360 (Panjabi-Massand et al. 2010) tightly linked to the resistance phenotype and negative selection by IP marker At2g44970 on one side and SSR marker At2g34700 on the other side. Detailed information on the markers is available in previous studies (Panjabi-Massand et al. 2008; Dhaka et al. 2017; Rout et al. 2018). The entire scheme took five generations of backcrossing followed by selfing to develop homozygous NILs.
2. **BAC Isolation, Sequencing, and Annotation:** A 5× pooled and amplified Hind III BAC library was developed using genomic DNA extracted from young leaves of Donskaja-IV. Partially digested, size-selected DNA fragments (150–350 kb) were ligated into Hind III digested vector pIndigoBAC-5 and transformed into *E. coli* DH10B cells. Colonies were distributed into 1 × 96 well plate (Pooled BAC library). Each well contained about 573 primary clones (total clones > 55,000). The BAC library was screened using PCR-based IP marker At2g36360. Two additional IP markers (At2g36480 and At2g36310) flanking At2g36360 were developed using sequence information of *B. rapa* line Chiifu-401 obtained from the BRAD database (Cheng et al. 2011). BAC clones giving amplifications with all the three markers were sequenced on a PacBio RSII instrument using single-molecule real-time (SMRT) technology followed by assembly of the reads using a hierarchical genome assembly process.
   BAC sequences were annotated for ORFs using Augustus (Stanke et al. 2004), a web server for gene prediction in eukaryotes. ORFs were screened for conserved domains using NCBI Batch CD-Search (Marchler-Bauer and Bryant 2004; Marchler-Bauer et al. 2011). The domains of BjuWRR1 protein were also predicted using an online tool InterPro (Finn et al. 2017) and COILS (ExPASy Bioinformatics Resource portal). The conserved motifs of the NB domain were identified following Meyers et al. (2003). Molecular weight and the isoelectric point of the protein were computed by Compute pI/MW online (ExPASy Bioinformatics Resource portal: https://web.expasy.org/compute_pi/).
3. **Vector Construction and Genetic Transformation of *B. juncea*:** PCR amplification of the candidate gene using BAC clone as a template was carried out with Q5 High-Fidelity DNA polymerase. The amplified product was cloned into a modified binary vector pCGMCP22 (Arumugam et al. 2007) containing the *bar* gene, conferring resistance to phosphinothricin (PPT), as a selectable marker between the *loxP* sites. The *bar* gene was driven by the CaMV35S promoter with an AMV leader sequence (Jobling and Gehrke 1987) and ocspA as the terminator. The modified vector was digested with restriction enzymes HindIII and PstI and amplified product was cloned using In-Fusion HD Cloning Kit followed by

sequencing on a 3730 DNA Analyzer. The binary vector containing the cloned gene was transferred into disarmed *Agrobacterium tumefaciens* strain GV3101 by electroporation. Genetic transformation of *B. juncea* line Varuna was carried out using hypocotyl explants following a protocol described by Mehra et al. (2000). Positive transformants were selected in vitro on a medium containing PPT at a concentration of 10 mg L^{-1}. For segregation analysis of the progeny of the T0 transgenics for resistance and susceptibility to PPT, a concentration of 200 mg L^{-1} was used in the field. For Southern blot analysis, genomic DNA was isolated from well-expanded leaves collected from the field grown transgenic lines (T0) following the protocol of Rogers and Bendich (1994). Six micrograms of DNA was digested with restriction enzyme EcoRI. Digested DNA was separated on 0.8% agarose gel and blotted onto Hybond-N+ nylon membrane. The blots were hybridized with probes that were labeled using PCR DIG Probe Synthesis Kit, and hybridization was detected with DIG Luminescent Detection Kit following the manufacturer's instructions.

4. **Gene Expression and cDNA Analysis:** For the gene expression studies, RNA was extracted from the cotyledons of 1-week-old seedlings of Donskaja-IV, Varuna, Heera, and Tumida using Spectrum Plant Total RNA Kit. RNA was treated with RNase free DNaseI (Qiagen) to remove any contaminating genomic DNA. Two micrograms of total RNA was used to synthesize the first-strand cDNA using High-Capacity cDNA Reverse Transcription Kit according to the manufacturer's instructions. For studying changes in transcript accumulation in response to pathogen infection, cotyledons of 1-week-old seedlings of Donskaja-IV were inoculated with 10 μl of sporangial suspension (2×10^4 sporangia ml^{-1} in water) of *A. candida* isolate AcB1 using the drop inoculation method. The cotyledon samples inoculated with the pathogen and distilled sterile water (mock) were collected 4 h, 48 h, 72 h, and 14 days after the inoculations—with four seedlings for each sample. Total RNA was extracted from the collected samples, and first-strand cDNA was synthesized as mentioned above. Real-time quantitative reverse transcription-PCR (qRT-PCR) analyses were carried out using a Power Up SYBR Green Master Mix on Quant Studio 6 Flex Real-Time PCR System. Comparisons of relative gene expression were made between the pathogen and mock-inoculated samples collected at the same time points. The ubiquitin gene UBQ9 (Chandna et al. 2012) was used as an internal control. The $2^{-\Delta\Delta C}T$ method (Livak and Schmittgen 2001) was followed in evaluating the relative gene expression levels. To obtain a cDNA sequence, RT-PCR was carried out using total RNA isolated from 1-week-old cotyledons of Donskaja-IV. The first-strand cDNA was synthesized as described above and was used as a template for PCR using Phusion High-Fidelity DNA Polymerase with a gene-specific primer pair. The PCR product was cloned in the pGEM-T Easy vector and sequenced on a 3730 DNA analyzer.
5. **Sequence Information of Syntenic Regions in *B. juncea* Lines and Sequenced Brassicaceae Genomes:** The sequence of the syntenic region in line Heera was obtained by screening a genomic DNA BAC library of the line reported earlier (Sharma et al. 2014). Identified BAC clone was sequenced by the procedure

reported for the Donskaja-IV BAC. The sequence of line Varuna was obtained from an SMRT-based genome assembly of the line. Sequences present in Cutlass, Skorospieka, Kranti, *B. nigra* (Sangam), and the B-genome of *B. juncea* (Donskaja-IV) were obtained from Illumina short-read sequencing of the genomes of these lines, the sequence of Tumida was downloaded from the BRAD database (Wang et al. 2015). Sequence information of other members of the Brassicaceae family was obtained from the BRAD database. Multiple sequence alignment of the alleles was done with Clustal W using MegAlign software (Arora et al. 2019).

8.2.3 Proteome Analysis of *Brassica-Albugo* Pathosystem

1. **Protein Extraction:** For protein extraction, cotyledon tissue (1 g) was ground to a fine powder in liquid nitrogen before homogenizing in 10 ml cold 20% (w/v) trichloroacetic acid in acetone containing 0.2% (w/v) dithiothreitol (DTT). Supernatants were incubated overnight at −20 °C to precipitate the protein completely before collecting by centrifugation at 30,000 × *g* for 20 min at 4 °C. The pellet was washed three times with ice-cold acetone containing 0.2% (w/v) DTT, dried at room temperature for 30 min, and resuspended in 1 ml of rehydration buffer [2 M thio-urea, 7 M urea, 2% (w/v) 3-[(3-cholamidopropyl) dimethylammonia]-1-propanesulphonate hydrate (CHAPS), 10% (w/v) phenylmethylsulphonyl fluoride (PMSF), and 0.2% (w/v) DTT]. The sample was mixed using a vortex mixer and incubated with orbital shaking for 16 h at 4 °C to solubilize the protein completely. The insoluble material was removed by centrifugation as above. The concentration of the protein in the supernatant was determined (Peterson 1983) and samples were stored at −80 °C. Extracts from pooled cotyledon tissue were prepared from at least three independent inoculation experiments.
2. **Two-Dimensional Gel Electrophoresis:** The method described by Yajima et al. (2004) for 2D separation of proteins was used with some modifications. A solution containing 500 μg of total protein in 0.04% (w/v) ampholytes in a total volume of 200 μl was used for passive rehydration of 11 cm immobilized pH gradient strips (IPG). The IPG strips were exposed to a total of 30 kVh using the IEF program: 300 V in linear mode for 1 h, 8 kV in linear mode for 3 h, and 8 kV in rapid mode until 30 kVh were reached, before storing at −20 °C overnight. Proteins in the IPG strips were reduced by gently agitating the strips in 6 M urea, 2% (w/v) SDS, 0.375 M TRISHCl, pH 8.8, 20% (v/v) glycerol, and 130 mM DTT at ambient temperature for 30 min, replacing with fresh solution once. Proteins were alkylated in 6 M urea, 2% SDS, 0.375 M TRISHCl, pH 8.8, 20% (v/v) glycerol, and 135 mM iodoacetamide, with gentle agitation at ambient temperature for 30 min, replacing the solution once. Each IPG strip was embedded at the top of an SDS-PAGE in 0.5% (w/v) low-melting point agarose in 13 TRIS-Glycine-SDS (TGS) buffer (Aldrich, USA). Precision wide-range standard proteins were used for molecular mass determinations. Proteins were separated

by electrophoresis in 13 TGS buffer at 150–200 V constant current. Gels were stained for at least 3 h with Coomassie Brilliant Blue R250 and destained according to the manufacturer's instructions.

3. ***2D* Gel Analyses and Protein Identification:** Images of the *2D* gels were acquired and recorded using a red filter (595–750 nm) and a resolution of 36.3336.3 lm before analysis. The software allowed the creation of match sets where replicate gels could be compared with corresponding gels from other treatments. The analysis was reevaluated by manual validation to include missing spots and to eliminate artefacts. The mean value for the normalized spot volume and the standard deviation (SD) were determined for each spot. Each set of 2D gels was analyzed using the Student's t test tool of the software to determine the statistically significant differences ($P = 0.05$) in protein amount between pathogen-inoculated and mock-inoculated cotyledons. Among the differentially staining spots, only those protein spots that were reproducibly unique to only one variety (susceptible or resistant) or that showed reproducibly altered staining intensities compared with the control gel images in response to pathogen infection were quantified.
   Protein spots were removed from the gels, destained, trypsin digested, and analyzed by mass spectrometry (MS) according to the method of Bringans et al. (2008). Mass spectra were analyzed using peptide sequence matching software. For the PR-5 protein, MS de novo sequence analysis was performed by further analysis of the spectra (PEAKS Studio, Version 4.5, SP2 program). The ion score automatically calculated by the software was $-10 \times \log (P)$ where P was the probability that the observed match was a random event. Ion scores exceeding the threshold value for a random event indicated sequence identity or extensive homology ($P < 0.05$) (Sharma et al. 2007). The identity of the spot was selected as the protein that produced the highest score and, consequently, the best match with its peptide sequence. For the MS peptides identified, database searches were used to create protein sequence alignments. Searches were performed over the full molecular weight and pI range and no species restriction was applied. Since the full genome sequence of *B. juncea* is not yet available, only two sets of peptides could be directly matched to *B. juncea* protein accessions. The rest of the peptide sets were matched to the closest protein sequence in another species.
4. **Quantitative Real-Time PCR (qRT-PCR) Primer Design:** Where available, *B. juncea* cDNA sequences were used for primer design. When *B. juncea* peptides could only be matched to homologues in related *Brassica* species, it was assumed that the corresponding cDNA sequences would have sufficient identity with those from *B. juncea* to be useful in qRT-PCR (Sharma et al. 2007). To ensure maximum sequence complementarily, primers were placed in those regions of the cDNA that matched the *B. juncea* peptide sequences. Primers were designed to generate amplicons of approximately 60 bp using the default program settings. Primer sequences were cross-checked by database searches (National Center for Biotechnology Information) to ensure that only the target of interest would be amplified in PCR reactions. All primers were tested on

B. juncea cDNA and the products analyzed by melt-curve analysis and agarose gel electrophoresis to ensure that a single PCR product was formed. The PCR efficiency of each primer pair was calculated (Ramakers et al. 2003).

5. **qRT-PCR:** Total RNA was extracted from pooled cotyledon samples of the inoculated and mock-inoculated plants and treated with DNase I (RNase-free DNase I, Qiagen) to ensure the absence of contaminating genomic DNA according to the manufacturer's instructions. First strand cDNA was synthesized using oligo (dT) primers according to the manufacturer's directions. qRT-PCR was performed using SYBR green following the manufacturer's standard protocol. The relative abundance of mRNAs was expressed as 40− ΔCt, with ΔCt being the difference in the threshold cycle (Ct) number of the gene-of-interest and a reference gene with 40 being randomly chosen since it represents the highest possible Ct value (Bari et al. 2006). The actin gene (*B. napus*) (GenBank accession no. AF111812) (Yang et al. 2007) was used as the reference gene in all experiments (Kaur et al. 2010).

8.2.4 Construction of a Linkage Map and Mapping of the Resistance Trait to *Albugo*

Genetic mapping was carried out using IP (Intron Polymorphism), genic SSR (simple sequence repeats), and genic SNP markers developed earlier for different *B. juncea* mapping populations in the laboratory (Panjabi-Massand et al. 2008; Paritosh et al. 2014; Dhaka et al. 2017). Further, GBS-based markers described (Paritosh et al. 2019) were added to the map. In the first instance, an anchor map was developed with the IP, SSR, and genic SNP markers using the mapping software JoinMap 4.1 (Van Ooijen 2006) at an LOD threshold value greater than 5.0 using regression mapping. The Kosambi mapping function (Kosambi 1944) was used to convert recombination frequencies into the genetic distance. In the second step, a high-density linkage map was developed by combining the anchor markers with GBS-based SNPs using the ASMap package (Taylor and Butler 2017) which uses MSTmap algorithm (Wu et al. 2008) for linkage group construction. The package was used with the default parameters except for a few changes—the probability value (*p*. value) was set to 1*e*−15, and the missing threshold (miss thresh) value to 0.3. A total of 18 linkage groups (LGs) referred to as BjuA1–BjuA10 and BjuB1–BjuB8, as described previously (Paritosh et al. 2019), were identified. Mapchart 2.2 (Voorrips 2002) was used for a graphical representation of LG BjuA6.

QTL mapping was performed using the software package MapQTL6 (Van Ooijen and Kyazma 2009) using Kruskal-Wallis, interval mapping, and Multiple QTL Mapping (MQM) algorithms. LOD thresholds for QTL significance were determined for each LG by taking the number of permutations to be 1000 and a significance level of 0.05. The nonparametric mapping using Kruskal-Wallis and interval mapping was performed prior to MQM mapping to identify closely linked markers. The markers which were close to the QTL and were identified both in the Kruskal-Wallis test and interval mapping were used as cofactors in MQM mapping. The marker closest to the peak LOD was identified and the corresponding flanks were determined subsequently.

1. **Physical Mapping, Gene Characterization, and Synteny Analysis:** Genomic sequences between the flanking markers in the mapped interval were downloaded from the BRAD database (Cheng et al. 2011). Genes annotated in *B. juncea* Tumida were taken from the genome assembly of the line (Yang et al. 2016). BRAD database was used for obtaining the sequence information of the syntenic regions in *A. thaliana, B. rapa*, and other sequenced *Brassica* species. The sequences of the genes present in Varuna were obtained from an SMRT-based gene assembly of *B. juncea* Varuna (Paritosh et al. 2019). The nucleotide sequences of the resistant/susceptible alleles were compared using the CLUSTALW algorithm of MegAlign. The gene structure and the encoded protein sequences were represented by Snap gene viewer. The structure of allelic variants in different lines of *B. juncea* and the synteny analysis was plotted using R programming for expression analysis, and RNA was extracted from the plant tissues with "Spectrum Plant Total RNA Kit" and treated with DNaseI to remove any genomic DNA contamination. About 2 μg of total RNA was used to synthesize the cDNA with "High-Capacity cDNA Reverse Transcription Kit." The ubiquitin gene UBQ9 (Chandna et al. 2012) was used as an internal control. cDNA was amplified with "Phusion High-Fidelity DNA Polymerase." Genomic DNA and cDNA were sequenced on an ABI PRISM sequencer.
2. **Phylogenetic Analysis**: Protein sequences of NBS type genes present in *A. thaliana* and *B. rapa* that belonged to the CNL clade (Yu et al. 2014) were downloaded from TAIR database (Rhee et al. 2003) and BRAD database. For the alignment of the NBS domain, complete predicted protein sequences were trimmed at ten amino acids N terminal to the first Gly in the P-loop motif and 30 amino acids beyond the MHDV motif. Sequences were then aligned with the MUSCLE program, available in MEGA7 (Kumar et al. 2016) using default options. A phylogenetic tree was constructed using the Maximum Likelihood method based on the JTT matrix-based model (Jones et al. 1992). All positions with less than 70% site coverage were eliminated, and the reliability of the interior nodes was assessed using 1000 bootstrap replicates (Bhayana et al. 2020).

8.3 *Brassica-Alternaria*: Molecular Techniques

8.3.1 Molecular Characterization of *Alternaria* Gene Showing Fungicidal Resistance

1. **Strains and Media:** All strains used were isolated from radish or cabbage during quality controls of commercial seed lots (French isolates) or seeds from small local markets (Romanian isolates). All strains were purified by mono spore isolation and maintained in tubes of malt agar medium (malt 20 g, agar 20 g in 1 l water) at 4 °C. The laboratory-resistant Abra20MLR and Abra43MLR strains

were obtained after transferring, onto selective medium, mycelial tips from fast-growing sectors observed with some sensitive isolates on iprodione-amended media. Fresh subcultures were made by transferring hyphal plugs to dishes of potato dextrose agar (PDA) medium amended or not with iprodione (dicarboximide) or fludioxonil (phenylyprrole) at a final concentration of 100 mg l^{-1}. The effects of fungicides on spore germination and germ tube elongation were tested as described by Iacomi-Vasilescu et al. (2004). Mycelial plugs were transferred onto PDA containing 2–4% NaCl (w/v) in order to test sensitivity to osmotic stress.

2. **DNA Procedures:** Genomic DNA was extracted from mycelium according to Adachi et al. (1993). For Southern analysis, DNA fragments resulting from genomic DNA digestion with Eco RI and Hind III were separated on 1% agarose gels and vacuum-transferred to Hybond N membranes. Blots were then probed with PCR fragments purified with Gene clean II and labeled with a random primer labeling kit.
Fungal RNA was extracted, DNAse-treated, and reverse-transcribed as described by Guillemette et al. (2004). PCR reactions using degenerated primers similar to those designed by Alex et al. (1996) and based on the conserved N and H boxes (N1A, 5¢-GARAAY TTI AII GCR TT-3¢; H1A, 5¢-CAYGAI HTI MGI ACI CC-3¢) were used to amplify a putative histidine kinase gene fragment from Abra43FS genomic DNA. The amplification was performed in 50 mM KCl, 1.5 mM $MgCl_2$, 10 mM Tris-HCl (pH 9.0), 0.1% Triton X-100, containing 0.2 mmol of dATP, dCTP, dTTP, and dGTP, 0.4 lmol of each primer, 50 ng of DNA, and 1 unit of Taq polymerase in a final volume of 50 ll. PCR was carried out in a Thermojet with 35 cycles at 95 °C (1 min), 40 °C (1 min), and 72 °C (1 min), followed by a 30-min extension at 72 °C. Amplification products were separated on 1.2% agarose gel, purified from gel by Gene-Clean II, cloned into a pGEM-T vector (Promega), and subsequently sequenced. Specific primers were designed from this sequence to perform PCR-walking according to the method described by Siebert et al. (1995). PCR amplifications (initial denaturation for 2 min at 94 °C, then 35 cycles of 1 min at 94 °C, 1 min at 55 °C, 1–4 min at 72 °C, with a final extension for 10 min at 72 °C) with specific primers were also performed on genomic DNA and first-strand cDNA in order to verify the presence of introns in coding sequences and to confirm the physical linkage between the 5¢ and 3¢ amplified parts of the histidine kinase gene. A set of 11 primer pairs (Table 8.1) that span the whole coding sequence of the AbNIK1 gene was designed for PCR amplification of corresponding portions of this gene using genomic DNA from three fungicide-sensitive strains (Abra43FS, AbraCMFS, Abra20FS) and from the seven resistant mutants. Nucleotide sequences were obtained directly from these PCR products by Genome Express (France) and deposited in GenBank (accessions AY700092–AY00098). Sequences were compared using the Multalin program (Corpet 1988).
3. **PCR-RFLP and Heteroduplex Mobility Assays:** Portions of the AbNIK1 gene potentially carrying mutations were PCR-amplified using relevant primers and either directly digested with different restriction enzymes (Promega) according to

Table 8.1 Primer sets used for amplification of the AbNIK1 gene from *Alternaria brassicicola* genomic DNA

Primer set	5¢ primer	3¢ primer	Amplified DNA fragment
1	5¢-TGCCAAAGGTTCTTCAGTCC-3¢	5¢-CAGCCCAAATACCAGCAACT-3¢	+154 to +995
2	5¢-TTGCCAGTCAAGTCACATTCC-3¢	5¢-CCATGGCGTTTACATTGACA-3¢	+900 to +1350
3	5¢-AGAGATGTCGGAACTGAAGGCAAGC-3¢	5¢-ATTTCTCGCACCTGTGTTGTG-3¢	+1254 to +1657
4	5¢-CGAAGTCACAAAGATTGCGCGAG-3¢	5¢-TACTTCGTACTTGCAGCGTGAG-3¢	+1511 to +1932
5	5¢-TCGCACAAGAAGTCAGCAAGGTC-3¢	5¢-TGAATTTGCCCTCCATTGCAGC-3¢	+1780 to +2247
6	5¢-CGACATGGTGATGAGACTGGAT-3¢	5¢-GTTGCGCATGCGTCAGATCGGTG-3¢	+2030 to +2487
7	5¢-CGCAACGAGAGATGTTGAC-3¢	5¢-GATGAGGTTAAGCTTATCG-3¢	+2482 to +2906
8	5¢-CCAGATCATTATGTCTTCCA-3¢	5¢-GCACCTCCTCAACCGAGTGA-3¢	+2832 to +3237
9	5¢-CGGATGAGATTACTGAGCAC-3¢	5¢-ACTTCTCCAGAATCCTAC-3¢	+3166 to +3573
10	5¢-CGTTCCAGATTCTTCTGGC-3¢	5¢-GAGAGCCGAATTAGTAGGG-3¢	+3505 to +3911
11	5¢-CGTCATTGCACTAACGGCGCAC-3¢	5¢-TCAGCTACTGGACTCCCAGC-3¢	+3794 to +4304

Fragment-numbering corresponds to nucleotide positions in the AbNIK1 gene, with the first nucleotide of the start codon considered as position +1 (Avenot et al. 2005)

the manufacturer's instructions or mixed for duplex formation as described by Sutra et al. (2001). Digestion products or mixtures of heteroduplexes and homoduplexes were then separated on non-denaturing polyacrylamide gels (Avenot et al. 2005).

8.3.2 Analysis of Molecular and Biochemical Mechanisms of Resistance in *Brassica* to *Alternaria*

1. **Plant Growth, Testing, and Sampling:** Twenty seedlings per each tested cultigen (*S. alba*, "Kamienna Głowa," PGH05I, and PGH09K) were grown from seeds for 45 days, and then inoculated with suspension of mixed *A. brassicicola* isolates and tested as per the optimized protocol. Each day, samples were collected from two plants per cultigen, at the same time (~11 am). At 3 dpi, Kamienna Głowa' started showing mild Alternaria dark spot symptoms, which exacerbated at 4 dpi; hence, sampling was stopped at this time. Collected samples were snap frozen in liquid nitrogen and stored at −80 °C until extracted.
2. **Biochemical Assays:** The homogenized samples were processed as in Nowicki et al. (2016). Extraction of soluble proteins (supernatant) was done with 5× (v: w) 50 mM potassium phosphate, pH 7.2, whereas the reactive oxygen species (ROS) and free phenols were assayed from the methanolic extracts (18,000 × *g* for 15 min at 4 °C each, with 10:1 (v: w) of 80% methanol in water), supernatants combined, and pellets were used for lignin assessment.
3. **Assessment of ROS Generation (H_2O_2, HO*, and O_2^-):** Generation of either ROS species followed the published methods (Loreto et al. 2001; Loreto and Velikova 2001; Habu and Ibeh 2015; Moukette et al. 2015). Methanolic extracts were assayed at A390, A560, and A530 for H_2O_2, HO*, and O_2−, respectively. The H_2O_2 as one of the standards; in the HO* and O_2^- assays, the absorbance readout was recalculated per mg of fresh tissue.
4. **Assessment of Plant Red-ox Enzymic Activities:** To analyze the dynamics of polyphenol oxidase (PPO), peroxidase (POX), superoxide dismutase (SOD), polygalacturonase (PGA), and catalase (CAT), standard methodology published elsewhere were used (Korgan et al. 2011; Moukette et al. 2015; Sheehy et al. 1988). Buffer extracted samples (as above) were assayed for protein using the BCA kit, and analyzed for linear assay range on a randomly chosen sample (0 dpi). Proper assays were run in micro-titer plates for 3–5 min, respectively, whereas absorbance readouts were recorded every minute in the Epoch2T reader. Protein activity was calculated as change in absorbance per minute and per mg of protein (PPO) or metabolized substrate/min/mg protein (POX, SOD, PGA, CAT).
5. **Accumulation of Free Phenols and Deposition of Lignins:** Levels of lignins deposition and free phenolics were assessed with published spectrophotometric methods (Campbell and Ellis 1992; Cvikrova et al. 1992). Methanolic extracts (see above) and pellets were used, and the assays used technical lignins and gallic acid as the respective standards (Nowakowska et al. 2019).

8.3.3 Genome-wide Identification of Defensin Genes in *Brassica juncea* and *Camelina sativa*

The available Arabidopsis thaliana defensin genes were downloaded from TAIR database (https://www.arabidopsis.org/), and their coding sequences were identified using ORFfinder program (https://www.ncbi.nlm.nih.gov/orffinder/). By using the ExPASy translate tool (https:// web. expasy. org/translate/) the coding sequences were translated to peptide sequences. The above peptide sequences of *A. thaliana* defensin genes were queried against *B. juncea* and *C. sativa* peptide sequences in BRAD database (http://brassicadb.org/brad/) to find putative defensins in *B. juncea* genome using blastp. From the blast output the peptide sequences showing >100 bit score were selected and redundant hits were removed to select unique sequences.

1. **Protein Structure, Conserved Domain Identification, and Computation of Physical and Chemical Parameters:** The NCBI-CDD database (https://www.ncbi.nlm.nih.gov/Structure/cdd/wrpsb.cgi) was used to analyze the conserved domain of all non-redundant sequence and SMART (http://smart.embl-heidelberg.de/) and Pfam (https://pfam.xfam.org/) were used to confirm that identified genes were members of defensin family. Those protein containing gamma-thionin domain under Knot1 super family were defined to be belonging to the defensin family (previously known as gamma-thionin family). Full length amino acid sequences of the selected defensin proteins were compiled and aligned using Clustal W and the alignment was visualized by Bio Edit V 7.0.4. software to find conserved regions, and the three-dimensional secondary structure of defensin proteins were constructed by a ProMod3 Version 1.3.0. homology modeling method using SWISS-MODEL server (https://swissmodel.expasy.org/interactive) to confirm that the selected proteins have structures similar to plant defensins, as they are generally defined by their conserved cysteine scaffold with α-helix and triple strand antiparallel β-sheets connected to the scaffold. Computer analysis of the amino acid sequence to compute chemical and physical parameters were performed with the Prot Param tool on the ExPASy server (https://web.expasy.org/protparam/) and possible disulfide bridges were determined using the DISUFIND server (http://disulfind.dsi.unifi.it/). The subcellular localization of each defensin protein of *B. juncea* and *C. sativa* was predicted using Plant-mPLoc (http://www.csbio.sjtu.edu.cn/bioinf/plant-multi/).
2. **Gene Structure Construction and Prediction of *Cis*-Acting Elements:** The defensin sequences retrieved from BRAD database were employed to identify respective genes from *B. juncea* and *C. sativa* genome using local blast in Bio Edit V 7.0.4 software. The exon intron structure of *B. juncea* defensin-encoding genes were determined based on alignment of coding sequences with corresponding genome sequences and graphical display was created using online Gene Structure Display Server (http://gsds.cbi.pku.edu.cn/). Conserved *cis*-acting regulatory elements in the promoter region of the putative defensin genes were analyzed by plant care database (http://bioinformatics.psb.ugent.be/webtools/plantcare/html/).

3. **Phylogenetic Analysis of Defensin Proteins of *B. juncea, C. sativa*, and Related Species:** To compare the composition of the defensin gene family in *Brassica*, defensin genes were identified in three other *Brassica* species, *B. rapa, B. napus,* and *B. oleracea* and also the wild-type *C. sativa* (http://brassicadb.org/brad/) using the same method. To study the evolutionary relationships among *B. juncea, B. rapa, B. oleracea, B. napus,* and *A. thaliana* defensins, an unrooted neighbor joining phylogenetic tree was constructed using MEGA 7.0 software based on the amino acid sequences of the defensin proteins of above genotype and a bootstrap test with 1000 replicates was performed.
4. **Expression Analysis of Defensin Genes in *B. juncea* and *C. sativa* in Response to SA, JA, and *A. brassicae* Infection and Wounding:** *B. juncea* and *C. sativa* plants were raised from seeds in pots and were maintained at 25 °C for 16 h light/ 8 h dark in glass house at National Phytotron Facility, Indian Agriculture Research Institute (IARI). *A. brassicae* isolated from disease infected field grown plants was cultured on Radish Dextrose Agar medium and was identified by Indian Type Culture Collection, IARI (ID No. 81651). Conidial suspensions were prepared by scraping sporulated mycelium from 21-day-old cultures and suspending in sterilized distilled water. Conidial concentration was adjusted to 5×10^3 conidia/ml using hemocytometer. Forty-five-days-old healthy plants with six leaves were treated with conidial suspensions using a hypodermic needle and control plants were treated same way with sterilized distilled water. To study the effect of hormone on defensin gene expression, plants were sprayed with 100 μM MeJA, and 2 mM SA solutions separately and control plants were sprayed with sterilized distilled water. Small puncture wounds were made on the leaves using a hypodermic needle filled with sterilized distilled water, to analyze the effect of wounding on the expression of defensin genes. Leaf samples were collected at 3, 6, 12, 24, 48, 72, and 96 h post treatment. Total RNA was isolated from leaf tissues using TRIzol reagent and first strand cDNA synthesis was performed from 2 μg total RNA using Superscript III cDNA synthesis kit following the manufacturer's instructions. Quantitative real-time PCR was performed using SYBR Premix Ex Taq Kit run at 95 °C for 10 min followed by 40 cycles of 95 °C for 10 s and 55 °C for 30 s with F-5′ CATGAAGCTCTCTATGCG 3′ and R-5′ CGATGGATCAGCGATTTCTGG 3′ primers. β-tubulin was used as the reference gene for mRNA relative expression pattern analysis. The reactions were performed with three biological replicates and three technical replicates per sample. The relative quantification method (ΔΔCT) was used to evaluate quantitative variation between the samples (Chaturani et al. 2019).

8.3.4 Genome-Wide Identification and Distribution of Chitinase Genes in *Brassica juncea* and *Camelina sativa* in Response to *Alternaria brassicae*

The amino acid sequences of *B. juncea* and *C. sativa* genome were retrieved from the *Brassica* database (BRAD) (http://brassicadb.org/brad/). Protcin sequences local database of *B. juncea* and *C. sativa* genes were created in Bioedit ver. 7.2.5. The

24 chitinase genes of *A. thaliana* downloaded from the *Arabidopsis* Information Resource (TAIR) database (https://www.arabidopsis.org/) were used as query to identify putative orthologs of chitinase genes in *B. juncea* and *C. sativa* local database using BLASTp. To identify high scoring pairs (HSPs), the initial cut-off for an *e*-value of 10^{-5} was kept. The HSPs showing *e*-value cut-off 10^{-5} were selected based on tabulated blast output. Finally, unique sequences were selected by removing redundant hits with highest similarity for the further analysis. The entire set of unique sequences was submitted to Pfam and SMART (http://smart.emblheidelberg.de/) to confirm the presence of the chitin binding domain (CBD). To remove sequences that lack CBD, multiple sequence alignment (MSA) was performed.

1. **Phylogenetic Analysis and Multiple Sequence Alignment:** The CLUSTALW alignment function in MEGA 7.0 was used to align chitinase sequences. Maximum likelihood method was used to construct the phylogenetic tree, and 1000 bootstraps were used to measure the stability of the branch node. The chitinases were grouped in accordance with different chitinase classes (I–V) of *A. thaliana*. The phylogenetic tree was constructed using *B. juncea* and *C. sativa* chitinase sequences and two separate trees were constructed using *B. juncea* and *C. sativa* chitinase sequences, respectively, with the common 24 chitinase sequences of *Arabidopsis*.
2. **Gene Structure and Chromosomal Localization:** The gene and CDS sequence of chitinases identified for *B. juncea* and *C. sativa* genome were retrieved from BRAD database (http://brassicadb.org/brad/). The exon-intron structures of chitinase-encoding genes of both crop species were determined based on alignment of CDS sequences with corresponding genomic sequences, and graphical display was created using the online Gene Structure Display Server (2.071) (http://gsds.cbi.pku.edu.cn/). The positions of the putative chitinase genes were visualized by mapping to the 18 and 20 chromosomes of *B. juncea* and *C. sativa*, respectively, using base pair start positions in Mapchart 2.2.
3. **Distribution of Conserved Motifs and 3D Structure of Chis Genes:** The upstream region (1.5 kb) of randomly selected five chitinase genes each from *B. juncea* and *C. sativa* were analyzed for conserved *cis*-regulatory elements involved in multiple stresses using Plant Care Database (http://bioinformatics.psb.ugent.be/webtools/plantcare/html/). For final quality check three-dimensional (3D) models of the same chitinase proteins for both crop species were constructed using Phyre2 server. For structural models the quality control and thresholds are as follows: alignment coverage >65% and confidence = 100%. MEMSAT-SVM prediction method available in Phyre2 server was used to predict transmembrane helix and topology of the chitinases.
4. **Genome Synteny and Gene Duplication:** The syntenic information of *A. thaliana, B. rapa, B. oleracea,* and *C. sativa* was downloaded from the BRAD database (http://brassicadb.org/brad/), whereas the same information is not available for *B. juncea* and therefore it is not included in synteny analysis. Chitinase genes were mapped to the syntenic blocks for inter-genomic

comparison. The syntenic diagram was drawn using Circos software version 0.63. Tandem duplications of chitinase genes in the C. sativa genome were identified based on their physical locations on individual chromosomes. Genes having an adjacent homologous ID gene on the same *C. sativa* chromosome with no more than one intervening gene was considered to be tandemly duplicated.

5. **Culture and Inoculation with *A. brassicae*:** To examine the induction of *B. juncea* and *C. sativa* chitinase genes, 40-days-old plants of both crop species were infected with *A. brassicae* strain. The *A. brassicae* culture was collected and was cultured at 22 °C for 20 days on radish dextrose agar. The conidia were taken from *A. brassicae* and suspended in sterile distilled water, and the muslin cloth was used for filtering, and diluted to 5×10^3 conidia/ml. Spore suspensions (4–6 drops) of *A. brassicae* (5×10^3 spores cm^{-3}) were inoculated on four different selected spots of the leaf surface of 40 days old *B. juncea* and *C. sativa* plants, and were then incubated in a chamber at 25 °C, with relative humidity 100%. Control plants for each treatment were treated with sterile distilled water. For RNA isolation leaf samples were harvested from control and infected plants after 6, 12, 24, and 48 h post-inoculation (hpi), and after flash frozen in liquid nitrogen were stored at −80 °C. Three different plants of both *B. juncea* and *C. sativa* were infected with *A. brassicae* on separate occasions to provide biological replicates for qRT-PCR analysis.
6. **DAB Staining (3, 3-Diaminobenzidine):** Leaves of 25-days-old *B. juncea* and *C. sativa* were used for DAB staining described by Thordal-Christensen et al. (1997). Leaves were stained with DAB (3,3′-Diaminobenzidine) for H_2O_2 production and were kept in dark for 8 h after incubation leaves were destained using bleach solution (ethanol:acetic acid:glycerol = 3:1:1) at 100 °C.
7. **Enzyme Assays:** Enzyme assays were carried out in *B. juncea* and *C. sativa* leaf extracts as per the method of Smith et al. (1988). The activity of SOD was determined through its ability to prevent nitro blue tetrazolium (NBT) photochemical reaction via the method of Beauchamp and Fridovich (1971). In 5 ml of 0.1 M Sodium Phosphate buffer (pH = 7) the 1 g of leaf tissue was ground and centrifuged at 12,000 rpm to get supernatant that was further used for enzyme assay's to the total reaction 1.110 ml of 50 mM phosphate buffer (pH 7.4), 0.040 ml of 1% (v/v) Triton X-100, 0.075 ml of 20 mM L-methionine, 0.075 ml of 10 mM hydroxylamine hydrochloride and 0.1 ml of 50 μM EDTA reaction was started by adding 100 μl of enzyme extract and 80 μl of riboflavin (50 μM). It was kept under illuminated W-Philips fluorescent lamps (60 μmol m^{-2} s^{-1}) for 10 min and the extract of peroxidase enzyme assay was prepared by homogenizing tissues in 50 mM potassium phosphate buffer (pH 7.8). The final volume of reaction contains 3 ml of 0.25% (v/v) guaiacol in potassium phosphate buffer (pH −7) containing 10 mM hydrogen peroxide reaction was started by adding 100 μl crude enzyme extract to the reaction mixture, which was measured spectrophotometrically at 470 nm. The last component to be added was H_2O_2 and the reaction was monitored by the decrease in absorbance at 290 nm (extinction coefficient of 2.8 mM^{-1} cm^{-1}) up to 5 min. The infected leaves were crushed in 50 mM Tris–NaOH buffer (pH 7.0) to prepare

catalase enzyme assay. In the final volume of 3 ml assay mixture contained 50 mM H_2O_2, 100 mM potassium phosphate buffer (pH 7.0), and 200 μl enzyme extract. The H_2O_2 decomposition was followed at 240 nm (extinction coefficient of 0.036 mM^{-1} cm^{-1}) by a decrease in absorbance for 5 min at 25 °C. Anderson (1996) method was followed for the assessment of GR activity, and catalase activity was expressed as μmol of H_2O_2 oxidized min^{-1} mg^{-1} protein. The chilled mortar and pestle were used to homogenize the leaf samples (0.1 g) in 5 ml of 50 mM Tris-HCl buffer (pH 7.6). At a frequency of 15,000 × *g* homogenate was centrifuged for 30 min at 4 °C and the supernatant obtained was further used for enzymatic reactions. Reaction was set up and 50 mM Tris-HCl buffer (pH 7.6), 1 ml NADPH (0.15 mM), 100 μl oxidized glutathione (1 mM GSSG), 3 mM $MgCl_2$, and 0.3 ml enzyme extract was added in a 2 ml centrifuge tube. Activity of GR was determined by measuring absorbance decline of NADPH at 340 nm and enzyme activity was expressed as NADPH oxidized μmol min^{-1} mg^{-1} protein.

8. **RNA Isolation and qRT-PCR Analysis:** A leaf sample of 100 mg was used for the extraction of total RNA needed of both control and treated *B. juncea* and *C. sativa* seedlings using PureLink RNA Mini Kit. RNA concentration and quality were measured by Nanodrop spectrophotometer. Superscript III cDNA synthesis kit according to the manufacturer's protocol was used to generate first strand cDNA from 2 μg of DNaseI-treated total RNA in 20 μl reaction volume. Primers of *B. juncea* and *C. sativa* chitinase genes as well as alpha-tubulin were designed using Oligoanalyzer software. qRT-PCR mixture was run for 5 min at 95 °C, followed by 40 cycles of 30 s at 94 °C, 30 s at 60 °C, and 30 s at 72 °C, and this mixture contained 2 μl of cDNA, 5 μl of SYBR green qRT-PCR master mix and 0.5 μl (10 picomol) of each primer. Each reaction was carried out in triplicates, and was replicated in three biological repeats. For internal control alpha-tubulin was used as house-keeping gene in all the experiments, and delta CT method was used to check relative expression levels of each gene. Significant level of expression was considered at fold changes with p values <0.05 (Mir et al. 2020).

8.4 *Brassica-Erysiphe*: Molecular Techniques

8.4.1 Identification of Molecular Markers Linked to Powdery Mildew R-Genes

1. **Collection of Plant Material:** The experimental material comprised of four females, viz., GM-1, GM-2, GM-3, and GDM-4 (Brassica juncea L.), and three males, viz., NRCY-05-02 (*B. rapa* L.), Neelam (*B. napus L.*), and Pusa Swarnim (*B. carinata* L).
2. **Generation of Hybrid Population:** Twelve F1s were developed by crossing four females with three males in an all possible crosses made (Excluding reciprocal). The seeds of all F1 hybrids were produced by hand emasculation and pollination

while the parental seeds were obtained through selfing. Subsequently, in the next year, a part of the F1 seeds were utilized for raising the F1 plants. The F1s were screened for powdery mildew incidence and selfed seeds of all the 12 hybrids were collected as F2 seeds. However, only one cross, that is, GM-3 × Pusa Swarnim was advanced to F2 generation due to its lowest disease score

3. **Development and Identification of Transgressive Segregants (F2):** The F2 seeds of the cross GM-3 × Pusa Swarnim was preserved for raising the F2 generation. F2 evaluation trial for powdery mildew resistance was conducted along with the parents, viz., GM-3 and Pusa Swarnim. The resistance plants were selected from the F2 population and the selected plants were used further for molecular analysis.
4. **Identification of DNA Markers:** A set of 165 A-, B-, and C-genome-specific SSR primers including functional SSR primer pairs were employed for the parental screening to detect the polymorphism between the resistant parent Pusa Swarnim and susceptible parent, GM-3. The SSR primers were custom synthesized from Eurofins Bengaluru. The genomic DNA of the two parents was subjected to PCR amplification with individual DNA markers. The amplicons scored as codominant-differences in size of the amplicon corresponding to the target loci was noted. The polymorphic markers were noted and later used in Bulk Segregant Analysis (BSA).
5. **DNA Extraction:** DNA from the F1 of the cross GM-3 × Pusa Swarnim along with the respective parents were extracted. The F1 was advanced to F2 generation and 210 plants were selected randomly for further molecular analysis (BSA). Leaves from 3 to 4 weeks old plants were collected from field, surface sterilized with 70% alcohol and freshly used for DNA isolation. The genomic DNA was extracted following Doyle and Doyle (1990) method with a few minor modifications.
6. **Assessment of Disease Infection:** Observations of powdery mildew severity was recorded at 15 days before maturity of randomly selected ten plants from each plot/row of F2 population. In each F2 individual plants were randomly selected and powdery mildew severity was recorded by referring the 0–9 scale (Mayee and Datar 1986).
7. **Bulked Segregant Analysis (BSA):** F2 population of the cross GM-3 × Pusa Swarnim was rose which had individuals segregating for disease resistance and susceptibility. Percent Disease Index (PDI) was calculated with the above scales using the formula given by Wheeler (1969). The PDI for individual plants were plotted each from two extremes, that is, resistant and susceptible was selected and their DNA was bulked. This was done with a view of gene interaction prevailing and thus resulting genotype-phenotype deviation expectable as a deviation from simple Mendelian situation (Govindaraj et al. 2005). The SSR marker found polymorphic between parents were screened across all the bulks and markers resulting in expected amplicon patterns were further shortlisted. Shortlisted markers were tested in the respective constituents of the bulks and selected ones were genotyped across all the F2 individual plants of a cross.

8. **Validation of Molecular Markers:** The 210 F2 individuals' plants of the cross GM-3 × Pusa Swarnim and parents were genotyped with shortlisted SSR markers through BSA. Marker segregation pattern was scored and recorded. The SSRs were scored as codominant marker type. After scoring the individual progenies the data was entered in Microsoft Excel spreadsheet and tested for goodness of fit through *Chi-square* analysis. The calculated $\chi 2$ values were compared with table values at one degree of freedom for each marker locus.

$$X2 = \sum_{i=n}^{k} \frac{(Oi - Ei)2}{Ei}$$

where $i = n = 1, 2, 3\ldots 210$ and $k = 1, 2, 3\ldots$ (Class of frequency), $O =$ Observed frequency of kth class, and $E =$ Expected frequency of kth class. The data sheet was saved in a format suitable for Single Marker Analysis (SMA). A similar approach has also been adapted previously in mustard and other crops where gene interaction prevailed (Pushpa et al. 2016; Govindaraj et al. 2005; Zhang et al. 2009; Gemenet et al. 2010; Milad et al. 2011; Kapadia et al. 2019).

8.4.2 DNA Sequence Analysis

DNA from the MGH isolate was prepared by a slight modification of the method of Pfister (1997). Briefly, conidia (about 0.1 ml dry vol) were ground in a microcentrifuge tube with a plastic pestle. The ground conidia were resuspended in 0.4 ml lysis buffer (200 mM Tris/HCl, pH 8.5, 250 mM NaCl, 25 mM EDTA, 0.5% SDS), and the suspension was treated as described by Pfister (1997). About 1 μl of the resulting DNA preparation was used for amplification by the polymerase chain reaction (PCR). The region of the ribosomal DNA containing the internal transcribed pacers (ITS) 1 and 2 was amplified using the primers ITS4 and NS7 (White et al. 1990), and the reaction conditions recommended by White et al. (1990) with an annealing temperature of 55 °C. The PCR products were sequenced without cloning utilizing ABI PRISM Dye Terminator chemistries, and ABI 377 DNA Sequencers. The primers ITS4, ITS2, NS8, NS7, ITS5, and ITS3 were used for DNA sequencing (White et al. 1990). Products from three independent PCR amplifications, one of which was performed on an independent DNA preparation, were sequenced to ensure that no PCR errors were obtained in the sequence. Analysis of the sequence data was carried out using Genetics Computer Group software, version 8 (Genetics Computer Group, Madison, Wisconsin), and the sequences of the ITS regions of the MGH isolate, and the *E. cichoracearum* UCSC isolate were compared using the GAOP algorithm (Plotnikova et al. 1998).

8.4.3 Embryo Rescue Technique to Transfer Powdery Mildew Resistance

Fully developed flower buds were emasculated prior to anthesis, and allowed to bloom. Fresh pollen grains from newly opened flowers were used to pollinate emasculated flowers. For embryo rescue, pistils were harvested 5–7 days after pollination, washed in 70% ethanol for 1 min, washed in 10% sodium hypochlorite solution for 10 min, and then rinsed twice with sterile distilled water. Ovaries were cultured for 7 days in hormone-free MS medium (Murashige and Skoog 1962) supplemented with 5% sucrose, 0.8% agar, and 500 mg/l casein hydrolysate (MS1) under continuous illumination with fluorescent lights (115 lmol/m^2/s). Pistils were cut open aseptically, and ovules were transferred to MS medium with 3% sucrose, and 0.8% agar without casein hydrolysate (MS2). All ovules were incubated under fluorescent lights with 16 h photoperiod, and developed ovules were transferred to individual plant boxes. Well-established seedlings were transferred to plastic pots containing Cornell mix (Boodley and Sheldrake 1972), covered with plastic bags, and placed in growth chambers for acclimatization with 6 h photoperiod day/night for two cycles per day with fluorescent lights. Fully developed plants were transferred to a greenhouse where they were grown to maturity (Tonguc and Griffiths 2004).

8.4.4 Identification of Molecular Markers Linked to Powdery Mildew R-Genes

1. **Collection of Plant Material:** The experimental material was comprised of four females, viz., GM-1, GM-2, GM-3, and GDM-4 (Brassica juncea L.), and three males, viz., NRCY-05-02 (*B. rapa* L.), Neelam (*B. napus L.*), and Pusa Swarnim (*B. carinata* L.).
2. **Generation of Hybrid Population:** Twelve F1s were developed by crossing four females with three males in an all possible crosses made (Excluding reciprocal). The seeds of all F1 hybrids were produced by hand emasculation and pollination while the parental seeds were obtained through selfing. Subsequently, in the next year, a part of the F1 seeds were utilized for raising the F1 plants. The F1s were screened for powdery mildew incidence and selfed seeds of all the 12 hybrids were collected as F2 seeds. However, only one cross, that is, GM-3 × Pusa Swarnim was advanced to F2 generation due to its lowest disease score
3. **Development and Identification of Transgressive Segregants (F2):** The F2 seeds of the cross GM-3 × Pusa Swarnim was preserved for raising the F2 generation. F2 evaluation trial for powdery mildew resistance was conducted along with the parents, viz., GM-3 and Pusa Swarnim. The resistance plants were selected from the F2 population and the selected plants were used further for molecular analysis.
4. **Identification of DNA Markers:** A set of 165 A-, B-, and C-genome-specific SSR primers including functional SSR primer pairs werc employed for the

parental screening to detect the polymorphism between the resistant parent Pusa Swarnim and susceptible parent, GM-3. The SSR primers were custom synthesized from Eurofins, Bengaluru, India. The genomic DNA of the two parents was subjected to PCR amplification with individual DNA markers. The amplicons scored as codominant-differences in size of the amplicon corresponding to the target loci was noted. The polymorphic markers were noted and later used in Bulk Segregant Analysis (BSA).

5. **DNA Extraction:** DNA from the F1 of the cross GM-3 × Pusa Swarnim along with the respective parents were extracted. The F1 was advanced to F2 generation and 210 plants were selected randomly for further molecular analysis (BSA). Leaves from 3 to 4 weeks old plants were collected from field, surface sterilized with 70% alcohol, and freshly used for DNA isolation. The genomic DNA was extracted following Doyle and Doyle (1990) method with a few minor modifications.
6. **Assessment of Disease Infection:** Observations of powdery mildew severity was recorded at 15 days before maturity of randomly selected ten plants from each plot/row of F2 population. In each F2 individual plants were randomly selected and powdery mildew severity was recorded by referring the 0–9 scale (Mayee and Datar 1986).
7. **Bulked Segregant Analysis (BSA):** F2 population of the cross GM-3 × Pusa Swarnim was raised which had individuals segregating for disease resistance and susceptibility. Percent Disease Index (PDI) was calculated with the above scales using the formula given by Wheeler (1969). The PDI for individual plants were plotted each from two extremes, that is, resistant and susceptible was selected and their DNA was bulked. This was done with a view of gene interaction prevailing and thus resulting genotype-phenotype deviation expectable as a deviation from simple Mendelian situation (Govindaraj et al. 2005). The SSR marker found polymorphic between parents were screened across all the bulks and markers resulting in expected amplicon patterns were further shortlisted. Shortlisted markers were tested in the respective constituents of the bulks and selected ones were genotyped across all the F2 individual plants of a cross.
8. **Validation of Molecular Markers:** The 210 F2 individuals' plants of the cross GM-3 × Pusa Swarnim and parents were genotyped with shortlisted SSR markers through BSA. Marker segregation pattern was scored and recorded. The SSRs were scored as codominant marker type. After scoring the individual progenies the data was entered in Microsoft Excel spreadsheet and tested for goodness of fit through *Chi-square* analysis. The calculated $\chi 2$ values were compared with table values at one degree of freedom for each marker locus.

$$X2 = \sum_{i=n}^{k} \frac{(Oi - Ei)2}{Ei}$$

where $i = n = 1, 2, 3...210$ and $k = 1, 2, 3...$ (Class of frequency), $O =$ Observed frequency of kth class and $E =$ Expected frequency of kth class. The data sheet was saved in a format suitable for Single Marker Analysis (SMA). A similar approach has also been adapted previously in mustard and other crops where gene interaction prevailed (Pushpa et al. 2016; Govindaraj et al. 2005; Zhang et al. 2009; Gemenet et al. 2010; Milad et al. 2011; Kapadia et al. 2019).

8.4.5 Molecular Identification of Anamorphic Powdery Mildews (Erysiphales)

The powdery mildew fungi (Erysiphales) are a common group of obligate plant pathogens that can be extremely difficult to identify, depending on the reproductive stage encountered. The ribosomal DNA Internal Transcribed Spacer (ITS) region was assessed for its usefulness to link anamorphic Erysiphales with their respective teleomorphs. PCR primers for the rDNA ITS region were designed, and found to have enhanced specificity for the 12 genera tested, even in the presence of contaminating fungi. Recent herbarium specimens of both anamorphic and teleomorphic material yielded sufficient DNA for amplification using a Chelex-based extraction method. The ITS regions from 25 anamorphic specimens were sequenced, and compared with the ITS sequences of their suspected teleomorphs. In most cases, an ITS region sequence similarity of above 99% indicated that a correct match had been made. Although this technique will not always unambiguously identify an anamorphic specimen, it will provide valuable information to use in conjunction with morphological, and host range data to aid in the final identification (Cunnington et al. 2003).

8.5 *Brassica-Hyaloperonospora*: Molecular Techniques

8.5.1 Assessment of Small RNA Role in *Brassica* to *Hyaloperonospora*

1. **Plant Lines, Pathogen Isolates, and Propagation:** *H. arabidopsidis* isolates Emoy2 and Cala2 were maintained on *A. thaliana* Ws-eds1 (Parker et al. 1996). Preparation of inoculum for experiments was performed as described previously (Tor et al. 2002). Sporulation was assessed 7 dpi, when the *Hpa* life cycle had been completed. To quantify sporulation, ten infected seedlings from each replicate were taken and placed into an Eppendorf tube containing 250 μl H_2O. Samples were vortexed and conidiospores were counted using a hemocytometer.
2. **sRNA and DNA Oligonucleotide Synthesis:** The *H. parasitica-CesA3* gene (*HpaG810051*) was used as the target gene in this method. Sense and antisense sRNAs designed at various lengths were *Hpa*_CesA3_RNA_AS_24 5-′-GCCGCAUCGCACGUACCUCAGUAC-3′, *Hpa*_CesA3_RNA_AS_25 5′-GCCGC AUCGCACGUACCUCAGUACG-3′, *Hpa*_CesA3_RNA_S_25 5-′-CGUACUGAGGUACGUGCGAUGCGGC-3′, and Hpa_CesA3_RNA_S_30 5′-GUCGUACUGAGGUACGUGCGAUGCGGCACU-3′. *Hpa*_CesA3_

RNA_ AS_30 5′-AGUGCCGCAUCGCACGUACCUCAGUACGAC-3′ and double-stranded sRNA were generated by mixing equal volume of *Hpa*_CesA3_RNA_S_30 and *Hpa*_CesA3_RNA_AS_30 oligos, heating at 95 °C for 5 min and allowing to anneal at room temperature for 20 min. Similarly, sense and antisense DNA oligonucleotides were also designed to the same region. These were Hpa_CesA3_DNA_S_25 5-′-GCCGCATCGCACGTACCTCAGTACG-3′ and *Hpa*_CesA3_DNA_AS_25 5′-CGTCT GAGGTACGTGCGATGCG GC-3′. These were obtained as synthesized deoxyribonucleotides or ribonucleotides from Sigma or Eurofin.

3. **Application of sRNAs to Pathogen Spores and Plant Inoculations:** *Hpa* spores were collected from infected *A. thaliana* Ws-eds1 seedlings, washed twice in sterile distilled water, and the spore concentration was adjusted to 5 × 10^4/ml using a hemocytometer. sRNAs were added to the spore suspension at a final concentration of 5, 10, and 20 μM and spores were subsequently drop inoculated onto 7-day-old seedlings. As a control, seedlings were also inoculated with spores in the same way with 20 μM DNA oligo-nucleotides or without sRNAs or DNA oligonucleotides. Seedlings were inspected from 3 dpi for sporulation. Sporulation was quantified as described above.
4. **Spore Germination Assays:** MS medium (Murashige and Skoog 1962) was prepared with 4.3 g/l MS basal salt mixture powder, agar (1.5%), sucrose (10 g/l), and distilled water. MS powder was dissolved in sterile distilled H_2O_2 the pH was adjusted to 5.7 using 1 M NaOH/HCl, and agar and sugar were added. The medium was sterilized by autoclaving at 15 psi and 121 °C for 15 min. Approximately 20 ml of the medium was aliquoted into each sterile Petri dish in a laminar airflow unit.
   Cellophane strips, 1.5 cm in length, were cut from plain transparent florists' cellophane and autoclaved in distilled water. After autoclaving, cellophane strips were placed onto the MS medium in Petri dishes under a laminar airflow and dishes were kept in the fridge for long-term storage. *Hpa* spores were collected, washed twice in sterile distilled water, and the spore concentration was adjusted to 5 × 10^4 spores/ml using a hemocytometer. Approximately 10 μl spore suspension, with 0 or 20 μM antisense sRNA, was dropped onto each piece of cellophane. Plates were incubated with a 12 h light/12 h dark regime at 16 °C. Spores were examined under a light microscope 48 h after incubation and germinated spores were counted.
5. **Staining Plant Tissues:** Seedlings of infected and non-inoculated controls were stained with a solution of phenol, lactic acid, glycerol, and water (1:1:1:1) supplemented with 1 mg/ml trypan blue, decolorized in chloral hydrate and visualized under a compound microscope as described in Woods-Tor et al. (2018).
6. **Statistical Analysis:** For statistical analysis, paired Student's *t*-tests were performed on data obtained from plant infection assays.
7. **Bioinformatics:** IICB Genomics and Transcriptomics Resources (http://eumicrobedb.org) and the Ensembl Protist (http://protists.ensembl.org) database were used for information on Hpa. Web servers including InterPro (Quevillon

et al. 2005) (http://www.ebi.ac.uk/interpro/) and Pfam (Punta et al. 2012) (http://pfam.wustl.edu/) were used for the analysis of Hpa-CesA3. BLAST (Altschul et al. 1997) was used to perform similarity-search of nucleotide and amino acid sequences of *Hpa-CesA3* against oomycete and *Arabidopsis* sequences. Primer design was performed using Geneious v. 10.0 (Kearse et al. 2012; Bilir et al. 2019).

8.5.2 Identification of R-Genes Overexpression in *Brassica-Hyaloperonospora* Pathosystem

1. **Plant Materials:** A mapping population was derived from a cross between BY and MM. BY is a Chinese cabbage in-bred line (*B. rapa* ssp. *pekinensis*) that was self-fertilized for eight generations. MM is a European turnip line (*B. rapa* L. ssp. *rapifera*) that was self-fertilized for six generations. The resulting F1 plants were subjected to self-pollination and microspore culture to produce an F2 segregating population and a double haploid (DH) population with 80 DH lines. The method of embryogenesis and doubling by isolated microspore culture in Chinese cabbage was described in Wang et al. (2010).
2. **Pathogen Inoculation and Evaluation of Resistance and Susceptible Phenotypes in the Mapping Populations**: The seeds of the F2 and DH mapping population were germinated on wet filter paper at 25 °C for 24–48 h. The seedlings were then transferred into soil and were grown at 25 °C with 16 h light followed by 20 °C in an 8 h dark cycle. For the DH population, three inoculation replicates were conducted, with 10 plants per replicate ($n = 30$). Two-week-old seedlings were used for pathogen inoculation. The pathogen isolate used in this study and the method of inoculum maintained are described in Yu et al. (2009). As described previously (Yu et al. 2009), about 2×10^5 spores μl^{-1} of conidial suspension of the *P. parasitica* isolate were sprayed on the abaxial side of leaves. After inoculation, the seedlings needed dark conditions and high humidity (nearly 100%) during the first 24 h and high humidity with 16 h light/8 h dark for subsequent 3–4 days for pathogen invasion. Inoculated plants were grown under 100% humidity and dark conditions at 20 °C for 24 h. After 1 week high humidity treatment, phenotypic data were collected. For the F2 population, the susceptible phenotype was evaluated according to the rating system described by Yu et al. (2009). Plants in classes 0, 1, and 3 with no sporulation observed were considered resistant (R), and those in classes 5, 7, and 9 with obvious sporulation were classified as susceptible (S). For the DH population, the disease indices (DIs) for each line were calculated as described in Yu et al. (2009). For further molecular marker analysis, young leaves were collected and DNA was subsequently extracted as described by Murray and Thompson (1980), with minor modifications.
3. **Molecular Marker Analysis:** Genomic DNA of the parental lines was re-sequenced and the results were compared to the reference sequence of Chiifu-401-42 (v 1.5). Sequence differences between the parental line in the

target QTL region were used to develop SNP markers, designed and synthesized in Britain for genotyping using the Kompetitive Allele-Specific PCR (KASP) method, as described by Saxena et al. (2014). For each SNP, two allele-specific forward primers and one common reverse primer were designed and tested by LGC. The PCR reactions of 20 μl contained 2 μl PCR buffer, 0.8 μl dNTPs (2.5 mM each), 1.0 μl oligonucleotide primers each (10 mM), 0.2 μl Taq DNA polymerase (2.5 U/μl), 2 μl DNA template (50 ng/μl), and 13 μl ddH_2O. The amplifications were performed in a PTC-200 Thermal-cycler using the following program: 95 °C for 5 min followed by 35 cycles of 95 °C for 30 s, 55 °C for 30 s, and 72 °C for 30 s, with a final extension at 72 °C for 5 min. Fluorescent detection of the reaction products was performed using an Omega Fluorostar scanner, and the data were analyzed using KlusterCaller 1.1 software (KBioscience). A total of 14 SNP markers were screened to combine with the phenotypes of F2 population and locate the *Br-DM04* to a 160-kb region.

4. **Linkage Map Construction and QTL Analysis:** The linkage map was constructed with Join Map 4.0 software (Van Ooijen 2006). The details of linkage map construction were identical to Su et al. (2018). The Kosambi mapping function was used to calculate genetic linkage distances (Kosambi 1944). Logarithm of odds difference (LOD) thresholds for determining significant QTLs were estimated from 1000 permutation tests (Churchill and Doerge 1994). The genome-wide threshold for downy mildew resistance was 3.0, and interval mapping was used to identify putative QTLs (Lander and Botstein 1989; Young 1996). Subsequently, molecular markers coinciding with, or closely flanking, the maximum QTL LOD were used as cofactors in multiple QTL analysis. All calculations employed Map QTL 3.0 software (Van Ooijen et al. 2002). In order to provide visual images of marker loci genomic positions, integrated markers and QTL were plotted using Map chart 2.2 (Voorrips 2002).
5. **Gene Cloning and DNA Sequence Analysis:** The primers used for amplification of genomic DNA sequences of *BrRLP47, BrRLP48, BrLRR1*, and *BrLRR2* were designed using reference sequences identified from the *Brassica* database1. The PCR products were then cloned into the pCR8/GM/TOPOTM entry vector and sequenced. Sequence similarities were calculated using DNAMAN version 6. The coding sequences of *BrRLP47, BrRLP48, BrLRR1*, and *BrLRR2* were cut by using Hind III and Kpn I restriction sites at 50- and 30-terminals, respectively. Subsequently, they were cloned into the pSuper1300 vector fused with a GFP tag at the C-terminus (Gong et al. 2002).
The pYBA1311 vector was used for the RNA interference (RNAi) analysis (Yan et al. 2012). In *BrRLP47* CDS, a highly specific fragment containing 169 bp was amplified with two pairs of adaptors (Kpn I/Xho I and Cla I/BamH I). To generate hairpin RNA, the fragment was inserted into Kpn I/Xho I sites and Cla I/BamHI sites of pYBA1311 to form a pair of reverse complement sequences along the transcript direction. A 153 bp fragment in the CDS sequence of *BrRLP48* was constructed into pYBA1311 vector in the same way. Then these recombinant plasmids were used to transform into *Agrobacterium tumefaciens* strain GV3101 for further agro-infiltration.

6. **RNA Extraction and Real-Time PCR Analysis:** Total RNA was isolated from leaves with the plant RNeasy kit. RNA was reverse transcribed into cDNA using a Prime Script TM RT Reagent Kit. Real-time PCR reactions were performed using the SYBR Green I Master Mix and were quantified with the Light Cycler 480 II. Reaction conditions were as follows: an initial denaturation at 95 °C for 3 min, followed by 40 cycles of 95 °C for 15 s (denaturation), 60 °C for 30 s (annealing), and 72 °C for 45 s (extension). PCR amplification was followed by heating for 1 min at 60–95 °C for melting curve analysis. Each sample reaction was performed with three replicates using 5 ml of Master Mix, 0.25 mM of each primer, 1 ml of diluted cDNA, and DNase free water to a final volume of 10 ml. The PCR products were sequenced to confirm the gene-specific amplifications.
7. **Expression Analysis of Candidate-Resistant Genes After Inoculation**: Two-week-old seedlings of disease resistant and susceptible parents MM and BY were used for pathogen inoculation. Thirty plants of MM and BY were used ($n = 30$). Under the same high humidity condition, the leaves with inoculated and non-inoculated leaves were sampled at 0, 2, 6, 12, 24, and 48 h after inoculation. Ten to 15 leaves were mixed to one sample. Then RNA of these samples was extracted for qRTPCR analysis. Three technical replicates were performed. The primers of *BrRLP47*, *BrRLP48*, *BrLRR1*, and *BrLRR2* for qRTPCR analysis were designed by Primer Premier 5 (Lalitha 2000).
8. **Agroinfiltration:** *Agrobacterium tumefaciens* strain GV3101 was used for transient expression studies. The cotyledons of 7-day-old seedlings were used in the Agroinfiltration experiment. Methodology was employed following Liu et al. (2010) with minor modifications. The *Agrobacterium* strains including gene silencing suppressor p19 were grown at 28 °C to an OD600 of approximately 2.5. Then the bacteria pellets were resuspended in 10 mM $MgCl_2$. The p19 and each strain were mixed with to final OD600 of 0.1 and 0.2. Final concentration of 10 mM Acetosyringone and 200 mM 2-(*N*-morpholine)-ethanesulfonic acid (MES, pH 5.6) were added to the mixture, which was kept at room temperature for at least 3 h without shaking before infiltration. Cotyledon injection was performed by depressing a 2 ml syringe to the back of leaf till the mixture was infiltrated to the whole leaf. Cotyledons without injection and with empty vector infiltrated were conducted as the controls. Overexpression vector with GFP tag was infiltrated into the cotyledons of susceptible parent BY and RNAi vector was injected into the cotyledons of resistant parent MM. Three days after infiltration, GFP fluorescence was observed under the confocal microscope and 5–10 injected cotyledons were mixed to each sample for qRT-PCR analysis. The RNA samples in three biological replicates were employed for qRT-PCR analysis.

   Subsequently, pathogen was inoculated into these 3-day infiltrated cotyledons. Keeping high humidity for 5–7 days after inoculation, the disease-resistant phenotype of the infiltrated cotyledons was investigated. The infected cotyledon, of which more than a half leaf area with observed sporulation, was considered to be the susceptible cotyledon. The expression levels of injected genes and the percent of susceptible cotyledons were derived from four independent experiments.

9. **SA and JA Treatments:** Eighteen-day-old BY and MM seedlings with two or three fully expanded leaves were subjected to SA and JA treatments. SA at 1.5 mM and 1 mM Me-JA dissolved in water were sprayed on the abaxial side of the leaves. Seedlings sprayed with water were designated as the controls. The leaves were sampled at 0, 2, 4, 6, 8, 12, 24, 48, 72, 96, and 120 h for detection of gene expression. Five to eight leaves were pooled in each sample. Three parallel experiments were performed as the biological replicates in further qRT-PCR analysis. The ratio of gene expression under phytohormone treatment to that under water treatment was considered to be the gene induction level by the phytohormone.
10. **Phytohormone Measurement:** The leaves of 20-day-old BY and MM seedlings were placed under high humidity and dark condition for 6 h and then pathogen inoculation was performed. The samples were collected at 0, 4, and 12 h after pathogen inoculation (the inoculation method is described above). For each sample for SA and JA measurements, approximately 50 mg of leaf tissue from 5 to 10 seedlings was needed. Three biological repetitions were made at each time point, each consisting of a mix of five seedlings. After freezing in liquid nitrogen, the samples were stored at −80 °C until further analysis. Each sample was ground to a fine powder in liquid nitrogen. The 50 mg powder was extracted in 15 ml cold 100% methanol followed by sonication in ultrasonic water bath for 30 min at frequency 40 kHz (25 °C). Then the mixture was centrifuged at 18,000 × *g* for 10 min at 25 °C. The supernatant was filtered through 0.2 μm membrane and dried under vacuum condition. The residue was dissolved in 400 μl of H_2O/methanol (1:1, v/v). Samples were centrifuged (10 min at 18,000 × *g*), and 10 μl of the aqueous phase was retained for analysis. The external standard method was used for quantification, in which internal standard SA (CAS: 69-72-7) and JA (CAS: 3572-66-5) were diluted in methanol to the concentrations of 1, 2, 5, 10, 20, 50, and 100 ng/ml to draw standard curve. The contents of free SA and JA were measured using a Shimadzu UHPLC system equipped a LC-30AD solvent delivery system, a SIL-30AC auto sampler, a CTO-30A column oven, a DGU-20A3 degasser, and a CBM-20A controller (Zhang et al. 2018).

8.5.3 CDNA-AFLP Analysis to Reveal Gene Expression

1. **Plant Material, Inoculum, and Pathogen Infection:** Two non-heading Chinese cabbage inbred lines, “Suzhou Qing” (resistant to *H. parasitica*) and “Aijiao Huang” (susceptible to *H. parasitica*), were used. The transcript-derived fragments (TDFs) were obtained from interaction between “Suzhou Qing” and downy mildew (*H. parasitica*). “Aijiao Huang” was used for the comparison of expression patterns of the four genes related with fungal resistance between resistant and susceptible line. Plants were grown in plastic nurseries (inner size:

45 × 45 mm; height: 57 mm) and transferred to a growth chamber under 25 °C day/20 °C night temperature with 85 ± 5% relative humidity and a 12 h light/12 h dark after germination for 36 h under dark. *H. parasitica* was isolated from leaves of susceptible line "Aijiao Huang." Conidial suspensions were adjusted to 1 × 10^5 spores per ml and Tween-20 was added as a surfactant to a final concentration of 0.1%. One hundred of 3-week-old seedlings (with four true-leaves) were sprayed with 50 ml pathogen suspension and distilled water (as control), respectively. After inoculation, the seedlings were covered with plastic film separately and transferred to a growth chamber under 20 °C, 100% relative humidity in the dark for the first 24 h to promote sporulation, then moved back to the initiatory conditions. Both control and treated third leaf of five plants were harvested and pooled at 0, 24, 48, and 72 hours post inoculation (hpi), immediately frozen in liquid nitrogen and stored at −70 °C until use.

2. **RNA Isolation and cDNA-AFLP Analysis:** Total RNAs were extracted using the RNAeasy Plant Mini kit (Qiagen; https://www.qiagen.com/cn/shop/sample-technologies/rna/rna-preparation/rneasymini-kit#orderinginformation) and synthesis of the first strand of cDNA is made using the M-MLV reverse transcriptase. To synthesize the second strand, the following components were added to the first-strand solution. A volume of 30 μl, 5 × second strand synthesis buffer, 3 μl dNTP mixture, 89 μl RNase-free H_2O, 2 μl *Escherichia coli* DNA polymerase I, 2 μl *E. coli* RNase H/*E. coli* DNA ligase mixture, and 4 μl T4 DNA polymerase in a final volume of 150 μl. The components were gently mixed and incubated at 16 °C for 2 h. Double-stranded cDNA was purified using the DNA Fragment Purification Kit Ver.2.0. Second-strand cDNA was digested by two restriction enzymes Tag I (restriction site TCGA) and Ase I, and then ligated to Tag I and Ase I double-strand adaptors. The AFLP adaptor primers 5′-GACGATGAGTCCTGAC-3′, 5′-CGGTCAGGACTCAT-3′ (Taq I-adaptor primers) and 5′-GCGTAGACTGCGTACC-3′, 5′-TAGGTACGCAGTC-3′ (Ase I-adaptor primers) were ligated onto the restriction fragments: Taq I pre-amplification primer, 5′-GACGATGAGTCCTGACCGA-3′; Ase I pre-amplification primer, 5-CTCGTAGACTGCGTACCTAAT-3′; Taq I selective amplification primer, 5′-GATGAGTCCAGACCGA+NN-3′; Ase I selective amplification primer, 5′-GACTGCGTACCTAAT+NN-3′ (indicated by N, representing an A, C, G, or T). The initial small-scale screen using 96 AFLP primer combinations were done using six Taq I forward selective amplification primers (extension CG, CA, CT, CC, GA, or GT) in combination with 16 Ase I reverse selective amplification primers (extension NN), respectively. Pre-amplification PCR was carried out with one-tenth volume of the restriction/ligation mix, the pre-amplification PCR was carried out as follows: 94 °C, 3 min; 94 °C, 30 s, 55 °C, 30 s, 72 °C, 60 s, 25 cycles; and 72 °C, 5 min. The products of pre-amplification was diluted tenfold, and the selective amplification PCR was carried out as follows: 94 °C, 30 s; 94 °C, 30 s, 65 °C, 30 s (−0.7 °C per cycle), 72 °C, 60 s, 12 cycles; 94 °C, 30 s, 56 °C, 30 s, 72 °C, 60 s, 24 cycles; and 72 °C, 5 min.

Selective amplification products were separated on a 6% polyacrylamide gels running at 60 W for 2 h and visualized by silver staining. Differential bands were excised from the polyacrylamide gel electrophoresis gels based on the alignment between films and markers on the gels, and incubated in 30 μl of water and then at 95 °C for 30 min. The TDFs were then re-amplified by PCR using same primers under the similar conditions. The amplified fragments were retrieved from a 1% agarose gel with the Sephaglas BandPrep kit, cloned into pGEM-T Easy vector (http://en.cellfood.com.cn/culture.aspx), and sequence information was BLASTed in the *Brassica* database (http://brassicadb.org/brad/).

3. **Quantitative Real-Time PCR:** The single-strand cDNA of resistant line "Suzhou Qing" and susceptible line "Aijiao Huang" were diluted to 30 ng μl^{-1}, and were used for quantitative reverse transcription PCR (qRT-PCR) analyses. Primers were designed by the Primers 3 (http://frodo.wi.mit.edu/primer3/) based on the interested cDNA sequence. The qRT-PCR reaction mixtures contained 12.5 μl, 2 × SYBR Green PCR Master Mix, 10 pM of each primer, 2 μl template and sterile distilled water to total volume of 25 μl, as well as also performed on CFX96 Real-Time System. Thermal conditions were 2 min of denaturation at 95 °C, followed by 45 cycles of 95 °C for 10 s, annealing at 55 °C for 20 s, and extension at 72 °C for 20 s and 72 °C for 5 min. Three technical replicates were analyzed for each biological replicate. All the cycle threshold (Ct) values from one gene were determined at the same threshold fluorescence value of 0.2 using the ΔΔCt method (Xiao et al. 2016).

8.6 *Brassica-Leptosphaeria*: Molecular Techniques

8.6.1 Phylogenetic Relationship of R-Loci in *Brassica* to *Leptosphaeria*

1. **Sample Collection, DNA Extraction, and PCR Amplification:** Forty-five plant materials were used in this analysis. The samples comprised 15 accessions each of *B. rapa* and *B. oleracea* var. *virids* from the USDA germplasm collections, and 15 cultivars of *B. napus* from the National Winter Canola Variety Trials. More detailed information on the accessions is available at the USDA-ARS Germplasm Resources Information Network available online at (http://www.arsgrin.gov/npgs). Genomic DNA was extracted from leaves of 5-week-old seedlings according to Edwards et al. (1991). Thirteen polymorphic RAPD primers, OPB01, OPB05, OPC08, OPE03, OPE11, OPE12, OPE14, OPE16, OPF02, OPF10, OPG02, OPT01, and OPI01 mapped to Chromosome 4 of *B. nigra* (Chevre et al. 1997), were selected for this study. The selected markers were obtained from MWG-Biotech PCR reactions were carried out in 25 μl volumes comprising the following: 0.4 mM of each dNTP, 1.5 mM $MgCl_2$, 1.25 units of gold Taq DNA polymerase, 0.4 pmol of each primer, 1× PCR buffer, and 25 ng DNA template. The amplifications were performed on a Peltier Thermal Cycler. Amplifications comprised 45 cycles each having the following repeated

sequence: Denaturation at 94 °C for 1 min, annealing at between 32 and 34 °C for 1 min depending on the marker, extension at 72 °C for 2 min. A final extension at 72 °C for 7 min was applied to all reactions. Amplified PCR products, mixed with a loading dye (0.25% bromophenol blue, 0.25% xylene cynol FF, 15% Ficoll) to increase its density and aid visualization, were loaded onto 1% agarose. A Bench-top 1 kb DNA ladder was loaded alongside the PCR products for size calling. The fragments were separated under electric current for 4 h in a standard horizontal electrophoresis unit, followed by a 15 min ethidium bromide staining. Gels were de-stained and visualized under UV light. Gel images were captured using AlphaImager v5.5 2000.

2. **Statistical Data Analyses:** Amplified DNA fragments were evaluated for size in base pairs using Alpha Ease FC version 4.0.1. Polymorphic information content (PIC) values were calculated for each primer as demonstrated by Botstein et al. (1980) and also by using the online resource, Polymorphism Information Content Calculator (http://www.agri.huji.ac.il/~weller/Hayim/parent/PIC.htm). PIC compares the polymorphism levels across markers (Hongtrakul et al. 1977; Garcia-Mas et al. 2000) and is used to determine the usefulness of markers for specific studies. Polymorphic bands were scored as binary data (present = 1; absent = 0) across the genotypes to generate a binary data matrix. The matrix was then used to generate a Genetic similarity (GS) matrix based on Jaccard's coefficient of similarity (Jaccard 1908). So, GS (ij) = $2a/(2a + b + c)$, where GS (ij) is the measure of genetic similarity between individuals i and j, a is the number of polymorphic bands that are shared by i and j, b is the number of bands present in i and absent in j, and c is the number of bands present in j and absent in i. The matrix of Genetic similarity was used to construct a dendrogram using the unweighted pair group method with arithmetic averages (UPGMA) algorithm (Ananga et al. 2008).

8.6.2 Transcriptome Analysis of *Brassica-Leptosphaeria* Pathosystem

1. **Plant Growth and Inoculation:** Susceptible Topas DH16516 and the *Rlm2* transgenic line NLA51-1 (Larkan et al. 2015) were grown in Sunshine mix #4 soil in controlled environments with a 16-h photoperiod (21 °C light, 16 °C dark). Seven-day-old cotyledons were point inoculated with 10 μl of a *L. maculans* isolate 00-100 (*AvrLm2*; Ghanbarnia et al. 2015), pycnidia suspension (106), or sterile water (mock). All inoculations and tissue collections were performed between 11:00 AM and 2:00 PM to minimize time of day effects. To explore the role of calcium in resistance, 10 μl of 5 mM EGTA was added to the inoculation site as an additional treatment and allowed to dry before inoculation.
2. **Tissue Processing:** Inoculated cotyledons were processed for LMD at 1 and 3 dpi per the methods of Becker et al. (2017b). Inoculation sites were excised by making two cuts parallel to the cotyledon petiole-like structure. A minimum of 30 inoculation sites per biological replicate (three biological replicates per

treatment) were collected and fixed overnight in 3:1 (v:v) ethanol:acetic acid at 4 °C. Tissues were then dehydrated in a graded ethanol series (30 min each: 75%, 85%, 95%, 100%, 100%) followed by xylenes infiltration (1 h each: 3:1, 1:1, 1:3 ethanol: xylenes (v: v), 100% xylenes) at 4 °C. Ten paraffin chips were added to the xylenes and kept at 4 °C overnight. Following, samples were brought to room temperature before incubating at 42 °C for 30 min and then at 58 °C for 1 h. The mixture was replaced with 100% melted paraffin every hour for 4 h before embedding.

3. **Sectioning and Laser Micro-dissection:** Embedded tissues were sectioned at 10 μm using a Leica RM2125 rotary microtome in RNase-free conditions and mounted on PEN membrane slides as per Millar et al. (2015). Slides were washed for 1 min in xylenes to deparaffinize before proceeding to LMD on a Leica LMD6000. Sections of 100 μm proximal to the inoculation site were collected in 60 μl of lysis buffer. RNA was isolated from at least 20,000,000 μm^2 of collected tissue per biological replicate. RNA extractions were performed exactly as reported in Belmonte et al. (2013). Following, RNA quality was assessed using the Agilent 2100 Bio-analyzer, and were of sufficient quality for qRT-PCR and RNA-seq as described in Millar et al. (2015).
4. **Library Construction:** First and second strand cDNA synthesis was performed per the alternative HTR (C2) described in Kumar et al. (2012) with the following modifications: Input RNA decreased to 50–75 ng; no Poly (A) bead isolation was performed; anchored oligo (dT20) primers were used during first strand synthesis instead of random hexamers; half reaction volumes for all steps; cDNA sheared via sonication instead of enzymatic digestion. Libraries were prepared using the NEB Next Ultra II kit per the manufacturer's instructions. Quality and distribution of libraries was assessed with the Agilent 2100 Bioanalyzer, and libraries were quantified with the NEBNext Library Quant Kit for Illumina and the NanoDrop 3300 fluorospectrometer. Library size selection of the 300–500 bp regions was performed with the E-gel size select system. Sequencing was performed on the HiSeq 2500 at Genome Prairie, SK, Canada.
5. **Data Analyses:** RNA sequencing of libraries produced ~423 million paired end reads, with an average of 17.6 million reads/sample (PRJNA477556). Adapter sequences, low-quality bases ($Q < 25$), and low-quality reads (avg. $Q < 30$) were removed using the Trimmomatic v0.36 software (Bolger et al. 2014). Paired-end reads passing quality control were aligned to the *B. napus* genome v4.1 (Chalhoub et al. 2014) with HISAT2 in cufflinks compatibility mode, allowing no more than two mismatches. Output SAM files were converted to BAM and sorted with Samtools. Raw counts were obtained from BAM files using the HTSeq Python Framework (Anders et al. 2015). Dendrograms and similarity matrices were generated from raw counts using DESeq (Love et al. 2014). From BAM files, Cuffquant, Cuffnorm, and Cuffdiff (Trapnell et al. 2012) were used to generate normalized counts in FPKM and to identify significantly differentially expressed genes (DEGs; $P < 0.05$; false discovery rate $= 0.05$). Differentially expressed gene sets were input into SeqEnrich (Becker et al. 2017a) to identify enriched metabolic pathways, GO terms, promoter TF binding motifs, and

predictive transcription factor networks. Networks were visualized in Cytoscape 3.6.0 (www.cytoscape.org). *L. maculans* transcriptome analyses were performed per Haddadi et al. (2016) and all reads were trimmed as above and mapped to the *L. maculans* isolate v23.1.3 genome (Rouxel et al. 2011). Transcript abundance was measured as Log2 RPKM (Reads Per Kilobase of exon per Million mapped reads). DEGs ($P < 0.05$; false discovery rate $= 0.05$) were identified with DESeq2 (Love et al. 2014). Identification of secretory protein signals was performed according to Haddadi et al. (2016).

6. **Real-Time qPCR Experiments:** RNA was converted to cDNA with the qScript cDNA Supermix (Quantabio). Following, qPCR was performed using the CFX Connect Real-Time System and PerfeCTa SYBR-Green Supermix as per manufacturer's instructions in a 15 μl reaction volume. The ΔΔCt method (Rieu and Powers 2009) was used to compare relative mRNA abundance, normalized to the endogenous housekeeping gene *Actin* (GenBank AF111812.1). Statistics were performed as described in McLoughlin et al. (2019).
7. **Jasmonic Acid Measurements:** Inoculation sites (30 per bio-replicate, and five bio-replicates per treatment) were hand-dissected and immediately flash-frozen and pulverized in liquid nitrogen. Following, 5 ml of extraction solvent (2-propanol:H_2O:HCl, 2:1:0.002, pH 2.4) and 25 ng of dihydro jasmonic acid internal standard were added to each sample, which were then shaken at 4 °C for 30 min. Next, 7 ml of dichloromethane was added, and samples shaken again at 4 °C for 30 min. After centrifugation at 5300 rpm for 5 min, the lower organic phase was transferred and dried on ice in a nitrogen evaporator system. The dried sample was then dissolved in methanol, mixed aggressively, and filtered through 0.2 μM PTFE before ultra performance liquid chromatography (UPLC). JA standards were prepared with internal standards to generate calibration curves and response factors. Samples were analyzed on a Waters Acquity UPLC/MS with Micromass Quattromicro API, as described by Henriquez et al. (2016) and (Becker et al. 2019).

8.6.3 Cloning and Transformation of *Leptosphaeria* Avirulence Gene

1. **Fungal Material, Mapping Population, and Inoculum Preparation:** The fungal F1 population was generated by crossing Lm isolate v23.1.3 [pathotype *AvrLm1*–4-5-6-7–8-(10)-11-J1-S, *AvrLep1*–3-4] to isolate 00–100 [*AvrLm2*–3-(5)-6-(8)-9-(10)-J1-S, *AvrLep1*–2-4] (Rouxel et al. 2011) by pairing single pycnidiospores isolates as described by Mengistu et al. (1995). Culturing and harvesting of the pycnidiospores were performed as described previously by Ghanbarnia et al. (2012).
2. **Plant Material and Virulence Phenotyping:** Parental isolates, progenies of the mapping population, and Lm transgenic lines were patho typed for *AvrLm9* by pathology test on *B. napus* line Goeland, which carries the *Rlm9* resistance gene

(Balesdent et al. 2005). Susceptible *B. napus* lines Topas DH16516 and Westar were used as controls. Pathology tests were performed using a 0–9 rating scale as described previously (Larkan et al. 2013). In addition, to show the specificity of response, additional *B. napus* differential lines, including Topas differential lines with single R-genes (Larkan et al. 2016b), *B. juncea* cv. *Vulcan* (*Rlm6*), and *B. napus* cv. *Roxet* (*Rlm7*), were included.

3. **DNA Extraction and Sequencing:** DNA extraction and re-sequencing of *L. maculans* isolates have been described elsewhere (Ghanbarnia et al. 2015). Sequence reads for each isolate were mapped to the reference genome v23.1.3 (Rouxel et al. 2011) and AAFC reference genome 00–100 using CLC.
4. **Mapping and Marker Development:** The genome sequences of the parental isolates v23.1.3 and 00–100 were utilized to generate SNP targets spanning the Lm genome for use in marker development. KASP assays (He et al. 2014) targeting individual SNPs were developed using Primer Picker software. Assays were first tested for clear detection of polymorphic alleles using DNA from the parental isolates before being applied to the entire F1 population. Assays were run and scored using a CFX96 Real-Time PCR System. Map distances (in cM) were calculated manually as described previously (Larkan et al. 2014).
5. **Cloning and Transformation of Candidate Gene:** Genomic clones were generated by amplification of an 1813-bp fragment containing the avirulent (00–100) allele, including native promoter and 30 untranslated region (UTR), via PCR using Q5 High-Fidelity Master Mix. Cloning was conducted using Gateway technology by adding attB sites to *AvrLmJ1* primers. These were transferred to the fungal transformation vector pNL11 (Larkan et al. 2015) prior to *Agrobacterium*-medi (Lucas et al. 1988) and transformation (Utermark and Karlovsky 2008). SNP alleles (cDNAs only) were synthesized and cloned into the pLM4 vector, in which expression of the construct was driven by the *AvrLm1* promoter (Ghanbarnia et al. 2015). Positive transformants were tested on *B. napus* cvs Goeland and Darmor which carry the *Rlm9* resistance gene. In addition, *B. napus* introgression lines (Larkan et al. 2016a, b) were used to further confirm that the interaction of *AvrLm9* was specific to *Rlm9*. These lines were Topas-*Rlm1*, Topas-*Rlm2*, Topas-*Rlm3*, Topas-*Rlm4*, Topas-*LepR1*, Topas-*LepR2*, and Topas-*LepR3* (Larkan et al. 2016a, b).
6. **RNA Extraction, Reverse Transcription, and Quantitative Reverse Transcription-Polymerase Chain Reaction (RT-PCR) Analysis:** Total RNA was extracted from infected cotyledons of the susceptible cultivar *B. napus* Topas DH16516 using the TRIzol LS reagent, and subsequently purified with an RNAeasy Mini Kit according to the manufacturer's protocol. DNA was removed by on-column treatment with RNase-free DNase in solution. All samples were adjusted to 1 lg of RNA, and cDNA was synthesized using a SuperScript III First-Strand Synthesis Super Mix Kit according to the manufacturer's protocol. qRT-PCR was performed using a CFX96 Touch Real-Time PCR System (*Bio-RAD*) and Sso Advanced Universal SYBR Green Super mix. RT-PCR was performed for the *AvrLm4–7* and *AvrLm9* genes with three biological samples.

Ct values were analyzed using the Q-gene program (Muller et al. 2002; Simon 2003). The *L. maculans* actin gene served as a reference.

7. **RNA Sequencing Analysis:** *L. maculans* genome-wide transcriptomic analyses were performed according to the methods described previously (Haddadi et al. 2016). Briefly, it was applied deep RNA sequencing using an Illumina HiSeq-2500 platform to assess genome-wide expression profiling of two natural *L. maculans* isolates harboring (v23.1.3/JN3) or lacking (00–100) *AvrLm4–7* during cotyledon infection at different time points (0, 2, 4, 6, and 8 dpi). All reads were mapped to the genome sequence of *L. maculans* isolate v23.1.3 (Rouxel et al. 2011). Differentially expressed genes (DEGs) were detected using the edgeR (CLC) and DESeq2 packages (Love et al. 2014; Robinson et al. 2010). The RNA sequencing data corresponding to 00–100 were presented as log2 Reads Per Kilobase per Million mapped reads (RPKM) of the average of three biological replicates, whereas the data corresponding to v23.1.3/JN3 were presented as log2(RPKM) of the merged three technical replicates.
8. **Yeast Two-Hybrid Assay:** For the yeast two-hybrid assay, the *AvrLm3* open reading frame (ORF) lacking signal peptide sequences was amplified using primers DspAvrLm3-NcoRI-F and DspAvrLm3-EcoRI-R from the *L. maculans–B. napus* cDNA library. Similarly, the *AvrLm4–7* ORF lacking signal peptide sequences was amplified using the primer set DspAvrLm4–7-NdeI-F and DspAvrLm4–7-PstI-R. PCR products were cloned into pGBKT7 vector as bait constructs. *AvrLm9* lacking its signal peptide was amplified using primers DspAvrLm9-NdeI-F and DspAvrLm9-EcoRI-R and sub-cloned into the vector pGADT7 as a prey construct (Clontech). The matchmaker GLA4 two-hybrid system and yeast strain Y2HGold (Clontech) was used. The yeast strain Y2HGold was co-transformed with bait and prey plasmid combinations using lithium acetate and polyethylene glycol 3350, as described previously (Gietz and Woods 2002). Transformants harboring both bait and prey plasmids were selected on plates containing minimal medium lacking leucine (Leu) and tryptophan (Trp) (MM-WL). For controls, empty prey vector pGBKT7 or pGADT7 was used as bait or prey. Next, one colony per combination was picked up from MM-WL plates to inoculate 1 ml of MM-WL culture. After 36 h of growth, the culture was centrifuged, the cells were resuspended in 25 μl of 0.9% NaCl from $OD_{600} = 1$ to $OD_{600} = 0.00001$, and then spotted on MM-WL and MM-AHWL plates supplemented with 40 μg/ml (X-α-D-Galactoside) (Clontech) and 200 ng/ml Aureobasidin A (Clontech). After 3 days of incubation, observations were recorded (Ghanbarnia et al. 2018).

8.6.4 Identification of QTLs in *Brassica* to *Leptosphaeria*

1. **Plant Populations:** F1 seedlings, produced by crossing the Swedish spring type, blackleg-susceptible *B. napus* variety Topas with the Australian blackleg-resistant varieties AG-Castle and AV-Sapphire, were used to produce DH populations via microspore culture by the method of Ferreira et al. (1995) using

modified Lichter medium (NLN) with 13% sucrose. The Topas/AG-Castle DH population (TC) utilized consisted of a total of 242 DH lines from two subpopulations of 139 and 103 DH lines, with each subpopulation produced from a single F1 plant. The Topas/AV-Sapphire DH population (TS) contained a total of 109 lines, also consisting of two subpopulations of 66 and 43 DH lines. Seeds of parental and DH lines used in the QTL trials were produced in tented field plots in Saskatoon, Canada, during the 2007 (TS) and 2008 (TC) summer growing seasons, packaged into replicates of 100 seeds per line and sent to Australia for testing field resistance to blackleg. The Topas *B. napus* line was sourced from the AAFC Saskatoon collection. AG-Castle and AV-Sapphire lines were provided by DEDJTR Victoria.

2. **Field Characterization of DH Populations:** The parental lines and their respective DH populations, along with several *B. napus* variety checks, were grown from May through November in blackleg disease nurseries situated at Wagga Wagga (New South Wales) and two sites (Wonwondah and Green Lake) south of Horsham (Victoria). Rainfall differed by an average of 103.2 mm/year over the 3 trial years at each site, with the Wagga Wagga site receiving an average 38% more precipitation. Each trial was planted in a randomized block design with three to four replicates per line. The TC population was tested in three separate environments: both Wagga Wagga and Horsham (Wonwondah site) and once more at Wagga Wagga. The TS population was tested in four environments: Horsham (Wonwondah site), and Wagga Wagga, as three-replicate trials and Wagga Wagga, and Horsham (Green Lake site), as four-replicate trials. Emergence counts were made 4–5 weeks after sowing. For trials at the Horsham sites, local diseased canola stubble, collected in the previous year, was spread as the source of *L. maculans* inoculum, while at Wagga Wagga the trials were sown into in situ canola stubble from the preceding season as previously described. At the end of the growing season trials were assessed for blackleg disease using both survival percent (S %) and internal infection (II) metrics. Survival percent was calculated for each row as (Standing plants at end of season/ Emergence count)/ 100, Internal infection was determined by examining 5–20 random plants per row for internal infection after cutting the stem at the crown and rating fungal discoloration on a 0–5 scale (0 = no infection, 1 = 1–25% infection, 2 = 26–50% infection, 3 = 51–75% infection, 4 = 76–99% infection, and 5 = 100% infection/missing plant). Median scores were calculated for each entry and metric using Microsoft Excel. Statistical analysis of populations and plotting of distribution histograms were performed using R software. Selected lines representative of the three major QTL identified from the TC DH population, both singularly and in combination were also entered into Saskatchewan disease nurseries at Melfort and Saskatoon in attempt to test the efficacy of the QTL resistance against Canadian populations of *L. maculans*. However, neither site produced sufficient disease pressure to assess the resistance, with Westar control entries presenting average internal infections of less than 25% at the end of the growing season.

3. **Single-Isolate Characterization:** Each DH line in the TC and TS populations were characterized for the presence or absence of hypersensitive response during infection by the single spore-derived *L. maculans* isolates WA30 or v23.1.3, respectively, via cotyledon infection tests. Four plants for each line were scored at 14 days post-infection on a 0–9 scale as described by Larkan et al. (2013). Further differential pathology testing, using isolates of *L. maculans* varying in their avirulence gene profiles (00–100, 98–15, 99–53, 89–12, 2354-Rimmer Collection), was performed with the parental lines and the R-gene control lines Topas DH16516 (a DH line of Topas with no effective blackleg resistance), Columbus (*Rlm1*, *Rlm3*), Quantum (*Rlm3*), Jet Neuf (*Rlm4*), Roxet (*Rlm7*), and Goeland (*Rlm9*) in order to produce a hypothesis for the likely R-gene content of each population. Inoculum for the tests was produced as described previously. Spores were diluted to 2×10^7spores/ml and 10 μl was used to inoculate each of four small wounds on each plant (one wound at the center of each cotyledon lobe). The transgenic isolate 3R11: AvrLm4–7 was produced by *Agrobacterium* transformation (Utermark and Karlovsky 2008). A 2205 bp amplicon was produced from the isolate v23.1.3 (primers: GW-AvrLm47G-Fc; GGGGACAAGTTTGTACAAAAAAGCAGGCTTCCTCTGGTAAGGAAGG-TTTACCAATTATA CACCT,GW-*AvrLm47G*-Rb; GGGGACCACTTTGTACAAGAAAGCTGGGTCGGCGGTAGATTTGCTA-CTAAAAG GTAACTTT) which spanned 1213 bp upstream and 514 bp downstream of the *L. maculans AvrLm4–7* CDS and inserted into the Gateway compatible fungal transformation vector pNL11 prior to transformation of the *L. maculans* isolate 3R11 as previously described (Parlange et al. 2009; Larkan et al. 2013, 2016a, b).
4. **Map Construction and QTL Identification:** The TC and TS populations were characterized using 225 and 212 simple sequence repeat (SSR) markers, respectively, as described previously. A subset of the TC population had previously been characterized using 279 DArT markers as part of a consensus map of *B. napus*, and this data was also incorporated into the TC map to provide additional resolution and to allow orientation to other maps (503 markers total, grouped into 360 bins). Linkage map construction and inclusive composite interval mapping (ICIM) was performed using QTL IciMapping version 3.3 (http://www.isbreeding.net). Prior to final map construction severely distorted markers (Chi-square Test, $P < 0.001$) were excluded. A threshold of LOD 6 was applied for initial grouping of markers. LGs consisting of >3 markers were anchored to chromosomes based on BLAST-like alignment tool (BLAT) analysis of SSR marker sequences to the published *B. napus* genome before recalculating LGs at LOD 3.

   Marker ordering was performed using nearest neighbor algorithm and rippling was performed once using the sum of adjacent recombination fractions (SARF) criterion. For QTL identification, the "ICIM-ADD" and "ICIM-EPI" functions of the software were utilized to investigate both additive and digenic epistatic QTL expressed in each environment and for each scoring metric (S and II). Permutation tests (1000 permutations, 95% confidence level, 1 cM interval) were performed for each scoring metric in each environment to determine significant LOD thresholds. The precision of QTL positions were then improved by re-analyzing the data using the determined LOD thresholds and a smaller (0.5 cM) scan interval. The "Multi-Environment Trials" (MET) function of the software was also utilized to determine the consensus positions for the major QTL and to assess the genotype by environment (GxE) interactions. QTL identified from individual environments were considered significant if they exceeded the LOD significance threshold and accounted for >5% of the variance. MET QTL were considered significant if they accounted for >5% of the variance and had a heritability ($h2 = \sigma 2$ (Additive)/$\sigma 2$ (Total)) > 0.5. QTLs positions were defined by both support and peak marker intervals, with support intervals representing the map interval in which the LOD exceeded the calculated LOD threshold (for single-environment QTL), or peak LOD-1 (for multi-environment analyses). The data sets for each population were analyzed both as total populations and as their separate component subpopulations. Final map figures were produced using Map chart 2.2 and Microsoft PowerPoint software. To determine the location of the major QTL relative to the *B. napus* reference Darmor-bzh genome the target sequences for the flanking SSR markers were obtained from the AAFC *Brassica* MAST Database (http://aafc-aac.usask.ca/BrassicaMAST/) and used to search the public *B. napus* genome database (http://www.genoscope.cns.fr) using default BLAT search Settings (Li et al. 2008; Zhang et al. 2012; Voorrips 2002; Larkan et al. 2016a, b)

8.6.5 Molecular Mapping of R-Genes in *Brassica-Leptosphaeria*

1. **Plant Material:** The DH population, SASDH, used for the genetic mapping of blackleg resistance loci consisted of 186 lines derived from a cross between the blackleg resistant cultivars Skipton [Barossa (*Rlm4*)/BLN356-3///58410K/ Shiralee(*Rlm4*)//Cobra) and Ag-Spectrum (Eureka (*Rlm4*)/ZE6]. This population was generated via microspore culture and showed segregation for several traits of agronomic importance such as blackleg resistance, flowering time, and carbon isotope discrimination (Luckett et al. 2011; Raman et al. 2011).
2. **Inoculum Preparation:** Eleven single-spore isolates were acquired from the national blackleg isolate collection. Inoculum was prepared by sub-culturing the *L. maculans* isolates on 10% V8 agar plates. Approximately 2 weeks later, pycnidiospores were collected in 10 ml of sterile water by dislodging the pycnidia. The spore suspension was filtered through muslin cloth and spore concentration adjusted to 1×10^6 per ml using a hemocytometer. All isolates were also screened for the presence of the three cloned Avr genes *AvrLm1*, *AvrLm4*, and *AvrLm6* using either PCR-based markers and or whole-genome sequence data.
3. **Single-Spore Isolate Screen:** The parental lines of SASDH along with other check lines were screened for resistance at the cotyledon and adult plant stages against the 11 single-spore *L. maculans* isolates. The two isolates, 04MGPS021

and 06MGPP041, which were each found to be virulent and avirulent against one of the parental canola lines, were selected for evaluating the DH lines for resistance to *L. maculans*. Twelve seeds of each genotype from SASDH and parental lines were sown in plastic pots (20 cm diameter) containing a commercial potting mix. The plants were thinned to four per pot, along with one Q2 plant (the susceptible control) in the center of each pot. Each genotype had two replicates. Both cotyledons of each seedling were punctured with a pair of bent tweezers and both lobes were inoculated with 10 μl of a suspension containing 10^6 pycnidiospore/ml. Plants were placed in a dew chamber at 100% relative humidity, at 20 °C, for 48 h, and then returned to a shade house. Seventeen days after inoculation, each inoculation point on the cotyledons was scored for resistance using the rating system of Koch et al. (1991), where 0 = no darkening around wounds and 9 = large gray-green lesions with profuse sporulation. The same plants were then allowed to develop to maturity and scored for resistance by assessing plant mortality and internal infection of the crown. Plants were severed at the crown with a pair of secateurs to enable the crown to be visually inspected for blackleg symptoms. Plants were scored for basal internal infection (0–100% area discoloration). Average internal infection was rated as 0–35% = resistant; 36–49% = intermediate, and 50–100% = susceptible. A spatially optimized incomplete block design with a nested blocking structure was employed to estimate different variance components. This design was made using the spatial design search program DiGGer (Coombes 2002) assuming positive correlation between neighboring pots in rows and columns and allowing for random row and column effects within a column-pair of benches. Twenty-five of the DH lines were duplicated and 137 lines evaluated by a single replicate.

4. **Field Screen:** Parental and SASDH lines were screened, in blackleg nurseries. The blackleg nurseries consisted of 6-month-old canola stubble from the previous year's crop. Stubble sourced from a mixture of triazine-tolerant varieties was scattered in the field nursery prior to sowing to increase pathogen pressure. Each line was sown into a 2-m row containing up to 30 individual plants. Irrigation was used to promote development of the epidemic. Up to ten mature plants from each DH line were cut with secateurs and scored for percentage of internal infection at physiological maturity (November/December).
   The experimental design for trial consisted of 177 DH lines, each parent, and Karoo, as a resistant variety, in a two-replicate design arranged in 30 rows by 12 columns of plots. Due to shortage of seed of some of the DH lines, only 154 DH lines of the population were screened for blackleg resistance, along with the parental lines. This was a partially replicated (p-rep) design (Cullis et al. 2006) with two replicates of 115 DH lines and one plot of 39 DH lines, with the parents replicated seven times. The control cultivars in 2009 were Karoo (resistant), Trigold (susceptible), and Hyola50 (resistant).
5. **Construction of Framework Map:** DNA was isolated from approximately 10-week-old, shade house-grown seedlings using a standard phenol–chloroform method. Six hundred and eighty-four simple sequence repeat (SSR), sequence-related amplified polymorphism (SRAP), sequence characterized amplified

region (SCAR), and candidate gene-based markers, originating from *B. rapa, B. oleracea, B. napus,* and *B. juncea* were collated from sources in the public domain (Cheng et al. 2009; Choi et al. 2007; Hopkins et al. 2007; http://ukcrop.net/perl/ace/search/BrassicaDB; Li and Quiros 2001; Long et al. 2007; Lowe et al. 2002, 2004; Piquemal et al. 2005; Sun et al. 2007; Suwabe et al. 2002, 2006; Tsuro et al. 2005) and investigated for polymorphism. The SSR primer pairs were synthesized. The forward primers of each primer pair were tagged with a 19-bp M13 sequence and PCR amplifications and allele sizing were carried out as described by Raman et al. (2005). A SCAR marker (BN204) derived from a region showing 92% amino acid identity with the defense-related gene serine threonine 20 (ste-20) protein kinase of Arabidopsis thaliana was also used. This marker was completely linked with the Rpg3Dun gene in an F2 population from Westar/Dunkeld (Dusabenyagasani and Fernando 2008). The genetic linkage map was produced using Map Manager version QTL20b (Manly et al. 2001) using the Kosambi mapping function at a probability of 0.01, as described previously (Raman et al. 2009). Accuracy of the marker order within linkage groups was checked using the R/qtl statistical analysis package (Broman et al. 2003), RECORD computer package (van Os et al. 2005), and compared with previously published maps (Choi et al. 2007; Lowe et al. 2004; Piquemal et al. 2005; Suwabe et al. 2006, 2008). The linkage data were exported into the Map Chart package (Voorrips 2002) to display the trait-marker data graphically.

6. **QTL Detection and Validation of *RlmSkipton* Linkage:** An integrated map consisting of 216 SSR, SRAP, SCAR, and EST-SSR markers covering 24 linkage groups, representing at least 17 chromosomes, was subsequently employed for the QTL analysis for blackleg resistance using the whole-genome average interval mapping approach (Verbyla et al. 2006), which simultaneously models genetic and environmental variation. Environmental variation was accounted for by including terms relating to design factors such as replicates, columns, rows, and scorer effects. Putative QTLs with a LOD score C2.0 have been reported. All QTL analyses were conducted using the ASREML-R package (Butler et al. 2007) using original disease scores (0–9) rather than using arbitrary thresholds.
   Subsequently, leaf lesion scores based on the cotyledon reaction were classified into two groups in order to map blackleg resistance precisely onto a genetic map of the SASDH population. A disease score of 1 was rated as resistant and a score of 4 was rated as susceptible. Standard Chi-squared (v2) tests for "goodness-of-fit" were used to test the validity of Mendelian ratios with observed data. Linkage between phenotypic and marker alleles was determined by Map Manager version QTX20b (Manly et al. 2001), at a threshold of $P = 0.001$. In order to confirm the location of the markers Xbrms075 and Xcb10439, that was found to be linked with the RlmSkipton resistance locus on A7. The sequence of 12 of the SSR markers was compared with the assembled genome sequence of *B. rapa* using BLAST (Altschul et al. 1990).

   Genetic control of blackleg resistance and *RlmSkipton*-SSR marker linkage was verified in an F2 population comprising 101 plants derived from Skipton/Ag-

Spectrum. Disease expression was tested by inoculating both lobes of cotyledons with the single-spore isolate 04MGPS021 under shade house conditions, as described earlier. After phenotyping, leaf tissue was harvested for DNA analysis. Five flanking markers to the *Rlm-Skipton* locus, *Xol09-a06*, *Xcb10278a*, *Xcb10439*, *Xbrms075*, and *Xbn204*, were verified to determine whether the markers discriminate between corresponding alleles associated with resistance and/or susceptibility. An integrated linkage map including *Rlm-Skipton* and molecular marker loci in an F2 population was generated with the segregation data using Map Manager QT X 017b (Manly et al. 2001). Linkage analysis and Chi-square tests were performed at a threshold of $P = 0.001$. The *Xcb10278a* and *Xbn204* markers exhibited segregation distortion and, therefore, were not used in map construction. To test the effectiveness of the SSR alleles in predicting blackleg resistance, the allele diversity of two validated markers (*Xcb10439* and *Xbrms075*), flanking the *Rlm-Skipton* locus in the DH and F2 populations from Skipton/Ag-Spectrum, as well as the *Xbn204* marker linked with *Rpg3Dun* (Dusabenyagasani and Fernando 2008) were compared in a set of 15 canola genotypes used as parents in Australian canola breeding programs. These genotypes were also evaluated for cotyledon resistance to isolate 04MGPS021, as described previously. At least five seedlings (20 lobes) of each genotype were inoculated as described previously, and experiments were repeated twice.

7. **Nomenclature of Chromosomes, Qualitative Genes, and QTL:** Standard nomenclature endorsed by the Multinational *Brassica* Genome Project steering committee was adopted to name the linkage groups of *B. napus* (N1–N10 correspond to A1–A10, respectively, and N11–N19 correspond to C1–C9, respectively), as described previously (http://www.brassica.info/resource/maps/lg-assignments.php). The pragmatic approach to assign linkage groups to endorsed nomenclature; a linkage group with at least two markers that have been mapped previously in *B. rapa, B. oleracea*, and/or *B. napus* were designated accordingly. QTLs identified were named using a standard "designation" system adopted by the international wheat community (Mcintosh et al. 2003). The "Q" indicates a QTL or a genomic region associated with the trait (in this case resistance to *L. maculans*) detected through QTL mapping, which is followed by an abbreviation of the laboratory designator (wwai), a hyphen (-), and the symbol for the chromosome in which the QTL is located. The symbols were "*ii*" and "*s*" for QTLs identified in field conditions using internal infection (canker development) and percent plant survival, respectively, as measures for blackleg resistance. *QRlm* (ii*).wwai-A1* represents a QTL associated with resistance to *L. maculans* (*Rlm*) identified using internal infection that was mapped at Wagga Wagga on chromosome A1. An additional suffix (a, b, c, d, and e) was used if either more than one QTL affecting the trait was identified on the same chromosome or if multiple segregating loci were detected by a primer pair (Raman et al. 2012a).

8.6.6 Identification of NBS-Encoding Genes in *Brassica napus*, *Brassica rapa*, and *Brassica oleracea*

The entire genome sequences and annotation data for *B. rapa* (Wang et al. 2011), *B. oleracea* (Liu et al. 2014; Parkin et al. 2014), and *B. napus* (Chalhoub et al. 2014) were downloaded from the BRAD database1. NBS-encoding genes in the three species were identified by using the amino acid sequence of the Pfam NB-ARC domain (PF00931) as a "blastp" query against all known protein sequences, via the HMMER V3.0 program with "trusted cut-off" as the threshold (Finn et al. 2011). The obtained hits were further submitted to the Pfam website2 to verify the presence of the NBARC domain. Those proteins with verified NB-ARC domains were further subdivided into groups based on the structure of the N-terminal and C-terminal domains of the protein. By using the Pfam database3, SMART protein motif analyses4, and the COILS program5, the presence or absence of TIR, CC, LRR domains of NBS-encoding genes was identified, which was used to classify the NBS-encoding genes into different groups.

1. **Location Anchoring and Gene Cluster Analysis of NBS-Encoding Genes in *B. rapa, B. oleracea*, and *B. napus***: The chromosome size and physical position of NBS-encoding genes were downloaded from the BRAD database6 for location anchoring of NBS-encoding genes. The visualization of NBS-encoding genes on 10, 9, and 19 chromosomes (assembled pseudomolecules) of *B. rapa, B. oleracea*, and *B. napus* was drawn by "Pheno Gram Plot." According to the gene cluster definition proposed by Richly et al. (2002) and Meyers et al. (2003), two or more genes located within eight ORFs of each other were treated as a gene cluster.
2. **Detection of Duplicated Genes Within Each Species and Identification of Homologous Gene Pairs Between *B. rapa, B. oleracea*, and *B. napus***: To detect the duplicated NBS-encoding genes within individual species and to identify homologous gene pairs between species, the BLASTP program of Blast2GO (Conesa et al. 2005) was employed using protein sequences with stringent parameters, that is, *E*-value cut-off ($e = 0.0$) (Ariyarathna and Francki 2016). The gene pairs across genomes were visualized using the Circos software (Krzywinski et al. 2009). If no homologous gene was found for a particular gene using this stringent criterion, the top hit was selected and designated as the most similar gene.
3. **Multiple Alignments and Phylogenetic Analysis of NBS-Encoding Genes:** Multiple alignments of amino acid sequences were performed using the program "Muscle" with default options (Thompson et al. 1994). The software RA x ML was then used to construct phylogenetic trees based on the Maximum Likelihood (ML) method (Stamatakis 2015).
4. **Non-synonymous/Synonymous Substitution (Ka/Ks) Ratios of Homologous Gene Pairs:** Protein sequence variation between each homologous pair was firstly calculated via sequence alignment using the program Muscle with default options (Thompson et al. 1994). Based on the alignment results, the number of

nucleotide variants was then calculated by reverse translation of the protein sequence to nucleotides. The ratio of nucleotide variants was calculated by the formula: ratio of nucleotide variants = number of variant nucleotides/the number of aligned nucleotides (gaps were not considered). To investigate selective pressure on NBS-encoding genes, the ratio of non-synonymous substitutions to synonymous substitutions (Ka/Ks) was calculated. Firstly, the protein sequences of NBS-encoding genes in each homologous gene family were aligned using the program Muscle with default options (Thompson et al. 1994). Then, non-synonymous substitutions (Ka) and synonymous substitutions (Ks), and the ratio between them (Ka/Ks) were calculated in each homologous gene family using the tool yn00 in the paml package (Yang 1997). Positive and negative selective pressure were justified as a Ka/Ks ratio > 1 and Ka/Ks ratio < 1, respectively, while a ratio of 1 indicated neutral evolution (Yang 2007). Ka/Ks ratio cut-offs of >1.2 and <0.8 were chosen to identify positively and negatively selected genes, respectively, with slight modification of the method of Mayer et al. (2011).

5. **RNA-seq Data Analysis of NBS-Encoding Genes:** In order to analyze the expression of the NBS-encoding genes, RNA-seq data of NBS-encoding genes of *B. rapa, B. oleracea,* and *B. napus* that was generated earlier and submitted to the GEO database was downloaded. RNA-seq data of five tissues (stem transcripts, root transcripts, leaf transcripts, flower transcripts, and seed transcripts) of *B. napus, B. rapa*, and *B. oleracea* were obtained from Tong et al. (2013); Liu et al. (2014), and Miao et al. (2016), respectively. Fragments per kilobase of exon model per million mapped reads (FPKM) were calculated for NBS-encoding genes. The FPKM values were log2 transformed, and a hierarchical cluster was created using the Genesis software using the hierarchical clustering algorithm (Eisen et al. 1998).
6. **Co-localization of NBS-Encoding Genes and Mapped Quantitative Resistance Loci Against Three Major Diseases of *B. napus*:** Resistance quantitative trait loci for blackleg (Leptosphaeria maculans and *L. biglobosa*), clubroot (*Plasmodiophora brassicae*), and Sclerotinia stem rot (Sclerotinia sclerotiorum) were collected from the literature (Hirai et al. 2004; Suwabe et al. 2006; Zhao et al. 2006; Delourme et al. 2008; Sakamoto et al. 2008; Kaur et al. 2009; Nagaoka et al. 2010; Raman et al. 2012b; Jian et al. 2013; Kato et al. 2013; Mei et al. 2013; Wei et al. 2014, 2016; Larkan et al. 2016a, b; Li et al. 2016; Sanjaya et al. 2016; Wu et al. 2016; Kumar et al. 2018; Pang et al. 2018). The QTL intervals were aligned to the reference genomes of *B. napus*[9] by BLAST analysis of the sequences of SSR primers or probe sequences linked to the QTLs, using the method of Cai et al. (2012) with slight modifications. The alignment criteria were set to allow three mismatches and one gap for a given primer pair. When a marker had multiple amplification loci on the same chromosome, an accurate position for a particular locus was determined manually by referring to the physical positions of its upstream and downstream markers (Fu et al. 2019).

8.7 *Brassica-Plasmodiophora*: Molecular Techniques

8.7.1 Genetics and Molecular Mapping of R-Genes to *Plasmodiophora* Pathotypes

1. **Plant Material:** The parent materials included a clubroot-resistant rutabaga inbred line Rutabaga-BF and a susceptible spring *B. napus* canola cultivar A07-29NI (Rahman 2016). The line Rutabaga-BF was developed from the cultivar Brookfield through self-pollination of a single clubroot-resistant plant. Seeds of Brookfield were received, and A07-29NI a canola cultivar was developed. The line Rutabaga-BF showed complete resistance to Canadian *P. brassicae* pathotypes 2, 3, 5, 6, and 8.
2. **Crossing and Population Development:** The parent Rutabaga-BF was vernalized for 8 weeks at 4 °C with an 8/16 h (day/night) photoperiod, and crossed to the susceptible parent A07-29NI. The F1 plants were vernalized for 8 weeks, and F2 seeds were produced through self-pollination of the F1 plants. Backcross (BC1) seeds were produced by crossing the F1 to A07-29NI. The F2 and BC1 seeds were harvested from the individual F1 plants and maintained as separate families.
3. **Doubled Haploid (DH) Lines Production:** Doubled haploid lines were produced from five F1 plants through application of microspore culture technique (Coventry et al. 1988). For this, flower buds of 3.5–4.5 mm in length, where the size of the petals was about two-thirds of the length of the anther, were used for microspore isolation. The buds were collected in a falcon tube and were placed on a bed of ice to prevent microspore degeneration. Prior to microspore isolation, the buds were sterilized by treating with 6% calcium hypochlorite (Ca $(OCl)_2$) solution for 15 min, and then rinsed three times with cold sterile milli-Q Millipore water. About 15–20 sterilized flower buds were homogenized in 25 ml liquid B5 medium (Gamborg et al. 1968) containing 13% sucrose using a sterile mortar and pestle. The homogenate was poured through two layers of nested sterile filters (64 μm top (NT X 64) and 41 μm bottom (NTX41)) into a sterile falcon tube. The falcon tube was centrifuged for 5 min at 1000 rpm, and the dark green supernatant was decanted leaving the pellet of yellow spores at the bottom of the tube. Microspores were washed three times with B5 liquid medium. After washing, the clean microspore pellet was re-suspended in 30 ml NLN 13 medium (Lichter 1982) containing 50 mg/l colchicine and poured into sterile Petri plates. The Petri plates were sealed with parafilm and incubated in dark in an incubator at 30 °C for 24 h. After 24 h, the suspension was transferred to a falcon tube and centrifuged at 1000 rpm for 5 min. The supernatant was decanted and 25 ml of NLN 13 medium (without colchicine) was added to the microspores and cultured in Petri plates for 14 days at 30 °C. After 14 days, the plates were transferred on a shaker rotating at 59 rpm to provide enough air to the developing embryos. The torpedo-shaped embryos were transferred to solid B5 medium (liquid B5 with 20% sucrose and 2% agar) at a density of 20 embryos per Petri plate. The plates were kept at 4 °C for 14 days, and then

transferred to room temperature under light (16/8 h day/night) for germination of the embryos and development of plantlets. Seedlings with well-developed roots and true leaves were transferred to soil-free growth media and covered with transparent cups for about a week to maintain high humidity.

At flowering, the plants producing visible pollen were considered to be doubled haploids, while the sterile plants with smaller size flower/petal and lacking pollen grains were considered as haploids. The haploid plants were treated with 0.35% aqueous solution of colchicines for chromosome doubling. For this, roots were washed and immersed in colchicine solution for 1 h. After this, roots were washed thoroughly under running tap water and the shoots were clipped off leaving only a few auxiliary buds. The treated plants were transplanted to soil-free growth medium in pots. Fertile plants with visible pollen grains were self-pollinated by bag isolation and harvested as doubled haploids.

4. **Pathogen Isolate:** Single-spore derived isolates of *P. brassicae*, classified as pathotypes 2, 3, 5, 6, and 8 based on Williams (1966) differentials, were used. These isolates were obtained from University of Alberta, in the form of galls. Resting spore suspensions (inoculum) from the preserved galls were prepared following the protocol described by Strelkov et al. (2007), and concentration of the suspension was adjusted to 1×10^7 to 1×10^8 resting spores/ml.
5. **Resistance Test:** Four F2 and their corresponding BC1 families were evaluated for resistance to *P. brassicae* pathotype 3, and the DH population was evaluated for resistance to *P. brassicae* pathotypes 2, 3, 5, 6, and 8. Seedlings germinated on moistened Whatman filter paper No. 1, at the age of 7 days after seeding, were inoculated by dipping the roots in resting spore suspension (Nieuwhof and Wiering 1961). The inoculated seedlings were transplanted to 3 cm × 3 cm × 5 cm ($L \times W \times D$) cells filled with Sunshine Professional Natural and Organic Mixes #4, placed in a tray (2 m × 4 m), and grown in a greenhouse at 21 ± 2 °C with a 16 h photoperiod. After transplanting, 2 ml inoculum was pipetted to each cell to ensure successful inoculation. During the first 7 days, the cells were kept saturated with water to ensure sufficient moisture for development of the pathogen. To ensure acidic condition of the soil, HCl solution (10% *v/v*) at 20 ml/tray (2 m × 4 m) was added each day during this time. Starting the second week, watering was done once a day. Seedlings were evaluated for clubroot resistance 42–45 days after inoculation, and the severity of gall development was rated on a 0–3 scale, where 0 = no galling, 1 = one or few tiny galls on lateral roots, 2 = moderate galling on lateral roots, and 3 = severe galling on lateral or main root. The details of inoculum preparation, inoculation technique, and scoring are described (Rahman et al. 2011; Hasan et al. 2012). Based on disease score, plants with scores of 0 or 1 were classified as resistant, while the plants with scores of 2 or 3 were classified as susceptible.
6. **DNA Extraction:** Leaf samples of the parents and the DH population (3–5 weeks old plants) grown in greenhouse were collected in aluminum foil and kept at −80 °C until use. DNA was extracted using a Wizard Genomic DNA purification kit following instructions of the manufacturer. The DNA pellet

was air-dried for 60 min and rehydrated in 200 μl DNA rehydration solution (TE buffer) overnight at 4 °C. The quality and quantity of the DNA was estimated by spectrometry using 260/280 nm absorbance ratio, and the concentration of DNA was estimated at 260 nm with an ND-2000 Nano Drop spectrophotometer.

7. **Molecular Markers:** Simple sequence repeat (SSR) primer sequences were obtained from Agriculture and Agri-Food Canada (AAFC), Saskatoon and the SSR primers published by different researchers, such as Lowe et al. (2002, 2004), Suwabe et al. (2002, 2003), Piquemal et al. (2005), Saito et al. (2006), Cheng et al. (2009), and Kato et al. (2012), and SSR markers designed based on *B. rapa* cultivar Chiifu-401 whole-genome sequence assembly version 1.5 (Cheng et al. 2011) were used. To design SSR primers from the *B. rapa* genome sequence, an open access software program, QDD version 3, was used to detect microsatellites (SSRs) in the whole genome and primers were designed from the targeted regions (Meglecz et al. 2010).
8. **Polymerase Chain Reaction (PCR):** Polymerase chain reaction was performed in a total volume of 14 μl containing 45 ng (15 ng/μl × 3 μl) of DNA, 2.5 μl of 1× PCR buffer, 1.0 μl of 25 mmol/l $MgCl_2$, 0.35 μl of 0.4 mmol/l dNTPs, 0.5 μmol/l of each of the forward and reverse SSR primers, 0.25 pmol/l of labeling dye, 0.125 U of GoTaq DNA Polymerase enzyme, and 6.995 μl of ddH_2O. Amplification reactions were carried out using a Gene-Amp PCR system 9700 starting with 95 °C for 2 min, followed by 35 cycles of denaturation at 95 °C for 30 s, annealing at 56 °C for 30 s, and extension at 72 °C for 45 s, with a final extension at 72 °C for 15 min. The PCR products were labeled with an M13 tailing technique as described by Schuelke (2000). For this, each forward SSR primer was attached to a universal M13 primer sequence 5=-CACGACGTTGTAAAACGAC-3= fluorescently labeled with infrared FAM, VIC, NED, and PET dyes. Electrophoretic separation and detection of the amplification products was performed using a capillary ABI sequencer No. 3730 and was scored and analyzed using Genemapper software version 3.7.
9. **ix. Map Construction and Identification of the Locus Controlling Clubroot Resistance:** All primer pairs that showed polymorphism between the two parents were used to genotype the DH population. Based on segregation in F2, BC1, and DH populations, it was evident that a major gene is involved in the control of resistance to *P. brassicae* pathotypes in Rutabaga-BF. Therefore, in the first step, single marker analysis was performed with the DH population using the polymorphic SSR markers to identify the chromosome(s) carrying the resistance gene(s). This analysis indicated that SSR markers from chromosome A8 were associated with resistance.

   Therefore, a genetic linkage map of this chromosome was constructed to identify the genomic region associated with resistance as well as to identify markers from this region. The linkage map was constructed using the program Joinmap 4.0 (Van Ooijen 2006). For this, a LOD threshold of 3.0 and a recombination frequency of 0.40 were used to group the marker loci. The recombination frequencies were transformed into centi Morgans (cM) using the Kosambi mapping function (Kosambi 1943). Composite interval mapping analysis was performed by using the linkage map to identify the genomic region carrying the clubroot resistance gene as well as to identify the linked markers for use in marker-assisted selection. For this, genotypic and phenotypic data were subjected to Inclusive Composite Interval Mapping (ICIM) analysis using the software program QTL Ici Mapping version 4.0 (http://www.isbreeding.net), and the linkage map was scanned for association with phenotypic variation for resistance to *P. brassicae* pathotypes 2, 3, 5, 6, and 8 at every 0.1 cM interval. Empirical threshold levels, which confirmed the existence of the locus at a significance level of 0.05, were obtained by analyzing 1000 permutations of the data set following the method described by Churchill and Doerge (1994). The location of the resistance locus was declared at the position where the LOD score exceeded the threshold significance level of 3.0.

10. **Candidate Gene Prediction:** To identify the physical location of the clubroot resistance gene in the *Brassica* genome, SSR marker sequences flanking the resistance locus were mapped on *B. rapa* cultivar Chiifu-401 genome sequence assembly version 1.5 (Cheng et al. 2011; www.brassicadb.org) by using electronic-PCR (e-PCR) (Schuler 1997; Rotmistrovsky et al. 2004). e-PCR is a computational procedure that searches DNA sequences that closely match the PCR primers in context of correct order, orientation, and spacing, thereby identifying sequence(s) within the genome whose size is similar to the PCR product to be generated by use of these primer pairs. When the primer sequences matched with a fragment of DNA sequence from the target chromosome and the length of the deduced fragment to be amplified by the primers is similar to the actual size of the fragment amplified by these primers, the sequences were considered to be the marker sequence. The sequence of the A8 chromosome segment of *B. rapa* cultivar Chiifu-401 (genome sequence assembly version 1.5), flanked by the SSR markers associated with resistance, was scanned using Augustus, a server-based software program that predicts the ab initio gene in the eukaryotic genomic sequence (Stanke et al. 2004; http://www.augustus.gobics.de), for detection of the gene(s) in this genomic region to be associated with resistance. The protein sequences, predicted based on the gene sequences, were scanned for similarity with Arabidopsis thaliana protein(s) of known function (https://www.arabidopsis.org) using the software program WU-BLAST2 (Gish 2003; https://www.arabidopsis.org/wublast/index2.jsp; Hasan and Rahman 2016).

8.7.2 Mapping of R-QTLs in *Brassica* to *Plasmodiophora*

1. **Collection of *Plasmodiophora brassicae* Pathotypes**: Infected clubroot galls were collected by Kim et al. (2016) and Jo et al. (2011). The Seosan isolate was assigned to pathotype group 4. The Seosan isolate was used to screen resistance and susceptibility of Chinese cabbage lines. The screening procedure was essentially the same as that previously described by Kim et al. (2016). A

susceptible Chinese cabbage line (CC-F920) and a resistant turnip line (SCNU-T2016) were selected to produce the segregating population.

2. **Production of the Segregating Population:** The resistant turnip line (SCNU-T2016) was used as a donor parent to transfer resistance-related genes to the susceptible Chinese cabbage line (CC-F920). F1 plants were subsequently selfed to obtain the F2 population.
3. **Spore Collection from Seosan Clubs:** Spores were collected from frozen club root-infected galls according to the protocol described by Feng et al. (2013). A 10% (w/v) sucrose solution was used to homogenize galls with a mortar and pestle. The liquid mixture of sucrose solution with spores was passed through a neight-layered cheesecloth. The suspension was centrifuged at 50 × *g* for 5 min. The supernatant was separated and centrifuged at 2000 × *g* for 5 min. The second centrifugation process produced two layers: the upper layer containing spores with a white to brownish appearance was distinctly separated from supernatant. This upper layer was initially suspended in 5 ml of double-distilled water with gentle pipetting and then more water was added to a final volume of 40. The spore concentration was adjusted to 1×10^7 ml^{-1}.
4. **Procedure for Infecting Chinese Cabbage with the Seosan Isolate and Scoring Disease:** Seedlings from parental, F1, and F2 lines were inoculated with concentrated spore suspension (1×10^7 ml^{-1}) of the Seosan isolate at 10 days after sowing. There were three replicates for each line. About 5.0 mm root tip was excised from each seedling and roots were dipped into 5.0 ml spore solution. Each of the inoculated seedlings was transplanted into a single pot. The seedlings were grown in a plant culture room at 25 ± 1 °C and a 12 h day and 12 h night cycle. Each individual pot was covered by a perforated, transparent polythene bag to maintain high internal relative humidity. Club formation was investigated at 1-week intervals. At 35 days after inoculation, infection was scored on a disease index (DI) following a modified method with a scale of 0–4, where 0 = no gall, 1 = a single club or a few small clubs of less than 5 mm in diameter formed on the lateral roots, 2 = a few clubs larger than 5 mm on the lateral roots or main root, 3 = a few medium size club 5–10 mm on the main root, and 4 = severe clubbing larger than 10 mm on either the main root or lateral roots. Plants with an average DI of 0.0 were classified as resistant, and plants with scores higher than 1.0 were classified as susceptible.
5. **Extraction of DNA from Parental and F2 Population Lines:** To extract genomic DNA (gDNA), young leaves from the F2 lines were collected, immediately snap frozen with liquid nitrogen, and stored at 80 °C. Genomic DNA was extracted using the QIAGEN DNeasy plant mini kit.
6. **Double Digest RAD-Seq Analysis:** A total of 45 F2 lines from each of the resistant and susceptible phenotypes from a cross between the clubroot-susceptible line (CC-F8920) and the CR line (SCNU-T2016) as well as the parental lines were selected for ddRAD-seq analysis. A set of ddRAD-Seq libraries were constructed for selected 96 lines. The double digest RAD-Seq (ddRAD-- Seq) analysis was executed using the procedures described by Shirasawa et al. (2016). Double digestion of 250 ng genomic DNA from each

line was accomplished using two restriction enzymes PstI and MspI. The double-digested genomic DNA was then ligated to adapters using the Liga Fast Rapid DNA Ligation System, and purified using Agencourt AMPure XP to eliminate short DNA fragments (<300 bp). Purified restriction-digested DNA was diluted with water and amplified with PCR using indexed primers. The PCR mixture (50 μl) contained 0.4 ng DNA, 0.2 μM of each indexed primer (one pair per mixture), 1× PCR buffer of KOD –plus– Ver. 2, 160 μM dNTPs, 1 mM $MgSO_4$, and 1 U DNA polymerase (KOD–plus–; Toyobo). Thermal cycling conditions were as follows: a 3 min initial denaturation at 95 °C; 20 cycles of 30 s denaturation at 94 °C, 30 s annealing at 55 °C, and 60 s extension at 72 °C; and a final 3 min extension at 72 °C. Amplicons were pooled and separated on a Blue Pippin 1.5% agarose cassette. Fragments of 300–900 bp were gel purified using the QIAGEN Mini Elute Kit.

The KAPA Library Quantification Kit was used for measuring concentrations of the DNA libraries on an ABI-7900HT real-time PCR system. The libraries (one for the F2 mapping population and another for the parental lines) with index tags distinguishing each line were pooled and sequenced on an Illumina HiSeq (in paired-end, 93 bp mode), yielding an average of 2.8 million reads per line. The reagent used in sequencing was TruSeq SBS Kit v3-HS 200 Cycles.

7. **Computational Processing of ddRAD-Seq Data:** The FASTX-Toolkit (http://hannonlab.cshl.edu/fastx_toolkit) was used to remove low-quality sequences and to trim adapters by using PRINSEQ (−trim_right 1 -trim_qual_right 10 -derep) and fastx_clipper (−a AGATCGG AAGAGC -M 10 -n) programs. About 1.8 million high-quality reads were obtained for each line after removing low-quality sequences and trimming of adapters. The filtered reads were mapped onto the *B. rapa* genome sequence (IVFCAASv1, using Bowtie 2 (version 2.1.0; parameters: --minins 100 --no-mixed) (NCBI accession: SRR7772575-SRR7772670) (Langmead and Salzberg 2012). This *B. rapa* genome database is similar to another *B. rapa* genome database, BRAD. The resulting sequence alignment/ map format (SAM) files were converted to binary sequence alignment/map format (BAM) files and subjected to single nucleotide polymorphism (SNP) calling using the mpileup option from SAM tools (version 0.1.19; parameters: -Duf) and the view option from BCF tools (parameters: -vcg). Furthermore, variant call format (VCF) files were filtered with VCF tools (version 0.1.11; parameters: --minQ 20--minDP 10 --max-missing 0.5 –maf 0.2 --min-alleles 2 --max-alleles 2 --remove-indels for the F2 data), in which parents must be homozygous and different genotypes. Any missing value was attributed using Beagle4 (Laila et al. 2019).
8. **Linkage Map Construction:** The linkage map was constructed using QTL Ici Mapping software version 4.1.0.0. To construct the linkage map with 4230 SNP markers, 2154 markers were randomly selected including at least 200 makers from each chromosome. Among the 2154 SNP markers, 2152 markers appropriately anchored with the 10 chromosomes in *B. rapa*. QTL mapping was conducted using Windows QTL cartographer version 2.5_011. During QTL mapping, 1000 permutations and a 0.05 significance level were applied. Both

interval mapping (IM) and composite interval mapping (ICM) functions were used in Ici Mapping (Lei et al. 2015).

9. **Probe and Primer Design, and HRM Validation:** For designing probes and primers to validate each of the 17 SNPs with high LOD scores (over 20), 300 bp nucleotide sequences of *B. rapa* genome were retrieved from the Ensemble Plants database (IVFCAASv1) with the target SNP positioned as the 150th nucleotide. Each of the 17 probes was designed with a phosphorylated 3′ end. The probes were synthesized. All probes were designed to be intergenic. A total of 17 forward and reverse primer sets were also developed to amplify the designed probe sequences. A real-time PCR thermocycler was used for HRM analysis. The HRM conditions, composition of the HRM-PCR mixture, and HRM curve analyses, were done according to the protocol used by Laila et al. (2017). A PCR mixture for HRM was prepared by mixing 1 μl genomic DNA at 5 ng μl^{-1}, 0.1 μl forward and 0.5 μl reverse primers at 10 pmol l^{-1}, 1 μl probe, 7.4 μl ultra-pure water, and 10 μl of Quanti Speed HRM Kit master mix.
   The PCR was conducted with the following conditions: pre-incubation for 10 min at 95 °C, followed by 50 cycles of amplification including denaturation for 20 s at 95 °C, annealing for 20 s between 63 and 55 °C under touchdown command, and melting for 20 s at 72 °C. The amplified PCR products were then subjected to the following HRM cycle: 60 s at 95 °C, 120 s at 40 °C, and 1 s at 83 °C. The fluorescence value ($-dF/dT$) was recorded during the final step with five readings per °C. The Light Cycler 96 software was used to analyze the HRM curves. Each of the samples was appropriately labelled in sample editor. Both delta Tm discrimination and curve shape discrimination were set at 75%, whereas a 0.2 positive/negative threshold level was set to obtain the final output (Laila et al. 2019).
10. **PCR Confirmation of Novelty for the New Resistance Locus:** To confirm that the newly identified locus is novel and different from the adjacent resistance locus *Crr1*, a set of11 samples were tested with a *Crr1*-specific primer (Forward sequence: GATTACCACTATGTACTGAACT and reverse sequence: CTTTCAAAAACGATTGAAATTTCAT). The samples were Akimeki (*Crr1*-resistant, homozygous), Cheonghajanggun (*Crr1*-resistant, heterozygous), Bulam-3-ho (*Crr1*-susceptible, homozygous), Seosan-resistant parent (SCNU-T2016, two samples), Seosan-susceptible parent (CC-F920, two samples), and two samples each from the resistant and susceptible F2 lines (one homozygous and one heterozygous from both types). DNA from the Chinese cabbage lines, Akimeki, Cheonghajanggun, and Bulam-3-ho, was extracted following the protocol utilized for F2 lines. PCR amplification of DNA was conducted following previously described methods. Similarly, a separate PCR amplification was conducted with six samples (two Akimeki, and two lines from each resistant, heterozygous-resistant and susceptible F2 genotypes) to show that the candidate *Rcr9* gene (Bra020936, Forward sequence: AGTGGAGAA CCAAAGCCAAA and reverse sequence: CAATTGTG GCCACCTTCTTC) is not CRs (Laila et al. 2019).

8.7.3 SNP Array, Mapping, Population Structure, and Linkage Disequilibrium Analysis Techniques

The young leaves from the representative plants of all accessions were collected, and extracted their genomic DNAs using a modified cetyltri-methyl ammonium bromide (CTAB, Murray and Thompson 1980) method. SNP genotyping of 472 accessions was performed using *Brassica* 60K Illumina Infinium R HD Assay SNP arrays according to the manufacturer's protocol 2. The quality control process of SNP array genotyping data was the same to previous study by Li et al. (2014). The LD was calculated using the software TASSEL 3.0 (Bradbury et al. 2007). The principal component analysis (PCA), population structure and kinship were analyzed by Li et al. (2014). To confirm the physical position of each SNP, the probe sequences of 26,841 high-quality SNPs selected (Li et al. 2014) were used to perform a BlastN search against the *B. napus* (*Darmor-bzh*) reference genome by Li et al. (2016). Only the top blast hits were considered to be mapped in the genome with an *E*-value threshold of 1*E*–15, and blast matches to multiple loci with the same top *E*-value were not considered to be mapped successfully. Totally, 19,945 SNPs were assigned to *B. napus* chromosomes (Li et al. 2016).

8.7.4 Genome-Wide Association Study

To evaluate the effects of population structure (Q, PC), the trait-SNP association analysis was performed using TASSEL 4.0 with six models: (1) GLM, controlling for nothing, (2) Q model, controlling for Q, (3) PCA model, controlling for PC, (4) K model, controlling for K, (5) Q+K model, controlling for both Q and K, (6) P+K model, controlling for both PC and K. The GLM, Q, K, and PCA models were performed using a general linear model with the following equation: $y = X\alpha + e$. The Q+K and P+K models were performed using a mixed linear model with the following equation: $y = X\alpha + K\mu + e$. In above equations, y represented phenotype, X represented genotype, α was a vector containing fixed effects, K was the relative kinship matrix, μ was a vector of random additive genetic effects, and *e* was the unobserved vector of random residual. The distribution of observed $-\log_{10}(p)$ for each SNP from marker-trait associations was compared with the expected distribution in a quantile-quantile plot(Q-Q plot). The significance of associations between SNPs and traits was based on a threshold $P < 5.01 \times 10^{-5}$ [$P = 1/N$, where $N =$ the number of SNPs used, $-\log_{10}(1/19{,}945) = 4.3$]. Meanwhile, to avoid ignoring the effects of minor loci, the threshold *P* was lowered to $2/N = 1.00 \times 10^{-4}$. Q-Q plots and Manhattan plots were drawn using the R package qqman[3] (Li et al. 2016).

8.7.5 QTL Alignment and Candidate Gene Prediction

To compare the QTLs detected in a study with previous studies on CR, the sequences of QTLs/genes which have been reported in previous studies were collected and performed BlastN or e-PCR on the *B. napus* reference genome (*Darmor-bzh*) with a threshold of 1*E*–10. We could ensure the homoeologous fragments' physical positions of QTLs/genes reported in studies on *B. napus* genome. It was regarded as an LD blocks that the region containing all SNPs which $r^2 > 0.4$ with the most significant SNP. To ascertain regions that potential candidate genes of each QTL identified in may be located by considering the flanking markers that outside and adjacent to the LD lock as the candidate region's boundary. To further identify the candidate genes and the participated pathways of the identified loci, the gene function annotation and enrichment analysis were carried out. All *B. napus* genes which located in the candidate regions were searched against the NCBI nonredundant (Nr) protein database using Blast P with an *E*-value ≤1*E*–05. The gene ontology (GO) terms of the *B. napus* genes were annotated by merging the Blast 2GO and Inter Pro annotation results (Wu et al. 2016). GO enrichment analysis provided all of the GO terms that were significantly enriched in candidate genes compared with the genome back ground using Blast 2GO with a false discovery rate (FDR) ≤0.01. The published differential expressed genes (DEGs) information in *B. rapa* were downloaded (Chen et al. 2015), and searched for the homoeologous genes in *B. napus* with at threshold of 1*E*–50 (Li et al. 2016).

8.8 *Brassica-Sclerotinia*: Molecular Techniques

8.8.1 Transcriptome Analysis of *Brassica-Sclerotinia* Pathosystem

1. **Plant Material and Fungal Strain:** Two *B. napus* doubled haploid lines differing in their susceptibility to *S. Sclerotiorum* were used. Both lines, NEP32 (susceptible) and NEP63 (resistant), were developed via microspore culture and were evaluated for their reaction to SSR. The susceptible line NEP32 was derived from spring type canola variety Helga (PI649136), and the resistant line NEP63 was developed from a cross between Helga and winter type canola accession PI458940. When inoculated using the petiole inoculation technique, the susceptible line develops lesions on the stem that expand to >4 cm in length with 100% girdling within 8 days post inoculation, eventually leading to wilting and death. Within the same period, the average lesion size on the resistant line is limited to <1 cm in length and <40% stem girdling, typically surrounded by purple margins suggesting accumulation of anthocyanins. The highly aggressive strain 1980 of *S. sclerotiorum* was used for inoculations and culture controls. Strain 1980 was chosen as it is the genome sequenced strain (Derbyshire et al. 2017; Kabbage et al. 2015).
2. **Plant Growth and Inoculation**: The seeds of NEP32 and NEP63 were surface sterilized by soaking them in 3% NaOCl for 1 min followed by immersion in 70% EtOH for 1 min and were rinsed three times in sterilized deionized water for 1 min. Seeds were planted in SunGro Sunshine mix #1 in 4 × 10 plastic plots with one seed per pot and plants were grown for 4 weeks in a growth chamber with a 16 h photoperiod, 21 °C/16 °C day/night temperature, and 60% relative humidity.

Once germinated, seedlings were watered as necessary and fertilized once a week with 20-20-20 fertilizer. Four weeks after planting, plants were infected following an established petiole inoculation technique. The petiole of one leaf per plant was cut approximately 2.5 cm away from the stem and the stump was capped with PDA plugs containing hyphal tips of an actively growing 2-days-old, *S. sclerotiorum* colony using 1 ml pipette tip. Inoculated petioles (≈2.5 cm) were harvested at 8, 16, 24, and 48 h post inoculation (hpi), flash frozen in liquid nitrogen and stored at −80 °C until RNA extraction. Each treatment-time point consisted of three biological replicates; petioles collected from five individual plants were pooled to constitute one biological replicate (Zhao et al. 2004).

3. **RNA Extraction, Library Preparation, and Sequencing:** Total RNA was extracted from inoculated petioles using RNeasy mini kit, and mRNA was isolated using Oligotex mRNA mini kit following manufacturers' instructions. cDNA libraries were constructed using NEB Next mRNA sample prep kit and NEB Next Multiplex Oligos kit. Similar procedure was followed to construct cDNA libraries from *S. sclerotiorum* mycelia grown on PDB in 10 cm Petri plates for 48 h under the same conditions described above. For sequencing, RNA from 8 and 16 hpi were pooled and referred as early infection stage (T1). Similarly, RNA from 24 and 48 hpi were pooled and referred as late infection stage (T2). mRNA libraries were sequenced at the University of Minnesota Biomedical Genomics Center. Of the three biological replicates, two replicates per each treatment-time point were sequenced on Illumina GA-IIx platform (1× 76 bp Single Reads), while one replicate was sequenced on Illumina HiSeq 2500 platform (2× 50 bp Paired End).
4. **Quality Control and Read Mapping to the Reference Genome:** Quality checks for the raw fastq files were conducted through a pipeline consisting of FastQC and FastX toolkit (http://hannonlab.cshl.edu/fastxtoolkit/) to retain reads that were at least 50 bp long with a minimum quality score of 30. Reference genomes and the corresponding annotations for *S. sclerotiorum* and *B. napus* were downloaded from ensemble fungi (ftp://ftp.ensemblgenomes.org/pub/release-42/fungi/) and genoscope (http://www.genoscope.cns.fr/brassicanapus/data/), respectively. The most recent full genome sequence of *S. sclerotiorum* (Sclerotinia sclerotiorum_1980_uf_70_gca_001857865) was used. Indexes for each reference genome were built and quality filter-passed reads were mapped to the reference genomes using HiSat2. Raw read counts mapped to each gene from the HiSat2 generated alignments were obtained using the feature Counts command of the Sub-read package (Kim et al. 2015; Liao et al. 2013).
5. **Differential Gene Expression Analysis:** Differential gene expression (DGE) analysis was conducted using edge R package. The raw count data were normalized with trimmed mean of means (TMM) normalization method implemented in edge R. Principal component analysis (PCA) was conducted to determine relatedness of the biological replicates. PCA plots were generated using scatter plot 3d package in R. Statistical analysis was performed using negative binomial distribution extended to generalized linear models. Pairwise contrasts were performed following quasi-likelihood F tests. A false discovery

rate (FDR) cut-off of 0.05 was applied to account for multiple testing corrections. A gene was considered as differentially expressed when the change in expression level was ≥2fold (absolute value of log 2-fold change (l2fc) ≥ 1), with an FDR-adjusted *p* value <0.05 (Lun et al. 2016).

6. **Functional Classification and Enrichment Analyses of DEGs:** Gene ontology (GO) enrichment analysis was performed in Blast2GO. GO terms were assigned to *S. sclerotiorum* total genes list in Blast2GO, which were used as the background list for enrichment analysis. Fishers' exact test implemented in Blast2GO was used to identify significantly enriched GO categories. A GO category was considered significantly enriched only when the *P*-value for that category was <0.05 after applying FDR correction (Conesa and Stefan 2008).
7. **Differential Gene Expression Data Validation:** The RNA-seq differential gene expression data was validated by performing qRT-PCR on six selected genes. Primers for qRT-PCR were designed using Primers-Blast. QuantiTect reverse transcription kit was used to synthesize cDNA from total RNA. Real-time quantification was performed using two technical replications in a BioRad CFX96 Real-Time system using iTaq Universal SYBR Green Super mix with the following cycling conditions: 95 °C for 30 s followed by 40 cycles of 95 °C for 10 s and 60 °C for 30 s. Expression levels of the DEGs were normalized against the *S. sclerotiorum* actin gene (sscle_14g099090) and the relative expression levels were calculated using the $2^{-\Delta\Delta Ct}$ method (Chittem et al. 2020).

8.8.2 Identification and Pyramiding of R-QTLs in *Brassica* to *Sclerotinia*

1. **Plant Materials and Breeding Process:** A schematic diagram of the breeding process is presented in Fig. 8.1. Interspecific hexaploid hybrid (AnAnCnCnCoCo) was generated between a Chinese semi-winter rapeseed variety "Zhongshuang 9" (female parent) which possesses partial resistance to *S. sclerotiorum* and a wild *B. oleracea* genotype "C01" (*B. incana*, male parent) which exhibited high resistance to *S. sclerotiorum* (Mei et al. 2015). The hexaploid progeny was backcrossed with Zhongshuang 9 for one round and then selfed until BC1F8. The selection in BC1F1 and BC1F2 was reported previously (Mei et al. 2015), and two elite BC1F2 individual plants, that is, BC1F2-1 and BC1F2-3, were selected to develop the following generations.
2. **Marker Assisted Selection:** According to the resistant QTLs identified from the wild *B. oleracea* "C01" (Mei et al. 2013), 28 simple sequence repeat (SSR) markers located within and adjacent to all the QTL intervals were screened for polymorphism between wild *B. oleracea* and Zhongshuang 9. Seven polymorphic markers, which were linked to one major QTL for leaf resistance on linkage group (LG) C01 and two major QTLs on LG C09 for both stem and leaf resistance, were finally used to trace the three resistance loci. For one locus, plants exhibited the same genotypes (bands) as the wild *B. oleracea* "C01" in all the markers linked to the QTL was considered to carry this resistant locus.

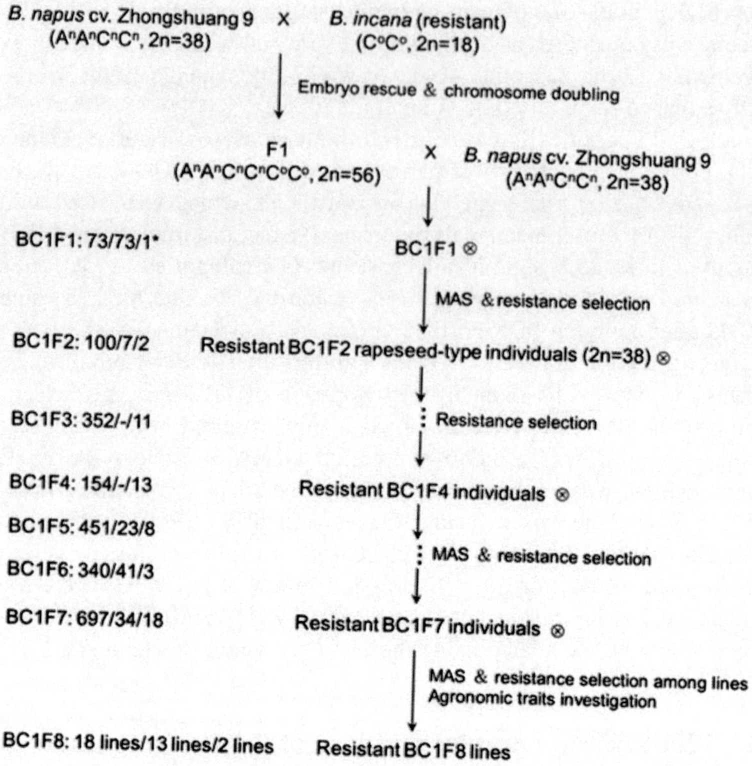

Fig. 8.1 A diagram of breeding process of *Sclerotinia*-resistant Brassica napus lines by using *Brassica incana* as the donor of resistance. Asterisk represents the number of plants for selection/the number of plants selected by MAS/the number of plants planted for the next generation (Mei et al. 2020)

3. **Resistance Evaluation:** The detached inoculation technique was conducted under controlled environment according to Mei et al. (2013) by using the same *S. sclerotiorum* isolate at ~80 days after sowing (DAS) (around nine-leaf stage) for leaf resistance and ~205 DAS (around 2-week post end flowering) for stem resistance. The infection temperature was maintained at 22 ± 1 °C. Lesion size/length on detached leaves/stems was recorded at 2/3 days after inoculation. In order to evaluate the resistance level of offspring, the susceptibility (*S*) relative to the rapeseed parental line Zhongshuang 9 was calculated with the equation $S = V/V_{\text{Zhongshuang 9}}$, where *V* is the lesion size (leaf) or lesion length (stem) in the progenies, and $V_{\text{Zhongshuang 9}}$ is that of Zhongshuang 9.
   Natural field resistance was investigated in BC1F8 generation. Lines were arranged in a randomized complete block design with three replications. Plot size was 2 by 2.4 m, with 60 plants in six rows. Plots were naturally infested with *S. sclerotiorum*, and disease incidence (percentage of infected plants in total

plants) was recorded at 210 DAS (3-week post end flowering). Disease index (*I*) was calculated by the formula

$$I = \frac{\Sigma\ (di \times li)}{L \times 4} \times 100$$

where *di, li,* and *L* represent the disease grade (classified on a scale of 0–3) of each plant, the number of plants in each grade and the total number of plants, respectively (China 2011). Disease severity was classified on a scale of 1–5 based on disease incidence and disease index according to the agricultural trade standard on investigation of rape Sclerotiniose (China 2011).

4. **Seed Yield and Seed Quality:** Seed setting (seeds per pod) was investigated in five plants per line, with 30 continuous siliques sampled from the main inflorescence of each plant. Seed yield per single plant was measured from ten open-pollinated plants in each plot. The 1000-seeds weight was measured from five open-pollinated plants. Seed quality was determined by near-infrared reflectance spectroscopy (NIRS) from 5 to 10 self-pollinated plants in each line, with three technical replicates.
5. **Data Analysis:** Statistical analyses such as analysis of variance (ANOVA), *F*, and *t* tests were performed using the SAS software, version 6.07 (SAS Institute 1992) (Mei et al. 2020).

8.8.3 Identification of Genome-Wide Associated R-Loci in *Brassica* to *Sclerotinia*

1. **Plant Materials:** The association panel comprised 448 rapeseed inbred lines, including 420 accessions from Asia, 18 accessions from Europe, 5 accessions from North America, and 5 accessions from Australia (Liu et al. 2016). Population structure and genetic relatedness analyses showed that most of the accessions included were distantly related (Liu et al. 2016).
2. **Field Experiments:** All 448 accessions were grown in two consecutive years. All field experiments followed a randomized complete block design with two replications. Each accession was grown in a plot with 10–12 plants in each of two rows, with a distance of 21 cm between plants within each row and 30 cm between rows. The field management regime followed essentially regular breeding practices.
3. **Assessment of Resistance to *S. sclerotiorum*:** The *S. sclerotiorum* isolate SS-1 was maintained and cultured on potato dextrose agar (PDA, 25% potato, 2.5% dextrose and 1.5% agar, pH 5.8; Wu et al. 2013). The isolate was maintained at 4 °C in darkness and cultured twice prior to inoculation at 23 °C in darkness. Mycelial agar plugs (7 mm in diameter) punched from the margin of a 2-day-old culture of *S. sclerotiorum* grown on PDA were used as the inoculum. Sclerotinia stem rot resistance was assessed using the detached stem inoculation assay according to Mei et al. (2012) with some modifications. The stems of 3–6

individuals of each accession in each replicate with a length of approximately 30 cm were cut using a sharp knife at a height of 20 cm above the ground, and the cut ends were immediately wrapped with polyethylene to retain freshness. Subsequently, the detached stems were placed in plastic trays (56 × 38 × 15 cm) with wet gauze at the bottom of the trays. Each 30-cm stem was inoculated with a mycelial agar plug at two points at an interval of 10–13 cm. The inoculated stems were sprayed with a fine mist of water, and the plastic tray was covered with plastic film to maintain a high level of relative humidity. The plastic trays with inoculated stems were maintained at 22 ± 2 °C in darkness. The length of lesions along the stems was measured after 5 days post inoculation (dpi). This assessment was conducted with two replications at the stage of flowering termination.

4. **Statistical Analysis:** Broad-sense heritability was calculated as $h^2 = \alpha^2_g/(\alpha^2_g + \alpha^2_{ge}/n + \alpha^2_e/nr)$, where s^2_g is the genetic variance; s^2_{ge} is the interaction variance of the genotype with the environment; s^2_e is the error variance; n is the number of environments; and r is the number of replications. The estimates of s^2_g, s^2_{ge}, and s^2_e were obtained through ANOVA (analysis of variance) using SAS9.3. Best line a run biased predictions (BLUPs) for each line across the two environments were calculated using a MIXED procedure in SAS 9.3 as phenotypes for the association analysis.
5. **Genome-wide Association Analysis:** The association panel was genotyped using *Brassica* 60K Illumina Infinium R HD Assay SNP arrays (Clarke et al. 2016) according to the manufacturer's protocol. Quality control and mapping of the SNPs to *B. napus* "Darmor-bzh" reference genome (Chalhoub et al. 2014) were performed and 25,573 SNPs retained for further analysis as described earlier (Liu et al. 2016). Principal component analysis (PCA) based on all the 25,573 SNPs was performed using the GCTA tool (Yang et al. 2011). The relative kinship matrix of 448 *B. napus* lines was generated using the SPAGeDi software package (Hardy and Vekemans 2002). Negative values between two accessions were set to 0. Trait-SNP association was performed using the PCACK model implemented in TASSEL 4.0 (Bradbury et al. 2007). The significance of the associations between SNPs and the trait was based on a single threshold of $P < 3.91 \times 10^{-5}$ ($P = 1/n$, where $n =$ the number of markers used; $-\log_{10}$ $(1/25{,}573) = 4.4$). The linkage disequilibrium (LD) measurement parameter r^2 was employed to estimate the LD between SNPs using the software package TASSEL 4.0 (Bradbury et al. 2007). Only homozygous SNPs were used, and heterozygous SNPs were set to missing. Marker haplo types and LD blocks were defined for each associated locus when flanking markers showed strong LD ($r^2 > 0.4$) with a lead SNP, as described by Hatzig et al. (2015). The interval of the LD block was considered for candidate gene disclosure.
6. **Alignment of SSR Resistance QTLs to the *B. napus* Genome:** Quantitative trait locus for SSR resistance were collected from previous bi-parental QTL mapping studies (Zhao et al. 2006; Wu et al. 2013; Wei et al. 2014) and a GWAS (Wei et al. 2015). All QTLs were aligned to the *B. napus* "Darmor-bzh" reference genome (Chalhoub et al. 2014) based on the physical positions of markers in confidence intervals or LD blocks. The physical localization of the

markers of RFLP (restricted fragment length polymorphism), SSR (single sequence repeat), and SNP (single nucleotide polymorphism) was determined using electronic PCR [e-PCR, (Schuler 1997)] or Blast N (Altschul et al. 1997) employing the marker primer or probe sequences. For SSR markers, e-PCR was performed with the primer sequences, using the genome sequences as templates (Cai et al. 2012). The parameters for e-PCR were set to allow three mismatches and one gap for a given primer pair. If a pair of primers had multiple amplicons, subsequently the amplicon for the primers were selected according the linkage group of the QTL (Cai et al. 2012). For RFLP markers, the probe sequences were used to perform BlastN1 http://www.illumina.com/searches against the reference genome sequences with an *E*-value of 1*E*-20.If the probe aligned to multiple loci, subsequently hits were selected according to the linkage group of the QTL from the top several blast hits. The physical localization of the SNP markers used in the GWAS was performed (Wei et al. 2015; Liu et al. 2016; Wu et al. 2016).

8.8.4 Marker Assisted Introgression of Resistance into *Brassica juncea* Against *Sclerotinia*

1. **Plant Materials:** A total of 96 *B. juncea–E. cardaminoides* ILs were developed previously through wide hybridization between *B. juncea* and *E. cardaminoides* as described.
2. **Chromosome Preparation and In Situ Hybridization:** The method for preparation of chromosome spreads was used as described by Rana et al. (2017). To prepare species-specific genomic in situ hybridization (GISH) probes, purified DNA of *B. cardaminoides, B. nigra,* and *B. juncea* were extracted using the DNeasy plant mini kit (Qiagen) as per manufacturer's specifications. Genomic DNA of *E. cardaminoides* was labeled with rhodamine-5-dUTP dye (red color). DNA of *B. nigra* was labeled with fluorescein-12-dUTP dye (green) using a nick translation kit. Two-step hybridization to perform GISH as described earlier (Rana et al. 2017). Visualization was carried out with Zeiss fluorescent microscope (Imager Z2A X 10). Digital images were captured using IsisR software. Images were cropped and optimized with Image J using only functions affecting the whole image.
3. **Screening of ILs for Resistance Responses to *S. sclerotiorum* Inoculation:** For field evaluations, ILs were planted in a randomized complete block design, with 20 plants in paired rows of 2 m length with inter-row spacing 30 cm. The phenotypic screening was carried out in two replications. Foggers (with one fogger per 9 m^2) were used to maintain high humidity for disease development, operating 2–3 times a day for 15 min each. Phenotyping of ILs for their resistance responses was carried out. Stem inoculations were carried out by using isolate PAU-4, an aggressive and virulent strain of *S. sclerotiorum* collected from infested *Brassica* fields. PAU-4 is a known highly virulent isolate on *B. juncea* (Rana et al. 2017). *Sclerotium* was first surface sterilized, cultured, and subcultured using standard procedures (Clarkson et al. 2003; Li et al. 2006). The

parents, viz., the recipient (*B. juncea* cv. RLC-1) as well as the donor (*E. cardaminoides*), were used as susceptible and resistant checks, respectively. The stem inoculation method (Buchwaldt et al. 2005; Bradley et al. 2006) was used for disease inoculation. The ILs were evaluated for disease incidence and development at 4–5 weeks post-inoculation. Ten plants per genotype per replication were scored for disease incidence by measuring stem lesion lengths according to the method described by Li et al. (2006, 2007). The resistance responses of ILs were categorized into five classes, based on stem lesion length, viz., highly resistant (HR), (0 < 2.5 cm); resistant (R), (2.5 < 5.0 cm); moderately resistant (MR), (5.0 < 7.5 cm); susceptible (S), (7.5 < 10.0 cm); and highly susceptible (HS), (>10.0 cm) as categorized earlier by Garg et al. (2010). Analysis of variance (ANOVA) was undertaken and standard deviation and standard errors were estimated using SAS software.

4. **DNA Extraction:** A total of 96 ILs comprising all five disease expression categories (HR, R, MR, S, and HS) were used for molecular profiling. DNA was harvested from young leaves with minor modifications of CTAB (cetyltrimethylammonium bromide) standard procedure (Doyle and Doyle 1990). Isolated DNA was further digested with RNase at 37 °C for 1 h. DNA was quantified by NanoDropR 2000 spectrophotometer and for purity by assessing the OD260/OD280 ratio and Qubit^R 2.0 fluorometer quantitation to measure DNA concentration. DNA samples with OD260/OD280 ratio of 1.8–2.0 and total amount >1.5 μg were qualified for library construction.
5. **SSR Genotyping:** A total of 96 ILs, along with resistant and susceptible parents, were first genotyped by using 100 polymorphic SSR markers. These markers were identified as transferable based on studies involving over 650 A- and B-genome-specific markers (Kim et al. 2009; Lowe et al. 2004). Sequences of 48 B-genome specific SSRs were obtained under MTA from Isobel Parkin. PCR reactions were performed in a 384-well Applied Biosystems thermocycler. Amplified DNA product was fractioned using an automated high-throughput electrophoresis system. Allelic polymorphism of all markers was recorded and mapping positions were inferred from published data (Lowe et al. 2002, 2004; Kim et al. 2009).
6. **Genotyping by Sequencing:** A smaller set of 84 ILs was genotyped by sequencing (GBS) (Elshire et al. 2011). Genomic services were outsourced. For this, high quality DNA of each sample was digested with appropriate combination of restriction enzymes based on in silico evaluation. This step was followed by several rounds of PCR amplification. Samples were then individually pooled and size-selected for the required fragments to complete the library construction. High quality libraries with appropriate insert sizes were then used for pair-end sequencing on Illumina^R HiSeq platform, with the read length of 150 bp at each end. The sequences and corresponding sequencing quality information were stored in a FASTQ file. The adapter sequences were removed from raw reads using the software Cutadapt (Martin 2011). The available genome sequence of *B. juncea* v1.51 (Yang et al. 2016) was used for reference-based alignments of whole genome sequences (25×) from the four

most prominent ILs (>25×), using software bowtie2 (Langmead and Salzberg 2012). Initially, one introgression line was aligned into the reference genome and SNP called using the NGSEP-GBS pipeline (Duitama et al. 2014). Total SNPs were replaced in genome reference using a perl script, pseudomaker.pl implemented in SEG-Map (Zhao et al. 2010) to construct the first step mock-up pseudomolecules, which were then used as a reference for next ILs. This process was repeated four times to construct final mock-up reference for alignment of sequence tags. Identification of SNPs was carried out by using NGSEP-GBS pipeline (Duitama et al. 2014) after aligning the paired end reads of 84 introgressed lines on final mockup reference genome. The resulting marker dataset comprised 30,863,034 SNPs. These were then filtered to include only quality SNPs for further analysis. Filtering parameters such as minimum mapping quality, minor allele frequency (0.1), only bi-allelic SNPs, minimum number of samples genotyped, maximum observed heterozygosity, and maximum missing calls (30%) were used for finding putative SNPs. After filtering, 78,578 SNPs were identified and imputed using fcgene and Beagle (Browning and Browning 2016) software.

7. **Association Mapping Based on SSR Genotyping:** The normalization of phenotypic data was done by using PBTools software2 in R-Package version 1.5 (R Core Team 2013). The marker trait associations (MTAs) were identified by using two models executed in TASSEL version 2.1 (Bradbury et al. 2007). GLM (generalized linear model) and MLM (mixed linear model) models were used. Bayesian model-based software, STRUCTURE version 2.2 (Pritchard et al. 2000), was used to determine the population structure by using multi-locus (SSR) genotypic data. Resultant Q-matrix was used as a covariate during association mapping analysis to reduce the bias from population structure. Association mapping was implemented in TASSEL software version 2.1 (Bradbury et al. 2007), measuring the nonrandom association between marker alleles from different loci (Zhu et al. 2008). Squared correlation coefficients between marker trait data ($R2$ values) and associated probabilities were calculated and converted into $-\log_{10}(P)$ values. The associated values are calculated with a false discovery rate (FDR) of 0.09 (Storey and Tibshirani 2003) to reduce the false marker-trait associations. LD, also known as gametic phase disequilibrium, was established between markers (Flint-Garcia et al. 2003). Annotation, or gene prediction study, of significant markers was carried out using MEGANTE software (Numa and Itoh 2014).
8. **Genome-wide Association Analysis (GWAS) Based on SNP Genotyping:** For genome-wide association analysis (GWAS), resistance responses (in terms of lesion length) were first converted into rank data and then transformed to log (x) for three crop seasons. These were also pooled over seasons. A principal component analysis (PCA) was performed across the introgressed lines to identify population stratification by MVP-GWAS tool. The imputed dataset of 78,578 SNP markers were used to calculate the PCs. First three components showed maximum variance. MVP tool5 was used for marker trait association with two different models, MLM and Farm CPU. GWAS were performed using

MLM association accounting for kinship, and GLM and Farm CPU were selected with PCs as covariate in MVP tool. R software package "adegenet" was used for applying discriminant analysis of principal components (DAPC) in the association analysis. After DAPC correction, first three discriminant functions were used as covariate in GLM model and kinship for MLM model for association analysis by software TASSELv5.2 (Bradbury et al. 2007). Manhattan plots were generated with multi-model plotting using MVP tools.

9. **SNP Validation:** Six peak SNPs associated with trait variation were identified for validation. Primers were designed from flanking region of SNP using Primer 3 software and their thermodynamic properties were confirmed by Vector NTI. Eight ILs were selected that differed for resistance responses. Genomic DNA from these test genotypes were amplified using designed primers. The PCR products were then purified and used for Sanger sequencing.
10. **In Silico Prediction of Candidate Genes:** The 25 kb flanking regions on each side of resistance associated peak SNP found to predict candidate genes using *B. juncea* pseudomolecules as a reference. The predicted genes and their orthologous sequences were then annotated by BLAST run against the *A. thaliana* database using Blast2GO v5.2.5 tool (Gotz et al. 2008). *Arabidopsis* protein database was used for gene finding as well as blast search. Protein IDs generated allowed annotation against all flowering plant databases (NCBI). These were further enriched by the biological functions inferred from the putative *Arabidopsis* orthologs. The gene ontologies of *Arabidopsis* orthologs was used for all analysis because they are far better curated than for *B. juncea*. Positions of the predicted candidate genes w.r.t. the SNPs were detected by blast searching sequences from the predicted genes against *B. juncea* mock-up pseudomolecules. Functions of the predicted candidate genes were verified from literature to determine their relevance for the trait in question (Rana et al. 2019).

8.9 *Brassica*-Turnip Mosaic Virus (*TuMV*): Molecular Techniques

8.9.1 TuMV Detection, Preservation, and Identification by ELISA

1. **Collection of Plant Samples:** Initially, fresh field specimens were tested for viruses by enzyme-linked immunosorbent assays (ELISAs), using purified anti-virus immunoglobulins, alkaline phosphatase conjugates, and ELISA protocols from Agdia, Inc. Infection of fresh leaves of *B. napus* differential lines inoculated with TuMV isolates was identified by ELISA with TuMV monoclonal antibody EMA67 (Jenner et al. 1999). However, the majority of testing of field specimens and isolates was done on stored petiole and stem blots on nitrocellulose membranes (Schwinghamer et al. 2010) by tissue blot immunoassays (TBIAs). These TBIAs were indirect procedures that involved incubation with antigen-specific primary antibodies followed by alkaline-phosphatase conjugated

secondary antibodies (anti-rabbit or anti-mouse), and finally chromogenic substrate (nitroblue tetrazolium plus 5-bromo-4-chloro-3-indolyl phosphate) to detect virus in blots or tissue prints on nitrocellulose membranes. Primary antibodies for TuMV TBIA and potyvirus TBIA were, respectively, TuMV polyclonal and general potyvirus monoclonal PTY 1, both from ELISA kits supplied by Bio-Rad Phyto-Diagnostics, INRA, France. Primary antibodies for *CaMV* and Pea seed-borne mosaic virus (*PSbMV*) TBIAs were polyclonals from DSMZ (Deutsche Sammlung von Mikroorganismen und Zellkulturen, Germany). The TBIA procedure of Lin et al. (1990) and Hsu and Lawson (1991) were modified as follows: Tris-buffered saline (TBS) (0.02 M Tris–HCl, pH 7.5, 0.15 M NaCl) was used rather than phosphate buffered saline as basic buffer for antibody dilution and membrane washing (Hammond and Jordan 1990); 1 μg/ml polyvinyl alcohol (Sigma MW 30,000–70,000) was used as pre-incubation blocking agent (Makkouk and Comeau 1994); 2% polyvinyl pyrrolidone (Sigma PVP-40) and 0.2% egg albumin were added to basic buffer for dilution and incubation of monoclonal antibodies and alkaline phosphatase conjugated antibodies.

For TBIAs that used polyclonals as primary antibodies, including TuMV, Cauliflower mosaic virus (*CaMV*), and other TBIAs, binding to host antigens was eliminated by dilution and pre-incubation at 37 °C for 2 h in strained leaf extract (1 g leaf per 50 ml TBS) of uninfected analogous host, usually wild radish, before adding the antibody mixture to membranes. Sensitivity of Potyvirus TBIA was maximized by heating membranes at 60 °C in 0.05 M Tris pH 9.5 (Jordan and Hammond 1991) for 5 min to expose antigenic determinants before blocking and incubating. Voucher specimens representing TuMV-positive hosts were lodged in the NSW Plant Pathology Herbarium (reference numbers prefaced by "DAR") and comprised one or more of the following: pressed leaves, diced leaves (desiccated over $CaCl_2$ or frozen at −80 °C), blots on replicated nitrocellulose membranes, or isolates transmitted from field specimens and stored as desiccated or frozen diced leaves.

2. **Isolates:** There were six isolates of TuMV transmitted first by aphids (*Lipaphis pseudobrassicae* or *Brevicoryne brassicae*) or rub inoculation and propagated by 1–12 serial rub inoculations of hosts including Chinese cabbage, turnip weed, mustard, and Buchan weed. All six TuMV isolates produced clear symptoms in these hosts and were positive in TuMV TBIA, positive in potyvirus TBIA, and negative in CaMV TBIA. Source plants and/or isolates except for 4297Awere also tested for a range of other mechanically transmissible viruses by species-specific TBIAs (AMV, BYMV, CMV, TSWV, and PSbMV) and found to be negative. The attempt to isolate single variants of TuMV that might occur within isolates was done by means of single lesion isolations. For TuYV, used one isolate (3434B). This was maintained in canola by aphid (*M. persicae*) transmission and determined to be BWYV-like by reaction with polyclonal antiserum to BWYV. Identification as TuYV rather than BWYV was based on infectivity in canola (Graichen and Rabenstein 1996). The experiment was conducted in a glasshouse and included seven host genotypes (six mustard and one canola) and

three inoculum classes: aphids (*M. persicae*) fed on *TuMV*-infected turnip weed, aphids fed on TuYV-infected canola, and non-acquisition fed aphids (mock inoculum) from a colony maintained on virus-free canola. There were seven replicates (49 individually potted plants) for TuMV and TuYV and six replicates (42 plants) for the mock inoculations. Field observations in previous years suggested that one mustard line (887–1) was particularly susceptible to TuMV, one line (J397) was relatively resistant, and the other four were intermediate.

3. **Glasshouse Inoculations:** Rub inoculations of *B. napus* differential lines S6, S1, R4, and 165 were conducted as described by Jenner and Walsh (1996). Five plants of each line were inoculated for each of the six NSW grain belt TuMV isolates. Inoculations of canola cv. Outback with TuMVisolates UK 1, CDN 1, and CZE 1 (Jenner and Walsh 1996) were done the same way.
   Plants were inoculated sown in pinebark/sand based potting mixture in 14 cm diameter × 14.5 cm high black plastic pots and thinned to either one to three plants per pot. Freshly harvested virus infected leaves and uninfected leaves were used for inocula and mock inocula, respectively. Extracts for rub inoculations were prepared by crushing leaves with a mortar and pestle in 0.07 M sodium phosphate buffer, pH 7.4, with 0.02 M Na_2SO_3 added immediately before use. The pestles wetted with freshly prepared extracts were used to inoculate test plants by gently rubbing cotyledons or first or second true leaves after pre-dusting with carborundum, using the forefinger of the other hand as support. Aphid transmissions from field plant specimens were achieved by transferring colonizing aphids directly to test seedlings. Aphid transmission for the mustard pathogenicity experiment involved an acquisition access period of 48 h and inoculation access period (IAP) of 72 h, using six *Myzus persicae* per inoculated plant, inoculated turnip weed as source for *TuMV*, and inoculated canola as source for TuYV. Inoculations were done and plants were grown year round in a glasshouse with evaporative cooling (10–28 °C) fitted with shade cloth from October to March. From flowering onward, nutrient status was maintained with inorganic fertilizer (Thrive®). Insects were eliminated by screens and regular monitoring. Occasional occurrences of fungal foliar diseases were controlled with application of commercially recommended foliar fungicides.
4. **Identification by ELISA:** Initially, fresh field specimens were tested for viruses by enzyme-linked immunosorbent assays (ELISAs), using purified anti-virus immunoglobulins, alkaline phosphatase conjugates, and ELISA protocols from Agdia, Inc. Infection of fresh leaves of B. napus differential lines inoculated with TuMV isolates was identified by ELISA with TuMV monoclonal antibody EMA67 (Jenner et al. 1999). However, the majority of testing of field specimens and isolates was done on stored petiole and stem blots on nitrocellulose membranes (Schwinghamer et al. 2010) by tissue blot immunoassays (TBIAs). These TBIAs were indirect procedures that involved incubation with antigen-specific primary antibodies followed by alkaline-phosphatase conjugated secondary antibodies (anti-rabbit or anti-mouse), and finally chromogenic substrate (nitroblue tetrazolium plus 5-bromo-4-chloro-3-indolyl phosphate) to detect virus in blots or tissue prints on nitrocellulose membranes.

Primary antibodies for TuMV TBIA and potyvirus TBIA were, respectively, TuMV polyclonal and general potyvirus monoclonal PTY 1, both from ELISA kits supplied by Bio-Rad Phyto-Diagnostics, INRA (Institut National de la Recherche Agronomique, France). Primary antibodies for CaMV and Pea seed-borne mosaic virus (PSbMV) TBIAs were polyclonals from DSMZ (Deutsche Sammlung von Mikroorganismen und Zellkulturen, Germany). The TBIA procedure of Lin et al. (1990) and Hsu and Lawson (1991) was modified as follows: Tris-buffered saline (TBS) (0.02 M Tris–HCl, pH 7.5, 0.15 M NaCl) was used rather than phosphate buffered saline as basic buffer for antibody dilution and membrane washing (Hammond and Jordan 1990); 1 μg/ml polyvinyl alcohol (Sigma MW 30,000–70,000) was used as pre-incubation blocking agent (Makkouk and Comeau 1994); 2% polyvinylpyrrolidone (Sigma PVP-40) and 0.2% egg albumin were added to basic buffer for dilution and incubation of monoclonal antibodies and alkaline phosphatase conjugated antibodies. For TBIAs that used polyclonals as primary antibodies, including TuMV, Cauliflower mosaic virus (CaMV), and other TBIAs, binding to host antigens was eliminated by dilution and pre-incubation at 37 °C for 2 h in strained leaf extract (1 g leaf per 50 ml TBS) of uninfected analogous host, usually wild radish, before adding the antibody mixture to membranes. Sensitivity of Potyvirus TBIA was maximized by heating membranes at 60 °C in 0.05 M Tris pH 9.5 (Jordan and Hammond 1991) for 5 min to expose antigenic determinants before blocking and incubating (Schwinghamer et al. 2014).

8.9.2 TuMV Biological and Serological Detection

In order to confirm the presence/absence of the most commonly occurring oilseed rape viruses, comprehensive characterization of herbaceous plant responses was performed. Mechanically inoculated TuMV assay hosts *N. tabacum* and *Ch. amaranticolor* showed chlorotic local lesions up to 10 days post inoculation (dpi). Chlorotic lesions regularly turned necrotic, but no systemic symptoms appeared. On the other hand, the local lesions that appeared in *C. quinoa* and *B. rapa* 10 dpi preceded systemic symptoms that were intensely expressed as vein clearing, mosaic, distortion, and stunting in *B. rapa* 24 dpi. The latter host was also suitable for the maintenance of the virus. *N. benthamiana, N. glutinosa,* and *N. megalosiphon* reacted with mild local chlorotic spots and systemic chlorosis of the veins with mild leaf crinkling, but *N. clevelandii* showed no signs of infection. *Pisum sativum* reacted strongly to the virus. The plants wilted and succumbed to the severe systemic necroses within 21 dpi, whereas *Phaseoulus vulgaris*, known to be the nonhost for TuMV (Plant Viruses Online, http://pvo.bio-mirror.cn/descr855.htm), showed no sign of infection. Symptoms indicating the presence of other viruses (CMV, TYMV, RaMV) commonly found in rapeseed in single or mixed infections (Rimmer et al. 2007) were not observed within the course of these experiments. The double-radial immune diffusion test with polyclonal antisera confirmed the TuMV presence, the RaMV and TYMV absence in the field oilseed rape leaf and floral tissue, as well as in the experimentally infected *B. rapa* and *N. benthamiana* plants (Seruga Music et al. 2014).

8.9.3 Plants, Chemical Treatments, and Virus Inoculation Method

Leafy mustard (*B. juncea* var. *foliosa Bailey*) plants were grown in a greenhouse at 23 °C and 20 °C (day and night air temperature, respectively) with a 16-h photoperiod. The four to five leaf-stage seedlings were subjected to chemical pretreatment for 4 consecutive days, and then the lower fully developed leaves were used for inoculation with TuMV. Six days later, the upper third fully developed leaf of each plant was collected for further study. Stock solutions of AA and SHAM were dissolved in deionized water. Mustard plants were sprayed with water (control plants) or freshly diluted solution of either 0.1 mM AA or 3 mM SHAM on both the adaxial and abaxial surfaces. SHAM was sprayed daily for 10 days.

Isolate CHZH33 in B strain of TuMV was procured for inoculation. This strain infects a wide range of plant species, mostly of the family Brassicaceae and contains a single copy of the genome, which is a single-stranded positive sense RNA (Ma et al. 2010). The main visible symptoms of systemic infection are mosaic, chlorosis, and deformity. Viral inoculum was prepared in 5 mM potassium phosphate buffer with pH 7.2 according to the method described by Tan et al. (2004). For inoculation, the plants were first sprayed with carborundum and then inoculated by rubbing 100 μl of viral inoculum onto each leaf with cotton swabs. Mock-inoculation controls were prepared by following the same procedure using only buffer. The upper newly developed, un-inoculated leaves were collected at different times for determining leaf rate of respiration and quantitative reverse-transcription polymerase chain reaction (qRT-PCR) measurements. Experiments were replicated three times.

Pot positions were randomized within each of seven replicate blocks on separate benches. Plants were inoculated (IAP initiated) 18 days after sowing, maintained and watered until maturity 105 days after inoculation, and pods and seed harvested after plants had desiccated. Infection status at 21 days post inoculation was determined by TBIA and/or characteristic symptoms for TuMV, and solely by TBIA for TuYV which was symptomless. The TBIAs showed that there was no unintended spread of the viruses between inoculum classes, indicating that aphids were eliminated successfully after the IAP. Data including plant height, seed weight, pod count, and days to death were analyzed using generalized linear models (GLM) (McCullagh and Nelder 1989) to test the effects of infection (TuMV, TuYV, nil) and genotype on the mustard plants. All GLM analyses were conducted using ASReml, a statistical package that fits linear mixed models using Residual Maximum Likelihood (Gilmour et al. 2008). Eighteen mustard plants that failed to become infected were combined with plants of their respective genotypes in the "non-viruliferous" inoculum class. Pairwise comparisons were conducted using least significant differences at $P < 0.05$. Canola was not included in the analysis as none of the seven replicate TuMV-inoculated canola plants became infected.

8.10 *Brassica* Species: Molecular Techniques

8.10.1 RNA Sequencing of *Brassica napus*

1. **RNA Extraction:** Frozen roots were placed in 15 ml Falcon tubes and ground using 5–7 mm diameter steel beads at 1100 to 1200 rpm (3–4 rounds of 1 min each) in a Genogrinder, followed by additional grinding in a mortar with a pestle in the presence of liquid nitrogen. Total RNA was extracted with a combination of TRIzol reagent and the RNeasy kit to increase the quantity and purity of the RNA. DNase treatment was performed on-column with the RNase-Free DNase set. The quantity of RNA was assessed with a Nanodrop 2000c spectrophotometer and quality was checked in a 2200 Tape Station.
2. **RNA-seq:** Three micrograms of RNA per pooled sample corresponding to each biological replicate with RNA Integrity Numbers (RIN) > 9 were sent for sequencing at Oklahoma State Genomics (Stillwater, OK, USA). Libraries were prepared with a TruSeq Stranded mRNA library preparation kit for 48 samples and samples were sequenced using TG Nextseq 500/550 High Output Kit v2 (75 cycles), in a NextSeq500 equipment. The raw data reads were filtered using the bcl2fastq conversion software provided by Illumina. This software uses read indexes to assign each read to the corresponding sample and then removes any adapters, barcodes, or primers using a trimmomatic-like algorithm. Any reads with an average Qscore <30 were removed and trimming of read ends was performed for bins of 15 bp that had a *Q*-score <30. Sequencing reads were deposited in the NCBI Sequence Read Archive (SRA) under accession number: PRJNA597078.

   The reads from all treatments were mapped to the *B. napus* reference genome (AST_PRJEB5043_v1) using the genome and gene annotation deposited in Ensembl (Chalhoub et al. 2014; Ensembl 2018; Kersey et al. 2018). Genome indexes were generated with Bowtie2 v2.3.3.1 (Langmead and Salzberg 2012) and transcriptome indexes with TopHat v2.1.1 (Trapnell et al. 2009). Reads mapped with TopHat were used along gene models (obtained from the Ensembl website) as input for Cufflinks v.2.2.1 to generate all potential transcripts in each sample. Cufflinks was run using the GTF-guide option, multi-read correction and fragment bias correction (multi-read correction and fragment bias correction were also used in the subprograms downstream). GTF files generated from all treatments and replicates were used to generate a consensus merged transcript gtf file with Cuffmerge. This consensus file was employed as a reference for re-mapping reads for quantification. Cuffquant was run with the merged gtf file to create ready-to-use files with quantified transcripts, to perform the necessary comparisons of differential expression among treatments. Cuffdiff was used to compare cuffquant files, contrasting controls with inoculated plants at each time-point and for each genotype separately. Expression levels were quantified as Fragments Per Kilobase of transcript per Million mapped reads (FPKM) and significant differential expression for each evaluated transcript was assessed after multiple comparison testing with the Benjamini-Hochberg correction (Benjamini and Hochberg 1995). The number of mapped reads from all treatments was compared using a one-way ANOVA implemented in R. FPKM values of all genes from treated and inoculated replicates were log-transformed to build PCA plots using the R-package ggfortify (https://github.com/sinhrks/ggfortify) (Son et al. 2018).

3. **cDNA Synthesis and** Quantitative **Reverse** Transcription **PCR (qRT-PCR):** Five hundred nanograms of RNA per sample were subjected to DNAse treatment using RNase-free-DNase I. cDNA was synthesized using the Revert Aid H Minus Reverse transcriptase with oligo dT following manufacturers' specifications. To test for residual DNA contamination, a 1:100 dilution of cDNA was used to perform PCR with primers corresponding to a constitutively expressed Clathrin Adaptor Complex (CAC) gene, producing 125 and 288 bp bands on cDNA and DNA, respectively. PCR analysis was conducted with 1× Buffer + KCl, 2.5 mM $MgCl_2$, 0.2 mM of each dNTP, 0.2 μM or each primer, and 1 unit of high fidelity Taq polymerase. Cycling conditions were 94 °C for 3 min, followed by 35 cycles of 94 °C for 30 s, 60 °C for 30 s, and 72 °C for 1 min, finishing with an extension step of 72 °C for 10 min, and a 4 °C step until samples were recovered from the thermocycler. To validate the differential expression detected by RNA-seq analysis, 15 primer pairs from genes showing significant differential expression in at least one time-point or host genotype were evaluated for efficiency (range of 90–110%) using a 5-point dilution series (1:4 serial dilutions) of a mix of all cDNA samples. Ten target primers within the expected efficiency range were used for final RNA-seq validation. Five reference gene primers with valid efficiencies were tested for stability across all samples using Best Keeper (Pfaffl et al. 2004), and the three genes showing the best correlation and lower standard deviation were chosen for relative quantification. Primers were designed using Primer3 (NIH 2012; Untergasser et al. 2012) with the following parameters: Na^+ concentration 50 mM, Mg^{++} concentration 3 mM, dNTPs concentration 0.8 mM, oligo length: min 18, opt 22, max 30 bp, melting temperature: min 60, opt 62, max 64 °C, GC content: min 35, opt 50, max 65%, amplicon length: min 70, opt 100, max 250 bp. Relative quantification of expression was performed using the $2^{-\Delta\Delta}$CT method (Livak and Schmittgen 2001). All qRT-PCR experiments were performed with four biological replicates and three technical replicates per biological replicate. The experiments were run on a ViiA7 Real-Time PCR system. Reactions were performed in 10 μl with 5 μl of an in-house made SYBR-green reagent mix, 2.5 μl of the pair of mixed primers (3.2 μM), and 2.5 μl of a 1:100 dilution of cDNA. Cycling conditions were 95 °C for 2 min followed by 40 cycles of 95 °C for 30 s and 60 °C for 1 min. A melting curve stage was added to verify product specificity: 95 °C for 15 s, 60 °C for 1 min, and 95 °C for 15 s. Log_2-fold change values from the qRT-PCR analysis were compared with the RNA-seq values of the respective genes to validate the RNA-seq data. Correlation coefficients were calculated using the CORREL function from Microsoft Excel for each day and host genotype using the ten target genes (Son et al. 2018).

8.10.2 Identification of *Brassica oleracea* Genes That Encode NBS Domain and NBS-Associated Conserved Domains

1. In the draft genome of *B. oleracea*, NBS-encoding genes were identified through Hidden Markov Model (HMM) profile corresponding to the Pfam NBS (NB-ARC) family PF00931 domain using HMMER V3.0 programme with "trusted cut-off" as threshold. From the selected protein sequences screened through NBS domain, high quality sequences were aligned through CLUSTALW and used to construct *B. oleracea* specific NBS profile using the "hmmbuild" module by HMMER V3.0 programme. With this model final set of NBS-encoding proteins were identified and only 157 proteins were selected as NBS candidate genes with stringent parameters. The NBS R-gene family is subdivided into different groups based on the structure of the N-terminal and C-terminal domains of the protein. For the identification of N-terminal and C-terminal domains of NBS-encoding genes, HMM Pfam and HMM Smart was used for detection. The employed PAIRCOIL2 (*P* score cut-off of 0.025) and MARCOIL programs with a threshold probability of 90 confirmed Coiled-Coil (CC) motif. From the result generated by these programs, the selected overlapping sequences as candidate genes with CC motif. Used same procedures to identify genes that contain TIR domain only and excluded the NBS-encoding genes as TIR-X genes. NBS-encoding genes in *A. thaliana* and *B. rapa* have been reported earlier but in order to get the latest NBS-encoding genes in these two species for comparative analysis, the same procedures were followed to screen NBS candidate genes in *B. rapa* and *A. thaliana* for consistency (Finn et al. 2011; Thompson et al. 1994; McDonnell et al. 2006).
2. **Assigning the Location of NBS-Encoding Genes to *B. oleracea* and *B. rapa* Genome:** The physical position of NBS-encoding genes was mapped to the 9 and 10 pseudo-molecular chromosomes of *B. oleracea* and *B. rapa* using GFF file which was downloaded from Bolbase and BRAD database, respectively. After that, in-house perl script to draw graphic potryl of NBS-encoding genes on pseudomolecular chromosomes with SVG module was used (Cheng et al. 2011; Yu et al. 2013; Altschul et al. 1997).
3. **Identification of Tandem Duplicated Arrays:** To detect the generated mechanism of NBS-encoding genes, BLASTP program was employed to identify the tandem duplicated genes using protein sequences with *E*-value cut-off $\leq 1e\text{-}20$, and one unrelated gene was allowed within a tandem array (Altschul et al. 1997).
4. **Alignment and Phylogenetic Analysis of NBS-Encoding Genes:** According to location of conserved domains for NBS (Nucleotide-binding Site) in complete predicted NBS protein sequences, conserved domain sequences of NBS-encoding genes were extracted and aligned using the programme Clustal W with default options for the phylogenetic analysis among three species. The poor alignment sequences were excluded by manually curation using Jalview. The resulting sequences were used to construct a phylogenetic tree using Maximum Likelihood (ML) method in MEGA 5.0 (Thompson et al. 1994; Clamp et al. 2004).

5. **Orthologous Gene Pairs Between *B. rapa, A. thaliana,* and *B. oleracea*:** Orthologous gene pairs provide information about the evolutionary relationship between different species. Two steps were used to detect gene pairs precisely. First, MCscan programme was employed to identify orthologous regions with the parameters ($e = 1e\text{-}20$, $u = 1$ and $s = 5$. Parameter of $s = 5$) between *B. rapa, A. thaliana,* and *B. oleracea* genomes. Second, after extracting orthologous regions that contained NBS-encoding genes, orthologous gene pairs of NBS-encoding genes were extracted (Tang et al. 2008).
6. **Non-synonymous/Synonymous Substitution (Ka/Ks) Ratios of Gene Pairs Between *B. rapa, A. thaliana,* and *B. oleracea*:** For the estimation of selection mode for the NBS-encoding genes among *B. oleracea, B. rapa,* and *A. thaliana,* the ratio of the rates of non-synonymous to synonymous substitutions (Ka/Ks) of all orthologous gene pairs were calculated for each branch of the phylogenetic tree using PAML software. For each sub-tree of NBS orthologous gene pairs among three species, model 1 with a free Ka/Ks ratio was calculated separately for each branch. The Ka/Ks values associated with terminal branches between modern species and their most recent reconstructed ancestors were employed in the subsequent analyses. In order to detect selection pressure, Ka/Ks ratio greater than 1, less than 1, and equal to 1 represents positive selection, negative or stabilizing selection, and neutral selection, respectively (Yang et al. 2007).
7. **RNA-seq Data Analysis of NBS-Encoding Genes:** For expression profiling of NBS-encoding genes, RNA-seq data that was generated earlier and submitted into GEO database was used. Transcript abundance is calculated by fragments per kilobase of exon model per million mapped reads (FPKM) and the FPKM values were log2 transformed. A hierarchical cluster was created using the Cluster 3.0 and heat map generated using Tree View Version 1.60 software (Eisen et al. 1998; Yu et al. 2014).

References

Adachi Y, Watanabe H, Tanabe K, Doke N, Nishimura S, Tsuge T (1993) Nuclear ribosomal DNA as a probe for genetic variability in the Japanese pear pathotype of *Alternaria alternata*. Appl Environ Microbiol 59:3197–3205

Alex LA, Borkovich KA, Simon MI (1996) Hyphal development in *Neurospora crassa*: involvement of a two-component histidine kinase. Proc Natl Acad Sci U S A 93(8):3416–3421

Altschul SF, Gish W, Miller W, Myers EW, Lipman DJ (1990) Basic local alignment search tool. J Mol Biol 215:403–410

Altschul SF, Madden TL, Schaffer AA, Zhang J, Zhang Z, Miller W, Lipman DJ (1997) Gapped BLAST and PSI-BLAST: a new generation of protein database search programs. Nucleic Acids Res 25:3389–3402

Ananga AO, Cebert E, Soliman K, Kantety R, Konan K, Ochieng JW (2008) Phylogenetic relationships within and among *Brassica* species from RAPD loci associated with blackleg resistance. Afr J Biotechnol 7(9):1287–1293

Anders S, Pyl PT, Huber W (2015) HTSeq—a Python framework to work with high-throughput sequencing data. Bioinformatics 31(2):166–169. https://doi.org/10.1093/bioinformatics/btu638

Anderson M (1996) Glutathione in free radicals. In: Punchard NA, Kelly FJ (eds) A practical approach. Oxford University Press, New York, p 213

Ariyarathna HCK, Francki MG (2016) Phylogenetic relationships and protein modeling revealed two distinct subfamilies of group II HKT genes between crop and model grasses. Genome 59:509–517. https://doi.org/10.1139/gen-2016-0035

Arora H, Padmaja KL, Paritosh K, Mukhi N, Tewari A, Mukhopadhyay A, Gupta V, Pradhan AK, Pental D (2019) BjuWRR1, a CC-NB-LRR gene identified in *Brassica juncea*, confers resistance to white rust caused by *Albugo candida*. Theor Appl Genet 132:2223–2236. https://doi.org/10.1007/s00122-019-03350-z

Arumugam N, Gupta V, Jagannath A, Mukhopadhyay A, Pradhan AK, Burma PK, Pental D (2007) A passage through in vitro culture leads to efficient production of marker-free transgenic plants in *Brassica juncea* using the Cr–*loxP* system. Transg Res 16:703–712

Avenot HF, Simoneau P, Iacomi B, Bataille-Simoneau N (2005) Characterization of mutations in the two-component histidine kinase gene AbNIK1 from *Alternaria brassicicola* that confer high dicarboximide and phenylpyrrole resistance. Curr Genet 47(4):234–243. https://doi.org/10.1007/s00294-005-0568-2

Balesdent MH, Barbetti M, Hua L, Sivasithamparam K, Gout L, Rouxel T (2005) Analysis of *Leptosphaeria maculans* race structure in a worldwide collection of isolates. Phytopathology 95:1061–1071

Bari R, Pant BD, Stitt M, Scheible WR (2006) PHO2, microRNA399, and PHR1 define a phosphate-signaling pathway in plants. Plant Physiol 141:988–999

Beauchamp C, Fridovich I (1971) Superoxide dismutase: improved assays and an assay applicable to acrylamide gels. Anal Biochem 44(1):276–287

Becker MG, Walker PL, Pulgar-Vidal NC, Belmonte MF (2017a) SeqEnrich: a tool to predict transcription factor networks from co-expressed *Arabidopsis* and *Brassica napus* gene sets. PLoS One 12(6):1–13. https://doi.org/10.1371/journal.pone.0178256

Becker MG, Zhang X, Walker PL, Wan JC, Millar JL, Khan D, Granger MJ, Cavers JD, Chan AC, Fernando DW, Belmonte MF (2017b) Transcriptome analysis of the *Brassica napus*-*Leptosphaeria maculans* pathosystem identifies receptor, signaling and structural genes underlying plant resistance. Plant J 90(3):573–586. https://doi.org/10.1111/tpj.13514

Becker MG, Haddadi P, Wan J, Adam L, Walker P, Larkan NJ, Daayf F, Borhan MH, Belmonte MF (2019) Transcriptome analysis of *Rlm2*-mediated host immunity in the *Brassica napus*–*Leptosphaeria maculans* pathosystem. Mol Plant-Microbe Interact 32(8):1001–1012. https://doi.org/10.1094/MPMI-01-19-0028-R

Belmonte MF, Kirkbride RC, Stone SL, Pelletier JM, Bui AQ, Yeung EC, Hashimoto M, Fci J, Harada CM, Munoz MD, Le BH, Drews GN, Brady SM, Goldberg RB, Harada JJ (2013) Comprehensive developmental profiles of gene activity in regions and subregions of the Arabidopsis seed. Proc Natl Acad Sci USA 110(5):E435–E444. https://doi.org/10.1073/pnas.1222061110

Benjamini Y, Hochberg Y (1995) Controlling the false discovery rate: a practical and powerful approach to multiple testing. J R Statist Soc B 57(1):289–300

Bhayana L, Paritosh K, Arora H, Yadava SK, Singh P, Nadan D, Mukhopadhyay A, Gupta V, Pradhan AK, Pental D (2020) A mapped locus on LG A6 of *Brassica juncea* line tumida conferring resistance to white rust contains a CNL type R gene. Front Plant Sci 10:1690. https://doi.org/10.3389/fpls.2019.01690

Bilir O, Telli O, Norman C, Budak H, Hong Y, Tor M (2019) Small RNA inhibits infection by downy mildew pathogen *Hyaloperonospora arabidopsidis*. Mol Plant Pathol 20(11):1523–1534. https://doi.org/10.1111/mpp.12863

Bolger AM, Lohse M, Usadel B (2014) Trimmomatic: a flexible trimmer for illumina sequence data. Bioinformatics 30:2114–2120

Boodley JW, Sheldrake R Jr (1972) Cornell peat-lite mixes for commercial plant growing. Cornell Univ Coop Exten Info Bull 43:1–8

Botstein D, White RL, Skolnick M, David R (1980) Construction of a genetic linkage map in man using restriction fragment length polymorphisms. Am J Hum Genet 32:314–331

Bradbury PJ, Zhang Z, Kroon DE, Casstevens TM, Ramdoss Y, Buckler ES (2007) TASSEL: software for association mapping of complex traits in diverse samples. Bioinformatics 23:2633–2635

Bradley CA, Henson RA, Porter PM, LeGare DG, Del Rio LE, Khot SD (2006) Response of canola cultivars to *Sclerotinia sclerotiorum* in controlled and field environments. Plant Dis 90:215–219

Bringans S, Eriksen S, Kendrick T, Gopalakrishnakone P, Livk A, Lock R, Lipscombe R (2008) Proteomic analyses of the venom of *Heterometrus longimanus* (Asian black scorpion). Proteomics 8:1081–1096

Broman KW, Wu H, Sen S, Churchill GA (2003) R/qtl: QTL mapping in experimental crosses. Bioinformatics 19:889–890. https://doi.org/10.1093/bioinformatics/btg112

Browning BL, Browning SR (2016) Genotype imputation with millions of reference samples. Am J Hum Genet 98:116–126. https://doi.org/10.1016/j.ajhg.2015.11.020

Buchwaldt L, Li R, Hegedus DD, Rimmer S (2005) Pathogenesis of *Sclerotinia sclerotiorum* in relation to screening for resistance. In: Proceedigns of 13th Intel Sclerotinia Workshop, Monterey, CA, p 22

Butler DG, Cullis BR, Gilmour AR, Gogel BJ (2007) ASReml-R reference manual. Release 2.0. Technical report. Queensland Department of Primary Industries, Australia

Cai G, Yang Q, Yang Q, Zhao Z, Chen H, Wu J, Fan C, Zhou Y (2012) Identification of candidate genes of QTLs for seed weight in *Brassica napus* through comparative mapping among *Arabidopsis*and *Brassica* species. BMC Genet 13:105. https://doi.org/10.1186/1471-2156-13-105

Campbell MM, Ellis BE (1992) Fungal elicitor-mediated responses in pine cell cultures: cell wall-bound phenolics. Phytochemistry 31:737–742

Chalhoub B, Denoeud F, Liu S, Parkin IA, Tang H, Wang X, Chiquet J, Belcram H, Tong C, Samans B, Corréa M, Da Silva C, Just J, Falentin C, Koh CS, Le Clainche I, Bernard M, Bento P, Noel B, Labadie K, Alberti A, Charles M, Arnaud D, Guo H, Daviaud C, Alamery S, Jabbari K, Zhao M, Edger PP, Chelaifa H, Tack D, Lassalle G, Mestiri I, Schnel N, Le Paslier MC, Fan G, Renault V, Bayer PE, Golicz AA, Manoli S, Lee TH, Thi VH, Chalabi S, Hu Q, Fan C, Tollenaere R, Lu Y, Battail C, Shen J, Sidebottom CH, Wang X, Canaguier A, Chauveau A, Bérard A, Deniot G, Guan M, Liu Z, Sun F, Lim YP, Lyons E, Town CD, Bancroft I, Wang X, Meng J, Ma J, Pires JC, King GJ, Brunel D, Delourme R, Renard M, Aury JM, Adams KL, Batley J, Snowdon RJ, Tost J, Edwards D, Zhou Y, Hua W, Sharpe AG, Paterson AH, Guan C, Wincker P (2014) Early allopolyploid evolution in the post-Neolithic *Brassica napus* oilseed genome. Science 345(6199):950–953

Chandna R, Augustine R, Bisht NC (2012) Evaluation of candidate reference genes for gene expression normalization in *Brassica juncea* using real time quantitative RT-PCR. PLoS One 7:e36918

Chaturani GDG, Mir ZA, Mishra P, Grover A (2019) Identification and characterization of defensin genes in *Brassica juncea* and *Camelina sativa* and analysis of their expression in response to biotic and abiotic stress. Acta Sci Agric 3(12):96–112

Chen J, Pang W, Chen B, Zhang C, Piao Z (2015) Transcriptome analysis of *Brassica rapa* near-isogenic lines carrying clubroot-resistant and -susceptible alleles in response to *Plasmodiophora brassicae* during early infection. Front Plant Sci 6:1183. https://doi.org/10.3389/fpls.2015.01183

Cheng X, Xu J, Xia S, Gu J, Yang Y, Fu J, Qian X, Zhang S, Wu J, Liu K (2009) Development and genetic mapping of microsatellite markers from genome survey sequences in *Brassica napus*. Theor Appl Genet 118(6):1121–1131

Cheng F, Liu S, Wu J, Fang L, Sun S, Liu B, Li P, Hua W, Wang X (2011) BRAD, the genetics and genomics database for *Brassica* plants. BMC Plant Biol 11:136. https://doi.org/10.1186/1471-2229-11-136

Chevre AM, Barret P, Eber F, Dupuy P, Brun H, Tanguy X, Renard M (1997) Selection of stable *Brassica napus-B.juncea* recombinant lines resistant to blackleg (*Leptosphaeria maculans*). I. Identification of molecular markers, chromosomal and genomic origin of the introgression. Theor Appl Genet 95:1104–1111

China MoAotPsRo (2011) Rules for investigation and forecast technology of rape sclerotiniose [*Sclerotinia sclerotiorum* (Lib.) de Bary]. Standards Press of China

Chittem K, Yajima WR, Goswami RS, del Rio Mendoza LE (2020) Transcriptome analysis of the plant pathogen *Sclerotinia sclerotiorum* interaction with resistant and susceptible canola (*Brassica napus*) lines. PLoS One. https://doi.org/10.1371/journal.pone.0229844

Choi S, Teakle G, Plaha P, Kim J, Allender C, Beynon E, Piao Z, Soengas P, Han T, King G, Barker G, Hand P, Lydiate D, Batley J, Edwards D, Koo D, Bang J, Park B-S, Lim Y (2007) The reference genetic linkage map for the multinational *Brassica rapa* genome sequencing project. Theor Appl Genet 115:777–792

Churchill GA, Doerge RW (1994) Empirical threshold values for quantitative trait mapping. Genet 138:963–971

Clamp M, Cuff J, Searle SM, Barton GJ (2004) The Jalview Java alignment editor. Bioinformatics 20(3):426–427. https://doi.org/10.1093/bioinformatics/btg430

Clarke WE, Higgins EE, Plieske J, Wieseke R, Sidebottom C, Khedikar Y, Batley J, Edwards D, Meng J, Li R, Lawley CT, Pauquet J, Laga B, Cheung W, Iniguez-Luy F, Dyrszka E, Rae S, Stich B, Snowdon RJ, Sharpe AG, Ganal MW, Parkin I (2016) A high-density SNP genotyping array for *Brassica napus* and its ancestral diploid species based on optimised selection of single-locus markers in the allotetraploid genome. Theor Appl Genet 129:1887–1899

Clarkson JP, Staveley J, Phelps K, Young CS, Whipps JM (2003) Ascospore release and survival in *Sclerotinia sclerotiorum*. Mycol Res 107:213–222. https://doi.org/10.1017/s0953756203007159

Conesa A, Stefan G (2008) Blast2GO: a comprehensive suite for functional analysis in plant genomics. Int J Plant Genom 2008:619832. https://doi.org/10.1155/2008/619832

Conesa A, Gotz S, Garcia-Gomez JM, Terol J, Talon M, Robles M (2005) Blast2GO: a universal tool for annotation, visualization and analysis in functional genomics research. Bioinformatics 21:3674–3676. https://doi.org/10.1093/bioinformatics/bti610

Coombes NE (2002) The reactive tabu search for efficient correlated experimental designs. PhD thesis, John Moores University, Liverpool

Corpet F (1988) Multiple sequence alignment with hierarchical clustering. Nucleic Acids Res 16:10881–10890

Coventry J, Kott L, Beversdorf WD (1988) Manual for microspore culture technique for *Brassica napus*. University of Guelph, Guelph

Cullis BR, Smith AB, Coombes NE (2006) On the design of early generation variety trials with correlated data. J Agric Biol Environ Stat 11:381–393

Cunnington JH, Takamatsu S, Lawrie AC, Pascoe IG (2003) Molecular identification of anamorphic powdery mildews (*Erysiphales*). Aust Plant Pathol 32:421–428

Cvikrova M, Binarova P, Eder J, Nedelnik J (1992) Accumulation of phenolic acids in filtrate-treated alfalfa cell cultures derived from genotypes with different susceptibility to *Fusarium oxysporum*. J Plant Physiol 140:21–27

Delourme R, Brun H, Ermel M, Lucas MO, Vallee P, Domin C, Walton G, Li H, Sivasithamparam K, Barbetti MJ (2008) Expression of resistance to *Leptosphaeria maculans* in *Brassica napus* double haploid lines in France and Australia is influenced by location. Ann Appl Biol 153:259–269

Derbyshire M, Denton-Giles M, Hegedus D, Seifbarghy S, Rollins J, van Kan J, Seidl MF, Faino L, Mbengue M, Navaud O, Raffaele S, Hammond-Kosack K, Heard S, Oliver R (2017) The complete genome sequence of the phytopathogenic fungus *Sclerotinia sclerotiorum* reveals insights into the genome architecture of broad host range pathogens. Genome Biol Evol 9:593–618

Dhaka N, Mukhopadhyay A, Paritosh K, Gupta V, Pental D, Pradhan AK (2017) Identification of genic SSRs and construction of a SSR-based linkage map in *Brassica juncea*. Euphytica 213:15

Doyle J, Doyle J (1990) Isolation of plant DNA from fresh tissue. Focus 12:13–15

Duitama J, Quintero JC, Cruz DF, Quintero C, Hubmann G, Foulquie-Moreno MR, Verstrepen KJ, Thevelein JM, Tohme J (2014) An integrated framework for discovery and genotyping of genomic variants from high-throughput sequencing experiments. Nucleic Acids Res 42:e44

Dusabenyagasani M, Fernando WGD (2008) Development of a SCAR marker to track canola resistance against blackleg caused by *Leptosphaeria maculans* pathogenicity group 3. Plant Dis 92:903–908

Edwards K, Johnstone C, Thompson C (1991) A simple and rapid method for the preparation of plant genomic DNA for PCR analysis. Nucl Acids Res 19:1349

Eisen MB, Spellman PT, Brown PO, Botstein D (1998) Cluster analysis and display of genome-wide expression patterns. Proc Natl Acad Sci U S A 95(25):14863–14868. https://doi.org/10.1073/pnas.95.25.14863

Elshire RJ, Glaubitz JC, Sun Q, Poland JA, Kawamoto K, Buckler ES et al (2011) A Robust, simple genotyping-by-sequencing (GBS) approach for high diversity species. PLoS One 6:e19379. https://doi.org/10.1371/journal.pone.0019379

Ensembl (2018) Ensembl Plants. *Brassica napus* (AST_PRJEB5043_v1). http://plants.ensembl.org/Brassica_napus/Info/Index.

Feng J, Hwang SF, Strelkov SE (2013) Studies into primary and secondary infection processes by *Plasmodiophora brassicae* on canola. Plant Pathol 62:177–183. https://doi.org/10.1111/j.1365-3059.2012.02612.x

Ferreira ME, Rimmer SR, Williams PH, Osborn TC (1995) Mapping loci controlling *Brassica napus* resistance to *Leptosphaeria maculans* under different screening conditions. Phytopathology 85:213–217

Finn RD, Clements J, Eddy SR (2011) HMMER web server: interactive sequence similarity searching. Nucleic Acids Res 39(Web Server issue):W29–W37. https://doi.org/10.1093/nar/gkr367

Finn RD, Attwood TK, Babbitt PC, Bateman A, Bork P, Bridge AJ, Chang HY, Dosztányi Z, El-Gebali S, Fraser M, Gough J, Haft D, Holliday GL, Huang H, Huang X, Letunic I, Lopez R, Lu S, Marchler-Bauer A, Mi H, Mistry J, Natale DA, Necci M, Nuka G, Orengo CA, Park Y, Pesseat S, Piovesan D, Potter SC, Rawlings ND, Redaschi N, Richardson L, Rivoire C, Sangrador-Vegas A, Sigrist C, Sillitoe I, Smithers B, Squizzato S, Sutton G, Thanki N, Thomas PD, Tosatto SC, Wu CH, Xenarios I, Yeh LS, Young SY, Mitchell AL (2017) InterPro in 2017-beyond protein family and domain annotations. Nucleic Acids Res 45:D190–D199

Flint-Garcia SA, Thornsberry JM, Buckler ES (2003) Structure of linkage disequilibrium in plants. Annu Rev Plant Biol 54:357–374

Fu Y, Zhang Y, Mason AS, Lin B, Zhang D, Yu H, Fu D (2019) NBS-encoding genes in *Brassica napus* evolved rapidly after allopolyploidization and co-localize with known disease resistance loci. Front Plant Sci 10:26. https://doi.org/10.3389/fpls.2019.00026

Gamborg OL, Miller RA, Ojima K (1968) Nutrient requirements of suspension cultures of soybean root cells. Exp Cell Res 50:151–158

Garcia-Mas J, Oliver M, Gomez-Paniagua H, de Vicente MC (2000) Comparing AFLP, RAPD and RFLP markers for measuring genetic diversity in melon. Theor Appl Genet 101:860–864

Garg H, Atri C, Sandhu PS, Kaur B, Renton M, Banga SK, Singh H, Singh C, Barbetti MJ, Banga SS (2010) High level of resistance to Sclerotinia sclerotiorum in introgression lines derived from hybridization between wild crucifers and the crop Brassica species B. napus and B. juncea. Field Crops Res 117:51–58. https://doi.org/10.1016/j.fcr.2010.01.013

Gemenet DC, Wachira FN, Pathak RS, Munyiri SW (2010) Identification of molecular markers linked to drought tolerance using bulked segregant analysis in Kenyan maize (*Zea mays* L.) landraces. J Anim Plant Sci 9(1):1122–1134

Ghanbarnia K, Lydiate DJ, Rimmer SR, Li G, Kutcher HR, Larkan NJ, McVetty PBE, Fernando WGD (2012) Genetic mapping of the *Leptosphaeria maculans* avirulence gene corresponding to the *LepR1* resistance gene of *Brassica napus*. Theor Appl Genet 124:505–513

Ghanbarnia K, Fudal I, Larkan NJ, Links MG, Balesdent MH, Profotova B, Fernando DWG, Rouxel T, Borhan H (2015) Rapid identification of the *Leptosphaeria maculans* avirulence gene *AvrLm2* using an intraspecific comparative genomics approach. Mol Plant Pathol 16:699–709

Ghanbarnia K, Ma L, Larkan NJ, Haddadi P, Fernando WGD, Borhan MH (2018) *Leptosphaeria maculans* AvrLm9: a new player in the game of hides and seek with AvrLm4-7. Mol Plant Pathol 19(7):1754–1764

Gietz RD, Woods RA (2002) Transformation of yeast by lithium acetate/single-stranded carrier DNA/polyethylene glycol method. Methods Enzymol 350:87–96

Gilmour AR, Gogel BJ, Cullis BR, Thompson R (2008) ASReml User Guide Release 3.0 VSN International Ltd, Hemel Hempstead, HP1 1ES, UK. www.vsni.co.uk

Gish W (2003) WU BLAST. Available from http://blast.wustl.edu

Gong ZZ, Lee H, Xiong LM, Jagendorf A, Stevenson B, Zhu JK (2002) RNA helicase-like protein as an early regulator of transcription factors for plant chilling and freezing tolerance. Proc Natl Acad Sci U S A 99:11507–11512. https://doi.org/10.1073/pnas.172399299

Gotz S, Garcia-Gomez JM, Terol J, Williams TD, Nagaraj SH, Nueda MJ, Robles M, Talon M, Dopazo J, Conesa A (2008) High–throughput functional annotation and data mining with the Blast2GO suite. Nucleic Acids Res 36:3420–3435. https://doi.org/10.1093/nar/gkn176

Govindaraj P, Arumugachamy S, Maheswaran M (2005) Bulked segregant analysis to detect main effect QTL associated with grain quality parameters in Basmati 370/ASD16 cross in rice (*Oryza sativa* L.) using SSR markers. Euphytica 144:61–68

Graichen K, Rabenstein F (1996) European isolates of beet western yellows virus (BMYV) from oilseed rape (*Brassica napus* L. ssp. *napus*) are non-pathogenic on sugar beet (*Beta vulgaris* L. *altissima*) but represent isolates of turnip yellows virus (TuYV). J Plant Dis Protect 103 (3):233–245

Guillemette T, Sellam A, Simoneau P (2004) Analysis of a non-ribosomal peptide synthetase gene from *Alternaria brassicae* and flanking genomic sequences. Curr Genet 45:214–224

Habu JB, Ibeh BO (2015) In vitro antioxidant capacity and free radical scavenging evaluation of active metabolite constituents of *Newbouldia laevis* ethanolic leaf extract. Biol Res 48:16

Haddadi P, Ma L, Wang H, Borhan MH (2016) Genome-wide transcriptome analyses provides insights into the lifestyle transition and effector repertoire of *Leptosphaeria maculans* during colonization of *Brassica napus* seedlings. Mol Plant Pathol 17(8):1196–1210. https://doi.org/10.1111/mpp.12356

Hammond J, Jordan RL (1990) Dot blots (viruses) and colony screening. In: Hampton RO, Ball EM, De Boer SH (eds) Serological methods for detection and identification of viral and bacterial plant pathogens. APS Press, St. Paul, pp 237–248

Hardy OJ, Vekemans X (2002) SPAGeDi: a versatile computer program to analyse spatial genetic structure at the individual or population levels. Mol Ecol Notes 2:618–620. https://doi.org/10.1046/j.1471-8286.2002.00305.x

Hasan MJ, Rahman H (2016) Genetics and molecular mapping of resistance to *Plasmodiophora brassicae* pathotypes 2, 3, 5, 6, and 8 in rutabaga (*Brassica napus* var. *napobrassica)*. Genome 59:805–815. https://doi.org/10.1139/gen-2016-0034

Hasan M, Strelkov SE, Howard RJ, Rahman H (2012) Screening of *Brassica* germplasm for resistance to *Plasmodiophora brassicae* pathotypes prevalent in Canada for broadening diversity in clubroot resistance. Can J Plant Sci 92:501–515

Hatzig SV, Frisch M, Breuer F, Nesi N, Ducournau S, Wagner MH, Leckband G, Abbadi A, Snowdon RJ (2015) Genome-wide association mapping unravels the genetic control of seed germination and vigor in *Brassica napus*. Front Plant Sci 6:221. https://doi.org/10.3389/fpls.2015.00221

He C, Holme J, Anthony J (2014) SNP genotyping: the KASP assay. In: Fleury D, Whitford R (eds) Crop breeding. Methods in molecular biology (methods and protocols), vol 1145. Humana Press, New York

Henriquez MA, Soliman A, Li G, Hannoufa A, Ayele BT, Daayf F (2016) Molecular cloning, functional characterization and expression of potato (*Solanum tuberosum*) 1-deoxy-d-xylulose 5-phosphate synthase 1 (StDXS1) in response to *Phytophthora infestans*. Plant Sci 243:71–83

Hirai M, Harada T, Kubo N, Tsukada M, Suwabe K, Matsumoto S (2004) A novel locus for clubroot resistance in *Brassica rapa* and its linkage markers. Theor Appl Genet 108:639–643

Hongtrakul V, Huestis GM, Knapp SJ (1977) Amplified fragment length polymorphisms as a tool for DNA fingerprinting sunflower germplasm: genetic diversity among oilseed inbred lines. Theor Appl Genet 95:400–407

Hopkins CJ, Cogan NOI, Hand M, Jewell E, Kaur J, Li X, Lim GAC, Ling AE, Love C, Mountford H, Todorovic M, Vardy M, Spangenberg GC, Edwards D, Batley J (2007) Sixteen new simple sequence repeat markers from *Brassica juncea* expressed sequences and their cross-species amplification. Mol Ecol Notes 7:697–700

Hsu HT, Lawson RH (1991) Direct tissue blotting for detection of tomato spotted wilt virus in Impatiens. Plant Dis 75:292–295

Hudspeth DSS, Nadler SA, Hudspeth MES (2000) A cox2 phylogeny of the Peronosporomycestes (Oomycetes). Mycologia 92:674–684

Iacomi-Vasilescu B, Avenot H, Bataille-Simoneau N, Laurent E, Guenard M, Simoneau P (2004) In vitro fungicide sensitivity of *Alternaria* species pathogenic to crucifers and identification of *Alternaria brassicicola* field isolates highly resistant to both di-carboximides and phenylpyrroles. Crop Prot 23:481–488

Jaccard P (1908) Nouvelles recherces sur la distribution florale. Bull Soc Vaudoise Sci Nat 44:223–270

Jenner CE, Walsh JA (1996) Pathotypic variation in Turnip mosaic virus with special reference to European isolates. Plant Pathol 45:848–856

Jenner CE, Keane GJ, Jones JE, Walsh JA (1999) Serotypic variation in Turnip mosaic virus. Plant Pathol 48:101–108

Jian W, Cai G, Tu J, Li L, Sheng L, Luo X, Zhou L, Fan C, Zhou Y (2013) Identification of QTLS for resistance to *Sclerotinia* Stem Rot and *Bnac. IGMT5.a* as a candidate gene of the major resistant QTL *SRC6* in *Brassica napus*. PLoS One 8:e67740. https://doi.org/10.1371/journal.pone.0067740

Jo SJ, Shim SA, Jang KS, Choi YH, Kim JC, Choi GJ (2011) Resistance of cultivars of Chinese cabbage to *Plasmodiophora brassicae* isolates of several races collected in Korea. Korean J Hort Sci Technol 29:610–616

Jobling SA, Gehrke L (1987) Enhanced translation of chimaeric messenger RNAs containing a plant viral untranslated leader sequence. Nat 325:622–625

Jones DT, Taylor WR, Thornton JM (1992) The rapid generation of mutation data matrices from protein sequences. Comput. Appl Biosci 8:275–282. https://doi.org/10.1093/bioinformatics/8.3.275

Jordan R, Hammond J (1991) Comparison and differentiation of potyvirus isolates and identification of strain-, virus-, subgroup-specific and potyvirus group-common epitopes using monoclonal antibodies. J Gen Virol 72:25–36

Kabbage M, Yarden O, Dickman MB (2015) Pathogenic attributes of *Sclerotinia sclerotiorum*: switching from a biotrophic to necrotrophic lifestyle. Plant Sci 233:53–60

Kapadia VN, Sasidharan N, Patel DA, Kalyanrao (2019) Identification of the SSR marker linked to powdery mildew resistant gene(s) in *Brassica* spp. J Pharmacog Phytochem 8(3):388–395

Kato T, Hatakayema K, Fukino N, Matsumoto S (2012) Identification of a clubroot resistance locus conferring resistance to a *Plasmodiophora brassicae* classified into pathotype group 3 in Chinese cabbage (*Brassica rapa* L.). Breed Sci 62:282–287. https://doi.org/10.1270/jsbbs.62.282

Kato T, Hatakeyama K, Fukino N, Matsumoto S (2013) Fine mapping of the Clubroot resistance gene *CRb* and development of a useful selectable marker in *Brassica rapa*. Breed Sci 63:116–124. https://doi.org/10.1270/jsbbs.63.116

Kaur S, Cogan NO, Ye G, Baillie RC, Hand ML, Ling AE, McGearey AK, Kaur J, Hopkins CJ, Todorovic M, Mountford H, Edwards D, Batley J, Burton W, Salisbury P, Gororo N, Marcroft S, Kearney G, Smith KF, Forster JW, Spangenberg GC (2009) Genetic map construction and QTL mapping of resistance to blackleg (*Leptosphaeria maculans*) disease in Australian canola (*Brassica napus* L.) cultivars. Theor Appl Genet 120:71–83. https://doi.org/10.1007/s00122-009-1160-9

Kaur P, Jost R, Sivasithamparam K, Barbetti MJ (2010) Proteome analysis of the *Albugo candida–Brassica juncea* pathosystem reveals that the timing of the expression of defence-related genes is a crucial determinant of pathogenesis. J Exp Bot 62(3):1285–1298. https://doi.org/10.1093/jxb/erq365

Kearse M, Moir R, Wilson A, Stones-Havas S, Cheung M, Sturrock S, Buxton S, Cooper A, Markowitz S, Duran C, Thierer T, Ashton B, Meintjes P, Drummond A (2012) Geneious basic: an integrated and extendable desktop software platform for the organization and analysis of sequence data. Bioinformatics 28:1647–1649

Kersey PJ, Allen JE, Allot A, Barba M, Boddu S, Bolt BJ, Carvalho-Silva D, Christensen M, Davis P, Grabmueller C, Kumar N, Liu Z, Maurel T, Moore B, McDowall MD, Maheswari U, Naamati G, Newman V, Ong CK, Paulini M, Pedro H, Perry E, Russell M, Sparrow H, Tapanari E, Taylor K, Vullo A, Williams G, Zadissia A, Olson A, Stein J, Wei S, Tello-Ruiz M, Ware D, Luciani A, Potter S, Finn RD, Urban M, Hammond-Kosack KE, Bolser DM, De Silva N, Howe KL, Langridge N, Maslen G, Staines DM, Yates A (2018) Ensembl Genomes 2018: an integrated omics infrastructure for non-vertebrate species. Nucleic Acid Res 46:D802–D808. https://doi.org/10.1093/nar/gkx1011

Kim H, Choi SR, Bae J (2009) Sequenced BAC anchored reference genetic map that reconciles the ten individual chromosomes of *Brassica rapa*. BMC Genomics 10:432. https://doi.org/10.1186/1471-2164-10-432

Kim D, Langmead B, Salzberg SL (2015) HISAT: a fast spliced aligner with low memory requirements. Nat Methods 12(4):357–360. pmid:25751142

Kim H, Jo EJ, Choi YH, Jang KS, Choi GJ (2016) Pathotype classification of *Plasmodiophora brassicae* isolates using Clubroot-resistant cultivars of Chinese cabbage. Plant Pathol J 32:423. https://doi.org/10.5423/PPJ.OA.04.2016.0081

Koch E, Cox R, Williams PH (1991) Infection of *Arabidopsis thaliana* by *Plasmodiophora brassicae*. J Phytopathol 132:99–104

Korgan S, Wolski EA, Cicore P, Suarez P, Capezio S, Huarte MA, Andreu AB (2011) *Solanum tarijense* reaction to *Phytophthora infestans* and the role of plant defense molecules. Plant Breed 130:231–236

Kosambi DD (1943) The estimation of map distances from recombination values. Ann Eugen 12:172–175

Kosambi DD (1944) The estimation of map distances from recombination values. Ann Eugen 12:172–175

Krzywinski M, Schein J, Birol İ, Connors J, Gascoyne R, Horsman D, Jones SJ, Marra MA (2009) Circos: an information aesthetic for comparative genomics. Genome Res 19(9):1639–1645. https://doi.org/10.1101/gr.092759.109

Kumar R, Ichihashi Y, Kimura S, Chitwood DH, Headland LR, Peng J, Maloof JN, Sinha NR (2012) A high-throughput method for illumina RNA-Seq library preparation. Front Plant Sci 3 (8):202. https://doi.org/10.3389/fpls.2012.00202

Kumar R, Megha S, Ha M, Basu U, Kav NNV (2016) A global study of transcriptome dynamics in canola (*Brassica napus* L) responsive to *Sclerotinia sclerotiorum* infection using RNA-seq. Gene 590:57–67. https://doi.org/10.1016/j.gene.2016.06.003

Kumar V, Paillard S, Fopa-Fomeju B, Falentin C, Deniot G, Baron C, Vallee P, Manzanares-Dauleux MJ, Delourme R (2018) Multi-year linkage and association mapping confirm the high number of genomic regions involved in oilseed rape quantitative resistance to blackleg. Theor Appl Genet 131:1627–1643

Laila R, Robin AHK, Yang K, Choi GJ, Park JI, Nou IS (2017) Detection of ribosomal DNA sequence polymorphisms in the Protist *Plasmodiophora brassicae* for the identification of geographical isolates. Int J Mol Sci 18:84. https://doi.org/10.3390/ijms18010084

Laila R, Park JI, Robin AHK, Natarajan S, Vijayakumar H, Shirasawa K, Isobe S, Kim HT, Nou IS (2019) Mapping of a novel clubroot resistance QTL using ddRAD-seq in Chinese cabbage (*Brassica rapa* L.). BMC Plant Biol 19:13. https://doi.org/10.1186/s12870-018-1615-8

Lalitha S (2000) Primer premier 5. Biotech Softw Internet Rep 1:270–272. https://doi.org/10.1089/152791600459894

Lander ES, Botstein D (1989) Mapping mendelian factors underlying quantitative traits using RFLP linkage maps. Genet 121:185–199

Langmead B, Salzberg SL (2012) Fast gapped-read alignment with Bowtie 2. Nat Methods 9:357–360. https://doi.org/10.1038/nmeth.1923

Larkan NJ, Lydiate DJ, Parkin IA, Nelson MN, Epp DJ, Cowling WA, Rimmer SR, Borhan MH (2013) The *Brassica napus* blackleg resistance gene LepR3 encodes a receptor-like protein triggered by the *Leptosphaeria maculans* effector AVRLM1. New Phytol 197:595–605. https://doi.org/10.1111/nph.12043

Larkan NJ, Lydiate DJ, Yu F, Rimmer SR, Borhan MH (2014) Co-localization of the blackleg resistance genes *Rlm2* and *LepR3* on *Brassica napus* chromosome A10. BMC Plant Biol 14:387. https://doi.org/10.1186/s12870-014-0387-z

Larkan NJ, Ma L, Borhan H (2015) The *Brassica napus* receptor-like protein RLM2 is encoded by a second allele of the *LepR3/Rlm2* blackleg resistance locus. Plant Biotech J 13:983–992

Larkan NJ, Raman H, Lydiate DJ, Robinson SJ, Yu F, Barbulescu DM, Raman R, Luckett DJ, Burton W, Wratten N, Salisbury PA, Rimmer SR, Borhan MH (2016a) Multi-environment QTL studies suggest a role for cysteine-rich protein kinase genes in quantitative resistance to blackleg disease in *Brassica napus*. BMC Plant Biol 16:183. https://doi.org/10.1186/s12870-016-0877-2

Larkan NJ, Yu F, Lydiate DJ, Rimmer SR, Borhan MH (2016b) Single R gene introgression lines for accurate dissection of the *Brassica-Leptosphaeria* pathosystem. Front Plant Sci 7:1771. https://doi.org/10.3389/fpls.2016.01771

Lee SB, Taylor JW (1990) Isolation of DNA from fungal mycelia and single spores. Chapter 34. In: Innis M, Gelfand D, Sninsky J, White T (eds) PCR protocols: a guide to methods and applications. Academic Press, Orlando

Lei M, Huihui L, Luyan Z, Jiankang W (2015) QTL IciMapping: integrated software for genetic linkage map construction and quantitative trait locus mapping in biparental populations. The Crop J 3(3):269–283. https://doi.org/10.1016/j.cj.2015.01.001

Li G, Quiros C (2001) Sequence-related amplified polymorphism (SRAP), a new marker system based on a simple PCR reaction: its application to mapping and gene tagging in *Brassica*. Theor Appl Genet 103:455–461

Li CX, Li H, Sivasithamparam K, Fu TD, Li YC, Liu SY, Barbetti MJ (2006) Expression of field resistance under Western Australian conditions to *Sclerotinia sclerotiorum* in Chinese and Australian *Brassica napus* and *Brassica juncea* germplasm and its relation with stem diameter. Aust J Agric Res 57:1131–1135

Li CX, Li H, Siddique AB, Sivasithamparam K, Salisbury P, Banga SS, Banga S, Chattopadhyay C, Kumar A, Singh R, Singh D, Agnihotri A, Liu SY, Li YC, Tu J, Fu TD, Wang YF, Barbetti MJ (2007) The importance of the type and time of inoculation and assessment in the determination of resistance in *Brassica napus* and *B. juncea* to *Sclerotinia sclerotiorum*. Aust J Agric Res 58:1198–1203. https://doi.org/10.1071/AR07094

Li CX, Sivasithamparam K, Walton G, Fels P, Barbetti MJ (2008) Both incidence and severity of white rust disease reflect host resistance in *Brassica juncea* germplasm from Australia, China and India. Field Crops Res 106:1–8

Li F, Chen B, Xu K, Wu J, Song W, Bancroft I et al (2014) Genome-wide association study dissects the genetic architecture of seed weight and seed quality in rapeseed (*Brassica napus* L.). DNA Res 21:355–367. https://doi.org/10.1093/dnares/dsu002

Li L, Luo Y, Chen B, Xu K, Zhang F, Li H, Huang Q, Xiao X, Zhang T, Hu J, Li F, Wu X (2016) A genome-wide association study reveals new loci for resistance to clubroot disease in *Brassica napus*. Front Plant Sci 7:1483. https://doi.org/10.3389/fpls.2016.01483

Liao Y, Smyth G, Shi W (2013) featureCounts: an efficient general-purpose program for assigning sequence reads to genomic features. Bioinformatics 30(7):923–930

Lichter R (1982) Induction of haploid plants from isolated pollen of *Brassica napus*. Zeitschrift für Pflanzenphysiologie 105(5):427–434. https://doi.org/10.1016/j.bbr.2011.03.031

Lin NS, Hsu YH, Hsu HT (1990) Immunological detection of plant viruses and a mycoplasmalike organism by direct tissue blotting on nitrocellulose membranes. Phytopathology 80:824–828

Liu LJ, Zhang YY, Tang SY, Zhao QZ, Zhang ZH, Zhang HW, Dong L, Guo H, Xie Q (2010) An efficient system to detect protein ubiquitination by agroinfiltration in *Nicotiana benthamiana*. Plant J 61:893–903. https://doi.org/10.1111/j.1365-313X.2009.04109.x

Liu S, Liu Y, Yang X, Tong C, Edwards D, Parkin IA, Zhao M, Ma J, Yu J, Huang S, Wang X, Wang J, Lu K, Fang Z, Bancroft I, Yang TJ, Hu Q, Wang X, Yue Z, Li H, Yang L, Wu J, Zhou Q, Wang W, King GJ, Pires JC, Lu C, Wu Z, Sampath P, Wang Z, Guo H, Pan S, Yang L, Min J, Zhang D, Jin D, Li W, Belcram H, Tu J, Guan M, Qi C, Du D, Li J, Jiang L, Batley J, Sharpe AG, Park BS, Ruperao P, Cheng F, Waminal NE, Huang Y, Dong C, Wang L, Li J, Hu Z, Zhuang M, Huang Y, Huang J, Shi J, Mei D, Liu J, Lee TH, Wang J, Jin H, Li Z, Li X, Zhang J, Xiao L, Zhou Y, Liu Z, Liu X, Qin R, Tang X, Liu W, Wang Y, Zhang Y, Lee J, Kim HH, Denoeud F, Xu X, Liang X, Hua W, Wang X, Wang J, Chalhoub B, Paterson AH (2014) The *Brassica oleracea* genome reveals the asymmetrical evolution of polyploid genomes. Nat Commun 5:3930. https://doi.org/10.1038/ncomms4930

Liu Y, Huang X, Li M, He P, Zhang Y (2016) Loss-of-function of *Arabidopsis* receptor-like kinase BIR1 activates cell death and defense responses mediated by BAK1 and SOBIR1. New Phytol 212:637–645

Livak KJ, Schmittgen TD (2001) Analysis of relative gene expression data using real-time quantitative PCR and the $2^{-\Delta\Delta C_T}$ method. Methods 25:402–408

Long Y, Shi J, Qiu D, Li R, Zhang C, Wang J, Hou J, Zhao J, Shi L, Park B-S, Choi SR, Lim YP, Meng J (2007) Flowering time quantitative trait loci analysis of oilseed *Brassica* in multiple environments and genomewide alignment with *Arabidopsis*. Genetics 177:2433–2444

Loreto F, Velikova V (2001) Isoprene produced by leaves protects the photosynthetic apparatus against ozone damage, quenches ozone products, and reduces lipid peroxidation of cellular membranes. Plant Physiol 127:1781–1787

Loreto F, Mannozzi M, Maris C, Nascetti P, Ferranti F, Pasqualini S (2001) Ozone quenching properties of isoprene and its antioxidant role in leaves. Plant Physiol 126:993–1000

Love MI, Huber W, Anders S (2014) Moderated estimation of fold change and dispersion for RNA-Seq data with DESeq2. Genome Biol 15:550. https://doi.org/10.1101/002832

Lowe AJ, Jones AE, Raybould AF, Trick M, Moule CJ, Edwards KJ (2002) Transferability and genome specificity of a new set of microsatellite primers among *Brassica* species of the U triangle. Mol Ecol Notes 2:7–11. https://doi.org/10.1046/j.1471-8286.2002.00126.x

Lowe A, Moule C, Trick M, Edwards K (2004) Efficient large scale development of microsatellites for marker and mapping applications in *Brassica* crop species. Theor Appl Genet 108:1103–1112. https://doi.org/10.1007/s00122-003-1522-7

Lucas JA, Crute IR, Sherriff C, Gordon PL (1988) The identification of a gene for race-specific resistance to *Peronospora parasitica* (downy mildew) in *Brassica napus var. oleifera* (Oilseed rape). Plant Pathol 37:538–545

Luckett DJ, Cowley R, Moroni S, Raman H (2011) Improving water-use efficiency and drought tolerance in canola—potential contribution from improved carbon isotope discrimination (CID). In: Proceedings of 13th Intel Rapeseed Congress, Prague

Lun AT, Chen Y, Smyth GK (2016) It's DE-licious: a recipe for differential expression analyses of RNA-seq experiments using quasi-likelihood methods in edgeR. Methods Mol Biol 1418:391–416. pmid:27008025

Ma JF, Hou XL, Xiao D, Qi L, Wang F, Sun F, Wang Q (2010) Cloning and characterization of the BcTUR3, gene related to resistance to Turnip mosaic virus (TuMV) from non-heading Chinese cabbage. Plant Mol Biol Rep 28:588–596

Makkouk K, Comeau A (1994) Evaluation of various methods for the detection of barley yellow dwarf virus by the tissue-blot immunoassay and its use for virus detection in cereals inoculated at different growth stages. Eur J Plant Pathol 100:71–80

Manly KF, Cudmore JRH, Meer JM (2001) Map Manager QTX, cross platform software for genetic mapping. Mamm Genome 12:930–932

Marchler-Bauer A, Bryant SH (2004) CD-Search: protein domain annotations on the fly. Nucleic Acids Res 32:W327–W331

Marchler-Bauer A, Lu S, Anderson JB, Chitsaz F, Derbyshire MK, DeWeese-Scott C, Fong JH, Geer LY, Geer RC, Gonzales NR, Gwadz M, Hurwitz DI, Jackson JD, Ke Z, Lanczycki CJ, Lu F, Marchler GH, Mullokandov M, Omelchenko MV, Robertson CL, Song JS, Thanki N, Yamashita RA, Zhang D, Zhang N, Zheng C, Bryant SH (2011) CDD: a conserved domain database for the functional annotation of proteins. Nucleic Acids Res 39:D225–D229

Martin M (2011) Cutadapt removes adapter sequences from high-throughput sequencing reads. EMBnet J 17:10–12

Mayee CD, Datar VV (1986) Phytopathometry bulletin-1. Marathawada Agricultural University, Parbhani, p 46

Mayer KF, Martis M, Hedley PE, Simkova H, Liu H, Morris JA, Steuernagel B, Taudien S, Roessner S, Gundlach H, Kubalakova M, Suchankova P, Murat F, Felder M, Nussbaumer T, Graner A, Salse J, Endo T, Sakai H, Tanaka T, Itoh T, Sato K, Platzer M, Matsumoto T, Scholz U, Dolezel J, Waugh R, Stein N (2011) Unlocking the barley genome by chromosomal and comparative genomics. Plant Cell 23:1249–1263. https://doi.org/10.1105/tpc.110.082537

McCullagh P, Nelder JA (1989) Generalized linear models, volume 37 of monographs on statistics and applied probability, 2nd edn. Chapman and Hall, London

McDonnell AV, Jiang T, Keating AE, Berger B (2006) Paircoil2: improved prediction of coiled coils from sequence. Bioinformatics 22(3):356–358. https://doi.org/10.1093/bioinformatics/bti797

Mcintosh RA, Yamazaki Y, Devos KM, Dubcovsky J, Rogers WJ, Appels R (2003) Catalogue of gene symbols for wheat. In: Proceedings of tenth international wheat genetics symposium, Paestum, Italy, 1–6 Sept 2003

McLoughlin KJ, Pedrini E, MacMahon M, Guduric-Fuchs J, Medina RJ (2019) Selection of a real-time PCR housekeeping gene panel in human endothelial colony forming cells for cellular senescence studies. Front Med 6:33. https://doi.org/10.3389/fmed.2019.00033

Meglecz E, Costedoat C, Dubut V, Gilles A, Malausa T, Pech N, Martin JF (2010) QDD: a user-friendly program to select microsatellite markers and design primers from large sequencing projects. Bioinformatics 26(3):403–404. https://doi.org/10.1093/bioinformatics/btp670

Mehra S, Pareek A, Bandyopadhyay P, Sharma P, Burma P, Pental D (2000) Development of transgenics in Indian oilseed mustard (*Brassica juncea*) resistant to herbicide phosphinothricin. Curr Sci 78:1358–1364

Mei J, Wei D, Disi JO, Ding Y, Liu Y, Qian W (2012) Screening resistance against *Sclerotinia sclerotiorum* in *Brassica* crops with use of detached stem assay under controlled environment. Eur J Plant Pathol 134:599–604. https://doi.org/10.1007/s10658-012-0040-3

Mei J, Ding Y, Lu K, Wei D, Liu Y, Disi JO, Li J, Liu L, Liu S, McKay J, Qian W (2013) Identification of genomic regions involved in resistance against *Sclerotinia sclerotiorum* from wild *Brassica oleracea*. Theor Appl Genet 126:549–556. https://doi.org/10.1007/s00122-012-2000-x

Mei J, Liu Y, Wei D, Wittkop B, Ding Y, Li Q, Li J, Wan H, Li Z, Ge X, Frauen M, Snowdon RJ, Qian W, Friedt W (2015) Transfer of *Sclerotinia* resistance from wild relative of *Brassica oleracea* into *Brassica napus* using a hexaploidy step. Theor Appl Genet 128:639–644

Mei J, Shao C, Yang R, Feng Y, Gao Y, Ding Y, Li J, Qian W (2020) Introgression and pyramiding of genetic loci from wild *Brassica oleracea* into *B. napus* for improving *Sclerotinia* resistance of rapeseed. Theor Appl Genet 133:1313–1319. https://doi.org/10.1007/s00122-020-03552-w

Mengistu A, Hill CB, Williams PH (1995) A toothpick method for mating *Leptosphaeria maculans*, the causal agent of blackleg of crucifers. Plant Dis 79:755–756

Meyers BC, Kozik A, Griego A, Kuang H, Michelmore RW (2003) Genome-wide analysis of NB-LRR-encoding genes in *Arabidopsis*. Plant Cell 15:809–834

Miao L, Zhang L, Raboanatahiry N, Lu G, Zhang X, Xiang J, Gan J, Fu C, Li M (2016) Transcriptome analysis of stem and globally comparison with other tissues in *Brassica napus*. Front Plant Sci 7:1403. https://doi.org/10.3389/fpls.2016.01403

Milad SI, Wahba LE, Barakat MN (2011) Identification of RAPD and ISSR markers associated with flag leaf senescence under water stressed conditions in wheat (*Triticum aestivum* L.). Aus J Crop Sci 5(3):337–343

Millar JL, Becker MG, Belmonte MF (2015) Laser micro-dissection of plant tissues. In: Plant micro-techniques and protocols. Springer, New York, pp 337–350. https://doi.org/10.1007/978-3-319-19944-3

Mir ZA, Ali S, Shivaraj SM, Bhat JA, Singh A, Yadav P, Rawat S, Paplao PK, Grover A (2020) Genome-wide identification and characterization of Chitinase gene family in *Brassica juncea* and *Camelina sativa* in response to *Alternaria brassicae*. Genomics 112(1):749–763

Moukette BM, Pieme CA, Njimou JR, Biapa CPN, Marco B, Ngogang JY (2015) *In vitro* antioxidant properties, free radicals scavenging activities of extracts and polyphenol composition of a non-timber forest product used as spice: *Monodora myristica*. Biol Res 48:15

Muller PY, Janovjak H, Miserez AR, Dobbie Z (2002) Processing of gene expression data generated by quantitative real-time RT–PCR. Biotechniques 32:1372–1379

Murashige T, Skoog F (1962) A Revised medium for rapid growth and bio assays with tobacco tissue cultures. Physiol Plant 15:473–497

Murray MG, Thompson WF (1980) Rapid isolation of high molecular weight plant DNA. Nucl Acids Res 8:4321–4326

Nagaoka T, Doullah MA, Matsumoto S, Kawasaki S, Ishikawa T, Hori H, Okazaki K (2010) Identification of QTLs that control clubroot resistance in *Brassica oleracea* and comparative analysis of clubroot resistance genes between *B. rapa* and *B. oleracea*. Theor Appl Genet 120:1335–1346. https://doi.org/10.1007/s00122-010-1259-z

Nieuwhof M, Wiering D (1961) Testing cabbage plants for clubroot resistance. Euphytica 10 (2):191–200. https://doi.org/10.1007/BF00022210

NIH (2012) Primer3. http://bioinfo.ut.ee/primer3-0.4.0/

Nowakowska M, Wrzesińska M, Kaminski P et al (2019) *Alternaria brassicicola* – Brassicaceae pathosystem: insights into the infection process and resistance mechanisms under optimized artificial bio-assay. Eur J Plant Pathol 153:131–151. https://doi.org/10.1007/s10658-018-1548-y

Nowicki M, Klosinska U, Szczechura W, Markiewicz M, Sowik I, Michalczuk L, Kozik E U (2016) Cellular and biochemical mechanisms of cucumber resistance against *Pseudoperonospora cubensis*. In: Cucurbitaceae 2016, XIth Eucarpia meeting on cucurbit genetics & breeding, Warsaw, Poland, 24–28 July 2016, pp 128–131. Cucurbitaceae 2016 Organizing Committee.

Numa H, Itoh T (2014) MEGANTE: a Web-based system for integrated plant genome annotation. Plant Cell Physiol 55:e2. https://doi.org/10.1093/pcp/pct157

van Os H, Stam P, Visser RGF, van Eck HJ (2005) RECORD: a novel method for ordering loci on a genetic linkage map. Theor Appl Genet 112:30–40

Pang W, Fu P, Li X, Zhan Z, Yu S, Piao Z (2018) Identification and mapping of the clubroot resistance gene CRd in Chinese cabbage (*Brassica rapa* ssp. *pekinensis*). Front Plant Sci 9:653. https://doi.org/10.3389/fpls.2018.00653

Panjabi-Massand P, Jagannath A, Bisht NC, Padmaja KL, Sharma S, Gupta V, Pradhan AK, Pental D (2008) Comparative mapping of *Brassica juncea* and *Arabidopsis thaliana* using Intron Polymorphism (IP) markers: homoeologous relationships, diversification and evolution of the A, B and C *Brassica* genomes. BMC Genom 9:113. https://doi.org/10.1186/1471-2164-9-113

Panjabi-Massand P, Yadava SK, Sharma P, Kaur A, Kumar A, Arumugam N, Sodhi YS, Mukhopadhyay A, Gupta V, Pradhan AK, Pental D (2010) Molecular mapping reveals two independent loci conferring resistance to *Albugo candida* in the east European germplasm of oilseed mustard *Brassica juncea*. Theor Appl Genet 121:137–145. https://doi.org/10.1007/s00122-010-1297-6

Paritosh K, Gupta V, Yadava SK, Singh P, Pradhan AK, Pental D (2014) RNA-seq based SNPs for mapping in *Brassica juncea* (AABB): synteny analysis between the two constituent genomes A (from *B. rapa*) and B (from *B. nigra*) shows highly divergent gene block arrangement and unique block fragmentation patterns. BMC Genomics 15:396. https://doi.org/10.1186/1471-2164-15-396

Paritosh K, Yadava SK, Singh P, Bhayana L, Mukhopadhyay A, Gupta V, Bisht NC, Zhang J, Kudrna D, Copetti D, Wing RA, Lachagari VBR, Pradhan AK, Pental D (2019) A chromosome-scale assembly of allotetraploid *Brassica juncea* (AABB) elucidates comparative architecture of the A and B genomes. Plant Biotechnol. https://doi.org/10.1101/681080

Parker JE, Holub EB, Frost LN, Falk A, Gunn ND, Daniels MJ (1996) Characterization of *eds1*, a mutation in *Arabidopsis* suppressing resistance to *Peronospora parasitica* specified by several different RPP genes. Plant Cell 8:2033–2046

Parkin IAP, Koh C, Tang H, Robinson SJ, Kagale S, Clarke WE, Town CD, Nixon J, Krishnakumar V, Bidwell SL, Denoeud F, Belcram H, Links MG, Just J, Clarke C, Bender T, Huebert T, Mason AS, Pires JC, Barker G, Moore J, Walley PG, Manoli S, Batley J, Edwards D, Nelson MN, Wang X, Paterson AH, King G, Bancroft I, Chalhoub B, Sharpe AG (2014) Transcriptome and methylome profiling reveals relics of genome dominance in the mesopolyploid *Brassica oleracea*. Genome Biol 15(6):R77

Parlange F, Daverdin G, Fudal I, Kuhn ML, Balesdent MH, Blaise F, Grezes-Besset B, Rouxel T (2009) *Leptosphaeria maculans* avirulence gene AvrLm4-7 confers a dual recognition specificity by the Rlm4 and Rlm7 resistance genes of oilseed rape, and circumvents Rlm4-mediated recognition through a single amino acid change. Mol Microbiol 71:851–863. https://doi.org/10.1111/j.1365-2958.2008.06547.x

Peterson GL (1983) Determination of total protein. Methods Enzymol 91:95–119

Pfaffl MW, Tichopad A, Prgomet C, Neuvians TP (2004) Determination of stable housekeeping genes, differentially regulated target genes and sample integrity: best Keeper–Excel-based tool using pair-wise correlations. Biotechnol Lett 26:509–515. https://doi.org/10.1023/B:BILE.0000019559.84305.47

Pfister DH (1997) Castor, pollux, and life histories of fungi. Mycologia 89:1–23

Piquemal J, Cinquin E, Couton F, Rondeau C, Seignoret E, Doucet I, Perret D, Villeger M, Vincourt P, Blanchard P (2005) Construction of an oilseed rape (*Brassica napus* L.) genetic map with SSR markers. Theor Appl Genet 111:1514–1523

Plotnikova JM, Reuber TL, Ausubel FM, Pfister DH (1998) Powdery mildew pathogenesis of *Arabidopsis thaliana*. Mycologia 90:1009–1016

Pritchard JK, Stephens M, Donnelly P (2000) Inference of population structure using multilocus genotype data. Genetics 155:945–959

Punta M, Coggill PC, Eberhardt RY, Mistry J, Tate J, Boursnell C, Pang N, Forslund K, Ceric G, Clements J, Heger A, Holm L, Sonnhammer ELL, Eddy SR, Bateman A, Finn RD (2012) The Pfam protein families database. Nucleic Acids Res 40:D290–D301

Pushpa HD, Devendra K, Yadava, Naveen S, Vasudev S, Saini N, Muthusamy V, Prabhu K (2016) Validation of molecular markers linked to low glucosinolate QTLs for marker assisted selection in Indian mustard [*Brassica juncea* (L.) Czern & Coss.]. Indian J Genetic 76(1):64–68. https://doi.org/10.5958/0975-6906.2016.00009.2

Quevillon E, Silventoinen V, Pillai S, Harte N, Mulder N, Apweiler R, Lopez R (2005) InterProScan: protein domains identifier. Nucleic Acids Res 33:W116–W120

R Core Team (2013) *R:* a language and environment for statistical computing. R Foundation for Statistical Computing, Vienna

Rahman H (2016) UA AlfaGold Clearfield herbicide-tolerant spring Brassica napus canola developed from winter × spring canola cross. Can J Plant Sci 97(1):144–146. https://doi.org/10.1139/cjps-2016-0028

Rahman H, Shakir A, Jakir Hasan M (2011) Breeding for clubroot resistant spring canola (*Brassica napus* L.) for the Canadian prairies: can the European winter canola cv. mendel be used as a source of resistance? Can J Plant Sci 91:447–458. https://doi.org/10.1093/bioinformatics/btq033

Ramakers C, Ruijter JM, Deprez RH, Moorman AF (2003) Assumption-free analysis of quantitative real-time polymerase chain reaction (PCR) data. Neuroscience Lett 339:62–66

Raman R, Raman H, Johnstone K, Lisle C, Smith A, Martin P, Allen H (2005) Genetic and in silico comparative mapping of the polyphenol oxidase gene in bread wheat (*Triticum aestivum* L.). Funct Integr Genomics 5:185–200

Raman R, Allen H, Diffey S, Raman H, Martin P, McKelvie K (2009) Localization of quantitative trait loci for quality attributes in a doubled haploid population of wheat (*Triticum aestivum* L.). Genome 52:701–715

Raman H, Raman R, Prangnell R, Eckermann P, Edwards D, Batley J, Coombes N, Taylor B, Wratten N, Luckett D, Dennis L (2011) Genetic dissection of natural variation for flowering time in rapeseed. In: Proceedings of Intel Rapeseed congress (Abstr) Prague, Czech Republic, p 139

Raman R, Taylor B, Lindbeck K, Coombes N, Barbulescu D, Salisbury P, Raman H (2012a) Molecular mapping and validation of *Rlm1* genes for resistance to *Leptosphaeria maculans* in canola (*Brassica napus* L). Crop Pasture Sci 63:1007–1017. https://doi.org/10.1071/CP12255

Raman R, Taylor B, Marcroft S, Stiller J, Eckermann P, Coombes N, Rehman A, Lindbeck K, Luckett D, Wratten N, Batley J, Edwards D, Wang X, Raman H (2012b) Molecular mapping of qualitative and quantitative loci for resistance to *Leptosphaeria maculans* causing blackleg disease in Canola (*Brassica napus* L.). Theor Appl Genet 125:405–418. https://doi.org/10.1007/s00122-012-1842-6

Rana K, Atri C, Gupta M, Akhatar J, Sandhu PS, Kumar N, Jaswal R, Barbetti MJ, Banga SS (2017) Mapping resistance responses to *Sclerotinia* infestation in introgression lines of *Brassica juncea* carrying genomic segments from wild *Brassicaceae B. fruticulosa.* Sci Rep 7:5904

Rana K, Atri C, Akhatar J, Kaur R, Goyal A, Singh MP, Kumar N, Sharma A, Sandhu PS, Kaur G, Barbetti MJ, Banga SS (2019) Detection of first marker trait associations for resistance against *Sclerotinia sclerotiorum* in *Brassica juncea–Erucastrum cardaminoides* introgression lines. Front Plant Sci 10:1015. https://doi.org/10.3389/fpls.2019.01015

Rhee SY, Beavis W, Berardini TZ, Chen G, Dixon D, Doyle A, Garcia-Hernandez M, Huala E, Lander G, Montoya M, Miller N, Mueller LA, Mundodi S, Reiser L, Tacklind J, Weems DC, Wu Y, Xu I, Yoo D, Yoon J, Zhang P (2003) The Arabidopsis information resource (TAIR): a model organism database providing a centralized, curated gateway to *Arabidopsis* biology, research materials and community. Nucleic Acids Res 31:224–228. https://doi.org/10.1093/nar/gkg076

Richly E, Kurth J, Leister D (2002) Mode of amplification and reorganization of resistance genes during recent *Arabidopsis thaliana* evolution. Mol Biol Evol 19(1):76–84. https://doi.org/10.1093/oxfordjournals.molbev.a003984

Rieu I, Powers SJ (2009) Real-time quantitative RT-PCR: design, calculations, and statistics. Plant Cell Online 21(4):1031–1033. https://doi.org/10.1105/tpc.109.066001

Rimmer SR, Shattuck VI, Buchwaldt L (2007) Compendium of Brassica diseases. APS Press, St. Paul

Robinson MD, McCarthy DJ, Smyth GK (2010) edgeR: a Bio-conductor package for differential expression analysis of digital gene expression data. Bioinformatics 26:139–140

Rogers SO, Bendich AJ (1994) Extraction of total cellular DNA from plants, algae and fungi. Plant Mol Biol Manual D1:1–8

Rotmistrovsky K, Jang W, Schuler GD (2004) A web server for performing electronic PCR. Nucleic Acids Res 32(Web server issue):108–112. https://doi.org/10.1093/nar/gkh450. PMID: 15215361

Rout K, Yadav BG, Yadava SK, Mukhopadhyay A, Gupta V, Pental D, Pradhan AK (2018) QTL landscape for oil content in *Brassica juncea*: analysis in multiple bi-parental populations in high and "0" erucic background. Front Plant Sci 9:1448

Rouxel T, Grandaubert J, Hane JK, Hoede C, Van de Wouw AP, Couloux A, Dominguez V, Anthouard V, Bally P, Bourras S, Cozijnsen AJ, Ciuffetti LM, Degrave A, Dilmaghani A, Duret L, Fudal I, Goodwin SB, Gout L, Glaser N, Linglin J, Kema GH, Lapalu N, Lawrence CB, May K, Meyer M, Ollivier B, Poulain J, Schoch CL, Simon A, Spatafora JW, Stachowiak A, Turgeon BG, Tyler BM, Vincent D, Weissenbach J, Amselem J, Quesneville H, Oliver RP, Wincker P, Balesdent MH, Howlett BJ (2011) Effector diversification within compartments of the *Leptosphaeria maculans* genome affected by repeat-induced point mutations. Nat Commun 2:202

Saito M, Kubo N, Matsumoto S, Suwabe K, Tsukada M, Hirai M (2006) Fine mapping of the clubroot resistance gene, Crr3, in *Brassica rapa*. Theor Appl Genet 114:81–91. https://doi.org/10.1007/s00122-006-0412-1

Sakamoto K, Saito A, Hayashida N, Taguchi G, Matsumoto E (2008) Mapping of isolate-specific QTLs for clubroot resistance in Chinese cabbage (*Brassica rapa* L. ssp. *pekinensis*). Theor Appl Genet 117:759–767. https://doi.org/10.1007/s00122-008-0817-0

Sanjaya G, Myrtle H, Jonathan D, Kyla H, Parkin IAP, Hegedus DD, Bekkaoui D, Buchwaldt L (2016) Microsatellite markers used for genome-wide association mapping of partial resistance to *Sclerotinia sclerotiorum* in a world collection of *Brassica napus*. Mol Breed 36:1–13

Saxena RK, von Wettberg E, Upadhyaya HD, Sanchez V, Songok S, Saxena K, Kimurto P, Varshney RK (2014) Genetic diversity and demographic history of Cajanus spp. Illustrated from genome-wide SNPs. PLoS One 9:e88568. https://doi.org/10.1371/journal.pone.0088568

Schuelke M (2000) An economic method for the fluorescent labeling of PCR fragments. Nat Biotechnol 18:233–234

Schuler GD (1997) Sequence mapping by electronic PCR. Genome Res. 7(5):541–550. https://doi.org/10.1101/gr.7.5.541. PMID: 9149949

Schwinghamer MW, Thomas JE, Schilg MA, Parry JN, Dann EK, Moore K, Kumari SG (2010) Mastreviruses in chickpea (Cicer arietinum) and other dicotyledonous crops and weeds in Queensland and northern New South Wales, Australia. Australas Plant Pathol 39(6):551–561. https://doi.org/10.1071/AP10032

Schwinghamer MW, Schilg MA, Walsh JA, Bambach RW, Cossu RM, Bambridge JM, Hind-Lanoiselet TL, McCorkell BE, Cross P (2014) Turnip mosaic virus: potential for crop losses in the grain belt of New South Wales, Australia. Australasian Plant Pathol 43:663–678. https://doi.org/10.1007/s13313-014-0304-9

Seruga Music M, Duc Nguyen H, Cerni S, Mamula D, Ohshima K, Skoric D (2014) Multilocus sequence analysis of 'Candidatus Phytoplasma asteris' strain and the genome analysis of Turnip mosaic virus co-infecting oilseed rape. J Appl Microbiol 117:774–785

Sharma N, Rahman MH, Strelkov S, Thiagarajah M, Bansal VK, Kav NNV (2007) Proteome-level changes in two *Brassica napus* varieties exhibiting differential responses to the fungal pathogen *Alternaria brassicae*. Plant Sci 172:95–110

Sharma S, Padmaja KL, Gupta V, Paritosh K, Pradhan AK, Pental D (2014) Two plastid DNA lineages—Rapa/Oleracea and Nigra—within the tribe Brassiceae can be best explained by reciprocal crosses at hexaploidy: evidence from divergence times of the plastid genomes and R-block Genes of the A and B genomes of *Brassica juncea*. PLoS One 9:e93260

Sheehy RE, Kramer M, Hiatt WR (1988) Reduction of polygalacturonase activity in tomato fruit by antisense RNA. Proc Natl Acad Sci 85:8805–8809

Shirasawa K, Hirakawa H, Isobe S (2016) Analytical workflow of double-digest restriction site-associated DNA sequencing based on empirical and *in silico* optimization in tomato. DNA Res 23:145–153. https://doi.org/10.1093/dnares/dsw004

Siebert PD, Chenchik A, Kellogg DE, Lukyanov KA, Lukyanov SA (1995) An improved PCR method for walking in uncloned genomic DNA. Nucleic Acids Res 23(6):1087–1088. https://doi.org/10.1093/nar/23.6.1087

Simon P (2003) Q-Gene: processing quantitative real-time RT–PCR data. Bioinformatics 19:1439–1440

Smith IK, Vierheller TL, Thorne CA (1988) Assay of glutathione reductase in crude tissue homogenates using 5, 5′-dithiobis (2-nitrobenzoic acid). Anal Biochem 175(1):408–413

Son K, Yu S, Shin W, Han K, Kang K (2018) A simple guideline to assess the characteristics of RNA-Seq data. BioMed Res Int 2018:2906292. https://doi.org/10.1155/2018/2906292

Stamatakis A (2015) Using RAxML to infer phylogenies. Curr Protoc Bioinformatics 51:6–14. https://doi.org/10.1002/0471250953.bi0614s51

Stanke M, Steinkamp R, Waack S, Morgenstern B (2004) AUGUSTUS: a web server for gene finding in eukaryotes. Nucleic Acids Res 32:W309–W312

Storey JD, Tibshirani R (2003) R: statistical significance for genome-wide experiments. Proc Natl Acad Sci U S A 100:9440–9445

Strelkov SE, Manolii VP, Cao T, Xue S, Hwang SF (2007) Pathotype classification of *Plasmodiophora brassicae* and its occurrence in *Brassica napus* in Alberta, Canada. J Phytopathol 155:706–712. https://doi.org/10.1111/j.1439-0434.2007.01303.x

Su TB, Wang WH, Li PR, Zhang B, Li P, Xin XY, Sun H, Yu Y, Zhang D, Zhao X, Wen C, Zhou G, Wang Y, Zheng H, Yu S, Zhang F (2018) A genomic variation map provides insights into the genetic basis of spring Chinese Cabbage (*Brassica rapa* ssp. pekinensis) selection. Mol Plant 11:1360–1376. https://doi.org/10.1016/j.molp.2018.08.006

Sun Z, Wang Z, Tu J, Zhang J, Yu F, McVetty P, Li G (2007) An ultradense genetic recombination map for *Brassica napus* consisting of 13551 SRAP markers. Theor Appl Genet 114:1305–1317

Sutra L, Simoneau P, Gardan L (2001) Genetic diversity of *Pseudomonas syringae* pathovars and related species assessed by heteroduplex mobility assay. In: De Boer SH (ed) Plant pathogenic bacteria. Kluwer, Dordrecht, pp 128–130

Suwabe K, Iketani H, Nunome T, Kage T, Hirai M (2002) Isolation and characterization of microsatellites in *Brassica rapa* L. Theor Appl Genet 104:1092–1098

Suwabe K, Tsukazaki H, Iketani H, Hatakeyama K, Fujimura M, Nunome T, Fukuoka H, Matsumoto S, Hirai M (2003) Identification of two loci for resistance to clubroot (*Plasmodiophora brassicae* Woronin) in *Brassica rapa* L. Theor Appl Genet 107:997–1002

Suwabe K, Tsukazaki H, Iketani H, Hatakeyama K, Kondo M, Fujimura M, Nunome T, Fukuoka H, Hirai M, Matsumoto S (2006) Simple sequence repeat-based comparative genomics between *Brassica rapa* and *Arabidopsis thaliana*: the genetic origin of clubroot resistance. Genetics 173:309–319

Suwabe K, Morgan C, Bancroft I (2008) Integration of *Brassica* A genome genetic linkage map between *Brassica napus* and *B. rapa*. Genome 41:169–176

Tan ZY, Wada Y, Chen JS, Ohshima K (2004) Inter- and intralineage recombinants are common in natural populations of Turnip mosaic virus. J Gen Virol 85:2683–2696

Tang H, Wang X, Bowers JE, Ming R, Alam M, Paterson AH (2008) Unraveling ancient hexaploidy through multiply-aligned angiosperm gene maps. Genome Res 18(12):1944–1954. https://doi.org/10.1101/gr.080978.108

Taylor J, Butler D (2017) R package ASMap: efficient genetic linkage map construction and diagnosis. J Stat Softw 79(6):1–28

Thompson JD, Higgins DG, Gibson TJ, Clustal W (1994) improving the sensitivity of progressive multiple sequence alignment through sequence weighting, position-specific gap penalties and weight matrix choice. Nucleic Acids Res 22(22):4673–4680. https://doi.org/10.1093/nar/22.22.4673

Thordal-Christensen H, Zhang ZG, Wei YD, Collinge DB (1997) Subcellular localization of H_2O_2 in plants. H2O2 accumulation in papillae and hypersensitive response during the barley-powdery mildew interaction. Plant J 11:1187–1194

Tong C, Wang X, Yu J, Wu J, Li W, Huang J, Dong C, Hua W, Liu S (2013) Comprehensive analysis of RNA-seq data reveals the complexity of the transcriptome in *Brassica rapa*. BMC Genomics 14:689. https://doi.org/10.1186/1471-2164-14-689

Tonguc M, Griffiths PD (2004) Transfer of powdery mildew resistance from *Brassica carinata* to *Brassica oleracea* through embryo rescue. Plant Breed 123:587–589

Tor M, Gordon P, Cuzick A, Eulgem T, Sinapidou E, Mert-Turk F, Can C, Dangl JL, Holub EB (2002) *Arabidopsis* SGT1b is required for defense signaling conferred by several downy mildew resistance genes. Plant Cell 14:993–1003

Trapnell C, Pachter L, Salzberg SL (2009) TopHat: discovering splice junctions with RNA-Seq. Bioinformatics 25:1105–1111. https://doi.org/10.1093/bioinformatics/btp120

Trapnell C, Roberts A, Goff L, Pertea G, Kim D, Kelley DR, Pimentel H, Salzberg SL, Rinn JL, Pachter L (2012) Differential gene and transcript expression analysis of RNA-seq experiments with TopHat and Cufflinks. Nat Prot 7(3):562–578. https://doi.org/10.1038/nprot.2012.016

Tsuro M, Suwabe K, Kubo N, Matsumoto S, Hirai M (2005) Construction of a molecular linkage map of radish (*Raphanus sativus* L.), based on AFLP and *Brassica*-SSR markers. Breed Sci 55:107–111

Untergasser A, Cutcutache I, Koressaar T, Ye J, Faircloth BC, Remm M, Rozen SG (2012) Primer3–new capabilities and interfaces. Nucleic Acids Res 40:e115. https://doi.org/10.1093/nar/gks596

Utermark J, Karlovsky P (2008) Genetic transformation of filamentous fungi by *Agrobacterium tumefaciens*. Nat Prot Exch 427:1–18

Van Ooijen J (2006) JoinMap^{R} 4, Software for the calculation of genetic linkage maps in experimental populations. Kyazma BV, Wageningen, p 33

Van Ooijen J, Kyazma B (2009) MapQTL 6: software for the mapping of quantitative trait loci in experimental populations of diploid species. Kyazma BV, Wageningen

Van Ooijen JW, Boer MP, Jansen RC, Maliepaard C (2002) MapQTL Version^{R}4.0 software for the calculation of QTL positions on genetic maps. Plant Res Intel Wageningen, Wageningen

Verbyla AP, Cullis BR, Thompson R (2006) The analysis of QTL by simultaneous use of the full linkage map. Theor Appl Genet 116:95–111

Voorrips RE (2002) MapChart: software for the graphical presentation of linkage maps and QTLs. J Hered 93:77–78

Wang Y, Li J, Hou S, Wang X, Li Y, Ren D, Chen S, Tang X, Zhou JM (2010) A *Pseudomonas syringae* ADP-ribosyl transferase inhibits *Arabidopsis* mitogen-activated protein kinase kinases. Plant Cell 22:2033–2044

Wang X, Wang H, Wang J, Sun R, Wu J, Liu S, Bai Y, Mun JH, Bancroft I, Cheng F, Huang S, Li X, Hua W, Wang J, Wang X, Freeling M, Pires JC, Paterson AH, Chalhoub B, Wang B, Hayward A, Sharpe AG, Park BS, Weisshaar B, Liu B, Li B, Liu B, Tong C, Song C, Duran C, Peng C, Geng C, Koh C, Lin C, Edwards D, Mu D, Shen D, Soumpourou E, Li F, Fraser F, Conant G, Lassalle G, King GJ, Bonnema G, Tang H, Wang H, Belcram H, Zhou H, Hirakawa H, Abe H, Guo H, Wang H, Jin H, Parkin IA, Batley J, Kim JS, Just J, Li J, Xu J, Deng J, Kim JA, Li J, Yu J, Meng J, Wang J, Min J, Poulain J, Wang J, Hatakeyama K, Wu K, Wang L, Fang L, Trick M, Links MG, Zhao M, Jin M, Ramchiary N, Drou N, Berkman PJ, Cai Q, Huang Q, Li R, Tabata S, Cheng S, Zhang S, Zhang S, Huang S, Sato S, Sun S, Kwon SJ, Choi SR, Lee TH, Fan W, Zhao X, Tan X, Xu X, Wang Y, Qiu Y, Yin Y, Li Y, Du Y, Liao Y, Lim Y, Narusaka Y, Wang Y, Wang Z, Li Z, Wang Z, Xiong Z, Zhang Z (2011) *Brassica rapa* genome sequencing project consortium. The genome of the mesopolyploid crop species *Brassica rapa*. Nat Genet 43:1035–1039. https://doi.org/10.1038/ng.919

Wang X, Wu J, Liang J, Cheng F, Wang X (2015) *Brassica* database (BRAD) version 2.0: integrating and mining Brassicaceae species genomic resources. Database 2015:bav093. https://doi.org/10.1093/database/bav093

Wei D, Mei J, Fu Y, Disi JO, Li J, Qian W (2014) Quantitative trait loci analyses for resistance to *Sclerotinia sclerotiorum* and flowering time in *Brassica napus*. Mol Breed 34:1797–1804

Wei L, Jian H, Lu K, Filardo F, Yin N, Liu L, Qu C, Li W, Du H, Li J (2015) Genome-wide association analysis and differential expression analysis of resistance to Sclerotinia stem rot in *Brassica napus*. Plant Biotechnol J. https://doi.org/10.1111/pbi.12501

Wei L, Jian H, Lu K, Filardo F, Yin N, Liu L, Qu C, Li W, Du H, Li J (2016) Genome-wide association analysis and differential expression analysis of resistance to Sclerotinia stem rot in *Brassica napus*. Plant Biotechnol J 14:1368–1380. https://doi.org/10.1111/pbi.12501

Wheeler BEJ (1969) An introduction to plant disease. Wiley, London, p 301

White TJ, Bruns T, Lee S, Taylor J (1990) Amplification and direct sequencing of fungal ribosomal RNA genes for phylogenetics. In: Innis MA, Gelfand DH, Sninsky JJ, White TJ (eds) PCR Protocols: a guide to methods and applications. Academic Press, San Diego, pp 315–322

Williams PH (1966) A system for the determination of races of *Plasmodiophora brassicae* that infect cabbage and rutabaga. Phytopathology 56:624–626

Woods-Tor A, Studholme DJ, Çevik V, Telli O, Holub EB, Tor M (2018) A suppressor/avirulence gene combination in *Hyaloperonospora arabidopsidis* determines race specificity in *Arabidopsis thaliana*. Front Plant Sci 9:1957

Wu Y, Bhat PR, Close TJ, Lonardi S (2008) Efficient and accurate construction of genetic linkage maps from the minimum spanning tree of a graph. PloS Genet 4:e1000212. https://doi.org/10.1371/journal.pgen.1000212

Wu J, Cai G, Tu J, Li L, Liu S, Luo X, Zhou L, Fan C, Zhou Y (2013) Identification of QTLs for resistance to Sclerotinia stem rot and BnaC.IGMT5.a as a candidate gene of the major resistant QTL SRC6 in *Brassica napus*. PLoS One 8:e67740. https://doi.org/10.1371/journal.pone.0067740

Wu J, Zhao Q, Liu S, Shahid M, Lan L, Cai G, Zhang C, Fan C, Wang Y, Zhou Y (2016) Genome-wide association study identifies new loci for resistance to Sclerotinia stem rot in *Brassica napus*. Front Plant Sci 7:1418. https://doi.org/10.3389/fpls.2016.01418

Xiao D, Liu ST, Wei YP, Zhou DY, Hou XL, Li Y, Hu CM (2016) c-DNA-AFLP analysis reveals differential gene expression in incompatible interaction between infected non-heading Chinese cabbage and *Hyaloperonospora parasitica*. Hort Res 3:1–16

Yajima W, Hall JC, Kav NNV (2004) Proteome-level differences between auxinic herbicide-susceptible and -resistant wild mustard (*Sinapis arvensis* L.). J Agric Food Chem 52:5063–5070

Yan X, Wang H, Ye Y, Zeng G, Ma R, Mi F, Yao L (2012) pYBA100: an ease-of-use binary vector with LoxP-FRT recombinase site for plant transformation. Mol Plant Breed 10:371–379

Yang Z (1997) PAML: a program package for phylogenetic analysis by maximum likelihood. Comput Appl Biosci 13:555–556. http://abacus.gene.ucl.ac.uk/software/paml.html

Yang Z (2007) PAML 4: phylogenetic analysis by maximum likelihood. Mol Biol Evol 24 (8):1586–1591. https://doi.org/10.1093/molbev/msm088

Yang B, Srivastava S, Deyholos MK, Kav NNV (2007) Transcriptional profiling of canola (*Brassica napus* L.) responses to the fungal pathogen *Sclerotinia sclerotiorum*. Plant Sci 173:156–171

Yang J, Weedon MN, Purcell S, Lettre G, Estrada K, Willer CJ, Smith AV, Ingelsson E, O'Connell JR, Mangino M, Mägi R, Madden PA, Heath AC, Nyholt DR, Martin NG, Montgomery GW, Frayling TM, Hirschhorn JN, McCarthy MI, Goddard ME, Visscher PM, GIANT Consortium (2011) Genomic inflation factors under polygenic inheritance. Eur J Hum Genet 19(7):807–812

Yang J, Liu D, Wang X, Ji C, Cheng F, Liu B, Hu Z, Chen S, Pental D, Ju Y, Yao P, Li X, Xie K, Zhang J, Wang J, Liu F, Ma W, Shopan J, Zheng H, Mackenzie SA, Zhang M (2016) The genome sequence of allopolyploid *Brassica juncea* and analysis of differential homoeolog gene expression influencing selection. Nat Genet 48:1225–1232. https://doi.org/10.1038/ng.3657

Young ND (1996) QTL mapping and quantitative disease resistance in plants. Annu Rev Phytopathol 34:479–501. https://doi.org/10.1146/annurev.phyto.34.1.479

Yu S, Zhang F, Yu R, Zou Y, Qi J, Zhao X, Yu Y, Zhang D, Li L (2009) Genetic mapping and localization of a major QTL for seedling resistance to downy mildew in Chinese cabbage

(*Brassica rapa* ssp. *pekinensis*). Mol Breed 23:573–590. https://doi.org/10.1007/s11032-009-9257-z

Yu F, Gugel RK, Kutcher HR, Peng G, Rimmer SR (2013) Identification and mapping of a novel blackleg resistance locus LepR4 in the progenies from *Brassica napus* × *B. rapa* subsp. *sylvestris*. Theor Appl Genet 126:307–315. https://doi.org/10.1007/s00122-012-1919-2

Yu J, Tehrim S, Zhang F, Tong C, Huang J, Cheng X, Dong C, Zhou Y, Qin R, Hua W, Liu S (2014) Genome-wide comparative analysis of NBS-encoding genes between *Brassica* species and *Arabidopsis thaliana*. BMC Genomics 15:3. https://doi.org/10.1186/1471-2164-15-3

Zhang XW, Yuan YX, Wang XW (2009) QTL mapping for TuMV resistance in Chinese cabbage [*Brassica campestris* L. ssp. *pekinensis* (Lour.)Olssom]. Acta Hortic Sin 36:731–736

Zhang L, Li H, Wang J (2012) The statistical power of inclusive composite interval mapping in detecting digenic epistasis showing common F_2 segregation ratios. J Integr Plant Biol 54 (4):270–279

Zhang B, Li P, Su T, Li P, Xin X, Wang W, Zhao X, Yu Y, Zhang D, Yu S, Zhang F (2018) BrRLP48, encoding a receptor-like protein, involved in downy mildew resistance in *Brassica rapa*. Front Plant Sci. https://doi.org/10.3389/fpls.2018.01708

Zhao J, Peltier AJ, Meng J, Osborn TC, Grau CR (2004) Evaluation of Scleroinia stem rot resistance in oilseed *Brassica napus* using an inoculation technique under green house conditions. Plant Dis 88:1033–1039

Zhao J, Udall JA, Quijada PA, Grau CR, Meng J, Osborn TC (2006) Quantitative trait loci for resistance to *Sclerotinia sclerotiorum* and its association with a homeologous non-reciprocal transposition in *Brassica napus* L. Theo Appl Genet 112:509–516

Zhao Q, Huang X, Lin Z, Han B (2010) SEG-Map: a novel software for genotype calling and genetic map construction from next-generation sequencing. Rice 3:98–102. https://doi.org/10.1007/s12284-010-9051-x

Zhu C, Gore M, Buckler ES, Yu J (2008) Status and prospects of association mapping in plants. Plant Genome 1:5–20

Future Research Priorities

9

Abstract

With the advances made in genomics of crucifers' host pathosystem using multiomics approaches, significant understanding has been achieved on the molecular mechanisms of host resistance. However, the use of next-generation sequencing (NGS) technology will help sequencing of entire genomes or transcriptomes of *Brassica*–host pathosystem more efficiently for better comprehension and molecular mechanisms operation and implementation for practical use. In nature, the coevolution of host and pathogen is a continuous process, and some of the pathogens adopt specific host, subvert innate immunity of the host through evolution of new virulent pathotypes. To fortify host immunity against the notorious pathogens, there is a need to be more vigilant through filling the gaps in our molecular host resistance mechanisms on the following areas of research priorities.

Keywords

Calmodulin binding transcription activators · Chitin-binding gene · Complex stress-response system · Durability of R-genes · Effector proteins · Epigenetics affects plant responses · Host defense R-genes · Interaction genes · Marker-linked stable QTLs · Mechanisms of multipathoge · Metabolomics · Molecular mechanisms of host resistance · Multiplex gene knock-out · Next-generation sequencing (NGS) · Proteomics · Transcriptomes of *Brassica*–host pathosystem · Transcriptomics

G. S. Saharan et al., *Molecular Mechanism of Crucifer's Host-Resistance*, https://doi.org/10.1007/978-981-16-1974-8_9

9.1 Introduction

With the advances made in genomics of crucifers' host pathosystem using multiomics approaches, significant understanding has been achieved on the molecular mechanisms of host resistance. However, the use of next-generation sequencing (NGS) technology will help sequencing of entire genomes or transcriptomes of *Brassica*–host pathosystem more efficiently for better comprehension and molecular mechanisms operation and implementation for practical use. In nature, the coevolution of host and pathogen is a continuous process, and some of the pathogens adopt specific host, subvert innate immunity of the host through evolution of new virulent pathotypes. To fortify host immunity against the notorious pathogens, there is a need to be more vigilant through filling the gaps in our molecular host resistance mechanisms (Saharan et al. 2016, 2018, 2019, 2021a, b) on the following areas of research priorities.

9.2 Research Priorities

1. Except for the four R-genes that have been cloned, the QTLs for the major diseases in *Brassica* have been identified, but the causal genes remain unknown. Cloning of R-genes has historically been undertaken using traditional methods, such as map-based cloning. However, the recent genomics revolution has provided significant insights into the presence and absence of expressed genes in the *Brassica* A- and C-genomes using a pan-transcriptome approach. This, along with the recent pan-genome research readily allows cloning of R-genes in crops such as *Brassica* that have complex genomes. More R-gene cloning work in *Brassica* will help us better understand how the plant defends itself against these pathogens, and we can use this knowledge to exploit novel R-genes in a quicker and more effective manner for breeding purposes.
2. The impact of climate change on crop diseases has been increasingly addressed over recent decades, making the characterization of new R-genes more crucial than ever. MutRenSeq technology has been used to clone resistance genes. This technique allows more rapid and precise cloning. Such methodologies can now be similarly applied for precise alteration of disease resistance genes in *Brassica*.
3. Mapping methods based on high-throughput sequencing data have developed rapidly, especially SNP-based methods, such as KASP markers and microarrays, BSA/BSR, and GWAS. Based on whole genome level mutations and their association with trait values, GWAS enables fast and accurate target trait gene characterization using natural populations. In addition, KASP technology possesses high assay robustness and accuracy and allows notable savings in terms of cost and time. These approaches should be explored for *Brassica*–hemibiotrophs and necrotrophs pathosystem.
4. Current molecular and omics methods, including transcriptomics, proteomics, and metabolomics, provide new opportunities for mining genes in the resistance-regulating network, which could be either directly used in resistance breeding or used in indirect prebreeding studies to promote understanding of

Brassica–pathogen interactions. In addition, the sequenced *Brassica* accessions do not contain all R-genes due to variations between individuals, whereas the establishment of the pangenomes could facilitate gene mining from a wider platform.

5. Achieving expanded knowledge of the molecular host–pathogen interactions could provide evidence for disease control. Protein–protein interactions in the *Brassica*–Turnip mosaic virus system are a heavily researched topic. Previous studies have shown that the cytoplasmic inclusion protein determines the viral avirulence for TuRB01/01b/04, whereas P3 determines the avirulence of TuRB03/05. There is need to search these areas in depth.
6. In *Brassica* crops, although over ten R-genes have been characterized to date, most avirulence or interaction genes in the pathogens have not yet been thoroughly characterized. Thus, the next emphasis should be on the establishment of global collections of pathogen isolates for the identification of important avirulence or interaction genes. In addition, for the host, the development of a series of ILs through recurrent backcrossing to "Mendalize" the quantitative loci would be beneficial.
7. As many R-genes have been cloned in *Arabidopsis*, the orthologous genes in *Brassica* can be investigated, which will greatly facilitate the cloning of these genes and the clarification of their molecular functions.
8. Studies have shown the role of certain effector proteins, but most have still not been characterized and their molecular targets and function are still unknown. Most of the effector/avirulence genes in the postulated gene-for-gene model have not yet been identified molecularly.
9. In future, the continuous advances in molecular and biological investigation in exploring the function of different isoform of single gene should facilitate the findings of the candidate isoforms in crucifer's pathogen defense.
10. There is need to fully understand the complex stress-response system of the plants to minimize differences in the tolerance capacity of transgenic plants between laboratories and crop fields.
11. The molecular mechanisms of seedling and adult plant resistance needs in-depth data generation.
12. Beyond the gene level, a wider perspective on how epigenetics affects plant responses toward pathogen attack should be considered to enhance understanding about the resistance mechanisms in *Brassica*. TEs, which are associated with DNA methylation, might contribute toward the resistance and susceptibility of the *Brassica* crop. How epigenetic variability is linked to phenotypic responses towards plant pathogens, as in case of *Arabidopsis*–*P. brassicae* pathosystem, where DNA methylation contributes toward quantitative resistance to clubroot, based on epigenotyped epigenetic recombinant inbred lines needs to be explored for better breeding programs.
13. Functional analysis of candidate genes in the *Brassica*–pathogen system can be performed using the clustered, regularly interspaced, palindromic repeats (CRISPR)–Cas9 (CRISPR-associated protein 9) system. In *B. napus*, two genes, *BnWRKY11* and *BnWRKY70*, that encode for WRKY transcription factors were successfully knocked out using CRISPR-Cas9, showing the latter

gene regulates disease resistance to *S. sclerotiorum*; more data need to be generated. This multiplex gene knock-out study still requires more information which is very useful for gene functional studies in polyploid crops like *Brassica* that contain many copies of the same genes as homeologs within the genome.

14. The CRISPR-Cas9 system was also successfully applied in *B. rapa* for early-flowering trait genome editing. In *L. maculans*, the chitin-binding gene LmCBP1, which was highly expressed during infection in *B. napus* and functional analysis using CRISPR-Cas9, was shown to play a role in enhancing cell death for pathogen growth and also to be involved in tolerance toward H_2O_2 during the plant immune response.
15. The role of WRKY genes and calmodulin binding transcription activators (CAMTA) in providing resistance to *Brassica* species against different pathogens needs further characterization.
16. MAPKs molecular are major players in molecular mechanism of host resistance. However, their role is poorly known against fungal effectors.
17. Identification of marker-linked stable QTLs will be very valuable for breeding effective R-cvs. to *L. maculans* in different environments to contribute for sustainable management of the disease.
18. There is a need to analyze huge data accumulated over the years on expression of molecular mechanisms of host–pathogen interaction and characterization of candidate's gene to engineer disease-resistant *Brassica* crops.
19. There is a need to study environmental and soil factors including nutrition supply under different agro-ecosystems which modulates durability of R-genes in *Brassica* cultivars.
20. There is a need to strengthen molecular mechanisms of resistance against *Xanthomonas* and Turnip mosaic virus (TuMV) diseases of crucifer.
21. Mechanisms of multipathogen and multipathotypes host defense R-genes and their functional nature including signaling molecules can help breeding durable resistant cultivars of *Brassica* crops.

Reference

Saharan GS, Mehta N, Meena PD (2016) Alternaria blight of crucifers: biology, ecology and management. Springer Verlag, Singapore, p 326

Saharan GS, Mehta N, Meena PD (2018) Downy mildew disease of crucifers: biology, ecology and disease management, vol LVI. Springer Verlag, Singapore, p 357

Saharan GS, Mehta N, Meena PD (2019) Powdery mildew disease of crucifers: biology, ecology and disease management. Springer Verlag, Singapore, p 362

Saharan GS, Mehta N, Meena PD (2021a) Club root disease of crucifers: biology, ecology and disease management. Springer Verlag, Singapore, p 756. https://doi.org/10.1007/978-981-16-2132-8

Saharan GS, Mehta N, Meena PD (2021b) Genomics of crucifer's host-resistance. Springer Nature Singapore Pvt. Ltd, Singapore, p 784. https://doi.org/10.1007/978-981-16-0862-9

Index

G. S. Saharan et al., *Molecular Mechanism of Crucifer's Host-Resistance*,
https://doi.org/10.1007/978-981-16-1974-8